The secret of KOTA
now at your Doorstep

DPP Daily Practice Problem Sheets

for NEET/AIIMS/JIPMER

PHYSICS

DPP BOOKLET

60 + 28
Topic-wise Chapter-wise
Tests for Concept Checking & Speed Building

Improves your learning by at least 20%

- Collection of 3100 + MCQs of all variety of questions
- Unique & innovative way of learning
- Detailed solutions to Topic-wise & Chapter-wise practice sheets
- Covers all important concepts of each topic
- As per latest pattern & syllabus

• **Corporate Office** : 45, 2nd Floor, Maharishi Dayanand Marg, Corner Market,
Malviya Nagar, New Delhi-110017
Tel. : 011-49842349 / 49842350

Typeset by Disha DTP Team

For further information about books from DISHA,

Log on to www.dishapublication.com or email to info@dishapublication.com

Daily Practice Problem (DPP) Sheets PHYSICS for NEET/AIIMS/JIPMER (Kota's formula to Success)

PREPARE ⟶ ASSESS ⟶ IMPROVE

Assessment is the most integral part of a student's preparation but still most of them avoid it. Only assessment can tell where you stand and how you can improve from that point. So it is very important that you take the right assessment, which is on the correct pattern, has the same level of difficulty as the actual exam and covers all the important concepts of the subject.

Disha Publication launches a first of its kind product which changed the way coaching was conducted in KOTA - the hub of Engineering and Medical Entrance education in India.

The book "Daily Practice Problem (DPP) Sheets for NEET/AIIMS" is precise, apt and tuned to all the requirements of a NEET/AIIMS aspirant.

KEY DIFFERENTIATING FEATURES OF THE DPP SHEETS

- Part A provides 60 DPP's with division of the complete NEET syllabus of Physics into 60 most important Topics. Each of the chapter has been broken into 2 or more topics.
- Part B consist of — Chapter-wise tests based on NCERT and NEET syllabus.
- Time Limit and Maximum Marks have been provided for each DPP Sheet/ topic. You must attempt each Sheet in test like conditions following the time limits. Further to achieve perfect preparation in a topic or chapter one has to score atleast 135 marks.
- Ultimate tool for Concept Checking & Speed Building.
- Collection of 3100 Standardised MCQ's of all variety of NEW pattern questions – MCQ only one correct option and Assertion-Reason.
- Unique & innovative way of learning. Whenever you have prepared a topic(Part A) or a chapter (Part B) just attempt that worksheet.
- Do not refer the Solution Booklet until and unless you have made all the efforts to solve the DPP Sheets.
- Covers all important Concepts of each Topic in the form of different Questions in the DPP Sheets.
- As per latest pattern & syllabus of NEET/AIIMS JIPMER exam.
- Compliant to all boards of education.

No matter where you PREPARE from – a coaching or NCERT books or any other textbook/ Guide - Daily Practice Problem Sheets provides you the right ASSESSMENT on each topic. Your performance provides you the right clues to IMPROVE your concepts so as to perform better in the final examination.

It is to be noted here that these are not tests but act as a checklist of student's learning and ability to apply concepts to different problems. Do proper analysis after you attempt each DPP sheet and try to locate your weak areas.

It is our strong belief that if an aspirant works hard on the clues provided through each of the DPP sheets he/ she can **improve his/ her learning and finally the SCORE by at least 20%.**

Disha Experts

The book comprises of following two parts

TOPIC-WISE DPP SHEETS WITH SOLUTIONS

INDEX/SYLLABUS

DPP - Daily Practice Problems

Name :

Date :

Start Time :

End Time :

PHYSICS 01

SYLLABUS : Physical World, Units & Dimensions

Max. Marks : 120 | **Time : 60 min.**

GENERAL INSTRUCTIONS

- The Daily Practice Problem Sheet contains 30 MCQ's. For each question only one option is correct. Darken the correct circle/ bubble in the Response Grid provided on each page.
- You have to evaluate your Response Grids yourself with the help of solution booklet.
- Each correct answer will get you 4 marks and 1 mark shall be deduced for each incorrect answer. No mark will be given/ deducted if no bubble is filled. Keep a timer in front of you and stop immediately at the end of 60 min.
- The sheet follows a particular syllabus. Do not attempt the sheet before you have completed your preparation for that syllabus. Refer syllabus sheet in the starting of the book for the syllabus of all the DPP sheets.
- After completing the sheet check your answers with the solution booklet and complete the Result Grid. Finally spend time to analyse your performance and revise the areas which emerge out as weak in your evaluation.

DIRECTIONS (Q.1-Q.21) : There are 21 multiple choice questions. Each question has 4 choices (a), (b), (c) and (d), out of which **ONLY ONE** choice is correct.

Q.1 If L, C and R represent inductance, capacitance and resistance respectively, then which of the following does not represent dimensions of frequency?

(a) $\frac{1}{RC}$ (b) $\frac{R}{L}$

(c) $\frac{1}{\sqrt{LC}}$ (d) $\frac{C}{L}$

Q.2 Number of particles crossing unit area perpendicular to X-axis in unit time is given by $n = -D\frac{n_2 - n_1}{x_2 - x_1}$, where n_1 and n_2 are number of particles per unit volume in the position x_1 and x_2. Find dimensions of D called as diffusion constant.

(a) $[M^0 L T^2]$ (b) $[M^0 L^2 T^{-4}]$

(c) $[M^0 L T^{-3}]$ (d) $[M^0 L^2 T^{-1}]$

Q.3 $X = 3YZ^2$ find dimensions of Y in (MKSA) system, if X and Z are the dimensions of capacity and magnetic field respectively

(a) $[M^{-3}L^{-2}T^{-4}A^{-1}]$ (b) $[ML^{-2}]$

(c) $[M^{-3}L^{-2}T^4A^4]$ (d) $[M^{-3}L^{-2}T^8A^4]$

Q.4 In the relation $P = \frac{\alpha}{\beta}e^{-\frac{\alpha Z}{k\theta}}$, P is pressure, Z is the distance, k is Boltzmann constant and θ is the temperature. The dimensional formula of β will be

(a) $[M^0L^2T^0]$ (b) $[M^1L^2T^1]$

(c) $[M^1L^0T^{-1}]$ (d) $[M^0L^2T^{-1}]$

RESPONSE GRID	1. ⓐⓑⓒⓓ	2. ⓐⓑⓒⓓ	3. ⓐⓑⓒⓓ	4. ⓐⓑⓒⓓ

Space for Rough Work

Q.5 The frequency of vibration of string is given by $\nu = \frac{P}{2l}\left[\frac{F}{m}\right]^{1/2}$. Here P is number of segments in the string and l is the length. The dimensional formula for m will be

(a) $[M^0LT^{-1}]$ (b) $[ML^0T^{-1}]$
(c) $[ML^{-1}T^0]$ (d) $[M^0L^0T^0]$

Q.6 What is the relationship between dyne and newton of force?

(a) 1 dyne = 10^{-5} newton (b) 1 dyne = 10^{-7} newton
(c) 1 dyne = 10^5 newton (d) 1 dyne = 10^7 newton

Q.7 The speed of light (c), gravitational constant (G) and Planck's constant (h) are taken as the fundamental units in a system. The dimensions of time in this new system should be

(a) $G^{1/2}h^{1/2}c^{-5/2}$ (b) $G^{-1/2}h^{1/2}c^{1/2}$
(c) $G^{1/2}h^{1/2}c^{-3/2}$ (d) $G^{1/2}h^{1/2}c^{1/2}$

Q.8 If the constant of gravitation (G), Planck's constant (h) and the velocity of light (c) be chosen as fundamental units. The dimensions of the radius of gyration is

(a) $h^{1/2}c^{-3/2}G^{1/2}$ (b) $h^{1/2}c^{3/2}G^{1/2}$
(c) $h^{1/2}c^{-3/2}G^{-1/2}$ (d) $h^{-1/2}c^{-3/2}G^{1/2}$

Q.9 The magnitude of any physical quantity

(a) depends on the method of measurement
(b) does not depend on the method of measurement
(c) is more in SI system than in CGS system
(d) directly proportional to the fundamental units of mass, length and time

Q.10 The unit of Stefan's constant σ is

(a) $Wm^{-2}K^{-1}$ (b) Wm^2K^{-4}
(c) $Wm^{-2}K^{-4}$ (d) $Wm^{-2}K^4$

Q.11 In $S = a + bt + ct^2$, S is measured in metres and t in seconds. The unit of c is

(a) ms^{-2} (b) m
(c) ms^{-1} (d) None

Q.12 Wavelength of ray of light is 0.00006 m. It is equal to

(a) 6 microns (b) 60 microns
(c) 600 microns (d) 0.6 microns

Q.13 SI unit of permittivity is

(a) $C^2m^2N^{-2}$ (b) $C^{-1}m^2N^{-2}$
(c) $C^2m^2N^2$ (d) $C^2m^{-2}N^{-1}$

Q.14 The dimensions of $\frac{1}{2}\varepsilon_0 E^2$ (ε_0 = permittivity of free space and E = electric field) are

(a) MLT^{-1} (b) ML^2T^{-2}
(c) $ML^{-1}T^{-2}$ (d) ML^2T^{-1}

Q.15 Which of the following pairs is wrong?

(a) Pressure-Baromter
(b) Relative density-Pyrometer
(c) Temperature-Thermometer
(d) Earthquake-Seismograph

Q.16 A physical quantity x depends on quantities y and z as follows: $x = Ay + B \tan Cz$, where A, B and C are constants. Which of the following do not have the same dimensions?

(a) x and B (b) C and z^{-1}
(c) y and B/A (d) x and A

Q.17 If the time period (T) of vibration of a liquid drop depends on surface tension (S), radius (r) of the drop and density (ρ) of the liquid, then the expression of T is

(a) $T = k\sqrt{\rho r^3/S}$ (b) $T = k\sqrt{\rho^{1/2}r^3/S}$
(c) $T = k\sqrt{\rho r^3/S^{1/2}}$ (d) None of these

Q.18 The dimensional formula for Planck's constant (h) is

(a) $[ML^{-2}T^{-3}]$ (b) $[M^0L^2T^{-2}]$
(c) $[M^0L^2T^{-1}]$ (d) $[ML^{-2}T^{-2}]$

Q.19 What are the dimensions of permeability (μ_0) of vaccum?

(a) $MLT^{-2}I^2$ (b) $MLT^{-2}I^{-2}$
(c) $ML^{-1}T^{-2}I^2$ (d) $ML^{-1}T^{-2}I^{-2}$

RESPONSE GRID					
	5. ⓐⓑⓒⓓ	6. ⓐⓑⓒⓓ	7. ⓐⓑⓒⓓ	8. ⓐⓑⓒⓓ	9. ⓐⓑⓒⓓ
	10. ⓐⓑⓒⓓ	11. ⓐⓑⓒⓓ	12. ⓐⓑⓒⓓ	13. ⓐⓑⓒⓓ	14. ⓐⓑⓒⓓ
	15. ⓐⓑⓒⓓ	16. ⓐⓑⓒⓓ	17. ⓐⓑⓒⓓ	18. ⓐⓑⓒⓓ	19. ⓐⓑⓒⓓ

Space for Rough Work

Q.20 A small steel ball of radius r is allowed to fall under gravity through a column of a viscous liquid of coefficient of viscosity η. After some time the velocity of the ball attains a constant value known as terminal velocity v_T. The terminal velocity depends on (i) the mass of the ball m, (ii) η, (iii) r and (iv) acceleration due to gravity g. Which of the following relations is dimensionally correct?

(a) $v_T \propto \frac{mg}{\eta r}$ (b) $v_T \propto \frac{\eta r}{mg}$

(c) $v_T \propto \eta rmg$ (d) $v_T \propto \frac{mgr}{\eta}$

Q.21 The equation of state of some gases can be expressed as $\left(P+\frac{a}{V^2}\right)(V-b)=RT$. Here P is the pressure, V is the volume, T is the absolute temperature and a, b and R are constants. The dimensions of 'a' are

(a) ML^5T^{-2} (b) $ML^{-1}T^{-2}$

(c) $M^0L^3T^0$ (d) $M^0L^6T^0$

DIRECTIONS (Q.22-Q.24) : In the following questions, more than one of the answers given are correct. Select the correct answers and mark it according to the following codes:

Codes :

(a) 1, 2 and 3 are correct **(b)** 1 and 2 are correct

(c) 2 and 4 are correct **(d)** 1 and 3 are correct

Q.22 The frequency of vibration f of a mass m suspended from a spring of spring constant k is given by a relation of the type $f = c\, m^x k^y$, where c is a dimensionless constant. The values of x and y are

(1) $x=\frac{1}{2}$ (2) $x=-\frac{1}{2}$

(3) $y=-\frac{1}{2}$ (4) $y=\frac{1}{2}$

Q.23 P represents radiation pressure, c represents speed of light and S represents radiation energy striking unit area per sec. The non zero integers x, y, z such that $P^x S^y c^z$ is dimensionless are

(1) x = 1 (2) y = – 1

(3) z = 1 (4) x = – 1

Q.24 Which of the following pairs have same dimensions?

(1) Angular momentum and work

(2) Torque and work

(3) Energy and Young's modulus

(4) Light year and wavelength

DIRECTION (Q.25-Q.27) : Read the passage given below and answer the questions that follows :

Three of the fundamental constants of physics are the universal gravitational constant, $G = 6.7 \times 10^{-11} m^3 kg^{-1} s^{-2}$, the speed of light, $c = 3.0 \times 10^8$ m/s, and Planck's constant, $h = 6.6 \times 10^{-34}$ kg m^2 s^{-1}.

Q.25 Find a combination of these three constants that has the dimensions of time. This time is called the Planck time and represents the age of the universe before which the laws of physics as presently understood cannot be applied.

(a) $\sqrt{\frac{hG}{c^4}}$ (b) $\sqrt{\frac{hG}{c^3}}$

(c) $\sqrt{\frac{hG}{c}}$ (d) $\sqrt{\frac{hG}{c^5}}$

Q.26 Find the value of Planck time in seconds

(a) 1.3×10^{-33} s (b) 1.3×10^{-43} s

(c) 2.3×10^{-13} s (d) 0.3×10^{-23} s

Q.27 The energy of a photon is given by $E = \frac{hc}{\lambda}$.

If $\lambda = 4 \times 10^{-7} m$, the energy of photon is

(a) 3.0 eV (b) 4.5 eV

(c) 2.10 eV (d) 3.95 eV

RESPONSE GRID					
	20. ⓐⓑⓒⓓ	21. ⓐⓑⓒⓓ	22. ⓐⓑⓒⓓ	23. ⓐⓑⓒⓓ	24. ⓐⓑⓒⓓ
	25. ⓐⓑⓒⓓ	26. ⓐⓑⓒⓓ	27. ⓐⓑⓒⓓ		

Space for Rough Work

DIRECTIONS (Q.28-Q.30) : Each of these questions contains two statements: Statement-1 (Assertion) and Statement-2 (Reason). Each of these questions has four alternative choices, only one of which is the correct answer. You have to select the correct choice.

(a) Statement-1 is True, Statement-2 is True; Statement-2 is a correct explanation for Statement-1.
(b) Statement-1 is True, Statement-2 is True; Statement-2 is NOT a correct explanation for Statement-1.
(c) Statement -1 is False, Statement-2 is True.
(d) Statement -1 is True, Statement-2 is False.

Q.28 Statement -1 : Unit of Rydberg constant R is m^{-1}.

Statement -2 : It follows from Bohr's formula $\bar{v} = R\left(\frac{1}{n_1^2} - \frac{1}{n_2^2}\right)$, where the symbols have their usual meaning.

Q.29 Statement -1: The time period of a pendulum is given by the formula, $T = 2\pi\sqrt{g/l}$.

Statement -2: According to the principle of homogeneity of dimensions, only that formula is correct in which the dimensions of L.H.S. is equal to dimensions of R.H.S.

Q.30 Statement -1: L/R and CR both have same dimensions.

Statement -2: L/R and CR both have dimension of time.

RESPONSE GRID	28. ⓐⓑⓒⓓ	29. ⓐⓑⓒⓓ	30. ⓐⓑⓒⓓ

DAILY PRACTICE PROBLEM SHEET 1 - PHYSICS			
Total Questions	30	Total Marks	120
Attempted		Correct	
Incorrect		Net Score	
Cut-off Score	30	Qualifying Score	50
Success Gap = Net Score – Qualifying Score			
Net Score = (Correct × 4) – (Incorrect × 1)			

Space for Rough Work

DPP - Daily Practice Problems

Name : Date :

Start Time : End Time :

PHYSICS 02

SYLLABUS : Measurements (Errors)

Max. Marks : 120 **Time : 60 min.**

GENERAL INSTRUCTIONS

- The Daily Practice Problem Sheet contains 30 MCQ's. For each question only one option is correct. Darken the correct circle/ bubble in the Response Grid provided on each page.
- You have to evaluate your Response Grids yourself with the help of solution booklet.
- Each correct answer will get you 4 marks and 1 mark shall be deduced for each incorrect answer. No mark will be given/ deducted if no bubble is filled. Keep a timer in front of you and stop immediately at the end of 60 min.
- The sheet follows a particular syllabus. Do not attempt the sheet before you have completed your preparation for that syllabus. Refer syllabus sheet in the starting of the book for the syllabus of all the DPP sheets.
- After completing the sheet check your answers with the solution booklet and complete the Result Grid. Finally spend time to analyse your performance and revise the areas which emerge out as weak in your evaluation.

Q.1 A wire has a mass 0.3 ± 0.003g, radius 0.5 ± 0.005 mm and length 6 ± 0.06 cm. The maximum percentage error in the measurement of its density is

(a) 1 (b) 2 (c) 3 (d) 4

Q.2 If 97.52 is divided by 2.54, the correct result in terms of significant figures is

(a) 38.4 (b) 38.3937

(c) 38.394 (d) 38.39

Q.3 A physical quantity A is related to four observable a, b, c and d as follows, $A = \frac{a^2b^3}{c\sqrt{d}}$ the percentage errors of measurement in a, b, c and d are 1%, 3%, 2% and 2% respectiely. What is the percentage error in the quantity A

(a) 12% (b) 7%

(c) 5% (d) 14%

Q.4 A physical quantity is given by $X = M^aL^bT^c$. The percentage error in measurement of M, L and T are α, β and γ respectively. Then maximum percentage error in the quantity X is

(a) $a\alpha + b\beta + c\gamma$ (b) $a\alpha + b\beta - c\gamma$

(c) $\frac{a}{\alpha}+\frac{b}{\beta}+\frac{c}{\gamma}$ (d) None of these

RESPONSE GRID	1. ⓐⓑⓒⓓ	2. ⓐⓑⓒⓓ	3. ⓐⓑⓒⓓ	4. ⓐⓑⓒⓓ

Q.5 If the length of rod A is 3.25 ± 0.01 cm and that of B is 4.19 ± 0.01 cm then the rod B is longer than rod A by
(a) 0.94 ± 0.00 cm (b) 0.94 ± 0.01 cm
(c) 0.94 ± 0.02 cm (d) 0.94 ± 0.005 cm

Q.6 If L = 2.331 cm, B = 2.1 cm, then L + B =
(a) 4.431 cm (b) 4.43 cm
(c) 4.4 cm (d) 4 cm

Q.7 The number of significant figures in all the given numbers 25.12, 2009, 4.156 and 1.217×10^{-4} is
(a) 1 (b) 2 (c) 3 (d) 4

Q.8 In an experiment, the following observation's were recorded: L = 2.820 m, M = 3.00 kg, l = 0.087 cm, Diameter D = 0.041 cm. Taking $g = 9.81 \text{ m/s}^2$ using the formula, $Y = \frac{4MgL}{\pi D^2 l}$, the maximum percentage error in Y is
(a) 7.96% (b) 4.56% (c) 6.50% (d) 8.42%

Q.9 A physical parameter a can be determined by measuring the parameters b, c, d and e using the relation $a = \frac{b^\alpha c^\beta}{d^\gamma e^\delta}$.
If the maximum errors in the measurement of b, c, d and e are b_1%, c_1%, d_1%, and e_1%, then the maximum error in the value of a determined by the experiment is
(a) $(b_1 + c_1 + d_1 + e_1)$% (b) $(b_1 + c_1 - d_1 - e_1)$%
(c) $(\alpha b_1 + \beta c_1 - \gamma d_1 - \delta e_1)$% (d) $(\alpha b_1 + \beta c_1 + \gamma d_1 + \delta e_1)$%

Q.10 The period of oscillation of a simple pendulum is given by $T = 2\pi\sqrt{\frac{l}{g}}$ where l is about 100 cm and is known to have 1mm accuracy. The period is about 2s. The time of 100 oscillations is measured by a stopwatch of least count 0.1s. The percentage error in g is
(a) 0.1% (b) 1%
(c) 0.2% (d) 0.8%

Q.11 The mean time period of second's pendulum is 2.00s and mean absolute error in the time period is 0.05s. To express maximum estimate of error, the time period should be written as
(a) (2.00 ± 0.01)s (b) (2.00 + 0.025) s
(c) (2.00 ± 0.05) s (d) (2.00 ± 0.10) s

Q.12 Error in the measurement of radius of a sphere is 1%. The error in the calculated value of its volume is
(a) 1% (b) 3%
(c) 5% (d) 7%

Q.13 The relative density of material of a body is found by weighing it first in air and then in water. If the weight in air is (5.00 ± 0.05) newton and weight in water is (4.00 ± 0.05) newton. Then the relative density along with the maximum permissible percentage error is
(a) 5.0 ± 11% (b) 5.0 ± 1%
(c) 5.0 ± 6% (d) 1.25 ± 5%

Q.14 The resistance $R = \frac{V}{i}$ where V = 100 ± 5 volts and i = 10 ± 0.2 amperes. What is the total error in R ?
(a) 5% (b) 7% (c) 5.2% (d) $\frac{5}{2}$%

Q.15 The length of a cylinder is measured with a meter rod having least count 0.1 cm. Its diameter is measured with vernier calipers having least count 0.01 cm. Given that length is 5.0 cm. and radius is 2.0 cm. The percentage error in the calculated value of the volume will be
(a) 1% (b) 2%
(c) 3% (d) 4%

Q.16 According to Joule's law of heating, heat produced $H = I^2Rt$, where I is current, R is resistance and t is time. If the errors in the measurements of I, R. and t are 3%, 4% and 6% respectively then error in the measurement of H is
(a) ± 17% (b) ± 16%
(c) ± 19% (d) ± 25%

RESPONSE GRID					
	5. ⓐⓑⓒⓓ	6. ⓐⓑⓒⓓ	7. ⓐⓑⓒⓓ	8. ⓐⓑⓒⓓ	9. ⓐⓑⓒⓓ
	10. ⓐⓑⓒⓓ	11. ⓐⓑⓒⓓ	12. ⓐⓑⓒⓓ	13. ⓐⓑⓒⓓ	14. ⓐⓑⓒⓓ
	15. ⓐⓑⓒⓓ	16. ⓐⓑⓒⓓ			

Space for Rough Work

Q.17 A physical quantity P is given by $P = \frac{A^3 B^{\frac{1}{2}}}{C^{-4} D^{\frac{3}{2}}}$. The quantity which brings in the maximum percentage error in P is

(a) A (b) B
(c) C (d) D

Q.18 If there is a positive error of 50% in the measurement of velocity of a body, then the error in the measurement of kinetic energy is

(a) 25% (b) 50%
(c) 100% (d) 125%

Q.19 The random error in the arithmetic mean of 100 observations is x; then random error in the arithmetic mean of 400 observations would be

(a) $4x$ (b) $\frac{1}{4}x$ (c) $2x$ (d) $\frac{1}{2}x$

Q.20 The percentage errors in the measurement of mass and speed are 2% and 3% respectively. How much will be the maximum error in the estimation of the kinetic energy obtained by measuring mass and speed?

(a) 11% (b) 8%
(c) 5% (d) 1%

Q.21 The unit of percentage error is

(a) Same as that of physical quantity
(b) Different from that of physical quantity
(c) Percentage error is unitless
(d) Errors have got their own units which are different from that of physical quantity measured

DIRECTIONS (Q.22-Q.24) : In the following questions, more than one of the answers given are correct. Select the correct answers and mark it according to the following codes:

Codes :

(a) 1, 2 and 3 are correct **(b)** 1 and 2 are correct
(c) 2 and 4 are correct **(d)** 1 and 3 are correct

Q.22 In the context of accuracy of measurement and significant figures in expressing results of experiment, which of the following is/are correct?

1. Out of the two measurements 50.14 cm and 0.00025 ampere, the first one has greater accuracy
2. Out of the two measurements 50.14 cm and 0.00025 ampere, the second has greater accuracy.
3. If one travels 478 km by rail and 397 m by road, the total distance travelled is 875 km.
4. If one travels 697 m by rail and 478 km by road, the total distance is 478 km.

Q.23 A thin copper wire of length l metre increases in length by 2% when heated through 10°C. Which is not the percentage increase in area when a square copper sheet of length l metre is heated through 10°C

(1) 12% (2) 8% (3) 16% (4) 4%

Q.24 A body travels uniformly a distance of (13.8 ± 0.2) m in a time (4.0 ± 0.3) s.

1. Its velocity with error limit is $(3.5 \pm 0.31)\,\text{ms}^{-1}$
2. Its velocity with error limit is $(3.5 \pm 0.11)\,\text{ms}^{-1}$
3. Percentage error in velocity is $\pm 4\%$
4. Percentage error in velocity is $\pm 9\%$

DIRECTION (Q.25-Q.27) : Read the passage given below and answer the questions that follows :

The internal radius of a 1m long resonance tube is measured as 3 cm. A tuning fork of frequency 2000 Hz is used. The first resonating length is measured as 4.6 cm and the second resonating length is measured as 14.0 cm.

Q.25 Calculate the maximum percentage error in measurement of e.

(a) 3.33% (b) 2.23% (c) 4.33% (d) 5.33%

Q.26 Calculate the speed of sound at the room temperature.

(a) 275 m/s (b) 376 m/s
(c) 356 m/s (d) 330 m/s

RESPONSE GRID					
	17. ⓐⓑⓒⓓ	18. ⓐⓑⓒⓓ	19. ⓐⓑⓒⓓ	20. ⓐⓑⓒⓓ	21. ⓐⓑⓒⓓ
	22. ⓐⓑⓒⓓ	23. ⓐⓑⓒⓓ	24. ⓐⓑⓒⓓ	25. ⓐⓑⓒⓓ	26. ⓐⓑⓒⓓ

Space for Rough Work

Q.27 Calculate the end correction.

(a) 0.2 cm (b) 0.3 cm

(c) 0.1 cm (d) 0.4 cm

DIRECTIONS (Q. 28-Q.30) : Each of these questions contains two statements: Statement-1 (Assertion) and Statement-2 (Reason). Each of these questions has four alternative choices, only one of which is the correct answer. You have to select the correct choice.

(a) Statement-1 is True, Statement-2 is True; Statement-2 is a correct explanation for Statement-1.

(b) Statement-1 is True, Statement-2 is True; Statement-2 is NOT a correct explanation for Statement-1.

(c) Statement-1 is False, Statement-2 is True.

(d) Statement-1 is True, Statement-2 is False.

Q.28 **Statement-1:** Number of significant figures in 0.005 is one and that in 0.500 is three.

Statement-2 : This is because zero is not significant.

Q.29 **Statement-1:** Out of three measurements l = 0.7 m; l = 0.70 m and l = 0.700 m, the last one is most accurate.

Statement-2: In every measurement, only the last significant digit is not accurately known.

Q.30 **Statement-1:** Parallex method cannot be used for measuring distances of stars more than 100 light years away.

Statement-2: Because parallex angle reduces so much that it cannot be measured accurately.

RESPONSE GRID	27. ⓐⓑⓒⓓ	28. ⓐⓑⓒⓓ	29. ⓐⓑⓒⓓ	30. ⓐⓑⓒⓓ

DAILY PRACTICE PROBLEM SHEET 2 - PHYSICS			
Total Questions	30	Total Marks	120
Attempted		Correct	
Incorrect		Net Score	
Cut-off Score	28	Qualifying Score	46
Success Gap = Net Score – Qualifying Score			
Net Score = (Correct × 4) – (Incorrect × 1)			

Space for Rough Work

DPP - Daily Practice Problems

Name :   Date :

Start Time :   End Time :

PHYSICS 03

SYLLABUS : MOTION IN A STRAIGHT LINE 1 (Distance, Displacement, Uniform & Non-uniform motion)

Max. Marks : 116 **Time : 60 min.**

GENERAL INSTRUCTIONS

- The Daily Practice Problem Sheet contains 29 MCQ's. For each question only one option is correct. Darken the correct circle/ bubble in the Response Grid provided on each page.
- You have to evaluate your Response Grids yourself with the help of solution booklet.
- Each correct answer will get you 4 marks and 1 mark shall be deduced for each incorrect answer. No mark will be given/ deducted if no bubble is filled. Keep a timer in front of you and stop immediately at the end of 60 min.
- The sheet follows a particular syllabus. Do not attempt the sheet before you have completed your preparation for that syllabus. Refer syllabus sheet in the starting of the book for the syllabus of all the DPP sheets.
- After completing the sheet check your answers with the solution booklet and complete the Result Grid. Finally spend time to analyse your performance and revise the areas which emerge out as weak in your evaluation.

DIRECTIONS (Q.1-Q.20) : There are 20 multiple choice questions. Each question has 4 choices (a), (b), (c) and (d), out of which **ONLY ONE** choice is correct.

Q.1 A particle moving in a straight line covers half the distance with speed of 3 m/s. The other half of the distance is covered in two equal time intervals with speed of 4.5 m/s and 7.5 m/s respectively. The average speed of the particle during this motion is

(a) 4.0 m/s (b) 5.0 m/s
(c) 5.5 m/s (d) 4.8 m/s

Q.2 The acceleration of a particle is increasing linearly with time t as bt. The particle starts from the origin with an initial velocity v_0. The distance travelled by the particle in time t will be

(a) $v_0 t + \frac{1}{3}bt^2$ (b) $v_0 t + \frac{1}{3}bt^3$
(c) $v_0 t + \frac{1}{6}bt^3$ (d) $v_0 t + \frac{1}{2}bt^2$

Q.3 The motion of a body is given by the equation $\frac{dv(t)}{dt} = 6.0 - 3v(t)$, where v(t) is speed in m/s and t in sec.

If body was at rest at t = 0

(a) The terminal speed is 4 m/s
(b) The speed varies with the time as $v(t) = 2(1 - e^{-5t})$m/s
(c) The speed is 0.1m/s when the acceleration is half the initial value
(d) The magnitude of the initial acceleration is 6.0 m/s^2

RESPONSE GRID	1. ⓐⓑⓒⓓ	2. ⓐⓑⓒⓓ	3. ⓐⓑⓒⓓ

Space for Rough Work

Q.4 A particle of mass m moves on the x-axis as follows: it starts from rest at t = 0 from the point x = 0 and comes to rest at t = 1 at the point x = 1. No other information is available about its motion at intermediate time (0 < t < 1). If α denotes the instantaneous acceleration of the particle, then

(a) α cannot remain positive for all t in the interval $0 \le t \le 1$
(b) | α | cannot exceed 2 at any point in its path
(c) | α | must be > 4 at some point or points in its path
(d) | α | = 2 at any point in its path.

Q.5 A particle starts from rest. Its acceleration (a) versus time (t) graph is as shown in the figure. The maximum speed of the particle will be

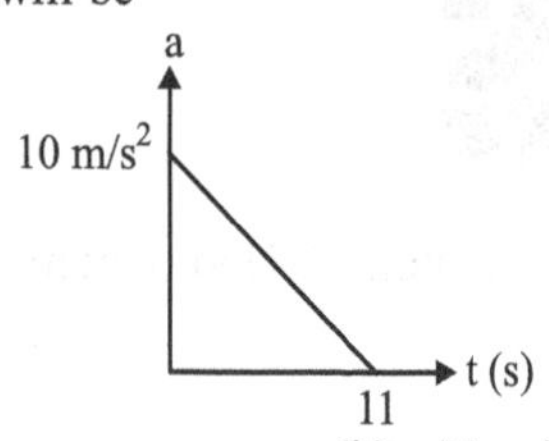

(a) 110 m/s (b) 55 m/s
(c) 550 m/s (d) 660 m/s

Q.6 A car accelerates from rest at a constant rate α for some time, after which it decelerates at a constant rate β and comes to rest. If the total time elapsed is t, then the maximum velocity acquired by the car is

(a) $\left(\frac{\alpha^2+\beta^2}{\alpha\beta}\right)t$ (b) $\left(\frac{\alpha^2-\beta^2}{\alpha\beta}\right)t$

(c) $\frac{(\alpha+\beta)t}{\alpha\beta}$ (d) $\frac{\alpha\beta t}{\alpha+\beta}$

Q.7 A small block slides without friction down an inclined plane starting from rest. Let S_n be the distance travelled from time t = n – 1 to t = n. Then $\frac{S_n}{S_{n+1}}$ is

(a) $\frac{2n-1}{2n}$ (b) $\frac{2n+1}{2n-1}$ (c) $\frac{2n-1}{2n+1}$ (d) $\frac{2n}{2n+1}$

Q.8 A particle starts moving from the position of rest under a constant acc. If it covers a distance x in t second, what distance will it travel in next t second?

(a) x (b) 2 x (c) 3 x (d) 4 x

Q.9 What will be the ratio of the distances moved by a freely falling body from rest in 4th and 5th seconds of journey?

(a) 4 : 5 (b) 7 : 9 (c) 16 : 25 (d) 1 : 1

Q.10 If a ball is thrown vertically upwards with speed u, the distance covered during the last t seconds of its ascent is

(a) (u+gt)t (b) ut (c) $\frac{1}{2}gt^2$ (d) $ut - \frac{1}{2}gt^2$

Q.11 If the displacement of a particle is $(2t^2 + t + 5)$ meter then, what will be acc. at t = 5 second?

(a) 21 m/s² (b) 20 m/s²
(c) 4 m/s² (d) 10 m/s²

Q.12 A particle moves along x-axis with acceleration $a = a_0(1 - t/T)$ where a_0 and T are constants if velocity at t = 0 is zero then find the average velocity from t = 0 to the time when a = 0.

(a) $\frac{a_0T}{3}$ (b) $\frac{a_0T}{2}$ (c) $\frac{a_0T}{4}$ (d) $\frac{a_0T}{5}$

Q.13 A point moves with uniform acceleration and v_1, v_2 and v_3 denote the average velocities in the three successive intervals of time t_1, t_2 and t_3. Which of the following relations is correct ?

(a) $(v_1-v_2):(v_2-v_3)=(t_1-t_2):(t_2+t_3)$
(b) $(v_1-v_2):(v_2-v_3)=(t_1+t_2):(t_2+t_3)$
(c) $(v_1-v_2):(v_2-v_3)=(t_1-t_2):(t_2-t_3)$
(d) $(v_1-v_2):(v_2-v_3)=(t_1-t_2):(t_2-t_3)$

Q.14 The position of a particle moving in the xy-plane at any time t is given by $x = (3t^2 - 6t)$ metres, $y = (t^2 - 2t)$ metres. Select the correct statement about the moving particle from the following

(a) The acceleration of the particle is zero at t = 0 second
(b) The velocity of the particle is zero at t = 0 second
(c) The velocity of the particle is zero at t = 1 second
(d) The velocity and acceleration of the particle are never zero

RESPONSE GRID				
4. ⓐⓑⓒⓓ	5. ⓐⓑⓒⓓ	6. ⓐⓑⓒⓓ	7. ⓐⓑⓒⓓ	8. ⓐⓑⓒⓓ
9. ⓐⓑⓒⓓ	10. ⓐⓑⓒⓓ	11. ⓐⓑⓒⓓ	12. ⓐⓑⓒⓓ	13. ⓐⓑⓒⓓ
14. ⓐⓑⓒⓓ				

Space for Rough Work

Q.15 Two cars A and B are travelling in the same direction with velocities v_1 and v_2 ($v_1 > v_2$). When the car A is at a distance d ahead of the car B, the driver of the car A applied the brake producing a uniform retardation a. There will be no collision when

(a) $d < \frac{(v_1 - v_2)^2}{2a}$ (b) $d < \frac{v_1^2 - v_2^2}{2a}$

(c) $d > \frac{(v_1 - v_2)^2}{2a}$ (d) $d > \frac{v_1^2 - v_2^2}{2a}$

Q.16 A body travels for 15 second starting from rest with constant acceleration. If it travels distances S_1, S_2 and S_3 in the first five seconds, second five seconds and next five seconds respectively the relation between S_1, S_2 and S_3 is

(a) $S_1 = S_2 = S_3$ (b) $5S_1 = 3S_2 = S_3$

(c) $S_1 = \frac{1}{3}S_2 = \frac{1}{5}S_3$ (d) $S_1 = \frac{1}{5}S_2 = \frac{1}{3}S_3$

Q.17 The position of a particle moving along the x-axis at certain times is given below

$t(s)$	0	1	2	3
$x(m)$	–2	0	6	16

Which of the following describes the motion correctly?

(a) Uniform, accelerated
(b) Uniform, decelerated
(c) Non-uniform, accelerated
(d) There is not enough data for generalization

Q.18 A body A moves with a uniform acceleration a and zero initial velocity. Another body B, starts from the same point moves in the same direction with a constant velocity v. The two bodies meet after a time t. The value of t is

(a) $\frac{2v}{a}$ (b) $\frac{v}{a}$

(c) $\frac{v}{2a}$ (d) $\sqrt{\frac{v}{2a}}$

Q.19 A particle moves along x-axis as $x = 4(t - 2) + a(t - 2)^2$ Which of the following is true?

(a) The initial velocity of particle is 4
(b) The acceleration of particle is 2a
(c) The particle is at origin at t = 0
(d) None of these

Q.20 The displacement x of a particle varies with time t, $x = ae^{-\alpha t} + be^{\beta t}$, where a, b, α and β are positive constants. The velocity of the particle will

(a) Go on decreasing with time
(b) Be independent of α and β
(c) Drop to zero when $\alpha = \beta$
(d) Go on increasing with time

DIRECTIONS (Q.21-Q.23) : In the following questions, more than one of the answers given are correct. Select the correct answers and mark it according to the following codes:

Codes :

(a) 1, 2 and 3 are correct **(b)** 1 and 2 are correct
(c) 2 and 4 are correct **(d)** 1 and 3 are correct

Q.21 A particle moves as such acceleration is given by a = 3 sin 4t, then :

(1) the acceleration of the particle becomes zero after each interval of $\frac{\pi}{4}$ second
(2) the initial velocity of the particle must be zero
(3) the particle comes at its initial position after sometime
(4) the particle must move on a circular path

Q.22 A reference frame attached to the earth :

(1) is an inertial frame by definition
(2) cannot be an inertial frame because the earth is revolving round the sun
(3) is an inertial frame because Newton's laws are applicable in this frame
(4) cannot be an inertial frame because the earth is rotating about its own axis

RESPONSE GRID					
	15. ⓐⓑⓒⓓ	16. ⓐⓑⓒⓓ	17. ⓐⓑⓒⓓ	18. ⓐⓑⓒⓓ	19. ⓐⓑⓒⓓ
	20. ⓐⓑⓒⓓ	21. ⓐⓑⓒⓓ	22. ⓐⓑⓒⓓ		

Space for Rough Work

Q.23 If a particle travels a linear distance at speed v_1 and comes back along the same track at speed v_2.
(1) Its average speed is arithmetic mean $(v_1 + v_2)/2$
(2) Its average speed is harmonic mean $2 v_1 v_2/(v_1 + v_2)/2$
(3) Its average speed is geometric mean $\sqrt{v_1 v_2}$
(4) Its average velocity is zero

DIRECTION (Q.24-Q.26) : Read the passage given below and answer the questions that follows :

A particle moves along x-axis and its acceleration at any time t is $a = 2 \sin(\pi t)$, where t is in seconds and a is in m/s^2. The initial velocity of particle (at time t = 0) is u = 0.

Q.24. The distance travelled (in meters) by the particle from time to t = 0 to t = 1s will be –
(a) $\frac{2}{\pi}$ (b) $\frac{1}{\pi}$
(c) $\frac{4}{\pi}$ (d) None of these

Q.25 The distance travelled (in meters) by the particle from time t = 0 to t = t will be –
(a) $\frac{2}{\pi^2}\sin \pi t - \frac{2t}{\pi}$ (b) $-\frac{2}{\pi^2}\sin \pi t + \frac{2t}{\pi}$
(c) $\frac{2t}{\pi}$ (d) None of these

Q.26 The magnitude of displacement (in meters) by the particle from time t = 0 to t = t will be –
(a) $\frac{2}{\pi^2}\sin \pi t - \frac{2t}{\pi}$ (b) $-\frac{2}{\pi^2}\sin \pi t + \frac{2t}{\pi}$
(c) $\frac{2t}{\pi}$ (d) None of these

DIRECTIONS (Qs. 27-Q.29) : Each of these questions contains two statements: Statement-1 (Assertion) and Statement-2 (Reason). Each of these questions has four alternative choices, only one of which is the correct answer. You have to select the correct choice.

(a) Statement-1 is True, Statement-2 is True; Statement-2 is a correct explanation for Statement-1.
(b) Statement-1 is True, Statement-2 is True; Statement-2 is NOT a correct explanation for Statement-1.
(c) Statement -1 is False, Statement-2 is True.
(d) Statement -1 is True, Statement-2 is False.

Q.27 Statement-1 : The position-time graph of a uniform motion in one dimension of a body can have negative slope.
Statement-2 : When the speed of body decreases with time, the position-time graph of the moving body has negative slope.

Q.28 Statement-1 : A body having non-zero acceleration can have a constant velocity.
Statement-2 : Acceleration is the rate of change of velocity.

Q.29 Statement-1 : Displacement of a body may be zero when distance travelled by it is not zero.
Statement-2 : The displacement is the longest distance between initial and final position.

RESPONSE GRID					
	23. ⓐⓑⓒⓓ	24. ⓐⓑⓒⓓ	25. ⓐⓑⓒⓓ	26. ⓐⓑⓒⓓ	27. ⓐⓑⓒⓓ
	28. ⓐⓑⓒⓓ	29. ⓐⓑⓒⓓ			

DAILY PRACTICE PROBLEM SHEET 3 - PHYSICS			
Total Questions	29	Total Marks	116
Attempted		Correct	
Incorrect		Net Score	
Cut-off Score	28	Qualifying Score	48
Success Gap = Net Score – Qualifying Score			
Net Score = (Correct × 4) – (Incorrect × 1)			

Space for Rough Work

DPP - Daily Practice Problems

Name :

Date :

Start Time :

End Time :

PHYSICS 04

SYLLABUS : MOTION IN A STRAIGHT LINE 2 (Relative Motion & Motion Under Gravity)

Max. Marks : 112 **Time : 60 min.**

GENERAL INSTRUCTIONS

- The Daily Practice Problem Sheet contains 28 MCQ's. For each question only one option is correct. Darken the correct circle/ bubble in the Response Grid provided on each page.
- You have to evaluate your Response Grids yourself with the help of solution booklet.
- Each correct answer will get you 4 marks and 1 mark shall be deduced for each incorrect answer. No mark will be given/ deducted if no bubble is filled. Keep a timer in front of you and stop immediately at the end of 60 min.
- The sheet follows a particular syllabus. Do not attempt the sheet before you have completed your preparation for that syllabus. Refer syllabus sheet in the starting of the book for the syllabus of all the DPP sheets.
- After completing the sheet check your answers with the solution booklet and complete the Result Grid. Finally spend time to analyse your performance and revise the areas which emerge out as weak in your evaluation.

DIRECTIONS (Q.1-Q.19) : There are 19 multiple choice questions. Each question has 4 choices (a), (b), (c) and (d), out of which **ONLY ONE** choice is correct.

Q.1 A stone is dropped from a minar of height h and it reaches after t seconds on earth. From the same minar if two stones are thrown (one upwards and other downwards) with the same velocity u and they reach the earth surface after t_1 and t_2 seconds respectively, then

(a) $t = t_1 - t_2$ (b) $t = \frac{t_1 + t_2}{2}$

(c) $t = \sqrt{t_1 t_2}$ (d) $t = t_1^2 \, t_2^2$

Q.2 A ball is projected upwards from a height h above the surface of the earth with velocity v. The time at which the ball strikes the ground is

(a) $\frac{v}{g} + \frac{2hg}{\sqrt{2}}$ (b) $\frac{v}{g}\left[1 - \sqrt{1 + \frac{2h}{g}}\right]$

(c) $\frac{v}{g}\left[1 + \sqrt{1 + \frac{2gh}{v^2}}\right]$ (d) $\frac{v}{g}\left[1 + \sqrt{v^2 + \frac{2g}{h}}\right]$

Q.3 A man throws balls with the same speed vertically upwards, one after the other at an interval of 2 seconds. What should be the speed of the throw so that more than two balls are in the sky at any time? (Given g = 9.8 m/s^2)

(a) At least 0.8 m/s

(b) Any speed less than 19.6 m/s

(c) Only with speed 19.6 m/s

(d) More than 19.6 m/s

RESPONSE GRID	1. ⓐⓑⓒⓓ	2. ⓐⓑⓒⓓ	3. ⓐⓑⓒⓓ

Space for Rough Work

Q.4 If a ball is thrown vertically upwards with speed u, the distance covered during the last t second of its ascent is

(a) $\frac{1}{2}gt^2$ (b) $ut - \frac{1}{2}gt^2$ (c) (u – gt)t (d) ut d

Q.5 A ball is thrown vertically upwards. Which of the following graphs represent velocity-time graph of the ball during its flight? (air resistance is neglected)

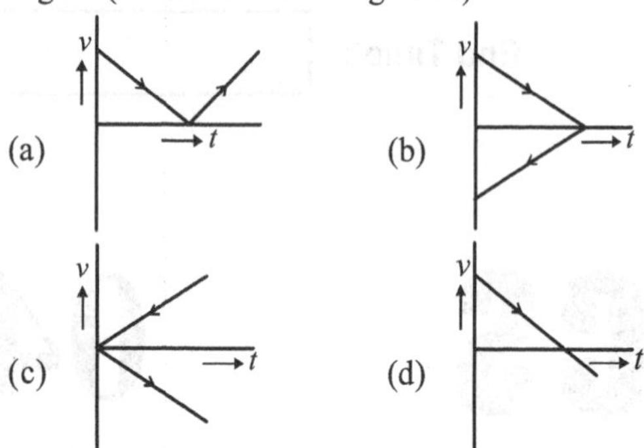

Q.6 A ball is dropped vertically from a height d above the ground. It hits the ground and bounces up vertically to a height d/2. Neglecting subsequent motion and air resistance, its velocity v varies with the height h above the ground is

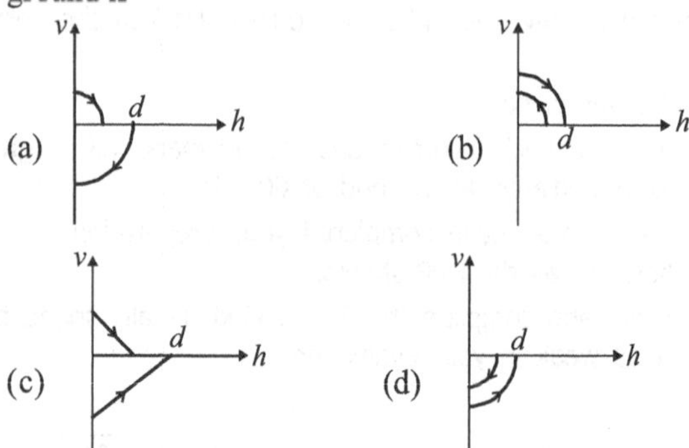

Q.7 P, Q and R are three balloons ascending with velocities U, 4U and 8U respectively. If stones of the same mass be dropped from each, when they are at the same height, then

(a) They reach the ground at the same time
(b) Stone from P reaches the ground first
(c) Stone from R reaches the ground first
(d) Stone from Q reaches the ground first

Q.8 A body is projected up with a speed '*u*' and the time taken by it is *T* to reach the maximum height *H*. Pick out the correct statement

(a) It reaches *H*/2 in *T*/2 sec
(b) It acquires velocity *u*/2 in *T*/2 sec
(c) Its velocity is *u*/2 at *H*/2
(d) Same velocity at 2*T*

Q.9 Time taken by an object falling from rest to cover the height of h_1 and h_2 is respectively t_1 and t_2 then the ratio of t_1 to t_2 is

(a) $h_1 : h_2$ (b) $\sqrt{h_1} : \sqrt{h_2}$ (c) $h_1 : 2h_2$ (d) $2h_1 : h_2$

Q.10 Three different objects of masses m_1, m_2 and m_3 are allowed to fall from rest and from the same point 'O' along three different frictionless paths. The speeds of the three objects, on reaching the ground, will be in the ratio of

(a) $m_1 : m_2 : m_3$ (b) $m_1 : 2m_2 : 3m_3$
(c) 1 : 1 : 1 (d) $\frac{1}{m_1} : \frac{1}{m_2} : \frac{1}{m_3}$

Q.11 From the top of a tower, a particle is thrown vertically downwards with a velocity of 10 m/s. The ratio of the distances, covered by it in the 3rd and 2nd seconds of the motion is (Take g = 10m/s^2)

(a) 5 : 7 (b) 7 : 5 (c) 3 : 6 (d) 6 : 3

Q.12 A body falls from a height h = 200 m. The ratio of distance travelled in each 2 second during t = 0 to t = 6 second of the journey is

(a) 1 : 4 : 9 (b) 1 : 2 : 4
(c) 1 : 3 : 5 (d) 1 : 2 : 3

Q.13 The effective acceleration of a body, when thrown upwards with acceleration a will be :

(a) $\sqrt{a - g^2}$ (b) $\sqrt{a^2 + g^2}$
(c) $(a - g)$ (d) $(a + g)$

Q.14 An aeroplane is moving with a velocity u. It drops a packet from a height h. The time t taken by the packet in reaching the ground will be

(a) $\sqrt{\left(\frac{2g}{h}\right)}$ (b) $\sqrt{\left(\frac{2u}{g}\right)}$ (c) $\sqrt{\left(\frac{h}{2g}\right)}$ (d) $\sqrt{\left(\frac{2h}{g}\right)}$

RESPONSE GRID					
	4. ⓐⓑⓒⓓ	5. ⓐⓑⓒⓓ	6. ⓐⓑⓒⓓ	7. ⓐⓑⓒⓓ	8. ⓐⓑⓒⓓ
	9. ⓐⓑⓒⓓ	10. ⓐⓑⓒⓓ	11. ⓐⓑⓒⓓ	12. ⓐⓑⓒⓓ	13. ⓐⓑⓒⓓ
	14. ⓐⓑⓒⓓ				

Space for Rough Work

Q.15 Two trains, each 50 m long are travelling in opposite direction with velocity 10 m/s and 15 m/s. The time of crossing is

(a) 2s (b) 4s

(c) $2\sqrt{3}s$ (d) $4\sqrt{3}\,s$

Q.16 A train of 150 metre length is going towards north direction at a speed of 10 m/s. A parrot flies at the speed of 5 m/s towards south direction parallel to the railway track. The time taken by the parrot to cross the train is

(a) 12 sec (b) 8 sec

(c) 15 sec (d) 10 sec

Q.17 The distance between two particles is decreasing at the rate of 6 m/sec. If these particles travel with same speeds and in the same direction, then the separation increase at the rate of 4 m/s. The particles have speeds as

(a) 5 m/sec; 1 m/sec (b) 4 m/sec; 1 m/sec

(c) 4 m/sec; 2 m/sec (d) 5 m/sec; 2 m/sec

Q.18 A train is moving towards east and a car is along north, both with same speed. The observed direction of car to the passenger in the train is

(a) East-north direction (b) West-north direction

(c) South-east direction (d) None of these

Q.19 An express train is moving with a velocity v_1. Its driver finds another train is moving on the same track in the same direction with velocity v_2. To escape collision, driver applies a retardation a on the train, the minimum time of escaping collision will be

(a) $t = \frac{v_1 - v_2}{a}$ (b) $t_1 = \frac{v_1^2 - v_2^2}{2}$

(c) Both (a) and (b) (d) None of these

DIRECTIONS (Q.20-Q.22) : In the following questions, more than one of the answers given are correct. Select the correct answers and mark it according to the following codes:

Codes :

(a) 1, 2 and 3 are correct **(b)** 1 and 2 are correct

(c) 2 and 4 are correct **(d)** 1 and 3 are correct

Q.20 Two particles move simultaneously from two points A and B, 300m apart. The particle at A, starts towards B with a velocity of 25 m/s and that at B, moves normal to the former with a velocity of 20 m/s.

(1) The relative velocity of the particle at A, w.r.t. that at B is 32.02 m/s

(2) The relative velocity of the particle at A, w.r.t. that at B is 12.04 m/s

(3) They are closest to each other after 7.32 sec.

(4) They are closest to each other after 4.25 sec.

Q.21 A plane is to fly due north. The speed of the plane relative to the air is 200 km/h, and the wind is blowing from west to east at 90 km/h.

(1) The plane should head in a direction of $\sin^{-1}(0.45)$

(2) The plane should head in a direction of $\sin^{-1}(0.60)$

(3) The relative velocity of plane w.r.t. ground is 179 km/h

(4) The relative velocity of plane w.r.t. ground is 149 km/h

Q.22 From the top of a multi-storeyed building 40m tall, a boy projects a stone vertically upwards with an initial velocity of 10 ms^{-1} such that it eventually falls to the ground.

(1) After 4 s the stone will strike the ground

(2) After 2 s the stone will pass through the point from where it was projected

(3) Its velocity when it strikes the ground is 30 m/s

(4) Its velocity when it strikes the ground is 40 m/s

DIRECTIONS (Q.23-Q.25) : Read the passage given below and answer the questions that follows :

When an airplane flies, its total velocity with respect to the ground is $v_{total} = v_{plane} + v_{wind}$, where v_{plane} denotes the plane's velocity through motionless air, and v_{wind} denotes the wind's velocity. Crucially, all the quantities in this equation are vectors. The magnitude of a velocity vector is often called the "speed."

Consider an airplane whose speed through motionless air is 100 meters per second (m/s). To reach its destination, the plane must fly east.

The "heading" of a plane is in the direction in which the nose of the plane points. So, it is the direction in which the engines propel the plane.

RESPONSE GRID					
	15. ⓐⓑⓒⓓ	16. ⓐⓑⓒⓓ	17. ⓐⓑⓒⓓ	18. ⓐⓑⓒⓓ	19. ⓐⓑⓒⓓ
	20. ⓐⓑⓒⓓ	21. ⓐⓑⓒⓓ	22. ⓐⓑⓒⓓ		

Space for Rough Work

Q.23 If the plane has an eastward heading, and a 20 m/s wind blows towards the southwest, then the plane's speed is –

(a) 80 m/s
(b) more than 80 m/s but less than 100 m/s
(c) 100 m/s
(d) more than 100 m/s

Q.24 The pilot maintains an eastward heading while a 20 m/s wind blows northward. The plane's velocity is deflected from due east by what angle?

(a) $\sin^{-1}\frac{20}{100}$ (b) $\cos^{-1}\frac{20}{100}$
(c) $\tan^{-1}\frac{20}{100}$ (d) none

Q.25 Let ϕ denote the answer of above question. The plane has what speed with respect to the ground ?

(a) $(100\text{ m/s})\sin\phi$ (b) $(100\text{ m/s})\cos\phi$
(c) $\frac{100\text{ m/s}}{\sin\phi}$ (d) $\frac{100\text{ m/s}}{\cos\phi}$

DIRECTIONS (Qs. 26-Q.28) : Each of these questions contains two statements: Statement-1 (Assertion) and Statement-2 (Reason). Each of these questions has four alternative choices, only one of which is the correct answer. You have to select the correct choice.

(a) Statement-1 is True, Statement-2 is True; Statement-2 is a correct explanation for Statement-1.
(b) Statement-1 is True, Statement-2 is True; Statement-2 is NOT a correct explanation for Statement-1.
(c) Statement -1 is False, Statement-2 is True.
(d) Statement -1 is True, Statement-2 is False.

Q.26 Statement-1 : The magnitude of velocity of two boats relative to river is same. Both boats start simultaneously from same point on the bank may reach opposite bank simultaneously moving along different paths.

Statement-2 : For boats to cross the river in same time. The component of their velocity relative to river in direction normal to flow should be same.

Q.27 Statement-1 : The acceleration of a body of mass 2 kg thrown vertically upwards is always constant.

Statement-2 : A body of all mass group travels under constant acceleration when only gravity acts on it.

Q.28 Statement-1 : The velocity of a body A relative to the body B is the sum of the velocities of bodies A and B if both travel in opposite direction on a straight line.

Statement-2 : The velocity of a body A relative to the body B is the difference of the velocities of bodies A and B if both travel in opposite direction on a straight line.

RESPONSE GRID	
	23. ⓐⓑⓒⓓ 24. ⓐⓑⓒⓓ 25. ⓐⓑⓒⓓ 26. ⓐⓑⓒⓓ 27. ⓐⓑⓒⓓ 28. ⓐⓑⓒⓓ

DAILY PRACTICE PROBLEM SHEET 4 - PHYSICS			
Total Questions	28	Total Marks	112
Attempted		Correct	
Incorrect		Net Score	
Cut-off Score	28	Qualifying Score	44
Success Gap = Net Score – Qualifying Score			
Net Score = (Correct × 4) – (Incorrect × 1)			

Space for Rough Work

DPP - Daily Practice Problems

Name :

Date :

Start Time :

End Time :

PHYSICS 05

SYLLABUS : Vectors

Max. Marks : 116 **Time : 60 min.**

GENERAL INSTRUCTIONS

- The Daily Practice Problem Sheet contains 29 MCQ's. For each question only one option is correct. Darken the correct circle/ bubble in the Response Grid provided on each page.
- You have to evaluate your Response Grids yourself with the help of solution booklet.
- Each correct answer will get you 4 marks and 1 mark shall be deduced for each incorrect answer. No mark will be given/ deducted if no bubble is filled. Keep a timer in front of you and stop immediately at the end of 60 min.
- The sheet follows a particular syllabus. Do not attempt the sheet before you have completed your preparation for that syllabus. Refer syllabus sheet in the starting of the book for the syllabus of all the DPP sheets.
- After completing the sheet check your answers with the solution booklet and complete the Result Grid. Finally spend time to analyse your performance and revise the areas which emerge out as weak in your evaluation.

DIRECTIONS (Q.1-Q.21) : There are 21 multiple choice questions. Each question has 4 choices (a), (b), (c) and (d), out of which **ONLY ONE** choice is correct.

Q.1 The length of second's hand in watch is 1 cm. The change in velocity of its tip in 15 seconds is

(a) zero (b) $\frac{\pi}{30\sqrt{2}}$ cm/sec

(c) $\frac{\pi}{30}$ cm/sec (d) $\frac{\pi\sqrt{2}}{30}$ cm/sec

Q.2 A particle moves towards east with velocity 5 m/s. After 10 seconds its direction changes towards north with same velocity. The average acceleration of the particle is

(a) zero (b) $\frac{1}{\sqrt{2}}$ m/s^2 N-W

(c) $\frac{1}{\sqrt{2}}$ m/s^2 N-E (d) $\frac{1}{\sqrt{2}}$ m/s^2 S-W

Q.3 A force $\vec{F} = -K(y\hat{i} + x\hat{j})$ (where K is a positive constant) acts on a particle moving in the x-y plane. Starting from the origin, the particle is taken along the positive x-axis to the point $(a, 0)$ and then parallel to the y-axis to the point (a, a). The total work done by the forces $\vec{F}$ on the particle is

(a) $-2Ka^2$ (b) $2Ka^2$

(c) $-Ka^2$ (d) Ka^2

RESPONSE GRID	1. ⓐⓑⓒⓓ	2. ⓐⓑⓒⓓ	3. ⓐⓑⓒⓓ

Space for Rough Work

Q.4 A metal sphere is hung by a string fixed to a wall. The sphere is pushed away from the wall by a stick. The forces acting on the sphere are shown in the second diagram. Which of the following statements is wrong?

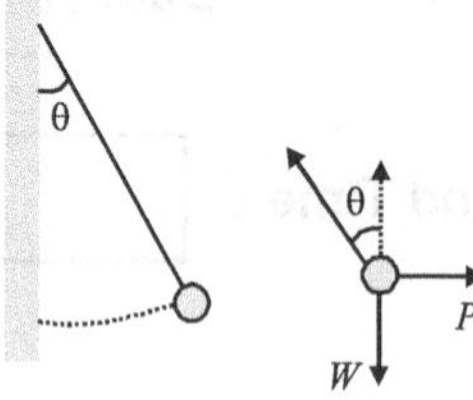

(a) $P = W \tan \theta$ (b) $\vec{T}+\vec{P}+\vec{W}=0$

(c) $T^2 = P^2 + W^2$ (d) $T = P + W$

Q.5 The speed of a boat is 5km/h in still water. It crosses a river of width 1 km along the shortest possible path in 15 minutes. The velocity of the river water is

(a) 1 km/h (b) 3 km/h

(c) 4 km/h (d) 5 km/h

Q.6 A man crosses a 320 m wide river perpendicular to the current in 4 minutes. If in still water he can swim with a speed 5/3 times that of the current, then the speed of the current in m/min is

(a) 30 (b) 40

(c) 50 (d) 60

Q.7 *P*, *Q* and *R* are three coplanar forces acting at a point and are in equilibrium. Given $P = 1.9318$ kg *wt*, $\sin \theta_1 = 0.9659$, the value of *R* is (in kg *wt*)

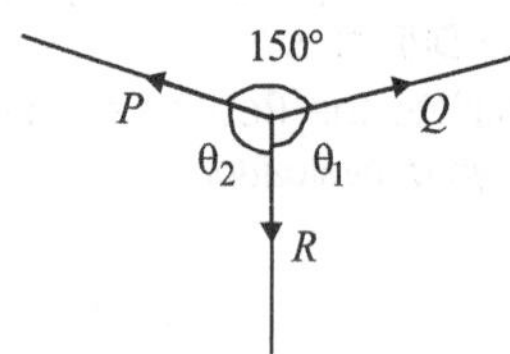

(a) 0.9659 (b) 2

(c) 1 (d) $\frac{1}{2}$

Q.8 As shown in figure the tension in the horizontal cord is 30 N. The weight *W* and tension in the string *OA* in newton are

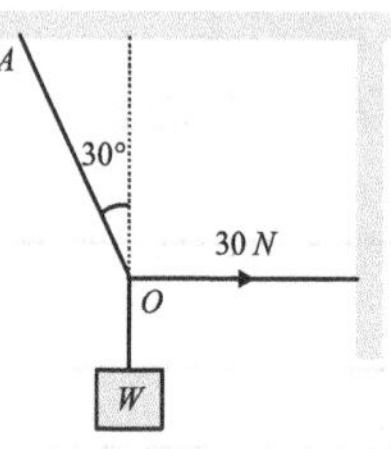

(a) $30\sqrt{3}, 30$ (b) $30\sqrt{3}, 60$

(c) $60\sqrt{3}, 30$ (d) None of these

Q.9 A boat is moving with a velocity $3\hat{i}+4\hat{j}$ with respect to ground. The water in the river is moving with a velocity $-3\hat{i}-4\hat{j}$ with respect to ground. The relative velocity of the boat with respect to water is

(a) $8\hat{j}$ (b) $-6\hat{i}-8\hat{j}$ (c) $6\hat{i}+8\hat{j}$ (d) $5\sqrt{2}\hat{i}$

Q.10 A person aiming to reach the exactly opposite point on the bank of a stream is swimming with a speed of 0.5 m/s at an angle of 120° with the direction of flow of water. The speed of water in the stream is

(a) 1 m/s (b) 0.5 m/s (c) 0.25 m/s (d) 0.433 m/s

Q.11 A man can swim with velocity *v* relative to water. He has to cross a river of width *d* flowing with a velocity *u* ($u > v$). The distance through which he is carried down stream by the river is *x*. Which of the following statements is correct?

(a) If he crosses the river in minimum time $x = \frac{du}{v}$

(b) *x* cannot be less than $\frac{du}{v}$

(c) For *x* to be minimum he has to swim in a direction making an angle of $\frac{\pi}{2} - \sin^{-1}\left(\frac{v}{u}\right)$ with the direction of the flow of water.

(d) *x* will be maximum if he swims in a direction making an angle of $\frac{\pi}{2} + \sin^{-1}\left(\frac{v}{u}\right)$ with direction of the flow of water.

RESPONSE GRID					
	4. ⓐⓑⓒⓓ	5. ⓐⓑⓒⓓ	6. ⓐⓑⓒⓓ	7. ⓐⓑⓒⓓ	8. ⓐⓑⓒⓓ
	9. ⓐⓑⓒⓓ	10. ⓐⓑⓒⓓ	11. ⓐⓑⓒⓓ		

Space for Rough Work

Q.12 A 120 m long train is moving towards west with a speed of 10 m/s. A bird flying towards east with a speed of 5 m/s crosses the train. The time taken by the bird to cross the train will be
(a) 16 sec (b) 12 sec (c) 10 sec (d) 8 sec

Q.13 What is the value of linear velocity, if $\vec{\omega} = 3\hat{i} - 4\hat{j} + \hat{k}$ and $\vec{r} = 5\hat{i} - 6\hat{j} + 6\hat{k}$
(a) $6\hat{i} - 2\hat{j} + 3\hat{k}$ (b) $6\hat{i} - 2\hat{j} + 8\hat{k}$
(c) $4\hat{i} - 13\hat{j} + 6\hat{k}$ (d) $-18\hat{i} - 13\hat{j} + 2\hat{k}$

Q.14 If $|\vec{A} \times \vec{B}| = \sqrt{3}\,\vec{A}.\vec{B}$, then the value of $|\vec{A} + \vec{B}|$ is
(a) $\left(A^2 + B^2 + \frac{AB}{\sqrt{3}}\right)^{1/2}$ (b) $A + B$
(c) $(A^2 + B^2 + \sqrt{3}AB)^{1/2}$ (d) $(A^2 + B^2 + AB)^{1/2}$

Q.15 Find the torque of a force $\vec{F} = -3\hat{i} + \hat{j} + 5\hat{k}$ acting at a point $\vec{r} = 7\hat{i} + 3\hat{j} + \hat{k}$
(a) $14\hat{i} - 38\hat{j} + 16\hat{k}$ (b) $4\hat{i} + 4\hat{j} + 6\hat{k}$
(c) $21\hat{i} + 4\hat{j} + 4\hat{k}$ (d) $-14\hat{i} + 34\hat{j} - 16\hat{k}$

Q.16 If $|\vec{A} \times \vec{B}| = |\vec{A}.\vec{B}|$, then angle between $\vec{A}$ and $\vec{B}$ will be
(a) 30° (b) 45° (c) 60° (d) 90°

Q.17 The vector $\vec{P} = a\hat{i} + a\hat{j} + 3\hat{k}$ and $\vec{Q} = a\hat{i} - 2\hat{j} - \hat{k}$ are perpendicular to each other. The positive value of a is
(a) 3 (b) 4
(c) 9 (d) 13

Q.18 A particle moves from position $3\hat{i} + 2\hat{j} - 6\hat{k}$ to $14\hat{i} + 13\hat{j} + 9\hat{k}$ due to a uniform force of $(4\hat{i} + \hat{j} + 3\hat{k})N$. If the displacement in metres then work done will be
(a) 100 J (b) 200 J (c) 300 J d) 250 J

Q.19 The three vectors $\vec{A} = 3\hat{i} - 2\hat{j} + \hat{k}$, $\vec{B} = \hat{i} - 3\hat{j} + 5\hat{k}$ and $\vec{C} = 2\hat{i} + \hat{j} - 4\hat{k}$ form
(a) an equilateral triangle (b) isosceles triangle
(c) a right angled triangle (d) no triangle

Q.20 Two forces $\vec{F_1} = 5\hat{i} + 10\hat{j} - 20\hat{k}$ and $\vec{F_2} = 10\hat{i} - 5\hat{j} - 15\hat{k}$ act on a single point. The angle between $\vec{F_1}$ and $\vec{F_2}$ is nearly
(a) 30° (b) 45°
(c) 60° (d) 90°

Q.21 With respect to a rectangular cartesian coordinate system, three vectors are expressed as
$\vec{a} = 4\hat{i} - \hat{j}$, $\vec{b} = -3\hat{i} + 2\hat{j}$, and $\vec{c} = -\hat{k}$
where $\hat{i}, \hat{j}, \hat{k}$ are unit vectors, along the *X*, *Y* and *Z*-axis respectively. The unit vectors $\hat{r}$ along the direction of sum of these vector is
(a) $\hat{r} = \frac{1}{\sqrt{3}}(\hat{i} + \hat{j} - \hat{k})$ (b) $\hat{r} = \frac{1}{\sqrt{2}}(\hat{i} + \hat{j} - \hat{k})$
(c) $\hat{r} = \frac{1}{3}(\hat{i} - \hat{j} + \hat{k})$ (d) $\hat{r} = \frac{1}{\sqrt{2}}(\hat{i} + \hat{j} + \hat{k})$

DIRECTIONS (Q.22-Q.24) : In the following questions, more than one of the answers given are correct. Select the correct answers and mark it according to the following codes:

Codes :
(a) 1, 2 and 3 are correct **(b)** 1 and 2 are correct
(c) 2 and 4 are correct **(d)** 1 and 3 are correct

Q.22 A boy walks uniformally along the sides of a rectangular park of size 400 m × 300 m, starting from one corner to the other corner diagonally opposite. Which of the following statements is correct?
(1) He has travelled a distance of 700 m
(2) His displacement is 500 m
(3) His velocity is not uniform throughout the walk
(4) His displacement is 700 m

Q.23 The three vectors $\vec{A} = 3\hat{i} - 2\hat{j} - \hat{k}$, $\vec{B} = \hat{i} - 3\hat{j} + 5\hat{k}$ and $\vec{C} = 2\hat{i} - \hat{j} - 4\hat{k}$ does not form
(1) an equilateral triangle (2) isosceles triangle
(3) a right angled triangle (4) no triangle

RESPONSE GRID					
	12. ⓐⓑⓒⓓ	13. ⓐⓑⓒⓓ	14. ⓐⓑⓒⓓ	15. ⓐⓑⓒⓓ	16. ⓐⓑⓒⓓ
	17. ⓐⓑⓒⓓ	18. ⓐⓑⓒⓓ	19. ⓐⓑⓒⓓ	20. ⓐⓑⓒⓓ	21. ⓐⓑⓒⓓ
	22. ⓐⓑⓒⓓ	23. ⓐⓑⓒⓓ			

Space for Rough Work

Q.24 If for two vectors $\vec{A}$ and $\vec{B}$, $\vec{A}\times\vec{B}=0$, which of the following is not correct?
(1) They are perpendicular to each other
(2) They act at an angle of 60°
(3) They act at an angle of 30°
(4) They are parallel to each other

DIRECTIONS (Q.25-Q.27) : Read the passage given below and answer the questions that follows :

$\vec{A}=2\hat{i}+\hat{j}+\hat{k}$ and $\vec{B}=\hat{i}+\hat{j}+\hat{k}$ are two vectors.

Q.25 The unit vector perpendicular to $\vec{A}$ is
(a) $\frac{-\hat{j}+\hat{k}}{\sqrt{2}}$ (b) $\frac{-\hat{j}-\hat{k}}{\sqrt{2}}$ (c) $\frac{\hat{i}+\hat{k}}{2}$ (d) $\frac{\hat{i}-\hat{k}}{2}$

Q.26 The unit vector parallel to $\vec{A}$ is
(a) $\frac{2\hat{i}-\hat{j}+3\hat{k}}{\sqrt{2}}$ (b) $\frac{2\hat{i}+\hat{j}+\hat{k}}{\sqrt{6}}$
(c) $\frac{2\hat{i}-\hat{j}-\hat{k}}{\sqrt{5}}$ (d) $\frac{2\hat{i}+\hat{j}-2\hat{k}}{\sqrt{6}}$

Q.27 The unit vector perpendicular to $\vec{B}$ is
(a) $\frac{-\hat{j}-\hat{k}}{\sqrt{3}}$ (b) $\frac{-\hat{j}+\hat{k}}{\sqrt{2}}$ (c) $\frac{\hat{i}-\hat{k}}{3}$ (d) $\frac{\hat{i}+\hat{k}}{2}$

DIRECTIONS (Q.28-Q.29) : Each of these questions contains two statements: Statement-1 (Assertion) and Statement-2 (Reason). Each of these questions has four alternative choices, only one of which is the correct answer. You have to select the correct choice.

(a) Statement-1 is True, Statement-2 is True; Statement-2 is a correct explanation for Statement-1.
(b) Statement-1 is True, Statement-2 is True; Statement-2 is NOT a correct explanation for Statement-1.
(c) Statement -1 is False, Statement-2 is True.
(d) Statement -1 is True, Statement-2 is False.

Q.28 **Statement-1:** If $|\vec{A}+\vec{B}|=|\vec{A}-\vec{B}|$, then angle between $\vec{A}$ and $\vec{B}$ is 90°

Statement-2 : $\vec{A}+\vec{B}=\vec{B}+\vec{A}$

Q.29 **Statement-1 :** The sum of two vectors can be zero.

Statement-2 : Two vectors cancel each other, when they are equal and opposite.

RESPONSE GRID	24. ⓐⓑⓒⓓ	25. ⓐⓑⓒⓓ	26. ⓐⓑⓒⓓ	27. ⓐⓑⓒⓓ	28. ⓐⓑⓒⓓ
	29. ⓐⓑⓒⓓ				

DAILY PRACTICE PROBLEM SHEET 5 - PHYSICS			
Total Questions	29	Total Marks	116
Attempted		Correct	
Incorrect		Net Score	
Cut-off Score	30	Qualifying Score	44
Success Gap = Net Score – Qualifying Score			
Net Score = (Correct × 4) – (Incorrect × 1)			

Space for Rough Work

DPP - Daily Practice Problems

Name :

Date :

Start Time :

End Time :

PHYSICS 06

SYLLABUS : MOTION IN A PLANE-1 (Projectile Motion)

Max. Marks : 112 **Time : 60 min.**

GENERAL INSTRUCTIONS

- The Daily Practice Problem Sheet contains 28 MCQ's. For each question only one option is correct. Darken the correct circle/ bubble in the Response Grid provided on each page.
- You have to evaluate your Response Grids yourself with the help of solution booklet.
- Each correct answer will get you 4 marks and 1 mark shall be deduced for each incorrect answer. No mark will be given/ deducted if no bubble is filled. Keep a timer in front of you and stop immediately at the end of 60 min.
- The sheet follows a particular syllabus. Do not attempt the sheet before you have completed your preparation for that syllabus. Refer syllabus sheet in the starting of the book for the syllabus of all the DPP sheets.
- After completing the sheet check your answers with the solution booklet and complete the Result Grid. Finally spend time to analyse your performance and revise the areas which emerge out as weak in your evaluation.

DIRECTIONS (Q.1-Q.20) : There are 20 multiple choice questions. Each question has 4 choices (a), (b), (c) and (d), out of which **ONLY ONE** choice is correct.

Q.1 The path followed by a body projected along y axis is given by $y = \sqrt{3}\,x - (1/2)\,x^2$. If $g = 10\ m/s^2$, then the initial velocity of projectile will be – (x and y are in m)

(a) $3\sqrt{10}$ m/s (b) $2\sqrt{10}$ m/s

(c) $10\sqrt{3}$ m/s (d) $10\sqrt{2}$ m/s

Q.2 When the angle of elevation of a gun are 60° and 30° respectively, the height it shoots are h_1 and h_2 respectively, h_1/h_2 equal to –

(a) 3/1 (b) 1/3 (c) 1/2 (d) 2/1

Q.3 If t_1 be the time taken by a body to clear the top of a building and t_2 be the time spent in air, then $t_2 : t_1$ will be –

(a) 1 : 2 (b) 2 : 1 (c) 1 : 1 (d) 1 : 4

Q.4 The co-ordinates of a moving particle at any time t are given by $x = ct^2$ and $y = bt^2$. The speed of the particle is

(a) $2t(c + b)$ (b) $2t\sqrt{c^2 - b^2}$

(c) $t\sqrt{c^2 + b^2}$ (d) $2t\sqrt{c^2 + b^2}$

Q.5 The height y and the distance x along the horizontal at plane of the projectile on a certain planet (with no surrounding atmosphere) are given by $y = (8t - 5t^2)$ metre and $x = 6t$ metre where t is in second. The velocity with which the projectile is projected is

(a) 8 m/s (b) 6 m/s

(c) 10 m/s (d) Data is insufficient

RESPONSE GRID	1. ⓐⓑⓒⓓ	2. ⓐⓑⓒⓓ	3. ⓐⓑⓒⓓ	4. ⓐⓑⓒⓓ	5. ⓐⓑⓒⓓ

Space for Rough Work

Q.6 A body is thrown at an angle 30° to the horizontal with the velocity of 30 m/s. After 1 sec, its velocity will be (in m/s) $(g = 10\ m/s^2)$

(a) $10\sqrt{7}$ (b) $700\sqrt{10}$ (c) $100\sqrt{7}$ (d) $\sqrt{10}$

Q.7 A particle is moving in a plane with a velocity given by, $\vec{u} = u_0\,\hat{i} + (\omega a \cos \omega t)\,\hat{j}$, where $\hat{i}$ and $\hat{j}$ are unit vectors along x and y-axes respectively. If the particle is at the origin at t = 0, then its distance from the origin at time $t = 3\pi/2\omega$ will be

(a) $\sqrt{\left[\left(\frac{3\pi u_0}{2\omega}\right)^2 + a^2\right]}$ (b) $\sqrt{\left[\left(\frac{3\pi u_0}{2\omega}\right) + a^2\right]}$

(c) $\sqrt{\left[\left(\frac{3\pi u_0}{2\omega}\right)^2 + a\right]}$ (d) $\sqrt{\left[\left(\frac{4\pi u_0}{2\omega}\right)^2 + a^2\right]}$

Q.8 A ball thrown by one player reaches the other in 2 sec. The maximum height attained by the ball above the point of projection will be about-

(a) 2.5 m (b) 5 m (c) 7.5 m (d) 10 m

Q.9 Rishabh and Bappy are playing with two different balls of masses m and 2m respectively. If Rishabh throws his ball vertically up and Bappy at an angle θ, both of them stay in our view for the same period. The height attained by the two balls are in the ratio of

(a) 2 : 1 (b) 1 : 1 (c) 1 : cos θ (d) 1 : sec θ

Q.10 A projectile is thrown at an angle θ and $(90^0 - \theta)$ from the same point with same velocity 98 m/s. The heights attained by them, if the difference of heights is 50 m will be (in m)

(a) 270, 220 (b) 300, 250
(c) 250, 200 (d) 200, 150

Q.11 A particle is projected with a velocity u so that its horizontal range is twice the maximum height attained. The horizontal range is

(a) u^2/g (b) $2u^2/3g$ (c) $4u^2/5g$ (d) $u^2/2g$

Q.12 Mr C.P. Nawani kicked off a football with an initial speed 19.6 m/s at a projection angle 45°. A receiver on the goal line 67.4 m away in the direction of the kick starts running to meet the ball at that instant. What must be his speed so that he could catch the ball before hitting the ground ?

(a) 2.82 m/s (b) $2/\sqrt{2}$ m/s
(c) 39.2 m/s (d) 10 m/s

Q.13 A ball is thrown from ground level so as to just clear a wall 4 metres high at a distance of 4 metres and falls at a distance of 14 metres from the wall. The magnitude of velocity of the ball will be

(a) $\sqrt{182}$ m/s (b) $\sqrt{181}$ m/s
(c) $\sqrt{185}$ m/s (d) $\sqrt{186}$ m/s

Q.14 A ball is projeced from O with an initial velocity 700 cm/s in a direction 37° above the horizontal. A ball B, 500 cm away from O on the line of the initial velocity of A, is released from rest at the instant A is projected. The height through which B falls, before it is hit by A and the direction of the velocity A at the time of impact will respectively be [given $g = 10\ m/s^2$, sin 37° = 0.6 and cos 37° = 8.0]

(a) 250 cm, 28° 42' (b) 255 cm, 27° 43'
(c) 245 cm, 20° 44' (d) 300 cm, 27° 43'

Q.15 A ball is thrown horizontally from a height of 20 m. It hits the ground with a velocity three times its initial velocity. The initial velocity of ball is

(a) 2 m/s (b) 3 m/s (c) 5 m/s (d) 7 m/s

Q.16 A projectile thrown from a height of 10 m with velocity of $\sqrt{2}$ m/s, the projectile will fall, from the foot of projection, at distance-$(g = 10\ m/s^2)$

(a) 1 m (b) 2 m (c) 3m (d) $\sqrt{2}$ m

Q.17 Savita throws a ball horizontally with a velocity of 8 m/s from the top of her building. The ball strikes to her brother Sudhir playing at 12 m away from the building. What is the height of the building ?

(a) 11m (b) 10 m (c) 8 m (d) 7 m

Q.18 A body is projected downwards at an angle of 30° to the horizontal with a velocity of 9.8 m/s from the top of a tower 29.4 m high. How long will it take before striking the ground?

(a) 1s (b) 2s (c) 3s (d) 4s

RESPONSE GRID					
	6. ⓐⓑⓒⓓ	7. ⓐⓑⓒⓓ	8. ⓐⓑⓒⓓ	9. ⓐⓑⓒⓓ	10. ⓐⓑⓒⓓ
	11. ⓐⓑⓒⓓ	12. ⓐⓑⓒⓓ	13. ⓐⓑⓒⓓ	14. ⓐⓑⓒⓓ	15. ⓐⓑⓒⓓ
	16. ⓐⓑⓒⓓ	17. ⓐⓑⓒⓓ	18. ⓐⓑⓒⓓ		

Space for Rough Work

Q.19 A ball is thrown from the top of a tower with an initial velocity of 10 m/s at an angle of 30° above the horizontal. It hits the ground at a distance of 17.3 m from the base of the tower. The height of the tower ($g = 10\ m/s^2$) will be

(a) 10m (b) 12m (c) 110m (d) 100m

Q.20 A ball 'A' is projected from origin with an initial velocity $v_0 = 700$ cm/sec in a direction 37° above the horizontal as shown in fig. Another ball 'B' 300 cm from origin on a line 37° above the horizontal is released from rest at the instant A starts. How far will B have fallen when it is hit by A ?

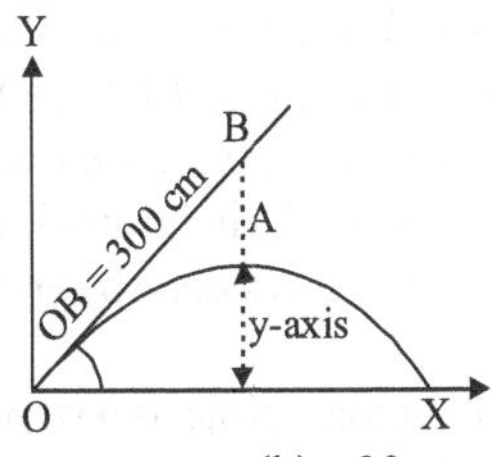

(a) 9 cm (b) 90 cm
(c) 0.9 cm (d) 900 cm

DIRECTIONS (Q.21-Q.23) : In the following questions, more than one of the answers given are correct. Select the correct answers and mark it according to the following codes:

Codes :

(a) 1, 2 and 3 are correct **(b)** 1 and 2 are correct
(c) 2 and 4 are correct **(d)** 1 and 3 are correct

Q.21 Choose the correct options

(1) A ball is dropped from the window of a moving train on horizontal rails, the path followed by the ball as observed by the observer on the ground is parabolic path.

(2) If T be the total time of flight of a current of water and H be the maximum height attained by it from the point of projection, then H/T will be (1/4) u sinθ
(u = projection velocity and θ = projection angle)

(3) A hunter aims his gun and fires a bullet directly at a monkey on a tree. At the instant bullet leaves the gun, monkey drops, the bullet misses to hit the monkey.

(4) If a baseball player can throw a ball at maximum distance = d over a ground, the maximum vertical height to which he can throw it, will be d
(Ball have same initial speed in each case)

Q.22 A ball projected with speed 'u' at an angle of projection 15° has range R. The other angle of projection at which the range will not be same with same initial speed 'u' is

(1) 45° (2) 35°
(3) 90° (4) 75°

Q.23 A projectile can have the same range R for two angles of projections. If t_1 and t_2 be the times of flight in two cases, then choose the incorrect relations –

(1) $t_1 t_2 \propto 1/R^2$ (2) $t_1 t_2 \propto R^2$
(3) $t_1 t_2 \propto 1/R$ (4) $t_1 t_2 \propto R$

DIRECTIONS (Q.24-Q.26) : Read the passage given below and answer the questions that follows :

Velocity at a general point P(x, y) for a horizontal projectile motion is given by

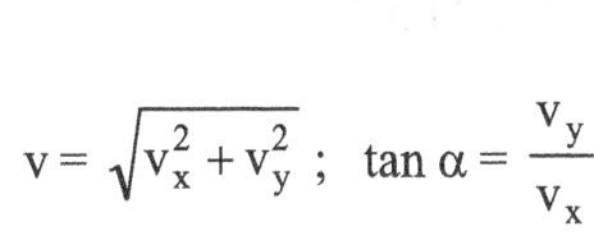

$$v = \sqrt{v_x^2 + v_y^2}\ ;\ \tan\alpha = \frac{v_y}{v_x}$$

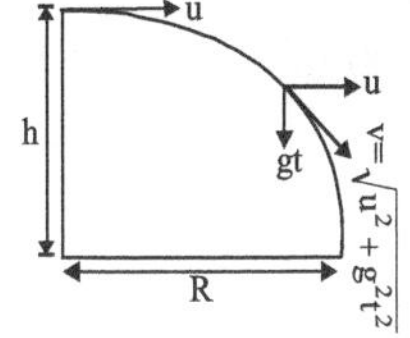

α is angle made by v with horizontal in clockwise direction

Trajectory equation for a horizontal projectile motion is given by $x = v_x t = ut$

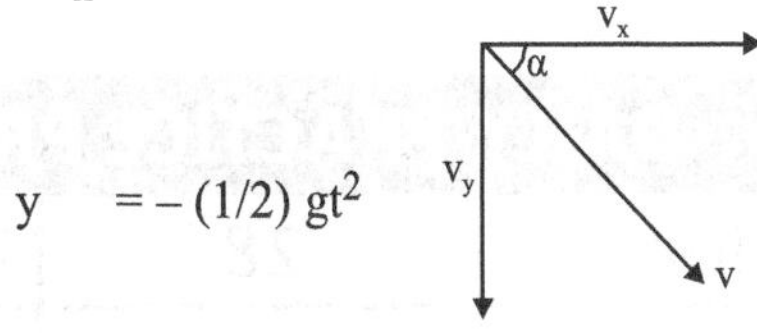

$$y = -(1/2)\,gt^2$$

eliminating t, we get $y = -(1/2)\dfrac{gx^2}{u^2}$

RESPONSE GRID	19. ⓐⓑⓒⓓ	20. ⓐⓑⓒⓓ	21. ⓐⓑⓒⓓ	22. ⓐⓑⓒⓓ	23. ⓐⓑⓒⓓ

Space for Rough Work

Q.24 A ball rolls off top of a stair way with a horizontal velocity u m/s. If the steps are h m high and b meters wide, the ball will just hit the edge of nth step if n equals to

(a) $\frac{hu^2}{gb^2}$ (b) $\frac{u^2g}{gb^2}$

(c) $\frac{2hu^2}{gb^2}$ (d) $\frac{2u^2g}{hb^2}$

Q.25 An aeroplane is in a level flying at an speed of 144 km/hr at an altitude of 1000 m. How far horizontally from a given target should a bomb be released from it to hit the target ?

(a) 571.43 m (b) 671.43 m

(c) 471.34 m (d) 371.34 m

Q.26 An aeroplane is flying horizontally with a velocity of 720 km/h at an altitude of 490 m. When it is just vertically above the target a bomb is dropped from it. How far horizontally it missed the target?

(a) 1000 m (b) 2000 m

(c) 100 m (d) 200 m

DIRECTIONS (Q.27-Q.28) : Each of these questions contains two statements: Statement-1 (Assertion) and Statement-2 (Reason). Each of these questions has four alternative choices, only one of which is the correct answer. You have to select the correct choice.

(a) Statement-1 is True, Statement-2 is True; Statement-2 is a correct explanation for Statement-1.
(b) Statement-1 is True, Statement-2 is True; Statement-2 is NOT a correct explanation for Statement-1.
(c) Statement -1 is False, Statement-2 is True.
(d) Statement -1 is True, Statement-2 is False.

Q.27 Statement -1 : Two projectiles are launched from the top of a cliff with same initial speed with different angles of projection. They reach the ground with the same speed.
Statement -2 : The work done by gravity is same in both the cases.

Q.28 Statement-1 : A man projects a stone with speed *u* at some angle. He again projects a stone with same speed such that time of flight now is different. The horizontal ranges in both the cases may be same. (Neglect air friction)
Statement-2 : The horizontal range is same for two projectiles projected with same speed if one is projected at an angle θ with the horizontal and other is projected at an angle (90° – θ) with the horizontal. (Neglect air friction)

RESPONSE GRID	24. ⓐⓑⓒⓓ	25. ⓐⓑⓒⓓ	26. ⓐⓑⓒⓓ	27. ⓐⓑⓒⓓ	28. ⓐⓑⓒⓓ

DAILY PRACTICE PROBLEM SHEET 6 - PHYSICS			
Total Questions	28	Total Marks	112
Attempted		Correct	
Incorrect		Net Score	
Cut-off Score	28	Qualifying Score	42
Success Gap = Net Score – Qualifying Score			
Net Score = (Correct × 4) – (Incorrect × 1)			

Space for Rough Work

DPP - Daily Practice Problems

Name :

Date :

Start Time :

End Time :

PHYSICS 07

SYLLABUS : MOTION IN A PLANE-2 (Horizontal Circular Motion)

Max. Marks : 120

Time : 60 min.

GENERAL INSTRUCTIONS

- The Daily Practice Problem Sheet contains 30 MCQ's. For each question only one option is correct. Darken the correct circle/ bubble in the Response Grid provided on each page.
- You have to evaluate your Response Grids yourself with the help of solution booklet.
- Each correct answer will get you 4 marks and 1 mark shall be deduced for each incorrect answer. No mark will be given/ deducted if no bubble is filled. Keep a timer in front of you and stop immediately at the end of 60 min.
- The sheet follows a particular syllabus. Do not attempt the sheet before you have completed your preparation for that syllabus. Refer syllabus sheet in the starting of the book for the syllabus of all the DPP sheets.
- After completing the sheet check your answers with the solution booklet and complete the Result Grid. Finally spend time to analyse your performance and revise the areas which emerge out as weak in your evaluation.

DIRECTIONS (Q.1-Q.21) : There are 21 multiple choice questions. Each question has 4 choices (a), (b), (c) and (d), out of which **ONLY ONE** choice is correct.

Q.1 A particle completes 1.5 revolutions in a circular path of radius 2 cm. The angular displacement of the particle will be – (in radian)

(a) 6π (b) 3π (c) 2π (d) π

Q.2 A particle revolving in a circular path completes first one third of circumference in 2 sec, while next one third in 1 sec. The average angular velocity of particle will be – (in rad/sec)

(a) $2\pi/3$ (b) $\pi/3$ (c) $4\pi/3$ (d) $5\pi/3$

Q.3 The ratio of angular speeds of minute hand and hour hand of a watch is -

(a) 1 : 12 (b) 6 : 1 (c) 12 : 1 (d) 1 : 6

Q.4 The angular displacement of a particle is given by

$\theta = \omega_0 t + \frac{1}{2}\alpha t^2$, where ω_0 and α are constant and

$\omega_0 = 1$ rad/sec, $\alpha = 1.5$ rad/sec^2. The angular velocity at time, t = 2 sec will be (in rad/sec) -

(a) 1 (b) 5

(c) 3 (d) 4

Q.5 The magnitude of the linear acceleration of the particle moving in a circle of radius 10 cm with uniform speed completing the circle in 4 s, will be -

(a) 5π cm/s^2 (b) 2.5π cm/s^2

(c) $5\pi^2$ cm/s^2 (d) $2.5\pi^2$ cm/s^2

RESPONSE GRID	1. ⓐⓑⓒⓓ	2. ⓐⓑⓒⓓ	3. ⓐⓑⓒⓓ	4. ⓐⓑⓒⓓ	5. ⓐⓑⓒⓓ

Space for Rough Work

Q.6 A cane filled with water is revolved in a vertical circle of radius 4 m and water just does not fall down. The time period of revolution will be –
(a) 1 s (b) 10 s
(c) 8 s (d) 4 s

Q.7 The length of second's hand in a watch is 1 cm. The change in velocity of its tip in 15 second is -
(a) 0 (b) $\frac{\pi}{30\sqrt{2}}$ cm/s
(c) $\frac{\pi}{30}$ cm/s (d) $\frac{\pi\sqrt{2}}{30}$ cm/s

Q.8 An electron is moving in a circular orbit of radius 5.3×10^{-11} metre around the atomic nucleus at a rate of 6.6×10^{15} revolutions per second. The centripetal force acting on the electron will be -
(The mass of the electron is 9.1×10^{-31}kg)
(a) 8.3×10^{-8}N (b) 3.8×10^{-8}N
(c) 4.15×10^{-8}N (d) 2.07×10^{-8}N

Q.9 An air craft executes a horizontal loop of radius 1 km with a steady speed of 900 km/h. The ratio of centripetal acceleration to that gravitational acceleration will be-
(a) 1 : 6.38 (b) 6. 38 : 1
(c) 2.25 : 9.8 (d) 2.5 : 9.8

Q.10 A car driver is negotiating a curve of radius 100 m with a speed of 18 km/hr. The angle through which he has to lean from the vertical will be -
(a) $\tan^{-1}\frac{1}{4}$ (b) $\tan^{-1}\frac{1}{40}$
(c) $\tan^{-1}\left(\frac{1}{2}\right)$ (d) $\tan^{-1}\left(\frac{1}{20}\right)$

Q.11 A particle moves in a circle of radius 20cm with a linear speed of 10m/s. The angular velocity will be -
(a) 50 rad/s (b) 100 rad/s
(c) 25 rad/s (d) 75 rad/s

Q.12 The angular velocity of a particle is given by $\omega = 1.5\,t - 3t^2 + 2$, the time when its angular acceleration decreases to be zero will be -
(a) 25 sec (b) 0.25 sec
(c) 12 sec (d) 1.2 sec

Q.13 A particle is moving in a circular path with velocity varying with time as $v = 1.5t^2 + 2t$. If the radius of circular path is 2 cm, the angular acceleration at t = 2 sec will be -
(a) 4 rad/sec^2 (b) 40 rad/sec^2
(c) 400 rad/sec^2 (d) 0.4 rad/sec^2

Q.14 A grind stone starts from rest and has a constant-angular acceleration of 4.0 rad/sec^2.The angular displacement and angular velocity, after 4 sec. will respectively be -
(a) 32 rad, 16 rad/s (b) 16 rad, 32 rad/s
(c) 64 rad, 32 rad/s (d) 32 rad, 64 rad/s

Q.15 The shaft of an electric motor starts from rest and on the application of a torque, it gains an angular acceleration given by $\alpha = 3t - t^2$ during the first 2 seconds after it starts after which $\alpha = 0$. The angular velocity after 6 sec will be -
(a) 10/3 rad/sec (b) 3/10 rad/sec
(c) 30/4 rad/sec (d) 4/30 rad/sec

Q.16 Using rectangular co-ordinates and the unit vectors **i** and **j**, the vector expression for the acceleration a will be (r is a position vector) -
(a) ωr^2 (b) $-\omega^2 r/2$
(c) $-2\omega r^2$ (d) $-\omega^2 r$

Q.17 The vertical section of a road over a canal bridge in the direction of its length is in the form of circle of radius 8.9 metre. Find the greatest speed at which the car can cross this bridge without losing contact with the road at its highest point, the center of gravity of the car being at a height h = 1.1 metre from the ground. (Take g = 10 m/sec^2)
(a) 5 m/s (b) 7 m/s
(c) 10 m/s (d) 13 m/s

Q.18 The maximum speed at which a car can turn round a curve of 30 metre radius on a level road if the coefficient of friction between the tyres and the road is 0.4, will be -
(a) 10.84 m/s (b) 17.84 m/s
(c) 11.76 m/s (d) 9.02 m/s

Q.19 The angular speed with which the earth would have to rotate on its axis so that a person on the equator would weigh $(3/5)^{th}$ as much as present, will be:
(Take the equatorial radius as 6400 km)
(a) 8.7×10^4 rad/sec (b) 8.7×10^3 rad/sec
(c) 7.8×10^4 rad/sec (d) 7.8×10^3 rad/sec

RESPONSE GRID					
	6. ⓐⓑⓒⓓ	7. ⓐⓑⓒⓓ	8. ⓐⓑⓒⓓ	9. ⓐⓑⓒⓓ	10. ⓐⓑⓒⓓ
	11. ⓐⓑⓒⓓ	12. ⓐⓑⓒⓓ	13. ⓐⓑⓒⓓ	14. ⓐⓑⓒⓓ	15. ⓐⓑⓒⓓ
	16. ⓐⓑⓒⓓ	17. ⓐⓑⓒⓓ	18. ⓐⓑⓒⓓ	19. ⓐⓑⓒⓓ	

Space for Rough Work

Q.20 A smooth table is placed horizontally and a spring of unstreched length ℓ_0 and force constant k has one end fixed to its centre. To the other end of the spring is attached a mass m which is making n revolution per second around the centre. Tension in the spring will be
(a) $4\pi^2 m k \ell_0 n^2/(k - 4\pi^2 m n^2)$
(b) $4\pi^2 m k \ell_0 n^2/(k + 4\pi^2 m n^2)$
(c) $2\pi^2 m k \ell_0 n^2/(k - 4\pi^2 m n^2)$
(d) $2\pi m k \ell_0 n^2/(k - 4\pi^2 m n^2)$

Q.21 A motor car is travelling at 30 m/s on a circular road of radius 500 m. It is increasing its speed at the rate of 2 m/s^2. Its net acceleration is (in m/s^2) –
(a) 2 (b) 1. 8
(c) 2.7 (d) 0

DIRECTIONS (Q.22-Q.24) : In the following questions, more than one of the answers given are correct. Select the correct answers and mark it according to the following codes:

Codes :
(a) 1, 2 and 3 are correct **(b)** 1 and 2 are correct
(c) 2 and 4 are correct **(d)** 1 and 3 are correct

Q.22 Three identical particles are connected by three strings as shown in fig. These particles are revolving in a horizontal circle. The velocity of outer most particle is v, then choose correct relation for T_1, T_2 and T_3
(where T_1 is tension in the outer most string etc.)

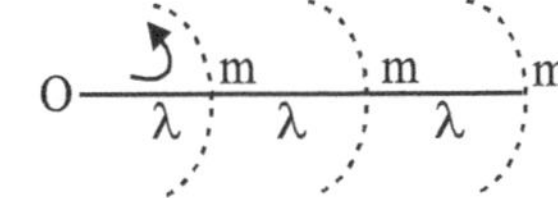

(1) $T_1 = \frac{mv_A^2}{3\ell}$ (2) $T_2 = \frac{5mv_A^2}{9\ell}$
(3) $T_3 = \frac{6mv_A^2}{9\ell}$ (4) $T_3 = \frac{5mv_A^2}{9\ell}$

Q.23 A particle describes a horizontal circle on the smooth surface of an inverted cone. The height of the plane of the circle above the vertex is 9.8 cm, then choose the correct options
(1) The speed of the particle will be 0.98 m/s
(2) $\tan\theta = \frac{rg}{v^2}$ (θ is semi-apex angle)
(3) The speed of the particle will be 98 m/s
(4) $\tan\theta = \frac{rg}{v}$ (θ is semiapex angle)

Q.24 Choose the correct statements
(1) Centripetal force is not a real force. It is only the requirement for circular motion.
(2) Work done by centripetal force may or may not be zero.
(3) Work done by centripetal force is always zero.
(4) Centripetal force is a fundamental force.

DIRECTIONS (Q.25-Q.27) : Read the passage given below and answer the questions that follows :

The velocity of the particle changes moving on the curved path, this change in velocity is brought by a force known as centripetal force and the acceleration so produced in the body is known as centripetal acceleration. The direction of centripetal force or acceleration is always towards the centre of circular path.

Q.25 A ball is fixed to the end of a string and is rotated in a horizontal circle of radius 5 m with a speed of 10 m/sec. The acceleration of the ball will be -
(a) 20 m/s^2 (b) 10 m/s^2
(c) 30 m/s^2 (d) 40 m/s^2

Q.26 A body of mass 2 kg lying on a smooth surface is attached to a string 3 m long and then whirled round in a horizontal circle making 60 revolution per minute. The centripetal acceleration will be-
(a) 118.4 m/s^2 (b) 1.18 m/s^2
(c) 2.368 m/s^2 (d) 23.68 m/s^2

Q.27 A body of mass 0.1 kg is moving on circular path of diameter 1.0 m at the rate of 10 revolutions per 31.4 seconds. The centripetal force acting on the body is -
(a) 0.2 N (b) 0.4 N
(c) 2 N (d) 4 N

RESPONSE GRID					
	20. ⓐⓑⓒⓓ	21. ⓐⓑⓒⓓ	22. ⓐⓑⓒⓓ	23. ⓐⓑⓒⓓ	24. ⓐⓑⓒⓓ
	25. ⓐⓑⓒⓓ	26. ⓐⓑⓒⓓ	27. ⓐⓑⓒⓓ		

Space for Rough Work

DIRECTIONS (Q.28-Q.30) : Each of these questions contains two statements: Statement-1 (Assertion) and Statement-2 (Reason). Each of these questions has four alternative choices, only one of which is the correct answer. You have to select the correct choice.

(a) Statement-1 is True, Statement-2 is True; Statement-2 is a correct explanation for Statement-1.
(b) Statement-1 is True, Statement-2 is True; Statement-2 is NOT a correct explanation for Statement-1.
(c) Statement -1 is False, Statement-2 is True.
(d) Statement -1 is True, Statement-2 is False.

Q.28 Statement - 1 : In non-uniform circular motion, velocity vector and acceleration vector are not perpendicular to each other.
Statement - 2 : In non-uniform circular motion, particle has normal as well as tangential acceleration.

Q.29 Statement - 1 : A cyclist is cycling on rough horizontal circular track with increasing speed. Then the frictional force on cycle is always directed towards centre of the circular track.
Statement - 2 : For a particle moving in a circle, radial component of net force should be directed towards centre.

Q.30 Statement - 1 : If net force $\vec{F}$ acting on a system is changing in direction only, the linear momentum $(\vec{p})$ of system changes in direction.
Statement - 2 : In case of uniform circular motion, magnitude of linear momentum is constant but direction of centripetal force changes at every instant.

RESPONSE GRID	28. ⓐⓑⓒⓓ	29. ⓐⓑⓒⓓ	30. ⓐⓑⓒⓓ

DAILY PRACTICE PROBLEM SHEET 7 - PHYSICS			
Total Questions	30	Total Marks	120
Attempted		Correct	
Incorrect		Net Score	
Cut-off Score	30	Qualifying Score	48
Success Gap = Net Score – Qualifying Score			
Net Score = (Correct × 4) – (Incorrect × 1)			

Space for Rough Work

DPP - Daily Practice Problems

Name :

Date :

Start Time :

End Time :

PHYSICS 08

SYLLABUS : MOTION IN A PLANE-3 (Vertical Circular Motion, Relative Motion)

Max. Marks : 112 **Time : 60 min.**

GENERAL INSTRUCTIONS

- The Daily Practice Problem Sheet contains 28 MCQ's. For each question only one option is correct. Darken the correct circle/ bubble in the Response Grid provided on each page.
- You have to evaluate your Response Grids yourself with the help of solution booklet.
- Each correct answer will get you 4 marks and 1 mark shall be deduced for each incorrect answer. No mark will be given/ deducted if no bubble is filled. Keep a timer in front of you and stop immediately at the end of 60 min.
- The sheet follows a particular syllabus. Do not attempt the sheet before you have completed your preparation for that syllabus. Refer syllabus sheet in the starting of the book for the syllabus of all the DPP sheets.
- After completing the sheet check your answers with the solution booklet and complete the Result Grid. Finally spend time to analyse your performance and revise the areas which emerge out as weak in your evaluation.

DIRECTIONS (Q.1-Q.19) : There are 19 multiple choice questions. Each question has 4 choices (a), (b), (c) and (d), out of which **ONLY ONE** choice is correct.

Q.1 A man whirls a stone round his head on the end of a string 4.0 metre long. Can the string be in a horizontal, plane? If the stone has a mass of 0.4 kg and the string will break, if the tension in it exceeds 8 N. The smallest angle the string can make with the horizontal and the speed of the stone will respectively be (Take g = 10 m/sec^2)

(a) 30º, 7.7 m/s (b) 60º, 7.7 m/s
(c) 45º, 8.2 m/s (d) 60º, 8.7 m/s

Q.2 In figure ABCDE is a channel in the vertical plane, part BCDE being circular with radius r. A ball is released from A and slides without friction and without rolling. It will complete the loop path when

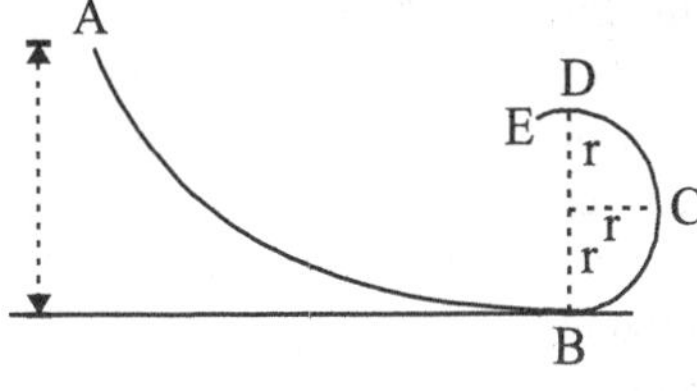

(a) $h > 5\ r/2$ (b) $h < 5\ r/2$
(c) $h < 2r/5$ (d) $h > 2r/5$

Q.3 An aircraft loops the loop of radius R = 500 m with a constant velocity v = 360 km/h. The weight of the flyer of mass m = 70 kg in the lower, upper and middle points of the loop will respectively be-

(a) 210 N, 700 N, 1400 N (b) 1400 N, 700 N, 2100 N
(c) 700 N, 1400 N, 210 N (d) 2100 N, 700 N, 1400 N

RESPONSE GRID	1. ⓐⓑⓒⓓ	2. ⓐⓑⓒⓓ	3. ⓐⓑⓒⓓ

Space for Rough Work

Q.4 A particle of mass 3 kg is moving under the action of a central force whose potential energy is given by U(r) = $10r^3$ joule. For what energy and angular momentum will the orbit be a circle of radius 10 m ?
(a) 2.5×10^4 J, 3000 kgm^2/sec
(b) 3.5×10^4 J, 2000 kgm^2/sec
(c) 2.5×10^3 J, 300 kgm^2/sec
(d) 3.5×10^3 J, 300 kgm^2/sec

Q.5 A string of length 1 m is fixed at one end and carries a mass of 100 gm at the other end. The string makes $2/\pi$ revolutions per second about a vertical axis through the fixed end. The angle of inclination of the string with the vertical, and the linear velocity of the mass will respectively be - (in M.K.S. system)
(a) 52°14', 3.16 (b) 50°14', 1.6
(c) 52°14', 1.6 (d) 50°14', 3.16

Q.6 A particle of mass m is moving in a circular path of constant radius r such that its centripetal acceleration a_c is varying with time t as $a_c = k^2rt^2$, where k is a constant. The power delivered to the particle by the force acting on it will be -
(a) mk^2t^2r (b) $mk^2r^2t^2$ (c) $m^2k^2t^2r^2$ (d) mk^2r^2t

Q.7 A car is moving in a circular path of radius 100 m with velocity of 200 m/sec such that in each sec its velocity increases by 100 m/s, the net acceleration of car will be - (in m/sec)
(a) $100\sqrt{17}$ (b) $10\sqrt{7}$ (c) $10\sqrt{3}$ (d) $100\sqrt{3}$

Q.8 A 4 kg balls is swing in a vertical circle at the end of a cord 1 m long. The maximum speed at which it can swing if the cord can sustain maximum tension of 163.6 N will be -
(a) 6 m/s (b) 36 m/s (c) 8 m/s (d) 64 m/s

Q.9 The string of a pendulum is horizontal. The mass of the bob is m. Now the string is released. The tension in the string in the lowest position is -
(a) 1 mg (b) 2 mg (c) 3 mg (d) 4 mg

Q.10 A swimmer can swim in still water at a rate 4 km/h. If he swims in a river flowing at 3 km/h and keeps his direction (w.r.t. water) perpendicular to the current. Find his velocity w.r.t. the ground.
(a) 3 km/hr (b) 5 km/hr
(c) 4 km/hr (d) 7 km/hr

Q.11 The roadway bridge over a canal is the form of an arc of a circle of radius 20 m. What is the minimum speed with which a car can cross the bridge without leaving contact with the ground at the highest point (g = 9.8 m/s^2)
(a) 7 m/s (b) 14 m/s (c) 289 m/s (d) 5 m/s

Q.12 A cane filled with water is revolved in a vertical circle of radius 0.5 m and the water does not fall down. The maximum period of revolution must be -
(a) 1.45 (b) 2.45 (c) 14.15 (d) 4.25

Q.13 A particle of mass m slides down from the vertex of semi-hemisphere, without any initial velocity. At what height from horizontal will the particle leave the sphere-
(a) $\frac{2}{3}R$ (b) $\frac{3}{2}R$ (c) $\frac{5}{8}R$ (d) $\frac{8}{5}R$

Q.14 A body of mass m tied at the end of a string of length ℓ is projected with velocity $\sqrt{4\ell g}$, at what height will it leave the circular path -
(a) $\frac{5}{3}\ell$ (b) $\frac{3}{5}\ell$ (c) $\frac{1}{3}\ell$ (d) $\frac{2}{3}\ell$

Q.15 A string of length L is fixed at one end and carries a mass M at the other end. The string makes $2/\pi$ revolutions per second around the vertical axis through the fixed end as shown in the figure, then tension in the string is
(a) ML
(b) 2 ML
(c) 4 ML
(d) 16 ML

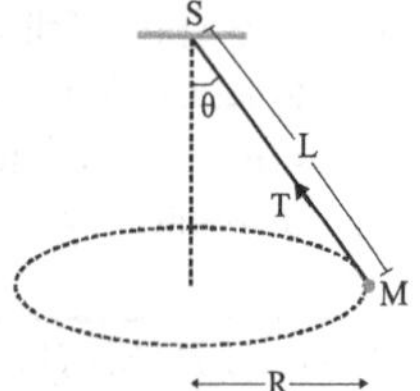

Q.16 A train has to negotiate a curve of radius 400 m. By how much should the outer rail be raised with respect to inner rail for a speed of 48 km/hr. The distance between the rails is 1 m.
(a) 12 m (b) 12 cm
(c) 4.5 cm (d) 4.5 m

RESPONSE GRID					
	4. ⓐⓑⓒⓓ	5. ⓐⓑⓒⓓ	6. ⓐⓑⓒⓓ	7. ⓐⓑⓒⓓ	8. ⓐⓑⓒⓓ
	9. ⓐⓑⓒⓓ	10. ⓐⓑⓒⓓ	11. ⓐⓑⓒⓓ	12. ⓐⓑⓒⓓ	13. ⓐⓑⓒⓓ
	14. ⓐⓑⓒⓓ	15. ⓐⓑⓒⓓ	16. ⓐⓑⓒⓓ		

Space for Rough Work

Q.17 A ship is steaming towards east at a speed of 12 ms^{-1}. A woman runs across the deck at a speed of 5 ms^{-1} in the direction at right angles to the direction of motion of the ship i.e. towards north. What is the velocity of the woman relative to sea ?

(a) 13 m/s (b) 5 m/s (c) 12 m/s (d) 17 m/s

Q.18 A man is walking on a level road at a speed of 3 km/h. Raindrops fall vertically with a speed of 4 km/h. Find the velocity of raindrops with respect to the men.

(a) 3 km/hr (b) 4 km/hr (c) 5 km/hr (d) 7 km/hr

Q.19 A stone of mass 1 kg tied to a light inextensible string of length $L = \frac{10}{3}$m is whirling in a circular path of radius L in a vertical plane. If the ratio of the maximum tension in the string to the minimum tension in the string is 4 and if g is taken to be $10m / sec^2$, the speed of the stone at the highest point of the circle is

(a) 20 m/sec (b) $10\sqrt{3}$m/sec

(c) $5\sqrt{2}$m/sec (d) 10 m/sec

DIRECTIONS (Q.20-Q.22) : In the following questions, more than one of the answers given are correct. Select the correct answers and mark it according to the following codes:

Codes :

(a) 1, 2 and 3 are correct **(b)** 1 and 2 are correct

(c) 2 and 4 are correct **(d)** 1 and 3 are correct

Q.20 Two bodies P and Q are moving along positive x-axis their position-time graph is shown below. If $\vec{V}_{PQ}$ is velocity of P w.r.t. Q and $\vec{V}_{QP}$ is velocity of Q w.r.t P, then

(1) $|\vec{V}_{PQ}| = |\vec{V}_{QP}|$ = constant

(2) $\vec{V}_{PQ}$ towards origin

(3) $\vec{V}_{QP}$ towards origin

(4) $|\vec{V}_{PQ}| \neq |\vec{V}_{QP}|$ = constant

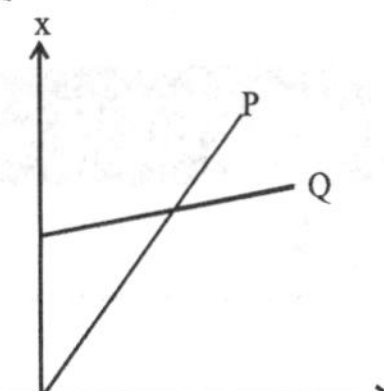

Q.21 A swimmer who can swim in a river with speed mv (with respect to still water) where v is the velocity of river current, jumps into the river from one bank to cross the river. Then

(1) If m < 1 he cannot cross the river

(2) If m ≤ 1 he cannot reach a point on other bank directly opposite to his starting point.

(3) If m > 1 he can reach a point on other bank

(4) He can reach the other bank at some point, whatever be the value of m.

Q.22 Consider two children riding on the merry-go-round Child 1 sits near the edge, Child 2 sits closer to the centre. Let v_1 and v_2 denote the linear speed of child 1 and child 2, respectively. Which of the following is/are wrong ?

(1) We cannot determine v_1 & v_2 without more information

(2) $v_1 = v_2$

(3) $v_1 < v_2$

(4) $v_1 > v_2$

DIRECTIONS (Q.23-Q.25) : Read the passage given below and answer the questions that follows :

Three of the fundamental constants of physics are the universal gravitational constant, $G = 6.7 \times 10^{-11} m^3kg^{-1}s^{-2}$, the speed of light, $c = 3.0 \times 10^8$ m/s, and Planck's constant, $h = 6.6 \times 10^{-34}$ Js^{-1}. Two particles A and B are projected in the vertical plane with same initial velocity u_0 from part (0, 0) and (ℓ, –h) towards each other as shown in figure at t = 0.

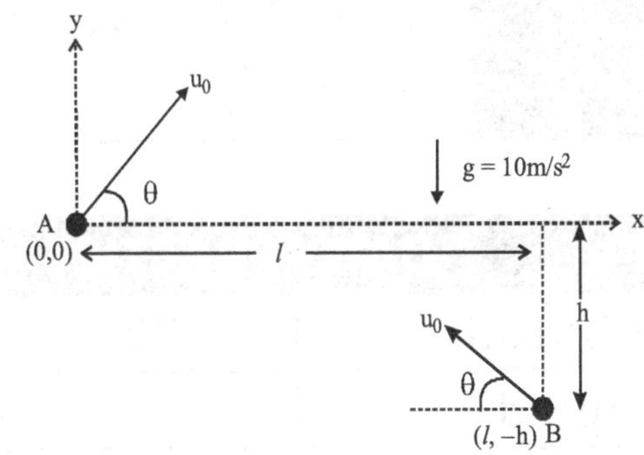

RESPONSE GRID					
	17. ⓐⓑⓒⓓ	18. ⓐⓑⓒⓓ	19. ⓐⓑⓒⓓ	20. ⓐⓑⓒⓓ	21. ⓐⓑⓒⓓ
	22. ⓐⓑⓒⓓ				

Space for Rough Work

Q.23 The path of particle A with respect to particle B will be –
(a) parabola
(b) straight line parallel to x-axis
(c) straight line parallel to y-axis
(d) None of these

Q.24 Minimum distance between particle A and B during motion will be –
(a) ℓ (b) h
(c) $\sqrt{\ell^2+h^2}$ (d) $\ell+h$

Q.25 The time when separation between A and B is minimum is
(a) $\frac{x}{u_0 \cos\theta}$ (b) $\sqrt{\frac{2h}{g}}$
(c) $\frac{\ell}{2u_0 \cos\theta}$ (d) $\frac{2\ell}{u_0 \cos\theta}$

DIRECTIONS (Qs. 26-Q.28) : Each of these questions contains two statements: Statement-1 (Assertion) and Statement-2 (Reason). Each of these questions has four alternative choices, only one of which is the correct answer. You have to select the correct choice.

(a) Statement-1 is True, Statement-2 is True; Statement-2 is a correct explanation for Statement-1.
(b) Statement-1 is True, Statement-2 is True; Statement-2 is NOT a correct explanation for Statement-1.
(c) Statement-1 is False, Statement-2 is True.
(d) Statement-1 is True, Statement-2 is False.

Q.26 Statement-1 : The relative velocity between any two bodies moving in opposite direction is equal to sum of the velocities of two bodies.

Statement-2 : Sometimes relative velocity between two bodies is equal to difference in velocities of the two.

Q.27 Statement-1: A river is flowing from east to west at a speed of 5m/min. A man on south bank of river, capable of swimming 10 m/min in still water, wants to swim across the river in shortest time. He should swim due north.

Statement-2 : For the shortest time the man needs to swim perpendicular to the bank.

Q.28 Statement-1 : Rain is falling vertically downwards with velocity 6 km/h. A man walks with a velocity of 8 km/h. Relative velocity of rain w.r.t. the man is 10 km/h.
Statement-2 : Relative velocity is the ratio of two velocities.

RESPONSE GRID	
	23. ⓐⓑⓒⓓ 24. ⓐⓑⓒⓓ 25. ⓐⓑⓒⓓ 26. ⓐⓑⓒⓓ 27. ⓐⓑⓒⓓ
	28. ⓐⓑⓒⓓ

DAILY PRACTICE PROBLEM SHEET 8 - PHYSICS			
Total Questions	28	Total Marks	112
Attempted		Correct	
Incorrect		Net Score	
Cut-off Score	28	Qualifying Score	44
Success Gap = Net Score – Qualifying Score			
Net Score = (Correct × 4) – (Incorrect × 1)			

Space for Rough Work

DPP - Daily Practice Problems

Name : 　　　　　　　　　　　　　　　　　　　　　　Date :

Start Time : 　　　　　　　　　　　　　　　　　　　End Time :

PHYSICS 09

SYLLABUS : LAWS OF MOTION-1 (Newton's laws, momentum, pseudo force concept)

Max. Marks : 116 　　　　　　　　　　　　　　　　　　　　**Time : 60 min.**

GENERAL INSTRUCTIONS

- The Daily Practice Problem Sheet contains 29 MCQ's. For each question only one option is correct. Darken the correct circle/ bubble in the Response Grid provided on each page.
- You have to evaluate your Response Grids yourself with the help of solution booklet.
- Each correct answer will get you 4 marks and 1 mark shall be deduced for each incorrect answer. No mark will be given/ deducted if no bubble is filled. Keep a timer in front of you and stop immediately at the end of 60 min.
- The sheet follows a particular syllabus. Do not attempt the sheet before you have completed your preparation for that syllabus. Refer syllabus sheet in the starting of the book for the syllabus of all the DPP sheets.
- After completing the sheet check your answers with the solution booklet and complete the Result Grid. Finally spend time to analyse your performance and revise the areas which emerge out as weak in your evaluation.

DIRECTIONS (Q.1-Q.20) : There are 20 multiple choice questions. Each question has 4 choices (a), (b), (c) and (d), out of which **ONLY ONE** choice is correct.

Q.1 A boy standing on a weighing machine observes his weight as 200 N. When he suddenly jumpes upwards, his friend notices that the reading increased to 400 N. The acceleration by which the boy jumped will be-

(a) 9.8 m/s^2 　　(b) 29.4 m/s^2
(c) 4.9 m/s^2 　　(d) 14.7 m/s^2

Q.2 A force of $(6\hat{i} + 8\hat{j})$ N acted on a body of mass 10 kg. The displacement after 10 sec, if it starts from rest, will be -

(a) 50 m along $\tan^{-1} 4/3$ with x axis
(b) 70 m along $\tan^{-1} 3/4$ with x axis
(c) 10 m along $\tan^{-1} 4/3$ with x axis
(d) None

Q.3 A boat of mass 1000 kg is moving with a velocity of 5 m/s. A person of mass 60 kg jumps into the boat. The velocity of the boat with the person will be -

(a) 4.71 m/s 　　(b) 4.71 cm/s
(c) 47.1 m/s 　　(d) 47.1 cm/s

Q.4 A disc of mass 10 gm is kept horizontally in air by firing bullets of mass 5 g each at the rate of 10/s. If the bullets rebound with same speed. The velocity with which the bullets are fired is -

(a) 49 cm/s (b) 98 cm/s (c) 147 cm/s (d) 196 cm/s

Q.5 A fire man has to carry an injured person of mass 40 kg from the top of a building with the help of the rope which can withstand a load of 100 kg. The acceleration of the fireman if his mass is 80 kg, will be-

(a) 8.17 m/s^2 　　(b) 9.8 m/s^2
(c) 1.63 m/s^2 　　(d) 17.97 m/s^2

RESPONSE GRID	1. ⓐⓑⓒⓓ	2. ⓐⓑⓒⓓ	3. ⓐⓑⓒⓓ	4. ⓐⓑⓒⓓ	5. ⓐⓑⓒⓓ

Space for Rough Work

Q.6 A body of mass 0.02 kg falls from a height of 5 metre into a pile of sand. The body penetrates the sand a distance of 5 cm before stoping. What force has the sand exerted on the body ?

(a) 1.96 N (b) –19.6 N
(c) –0.196 N (d) 0.0196 N

Q.7 The magnitude of the force (in newton) acting on a body varies with time t (in microsecond) as shown in fig. AB, BC, and CD are straight line segments. The magnitude of the total impulse of the force on the body from t = 4 μs to t = 16 μs is

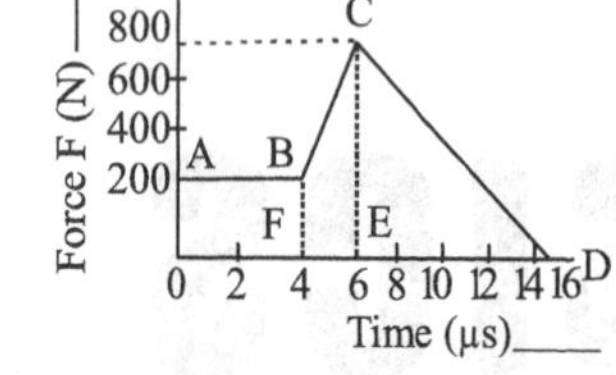

(a)  5×10^{-4} N.s
(b) 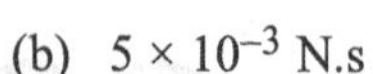 5×10^{-3} N.s
(c) 5×10^{-5} N.s
(d) 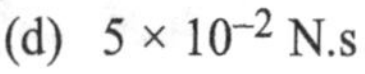 5×10^{-2} N.s

Q.8 The total mass of an elevator with a 80 kg man in it is 1000 kg. This elevator moving upward with a speed of 8 m/sec, is brought to rest over a distance of 16m. The tension T in the cables supporting the elevator and the force exerted on the man by the elevator floor will respectively be-

(a) 7800 N, 624 N (b) 624 N, 7800 N
(c) 78 N, 624 N (d) 624 N, 78 N

Q.9 In the arrangement shown in fig. the ends P and Q of an unstretchable string move downwards with a uniform speed U. Pulleys A and B are fixed. Mass M moves upwards with a speed of

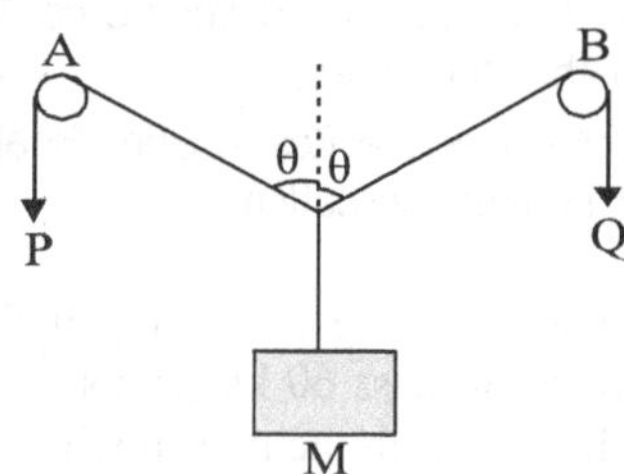

(a) 2U cos θ (b) U cos θ
(c) 2U/cos θ (d) U/cos θ

Q.10 An engine of mass 5×10^4 kg pulls a coach of mass 4×10^4 kg. Suppose that there is a resistance of 1 N per 100 kg acting on both coach and engine, and that the driving force of engine is 4500 N. The acceleration of the engine and tension in the coupling will respectively be-

(a) 0.04 m/s^2, 2000 N (b) 0.4 m/s^2, 200 N
(c) 0.4 m/s^2, 20 N (d) 4 m/s^2, 200 N

Q.11 A body whose mass 6 kg is acted upon by two forces $(8\hat{i}+10\hat{j})$ N and $(4\hat{i}+8\hat{j})$ N. The acceleration produced will be (in m/s^2) –

(a) $(3\hat{i}+2\hat{j})$ (b) $12\hat{i}+18\hat{j}$
(c) $\frac{1}{3}(\hat{i}+\hat{j})$ (d) $2\hat{i}+3\hat{j}$

Q.12 A car of 1000 kg moving with a velocity of 18 km/hr is stopped by the brake force of 1000 N. The distance covered by it before coming to rest is -

(a) 1 m (b) 162 m (c) 12.5 m (d) 144 m

Q.13 A block of metal weighing 2 kg is resting on a frictionless plane. It is struck by a jet releasing water at a rate of 1 kg/s and at a speed of 5 m/s. The initial acceleration of the block will be –

(a) 2.5 m/s^2 (b) 5 m/s^2 (c) 0.4 m/s^2 (d) 0

Q.14 A man fires the bullets of mass m each with the velocity v with the help of machine gun, if he fires n bullets every sec, the reaction force per second on the man will be -

(a) $\frac{m}{v}n$ (b) m n v (c) $\frac{mv}{n}$ (d) $\frac{vn}{m}$

Q.15 A body of mass 15 kg moving with a velocity of 10 m/s is to be stopped by a resistive force in 15 sec, the force will be -

(a) 10 N (b) 5 N
(c) 100 N (d) 50 N

Q.16 A cricket ball of mass 250 gm moving with a velocity of 24 m/s is hit by a bat so that it acquires a velocity of 28 m/s in the opposite direction. The force acting on the ball, if the contact time is 1/100 of a second, will be -

(a) 1300 N in the final direction of ball
(b) 13 N in the initial direction of ball
(c) 130 N in the final direction of ball
(d) 1.3 N in the initial direction of ball

RESPONSE GRID					
	6. ⓐⓑⓒⓓ	7. ⓐⓑⓒⓓ	8. ⓐⓑⓒⓓ	9. ⓐⓑⓒⓓ	10. ⓐⓑⓒⓓ
	11. ⓐⓑⓒⓓ	12. ⓐⓑⓒⓓ	13. ⓐⓑⓒⓓ	14. ⓐⓑⓒⓓ	15. ⓐⓑⓒⓓ
	16. ⓐⓑⓒⓓ				

Space for Rough Work

Q.17 A force of 2 N is applied on a particle for 2 sec, the change in momentum will be -

(a) 2 Ns (b) 4 Ns (c) 6 Ns (d) 3 Ns

Q.18 A body of mass 2 kg is moving along x-direction with a velocity of 2 m/sec. If a force of 4 N is applied on it along y-direction for 1 sec, the final velocity of particle will be -

(a) $2\sqrt{2}$ m/s (b) $\sqrt{2}$ m/s

(c) $1/\sqrt{2}$ m/s (d) $1/2\sqrt{2}$ m/s

Q.19 A cricket ball of mass 150 g is moving with a velocity of 12m/sec and is hit by a bat so that the ball is turned back with a velocity of 20 m/sec, the force on the ball acts for 0.01 sec, then the average force exerted by the bat on the ball will be

(a) 48 N (b) 40 N (c) 480 N (d) 400 N

Q.20 A body of mass 20 kg moving with a velocity of 3 m/s, rebounds on a wall with same velocity. The impulse on the body is -

(a) 60 Ns (b) 120 Ns (c) 30 Ns (d) 180 Ns

DIRECTIONS (Q.21-Q.23) : In the following questions, more than one of the answers given are correct. Select the correct answers and mark it according to the following codes:

Codes :

(a) 1, 2 and 3 are correct **(b)** 1 and 2 are correct

(c) 2 and 4 are correct **(d)** 1 and 3 are correct

Q.21 A mass of 60 kg is on the floor of a lift moving down. The lift moves at first with an acceleration of 3 m/sec^2, then with constant velocity and finally with a retardation of 3m/sec^2. Choose the correct options related to possible reactions exerted by the lift on the body in each part of the motion –

(1) 408 N (2) 588 N (3) 768 N (4) 508 N

Q.22 A mass of 10 kg is hung to a spring balance in lift. If the lift is moving with an acceleration g/3 in upward & downward directions, choose the correct options related to the reading of the spring balance.

(1) 13.3 kg (2) 6.67 kg (3) 32.6 kg (4) 0

Q.23 Choose the correct options

(1) A reference frame in which Newton's first law is valid is called an inertial reference frame.

(2) Frame moving at constant velocity relative to a known inertial frame is also an inertial frame.

(3) Idealy, no inertial frame exists in the universe for practical purpose, a frame of reference may be considered as inertial if its acceleration is negligible with respect to the acceleration of the object to be observed.

(4) To measure the acceleration of a falling apple, earth cannot be considered as an inertial frame.

DIRECTIONS (Q.24-Q.26) : Read the passage given below and answer the questions that follows :

Pseudo force is an imaginary force which is recognised only by a non-inertial observer to explain the physical situation according to newton's laws. Magnitude of pseudo force F_p is equal to the product of the mass m of the object and the acceleration a of the frame of reference. The direction of the force is opposite to the direction of acceleration, $F_p = -ma$

Q.24 A spring weighing machine inside a stationary lift reads 50 kg when a man stand on it. What would happen to the scale reading if the lift is moving upward with (i) constant velocity (ii) constant acceleratioin ?

(a) 50 kg wt, $\left(50+\frac{50a}{g}\right)$ kg wt

(b) 50 kg wt, $\left(50+\frac{50g}{a}\right)$ kg wt

(c) 50 kg wt, $\left(\frac{50a}{g}\right)$ kg wt

(d) 50 kg wt, $\left(\frac{50g}{a}\right)$ kg wt

Q.25 A 25 kg lift is supported by a cable. The acceleration of the lift when the tension in the cable is 175 N, will be -

(a) 2.8 m/s^2 (b) 16.8 m/s^2

(c) – 9.8 m/s^2 (d) 14 m/s^2

RESPONSE GRID					
	17. ⓐⓑⓒⓓ	18. ⓐⓑⓒⓓ	19. ⓐⓑⓒⓓ	20. ⓐⓑⓒⓓ	21. ⓐⓑⓒⓓ
	22. ⓐⓑⓒⓓ	23. ⓐⓑⓒⓓ	24. ⓐⓑⓒⓓ	25. ⓐⓑⓒⓓ	

Space for Rough Work

Q.26 A body is suspended by a string from the ceiling of an elevator. It is observed that the tension in the string is doubled when the elevator is accelerated. The acceleration will be -

(a) 4.9 m/s^2 (b) 9.8 m/s^2
(c) 19.6 m/s^2 (d) 2.45 m/s^2

DIRECTIONS (Q. 27-Q.29) : Each of these questions contains two statements: Statement-1 (Assertion) and Statement-2 (Reason). Each of these questions has four alternative choices, only one of which is the correct answer. You have to select the correct choice.

(a) Statement-1 is True, Statement-2 is True; Statement-2 is a correct explanation for Statement-1.
(b) Statement-1 is True, Statement-2 is True; Statement-2 is NOT a correct explanation for Statement-1.
(c) Statement-1 is False, Statement-2 is True.
(d) Statement-1 is True, Statement-2 is False.

Q.27 Statement-1 : A cloth covers a table. Some dishes are kept on it. The cloth can be pulled out without dislodging the dishes from the table.
Statement-2 : To every action there is an equal and opposite reaction.

Q.28 Statement-1 : If the net external force on the body is zero then its acceleration is zero.
Statement-2 : Acceleration does not depend on force.

Q.29 Statement-1 : The slope of momentum versus time graph give us the acceleration.
Statement-2 : Force is given by the rate of change of momentum.

RESPONSE GRID	26. ⓐⓑⓒⓓ	27. ⓐⓑⓒⓓ	28. ⓐⓑⓒⓓ	29. ⓐⓑⓒⓓ

DAILY PRACTICE PROBLEM SHEET 9 - PHYSICS			
Total Questions	29	Total Marks	116
Attempted		Correct	
Incorrect		Net Score	
Cut-off Score	30	Qualifying Score	44
Success Gap = Net Score – Qualifying Score			
Net Score = (Correct × 4) – (Incorrect × 1)			

Space for Rough Work

DPP - Daily Practice Problems

Name :

Date :

Start Time :

End Time :

PHYSICS 10

SYLLABUS : LAWS OF MOTION-2 (Blocks in contact, connected by string, pulley arrangement)

Max. Marks : 112 **Time : 60 min.**

GENERAL INSTRUCTIONS

- The Daily Practice Problem Sheet contains 28 MCQ's. For each question only one option is correct. Darken the correct circle/ bubble in the Response Grid provided on each page.
- You have to evaluate your Response Grids yourself with the help of solution booklet.
- Each correct answer will get you 4 marks and 1 mark shall be deduced for each incorrect answer. No mark will be given/ deducted if no bubble is filled. Keep a timer in front of you and stop immediately at the end of 60 min.
- The sheet follows a particular syllabus. Do not attempt the sheet before you have completed your preparation for that syllabus. Refer syllabus sheet in the starting of the book for the syllabus of all the DPP sheets.
- After completing the sheet check your answers with the solution booklet and complete the Result Grid. Finally spend time to analyse your performance and revise the areas which emerge out as weak in your evaluation.

DIRECTIONS (Q.1-Q.20) : There are **20** multiple choice questions. Each question has 4 choices (a), (b), (c) and (d), out of which **ONLY ONE** choice is correct.

Q.1 A block of mass M is pulled along a horizontal frictionless surface by a rope of mass m. If a force P is applied at the free end of the rope, the force exerted by the rope on the block will be -

(a) P (b) $\frac{Pm}{M+m}$ (c) $\frac{MP}{M+m}$ (d) $\frac{mP}{M+m}$

Q.2 A body of mass 50 kg is pulled by a rope of length 8 m on a surface by a force of 108N applied at the other end. The force that is acting on 50 kg mass, if the linear density of rope is 0.5 kg/m will be -

(a) 108 N (b) 100 N (c) 116 N (d) 92 N

Q.3 A rope of length 15 m and linear density 2 kg/m is lying length wise on a horizontal smooth table. It is pulled by a force of 25 N. The tension in the rope at the point 7 m away from the point of application, will be -

(a) 11.67 N (b) 13.33 N
(c) 36.67 N (d) None of these

Q.4 A force of 100 N acts in the direction as shown in figure on a block of mass 10 kg resting on a smooth horizontal table. The speed acquired by the block after it has .moved a distance of 10 m, will be -
(in m/sec) (g = 10 m/sec^2)

(a) 17 m/sec
(b) 13.17 m/sec
(c) 1.3 m/sec
(d) 1.7 m/sec

100N
30°
10kg

Q.5 In the above example, the velocity after 2 sec will be - (in m/sec)

(a) $10\sqrt{3}$ (b) $5\sqrt{3}$ (c) 10 (d) 5

RESPONSE GRID	1. ⓐⓑⓒⓓ	2. ⓐⓑⓒⓓ	3. ⓐⓑⓒⓓ	4. ⓐⓑⓒⓓ	5. ⓐⓑⓒⓓ

Space for Rough Work

Q.6 Two blocks of mass m = 1 kg and M = 2 kg are in contact on a frictionless table. A horizontal force F(= 3N) is applied to m. The force of contact between the blocks, will be-

(a) 2N (b) 1N (c) 4N (d) 5N

Q.7 A force produces an acceleration of 5 m/s^2 in a body and same force an acceleration of 15 m/s^2 in another body. The acceleration produced by the same force when applied to the combination of two bodies will be -

(a) 3.75 m/s^2 (b) 20 m/s^2
(c) 10 m/s^2 (d) 0.667 m/s^2

Q.8 What is the tension in a rod of length L and mass M at a distance y from F_1 when the rod is acted on by two unequal forces F_1 and F_2 ($<F_1$) as shown in fig

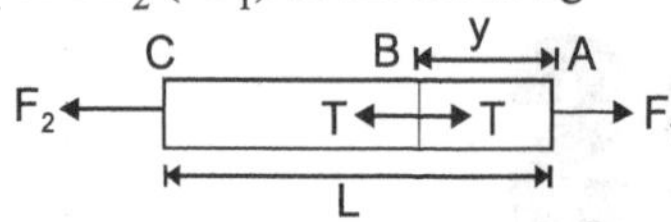

(a) $F_1\left(1-\frac{y}{L}\right)+F_2\left(\frac{y}{L}\right)$ (b) $\frac{M}{L}y\left(\frac{F_1-F_2}{M}\right)$

(c) $F_1\left(1+\frac{y}{L}\right)+F_2\left(\frac{y}{L}\right)$ (d) $\frac{M}{L}y\left(\frac{F_1+F_2}{M}\right)$

Q.9 A force produces acceleration $1, \frac{1}{2}, \frac{1}{3}, \frac{1}{4}, \ldots\ldots\ldots$ (all in m/s^2), applied separetly to n bodies. If these bodies are combined to form single one, then the acceleration of the system will be, if same force is taken into account.

(a) $\frac{n}{2}$ (b) $\frac{2}{n(n+1)}$ (c) $\frac{n^2}{2}$ (d) $\frac{n^2(n+1)}{2}$

Q.10 Two blocks of masses 6 kg and 4 kg connected by a rope of mass 2 kg are resting on frictionless floor as shown in fig. If a constant force of 60 N is applied to 6 kg block, tension in the rope at A, B, and C will respectively be -

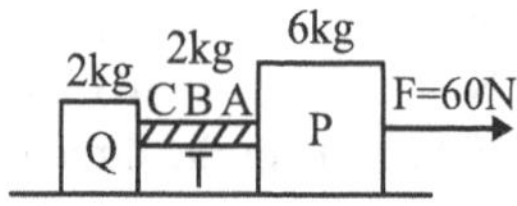

(a) 30 N, 25 N, 20 N (b) 25 N, 30 N, 20 N
(c) 20 N, 30 N, 25 N (d) 30 N, 20 N, 25 N

Q.11 The pulley arrangements of fig (a) and (b) are identical. The mass of the rope is negligible. In (a) the mass m is lifted up by attaching a mass 2 m to the other end of the rope. In (b) m is lifted up by pulling the other end of the rope with a constant downward force F = 2 mg. Which of the following is correct?

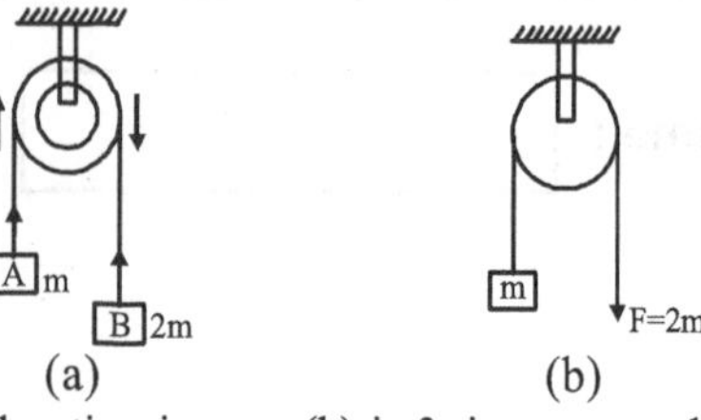

(a) Acceleration in case (b) is 3 times more than that in case (a)
(b) In case (a) acceleration is g, while in case (b) it is 2g
(c) In both the cases, acceleration is same
(d) None of the above

Q.12 Three equal weights of mass m each are hanging on a string passing over a fixed pulley as shown in fig. The tensions in the string connecting weights A to B and B to C will respectively be -

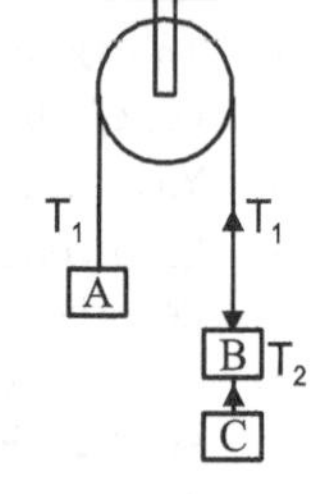

(a) $\frac{2}{3}$ mg, $\frac{2}{3}$ mg
(b) $\frac{2}{3}$ mg, $\frac{4}{3}$ mg
(c) $\frac{4}{3}$ mg, $\frac{2}{3}$ mg
(d) $\frac{3}{2}$ mg, $\frac{3}{4}$ mg

Q.13 In the situation shown in figure, both the pulleys and the strings are light and all the surfaces are frictionless. The acceleration of mass M, tension in the string PQ and force exerted by the clamp on the pulley, will respectively be -

(a) (2/3)g, (1/3)Mg, ($\sqrt{2}$/3)Mg
(b) (1/3)g, (1/3)Mg, ($\sqrt{2}$/3)Mg
(c) (1/3)g, (2/3)Mg, $\sqrt{3}$ Mg
(d) 2g, (1/2)g, $\sqrt{2}$ Mg

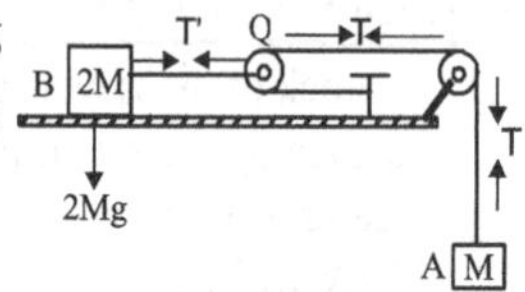

RESPONSE GRID	
	6. ⓐⓑⓒⓓ 7. ⓐⓑⓒⓓ 8. ⓐⓑⓒⓓ 9. ⓐⓑⓒⓓ 10. ⓐⓑⓒⓓ
	11. ⓐⓑⓒⓓ 12. ⓐⓑⓒⓓ 13. ⓐⓑⓒⓓ

Space for Rough Work

Q.14 A body of mass 50kg resting on a smooth inclined plane is connected by a massless inextensible string passing over a smooth pulley, at the top of the inclined plane have another mass of 40 kg as shown in the figure. The distance through which 50 kg mass fall in 4 sec will be -
(The angle of the inclined plane is 30º)

(a) 13.04 m
(b) 1.63m
(c) 1.304 m
(d) 16.3m

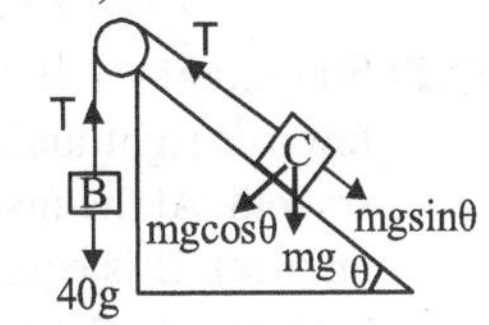

Q.15 A bob is hanging over a pulley inside a car through a string. The second end of the string is in the hand of a person standing in the car. The car is moving with constant acceleration 'a' directed horizontally as shown in figure. Other end of the string is pulled with constant acceleration 'a' vertically. The tension in the string is –

(a) $m\sqrt{g^2+a^2}$

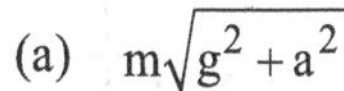

(b) $m\sqrt{g^2+a^2}-ma$
(c) $m\sqrt{g^2+a^2}+ma$
(d) m (g + a)

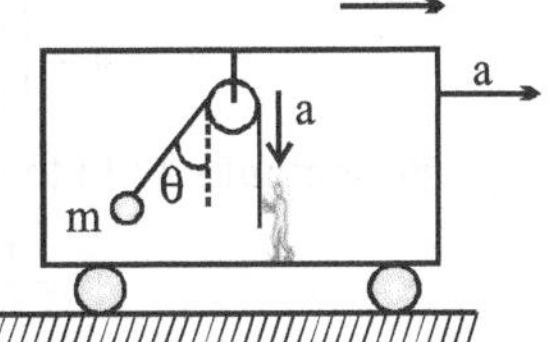

Q.16 In the fig shown, the velocity of each particle at the end of 4 sec will be -
(a) 0.872 m/s
(b) 8.72 m/s
(c) 0.218 m/s
(d) 2.18 m/s

Q.17 In the above example, the height ascended or descended, as the case may be, during that time i.e. 4 sec will be -
(a) 1.744 m (b) 17.44 m
(c) 0.1744 m (d) None of these

Q.18 In the above question, if at the end of 4 sec, the string be cut, the position of each particle in next 2 seconds will respectively be -
(a) 17.856 m, 21.344 m (b) –21.344 m, 17.856 m
(c) –17.856 m, 21.344 m (d) –17.856 m, –21.344 m

Q.19 Consider the double Atwood's machine as shown in the figure. What is acceleration of the masses ?
(a) g/3
(b) g/2
(c) g
(d) g/4

Q.20 In above question, what is the tension in each string ?
(a) mg/3 (b) 4mg/3 (c) 2mg/3 (d) 5mg/3

DIRECTIONS (Q.21-Q.22) : In the following questions, more than one of the answers given are correct. Select the correct answers and mark it according to the following codes:

Codes :
(a) 1, 2 and 3 are correct **(b)** 1 and 2 are correct
(c) 2 and 4 are correct **(d)** 1 and 3 are correct

Q.21 Choose the correct options –
(1) Inertia ∝ mass (2) 1 newton = 10^5 dyne
(3) Thrust on rocket $\vec{F} = \frac{\Delta M}{\Delta t}\vec{v} - M\vec{g}$
(4) Apparent weight of a body in the accelerated lift is W = m (g + a).

Q.22 Choose the correct statements –
(1) For equilibrium of a body under the action of concurrent forces $\vec{F}_1+\vec{F}_2+\vec{F}_3+.....\vec{F}_n=0$
(2) If the downward acceleration of the lift is a = g, then the body will experience weightlessness.
(3) If the downward acceleration of the body is a > g, then the body will rise up to the ceiling of lift
(4) If the downward acceleration of the lift is a > g, then the body will experience weightlessness.

DIRECTIONS (Q.23-Q.25) : Read the passage given below and answer the questions that follows :

A bead of mass m is attached to one end of a spring of natural length R and spring constant $K = \frac{(\sqrt{3}+1)\,mg}{R}$. The other end of the spring is fixed at point A on a smooth ring of radius R as shown in figure. When bead is released to move then

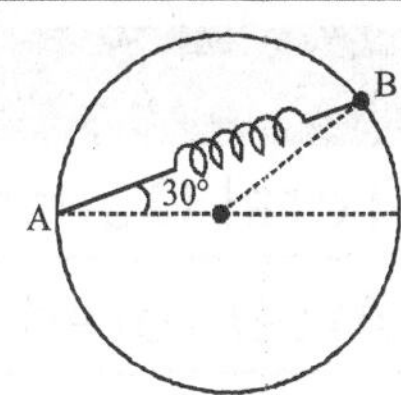

RESPONSE GRID					
	14. ⓐⓑⓒⓓ	15. ⓐⓑⓒⓓ	16. ⓐⓑⓒⓓ	17. ⓐⓑⓒⓓ	18. ⓐⓑⓒⓓ
	19. ⓐⓑⓒⓓ	20. ⓐⓑⓒⓓ	21. ⓐⓑⓒⓓ	22. ⓐⓑⓒⓓ	

Space for Rough Work

Q.23 Initial elongation in the spring is –

(a) R (b) 2R (c) $\sqrt{2}R$ (d) $\sqrt{3}R$

Q.24 The normal reaction force at B is –

(a) $\frac{mg}{2}$ (b) $\sqrt{3}mg$ (c) $3\sqrt{3}mg$ (d) $\frac{3\sqrt{3}mg}{2}$

Q.25 Tangential acceleration of bead just after it is released.

(a) $\frac{g}{2}$ (b) $\frac{3}{4}g$ (c) $\frac{g}{4}$ (d) $\frac{2}{3}g$

DIRECTIONS (Qs. 26-Q.28) : Each of these questions contains two statements: Statement-1 (Assertion) and Statement-2 (Reason). Each of these questions has four alternative choices, only one of which is the correct answer. You have to select the correct choice.

(a) Statement-1 is True, Statement-2 is True; Statement-2 is a correct explanation for Statement-1.
(b) Statement-1 is True, Statement-2 is True; Statement-2 is NOT a correct explanation for Statement-1.
(c) Statement -1 is False, Statement-2 is True.
(d) Statement -1 is True, Statement-2 is False.

Q.26 Statement–1 In fig the ground is smooth and the masses of both the blocks are different. Net force acting on each of block is not same.

Statement– 2 Acceleration of the both blocks will be different.

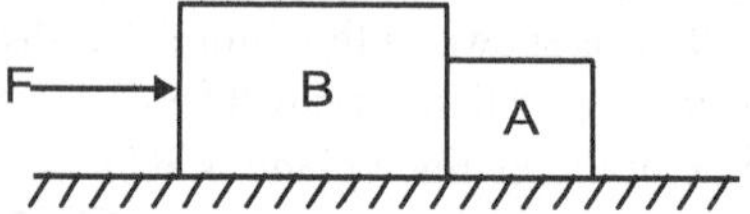

Q.27 Statement– 1 Block A is moving on horizontal surface towards right under the action of force F. All surfaces are smooth. At the instant shown the force exerted by block A on block B is equal to net force on block B.

Statement– 2 From Newton's third law of motion, the force exerted by block A on B is equal in magnitude to force exerted by block B on A.

Q.28 Statement– 1 : In the given fig.

$$a = \left(\frac{m_2 - m_1}{m_1 + m_2}\right)g$$

Statement– 2 : In the given fig.,

$$T = \frac{m_1 + m_2}{2m_1 m_2}g$$

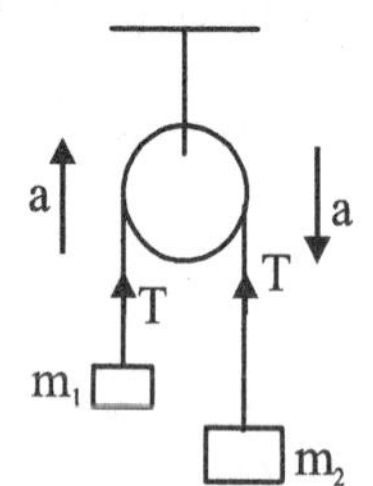

RESPONSE GRID	
	23. ⓐⓑⓒⓓ 24. ⓐⓑⓒⓓ 25. ⓐⓑⓒⓓ 26. ⓐⓑⓒⓓ 27. ⓐⓑⓒⓓ
	28. ⓐⓑⓒⓓ

DAILY PRACTICE PROBLEM SHEET 10 - PHYSICS			
Total Questions	28	Total Marks	112
Attempted		Correct	
Incorrect		Net Score	
Cut-off Score	28	Qualifying Score	42
Success Gap = Net Score – Qualifying Score			
Net Score = (Correct × 4) – (Incorrect × 1)			

Space for Rough Work

DPP - Daily Practice Problems

Name :

Date :

Start Time :

End Time :

PHYSICS 11

SYLLABUS : LAWS OF MOTION-3 (Friction)

Max. Marks : 112 **Time : 60 min.**

GENERAL INSTRUCTIONS

- The Daily Practice Problem Sheet contains 28 MCQ's. For each question only one option is correct. Darken the correct circle/ bubble in the Response Grid provided on each page.
- You have to evaluate your Response Grids yourself with the help of solution booklet.
- Each correct answer will get you 4 marks and 1 mark shall be deduced for each incorrect answer. No mark will be given/ deducted if no bubble is filled. Keep a timer in front of you and stop immediately at the end of 60 min.
- The sheet follows a particular syllabus. Do not attempt the sheet before you have completed your preparation for that syllabus. Refer syllabus sheet in the starting of the book for the syllabus of all the DPP sheets.
- After completing the sheet check your answers with the solution booklet and complete the Result Grid. Finally spend time to analyse your performance and revise the areas which emerge out as weak in your evaluation.

DIRECTIONS (Q.1-Q.20) : There are 20 multiple choice questions. Each question has 4 choices (a), (b), (c) and (d), out of which **ONLY ONE** choice is correct.

Q.1 A body of mass 400 g slides on a rough horizontal surface. If the frictional force is 3.0 N, the angle made by the contact force on the body with the vertical will be

(a) 37° (b) 53° (c) 63° (d) 27°

Q.2 In the above question, the magnitude of the contact force is
($g = 10\ m/s^2$)

(a) 3.0 N (b) 4.0 N (c) 5.0 N (d) 7.0 N

Q.3 The coefficient of static friction between a block of mass m and an inclined plane is $\mu_s = 0.3$. What can be the maximum angle θ of the inclined plane with the horizontal so that the block does not slip on the plane?

(a) $\tan^{-1}(0.1)$ (b) $\tan^{-1}(0.2)$
(c) $\tan^{-1}(0.3)$ (d) $\tan^{-1}(0.4)$

Q.4 The coefficient of static friction between the two blocks shown in figure is μ and the table is smooth. What maximum horizontal force F can be applied to the block of mass M so that the blocks move together?

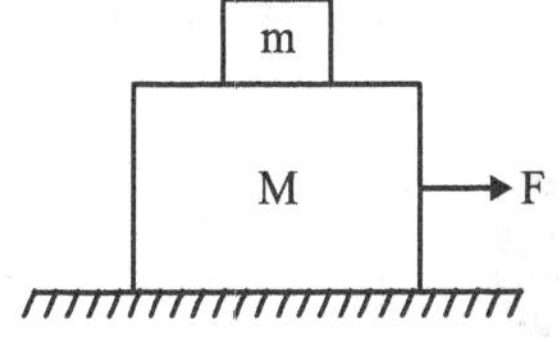

(a) $\mu g(M+m)$ (b) $\mu g(M-m)$
(c) $2\mu g(M+m)$ (d) $\mu g(M+2m)$

RESPONSE GRID 1. ⓐⓑⓒⓓ 2. ⓐⓑⓒⓓ 3. ⓐⓑⓒⓓ 4. ⓐⓑⓒⓓ

Space for Rough Work

Q.5 Block A weighs 4 N and block B weighs 8 N. The coefficient of kinetic friction is 0.25 for all surfaces. Find the force F to slide B at a constant speed when A rests on B and moves with it.

(a) 2N (b) 3N (c) 1N (d) 5N

Q.6 In the above question, find the force F to slide B at a constant speed when A is held at rest.

(a) 2N (b) 3N (c) 1N (d) 4N

Q.7 In the above question, find the force F to slide B at a constant speed when A and B are connected by a light cord passing over a smooth Pulley.

(a) 2N (b) 3N (c) 1N (d) 5N

Q.8 Find the maximum value of M/m in the situation shown in figure so that the system remains at rest. Friction coefficient at both the contacts is μ.

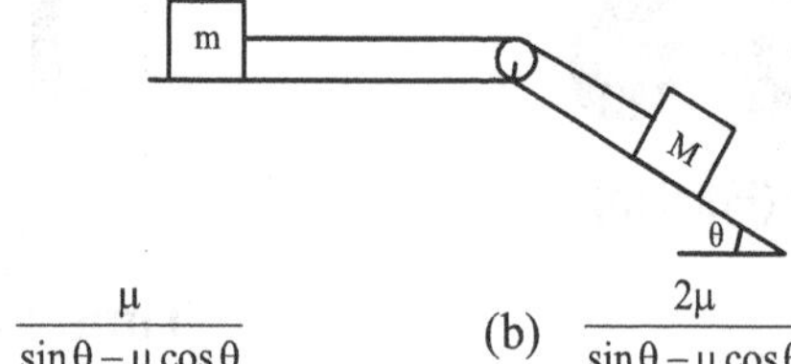

(a) $\frac{\mu}{\sin\theta-\mu\cos\theta}$ (b) $\frac{2\mu}{\sin\theta-\mu\cos\theta}$

(c) $\frac{\mu}{\sin\theta+\mu\cos\theta}$ (d) $\frac{\mu}{\cos\theta-\mu\sin\theta}$

Q.9 A block placed on a horizontal surface is being pushed by a force F making an angle θ with the vertical, if the coefficient of friction is μ, how much force is needed to get the block just started?

(a) $\frac{\mu}{\sin\theta-\mu\cos\theta}$ (b) $\frac{2\mu}{\sin\theta-\mu\cos\theta}$

(c) $\frac{\mu}{\sin\theta+\mu\cos\theta}$ (d) $\frac{\mu}{\cos\theta-\mu\sin\theta}$

Q.10 Assuming the length of a chain to be L and coefficient of static friction μ. Compute the maximum length of the chain which can be held outside a table without sliding.

(a) $\frac{2\mu L}{1+\mu}$ (b) $\frac{\mu L}{1-\mu}$ (c) $\frac{\mu L}{1+\mu}$ (d) $\frac{3\mu L}{1+\mu}$

Q.11 If the coefficient of friction between an insect and bowl is μ and the radius of the bowl is r, find the maximum height to which the insect can crawl in the bowl.

(a) $R\left[1+\frac{1}{\sqrt{(\mu^2+1)}}\right]$ (b) $R\left[1-\frac{1}{\sqrt{(\mu^2-1)}}\right]$

(c) $R\left[1-\frac{2}{\sqrt{(\mu^2+1)}}\right]$ (d) $R\left[1-\frac{1}{\sqrt{(\mu^2+1)}}\right]$

Q.12 A body of mass m is released from the top of a rough inclined plane as shown in figure. If the frictional force be F, then body will reach the bottom with a velocity

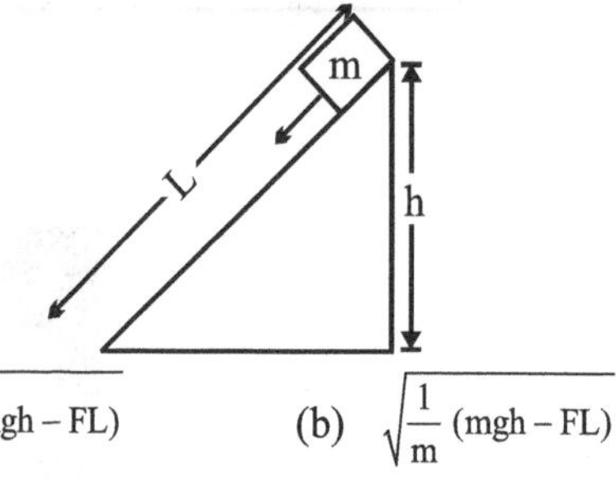

(a) $\sqrt{\frac{2}{m}(mgh-FL)}$ (b) $\sqrt{\frac{1}{m}(mgh-FL)}$

(c) $\sqrt{\frac{2}{m}(mgh+FL)}$ (d) None of these

Q.13 A block of mass 2 kg is placed on the floor. The coefficient of static friction is 0.4. A force F of 2.5 N is applied on the block, as shown. Calculate the force of friction between the block and the floor. ($g = 9.8\ ms^{-2}$)

(a) 2.5 N (b) 25 N (c) 7.84 N (d) zero

Q.14 A block is kept on a horizontal table. The table is undergoing simple harmonic motion of frequency 3 Hz in a horizontal plane. The coefficient of static friction between the block and the table surface is 0.72. Find the maximum amplitude of the table at which the block does not slip on the surface ($g = 10\ ms^{-2}$)

(a) 0.01 m (b) 0.02 m (c) 0.03 m (d) 0.04 m

Q.15 Two cars of unequal masses use similar tyres. If they are moving at the same initial speed, the minimum stopping distance -

(a) is smaller for the heavier car
(b) is smaller for the lighter car
(c) is same for both cars
(d) depends on the volume of the car

RESPONSE GRID				
5. ⓐⓑⓒⓓ	6. ⓐⓑⓒⓓ	7. ⓐⓑⓒⓓ	8. ⓐⓑⓒⓓ	9. ⓐⓑⓒⓓ
10. ⓐⓑⓒⓓ	11. ⓐⓑⓒⓓ	12. ⓐⓑⓒⓓ	13. ⓐⓑⓒⓓ	14. ⓐⓑⓒⓓ
15. ⓐⓑⓒⓓ				

Space for Rough Work

Q.16 Consider the situation shown in figure. The wall is smooth but the surfaces of A and B in contact are rough in equilibrium the friction on B due to A –

(a) is upward
(b) is downward
(c) is zero
(d) the system cannot remain in equilibrium

Q.17 A block is placed on a rough floor and a horizontal force F is applied on it. The force of friction f by the floor on the block is measured for different values of F and a graph is plotted between them –

(i) The graph is a straight line of slope 45°
(ii) The graph is straight line parallel to the F axis
(iii) The graph is a straight line of slope 45° for small F and a straight line parallel to the F-axis for large F.
(iv) There is small kink on the graph

(a) iii, iv (b) i, iv (c) i, ii (d) i, iii

Q.18 The contact force exerted by a body A on another body B is equal to the normal force between the bodies. We conclude that -

(i) the surfaces must be smooth
(ii) force of friction between two bodies may be equal to zero
(iii) magnitude of normal reaction is equal to that of friction
(iv) bodies may be rough

(a) ii, iv (b) i, ii (c) iii, iv (d) i, iv

Q.19 It is easier to pull a body than to push, because -

(a) the coefficient of friction is more in pushing than that in pulling
(b) the friction force is more in pushing than that in pulling
(c) the body does not move forward when pushed
(d) None of these

Q.20 A block of metal is lying on the floor of a bus. The maximum acceleration which can be given to the bus so that the block may remain at rest, will be -

(a) μg (b) μ/g (c) $\mu^2 g$ (d) μg^2

DIRECTIONS (Q.21-Q.23) : In the following questions, more than one of the answers given are correct. Select the correct answers and mark it according to the following codes:

Codes :

(a) 1, 2 and 3 are correct **(b)** 1 and 2 are correct
(c) 2 and 4 are correct **(d)** 1 and 3 are correct

Q.21 Choose the correct statements –

(1) Kinetic friction is lesser than limiting friction.
(2) In rolling the surfaces at contact do not rub each other.
(3) If a body is at rest and no pulling force is acting on it, force of friction on it is zero.
(4) Kinetic friction is greater than limiting friction.

Q.22 Choose the correct statements –

(1) Force of friction is partically independent of microscopic area of surface in contact and relative velocity between them. (if it is not high)
(2) Normally with increase in smoothness friction decreases. But if the surface area are made too smooth by polishing and cleaning the bonding force of adhesion will increase and so the friction will increase resulting in 'Cold welding'
(3) Friction is a non conservative force, i.e. work done against friction is path dependent.
(4) Force of fricton depends on area

Q.23 Choose the correct options –

(1) Friction always opposes the motion
(2) Friction may opposes the motion
(3) If the applied force is increased the force of static friction remains constant.
(4) If the applied force is increased the force of static friction also increases upto limiting friction.

DIRECTIONS (Q.24-Q.26) : Read the passage given below and answer the questions that follows :

A block of mass 1 kg is placed on a rough horizontal surface. A spring is attached to the block whose other end is joined to a rigid wall, as shown in the figure. A horizontal force is applied on the block so that it remains at rest while the spring is elongated by x.

RESPONSE GRID					
	16. ⓐⓑⓒⓓ	17. ⓐⓑⓒⓓ	18. ⓐⓑⓒⓓ	19. ⓐⓑⓒⓓ	20. ⓐⓑⓒⓓ
	21. ⓐⓑⓒⓓ	22. ⓐⓑⓒⓓ	23. ⓐⓑⓒⓓ		

Space for Rough Work

$x \geq \frac{\mu mg}{k}$. Let F_{max} and F_{min} be the maximum and minimum values of force F for which the block remains a equilibrium. For a particular x,

$$F_{max} - F_{min} = 2N.$$

Also shown is the variation of $F_{max} + F_{min}$ versus x, the elongation of the spring.

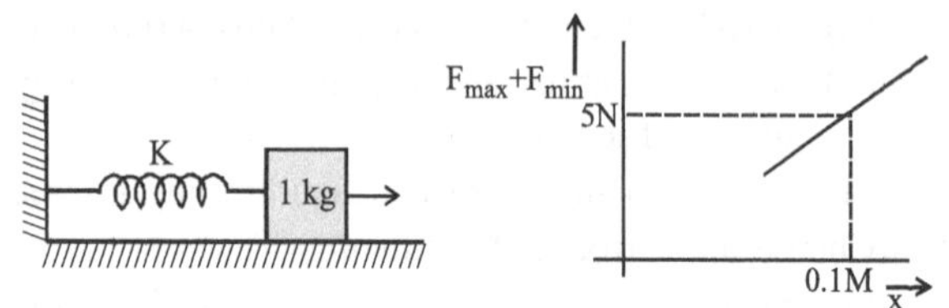

Q.24 The coefficient of friction between the block and the horizontal surface is –
(a) 0.1 (b) 0.2
(c) 0.3 (d) 0.4

Q.25 The spring constant of the spring is –
(a) 25 N/m (b) 20 N/m
(c) 2.5 N/m (d) 50 N/m

Q.26 The value of F_{min}, if x = 3 cm. is –
(a) 0 (b) 0.2 N
(c) 5N (d) 1N

DIRECTIONS (Q. 27-Q.28) : Each of these questions contains two statements: Statement-1 (Assertion) and Statement-2 (Reason). Each of these questions has four alternative choices, only one of which is the correct answer. You have to select the correct choice.

(a) Statement-1 is True, Statement-2 is True; Statement-2 is a correct explanation for Statement-1.
(b) Statement-1 is True, Statement-2 is True; Statement-2 is NOT a correct explanation for Statement-1.
(c) Statement -1 is False, Statement-2 is True.
(d) Statement -1 is True, Statement-2 is False.

Q.27 A solid sphere and a hollow sphere of same mass *M* and same radius *R* are released from the top of a rough inclined plane. Friction coefficient is same for both the bodies. If both bodies perform imperfect rolling, then

Statement - 1 : Work done by friction for the motion of bodies from top of incline to the bottom will be same for both the bodies.

Statement - 2 : Force of friction will be same for both the bodies.

Q.28 Statement - 1 : Maximum value of friction force between two surfaces is μ × normal reaction.
where μ = coefficient of friction between surfaces.

Statement - 2 : Friction force between surfaces of two bodies is always less than or equal to externally applied force.

RESPONSE GRID	24.ⓐⓑⓒⓓ	25.ⓐⓑⓒⓓ	26.ⓐⓑⓒⓓ	27.ⓐⓑⓒⓓ	28.ⓐⓑⓒⓓ

DAILY PRACTICE PROBLEM SHEET 11 - PHYSICS			
Total Questions	28	Total Marks	112
Attempted		Correct	
Incorrect		Net Score	
Cut-off Score	26	Qualifying Score	42
Success Gap = Net Score – Qualifying Score			
Net Score = (Correct × 4) – (Incorrect × 1)			

DPP - Daily Practice Problems

Name :

Date :

Start Time :

End Time :

PHYSICS 12

SYLLABUS : Work, Energy and Power-1 (Work by constant and variable forces, kinetic and potential energy, work energy theorem)

Max. Marks : 112 **Time : 60 min.**

GENERAL INSTRUCTIONS

- The Daily Practice Problem Sheet contains 28 MCQ's. For each question only one option is correct. Darken the correct circle/ bubble in the Response Grid provided on each page.
- You have to evaluate your Response Grids yourself with the help of solution booklet.
- Each correct answer will get you 4 marks and 1 mark shall be deduced for each incorrect answer. No mark will be given/ deducted if no bubble is filled. Keep a timer in front of you and stop immediately at the end of 60 min.
- The sheet follows a particular syllabus. Do not attempt the sheet before you have completed your preparation for that syllabus. Refer syllabus sheet in the starting of the book for the syllabus of all the DPP sheets.
- After completing the sheet check your answers with the solution booklet and complete the Result Grid. Finally spend time to analyse your performance and revise the areas which emerge out as weak in your evaluation.

DIRECTIONS (Q.1-Q.20) : There are 20 multiple choice questions. Each question has 4 choices (a), (b), (c) and (d), out of which **ONLY ONE** choice is correct.

Q.1 A body is acted upon by a force $\vec{F} = -\hat{i} + 2\hat{j} + 3\hat{k}$. The work done by the force in displacing it from (0,0,0) to (0,0,4m) will be -

(a) 12 J (b) 10 J
(c) 8 J (d) 6 J

Q.2 The work done in pulling a body of mass 5 kg along an inclined plane (angle 60°) with coefficient of friction 0.2 through 2 m, will be -

(a) 98.08 J (b) 94.08 J
(c) 90.08 J (d) 91.08 J

Q.3 A force $\vec{F} = (7 - 2x + 3x^2)$ N is applied on a 2 kg mass which displaces it from x = 0 to x = 5 m. Work done in joule is -

(a) 70 (b) 270 (c) 35 (d) 135

Q.4 An automobile of mass m accelerates from rest. If the engine supplies a constant power P, the velocity at time t is given by -

(a) $v = \frac{Pt}{m}$ (b) $v = \frac{2Pt}{m}$ (c) $\sqrt{\frac{Pt}{m}}$ (d) $\sqrt{\frac{2Pt}{m}}$

Q.5 In the above question, the position (s) at time (t) is given by -

(a) $\left(\frac{2Pt}{m}\right)t$ (b) $\left(\frac{8P}{9m}\right)^{1/2} t^{3/2}$

(c) $\left(\frac{9P}{8m}\right)^{1/2} t^{1/2}$ (d) $\left(\frac{8P}{9m}\right)^{1/2} t$

RESPONSE GRID	1. ⓐⓑⓒⓓ	2. ⓐⓑⓒⓓ	3. ⓐⓑⓒⓓ	4. ⓐⓑⓒⓓ	5. ⓐⓑⓒⓓ

Space for Rough Work

Q.6 A particle moving in a straight line is acted by a force, which works at a constant rate and changes its velocity from u to v in passing over a distance x. The time taken will be -

(a) $x = \frac{v-u}{v^2+u^2}$ (b) $x\left(\frac{v+u}{v^2+u^2}\right)$

(c) $\frac{3}{2}(x)\left(\frac{v^2-u^2}{v^3-u^3}\right)$ (d) $x\left(\frac{v}{u}\right)$

Q.7 A chain of linear density 3 kg /m and length 8 m is lying on the table with 4 m of chain hanging from the edge. The work done in lifting the chain on the table will be -

(a) 117.6 J (b) 235.2 J (c) 98 J (d) 196 J

Q.8 The work done in lifting water from a well of depth 6 m using a bucket of mass 0.5 kg and volume 2 litre, will be-

(a) 73.5 J (b) 147 J (c) 117. 6 J (d) 98 J

Q.9 An object of mass 5 kg falls from rest through a vertical distance of 20 m and reaches a velocity of 10 m/s. How much work is done by the push of the air on the object ? ($g = 10\ m/s^2$).

(a) 350 J (b) 750 J (c) 200 J (d) 300 J

Q.10 A boy pulls a 5 kg block 20 metres along a horizontal surface at a constant speed with a force directed 45° above the horizontal. If the coefficient of kinetic friction is 0.20, how much work does the boy do on the block?

(a) 163.32 J (b) 11.55 J
(c) 150 J (d) 115 J

Q.11 A uniform chain is held on a frictionless table with one-fifth of its length hanging over the edge. If the chain has a length ℓ and a mass m, how much work is required to pull the hanging part back on the table ?

(a) $mg\ \ell / 10$ (b) $mg\ \ell / 5$
(c) $mg\ \ell / 50$ (d) $mg\ \ell / 2$

Q.12 A bus of mass 1000 kg has an engine which produces a constant power of 50 kW. If the resistance to motion, assumed constant is 1000 N. The maximum speed at which the bus can travel on level road and the acceleration when it is travelling at 25 m/s, will respectively be -

(a) 50 m/s, $1.0\ m/s^2$ (b) 1.0 m/s, $50\ m/s^2$
(c) 5.0 m/s, $10\ m/s^2$ (d) 10 m/s, $5 m/s^2$

Q.13 The power output of a $_{92}U^{235}$ reactor if it takes 30 days to use up 2 kg of fuel and if each fission gives 185 MeV of energy (Avogadro number = 6×10^{23}/mole) will be -

(a) 58.4 MW (b) 5.84 MW
(c) 584 W (d) 5840 MW

Q.14 The stopping distance for a vehicle of mass M moving with a speed v along a level road, will be -
(μ is the coefficient of friction between tyres and the road)

(a) $\frac{v^2}{\mu g}$ (b) $\frac{2v^2}{\mu g}$ (c) $\frac{v^2}{2\mu g}$ (d) $\frac{v}{\mu g}$

Q.15 The earth circles the sun once a year. How much work would have to be done on the earth to bring it to rest relative to the sun, (ignore the rotation of earth about - its own axis) Given that mass of the earth is 6×10^{24} kg and distance between the sun and earth is 1.5×10^8 km-

(a) 2.7×10^{33} (b) 2.7×10^{24}
(c) 1.9×10^{23} (d) 1.9×10^{24}

Q.16 A particle of mass m is moving in a horizontal circle of radius r, under a centripetal force equal to $(-k/r^2)$, where k is a constant. The total energy of the particle is -

(a) k/2r (b) –k/2r
(c) kr (d) –k/r

Q.17 The work done by a person in carrying a box of mass 10 kg through a vertical height of 10 m is 4900 J. The mass of the person is -

(a) 60 kg (b) 50 kg
(c) 40 kg (d) 130 kg

Q.18 A uniform rod of length 4 m and mass 20 kg is lying horizontal on the ground. The work done in keeping it vertical with one of its ends touching the ground, will be -

(a) 784 J (b) 392 J (c) 196 J (d) 98 J

Q.19 If g is the acceleration due to gravity on the earth's surface, the gain in the potential energy of an object of mass m raised from surface of the earth to a height equal to radius R of the earth is - [M = mass of earth]

(a) $\frac{GMm}{2R}$ (b) $\frac{GM}{R}$ (c) $\frac{GMm}{R}$ (d) $\frac{GM}{2R}$

RESPONSE GRID					
	6. ⓐⓑⓒⓓ	7. ⓐⓑⓒⓓ	8. ⓐⓑⓒⓓ	9. ⓐⓑⓒⓓ	10. ⓐⓑⓒⓓ
	11. ⓐⓑⓒⓓ	12. ⓐⓑⓒⓓ	13. ⓐⓑⓒⓓ	14. ⓐⓑⓒⓓ	15. ⓐⓑⓒⓓ
	16. ⓐⓑⓒⓓ	17. ⓐⓑⓒⓓ	18. ⓐⓑⓒⓓ	19. ⓐⓑⓒⓓ	

Space for Rough Work

Q.20 The potential energy between two atoms in a molecule is given by, $U_{(x)} = \frac{a}{x^{12}} - \frac{b}{x^6}$, where a and b are positive constant and x is the distance between the atoms. The atoms is an stable equilibrium, when-

(a) $x = 0$ (b) $x = \left(\frac{a}{2b}\right)^{1/6}$

(c) $x = \left(\frac{2a}{b}\right)^{1/6}$ (d) $x = \left(\frac{11a}{5b}\right)^{1/6}$

DIRECTIONS (Q.21-Q.22) : In the following questions, more than one of the answers given are correct. Select the correct answers and mark it according to the following codes:

Codes :

(a) 1, 2 and 3 are correct

(b) 1 and 2 are correct

(c) 2 and 4 are correct

(d) 1 and 3 are correct

Q.21 A man pushes a wall and fails to displace it. Choose incorrect statements related to his work

(1) Negative work

(2) Positive but not maximum work

(3) Maximum work

(4) No work at all

Q.22 Choose the correct options –

(1) The work done by forces may be equal to change in kinetic energy

(2) The work done by forces may be equal to change in potential energy

(3) The work done by forces may be equal to change in total energy

(4) The work done by forces must be equal to change in potential energy.

DIRECTIONS (Q.23-Q.25) : Read the passage given below and answer the questions that follows :

In the figure shown, the system is released from rest with both the springs in unstretched positions. Mass of each block is 1 kg and force constant of each spring is 10 N/m.

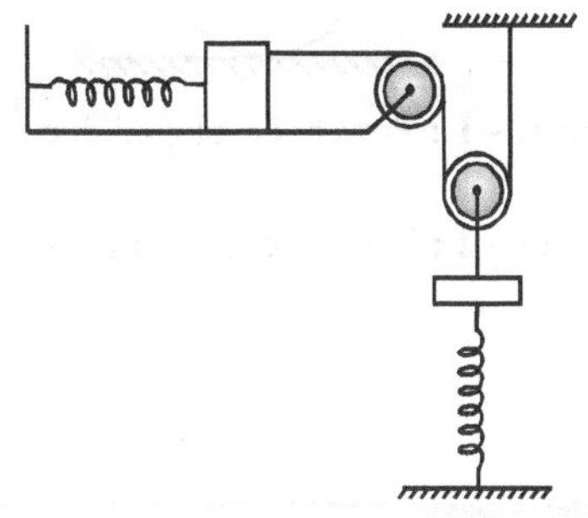

Q.23 Extension of horizontal spring in equilibrium is:

(a) 0.2 m (b) 0.4 m

(c) 0.6 m (d) 0.8 m

Q.24 Extension of vertical spring in equilibrium is

(a) 0.4 m (b) 0.2 m

(c) 0.6 m (d) 0.8 m

Q.25 Maximum speed of the block placed horizontally is:

(a) 3.21 m/s (b) 2.21 m/s

(c) 1.93 m/s (d) 1.26 m/s

DIRECTIONS (Qs. 26-Q.28) : Each of these questions contains two statements: Statement-1 (Assertion) and Statement-2 (Reason). Each of these questions has four alternative choices, only one of which is the correct answer. You have to select the correct choice.

(a) Statement-1 is True, Statement-2 is True; Statement-2 is a correct explanation for Statement-1.

(b) Statement-1 is True, Statement-2 is True; Statement-2 is NOT a correct explanation for Statement-1.

(c) Statement -1 is False, Statement-2 is True.

(d) Statement -1 is True, Statement-2 is False.

RESPONSE GRID	
20. ⓐⓑⓒⓓ 21. ⓐⓑⓒⓓ 22. ⓐⓑⓒⓓ 23. ⓐⓑⓒⓓ 24. ⓐⓑⓒⓓ	
25. ⓐⓑⓒⓓ	

Q.26 As shown in the figure, a uniform sphere is rolling on a horizontal surface without slipping, under the action of a horizontal force F.

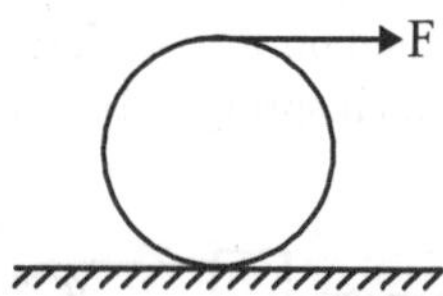

Statement - 1 : Power developed due to friction force is zero.

Statement - 2 : Power developed by gravity force is non-zero.

Q.27 Statement - 1 : Sum of work done by the Newton's 3rd law pair internal forces, acting between two particles may be zero.

Statement - 2 : If two particles undergo same displacement then work done by Newton's 3rd law pair forces on them is of opposite sign and equal magnitude.

Q.28 Statement - 1: A particle moves along a straight line with constant velocity. Now a constant non-zero force is applied on the particle in direction opposite to its initial velocity. After the force is applied, the net work done by this force may be zero in certain time intervals.

Statement - 2 : The work done by a force acting on a particle is zero in any time interval if the force is always perpendicular to velocity of the particle.

RESPONSE GRID	26. ⓐⓑⓒⓓ 27. ⓐⓑⓒⓓ 28. ⓐⓑⓒⓓ

DAILY PRACTICE PROBLEM SHEET 12 - PHYSICS			
Total Questions	28	Total Marks	112
Attempted		Correct	
Incorrect		Net Score	
Cut-off Score	28	Qualifying Score	44
Success Gap = Net Score – Qualifying Score			
Net Score = (Correct × 4) – (Incorrect × 1)			

Space for Rough Work

DPP - Daily Practice Problems

Name :

Date :

Start Time :

End Time :

PHYSICS 13

SYLLABUS : Work, Energy and Power-2 (Conservation of momentum and energy, collision, rocket case)

Max. Marks : 112 **Time : 60 min.**

GENERAL INSTRUCTIONS

- The Daily Practice Problem Sheet contains 28 MCQ's. For each question only one option is correct. Darken the correct circle/ bubble in the Response Grid provided on each page.
- You have to evaluate your Response Grids yourself with the help of solution booklet.
- Each correct answer will get you 4 marks and 1 mark shall be deduced for each incorrect answer. No mark will be given/ deducted if no bubble is filled. Keep a timer in front of you and stop immediately at the end of 60 min.
- The sheet follows a particular syllabus. Do not attempt the sheet before you have completed your preparation for that syllabus. Refer syllabus sheet in the starting of the book for the syllabus of all the DPP sheets.
- After completing the sheet check your answers with the solution booklet and complete the Result Grid. Finally spend time to analyse your performance and revise the areas which emerge out as weak in your evaluation.

DIRECTIONS (Q.1-Q.20) : There are 20 multiple choice questions. Each question has 4 choices (a), (b), (c) and (d), out of which **ONLY ONE** choice is correct.

Q.1 A rifle man, who together with his rifle has a mass of 100 kg, stands on a smooth surface fires 10 shots horizontally. Each bullet has a mass 10 gm and a muzzle velocity of 800 m/s. What velocity does rifle man acquire at the end of 10 shots

(a) 0.8 m/s (b) 0.5 m/s (c) 0.3 m/s (d) 1.2 m/s

Q.2 A bullet of mass 10 g travelling horizontally with a velocity of 300 m/s strikes a block of wood of mass 290 g which rests on a rough horizontal floor. After impact the block and the bullet move together and come to rest when the block has travelled a distance of 15 m. The coefficient of friction between the block and the floor will be - (Duration of impact is very short)

(a) $\frac{1}{2}$ (b) $\frac{2}{3}$ (c) $\frac{1}{3}$ (d) $\frac{3}{4}$

Q.3 A 20 g bullet pierces through a plate of mass $m_1 = 1$ kg and then comes to rest inside a second plate of mass $m_2 = 2.98$ kg. It is found that the two plates, initially at rest, now move with equal velocities. The percentage loss in the initial velocity of bullet when it is between m_1 and m_2. (Neglect any loss of material of the bodies, due to action of bullet.) will be -

(a) 20% (b) 25%
(c) 30% (d) 45%

RESPONSE GRID	1. ⓐⓑⓒⓓ	2. ⓐⓑⓒⓓ	3. ⓐⓑⓒⓓ

Space for Rough Work

Q.4 A bullet of mass 20 g hits a block of mass 1.98 kg suspended from a massless string of length 100 cm and sticks to it. The bullet flies down at an angle of 30° to the horizontal with a velocity of 200 m/s. Through what height the block will rise-

(a) 0.15 m

(b) 0.30 m

(c) 0.45 m

(d) 0.75 m

Q.5 A bullet of mass 0.01 kg travelling at a speed of 500 m/s strikes a block of mass 2 kg, which is suspended by a string of length 5 m. The centre of gravity of the block is found to rise a vertical distance of 0.1 m. The speed of the bullet after it emerges from the block will be -

(a) 1.4 m/s

(b) 110 m/s

(c) 220 m/s

(d) 14 m/s

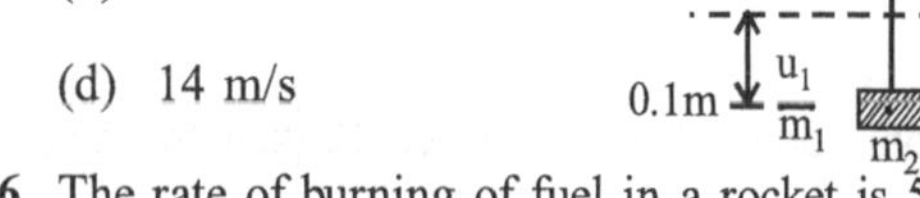

Q.6 The rate of burning of fuel in a rocket is 50 gm/sec. and comes out with and velocity 4×10^3 m/s. The force exerted by gas on rocket will be -

(a) 200 N (b) 250 N
(c) 2.5×10^6 N (d) 2.5×10^4 N

Q.7 A body of mass 1 kg strikes elastically with another body at rest and continues to move in the same direction with one fourth of its initial velocity. The mass of the other body is -

(a) 0.6 kg (b) 2.4 kg (c) 3 kg (d) 4 kg

Q.8 A ball moving with a speed of 9 m/s strikes with an identical stationary ball such that after the collision the direction of each ball makes an angle of 30° with the original line of motion. Find the speeds of the two balls after the collision. Is the kinetic energy conserved in this collision process ?

(a) $3\sqrt{3}$ m/s, no (b) $3\sqrt{3}$ m/s, no
(c) $6\sqrt{3}$ m/s, yes (d) 0, yes

Q.9 The mass of a rocket is 500 kg and the relative velocity of the gases ejecting from it is 250 m/s with respect to the rocket. The rate of burning of the fuel in order to give the rocket an initial acceleration 20 m/s^2 in the vertically upward direction (g = 10 m/s^2), will be -

(a) 30 kg/s (b) 60 kg/s
(c) 45 kg/s (d) 10 kg/s

Q.10 A slow moving electron collides elastically with a hydrogen atom at rest. The initial and final motions are along the same straight line. What fraction of electron's kinetic energy is transferred to the hydrogen atom? The mass of hydrogen atom is 1850 times the mass of electron.

(a) 0.217 % (b) 2.17 % (c) 0.0217 % (d) 21.7 %

Q.11 A particle of mass 4 m which is at rest explodes into three fragments, two of the fragments each of mass m are found to move each with a speed v making an angle 90° with each other. The total energy relased in this explosion is -

(a) $\frac{1}{2}mv^2$ (b) mv^2
(c) $\frac{3}{2}mv^2$ (d) $2mv^2$

Q.12 A body of mass M splits into two parts αM and (1 – α) M by an internal explosion, which generates kinetic energy T. After explosion if the two parts move in the same direction as before, their relative speed will be -

(a) $\sqrt{\frac{T}{(1-\alpha)M}}$ (b) $\sqrt{\frac{2T}{\alpha(1-\alpha)M}}$
(c) $\sqrt{\frac{T}{2(1-\alpha)M}}$ (d) $\sqrt{\frac{2T}{(1-\alpha)M}}$

Q.13 A body of mass 1 kg initially at rest explodes and breaks into three fragments of masses in the ratio 1 : 1 : 3. The two pieces of equal mass fly off perpendicular to each other with a speed of 30 m/sec each. What is the velocity of the heavier fragment ?

(a) $10\sqrt{2}$ m/s (b) $15\sqrt{2}$ m/s
(c) $5\sqrt{2}$ m/s (d) $20\sqrt{2}$ m/s

RESPONSE GRID					
	4. ⓐⓑⓒⓓ	5. ⓐⓑⓒⓓ	6. ⓐⓑⓒⓓ	7. ⓐⓑⓒⓓ	8. ⓐⓑⓒⓓ
	9. ⓐⓑⓒⓓ	10. ⓐⓑⓒⓓ	11. ⓐⓑⓒⓓ	12. ⓐⓑⓒⓓ	13. ⓐⓑⓒⓓ

Space for Rough Work

Q.14 A body of mass m moving with a velocity v_1 in the X-direction collides with another body of mass M moving in Y-direction with a velocity v_2. They coalasce into one body during collision. The magnitude and direction of the momentum of the final body, will be-

(a) $\sqrt{(mv_1)+(Mv_2)}$, $\tan^{-1}\left(\frac{Mv_2}{mv_1}\right)$

(b) $\sqrt{(mv_1)+(Mv_2)}$, $\tan^{-1}\left(\frac{Mv_1}{mv_2}\right)$

(c) $\sqrt{(mv_1)^2+(Mv_2)^2}$, $\tan^{-1}\left(\frac{Mv_2}{mv_1}\right)$

(d) $\sqrt{(mv_1)^2+(Mv_2)^2}$, $\tan^{-1}\left(\frac{Mv_1}{mv_2}\right)$

Q.15 A ball of mass m hits a wall with a speed v making an angle θ with the normal. If the coefficient of restitution is e, the direction and magnitude of the velocity of ball after reflection from the wall will respectively be -

(a) $\tan^{-1}\left(\frac{\tan\theta}{e}\right)$, $v\sqrt{\sin^2\theta+e^2\cos^2\theta}$

(b) $\tan^{-1}\left(\frac{e}{\tan\theta}\right)$, $\frac{1}{v}\sqrt{e^2\sin^2\theta+\cos^2\theta}$

(c) $\tan^{-1}(e\tan\theta)$, $\frac{v}{e}\tan\theta$

(d) $\tan^{-1}(e\tan\theta)$, $v\sqrt{\sin^2\theta+e^2}$

Q.16 A tennis ball dropped from a height of 2 m rebounds only 1.5 metre after hitting the ground. What fraction of energy is lost in the impact?

(a) 1/2 (b) 1/4 (c) 1/8 (d) 1/16

Q.17 A bullet is fired from the gun. The gun recoils, the kinetic energy of the recoil shall be-

(a) equal to the kinetic energy of the bullet
(b) less than the kinetic energy of the bullet
(c) greater than the kinetic energy of the bullet
(d) double that of the kinetic energy of the bullet

Q.18 Conservation of linear momentum is equivalent to-

(a) Newton's second law of motion
(b) Newton's first law of motion
(c) Newton's third law of motion
(d) Conservation of angular momentum.

Q.19 In an inelastic collision-

(a) momentum is conserved but kinetic energy is not conserved
(b) momentum is not conserved but kinetic energy is conserved
(c) neither momentum nor kinetic energy is conserved
(d) both the momentum and kinetic energy are conserved

Q.20 Inelastic collision is the-

(a) collision of ideal gas molecules with the walls of the container
(b) collision of electron and positron to an inhilate each other.
(c) collision of two rigid solid spheres lying on a frictionless table
(d) scattering of α-particles with the nucleus of gold atom

DIRECTIONS (Q.21-Q.23) : In the following questions, more than one of the answers given are correct. Select the correct answers and mark it according to the following codes:

Codes :

(a) 1, 2 and 3 are correct **(b)** 1 and 2 are correct
(c) 2 and 4 are correct **(d)** 1 and 3 are correct

Q.21 Which of the following statements is false for collisions-

(1) Momentum is conserved in elastic collisions but not in inelastic collisions.
(2) Total-kinetic energy is conserved in elastic collisions but momentum is not conserved.
(3) Total kinetic energy and momentum both are conserved in all types of collisions
(4) Total kinetic energy is not conserved in inelastic collisions but momentum is conserved

Q.22 Which of the following hold when two particles of masses m_1 and m_2 undergo elastic collision?

(1) When $m_1 = m_2$ and m_2 is stationary, there is maximum transfer of kinetic energy in head on collision
(2) When $m_1 = m_2$ and m_2 is stationary, there is maximum transfer of momentum in head on collision
(3) When $m_1 >> m_2$ and m_2 is stationary, after head on collision m_2 moves with twice the velocity of m_1.
(4) When the collision is oblique and $m_1 = m_2$ with m_2 stationary, after the collision the particle move in opposite directions.

RESPONSE GRID					
	14. ⓐⓑⓒⓓ	15. ⓐⓑⓒⓓ	16. ⓐⓑⓒⓓ	17. ⓐⓑⓒⓓ	18. ⓐⓑⓒⓓ
	19. ⓐⓑⓒⓓ	20. ⓐⓑⓒⓓ	21. ⓐⓑⓒⓓ	22. ⓐⓑⓒⓓ	

Space for Rough Work

Q.23 Two balls at the same temperature collide inelastically. Which of the following is not conserved?
(1) Kinetic energy (2) Velocity
(3) Temperature (4) Momentum

DIRECTIONS (Q.24-Q.26) : Read the passage given below and answer the questions that follows :

A small particle of mass m/10 is moving horizontally at a height of $3R/2$ from ground with velocity 10 m/s. A perfectly inelastic collision occurs at point P of sphere of mass m placed on smooth horizontal surface. The radius of sphere is R. (m = 10 kg and R = 0.1 m) (Assume all surfaces to be smooth).

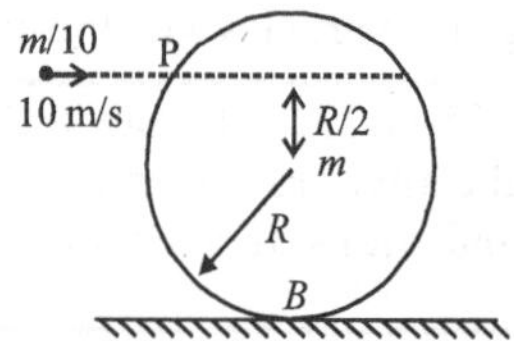

Q.24 Speed of particle just after collision is
(a) approx 5.0 m/s (b) approx 10 m/s
(c) approx. 15.0 m/s (d) approx 20.0 m/s

Q.25 Speed of sphere just after collision is
(a) 27/43 m/s (b) 30/43 m/s
(c) 35/43 m/s (d) 40/43 m/s

Q.26 Angular speed of sphere just after collision is
(a) zero (b) 2 rad/sec
(c) 2.5 rad/sec (d) 3 rad/sec

DIRECTIONS (Q. 27-Q.28) : Each of these questions contains two statements: Statement-1 (Assertion) and Statement-2 (Reason). Each of these questions has four alternative choices, only one of which is the correct answer. You have to select the correct choice.

(a) Statement-1 is True, Statement-2 is True; Statement-2 is a correct explanation for Statement-1.
(b) Statement-1 is True, Statement-2 is True; Statement-2 is NOT a correct explanation for Statement-1.
(c) Statement -1 is False, Statement-2 is True.
(d) Statement -1 is True, Statement-2 is False.

Q.27 A particle of mass m strikes a wedge of mass M horizontally as shown in the figure.

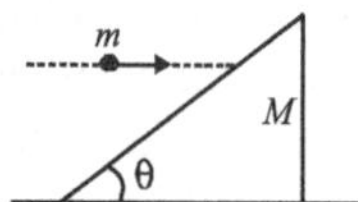

Statement - 1 : If collision is perfectly inelastic then, it can be concluded that the particle sticks to the wedge.
Statement - 2 : In perfectly inelastic collision velocity of both bodies is same along common normal just after collision.

Q.28 Statement - 1 : In an elastic collision in one dimension between two bodies, total momentum remains the same before, during and after the collision.
Statement - 2 : In an elastic collision in one dimension between two bodies, total kinetic energy remains the same before, during and after the collision.
[Assume external forces are absent in both the above statements].

RESPONSE GRID	
	23. ⓐⓑⓒⓓ 24. ⓐⓑⓒⓓ 25. ⓐⓑⓒⓓ 26. ⓐⓑⓒⓓ 27. ⓐⓑⓒⓓ 28. ⓐⓑⓒⓓ

DAILY PRACTICE PROBLEM SHEET 13 - PHYSICS			
Total Questions	28	Total Marks	112
Attempted		Correct	
Incorrect		Net Score	
Cut-off Score	28	Qualifying Score	44
Success Gap = Net Score – Qualifying Score			
Net Score = (Correct × 4) – (Incorrect × 1)			

Space for Rough Work

DPP - Daily Practice Problems

Name :

Date :

Start Time :

End Time :

PHYSICS 14

SYLLABUS : Centre of mass and its motion

Max. Marks : 112 **Time : 60 min.**

GENERAL INSTRUCTIONS

- The Daily Practice Problem Sheet contains 28 MCQ's. For each question only one option is correct. Darken the correct circle/ bubble in the Response Grid provided on each page.
- You have to evaluate your Response Grids yourself with the help of solution booklet.
- Each correct answer will get you 4 marks and 1 mark shall be deduced for each incorrect answer. No mark will be given/ deducted if no bubble is filled. Keep a timer in front of you and stop immediately at the end of 60 min.
- The sheet follows a particular syllabus. Do not attempt the sheet before you have completed your preparation for that syllabus. Refer syllabus sheet in the starting of the book for the syllabus of all the DPP sheets.
- After completing the sheet check your answers with the solution booklet and complete the Result Grid. Finally spend time to analyse your performance and revise the areas which emerge out as weak in your evaluation.

DIRECTIONS (Q.1-Q.20) : There are 20 multiple choice questions. Each question has 4 choices (a), (b), (c) and (d), out of which **ONLY ONE** choice is correct.

Q.1 In the *HCl* molecule, the separation between the nuclei of the two atom is about 1.27 A° (1A° = 10^{-10} m). The approximate location of the centre of mass from the hydrogen atom, assuming the chlorine atom to be about 35.5 times massive as hydrogen is

(a) 1 Å (b) 2.5 Å (c) 1.24 Å (d) 1.5 Å

Q.2 A 2 kg body and a 3 kg body are moving along the x-axis. At a particular instant the 2 kg body has a velocity of 3 ms^{-1} and the 3 kg body has the velocity of $2ms^{-1}$. The velocity of the centre of mass at that instant is

(a) 5 ms^{-1} (b) 1 ms^{-1}
(c) 0 (d) None of these

Q.3 The distance between the carbon atom and the oxygen atom in a carbon monoxide molecule is 1.1 Å. Given, mass of carbon atom is 12 a.m.u. and mass of oxygen atom is 16 a.m.u., calculate the position of the centre of mass of the carbon monoxide molecule

(a) 6.3 Å from the carbon atom
(b) 1 Å from the oxygen atom
(c) 0.63 Å from the carbon atom
(d) 0.12 Å from the oxygen atom

Q.4 The velocities of three particles of masses 20g, 30g and 50g are $10\hat{i}, 10\hat{j}$ and $10\hat{k}$ respectively. The velocity of the centre of mass of the three particles is

(a) $2\hat{i}+3\hat{j}+5\hat{k}$ (b) $10(\hat{i}+\hat{j}+\hat{k})$
(c) $20\hat{i}+30\hat{j}+5\hat{k}$ (d) $2\hat{i}+30\hat{j}+50\hat{k}$

RESPONSE GRID	1. ⓐⓑⓒⓓ	2. ⓐⓑⓒⓓ	3. ⓐⓑⓒⓓ	4. ⓐⓑⓒⓓ

Space for Rough Work

Q.5 The centre of mass of a triangle shown in figure has coordinates

(a) $x=\frac{h}{2}, y=\frac{b}{2}$

(b) $x=\frac{b}{2}, y=\frac{h}{2}$

(c) $x=\frac{b}{3}, y=\frac{h}{3}$

(d) $x=\frac{h}{3}, y=\frac{b}{3}$

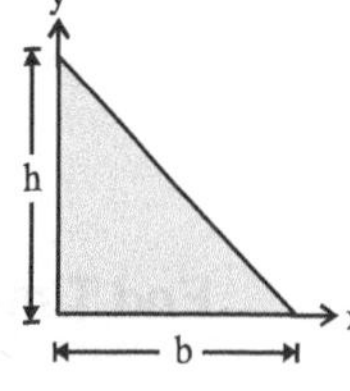

Q.6 Two bodies of masses 2 kg and 4 kg are moving with velocities 2 m/s and 10 m/s respectively along same direction. Then the velocity of their centre of mass will be

(a) 8.1 m/s (b) 7.3 m/s
(c) 6.4 m/s (d) 5.3 m/s

Q.7 Four particles of masses *m*, 2*m*, 3*m* and 4*m* are arranged at the corners of a parallelogram with each side equal to *a* and one of the angle between two adjacent sides is 60°. The parallelogram lies in the *x*-*y* plane with mass *m* at the origin and 4m on the *x*-axis. The centre of mass of the arrangement will be located at

(a) $\left(\frac{\sqrt{3}}{2}a, 0.95a\right)$ (b) $\left(0.95a, \frac{\sqrt{3}}{4}a\right)$

(c) $\left(\frac{3a}{4}, \frac{a}{2}\right)$ (d) $\left(\frac{a}{2}, \frac{3a}{4}\right)$

Q.8 Three identical metal balls each of radius *r* are placed touching each other on a horizontal surface such that an equilateral triangle is formed, when centres of three balls are joined. The centre of the mass of system is located at

(a) Horizontal surface
(b) Centre of one of the balls
(c) Line joining centres of any two balls
(d) Point of intersection of the medians

Q.9 2 bodies of different masses of 2 kg and 4 kg are moving with velocities 20 m/s and 10 m/s towards each other due to mutual gravitational attraction. What is the velocity of their centre of mass?

(a) 5 m/s (b) 6 m/s
(c) 8 m/s (d) Zero

Q.10 Two particles of masses m_1 and m_2 initially at rest start moving towards each other under their mutual force of attraction. The speed of the centre of mass at any time *t*, when they are at a distance *r* apart, is

(a) zero (b) $\left(G\frac{m_1 m_2}{r^2}.\frac{1}{m_1}\right)t$

(c) $\left(G\frac{m_1 m_2}{r^2}.\frac{1}{m_2}\right)t$ (d) $\left(G\frac{m_1 m_2}{r^2}.\frac{1}{m_1+m_2}\right)t$

Q.11 A 'T' shaped object, dimensions shown in the figure, is lying on a smooth floor. A force '$\vec{F}$' is applied at the point P parallel to AB, such that the object has only the translational motion without rotation. Find the location of P with respect to C

(a) $\frac{4}{3}l$

(b) l

(c) $\frac{2}{3}l$

(d) $\frac{3}{2}l$

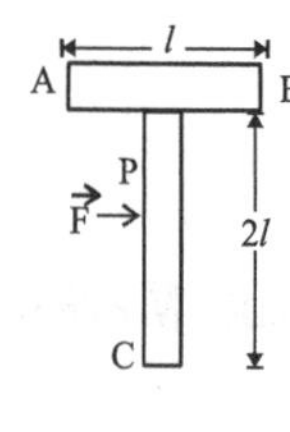

Q.12 Two spheres of masses 2*M* and *M* are initially at rest at a distance *R* apart. Due to mutual force of attraction, they approach each other. When they are at separation *R*/2, the acceleration of the centre of mass of spheres would be

(a) 0 m/s^2 (b) g m/s^2
(c) 3 g m/s^2 (d) 12 g m/s^2

Q.13 Masses 8 kg, 2 kg, 4 kg and 2 kg are placed at the corners A, B, C, D respectively of a square ABCD of diagonal 80 cm. The distance of centre of mass from A will be

(a) 20 cm (b) 30 cm
(c) 40 cm (d) 60 cm

Q.14 If linear density of a rod of length 3m varies as $\lambda = 2 + x$, them the position of the centre of gravity of the rod is

(a) $\frac{7}{3}$m (b) $\frac{12}{7}$m

(c) $\frac{10}{7}$m (d) $\frac{9}{7}$m

RESPONSE GRID					
	5. ⓐⓑⓒⓓ	6. ⓐⓑⓒⓓ	7. ⓐⓑⓒⓓ	8. ⓐⓑⓒⓓ	9. ⓐⓑⓒⓓ
	10. ⓐⓑⓒⓓ	11. ⓐⓑⓒⓓ	12. ⓐⓑⓒⓓ	13. ⓐⓑⓒⓓ	14. ⓐⓑⓒⓓ

Space for Rough Work

Q.15 Four bodies of equal mass start moving with same speed as shown in the figure. In which of the following combination the centre of mass will remain at origin?

(a) *c* and *d*

(b) *a* and *b*

(c) *a* and *c*

(d) *b* and *d*

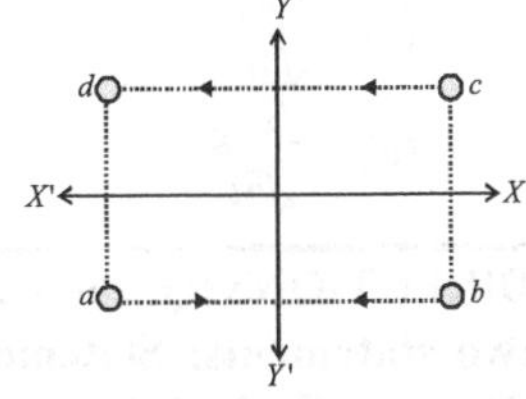

Q.16 Three identical spheres, each of mass 1 kg are kept as shown in figure, touching each other, with their centres on a straight line. If their centres are marked *P*, *Q*, *R* respectively, the distance of centre of mass of the system from *P* is

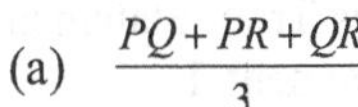

(a) $\frac{PQ+PR+QR}{3}$

(b) $\frac{PQ+PR}{3}$

(c) $\frac{PQ+QR}{3}$

(d) $\frac{PR+QR}{3}$

Q.17 A ladder is leaned against a smooth wall and it is allowed to slip on a frictionless floor. Which figure represents trace of motion of its centre of mass

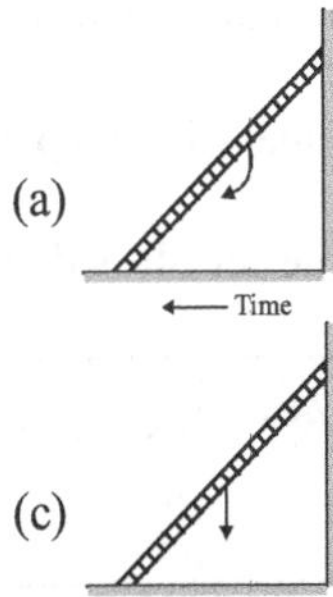

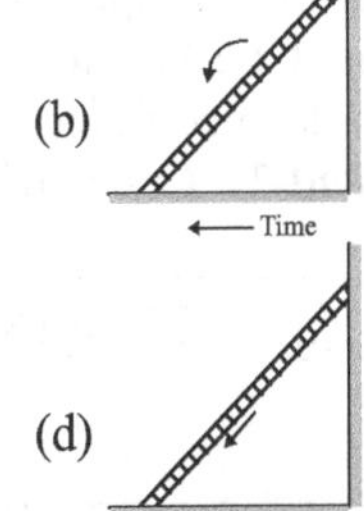

Q.18 The two particles *X* and *Y*, initially at rest, start moving towards each other under mutual attraction. If at any instant the velocity of *X* is *V* and that of *Y* is 2*V*, the velocity of their centre of mass will be

(a) 0 (b) *V* (c) 2*V* (d) *V*/2

Q.19 A cricket bat is cut at the location of its centre of mass as shown in the fig. Then

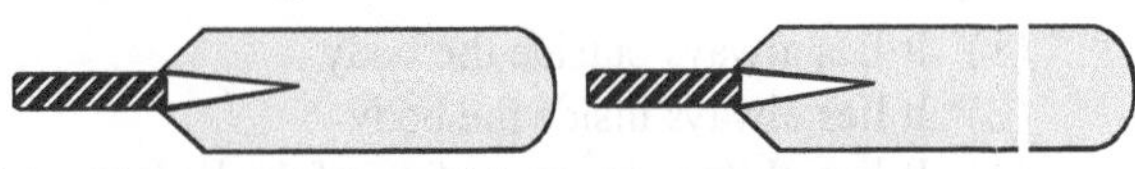

(a) The two pieces will have the same mass
(b) The bottom piece will have larger mass
(c) The handle piece will have larger mass
(d) Mass of handle piece is double the mass of bottom piece

Q.20 Consider a system of two particles having mass m_1 and m_2. If the particle of mass m_1 is pushed towards the centre of mass of particles through a distance *d*, by what distance would be particle of mass m_2 move so as to keep the centre of mass of particles at the original position?

(a) $\frac{m_1}{m_1+m_2}d$ (b) $\frac{m_1}{m_2}d$ (c) d (d) $\frac{m_2}{m_1}d$

DIRECTIONS (Q.21-Q.22) : In the following questions, more than one of the answers given are correct. Select the correct answers and mark it according to the following codes:

Codes :

(a) 1, 2 and 3 are correct **(b)** 1 and 2 are correct
(c) 2 and 4 are correct **(d)** 1 and 3 are correct

Q.21 Choose the wrong statements about the centre of mass (CM) of a system of two particles

(1) The CM lies on the line joining the two particles midway between them
(2) The CM lies on the line joining them at a point whose distance from each particle is proportional to the square of the mass of that particle
(3) The CM is on the line joining them at a point whose distance from each particle is proportional to the mass of that particle
(4) The CM lies on the line joining them at a point whose distance from each particle is inversely proportional to the mass of that particle

RESPONSE GRID					
	15. ⓐⓑⓒⓓ	16. ⓐⓑⓒⓓ	17. ⓐⓑⓒⓓ	18. ⓐⓑⓒⓓ	19. ⓐⓑⓒⓓ
	20. ⓐⓑⓒⓓ	21. ⓐⓑⓒⓓ			

Space for Rough Work

Q.22 Choose the wrong statements about the centre of mass of a body
(1) It lies always outside the body
(2) It lies always inside the body
(3) It lies always on the surface of the body
(4) It may lie within, outside or on the surface of the body

DIRECTIONS (Q.23-Q.25) : Read the passage given below and answer the questions that follows :

A system consists of block A and B each of mass m connected by a light spring as shown in the figure with block B in contact with a wall. The block A compresses the spring by $3mg/k$ from natural length of spring and then released from rest. Neglect friction anywhere.

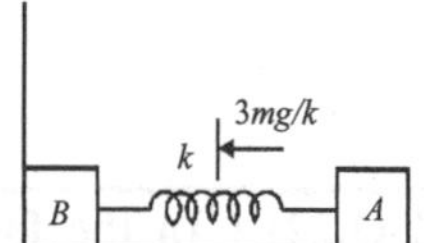

Q.23 Acceleration of centre of mass of system comprising A and B just after A is released is
(a) 0 (b) $3g/2$
(c) $3g$ (d) None of these

Q.24 Velocity of centre of mass of system comprising A and B when block B just loses contact with the wall
(a) $3g\sqrt{\frac{m}{k}}$ (b) $\frac{3g}{2}\sqrt{\frac{m}{k}}$
(c) $2g\sqrt{\frac{m}{k}}$ (d) None of these

Q.25 Maximum extension in the spring after system loses contact with wall
(a) $\frac{3mg}{\sqrt{2}k}$ (b) $\frac{\sqrt{3}mg}{2k}$
(c) $\frac{\sqrt{3}mg}{\sqrt{2}k}$ (d) None of these

DIRECTIONS (Q. 26-Q.28) : Each of these questions contains two statements: Statement-1 (Assertion) and Statement-2 (Reason). Each of these questions has four alternative choices, only one of which is the correct answer. You have to select the correct choice.

(a) Statement-1 is True, Statement-2 is True; Statement-2 is a correct explanation for Statement-1.
(b) Statement-1 is True, Statement-2 is True; Statement-2 is NOT a correct explanation for Statement-1.
(c) Statement -1 is False, Statement-2 is True.
(d) Statement -1 is True, Statement-2 is False.

Q.26 Statement-1 : The centre of mass of a system of n particles is the weighted average of the position vector of the n particles making up the system.
Statement-2 : The position of the centre of mass of a system is independent of coordinate system.

Q.27 Statement-1 : The centre of mass of a proton and an electron, released from their respective positions remains at rest.
Statement-2 : The centre of mass remains at rest, if no external force is applied.

Q.28 Statement-1 : Position of centre of mass is independent of the reference frame.
Statement-2 : Centre of mass is same for all bodies.

RESPONSE GRID					
	22. ⓐⓑⓒⓓ	23. ⓐⓑⓒⓓ	24. ⓐⓑⓒⓓ	25. ⓐⓑⓒⓓ	26. ⓐⓑⓒⓓ
	27. ⓐⓑⓒⓓ	28. ⓐⓑⓒⓓ			

DAILY PRACTICE PROBLEM SHEET 14 - PHYSICS			
Total Questions	28	Total Marks	112
Attempted		Correct	
Incorrect		Net Score	
Cut-off Score	28	Qualifying Score	46
Success Gap = Net Score – Qualifying Score			
Net Score = (Correct × 4) – (Incorrect × 1)			

Space for Rough Work

DPP - Daily Practice Problems

Name : 　　　　Date :

Start Time : 　　　　End Time :

PHYSICS 15

SYLLABUS : Rotational Motion – 1: Basic concepts of rotational motion, moment of a force, torque, angular momentum and its conservation with application

Max. Marks : 112 　　　　**Time : 60 min.**

GENERAL INSTRUCTIONS

- The Daily Practice Problem Sheet contains 28 MCQ's. For each question only one option is correct. Darken the correct circle/ bubble in the Response Grid provided on each page.
- You have to evaluate your Response Grids yourself with the help of solution booklet.
- Each correct answer will get you 4 marks and 1 mark shall be deduced for each incorrect answer. No mark will be given/ deducted if no bubble is filled. Keep a timer in front of you and stop immediately at the end of 60 min.
- The sheet follows a particular syllabus. Do not attempt the sheet before you have completed your preparation for that syllabus. Refer syllabus sheet in the starting of the book for the syllabus of all the DPP sheets.
- After completing the sheet check your answers with the solution booklet and complete the Result Grid. Finally spend time to analyse your performance and revise the areas which emerge out as weak in your evaluation.

DIRECTIONS (Q.1-Q.20) : There are 20 multiple choice questions. Each question has 4 choices (a), (b), (c) and (d), out of which **ONLY ONE** choice is correct.

Q.1 A thin circular ring of mass M and radius r is rotating about its axis with a constant angular velocity ω. Four objects each of mass m, are kept gently to the opposite ends of two perpendicular diameters of the ring. The angular velocity of the ring will be

(a) $\frac{M\omega}{M+4m}$ (b) $\frac{(M+4m)\omega}{M}$ (c) $\frac{(M-4m)\omega}{M+4m}$ (d) $\frac{M\omega}{4m}$

Q..2 The angular momentum of a system of particles is conserved

(a) When no external force acts upon the system
(b) When no external torque acts upon the system
(c) When no external impulse acts upon the system
(d) When axis of rotation remains same

Q.3 Two rigid bodies A and B rotate with rotational kinetic energies E_A and E_B respectively. The moments of inertia of A and B about the axis of rotation are I_A and I_B respectively. If $I_A = I_B/4$ and $E_A = 100\,E_B$, the ratio of angular momentum (L_A) of A to the angular momentum (L_B) of B is

(a) 25 (b) 5/4 (c) 5 (d) 1/4

Q.4 A uniform heavy disc is rotating at constant angular velocity ω about a vertical axis through its centre and perpendicular to the plane of the disc. Let L be its angular momentum. A lump of plasticine is dropped vertically on the disc and sticks to it. Which of the following will be constant?

(a) ω (b) ω and L both
(c) L only (d) Neither ω nor L

RESPONSE GRID	1. ⓐⓑⓒⓓ	2. ⓐⓑⓒⓓ	3. ⓐⓑⓒⓓ	4. ⓐⓑⓒⓓ

Space for Rough Work

Q.5 Two discs of moment of inertia I_1 and I_2 and angular speeds ω_1 and ω_2 are rotating along collinear axes passing through their centre of mass and perpendicular to their plane. If the two are made to rotate combindly along the same axis the rotational KE of system will be

(a) $\frac{I_1\omega_1+I_2\omega_2}{2(I_1+I_2)}$ (b) $\frac{(I_1+I_2)(\omega_1+\omega_2)^2}{2}$

(c) $\frac{(I_1\omega_1+I_2\omega_2)^2}{2(I_1+I_2)}$ (d) None of these

Q.6 A particle performs uniform circular motion with an angular momentum L. If the frequency of a particle's motion is doubled and its kinetic energy is halved, the angular momentum becomes.

(a) 2 L (b) 4 L (c) L/2 (d) L/4

Q.7 A round disc of moment of inertia I_2 about its axis perpendicular to its plane and passing through its centre is placed over another disc of moment of inertia I_1 rotating with an angular velocity ω about the same axis. The final angular velocity of the combination of discs is

(a) $\frac{I_2\omega}{I_1+I_2}$ (b) ω (c) $\frac{I_1\omega}{I_1+I_2}$ (d) $\frac{(I_1+I_2)\omega}{I_1}$

Q.8 Calculate the angular momentum of a body whose rotational energy is 10 joule. If the angular momentum vector coincides with the axis of rotation and its moment of inertia about this axis is 8×10^{-7} kg m^2

(a) 4×10^{-3} kg m^2/s (b) 2×10^{-3} kg m^2/s

(c) 6×10^{-3} kg m^2/s (d) None of these

Q.9 If the earth is treated as a sphere of radius R and mass M. Its angular momentum about the axis of rotation with period T is

(a) $\frac{\pi MR^3}{T}$ (b) $\frac{MR^2\pi}{T}$ (c) $\frac{2\pi MR^2}{5T}$ (d) $\frac{4\pi MR^2}{5T}$

Q.10 If the earth is a point mass of 6×10^{24}kg revolving around the sun at a distance of 1.5×10^8 km and in time T = 3.14 $\times10^7$s. then the angular momentum of the earth around the sun is

(a) 1.2×10^{18} kg m^2/s (b) 1.8×10^{29} kg m^2/s

(c) 1.5×10^{37} kg m^2/s (d) 2.7×10^{40} kg m^2/s

Q.11 An automobile engine develops 100 kW when rotating at a speed of 1800 rev/min. What torque does it deliver

(a) 350 N-m (b) 440 N-m

(c) 531 N-m (d) 628 N-m

Q.12 A constant torque acting on a uniform circular wheel changes its angular momentum from A_0 to $4A_0$ in 4 seconds. The magnitude of this torque is

(a) $\frac{3A_0}{4}$ (b) A_0

(c) $4A_0$ (d) $12A_0$

Q.13 A wheel having moment of inertia 2 kg -m^2 about its vertical axis, rotates at the rate of 60 rpm about this axis. The torque which can stop the wheel's rotation in 1 minute would be

(a) $\frac{2\pi}{15}$Nm (b) $\frac{\pi}{12}$Nm

(c) $\frac{\pi}{15}$Nm (d) $\frac{\pi}{18}$Nm

Q.14 Find the torque of a force $\vec{F}=-3\hat{i}+\hat{j}+5\hat{k}$ acting at the point $\vec{r}=7\hat{i}+3\hat{j}+\hat{k}$

(a) $14\hat{i}-38\hat{j}+16\hat{k}$ (b) $4\hat{i}+4\hat{j}+6\hat{k}$

(c) $-14\hat{i}+38\hat{j}-16\hat{k}$ (d) $-21\hat{i}+3\hat{j}+5\hat{k}$

Q.15 A constant torque of 1000 N -m, turns a wheel of moment of inertia 200 kg -m^2 about an axis passing through the centre. Angular velocity of the wheel after 3 s will be

(a) 15 rad/s (b) 10 rad/s (c) 5 rad/s (d) 1 rad/s

Q.16 A torque of 30 N-m is applied on a 5 kg wheel whose moment of inertia is 2kg-m^2 for 10 sec. The angle covered by the wheel in 10 sec will be

(a) 750 rad (b) 1500 rad

(c) 3000 rad (d) 6000 rad

RESPONSE GRID					
	5. ⓐⓑⓒⓓ	6. ⓐⓑⓒⓓ	7. ⓐⓑⓒⓓ	8. ⓐⓑⓒⓓ	9. ⓐⓑⓒⓓ
	10. ⓐⓑⓒⓓ	11. ⓐⓑⓒⓓ	12. ⓐⓑⓒⓓ	13. ⓐⓑⓒⓓ	14. ⓐⓑⓒⓓ
	15. ⓐⓑⓒⓓ	16. ⓐⓑⓒⓓ			

Space for Rough Work

Q.17 A horizontal force F is applied such that the block remains stationary, then which of the following statement is false

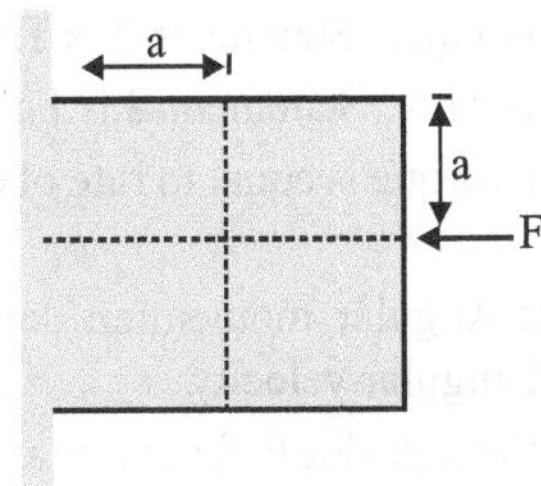

(a) f = mg [where f is the friction force]
(b) F = N [where N is the normal reaction]
(c) F will not produce torque
(d) N will not produce torque

Q.18 In a bicycle, the radius of rear wheel is twice the radius of front wheel. If r_F and r_r are the radius, v_F and v_r are speeds of top most points of wheel, then

(a) $v_r = 2\,v_F$ (b) $v_F = 2\,v_r$
(c) $v_F = v_r$ (d) $v_F > v_r$

Q.19 The wheel of a car is rotating at the rate of 1200 revolutions per minute. On pressing the accelerator for 10 seconds, it starts rotating at 4500 revolutions per minute. The angular acceleration of the wheel is

(a) 30 rad/sec^2 (b) 1880 degree/sec^2
(c) 40 rad/sec^2 (d) 1980 degree/sec^2

Q.20 A wheel rotates with a constant acceleration of 2.0 radian/sec^2. It the wheel starts from rest, the number of revolutions it makes in the first ten seconds will be approximately

(a) 8 (b) 16 (c) 24 (d) 32

DIRECTIONS (Q.21-Q.22) : In the following questions, more than one of the answers given are correct. Select the correct answers and mark it according to the following codes:

Codes :

(a) 1, 2 and 3 are correct **(b)** 1 and 2 are correct
(c) 2 and 4 are correct **(d)** 1 and 3 are correct

Q.21 A child is standing with folded hands at the centre of a platform rotating about its central axis. The kinetic energy of the system is K. The child now stretches his arms so that the moment of inertia of the system doubles. The kinetic energy of the system now is

(1) less than 2K (2) equal to K/2
(3) more thanK/4 (4) equal to 4K

Q.22 Two uniforms discs of equal mass but unequal radii are mounted on fixed horizontal axiles. Light strings are wrapped on each of the discs. The strings are pulled by constant equal forces F for same amount of time as shown in the figure.

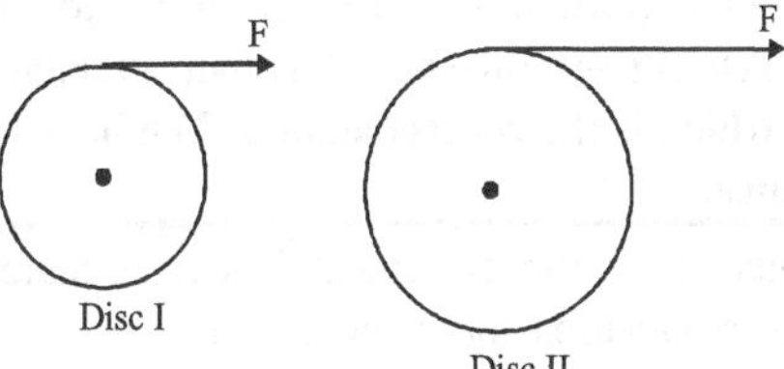

Angular momenta of discs are L_1 and L_2 and their kinetic energies are K_1 and K_2. Which of the following statements true –

(1) $L_1 = L_2$ (2) $L_1 < L_2$ (3) $K_1 > K_2$ (4) $K_1 = K_2$

DIRECTIONS (Q.23-Q.25) : Read the passage given below and answer the questions that follows :

Consider a cylinder of mass M = 1kg and radius R = 1 m lying on a rough horizontal plane. It has a plank lying on its top as shown in the figure.

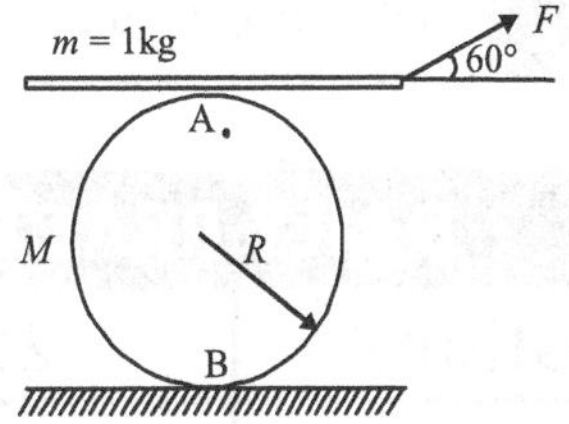

A force F = 55 N is applied on the plank such that the plank moves and causes the cylinder to roll. The plank always remains horizontal. There is no slipping at any point of contact.

RESPONSE GRID	17. ⓐⓑⓒⓓ	18. ⓐⓑⓒⓓ	19. ⓐⓑⓒⓓ	20. ⓐⓑⓒⓓ	21. ⓐⓑⓒⓓ
	22. ⓐⓑⓒⓓ				

Space for Rough Work

Q.23 Calculate the acceleration of cylinder.

(a) 20 m/s^2 (b) 10 m/s^2
(c) 5 m/s^2 (d) None of these

Q.24 Find the value of frictional force at *A*

(a) 7.5 N (b) 5.0 N
(c) 2.5 N (d) None of these

Q.25 Find the value of frictional force at *B*

(a) 7.5 N (b) 5.0 N
(c) 2.5 N (d) None of these

DIRECTIONS (Q.26-Q.28) : Each of these questions contains two statements: Statement-1 (Assertion) and Statement-2 (Reason). Each of these questions has four alternative choices, only one of which is the correct answer. You have to select the correct choice.

(a) Statement-1 is True, Statement-2 is True; Statement-2 is a correct explanation for Statement-1.
(b) Statement-1 is True, Statement-2 is True; Statement-2 is NOT a correct explanation for Statement-1.
(c) Statement -1 is False, Statement-2 is True.
(d) Statement -1 is True, Statement-2 is False.

Q.26 **Statement -1:** Torque is equal to rate of change of angular momentum.

Statement -2: Angular momentum depends on moment of inertia and angular velocity.

Q.27 **Statement -1:** Torque due to force is maximum when angle between $\vec{r}$ and $\vec{F}$ is 90°.

Statement -2: The unit of torque is newton- meter.

Q.28 **Statement -1:** It is harder to open and shut the door if we apply force near the hinge.

Statement -2: Torque is maximum at hinge of the door.

RESPONSE GRID	
	23. ⓐⓑⓒⓓ 24. ⓐⓑⓒⓓ 25. ⓐⓑⓒⓓ 26. ⓐⓑⓒⓓ 27. ⓐⓑⓒⓓ
	28. ⓐⓑⓒⓓ

DAILY PRACTICE PROBLEM SHEET 15 - PHYSICS			
Total Questions	28	Total Marks	112
Attempted		Correct	
Incorrect		Net Score	
Cut-off Score	28	Qualifying Score	44
Success Gap = Net Score – Qualifying Score			
Net Score = (Correct × 4) – (Incorrect × 1)			

Space for Rough Work

DPP - Daily Practice Problems

Name :

Date :

Start Time :

End Time :

PHYSICS 16

SYLLABUS : Rotational Motion-2 : Moment of inertia, radius of gyration, (values of moments of inertia simple geometrical objects)

Max. Marks : 120 **Time : 60 min.**

GENERAL INSTRUCTIONS

- The Daily Practice Problem Sheet contains 30 MCQ's. For each question only one option is correct. Darken the correct circle/ bubble in the Response Grid provided on each page.
- You have to evaluate your Response Grids yourself with the help of solution booklet.
- Each correct answer will get you 4 marks and 1 mark shall be deduced for each incorrect answer. No mark will be given/ deducted if no bubble is filled. Keep a timer in front of you and stop immediately at the end of 60 min.
- The sheet follows a particular syllabus. Do not attempt the sheet before you have completed your preparation for that syllabus. Refer syllabus sheet in the starting of the book for the syllabus of all the DPP sheets.
- After completing the sheet check your answers with the solution booklet and complete the Result Grid. Finally spend time to analyse your performance and revise the areas which emerge out as weak in your evaluation.

DIRECTIONS (Q.1-Q.21) : There are 21 multiple choice questions. Each question has 4 choices (a), (b), (c) and (d), out of which **ONLY ONE** choice is correct.

Q.1 Five particles of mass 2 kg are attached to the rim of a circular disc of radius 0.1 m & negligible mass. Moment of inertia of the system about an axis passing through the centre of the disc & perpendicular to its plane is

(a) 1 kg-m^2 (b) 0.1kg-m^2 (c) 2 kg-m^2 (d) 0.2 kg-m^2

Q.2 Two discs of the same material and thickness have radii 0.2 m and 0.6 m. Their moments of inertia about their axes will be in the ratio of

(a) 1 : 81 (b) 1: 27 (c) 1 : 9 (d) 1 : 3

Q.3 A cylinder of 500 g and radius 10 cm has moment of inertia (about its natural axis)

(a) 2.5×10^{-3} kg–m^2 (b) 2×10^{-3} kg–m^2
(c) 5×10^{-3} kg–m^2 (d) 3.5×10^{-3} kg–m^2

Q.4 A constant torque of 31.4 N–m is exerted on a pivoted wheel. If angular acceleration of wheel is 4π rad/sec^2, then the moment of inertia of the wheel is

(a) 2.5 kg–m^2 (b) 2.5 kg–m^2
(c) 4.5 kg–m^2 (d) 5.5 kg–m^2

Q.5 From a uniform wire, two circular loops are made (i) P of radius r and (ii) Q of radius nr. If the moment of inertia of loop Q about an axis passing through its centre and perpendicular to its plane is 8 times that of P about a similar axis, the value of *n* is (diameter of the wire is very much smaller than r or nr)

(a) 8 (b) 6
(c) 4 (d) 2

RESPONSE GRID	1. ⓐⓑⓒⓓ	2. ⓐⓑⓒⓓ	3. ⓐⓑⓒⓓ	4. ⓐⓑⓒⓓ	5. ⓐⓑⓒⓓ

Space for Rough Work

Q.6 The moment of inertia of a sphere of mass M and radius R about an axis passing through its centre is $\frac{2}{5}MR^2$. The radius of gyration of the sphere about a parallel axis to the above and tangent to the sphere is

(a) $\frac{7}{5}R$ (b) $\frac{3}{5}R$ (c) $\left(\sqrt{\frac{7}{5}}\right)R$ (d) $\left(\sqrt{\frac{3}{5}}\right)R$

Q.7 Four particles each of mass m are placed at the corners of a square of side length ℓ. The radius of gyration of the system about an axis perpendicular to the square and passing through its centre is

(a) $\frac{\ell}{\sqrt{2}}$ (b) $\frac{\ell}{2}$ (c) ℓ (d) $(\sqrt{2})\ell$

Q.8 The radius of gyration of a disc of mass 50 g and radius 2.5 cm, about an axis passing through its centre of gravity and perpendicular to the plane is

(a) 0.52 cm (b) 1.76 cm (c) 3.54 cm (d) 6.54 cm

Q.9 Moment of inertia of a ring of mass m = 3 gm and radius $r = 1$ cm about an axis passing through its edge and parallel to its natural axis is

(a) 10 gm–cm^2 (b) 100 gm–cm^2
(c) 6 gm–cm^2 (d) 1 gm–cm^2

Q.10 A disc is of mass M and radius r. The moment of inertia of it about an axis tangential to its edge and in plane of the disc or parallel to its diameter is

(a) $\frac{5}{4}Mr^2$ (b) $\frac{Mr^2}{4}$ (c) $\frac{3}{2}Mr^2$ (d) $\frac{Mr^2}{2}$

Q.11 Two spheres each of mass M and radius R/2 are connected with a massless rod of length 2R as shown in the figure.

The moment of inertia of the system about an axis passing through the centre of one of the spheres and perpendicular to the rod will be

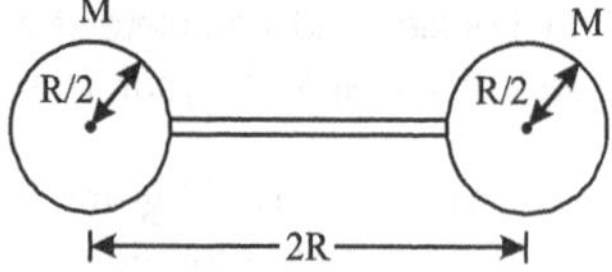

(a) $\frac{21}{5}Mr^2$ (b) $\frac{2}{5}Mr^2$ (c) $\frac{5}{2}Mr^2$ (d) $\frac{5}{21}Mr^2$

Q.12 Three point masses m_1, m_2, m_3 are located at the vertices of an equilateral triangle of length 'a'. The moment of inertia of the system about an axis along the altitude of the triangle passing through m_1, is

(a) $(m_2 + m_3)\frac{a^2}{4}$ (b) $(m_1 + m_2 + m_3)a^2$
(c) $(m_1 + m_2)\frac{a^2}{4}$ (d) $(m_2 + m_3)a^2$

Q.13 Three rods each of length L and mass M are placed along X, Y and Z axis in such a way that one end of each of the rod is at the origin. The moment of inertia of this system about Z axis is

(a) $\frac{2ML^2}{3}$ (b) $\frac{4ML^2}{3}$ (c) $\frac{5ML^2}{3}$ (d) $\frac{ML^2}{3}$

Q.14 ABC is a triangular plate of uniform thickness. The sides are in the ratio shown in the figure. I_{AB}, I_{BC}, I_{CA} are the moments of inertia of the plate about AB, BC, CA respectively. For this arrangement which one of the following relation is correct?

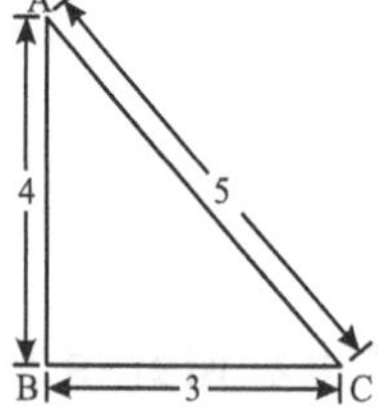

(a) I_{CA} is maximum (b) $I_{BC} > I_{AB}$
(c) $I_{BC} > I_{AB}$ (d) $I_{AB} + I_{BC} = I_{CA}$

Q.15 A 1m long rod has a mass of 0.12 kg. The moment of inertia about an axis passin through the centre and perpendicular to the length of rod will be

(a) 0.01kg-m^2 (b) 0.001 kg-m^2
(c) 1 kg-m^2 (d) 10 kg-m^2

Q.16 Two rings of the same radius and mass are placed such that their centres are at a common point and their planes are perpendicular to each other. The moment of inertia of the system about an axis passing through the centre and perpendicular to the plane of one of the rings is (mass of the ring = m and radius = r)

(a) $\frac{1}{2}mr^2$ (b) mr^2
(c) $\frac{3}{2}mr^2$ (d) $2mr^2$

RESPONSE GRID					
	6. ⓐⓑⓒⓓ	7. ⓐⓑⓒⓓ	8. ⓐⓑⓒⓓ	9. ⓐⓑⓒⓓ	10. ⓐⓑⓒⓓ
	11. ⓐⓑⓒⓓ	12. ⓐⓑⓒⓓ	13. ⓐⓑⓒⓓ	14. ⓐⓑⓒⓓ	15. ⓐⓑⓒⓓ
	16. ⓐⓑⓒⓓ				

Space for Rough Work

Q.17 One quarter sector is cut from a uniform circular disc of radius R. This sector has mass M. It is made to rotate about a line perpendicular to its plane and passing through the centre of the original disc. Its moment of inertia about the axis of rotation is

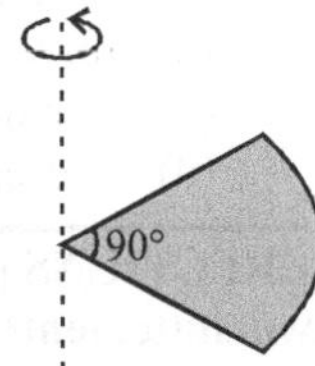

(a) $\frac{1}{2}MR^2$ (b) $\frac{1}{4}MR^2$ (c) $\frac{1}{8}MR^2$ (d) $\sqrt{2}MR^2$

Q.18 A thin wire of length L and uniform linear mass density ρ is bent into a circular loop with centre at O as shown in figure. The moment of inertia of the loop about the axis XX′ is

(a) 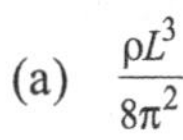 $\frac{\rho L^3}{8\pi^2}$

(b) $\frac{\rho L^3}{16\pi^2}$

(c) $\frac{5\rho L^3}{16\pi^2}$

(d) $\frac{3\rho L^3}{8\pi^2}$

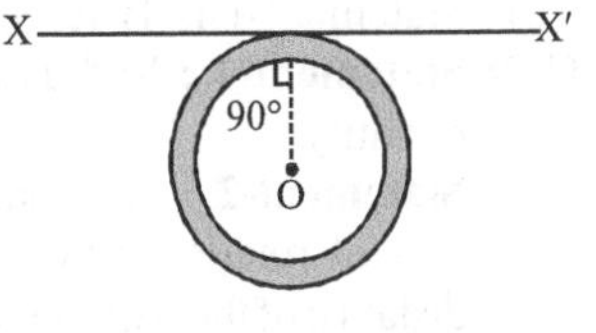

Q.19 Two disc of same thickness but of different radii are made of two different materials such that their masses are same. The densities of the materials are in the ratio 1 : 3. The ratio of the moments of inertia of these discs about the respective axes passing through their centres and perpendicular to their planes will be in

(a) 1 : 3 (b) 3 : 1 (c) 1 : 9 (d) 9 : 1

Q.20 A circular disc of radius R and thickness $\frac{R}{6}$ has moment of inertia I about an axis passing through its centre and perpendicular to its plane. It is melted and recasted into a solid sphere. The moment of inertia of the sphere about one of its diameter as an axis of rotation will be

(a) I (b) $\frac{2I}{8}$ (c) $\frac{I}{5}$ (d) $\frac{I}{10}$

Q.21 Three rings each of mass M and radius R are arranged as shown in the figure. The moment of inertia of the system about YY' will be

(a) $3\ MR^2$

(b) $\frac{3}{2}MR^2$

(c) $5\ MR^2$

(d) $\frac{7}{2}MR^2$

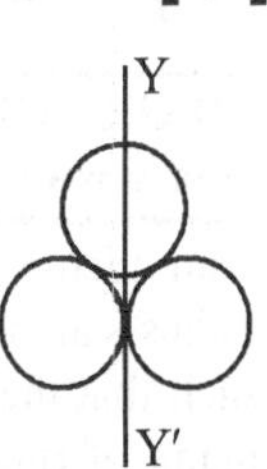

DIRECTIONS (Q.22-Q.24) : In the following questions, more than one of the answers given are correct. Select the correct answers and mark it according to the following codes:

Codes :

(a) 1, 2 and 3 are correct **(b)** 1 and 2 are correct
(c) 2 and 4 are correct **(d)** 1 and 3 are correct

Q.22 The density of a rod AB increases linearly from A to B. Its midpoint is O and its centre of mass is at C. Four axes pass through A, B, O and C, all perpendicular to the length of the rod. The moments of inertia of the rod about these axes are I_A, I_B, I_O and I_C respectively then:.

(1) $I_A > I_B$ (2) $I_A < I_B$
(3) $I_O > I_C$ (4) $I_O < I_C$

Q.23 The moment of inertia of a thin square plate ABCD of uniform thickness about an axis passing through the centre O and perpendicular to the plane of the plate is

(1) $I_1 + I_2$
(2) $I_3 + I_4$
(3) $I_1 + I_3$
(4) $I_1 + I_2 + I_3 + I_4$

where I_1, I_2, I_3 and I_4 are respectively moments of inertia about axes 1, 2, 3 and 4 which are in the plane of the plate.

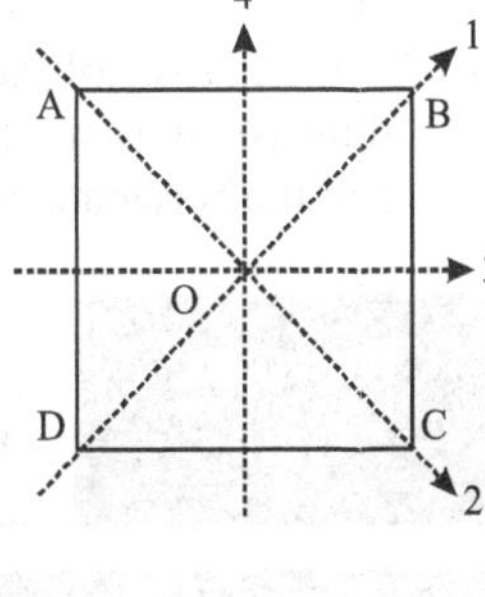

Q.24 Moment of inertia doesn't depend on

(1) distribution of particles
(2) mass
(3) position of axis of rotation
(4) None of these

RESPONSE GRID				
17. ⓐⓑⓒⓓ	18. ⓐⓑⓒⓓ	19. ⓐⓑⓒⓓ	20. ⓐⓑⓒⓓ	21. ⓐⓑⓒⓓ
22. ⓐⓑⓒⓓ	23. ⓐⓑⓒⓓ	24. ⓐⓑⓒⓓ		

Space for Rough Work

DIRECTIONS (Q.25-Q.27) : Read the passage given below and answer the questions that follows :

Four identical spheres having mass M and radius R are fixed tightly within a massless ring such that the centres of all spheres lie in the plane of ring. The ring is kept on a rough horizontal table as shown. The string is wrapped around the ring can roll without slipping.

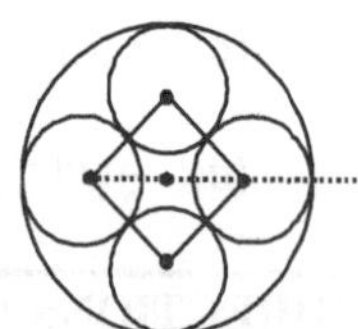

The other end of the string is passed over a massless frictionless pulley to a block of mass M. A force F is applied horizontally on the ring, at the same level as the centre, so that the system is in equilibrium.

Q.25 The moment of inertia of the combined ring system about the centre of ring will be

(a) $\frac{12}{5}MR^2$

(b) $\frac{48}{15}MR^2$

(c) $\frac{24}{5}MR^2$

(d) $\frac{48}{5}MR^2$

Q.26 The magnitude of F is

(a) Mg (b) $2Mg$

(c) $\frac{Mg}{2}$ (d) None of these

Q.27 If the masses of the spheres were doubled keeping their dimensions same, the force of friction between the ring and the horizontal surface would

(a) be doubled

(b) increase but be less than double

(c) remain the same

(d) decrease

DIRECTIONS (Q. 28-Q.30) : Each of these questions contains two statements: Statement-1 (Assertion) and Statement-2 (Reason). Each of these questions has four alternative choices, only one of which is the correct answer. You have to select the correct choice.

(a) Statement-1 is True, Statement-2 is True; Statement-2 is a correct explanation for Statement-1.

(b) Statement-1 is True, Statement-2 is True; Statement-2 is NOT a correct explanation for Statement-1.

(c) Statement-1 is False, Statement-2 is True.

(d) Statement-1 is True, Statement-2 is False.

Q.28 Statement-1 : Radius of gyration of a body is a constant quantity.

Statement-2 : The radius of gyration of a body about an axis of rotation may be defined as the root mean square distance of the particles of the body from the axis of rotation.

Q.29 Statement-1 : Moment of inertia of a particle is same, whatever be the axis of rotation.

Statement-2 : Moment of inertia depends on mass and perpendicular distance of the particle from its axis of rotation.

Q.30 Statement-1 : If earth shrink (without change in mass) to half of its present size, length of the day would become 6 hours.

Statement-2 : When the size of the earth will change, its moment of inertia will also change.

RESPONSE GRID	25. ⓐⓑⓒⓓ 26. ⓐⓑⓒⓓ 27. ⓐⓑⓒⓓ 28. ⓐⓑⓒⓓ 29. ⓐⓑⓒⓓ 30. ⓐⓑⓒⓓ

DAILY PRACTICE PROBLEM SHEET 16 - PHYSICS			
Total Questions	30	Total Marks	120
Attempted		Correct	
Incorrect		Net Score	
Cut-off Score	32	Qualifying Score	50
Success Gap = Net Score – Qualifying Score			
Net Score = (Correct × 4) – (Incorrect × 1)			

Space for Rough Work

DPP - Daily Practice Problems

Name : 　　　　Date :

Start Time : 　　　　End Time :

PHYSICS 17

SYLLABUS : Rotational Motion - 3 : Rolling Motion, Parallel and perpendicular theorems and their applications, Rigid body rotation, equations of rotational motion

Max. Marks : 116 　　　　**Time : 60 min.**

GENERAL INSTRUCTIONS

- The Daily Practice Problem Sheet contains 30 MCQ's. For each question only one option is correct. Darken the correct circle/ bubble in the Response Grid provided on each page.
- You have to evaluate your Response Grids yourself with the help of solution booklet.
- Each correct answer will get you 4 marks and 1 mark shall be deduced for each incorrect answer. No mark will be given/ deducted if no bubble is filled. Keep a timer in front of you and stop immediately at the end of 60 min.
- The sheet follows a particular syllabus. Do not attempt the sheet before you have completed your preparation for that syllabus. Refer syllabus sheet in the starting of the book for the syllabus of all the DPP sheets.
- After completing the sheet check your answers with the solution booklet and complete the Result Grid. Finally spend time to analyse your performance and revise the areas which emerge out as weak in your evaluation.

DIRECTIONS (Q.1-Q.20) : There are 20 multiple choice questions. Each question has 4 choices (a), (b), (c) and (d), out of which **ONLY ONE** choice is correct.

Q.1 A disc is rolling (without slipping) on a horizontal surface. C is its centre and Q and P are two points equidistant from C. Let v_P, v_Q and v_C be the magnitude of velocities of points P, Q and C respectively, then

Q
C
P

(a) $v_Q > v_C > v_P$

(b) $v_Q < v_C < v_P$

(c) $v_Q = v_P, v_C = \frac{v_P}{2}$

(d) $v_Q < v_C > v_P$

Q.2 A uniform rod of length $2L$ is placed with one end in contact with the horizontal and is then inclined at an angle α to the horizontal and allowed to fall without slipping at contact point. When it becomes horizontal, its angular velocity will be

(a) $\omega = \sqrt{\frac{3g\sin\alpha}{2L}}$

(b) $\omega = \sqrt{\frac{2L}{3g\sin\alpha}}$

(c) $\omega = \sqrt{\frac{6g\sin\alpha}{L}}$

(d) $\omega = \sqrt{\frac{L}{g\sin\alpha}}$

RESPONSE GRID 　1. ⓐⓑⓒⓓ 　2. ⓐⓑⓒⓓ

Space for Rough Work

Q.3 According to the theorem of parallel axes $I = I_{cm} + Mx^2$, the graph between I and x will be

(a) 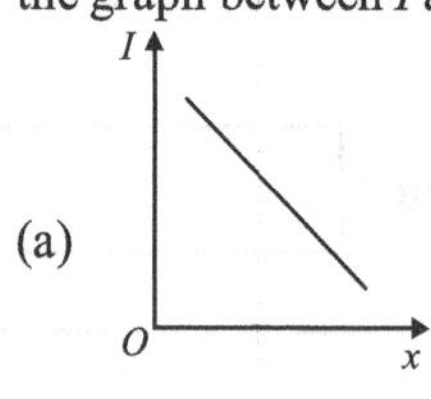

(b) 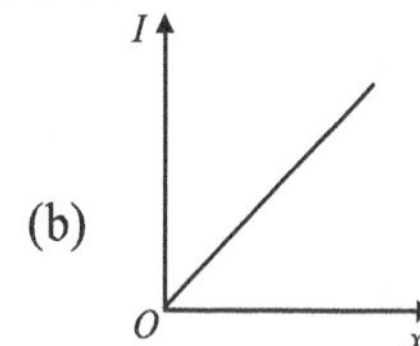

(c) 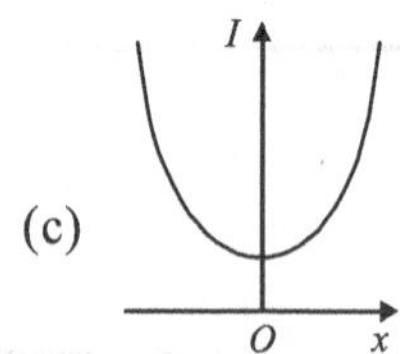

(d) 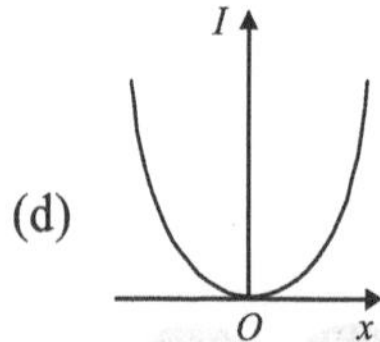

Q.4 A solid cylinder of mass M and radius R rolls without slipping down an inclined plane of length L and height h. What is the speed of its centre of mass when the cylinder reaches its bottom

(a) $\sqrt{\frac{3}{4}gh}$ (b) $\sqrt{\frac{4}{3}gh}$ (c) $\sqrt{4gh}$ (d) $\sqrt{2gh}$

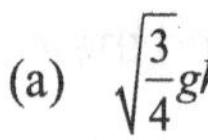

Q.5 An inclined plane makes an angle 30° with the horizontal. A solid sphere rolling down this inclined plane from rest without slipping has a linear acceleration equal to

(a) $\frac{g}{3}$ (b) $\frac{2g}{3}$ (c) $\frac{5g}{7}$ (d) $\frac{5g}{14}$

Q.6 A cord is wound round the circumference of wheel of radius r. The axis of the wheel is horizontal and moment of inertia about it is I. A weight mg is attached to the end of the cord and falls from the rest. After falling through a distance h, the angular velocity of the wheel will be

(a) $\sqrt{\frac{2gh}{I+mr}}$ (b) $\left[\frac{2mgh}{I+mr^2}\right]^{1/2}$

(c) $\left[\frac{2mgh}{I+2mr^2}\right]^{1/2}$ (d) $\sqrt{2gh}$

Q.7 A solid sphere, disc and solid cylinder all of the same mass and made up of same material are allowed to roll down (from rest) on an inclined plane, then

(a) Solid sphere reaches the bottom first
(b) Solid sphere reaches the bottom late
(c) Disc will reach the bottom first
(d) All of them reach the bottom at the same time

Q.8 A solid sphere is rolling on a frictionless surface, shown in figure with a transnational velocity v m/s. If sphere climbs up to height h then value of v should be

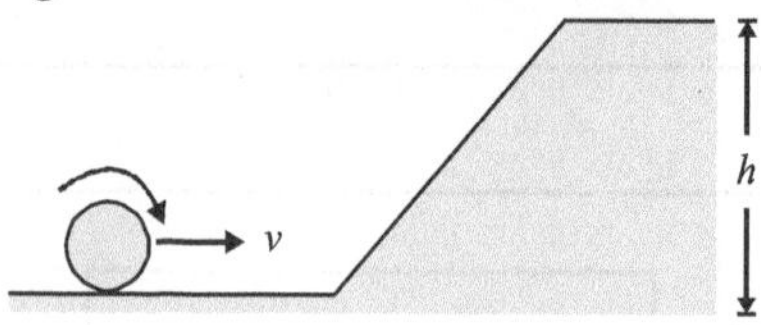

(a) $\geq \sqrt{\frac{10}{7}gh}$ (b) $\geq \sqrt{2gh}$ (c) $2gh$ (d) $\frac{10}{7}gh$

Q.9 Moment of inertia of a disc about its own axis is I. Its moment of inertia about a tangential axis in its plane is

(a) $\frac{5}{2}I$ (b) $3I$ (c) $\frac{3}{2}I$ (d) $2I$

Q.10 Three rings each of mass M and radius R are arranged as shown in the figure. The moment of inertia of the system about YY' will be

(a) $3MR^2$

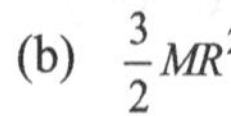

(b) $\frac{3}{2}MR^2$

(c) $5MR^2$

(d) $\frac{7}{2}MR^2$

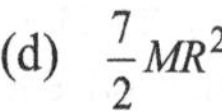

Q.11 One circular ring and one circular disc, both are having the same mass and radius. The ratio of their moments of inertia about the axes passing through their centres and perpendicular to their planes, will be

(a) 1 : 1 (b) 2 : 1
(c) 1 : 2 (d) 4 : 1

Q.12 From a disc of radius R, a concentric circular portion of radius r is cut out so as to leave an annular disc of mass M. The moment of inertia of this annular disc about the axis perpendicular to its plane and passing through its centre of gravity is

(a) $\frac{1}{2}M(R^2+r^2)$ (b) $\frac{1}{2}M(R^2-r^2)$

(c) $\frac{1}{2}M(R^4+r^4)$ (d) $\frac{1}{2}M(R^4-r^4)$

RESPONSE GRID					
	3. ⓐⓑⓒⓓ	4. ⓐⓑⓒⓓ	5. ⓐⓑⓒⓓ	6. ⓐⓑⓒⓓ	7. ⓐⓑⓒⓓ
	8. ⓐⓑⓒⓓ	9. ⓐⓑⓒⓓ	10. ⓐⓑⓒⓓ	11. ⓐⓑⓒⓓ	12. ⓐⓑⓒⓓ

Space for Rough Work

Q.13 The moment of inertia of a straight thin rod of mass M and length l about an axis perpendicular to its length and passing through its one end, is

(a) $\frac{M\ell^2}{12}$ (b) $\frac{M\ell^2}{3}$ (c) $\frac{M\ell^2}{2}$ (d) $M\ell^2$

Q.14 Four thin rods of same mass M and same length l, form a square as shown in figure. Moment of inertia of this system about an axis through centre O and perpendicular to its plane is

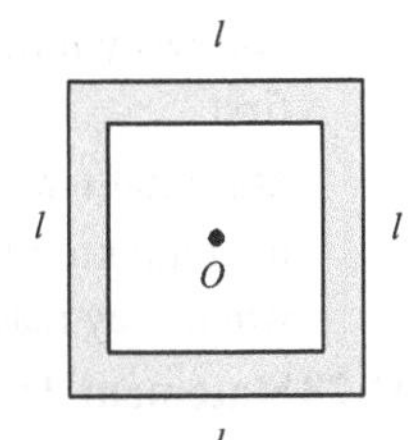

(a) $\frac{4}{3}Ml^2$

(b) $\frac{Ml^2}{3}$

(c) $\frac{Ml^2}{6}$

(d) $\frac{2}{3}Ml^2$

Q.15 The moment of inertia of a uniform circular ring, having a mass M and a radius R, about an axis tangential to the ring and perpendicular to its plane, is

(a) $2MR^2$ (b) $\frac{3}{2}MR^2$ (c) $\frac{1}{2}MR^2$ (d) MR^2

Q.16 The moment of inertia of uniform rectangular plate about an axis passing through its mid-point and parallel to its length l is (b = breadth of rectangular plate)

(a) $\frac{Mb^2}{4}$ (b) $\frac{Mb^3}{6}$ (c) $\frac{Mb^3}{12}$ (d) $\frac{Mb^2}{12}$

Q.17 The moment of inertia of a circular ring about an axis passing through its centre and normal to its plane is 200 gm × cm^2. Then moment of inertia about its diameter is

(a) 400 gm × cm^2 (b) 300 gm × cm^2
(c) 200 gm × cm^2 (d) 100 gm × cm^2

Q.18 From a circular disc of radius R and mass 9 M, a small disc of radius $R/3$ is removed from the disc. The moment of inertia of the remaining disc about an axis perpendicular to the plane of the disc and passing through O is

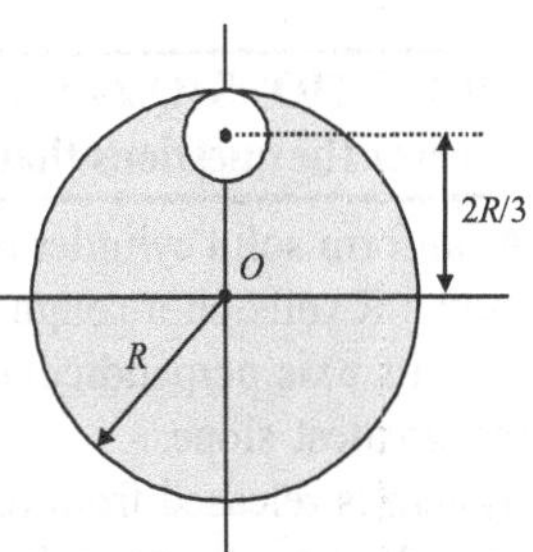

(a) $4MR^2$

(b) $\frac{40}{9}MR^2$

(c) $10MR^2$

(d) $\frac{37}{9}MR^2$

Q.19 The moment of inertia of a thin rod of mass M and length L about an axis perpendicular to the rod at a distance $L/4$ from one end is

(a) $\frac{ML^2}{6}$ (b) $\frac{ML^2}{12}$ (c) $\frac{7ML^2}{24}$ (d) $\frac{7ML^2}{48}$

Q.20 A wheel has a speed of 1200 revolutions per minute and is made to slow down at a rate of 4 radians /s^2. The number of revolutions it makes before coming to rest is

(a) 143 (b) 272 (c) 314 (d) 722

DIRECTIONS (Q.21-Q.23) : In the following questions, more than one of the answers given are correct. Select the correct answers and mark it according to the following codes:

Codes:

(a) 1, 2 and 3 are correct **(b)** 1 and 2 are correct
(c) 2 and 4 are correct **(d)** 1 and 3 are correct

Q.21 In pure rolling fraction of its total energy associated with rotation is α for a ring and β for a solid sphere. Then

(1) $\alpha = 1/2$ (2) $\beta = 2/7$ (3) $\beta = 2/5$ (4) $\alpha = 1/4$

Q.22 One solid sphere and a disc of same radius are falling along an inclined plane without slip. One reaches earlier than the other due to

(1) different size
(2) different radius of gyration
(3) different moment of inertia
(4) different friction

Q.23 A body is rolling down an inclined plane. Its translational and rotational kinetic energies are equal. The body is not a

(1) solid sphere (2) hollow sphere
(3) solid cylinder (4) hollow cylinder

RESPONSE GRID				
13. ⓐⓑⓒⓓ	14. ⓐⓑⓒⓓ	15. ⓐⓑⓒⓓ	16. ⓐⓑⓒⓓ	17. ⓐⓑⓒⓓ
18. ⓐⓑⓒⓓ	19. ⓐⓑⓒⓓ	20. ⓐⓑⓒⓓ	21. ⓐⓑⓒⓓ	22. ⓐⓑⓒⓓ
23. ⓐⓑⓒⓓ				

Space for Rough Work

DIRECTIONS (Q.24-Q.26) : Read the passage given below and answer the questions that follows :

A uniform solid cylinder of mass 2m and radius R rolls on a rough inclined plane with its axis perpendicular to the line of the greatest slope.

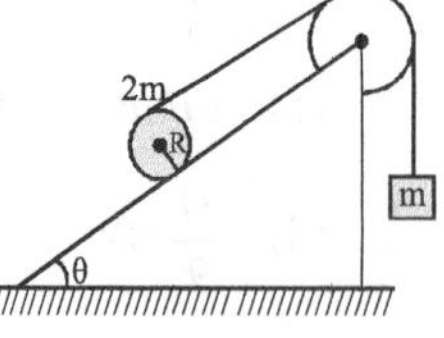

System is released from rest and as the cylinder rolls it winds up a light string which passes over a light pulley.

Q.24 The acceleration of block of mass m is -

(a) $\frac{2}{7}g\,(1-\cos\theta)$ (b) $\frac{4}{7}g\,(1-\sin\theta)$

(c) $\frac{2}{7}g\,(1-\sin\theta)$ (d) $\frac{2}{14}g\,(1+\sin\theta)$

Q.25 The tension in the string is –

(a) $\left(\frac{4+3\sin\theta}{7}\right)$ mg (b) $\left(\frac{3-4\sin\theta}{7}\right)$ mg

(c) $\left(\frac{3+4\sin\theta}{7}\right)$ mg (d) $\frac{2}{7}(1-\sin\theta)$ mg

Q.26 The frictional force acting on the cylinder is-

(a) $\frac{2}{7}(1-\sin\theta)$ mg (b) $\left(\frac{6-\sin\theta}{7}\right)$ mg

(c) $\left(\frac{1+6\cos\theta}{7}\right)$ mg (d) $\left(\frac{1+6\sin\theta}{7}\right)$ mg

DIRECTIONS (Q. 27-Q.29) : Each of these questions contains two statements: Statement-1 (Assertion) and Statement-2 (Reason). Each of these questions has four alternative choices, only one of which is the correct answer. You have to select the correct choice.

(a) Statement-1 is True, Statement-2 is True; Statement-2 is a correct explanation for Statement-1.

(b) Statement-1 is True, Statement-2 is True; Statement-2 is NOT a correct explanation for Statement-1.

(c) Statement -1 is False, Statement-2 is True.

(d) Statement -1 is True, Statement-2 is False.

Q.27 **Statement-1 :** Two cylinders, one hollow (metal) and the other solid (wood) with the same mass and identical dimensions are simultaneously allowed to roll without slipping down an inclined plane from the same height. The hollow cylinder will reach the bottom of the inclined plane first.

Statement-2 : By the principle of conservation of energy, the total kinetic energies of both the cylinders are identical when they reach the bottom of the incline.

Q.28 **Statement-1:** The force of frction in the case of a disc rolling without slipping down an inclined plane is 1/3 $g \sin\alpha$.

Statement-2: When the disc rolls without slipping, friction is required because for rolling condition velocity of point of contact is zero.

Q.29 **Statement-1:** If two different axes are at same distance from the centre of mass of a rigid body, then moment of inertia of the given rigid body about both the axes will always be the same.

Statement-2: From parallel axis theorem, $I = I_{cm} + md^2$, where all terms have usual meaning.

RESPONSE GRID	
	24.ⓐⓑⓒⓓ 25.ⓐⓑⓒⓓ 26.ⓐⓑⓒⓓ 27.ⓐⓑⓒⓓ 28. ⓐⓑⓒⓓ
	29.ⓐⓑⓒⓓ

DAILY PRACTICE PROBLEM SHEET 17 - PHYSICS			
Total Questions	29	Total Marks	116
Attempted		Correct	
Incorrect		Net Score	
Cut-off Score	28	Qualifying Score	44
Success Gap = Net Score – Qualifying Score			
Net Score = (Correct × 4) – (Incorrect × 1)			

Space for Rough Work

DPP - Daily Practice Problems

Name : Date :

Start Time : End Time :

PHYSICS 18

SYLLABUS : Gravitation - 1 (The Universal law of gravitation, Acceleration due to gravity and its variation with altitude and depth, Kepler's law of planetary motion)

Max. Marks : 116 **Time : 60 min.**

GENERAL INSTRUCTIONS

- The Daily Practice Problem Sheet contains 29 MCQ's. For each question only one option is correct. Darken the correct circle/ bubble in the Response Grid provided on each page.
- You have to evaluate your Response Grids yourself with the help of solution booklet.
- Each correct answer will get you 4 marks and 1 mark shall be deduced for each incorrect answer. No mark will be given/ deducted if no bubble is filled. Keep a timer in front of you and stop immediately at the end of 60 min.
- The sheet follows a particular syllabus. Do not attempt the sheet before you have completed your preparation for that syllabus. Refer syllabus sheet in the starting of the book for the syllabus of all the DPP sheets.
- After completing the sheet check your answers with the solution booklet and complete the Result Grid. Finally spend time to analyse your performance and revise the areas which emerge out as weak in your evaluation.

DIRECTIONS (Q.1-Q.20) : There are 20 multiple choice questions. Each question has 4 choices (a), (b), (c) and (d), out of which **ONLY ONE** choice is correct.

Q.1 A mass M splits into two parts m and (M – m), which are then separated by a certain distance. What ratio of (m / M) maximises the gravitational force between the parts. ?

(a) 2/3 (b) 3/4
(c) 1/2 (d) 1/3

Q.2 What would be the angular speed of earth, so that bodies lying on equator may experience weightlessness ? ($g = 10 m/s^2$ and radius of earth = 6400 km)

(a) 1.25×10^{-3} rad/sec (b) 1.25×10^{-2} rad/sec
(c) 1.25×10^{-4} rad/sec (d) 1.25×10^{-1} rad/sec

Q.3 The speed with which the earth have to rotate on its axis so that a person on the equator would weigh (3/5) th as much as present will be - (Take the equatorial radius as 6400 km.)

(a) 3.28×10^{-4} rad/sec (b) 7.826×10^{-4} rad/sec
(c) 3.28×10^{-3} rad/sec (d) 7.28×10^{-3} rad/sec

Q.4 On a planet (whose size is the same as that of earth and mass 4 times to the earth) the energy needed to lift a 2kg mass vertically upwards through 2m distance on the planet is ($g = 10 m/sec^2$ on the surface of earth) -

(a) 16 J (b) 32 J (c) 160 J (d) 320 J

Q.5 Two bodies of mass 10^2 kg and 10^3 kg are lying 1m apart. The gravitational potential at the mid-point of the line joining them is -

(a) 0 (b) –1.47 Joule/kg
(c) 1.47 Joule/kg (d) -147×10^{-7} Joule /kg

RESPONSE GRID	1. ⓐⓑⓒⓓ	2. ⓐⓑⓒⓓ	3. ⓐⓑⓒⓓ	4. ⓐⓑⓒⓓ	5. ⓐⓑⓒⓓ

Space for Rough Work

Q.6 If g is the acceleration due to gravity on the earth's surface, the gain in P.E. of an object of mass m raised from the surface of the earth to a height of the radius R of the earth is -

(a) mgR (b) 2mgR (c) $\frac{1}{2}$ mgR (d) $\frac{1}{4}$ mgR

Q.7 Four particles, each of mass m, are placed at the corners of square and moving along a circle of radius r under the influence of mutual gravitational attraction. The speed of each particle will be -

(a) $\sqrt{\frac{Gm}{r}(2\sqrt{2}+1)}$ (b) $\sqrt{\frac{Gm}{r}}$

(c) $\sqrt{\frac{Gm}{r}\left(\frac{2\sqrt{2}+1}{4}\right)}$ (d) $\sqrt{\frac{2\sqrt{2}Gm}{r}}$

Q.8 Three particles of equal mass m are situated at the vertices of an equilateral triangle of side l. What should be the velocity of each particle, so that they move on a circular path without changing l ?

(a) $\sqrt{\frac{GM}{2\ell}}$ (b) $\sqrt{\frac{GM}{\ell}}$ (c) $\sqrt{\frac{2GM}{\ell}}$ (d) $\sqrt{\frac{GM}{3\ell}}$

Q.9 What will be the acceleration due to gravity on the surface of the moon if its radius is 1/4 th the radius of the earth and its mass is 1/80 th the mass of the earth ?

(a) g/6 (b) g/5 (c) g/7 (d) g/8

Q.10 If the value of 'g' at a height h above the surface of the earth is the same as at a depth x below it, then (both x and h being much smaller than the radius of the earth)

(a) $x=h$ (b) $x=2h$ (c) $x=\frac{h}{2}$ (d) $x=h^2$

Q.11 At what height above the earth's surface the acceleration due to gravity will be 1/9 th of its value at the earth's surface? Radius of earth is 6400 km.

(a) 12800 km (b) 1280 km
(c) 128000 km (d) 128 km

Q.12 If the radius of the earth were to shrink by one percent, its mass remaining the same, the acceleration due to gravity on the earth's surface would -

(a) decrease (b) remain unchanged
(c) increase (d) None of these

Q.13 At what height above the earth's surface does the force of gravity decrease by 10% ? Assume radius of earth to be 6370 km.

(a) 350km. (b) 250km. (c) 150km. (d) 300km.

Q.14 A particle is suspended from a spring and it stretches the spring by 1 cm on the surface of earth. The same particle will stretches the same spring at a place 800 km above earth surface by -

(a) 0.79 cm (b) 0.1 cm
(c) $\pi/6$ rad/hr. (d) $2\pi/7$ rad/hr.

Q.15 The change in the value of acceleration of earth towards sun, when the moon comes from the position of solar eclipse to the position on the other side of earth in line with sun is (Mass of moon = 7.36×10^{22} kg, the orbital radius of moon 3.8×10^8m.

(a) 6.73×10^{-2} m/s^2 (b) 6.73×10^{-3} m/s^2
(c) 6.73×10^{-4} m/s^2 (d) 6.73×10^{-5} m/s^2

Q.16 The radius of the earth is R_e and the acceleration due to gravity at its surface is g. The work required in raising a body of mass m to a height h form the surface of the earth will be -

(a) $\frac{mgh}{\left(1-\frac{h}{R_e}\right)}$ (b) $\frac{mgh}{\left(1+\frac{h}{R_e}\right)^2}$ (c) $\frac{mgh}{\left(1+\frac{h}{R_e}\right)}$ (d) $\frac{mg}{\left(1+\frac{h}{R_e}\right)}$

Q.17 The masses and the radius of the earth and the moon are M_1, M_2 and R_1, R_2 respectively their centres are at distance d apart. The minimum speed with which a particle of mass m should be projected form a point midway between the two centres so as to escape to infinity will be -

(a) $2\sqrt{\frac{G}{d}(M_1+M_2)}$ (b) $\sqrt{\frac{G}{d}(M_1+M_2)}$

(c) $\sqrt{\frac{G}{2d}(M_1+M_2)}$ (d) $2\sqrt{\frac{G M_1}{d M_2}}$

Q.18 With what velocity must a body be thrown upward form the surface of the earth so that it reaches a height of 10 R_e? earth's mass $M_e = 6 \times 10^{24}$ kg, radius $R_e = 6.4 \times 10^6$ m and $G = 6.67 \times 10^{-11}$ N–m^2/kg^2.

(a) 10.7×10^4 m/s (b) 10.7×10^3 m/s
(c) 10.7×10^5 m/s (d) 1.07×10^4 m/s

RESPONSE GRID					
	6. ⓐⓑⓒⓓ	7. ⓐⓑⓒⓓ	8. ⓐⓑⓒⓓ	9. ⓐⓑⓒⓓ	10. ⓐⓑⓒⓓ
	11. ⓐⓑⓒⓓ	12. ⓐⓑⓒⓓ	13. ⓐⓑⓒⓓ	14. ⓐⓑⓒⓓ	15. ⓐⓑⓒⓓ
	16. ⓐⓑⓒⓓ	17. ⓐⓑⓒⓓ	18. ⓐⓑⓒⓓ		

Space for Rough Work

Q.19 Two concentric shells of uniform density having masses M_1 and M_2 are situated as shown in the figure. The force on the particle of mass m when it is located at $r = b$ is

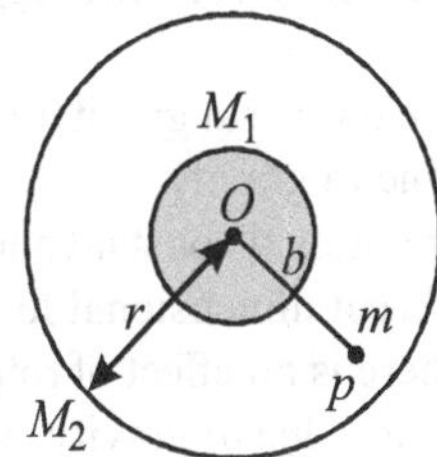

(a) $\frac{GM_1m}{b^2}$ (b) $\frac{GM_2m}{b^2}$

(c) $G\frac{(M_1+M_2)m}{b^2}$ (d) $G\frac{(M_1-M_2)m}{b^2}$

Q.20 What is the mass of the planet that has a satellite whose time period is T and orbital radius is r?

(a) $\frac{4\pi^2 r^3}{GT^2}$ (b) $\frac{3\pi^2 r^3}{GT^2}$ (c) $\frac{4\pi^2 r^3}{GT^3}$ (d) $\frac{4\pi^2 T}{GT^2}$

DIRECTIONS (Q.21-Q.23) : In the following questions, more than one of the answers given are correct. Select the correct answers and mark it according to the following codes:

Codes :

(a) 1, 2 and 3 are correct **(b)** 1 and 2 are correct
(c) 2 and 4 are correct **(d)** 1 and 3 are correct

Q.21 The gravitational force between two point masses m_1 and m_2 at separation r is given by $F = k\frac{m_1m_2}{r^2}$

The constant k doesn't

(1) depend on medium between masses
(2) depend on the place
(3) depend on time
(4) depend on system of units

Q.22 Which of the following statements about the gravitional constant are false ?

(1) It is a force
(2) It has no unit
(3) It has same value in all system of units
(4) It doesn't depend on the value of the masses

Q.23 Spot the correct statements :

The acceleration due to gravity 'g' decreases if

(1) We go down from the surface of the earth towards its centre
(2) We go up from the surface of the earth
(3) The rotational velocity of the earth is increased
(4) We go from the equator towards the poles on the surface of the earth

DIRECTIONS (Q.24-Q.26) : Read the passage given below and answer the questions that follows :

The orbit of Pluto is much more eccentric than the orbits of the other planets. That is, instead of being nearly circular, the orbit is noticeably elliptical. The point in the orbit nearest to the sun is called the perihelion and the point farthest from the sun is called the aphelion.

Q.24 At perihelion, the gravitational potential energy of Pluto in its orbit has

(a) its maximum value
(b) its minimum value
(c) the same value as at every other point in the orbit
(d) value which depends on sense of rotation

Q.25 At perihelion, the mechanical energy of Pluto's orbit has

(a) its maximum value
(b) its minimum value
(c) the same value as at every other point in the orbit
(d) value which depends on sense of rotation

Q.26 As Pluto moves from the perihelion to the aphelion, the work done by gravitational pull of Sun on Pluto is

(a) is zero
(b) is positive
(c) is negative
(d) depends on sense of rotation

RESPONSE GRID					
	19. ⓐⓑⓒⓓ	20. ⓐⓑⓒⓓ	21. ⓐⓑⓒⓓ	22. ⓐⓑⓒⓓ	23. ⓐⓑⓒⓓ
	24. ⓐⓑⓒⓓ	25. ⓐⓑⓒⓓ	26. ⓐⓑⓒⓓ		

Space for Rough Work

DIRECTIONS (Q.27-Q.29) : Each of these questions contains two statements: Statement-1 (Assertion) and Statement-2 (Reason). Each of these questions has four alternative choices, only one of which is the correct answer. You have to select the correct choice.

(a) Statement-1 is True, Statement-2 is True; Statement-2 is a correct explanation for Statement-1.
(b) Statement-1 is True, Statement-2 is True; Statement-2 is NOT a correct explanation for Statement-1.
(c) Statement -1 is False, Statement-2 is True.
(d) Statement -1 is True, Statement-2 is False.

Q.27 **Statement-1 :** Gravitational force between two particles is negligibly small compared to the electrical force.
Statement-2 : The electrical force is experienced by charged particles only.

Q.28 **Statement-1 :**The universal gravitational constant is same as acceleration due to gravity.
Statement-2 : Gravitional constant and acceleration due to gravity have different dimensional formula.

Q.29 **Statement-1 :**There is no effect of rotation of earth on the value of acceleration due to gravity at poles.
Statement-2 : Rotation of earth is about polar axis.

RESPONSE GRID	27. ⓐⓑⓒⓓ	28. ⓐⓑⓒⓓ	29. ⓐⓑⓒⓓ

DAILY PRACTICE PROBLEM SHEET 18 - PHYSICS			
Total Questions	29	Total Marks	116
Attempted		Correct	
Incorrect		Net Score	
Cut-off Score	30	Qualifying Score	48
Success Gap = Net Score – Qualifying Score			
Net Score = (Correct × 4) – (Incorrect × 1)			

Space for Rough Work

DPP - Daily Practice Problems

Name :

Date :

Start Time :

End Time :

PHYSICS 19

SYLLABUS : Gravitation - 2 (Gravitational potential energy, Gravitational potential, Escape velocity & Orbital velocity of a satellite, Geo-stationary satellites)

Max. Marks : 120 **Time : 60 min.**

GENERAL INSTRUCTIONS

- The Daily Practice Problem Sheet contains 30 MCQ's. For each question only one option is correct. Darken the correct circle/ bubble in the Response Grid provided on each page.
- You have to evaluate your Response Grids yourself with the help of solution booklet.
- Each correct answer will get you 4 marks and 1 mark shall be deduced for each incorrect answer. No mark will be given/ deducted if no bubble is filled. Keep a timer in front of you and stop immediately at the end of 60 min.
- The sheet follows a particular syllabus. Do not attempt the sheet before you have completed your preparation for that syllabus. Refer syllabus sheet in the starting of the book for the syllabus of all the DPP sheets.
- After completing the sheet check your answers with the solution booklet and complete the Result Grid. Finally spend time to analyse your performance and revise the areas which emerge out as weak in your evaluation.

DIRECTIONS (Q.1-Q.21) : There are 21 multiple choice questions. Each question has 4 choices (a), (b), (c) and (d), out of which **ONLY ONE** choice is correct.

Q.1 A body of mass 100 kg falls on the earth from infinity. What will be its energy on reaching the earth ? Radius of the earth is 6400 km and g = 9.8 m/s^2. Air friction is negligible.

(a) 6.27×10^9 J (b) 6.27×10^{10} J
(c) 6.27×10^{10} J (d) 6.27×10^7 J

Q.2 An artificial satellite of the earth is to be established in the equatorial plane of the earth and to an observer at the equator it is required that the satellite will move eastward, completing one round trip per day. The distance of the satellite from the centre of the earth will be- (The mass of the earth is 6.00×10^{24} kg and its angular velocity = 7.30×10^{-5} rad./sec.)

(a) 2.66×10^3 m. (b) 2.66×10^5 m.
(c) 2.66×10^6 m. (d) 2.66×10^7 m.

Q.3 Two satellites S_1 and S_2 revolve round a planet in the same direction in circular orbits. Their periods of revolutions are 1 hour and 8 hour respectively. The radius of S_1 is 10^4 km. The velocity of S_2 with respect to S_1 will be-

(a) $\pi \times 10^4$ km/hr (b) $\pi/3 \times 10^4$ km/hr
(c) $2\pi \times 10^4$ km/hr (d) $\pi/2 \times 10^4$ km/hr

Q.4 In the above example the angular velocity of S_2 as actually observed by an astronaut in S_1 is -

(a) $\pi/3$ rad/hr (b) $\pi/3$ rad/sec
(c) $\pi/6$ rad/hr (d) $2\pi/7$ rad/hr

RESPONSE GRID	1. ⓐⓑⓒⓓ	2. ⓐⓑⓒⓓ	3. ⓐⓑⓒⓓ	4. ⓐⓑⓒⓓ

Space for Rough Work

Q.5 The moon revolves round the earth 13 times in one year. If the ratio of sun-earth distance to earth-moon distance is 392, then the ratio of masses of sun and earth will be -

(a) 365 (b) 356
(c) 3.56×10^5 (d) 1

Q.6 Two planets of radii in the ratio 2 : 3 are made from the materials of density in the ratio 3 : 2. Then the ratio of acceleration due to gravity g_1/g_2 at the surface of two planets will be

(a) 1 (b) 2.25 (c) $\frac{4}{9}$ (d) 0.12

Q.7 A satellite of mass m is revolving in a circular orbit of radius r. The relation between the angular momentum J of satellite and mass m of earth will be -

(a) $J = \sqrt{G.Mm^2r}$ (b) $J = \sqrt{GMm}$
(c) $J = \sqrt{GMmr}$ (d) $J = \sqrt{\frac{mr}{M}}$

Q.8 A spaceship is launched into a circular orbit close to earth's surface. What additional velocity has now to be imparted to the spaceship in the orbit to overcome the gravitational pull? (Radius of earth = 6400 km, g = 9.8 m/sec^2)

(a) 3.285 km/sec (b) 32.85 m/sec
(c) 11.32 km/sec (d) 7.32 m/sec

Q.9 The ratio of the radius of the Earth to that of the moon is 10. The ratio of g on earth to the moon is 6. The ratio of the escape velocity from the earth's surface to that from the moon is approximately -

(a) 10 (b) 8 (c) 4 (d) 2

Q.10 Acceleration due to gravity on a planet is 10 times the value on the earth. Escape velocity for the planet and the earth are V_p and V_e respectively. Assuming that the radii of the planet and the earth are the same, then -

(a) $V_P = 10\,V_e$ (b) $V_P = \sqrt{10}\,V_e$
(c) $V_p = \frac{V_e}{\sqrt{10}}$ (d) $V_P = \frac{V_e}{10}$

Q.11 The Jupiter's period of revolution round the Sun is 12 times that of the Earth. Assuming the planetary orbits are circular, how many times the distance between the Jupiter and Sun exceeds that between the Earth and the sun.

(a) 5.242 (b) 4.242 (c) 3.242 (d) 2.242

Q.12 The mean distance of mars from sun is 1.524 times the distance of the earth from the sun. The period of revolution of mars around sun will be-

(a) 2.88 earth year (b) 1.88 earth year
(c) 3.88 earth year (d) 4.88 earth year

Q.13 The semi-major axes of the orbits of mercury and mars are respectively 0.387 and 1.524 in astronomical unit. If the period of Mercury is 0.241 year, what is the period of Mars.

(a) 1.2 years (b) 3.2 years
(c) 3.9 years (d) 1.9 years

Q.14 If a graph is plotted between T^2 and r^3 for a planet then its slope will be -

(a) $\frac{4\pi^2}{GM}$ (b) $\frac{GM}{4\pi^2}$
(c) 4π GM (d) 0

Q.15 The mass and radius of earth and moon are M_1, R_1 and M_2, R_2 respectively. Their centres are d distance apart. With what velocity should a particle of mass m be projected from the mid point of their centres so that it may escape out to infinity -

(a) $\sqrt{\frac{G(M_1+M_2)}{d}}$ (b) $\sqrt{\frac{2G(M_1+M_2)}{d}}$
(c) $\sqrt{\frac{4G(M_1+M_2)}{d}}$ (d) $\sqrt{\frac{GM_1M_2}{d}}$

Q.16 A satellite has to revolve round the earth in a circular orbit of radius 8×10^3 km. The velocity of projection of the satellite in this orbit will be -

(a) 16 km/sec (b) 8 km/sec
(c) 3 km/sec (d) 7.08 km/sec

Q.17 If the satellite is stopped suddenly in its orbit which is at a distnace = radius of earth from earth's surface and allowed to fall freely into the earth, the speed with which it hits the surface of earth will be -

(a) 7.919 m/sec (b) 7.919 km/sec
(c) 11.2 m/sec (d) 11.2 km/sec

RESPONSE GRID					
	5. ⓐⓑⓒⓓ	6. ⓐⓑⓒⓓ	7. ⓐⓑⓒⓓ	8. ⓐⓑⓒⓓ	9. ⓐⓑⓒⓓ
	10. ⓐⓑⓒⓓ	11. ⓐⓑⓒⓓ	12. ⓐⓑⓒⓓ	13. ⓐⓑⓒⓓ	14. ⓐⓑⓒⓓ
	15. ⓐⓑⓒⓓ	16. ⓐⓑⓒⓓ	17. ⓐⓑⓒⓓ		

Space for Rough Work

Q.18 A projectile is fired vertically upward from the surface of earth with a velocity K v_e m/s where v_e m/s is the escape velocity and K < 1. Neglecting air resistance, the maximum height to which it will rise measured from the centre of the earth is - (where R = radius of earth)

(a) $\frac{R}{1-K^2}$ (b) $\frac{R}{K^2}$ (c) $\frac{1-K^2}{R}$ (d) $\frac{K^2}{R}$

Q.19 A satellite is revolving in an orbit close to the earth's surface. Taking the radius of the earth as 6.4×10^6 metre, the value of the orbital speed and the period of revolution of the satellite will respectively be (g = 9.8 meter/sec^2)

(a) 7.2 km/sec., 84.6 minutes
(b) 2.7 km/sec., 8.6 minutes
(c) .72 km/sec., 84.6 minutes
(d) 7.2 km/sec., 8.6 minutes

Q.20 If the period of revolution of an artificial satellite just above the earth be T second and the density of earth be ρ, kg/m^3 then
($G = 6.67 \times 10^{-11}$ m^3/kg. second^2)

(a) ρT^2 is a universal constant
(b) ρT^2 varies with time
(c) $\rho T^2 = \frac{3\pi}{G}$
(d) Both (a) and (c)

Q.21 Two satellites P and Q of same mass are revolving near the earth surface in the equitorial plane. The satellite P moves in the direction of rotation of earth whereas Q moves in the opposite direction. The ratio of their kinetic energies with respect to a frame attached to earth will be -

(a) $\left(\frac{8363}{7437}\right)^2$ (b) $\left(\frac{7437}{8363}\right)^2$ (c) $\left(\frac{8363}{7437}\right)$ (d) $\left(\frac{7437}{8363}\right)$

DIRECTIONS (Q.22-Q.24) : In the following questions, more than one of the answers given are correct. Select the correct answers and mark it according to the following codes:

Codes :
(a) 1, 2 and 3 are correct **(b)** 1 and 2 are correct
(c) 2 and 4 are correct **(d)** 1 and 3 are correct

Q.22 Gas escapes from the surface of a planet because it acquires an escape velocity. The escape velocity will depend on which of the following factors:

(1) Mass of the planet
(2) Radius of the planet
(3) Mass of the particle escaping
(4) Temperature of the planet

Q.23 v_e and v_p denotes the escape velocity from the earth and another planet having twice the radius and the same mean density as the earth. Then which of the following is (are) wrong ?

(1) $v_e = v_p$ (2) $v_e = 2v_p$
(3) $v_e = v_p/4$ (4) $v_e = v_p/2$

Q.24 Select the wrong statements from the following

(1) The orbital velocity of a satellite increases with the radius of the orbit
(2) Escape velocity of a particle from the surface of the earth depends on the speed with which it is fired
(3) The time period of a satellite does not depend on the radius of the orbit
(4) The orbital velocity is inversely proportional to the square root of the radius of the orbit

DIRECTIONS (Q.25-Q.27) : Read the passage given below and answer the questions that follows :

It can be assumed that orbits of earth and mars are nearly circular around the sun. It is proposed to launch an artificial planet around the sun such that its apogee is at the orbit of mars while its perigee is at the orbit of earth. Let T_e and T_m be periods of revolution of earth and mars. Further the variables are assigned the meanings as follows.

$M_e \rightarrow$ Mass of earth
$M_m \rightarrow$ Mass of mars.
$L_e \rightarrow$ Angular momentum of earth around the sun.
$L_m \rightarrow$ Angular momentum of mars around the sun.
$R_e \rightarrow$ Semi major axis of earth's orbit.
$R_m \rightarrow$ Semi major axis of mars orbit.
$M \rightarrow$ Mass of the artificial planet.
$E_e \rightarrow$ Total energy of earth.
$E_m \rightarrow$ Total energy of mars.

RESPONSE GRID	18. ⓐⓑⓒⓓ	19. ⓐⓑⓒⓓ	20. ⓐⓑⓒⓓ	21. ⓐⓑⓒⓓ	22. ⓐⓑⓒⓓ
	23. ⓐⓑⓒⓓ	24. ⓐⓑⓒⓓ			

Space for Rough Work

Q.25 Time period of revolution of the artificial planet about sun will be (neglect gravitational effects of earth and mars)

(a) $\frac{T_e + T_m}{2}$ (b) $\sqrt{T_e T_m}$

(c) $\frac{2T_e T_m}{T_e + T_m}$ (d) $\left[\frac{T_e^{2/3} + T_m^{2/3}}{2}\right]^{3/2}$

Q.26 The total energy of the artificial planet's orbit will be

(a) $\frac{2M}{M_e}\left(\frac{R_e E_e}{R_e + R_m}\right)$ (b) $\frac{2M}{M_m}\left(\frac{R_e E_e}{R_e + R_m}\right)$

(c) $\frac{2E_e M}{M_m}\left(\frac{R_e + R_m}{R_m}\right)$ (d) $\frac{2E_e M}{M_e}\left(\frac{R_e + R_m}{\sqrt{R_e^2 + R_m^2}}\right)$

Q.27 Areal velocity of the artificial planet around the sun will be

(a) less than that of earth
(b) more than that of mars
(c) more than that of earth
(d) same as that of the earth

DIRECTIONS (Q. 28-Q.30) : Each of these questions contains two statements: Statement-1 (Assertion) and Statement-2 (Reason). Each of these questions has four alternative choices, only one of which is the correct answer. You have to select the correct choice.

(a) Statement-1 is True, Statement-2 is True; Statement-2 is a correct explanation for Statement-1.
(b) Statement-1 is True, Statement-2 is True; Statement-2 is NOT a correct explanation for Statement-1.
(c) Statement-1 is False, Statement-2 is True.
(d) Statement-1 is True, Statement-2 is False.

Q.28 **Statement-1** :The speed of revolution of an artificial satellite revolving very near the earth is $8kms^{-1}$.
Statement-2 : Orbital velocity of a satellite, become independent of height of near satellite.

Q.29 **Statement-1** :If an earth satellite moves to a lower orbit, there is some dissipation of energy but the speed of gravitational satellite increases.
Statement-2 :The speed of satellite is a constant quantity.

Q.30 **Statement-1** :Gravitational potential of earth at every place on it is negative.
Statement-2 :Every body on earth is bound by the attraction of earth.

RESPONSE GRID	25. ⓐⓑⓒⓓ 26. ⓐⓑⓒⓓ 27. ⓐⓑⓒⓓ 28. ⓐⓑⓒⓓ 29. ⓐⓑⓒⓓ 30. ⓐⓑⓒⓓ

DAILY PRACTICE PROBLEM SHEET 19 - PHYSICS			
Total Questions	30	Total Marks	120
Attempted		Correct	
Incorrect		Net Score	
Cut-off Score	32	Qualifying Score	50
Success Gap = Net Score – Qualifying Score			
Net Score = (Correct × 4) – (Incorrect × 1)			

Space for Rough Work

DPP - Daily Practice Problems

Name : 　　　　　　Date :

Start Time : 　　　　　　End Time :

PHYSICS 20

SYLLABUS : Mechanical Properties of solids

Max. Marks : 120 　　　　　　**Time : 60 min.**

GENERAL INSTRUCTIONS

- The Daily Practice Problem Sheet contains 30 MCQ's. For each question only one option is correct. Darken the correct circle/ bubble in the Response Grid provided on each page.
- You have to evaluate your Response Grids yourself with the help of solution booklet.
- Each correct answer will get you 4 marks and 1 mark shall be deduced for each incorrect answer. No mark will be given/ deducted if no bubble is filled. Keep a timer in front of you and stop immediately at the end of 60 min.
- The sheet follows a particular syllabus. Do not attempt the sheet before you have completed your preparation for that syllabus. Refer syllabus sheet in the starting of the book for the syllabus of all the DPP sheets.
- After completing the sheet check your answers with the solution booklet and complete the Result Grid. Finally spend time to analyse your performance and revise the areas which emerge out as weak in your evaluation.

DIRECTIONS (Q.1-Q.21) : There are 21 multiple choice questions. Each question has 4 choices (a), (b), (c) and (d), out of which **ONLY ONE** choice is correct.

Q.1 Length of a wire is doubled, when 20×10^8 N/m^2 stress is applied on it. Its Young's modulus of elasticity in N/m^2 will be

(a) 20×10^8 　(b) 20×10^9
(c) 20×10^{10} 　(d) 10×10^8

Q.2 A steel wire of uniform cross-sectional area 2mm^2 is heated upto 50°C and clamped rigidly at two ends. If the temperature of wire falls to 30° then change in tension in the wire will be, if coefficient of linear expansion of steel is 1.1×10^{-5} /°C and young's modulus of elasticity of steel is 2×10^{11} N/m^2.

(a) 44 N 　(b) 88 N
(c) 132 N 　(d) 22 N

Q.3 The work done in increasing the length of a one metre long wire of cross-sectional area 1mm^2 through 1mm will be ($Y = 2 \times 10^{11}$ N/m^2)

(a) 250 J 　(b) 10 J
(c) 5 J 　(d) 0.1 J

Q.4 A spring is stretched by 3cm when a load of 5.4×10^6 dyne is suspended from it. Work done will be-

(a) 8.1×10^6 J 　(b) 8×10^6 J
(c) 8.0×10^6 erg 　(d) 8.1×10^6 erg

RESPONSE GRID	1. ⓐⓑⓒⓓ	2. ⓐⓑⓒⓓ	3. ⓐⓑⓒⓓ	4. ⓐⓑⓒⓓ

Space for Rough Work

Q.5 A wire of length 1m and area of cross section 4×10^{-8} m^2 increases in length by 0.2 cm when a force of 16 N is applied. Value of Y for the material of the wire will be

(a) 2×10^6 N/m^2 (b) 2×10^{11} kg/m^2
(c) 2×10^{11} N/mm^2 (d) 2×10^{11} N/m^2

Q.6 The volume of a solid rubber ball when it is carried from the surface to the bottom of a 200 m deep lake decreases by 0.1%. The value for bulk modulus of elasticity for rubber will be

(a) 2×10^9 Pa (b) 2×10^6 Pa
(c) 2×10^4 Pa (d) 2×10^{-4} Pa

Q.7 A steel wire is 4.0 m long and 2 mm in diameter. Young's modulus of steel is 1.96×10^{11} N/m^2. If a mass of 20 kg is suspended from it the elongation produced will be -

(a) 2.54 mm (b) 1.27 mm
(c) 0.64 mm (d) 0.27 mm

Q.8 A brass rod is to support a load of 400 N. If its elastic limit is 4.0×10^8 N/m^2 its minimum diameter must be -

(a) 1.13 mm (b) 2.26 mm
(c) 3.71 mm (d) 4.52 mm

Q.9 A 4.0 m long copper wire of cross sectional area 1.2 cm^2 is stretched by a force of 4.8×10^3 N stress will be -

(a) 4.0×10^7 N/mm^2 (b) 4.0×10^7 KN/m^2
(c) 4.0×10^7 N/m^2 (d) None of these

Q.10 A copper rod 2m long is stretched by 1mm. Strain will be -

(a) 10^{-4}, volumetric (b) 5×10^{-4}, volumetric
(c) 5×10^{-4}, longitudinal (d) 5×10^{-3}, volumetric

Q.11 A wire of cross sectional area 3mm^2 is just stretched between two fixed points at a temperature of 20°C. Determine the tension when the temperature falls to 20°C. Coefficient of linear expansion $\alpha = 10^{-5}$ /°C and $Y = 2 \times 10^{11}$ N/m^2.

(a) 120 KN (b) 20 N (c) 120 N (d) 102 N

Q.12 The compressibility of water is 5×10^{-10} m^2/N. If it is subjected to a pressure of 15 MPa, the fractional decrease in volume will be -

(a) 3.3×10^{-5} (b) 5.6×10^{-4}
(c) 7.5×10^{-3} (d) 1.5×10^{-2}

Q.13 The Young's modulus of steel is 2×10^{11} N/m^2 and its coefficient of linear expansion is 1.1×10^{-5} per deg. The pressure to be applied to the ends of a steel cylinder to keep its length constant on raising its temperature by 100°C, will be -

(a) 5.5×10^4 N/m^2 (b) 1.8×10^6 N/m^2
(c) 2.2×10^8 N/m^2 (d) 2.0×10^{11} N/m^2

Q.14 For a given material, the Young's modulus is 2.4 times that of rigidity modulus. It's poisson's ratio is

(a) 1.2 (b) 1.02
(c) 0.2 (d) 2

Q.15 A wire of length 1m is stretched by a force of 10N. The area of cross-section of the wire is 2×10^{-6} m^2 and Y is 2×10^{11}N/m^2. Increase in length of the wire will be -

(a) 2.5×10^{-5} cm (b) 2.5×10^{-5} mm
(c) 2.5×10^{-5} m (d) None of these

Q.16 A stress of 1kg/mm^2 is applied on a wire. If the modulus of elasticity of the wire is 10^{10} dyne/cm^2, then the percentage increase in the length of the wire will be

(a) 0.007 (b) 0.0098 (c) 98 (d) 9.8

Q.17 A uniform steel wire of density 7800kg/m^3 is 2.5 m long and weighs 15.6×10^{-3} kg. It extends by 1.25 mm when loaded by 8kg. Calculate the value of young's modulus of elasticity for steel.

(a) 1.96×10^{11} N/m^2 (b) 19.6×10^{11} N/m^2
(c) 196×10^{11} N/m^2 (d) None of these

RESPONSE GRID					
	5. ⓐⓑⓒⓓ	6. ⓐⓑⓒⓓ	7. ⓐⓑⓒⓓ	8. ⓐⓑⓒⓓ	9. ⓐⓑⓒⓓ
	10. ⓐⓑⓒⓓ	11. ⓐⓑⓒⓓ	12. ⓐⓑⓒⓓ	13. ⓐⓑⓒⓓ	14. ⓐⓑⓒⓓ
	15. ⓐⓑⓒⓓ	16. ⓐⓑⓒⓓ	17. ⓐⓑⓒⓓ		

Space for Rough Work

Q.18 A metallic wire is suspended by suspending weight to it. If S is longitudinal strain and Y its young's modulus of elasticity then potential energy per unit volume will be

(a) $\frac{1}{2}Y^2S^2$ (b) $\frac{1}{2}Y^2S$

(c) $\frac{1}{2}YS^2$ (d) $2YS^2$

Q.19 The lengths and radii of two wires of same material are respectively L, 2L, and 2R, R. Equal weights are applied on then. If the elongations produced in them are l_1 and l_2 respectively then their ratio will be

(a) 2 : 1 (b) 4 : 1

(c) 8 : 1 (d) 1 : 8

Q.20 The ratio of radii of two wires of same material is 2 : 1. If these wires are stretched by equal forces, then the ratio of stresses produced in them will be

(a) 1 : 2 (b) 2 : 1

(c) 1 : 4 (d) 4 : 1

Q.21 A rod of length l and area of cross-section A is heated from 0°C to 100°C. The rod is so placed that it is not allowed to increase in length, then the force developed is proportional to

(a) l (b) l^{-1} (c) A (d) A^{-1}

DIRECTIONS (Q.22-Q.24) : In the following questions, more than one of the answers given are correct. Select the correct answers and mark it according to the following codes:

Codes :

(a) 1, 2 and 3 are correct **(b)** 1 and 2 are correct

(c) 2 and 4 are correct **(d)** 1 and 3 are correct

Q.22 Mark the correct statements

(1) Sliding of molecular layer is much easier than compression or expansion

(2) Reciprocal of bulk modulus of elasticity is called compressibility

(3) Hollow shaft is much stronger than a solid rod of same length and same mass

(4) It is difficult to twist a long rod as compared to small rod

Q.23 Which statements are false for a metal?

(1) $Y < \eta$ (2) $Y = \eta$

(3) $Y < 1/\eta$ (4) $Y > \eta$

Q.24 Which of the following relations are false

(1) $3Y = K(1-\sigma)$ (2) $\sigma = (6K+\eta)Y$

(3) $K = \frac{9\eta Y}{Y+\eta}$ (4) $\sigma = \frac{0.5Y-\eta}{\eta}$

DIRECTIONS (Q.25-Q.27) : Read the passage given below and answer the questions that follows :

A bar of cross section A is subjected to equal and opposite tensile forces F at its ends. Consider a plane through the bar making an angle θ with a plane at right angles to the bar as shown in figure.

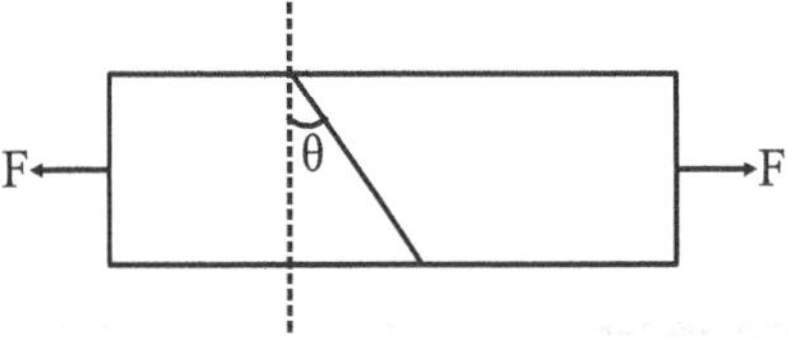

Q.25 The tensile stress at this plane in terms of F, A and θ is

(a) $\frac{F\cos^2\theta}{A}$ (b) $\frac{F}{A\cos^2\theta}$

(c) $\frac{F\sin^2\theta}{A}$ (d) $\frac{F}{A\sin^2\theta}$

Q.26 In the above problem, for what value of θ is the tensile stress maximum ?

(a) Zero (b) 90° (c) 45° (d) 30°

RESPONSE GRID					
	18. ⓐⓑⓒⓓ	19. ⓐⓑⓒⓓ	20. ⓐⓑⓒⓓ	21. ⓐⓑⓒⓓ	22. ⓐⓑⓒⓓ
	23. ⓐⓑⓒⓓ	24. ⓐⓑⓒⓓ	25. ⓐⓑⓒⓓ	26. ⓐⓑⓒⓓ	

Space for Rough Work

Q.27 The shearing stress at the plane, in terms of F, A and θ is

(a) $\frac{F\cos 2\theta}{2A}$ (b) $\frac{F\sin 2\theta}{2A}$

(c) $\frac{F\sin\theta}{A}$ (d) $\frac{F\cos\theta}{A}$

DIRECTIONS (Qs. 28-Q.30) : Each of these questions contains two statements: Statement-1 (Assertion) and Statement-2 (Reason). Each of these questions has four alternative choices, only one of which is the correct answer. You have to select the correct choice.

(a) Statement-1 is True, Statement-2 is True; Statement-2 is a correct explanation for Statement-1.

(b) Statement-1 is True, Statement-2 is True; Statement-2 is NOT a correct explanation for Statement-1.

(c) Statement -1 is False, Statement-2 is True.

(d) Statement -1 is True, Statement-2 is False.

Q.28 Statement -1 : Steel is more elastic than rubber.

Statement -2 : Under given deforming force, steel is deformed less than rubber.

Q.29 Statement -1 : Bulk modulus of elasticity (K) represents incompressibility of the material.

Statement -2 : Bulk modulus of elasticity is proportional to change in pressure.

Q.30 Statement -1 : The bridges declared unsafe after a long use.

Statement -2 : Elastic strength of bridges losses with time.

RESPONSE GRID	27. ⓐⓑⓒⓓ	28. ⓐⓑⓒⓓ	29. ⓐⓑⓒⓓ	30. ⓐⓑⓒⓓ

DAILY PRACTICE PROBLEM SHEET 20 - PHYSICS			
Total Questions	30	Total Marks	120
Attempted		Correct	
Incorrect		Net Score	
Cut-off Score	32	Qualifying Score	52
Success Gap = Net Score – Qualifying Score			
Net Score = (Correct × 4) – (Incorrect × 1)			

Space for Rough Work

DPP - Daily Practice Problems

Name :

Date :

Start Time :

End Time :

PHYSICS 21

SYLLABUS : Fluid Mechanics

Max. Marks : 112 **Time : 60 min.**

GENERAL INSTRUCTIONS

- The Daily Practice Problem Sheet contains 28 MCQ's. For each question only one option is correct. Darken the correct circle/ bubble in the Response Grid provided on each page.
- You have to evaluate your Response Grids yourself with the help of solution booklet.
- Each correct answer will get you 4 marks and 1 mark shall be deduced for each incorrect answer. No mark will be given/ deducted if no bubble is filled. Keep a timer in front of you and stop immediately at the end of 60 min.
- The sheet follows a particular syllabus. Do not attempt the sheet before you have completed your preparation for that syllabus. Refer syllabus sheet in the starting of the book for the syllabus of all the DPP sheets.
- After completing the sheet check your answers with the solution booklet and complete the Result Grid. Finally spend time to analyse your performance and revise the areas which emerge out as weak in your evaluation.

DIRECTIONS (Q.1-Q.20) : There are 20 multiple choice questions. Each question has 4 choices (a), (b), (c) and (d), out of which **ONLY ONE** choice is correct.

Q.1 The force required to separate two glass plates of area $10^{-2}m^{-2}$ with a film of water 0.05 m thick between them, is (Surface tension of water is 70×10^{-3} N/m)

(a) 28 N (b) 14 N (c) 50 N (d) 38 N

Q.2 A thin metal disc of radius r floats on water surface and bends the surface downwards along the perimeter making an angle θ with vertical edge of the disc. If the disc displaces a weight of water W and surface tension of water is T, then the weight of metal disc is

(a) $2\pi rT + W$ (b) $2\pi rT\cos\theta - W$

(c) $2\pi rT\cos\theta + W$ (d) $W - 2\pi rT\cos\theta$

Q.3 The amount of work done in blowing a soap bubble such that its diameter increases from d to D is (T = surface tension of the solution)

(a) $4\pi(D^2 - d^2)T$ (b) $8\pi(D^2 - d^2)T$

(c) $\pi(D^2 - d^2)T$ (d) $2\pi(D^2 - d^2)T$

Q.4 A film of water is formed between two straight parallel wires of length 10 cm each separated by 0.5 cm. If their separation is increased by 1 mm while still maintaining their parallelism, how much work will have to be done (Surface tension of water = 70×10^{-2} N/m)

(a) 7.22×10^{-6} Joule (b) 1.44×10^{-5} Joule

(c) 2.88×10^{-5} Joule (d) 5.76×10^{-5} Joule

RESPONSE GRID	1. ⓐⓑⓒⓓ	2. ⓐⓑⓒⓓ	3. ⓐⓑⓒⓓ	4. ⓐⓑⓒⓓ

Space for Rough Work

Q.5 The liquid meniscus in capillary tube will be convex, if the angle of contact is
(a) Greater than 90° (b) Less than 90°
(c) Equal to 90° (d) Equal to 0°

Q.6 Two soap bubbles of radii r_1 and r_2 equal to 4 cm and 5 cm are touching each other over a common surface S_1S_2 (shown in figure). Its radius will be

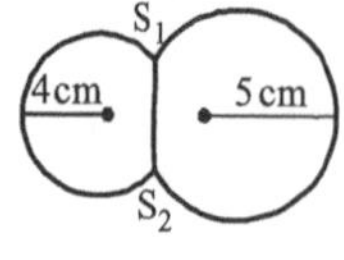

(a) 4 cm
(b) 20 cm
(c) 5 cm
(d) 4.5 cm

Q.7 The radii of two soap bubbles are r_1 and r_2. In isothermal conditions, two meet together in vaccum. Then the radius R of the resultant bubble is given by
(a) $R=(r_1+r_2)/2$ (b) $R=r_1(r_1r_2+r_2)$
(c) $R^2=r_1^2+r_2^2$ (d) $R=r_1+r_2$

Q.8 Two parallel glass plates are dipped partly in the liquid of density 'd' keeping them vertical. If the distance between the plates is 'x', surface tension for liquids is T and angle of contact is θ, then rise of liquid between the plates due to capillary will be
(a) $\dfrac{T\cos\theta}{xd}$ (b) $\dfrac{2T\cos\theta}{xdg}$ (c) $\dfrac{2T}{xdg\cos\theta}$ (d) $\dfrac{T\cos\theta}{xdg}$

Q.9 A capillary tube of radius R is immersed in water and water rises in it to a height H. Mass of water in the capillary tube is M. If the radius of the tube is doubled, mass of water that will rise in the capillary tube will now be
(a) M (b) $2M$ (c) $M/2$ (d) $4M$

Q.10 In a surface tension experiment with a capillary tube water rises upto 0.1 m. If the same experiment is repeated on an artificial satellite, which is revolving around the earth, water will rise in the capillary tube upto a heights of
(a) 0.1 m
(b) 0.2 m
(c) 0.98 m
(d) Full length of the capillary tube

Q.11 Which graph represents the variation of surface tension with temperature over small temperature ranges for water?

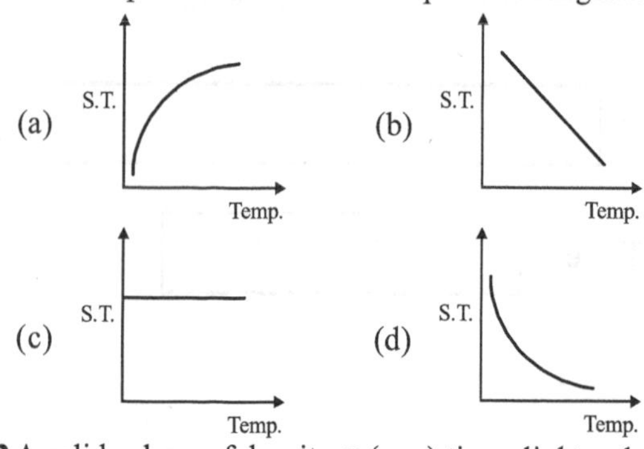

Q.12 A solid sphere of density $\eta\,(>1)$ times lighter than water is suspended in a water tank by a string tied to its base as shown in fig. If the mass of the sphere is m then the tension in the string is given by
(a) $\left(\dfrac{\eta-1}{\eta}\right)mg$
(b) ηmg
(c) $\dfrac{mg}{\eta-1}$
(d) $(\eta-1)mg$

Q.13 A candle of diameter d is floating on a liquid in a cylindrical container of diameter D $(D>>d)$ as shown in figure. If it is burning at the rate of 2 cm/hour then the top of the candle will

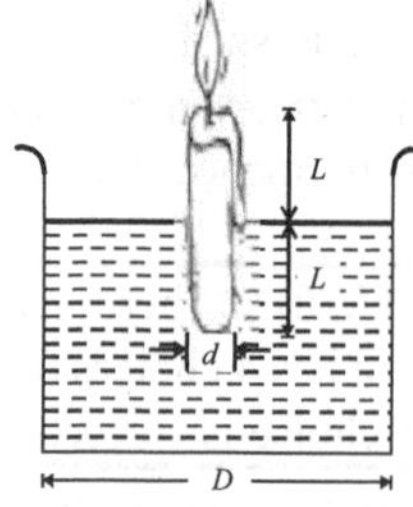

(a) Remain at the same height
(b) Fall at the rate of 1 cm/hour
(c) Fall at the rate of 2 cm/hour
(d) Go up the rate of 1 cm/hour

RESPONSE GRID					
	5. ⓐⓑⓒⓓ	6. ⓐⓑⓒⓓ	7. ⓐⓑⓒⓓ	8. ⓐⓑⓒⓓ	9. ⓐⓑⓒⓓ
	10. ⓐⓑⓒⓓ	11. ⓐⓑⓒⓓ	12. ⓐⓑⓒⓓ	13. ⓐⓑⓒⓓ	

Space for Rough Work

Q.14 A viscous fluid is flowing through a cylindrical tube. The velocity distribution of the fluid is best represented by the diagram

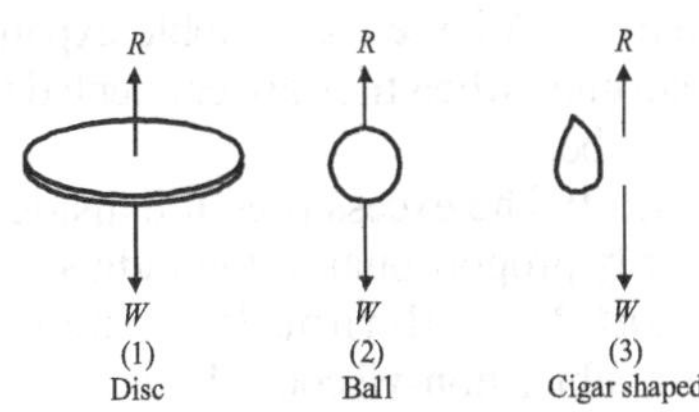

(d) None of these

Q.15 When a body falls in air, the resistance of air depends to a great extent on the shape of the body, 3 different shapes are given. Identify the combination of air resistances which truly represents the physical situation. (The cross sectional areas are the same).

(a) $1 < 2 < 3$ (b) $2 < 3 < 1$
(c) $3 < 2 < 1$ (d) $3 < 1 < 2$

Q.16 A homogeneous solid cylinder of length $L (L < H/2)$. Cross-sectional area A/5 is immersed such that it floats with its axis vertical at the liquid-liquid interface with length L/4 in the denser liquid as shown in the fig. The lower density liquid is open to atmosphere having pressure P_0. Then density of solid is given by

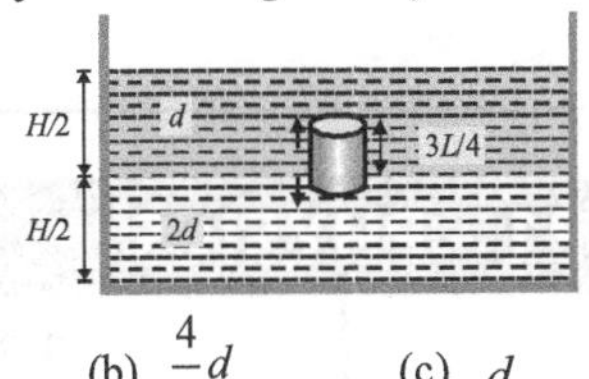

(a) $\frac{5}{4}d$ (b) $\frac{4}{5}d$ (c) d (d) $\frac{d}{5}$

Q.17 A large open tank has two holes in the wall. One is a square hole of side L at a depth y from the top and the other is a circular hole of radius R at a depth $4y$ from the top. When the tank is completely filled with water the quantities of water flowing out per second from both the holes are the same. Then R is equal to

(a) $2\pi L$ (b) $\frac{L}{\sqrt{2\pi}}$ (c) L (d) $\frac{L}{2\pi}$

Q.18 Water is filled in a cylindrical container to a height of 3m. The ratio of the cross-sectional area of the orifice and the beaker is 0.1. The square of the speed of the liquid coming out from the orifice is (g = 10m/s^2)

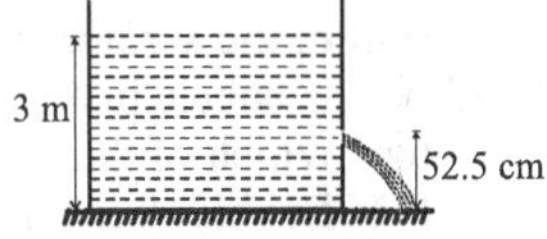

(a) $50\ m^2/s^2$
(b) $50.5\ m^2/s^2$
(c) $51\ m^2/s^2$
(d) $52\ m^2/s^2$

Q.19 An incompressible liquid flows through a horizontal tube as shown in the following fig. Then the velocity v of the fluid is

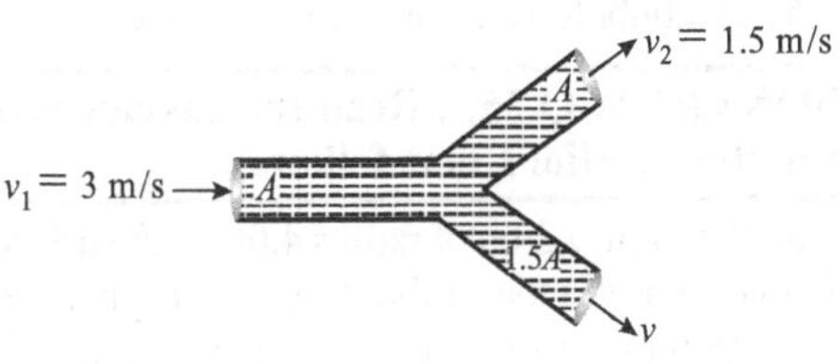

(a) 3.0 m/s (b) 1.5 m/s (c) 1.0 m/s (d) 2.25 m/s

Q.20 Radius of a capillary tube is 2×10^{-3}m. A liquid of weight 6.28×10^{-4} N may remain in the capillary tube then the surface tension of liquid will be

(a) 5×10^{-3} N/m (b) 5×10^{-2} N/m
(c) 5 N/m (d) 50 N/m

DIRECTIONS (Q.21-Q.23) : In the following questions, more than one of the answers given are correct. Select the correct answers and mark it according to the following codes:

Codes :

(a) 1, 2 and 3 are correct **(b)** 1 and 2 are correct
(c) 2 and 4 are correct **(d)** 1 and 3 are correct

RESPONSE GRID	14. ⓐⓑⓒⓓ	15. ⓐⓑⓒⓓ	16. ⓐⓑⓒⓓ	17. ⓐⓑⓒⓓ	18. ⓐⓑⓒⓓ
	19. ⓐⓑⓒⓓ	20. ⓐⓑⓒⓓ			

Space for Rough Work

Q.21 The temperature at which the surface tension of water is zero
(1) 370°C
(2) 0°C
(3) Slightly less than 647 K
(4) 277 K

Q.22 Which of the following statements are true in case when two water drops coalesce and make a bigger drop?
(1) Energy is released.
(2) Energy is absorbed.
(3) The surface area of the bigger drop is smaller than the sum of the surface areas of both the drops.
(4) The surface area of the bigger drop is greater than the sum of the surface areas of both the drops.

Q.23 An air bubble in a water tank rises from the bottom to the top. Which of the following statements are true?
(1) Bubble rises upwards because pressure at the bottom is greater than that at the top.
(2) As the bubble rises, its size increases.
(3) Bubble rises upwards because pressure at the bottom is less than that at the top.
(4) As the bubble rises, its size decreases.

DIRECTIONS (Q.24-Q.25) : Read the passage given below and answer the questions that follows :

There is a small mercury drop of radius 4.0mm. A surface P of area 1.0 mm^2 is placed at the top of the drop. Atmospheric pressure = 10^5 Pa. Surface tension of mercury = 0.465 N/m. Gravity effect is negligible.

Q.24 The force exerted by air on surface P is
(a) 0.1 N (b) 1.0023 N (c) 10^5 N (d) 1.0 N

Q.25 The force exerted by mercury drop on the surface P is
(a) 0.1 N (b) 1.0023 N
(c) 0.00023 N (d) 0.10023 N

DIRECTIONS (Qs. 26-Q.28) : Each of these questions contains two statements: Statement-1 (Assertion) and Statement-2 (Reason). Each of these questions has four alternative choices, only one of which is the correct answer. You have to select the correct choice.

(a) Statement-1 is True, Statement-2 is True; Statement-2 is a correct explanation for Statement-1.
(b) Statement-1 is True, Statement-2 is True; Statement-2 is NOT a correct explanation for Statement-1.
(c) Statement -1 is False, Statement-2 is True.
(d) Statement -1 is True, Statement-2 is False.

Q.26 **Statement-1 :** A large soap bubble expands while a small bubble shrinks, when they are connected to each other by a capillary tube.
Statement-2: The excess pressure inside bubble (or drop) is inverse ly proportional to the radius.

Q.27 **Statement-1 :** Bernoulli's theorem holds for incompressible, non-viscous fluids.
Statement-2 : The factor $\frac{v^2}{2g}$ is called velocity head.

Q.28 **Statement-1 :** The velocity increases, when water flowing in broader pipe enter a narrow pipe.
Statement-2 : According to equation of continuity, product of area and velocity is constant.

RESPONSE GRID					
	21. ⓐⓑⓒⓓ	22. ⓐⓑⓒⓓ	23. ⓐⓑⓒⓓ	24. ⓐⓑⓒⓓ	25. ⓐⓑⓒⓓ
	26. ⓐⓑⓒⓓ	27. ⓐⓑⓒⓓ	28. ⓐⓑⓒⓓ		

DAILY PRACTICE PROBLEM SHEET 21 - PHYSICS			
Total Questions	28	Total Marks	112
Attempted		Correct	
Incorrect		Net Score	
Cut-off Score	28	Qualifying Score	42
Success Gap = Net Score – Qualifying Score			
Net Score = (Correct × 4) – (Incorrect × 1)			

Space for Rough Work

DPP - Daily Practice Problems

Name : 　　　　　　　　Date :

Start Time : 　　　　　　　　End Time :

PHYSICS 22

SYLLABUS : Thermal Expansion, Calorimetry and Change of State

Max. Marks : 116 　　　　　　　　**Time : 60 min.**

GENERAL INSTRUCTIONS

- The Daily Practice Problem Sheet contains 29 MCQ's. For each question only one option is correct. Darken the correct circle/ bubble in the Response Grid provided on each page.
- You have to evaluate your Response Grids yourself with the help of solution booklet.
- Each correct answer will get you 4 marks and 1 mark shall be deduced for each incorrect answer. No mark will be given/ deducted if no bubble is filled. Keep a timer in front of you and stop immediately at the end of 60 min.
- The sheet follows a particular syllabus. Do not attempt the sheet before you have completed your preparation for that syllabus. Refer syllabus sheet in the starting of the book for the syllabus of all the DPP sheets.
- After completing the sheet check your answers with the solution booklet and complete the Result Grid. Finally spend time to analyse your performance and revise the areas which emerge out as weak in your evaluation.

DIRECTIONS (Q.1-Q.20) : There are 20 multiple choice questions. Each question has 4 choices (a), (b), (c) and (d), out of which **ONLY ONE** choice is correct.

Q.1 A glass flask is filled up to a mark with 50 cc of mercury at 18°C. If the flask and contents are heated to 38°C, how much mercury will be above the mark? (α for glass is 9×10^{-6}/°C and coefficient of real expansion of mercury is 180×10^{-6}/°C)

(a) 0.85 cc (b) 0.46 cc (c) 0.153 cc (d) 0.05 cc

Q.2 The coefficient of apparent expansion of mercury in a glass vessel is 153×10^{-6}/°C and in a steel vessel is 144×10^{-6}/°C. If α for steel is 12×10^{-6}/°C, then that of glass is

(a) 9×10^{-6}/°C (b) 6×10^{-6}/°C
(c) 36×10^{-6}/°C (d) 27×10^{-6}/°C

Q.3 An iron tyre is to be fitted on to a wooden wheel 1m in diameter. The diameter of tyre is 6mm smaller than that of wheel. The tyre should be heated so that its temperature increases by a minimum of (the coefficient of cubical expansion of iron is 3.6×10^{-5}/°C)

(a) 167°C (b) 334°C (c) 500°C (d) 1000°C

Q.4 A rod of length 20 cm is made of metal. It expands by 0.075 cm when its temperature is raised from 0°C to 100°C. Another rod of a different metal *B* having the same length expands by 0.045 cm for the same change in temperature. A third rod of the same length is composed of two parts, one of metal *A* and the other of metal *B*. This rod expands by 0.060 cm for the same change in temperature. The portion made of metal *A* has the length

(a) 20 cm (b) 10 cm (c) 15 cm (d) 18 cm

RESPONSE GRID	1. ⓐⓑⓒⓓ	2. ⓐⓑⓒⓓ	3. ⓐⓑⓒⓓ	4. ⓐⓑⓒⓓ

Space for Rough Work

Q.5 A glass flask of volume one litre at 0°C is filled, level full of mercury at this temperature. The flask and mercury are now heated to 100°C. How much mercury will spill out, if coefficient of volume expansion of mercury is 1.82×10^{-4}/°C and linear expansion of glass is 0.1×10^{-4}/°C?
(a) 21.2 cc (b) 15.2 cc (c) 1.52 cc (d) 2.12 cc

Q.6 The apparent coefficient of expansion of a liquid when heated in a copper vessel is C and when heated in a silver vessel is S. If A is the linear coefficient of expansion of copper, then the linear coefficient of expansion of silver is
(a) $\frac{C+S-3A}{3}$ (b) $\frac{C+3A-S}{3}$
(c) $\frac{S+3A-C}{3}$ (d) $\frac{C-S+3A}{3}$

Q.7 The coefficient of volumetric expansion of mercury is 18×10^{-5}/°C. A thermometer bulb has a volume 10^{-6} m^3 and cross section of stem is 0.004 cm^2. Assuming that bulb is filled with mercury at 0°C then the length of the mercury column at 100°C is
(a) 18.8 mm (b) 9.2 mm (c) 7.4 cm (d) 4.5 cm

Q.8 A piece of metal weight 46 gm in air, when it is immersed in the liquid of specific gravity 1.24 at 27°C it weighs 30 gm. When the temperature of liquid is raised to 42°C the metal piece weighs 30.5 gm, specific gravity of the liquid at 42°C is 1.20, then the linear expansion of the metal will be
(a) 3.316×10^{-5}/°C (b) 2.316×10^{-5}/°C
(c) 4.316×10^{-5}/°C (d) None of these

Q.9 2 kg of ice at – 20°C is mixed with 5 kg of water at 20°C in an insulating vessel having a negligible heat capacity. Calculate the final mass of water remaining in the container. It is given that the specific heats of water and ice are 1 kcal/kg /°C and 0.5 kcal/kg/°C while the latent heat of fusion of ice is 80 kcal/kg
(a) 7 kg (b) 6 kg (c) 4 kg (d) 2 kg

Q.10 A lead bullet at 27°C just melts when stopped by an obstacle. Assuming that 25% of heat is absorbed by the obstacle, then the velocity of the bullet at the time of striking (M.P. of lead = 327°C, specific heat of lead = 0.03 cal/gm°C, latent heat of fusion of lead = 6 cal/gm and J = 4.2 joule/cal)
(a) 410 m/sec (b) 1230 m/sec
(c) 307.5 m/sec (d) None of the above

Q.11 The temperature of equal masses of three different liquids *A*, *B* and *C* are 12°C, 19°C and 28°C respectively. The temperature when *A* and *B* are mixed is 16°C and when *B* and *C* are mixed is 23°C, The temperature when *A* and *C* are mixed is
(a) 18.2°C (b) 22°C (c) 20.2°C (d) 25.2°C

Q.12 50 gm of copper is heated to increase its temperature by 10°C. If the same quantity of heat is given to 10 gm of water, the rise in its temperature is (Specific heat of copper = 420 Joule-kg^{-1}°C^{-1})
(a) 5°C (b) 6°C (c) 7°C (d) 8°C

Q.13 A beaker contains 200 gm of water. The heat capacity of the beaker is equal to that of 20 gm of water. The initial temperature of water in the beaker is 20°C. If 440 gm of hot water at 92°C is poured in it, the final temperature (neglecting radiation loss) will be nearest to
(a) 58°C (b) 68°C (c) 73°C (d) 78°C

Q.14 One calorie is defined as the amount of heat required to raise temperature of 1g of water by 1°C and it is defined under which of the following condition
(a) From 14.5°C to 15.5°C at 760 mm of Hg
(b) From 98.5°C to 99.5°C at 760 mm of Hg
(c) From 13.5°C to 14.5°C at 76 mm of Hg
(d) From 3.5°C to 4.5°C at 76 mm of Hg

Q.15 A bullet moving with a uniform velocity *v*, stops suddenly after hitting the target and the whole mass melts be *m*, specific heat *S*, initial temperature 25°C, melting point 475°C and the latent heat *L*. Then *v* is given by
(a) $mL = mS(475-25)+\frac{1}{2}.\frac{mv^2}{J}$
(b) $mS(475-25)+mL=\frac{mv^2}{2J}$
(c) $mS(475-25)+mL=\frac{mv^2}{J}$
(d) $mS(475-25)-mL=\frac{mv^2}{2J}$

RESPONSE GRID					
	5. ⓐⓑⓒⓓ	6. ⓐⓑⓒⓓ	7. ⓐⓑⓒⓓ	8. ⓐⓑⓒⓓ	9. ⓐⓑⓒⓓ
	10. ⓐⓑⓒⓓ	11. ⓐⓑⓒⓓ	12. ⓐⓑⓒⓓ	13. ⓐⓑⓒⓓ	14. ⓐⓑⓒⓓ
	15. ⓐⓑⓒⓓ				

Space for Rough Work

Q.16 A stationary object at 4°C and weighing 3.5 kg falls from a height of 2000 m on a snow mountain at 0°C. If the temperature of the object just before hitting the snow is 0°C and the object comes to rest immediately? (g = 10m/s^2 and latent heat of ice = 3.5×10^5 joule/sec), then the object will melt

(a) 2 kg of ice (b) 200 gm of ice
(c) 20 gm of ice (d) 2 gm of ice

Q.17 Density of a substance at 0°C is 10 gm/cc and at 100°C, its density is 9.7 gm/cc. The coefficient of linear expansion of the substance will be

(a) 10^2 (b) 10^{-2} (c) 10^{-3} (d) 10^{-4}

Q.18 The real coefficient of volume expansion of glycerine is 0.000597 per °C and linear coefficient of expansion of glass is 0.000009 per°C. Then the apparent volume coefficient of expansion of glycerine is

(a) 0.000558 per°C (b) 0.00057 per°C
(c) 0.00027 per°C (d) 0.00066 per°C

Q.19 A constant volume gas thermometer shows pressure reading of 50 cm and 90 cm of mercury at 0°C and 100°C respectively. When the pressure reading is 60 cm of mercury, the temperature is

(a) 25°C (b) 40°C (c) 15°C (d) 12.5°C

Q.20 A student takes 50gm wax (specific heat = 0.6 kcal/kg°C) and heats it till it boils. The graph between temperature and time is as follows. Heat supplied to the wax per minute and boiling point are respectively.

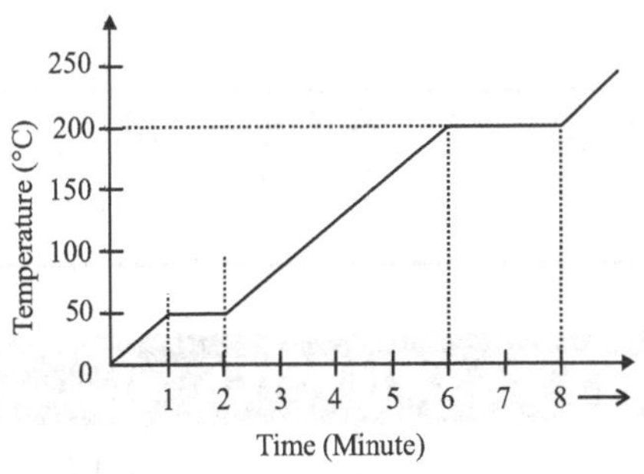

(a) 500 cal, 50°C (b) 1000 cal, 100°C
(c) 1500 cal, 200°C (d) 1000 cal, 200°C

DIRECTIONS (Q.21-Q.23) : In the following questions, more than one of the answers given are correct. Select the correct answers and mark it according to the following codes:

Codes :

(a) 1, 2 and 3 are correct **(b)** 1 and 2 are correct
(c) 2 and 4 are correct **(d)** 1 and 3 are correct

Q.21 Heat is supplied to a certain homogenous sample of matter, at a uniform rate. Its temperature is plotted against time, as shown. Which of the following conclusions can be drawn?

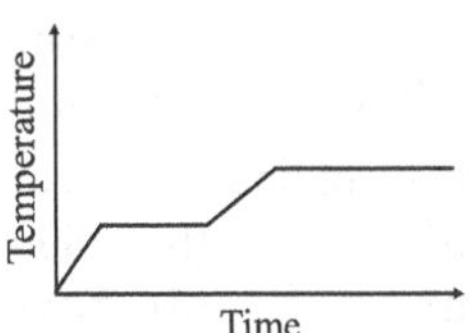

(1) Its specific heat capacity is greater in the liquid state than in the solid state
(2) Its latent heat of vaporization is greater than its latent heat of fusion
(3) Its specific heat capacity is greater in the solid state than in the liquid state
(4) Its latent heat of vaporization is smaller than its latent heat of fusion

Q.22 A bimetallic strip is formed out of two identical strips, one of copper and other of brass. The coefficients of linear expansion of the two metals are α_C and α_B. On heating, the temperature of the strip goes up by ΔT and the strip bends to form an arc of radius of curvature R. Then R is

(1) inversely proportional to ΔT
(2) proportional to $|\alpha_B - \alpha_C|$
(3) inversely proportional to $|\alpha_B - \alpha_C|$
(4) proportional to ΔT

Q.23 A bimetallic strip is heated. Choose wrong statements.

(1) does not bend at all
(2) gets twisted in the form of an helix
(3) bends in the form of an arc with the more expandable metal inside.
(4) bend in the form of an arc with the more expandable metal outside

RESPONSE GRID					
	16. ⓐⓑⓒⓓ	17. ⓐⓑⓒⓓ	18. ⓐⓑⓒⓓ	19. ⓐⓑⓒⓓ	20. ⓐⓑⓒⓓ
	21. ⓐⓑⓒⓓ	22. ⓐⓑⓒⓓ	23. ⓐⓑⓒⓓ		

DIRECTIONS (Q.24-Q.26) : Read the passage given below and answer the questions that follows :

In a thermally insulated tube of cross sectional area $4 cm^2$ a liquid of thermal expansion coefficient $10^{-3} K^{-1}$ is flowing. Its velocity at the entrance is 0.1 m/s. At the middle of the tube a heater of a power of 10kW is heating the liquid. The specific heat capacity of the liquid is 1.5 kJ/(kg K), and its density is $1500 kg/m^3$ at the entrance.

Q.24 The rise in temperature of the liquid as it pass through the tube is

(a) $\frac{1000}{9}$°C (b) $\frac{1}{9}$°C (c) $\frac{500}{9}$°C (d) None

Q.25 What is the density of liquid at the exit ?

(a) $1450 kg/m^3$ (b) $1400 kg/m^3$
(c) $1350 kg/m^3$ (d) None of these

Q.26 How much bigger is the volume rate of flow at the end of the tube than at the entrance in cubic meters ?

(a) 9×10^{-5} (b) $\frac{1}{3} \times 10^{-5}$
(c) $\frac{4}{9} \times 10^{-5}$ (d) None

DIRECTIONS (Q. 27-Q.29) : Each of these questions contains two statements: Statement-1 (Assertion) and Statement-2 (Reason). Each of these questions has four alternative choices, only one of which is the correct answer. You have to select the correct choice.

(a) Statement-1 is True, Statement-2 is True; Statement-2 is a correct explanation for Statement-1.
(b) Statement-1 is True, Statement-2 is True; Statement-2 is NOT a correct explanation for Statement-1.
(c) Statement -1 is False, Statement-2 is True.
(d) Statement -1 is True, Statement-2 is False.

Q.27 Statement-1 : Fahrenheit is the smallest unit measuring temperature.

Statement-2 : Fahrenheit was the first temperature scale used for measuring temperature.

Q.28 Statement-1 : A brass disc is just fitted in a hole in a steel plate. The system must be cooled to loosen the disc from the hole.

Statement-2 : The coefficient of linear expansion for brass is greater than the coefficient of linear expansion for steel.

Q.29 Statement-1 : Latent heat of fusion of ice is $336000 J kg^{-1}$.

Statement-2 : Latent heat refers to change of state without any change in temperature.

RESPONSE GRID	
	24. ⓐⓑⓒⓓ 25. ⓐⓑⓒⓓ 26. ⓐⓑⓒⓓ 27. ⓐⓑⓒⓓ 28. ⓐⓑⓒⓓ
	29. ⓐⓑⓒⓓ

DAILY PRACTICE PROBLEM SHEET 22 - PHYSICS			
Total Questions	29	Total Marks	116
Attempted		Correct	
Incorrect		Net Score	
Cut-off Score	28	Qualifying Score	42
Success Gap = Net Score – Qualifying Score			
Net Score = (Correct × 4) – (Incorrect × 1)			

Space for Rough Work

DPP - Daily Practice Problems

Name :

Date :

Start Time :

End Time :

PHYSICS 23

SYLLABUS : Heat transfer & Newton's law of cooling

Max. Marks : 120 **Time : 60 min.**

GENERAL INSTRUCTIONS

- The Daily Practice Problem Sheet contains 30 MCQ's. For each question only one option is correct. Darken the correct circle/ bubble in the Response Grid provided on each page.
- You have to evaluate your Response Grids yourself with the help of solution booklet.
- Each correct answer will get you 4 marks and 1 mark shall be deduced for each incorrect answer. No mark will be given/ deducted if no bubble is filled. Keep a timer in front of you and stop immediately at the end of 60 min.
- The sheet follows a particular syllabus. Do not attempt the sheet before you have completed your preparation for that syllabus. Refer syllabus sheet in the starting of the book for the syllabus of all the DPP sheets.
- After completing the sheet check your answers with the solution booklet and complete the Result Grid. Finally spend time to analyse your performance and revise the areas which emerge out as weak in your evaluation.

DIRECTIONS (Q.1-Q.21) : There are 21 multiple choice questions. Each question has 4 choices (a), (b), (c) and (d), out of which **ONLY ONE** choice is correct.

Q.1 Two rods (one semi-circular and other straight) of same material and of same cross-sectional area are joined as shown in the figure. The points A and B are maintained at different temperature. The ratio of the heat transferred through a cross-section of a semi-circular rod to the heat transferred through a cross section of the straight rod in a given time is

(a) $2 : \pi$

(b) $1 : 2$

(c) $\pi : 2$

(d) $3 : 2$

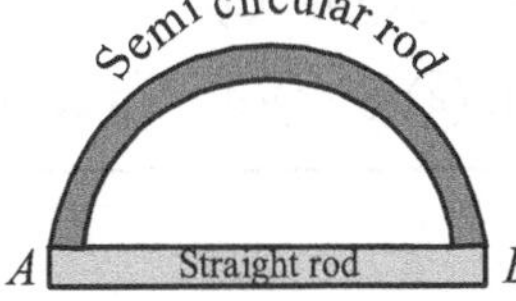

Q.2 A wall is made up of two layers A and B. The thickness of the two layers is the same, but materials are different. The thermal conductivity of A is double than that of B. In thermal equilibrium the temperature difference between the two ends is 36°C. Then the difference of temperature at the two surfaces of A will be

(a) 6°C (b) 12°C (c) 18°C (d) 24°C

Q.3 A room is maintained at 20°C by a heater of resistance 20 ohm connected to 200 volt mains. The temperature is uniform through out the room and heat is transmitted through a glass window of area 1m^2 and thickness 0.2 cm. What will be the temperature outside? Given that thermal conductivity K for glass is 0.2 cal/m/°C/sec and J = 4.2 J/cal

(a) 15.24°C (b) 15.00°C

(c) 24.15°C (d) None of these

RESPONSE GRID	1. ⓐⓑⓒⓓ	2. ⓐⓑⓒⓓ	3. ⓐⓑⓒⓓ

Space for Rough Work

Q.4 A composite metal bar of uniform section is made up of length 25 cm of copper, 10 cm of nickel and 15 cm of aluminium. Each part being in perfect thermal contact with the adjoining part. The copper end of the composite rod is maintained at 100°C and the aluminium end at 0°C. The whole rod is covered with belt so that there is no heat loss occurs at the sides. If $K_{Cu} = 2K_{Al}$ and $K_{Al} = 3K_{Ni}$, then what will be the temperatures of Cu – Ni and Ni – Al junctions respectively

Cu	Ni	Al

100° C 0° C

(a) 23.33°C and 78.8°C (b) 83.33°C and 20°C
(c) 50°C and 30°C (d) 30°C and 50°C

Q.5 Three rods of the same dimension have thermal conductivities 3*K*, 2*K* and *K*. They are arranged as shown in fig. with their ends at 100°C, 50°C and 20°C. The temperature of their junction is

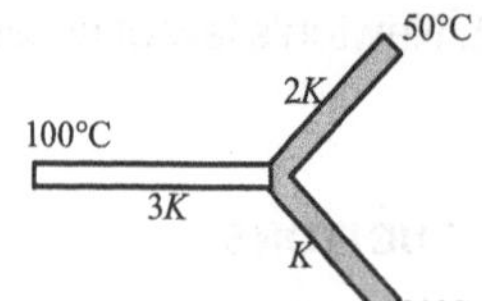

(a) 60°C
(b) 70°C
(c) 50°C
(d) 35°C

Q.6 A black body is at a temperature of 2880 K. The energy of radiation emitted by this object with wavelength between 499 nm and 500 nm is U_1, between 999 nm and 1000 nm is U_2 and between 1499 nm and 1500 nm is U_3. The Wein's constant $b = 2.88 \times 10^6$ nm K. Then
(a) $U_1 = 0$ (b) $U_3 = 0$ (c) $U_1 > U_2$ (d) $U_2 > U_1$

Q.7 A body initially at 80° C cools to 64° C in 5 minutes and to 52° C in 10 minutes. The temperature of the body after 15 minutes will be
(a) 42.7° C (b) 35° C (c) 47° (d) 40° C

Q.8 A 5 cm thick ice block is there on the surface of water in a lake. The temperature of air is – 10° C; how much time it will take to double the thickness of the block
(L = 80 cal/g, K_{ice} = 0.004 erg/s-k, d_{ice} = 0.92 g cm^{-3})
(a) 1 hour (b) 191 hours
(c) 19.1 hours (d) 1.91 hours

Q.9 A cylindrical rod with one end in a steam chamber and the other end in ice results in melting of 0.1 gm of ice per second. If the rod is replaced by another with half the length and double the radius of the first and if the thermal conductivity of material of second rod is $\frac{1}{4}$ that of first, the rate at which ice melts in gm/sec will be
(a) 3.2 (b) 1.6 (c) 0.2 (d) 0.1

Q.10 An ice box used for keeping eatable cold has a total wall area of 1 metre^2 and a wall thickness of 5.0 cm. The thermal conductivity of the ice box is $K = 0.01$ joule/metre °C. It is filled with ice at 0° C along with eatables on a day when the temperature is 30°C. The latent heat of fusion of ice is 334×10^3 joules/kg. The amount of ice melted in one day is (1 day = 86,400 seconds)
(a) 776 gm (b) 7760 gm
(c) 11520 gm (d) 1552 gm

Q.11 A solid copper sphere (density ρ and specific heat capacity *c*) of radius *r* at an initial temperature 200 K is suspended inside a chamber whose walls are at almost 0 K. The time required (in μ *s*) for the temperature of the sphere to drop to 100 K is
(a) $\frac{72}{7}\frac{r\rho c}{\sigma}$ (b) $\frac{7}{72}\frac{r\rho c}{\sigma}$ (c) $\frac{27}{7}\frac{r\rho c}{\sigma}$ (d) $\frac{7}{27}\frac{r\rho c}{\sigma}$

Q.12 Four rods of identical cross-sectional area and made from the same metal form the sides of square. The temperature of two diagonally opposite points are T and $\sqrt{2}T$ respectively in the steady state. Assuming that only heat conduction takes place, what will be the temperature difference between other two points
(a) $\frac{\sqrt{2}+1}{2}T$ (b) $\frac{2}{\sqrt{2}+1}T$
(c) 0 (d) None of these

Q.13 Consider two hot bodies B_1 and B_2 which have temperature 100° C and 80° C respectively at $t = 0$. The temperature of surroundings is 40° C. The ratio of the respective rates of cooling R_1 and R_2 of these two bodies at $t = 0$ will be
(a) $R_1 : R_2 = 3 : 2$ (b) $R_1 : R_2 = 5 : 4$
(c) $R_1 : R_2 = 2 : 3$ (d) $R_1 : R_2 = 4 : 5$

RESPONSE GRID	4. ⓐⓑⓒⓓ	5. ⓐⓑⓒⓓ	6. ⓐⓑⓒⓓ	7. ⓐⓑⓒⓓ	8. ⓐⓑⓒⓓ
	9. ⓐⓑⓒⓓ	10. ⓐⓑⓒⓓ	11. ⓐⓑⓒⓓ	12. ⓐⓑⓒⓓ	13. ⓐⓑⓒⓓ

Space for Rough Work

Q.14 A body cools from 60° C to 50° C in 10 minutes. If the room temperature is 25° C and assuming Newton's law of cooling to hold good, the temperature of the body at the end of the next 10 minutes will be
(a) 38.5° C (b) 40° C (c) 42.85° C (d) 45° C

Q.15 The rates of cooling of two different liquids put in exactly similar calorimeters and kept in identical surroundings are the same if
(a) The masses of the liquids are equal
(b) Equal masses of the liquids at the same temperature are taken
(c) Different volumes of the liquids at the same temperature are taken
(d) Equal volumes of the liquids at the same temperature are taken

Q.16 For cooking the food, which of the following type of utensil is most suitable
(a) High specific heat and low conductivity
(b) High specific heat and high conductivity
(c) Low specific heat and low conductivity
(d) Low specific heat and high conductivity

Q.17 Two rods A and B are of equal lengths. Their ends are kept between the same temperature and their area of cross-sections are A_1 and A_2 and thermal conductivities K_1 and K_2. The rate of heat transmission in the two rods will be equal, if
(a) $K_1A_2 = K_2A_1$ (b) $K_1A_1 = K_2A_2$
(c) $K_1 = K_2$ (d) $K_1A_1^2 = K_2A_2^2$

Q.18 While measuring the thermal conductivity of a liquid, we keep the upper part hot and lower part cool, so that
(a) Convection may be stopped
(b) Radiation may be stopped
(c) Heat conduction is easier downwards
(d) It is easier and more convenient to do so

Q.19 When fluids are heated from the bottom, convection currents are produced because
(a) Molecular motion of fluid becomes aligned
(b) Molecular collisions take place within the fluid
(c) Heated fluid becomes more dense than the cold fluid above it
(d) Heated fluid becomes less dense than the cold fluid above it

Q.20 If between wavelength λ and $\lambda + d\lambda$, e_λ and a_λ be the emissive and absorptive powers of a body and E_λ be the emissive power of a perfectly black body, then according to Kirchoff's law, which is true
(a) $e_\lambda = a_\lambda = E_\lambda$ (b) $e_\lambda E_\lambda = a_\lambda$
(c) $e_\lambda = a_\lambda E_\lambda$ (d) $e_\lambda a_\lambda E_\lambda$ = constant

Q.21 Two thermometers A and B are exposed in sunlight. The bulb of A is painted black, But that of B is not painted. The correct statement regarding this case is
(a) Temperature of A will rise faster than B but the final temperature will be the same in both
(b) Both A and B show equal rise in beginning
(c) Temperature of A will remain more than B
(d) Temperature of B will rise faster

DIRECTIONS (Q.22-Q.24) : In the following questions, more than one of the answers given are correct. Select the correct answers and mark it according to the following codes:

Codes :
(a) 1, 2 and 3 are correct **(b)** 1 and 2 are correct
(c) 2 and 4 are correct **(d)** 1 and 3 are correct

Q.22 Two bodies A and B have thermal emissivities of 0.01 and 0.81 respectively. The outer surface areas of the two bodies are the same. The two bodies emit total radiant power at the same rate. The wavelength λ_B corresponding to maximum spectral radiancy in the radiation from B is shifted from the wavelength corresponding to maximum spectral radiancy in the radiation from A, by 1.00 μm. If the temperature of A is 5802 K
(1) The temperature of B is 1934 K
(2) $\lambda_B = 1.5$ μm
(3) The temperature of B is 11604 K
(4) The temperature of B is 2901 K

Q.23 A cane is taken out from a refrigerator at 0°C. The atmospheric temperature is 25°C. If t_1 is the time taken to heat from 0°C to 5°C and t_2 is the time taken from 10°C to 15°C, then the wrong statements are
(1) $t_1 > t_2$ (2) $t_1 = t_2$
(3) There is no relation (4) $t_1 < t_2$

RESPONSE GRID					
	14. ⓐⓑⓒⓓ	15. ⓐⓑⓒⓓ	16. ⓐⓑⓒⓓ	17. ⓐⓑⓒⓓ	18. ⓐⓑⓒⓓ
	19. ⓐⓑⓒⓓ	20. ⓐⓑⓒⓓ	21. ⓐⓑⓒⓓ	22. ⓐⓑⓒⓓ	23. ⓐⓑⓒⓓ

Space for Rough Work

Q.24 The rate of loss of heat from a body cooling under conditions of forced convection is proportional to its
(1) surface area
(2) excess of temperature over that of surrounding
(3) heat capacity
(4) absolute temperature

DIRECTIONS (Q.25-Q.27) : Read the passage given below and answer the questions that follows :

A brass ball of mass 100g is heated to 100°C and then dropped into 200g of turpentine in a calorimeter at 15°C. The final temperature is found to be 23°C. Take specific heat of brass as 0.092 cal/g°C and water equivalent of calorimeter as 4g.

Q.25 The specific heat of turpentine is
(a) 0.42 cal/g°c (b) 0.96 cal/g°c
(c) 0.72 cal/g°c (d) 0.12 cal/g°c

Q.26 Heat lost by the ball is approximately
(a) 810 cal (b) 610 cal
(c) 710 cal (d) 510 cal

Q.27 Heat gained by turpentine and calorimeter is approximately
(a) 810 cal (b) 610 cal
(c) 710 cal (d) 510 cal

DIRECTIONS (Q.28-Q.30) : Each of these questions contains two statements: Statement-1 (Assertion) and Statement-2 (Reason). Each of these questions has four alternative choices, only one of which is the correct answer. You have to select the correct choice.

(a) Statement-1 is True, Statement-2 is True; Statement-2 is a correct explanation for Statement-1.
(b) Statement-1 is True, Statement-2 is True; Statement-2 is NOT a correct explanation for Statement-1.
(c) Statement -1 is False, Statement-2 is True.
(d) Statement -1 is True, Statement-2 is False.

Q.28 Statement-1 : The equivalent thermal conductivity of two plates of same thickness in contact (series) is less than the smaller value of thermal conductivity.
Statement-2 : For two plates of equal thickness in contact (series) the equivalent thermal conductivity is given by

$$\frac{2}{K}=\frac{1}{K_1}+\frac{1}{K_2}$$

Q.29 Statement-1 : The absorbance of a perfect black body is unity.
Statement-2 : A perfect black body when heated emits radiations of all possible wavelengths at that temperature.

Q.30 Statement-1 : As temperature of a black body is raised, wavelength corresponding to maximum energy reduces.
Statement-2 : Higher temperature would mean higher energy and hence higher wavelength.

RESPONSE GRID	
24. ⓐⓑⓒⓓ 25. ⓐⓑⓒⓓ 26. ⓐⓑⓒⓓ 27. ⓐⓑⓒⓓ 28. ⓐⓑⓒⓓ	
29. ⓐⓑⓒⓓ 30. ⓐⓑⓒⓓ	

DAILY PRACTICE PROBLEM SHEET 23 - PHYSICS			
Total Questions	30	Total Marks	120
Attempted		Correct	
Incorrect		Net Score	
Cut-off Score	30	Qualifying Score	48
Success Gap = Net Score – Qualifying Score			
Net Score = (Correct × 4) – (Incorrect × 1)			

Space for Rough Work

DPP - Daily Practice Problems

Name :

Date :

Start Time :

End Time :

PHYSICS 24

SYLLABUS : Thermodynamics-1 (Thermal equilibrium, zeroth law of thermodynamics, concept of temperature, Heat, work and internal energy, Different thermodynamic processes)

Max. Marks : 116 **Time : 60 min.**

GENERAL INSTRUCTIONS

- The Daily Practice Problem Sheet contains 29 MCQ's. For each question only one option is correct. Darken the correct circle/ bubble in the Response Grid provided on each page.
- You have to evaluate your Response Grids yourself with the help of solution booklet.
- Each correct answer will get you 4 marks and 1 mark shall be deduced for each incorrect answer. No mark will be given/ deducted if no bubble is filled. Keep a timer in front of you and stop immediately at the end of 60 min.
- The sheet follows a particular syllabus. Do not attempt the sheet before you have completed your preparation for that syllabus. Refer syllabus sheet in the starting of the book for the syllabus of all the DPP sheets.
- After completing the sheet check your answers with the solution booklet and complete the Result Grid. Finally spend time to analyse your performance and revise the areas which emerge out as weak in your evaluation.

DIRECTIONS (Q.1-Q.21) : There are 21 multiple choice questions. Each question has 4 choices (a), (b), (c) and (d), out of which **ONLY ONE** choice is correct.

Q.1 For an isothermal expansion of a perfect gas, the value of $\frac{\Delta P}{P}$ is

(a) $-\gamma^{1/2}\frac{\Delta V}{V}$ (b) $-\frac{\Delta V}{V}$ (c) $-\gamma\frac{\Delta V}{V}$ (d) $-\gamma^2\frac{\Delta V}{V}$

Q.2 When an ideal gas in a cylinder was compressed isothermally by a piston, the work done on the gas was found to be 1.5×10^4 Joule. During this process about

(a) 3.6×10^3 cal of heat flowed out from the gas
(b) 3.6×10^3 cal of heat flowed into the gas
(c) 1.5×10^4 cal of heat flowed into the gas
(d) 1.5×10^4 cal of heat flowed out from the gas

Q.3 The latent heat of vaporisation of water is 2240 J/gm. If the work done in the process of expansion of 1 g of water is 168 J, then increase in internal energy is

(a) 2408 J (b) 2240 J
(c) 2072 J (d) 1904 J

Q.4 One mole of an ideal gas expands at a constant temperature of 300 K from an initial volume of 10 litres to a final volume of 20 litres. The work done in expanding the gas is

$(R = 8.31 \text{ J/mole} - \text{K})$

(a) 750 Joules (b) 1728 Joules
(c) 1500 Joules (d) 3456 Joules

RESPONSE GRID	1. ⓐⓑⓒⓓ	2. ⓐⓑⓒⓓ	3. ⓐⓑⓒⓓ	4. ⓐⓑⓒⓓ

Space for Rough Work

Q.5 The pressure in the tyre of a car is four times the atmospheric pressure at 300 K. If this tyre suddenly bursts, its new temperature will be $(\gamma = 1.4)$

(a) $300(4)^{1.4/0.4}$ (b) $300\left(\frac{1}{4}\right)^{-0.4/1.4}$

(c) $300(2)^{-0.4/1.4}$ (d) $300(4)^{-0.4/1.4}$

Q.6 A monoatomic gas $(\gamma = 5/3)$ is suddenly compressed to $\frac{1}{8}$ of its original volume adiabatically, then the pressure of the gas will change to

(a) $\frac{24}{5}$ (b) 8

(c) $\frac{40}{3}$ (d) 32 times its initial pressure

Q.7 An ideal gas at 27°C is compressed adiabatically to $\frac{8}{27}$ of its original volume. If $\gamma = \frac{5}{3}$, then the rise in temperature is

(a) 450 K (b) 375 k (c) 225 K (d) 405 K

Q.8 A given system undergoes a change in which the work done by the system equals the decrease in its internal energy. The system must have undergone an

(a) Isothermal change (b) Adiabatic change

(c) Isobaric change (d) Isochoric change

Q.9 Helium at 27° has a volume of 8 litres. It is suddenly compressed to a volume of 1 litre. The temperature of the gas will be $[\gamma = 5/3]$

(a) 108°C (b) 9327°C (c) 1200°C (d) 927°C

Q.10 One mole of an ideal gas at an initial temperature of TK does 6 R joules of work adiabatically. If the ratio of specific heats of this gas at constant pressure and at constant volume is 5/3, the final temperature of gas will be

(a) $(T+2.4)K$ (b) $(T-2.4)K$

(c) $(T+4)K$ (d) $(T-4)K$

Q.11 For an adiabatic expansion of a perfect gas, the value of $\frac{\Delta P}{P}$ is equal to

(a) $-\sqrt{\gamma\frac{\Delta V}{V}}$ (b) $-\frac{\Delta V}{V}$

(c) $-\gamma\frac{\Delta V}{V}$ (d) $-\gamma^2\frac{\Delta V}{V}$

Q.12 If 300 ml of gas at 27°C is cooled to 7°C at constant pressure, then its final volume will be

(a) 540 ml (b) 350 ml

(c) 280 ml (d) 135 ml

Q.13 A sample of gas expands from volume V_1 to V_2. The amount of work done by the gas is greatest when the expansion is

(a) isothermal (b) isobaric

(c) adiabatic (d) equal in all cases

Q.14 How much work to be done in decreasing the volume of and ideal gas by an amount of $2.4\times10^{-4}\text{m}^3$ at normal temperature and constant normal pressure of $1\times10^5 N/\text{m}^2$

(a) 28 Joule (b) 27 Joule

(c) 25 Joule (d) 24 Joule

Q.15 One mole of a perfect gas in a cylinder fitted with a piston has a pressure P, volume V and temperature T. If the temperature is increased by 1 K keeping pressure constant, the increase in volume is

(a) $\frac{2V}{273}$ (b) $\frac{V}{91}$ (c) $\frac{V}{273}$ (d) V

Q.16 Work done by 0.1 mole of a gas at 27°C to double its volume at constant pressure is $(R = 2\text{ cal mol}^{-1\circ}\text{K}^{-1})$

(a) 54 cal (b) 600 cal (c) 60 cal (d) 546 cal

Q.17 When an ideal diatomic gas is heated at constant pressure, the fraction of the heat energy supplied which increases the internal energy of the gas, is

(a) $\frac{2}{5}$ (b) $\frac{3}{5}$ (c) $\frac{3}{7}$ (d) $\frac{5}{7}$

RESPONSE GRID					
	5. ⓐⓑⓒⓓ	6. ⓐⓑⓒⓓ	7. ⓐⓑⓒⓓ	8. ⓐⓑⓒⓓ	9. ⓐⓑⓒⓓ
	10. ⓐⓑⓒⓓ	11. ⓐⓑⓒⓓ	12. ⓐⓑⓒⓓ	13. ⓐⓑⓒⓓ	14. ⓐⓑⓒⓓ
	15. ⓐⓑⓒⓓ	16. ⓐⓑⓒⓓ	17. ⓐⓑⓒⓓ		

Space for Rough Work

Q.18 When heat is given to a gas in an isothermal change, the result will be
(a) external work done
(b) rise in temperature
(c) increase in internal energy
(d) external work done and also rise in temp.

Q.19 An ideal gas expands isothermally from a volume V_1 to V_2 and then compressed to original volume V_1 adiabatically. Initial pressure is P_1 and final pressure is P_3. The total work done is W. Then
(a) $P_3 > P_1, W > 0$ (b) $P_3 < P_1, W < 0$
(c) $P_3 > P_1, W < 0$ (d) $P_3 = P_1, W = 0$

Q.20 An ideal gas expands in such a manner that its pressure and volume can be related by equation PV^2 = constant. During this process, the gas is
(a) heated
(b) cooled
(c) neither heated nor cooled
(d) first heated and then cooled

Q.21 In the following $P-V$ diagram two adiabatics cut two isothermals at temperatures T_1 and T_2 (fig.). The value of $\frac{V_a}{V_d}$ will be

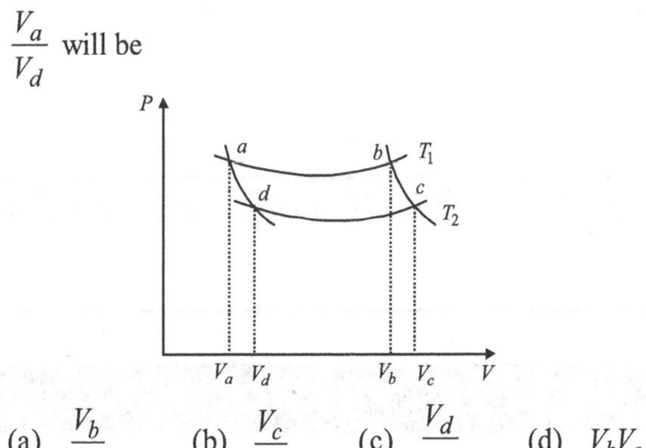

(a) $\frac{V_b}{V_c}$ (b) $\frac{V_c}{V_b}$ (c) $\frac{V_d}{V_a}$ (d) $V_b V_c$

DIRECTIONS (Q.22 - Q.23) : In the following questions, more than one of the answers given are correct. Select the correct answers and mark it according to the following codes:

Codes : **(a)** 1, 2 and 3 are correct **(b)** 1 and 2 are correct
(c) 2 and 4 are correct **(d)** 1 and 3 are correct

Q.22 During the melting of a slab of ice at 273 K and one atmospheric pressure
(1) Positive work is done on the ice-water system by the atmosphere
(2) Positive work is done by ice-water system on the atmosphere
(3) The internal energy of the ice-water system increases
(4) The internal energy of the ice-water system decreases

Q.23 One mole of an ideal monatomic gas is taken from A to C along the path ABC. The temperature of the gas at A is T_0. For the process ABC –

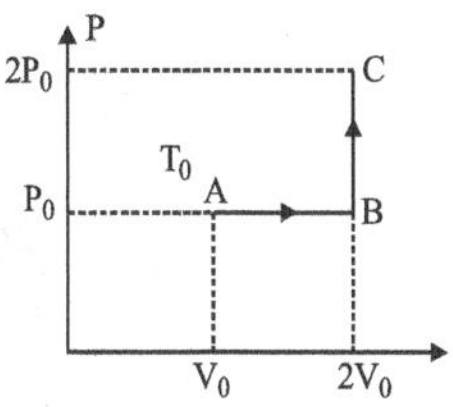

(1) Work done by the gas is RT_0
(2) Change in internal energy of the gas is $\frac{11}{2}RT_0$
(3) Heat absorbed by the gas is $\frac{11}{2}RT_0$
(4) Heat absorbed by the gas is $\frac{13}{2}RT_0$

DIRECTIONS (Q.24-Q.26) : Read the passage given below and answer the questions that follows :

In the figure n mole of a monoatomic ideal gas undergo the process ABC as shown in the P-V diagram. The process AB is isothermal and BC is isochoric. The temperature of the gas at A is T_0. Total heat given to the gas during the process ABC is measured to be Q.

RESPONSE GRID	18. ⓐⓑⓒⓓ	19. ⓐⓑⓒⓓ	20. ⓐⓑⓒⓓ	21. ⓐⓑⓒⓓ	22. ⓐⓑⓒⓓ
	23. ⓐⓑⓒⓓ				

Space for Rough Work

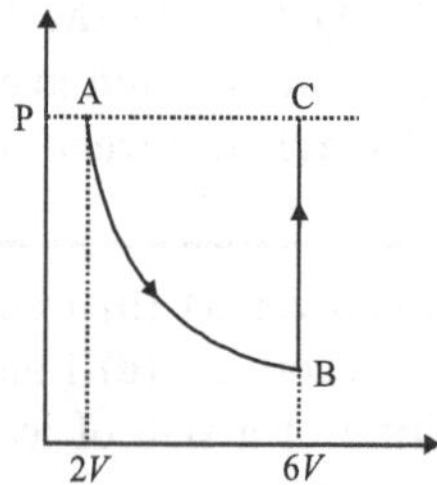

Q.24 Temperature of the gas at C is equal to

(a) T_0 (b) $3\,T_0$
(c) $6\,T_0$ (d) $2\,T_0$

Q.25 Heat absorbed by the gas in the process BC

(a) $3nRT_0$ (b) nRT_0
(c) $2nRT_0$ (d) $6nRT_0$

Q.26 The average molar heat capacity of the gas in process ABC is

(a) $\frac{Q}{nT_0}$ (b) $\frac{Q}{2nT_0}$
(c) $\frac{Q}{3nT_0}$ (d) $\frac{2Q}{nT_0}$

DIRECTIONS (Q. 27-Q.29) : Each of these questions contains two statements: Statement-1 (Assertion) and Statement-2 (Reason). Each of these questions has four alternative choices, only one of which is the correct answer. You have to select the correct choice.

(a) Statement-1 is True, Statement-2 is True; Statement-2 is a correct explanation for Statement-1.
(b) Statement-1 is True, Statement-2 is True; Statement-2 is NOT a correct explanation for Statement-1.
(c) Statement -1 is False, Statement-2 is True.
(d) Statement -1 is True, Statement-2 is False.

Q.27 **Statement-1 :** The isothermal curves intersect each other at a certain point.
Statement-2 : The isothermal change takes place slowly, so the isothermal curves have very little slope.

Q.28 **Statement-1 :** In adiabatic compression, the internal energy and temperature of the system get increased.
Statement-2 : The adiabatic compression is a slow process.

Q.29 **Statement-1 :** The specific heat of a gas in an adiabatic process is zero and in an isothermal process is infinite.
Statement-2 : Specific heat of a gas is directly proportional to change of heat in system and inversely proportional to change in temperature.

RESPONSE GRID	24. ⓐⓑⓒⓓ 25. ⓐⓑⓒⓓ 26. ⓐⓑⓒⓓ 27. ⓐⓑⓒⓓ 28. ⓐⓑⓒⓓ 29. ⓐⓑⓒⓓ

DAILY PRACTICE PROBLEM SHEET 24 - PHYSICS			
Total Questions	29	Total Marks	116
Attempted		Correct	
Incorrect		Net Score	
Cut-off Score	28	Qualifying Score	44
Success Gap = Net Score – Qualifying Score			
Net Score = (Correct × 4) – (Incorrect × 1)			

Space for Rough Work

DPP - Daily Practice Problems

Name :

Date :

Start Time :

End Time :

PHYSICS 25

SYLLABUS : Thermodynamics-2 (1st and 2nd laws of thermodynamics, Reversible & irreversible processes, Carnot engine and its efficiency)

Max. Marks : 112 **Time : 60 min.**

GENERAL INSTRUCTIONS

- The Daily Practice Problem Sheet contains 28 MCQ's. For each question only one option is correct. Darken the correct circle/ bubble in the Response Grid provided on each page.
- You have to evaluate your Response Grids yourself with the help of solution booklet.
- Each correct answer will get you 4 marks and 1 mark shall be deduced for each incorrect answer. No mark will be given/ deducted if no bubble is filled. Keep a timer in front of you and stop immediately at the end of 60 min.
- The sheet follows a particular syllabus. Do not attempt the sheet before you have completed your preparation for that syllabus. Refer syllabus sheet in the starting of the book for the syllabus of all the DPP sheets.
- After completing the sheet check your answers with the solution booklet and complete the Result Grid. Finally spend time to analyse your performance and revise the areas which emerge out as weak in your evaluation.

DIRECTIONS (Q.1-Q.19) : There are 19 multiple choice questions. Each question has 4 choices (a), (b), (c) and (d), out of which **ONLY ONE** choice is correct.

Q.1 Six moles of an ideal gas performs a cycle shown in figure. If the temperature $T_A = 600K, T_B = 800K, T_C = 2200K$ and $T_D = 1200K$, the work done per cycle is

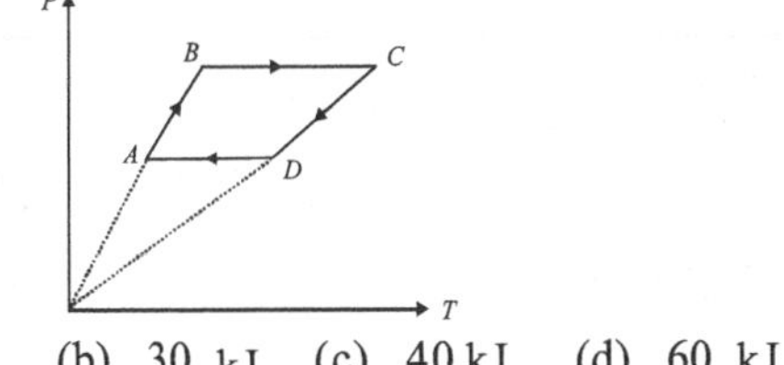

(a) 20 kJ (b) 30 kJ (c) 40 kJ (d) 60 kJ

Q.2 An ideal gas is taken from point A to the point B, as shown in the $P-V$ diagram, keeping the temperature constant. The work done in the process is

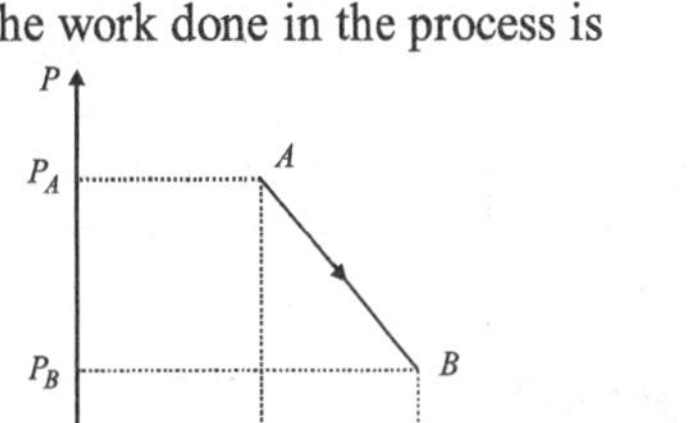

(a) $(P_A - P_B)(V_B - V_A)$ (b) $\frac{1}{2}(P_B - P_A)(V_B + V_A)$

(c) $\frac{1}{2}(P_B - P_A)(V_B - V_A)$ (d) $\frac{1}{2}(P_B + P_A)(V_B - V_A)$

RESPONSE GRID	1. ⓐⓑⓒⓓ	2. ⓐⓑⓒⓓ

Space for Rough Work

Q.3 In the diagrams (i) to (iv) of variation of volume with changing pressure is shown. A gas is taken along the path *ABCD*. The change in internal energy of the gas will be

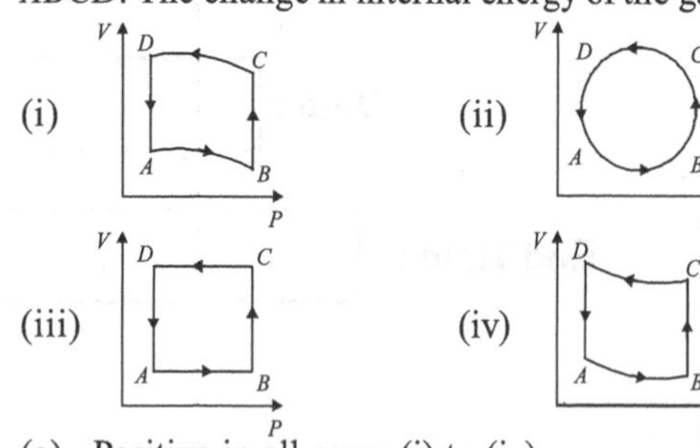

(a) Positive in all cases (i) to (iv)
(b) Positive in cases (i), (ii) and (iii) but zero in case (iv)
(c) Negative in cases (i), (ii) and (iii) but zero in case (iv)
(d) Zero in all cases

Q.4 A monoatomic ideal gas, initially at temperature T_1, is enclosed in a cylinder fitted with a frictionless piston. The gas is allowed to expand adiabatically to a temperature T_2 by releasing the piston suddenly. If L_1 and L_2 are the lengths of the gas column before and after expansion respectively, then T_1/T_2 is given by

(a) $\left(\frac{L_1}{L_2}\right)^{2/3}$ (b) $\frac{L_1}{L_2}$ (c) $\frac{L_2}{L_1}$ (d) $\left(\frac{L_2}{L_1}\right)^{2/3}$

Q.5 A gas mixture consists of 2 moles of oxygen and 4 moles argon at temperature T. Neglecting all vibrational modes, the total internal energy of the system is
(a) $4\,RT$ (b) $15\,RT$ (c) $9\,RT$ (d) $11\,RT$

Q.6 Two Carnot engines *A* and *B* are operated in succession. The first one, *A* receives heat from a source at $T_1 = 800K$ and rejects to sink at $T_2 K$. The second engine *B* receives heat rejected by the first engine and rejects to another sink at $T_3 = 300K$. If the work outputs of two engines are equal, then the value of T_2 will be
(a) 100K (b) 300K (c) 550K (d) 700 K

Q.7 A Carnot engine whose low temperature reservoir is at 7°C has an efficiency of 50%. It is desired to increase the efficiency to 70%. By how many degrees should the temperature of the high temperature reservoir be increased
(a) 840K (b) 280 K (c) 560 K (d) 380K

Q.8 An ideal heat engine working between temperature T_1 and T_2 has an efficiency η, the new efficiency of engine if both the source and sink temperature are doubled, will be

(a) $\frac{\eta}{2}$ (b) η (c) 2η (d) 3η

Q.9 Efficiency of a Carnot engine is 50% when temperature of outlet is 500 *K*. In order to increase efficiency up to 60% keeping temperature of intake the same what will be temperature of outlet
(a) 200 K (b) 400 K (c) 600 K (d) 800 K

Q.10 A scientist says that the efficiency of his heat engine which operates at source temperature 120°*C* and sink temperature 27°*C* is 26%, then
(a) It is impossible
(b) It is possible but less probable
(c) It is quite probable
(d) Data are incomplete

Q.11 The efficiency of Carnot's engine operating between reservoirs, maintained at temperatures 27°C and –123°C, is
(a) 50% (b) 24% (c) 0.75% (d) 0.4%

Q.12 The temperature of sink of Carnot engine is 27°C and Efficiency of engine is 25%. Then temperature of source is
(a) 227°C (b) 327°C (c) 127°C (d) 27°C

Q.13 In changing the state of thermodynamics from *A* to *B* state, the heat required is Q and the work done by the system is W. The change in its internal energy is

(a) $Q+W$ (b) $Q-W$ (c) Q (d) $\frac{Q-W}{2}$

Q.14 The first law of thermodynamics is concerned with the conservation of
(a) Momentum (b) Energy
(c) Mass (d) Temperature

Q.15 A system is given 300 calories of heat and it does 600 joules of work. The internal energy of the system change in this process is
(J = 4.18 Joule/cal)
(a) 654 Joule (b) 156.5 Joule
(c) –300 Joule (d) – 528.2 Joule

RESPONSE GRID					
	3. ⓐⓑⓒⓓ	4. ⓐⓑⓒⓓ	5. ⓐⓑⓒⓓ	6. ⓐⓑⓒⓓ	7. ⓐⓑⓒⓓ
	8. ⓐⓑⓒⓓ	9. ⓐⓑⓒⓓ	10. ⓐⓑⓒⓓ	11. ⓐⓑⓒⓓ	12. ⓐⓑⓒⓓ
	13. ⓐⓑⓒⓓ	14. ⓐⓑⓒⓓ	15. ⓐⓑⓒⓓ		

Space for Rough Work

Q.16 110 *J* of heat is added to a gaseous system, whose internal energy change is 40 *J,* then the amount of external work done is

(a) 150 J (b) 70 J (c) 110 J (d) 40 J

Q.17 For free expansion of the gas which of the following is true

(a) $Q = W = 0$ and $\Delta E_{int} = 0$

(b) $Q = 0, W > 0$ and $\Delta E_{int} = -W$

(c) $W = 0, Q > 0$, and $\Delta E_{int} = Q$

(d) $W > 0, Q < 0$ and $\Delta E_{int} = 0$

Q.18 In a given process for an ideal gas, $dW = 0$ and $dQ < 0$. Then for the gas

(a) The temperature will decrease

(b) The volume will increase

(c) The pressure will remain constant

(d) The temperature will increase

Q.19 The specific heat of hydrogen gas at constant pressure is $C_p = 3.4 \times 10^3 \text{cal/kg°C}$ and at constant volume is $C_V = 2.4 \times 10^3 \text{cal/kg°C}$. If one kilogram hydrogen gas is heated from 10°C to 20°C at constant pressure, the external work done on the gas to maintain it at constant pressure is

(a) 10^5 cal (b) 10^4 cal

(c) 10^3 cal (d) 5×10^3 cal

DIRECTIONS (Q.20-Q.22) : In the following questions, more than one of the answers given are correct. Select the correct answers and mark it according to the following codes:

Codes : **(a)** 1, 2 and 3 are correct **(b)** 1 and 2 are correct **(c)** 2 and 4 are correct **(d)** 1 and 3 are correct

Q.20 Which of the following processes are irreversible?

(1) Transfer of heat by radiation

(2) Electrical heating of a nichrome wire

(3) Transfer of heat by conduction

(4) Isothermal compression

Q.21 For a reversible process, unnecessary conditions are

(1) In the whole cycle of the system, the loss of any type of heat energy should be zero

(2) That the process should be too fast

(3) That the process should be slow so that the working substance should remain in thermal and mechanical equilibrium with the surroundings

(4) The loss of energy should be zero and it should be quasistatic

Q.22 One mole of an ideal gas is taken through the cyclic through the cyclic process shown in the V-T diagram, where V = volume and T = absolute temperature of the gas. Which of the following statements are correct

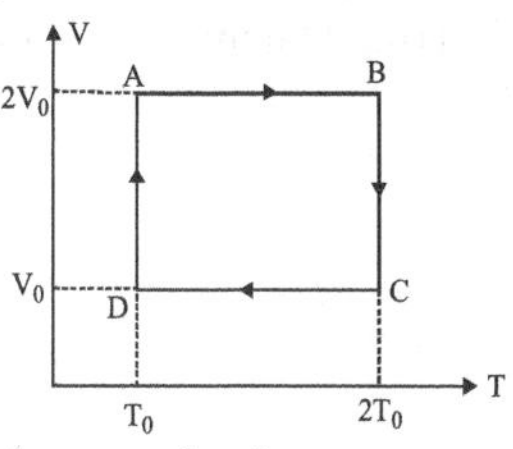

(1) Heat is given out by the gas

(2) Heat is absorbed by the gas

(3) The magnitude of the work done by the gas is RT_0 (ln 2)

(4) The magnitude of the work done by the gas is V_0T_0

DIRECTIONS (Q.23-Q.25) : Read the passage given below and answer the questions that follows :

V-T graph of a process of monoatomic ideal gas is as shown in figure.

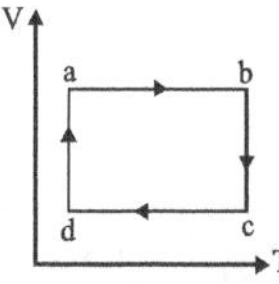

Q.23 Sum of work done by the gas in process abcd is –

(a) zero (b) positive

(c) negative (d) data is insufficient

Q.24 Heat is supplied to the gas in process(s) –

(a) *da, ab* and *bc* (b) *da* and *ab* only

(c) *da* only (d) *ab* and *bc* only

Q.25 Change in internal energy of the gas is zero in process(s) –

(a) *da, ab* and *bc* (b) *da* and *bc* only

(c) *da* only (d) *da* and *ab* only

RESPONSE GRID					
	16. ⓐⓑⓒⓓ	17. ⓐⓑⓒⓓ	18. ⓐⓑⓒⓓ	19. ⓐⓑⓒⓓ	20. ⓐⓑⓒⓓ
	21. ⓐⓑⓒⓓ	22. ⓐⓑⓒⓓ	23. ⓐⓑⓒⓓ	24. ⓐⓑⓒⓓ	25. ⓐⓑⓒⓓ

Space for Rough Work

DIRECTIONS (Q.26-Q.28) : Each of these questions contains two statements: Statement-1 (Assertion) and Statement-2 (Reason). Each of these questions has four alternative choices, only one of which is the correct answer. You have to select the correct choice.

(a) Statement-1 is True, Statement-2 is True; Statement-2 is a correct explanation for Statement-1.
(b) Statement-1 is True, Statement-2 is True; Statement-2 is NOT a correct explanation for Statement-1.
(c) Statement -1 is False, Statement-2 is True.
(d) Statement -1 is True, Statement-2 is False.

Q.26 Statement-1 : It is not possible for a system, unaided by an external agency to transfer heat from a body at lower temperature to another body at higher temperature.

Statement-2 : According to Clausius statement, "No process is possible whose sole result is the transfer of heat from a cooled object to a hotter object.

Q.27 Statement-1 : A room can be warmed by opening the door of a refrigerator in a closed room.

Statement-2 : Head flows from lower temperature (refrigerator) to higher temperature (room).

Q.28 Statement-1 : In isothermal process whole of the heat energy supplied to the body is converted into internal energy.

Statement-2 : According to the first law of themodynamics

$$\Delta Q = \Delta U + P\Delta V$$

RESPONSE GRID	26. ⓐⓑⓒⓓ	27. ⓐⓑⓒⓓ	28. ⓐⓑⓒⓓ

DAILY PRACTICE PROBLEM SHEET 25 - PHYSICS			
Total Questions	28	Total Marks	112
Attempted		Correct	
Incorrect		Net Score	
Cut-off Score	26	Qualifying Score	42
Success Gap = Net Score – Qualifying Score			
Net Score = (Correct × 4) – (Incorrect × 1)			

Space for Rough Work

DPP - Daily Practice Problems

Name : 　　　　　　　　　　　　Date :

Start Time : 　　　　　　　　　　End Time :

PHYSICS 26

SYLLABUS : Kinetic Theory

Max. Marks : 120 　　　　　　　　　　**Time : 60 min.**

GENERAL INSTRUCTIONS

- The Daily Practice Problem Sheet contains 30 MCQ's. For each question only one option is correct. Darken the correct circle/ bubble in the Response Grid provided on each page.
- You have to evaluate your Response Grids yourself with the help of solution booklet.
- Each correct answer will get you 4 marks and 1 mark shall be deduced for each incorrect answer. No mark will be given/ deducted if no bubble is filled. Keep a timer in front of you and stop immediately at the end of 60 min.
- The sheet follows a particular syllabus. Do not attempt the sheet before you have completed your preparation for that syllabus. Refer syllabus sheet in the starting of the book for the syllabus of all the DPP sheets.
- After completing the sheet check your answers with the solution booklet and complete the Result Grid. Finally spend time to analyse your performance and revise the areas which emerge out as weak in your evaluation.

DIRECTIONS (Q.1-Q.21) : There are 21 multiple choice questions. Each question has 4 choices (a), (b), (c) and (d), out of which **ONLY ONE** choice is correct.

Q.1 If pressure of a gas contained in a closed vessel is increased by 0.4% when heated by 1°C, the initial temperature must be

(a) 250 K (b) 250°C (c) 2500 K (d) 25°C

Q.2 To double the volume of a given mass of an ideal gas at 27°C keeping the pressure constant, one must raise the temperature in degree centigrade to

(a) 54 (b) 270 (c) 327 (d) 600

Q.3 Under which of the following conditions is the law $PV = RT$ obeyed most closely by a real gas?

(a) High pressure and high temperature
(b) Low pressure and low temperature
(c) Low pressure and high temperature
(d) High pressure and low temperature

Q.4 The pressure P, volume V and temperature T of a gas in the jar A and the other gas in the jar B at pressure $2P$, volume $\frac{V}{4}$ and temperature 2T, then the ratio of, the number of molecules in the jar A and B will be

(a) 1 : 1 (b) 1 : 2
(c) 2 : 1 (d) 4 : 1

Q.5 A flask is filled with 13 gm of an ideal gas at 27°C and its temperature is raised to 52°C. The mass of the gas that has to be released to maintain the temperature of the gas in the flask at 52°C and the pressure remaining the same is

(a) 2.5 g (b) 2.0 g (c) 1.5 g (d) 1.0 g

RESPONSE GRID	1. ⓐⓑⓒⓓ	2. ⓐⓑⓒⓓ	3. ⓐⓑⓒⓓ	4. ⓐⓑⓒⓓ	5. ⓐⓑⓒⓓ

Space for Rough Work

Q.6 The pressure and temperature of two different gases is P and T having the volume V for each. They are mixed keeping the same volume and temperature, the pressure of the mixture will be

(a) $P/2$ (b) P (c) $2P$ (d) $4P$

Q.7 The root mean square velocity of a gas molecule of mass m at a given temperature is proportional to

(a) m° (b) m (c) $\sqrt{m}$ (d) $\frac{1}{\sqrt{m}}$

Q.8 Which of the following statements is true?

(a) Absolute zero temperature is not zero energy temperature

(b) Two different gases at the same temperature and pressure have equal root mean square velocities

(c) The root mean square speed of the molecules of different ideal gases, maintained at the same temperature are the same

(d) Given sample of 1 cc of hydrogen and 1 cc of oxygen both at NTP; oxygen sample has a large number of molecules

Q.9 At room temperature, the *r.m.s.* speed of the molecules of certain diatomic gas is found to be 1930 m/s. The gas is

(a) H_2 (b) F_2 (c) O_2 (d) Cl_2

Q.10 Root mean square velocity of a particle is v at pressure P. If pressure is increased two times, then the *r.m.s.* velocity becomes

(a) $2\,v$ (b) $3\,v$ (c) $0.5\,v$ (d) v

Q.11 In the two vessels of same volume, atomic hydrogen and helium at pressure 1 atm and 2 atm are filled. If temperature of both the samples is same, then average speed of hydrogen atoms $<C_H>$ will be related to that of helium $<C_{He}>$ as

(a) $<C_H> = \sqrt{2} <C_{He}>$ (b) $<C_H> = <C_{He}>$

(c) $<C_H> = 2 <C_{He}>$ (d) $<C_H> = \frac{<C_{He}>}{2}$

Q.12 For a gas at a temperature T the root-mean-square velocity v_{rms}, the most probable speed v_{mp}, and the average speed v_{av} obey the relationship

(a) $v_{av} > v_{rms} > v_{mp}$ (b) $v_{rms} > v_{av} > v_{mp}$

(c) $v_{mp} > v_{av} > v_{rms}$ (d) $v_{mp} > v_{rms} > v_{av}$

Q.13 One mole of ideal monoatomic gas $(\gamma = 5/3)$ is mixed with one mole of diatomic gas $(\gamma = 7/5)$. What is γ for the mixture? $\left(\gamma = \frac{Cp}{Cv}\right)$

(a) 3/2 (b) 23/15 (c) 35/23 (d) 4/3

Q.14 The value of the gas constant (R) calculated from the perfect gas equation is 8.32 Joule/gm mol K, whereas its value calculated from the knowledge of C_p and C_V of the gas is 1.98 cal/gm mole K. From this data, the value of J is

(a) 4.16 J / cal (b) 4.18 J / cal

(c) 4.20 J / cal (d) 4.22 J / cal

Q.15 Gas at a pressure P_0 is contained in a vessel. If the masses of all the molecules are halved and their speeds are doubled, the resulting pressure P will be equal to

(a) $4P_0$ (b) $2P_0$ (c) P_0 (d) $\frac{P_0}{2}$

Q.16 The relation between the gas pressure P and average kinetic energy per unit volume E is

(a) $P = \frac{1}{2}E$ (b) $P = E$ (c) $P = \frac{3}{2}E$ (d) $P = \frac{2}{3}E$

Q.17 Mean kinetic energy (or average energy) per *gm* molecule of a monoatomic gas is given by

(a) $\frac{3}{2}RT$ (b) $\frac{1}{2}KT$ (c) $\frac{1}{2}RT$ (d) $\frac{3}{2}KT$

Q.18 At which of the following temperature would the molecules of a gas have twice the average kinetic energy they have at 20°C?

(a) 40°C (b) 80°C (c) 313°C (d) 586°C

RESPONSE GRID					
	6. ⓐⓑⓒⓓ	7. ⓐⓑⓒⓓ	8. ⓐⓑⓒⓓ	9. ⓐⓑⓒⓓ	10. ⓐⓑⓒⓓ
	11. ⓐⓑⓒⓓ	12. ⓐⓑⓒⓓ	13. ⓐⓑⓒⓓ	14. ⓐⓑⓒⓓ	15. ⓐⓑⓒⓓ
	16. ⓐⓑⓒⓓ	17. ⓐⓑⓒⓓ	18. ⓐⓑⓒⓓ		

Space for Rough Work

Q.19 The kinetic energy of one gram molecule of a gas at normal temperature and pressure is $(R = 8.31 \text{ J/mol} - K)$

(a) 0.56×10^4 J (b) 1.3×10^2 J

(c) 2.7×10^2 J (d) 3.4×10^3 J

Q.20 70 calories of heat are required to raise the temperature o f 2 moles of an ideal gas at constant pressure from 30°C to 35°C. The amount of heat required to raise the temperature of same gas through the same range (30°C to 35°C) at constant volume $(R = 2 \text{ cal/mol K})$

(a) 30 cal (b) 50 *cal* (c) 70 cal (d) 90 cal

Q.21 A vessel contains a mixture of one mole of oxygen and two moles of nitrogen at 300 K. The ratio of the average rotational kinetic energy per O_2 molecule to that per N_2 molecule is

(a) 1 : 1
(b) 1 : 2
(c) 2 : 1
(d) Depends on the moments of inertia of the two molecules

DIRECTIONS (Q.22-Q.24) : In the following questions, more than one of the answers given are correct. Select the correct answers and mark it according to the following codes:

Codes : **(a)** 1, 2 and 3 are correct **(b)** 1 and 2 are correct
(c) 2 and 4 are correct **(d)** 1 and 3 are correct

Q.22 From the following statements, concerning ideal gas at any given temperature T, select the correct one(s)

(1) The coefficient of volume expansion at constant pressure is same for all ideal gases
(2) In a gaseous mixture, the average translational kinetic energy of the molecules of each component is same
(3) The mean free path of molecules increases with the decrease in pressure
(4) The average translational kinetic energy per molecule of oxygen gas is 3KT (K being Boltzmann constant)

Q.23 Let $\bar{v}$, v_{rms} and v_p respectively denote the mean speed, root mean square speed and most probable speed of the molecules in an ideal monoatomic gas at absolute temperature T, the mass of a molecule is m. Then

(1) $v_p < \bar{v} < v_{rms}$
(2) The average kinetic energy of a molecule is $\frac{3}{4}mv_p^2$
(3) No molecule can have speed greater than $\sqrt{2}v_{rms}$
(4) No molecule can have speed less than $v_p/\sqrt{2}$

Q.24 A gas in container A is in thermal equilibrium with another gas in container B, both contain equal masses of the two gases in the respective containers. Which of the following can be true

(1) $P_A = P_B, V_A \neq V_B$ (2) $P_A V_A = P_B V_B$

(3) $PA \neq P_B, V_A = V_B$ (4) $\frac{P_A}{V_A} = \frac{P_B}{V_B}$

DIRECTIONS (Q.25-Q.27) : Read the passage given below and answer the questions that follows :

A diathermic piston divides adiabatic cylinder of volume V_0 into two equal parts as shown in the figure. Both parts contain ideal monoatomic gases. The initial pressure and temperature of gas in left compartment are P_0 and T_0 while that in right compartment are $2P_0$ and $2T_0$. Initially the piston is kept fixed and the system is allowed to acquire a state of thermal equilibrium.

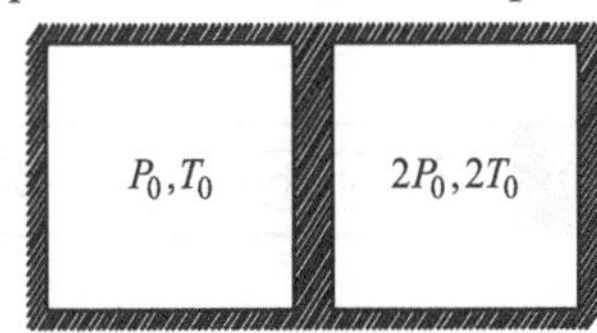

Q.25 The pressure in left compartment after thermal equilibrium is achieved is

(a) P_0 (b) $\frac{3}{2}P_0$

(c) $\frac{4}{3}P_0$ (d) None of these

RESPONSE GRID					
	19. ⓐⓑⓒⓓ	20. ⓐⓑⓒⓓ	21. ⓐⓑⓒⓓ	22. ⓐⓑⓒⓓ	23. ⓐⓑⓒⓓ
	24. ⓐⓑⓒⓓ	25. ⓐⓑⓒⓓ			

Space for Rough Work

Q.26 The heat that flown from right compartment to left compartment before thermal equilibrium is achieved is

(a) P_0V_0 (b) $\frac{3}{4}P_0V_0$ (c) $\frac{3}{8}P_0V_0$ (d) $\frac{2}{3}P_0V_0$

Q.27 If the pin which was keeping the piston fixed is removed and the piston is allowed to slide slowly such that a state of mechanical equilibrium is achieved. The volume of left compartment when piston is in equilibrium is

(a) $\frac{3}{4}V_0$ (b) $\frac{V_0}{4}$ (c) $\frac{V_0}{2}$ (d) $\frac{2}{3}V_0$

DIRECTIONS (Qs. 28-Q.30) : Each of these questions contains two statements: Statement-1 (Assertion) and Statement-2 (Reason). Each of these questions has four alternative choices, only one of which is the correct answer. You have to select the correct choice.

(a) Statement-1 is True, Statement-2 is True; Statement-2 is a correct explanation for Statement-1.
(b) Statement-1 is True, Statement-2 is True; Statement-2 is NOT a correct explanation for Statement-1.
(c) Statement -1 is False, Statement-2 is True.
(d) Statement -1 is True, Statement-2 is False.

Q.28 Statement-1 : Internal energy of an ideal gas does not depend upon volume of the gas.

Statement-2 : Internal energy of an ideal gas depends on temperature of gas.

Q.29 Statement-1 : Equal masses of helium and oxygen gases are given equal quantities of heat. There will be a greater rise in the temperature of helium compared to that of oxygen.

Statement-2 : The molecular weight of oxygen is more than the molecular weight of helium.

Q.30 Statement-1 : Maxwell speed distribution graph is asymmetric about most probable speed.

Statement-2 : *rms* speed of ideal gas, depends upon it's type (monoatomic, diatomic and polyatomic).

RESPONSE GRID	26. ⓐⓑⓒⓓ	27. ⓐⓑⓒⓓ	28. ⓐⓑⓒⓓ	29. ⓐⓑⓒⓓ	30. ⓐⓑⓒⓓ

DAILY PRACTICE PROBLEM SHEET 26 - PHYSICS			
Total Questions	30	Total Marks	120
Attempted		Correct	
Incorrect		Net Score	
Cut-off Score	30	Qualifying Score	48
Success Gap = Net Score – Qualifying Score			
Net Score = (Correct × 4) – (Incorrect × 1)			

Space for Rough Work

DPP - Daily Practice Problems

Name :

Date :

Start Time :

End Time :

PHYSICS 27

SYLLABUS : Oscillations-1 (Periodic motion - period, Frequency, Displacement as a function of time. Periodic functions, Simple harmonic motion and its equation, Energy in S.H.M. - kinetic and potential energies)

Max. Marks : 120 **Time : 60 min.**

GENERAL INSTRUCTIONS

- The Daily Practice Problem Sheet contains 30 MCQ's. For each question only one option is correct. Darken the correct circle/ bubble in the Response Grid provided on each page.
- You have to evaluate your Response Grids yourself with the help of solution booklet.
- Each correct answer will get you 4 marks and 1 mark shall be deduced for each incorrect answer. No mark will be given/ deducted if no bubble is filled. Keep a timer in front of you and stop immediately at the end of 60 min.
- The sheet follows a particular syllabus. Do not attempt the sheet before you have completed your preparation for that syllabus. Refer syllabus sheet in the starting of the book for the syllabus of all the DPP sheets.
- After completing the sheet check your answers with the solution booklet and complete the Result Grid. Finally spend time to analyse your performance and revise the areas which emerge out as weak in your evaluation.

DIRECTIONS (Q.1-Q.22) : There are 22 multiple choice questions. Each question has 4 choices (a), (b), (c) and (d), out of which **ONLY ONE** choice is correct.

Q.1 A simple harmonic motion is represented by $F(t)=10\sin(20t+0.5)$. The amplitude of the S.H.M. is

(a) $a=30$ cm (b) $a=20$ cm
(c) $a=10$ cm (d) $a=5$ cm

Q.2 A particle executes a simple harmonic motion of time period *T*. Find the time taken by the particle to go directly from its mean position to half the amplitude

(a) $T/2$ (b) $T/4$ (c) $T/8$ (d) $T/12$

Q.3 The periodic time of a body executing simple harmonic motion is 3 sec. After how much time from time $t=0$, its displacement will be half of its amplitude

(a) $\frac{1}{8}$sec (b) $\frac{1}{6}$sec (c) $\frac{1}{4}$sec (d) $\frac{1}{3}$sec

Q.4 If $x=a\sin\left(\omega t+\frac{\pi}{6}\right)$ and $x'=a\cos\omega t$, then what is the phase difference between the two waves?

(a) $\pi/3$ (b) $\pi/6$
(c) $\pi/2$ (d) π

Q.5 A body is executing S.H.M. when its displacement from the mean position is 4 cm and 5 cm, the corresponding velocity of the body is 10 cm/sec and 8 cm/sec. Then the time period of the body is

(a) 2πsec (b) $\pi/2$sec
(c) πsec (d) $3\pi/2$sec

RESPONSE GRID	1. ⓐⓑⓒⓓ	2. ⓐⓑⓒⓓ	3. ⓐⓑⓒⓓ	4. ⓐⓑⓒⓓ	5. ⓐⓑⓒⓓ

Space for Rough Work

Q.6 If a simple pendulum oscillates with an amplitude of 50 mm and time period of 2 sec, then its maximum velocity is
(a) 0.10 m/s (b) 0.15m/s
(c) 0.8 m/s (d) 0.26 m/s

Q.7 The maximum velocity and the maximum acceleration of a body moving in a simple harmonic oscillator are 2 m/s and 4 m/s^2. Then angular velocity will be
(a) 3 rad/sec (b) 0.5 rad/sec
(c) 1 rad/sec (d) 2 rad/sec

Q.8 The amplitude of a particle executing SHM is 4 cm. At the mean position the speed of the particle is 16 cm/sec. The distance of the particle from the mean position at which the speed of the particle becomes $8\sqrt{3}$ cm/s, will be
(a) $2\sqrt{3}$ cm (b) $\sqrt{3}$ cm
(c) 1 cm (d) 2 cm

Q.9 The amplitude of a particle executing S.H.M. with frequency of 60 Hz is 0.01 m. The maximum value of the acceleration of the particle is
(a) $144\pi^2 m/sec^2$ (b) $144 m/sec^2$
(c) $\frac{144}{\pi^2} m/sec^2$ (d) $288\pi^2 m/sec^2$

Q.10 A particle executes simple harmonic motion with an angular velocity and maximum acceleration of 3.5rad/sec and 7.5 m/s^2 respectively. The amplitude of oscillation is
(a) 0.28 m (b) 0.36 m (c) 0.53 m (d) 0.61 m

Q.11 What is the maximum acceleration of the particle doing the SHM $y = 2\sin\left[\frac{\pi t}{2} + \phi\right]$ where y is in cm?
(a) $\frac{\pi}{2} cm/s^2$ (b) $\frac{\pi^2}{2} cm/s^2$
(c) $\frac{\pi}{4} cm/s^2$ (d) $\frac{\pi}{4} cm/s^2$

Q.12 The total energy of a particle executing S.H.M. is proportional to
(a) Displacement from equilibrium position
(b) Frequency of oscillation
(c) Velocity in equilibrium position
(d) Square of amplitude of motion

Q.13 When the displacement is half the amplitude, the ratio of potential energy to the total energy is
(a) $\frac{1}{2}$ (b) $\frac{1}{4}$ (c) 1 (d) $\frac{1}{8}$

Q.14 A particle is executing simple harmonic motion with frequency f. The frequency at which its kinetic energy changes into potential energy is
(a) $f/2$ (b) f (c) $2f$ (d) $4f$

Q.15 A particle executes simple harmonic motion with a frequency f. The frequency with which its kinetic energy oscillates is
(a) $f/2$ (b) f (c) $2f$ (d) $4f$

Q.16 The kinetic energy of a particle executing S.H.M. is 16 J when it is in its mean position. If the amplitude of oscillations is 25 cm and the mass of the particle is 5.12 kg, the time period of its oscillation is
(a) $\frac{\pi}{5}$sec (b) 2πsec (c) 20πsec (d) 5πsec

Q.17 The displacement x (in metres) of a particle performing simple harmonic motion is related to time t (in seconds) as $x = 0.05\cos\left(4\pi t + \frac{\pi}{4}\right)$. The frequency of the motion will be
(a) 0.5 Hz (b) 1.0 Hz (c) 1.5 Hz (d) 2.0 Hz

Q.18 A particle executes simple harmonic motion [amplitude = A] between $x = -A$ and $x = +A$. The time taken for it to go from 0 to $A/2$ is T_1 and to go from $A/2$ to A is T_2. Then
(a) $T_1 < T_2$ (b) $T_1 > T_2$ (c) $T_1 = T_2$ (d) $T_1 = 2T_2$

RESPONSE GRID					
	6. ⓐⓑⓒⓓ	7. ⓐⓑⓒⓓ	8. ⓐⓑⓒⓓ	9. ⓐⓑⓒⓓ	10. ⓐⓑⓒⓓ
	11. ⓐⓑⓒⓓ	12. ⓐⓑⓒⓓ	13. ⓐⓑⓒⓓ	14. ⓐⓑⓒⓓ	15. ⓐⓑⓒⓓ
	16. ⓐⓑⓒⓓ	17. ⓐⓑⓒⓓ	18. ⓐⓑⓒⓓ		

Space for Rough Work

Q.19 A cylindrical piston of mass M slides smoothly inside a long cylinder closed at one end, enclosing a certain mass of gas. The cylinder is kept with its axis horizontal. If the piston is disturbed from its equilibrium position, it oscillates simple harmonically. The period of oscillation will be

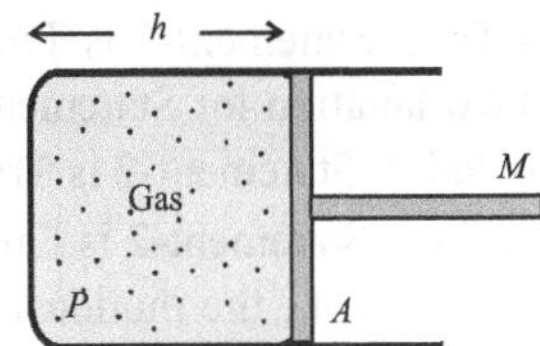

(a) $T = 2\pi\sqrt{\left(\frac{Mh}{PA}\right)}$ (b) $T = 2\pi\sqrt{\left(\frac{MA}{Ph}\right)}$

(c) $T = 2\pi\sqrt{\left(\frac{M}{PAh}\right)}$ (d) $T = 2\pi\sqrt{MPhA}$

Q.20 A particle is performing simple harmonic motion along x-axis with amplitude 4 cm and time period 1.2 sec. The minimum time taken by the particle to move from $x = 2$ cm to $x = +4$ cm and back again is given by

(a) 0.6 sec (b) 0.4 sec
(c) 0.3 sec (d) 0.2 sec

Q.21 A spring of force constant k is cut into two pieces such that one piece is double the length of the other. Then the long piece will have a force constant of

(a) (2/3)k (b) (3/2)k
(c) 3k (d) 6k

Q.22 A simple pendulum has time period T_1. The point of suspension is now moved upward according to equation $y = kt^2$ where $k = 1 m/sec^2$. If new time period is T_2 then ratio $\frac{T_1^2}{T_2^2}$ will be

(a) 2/3 (b) 5/6
(c) 6/5 (d) 3/2

DIRECTIONS (Q.23-Q.25) : In the following questions, more than one of the answers given are correct. Select the correct answers and mark it according to the following codes:

Codes : **(a)** 1, 2 and 3 are correct
(b) 1 and 2 are correct
(c) 2 and 4 are correct
(d) 1 and 3 are correct

Q.23 A particle constrained to move along the x-axis in a potential $V = kx^2$, is subjected to an external time dependent force $\vec{f}(t)$, here k is a constant, x the distance from the origin, and t the time. At some time T, when the particle has zero velocity at x = 0, the external force is removed. Choose the incorrect options –

(1) Particle executes SHM
(2) Particle moves along +x direction
(3) Particle moves along – x direction
(4) Particle remains at rest

Q.24 Three simple harmonic motions in the same direction having the same amplitude a and same period are superposed. If each differs in phase from the next by 45°, then –

(1) The resultant amplitude is $(1+\sqrt{2})\,a$
(2) The phase of the resultant motion relative to the first is 90°
(3) The energy associated with the resulting motion is $(3+2\sqrt{2})$ times the energy associated with any single motion
(4) The resulting motion is not simple harmonic

Q.25 For a particle executing simple harmonic motion, which of the following statements is correct?

(1) The total energy of the particle always remains the same
(2) The restoring force always directed towards a fixed point
(3) The restoring force is maximum at the extreme positions
(4) The acceleration of the particle is maximum at the equilibrium position

RESPONSE GRID					
	19. ⓐⓑⓒⓓ	20. ⓐⓑⓒⓓ	21. ⓐⓑⓒⓓ	22. ⓐⓑⓒⓓ	23. ⓐⓑⓒⓓ
	24. ⓐⓑⓒⓓ	25. ⓐⓑⓒⓓ			

Space for Rough Work

DIRECTIONS (Q.26-Q.27) : Read the passage given below and answer the questions that follows :

The differential equation of a particle undergoing SHM is given by $a\frac{d^2x}{dt^2}+bx=0$. The particle starts from the extreme position.

Q.26 The ratio of the maximum acceleration to the maximum velocity of the particle is –

(a) $\frac{b}{a}$ (b) $\frac{a}{b}$

(c) $\sqrt{\frac{a}{b}}$ (d) $\sqrt{\frac{b}{a}}$

Q.27 The equation of motion may be given by :

(a) $x = A\sin\left(\sqrt{\frac{b}{a}}\right)t$

(b) $x = A\cos\left(\sqrt{\frac{b}{a}}\right)t$

(c) $x = A\sin\left(\sqrt{\frac{b}{a}}t+\theta\right)$ where $\theta \neq \pi/2$

(d) None of these

DIRECTIONS (Q.28-Q.30) : Each of these questions contains two statements: Statement-1 (Assertion) and Statement-2 (Reason). Each of these questions has four alternative choices, only one of which is the correct answer. You have to select the correct choice.

(a) Statement-1 is True, Statement-2 is True; Statement-2 is a correct explanation for Statement-1.
(b) Statement-1 is True, Statement-2 is True; Statement-2 is NOT a correct explanation for Statement-1.
(c) Statement -1 is False, Statement-2 is True.
(d) Statement -1 is True, Statement-2 is False.

Q.28 Statement-1 : In S.H.M., the motion is 'to and fro' and periodic.
Statement-2 : Velocity of the particle $(v)=\omega\sqrt{k^2-x^2}$ (where x is the displacement and k is amplitude)

Q.29 Statement-1 : In simple harmonic motion, the velocity is maximum when acceleration is minimum.
Statement-2 : Displacement and velocity of S.H.M. differ in phase by $\pi/2$

Q.30 Statement-1 : The graph of total energy of a particle in SHM *w.r.t.*, position is a straight line with zero slope.
Statement-2 : Total energy of particle in SHM remains constant throughout its motion.

RESPONSE GRID	26. ⓐⓑⓒⓓ	27. ⓐⓑⓒⓓ	28. ⓐⓑⓒⓓ	29. ⓐⓑⓒⓓ	30. ⓐⓑⓒⓓ

DAILY PRACTICE PROBLEM SHEET 27 - PHYSICS			
Total Questions	30	Total Marks	120
Attempted		Correct	
Incorrect		Net Score	
Cut-off Score	30	Qualifying Score	45
Success Gap = Net Score – Qualifying Score			
Net Score = (Correct × 4) – (Incorrect × 1)			

Space for Rough Work

DPP - Daily Practice Problems

Name :

Date :

Start Time :

End Time :

PHYSICS 28

SYLLABUS : Oscillations-2 (Oscillations of a spring, simple pendulum, free, forced and damped oscillations, Resonance)

Max. Marks : 112 **Time : 60 min.**

GENERAL INSTRUCTIONS

- The Daily Practice Problem Sheet contains 28 MCQ's. For each question only one option is correct. Darken the correct circle/ bubble in the Response Grid provided on each page.
- You have to evaluate your Response Grids yourself with the help of solution booklet.
- Each correct answer will get you 4 marks and 1 mark shall be deduced for each incorrect answer. No mark will be given/ deducted if no bubble is filled. Keep a timer in front of you and stop immediately at the end of 60 min.
- The sheet follows a particular syllabus. Do not attempt the sheet before you have completed your preparation for that syllabus. Refer syllabus sheet in the starting of the book for the syllabus of all the DPP sheets.
- After completing the sheet check your answers with the solution booklet and complete the Result Grid. Finally spend time to analyse your performance and revise the areas which emerge out as weak in your evaluation.

DIRECTIONS (Q.1-Q.20) : There are 20 multiple choice questions. Each question has 4 choices (a), (b), (c) and (d), out of which **ONLY ONE** choice is correct.

Q.1 A particle of mass m is attached to three identical springs A, B and C each of force constant k as shown in figure. If the particle of mass m is pushed slightly against the spring A and released then the time period of oscillations is

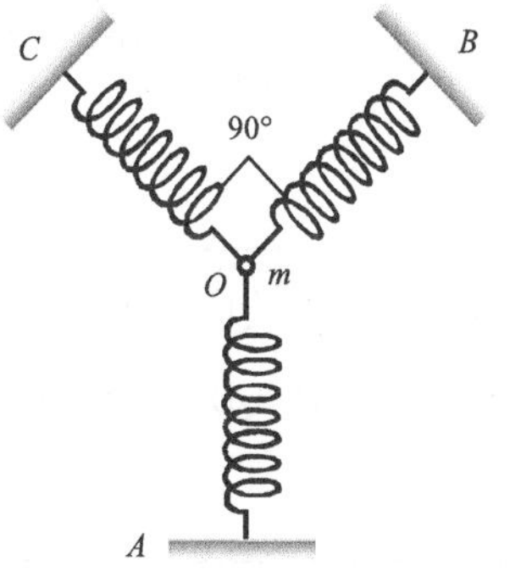

(a) $2\pi\sqrt{\frac{2m}{k}}$ (b) $2\pi\sqrt{\frac{m}{2k}}$ (c) $2\pi\sqrt{\frac{m}{k}}$ (d) $2\pi\sqrt{\frac{m}{3k}}$

Q.2 Three masses 700g, 500g, and 400g are suspended at the end of a spring as shown and are in equilibrium. When the 700g mass is removed, the system oscillates with a period of 3 seconds, when the 500 gm mass is also removed, it will oscillate with a period of

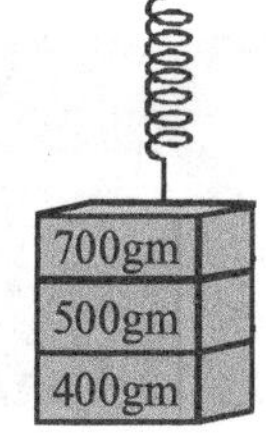

(a) 1 s
(b) 2 s
(c) 3 s
(d) $\sqrt{\frac{12}{5}}s$

RESPONSE GRID	1. ⓐⓑⓒⓓ	2. ⓐⓑⓒⓓ

Space for Rough Work

Q.3 The bob of a simple pendulum is displaced from its equilibrium position O to a position Q which is at height h above O and the bob is then released.
Assuming the mass of the bob to be m and time period of oscillations to be 2.0 sec, the tension in the string when the bob passes through O is

(a) $m(g+\pi\sqrt{2gh})$

(b) $m(g+\sqrt{\pi^2 gh})$

(c) $m\left(g+\sqrt{\frac{\pi^2}{2}gh}\right)$

(d) $m\left(g+\sqrt{\frac{\pi^2}{3}gh}\right)$

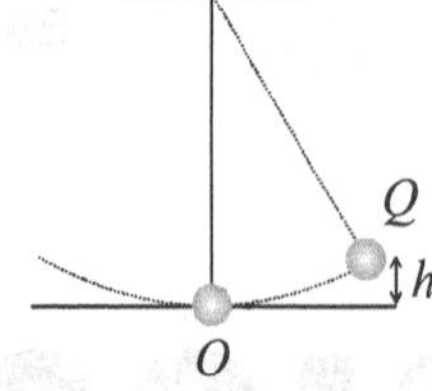

Q.4 A spring of force constant k is cut into two pieces such that one piece is double the length of the other. Then the long piece will have a force constant of
(a) $(2/3)k$ (b) $(3/2)k$ (c) $3k$ (d) $6k$

Q.5 A pendulum suspended from the ceiling of a train has a period T, when the train is at rest. When the train is accelerating with a uniform acceleration a, the period of oscillation will
(a) increase (b) decrease
(c) remain unaffected (d) become infinite

Q.6 A simple pendulum is set up in a trolley which moves to the right with an acceleration a on a horizontal plane. Then the thread of the pendulum in the mean position makes an angle θ with the vertical is

(a) $\tan^{-1}\frac{a}{g}$ in the forward direction

(b) $\tan^{-1}\frac{a}{g}$ in the backward direction

(c) $\tan^{-1}\frac{g}{a}$ in the backward direction

(d) $\tan^{-1}\frac{g}{a}$ in the forward direction

Q.7 The time period of a second's pendulum is 2 sec. The spherical bob which is empty from inside has a mass of 50 gm. This is now replaced by another solid bob of same radius but having different mass of 100 gm. The new time period will be
(a) 4 sec (b) 1 sec (c) 2 sec (d) 8 sec

Q.8 The length of a simple pendulum is increased by 1%. Its time period will
(a) Increase by 1% (b) Increase by 0.5%
(c) Decrease by 0.5% (d) Increase by 2%

Q.9 The bob of a pendulum of length l is pulled aside from its equilibrium position through an angle θ and then released. The bob will then pass through its equilibrium position with a speed v, where v equals
(a) $\sqrt{2gl(1-\sin\theta)}$ (b) $\sqrt{2gl(1+\cos\theta)}$
(c) $\sqrt{2gl(1-\cos\theta)}$ (d) $\sqrt{2gl(1+\sin\theta)}$

Q.10 A simple pendulum is executing simple harmonic motion with a time period T. If the length of the pendulum is increased by 21%, the percentage increase in the time period of the pendulum of is
(a) 10% (b) 21% (c) 30% (d) 50%

Q.11 A chimpanzee swinging on a swing in a sitting position, stands up suddenly, the time period will
(a) Become infinite (b) Remain same
(c) Increase (d) Decrease

Q.12 A simple pendulum consisting of a ball of mass m tied to a thread of length l is made to swing on a circular arc of angle θ in a vertical plane. At the end of this arc, another ball of mass m is placed at rest. The momentum transferred to this ball at rest by the swinging ball is

(a) Zero (b) $m\theta\sqrt{\frac{g}{l}}$ (c) $\frac{m\theta}{l}\sqrt{\frac{l}{g}}$ (d) $\frac{m}{l}2\pi\sqrt{\frac{l}{g}}$

Q.13 The time period of a simple pendulum of length L as measured in an elevator descending with acceleration $g/3$ is

(a) $2\pi\sqrt{\frac{3L}{g}}$ (b) $\pi\sqrt{\left(\frac{3L}{g}\right)}$ (c) $2\pi\sqrt{\left(\frac{3L}{2g}\right)}$ (d) $2\pi\sqrt{\frac{2L}{3g}}$

RESPONSE GRID					
	3. ⓐⓑⓒⓓ	4. ⓐⓑⓒⓓ	5. ⓐⓑⓒⓓ	6. ⓐⓑⓒⓓ	7. ⓐⓑⓒⓓ
	8. ⓐⓑⓒⓓ	9. ⓐⓑⓒⓓ	10. ⓐⓑⓒⓓ	11. ⓐⓑⓒⓓ	12. ⓐⓑⓒⓓ
	13. ⓐⓑⓒⓓ				

Space for Rough Work

Q.14 A mass m is suspended from the two coupled springs connected in series. The force constant for springs are k_1 and k_2. The time period of the suspended mass will be

(a) $T = 2\pi\sqrt{\left(\frac{m}{k_1+k_2}\right)}$ (b) $T = 2\pi\sqrt{\left(\frac{m}{k_1+k_2}\right)}$

(c) $T = 2\pi\sqrt{\left(\frac{m(k_1+k_2)}{k_1k_2}\right)}$ (d) $T = 2\pi\sqrt{\left(\frac{mk_1k_2}{k_1+k_2}\right)}$

Q.15 A spring having a spring constant k is loaded with a mass m. The spring is cut into two equal parts and one of these is loaded again with the same mass. The new spring constant is

(a) $k/2$ (b) k (c) $2k$ (d) k^2

Q.16 A mass $m = 100$ gm is attached at the end of a light spring which oscillates on a frictionless horizontal table with an amplitude equal to 0.16 metre and time period equal to 2 sec. Initially the mass is released from rest at $t = 0$ and displacement $x = -0.16$ metre. The expression for the displacement of mass at any time t is

(a) $x = 0.16\cos(\pi t)$ (b) $x = -0.16\cos(\pi t)$

(c) $x = 0.16\sin(\pi t + \pi)$ (d) $x = -0.16\sin(\pi t + \pi)$

Q.17 Two masses m_1 and m_2 are suspended together by a massless spring of constant k. When the masses are in equilibrium, m_1 is removed without disturbing the system. The amplitude of oscillations is

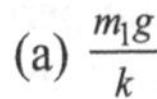

(a) $\frac{m_1 g}{k}$

(b) $\frac{m_2 g}{k}$

(c) $\frac{(m_1+m_2)g}{k}$

(d) $\frac{(m_1-m_2)g}{k}$

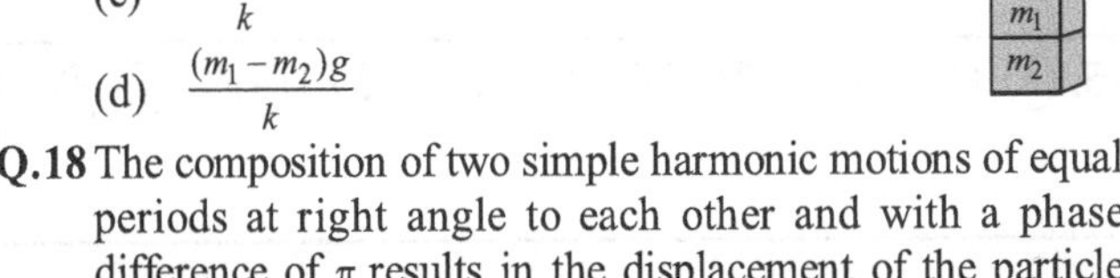

Q.18 The composition of two simple harmonic motions of equal periods at right angle to each other and with a phase difference of π results in the displacement of the particle along

(a) Straight line (b) Circle

(c) Ellipse (d) Figure of 8

Q.19 A particle with restoring force proportional to displacement and resisting force proportional to velocity is subjected to a force $F \sin \omega t$. If the amplitude of the particle is maximum for $\omega = \omega_1$ and the energy of the particle is maximum for $\omega = \omega_2$, then (where ω_0 natural frequency of oscillation of particle)

(a) $\omega_1 = \omega_0$ and $\omega_2 \neq \omega_0$ (b) $\omega_1 = \omega_0$ and $\omega_2 = \omega_0$

(c) $\omega_1 \neq \omega_0$ and $\omega_2 = \omega_0$ (d) $\omega_1 \neq \omega_0$ and $\omega_2 \neq \omega_0$

Q.20 Amplitude of a wave is represented by $A = \frac{c}{a+b-c}$

Then resonance will occur when

(a) $b = -c/2$ (b) $b = 0$ & $a = c$

(c) $b = -a/2$ (d) None

DIRECTIONS (Q.21-Q.23) : In the following questions, more than one of the answers given are correct. Select the correct answers and mark it according to the following codes:

Codes :

(a) 1, 2 and 3 are correct **(b)** 1 and 2 are correct

(c) 2 and 4 are correct **(d)** 1 and 3 are correct

Q.21 Two blocks A and B each of mass m are connected by a massless spring of natural length L and spring constant k. The blocks are initially resting on a smooth horizontal floor with the spring at its natural length. A third identical block C also of mass m moves on the floor with a speed v along the line joining A and B and collides with A. Then

(1) The kinetic energy of the A – B system at maximum compression of the spring is $mv^2/4$

(2) The maximum compression of the spring is $v\sqrt{m/2k}$

(3) The kinetic energy of the A-B system at maximum compression of the spring is zero

(4) The maximum compression of the spring is $v\sqrt{m/k}$

Q.22 A simple pendulum of length L and mass (bob) M is oscillating in a plane about a vertical line between angular limits $-\phi$ and $+\phi$. For an angular displacement $\theta(|\theta| < \phi)$, the tension in the string and the velocity of the bob are T and v respectively. The following relations hold good under the above conditions

(1) $T - Mg\cos\theta = \frac{Mv^2}{L}$

(2) $T\cos\theta = Mg$

(3) The magnitude of the tangential acceleration of the bob $|a_T| = g\sin\theta$

(4) $T = Mg\cos\theta$

RESPONSE GRID					
	14. ⓐⓑⓒⓓ	15. ⓐⓑⓒⓓ	16. ⓐⓑⓒⓓ	17. ⓐⓑⓒⓓ	18. ⓐⓑⓒⓓ
	19. ⓐⓑⓒⓓ	20. ⓐⓑⓒⓓ	21. ⓐⓑⓒⓓ	22. ⓐⓑⓒⓓ	

Space for Rough Work

Q.23 Identify wrong statements among the following

(1) The greater the mass of a pendulum bob, the shorter is its frequency of oscillation

(2) A simple pendulum with a bob of mass M swings with an angular amplitude of 40°. When its angular amplitude is 20°, the tension in the string is less than Mgcos20°.

(3) The fractional change in the time period of a pendulum on changing the temperature is independent of the length of the pendulum.

(4) As the length of a simple pendulum is increased, the maximum velocity of its bob during its oscillation will also decreases.

DIRECTIONS (Q.24-Q.25) : Read the passage given below and answer the questions that follows :

A particle performs linear SHM such that it is placed on platform & platform along with particles oscillate vertically up and down with amplitude A = 1cm. If the particle does not loose contact with platform anywhere and mass of particle is 1 kg, find :

Q.24 The minimum, possible time period (Take $\pi = \sqrt{g}$)

(a) 0.1 sec. (b) 0.2 sec.
(c) 0.3 sec. (d) 0.4 sec.

Q.25 For minimum time period condition average potential energy between t = 0 to t = 0.05 sec (Take g = 10 m/s^2)

(a) 0.025 Joule (b) 0.1 Joule
(c) 0.08 Joule (d) 0.06 Joule

DIRECTIONS (Q.26-Q.28) : Each of these questions contains two statements: Statement-1 (Assertion) and Statement-2 (Reason). Each of these questions has four alternative choices, only one of which is the correct answer. You have to select the correct choice.

(a) Statement-1 is True, Statement-2 is True; Statement-2 is a correct explanation for Statement-1.
(b) Statement-1 is True, Statement-2 is True; Statement-2 is NOT a correct explanation for Statement-1.
(c) Statement -1 is False, Statement-2 is True.
(d) Statement -1 is True, Statement-2 is False.

Q.26 Statement-1 : Consider motion for a mass spring system under gravity, motion of *M* is not a simple harmonic motion unless *Mg* is negligibly small.

Statement-2 : For simple harmonic motion acceleration must be proportional to displacement and is directed towards the mean position.

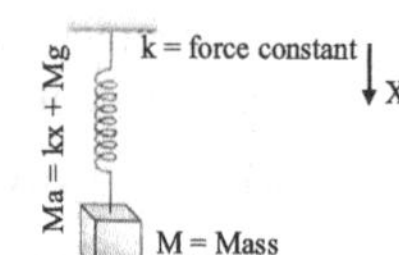

Q.27 Statement-1 : The periodic time of a hard spring is less as compared to that of a soft spring.

Statement-2 : The periodic time depends upon the spring constant, and spring constant is large for hard spring.

Q.28 Statement-1 : The percentage change in time period is 1.5%, if the length of simple pendulum increases by 3%

Statement-2: Time period is directly proportional to length of pendulum.

RESPONSE GRID	
	23. ⓐⓑⓒⓓ 24. ⓐⓑⓒⓓ 25. ⓐⓑⓒⓓ 26. ⓐⓑⓒⓓ 27. ⓐⓑⓒⓓ
	28. ⓐⓑⓒⓓ

DAILY PRACTICE PROBLEM SHEET 28 - PHYSICS			
Total Questions	28	Total Marks	112
Attempted		Correct	
Incorrect		Net Score	
Cut-off Score	28	Qualifying Score	42
Success Gap = Net Score – Qualifying Score			
Net Score = (Correct × 4) – (Incorrect × 1)			

Space for Rough Work

DPP - Daily Practice Problems

Name :

Date :

Start Time :

End Time :

PHYSICS 29

SYLLABUS : Waves-1 (Wave motion, longitudinal and transverse waves, speed of a wave, displacement relation for a progressive wave, principle of superposition of waves, reflection of waves)

Max. Marks : 116 **Time : 60 min.**

GENERAL INSTRUCTIONS

- The Daily Practice Problem Sheet contains 29 MCQ's. For each question only one option is correct. Darken the correct circle/ bubble in the Response Grid provided on each page.
- You have to evaluate your Response Grids yourself with the help of solution booklet.
- Each correct answer will get you 4 marks and 1 mark shall be deduced for each incorrect answer. No mark will be given/ deducted if no bubble is filled. Keep a timer in front of you and stop immediately at the end of 60 min.
- The sheet follows a particular syllabus. Do not attempt the sheet before you have completed your preparation for that syllabus. Refer syllabus sheet in the starting of the book for the syllabus of all the DPP sheets.
- After completing the sheet check your answers with the solution booklet and complete the Result Grid. Finally spend time to analyse your performance and revise the areas which emerge out as weak in your evaluation.

DIRECTIONS (Q.1-Q.20) : There are 20 multiple choice questions. Each question has 4 choices (a), (b), (c) and (d), out of which **ONLY ONE** choice is correct.

Q.1 A tuning fork makes 256 vibrations per second in air. When the velocity of sound is 330 m/s then wavelength of the tone emitted is

(a) 0.56 m (b) 0.89 m (c) 1.11 m (d) 1.29 m

Q.2 In a sinusoidal wave, the time required for a particular point to move from maximum displacement to zero displacement is 0.170 second. The frequency of the wave is

(a) 1.47Hz (b) 0.36 Hz (c) 0.73 Hz (d) 2.94 Hz

Q.3 A man is standing between two parallel cliffs and fires a gun. If he hears first and second echoes after 1.5 s and 3.5s respectively, the distance between the cliffs is (Velocity of sound in air = 340 ms^{-1})

(a) 1190 m

(b) 850 m

(c) 595 m

(d) 510 m

Q.4 v_1 and v_2 are the velocities of sound at the same temperature in two monoatomic gases of densities ρ_1 and ρ_2 respectively. If $\rho_1/\rho_2 = \frac{1}{4}$ then the ratio of velocities v_1 and v_2 will be

(a) 1 : 2 (b) 4 : 1 (c) 2 : 1 (d) 1 : 4

Q.5 A wave of frequency 500 Hz has velocity 360 m/sec. The distance between two nearest points 60° out of phase, is

(a) 0.6 cm (b) 12 cm (c) 60 cm (d) 120 cm

RESPONSE GRID	1. ⓐⓑⓒⓓ	2. ⓐⓑⓒⓓ	3. ⓐⓑⓒⓓ	4. ⓐⓑⓒⓓ	5. ⓐⓑⓒⓓ

Space for Rough Work

Q.6 Two waves are given by $y_1 = a\sin(\omega t - kx)$ and y_2 = a cos (ωt – kx). The phase difference between the two waves is

(a) $\frac{\pi}{4}$ (b) π (c) $\frac{\pi}{8}$ (d) $\frac{\pi}{2}$

Q.7 The relation between time and displacement for two particles is given by

$y = 0.06\sin 2\pi(0.04t + \phi_1), y_2 = 0.03\sin 2\pi(1.04t + \phi_2)$

The ratio of the intensities of the waves produced by the vibrations of the two particles will be

(a) 2 : 1 (b) 1 : 2 (c) 4 : 1 (d) 1 : 4

Q.8 A transverse wave is described by the equation $Y = Y_0 \sin 2\pi\left(ft - \frac{x}{\lambda}\right)$. The maximum particle velocity is four times the wave velocity if

(a) $\lambda = \frac{\pi Y_0}{4}$ (b) $\lambda = \frac{\pi Y_0}{2}$

(c) $\lambda = \pi Y_0$ (d) $\lambda = 2\pi Y_0$

Q.9 Which one of the following does not represent a travelling wave?

(a) $y = \sin(x - vt)$ (b) $y = y_m \sin k(x + vt)$

(c) $y = y_m \log(x - vt)$ (d) $y = f(x^2 - vt^2)$

Q.10 The path difference between the two waves $y_1 = a_1 \sin\left(\omega t - \frac{2\pi x}{\lambda}\right)$ and $y_2 = a_2 \cos\left(\omega t - \frac{2\pi x}{\lambda} + \phi\right)$ is

(a) $\frac{\lambda}{2\pi}\phi$ (b) $\frac{\lambda}{2\pi}\left(\phi + \frac{\pi}{2}\right)$

(c) $\frac{2\pi}{\lambda}\left(\phi - \frac{\pi}{2}\right)$ (d) $\frac{2\pi}{\lambda}\phi$

Q.11 A transverse wave is represented by the equation

$y = y_0 \sin\frac{2\pi}{\lambda}(vt - x)$, Here v = wave velocity

For what value of λ, the maximum particle velocity equal to two times the wave velocity

(a) $\lambda = 2\pi y_0$ (b) $\lambda = \pi y_0 / 3$

(c) $\lambda = \pi y_0 / 2$ (d) $\lambda = \pi y_0$

Q.12 The equation of a plane progressive wave is given by $y = 0.025\sin(100t + 0.25x)$. The frequency of this wave would be

(a) $\frac{50}{\pi}$Hz (b) $\frac{100}{\pi}$Hz (c) 100Hz (d) 50Hz

Q.13 A wave travelling in positive X-direction with A = 0.2m has a velocity of 360 m/sec. If λ = 60m, then correct expression for the wave is

(a) $y = 0.2\sin\left[2\pi\left(6t + \frac{x}{60}\right)\right]$ (b) $y = 0.2\sin\left[\pi\left(6t + \frac{x}{60}\right)\right]$

(c) $y = 0.2\sin\left[2\pi\left(6t - \frac{x}{60}\right)\right]$ (d) $y = 0.2\sin\left[\pi\left(6t - \frac{x}{60}\right)\right]$

Q.14 The equation of a wave is given as $y = 0.07\sin(12\pi x - 3000\pi t)$. where x is in metre and t in sec, then the correct statement is

(a) $\lambda = 1/6$m, $v = 250$ m/s (b) $a = 0.07$m, $v = 300$m/s

(c) $n = 1500, v = 200$ m/s (d) None

Q.15 The equation of a progressive wave is given by $y = 0.5\sin(20x - 400t)$ where x and y are in metre and t is in second. The velocity of the wave is

(a) 10 m/s (b) 20 m/s (c) 200 m/s (d) 400 m/s

Q.16 There is a destructive interference between the two waves of wavelength λ coming from two different paths at a point. To get maximum sound or constructive interference at that point, the path of one wave is to be increased by

(a) $\frac{\lambda}{4}$ (b) $\frac{\lambda}{2}$

(c) $\frac{3\lambda}{4}$ (d) λ

Q.17 If two waves of same frequency and same amplitude on superimposition produced a resultant disturbance of the same amplitude, the waves differ in phase by

(a) π (b) $2\pi/3$ (c) $\pi / 2$ (d) zero

RESPONSE GRID					
	6. ⓐⓑⓒⓓ	7. ⓐⓑⓒⓓ	8. ⓐⓑⓒⓓ	9. ⓐⓑⓒⓓ	10. ⓐⓑⓒⓓ
	11. ⓐⓑⓒⓓ	12. ⓐⓑⓒⓓ	13. ⓐⓑⓒⓓ	14. ⓐⓑⓒⓓ	15. ⓐⓑⓒⓓ
	16. ⓐⓑⓒⓓ	17. ⓐⓑⓒⓓ			

Space for Rough Work

Q.18 Equation of motion in the same direction is given by $y_1 = A\sin(\omega t - kx)$, $y_2 = A\sin(\omega t - kx - \theta)$. The amplitude of the medium particle will be

(a) $2A\cos\frac{\theta}{2}$ (b) $2A\cos\theta$

(c) $\sqrt{2}A\cos\frac{\theta}{2}$ (d) $\sqrt{2}A\cos\theta$

Q.19 The amplitude of a wave, represented by displacement equation $y = \frac{1}{\sqrt{a}}\sin\omega t \pm \frac{1}{\sqrt{b}}\cos\omega t$ will be

(a) $\frac{a+b}{ab}$ (b) $\frac{\sqrt{a}+\sqrt{b}}{ab}$ (c) $\frac{\sqrt{a}\pm\sqrt{b}}{ab}$ (d) $\sqrt{\frac{a+b}{ab}}$

Q.20 The displacement due to a wave moving in the positive x-direction is given by $y = \frac{1}{(1+x^2)}$ at time $t = 0$ and by $y = \frac{1}{[1+(x-1)^2]}$ at $t = 2$ seconds, where x and y are in metres. The velocity of the wave in *m/s* is

(a) 0.5 (b) 1 (c) 2 (d) 4

DIRECTIONS (Q.21-Q.23) : In the following questions, more than one of the answers given are correct. Select the correct answers and mark it according to the following codes:

Codes : **(a)** 1, 2 and 3 are correct
(b) 1 and 2 are correct
(c) 2 and 4 are correct
(d) 1 and 3 are correct

Q.21 P, Q and R are three particles of a medium which lie on the x-axis. A sine wave of wavelength λ is travelling through the medium in the x-direction. P and Q always have the same speed, while P and R always have the same velocity. The minimum distance between –

(1) P and Q is λ (2) P and Q is $\lambda/2$
(3) P and R is $\lambda/2$ (4) P and R is λ

Q.22 A wave represented by the given equation

$Y = A\sin\left(10\pi x + 15\pi t + \frac{\pi}{3}\right)$, where x is in meter and t is in second. The expression represents

(1) A wave travelling in the negative X direction with a velocity of 1.5 m/sec
(2) A wave travelling in the negative X direction with a wavelength of 0.2 m
(3) A wave travelling in the positive X direction with a velocity of 1.5 m/sec.
(4) A wave travelling in the positive X direction with a wavelength of 0.2 m

Q.23 It is usually more convenient to describe a sound wave in terms of pressure wave as compared to displacement wave because –

(1) Two waves of same intensity but different frequencies have different displacement amplitude but same pressure amplitude
(2) The human ear responds to the change in pressure and not to the displacement wave.
(3) The electronic detector (microphone) does respond to the change in pressure but not to the displacement.
(4) None of the above

DIRECTIONS (Q.24-Q.26) : Read the passage given below and answer the questions that follows :

Sound from a point isotropic source spreads equally in all directions in homogeneous medium. Therefore its intensity decreases with square of distance from the source. When distance between observer and the source changes, apart from changes in intensity, the observer listens sound of pitch higher or lower than actual pitch depending upon the fact that the distance between the observer and source is decreasing or increasing respectively. An observer O is at a distance 2R from centre of a circle of radius R. A point isotropic sound source S moves on the circle with uniform angular velocity $\omega = \pi/3$ rad/s. Initially observer, source and centre of the circle are in same line.

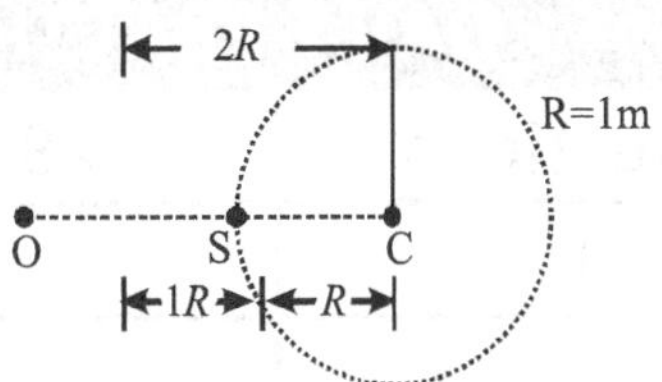

RESPONSE GRID	18. ⓐⓑⓒⓓ	19. ⓐⓑⓒⓓ	20. ⓐⓑⓒⓓ	21. ⓐⓑⓒⓓ	22. ⓐⓑⓒⓓ
	23. ⓐⓑⓒⓓ				

Space for Rough Work

Q.24 Starting from initial moment, the source moves through an angular displacement 180°. Intensity of the sound as observed by the observer decreases by a factor of –

(a) 2 (b) 3
(c) 4 (d) 9

Q.25 During a complete round trip of star on the circle, the observer listens a sound, whose –

(a) wavelength first decreases to a maximum value then increases to the original value
(b) wavelength first increases to a maximum value then decreases to the original value
(c) During the first half time wavelength increases then decreases to the original value
(d) None of the above is correct because in Doppler's effect, it is the pitch of sound which changes and not its wavelength, irrespective of motion of source or observer.

Q.26 Sound emitted by the source at two successive instants t_1 and t_2 has minimum and maximum observed pitch respectively, then –

(a) $t_1 = 1s, t_2 = 5s$
(b) $t_1 = 5s, t_2 = 7s$
(c) $t_1 = 7s, t_2 = 11s$
(d) $t_1 = 5s, t_2 = 11s$

DIRECTIONS (Q. 27-Q.29) : Each of these questions contains two statements: Statement-1 (Assertion) and Statement-2 (Reason). Each of these questions has four alternative choices, only one of which is the correct answer. You have to select the correct choice.

(a) Statement-1 is True, Statement-2 is True; Statement-2 is a correct explanation for Statement-1.
(b) Statement-1 is True, Statement-2 is True; Statement-2 is NOT a correct explanation for Statement-1.
(c) Statement -1 is False, Statement-2 is True.
(d) Statement -1 is True, Statement-2 is False.

Q.27 **Statement-1 :** Particle velocity and wave velocity both are independent of time.

Statement-2 : For the propagation of wave motion, the medium must have the properties of elasticity and inertia.

Q.28 **Statement-1 :** Speed of wave $= \frac{\text{Wavelength}}{\text{Time period}}$

Statement-2 : Wavelength is the distance between two nearest particles vibrating in phase.

Q.29 **Statement-1 :** Transverse waves are not produced in liquids and gases.

Statement-2 : Light waves are transverse waves.

RESPONSE GRID	
	24. ⓐⓑⓒⓓ 25. ⓐⓑⓒⓓ 26. ⓐⓑⓒⓓ 27. ⓐⓑⓒⓓ 28. ⓐⓑⓒⓓ
	29. ⓐⓑⓒⓓ

DAILY PRACTICE PROBLEM SHEET 29 - PHYSICS			
Total Questions	29	Total Marks	116
Attempted		Correct	
Incorrect		Net Score	
Cut-off Score	28	Qualifying Score	44
Success Gap = Net Score – Qualifying Score			
Net Score = (Correct × 4) – (Incorrect × 1)			

Space for Rough Work

DPP - Daily Practice Problems

Name :

Date :

Start Time :

End Time :

PHYSICS 30

SYLLABUS : Waves-2 (Standing waves in strings and organ pipes, Fundamental mode and harmonics, Beats, Doppler effect in sound)

Max. Marks : 120 **Time : 60 min.**

GENERAL INSTRUCTIONS

- The Daily Practice Problem Sheet contains 30 MCQ's. For each question only one option is correct. Darken the correct circle/ bubble in the Response Grid provided on each page.
- You have to evaluate your Response Grids yourself with the help of solution booklet.
- Each correct answer will get you 4 marks and 1 mark shall be deduced for each incorrect answer. No mark will be given/ deducted if no bubble is filled. Keep a timer in front of you and stop immediately at the end of 60 min.
- The sheet follows a particular syllabus. Do not attempt the sheet before you have completed your preparation for that syllabus. Refer syllabus sheet in the starting of the book for the syllabus of all the DPP sheets.
- After completing the sheet check your answers with the solution booklet and complete the Result Grid. Finally spend time to analyse your performance and revise the areas which emerge out as weak in your evaluation.

DIRECTIONS (Q.1-Q.22) : There are 22 multiple choice questions. Each question has 4 choices (a), (b), (c) and (d), out of which **ONLY ONE** choice is correct.

Q.1 A police car moving at 22 m/s, chases a motorcylist. The police man sounds his horn at 176 Hz, while both of them move towards a stationary siren of frequency 165 Hz. Calculate the speed of the motorcycle, if it is given that he does not observes any beats

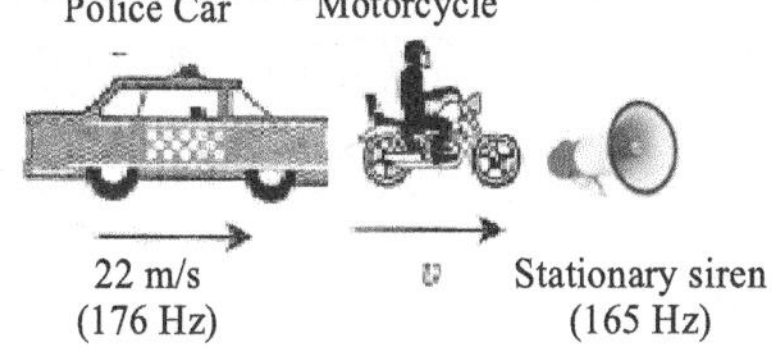

(a) 33 m/s (b) 22 m/s (c) Zero (d) 11 m/s

Q.2 A closed organ pipe of length L and an open organ pipe contain gases of densities ρ_1 and ρ_2 respectively. The compressibility of gases are equal in both the pipes. Both the pipes are vibrating in their first overtone with same frequency. The length of the open organ pipe is

(a) $\frac{L}{3}$ (b) $\frac{4L}{3}$ (c) $\frac{4L}{3}\sqrt{\frac{\rho_1}{\rho_2}}$ (d) $\frac{4L}{3}\sqrt{\frac{\rho_2}{\rho_1}}$

Q.3 Two whistles A and B produces notes of frequencies 660 Hz and 596 Hz respectively. There is a listener at the mid-point of the line joining them. Now the whistle B and the listener start moving with speed 30 m/s away from the whistle A. If speed of sound be 330 m/s, how many beats will be heard by the listener

(a) 2 (b) 4 (c) 6 (d) 8

RESPONSE GRID	1. ⓐⓑⓒⓓ	2. ⓐⓑⓒⓓ	3. ⓐⓑⓒⓓ

Space for Rough Work

Q.4 An open organ pipe is in resonance in its 2nd harmonic with tuning fork of frequency f_1. Now, it is closed at one end. If the frequency of the tuning fork is increased slowly from f_1 then again a resonance is obtained with a frequency f_2. If in this case the pipe vibrates n^{th} hamonics then

(a) $n = 3, f_2 = \frac{3}{4}f_1$ (b) $n = 3, f_2 = \frac{5}{4}f_1$

(c) $n = 5, f_2 = \frac{5}{4}f_1$ (d) $n = 5, f_2 = \frac{3}{4}f_1$

Q.5 The source producing sound and an observer both are moving along the direction of propagation of sound waves. If the respective velocities of sound, source and an observer are v, v_s and v_0, then the apparent frequency heard by the observer will be (n = frequency of sound)

(a) $\frac{n(v+v_0)}{v-v_0}$ (b) $\frac{n(v-v_0)}{v-v_s}$

(c) $\frac{n(v-v_0)}{v+v_s}$ (d) $\frac{n(v+v_0)}{v+v_s}$

Q.6 A whistle sends out 256 waves in a second. If the whistle approaches the observer with velocity 1/3 of the velocity of sound in air, the number of waves per second the observer will receive

(a) 384 (b) 192 (c) 300 (d) 200

Q.7 A source of sound emitting a note of frequency 200 Hz moves towards an observer with a velocity v equal to the velocity of sound. If the observer also moves away from the source with the same velocity v, the apparent frequency heard by the observer is

(a) 50 Hz (b) 100 Hz (c) 150 Hz (d) 200 Hz

Q.8 The speed of sound in air at a given temperature is 350 m/s. An engine blows whistle at a frequency of 1200 cps. It is approaching the observer with velocity 50 m/s. The apparent frequency in *cps* heard by the observer will be

(a) 600 (b) 1050 (c) 1400 (d) 2400

Q.9 A source of sound of frequency n is moving towards a stationary observer with a speed S. If the speed of sound in air is V and the frequency heard by the observer is n_1, the value of n_1 / n is

(a) (V + S)/V (b) V/ (V + S)

(c) (V – S)/V (d) V/ (V – S)

Q.10 An observer is moving away from source of sound of frequency 100 Hz. His speed is 33 m/s. If speed of sound is 330 m/s, then the observed frequency is

(a) 90 Hz (b) 100 Hz (c) 91 Hz (d) 110 Hz

Q.11 A whistle giving out 450 Hz approaches a stationary observer at a speed of 33 m/s. The frequency heard by the observer in Hz is

(a) 409 (b) 429 (c) 517 (d) 500

Q.12 Two sirens situated one kilometre apart are producing sound of frequency 330 Hz. An observer starts moving from one siren to the other with a speed of 2 m/s. If the speed of sound be 330 m/s, what will be the beat frequency heard by the observer

(a) 8 (b) 4 (c) 6 (d) 1

Q.13 A small source of sound moves on a circle as shown in the figure and an observer is standing on O. Let n_1, n_2 and n_3 be the frequencies heard when the source is at A, B and C respectively. Then

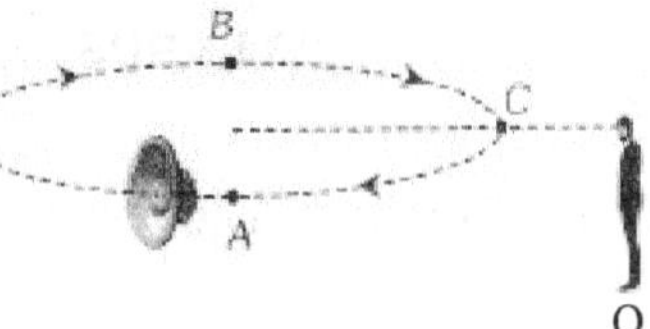

(a) $n_1 > n_2 > n_3$

(b) $n_2 > n_3 < n_1$

(c) $n_1 = n_2 > n_3$

(d) $n_2 > n_1 > n_3$

Q.14 A person carrying a whistle emitting continuously a note of 272 Hz is running towards a reflecting surface with a speed of 18 km/hour. The speed of sound in air is $345 ms^{-1}$. The number of beats heard by him is

(a) 4 (b) 6 (c) 8 (d) 3

Q.15 A source of sound of frequency 256 Hz is moving rapidly towards a wall with a velocity of 5m/s. The speed of sound is 330 m/s. If the observer is between the wall and the source, then the beats heard per second will be

(a) 7.8 Hz (b) 7.7 Hz

(c) 3.9 Hz (d) Zero

Q.16 The harmonics which are present in a pipe open at one end are

(a) odd harmonics

(b) even harmonics

(c) even as well as odd harmonics

(d) None of these

RESPONSE GRID					
	4. ⓐⓑⓒⓓ	5. ⓐⓑⓒⓓ	6. ⓐⓑⓒⓓ	7. ⓐⓑⓒⓓ	8. ⓐⓑⓒⓓ
	9. ⓐⓑⓒⓓ	10. ⓐⓑⓒⓓ	11. ⓐⓑⓒⓓ	12. ⓐⓑⓒⓓ	13. ⓐⓑⓒⓓ
	14. ⓐⓑⓒⓓ	15. ⓐⓑⓒⓓ	16. ⓐⓑⓒⓓ		

Space for Rough Work

Q.17 A source of sound placed at the open end of a resonance column sends an acoustic wave of pressure amplitude P_0 inside the tube. If the atmospheric pressure is P_A, then the ratio of maximum and minimum pressure at the closed end of the tube will be

(a) $\frac{(P_A+P_0)}{(P_A-P_0)}$ (b) $\frac{(P_A+2P_0)}{(P_A-2P_0)}$

(c) $\frac{P_A}{P_A}$ (d) $\frac{\left(P_A+\frac{1}{2}P_0\right)}{\left(P_A-\frac{1}{2}P_0\right)}$

Q.18 The frequency of fundamental tone in an open organ pipe of length 0.48 m is 320 Hz. Speed of sound is 320 m/sec. Frequency of fundamental tone in closed organ pipe will be

(a) 153.8 Hz (b) 160.0 Hz
(c) 320.0 Hz (d) 143.2 Hz

Q.19 A standing wave having 3 nodes and 2 antinodes is formed between two atoms having a distance 1.21Å between them. The wavelength of the standing wave is

(a) 1.21 Å (b) 2.42 Å (c) 6.05 Å (d) 3.63 Å

Q.20 A string on a musical instrument is 50 cm long and its fundamental frequency is 270 Hz. If the desired frequency of 1000 Hz, is to be produced, the required length of the string is

(a) 13.5 cm (b) 2.7 cm (c) 5.4 cm (d) 10.3 cm

Q.21 The loudness and the pitch of a sound depends on

(a) intensity and velocity
(b) frequency and velocity
(c) intensity and frequency
(d) frequency and number of harmonics

Q.22 If in an experiment for determination of velocity of sound by resonance tube method using a tuning fork of 512 Hz, first resonance was observed at 30.7 cm and second was obtainded at 63.2 cm, then maximum possible error in velocity of sound is (consider actual speed of sound in air is 332 m/s)

(a) 204 cm/sec (b) 110 cm/sec
(c) 58 cm/sec (d) 80 cm/sec

DIRECTIONS (Q.23-Q.25) : In the following questions, more than one of the answers given are correct. Select the correct answers and mark it according to the following codes:

Codes :

(a) 1, 2 and 3 are correct **(b)** 1 and 2 are correct
(c) 2 and 4 are correct **(d)** 1 and 3 are correct

Q.23 You are given four tuning forks, the lowest frequency of the fork is 300Hz. By striking two tuning forks at a time any of 1, 2, 3, 5, 7 & 8 Hz beat frequencies are heard. The possible frequencies of the other three forks –

(1) 301, 302 & 307 (2) 301, 303 & 308
(3) 300, 304 & 307 (4) 305, 307 & 308

Q.24 Doppler shift in frequency depends upon

(1) the frequency of the wave produced
(2) the velocity of the source
(3) the velocity of the observer
(4) distance from the source to the listener

Q.25 The (x, y) coordinates of the corners of a square plate are (0, 0), (L, 0), (L, L) and (0, L). The edges of the plate are clamped and transverse standing waves are set up in it. If $u(x, y)$ denotes the displacement of the plate at the point (x, y) at some instant of time, the possible expression(s) for u is (are) (a = positive constant)

(1) $a\sin\frac{\pi x}{L}\sin\frac{\pi y}{L}$ (2) $a\sin\frac{\pi x}{L}\sin\frac{2\pi y}{L}$

(3) $a\cos\frac{\pi x}{2L}\cos\frac{\pi y}{2L}$ (4) $a\cos\frac{2\pi x}{L}\cos\frac{\pi y}{L}$

DIRECTIONS (Q.26-Q.27) : Read the passage given below and answer the questions that follows :

A plate was cut from a quartz crystal and is used to control the frequency of an oscillating electrical circuit. Longitudinal standing waves are set up in the plate with displacement antinodes at opposite faces. The fundamental frequency of vibration is given by the equation $f_0 = \frac{2.87\times10^4}{s}$. Here s is thickness of the plate and density of quartz is 2658.76 kg/m^3.

RESPONSE GRID					
	17. ⓐⓑⓒⓓ	18. ⓐⓑⓒⓓ	19. ⓐⓑⓒⓓ	20. ⓐⓑⓒⓓ	21. ⓐⓑⓒⓓ
	22. ⓐⓑⓒⓓ	23. ⓐⓑⓒⓓ	24. ⓐⓑⓒⓓ	25. ⓐⓑⓒⓓ	

Space for Rough Work

Q.26 Young's modulus of elasticity for quartz is –

(a) 7×10^{11} N/m^2 (b) 8.76×10^{12} N/m^2
(c) 2×10^{12} N/m^2 (d) Information insufficient

Q.27 If the quartz plate is vibrating in 3rd harmonic while measuring the frequency of 1.2×10^6 Hz, then the thickness of the plateis

(a) 71.75cm (b) 7.175 cm
(c) 6.02 cm (d) 0.07 cm

DIRECTIONS (Q.28-Q.30) : Each of these questions contains two statements: Statement-1 (Assertion) and Statement-2 (Reason). Each of these questions has four alternative choices, only one of which is the correct answer. You have to select the correct choice.

Q.28 Statement-1 : Beats cannot be produced by light sources.

Statement-2: Light sources have constant phase difference.

Q.29 Statement-1 : In the case of a stationary wave, a person hear a loud sound at the nodes as compared to the antinodes.

Statement-2 : In a stationary wave all the particles of the medium vibrate in phase.

Q.30 Statement-1 : Velocity of particles, while crossing mean position (in stationary waves) varies from maximum at antinodes to zero at nodes.

Statement-2: Amplitude of vibration at antinodes is maximum and at nodes, the amplitude is zero, and all particles between two successive nodes cross the mean position together.

RESPONSE GRID	26. ⓐⓑⓒⓓ	27. ⓐⓑⓒⓓ	28. ⓐⓑⓒⓓ	29. ⓐⓑⓒⓓ	30. ⓐⓑⓒⓓ

DAILY PRACTICE PROBLEM SHEET 30 - PHYSICS			
Total Questions	30	Total Marks	120
Attempted		Correct	
Incorrect		Net Score	
Cut-off Score	30	Qualifying Score	48
Success Gap = Net Score – Qualifying Score			
Net Score = (Correct × 4) – (Incorrect × 1)			

Space for Rough Work

DPP - Daily Practice Problems

Name : 　　　　　　　　　　Date :

Start Time : 　　　　　　　　　　End Time :

PHYSICS 31

SYLLABUS : Practical Physics - 1

Max. Marks : 120 　　　　　　　　　　**Time : 60 min.**

GENERAL INSTRUCTIONS

- The Daily Practice Problem Sheet contains 30 MCQ's. For each question only one option is correct. Darken the correct circle/ bubble in the Response Grid provided on each page.
- You have to evaluate your Response Grids yourself with the help of solution booklet.
- Each correct answer will get you 4 marks and 1 mark shall be deduced for each incorrect answer. No mark will be given/ deducted if no bubble is filled. Keep a timer in front of you and stop immediately at the end of 60 min.
- The sheet follows a particular syllabus. Do not attempt the sheet before you have completed your preparation for that syllabus. Refer syllabus sheet in the starting of the book for the syllabus of all the DPP sheets.
- After completing the sheet check your answers with the solution booklet and complete the Result Grid. Finally spend time to analyse your performance and revise the areas which emerge out as weak in your evaluation.

DIRECTIONS (Q.1-Q.24) : There are 24 multiple choice questions. Each question has 4 choices (a), (b), (c) and (d), out of which **ONLY ONE** choice is correct.

Q.1 One cm on the main scale of a vernier callipers is divided into ten equal parts. If 20 divisions of vernier scale coincide with 8 small divisions of the main scale. What will be the least count of callipers ?

(a) 0.05 cm 　　(b) 0.06 cm
(c) 0.04 cm 　　(d) 0.01 cm

Q.2 The shape of stress vs strain graph within elastic limit is :

(a) parabolic 　　(b) curve line
(c) straight line 　　(d) ellipse

Q.3 In a vernier calliper N divisions of vernier scale coincides with N – 1 divisions of main scale (in which length of one division is 1 mm). The least count of the instrument should be

(a) N 　　(b) N – 1 　　(c) 1/10 N 　　(d) 1/N – 1

Q.4 The figure shows a situation when the jaws of vernier are touching each other. Each main scale division is of 1 mm. Find zero correction.

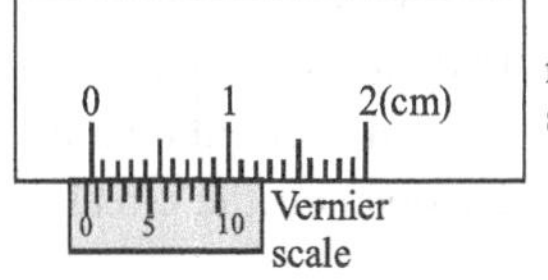

(a) – 0.5 mm
(b) + 0.5 mm
(c) – 0.4 mm
(d) + 0.4 mm

Q.5 In an experiment for measurement of young's modulus, following readings are taken. Load = 3.00 kg, length = 2.820 m, diameter = 0.041 cm and extension = 0.87. Determine the percentage error in the measurement of Y.

(a) ±5% 　　(b) ±6.5%
(c) ±5.5% 　　(d) ±15%

RESPONSE GRID	1. ⓐⓑⓒⓓ	2. ⓐⓑⓒⓓ	3. ⓐⓑⓒⓓ	4. ⓐⓑⓒⓓ	5. ⓐⓑⓒⓓ

Space for Rough Work

Q.6 When the zero of the circular scale of a screw gauge coincides with the zero of the main scale before A and B come in contact then the instrument has

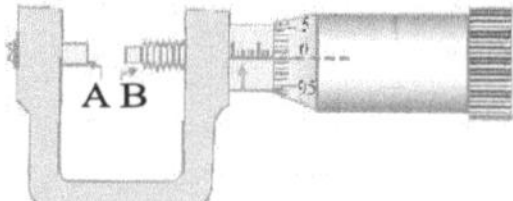

(a) positive zero error
(b) negative zero error
(c) no zero error
(d) can't be said anything

Q.7 If h be the elevation or depression of a spherical surface from the plane glass plate and c be the mean distance between two consecutive points corresponding to the impressions made by the three legs of a spherometer then the radius of curvature is

(a) $\frac{c^2}{6h} - \frac{h}{2}$ (b) $\frac{c^2}{6h} + \frac{h^2}{2}$ (c) $\frac{c^2}{6h} + \frac{h}{2}$ (d) $\frac{c^2}{6h} + \frac{2}{h}$

Q.8 The least count of a spherometer is given by
(a) pitch × no. of circular divisions
(b) $\frac{\text{pitch}}{\text{no. of circular divisions}}$
(c) $\frac{\text{no. of circular divisions}}{\text{pitch}}$
(d) $\frac{\text{pitch}}{\text{mean distance between two consecutive legs of the spherometer}}$

Q.9 The specific heat of a solid is determined by the method known as
(a) the method of fusion
(b) the method of mixture
(c) the method of vaporisation
(d) the method of cooling

Q.10 Which principle is involved in the experiment to determine the specific heat of a liquid by the method of mixture ?
(a) Heat gained by solid = Heat lost by calorimeter and liquid.
(b) Heat lost by solid = Heat gained by calorimeter and liquid.
(c) Heat lost by solid and liquid = Heat gained by calorimeter.
(d) Heat gained by solid and calorimeter = Heat lost by liquid.

Q.11 Two full turns of the circular scale of a screw gauge cover a distance of 1mm on its main scale. The total number of divisions on the circular scale is 50. Further, it is found that the screw gauge has a zero error of – 0.03 mm. While measuring the diameter of a thin wire, a student notes the main scale reading of 3 mm and the number of circular scale divisions in line with the main scale as 35. The diameter of the wire is
(a) 3.32 mm (b) 3.73 mm
(c) 3.67 mm (d) 3.38 mm

Q.12 In an experiment the angles are required to be measured using an instrument, 29 divisions of the main scale exactly coincide with the 30 divisions of the vernier scale. If the smallest division of the main scale is half- a degree (= 0.5°), then the least count of the instrument is :
(a) half minute (b) one degree
(c) half degree (d) one minute

Q.13 In a screw gauge, the zero of main scale coincides with fifth division of circular scale in figure (i). The circular divisions of screw gauge are 50. It moves 0.5 mm on main scale in one rotation. The diameter of the ball in figure (ii) is

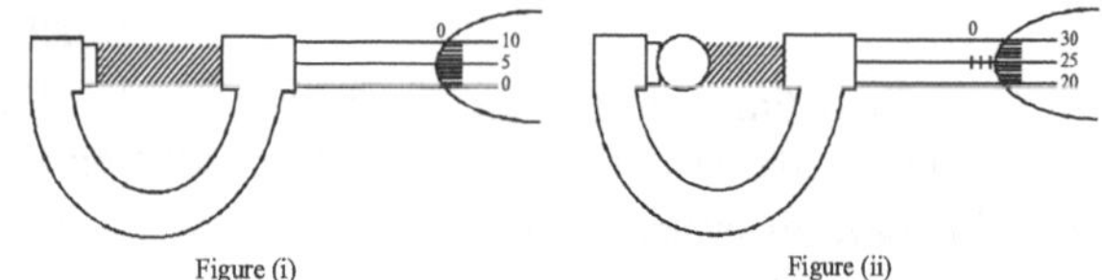

Figure (i) Figure (ii)

(a) 2.25 mm (b) 2.20 mm
(c) 1.20 mm (d) 1.25 mm

Q.14 A student performs an experiment for determination of $g\left(=\frac{4\pi^2\ell}{T^2}\right)$. The error in length ℓ is $\Delta\ell$ and in time T is ΔT and n is number of times the reading is taken. The measurement of g is most accurate for

	$\Delta\ell$	ΔT	n
(a)	5 mm	0.2 sec	10
(b)	5 mm	0.2 sec	20
(c)	5 mm	0.1 sec	10
(d)	1 mm	0.1 sec	50

RESPONSE GRID					
	6. ⓐⓑⓒⓓ	7. ⓐⓑⓒⓓ	8. ⓐⓑⓒⓓ	9. ⓐⓑⓒⓓ	10. ⓐⓑⓒⓓ
	11. ⓐⓑⓒⓓ	12. ⓐⓑⓒⓓ	13. ⓐⓑⓒⓓ	14. ⓐⓑⓒⓓ	

Space for Rough Work

Q.15 A student performs an experiment to determine the Young's modulus of a wire, exactly 2 m long, by Searle's method. In a particular reading, the student measures the extension in the length of the wire to be 0.8 mm with an uncertainty of ± 0.05 mm at a load of exactly 1.0 kg. The student also measures the diameter of the wire to be 0.4 mm with an uncertainty of ± 0.01 mm. Take $g = 9.8$ m/s^2 (exact). The Young's modulus obtained from the reading is

(a) $(2.0 \pm 0.3) \times 10^{11}$ N/m^2 (b) $(2.0 \pm 0.2) \times 10^{11}$ N/m^2
(c) $(2.0 \pm 0.1) \times 10^{11}$ N/m^2 (d) $(2.0 \pm 0.05) \times 10^{11}$ N/m^2

Q.16 Students I, II and III perform an experiment for measuring the acceleration due to gravity (g) using a simple pendulum. They use different lengths of the pendulum and /or record time for different number of oscillations. The observations are shown in the table.

Least count for length = 0.1 cm
Least count for time = 0.1 s

Student	Length of the pendulum (cm)	No. of oscillations (n)	Total time for (n) oscillations (s)	Time period (s)
I	64.0	8	128.0	16.0
II	64.0	4	64.0	16.0
III	20.0	4	36.0	9.0

If E_I, E_{II} and E_{III} are the percentage errors in g, i.e., $\left(\frac{\Delta g}{g} \times 100\right)$ for students I, II and III, respectively, then

(a) $E_I = 0$ (b) E_I is minimum
(c) $E_I = E_{II}$ (d) E_{II} is maximum

Q.17 If the terminal speed of a sphere of gold (density = 19.5 kg/m^3) is 0.2 m/s in a viscous liquid (density = 1.5 kg/m^3), find the terminal speed of a sphere of silver (density = 10.5 kg/m^3) of the same size in the same liquid

(a) 0.4 m/s (b) 0.133 m/s
(c) 0.1 m/s (d) 0.2 m/s

Q.18 A spherical solid ball of volume V is made of a material of density ρ_1. It is falling through a liquid of density ρ_2 ($\rho_2 < \rho_1$). Assume that the liquid applies a viscous force on the ball that is proportional to the square of its speed v, i.e., $F_{viscous} = -kv^2$ ($k > 0$). The terminal speed of the ball is

(a) $\sqrt{\frac{Vg(\rho_1 - \rho_2)}{k}}$ (b) $\frac{Vg\rho_1}{k}$
(c) $\sqrt{\frac{Vg\rho_1}{k}}$ (d) $\frac{Vg(\rho_1 - \rho_2)}{k}$

Q.19 A jar is filled with two non-mixing liquids 1 and 2 having densities ρ_1 and, ρ_2 respectively. A solid ball, made of a material of density ρ_3, is dropped in the jar. It comes to equilibrium in the position shown in the figure. Which of the following is true for ρ_1, ρ_2 and ρ_3?

Liquid 1: ρ_1
Liquid 2: ρ_2

(a) $\rho_3 < \rho_1 < \rho_2$
(b) $\rho_1 > \rho_3 > \rho_2$
(c) $\rho_1 < \rho_2 < \rho_3$
(d) $\rho_1 < \rho_3 < \rho_2$

Q.20 A capillary tube (A) is dipped in water. Another identical tube (B) is dipped in a soap-water solution. Which of the following shows the relative nature of the liquid columns in the two tubes?

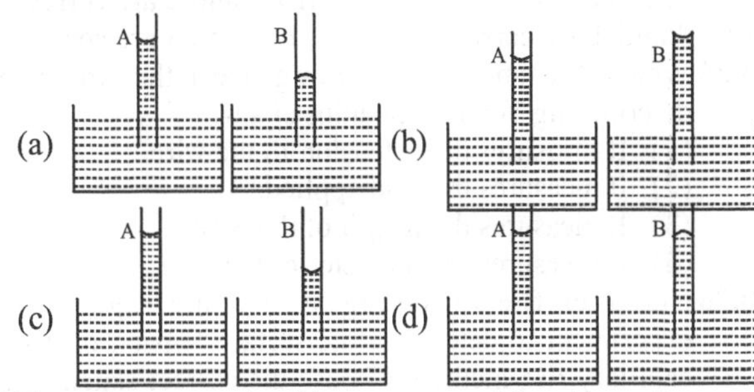

Q.21 Two wires are made of the same material and have the same volume. However wire 1 has cross-sectional area A and wire 2 has cross-sectional area $3A$. If the length of wire 1 increases by Δx on applying force F, how much force is needed to stretch wire 2 by the same amount of energy?

(a) 4 F (b) 6 F (c) 9 F (d) 1 F

Q.22 The vernier constant of two vernier callipers A and B are 0.01 cm and 0.01 mm respectively. Which one can measure the length of an object more accurately?

(a) Vernier A (b) Vernier B
(c) Accuracy in measurement does not depend on vernier constant
(d) Both A and B are equally accurate.

RESPONSE GRID					
	15. ⓐⓑⓒⓓ	16. ⓐⓑⓒⓓ	17. ⓐⓑⓒⓓ	18. ⓐⓑⓒⓓ	19. ⓐⓑⓒⓓ
	20. ⓐⓑⓒⓓ	21. ⓐⓑⓒⓓ	22. ⓐⓑⓒⓓ		

Space for Rough Work

Q.23 The acceleration due to gravity at a place can be determined with the help of a simple pendulum. For this purpose effective length of the pendulum is considered. If 'ℓ' be the length of the string and 'd' the diameter of the bob then the effective length is equal to

(a) $\ell + d$ (b) $\ell + \frac{d}{2}$ (c) $\ell - \frac{d}{2}$ (d) $\ell - d$

Q.24 If x, y, p and q represent the increase in length, the original length of the experimental wire, load applied to the wire and area of cross-section of the wire respectively then Young's modulus of the wire is given by

(a) $\frac{xy}{pq}$ (b) $\frac{xp}{yq}$ (c) $\frac{py}{xq}$ (d) $\frac{pq}{xy}$

DIRECTIONS (Q.25-Q.27) : In the following questions, more than one of the answers given are correct. Select the correct answers and mark it according to the following codes:

Codes :

(a) 1, 2 and 3 are correct **(b)** 1 and 2 are correct
(c) 2 and 4 are correct **(d)** 1 and 3 are correct

Q.25 What is the function of a screw gauge in the experiment of determining Young's modulus of a wire ?
(1) It measures the extension in the wire.
(2) It measures the load applied.
(3) It measures the length of the wire.
(4) It measures the diameter of the wire.

Q.26 Consider the following statements regarding the experiment to determine the surface tension of water by capillary rise method. Choose the correct statements.
(1) Capillary tube should be clean and liquid should be free from dirt and grease.
(2) Distilled water should be avoided.
(3) Distilled water should be added.
(4) Dirty liquid should be used.

Q.27 The temperature-time variation graphs, as obtained by four students 1, 2, 3 and 4 are as shown. The graphs, likely to be wrong are

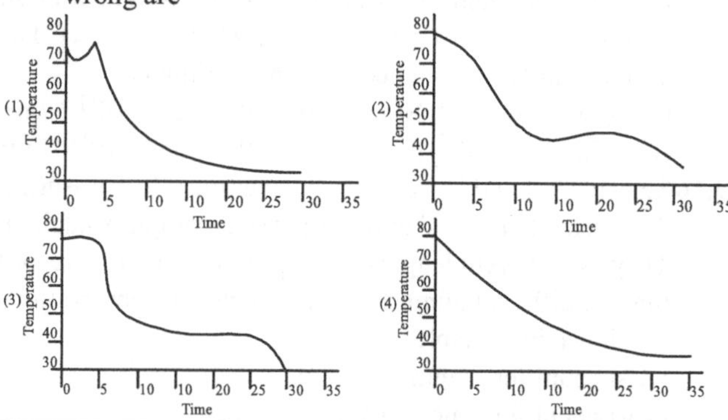

DIRECTIONS (Q.28-Q.30) : Read the passage given below and answer the questions that follows :

The internal radius of a 1m long resonance tube is measured as 3 cm. A tuning fork of frequency 2000 Hz is used. The first resonating length is measured as 4.6 cm and the second resonating length is measured as 14.0 cm.

Q.28 Calculate the maximum percentage error in measurement of e.
(a) 3.33% (b) 2.23% (c) 4.33% (d) 5.33%

Q.29 Calculate the speed of sound at the room temperature.
(a) 275 m/s (b) 376 m/s (c) 356 m/s (d) 330 m/s

Q.30 Calculate the end correction.
(a) 0.2 cm (b) 0.3 cm (c) 0.1 cm (d) 0.4 cm

RESPONSE GRID					
	23. ⓐⓑⓒⓓ	24. ⓐⓑⓒⓓ	25. ⓐⓑⓒⓓ	26. ⓐⓑⓒⓓ	27. ⓐⓑⓒⓓ
	28. ⓐⓑⓒⓓ	29. ⓐⓑⓒⓓ	30. ⓐⓑⓒⓓ		

DAILY PRACTICE PROBLEM SHEET 31 - PHYSICS			
Total Questions	30	Total Marks	120
Attempted		Correct	
Incorrect		Net Score	
Cut-off Score	30	Qualifying Score	48
Success Gap = Net Score – Qualifying Score			
Net Score = (Correct × 4) – (Incorrect × 1)			

Space for Rough Work

DPP - Daily Practice Problems

Name : ______________________ **Date :** __________

Start Time : __________ **End Time :** __________

PHYSICS 32

SYLLABUS : Electrostatics-1 (Coulomb's law, electric field, field lines, Gauss's law)

Max. Marks : 104 **Time : 60 min.**

GENERAL INSTRUCTIONS

- The Daily Practice Problem Sheet contains 26 MCQ's. For each question only one option is correct. Darken the correct circle/ bubble in the Response Grid provided on each page.
- You have to evaluate your Response Grids yourself with the help of solution booklet.
- Each correct answer will get you 4 marks and 1 mark shall be deduced for each incorrect answer. No mark will be given/ deducted if no bubble is filled. Keep a timer in front of you and stop immediately at the end of 60 min.
- The sheet follows a particular syllabus. Do not attempt the sheet before you have completed your preparation for that syllabus. Refer syllabus sheet in the starting of the book for the syllabus of all the DPP sheets.
- After completing the sheet check your answers with the solution booklet and complete the Result Grid. Finally spend time to analyse your performance and revise the areas which emerge out as weak in your evaluation.

DIRECTIONS (Q.1-Q.18) : There are 18 multiple choice questions. Each question has 4 choices (a), (b), (c) and (d), out of which **ONLY ONE** choice is correct.

Q.1 A total charge Q is broken in two parts Q_1 and Q_2 and they are placed at a distance R from each other. The maximum force of repulsion between them will occur, when

(a) $Q_2 = \frac{Q}{R}, Q_1 = Q - \frac{Q}{R}$ (b) $Q_2 = \frac{Q}{4}, Q_1 = Q - \frac{2Q}{3}$

(c) $Q_2 = \frac{Q}{4}, Q_1 = \frac{3Q}{4}$ (d) $Q_1 = \frac{Q}{2}, Q_2 = \frac{Q}{2}$

Q.2 Two small balls each having the charge $+Q$ are suspended by insulating threads of length L from a hook. This arrangement is taken in space where there is no gravitational effect, then the angle between the two suspensions and the tension in each thread will be

(a) $180°, \frac{1}{4\pi \in_0} \frac{Q^2}{(2L)^2}$ (b) $90°, \frac{1}{4\pi \in_0} \frac{Q^2}{L^2}$

(c) $180°, \frac{1}{4\pi \in_0} \frac{Q^2}{2L^2}$ (d) $180°, \frac{1}{4\pi \in_0} \frac{Q^2}{L^2}$

Q.3 Electric charges of $1\mu C, -1\mu C$ and $2\mu C$ are placed in air at the corners A,B and C respectively of an equilateral triangle ABC having length of each side 10 cm. The resultant force on the charge at C is

(a) 0.9 N (b) 1.8N (c) 2.7 N (d) 3.6 N

Q.4 An electron is moving round the nucleus of a hydrogen atom in a circular orbit of radius r. The coulomb force $\vec{F}$ between the two is (here $K = \frac{1}{4\pi \in_0}$)

(a) $-K\frac{e^2}{r^3}\hat{r}$ (b) $K\frac{e^2}{r^3}\vec{r}$ (c) $-K\frac{e^2}{r^3}\vec{r}$ (d) $K\frac{e^2}{r^2}\hat{r}$

Q.5 Equal charges q are placed at the four corners A, B, C, D of a square of length a. The magnitude of the force on the charge

RESPONSE GRID	1. ⓐⓑⓒⓓ	2. ⓐⓑⓒⓓ	3. ⓐⓑⓒⓓ	4. ⓐⓑⓒⓓ

Space for Rough Work

at B will be

(a) $\frac{3q^2}{4\pi \in_0 a^2}$ (b) $\frac{q^2}{\pi \in_0 a^2}$

(c) $\left(\frac{1+2\sqrt{2}}{2}\right)\frac{q^2}{4\pi \in_0 a^2}$ (d) $\left(2+\frac{1}{\sqrt{2}}\right)\frac{q^2}{4\pi \in_0 a^2}$

Q.6 The charges on two spheres are $+7\mu C$ and $-5\mu C$ respectively. They experience a force F. If each of them is given an additional charge of $-2\mu C$, the new force of attraction will be

(a) F (b) F/2 (c) $F/\sqrt{3}$ (d) 2F

Q.7 Electric lines of force about negative point charge are

(a) Circular, anticlockwise (b) Circular, clockwise
(c) Radial, inward (d) Radial, outward

Q.8 Figure shows the electric lines of force emerging from a charged body. If the electric field at A and B are E_A and E_B respectively and if the distance between A and B is r then

(a) $E_A > E_B$

(b) $E_A < E_B$

(c) $E_A = \frac{E_B}{r}$

(d) $E_A = \frac{E_B}{r^2}$

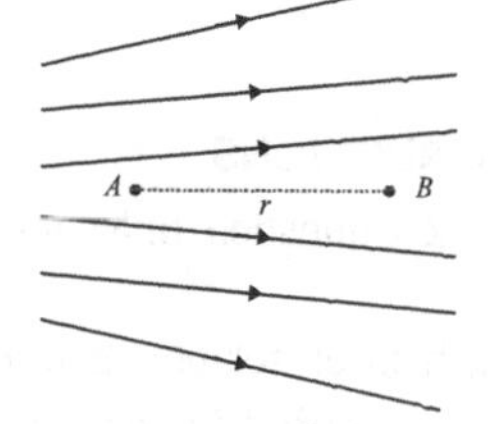

Q.9 The magnitude of electric field intensity E is such that, an electron placed in it would experience an electrical force equal to its weight is given by

(a) mge (b) $\frac{mg}{e}$ (c) $\frac{e}{mg}$ (d) $\frac{e^2}{m^2}g$

Q.10 A charge particle is free to move in an electric field. It will travel

(a) Always along a line of force
(b) Along a line of force, if its initial velocity is zero
(c) Along a line of force, if it has some initial velocity in the direction of an acute angle with the line of force
(d) None of the above

Q.11 Two point charges Q and $-3Q$ are placed at some distance apart. If the electric field at the location of Q is E then at the locality of $-3Q$, it is

(a) $-E$ (b) $E/3$ (c) $-3E$ (d) $-E/3$

Q.12 Charges q, $2q$, $3q$ and $4q$ are placed at the corners A, B, C and D of a square as shown in the following figure. The direction of electric field at the centre of the square is parallel to side.

(a) AB
(b) CB
(c) BD
(d) AC

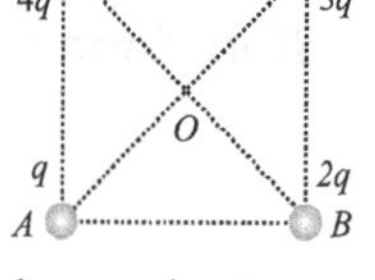

Q.13 Three infinitely long non-conducting charge sheets are placed as shown in figure. The electric field at point P is

(a) $\frac{2\sigma}{\varepsilon_0}\hat{k}$

(b) $-\frac{2\sigma}{\varepsilon_0}\hat{k}$

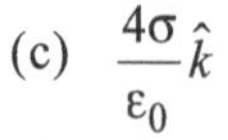
(c) $\frac{4\sigma}{\varepsilon_0}\hat{k}$

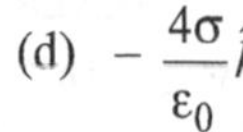
(d) $-\frac{4\sigma}{\varepsilon_0}\hat{k}$

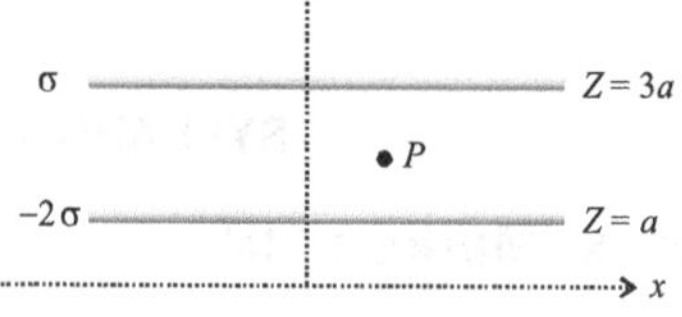

Q.14 Gauss's law is true only if force due to a charge varies as

(a) r^{-1} (b) r^{-2} (c) r^{-3} (d) r^{-4}

Q.15 The electric intensity due to an infinite cylinder of radius R and having charge q per unit length at a distance $r(r > R)$ from its axis is

(a) Directly proprotional to r^2
(b) Directly proprotional to r^3
(c) Inversely proprotional to r
(d) Inversely proprotional to r^2

Q.16 A sphere of radius R has a uniform distribution of electric charge in its volume. At a distance x from its centre, for $x < R$, the eletric field is directly proportional to

(a) $\frac{1}{x^2}$ (b) $\frac{1}{x}$ (c) x (d) x^2

Q.17 A charged ball B hangs from a silk thread S, which makes an angle θ with a large charged conducting sheet P, as shown

RESPONSE GRID					
	5. ⓐⓑⓒⓓ	6. ⓐⓑⓒⓓ	7. ⓐⓑⓒⓓ	8. ⓐⓑⓒⓓ	9. ⓐⓑⓒⓓ
	10. ⓐⓑⓒⓓ	11. ⓐⓑⓒⓓ	12. ⓐⓑⓒⓓ	13. ⓐⓑⓒⓓ	14. ⓐⓑⓒⓓ
	15. ⓐⓑⓒⓓ	16. ⓐⓑⓒⓓ			

Space for Rough Work

in the figure. The surface charge density σ of the sheet is proportional to

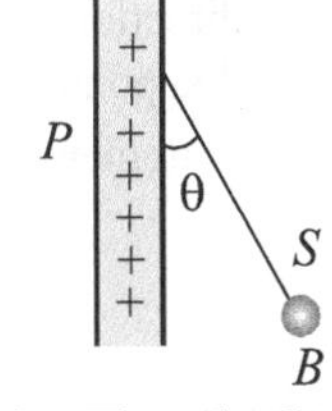

(a) $\sin\theta$
(b) $\tan\theta$
(c) $\cos\theta$
(d) $\cot\theta$

Q.18 A charge q is placed at the centre of a cube. Then the flux passing through one face of cube will be

(a) $\frac{q}{\epsilon_0}$ (b) $\frac{q}{2\epsilon_0}$ (c) $\frac{q}{4\epsilon_0}$ (d) $\frac{q}{6\epsilon_0}$

DIRECTIONS (Q.19-Q.20) : In the following questions, more than one of the answers given are correct. Select the correct answers and mark it according to the following codes:

Codes :
(a) 1, 2 and 3 are correct **(b)** 1 and 2 are correct
(c) 2 and 4 are correct **(d)** 1 and 3 are correct

Q.19 A solid sphere S_1 is connected to a charge reservoir through a heater H as shown in figure.

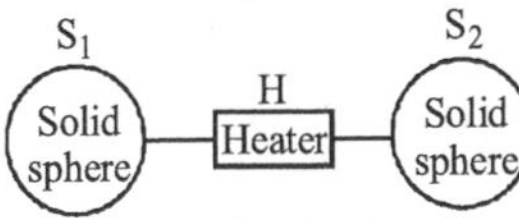

Flux through a closed spherical surface around S_1 is given by $\phi = \alpha t^2$ where α is a constant and t is time in seconds. If resistance of heater is R then select correct statements
(1) Power consumed by heater will be $4\alpha^2\varepsilon_0^2 Rt^2$.
(2) Electric flux through a closed spherical surface around S_2 will be $-\alpha t^2$.
(3) Rate of change of electric flux through a closed spherical surface around S_2 will be $-2\alpha t$
(4) All of the above are correct

Q.20 A simple pendulum has a time period T. The bob is now given some positive charge –
(1) If some positive charge is placed at the point of suspension, T will increases
(2) If some positive charge is placed at the point of suspension, T will not change
(3) If a uniform downward electric field is switched on, T will increase
(4) If a uniform downward electric field is switched on, T will decrease

DIRECTIONS (Q.21-Q.23) : Read the passage given below and answer the questions that follows :

A sphere of radius R contains charge density $\rho(r) = A(R - r)$, for $0 < r < R$. The total electric charge inside the sphere is Q.

Q.21 The value of A in terms of Q and R is

(a) $\frac{2Q^2}{\pi R^4}$ (b) $\frac{3Q}{\pi R^4}$ (c) $\frac{3Q^2}{\pi R^3}$ (d) $\frac{3Q}{\pi R}$

Q.22 The electric field inside the sphere is

(a) $\frac{3Q}{\epsilon_0 R^2}\left[\frac{1}{3}\left(\frac{r}{R}\right) - \frac{1}{4}\left(\frac{r}{R}\right)^2\right]$ (b) $\frac{12Q^2}{R^3}\left[\frac{1}{3}\left(\frac{r}{R}\right) - \frac{1}{4}\left(\frac{r}{R}\right)^2\right]$

(c) $\frac{120Q}{5\epsilon_0 R^2}\left[\frac{1}{4}\left(\frac{r}{R}\right) - \frac{1}{3}\left(\frac{r}{R}\right)^2\right]$ (d) $\frac{12}{R^2 Q}\left[\frac{1}{3}\left(\frac{r}{R}\right) - \frac{1}{4}\left(\frac{r}{R}\right)^2\right]$

Q.23 The electric field outside the sphere is $\left(k = \frac{1}{4\pi\epsilon_0}\right)$

(a) $\frac{kQ}{r}$ (b) $\frac{kQ}{r^2}$ (c) $\frac{kQ}{r^3}$ (d) $\frac{kQ^2}{r^2}$

DIRECTIONS (Q. 24-Q.26) : Each of these questions contains two statements: Statement-1 (Assertion) and Statement-2 (Reason). Each of these questions has four alternative choices, only one of which is the correct answer. You have to select the correct choice.

(a) Statement-1 is True, Statement-2 is True; Statement-2 is a correct explanation for Statement-1.
(b) Statement-1 is True, Statement-2 is True; Statement-2 is NOT a correct explanation for Statement-1.
(c) Statement -1 is False, Statement-2 is True.
(d) Statement -1 is True, Statement-2 is False.

Q.24 Statement-1 : Electric lines of force cross each other.
Statement-2 : Electric field at a point superimpose to give

RESPONSE GRID	17. ⓐⓑⓒⓓ	18. ⓐⓑⓒⓓ	19. ⓐⓑⓒⓓ	20. ⓐⓑⓒⓓ	21. ⓐⓑⓒⓓ
	22. ⓐⓑⓒⓓ	23. ⓐⓑⓒⓓ			

one resultant electric field.

Q.25 Statement-1 : A point charge is brought in an electric field. The field at a nearby point will increase, whatever be the nature of the charge.

Statement-2 : The electric field is dependent on the nature of charge.

Q.26 Statement-1 : Direction of electric field at a point signifies direction of force experienced by a point charge placed at that point.

Statement-2 : Electric field is a vector quantity.

RESPONSE GRID	24. ⓐⓑⓒⓓ	25. ⓐⓑⓒⓓ	26. ⓐⓑⓒⓓ

DAILY PRACTICE PROBLEM SHEET 32 - PHYSICS			
Total Questions	26	Total Marks	104
Attempted		Correct	
Incorrect		Net Score	
Cut-off Score	25	Qualifying Score	40
Success Gap = Net Score – Qualifying Score			
Net Score = (Correct × 4) – (Incorrect × 1)			

Space for Rough Work

DPP - Daily Practice Problems

Name : 　　　　　　Date :

Start Time : 　　　　　　End Time :

PHYSICS 33

SYLLABUS : Electrostatics-2 (Electric potential and potential difference, equipotential surfaces, electric dipole)

Max. Marks : 104 　　　　　　**Time : 60 min.**

GENERAL INSTRUCTIONS

- The Daily Practice Problem Sheet contains 26 MCQ's. For each question only one option is correct. Darken the correct circle/ bubble in the Response Grid provided on each page.
- You have to evaluate your Response Grids yourself with the help of solution booklet.
- Each correct answer will get you 4 marks and 1 mark shall be deduced for each incorrect answer. No mark will be given/ deducted if no bubble is filled. Keep a timer in front of you and stop immediately at the end of 60 min.
- The sheet follows a particular syllabus. Do not attempt the sheet before you have completed your preparation for that syllabus. Refer syllabus sheet in the starting of the book for the syllabus of all the DPP sheets.
- After completing the sheet check your answers with the solution booklet and complete the Result Grid. Finally spend time to analyse your performance and revise the areas which emerge out as weak in your evaluation.

DIRECTIONS (Q.1-Q.19) : There are 19 multiple choice questions. Each question has 4 choices (a), (b), (c) and (d), out of which ONLY ONE choice is correct.

Q1. In the electric field of a point charge q, a certain charge is carried from point A to B, C, D and E. Then the work done by electric force is

(a) least along the path AB

(b) least along the path AD

(c) zero along all the paths AB, AC, AD and AE

(d) least along AE

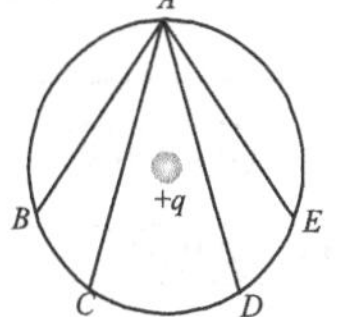

Q2. Four equal charges Q are placed at the four corners of a square of each side 'a'. Work done in removing a charge $-Q$ from its centre to infinity is

(a) 0 (b) $\frac{\sqrt{2}Q^2}{4\pi\epsilon_0 a}$ (c) $\frac{\sqrt{2}Q^2}{\pi\epsilon_0 a}$ (d) $\frac{Q^2}{2\pi\epsilon_0 a}$

Q3. A particle A has charge $+q$ and a particle B has charge $+4q$ with each of them having the same mass m. When allowed to fall from rest through the same electric potential difference, the ratio of their speed $\frac{v_A}{v_B}$ will become

(a) 2 : 1 (b) 1 : 2

(c) 1 : 4 (d) 4 : 1

RESPONSE GRID	1. ⓐⓑⓒⓓ	2. ⓐⓑⓒⓓ	3. ⓐⓑⓒⓓ

Space for Rough Work

Q4. In the figure the charge Q is at the centre of the circle. Work done (by electrostatic force on q) is maximum when another charge q is taken from point P to (consider both the charges to be positive)

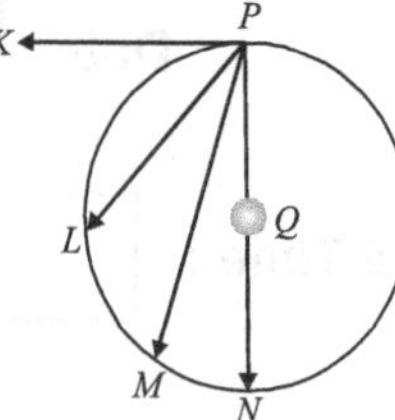

(a) K (b) L (c) M (d) N

Q5. How much kinetic energy will be gained by an α- particle in going from a point at 70V to another point at 50V ?
(a) 40 eV (b) 40 keV (c) 40 MeV (d) 0 eV

Q6. Ten electrons are equally spaced and fixed around a circle of radius R. Relative to $V = 0$ at infinity, the electrostatic potential V and the electric field E at the centre C are

(a) $V \neq 0$ and $\vec{E} \neq 0$ (b) $V \neq 0$ and $\vec{E} = 0$

(c) $V = 0$ and $\vec{E} = 0$ (d) $V = 0$ and $\vec{E} \neq 0$

Q7. The displacement of a charge Q in the electric field $\vec{E} = e_1\hat{i} + e_2\hat{j} + e_3\hat{k}$ is $\vec{r} = a\hat{i} + b\hat{j}$. The work done is

(a) $Q(ae_1 + be_2)$ (b) $Q\sqrt{(ae_1)^2 + (be_2)^2}$

(c) $Q(e_1 + e_2)\sqrt{a^2 + b^2}$ (d) $Q\left(\sqrt{e_1^2 + e_2^2}\right)(a + b)$

Q8. As shown in the figure, charges $+q$ and $-q$ are placed at the vertices B and C of an isosceles triangle. The potential at the vertex A is

(a) $\frac{1}{4\pi\varepsilon_0} \cdot \frac{2q}{\sqrt{a^2 + b^2}}$

(b) Zero

(c) $\frac{1}{4\pi\varepsilon_0} \cdot \frac{q}{\sqrt{a^2 + b^2}}$

(d) $\frac{1}{4\pi\varepsilon_0} \cdot \frac{(-q)}{\sqrt{a^2 + b^2}}$

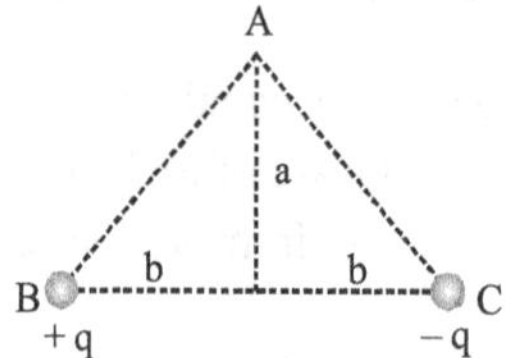

Q9. Two electric charges $12\mu C$ and $-6\mu C$ are placed 20 cm apart in air. If there will be a point P on the line joining these charges and outside the region between them, at which the electric potential is zero, then the distance of P from $-6\mu C$ charge is
(a) 0.10 m (b) 0.15 m (c) 0.20 m (d) 0.25 m

Q10. In the rectangle shown below, the two corners have charges $q_1 = -5\ \mu C$ and $q_2 = +2.0\ \mu C$. The work done by external agent in moving a charge $q = +3.0\ \mu C$ slowly from B to A is

(Take $1/4\pi\varepsilon_0 = 10^{10} Nm^2/C^2$)

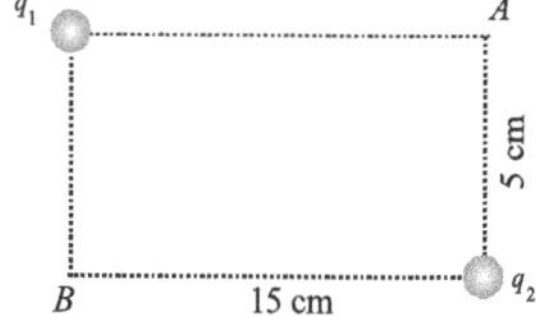

(a) 2.8 J
(b) 3.5 J
(c) 4.5 J
(d) 5.5 J

Q11. Electric charges q, q, $-2q$ are placed at the corners of an equilateral triangle ABC of side l. The magnitude of electric dipole moment of the system is

(a) ql (b) $2ql$ (c) $\sqrt{3}ql$ (d) $4ql$

Q12. A charge (–q) and another charge (+Q) are kept at two points A and B respectively. Keeping the charge (+Q) fixed at B, the charge (–q) at A is moved to another point C such that ABC forms an equilateral triangle of side l. The net work done by electrostatic field in moving the charge (–q) is

(a) $\frac{1}{4\pi\epsilon_0}\frac{Qq}{l}$ (b) $\frac{1}{4\pi\epsilon_0}\frac{Qq}{l^2}$

(c) $\frac{1}{4\pi\epsilon_0}Qql$ (d) zero

Q13. In an hydrogen atom, the electron revolves around the nucleus in an orbit of radius 0.53×10^{-10}m. Then the electrical potential produced by the nucleus at the position of the electron is
(a) –13.6V (b) –27.2V (c) 27.2V (d) 13.6V

Q14. Point charge $q_1 = 2\mu C$ and $q_2 = -1\mu C$ are kept at points $x = 0$ and $x = 6$ respectively. Electrical potential will be zero at points

(a) $x = 2$ and $x = 9$ (b) $x = 1$ and $x = 5$

(c) $x = 4$ and $x = 12$ (d) $x = -2$ and $x = 2$

RESPONSE GRID					
	4. ⓐⓑⓒⓓ	5. ⓐⓑⓒⓓ	6. ⓐⓑⓒⓓ	7. ⓐⓑⓒⓓ	8. ⓐⓑⓒⓓ
	9. ⓐⓑⓒⓓ	10. ⓐⓑⓒⓓ	11. ⓐⓑⓒⓓ	12. ⓐⓑⓒⓓ	13. ⓐⓑⓒⓓ
	14. ⓐⓑⓒⓓ				

Space for Rough Work

Q15. The distance between H^+ and Cl^- ions in HCl molecule is 1.28 Å. What will be the potential due to this dipole at a distance of 12 Å on the axis of dipole

(a) 0.13 V (b) 1.3 V (c) 13 V (d) 130 V

Q16. Two identical thin rings each of radius R metres are coaxially placed at a distance R metres apart. If Q_1 coulomb and Q_2 coulomb are respectively the charges uniformly spread on the two rings, the work done by external agent in moving a charge q slowly from the centre of ring with charge Q_1 to that of other is

(a) zero (b) $\frac{q(Q_2 - Q_1)(\sqrt{2}-1)}{\sqrt{2}.4\pi \epsilon_0 R}$

(c) $\frac{q\sqrt{2}(Q_1 + Q_2)}{4\pi \epsilon_0 R}$ (d) $\frac{q(Q_1 + Q_2)(\sqrt{2}+1)}{4\pi \epsilon_0 R}$

Q17. Identical point charges, each having $+q$ charge, are fixed at each of the points $x = x_0, x = 3x_0, x = 5x_0$ .... infinite, on the x-axis and a identical point charges, each having $-q$ charge, are fixed at each of the points $x = 2x_0, x = 4x_0, x = 6x_0$ .... infinite. Here x_0 is a positive constant. Potential at the origin due to the above system of charges is

(a) 0 (b) $\frac{q}{8\pi \epsilon_0 x_0 \ln 2}$

(c) ∞ (d) $\frac{q \ln 2}{4\pi \epsilon_0 x_0}$

Q18. A uniform electric field pointing in positive x - direction exists in a region. Let A be the origin, B be the point on the x-axis at $x = +1$ cm and C be the point on the y-axis at $y = +1$ cm. Then the potentials at the points A, B and C satisfy

(a) $V_A < V_B$ (b) $V_A > V_B$ (c) $V_A < V_C$ (d) $V_A > V_C$

Q19. A point Q lies on the perpendicular bisector of an electrical dipole of dipole moment p . If the distance of Q from the dipole is r (much larger than the size of the dipole), then electric field at Q is proportional to

(a) P^{-1} and r^{-2} (b) p and r^{-2}

(c) P^2 and r^{-3} (d) p and r^{-3}

DIRECTIONS (Q.20-Q.21) : In the following questions, more than one of the answers given are correct. Select the correct answers and mark it according to the following codes:

Codes :

(a) 1, 2 and 3 are correct **(b)** 1 and 2 are correct

(c) 2 and 4 are correct **(d)** 1 and 3 are correct

Q20. Consider a system of three charges $\frac{q}{3}, \frac{q}{3}$ and $-\frac{2q}{3}$ placed at point A, B and C respectively, as shown in the figure. Take O to be the centre of the circle of radius R and angle CAB = 60°. Choose the **incorrect** options

(1) The electric field at point O is $\frac{q}{8\pi\varepsilon_0 R^2}$ directed along the negative x-axis

(2) The potential energy of the system is zero

(3) The potential at point O is $\frac{q}{12\pi\varepsilon_0 R}$

(4) The magnitude of the force between the charges at C and B is $\frac{q^2}{54\pi\varepsilon_0 R^2}$

Q21. For spherical symmetrical charge distribution, variation of electric potential with distance from centre is given in diagram. Given that: $V = \frac{q}{4\pi\varepsilon_0 R_0}$ for $r \le R_0$ and $V = \frac{q}{4\pi\varepsilon_0 r}$ for $r \ge R_0$

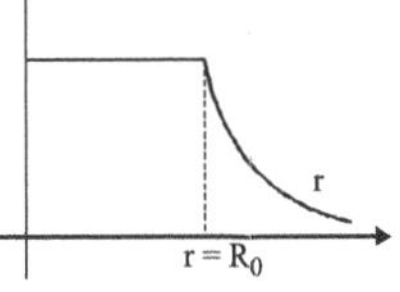

Then which option (s) are correct :

(1) Total charge within $2R_0$ is q

(2) Total electrostatic energy for $r \le R_0$ is non-zero

(3) At $r = R_0$ electric field is discontinuous

(4) There will be no charge anywhere except at $r < R_0$.

DIRECTIONS (Q.22-Q.23) : Read the passage given below and answer the questions that follows :

An electric dipole (AB) consisting of two particles of equal and opposite charge and same mass is released in an electric field. In the figure field lines are without considering effect of field of dipole.

RESPONSE GRID	15. ⓐⓑⓒⓓ	16. ⓐⓑⓒⓓ	17. ⓐⓑⓒⓓ	18. ⓐⓑⓒⓓ	19. ⓐⓑⓒⓓ
	20. ⓐⓑⓒⓓ	21. ⓐⓑⓒⓓ			

Space for Rough Work

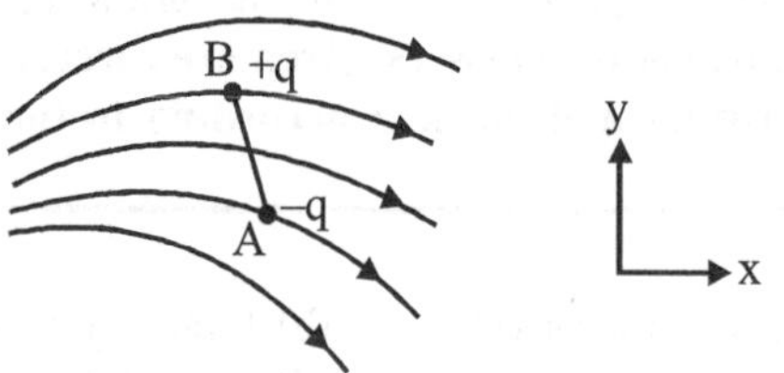

Q22. The centre of mass of the dipole
(a) Has no acceleration
(b) Has acceleration with positive x and y components
(c) Has acceleration with positive x component and negative y component
(d) Has acceleration with negative x component and positive y component

Q23. Angular acceleration of the dipole, immediately after it is released
(a) is zero
(b) is clockwise
(c) is anticlockwise
(d) cannot be determined from the given information.

DIRECTIONS (Q. 24-Q.26) : Each of these questions contains two statements: Statement-1 (Assertion) and Statement-2 (Reason). Each of these questions has four alternative choices, only one of which is the correct answer. You have to select the correct choice.

(a) Statement-1 is True, Statement-2 is True; Statement-2 is a correct explanation for Statement-1.
(b) Statement-1 is True, Statement-2 is True; Statement-2 is NOT a correct explanation for Statement-1.
(c) Statement -1 is False, Statement-2 is True.
(d) Statement -1 is True, Statement-2 is False.

Q24. Statement -1 : A bird perches on a high power line and nothing happens to the bird.
Statement -2 : The level of bird is very high from the ground.

Q25. Statement -1 : Electrons move away from a low potential to high potential region.
Statement- 2 : Because electrons have negative charge

Q26. Statement -1 : Surface of a symmetrical conductor can be treated as equipotential surface.
Statement -2 : Charges can easily flow in a conductor.

RESPONSE GRID	22. ⓐⓑⓒⓓ	23. ⓐⓑⓒⓓ	24. ⓐⓑⓒⓓ	25. ⓐⓑⓒⓓ	26. ⓐⓑⓒⓓ

DAILY PRACTICE PROBLEM SHEET 33 - PHYSICS			
Total Questions	26	Total Marks	104
Attempted		Correct	
Incorrect		Net Score	
Cut-off Score	26	Qualifying Score	42
Success Gap = Net Score – Qualifying Score			
Net Score = (Correct × 4) – (Incorrect × 1)			

Space for Rough Work

DPP - Daily Practice Problems

Name : 　　　　　　　　**Date :**

Start Time : 　　　　　　　　**End Time :**

PHYSICS 34

SYLLABUS : ELECTROSTATICS -3 (Electrostatic Potential energy, conductors)

Max. Marks : 96 　　　　　　　　**Time : 60 min.**

GENERAL INSTRUCTIONS

- The Daily Practice Problem Sheet contains 24 MCQ's. For each question only one option is correct. Darken the correct circle/ bubble in the Response Grid provided on each page.
- You have to evaluate your Response Grids yourself with the help of solution booklet.
- Each correct answer will get you 4 marks and 1 mark shall be deduced for each incorrect answer. No mark will be given/ deducted if no bubble is filled. Keep a timer in front of you and stop immediately at the end of 60 min.
- The sheet follows a particular syllabus. Do not attempt the sheet before you have completed your preparation for that syllabus. Refer syllabus sheet in the starting of the book for the syllabus of all the DPP sheets.
- After completing the sheet check your answers with the solution booklet and complete the Result Grid. Finally spend time to analyse your performance and revise the areas which emerge out as weak in your evaluation.

DIRECTIONS (Q.1-Q.16) : There are 16 multiple choice questions. Each question has 4 choices (a), (b), (c) and (d), out of which **ONLY ONE** choice is correct.

Q.1 Three charges Q, $+q$ and $+q$ are placed at the vertices of right-angled isosceles triangle as shown in the figure. The net electrostatic energy of the configuration is zero if Q is equal to

Q, +q, +q, a

(a) $\frac{-q}{1+\sqrt{2}}$ (b) $\frac{-2q}{2+\sqrt{2}}$

(c) $-2q$ (d) $+q$

Q.2 Three charges of equal value 'q' are placed at the vertices of an equilateral triangle. What is the net potential energy, if the side of equilateral Δ is l ?

(a) $\frac{1}{4\pi\epsilon_0}\frac{q^2}{l}$ (b) $\frac{1}{4\pi\epsilon_0}\frac{2q^2}{l}$

(c) $\frac{1}{4\pi\epsilon_0}\frac{3q^2}{l}$ (d) $\frac{1}{4\pi\epsilon_0}\frac{4q^2}{l}$

Q.3 If identical charges $(-q)$ are placed at each corner of a cube of side b, then electric potential energy of charge $(+q)$ which is placed at centre of the cube will be

(a) $\frac{8\sqrt{2}q^2}{4\pi\epsilon_0 b}$ (b) $\frac{-8\sqrt{2}q^2}{\pi\epsilon_0 b}$

(c) $\frac{-4\sqrt{2}q^2}{\pi\epsilon_0 b}$ (d) $\frac{-4q^2}{\sqrt{3}\pi\epsilon_0 b}$

RESPONSE GRID	1. ⓐⓑⓒⓓ	2. ⓐⓑⓒⓓ	3. ⓐⓑⓒⓓ

Space for Rough Work

Q.4 Two charges q_1 and q_2 are placed 30 cm apart, shown in the figure. A third charge q_3 is moved along the arc of a circle of radius 40 cm from C to D. The change in the potential energy of the system is $\frac{q_3}{4\pi \in_0} k$, here k is

(a) $8q_2$
(b) $8q_1$
(c) $6q_2$
(d) $6q_1$

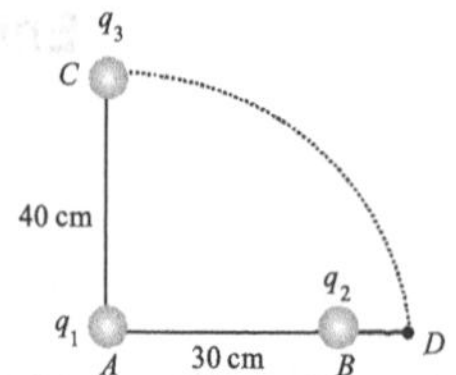

Q.5 Three particles, each having a charge of 10μC are placed at the corners of an equilateral triangle of side 10 cm. The electrostatic potential energy of the system is (Given $\frac{1}{4\pi \in_0} = 9 \times 10^9 \, Nm^2/C^2$)

(a) Zero (b) Infinite (c) 27 J (d) 100 J

Q.6 Two equal charges q are placed at a distance of $2a$ and a third charge $-2q$ is placed at the midpoint. The potential energy of the system is

(a) $\frac{q^2}{8\pi \in_0 a}$ (b) $\frac{6q^2}{8\pi \in_0 a}$

(c) $-\frac{7q^2}{8\pi \in_0 a}$ (d) $\frac{9q^2}{8\pi \in_0 a}$

Q.7 An electric dipole has the magnitude of its charge is q and its dipole moment is p. It is placed in a uniform electric field E. If its dipole moment is along the direction of the field, the force on it and its potential energy are respectively

(a) $2q.E$ and minimum (b) $q.E$ and $p.E$

(c) Zero and minimum (d) $q.E$ and maximum

Q8 In bringing an electron towards another electron, electrostatic potential energy of the system:

(a) decreases (b) increases
(c) remains unchanged (d) becomes zero

Q.9 Two identical charges are placed at the two corners of an equilateral triangle. The potential energy of the system is U. The work done in bringing an identical charge from infinity to the third vertex is

(a) U (b) 2U (c) 3U (d) zero

Q.10 Potential energy of two equal negative point charges 2μC held 1 m apart in air is

(a) 2 J (b) 2eV (c) 4 J (d) 0.036 J

Q.11 Four charges + q, –q, + q and –q are put together on four corners of a square as shown in figure. The work done by external agent in slowly assembling this configuration is

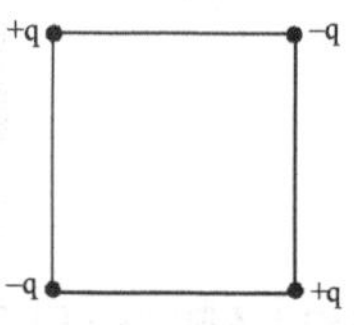

(a) zero (b) $-2.59kq^2/a$
(c) $+2.59kq^2/a$ (d) none of these

Q.12 As shown in figure a dust particle with mass $m = 5.0 \times 10^{-9}$ kg and charge $q_0 = 2.0$ nC starts from rest at point a and moves in a straight line to point b. What is its speed v at point b?

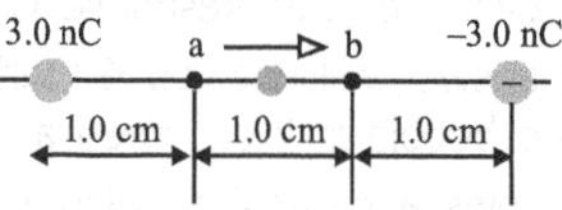

(a) 26 ms^{-1} (b) 34 ms^{-1} (c) 46 ms^{-1} (d) 14 ms^{-1}

Q.13 Charges –q, Q and –q are placed at equal distance on a straight line. If the total potential energy of the system of three charges is zero, then find the ratio Q/q.

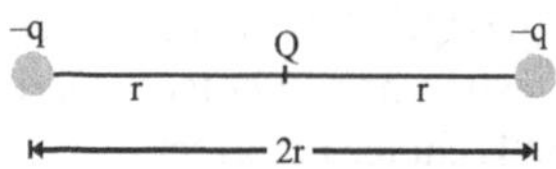

(a) 1/2 (b) 1/4 (c) 2/3 (d) 3/4

Q.14 When the separation between two charges is increased, the electric potential energy of the charges

(a) increases (b) decreases
(c) remains the same (d) may increase or decrease

RESPONSE GRID					
	4. ⓐⓑⓒⓓ	5. ⓐⓑⓒⓓ	6. ⓐⓑⓒⓓ	7. ⓐⓑⓒⓓ	8. ⓐⓑⓒⓓ
	9. ⓐⓑⓒⓓ	10. ⓐⓑⓒⓓ	11. ⓐⓑⓒⓓ	12. ⓐⓑⓒⓓ	13. ⓐⓑⓒⓓ
	14. ⓐⓑⓒⓓ				

Space for Rough Work

Q.15 A positive charge is moved from a low potential point A to high potential point B. Then the electric potential energy of the system
(a) increases
(b) decreases
(c) will remain the same
(d) nothing definite can be predicted

Q.16 If V and u are electric potential and energy density, respectively, at a distance r from a positive point charge, then which of the following graph is correct ?

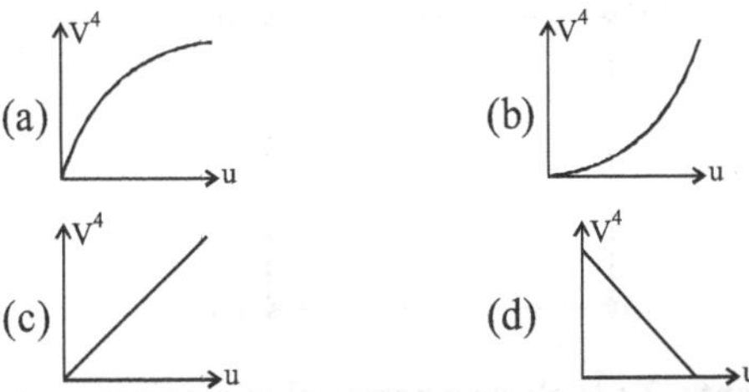

DIRECTIONS (Q.17-Q.18) : In the following questions, more than one of the answers given are correct. Select the correct answers and mark it according to the following codes:

Codes :
(a) 1, 2 and 3 are correct **(b)** 1 and 2 are correct
(c) 2 and 4 are correct **(d)** 1 and 3 are correct

Q.17 S is a solid neutral conducting sphere. A point charge q of 1×10^{-6} C is placed at point A. C is the centre of sphere and AB is a tangent. BC = 3m and AB = 4m.

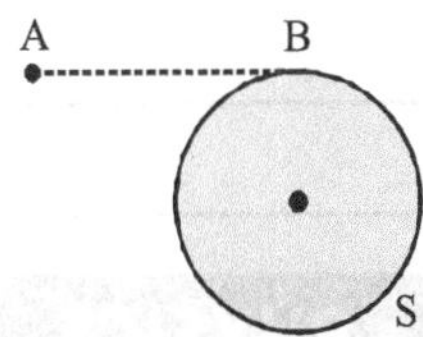

(1) The electric potential of the conductor is 1.8 kV
(2) The electric potential of the conductor is 2.25 kV
(3) The electric potential at B due to induced charges on the sphere is – 0.45 kV
(4) The electric potential at B due to induced charges on the sphere is 0.45 kV

Q.18 A proton moves a distance d in a uniform electric field $\vec{E}$ as shown in the figure. Then which of the following statements are correct ?

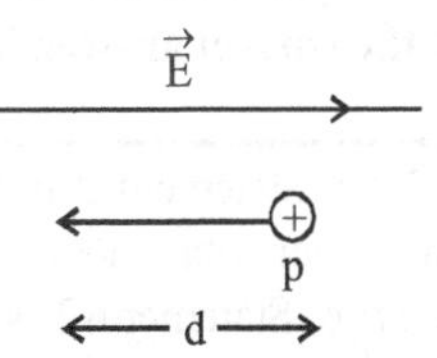

(1) Electric field do a negative work on the proton
(2) Electric potential energy of the proton increases
(3) Electric field do a positive work on the proton
(4) Electric potential energy of the proton decreases

DIRECTIONS (Q.19-Q.21) : Read the passage given below and answer the questions that follows :

Three concentric spherical conductors A, B and C of radii R, 2R and $4R$ respectively. A and C is shorted and B is uniformly charged.

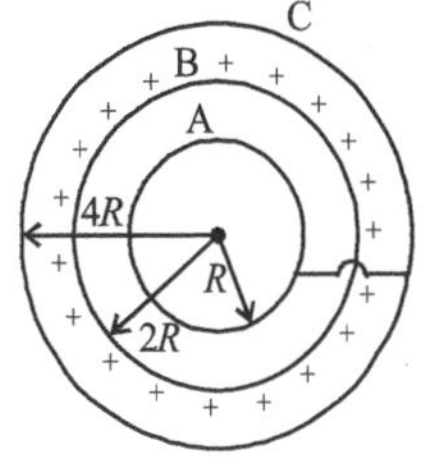

Q.19 Charge on conductor A is
(a) $Q/3$ (b) $-Q/3$
(c) $2Q/3$ (d) None of these

Q.20 Potential at A is
(a) $\frac{Q}{4\pi\varepsilon_0 R}$ (b) $\frac{Q}{16\pi\varepsilon_0 R}$
(c) $\frac{Q}{20\pi\varepsilon_0 R}$ (d) None of these

Q.21 Potential at B is
(a) $\frac{Q}{4\pi\varepsilon_0 R}$ (b) $\frac{Q}{16\pi\varepsilon_0 R}$
(c) $\frac{5Q}{48\pi\varepsilon_0 R}$ (d) None of these

RESPONSE GRID					
	15. ⓐⓑⓒⓓ	16. ⓐⓑⓒⓓ	17. ⓐⓑⓒⓓ	18. ⓐⓑⓒⓓ	19. ⓐⓑⓒⓓ
	20. ⓐⓑⓒⓓ	21. ⓐⓑⓒⓓ			

Space for Rough Work

DIRECTIONS (Q. 22-Q.24) : Each of these questions contains two statements: Statement-1 (Assertion) and Statement-2 (Reason). Each of these questions has four alternative choices, only one of which is the correct answer. You have to select the correct choice.

(a) Statement-1 is True, Statement-2 is True; Statement-2 is a correct explanation for Statement-1.
(b) Statement-1 is True, Statement-2 is True; Statement-2 is NOT a correct explanation for Statement-1.
(c) Statement -1 is False, Statement-2 is True.
(d) Statement -1 is True, Statement-2 is False.

Q.22 Statement-1 : No work is done in taking a small positive charge from one point to other inside a positively charged metallic sphere while outside the sphere work is done in taking the charge towards the sphere. Neglect induction due to small charge.

Statement-2 : Inside the sphere electric potential is same at each point, but outside it is different for different points.

Q.23 Statement-1 : Electric potential of earth is taken to be zero as a reference.

Statement-2 : The electric field produced by earth in surrounding space is zero.

Q.24 Statement - 1 : The electric potential and the electric field intensity at the centre of a square having four fixed point charges at their vertices as shown in figure are zero.

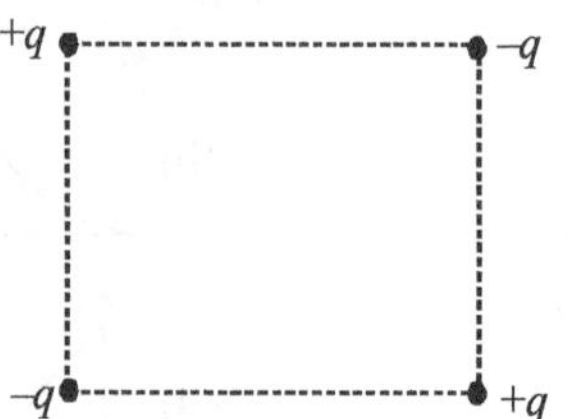

Statement - 2 : If electric potential at a point is zero then the magnitude of electric field at that point must be zero.

RESPONSE GRID	22. ⓐⓑⓒⓓ	23. ⓐⓑⓒⓓ	24. ⓐⓑⓒⓓ

DAILY PRACTICE PROBLEM SHEET 34 - PHYSICS			
Total Questions	24	Total Marks	96
Attempted		Correct	
Incorrect		Net Score	
Cut-off Score	22	Qualifying Score	39
Success Gap = Net Score – Qualifying Score			
Net Score = (Correct × 4) – (Incorrect × 1)			

Space for Rough Work

DPP - Daily Practice Problems

Name :

Date :

Start Time :

End Time :

PHYSICS 35

SYLLABUS : ELECTROSTATICS-4 (Capacitors, dielectrics)

Max. Marks : 96 **Time : 60 min.**

GENERAL INSTRUCTIONS

- The Daily Practice Problem Sheet contains 24 MCQ's. For each question only one option is correct. Darken the correct circle/ bubble in the Response Grid provided on each page.
- You have to evaluate your Response Grids yourself with the help of solution booklet.
- Each correct answer will get you 4 marks and 1 mark shall be deduced for each incorrect answer. No mark will be given/ deducted if no bubble is filled. Keep a timer in front of you and stop immediately at the end of 60 min.
- The sheet follows a particular syllabus. Do not attempt the sheet before you have completed your preparation for that syllabus. Refer syllabus sheet in the starting of the book for the syllabus of all the DPP sheets.
- After completing the sheet check your answers with the solution booklet and complete the Result Grid. Finally spend time to analyse your performance and revise the areas which emerge out as weak in your evaluation.

DIRECTIONS (Q.1-Q.15) : There are 15 multiple choice questions. Each question has 4 choices (a), (b), (c) and (d), out of which **ONLY ONE** choice is correct.

Q.1 A parallel plate capacitor is charged to a potential difference of 50V. It is discharged through a resistance. After 1 second, the potential difference between plates becomes 40V. Then

(a) Fraction of stored energy after 1 second is 16/25

(b) Potential difference between the plates after 2 seconds will be 30V

(c) Potential difference between the plates after 2 seconds will be $20V$

(d) Fraction of stored energy after 1 second is 4/5

Q.2 Five identical plates each of area A are joined as shown in the figure. The distance between the plates is d. The plates are connected to a potential difference of V volts. The charge on plates 1 and 4 will be respectively

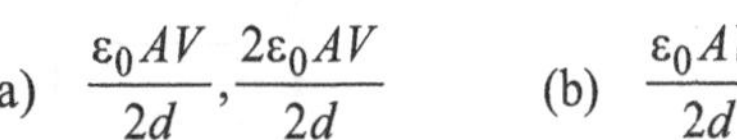

(a) $\frac{\varepsilon_0 AV}{2d}, \frac{2\varepsilon_0 AV}{2d}$ (b) $\frac{\varepsilon_0 AV}{2d}, \frac{2\varepsilon_0 AV}{2d}$

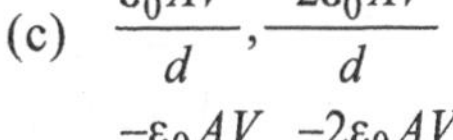

(c) $\frac{\varepsilon_0 AV}{d}, \frac{-2\varepsilon_0 AV}{d}$

(d) $\frac{-\varepsilon_0 AV}{d}, \frac{-2\varepsilon_0 AV}{d}$

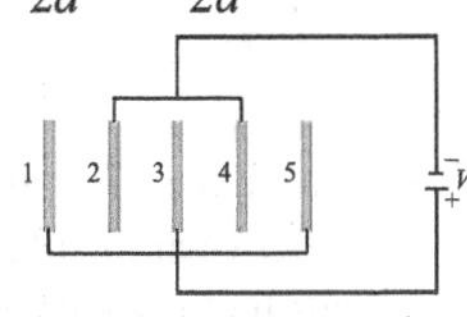

Q.3 Figure given below shows two identical parallel plate capacitors connected to a battery with switch S closed. The switch is now opened and the free space between the plates of capacitors is filled with a dielectric of dielectric constant 3. What will be the ratio of total electrostatic energy stored in both capacitors before and after the introduction of the dielectric?

(a) 3 : 1

(b) 5 : 1

(c) 3 : 5

(d) 5 : 3

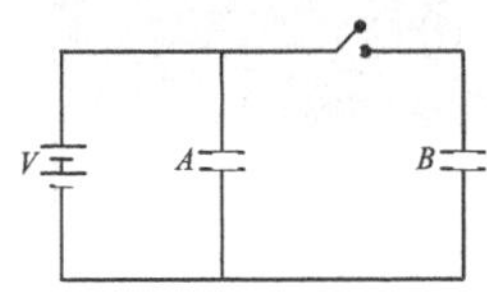

RESPONSE GRID 1. ⓐⓑⓒⓓ 2. ⓐⓑⓒⓓ 3. ⓐⓑⓒⓓ

Space for Rough Work

Q.4 All six capacitors shown are identical, Each can withstand maximum 200 volts between its terminals. The maximum voltage that can be safely applied between A and B is

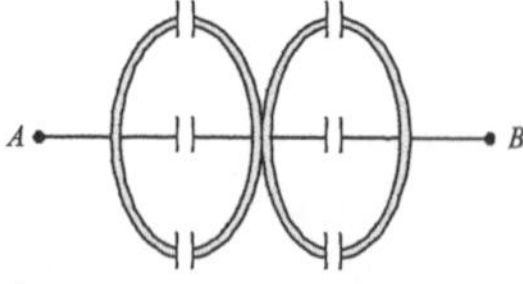

(a) 1200 V
(b) 400 V
(c) 800 V
(d) 200 V

Q.5 A capacitor of capacity C_1 is charged upto V volt and then connected to an uncharged capacitor of capacity C_2. Then final potential difference across each will be

(a) $\frac{C_2V}{C_1+C_2}$ (b) $\left(1+\frac{C_2}{C_1}\right)V$ (c) $\frac{C_1V}{C_1+C_2}$ (d) $\left(1-\frac{C_2}{C_1}\right)V$

Q.6 Two capacitors of capacitances 3μF and 6μF are charged to a potential of 12V each. They are now connected to each other, with the positive plate of each joined to the negative plate of the other. The potential difference across each will be

(a) 6 volt (b) 4 volt
(c) 3 volt (d) zero

Q.7 In the figure a capacitor is filled with dielectrics K_1, K_2 and K_3. The resultant capacitance is

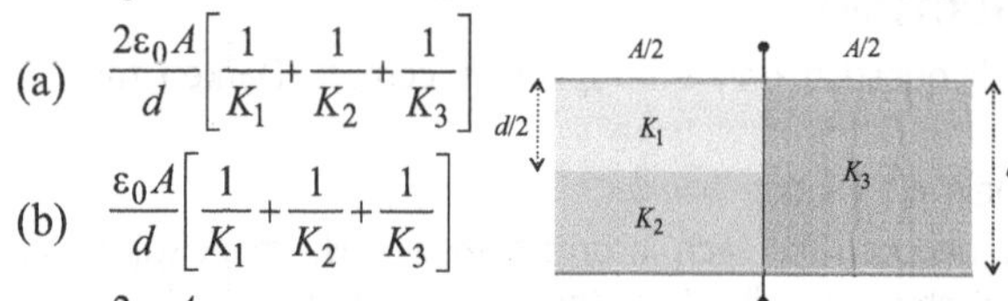

(a) $\frac{2\varepsilon_0 A}{d}\left[\frac{1}{K_1}+\frac{1}{K_2}+\frac{1}{K_3}\right]$

(b) $\frac{\varepsilon_0 A}{d}\left[\frac{1}{K_1}+\frac{1}{K_2}+\frac{1}{K_3}\right]$

(c) $\frac{2\varepsilon_0 A}{d}[K_1+K_2+K_3]$

(d) None of these

Q.8 The resultant capacitance of given circut is

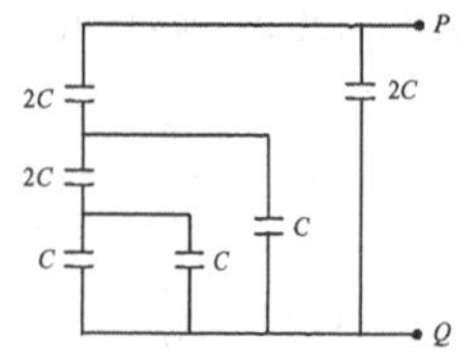

(a) $3C$
(b) $2C$
(c) C
(d) $\frac{C}{3}$

Q.9 Two dielectric slabs of constant K_1 and K_2 have been filled in between the plates of a capacitor as shown below. What will be the capacitance of the capacitor

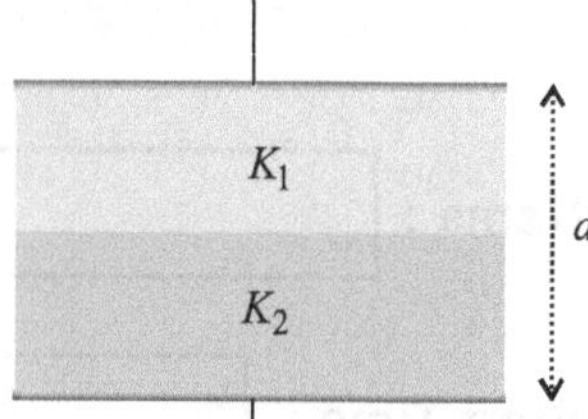

(a) $\frac{2\varepsilon_0 A}{d}(K_1+K_2)$

(b) $\frac{2\varepsilon_0 A}{d}\left(\frac{K_1+K_2}{K_1\times K_2}\right)$

(c) $\frac{4\varepsilon_0 A}{d}\left(\frac{K_1\times K_2}{K_1+K_2}\right)$

(d) $\frac{2\varepsilon_0 A}{d}\left(\frac{K_1\times K_2}{K_1+K_2}\right)$

Q.10 Eight drops of mercury of equal radii possessing equal charges combine to form a big drop. Then the capacitance of bigger drop compared to each individual small drop is

(a) 8 times (b) 4 times
(c) 2 times (d) 32 times

Q.11 Separation between the plates of a parallel plate capacitor is d and the area of each plate is A. When a slab of material of dielectric constant k and thickness $t(t<d)$ is introduced between the plates, its capacitance becomes

(a) $\frac{\varepsilon_0 A}{d+t\left(1-\frac{1}{k}\right)}$ (b) $\frac{\varepsilon_0 A}{d+t\left(1+\frac{1}{k}\right)}$

(c) $\frac{\varepsilon_0 A}{d-t\left(1-\frac{1}{k}\right)}$ (d) $\frac{\varepsilon_0 A}{d-t\left(1+\frac{1}{k}\right)}$

Q.12 There is an air filled 1pF parallel plate capacitor. When the plate separation is doubled and the space is filled with wax, the capacitance increases to 2pF. The dielectric constant of wax is

(a) 2 (b) 4 (c) 6 (d) 8

Q.13 Between the plates of a parallel plate condenser, a plate of thickness t_1 and dielectric constant k_1 is placed. In the rest of the space, there is another plate of thickness t_2 and dielectric constant k_2. The potential difference across the condenser will be

(a) $\frac{Q}{A\varepsilon_0}\left(\frac{t_1}{K_1}+\frac{t_2}{K_2}\right)$ (b) $\frac{\varepsilon_0 Q}{A}\left(\frac{t_1}{K_1}+\frac{t_2}{K_2}\right)$

(c) $\frac{Q}{A\varepsilon_0}\left(\frac{K_1}{t_1}+\frac{K_2}{t_2}\right)$ (d) $\frac{\varepsilon_0 Q}{A}(K_1t_1+K_2t_2)$

RESPONSE GRID					
	4. ⓐⓑⓒⓓ	5. ⓐⓑⓒⓓ	6. ⓐⓑⓒⓓ	7. ⓐⓑⓒⓓ	8. ⓐⓑⓒⓓ
	9. ⓐⓑⓒⓓ	10. ⓐⓑⓒⓓ	11. ⓐⓑⓒⓓ	12. ⓐⓑⓒⓓ	13. ⓐⓑⓒⓓ

Space for Rough Work

Q.14 A parallel plate capacitor is charged and the charging battery is then disconnected. If the plates of the capacitor are moved further apart by means of insulating handles, then
(a) The charge on the capacitor increases
(b) The voltage across the plates decreases
(c) The capacitance increases
(d) The electrostatic energy stored in the capacitor increases

Q.15 A parallel plate capacitor of plate area A and plates separation distance d is charged by applying a potential V_0 between the plates. The dielectric constant of the medium between the plates is K. What is the uniform electric field E between the plates of the capacitor ?

(a) $E = \epsilon_0 \frac{CV_0}{KA}$ (b) $E = \frac{V_0}{Kd}$

(c) $E = \frac{V_0}{KA}$ (d) $E = \frac{Kv_0 d}{\epsilon_0 A}$

DIRECTIONS (Q.16-Q.18) : In the following questions, more than one of the answers given are correct. Select the correct answers and mark it according to the following codes:

Codes :
(a) 1, 2 and 3 are correct **(b)** 1 and 2 are correct
(c) 2 and 4 are correct **(d)** 1 and 3 are correct

Q.16 A parallel plate air condenser is connected with a battery. Its charge, potenital difference, electric field and energy are Q_0, V_0, E_0, and U_0, respectively. In order to fill the complete space between the plates a dielectric slab is inserted, the battery is still connected. Now the corresponding values Q, V, E and U are in relation with the initially stated as
(1) $V > V_0$ (2) $Q > Q_0$ (3) $E > E_0$ (4) $U > U_0$

Q.17 The false statement are, on increasing the distance between the plates of a parallel plate condenser,
(1) The electric field intensity between the plates will decrease
(2) The electric field intensity between the plates will increase
(3) The P. D. between the plates will decrease
(4) The electric field intensity between the plates will remain unchanged

Q.18 The capacitance of a parallel plate condenser depends on
(1) Area of the plates
(2) Medium between the plates
(3) Distance between the plates
(4) Metal of the plates

DIRECTIONS (Q.19-Q.21) : Read the passage given below and answer the questions that follows :

Capacitor C_3 in the circuit is variable capacitor (its capacitance can be varied). Graph is plotted between potential difference V_1 (across capacitor C_1) versus C_3.
Electric potential V_1 approaches on asymptote of 10 volts as $C_3 \to \infty$.

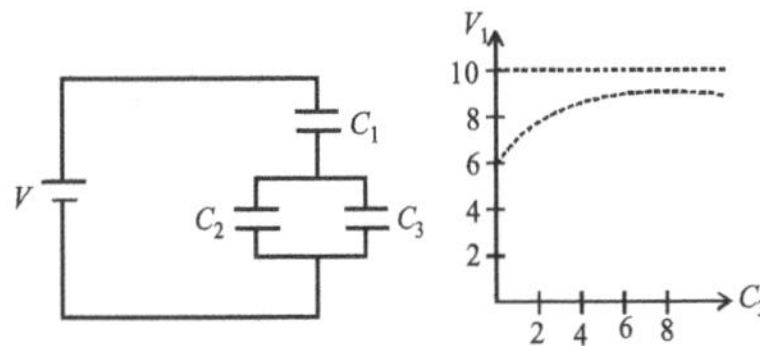

Q.19 The ratio of the capacitance $\frac{C_1}{C_2}$ will be
(a) 2 / 3 (b) 4 / 3 (c) 3 / 4 (d) 3 / 2

Q.20 The value of C_3 for which potential difference across C_1 will become 8V, is
(a) $1.5C_1$ (b) $2.5C_1$ (c) $3.5\,C_1$ (d) $4.5\,C_1$

Q.21 The ratio of energy stored in capacitor C_1 to that of total energy when $C_3 \to \infty$ is
(a) zero (b) 1/3
(c) 1 (d) Data insufficient

DIRECTIONS (Q. 22-Q.24) : Each of these questions contains two statements: Statement-1 (Assertion) and Statement-2 (Reason). Each of these questions has four alternative choices, only one of which is the correct answer. You have to select the correct choice.

(a) Statement-1 is True, Statement-2 is True; Statement-2 is a correct explanation for Statement-1.
(b) Statement-1 is True, Statement-2 is True; Statement-2 is NOT a correct explanation for Statement-1.
(c) Statement-1 is False, Statement-2 is True.
(d) Statement-1 is True, Statement-2 is False.

RESPONSE GRID					
	14. ⓐⓑⓒⓓ	15. ⓐⓑⓒⓓ	16. ⓐⓑⓒⓓ	17. ⓐⓑⓒⓓ	18. ⓐⓑⓒⓓ
	19. ⓐⓑⓒⓓ	20. ⓐⓑⓒⓓ	21. ⓐⓑⓒⓓ		

Space for Rough Work

Q.22 Statement-1 : The force with which one plate of a parallel plate capacitor is attracted towards the other plate is equal to square of surface density per $2 \in_0$ per unit area.

Statement-2 : The electric field due to one charged plate of the capacitor at the location of the other is equal to surface density per $2 \in_0$.

Q.23 Statement-1 : Circuit containing capacitors should be handled cautiously even when there is no current.

Statement-2 : The capacitors are very delicate and so quickly break down.

Q.24 Statement-1 : If the distance between parallel plates of a capacitor is halved and dielectric constant is made three times, then the capacitance becomes 6 times.

Statement-1 : Capacitance of the capacitor does not depend upon the nature of the material of plates.

RESPONSE GRID	22. ⓐⓑⓒⓓ	23. ⓐⓑⓒⓓ	24. ⓐⓑⓒⓓ

DAILY PRACTICE PROBLEM SHEET 35 - PHYSICS			
Total Questions	24	Total Marks	96
Attempted		Correct	
Incorrect		Net Score	
Cut-off Score	22	Qualifying Score	40
Success Gap = Net Score – Qualifying Score			
Net Score = (Correct × 4) – (Incorrect × 1)			

Space for Rough Work

DPP - Daily Practice Problems

Name :

Date :

Start Time :

End Time :

PHYSICS 36

SYLLABUS : CURRENT ELECTRICITY – 1 (Electric Current, drift velocity, Ohm's law, Electrical resistance, Resistances of different materials, V-I characteristics of Ohm and non-ohmic conductors, electrical energy and power, Electrical resistivity, Colour code of resistors, Temperature dependance of resistance)

Max. Marks : 92 **Time : 60 min.**

GENERAL INSTRUCTIONS

- The Daily Practice Problem Sheet contains 23 MCQ's. For each question only one option is correct. Darken the correct circle/ bubble in the Response Grid provided on each page.
- You have to evaluate your Response Grids yourself with the help of solution booklet.
- Each correct answer will get you 4 marks and 1 mark shall be deduced for each incorrect answer. No mark will be given/ deducted if no bubble is filled. Keep a timer in front of you and stop immediately at the end of 60 min.
- The sheet follows a particular syllabus. Do not attempt the sheet before you have completed your preparation for that syllabus. Refer syllabus sheet in the starting of the book for the syllabus of all the DPP sheets.
- After completing the sheet check your answers with the solution booklet and complete the Result Grid. Finally spend time to analyse your performance and revise the areas which emerge out as weak in your evaluation.

DIRECTIONS (Q.1-Q.14) : There are 14 multiple choice questions. Each question has 4 choices (a), (b), (c) and (d), out of which **ONLY ONE** choice is correct.

Q.1 In the following fig. the ratio of current in 3Ω and 1Ω resistances is–

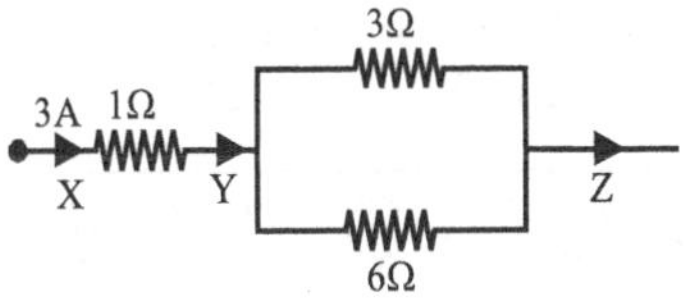

3Ω
3A 1Ω
X Y Z
6Ω

(a) 1/3
(b) 2/3
(c) 1
(d) 2

Q.2 The resultant resistance between the points A and B in the following fig. will be –

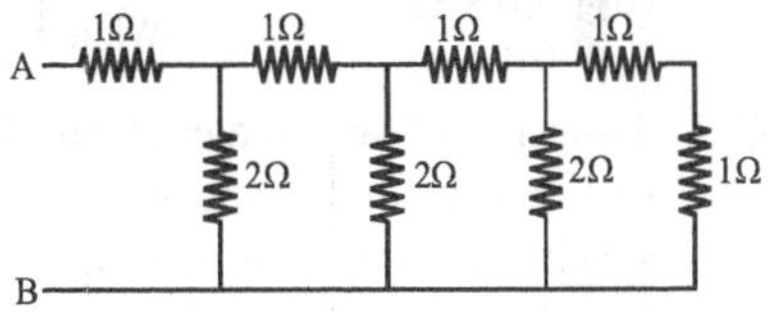

(a) $4\ \Omega$
(b) $8\ \Omega$
(c) $6\ \Omega$
(d) $2\ \Omega$

1Ω 1Ω 1Ω 1Ω
A
2Ω 2Ω 2Ω 1Ω
B

Q.3 How will reading in the ammeter A be affected if an other identical bulb Q is connected in parallel to P as shown in the fig. The voltage in the mains is maintained at constant value

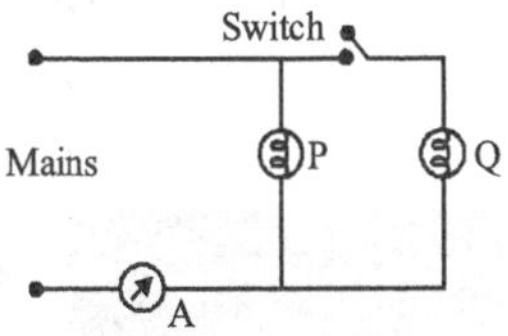

Switch
Mains
P Q
A

(a) the reading will be reduced to one half.
(b) the reading will be double of previous one.
(c) the reading will not be affected.
(d) the reading will increase four fold.

RESPONSE GRID	1. ⓐⓑⓒⓓ	2. ⓐⓑⓒⓓ	3. ⓐⓑⓒⓓ

Space for Rough Work

Q.4 In the circuit shown, the galvanometer G reads zero. If batteries have negligible internal resistances, the value of resistance X wil be –

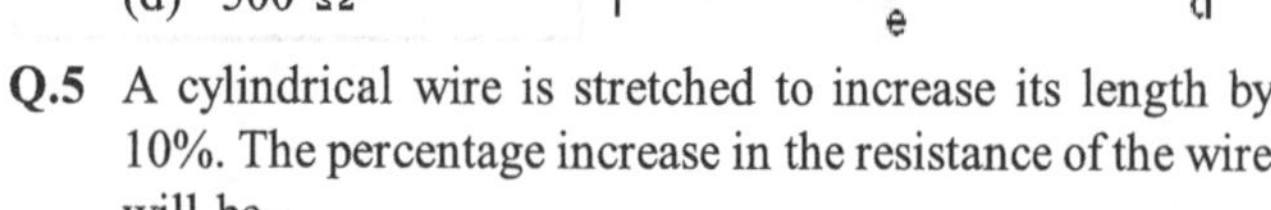

(a) 10 Ω
(b) 100 Ω
(c) 200 Ω
(d) 500 Ω

Q.5 A cylindrical wire is stretched to increase its length by 10%. The percentage increase in the resistance of the wire will be–
(a) 20% (b) 21% (c) 22% (d) 24%

Q.6 In the figure, the equivalent resistance between A and B is–

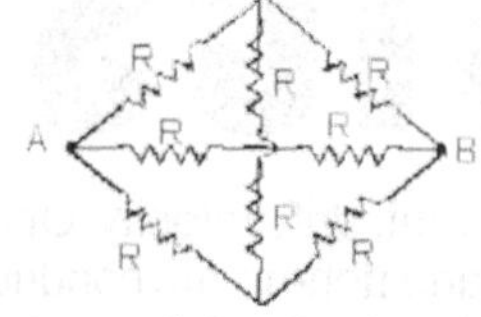

(a) 2R/3
(b) R/3
(c) R
(d) 3R

Q.7 In the adjoining network of resistors, each is of resistance r ohm, the equivalent resistance between points A and B is–

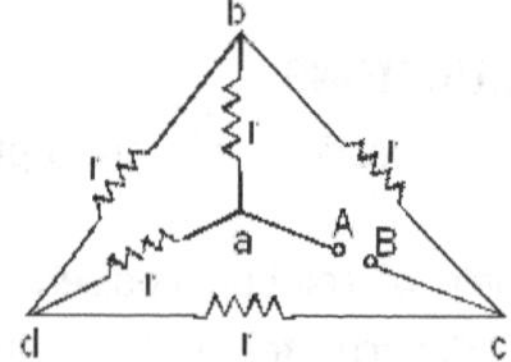

(a) 5r
(b) 2r/3
(c) r
(d) r/2.

Q.8 In the figure a carbon resistor has bands of different colours on its body as mentioned in the figure. The value of the resistance is

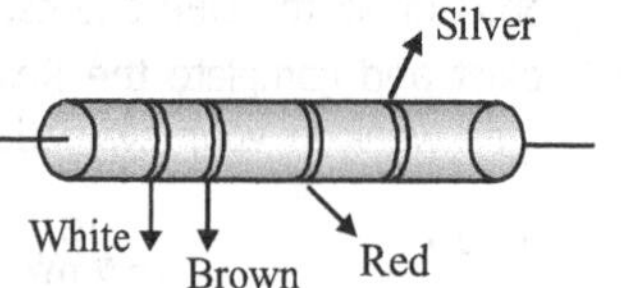

(a) 2.2 k Ω
(b) 3.3 k Ω
(c) 5.6 k Ω
(d) 9.1 k Ω

Q.9 Two wires of same material have length L and $2L$ and cross-sectional areas $4A$ and A respectively. The ratio of their specific resistance would be
(a) 1 : 2 (b) 8 : 1 (c) 1 : 8 (d) 1 : 1

Q.10 In the fig. shown, Calculate the current through 3 ohm resistor. The emf of battery is 2 volt and its internal resistance is 2/3 ohm.

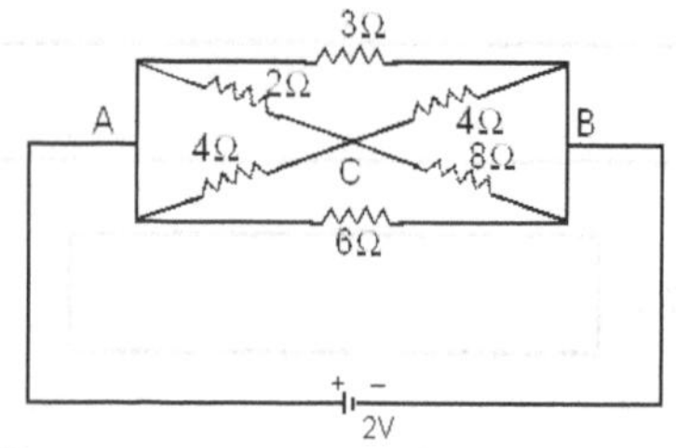

(a) 0.33 amp. (b) 0.44 amp.
(c) 1.22 amp. (d) 0.88 amp.

Q.11 The current in the given circuit will be

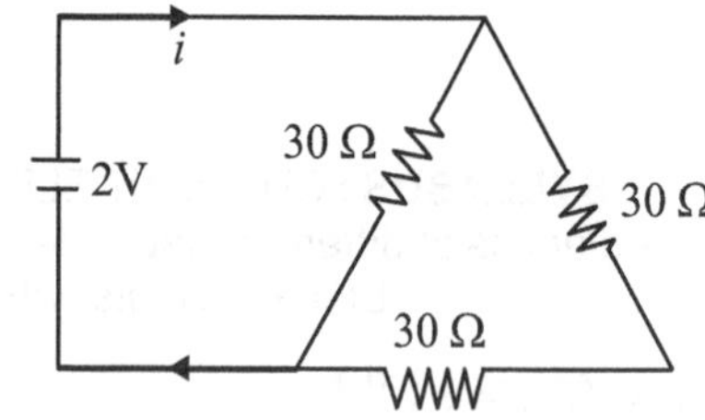

(a) 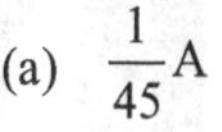 $\frac{1}{45}$ A
(b) 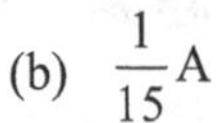 $\frac{1}{15}$ A
(c) 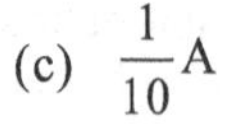 $\frac{1}{10}$ A
(d) $\frac{1}{5}$ A

Q.12 The equivalent resistance of the following infinite network of resistance is

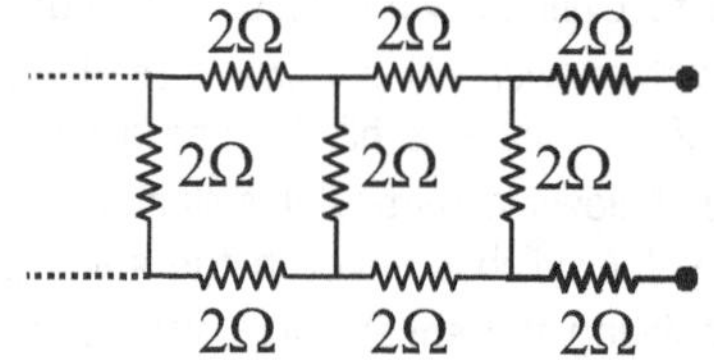

(a) Less than 4 Ω
(b) 4Ω
(c) More than 4Ω but less than 12Ω
(d) 12 Ω

Q.13 A heater coil connected to a supply of a 220 V is dissipating some power P_1. The coil is cut into half and the two halves are connected in parallel. The heater now dissipates a power P_2. The ratio of power $P_1 : P_2$ is
(a) 2 : 1 (b) 1 : 2 (c) 1 : 4 (d) 4 : 1

RESPONSE GRID					
	4. ⓐⓑⓒⓓ	5. ⓐⓑⓒⓓ	6. ⓐⓑⓒⓓ	7. ⓐⓑⓒⓓ	8. ⓐⓑⓒⓓ
	9. ⓐⓑⓒⓓ	10. ⓐⓑⓒⓓ	11. ⓐⓑⓒⓓ	12. ⓐⓑⓒⓓ	13. ⓐⓑⓒⓓ

Space for Rough Work

Q.14 An electric lamp is marked 60 *W*, 230 *V*. The cost of a 1 kWh of energy is ₹ 1.25. The cost of using this lamp 8 hrs a day for 30 days is (approximately)

(a) ₹ 10 (b) ₹ 16 (c) ₹ 18 (d) ₹ 20

DIRECTIONS (Q.15-Q.17) : In the following questions, more than one .of the answers given are correct. Select the correct answers and mark it according to the following codes:

Codes :

(a) 1, 2 and 3 are correct **(b)** 1 and 2 are correct
(c) 2 and 4 are correct **(d)** 1 and 3 are correct

Q.15 In the fig below the bulbs are identical, The bulbs, light most brightly are

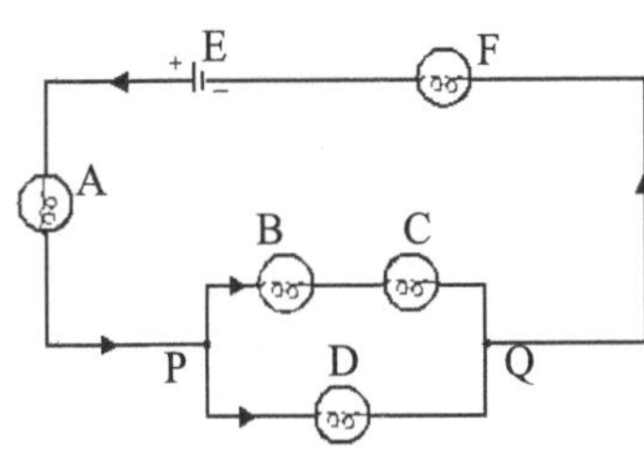

(1) A (2) B (3) F (4) D

Q.16 An electric kettle has two heating coils. When one of the coils is switched on, the water begins to boil in 6 minutes. When the other is switched on, the boiling begins in 8 minutes. The time when the boiling begin if both coils are switched on simultaneously is (i) in series (ii) in parallel

(1) 14 min in series (2) 3.43 min in parallel
(3) 3.43 min in series (4) 14 min in parallel

Q.17 For the circuit shown in the figure

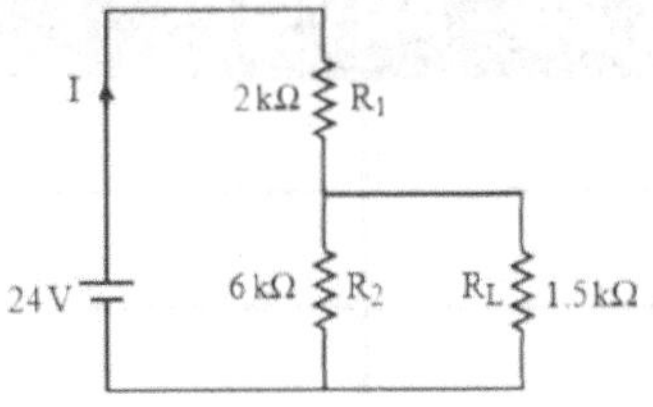

(1) The potential difference across R_L is 18 V
(2) The current I through the battery is 7.5 mA
(3) Ratio of powers dissipated in R_1 and R_2 is 3
(4) If R_1 and R_2 are interchanged magnitude of the power dissipated in R_L will decrease by a factor of 9

DIRECTIONS (Q.18-Q.20) : Read the passage given below and answer the questions that follows :

In the circuit shown in the figure,

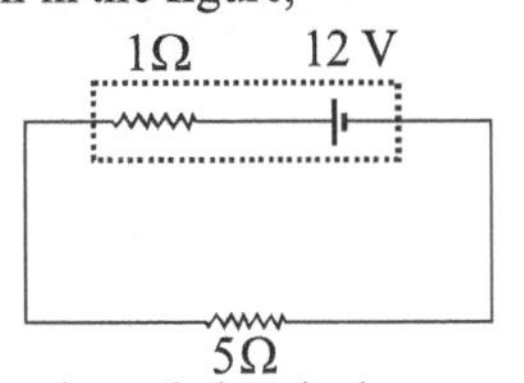

Q.18 Rate of conversion of chemical energy within the battery is

(a) 24 W (b) 20 W (c) 4 W (d) 14 W

Q.19 Rate of dissipation of electrical energy in battery is

(a) 24 W (b) 20 W (c) 4 W (d) 14 W

Q.20 Rate of dissipation of electrical energy in external resistor is

(a) 4 W (b) 20 W (c) 14 W (d) 24 W

DIRECTIONS (Q. 21-Q.23) : Each of these questions contains two statements': Statement-1 (Assertion) and Statement-2 (Reason). Each of these questions has four alternative choices, only one of which is the correct answer. You have to select the correct choice.

(a) Statement-1 is True, Statement-2 is True; Statement-2 is a correct explanation for Statement-1.
(b) Statement-1 is True, Statement-2 is True; Statement-2 is NOT a correct explanation for Statement-1.
(c) Statement -1 is False, Statement-2 is True.
(d) Statement -1 is True, Statement-2 is False.

Q.21 Statement-1 : The resistivity of a semiconductor decreases with temperature.

Statement-2 : The atoms of a semiconductor vibrate with larger amplitude at higher temperatures thereby increasing its resistivity.

RESPONSE GRID

14. ⓐⓑⓒⓓ 15. ⓐⓑⓒⓓ 16. ⓐⓑⓒⓓ 17. ⓐⓑⓒⓓ 18. ⓐⓑⓒⓓ
19. ⓐⓑⓒⓓ 20. ⓐⓑⓒⓓ 21. ⓐⓑⓒⓓ

Space for Rough Work

Q.22 Statement-1 : In a simple battery circuit the point of lowest potential is negative terminal of the battery.
Statement-2 : The current flows towards the point of the higher potential as it flows in such a circuit from the negative to the positive terminal.

Q.23 Statement-1 : The temperature coefficient of resistance is positive for metals and negative for p-type semiconductor.
Statement-2 : The effective charge carriers in metals are negatively charged whereas in p-type semiconductor they are positively charged.

RESPONSE GRID	22. ⓐⓑⓒⓓ	23. ⓐⓑⓒⓓ

DAILY PRACTICE PROBLEM SHEET 36 - PHYSICS			
Total Questions	23	Total Marks	92
Attempted		Correct	
Incorrect		Net Score	
Cut-off Score	24	Qualifying Score	40
Success Gap = Net Score – Qualifying Score			
Net Score = (Correct × 4) – (Incorrect × 1)			

Space for Rough Work

DPP - Daily Practice Problems

Name : 　　　　Date :

Start Time : 　　　　End Time :

PHYSICS 37

SYLLABUS : CURRENT ELECTRICITY – 2 Electrical cell and its internal resistance, Potential difference and E.M.F of a cell, Combination of cells in series and in parallel, Kirchoff's laws and their applications, RC transient circuit, Galvanometer, Ammeter, Voltmeter]

Max. Marks : 104 　　　　 **Time : 60 min.**

GENERAL INSTRUCTIONS

- The Daily Practice Problem Sheet contains 26 MCQ's. For each question only one option is correct. Darken the correct circle/ bubble in the Response Grid provided on each page.
- You have to evaluate your Response Grids yourself with the help of solution booklet.
- Each correct answer will get you 4 marks and 1 mark shall be deduced for each incorrect answer. No mark will be given/ deducted if no bubble is filled. Keep a timer in front of you and stop immediately at the end of 60 min.
- The sheet follows a particular syllabus. Do not attempt the sheet before you have completed your preparation for that syllabus. Refer syllabus sheet in the starting of the book for the syllabus of all the DPP sheets.
- After completing the sheet check your answers with the solution booklet and complete the Result Grid. Finally spend time to analyse your performance and revise the areas which emerge out as weak in your evaluation.

DIRECTIONS (Q.1-Q.18) : There are 18 multiple choice questions. Each question has 4 choices (a), (b), (c) and (d), out of which **ONLY ONE** choice is correct.

Q.1 The voltmeter shown in fig, reads 6V across the 60 Ω resistor. Then the resistance of the voltmeter is-

(a) 0 Ω
(b) ∞ Ω
(c) 200 Ω
(d) 300 Ω

Q.2 If only one hundredth part of total current flowing in the circuit is to be passed through a galvanometer of resistance GΩ, Then the value of shunt resistance required will be-

(a) G/10 (b) G/100 (c) G/99 (d) G/999

Q.3 The shunt required for 10% of main current to be sent through the moving coil galvanometer of resistance 99Ω will be-

(a) 0.9 Ω (b) 11 Ω (c) 90 Ω (d) 9.9 Ω

Q.4 The reading of voltmeter in the following circuit will be-

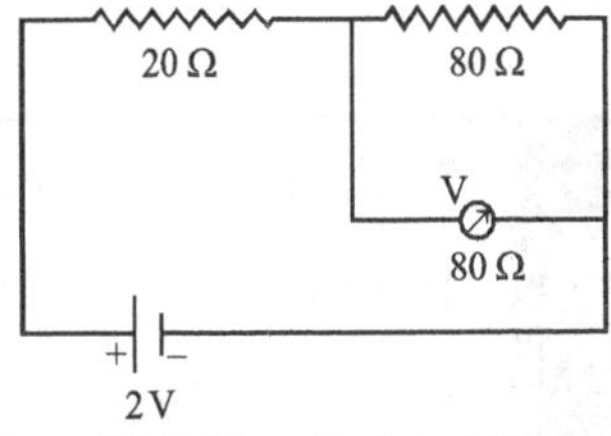

(a) 2 volt (b) 0.80 volt (c) 1.33 volt (d) 1.60 volt

RESPONSE GRID	1. ⓐⓑⓒⓓ	2. ⓐⓑⓒⓓ	3. ⓐⓑⓒⓓ	4. ⓐⓑⓒⓓ

Space for Rough Work

Q.5 The figure below shows currents in a part of electric circuit. The current i is
(a) 1.7 amp
(b) 3.7 amp
(c) 1.3 amp
(d) 1 amp

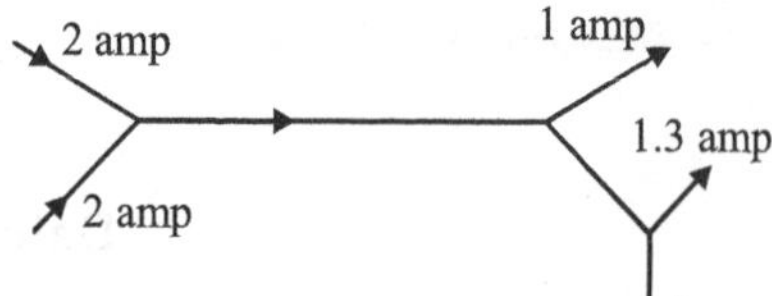

Q.6 A voltmeter can measure upto 25 volt and its resistance is 1000 Ω. The resistance required to add with voltmeter to measure upto 250 volt will be-
(a) 9000 Ω (b) 1000 Ω (c) 2500 Ω (d) 900 Ω

Q.7 When a Laclanche cell is connected to a 10Ω resistance then a current of 0.25 ampere flows in the circuit. If the resistance is reduced to 4Ω then current becomes 0.5ampere. The internal resistance of galvanometer will be-
(a) 1.5 Ω (b) 0.5 Ω (c) 1 Ω (d) 2 Ω

Q.8 Consider the circuit shown in the figure. The value of current I_3 is
(a) 5 A
(b) 3 A
(c) – 3 A
(d) – 5/6 A

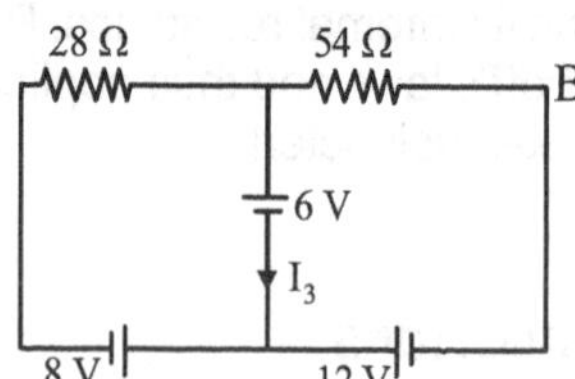

Q.9 If $V_B - V_A = 4$ V in the given figure, then resistance X will be
(a) 5 Ω
(b) 10 Ω
(c) 15 Ω
(d) 20 Ω

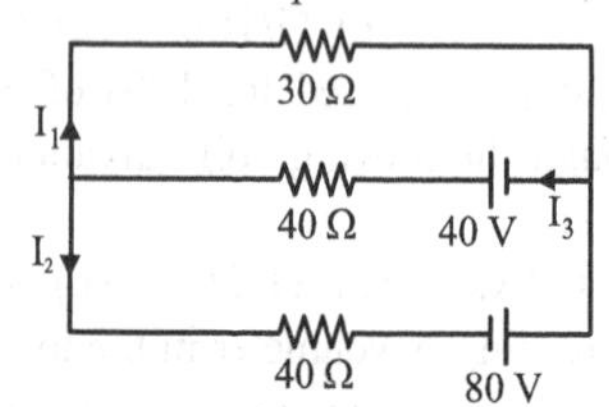

Q.10 In the given circuit the current I_1 is
(a) 0.4 A
(b) – 0.4 A
(c) 0.8 A
(d) – 0.8 A

Q.11 To get the maximum current from a parallel combination of n identical cells each of internal resistance r in an external resistance R,
(a) $R >> r$ (b) $R << r$ (c) $R > r$ (d) $R = r$

Q.12 In the circuit shown below, if the value of R is increased then what will be the effect on the reading of ammeter if the internal resistance of cell is negligible-

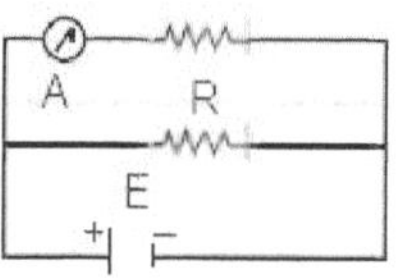

(a) The reading of ammeter will decrease
(b) The reading of ammeter will increase
(c) The reading of ammeter will remain unchanged
(d) The reading of ammeter will become zero.

Q.13 Twelve wires of equal length and same cross-section are connected in the form of a cube. If the resistance of each of the wires is R, then the effective resistance between the two diagonal ends would be
(a) 2 R
(b) 12 R
(c) $\frac{5}{6}$R
(d) 8 R

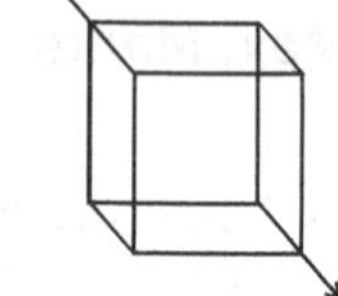

Q.14 The arrangement as shown in figure is called as
(a) Potential divider
(b) Potential adder
(c) Potential substracter
(d) Potential multiplier

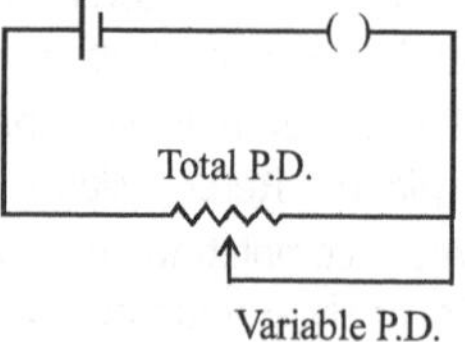

Q.15 When a cell of emf E and internal resistance r, is connected to the ends of a resistance R, then current through resistance is I. If the same cell is connected to the ends of a resistance R/2 then the current would be-
(a) less than I
(b) I
(c) greater then I but less than 2I
(d) greater than 2I

RESPONSE GRID				
5. ⓐⓑⓒⓓ	6. ⓐⓑⓒⓓ	7. ⓐⓑⓒⓓ	8. ⓐⓑⓒⓓ	9. ⓐⓑⓒⓓ
10. ⓐⓑⓒⓓ	11. ⓐⓑⓒⓓ	12. ⓐⓑⓒⓓ	13. ⓐⓑⓒⓓ	14. ⓐⓑⓒⓓ
15. ⓐⓑⓒⓓ				

Space for Rough Work

Q.16 The resistance of an ideal voltmeter is
(a) Zero (b) Very low
(c) Very large (d) Infinite

Q.17 An ammeter with internal resistance 90 Ω reads 1.85 A when connected in a circuit containing a battery and two resistors 700 Ω and 410 Ω in series. Actual current will be
(a) 1.85 A (b) Greater than 1.85 A
(c) Less than 1.85 A (d) None of these

Q.18 The figure shows a network of currents. The magnitude of currents is shown here. The current *I* will be

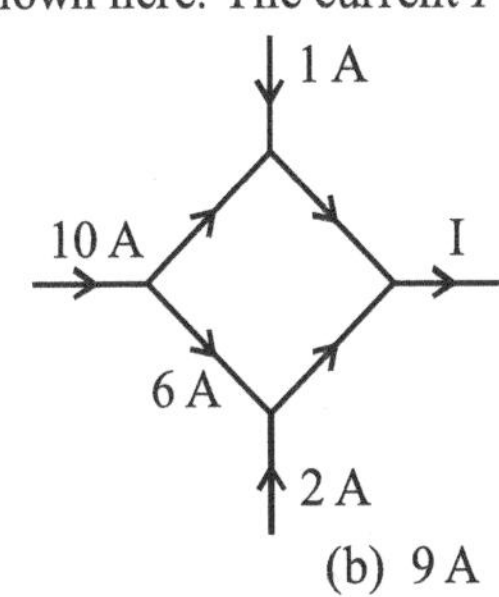

(a) 3 A (b) 9 A
(c) 13 A (d) 19 A

DIRECTIONS (Q.19-Q.21) : In the following questions, more than one of the answers given are correct. Select the correct answers and mark it according to the following codes:

Codes :
(a) 1, 2 and 3 are correct **(b)** 1 and 2 are correct
(c) 2 and 4 are correct **(d)** 1 and 3 are correct

Q.19 In the figure,

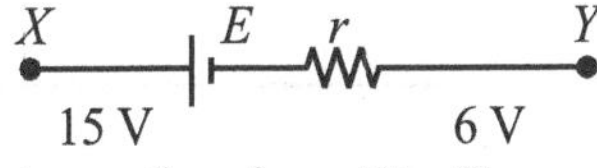

(1) current may flow from *X* to *Y*
(2) current may flow from *Y* to *X*
(3) current's direction depends on E
(4) current's direction depends on *r*

Q.20 Kirchoff's laws are based on conservation of
(1) charge (2) potential
(3) energy (4) mass

Q.21 A microammeter has a resistance of 100Ω and a full scale range of 50 μA. It can be used as a voltmeter or a higher range ammeter provided a resistance is added to it. Pick the correct range and resistance combination(s).
(1) 10V range with 200 kΩ resistance in series.
(2) 50V range with 10 kΩ resistance in series.
(3) 5 mA range with 1Ω resistance in parallel.
(4) 10 mA range with 1 kΩ resistance in parallel.

DIRECTIONS (Q.22-Q.23) : Read the passage given below and answer the questions that follows :

A 6V battery of negligible internal resistance is connected across a uniform wire *AB* of length 100cm. The positive terminal of another battery of emf 4V and internal resistance 1Ω is joined to the point *A* as shown in figure. Take the potential at *B* to be zero.

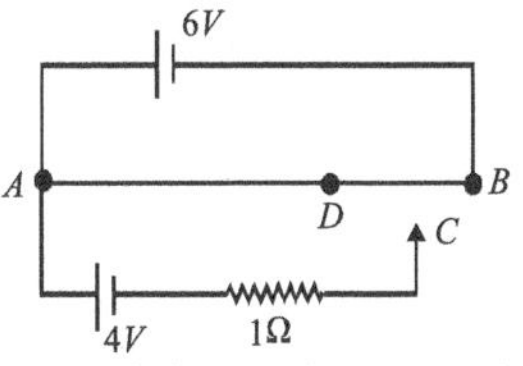

Q.22 What are the potentials at points *A* and *C* ?
(a) 6V, 2V (b) 8V, 4V
(c) 6V, 4V (d) 8V, 3V

Q.23 If the points *C* and *D* are connected by a wire, what will be the current through it ?
(a) zero (b) 1A
(c) 2A (d) 3A

DIRECTIONS (Qs. 24-Q.26) : Each of these questions contains two statements: Statement-1 (Assertion) and Statement-2 (Reason). Each of these questions has four alternative choices, only one of which is the correct answer. You have to select the correct choice.

(a) Statement-1 is True, Statement-2 is True; Statement-2 is a correct explanation for Statement-1.
(b) Statement-1 is True, Statement-2 is True; Statement-2 is NOT a correct explanation for Statement-1.
(c) Statement -1 is False, Statement-2 is True.
(d) Statement -1 is True, Statement-2 is False.

RESPONSE GRID					
	16. ⓐⓑⓒⓓ	17. ⓐⓑⓒⓓ	18. ⓐⓑⓒⓓ	19. ⓐⓑⓒⓓ	20. ⓐⓑⓒⓓ
	21. ⓐⓑⓒⓓ	22. ⓐⓑⓒⓓ	23. ⓐⓑⓒⓓ		

Space for Rough Work

Q.24 Statement -1 : Voltameter measures current more accurately than ammeter.

Statement -2 : Relative error will be small if measured from voltameter.

Q.25 Statement - 1 : A larger dry cell has higher emf.

Statement - 2 : The emf of a dry cell is independent of its size.

Q.26 Statement - 1 : In the circuit shown, V_{ab} or $V_a - V_b = 0$, if

$I = 2A$.

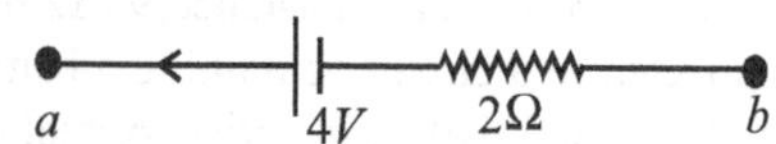

Statement - 2 : Potential difference across the terminals of a non ideal battery is less than its emf when a current flows through it.

RESPONSE GRID	24. ⓐⓑⓒⓓ	25. ⓐⓑⓒⓓ	26. ⓐⓑⓒⓓ

DAILY PRACTICE PROBLEM SHEET 37 - PHYSICS			
Total Questions	26	Total Marks	104
Attempted		Correct	
Incorrect		Net Score	
Cut-off Score	26	Qualifying Score	42
Success Gap = Net Score – Qualifying Score			
Net Score = (Correct × 4) – (Incorrect × 1)			

Space for Rough Work

DPP - Daily Practice Problems

Name :

Date :

Start Time :

End Time :

PHYSICS 38

SYLLABUS : CURRENT ELECTRICITY-3 : Wheatstone bridge, Meter bridge, Potentiometer-principle and its applications.

Max. Marks : 100 **Time : 60 min.**

GENERAL INSTRUCTIONS

- The Daily Practice Problem Sheet contains 25 MCQ's. For each question only one option is correct. Darken the correct circle/ bubble in the Response Grid provided on each page.
- You have to evaluate your Response Grids yourself with the help of solution booklet.
- Each correct answer will get you 4 marks and 1 mark shall be deduced for each incorrect answer. No mark will be given/ deducted if no bubble is filled. Keep a timer in front of you and stop immediately at the end of 60 min.
- The sheet follows a particular syllabus. Do not attempt the sheet before you have completed your preparation for that syllabus. Refer syllabus sheet in the starting of the book for the syllabus of all the DPP sheets.
- After completing the sheet check your answers with the solution booklet and complete the Result Grid. Finally spend time to analyse your performance and revise the areas which emerge out as weak in your evaluation.

DIRECTIONS (Q.1-Q.16) : There are **16** multiple choice questions. Each question has 4 choices (a), (b), (c) and (d), out of which **ONLY ONE** choice is correct.

Q.1 A Potentiometer wire of length 1 m is connected in series with 490 Ω resistance and 2V battery. If 0.2 mV/cm is the potential gradient, then resistance of the potentiometer wire is

(a) 4.9 Ω (b) 7.9 Ω (c) 5.9 Ω (d) 6.9 Ω

Q.2 Two resistances are connected in two gaps of a metre bridge. The balance point is 20 cm from the zero end. A resistance of 15 ohms is connected in series with the smaller of the two. The null point shifts to 40 cm. The value of the smaller resistance in ohm is

(a) 3 (b) 6 (c) 9 (d) 12

Q.3 In a potentiometer experiment the balancing with a cell is at length 240 cm. On shunting the cell with a resistance of 2 Ω, the balancing length becomes 120 cm. The internal resistance of the cell is

(a) 4 Ω (b) 2 Ω (c) 1 Ω (d) 0.5 Ω

Q.4 A potentiometer consists of a wire of length 4 m and resistance 10 Ω. It is connected to cell of emf 2 V. The potential difference per unit length of the wire will be

(a) 0.5 V/m (b) 10 V/m (c) 2 V/m (d) 5 V/m

Q.5 In given figure, the potentiometer wire AB has a resistance of 5 Ω and length 10 m. The balancing length AM for the emf of 0.4 V is

(a) 0.4 m
(b) 4 m
(c) 0.8 m
(d) 8 m

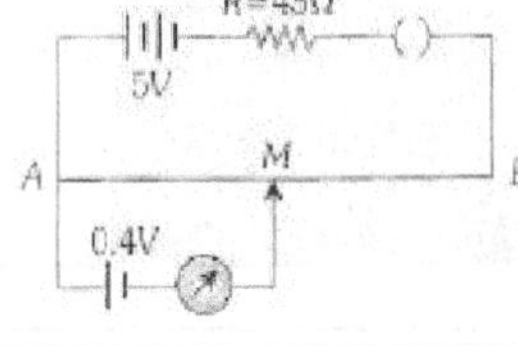

RESPONSE GRID	1. ⓐⓑⓒⓓ	2. ⓐⓑⓒⓓ	3. ⓐⓑⓒⓓ	4. ⓐⓑⓒⓓ	5. ⓐⓑⓒⓓ

Space for Rough Work

Q.6 In the circuit shown in the figure, the current flowing in 2Ω resistance

(a) 1.4 A
(b) 1.2 A
(c) 0.4 A
(d) 1.0 A

Q.7 For the post office box arrangement to determine the value of unknown resistance the unknown resistance should be connected between

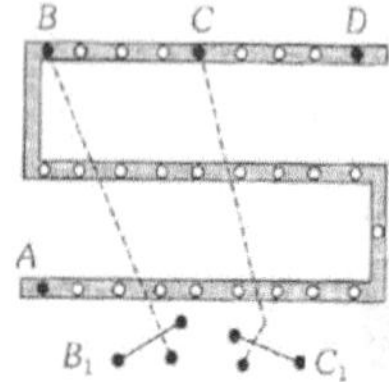

(a) B and C (b) C and D (c) A and D (d) B_1 and C_1

Q.8 The e.m.f. of a standard cell balances across 150 cm length of a wire of potentiometer. When a resistance of 2Ω is connected as a shunt with the cell, the balance point is obtained at 100 cm. The internal resistance of the cell is

(a) 0.1 Ω (b) 1Ω (c) 2Ω (d) 0.5Ω

Q.9 Five resistors are connected as shown in the diagram. The equivalent resistance between A and B is

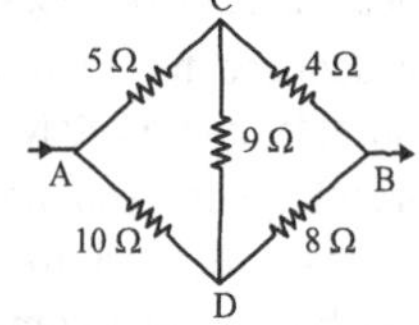

(a) 6 Ω
(b) 9 Ω
(c) 12 Ω
(d) 15 Ω

Q.10 A potentiometer has uniform potential gradient. The specific resistance of the material of the potentiometer wire is 10^{-7} ohm-meter and the current passing through it is 0.1 ampere; cross-section of the wire is $10^{-6}m^2$. The potential gradient along the potentiometer wire is

(a) 10^{-4} V/m (b) 10^{-6} V/m
(c) 10^{-2} V/m (d) 10^{-8} V/m

Q.11 Resistance in the two gaps of a meter bridge are 10 ohm and 30 ohm respectively. If the resistances are interchanged the balance point shifts by

(a) 33.3 cm (b) 66.67 cm (c) 25 cm (d) 50 cm

Q.12 A potentiometer has uniform potential gradient across it. Two cells connected in series (i) to support each other and (ii) to oppose each other are blanced over 6m and 2m respectively on the potentiometer wire. The e.m.f's of the cells are in the ratio of

(a) 1 : 2 (b) 1 : 1 (c) 3 : 1 (d) 2 : 1

Q.13 In a potentiometer experiment two cells of e.m.f E_1 and E_2 are used in series and in conjunction and the balancing length is found to be 58 cm of the wire. If the polarity of E_2 is reversed, then the balancing length becomes 29 cm.

The ratio $\frac{E_1}{E_2}$ of the e.m.f. of the two cells is

(a) 1:1 (b) 2:1 (c) 3:1 (d) 4:1

Q.14 The resistance of a 10 meter long potentiometer wire is 1 ohm/metre. A cell of e.m.f. 2.2 volts and a high resistance box are connected in series with this wire. The value of resistance taken from resistance box for getting potential gradient of 2.2 millivolt/metre will be

(a) 790 Ω (b) 810 Ω (c) 990 Ω (d) 1000 Ω

Q.15 In the shown arrangement of the experiment of the meter bridge if AC corresponding to null deflection of galvanometer is x, what would be its value if the radius of the wire AB is doubled

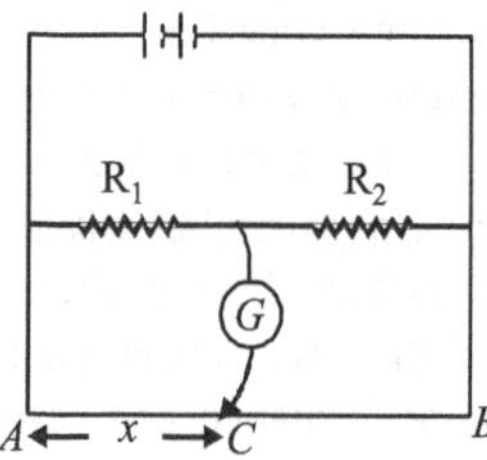

(a) x (b) $x/4$ (c) $4x$ (d) $2x$

Q.16 In meter bridge or Wheatstone bridge for measurement of resistance, the known and the unknown resistances are interchanged. The error so removed is

(a) End correction
(b) Index error
(c) Due to temperature effect
(d) Random error

RESPONSE GRID					
	6. ⓐⓑⓒⓓ	7. ⓐⓑⓒⓓ	8. ⓐⓑⓒⓓ	9. ⓐⓑⓒⓓ	10. ⓐⓑⓒⓓ
	11. ⓐⓑⓒⓓ	12. ⓐⓑⓒⓓ	13. ⓐⓑⓒⓓ	14. ⓐⓑⓒⓓ	15. ⓐⓑⓒⓓ
	16. ⓐⓑⓒⓓ				

Space for Rough Work

DIRECTIONS (Q.17-Q.19) : In the following questions, more than one of the answers given are correct. Select the correct answw]ers and mark it according to the following codes:

Codes :

(a) 1, 2 and 3 are correct **(b)** 1 and 2 are correct
(c) 2 and 4 are correct **(d)** 1 and 3 are correct

Q.17 Which of the following statements are correct?

(1) Voltmeter should have high resistance.
(2) Ammeter should have low resistance.
(3) Voltmeter is placed in parallel across the conductor in a circuit.
(4) Ammeter is placed in parallel across the conductor in a circuit.

Q.18 Which are correct statements?

(1) The Wheatstone bridge is most sensitive when all the four resistances are of the same order
(2) Kirchhoff's first law (for currents meeting at a junction in an electric circuit) expresses the conservtion of charge.
(3) The rheostat can be used as a potential divider.
(4) In a balanced Wheatstone bridge, interchanging the positions of galvanometer and cell affects the balance of the bridge.

Q.19 Figure shows a balanced Wheatstone's bridge

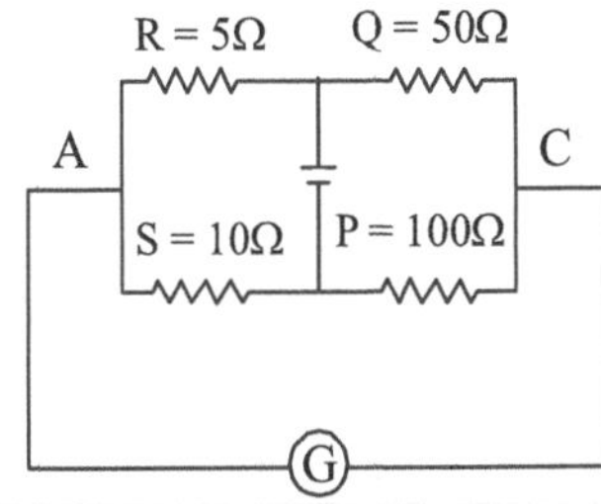

(1) If P is slightly increased, the current in the galvanometer flows from A to C.
(2) If P is slightly increased, the current in the galvanometer flows from C to A.
(3) If Q is slightly increased, the current in the galvanometer flows from C to A.
(4) If Q is slightly increased, the current in the galvanometer flows from A to C.

DIRECTIONS (Q.20-Q.22) : Read the passage given below and answer the questions that follows :

A battery is connected to a potentiometer and a balance point is obtained at 84 cm along the wire. When its terminals are connected by a 5Ω resistor, the balance point changes to 70 cm

Q.20 Calculate the internal resistance of the cell.

(a) 4 Ω (b) 2Ω (c) 5 Ω (d) 1 Ω

Q.21 Find the new position of the balance point when 5Ω resistance is replaced by 4Ω resistor.

(a) 26.5 cm (b) 52 cm (c) 67.2 cm (d) 83.3 cm

Q.22 How can we change a galvanometer with $R_e = 20.0\Omega$ and $I_{fs} = 0.00100$ A into a voltmeter with a maximum range of 10.0 V?

(a) By adding a resistance 9980 Ω in parallel with the galvanometer
(b) By adding a resistance 9980 Ω in series with the galvanometer
(c) By adding a resistance 8890 Ω in parallel with the galvanometer
(d) By adding a resistance 8890 Ω in series with the galvanometer

DIRECTIONS (Q. 23-Q.25) : Each of these questions contains two statements: Statement-1 (Assertion) and Statement-2 (Reason). Each of these questions has four alternative choices, only one of which is the correct answer. You have to select the correct choice.

(a) Statement-1 is True, Statement-2 is True; Statement-2 is a correct explanation for Statement-1.
(b) Statement-1 is True, Statement-2 is True; Statement-2 is NOT a correct explanation for Statement-1.
(c) Statement -1 is False, Statement-2 is True.
(d) Statement -1 is True, Statement-2 is False.

Q.23 Statement -1 : In meter bridge experiment, a high resistance is always connected in series with a galvanometer.

Statement -2 : As resistance increases current through the circuit increases.

RESPONSE GRID					
	17. ⓐⓑⓒⓓ	18. ⓐⓑⓒⓓ	19. ⓐⓑⓒⓓ	20. ⓐⓑⓒⓓ	21. ⓐⓑⓒⓓ
	22. ⓐⓑⓒⓓ	23. ⓐⓑⓒⓓ			

Space for Rough Work

Q.24 Statement -1: A potentiometer of longer length is used for accurate measurement.

Statement -2: The potential gradient for a potentiometer of longer length with a given source of *e.m.f.* becomes small.

Q.25 Statement -1: The e.m.f. of the driver cell in the potentiometer experiment should be greater than the e.m.f. of the cell to be determined.

Statement -2: The fall of potential across the potentiometer wire should not be less than the e.m.f. of the cell to be determined.

RESPONSE GRID	24. ⓐⓑⓒⓓ	25. ⓐⓑⓒⓓ

DAILY PRACTICE PROBLEM SHEET 38 - PHYSICS			
Total Questions	25	Total Marks	100
Attempted		Correct	
Incorrect		Net Score	
Cut-off Score	26	Qualifying Score	44
Success Gap = Net Score – Qualifying Score			
Net Score = (Correct × 4) – (Incorrect × 1)			

Space for Rough Work

DPP - Daily Practice Problems

Name : 　　　　　　　　　　　　**Date :**

Start Time : 　　　　　　　　　**End Time :**

PHYSICS 39

SYLLABUS : MAGNETIC EFFECTS OF CURRENT-1 (Magnetic field due to current carrying wires, Biot savart law)

Max. Marks : 108 　　　　　　　　　　　　**Time : 60 min.**

GENERAL INSTRUCTIONS

- The Daily Practice Problem Sheet contains 27 MCQ's. For each question only one option is correct. Darken the correct circle/ bubble in the Response Grid provided on each page.
- You have to evaluate your Response Grids yourself with the help of solution booklet.
- Each correct answer will get you 4 marks and 1 mark shall be deduced for each incorrect answer. No mark will be given/ deducted if no bubble is filled. Keep a timer in front of you and stop immediately at the end of 60 min.
- The sheet follows a particular syllabus. Do not attempt the sheet before you have completed your preparation for that syllabus. Refer syllabus sheet in the starting of the book for the syllabus of all the DPP sheets.
- After completing the sheet check your answers with the solution booklet and complete the Result Grid. Finally spend time to analyse your performance and revise the areas which emerge out as weak in your evaluation.

DIRECTIONS (Q.1-Q.19) : There are 19 multiple choice questions. Each question has 4 choices (a), (b), (c) and (d), out of which **ONLY ONE** choice is correct.

Q.1 The magnitude of magnetic field at a point having perpendicular distance 50 mm from a long straight conducting wire carrying a current of 3A is

(a) 0.12 G　　(b) 1.2 G

(c) 12 G　　(d) 0.012 G

Q.2 A circular arc of wire of radius of curvature r subtends an angle of $\pi/4$ radian at its centre. If i current is flowing in it then the magnetic induction at its centre is -

(a) $\frac{\mu_0 i}{8r}$　　(b) $\frac{\mu_0 i}{4r}$

(c) $\frac{\mu_0 i}{16r}$　　(d) 0

Q.3 A current i is flowing in a conductor PQRST shaped as shown in the figure. The radius of curved part QRS is r and length of straight portions PQ and ST is very large. The magnetic field at the centre O of the curved part is -

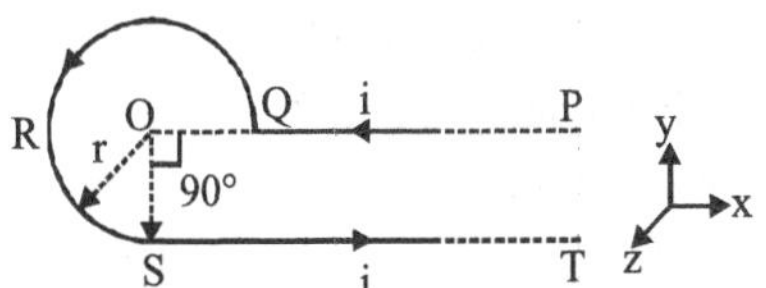

(a) $\frac{\mu_0 i}{4\pi r}\left[\frac{3\pi}{2}+1\right]\hat{k}$　　(b) $\frac{\mu_0 i}{4\pi r}\left[\frac{3\pi}{2}-1\right]\hat{k}$

(c) $\frac{\mu_0 i}{4\pi r}\left[\frac{3\pi}{2}+1\right](-\hat{k})$　　(d) $\frac{\mu_0 i}{4\pi r}\left[\frac{3\pi}{2}-1\right](-\hat{k})$

RESPONSE GRID 1. ⓐⓑⓒⓓ 2. ⓐⓑⓒⓓ 3. ⓐⓑⓒⓓ

Space for Rough Work

Q.4 Consider the loop PQRSP, carrying clockwise current i, shown in the figure. The magnitude of magnetic field at the centre O of the curved portion is

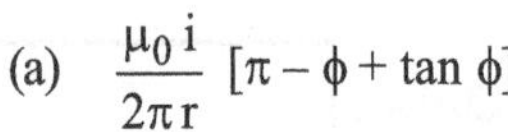

(a) $\frac{\mu_0 i}{2\pi r}[\pi - \phi + \tan\phi]$

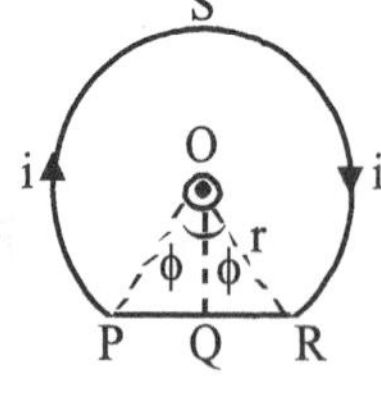

(b) $\frac{\mu_0 i}{2\pi r}$

(c) 0

(d) $\frac{\mu_0 i}{2\pi r}[\pi - \phi + \tan\phi]$

Q.5 A circular coil of 0.2 m diameter has 100 turns and carries a current of 0.1 ampere. The intensity of magnetic field at the centre of the coil is -

(a) 6.28×10^{-4} N/A.m (b) 62.8×10^{-4} N/A.m

(c) 6.28×10^{-5} N/A.m (d) 62.8×10^{-5} N/A.m

Q.6 For the arrangement of two current carrying identical coils shown in the figure, the magnetic field at the center O is (N and a represent number of turns and radius of each coil)-

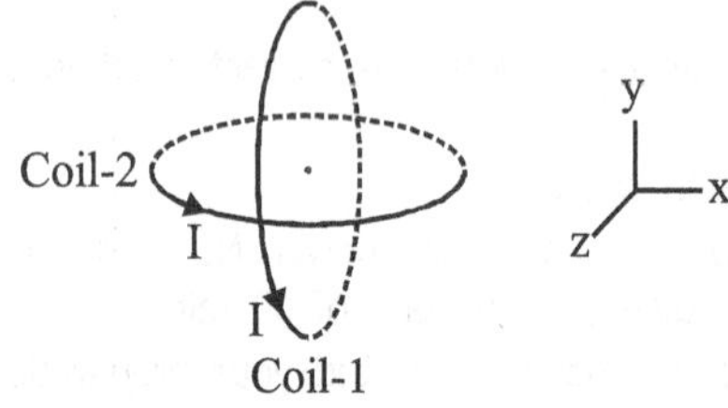

(a) $\frac{\mu_0 NI}{\sqrt{2}a}$ (b) $\frac{\mu_0 NI}{2\sqrt{2}a}$ (c) $\frac{\mu_0 NI}{2}$ (d) $\frac{\mu_0 NI}{2a}$

Q.7 A current is flowing through a conducting hollow pipe whose area of cross-section is shown in the fig. The value of magnetic induction will be zero at-

(a) Point P, Q and R

(b) Point R but not at P and Q

(c) Point Q but not at P and R

(d) Point P but not at Q and R

Q.8 Dimensional formula of μ_0 is-

(a) $MLT^{-2}A^{-2}$ (b) $MLT^{-2}A^{-2}$

(c) $MLT^{-2}A^{2}$ (d) $MLT^{2}A^{2}$

Q.9 A current of 1.0 ampere is flowing in the sides of an equilateral triangle of side 4.5×10^{-2} m. Find the magnetic field at the centroid of the triangle.
(Permeability constant $\mu_0 = 4\pi \times 10^{-7}$ V-s/A-m).

(a) 4.0×10^{-5} weber/m² (b) 6.0×10^{-8} weber/m²

(c) 2.0×10^{-5} weber/m² (d) 7.0×10^{-12} weber/m²

Q.10 An air-solenoid has 500 turns of wire in its 40 cm length. If the current in the wire be 1.0 ampere then the magnetic field on the axis inside the solenoid is -

(a) 15.7 gauss (b) 1.57 gauss

(c) 0.157 gauss (d) 0.0157 gauss

Q.11 A solenoid of length 0.2m has 500 turns on it. If 8.71×10^{-6} Weber/m² be the magnetic field at an end of the solenoid, then the current flowing in the solenoid is –

(a) $\frac{0.174}{\pi}$A (b) $\frac{0.0174}{\pi}$A (c) $\frac{17.4}{\pi}$A (d) $\frac{174}{\pi}$A

Q.12 A circular current carrying coil has a radius R. The distance from the centre of the coil on the axis where the magnetic induction will be $\frac{1}{8}th$ to its value at the centre of the coil, is

(a) $\frac{R}{\sqrt{3}}$ (b) $R\sqrt{3}$ (c) $2\sqrt{3}R$ (d) $\frac{2}{\sqrt{3}}R$

Q.13 The average radius of an air cored made toroid is 0.1 m and it has 500 turns. If it carries 0.5 ampere current, then the magnetic field inside it is :

(a) 5×10^{-4} tesla (b) 5×10^{-3} tesla

(c) 5×10^{-2} tesla (d) 2×10^{-3} tesla

Q.14 The straight long conductors AOB and COD are perpendicular to each other and carry current i_1 and i_2. The magnitude of the magnetic induction at point P at a distance a from the point O in a direction perpendicular to the plane ACBD is

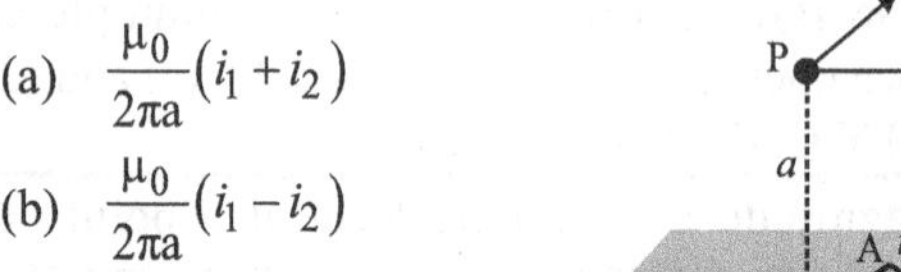

(a) $\frac{\mu_0}{2\pi a}(i_1 + i_2)$

(b) $\frac{\mu_0}{2\pi a}(i_1 - i_2)$

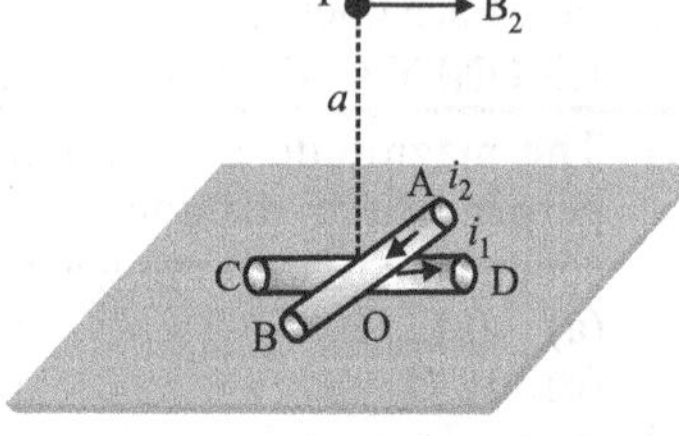

(c) $\frac{\mu_0}{2\pi a}\left(i_1^2 + i_2^2\right)^{1/2}$

(d) $\frac{\mu_0}{2\pi a}\frac{i_1 i_2}{(i_1 + i_2)}$

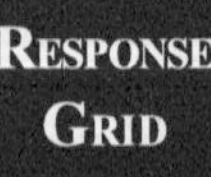

RESPONSE GRID				
4. ⓐⓑⓒⓓ	5. ⓐⓑⓒⓓ	6. ⓐⓑⓒⓓ	7. ⓐⓑⓒⓓ	8. ⓐⓑⓒⓓ
9. ⓐⓑⓒⓓ	10. ⓐⓑⓒⓓ	11. ⓐⓑⓒⓓ	12. ⓐⓑⓒⓓ	13. ⓐⓑⓒⓓ
14. ⓐⓑⓒⓓ				

Space for Rough Work

Q.15 A conducting circular loop of radius r carries a constant current i. It is placed in a uniform magnetic field $\vec{B}$, such that $\vec{B}$ is perpendicular to the plane of the loop. The magnetic force acting on the loop is

(a) $ir\vec{B}$ (b) $2\pi ri\vec{B}$ (c) zero (d) $\pi ri\vec{B}$

Q.16 The radius of a circular loop is r and a current i is flowing in it. The equivalent magnetic moment will be

(a) ir (b) $2\pi ir$ (c) $i\pi r^2$ (d) $\frac{1}{r^2}$

Q.17 A current of 30 A is flowing in a vertical straight wire. If the horizontal component of earth's magnetic field is 2×10^{-5} tesla then the distance of null point from wire is -

(a) 0.9 m (b) 0.3 mm (c) 0.3 cm (d) 0.3 m

Q.18 A charged particle is released from rest in a region of steady uniform electric and magnetic fields which are parallel to each other. The particle will move in a

(a) Straight line (b) Circle
(c) Helix (d) Cycloid

Q.19 A 6.28m long wire is turned into a coil of diameter 0.2m and a current of 1 amp is passed in it. The magnetic induction at its centre will be -

(a) 6.28×10^{-5} T (b) 0 T
(c) 6.28 T (d) 6.28×10^{-3} T

DIRECTIONS (Q.20-Q.21) : In the following questions, more than one of the answers given are correct. Select the correct answers and mark it according to the following codes:

Codes :

(a) 1, 2 and 3 are correct **(b)** 1 and 2 are correct
(c) 2 and 4 are correct **(d)** 1 and 3 are correct

Q.20 Two long straight parallel wires carry currents I_1 and I_2 respectively, in the same direction (as shown). The distance between the wires is R. The magnetic field at the centre of the two wires will be-

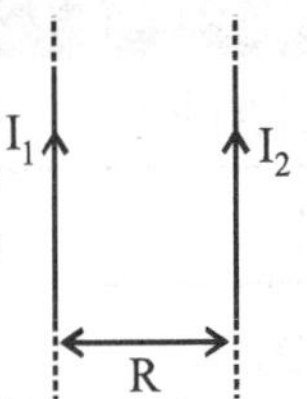

(1) $\frac{\mu_0 (I_1 - I_2)}{\pi R}$ into the plane of paper (If $I_1 > I_2$)

(2) $\frac{\mu_0 (I_2 - I_1)}{\pi R}$ out of the plane of paper (if $I_2 > I_1$)

(3) $\frac{\mu_0 (I_1 - I_2)}{\pi R^2}$ out of the plane of paper (if $I_2 > I_1$)

(4) $\frac{\mu_0 (I_2 - I_1)}{\pi R^2}$ into the plane of paper (if $I_1 > I_2$)

Q.21 A wire of length L carrying current I is bent into a circle of one turn. The field at the center of the coil is B_1. A similar wire of length L carrying current I is bent into a square of one turn. The field at its center is B_2. Then

(1) $B_1 > B_2$ (2) $B_1 = B_2$ (3) $\frac{B_1}{B_2} = 2$ (4) $B_1 < B_2$

DIRECTIONS (Q.22-Q.24) : Read the passage given below and answer the questions that follows :

A conducting wire is bent into a loop as shown in the figure. The segment AOB is parabolic given by the equation $y^2 = 2x$ while segment BA is a straight line parallel to the y-axis.

The magnetic field in the region is $\vec{B} = -8\hat{k}$ and the current in the wire is 2A.

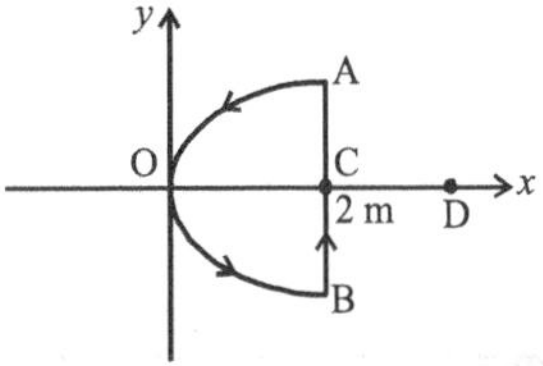

Q.22 The torque on the loop will be

(a) $16\sqrt{2}$ Nm (b) 16 Nm
(c) $18\sqrt{2}$ Nm (d) Zero

Q.23 The field created by the current in the loop at point C will be

(a) $-\frac{\mu_0}{4\pi}\hat{k}$ (b) $-\frac{\mu_0}{2\pi}\hat{k}$
(c) $-\frac{\mu_0\sqrt{2}}{\pi}\hat{k}$ (d) None of these

RESPONSE GRID					
	15. ⓐⓑⓒⓓ	16. ⓐⓑⓒⓓ	17. ⓐⓑⓒⓓ	18. ⓐⓑⓒⓓ	19. ⓐⓑⓒⓓ
	20. ⓐⓑⓒⓓ	21. ⓐⓑⓒⓓ	22. ⓐⓑⓒⓓ	23. ⓐⓑⓒⓓ	

Space for Rough Work

Q.24 Magnetic field at point D due to segment AO of the loop is directed parallel to

(a) $\hat{k}$ (b) $-\hat{k}$ (c) $\hat{i}$ (d) $\hat{j}$

DIRECTIONS (Q. 25-Q.27) : Each of these questions contains two statements: Statement-1 (Assertion) and Statement-2 (Reason). Each of these questions has four alternative choices, only one of which is] the correct answer. You have to select the correct choice.

(a) Statement-1 is True, Statement-2 is True; Statement-2 is a correct explanation for Statement-1.
(b) Statement-1 is True, Statement-2 is True; Statement-2 is NOT a correct explanation for Statement-1.
(c) Statement-1 is False, Statement-2 is True.
(d) Statement -1 is True, Statement-2 is False.

Q.25 Statement -1: Cyclotron does not accelerate electron.
Statement-2: Mass of the electron is very small.

Q.26 Statement-1: The ion cannot move with a speed beyond a certain limit in a cyclotron.
Statement-2: As velocity increases time taken by ion increases.

Q.27 Statement-1: If an electron, while coming vertically from outerspace, enter the earth's magnetic field, it is deflected towards west.
Statement-2: Electron has negative charge.

RESPONSE GRID	24. ⓐⓑⓒⓓ	25. ⓐⓑⓒⓓ	26. ⓐⓑⓒⓓ	27. ⓐⓑⓒⓓ

DAILY PRACTICE PROBLEM SHEET 39 - PHYSICS

Total Questions	27	Total Marks	108
Attempted		Correct	
Incorrect		Net Score	
Cut-off Score	28	Qualifying Score	44
Success Gap = Net Score – Qualifying Score			
Net Score = (Correct × 4) – (Incorrect × 1)			

Space for Rough Work

DPP - Daily Practice Problems

Name : ____________ **Date :** ________

Start Time : ________ **End Time :** ________

PHYSICS 40

SYLLABUS : MAGNETIC EFFECTS OF CURRENT-2 :
(Motion of charge particle in a magnetic field, force between current carrying wires.)

Max. Marks : 104 **Time : 60 min.**

GENERAL INSTRUCTIONS

- The Daily Practice Problem Sheet contains 26 MCQ's. For each question only one option is correct. Darken the correct circle/ bubble in the Response Grid provided on each page.
- You have to evaluate your Response Grids yourself with the help of solution booklet.
- Each correct answer will get you 4 marks and 1 mark shall be deduced for each incorrect answer. No mark will be given/ deducted if no bubble is filled. Keep a timer in front of you and stop immediately at the end of 60 min.
- The sheet follows a particular syllabus. Do not attempt the sheet before you have completed your preparation for that syllabus. Refer syllabus sheet in the starting of the book for the syllabus of all the DPP sheets.
- After completing the sheet check your answers with the solution booklet and complete the Result Grid. Finally spend time to analyse your performance and revise the areas which emerge out as weak in your evaluation.

DIRECTIONS (Q.1-Q.18) : There are 18 multiple choice questions. Each question has 4 choices (a), (b), (c) and (d), out of which **ONLY ONE** choice is correct.

Q.1 A proton, a deutron and an α-particle are accelerated through same potential difference and then they enter a normal uniform magnetic field. The ratio of their kinetic energies will be-
(a) 2 : 1 : 3 (b) 1 : 1 : 2 (c) 1 : 1 : 1 (d) 1 : 2 : 4

Q.2 A proton of energy 8eV is moving in a circular path in a uniform magnetic field. The energy of an α-particle moving in the same magnetic field and along the same path will be-
(a) 4eV (b) 2eV
(c) 8eV (d) 6eV

Q.3 An electron is revolving in a circular path of radius 2×10^{-10} m with a speed of 3×10^{6} m/s. The magnetic field at the centre of circular path will be-
(a) 1.2 T (b) 2.4 T (c) 0 (d) 3.6 T

Q.4 An α particle travels at an angle of 30° to a magnetic field 0.8 T with a velocity of 10^5 m/s. The magnitude of force will be-
(a) 1.28×10^{-14} N (b) $(1.28)\sqrt{3} \times10^{-4}$ N
(c) 1.28×10^{-4} N (d) $(12.8)\sqrt{3} \times10^{-4}$ N

Q.5 A beam of protons is moving horizontally towards you. As it approaches, it passes through a magnetic field directed downward. The beam deflects-
(a) to your left side
(b) to your right side
(c) does not deflect
(d) nothing can be said

RESPONSE GRID	1. ⓐⓑⓒⓓ	2. ⓐⓑⓒⓓ	3. ⓐⓑⓒⓓ	4. ⓐⓑⓒⓓ	5. ⓐⓑⓒⓓ

Space for Rough Work

Q.6 If a particle moves in a circular path in clockwise direction after entering into a downward vertical magnetic field. The charge on the particle is-

(a) positive (b) negative
(c) nothing can be said (d) neutral

Q.7 In the example above, after how much time, particle comes to the starting point for the first time. (mass of particle = m)

(a) $\frac{2\pi m}{3qB}$

(b) $\frac{2\pi m}{qB}$

(c) Never

(d) It will leave the circular path before coming to the starting point

Q.8 A current of 2.0 amp is flowing through a wire of length 50 cm. If this wire be placed at an angle of 60º with the direction of a uniform magnetic field of 5.0×10^{-4} N/Am the force on the wire will be-

(a) 4.33×10^{-4} N (b) 2.50×10^{-4} N
(c) 5.0×10^{-4} N (d) 2.33×10^{-4} N

Q.9 A particle of mass m and charge q moves with a constant velocity v along the positive x direction. It enters a region containing a uniform magnetic field B directed along the negative z direction, extending from $x = a$ to $x = b$. The minimum value of v required so that the particle can just enter the region $x < b$ is

(a) qb B/m (b) $q(b - a)$ B/m
(c) qa B/m (d) $q(b + a)$ B/2m

Q.10 A particle with charge q, moving with a momentum p, enters a uniform magnetic field normally. The magnetic field has magnitude B and is confined to a region of width d, where $d < \frac{p}{Bq}$, The particle is deflected by an angle θ in crossing the field. Then

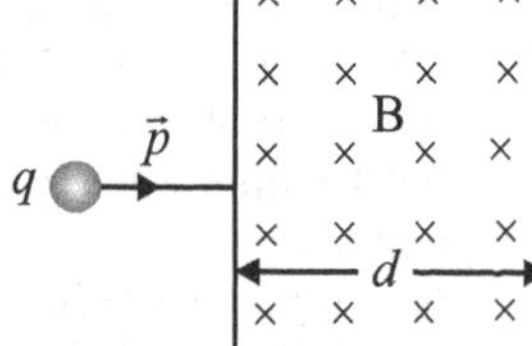

(a) $\sin\theta = \frac{Bqd}{p}$

(b) $\sin\theta = \frac{p}{Bqd}$

(c) $\sin\theta = \frac{Bp}{qd}$

(d) $\sin\theta = \frac{pd}{Bq}$

Q.11 An α particle is moving in a magnetic field of $(3\hat{i} + 2\hat{j})$ tesla with a velocity of $5 \times 10^5\ \hat{i}$ m/s. The magnetic force acting on the particle will be-

(a) 3.2×10^{-13} dyne (b) 3.2×10^{13} N
(c) 0 (d) 3.2×10^{-13} N

Q.12 If an α-particle moving with velocity 'v' enters perpendicular to a magnetic field then the magnetic force acting on it will be-

(a) evB (b) 2evB (c) 0 (d) 4evB

Q.13 What is the net force on the square coil ?

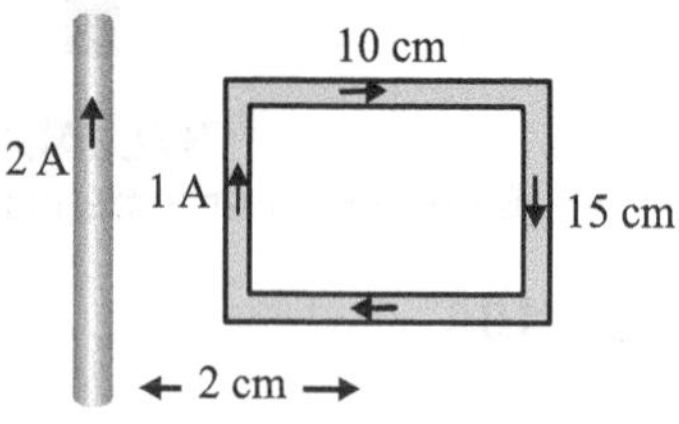

(a) 25×10^{-7} N towards wire
(b) 25×10^{-7} N away from wire
(c) 35×10^{-7} N towards wire
(d) 35×10^{-7} N away from wire

Q.14 A proton is to circulate the earth along the equator with a speed of 1.0×10^7 m/s. The minimum magnetic field which should be created at the equator for this purpose. (The mass of proton = 1.7×10^{-27} kg and radius of earth = 6.37×10^6 m.) will be (in Wb/m²)

(a) 1.6×10^{-19} (b) 1.67×10^{-8}
(c) 1.0×10^{-7} (d) 2×10^{-7}

Q.15 An α-particle is describing a circle of radius 0.45 m in a field of magnetic induction 1.2 weber/m². The potential difference required to accelerate the particle, (The mass of α-particle is 6.8×10^{-27} kg and its charge is 3.2×10^{-19} coulomb.) will be —

(a) 6×10^6 V (b) 2.3×10^{-12} V
(c) 7×10^6 V (d) 3.2×10^{-12} V

RESPONSE GRID					
	6. ⓐⓑⓒⓓ	7. ⓐⓑⓒⓓ	8. ⓐⓑⓒⓓ	9. ⓐⓑⓒⓓ	10. ⓐⓑⓒⓓ
	11. ⓐⓑⓒⓓ	12. ⓐⓑⓒⓓ	13. ⓐⓑⓒⓓ	14. ⓐⓑⓒⓓ	15. ⓐⓑⓒⓓ

Space for Rough Work

Q.16 An electron beam passes through a magnitic field of 2×10^{-3} weber/m^2 and an electric field of 1.0×10^4 volt/m both acting simultaneously. If the electric field is removed, what will be the radius of the electron path ?
(a) 1.43 cm. (b) 0.43 cm (c) 2.43 cm. (d) 3.43 cm.

Q.17 A straight horizontal copper wire carries a current i = 30 A. The linear mass density of the wire is 45 g/m. What is the magnitude of the magnetic field needed to balance its weight?
(a) 147 G (b) 441 G (c) 14.7 G (d) 0 G

Q.18 A 1m long conducting wire is lying at right angles to the magnetic field. A force of 1 kg. wt is acting on it in a magnetic field of 0.98 tesla. The current flowing in it will be-
(a) 100 A (b) 10 A (c) 1 A (d) 0

DIRECTIONS (Q.19-Q.20) : In the following questions, more than one of the answers given are correct. Select the correct answers and mark it according to the following codes:

Codes :
(a) 1, 2 and 3 are correct **(b)** 1 and 2 are correct
(c) 2 and 4 are correct **(d)** 1 and 3 are correct

Q.19 In the fig the two parallel wires PQ and ST are at 30 cm apart. The currents flowing in the wires are according to fig. The force acting over a length of 5m of the wires is-

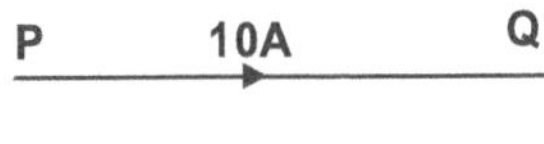

(1) 5×10^{-4} N (2) attraction
(3) 5×10^{-8} N (4) repulsion

Q.20 A beam of protons enters a uniform magnetic field of 0.3 tesla with a velocity of 4×10^5 m/s at an angle of 60° to the field. Then,
(Mass of the proton = 1.7×10^{-27} kg.)
(1) the radius of the helical path is 1.226×10^{-2} m
(2) the pitch of the helix is 4.45×10^{-2} m
(3) the radius of the helical path is 1.226×10^{-3} m
(4) the pitch of the helix is 4.45×10^{-4} m

DIRECTIONS (Q.21-Q.23) : Read the passage given below and answer the questions that follows :

A charge particles q enters in a magnetic field $\vec{B} = y\hat{i} + x\hat{j}$ with the velocity $\vec{v} = x\hat{i} + y\hat{j}$. Neglect any force other than magnetic force. Now answer the following question.

Q.21 When particle arrives at any point P (2, 2) then force acting on it, will be –
(a) Zero (b) $4\sqrt{2}q$ (c) 8q (d) $2\sqrt{2}q$

Q.22 Magnetic force F acting on charge is proportional to –
(a) $F \propto (x^2 - y^2)$
(b) $F \propto (x^2 + y^2)$
(c) $F \propto \sqrt{x^2 + y^2}$
(d) F does not depend on x or y co-ordinate

Q.23 Which of the following is true for the direction of magnetic force ?
(a) if x > y then force works along (–z) direction
(b) if x < y then force works along (+z) direction
(c) if x > y then force works along (+z) direction
(d) None of these

DIRECTIONS (Qs. 24-Q.26) : Each of these questions contains two statements: Statement-1 (Assertion) and Statement-2 (Reason). Each of these questions has four alternative choices, only one of which is the correct answer. You have to select the correct choice.

(a) Statement-1 is True, Statement-2 is True; Statement-2 is a correct explanation for Statement-1.
(b) Statement-1 is True, Statement-2 is True; Statement-2 is NOT a correct explanation for Statement-1.
(c) Statement -1 is False, Statement-2 is True.
(d) Statement -1 is True, Statement-2 is False.

Q.24 Statement -1 : If two long wires, hanging freely are connected to a battery in series, they come closer to each other.

Statement -2 : Force of repulsion acts between the two wires carrying current.

RESPONSE GRID					
	16. ⓐⓑⓒⓓ	17. ⓐⓑⓒⓓ	18. ⓐⓑⓒⓓ	19. ⓐⓑⓒⓓ	20. ⓐⓑⓒⓓ
	21. ⓐⓑⓒⓓ	22. ⓐⓑⓒⓓ	23. ⓐⓑⓒⓓ	24. ⓐⓑⓒⓓ	

Space for Rough Work

Q.25 Statement - 1 : For a charged particle to pass through a uniform electro-magnetic field without change in velocity, its velocity vector must be perpendicular to the magnetic field.

Statement - 2 : Net Lorentz force on the particle is given by $\vec{F} = q[\vec{E} + \vec{v} \times \vec{B}]$.

Q.26 Statement - 1 : If an electron is not deflected while passing through a certain region of space, then only possibility is that there is no magnetic region.

Statement - 2 : Magnetic force is directly proportional to the magnetic field applied.

RESPONSE GRID	25. ⓐⓑⓒⓓ	26. ⓐⓑⓒⓓ

DAILY PRACTICE PROBLEM SHEET 40 - PHYSICS			
Total Questions	26	Total Marks	104
Attempted		Correct	
Incorrect		Net Score	
Cut-off Score	26	Qualifying Score	46
Success Gap = Net Score – Qualifying Score			
Net Score = (Correct × 4) – (Incorrect × 1)			

Space for Rough Work

DPP - Daily Practice Problems

Name :

Date :

Start Time :

End Time :

PHYSICS 41

SYLLABUS : MAGNETIC EFFECTS OF CURRENT-3 (Magnetic dipole, Current carrying loop in magnetic field, Galvanometer)

Max. Marks : 116 **Time : 60 min.**

GENERAL INSTRUCTIONS

- The Daily Practice Problem Sheet contains 29 MCQ's. For each question only one option is correct. Darken the correct circle/ bubble in the Response Grid provided on each page.
- You have to evaluate your Response Grids yourself with the help of solution booklet.
- Each correct answer will get you 4 marks and 1 mark shall be deduced for each incorrect answer. No mark will be given/ deducted if no bubble is filled. Keep a timer in front of you and stop immediately at the end of 60 min.
- The sheet follows a particular syllabus. Do not attempt the sheet before you have completed your preparation for that syllabus. Refer syllabus sheet in the starting of the book for the syllabus of all the DPP sheets.
- After completing the sheet check your answers with the solution booklet and complete the Result Grid. Finally spend time to analyse your performance and revise the areas which emerge out as weak in your evaluation.

DIRECTIONS (Q.1-Q.20) : There are 20 multiple choice questions. Each question has 4 choices (a), (b), (c) and (d), out of which **ONLY ONE** choice is correct.

Q.1 A circular coil has radius 4 cm and 20 number of turns carries a current of 3 ampere. It is placed in a magnetic field of intensity 0.5 weber/m^2. The magnetic dipole moment of the coil is $\left(\text{Take } \pi = \frac{22}{7}\right)$

(a) 0.15 ampere m^2 (b) 0.3 ampere m^2
(c) 0.45 ampere m^2 (d) 0.6 ampere m^2

Q.2 A circular coil of radius 4 cm has 50 turns. In this coil a current of 2 A is flowing. It is placed in a magnetic field of 0.1 weber/m^2. The amount of work done in rotating it through 180° from its equilibrium position will be

(a) 0.1 J (b) 0.2 J (c) 0.4 J (d) 0.8 J

Q.3 The deflection in a moving coil galvanometer is

(a) directly proportional to the torsional constant
(b) directly proportional to the number of turns in the coil
(c) inversely proportional to the area of the coil
(d) inversely proportional to the current flowing

Q.4 A moving coil galvanometer has N number of turns in a coil of effective area A. It carries a current I. The magnetic field B is radial. The torque acting on the coil is

(a) NA^2B^2I (b) $NABI^2$ (c) N^2ABI (d) $NABI$

Q.5 A current carrying loop is free to turn in a uniform magnetic field. The loop will then come into equilibrium when its plane is inclined at

(a) 0° to the direction of the field
(b) 45° to the direction of the field
(c) 90° to the direction of the field
(d) 135° to the direction of the field

RESPONSE GRID	1. ⓐⓑⓒⓓ	2. ⓐⓑⓒⓓ	3. ⓐⓑⓒⓓ	4. ⓐⓑⓒⓓ	5. ⓐⓑⓒⓓ

Space for Rough Work

Q.6 A 100 turns coil shown in figure carries a current of 2 amp in a magnetic field B = 0.2Wb/m^2. The torque acting on the coil is

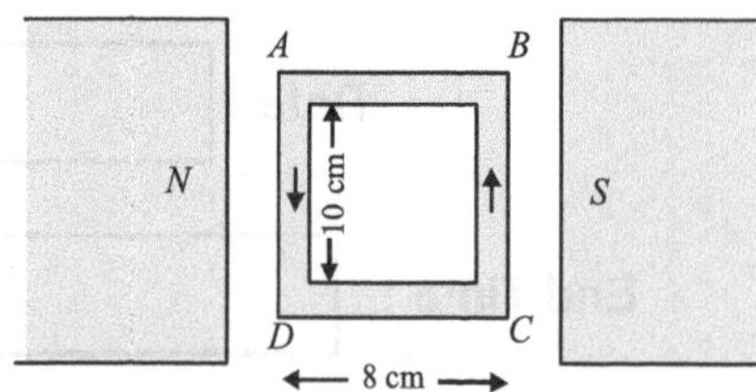

(a) 0.32 Nm tending to rotate the side *AD* out of the page
(b) 0.32 Nm tending to rotate the side *AD* into the page
(c) 0.0032 Nm tending to rotate the side *AD* out of the page
(d) 0.0032 Nm tending to rotate the side *AD* into the page

Q.7 A rectangular coil of size 20 cm × 20 cm has 100 turns and carries a current of 1 A. It is placed in a uniform magnetic field B = 0.5 T with the direction of magnetic field parallel to the plane of the coil. The magnitude of the torque required to hold this coil in this position is
(a) zero (b) 200 Nm
(c) 2 Nm (d) 10 Nm

Q.8 A circular loop of area 0.01m^2 carrying a current of 10 A, is held perpendicular to a magnetic field of intensity 0.1 T. The torque acting on the loop is
(a) zero (b) 0.01 Nm
(c) 0.001 Nm (d) 0.8Nm

Q.9 The magnetic moment of a current carrying circular coil is
(a) directly proportional to the length of the wire
(b) inversely proportional to the length of the wire
(c) directly proportional to the square of the length of the wire
(d) inversely proportional to the square of the length of the wire

Q.10 What is the shape of magnet in moving coil galvanometer to make the radial magnetic field?
(a) Concave cylindrical (b) Horse shoe magnet
(c) Convex cylindrical (d) None of these

Q.11 Current i is carried in a wire of length L. If the wire is turned into a circular coil, the maximum magnitude of torque in a given magnetic field B will be
(a) $\frac{LiB^2}{2}$ (b) $\frac{Li^2B}{2}$ (c) $\frac{L^2iB}{4\pi}$ (d) $\frac{Li^2B}{4\pi}$

Q.12 In ballistic galvanometer, the frame on which the coil is wound is non-metallic. It is
(a) to avoid the production of induced e.m.f.
(b) to avoid the production of eddy currents
(c) to increase the production of eddy currents
(d) to increase the production of induced e.m.f.

Q.13 A solenoid of length 0.4 m and having 500 turns of wire carries a current of 3 amp. A thin coil having 10 turns of wire and of radius 0.01 m carries a current of 0.4 amp. The torque (in Nm) required to hold the coil in the middle of the solenoid with its axis perpendicular to the axis of the solenoid is ($\mu_0 = 4\pi \times 10^{-7}$ V-s/A-m)
(a) 59.2×10^{-6} (b) 5.92×10^{-6}
(c) 0.592×10^{-6} (d) 0.592×10^{-4}.

Q.14 If an electron is moving with velocity v in an orbit of radius r in a hydrogen atom, then the equivalent magnetic moment is
(a) $\frac{\mu_0 e}{2r}$ (b) $\frac{ev}{r^2}$ (c) $\frac{ev \times 10^{-7}}{r^3}$ (d) $\frac{evr}{2}$

Q.15 In a moving coil galvanometer, the deflection of the coil θ is related to the electrical current i by the relation
(a) $i \propto \tan\theta$ (b) $i \propto \theta$
(c) $i \propto \theta^2$ (d) $i \propto \sqrt{\theta}$

Q.16 A thin circular wire carrying a current I has a magnetic moment M. The shape of the wire is changed to a square and it carries the same current. It will have a magnetic moment
(a) M (b) $\frac{4}{\pi^2}M$ (c) $\frac{4}{\pi}M$ (d) $\frac{\pi}{4}M$

Q.17 A ring of radius R, made of an insulating material carries a charge Q uniformly distributed on it. If the ring rotates about the axis passing through its centre and normal to plane of the ring with constant angular speed ω, then the magnitude of the magnetic moment of the ring is
(a) $Q\omega R^2$ (b) $\frac{1}{2}Q\omega R^2$ (c) $Q\omega^2 R$ (d) $\frac{1}{2}Q\omega^2 R$

RESPONSE GRID					
	6. ⓐⓑⓒⓓ	7. ⓐⓑⓒⓓ	8. ⓐⓑⓒⓓ	9. ⓐⓑⓒⓓ	10. ⓐⓑⓒⓓ
	11. ⓐⓑⓒⓓ	12. ⓐⓑⓒⓓ	13. ⓐⓑⓒⓓ	14. ⓐⓑⓒⓓ	15. ⓐⓑⓒⓓ
	16. ⓐⓑⓒⓓ	17. ⓐⓑⓒⓓ			

Space for Rough Work

Q.18 The $(\tau - \theta)$ graph for a current carrying coil placed in a uniform magnetic field is

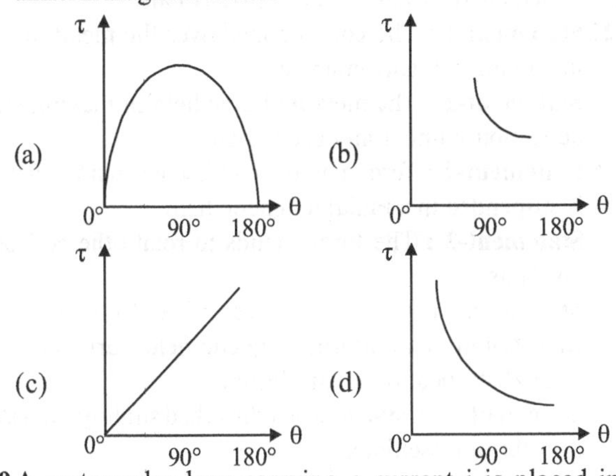

Q.19 A rectangular loop carrying a current i is placed in a uniform magnetic field B. The area enclosed by the loop is A. If there are n turns in the loop, the torque acting on the loop is given by

(a) $ni\vec{A}\times\vec{B}$ (b) $ni\vec{A}.\vec{B}$

(c) $\frac{1}{n}(i\vec{A}\times\vec{B})$ (d) $\frac{1}{n}(i\vec{A}.\vec{B})$

Q.20 The pole pieces of the magnet used in a pivoted coil galvanometer are

(a) plane surfaces of a bar magnet
(b) plane surfaces of a horse-shoe magnet
(c) cylindrical surfaces of a bar magnet
(d) cylindrical surfaces of a horse-shoe magnet

DIRECTIONS (Q.21-Q.23) : In the following questions, more than one of the answers given are correct. Select the correct answers and mark it according to the following codes:

Codes :

(a) 1, 2 and 3 are correct **(b)** 1 and 2 are correct
(c) 2 and 4 are correct **(d)** 1 and 3 are correct

Q.21 Four wires each of length 2.0 metre are bent into four loops *P, Q, R* and *S* and then suspended into uniform magnetic field. Same current is passed in each loop. Which statements are incorrect?

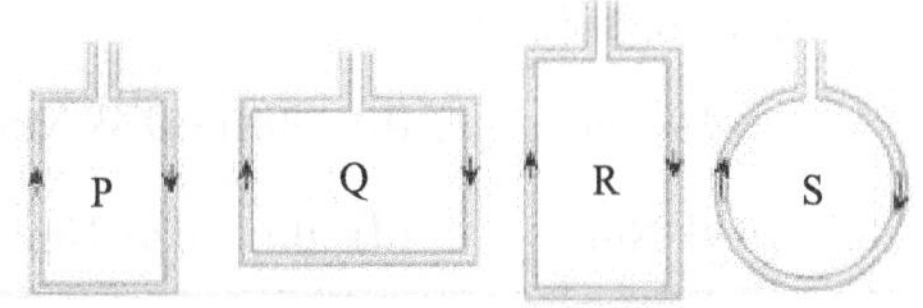

(1) Couple on loop *P* will be the highest
(2) Couple on loop *Q* will be the highest
(3) Couple on loop *R* will be the highest
(4) Couple on loop *S* will be the highest

Q.22 The sensitivity of a moving coil galvanometer can be increased by

(1) decreasing the couple per unit twist of the suspension
(2) increasing the number of turns in the coil
(3) decreasing the area of the coil
(4) decreasing the magnetic field

Q.23 A current carrying rectangular coil is placed in a uniform magnetic field. In which orientation, the coil will tend to rotate

(1) The magnetic field is parallel to the plane of the coil
(2) The magnetic field is at 45° with the plane of the coil
(3) In any orientation
(4) The magnetic field is perpendicular to the plane of the coil

DIRECTIONS (Q.24-Q.26) : Read the passage given below and answer the questions that follows :

A wire carrying a 10 A current is bent to pass through sides of a cube of side 10 cm as shown in figure. A magnetic field $\vec{B} = (2\hat{i} - 3\hat{j} + \hat{k})$T is present in the region. Then, find

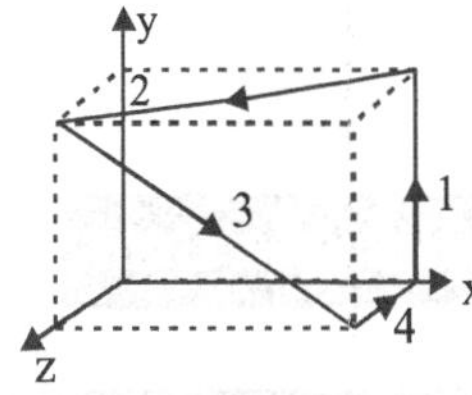

Q.24 The net force on the loop

(a) $\vec{F}_{net} = 0$ (b) $\vec{F}_{net} = (0.1\hat{i} - 0.2\hat{k})$N

(c) $\vec{F}_{net} = (0.3\hat{i} + 0.4\hat{k})$N (d) $\vec{F}_{net} = (0.36\hat{k})$N

Q.25 The magnetic moment vector of the loop.

(a) $(0.1\hat{i} + 0.05\hat{j} - 0.05\hat{k})Am^2$ (b) $(0.1\hat{i} + 0.05\hat{j} + 0.05\hat{k})Am^2$

(c) $(0.1\hat{i} - 0.05\hat{j} + 0.05\hat{k})Am^2$ (d) $(0.1\hat{i} - 0.05\hat{j} - 0.05\hat{k})Am^2$

RESPONSE GRID					
	18. ⓐⓑⓒⓓ	19. ⓐⓑⓒⓓ	20. ⓐⓑⓒⓓ	21. ⓐⓑⓒⓓ	22. ⓐⓑⓒⓓ
	23. ⓐⓑⓒⓓ	24. ⓐⓑⓒⓓ	25. ⓐⓑⓒⓓ		

Space for Rough Work

Q.26 The net torque on the loop.

(a) $-0.1\hat{i}+0.4\hat{k}$ Nm (b) $-0.1\hat{i}-0.4\hat{k}$ Nm

(c) $0.1\hat{i}-0.4\hat{k}$ Nm (d) $0.1\hat{i}-0.4\hat{k}$ Nm

DIRECTIONS (Q. 27-Q.29) : Each of these questions contains two statements: Statement-1 (Assertion) and Statement-2 (Reason). Each of these questions has four alternative choices, only one of which is the correct answer. You have to select the correct choice.

(a) Statement-1 is True, Statement-2 is True; Statement-2 is a correct explanation for Statement-1.
(b) Statement-1 is True, Statement-2 is True; Statement-2 is NOT a correct explanation for Statement-1.
(c) Statement-1 is False, Statement-2 is True.
(d) Statement-1 is True, Statement-2 is False.

Q.27 Statement-1 : The coil is bound over the metallic frame in moving coil galvanometer.
Statement-2 : The metallic frame help in making steady deflection without any oscillation.

Q.28 Statement-1 : Torque on the coil is maximum, when coil is suspended in a radial magnetic field.
Statement-2 : The torque tends to rotate the coil on its own axis.

Q.29 Statement-1 : A current carrying loop placed in equilibrium in a uniform magnetic field starts oscillating when disturbed from equilibrium.
Statement-2 : A system when disturbed slightly from stable equilibrium oscillates.

RESPONSE GRID	26. ⓐⓑⓒⓓ	27. ⓐⓑⓒⓓ	28. ⓐⓑⓒⓓ	29. ⓐⓑⓒⓓ

DAILY PRACTICE PROBLEM SHEET 41 - PHYSICS			
Total Questions	29	Total Marks	116
Attempted		Correct	
Incorrect		Net Score	
Cut-off Score	30	Qualifying Score	48
Success Gap = Net Score – Qualifying Score			
Net Score = (Correct × 4) – (Incorrect × 1)			

Space for Rough Work

DPP - Daily Practice Problems

Name : | Date :

Start Time : | End Time :

PHYSICS 42

SYLLABUS : MAGNETISM AND MATTER - 1 (Bar magnet as an equivalent solenoid, Magnetic field lines, Earth's magnetic field and magnetic elements)

Max. Marks : 120 **Time : 60 min.**

GENERAL INSTRUCTIONS

- The Daily Practice Problem Sheet contains 30 MCQ's. For each question only one option is correct. Darken the correct circle/ bubble in the Response Grid provided on each page.
- You have to evaluate your Response Grids yourself with the help of solution booklet.
- Each correct answer will get you 4 marks and 1 mark shall be deduced for each incorrect answer. No mark will be given/ deducted if no bubble is filled. Keep a timer in front of you and stop immediately at the end of 60 min.
- The sheet follows a particular syllabus. Do not attempt the sheet before you have completed your preparation for that syllabus. Refer syllabus sheet in the starting of the book for the syllabus of all the DPP sheets.
- After completing the sheet check your answers with the solution booklet and complete the Result Grid. Finally spend time to analyse your performance and revise the areas which emerge out as weak in your evaluation.

DIRECTIONS (Q.1-Q.21) : There are 21 mulltiple choice questions. Each question has 4 choices (a), (b), (c) and (d), out of which **ONLY ONE** choice is correct.

Q.1 A north pole of strength 50 Am and south pole of strength 100 Am are separated by a distance of 10 cm in air. Find the force between them.
(a) 50×10^{-3} N (b) 25×10^{-3} N
(c) 20×10^{-6} N (d) 30×10^{-18} N

Q.2 Calculate magnetic induction at a distance of 20 cm from a pole of strength 40 Am in air.
(a) 10^{-4} wb/m^2 (b) 10^{-8} wb/m^2
(c) 10^{-1} wb/m^2 (d) 10^{-12} wb/m^2

Q.3 A bar magnet of length 0.2 m and pole strength 5 Am is kept in a uniform magnetic induction field of strength 15 wb/m^{-2} making an angle of 30° with the field. Find the couple acting on it.
(a) 2.5 Nm (b) 5.5 Nm (c) 7.5 Nm (d) 9.0 Nm

Q.4 The force experienced by a pole of strength 100 Am at a distance of 0.2 m from a short magnet of length 5 cm and pole strength of 200 Am on its axial line will be
(a) 2.5×10^{-2} N (b) 2.5×10^{-3} N
(c) 5.0×10^{-2} N (d) 5.0×10^{-3} N

Q.5 A magnet of moment M is lying in a magnetic field of induction B. W_1 is the work done in turning it from 0° to 60° and W_2 is the work done in turning it from 30° to 90°. Then
(a) $W_2 = W_1$ (b) $W_2 = \frac{W_1}{2}$
(c) $W_2 = 2W_1$ (d) $W_2 = \sqrt{3}\, W_1$

RESPONSE GRID	1. ⓐⓑⓒⓓ	2. ⓐⓑⓒⓓ	3. ⓐⓑⓒⓓ	4. ⓐⓑⓒⓓ	5. ⓐⓑⓒⓓ

Space for Rough Work

Q.6 A bar magnet of magnetic moment 4.0 A-m^2 is free to rotate about a vertical axis through its centre. The magnet is released from rest from east-west position. Kinetic energy of the magnet in north-south position will be (Horizontal component of earth's magnetic field $B_H = 25\mu T$)
(a) 10^{-2} J (b) 10^{-4} J (c) 10^{-6} J (d) 0

Q.7 The length of a bar magnet is 10 cm and its pole strength is 10^{-3} Weber. It is placed in a magnetic field of induction $4\pi \times 10^{-3}$ Tesla in a direction making an angle 30° with the field direction. The value of torque acting on the magnet will be –
(a) $2\pi \times 10^{-7}$ N-m (b) $2\pi \times 10^{-5}$ N-m
(c) 0.5×10^{2} N-m (d) None of these

Q.8 At magnetic poles of earth, angle of dip is
(a) zero (b) 45° (c) 90° (d) 180°

Q.9 A short bar magnet is placed with its north pole pointing south. The neutral point is 10 cm away from the centre of the magnet. If H = 0.4 gauss, calculate magnetic moment of the magnet.
(a) 2 Am^2 (b) 1A m^2 (c) 0.1 A m^2 (d) 0.2 Am^2

Q.10 A bar magnet with its poles 25 cm apart and of pole strength 24.0 A-m rests with its centre on a frictionless pivot. A force F is applied on the magnet at a distance of 12 cm from the pivot, so that it is held in equilibrium at an angle of 30° with respect to a magnetic field of induction 0.25 T. The value of force F is
(a) 65.62 N (b) 2.56 N (c) 6.52 N (d) 6.25 N

Q.11 A small magnet of magnetic moment 4A-m^2 is placed on a deflection magnetometer in tan-B position at a distance of 20 cm from the compass needle. At what distance from compass needle should another small magnet of moment 0.5 A-m^2 be placed such that the deflection of the needle remains zero ?
(a) 12 cm (b) 10 cm (c) 20 cm (d) 30 cm

Q.12 The ratio of intensities of magnetic field, at distances x and 2x from the centre of magnet of length 2cm on its axis, will be
(a) 4 : 1 (b) 4 : 1 approx
(c) 8 : 1 (d) 8 : 1 approx

Q.13 Two magnets A and B are identical and these are arranged as shown in the figure. Their length is negligible in comparison to the separation between them. A magnetic needle is placed between the magnets at point P which gets deflected through an angle θ under the influence of magnets. The ratio of distances d_1 and d_2 will be
(a) $(2 \tan \theta)^{1/3}$
(b) $(2 \tan \theta)^{-1/3}$
(c) $(2 \cot \theta)^{1/3}$
(d) $(2 \cot \theta)^{-1/3}$

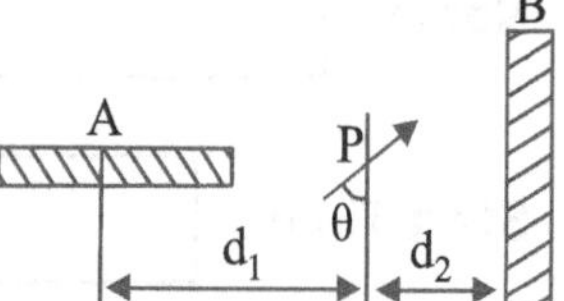

Q.14 The period of oscillation of a freely suspended bar magnet is 4 second. If it is cut into two equal parts length wise then the time period of each part will be
(a) 4 sec (b) 2 sec (c) 0.5 sec (d) 0.25 sec

Q.15 The length, breadth and mass of two bar magnets are same but their magnetic moments are 3M and 2M respectively. These are joined pole to pole and are suspended by a string. When oscillated in a magnetic field of strength B, the time period obtained is 5s. If the poles of either of the magnets are reverse then the time period of the combination in the same magnetic field will be –
(a) $3\sqrt{3}$ s (b) $2\sqrt{2}$ s (c) $5\sqrt{5}$ s (d) 1 s

Q.16 A thin magnetic needle oscillates in a horizontal plane with a period T. It is broken into n equals parts. The time period of each part will be
(a) T (b) $\frac{T}{n}$ (c) Tn^2 (d) $\frac{T}{n^2}$

Q.17 A bar magnet made of steel has a magnetic moment of 2.5 A-m^2 and a mass of 6.6×10^3 kg. If the density of steel is 7.9×10^9 kg/m^3, find the intensity of magnetization of the magnet.
(a) 3.0×10^6 A/m (b) 2.0×10^6 A/m
(c) 5.0×10^6 A/m (d) 1.2×10^6 A/m

Q.18 A short magnet of length 4 cm is kept at a distance of 20 cm to the east of a compass box such that is axis is perpendicular to the magnetic meridian. If the deflection produced is 45°, find the pole strength (H = 30 Am^{-1})
(a) 17.7 Am (b) 44.2 Am
(c) 27.7 Am (d) 37.7 Am

RESPONSE GRID					
	6. ⓐⓑⓒⓓ	7. ⓐⓑⓒⓓ	8. ⓐⓑⓒⓓ	9. ⓐⓑⓒⓓ	10. ⓐⓑⓒⓓ
	11. ⓐⓑⓒⓓ	12. ⓐⓑⓒⓓ	13. ⓐⓑⓒⓓ	14. ⓐⓑⓒⓓ	15. ⓐⓑⓒⓓ
	16. ⓐⓑⓒⓓ	17. ⓐⓑⓒⓓ	18. ⓐⓑⓒⓓ		

Space for Rough Work

Q.19 A 10 cm long bar magnet of magnetic moment 1.34 Am^2 is placed in the magnetic meridian with its south pole pointing geographical south. The neutral point is obtained at a distance of 15 cm from the centre of the magnet. Calculate the horizontal component of earth's magnetic field.
(a) 0.12×10^{-4} T (b) 0.21×10^{-4} T
(c) 0.34×10^{-4} T (d) 0.87×10^{-7} T

Q.20 A 30 cm long bar magnet is placed in the magnetic meridian with its north pole pointing south. The neutral point is obtained at a distance of 40 cm from the centre of the magnet. Pole strength of the magnet is (The horizontal component of earth's magnetic field is 0.34 Gauss)
(a) 26.7 Am (b) 16.7 Am (c) 12.7 Am (d) 15.2 Am

Q.21 A long straight horizontal cable carries a current of 2.5 A in the direction 10° south of west to 10° north of east. The magnetic meridian of the place happens to be 10° west of the geographic meridian. The earth's magnetic field at the location is 0.33 Gauss, and the angle of dip is zero. Distance of the line of neutral points from the cable is (Ignore the thickness of the cable).
(a) 1.5 cm (b) 2.5 cm (c) 3.5 cm (d) 2.0 cm

DIRECTIONS (Q.22-Q.24) : In the following questions, more than one of the answers given are correct. Select the correct answers and mark it according to the following codes:

Codes :
(a) 1, 2 and 3 are correct **(b)** 1 and 2 are correct
(c) 2 and 4 are correct **(d)** 1 and 3 are correct

Q.22 Which of the following is/are not the main difference between electric lines of force and magnetic lines of force ?
(1) Electric lines of force are closed curves whereas magnetic lines of force are open curves.
(2) Magnetic lines of force cut each other whereas electric lines of force do not cut.
(3) Electric lines of force cut each other whereas magnetic lines of force do not cut.
(4) Electric lines of force are open curves whereas magnetic lines of force are closed curves.

Q.23 The correct statements regarding the lines of force of the magnetic field B are
(1) Magnetic intensity is a measure of lines of force passing through unit area held normal to it
(2) Magnetic lines of force form a closed curve
(3) Due to a magnet magnetic lines of force never cut each other
(4) Inside a magnet, its magnetic lines of force move from north pole of a magnet towards its south pole

Q.24 A short bar magnet of magnetic moment 5.25×10^{-2} JT^{-1} is placed with its axis perpendicular to the earth's field direction. Magnitude of the earth's field at the place is given to be 0.42 G. Ignore the length of the magnet in comparison to the distance involved. Then
(1) the distance from the centre of the magnet on its normal bisector at which the resultant field is inclined at 45° with the earth's field is 5 cm
(2) the distance from the centre of the magnet on its axis at which the resultant field inclined at 45° with the earth's field is 6.3 cm
(3) the distance from the centre of the magnet on its normal bisector at which the resultant field inclined at 45° with the earth's field is 8.3 cm
(4) the distance from the centre of the magnet on its axis at which the resultant field inclined at 45° with the earth's field is 8 cm

DIRECTIONS (Q.25-Q.27) : Read the passage given below and answer the questions that follows :

A telephone cable at a place has four long, straight horizontal wires carrying a current of 1.0 A in the same direction east to west. The earth's magnetic field at the place is 0.39 Gauss, and the angle of dip is 35°. The magnetic declination is nearly zero. (cos 35° = 0.82, sin 35° = 0.57)

Q.25 The magnetic field produced by four current carrying straight cable wires at a distance 4 cm is
(a) 0.2 Gauss (b) 0.3 Gauss
(c) 0.4 Gauss (d) 0.5 Gauss

Q.26 The resultant magnetic field below at points 4cm and above the cable are
(a) 0.25, 0.56 Gauss (b) 0.14, 0.32 Gauss
(c) 0.23, 0.34 Gauss (d) 0.52, 0.62 Gauss

RESPONSE GRID				
19. ⓐⓑⓒⓓ	20. ⓐⓑⓒⓓ	21. ⓐⓑⓒⓓ	22. ⓐⓑⓒⓓ	23. ⓐⓑⓒⓓ
24. ⓐⓑⓒⓓ	25. ⓐⓑⓒⓓ	26. ⓐⓑⓒⓓ		

Space for Rough Work

Q.27 The angle that resultant makes with horizontal in case below and above the cable respectively, are

(a) 30°, 45° (b) $\tan^{-1} 1.8, \tan^{-1} 0.43$

(c) $\tan^{-1} 2, \tan^{-1} \sqrt{2}$ (d) $\sin^{-1} 0.7, \sin^{-1} 0.9$

DIRECTIONS (Q. 28-Q.30) : Each of these questions contains two statements: Statement-1 (Assertion) and Statement-2 (Reason). Each of these questions has four alternative choices, only one of which is the correct answer. You have to select the correct choice.

(a) Statement-1 is True, Statement-2 is True; Statement-2 is a correct explanation for Statement-1.

(b) Statement-1 is True, Statement-2 is True; Statement-2 is NOT a correct explanation for Statement-1.

(c) Statement-1 is False, Statement-2 is True.

(d) Statement-1 is True, Statement-2 is False.

Q.28 Statement-1 : Gauss theorem is not applicable in magnetism.

Statement-2 : Mono magnetic pole does not exist.

Q.29 Statement-1 : A compass needle when placed on the magnetic north pole of the earth cannot rotate in vertical direction.

Statement-2 : The earth has only horizontal component of its magnetic field at the north poles.

Q.30 Statement-1 : We cannot think of magnetic field configuration with three poles.

Statement-2 : A bar magnet does not exert a torque on itself due to its own field.

RESPONSE GRID	27. ⓐⓑⓒⓓ	28. ⓐⓑⓒⓓ	29. ⓐⓑⓒⓓ	30. ⓐⓑⓒⓓ

DAILY PRACTICE PROBLEM SHEET 42 - PHYSICS			
Total Questions	30	Total Marks	120
Attempted		Correct	
Incorrect		Net Score	
Cut-off Score	30	Qualifying Score	50
Success Gap = Net Score – Qualifying Score			
Net Score = (Correct × 4) – (Incorrect × 1)			

Space for Rough Work

DPP - Daily Practice Problems

Name :

Date :

Start Time :

End Time :

PHYSICS 43

SYLLABUS : MAGNETISM & MATTER-2 (Para, dia and ferro-magnetic substances, magnetic susceptibility and permeability, Hysteresis, Electromagnets and permanent magnets.)

Max. Marks : 120 **Time : 60 min.**

GENERAL INSTRUCTIONS

- The Daily Practice Problem Sheet contains 30 MCQ's. For each question only one option is correct. Darken the correct circle/ bubble in the Response Grid provided on each page.
- You have to evaluate your Response Grids yourself with the help of solution booklet.
- Each correct answer will get you 4 marks and 1 mark shall be deduced for each incorrect answer. No mark will be given/ deducted if no bubble is filled. Keep a timer in front of you and stop immediately at the end of 60 min.
- The sheet follows a particular syllabus. Do not attempt the sheet before you have completed your preparation for that syllabus. Refer syllabus sheet in the starting of the book for the syllabus of all the DPP sheets.
- After completing the sheet check your answers with the solution booklet and complete the Result Grid. Finally spend time to analyse your performance and revise the areas which emerge out as weak in your evaluation.

DIRECTIONS (Q.1-Q.24) : There are 24 multiple choice questions. Each question has 4 choices (a), (b), (c) and (d), out of which **ONLY ONE** choice is correct.

Q.1 Susceptibility of ferromagnetic substance is

(a) > 1 (b) < 1 (c) 0 (d) 1

Q.2 Among the following properties describing diamagnetism identify the property that is wrongly stated.

(a) Diamagnetic material do not have permanent magnetic moment

(b) Diamagnetism is explained in terms of electromagnetic induction.

(c) Diamagnetic materials have a small positive susceptibility

(d) The magnetic moment of individual electrons neutralize each other

Q.3 If the magnetic dipole moment of an atom of diamagnetic material, paramagnetic material and ferromagnetic material denoted by μ_d, μ_p, μ_f respectively then

(a) $\mu_d, \neq 0$ and $\mu_f \neq 0$ (b) $\mu_p = 0$ and $\mu_f \neq 0$

(c) $\mu_d = 0$ and $\mu_p \neq 0$ (d) $\mu_d \neq 0$ and $\mu_p = 0$

Q.4 When a piece of a ferromagnetic substance is put in a uniform magnetic field, the flux density inside it is four times the flux density away from the piece. The magnetic permeability of the material is

(a) 1 (b) 2

(c) 3 (d) 4

RESPONSE GRID	1. ⓐⓑⓒⓓ	2. ⓐⓑⓒⓓ	3. ⓐⓑⓒⓓ	4. ⓐⓑⓒⓓ

Space for Rough Work

Q.5 The given figure represents a material which is

(a) Paramagnetic

(b) Diamagnetic

(c) Ferromagnetic

(d) None of these

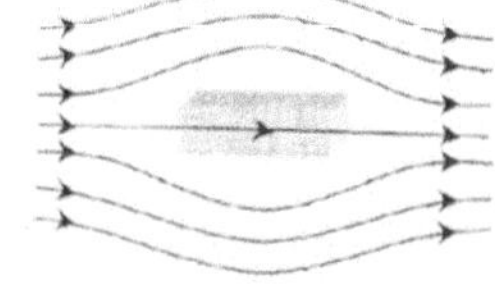

Q.6 Liquid oxygen remains suspended between two pole faces of a magnet because it is

(a) diamagnetic (b) paramagnetic

(c) ferromagnetic (d) antiferromagnetic

Q.7 A superconductor exhibits perfect

(a) ferrimagnetism (b) ferromagnetism

(c) paramagnetism (d) diamagnetism

Q.8 Which of the following is the most suitable for the core of electromagnets?

(a) Soft iron (b) Steel

(c) Copper-nickel alloy (d) Air

Q.9 The universal property of all substances is

(a) diamagnetism (b) ferromagnetism

(c) paramagnetism (d) all of these

Q.10 If a magnetic substance is kept in a magnetic field, then which of the following substance is thrown out ?

(a) Paramagnetic (b) Ferromagnetic

(c) Diamagnetic (d) Antiferromagnetic

Q.11 In the hysteresis cycle, the value of H needed to make the intensity of magnetisation zero is called

(a) Retentivity (b) Coercive force

(c) Lorentz force (d) None of these

Q.12 If a diamagnetic solution is poured into a *U*-tube and one arm of this *U*-tube placed between the poles of a strong magnet with the meniscus in a line with field, then the level of the solution will

(a) rise (b) fall

(c) oscillate slowly (d) remain as such

Q.13 The relative permeability is represented by μ_r and the susceptibility is denoted by χ for a magnetic substance. Then for a paramagnetic substance

(a) $\mu_r < 1, \chi < 0$ (b) $\mu_r < 1, \chi > 0$

(c) $\mu_r > 1, \chi < 0$ (d) $\mu_r > 1, \chi > 0$

Q.14 The use of study of hysteresis curve for a given material is to estimate the

(a) voltage loss (b) hysteresis loss

(c) current loss (d) all of these

Q.15 The magnetic moment of atomic neon is

(a) zero (b) μB/2

(c) μB (d) 3μB/2

Q.16 A ferromagnetic material is heated above its Curie temperature, then which one is a correct statement ?

(a) Ferromagnetic domains are perfectly arranged

(b) Ferromagnetic domains becomes random

(c) Ferromagnetic domains are not influenced

(d) Ferromagnetic material changes itself into diamagnetic material

Q.17 If a diamagnetic substance is brought near north or south pole of a bar magnet, it is

(a) attracted by the poles

(b) repelled by the poles

(c) repelled by the north pole and attracted by the south pole

(d) attracted by the north pole and repelled by the south pole

Q.18 The material of permanent magnet has

(a) high retentivity, low coercivity

(b) low retentivity, high coercivity

(c) low retentivity, low coercivity

(d) high retentivity, high coercivity

Q.19 Diamagnetic substances are

(a) feebly attracted by magnets

(b) strongly attracted by magnets

(c) feebly repelled by magnets

(d) strongly repelled by magnets

RESPONSE GRID					
	5. ⓐⓑⓒⓓ	6. ⓐⓑⓒⓓ	7. ⓐⓑⓒⓓ	8. ⓐⓑⓒⓓ	9. ⓐⓑⓒⓓ
	10. ⓐⓑⓒⓓ	11. ⓐⓑⓒⓓ	12. ⓐⓑⓒⓓ	13. ⓐⓑⓒⓓ	14. ⓐⓑⓒⓓ
	15. ⓐⓑⓒⓓ	16. ⓐⓑⓒⓓ	17. ⓐⓑⓒⓓ	18. ⓐⓑⓒⓓ	19. ⓐⓑⓒⓓ

Space for Rough Work

Q.20 For an isotropic medium B, μ, H and M are related as (where B, μ_0, H and M have their usual meaning in the context of magnetic material)

(a) $(B-M) = \mu_0 H$ (b) $M = \mu_0(H+M)$
(c) $H = \mu_0(H+M)$ (d) $B = \mu_0(H+M)$

Q.21 Relative permeability of iron is 5500, then its magnetic susceptibility will be

(a) 5500×10^7 (b) 5500×10^{-7}
(c) 5501 (d) 5499

Q.22 A magnetising field of 2×10^3 amp/m produces a magnetic flux density of 8π Tesla in an iron rod. The relative permeability of the rod will be

(a) 10^2 (b) 10^0 (c) 10^4 (d) 10^1

Q.23 The mass of a specimen of a ferromagnetic material is 0.6 kg. and its density is 7.8×10^3 kg/m^3. If the area of hysteresis loop of alternating magnetising field of frequency 50Hz is 0.722 MKS units then the hysteresis loss per second will be

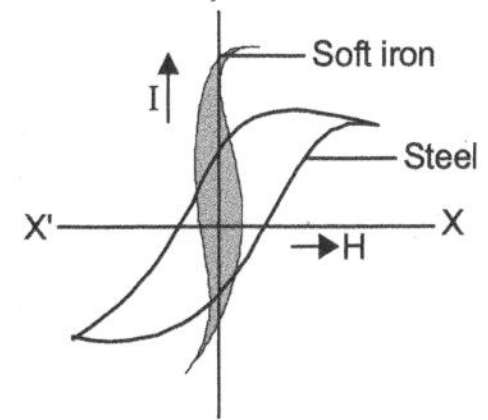

(a) 277.7×10^{-5} Joule
(b) 277.7×10^{-6} Joule
(c) 277.7×10^{-4} Joule
(d) 27.77×10^{-4} Joule

Q.24 A diamagnetic material in a magnetic field moves

(a) from weaker to the stronger parts of the field
(b) perpendicular to the field
(c) from stronger to the weaker parts of the field
(d) None of these

DIRECTIONS (Q.25-Q.27) : In the following questions, more than one of the answers given are correct. Select the correct answers and mark it according to the following codes:

Codes :

(a) 1, 2 and 3 are correct **(b)** 1 and 2 are correct
(c) 2 and 4 are correct **(d)** 1 and 3 are correct

Q.25 A magnetising field of 1600 Am^{-1} produces a magnetic flux of 2.4×10^{-5} weber in a bar of iron of area of cross-section 0.2 cm^2. Then,

(1) the magnetic permeability of the bar is 7.5×10^{-4} TA^{-1}m
(2) the susceptibility of the bar is 596.1
(3) the magnetic permeability of the bar is 4.1 Wbm^{-2}
(4) the susceptibility of the bar is 496.1

Q.26 Which of the following statements are correct about hysteresis?

(1) This effect is common to all ferromagnetic substances
(2) The hysteresis loop area is proportional to the thermal energy developed per unit volume of the material
(3) The shape of the hysteresis loop is characteristic of the material
(4) The hysteresis loop area is independent of the thermal energy developed per unit volume of the material

Q.27 Which of the following statments are false about the magnetic susceptibility χ_m of paramagnetic substance?

(1) Value of χ_m is directly proportional to the absolute temperature of the sample
(2) χ_m is negative at all temperature
(3) χ_m does not depend on the temperature of the sample
(4) χ_m is positive at all temperature

DIRECTIONS (Q. 28-Q.30) : Each of these questions contains two statements: Statement-1 (Assertion) and Statement-2 (Reason). Each of these questions has four alternative choices, only one of which is the correct answer. You have to select the correct choice.

(a) Statement-1 is True, Statement-2 is True; Statement-2 is a correct explanation for Statement-1.
(b) Statement-1 is True, Statement-2 is True; Statement-2 is NOT a correct explanation for Statement-1.
(c) Statement -1 is False, Statement-2 is True.
(d) Statement -1 is True, Statement-2 is False.

RESPONSE GRID					
	20. ⓐⓑⓒⓓ	21. ⓐⓑⓒⓓ	22. ⓐⓑⓒⓓ	23. ⓐⓑⓒⓓ	24. ⓐⓑⓒⓓ
	25. ⓐⓑⓒⓓ	26. ⓐⓑⓒⓓ	27. ⓐⓑⓒⓓ		

Space for Rough Work

Q.28 Statement-1 : The ferromagnetic substance do not obey Curie's law.

Statement-2 : At Curie point a ferromagnetic substance start behaving as a paramagnetic substance.

Q.29 Statement-1 : A paramagnetic sample displays greater magnetisation (for the same magnetising field) when cooled.

Statement-2 : The magnetisation does not depend on temperature.

Q.30 Statement-1 : The permeability of a ferromagnetic material dependent on the magnetic field.

Statement-2 : Permeability of a material is a constant quantity.

RESPONSE GRID	28. ⓐⓑⓒⓓ	29. ⓐⓑⓒⓓ	30. ⓐⓑⓒⓓ

DAILY PRACTICE PROBLEM SHEET 43 - PHYSICS			
Total Questions	30	Total Marks	120
Attempted		Correct	
Incorrect		Net Score	
Cut-off Score	28	Qualifying Score	46
Success Gap = Net Score – Qualifying Score			
Net Score = (Correct × 4) – (Incorrect × 1)			

Space for Rough Work

DPP - Daily Practice Problems

Name :   Date :

Start Time :   End Time :

PHYSICS 44

SYLLABUS : ELECTROMAGNETIC INDUCTION-1 (Magnetic flux, Faraday's law of electromagnetic induction, Lenz's law, motional e.m.f.)

Max. Marks : 116   **Time : 60 min.**

GENERAL INSTRUCTIONS

- The Daily Practice Problem Sheet contains 29 MCQ's. For each question only one option is correct. Darken the correct circle/ bubble in the Response Grid provided on each page.
- You have to evaluate your Response Grids yourself with the help of solution booklet.
- Each correct answer will get you 4 marks and 1 mark shall be deduced for each incorrect answer. No mark will be given/ deducted if no bubble is filled. Keep a timer in front of you and stop immediately at the end of 60 min.
- The sheet follows a particular syllabus. Do not attempt the sheet before you have completed your preparation for that syllabus. Refer syllabus sheet in the starting of the book for the syllabus of all the DPP sheets.
- After completing the sheet check your answers with the solution booklet and complete the Result Grid. Finally spend time to analyse your performance and revise the areas which emerge out as weak in your evaluation.

DIRECTIONS (Q.1-Q.20) : There are 20 multiple choice questions. Each question has 4 choices (a), (b), (c) and (d), out of which **ONLY ONE** choice is correct.

Q.1 A loop of wire is placed in a magnetic field $\vec{B} = 0.02\hat{i}$ T. Then the flux through the loop if its area vector $\vec{A} = 30\hat{i} + 16\hat{j} + 23\hat{k}\ \text{cm}^2$ is

(a) 60 μWb   (b) 32 μ Wb
(c) 46 μ Wb   (d) 138 μ Wb

Q.2 The magnetic flux passing perpendicular to the plane of the coil and directed into the paper is varying according to the relation $\phi = 3t^2 + 2t + 3$, where ϕ is in milliweber and t is in second. Then the magnitude of emf induced in the loop when t = 2 second is-

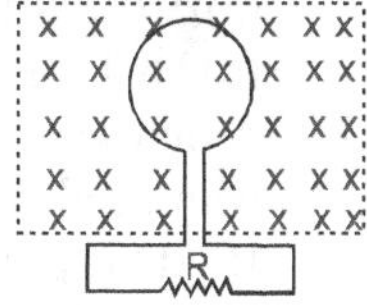

(a) 31 mV
(b) 19 mV
(c) 14 mV
(d) 6 mV

Q.3 A current carrying solenoid is approaching a conducting loop as shown in the figure. The direction of induced current as observed by an observer on the other side of the loop will be -

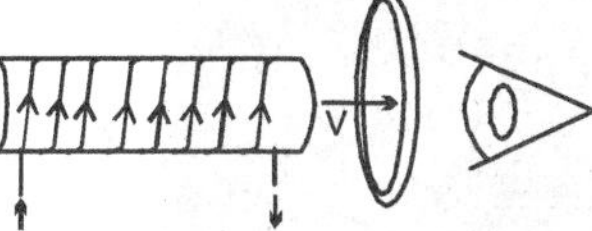

(a) anti-clockwise
(b) clockwise
(c) east
(d) west

RESPONSE GRID	1. ⓐⓑⓒⓓ	2. ⓐⓑⓒⓓ	3. ⓐⓑⓒⓓ

Space for Rough Work

Q.4 Consider the arrangement shown in figure in which the north pole of a magnet is moved away from a thick conducting loop containing capacitor. Then excess positive charge will arrive on

(a) plate a
(b) plate b
(c) both plates a and b
(d) neither a nor b plates

Q.5 An electron moves along the line AB, which lies in the same plane as a circular loop of conducting wires as shown in the diagram. What will be the direction of current induced if any, in the loop?

(a) No current will be induced
(b) The current will be clockwise
(c) The current will be anticlockwise
(d) The current will change direction as the electron passes by

Q.6 When a small piece of wire passes between the magnetic poles of a horse-shoe magnet in 0.1 sec, emf of 4×10^{-3} volt is induced in it. The magnetic flux between the poles is :

(a) 4×10^{-2} weber (b) 4×10^{-3} weber
(c) 4×10^{-4} weber (d) 4×10^{-6} weber

Q.7 The normal magnetic flux passing through a coil changes with time according to following equation

$$\phi = 10t^2 + 5t + 1$$

where ϕ is in milliweber and t is in second. The value of induced e.m.f. produced in the coil at t = 5s will be –

(a) zero (b) 1 V (c) 2 V (d) 0.105 V

Q.8 A bicycle wheel of radius 0.5 m has 32 spokes. It is rotating at the rate of 120 revolutions per minute, perpendicular to the horizontal component of earth's magnetic field $B_H = 4 \times 10^{-5}$ tesla. The emf induced between the rim and the centre of the wheel will be-

(a) 6.28×10^{-5} V (b) 4.8×10^{-5} V
(c) 6.0×10^{-5} V (d) 1.6×10^{-5} V

Q.9 A thin semicircular conducting ring of radius R is falling with its plane vertical in a horizontal magnetic induction B. At the position MNQ shown in the fig, the speed of the ring is V. The potential difference developed across the semicircular ring is

(a) Zero
(b) $B \nu R^2/2$ and M is at higher potential
(c) πRBV and Q is at higher potential
(d) 2RBV and Q is at higher potential

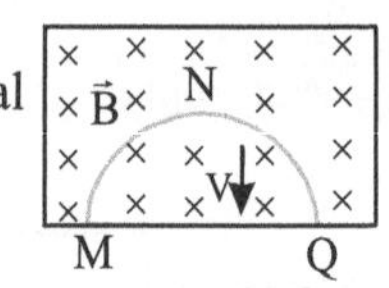

Q.10 An aeroplane having a distance of 50 metre between the edges of its wings is flying horizontally with a speed of 360km/hour. If the vertical component of earth's magnetic field is 4×10^{-4} weber/m^2, then the induced emf between the edges of its wings will be –

(a) 2 mV (b) 2 V (c) 0.2 V (d) 20 V

Q.11 At certain location in the northern hemisphere, the earth's magnetic field has a magnitude of 42 μT and points down ward at 57° to vertical. The flux through a horizontal surface of area 2.5 m^2 will be- (Given cos 33° = 0.839, cos 57° = 0.545)

(a) 42×10^{-6} Wb/m^2 (b) 42×10^{-6} Wb/m^2
(c) 57×10^{-6} Wb/m^2 (d) 57×10^{-6} Wb/m^2

Q.12 A square loop of side a is rotating about its diagonal with angular velocity ω in a perpendicular magnetic field as shown in the figure. If the number of turns in it is 10 then the magnetic flux linked with the loop at any instant will be–

(a) $10Ba^2 \cos \omega t$ (b) 10Ba
(c) $10Ba^2$ (d) $20Ba^2$

Q.13 Two identical coaxial circular loops carry current i each circulating in the clockwise direction. If the loops are approaching each other, then

(a) Current in each loop increases
(b) Current in each loop remains the same
(c) Current in each loop decreases
(d) Current in one-loop increases and in the other it decreases

Q.14 The distance between the ends of wings of an aeroplane is 3m. This aeroplane is descending down with a speed of 300 km/hour. If the horizontal component of earths magnetic field is 0.4 gauss then the value of e.m.f. induced in the wings of the plane will be –

(a) 1 V (b) 2 V (c) 0.01 V (d) 0.1 V

RESPONSE GRID					
	4. ⓐⓑⓒⓓ	5. ⓐⓑⓒⓓ	6. ⓐⓑⓒⓓ	7. ⓐⓑⓒⓓ	8. ⓐⓑⓒⓓ
	9. ⓐⓑⓒⓓ	10. ⓐⓑⓒⓓ	11. ⓐⓑⓒⓓ	12. ⓐⓑⓒⓓ	13. ⓐⓑⓒⓓ
	14. ⓐⓑⓒⓓ				

Space for Rough Work

Q.15 A gramophone disc of brass of diameter 30 cm rotates horizontally at the rate of 100/3 revolutions per minute. If the vertical component of the earth's magnetic field be 0.01 weber/metre^2, then the emf induced between the centre and the rim of the disc will be-

(a) 7.065×10^{-4} V (b) 3.9×10^{-4} V
(c) 2.32×10^{-4} V (d) None of the above

Q.16 A closed coil consists of 500 turns on a rectangular frame of area 4.0 cm^2 and has a resistance of 50 ohm. The coil is kept with its plane perpendicular to a uniform magnetic field of 0.2 weber/meter^2. The amount of charge flowing through the coil if it is turned over (rotated through 180°) will be -

(a) 1.6×10^{-19} C (b) 1.6×10^{-9} C
(c) 1.6×10^{-3} C (d) 1.6×10^{-2} C

Q.17 A copper disc of radius 0.1 m rotates about its centre with 10 revolution per second in 'a uniform magnetic field of 0.1 T. The emf induced across the radius of the disc is -

(a) $\pi/10$ V (b) $2\pi/10$ V (c) 10π mV (d) 20π mV

Q.18 Two rail tracks, insulated from each other and the ground, are connected to milli voltmeter. What is the reading of the milli voltmeter when a train passes at a speed of 180 km/hr along the track ? Given that – the horizontal component of earth's magnetic field B_H is 0.2×10^{-4} Wb/m^2 and rails are separated by 1 metre.

(a) 1 mV (b) 10 mV (c) 100 mV (d) 1 V

Q.19 The annular disc of copper, with inner radius a and outer radius b is rotating with a uniform angular speed ω, in a region where a uniform magnetic field B along the axis of rotation exists. Then, the emf induced between inner side and the outer rim of the disc is-

(a) Zero (b) $\frac{1}{2}B\omega a^2$
(c) $\frac{1}{2}B\omega b^2$ (d) $\frac{1}{2}B\omega(b^2 - a^2)$

Q.20 A conducting wire in the shape of Y with each side of length ℓ is moving in a uniform magnetic field B, with a uniform speed v as shown in fig. The induced emf at the two ends X and Y of the wire will be-

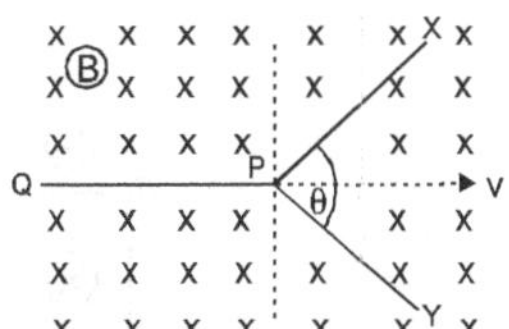

(a) zero
(b) $2B\ell v$
(c) $2B\ell v \sin(\theta/2)$
(d) $2B\ell v \cos(\theta/2)$

DIRECTIONS (Q.21-Q.23) : In the following questions, more than one of the answers given are correct. Select the correct answers and mark it according to the following codes:

Codes :
(a) 1, 2 and 3 are correct **(b)** 1 and 2 are correct
(c) 2 and 4 are correct **(d)** 1 and 3 are correct

Q.21 A rectangular coil of size 10 cm ×20 cm has 60 turns. It is rotating about one of its diagonals in magnetic field 0.5 Wb/m^2 with a rate of 1800 revolution per minute. The induced e.m.f. in the coil can be

(1) 111 V (2) 112 V (3) 113 V (4) 114 V

Q.22 A closed coil of copper whose area is 1m × 1m is free to rotate about an axis. The coil is placed perpendicular to a magnetic field of 0.10 Wb/m^2. It is rotated through 180° in 0.01 second. Then (The resistance of the coil is 2.0Ω)

(1) The induced e.m.f. in the coil is 20 V
(2) The induced current in the coil is 10 A
(3) The induced e.m.f. in the coil is 10 V
(4) The induced current in the coil is 20 A

Q.23 5.5×10^{-4} magnetic flux lines are passing through a coil of resistance 10 ohm and number of turns 1000. If the number of flux lines reduces to 5×10^{-5} in 0.1 sec. Then

(1) The electromotive force induced in the coil is 5V
(2) The electromotive force induced in the coil is 5×10^{-4} V
(3) The current induced in the coil is 0.5 A
(4) The current induced in the coil is 10 A

DIRECTIONS (Q.24-Q.26) : Read the passage given below and answer the questions that follows :

In the figure shown, the rod has a resistance *R*, the horizontal rails have negligible friction. Magnetic field of intensity B is directed perpendicular into the plane of paper. A cell of e.m.f. *E* and negligible internal resistance is connected between points *a* and *b*. The rod is initially at rest.

RESPONSE GRID					
	15. ⓐⓑⓒⓓ	16. ⓐⓑⓒⓓ	17. ⓐⓑⓒⓓ	18. ⓐⓑⓒⓓ	19. ⓐⓑⓒⓓ
	20. ⓐⓑⓒⓓ	21. ⓐⓑⓒⓓ	22. ⓐⓑⓒⓓ	23. ⓐⓑⓒⓓ	

Space for Rough Work

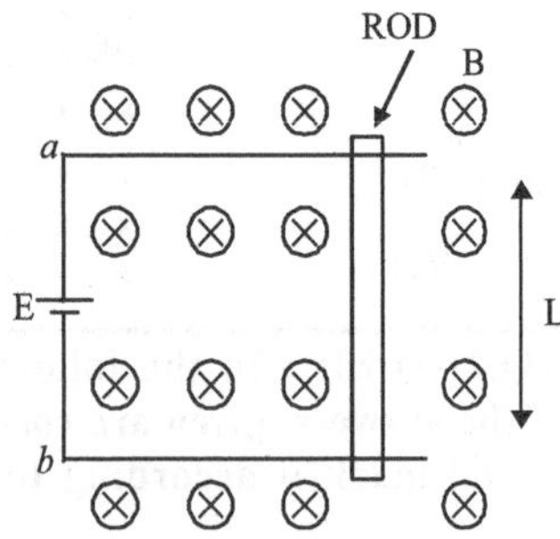

Q.24 The velocity of the rod as a function of time t (where $\tau = mR/B\ell^2$) is

(a) $\frac{E}{B\ell}(1-e^{-t/\tau})$ (b) $\frac{E}{B\ell}(1+e^{-t/\tau})$

(c) $\frac{3E}{2B\ell}(1-e^{-t/\tau})$ (d) $\frac{E}{2B\ell}(1-e^{-t/\tau})$

Q.25 After some time the rod will approach a terminal speed. The speed is

(a) $\frac{3E}{2B\ell}$ (b) $\frac{E}{2B\ell}$ (c) $\frac{E}{B\ell}$ (d) $\frac{2E}{B\ell}$

Q.26 The current when the rod attains its terminal speed is

(a) $\frac{2E}{R}$ (b) $\frac{E}{R}$ (c) $\frac{3E}{2R}$ (d) zero

DIRECTIONS (Q. 27-Q.29) : Each of these questions contains two statements: Statement-1 (Assertion) and Statement-2 (Reason). Each of these questions has four alternative choices, only one of which is the correct answer. You have to select the correct choice.

(a) Statement-1 is True, Statement-2 is True; Statement-2 is a correct explanation for Statement-1.
(b) Statement-1 is True, Statement-2 is True; Statement-2 is NOT a correct explanation for Statement-1.
(c) Statement -1 is False, Statement-2 is True.
(d) Statement -1 is True, Statement-2 is False.

Q.27 Statement-1 : The induced e.m.f. and current will be same in two identical loops of copper and aluminium, when rotated with same speed in the same magnetic field.
Statement-2 : Induced e.m.f. is proportional to rate of change of magnetic field while induced current depends on resistance of wire.

Q.28 Statement-1 : An aircraft flies along the meridian, the potential at the ends of its wings will be the same.
Statement-2 : Whenever there is change in the magnetic flux e.m.f. induces.

Q.29 Statement-1 : Lenz's law violates the principle of conservation of energy.
Statement-2 : Induced e.m.f. opposes the change in magnetic flux responsible for its production.

RESPONSE GRID	
	24. ⓐⓑⓒⓓ 25. ⓐⓑⓒⓓ 26. ⓐⓑⓒⓓ 27. ⓐⓑⓒⓓ 28. ⓐⓑⓒⓓ
	29. ⓐⓑⓒⓓ

DAILY PRACTICE PROBLEM SHEET 44 - PHYSICS			
Total Questions	29	Total Marks	116
Attempted		Correct	
Incorrect		Net Score	
Cut-off Score	26	Qualifying Score	46
Success Gap = Net Score – Qualifying Score			
Net Score = (Correct × 4) – (Incorrect × 1)			

Space for Rough Work

DPP - Daily Practice Problems

Name :   **Date :**

Start Time :   **End Time :**

PHYSICS 45

SYLLABUS : ELECTROMAGNETIC INDUCTION - 2 : Self inductance, mutual inductance, Growth and decay of current in L.R. circuit, Transformer, Electric motor, Generator

Max. Marks : 116   **Time : 60 min.**

GENERAL INSTRUCTIONS

- The Daily Practice Problem Sheet contains 29 MCQ's. For each question only one option is correct. Darken the correct circle/ bubble in the Response Grid provided on each page.
- You have to evaluate your Response Grids yourself with the help of solution booklet.
- Each correct answer will get you 4 marks and 1 mark shall be deduced for each incorrect answer. No mark will be given/ deducted if no bubble is filled. Keep a timer in front of you and stop immediately at the end of 60 min.
- The sheet follows a particular syllabus. Do not attempt the sheet before you have completed your preparation for that syllabus. Refer syllabus sheet in the starting of the book for the syllabus of all the DPP sheets.
- After completing the sheet check your answers with the solution booklet and complete the Result Grid. Finally spend time to analyse your performance and revise the areas which emerge out as weak in your evaluation.

DIRECTIONS (Q.1-Q.20) : There are 20 multiple choice questions. Each question has 4 choices (a), (b), (c) and (d), out of which **ONLY ONE** choice is correct.

Q.1 A current increases uniformly from zero to one ampere in 0.01 second, in a coil of inductance 10 mH. The induced e.m.f. will be -

(a) 1 V (b) 2 V (c) 3 V (d) 4 V

Q.2 The current in a coil varies with respect to time t as $I = 3t^2 + 2t$. If the inductance of coil be 10 mH, the value of induced e.m.f. at t = 2s will be-

(a) 0.14 V (b) 0.12 V (c) 0.11 V (d) 0.13 V

Q.3 Two circular coils can be arranged in any of the three situations shown in the figure. Their mutual inductance will be

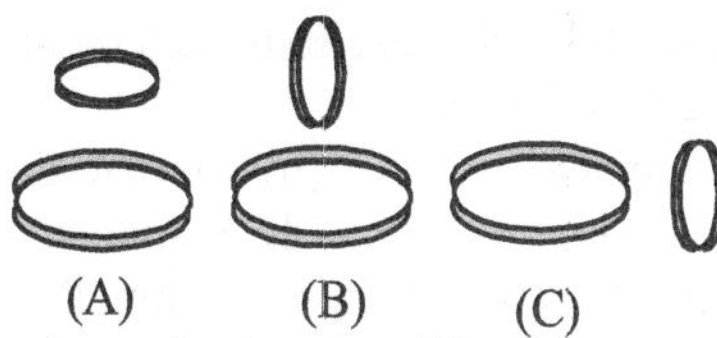

(a) Maximum in situation (A)
(b) Maximum in situation (B)
(c) Maximum in situation (C)
(d) The same in all situations

Q.4 A current of 10 A in the primary coil of a circuit is reduced to zero at a uniform rate in 10^{-3} second. If the coefficient of mutual inductance is 3H, the induced e.m.f. in the secondary coil will be-

(a) 3 kV (b) 30 kV (c) 2 kV (d) 20 kV

RESPONSE GRID	1. ⓐⓑⓒⓓ	2. ⓐⓑⓒⓓ	3. ⓐⓑⓒⓓ	4. ⓐⓑⓒⓓ

Space for Rough Work

Q.5 Three inductances are connected as shown below. Assuming no coupling, the resultant inductance will be- ($L_1 = 0.75$ H, $L_2 = L_3 = 0.5$ H)

(a) 0.25 H
(b) 0.75 H
(c) 0.01 H
(d) 1 H

Q.6 A solenoid has an inductance of 50 mH and a resistance of 0.025 Ω. If it is connected to a battery, how long will it take for the current to reach one half of its final equilibrium value?

(a) 1.34 s (b) 1.38 s (c) 1.38 ms (d) 0.23 s

Q.7 The current in the primary coil of a transformer (assuming no power loss) as shown in fig. will be –

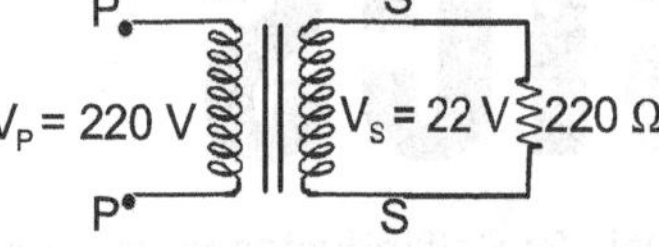

(a) 0.01 A
(b) 1.0 A
(c) 0.1 A
(d) 10^{-6} A.

Q.8 A current of 5A is flowing at 220V in the primary coil of a transformer. If the voltage produced in the secondary coil is 2200V and 50% of power is lost, then the current in the secondary coil will be –

(a) 2.5A (b) 5A (c) 0.25A (d) 0.025A

Q.9 An inductor (L = 100 mH), a resistor (R = 100Ω) and a battery (E = 100 V) are initially connected in series as shown in the figure. After a long time the battery is disconnected after short circuiting the points A and B. The current in the circuit 1 ms after the short circuit is

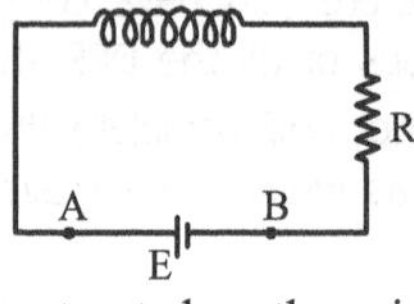

(a) e A
(b) 0.1 A
(c) 1 A
(d) 1/e A

Q.10 Which of the following is constructed on the principle of electromagnetic induction ?

(a) Galvanometer (b) Electric motor
(c) Generator (d) Voltmeter

Q.11 In the circuit, E = 10 volt, $R_1 = 5.0$ ohm, $R_2 = 10$ ohm and L = 5.0 henry. The current just after the switch S is pressed is.

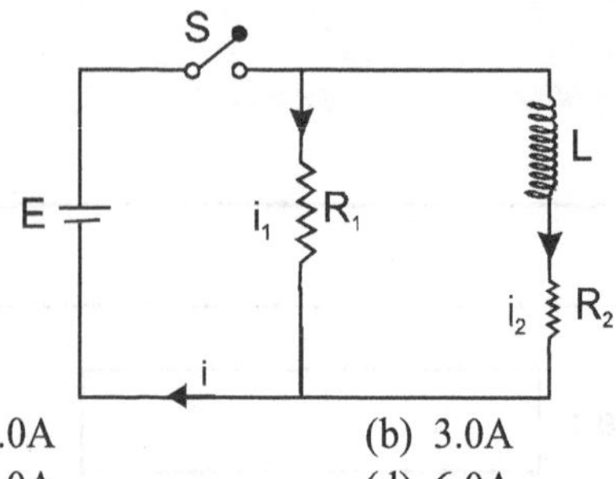

(a) 2.0A (b) 3.0A
(c) 5.0A (d) 6.0A

Q.12 Two inductors L_1 and L_2 are at a sufficient distance apart. Equivalent inductance when they are connected (i) in series (ii) in parallel are

(a) $L_1 + L_2, \frac{L_1L_2}{L_1 + L_2}$ (b) $L_1 - L_2, \frac{L_1L_2}{L_1 - L_2}$

(c) $L_1 L_2, \frac{L_1 + L_2}{L_1L_2}$ (d) None of these

Q.13 A small coil of N_1 turns, l_1 length is tightly wound over the centre of a long solenoid of length l_2, area of cross-section A and number of turns N_2. If a current I flows in the small coil, then the flux through the long solenoid is

(a) zero (b) $\frac{\mu_0 N_1^2 AI}{\ell_1}$

(c) inifinite (d) $\frac{\mu_0 N_1 N_2 AI}{\ell_2}$

Q.14 If the current in the primary coil is reduced from 3.0 ampere to zero in 0.001 second, the induced e.m.f in the secondary coil is 1500 volt. The mutual inductance of the two coils will be-

(a) 0.5 H (b) 0.05 H (c) 0.005 H (d) 0.0005 H

Q.15 A 50 Hz a.c. current of crest value 1A flows through the primary of a transformer. If the mutual inductance between the primary and secondary be 1.5 H, the crest voltage induced in secondary is-

(a) 75 V (b) 150 V (c) 471 V (d) 300 V

Q.16 In an inductor of inductance L = 100 mH, a current of I = 10A is flowing. The energy stored in the inductor is

(a) 5 J (b) 10 J (c) 100 J (d) 1000 J

RESPONSE GRID					
	5. ⓐⓑⓒⓓ	6. ⓐⓑⓒⓓ	7. ⓐⓑⓒⓓ	8. ⓐⓑⓒⓓ	9. ⓐⓑⓒⓓ
	10. ⓐⓑⓒⓓ	11. ⓐⓑⓒⓓ	12. ⓐⓑⓒⓓ	13. ⓐⓑⓒⓓ	14. ⓐⓑⓒⓓ
	15. ⓐⓑⓒⓓ	16. ⓐⓑⓒⓓ			

Space for Rough Work

Q.17 A step up transformer has transformation ratio 5 : 3. What is voltage in secondary if voltage in primary is 60 V

(a) 20 V (b) 60 V (c) 100 V (d) 180 V

Q.18 A transformer has turn ratio 100 : 1. If secondary coil has 4 amp current then current in primary coil is

(a) 4 A (b) 0.04 A (c) 0.4 A (d) 400 A

Q.19 A step-down transformer is used on a 1000 V line to deliver 20 A at 120 V at the secondary coil. If the efficiency of the transformer is 80%, the current drawn from the line is:

(a) 3 A (b) 30 A (c) 0.3 A (d) 2.4 A

Q.20 Energy stored in an inductor is proportional to (i = current in the inductor)

(a) i (b) $\sqrt{i}$ (c) i^2 (d) i^3

DIRECTIONS (Q.21-Q.23) : In the following questions, more than one of the answers given are correct. Select the correct answers and mark it according to the following codes:

Codes :

(a) 1, 2 and 3 are correct **(b)** 1 and 2 are correct
(c) 2 and 4 are correct **(d)** 1 and 3 are correct

Q.21 Voltage (r. m. s) in the secondary coil of a transformer depends upon

(1) voltage in the primary coil
(2) ratio of number of turns in the two coils
(3) frequency of the source
(4) time-period of the source

Q.22 Core of a transformer can't be made up of

(1) steel (2) alnico (3) iron (4) soft iron

Q.23 Large transformer, when used for some time, become hot and are cooled by circulating oil. The heating of transformer is due to

(1) heating effect of current
(2) hysteresis loss
(3) chemical effect of current
(4) magnetic effect of current

DIRECTIONS (Q.24-Q.26) : Read the passage given below and answer the questions that follows :

In Fig., there is a conducting loop ABCDEFA, of resistance λ per unit length placed near a long straight current-carrying wire. The dimensions are shown in the figure. The long wire lies in the plane of the loop. The current in the long wire varies as $I = I_0 t$.

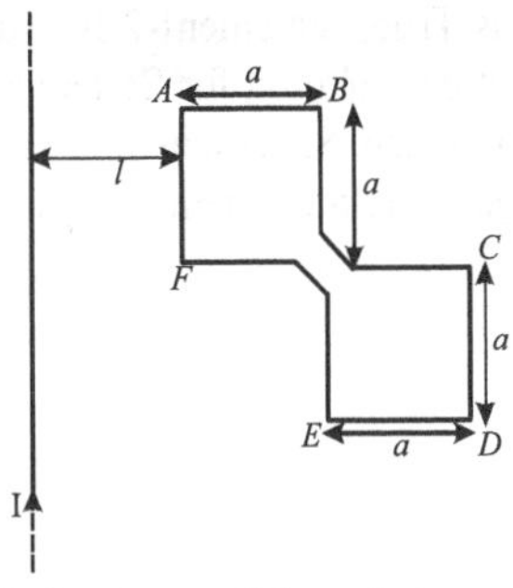

Q.24 The mutual inductance of the pair is

(a) $\frac{\mu_0 a}{2\pi}\ln\left(\frac{2a+\ell}{\ell}\right)$ (b) $\frac{\mu_0 a}{2\pi}\ln\left(\frac{2a-\ell}{\ell}\right)$

(c) $\frac{2\mu_0 a}{\pi}\ln\left(\frac{a+\ell}{\ell}\right)$ (d) $\frac{\mu_0 a}{\pi}\ln\left(\frac{a+\ell}{\ell}\right)$

Q.25 The e.m.f. induced in the closed loop is

(a) $\frac{\mu_0 I_0 a}{2\pi}\ln\left(\frac{2a+\ell}{\ell}\right)$ (b) $\frac{\mu_0 I_0 a}{2\pi}\ln\left(\frac{2a-\ell}{\ell}\right)$

(c) $\frac{2\mu_0 I_0 a}{\pi}\ln\left(\frac{a+\ell}{\ell}\right)$ (d) $\frac{\mu_0 I_0 a}{\pi}\ln\left(\frac{a+\ell}{\ell}\right)$

Q.26 The heat produced in the loop in time t is

(a) $\frac{\left[\frac{\mu_0}{2\pi}\ln\left(\frac{a+\ell}{\ell}\right)I_0\right]^2 at}{4\lambda}$ (b) $\frac{\left[\frac{\mu_0}{2\pi}\ln\left(\frac{2a+\ell}{\ell}\right)I_0\right]^2 at}{8\lambda}$

(c) $\frac{\left[\frac{2\mu_0}{\pi}\ln\left(\frac{a+\ell}{\ell}\right)I_0\right]^2 at}{3\lambda}$ (d) $\frac{\left[\frac{\mu_0}{2\pi}\ln\left(\frac{3a+\ell}{\ell}\right)I_0\right]^2 at}{6\lambda}$

RESPONSE GRID					
	17. ⓐⓑⓒⓓ	18. ⓐⓑⓒⓓ	19. ⓐⓑⓒⓓ	20. ⓐⓑⓒⓓ	21. ⓐⓑⓒⓓ
	22. ⓐⓑⓒⓓ	23. ⓐⓑⓒⓓ	24. ⓐⓑⓒⓓ	25. ⓐⓑⓒⓓ	26. ⓐⓑⓒⓓ

Space for Rough Work

DIRECTIONS (Qs. 27-Q.29) : Each of these questions contains two statements: Statement-1 (Assertion) and Statement-2 (Reason). Each of these questions has four alternative choices, only one of which is the correct answer. You have to select the correct choice.

(a) Statement-1 is True, Statement-2 is True; Statement-2 is a correct explanation for Statement-1.
(b) Statement-1 is True, Statement-2 is True; Statement-2 is NOT a correct explanation for Statement-1.
(c) Statement-1 is False, Statement-2 is True.
(d) Statement-1 is True, Statement-2 is False.

Q.27 Statement-1 : Soft iron is used as a core of transformer.

Statement-2 : Area of hysteresis loop for soft iron is small.

Q.28 Statement-1 : An electric motor will have maximum efficiency when back e.m.f. is equal to half of the applied e.m.f.

Statement-2 : Efficiency of electric motor depends only on magnitude of back e.m.f.

Q.29 Statement-1 : A transformer cannot work on dc supply.

Statement-2 : dc changes neither in magnitude nor in direction.

RESPONSE GRID	27. ⓐⓑⓒⓓ	28. ⓐⓑⓒⓓ	29. ⓐⓑⓒⓓ

DAILY PRACTICE PROBLEM SHEET 45 - PHYSICS			
Total Questions	29	Total Marks	116
Attempted		Correct	
Incorrect		Net Score	
Cut-off Score	28	Qualifying Score	48
Success Gap = Net Score – Qualifying Score			
Net Score = (Correct × 4) – (Incorrect × 1)			

Space for Rough Work

DPP - Daily Practice Problems

Name :

Date :

Start Time :

End Time :

PHYSICS 46

SYLLABUS : ALTERNATING CURRENT - 1 (Alternating currents, peak and rms value of alternating current/voltage; reactance and impedance, Pure circuits, LR, CR ac circuits.)

Max. Marks : 116 **Time : 60 min.**

GENERAL INSTRUCTIONS

- The Daily Practice Problem Sheet contains 29 MCQ's. For each question only one option is correct. Darken the correct circle/ bubble in the Response Grid provided on each page.
- You have to evaluate your Response Grids yourself with the help of solution booklet.
- Each correct answer will get you 4 marks and 1 mark shall be deduced for each incorrect answer. No mark will be given/ deducted if no bubble is filled. Keep a timer in front of you and stop immediately at the end of 60 min.
- The sheet follows a particular syllabus. Do not attempt the sheet before you have completed your preparation for that syllabus. Refer syllabus sheet in the starting of the book for the syllabus of all the DPP sheets.
- After completing the sheet check your answers with the solution booklet and complete the Result Grid. Finally spend time to analyse your performance and revise the areas which emerge out as weak in your evaluation.

DIRECTIONS (Q.1-Q.21) : There are 21 multiple choice questions. Each question has 4 choices (a), (b), (c) and (d), out of which **ONLY ONE** choice is correct.

Q.1 The resistance of a coil for dc is 5 ohm. In ac, the resistance will

(a) remain same (b) increase
(c) decrease (d) be zero

Q.2 If instantaneous current is given by $i = 4\cos(\omega t + \phi)$ amperes, then the r.m.s. value of current is

(a) 4 ampere (b) $2\sqrt{2}$ ampere
(c) $4\sqrt{2}$ ampere (d) zero ampere

Q.3 In an ac circuit I = 100 sin 200πt. The time required for the current to achieve its peak value will be

(a) $\frac{1}{100}$sec (b) $\frac{1}{200}$sec (c) $\frac{1}{300}$sec (d) $\frac{1}{400}$sec

Q.4 The frequency of ac mains in India is

(a) 30 c/s or Hz (b) 50 c/s or Hz
(c) 60 c/s or Hz (d) 120 c/s or Hz

Q.5 The peak value of an alternating e.m.f. E given by $E = E_0 \cos \omega t$ is 10 volts and its frequency is 50 Hz. At time $t = \frac{1}{600}$sec, the instantaneous e.m.f. is

(a) 10 V (b) $5\sqrt{3}\, V$ (c) 5 V (d) 1 V

RESPONSE GRID	1. ⓐⓑⓒⓓ	2. ⓐⓑⓒⓓ	3. ⓐⓑⓒⓓ	4. ⓐⓑⓒⓓ	5. ⓐⓑⓒⓓ

Space for Rough Work

Q.6 An alternating current is given by the equation $i = i_1 \cos \omega t + i_2 \sin \omega t$. The r.m.s. current is given by

(a) $\frac{1}{\sqrt{2}}(i_1 + i_2)$ (b) $\frac{1}{\sqrt{2}}(i_1 + i_2)^2$

(c) $\frac{1}{\sqrt{2}}(i_1^2 + i_2^2)^{1/2}$ (d) $\frac{1}{2}(i_1^2 + i_2^2)^{1/2}$

Q.7 In a circuit, the value of alternating current is measured by hot wire ammeter as 10 ampere. Its peak value will be

(a) 10 A (b) 20 A (c) 14.14 A (d) 7.07 A

Q.8 The frequency of an alternating voltage is 50 cycles/sec and its amplitude is 120*V*. Then the r.m.s. value of voltage is

(a) 101.3V (b) 84.8V (c) 70.7V (d) 56.5V

Q.9 A resistance of 20Ω is connected to a source of an alternating potential $V = 220 \sin(100\pi t)$. The time taken by the current to change from its peak value to r.m.s. value is

(a) 0.2 sec (b) 0.25 sec

(c) 25×10^{-3} sec (d) 2.5×10^{-3} sec

Q.10 An alternating current of frequency *f* is flowing in a circuit containing a resistor of resistance *R* and a choke of inductance *L* in series. The impedance of this circuit is

(a) $R + 2\pi f L$ (b) $\sqrt{R^2 + 4\pi^2 f^2 L^2}$

(c) $\sqrt{R^2 + L^2}$ (d) $\sqrt{R^2 + 2\pi f L}$

Q.11 An alternating voltage is connected in series with a resistance *R* and an inductance *L*. If the potential drop across the resistance is 200 V and across the inductance is 150 V, then the applied voltage is

(a) 350 V (b) 250 V

(c) 500 V (d) 300 V

Q.12 An inductive circuit contains resistance of 10Ω and an inductance of 20 *H*. If an ac voltage of 120 *V* and frequency 60 Hz is applied to this circuit, the current would be nearly

(a) 0.32 A (b) 0.016 A

(c) 0.48 A (d) 0.80 A

Q.13 A 20 volt ac is applied to a circuit consisting of a resistance and a coil with negligible resistance. If the voltage across the resistance is 12 *V*, the voltage across the coil is

(a) 16 volt (b) 10 volt

(c) 8 volt (d) 6 volt

Q.14 An alternating voltage $E = 200\sqrt{2} \sin(100t)$ is connected to a 1 microfarad capacitor through an ac ammeter. The reading of the ammeter will be

(a) 10 mA (b) 20 mA

(c) 40 mA (d) 80 mA

Q.15 A resistor and a capacitor are connected in series with an a.c. source. If the potential drop across the capacitor is 5 V and that across resistor is 12 V, applied voltage is

(a) 13 V (b) 17 V

(c) 5 V (d) 12 V

Q.16 A 120 volt ac source is connected across a pure inductor of inductance 0.70 henry. If the frequency of the ac source is 60 Hz, the current passing through the inductor is

(a) 4.55 amp (b) 0.355 amp

(c) 0.455 amp (d) 3.55 amp

Q.17 The instantaneous value of current in an A.C. circuit is I = 2 sin (100 πt + π/3)A. The current will be maximum for the first time at

(a) $t = \frac{1}{100} s$ (b) $t = \frac{1}{200} s$

(c) $t = \frac{1}{400} s$ (d) $t = \frac{1}{600} s$

Q.18 In an L – R circuit, the value of *L* is $\left(\frac{0.4}{\pi}\right)$ henry and the value of *R* is 30 ohm. If in the circuit, an alternating *e.m.f.* of 200 volt at 50 cycles per sec is connected, the impedance of the circuit and current will be

(a) 11.4Ω, 17.5*A* (b) 30.7Ω, 6.5*A*

(c) 40.4Ω, 5*A* (d) 50Ω, 4*A*

RESPONSE GRID					
	6. ⓐⓑⓒⓓ	7. ⓐⓑⓒⓓ	8. ⓐⓑⓒⓓ	9. ⓐⓑⓒⓓ	10. ⓐⓑⓒⓓ
	11. ⓐⓑⓒⓓ	12. ⓐⓑⓒⓓ	13. ⓐⓑⓒⓓ	14. ⓐⓑⓒⓓ	15. ⓐⓑⓒⓓ
	16. ⓐⓑⓒⓓ	17. ⓐⓑⓒⓓ	18. ⓐⓑⓒⓓ		

Space for Rough Work

Q.19 The voltage across a pure inductor is represented by the following diagram. Which one of the following diagrams will represent the current?

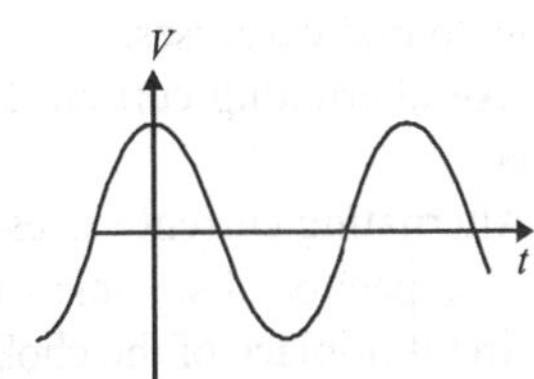

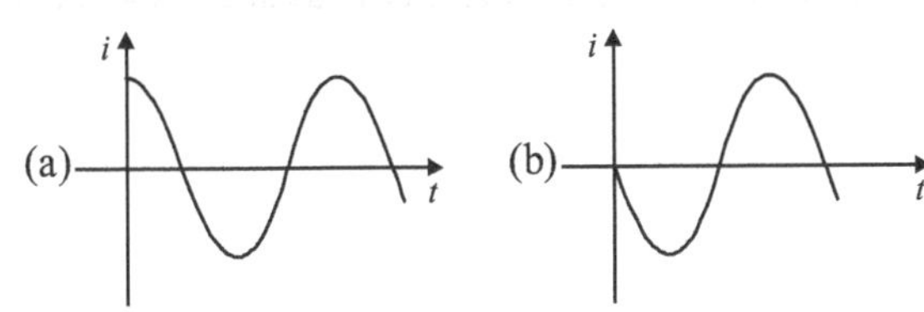

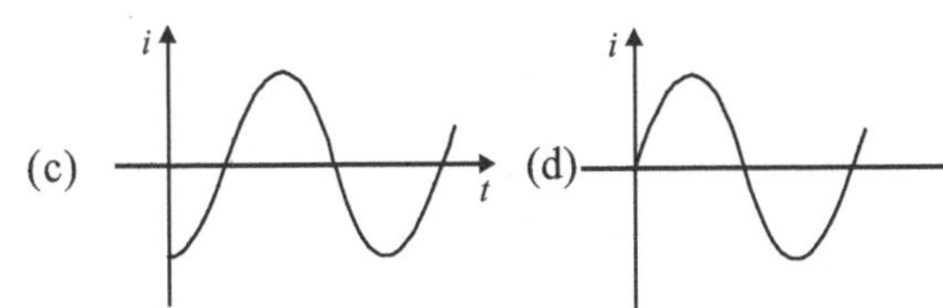

Q.20 One 10 V, 60 W bulb is to be connected to 100 V line. The required induction coil has self-inductance of value (f = 50 Hz)

(a) 0.052 H (b) 2.42 H
(c) 16.2 mH (d) 1.62 mH

Q.21 A resistance of 300Ω and an inductance of $\frac{1}{\pi}$ henry are connected in series to a ac voltage of 20 volt and 200 Hz frequency. The phase angle between the voltage and current is

(a) $\tan^{-1}\frac{4}{3}$ (b) $\tan^{-1}\frac{3}{4}$
(c) $\tan^{-1}\frac{3}{2}$ (d) $\tan^{-1}\frac{2}{5}$

DIRECTIONS (Q.22-Q.23) : In the following questions, more than one of the answers given are correct. Select the correct answers and mark it according to the following codes:

Codes :

(a) 1, 2 and 3 are correct
(b) 1 and 2 are correct
(c) 2 and 4 are correct
(d) 1 and 3 are correct

Q.22 If an alternating voltage is represented as E = 141 sin (628t), then

(1) the rms voltage is 141V
(2) the rms voltage is 100V
(3) the frequency is 50 Hz
(4) the frequency is 100 Hz

Q.23 The r.m.s. value of an ac of 50 Hz is 10 A.

(1) The time taken by the alternating current in reaching from zero to maximum value is 5×10^{-3} sec
(2) The time taken by the alternating current in reaching from zero to maximum value is 2×10^{-3} sec
(3) The peak current is 14.14 A
(4) The peak current is 7.07 A

DIRECTIONS (Q.24-Q.26) : Read the passage given below and answer the questions that follows :

If the voltage in an ac circuit is represented by the equation, $V = 220\sqrt{2}\sin(314t - \phi)$, then

Q.24 RMS value of the voltage is

(a) 220 V (b) 314 V
(c) $220\sqrt{2}$ V (d) $200/\sqrt{2}$ V

Q.25 Average voltage is

(a) 220 V (b) $622/\pi$ V
(c) $220\sqrt{2}$ V (d) $200/\sqrt{2}$ V

Q.26 Frequency of ac is

(a) 50 Hz (b) $50\sqrt{2}$ Hz
(c) $50\sqrt{2}$ Hz (d) 75 Hz

RESPONSE GRID					
	19. ⓐⓑⓒⓓ	20. ⓐⓑⓒⓓ	21. ⓐⓑⓒⓓ	22. ⓐⓑⓒⓓ	23. ⓐⓑⓒⓓ
	24. ⓐⓑⓒⓓ	25. ⓐⓑⓒⓓ	26. ⓐⓑⓒⓓ		

Space for Rough Work

DIRECTIONS (Q. 27-Q.29) : Each of these questions contains two statements: Statement-1 (Assertion) and Statement-2 (Reason). Each of these questions has four alternative choices, only one of which is the correct answer. You have to select the correct choice.

(a) Statement-1 is True, Statement-2 is True; Statement-2 is a correct explanation for Statement-1.
(b) Statement-1 is True, Statement-2 is True; Statement-2 is NOT a correct explanation for Statement-1.
(c) Statement -1 is False, Statement-2 is True.
(d) Statement -1 is True, Statement-2 is False.

Q27 **Statement-1:** The alternating current lags behind the em.f. by a phase angle of $\pi/2$, when ac flows through an inductor.
Statement-2 : The inductive reactance increases as the frequency of ac source decreases.

Q.28 **Statement-1 :** An alternating current does not show any magnetic effect.
Statement-2: Alternating current varies with time.

Q.29 **Statement-1 :** A capacitor of suitable capacitance can be used in an ac circuit in place of the choke coil.
Statement-2 : A capacitor blocks dc and allows ac only.

RESPONSE GRID	27. ⓐⓑⓒⓓ	28. ⓐⓑⓒⓓ	29. ⓐⓑⓒⓓ

DAILY PRACTICE PROBLEM SHEET 46 - PHYSICS			
Total Questions	29	Total Marks	116
Attempted		Correct	
Incorrect		Net Score	
Cut-off Score	26	Qualifying Score	44
Success Gap = Net Score – Qualifying Score			
Net Score = (Correct × 4) – (Incorrect × 1)			

Space for Rough Work

DPP - Daily Practice Problems

Name : 　　　　Date :

Start Time : 　　　　End Time :

PHYSICS 47

SYLLABUS : ALTERNATING CURRENT - 2 (LCR series circuit, resonance, quality factor, power in AC circuits, wattless and power current)

Max. Marks : 116 　　　　**Time : 60 min.**

GENERAL INSTRUCTIONS

- The Daily Practice Problem Sheet contains 29 MCQ's. For each question only one option is correct. Darken the correct circle/ bubble in the Response Grid provided on each page.
- You have to evaluate your Response Grids yourself with the help of solution booklet.
- Each correct answer will get you 4 marks and 1 mark shall be deduced for each incorrect answer. No mark will be given/ deducted if no bubble is filled. Keep a timer in front of you and stop immediately at the end of 60 min.
- The sheet follows a particular syllabus. Do not attempt the sheet before you have completed your preparation for that syllabus. Refer syllabus sheet in the starting of the book for the syllabus of all the DPP sheets.
- After completing the sheet check your answers with the solution booklet and complete the Result Grid. Finally spend time to analyse your performance and revise the areas which emerge out as weak in your evaluation.

DIRECTIONS (Q.1-Q.20) : There are 20 multiple choice questions. Each question has 4 choices (a), (b), (c) and (d), out of which **ONLY ONE** choice is correct.

Q.1 In a series *LCR* circuit capacitance is changed from *C* to 2 *C*. For the resonant frequency to remain unchanged, the inductance would be changed from *L* to

(a) $L/2$ (b) $2L$ (c) $4L$ (d) $L/4$

Q.2 The power factor of *LCR* circuit at resonance is

(a) 0.707 (b) 1 (c) Zero (d) 0.5

Q.3 An alternating current source of frequency 100 Hz is joined to a combination of a resistance, a capacitance and a inductance in series. The potential difference across the inductance, the resistance and the capacitor is 46, 8 and 40 volt respectively. The electromotive force of alternating current source in volt is

(a) 94 (b) 14 (c) 10 (d) 76

Q.4 A 10 ohm resistance, 5 mH inductance coil and 10 μF capacitor are joined in series. When a suitable frequency alternating current source is joined to this combination, the circuit resonates. If the resistance is halved, the resonance frequency

(a) is halved (b) is doubled

(c) remains unchanged (d) is quadrupled

Q.5 The phase difference between the current and voltage of LCR circuit in series combination at resonance is

(a) $0°$ (b) $\pi/2$

(c) π (d) $-\pi$

RESPONSE GRID	1. ⓐⓑⓒⓓ	2. ⓐⓑⓒⓓ	3. ⓐⓑⓒⓓ	4. ⓐⓑⓒⓓ	5. ⓐⓑⓒⓓ

Space for Rough Work

Q.6 The coefficient of induction of a choke coil is $0.1H$ and resistance is 12Ω. If it is connected to an alternating current source of frequency 60 Hz, then power factor is approximately
(a) 0.4 (b) 0.30 (c) 0.2 (d) 0.1

Q.7 The resonant frequency of a circuit is f. If the capacitance is made 4 times the initial values, then the resonant frequency will become
(a) f/2 (b) 2f (c) f (d) f/4

Q.8 In the non-resonant circuit, what will be the nature of the circuit for frequencies higher than the resonant frequency?
(a) Resistive (b) Capacitive
(c) Inductive (d) None of the above

Q.9 In a series LCR circuit, resistance $R = 10\Omega$ and the impedance $Z = 20\Omega$. The phase difference between the current and the voltage is
(a) 30° (b) 45° (c) 60° (d) 90°

Q.10 An alternating e.m.f. of frequency $\nu\left(=\frac{1}{2\pi\sqrt{LC}}\right)$ is applied to a series LCR circuit. For this frequency of the applied e.m.f.
(a) The circuit is at resonance and its impedance is made up only of a reactive part
(b) The current in the circuit is not in phase with the applied e.m.f. and the voltage across R equals this applied emf
(c) The sum of the p.d.'s across the inductance and capacitance equals the applied e.m.f. which is 180° ahead of phase of the current in the circuit
(d) The quality factor of the circuit is $\omega L/R$ or $1/\omega CR$ and this is a measure of the voltage magnification (produced by the circuit at resonance) as well as the sharpness of resonance of the circuit

Q.11 In a circuit L,C and R are connected in series with an alternating voltage source of frequency f. The current leads the voltage by 45°. The value of C is
(a) $\frac{1}{2\pi f(2\pi fL+R)}$ (b) $\frac{1}{\pi f(2\pi fL+R)}$
(c) $\frac{1}{2\pi f(2\pi fL-R)}$ (d) $\frac{1}{\pi f(2\pi fL-R)}$

Q.12 For the series LCR circuit shown in the figure, what is the resonance frequency and the amplitude of the current at the resonating frequency

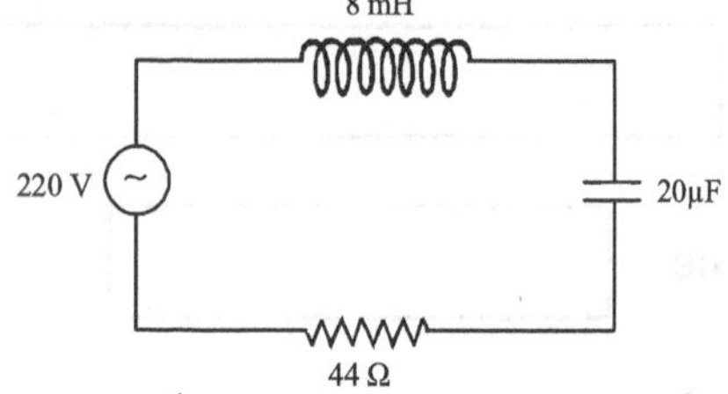

(a) $2500\,\text{rad s}^{-1}$ and $5\sqrt{2}$A (b) $2500\,\text{rad s}^{-1}$ and $5A$
(c) $2500\,\text{rad s}^{-1}$ and $\frac{5}{\sqrt{2}}$A (d) $25\,\text{rad s}^{-1}$ and $5\sqrt{2}$A

Q.13 In an ac circuit, V and I are given by $V = 100\sin(100\,t)$ volt, $I = \sin\left(100t+\frac{\pi}{3}\right)$ mA. The average power dissipated in circuit is
(a) 10^4 watt (b) 10 watt
(c) 0.025 watt (d) 2.5 watt

Q.14 For a series LCR circuit $R = X_L = 2X_C$. The impedance of the circuit and phase difference between V and I respectively will be
(a) $\frac{\sqrt{5}R}{2}, \tan^{-1}(2)$ (b) $\frac{\sqrt{5}R}{2}, \tan^{-1}(1/2)$
(c) $\sqrt{5}X_C, \tan^{-1}(2)$ (d) $\sqrt{5}R, \tan^{-1}(1/2)$

Q.15 If a current I given by $I_0\sin\left(\omega t-\frac{\pi}{2}\right)$ flows in an ac circuit across which an ac potential of $E = E_0\sin\omega t$ has been applied, then the average power consumption P in the circuit will be
(a) $P=\frac{E_0 I_0}{\sqrt{2}}$ (b) $P=\sqrt{2}E_0 I_0$
(c) $P=\frac{E_0 I_0}{2}$ (d) $P=0$

Q.16 An ac supply gives $30V$ r.m.s. which passes through a $10\,\Omega$ resistance. The power dissipated in it is
(a) $90\sqrt{2}$ W (b) 90 W (c) $45\sqrt{2}$ W (d) 45 W

RESPONSE GRID					
	6. ⓐⓑⓒⓓ	7. ⓐⓑⓒⓓ	8. ⓐⓑⓒⓓ	9. ⓐⓑⓒⓓ	10. ⓐⓑⓒⓓ
	11. ⓐⓑⓒⓓ	12. ⓐⓑⓒⓓ	13. ⓐⓑⓒⓓ	14. ⓐⓑⓒⓓ	15. ⓐⓑⓒⓓ
	16. ⓐⓑⓒⓓ				

Space for Rough Work

Q.17 The figure shows variation of R, X_L and X_C with frequency f in a series L, C, R circuit. Then for what frequency point, the circuit is inductive

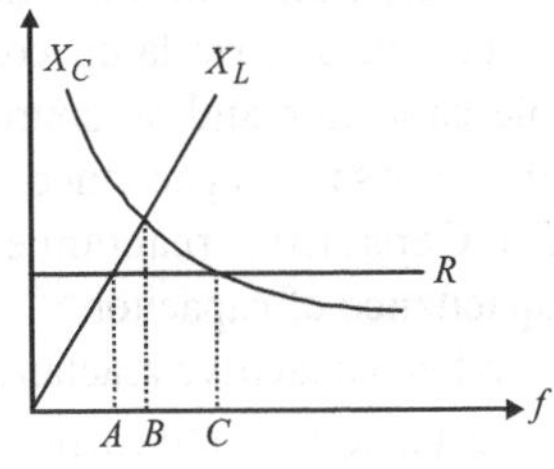

(a) A (b) B (c) C (d) All points

Q.18 An alternating e.m.f. of angular frequency ω is applied across an inductance. The instantaneous power developed in the circuit has an angular frequency

(a) $\frac{\omega}{4}$ (b) $\frac{\omega}{2}$ (c) ω (d) 2ω

Q.19 In the circuit shown in figure neglecting source resistance the voltmeter and ammeter reading will respectively, be

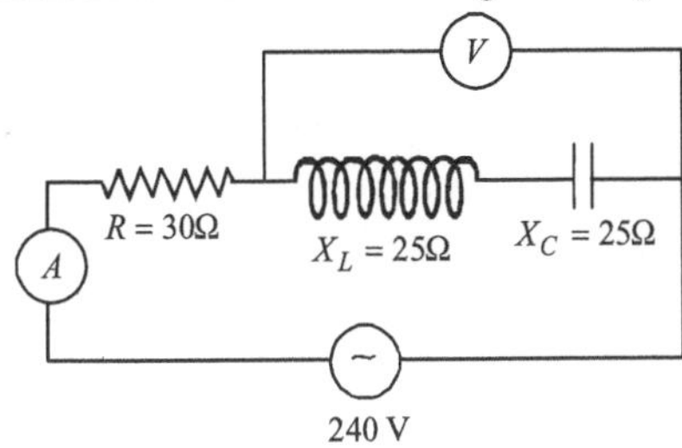

(a) $0V, 3A$ (b) $150V, 3A$
(c) $150V, 6A$ (d) $0V, 8A$

Q.20 In an LCR circuit, the sharpness of resonance depends on
(a) Inductance (L) (b) Capacitance (C)
(c) Resistance (R) (d) All of these

DIRECTIONS (Q.21-Q.23) : In the following questions, more than one of the answers given are correct. Select the correct answers and mark it according to the following codes:

Codes :
(a) 1, 2 and 3 are correct **(b)** 1 and 2 are correct
(c) 2 and 4 are correct **(d)** 1 and 3 are correct

Q.21 For series LCR circuit, correct statements are
(1) Applied e.m.f. and potential difference across resistance may be in phase
(2) Applied e.m.f. and potential difference at inductor coil have phase difference of $\pi/2$
(3) Potential difference across resitance and capacitor have phase difference of $\pi/2$
(4) Potential difference at capacitor and inductor have phase difference of $\pi/2$

Q.22 An ac source is connected to a resistive circuits. Which of the following statements are false?
(1) Current leads the voltage
(2) Current lags behind the voltage
(3) Any of (1) or (2) may be true depending upon the value of resistance
(4) Current and voltage are in same phase

Q.23 A series LCR arrangement with $X_L = 80\,\Omega, X_C = 50\,\Omega, R = 40\,\Omega$ is applied across a.c. source of 200 V. Choose the correct options.
(1) Wattless current = 3.2 A
(2) Power current = 3.2 A
(3) Power factor = 0.6
(4) Impedance of circuit = 50 Ω

DIRECTIONS (Q.24-Q.26) : Read the passage given below and answer the questions that follows :

A student constructs a series RLC circuit. While operating the circuit at a frequency f she uses an AC voltmeter and measures the potential difference across each device as $(\Delta V_R) = 8.8$ V, $(\Delta V_L) = 2.6$V and $(\Delta V_C) = 7.4$V.

Q.24 The circuit is constructed so that the inductor is next to the capacitor. What result should the student expect for a measurement of the combined potential difference $(\Delta V_L + \Delta V_C)$ across the inductor and capacitor ?
(a) 10.0 V (b) 7.8 V
(c) 7.4 V (d) 4.8 V

Q.25 What result should the student expect for a measurement of the amplitude E_m of the potential difference across the power supply ?
(a) 18.8 V (b) 13.6 V (c) 10.0 V (d) 4.0 V

RESPONSE GRID					
	17. ⓐⓑⓒⓓ	18. ⓐⓑⓒⓓ	19. ⓐⓑⓒⓓ	20. ⓐⓑⓒⓓ	21. ⓐⓑⓒⓓ
	22. ⓐⓑⓒⓓ	23. ⓐⓑⓒⓓ	24. ⓐⓑⓒⓓ	25. ⓐⓑⓒⓓ	

Space for Rough Work

Q.26 What will happen to the value of (ΔV_L) if the frequency is adjusted to increase the current through the circuit?
(a) (ΔV_L) will increase.
(b) (ΔV_L) will decrease.
(c) (ΔV_L) will remain the same regardless of any changes to f.
(d) There is not enough information to answer the question.

DIRECTIONS (Q. 27-Q.29) : Each of these questions contains two statements: Statement-1 (Assertion) and Statement-2 (Reason). Each of these questions has four alternative choices, only one of which is the correct answer. You have to select the correct choice.

(a) Statement-1 is True, Statement-2 is True; Statement-2 is a correct explanation for Statement-1.
(b) Statement-1 is True, Statement-2 is True; Statement-2 is NOT a correct explanation for Statement-1.
(c) Statement -1 is False, Statement-2 is True.
(d) Statement -1 is True, Statement-2 is False.

Q.27 Statement-1 : For an electric lamp connected in series with a variable capacitor and ac source, its brightness increases with increase in capacitance.
Statement-2 : Capacitive reactance decreases with increase in capacitance of capacitor.

Q.28 Statement-1 : When capacitive reactance is smaller than the inductive reactance in LCR current, e.m.f. leads the current.
Statement-2 : The phase angle is the angle between the alternating e.m.f. and alternating current of the circuit.

Q.29 Statement-1 : Choke coil is preferred over a resistor to adjust current in an ac circuit.
Statement-2 : Power factor for inductance is zero.

RESPONSE GRID	26. ⓐⓑⓒⓓ	27. ⓐⓑⓒⓓ	27. ⓐⓑⓒⓓ	28. ⓐⓑⓒⓓ	29. ⓐⓑⓒⓓ

DAILY PRACTICE PROBLEM SHEET 47 - PHYSICS			
Total Questions	29	Total Marks	116
Attempted		Correct	
Incorrect		Net Score	
Cut-off Score	26	Qualifying Score	46
Success Gap = Net Score – Qualifying Score			
Net Score = (Correct × 4) – (Incorrect × 1)			

Space for Rough Work

DPP - Daily Practice Problems

Name : 　　　　　　Date :

Start Time : 　　　　　　End Time :

PHYSICS 48

SYLLABUS : EM Waves

Max. Marks : 116 　　　　　　**Time : 60 min.**

GENERAL INSTRUCTIONS

- The Daily Practice Problem Sheet contains 29 MCQ's. For each question only one option is correct. Darken the correct circle/ bubble in the Response Grid provided on each page.
- You have to evaluate your Response Grids yourself with the help of solution booklet.
- Each correct answer will get you 4 marks and 1 mark shall be deduced for each incorrect answer. No mark will be given/ deducted if no bubble is filled. Keep a timer in front of you and stop immediately at the end of 60 min.
- The sheet follows a particular syllabus. Do not attempt the sheet before you have completed your preparation for that syllabus. Refer syllabus sheet in the starting of the book for the syllabus of all the DPP sheets.
- After completing the sheet check your answers with the solution booklet and complete the Result Grid. Finally spend time to analyse your performance and revise the areas which emerge out as weak in your evaluation.

DIRECTIONS (Q.1-Q.20) : There are 20 multiple choice questions. Each question has 4 choices (a), (b), (c) and (d), out of which **ONLY ONE** choice is correct.

Q.1 Light is an electromagnetic wave. Its speed in vacuum is given by the expression

(a) $\sqrt{\mu_0 \varepsilon_0}$ (b) $\sqrt{\dfrac{\mu_0}{\varepsilon_0}}$ (c) $\sqrt{\dfrac{\varepsilon_0}{\mu_0}}$ (d) $\sqrt{\dfrac{1}{\mu_0 \varepsilon_0}}$

Q.2 The range of wavelength of the visible light is

(a) 10 Å to 100 Å

(b) 4,000 Å to 8,000 Å

(c) 8,000 Å to 10,000 Å

(d) 10,000 Å to 15000 Å

Q.3 Which of the following radiations has the least wavelength?

(a) γ-rays (b) β-rays

(c) α-rays (d) X -rays

Q.4 A parallel plate capacitor with plate area A and seperation between the plates *d*, is charged by a constant current *i*. Consider a plane surface of area A/4 parallel to the plates and drawn symetrically between the plates, what is the displacement current through this area?

(a) i (b) 2i (c) i/4 (d) i/2

Q.5 The charging current for a capacitor is 1 A, then the displacement current is

(a) 1 A (b) 0.5 A

(c) 0 (d) 2 A

RESPONSE GRID	1. ⓐⓑⓒⓓ	2. ⓐⓑⓒⓓ	3. ⓐⓑⓒⓓ	4. ⓐⓑⓒⓓ	5. ⓐⓑⓒⓓ

Space for Rough Work

Q.6 If $\vec{E}$ and $\vec{B}$ be the electric and magnetic field of E.M. wave then the direction of propogation of E.M. wave is along the direction.

(a) $\vec{E}$ (b) $\vec{B}$

(c) $\vec{E} \times \vec{B}$ (d) $\vec{B} \times \vec{E}$

Q.7 Which of the following pairs of space and time varying E and B fields would generate a plane electromagnetic wave travelling in (–Z) direction ?

(a) E_x, B_y (b) E_y, B_x

(c) E_z, B_y (d) E_y, B_z

Q.8 Choose the wrong statement for E.M. wave. They-

(a) are transverse

(b) travel in vacuum with the speed of light

(c) are produced by accelerated charges

(d) travel with same speed in all medium

Q.9 The intensity of light from a source is $500/\pi$ W/m^2. Find the amplitude of electric field in this wave-

(a) $\sqrt{3} \times 10^2$ N/C (b) $2\sqrt{3} \times 10^2$ N/C

(c) $\frac{\sqrt{3}}{2} \times 10^2$ N/C (d) $2\sqrt{3} \times 10^1$ N/C

Q.10 A point source of 2 watt is radiating uniformly in all direction in vacuum. Find the amplitude of electric field at a distance 2m from it-

(a) 3×10^{-4} (b) $\sqrt{30}$

(c) $\sqrt{3} \times 10^{-4}$ (d) $\sqrt{3} \times 10^{-2}$

Q.11 In a EM wave the amplitude of electric field is 10 V/m. The frequency of wave is 5×10^4 Hz. The wave is propagating along Z-axis. Then the average energy density of magnetic field is-

(a) 2.21×10^{-10} J/m^3 (b) 2.21×10^{-8} J/m^3

(c) 2×10^{-8} J/m^3 (d) 2×10^{-10} J/m^3

Q.12 Elecromagnetic waves travel in a medium with a speed of 2×10^8 m/s. The relative permeability of the medium is 1. What is the relative permittivity of the medium ?

(a) 2.25 (b) 1.25 (c) 3.25 (d) 0.25

Q.13 A magnetic field of a plane electromagnetic wave is given by $B_y = 2 \times 10^{-7} \sin(0.5 \times 10^3 x + 1.5 \times 10^{11} t)$ T. Fequency of the wave is

(a) 23.9 Hz (b) 13.9 Hz

(c) 33.9 Hz (d) 12.9 Hz

Q.14 The electric field of a plane electromagnetic wave in vacuum is represented by

$E_x = 0$, $E_y = 0.5 \cos[2\pi \times 10^8 (t - x/c)]$ and $E_z = 0$.

Determine the wavelength of the wave.

(a) 4 m (b) 5 m (c) 3 m (d) 6 m

Q.15 A light beam travelling in the X-direction is described by the electric field $E_y = (300 \text{ V/m}) \sin\omega(t - x/c)$. An electron is constrained to move along the Y-direction with a speed of 2.0×10^7 m/s. Find the maximum magnetic force (in N) on the electron.

(a) 3.2×10^{-18} (b) 5.1×10^{-16}

(c) 6.5×10^{-11} (d) 7.8×10^{-12}

Q.16 Which of the following waves have minimum frequency ?

(a) Microwaves (b) Audible waves

(c) Ultrasonic waves (d) Radiowaves

Q.17 Electromagnetic waves travel in a medium which has relative permeability 1.3 and relative permittivity 2.14. Then the speed of the electromagnetic wave in the medium will be

(a) 13.6×10^6 m/s (b) 1.8×10^2 m/s

(c) 3.6×10^8 m/s (d) 1.8×10^8 m/s

Q.18 If λ_v, λ_x and λ_m represent the wavelength of visible light *x*-rays and microwaves respectively, then

(a) $\lambda_m > \lambda_x > \lambda_v$ (b) $\lambda_v > \lambda_m > \lambda_x$

(c) $\lambda_m > \lambda_v > \lambda_x$ (d) $\lambda_v > \lambda_x > \lambda_m$

RESPONSE GRID					
	6. ⓐⓑⓒⓓ	7. ⓐⓑⓒⓓ	8. ⓐⓑⓒⓓ	9. ⓐⓑⓒⓓ	10. ⓐⓑⓒⓓ
	11. ⓐⓑⓒⓓ	12. ⓐⓑⓒⓓ	13. ⓐⓑⓒⓓ	14. ⓐⓑⓒⓓ	15. ⓐⓑⓒⓓ
	16. ⓐⓑⓒⓓ	17. ⓐⓑⓒⓓ	18. ⓐⓑⓒⓓ		

Space for Rough Work

Q.19 Light wave is travelling along + y-direction. If the corresponding $\vec{E}$ vector at that time is along $+x$-direction, $\vec{B}$ vector must be directed along.

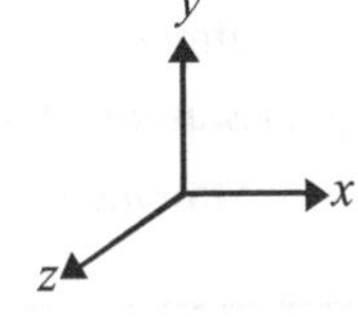

(a) y-axis

(b) x-axis

(c) + z-axis

(d) $-z$ axis

Q.20 A wave is propagating in a medium of dielectric constant 2 and relative magnetic permeability 50. The wave impedance of such a medium is

(a) 5Ω (b) 376.6Ω

(c) 1883 Ω (d) 3776 Ω

DIRECTIONS (Q.21-Q.23) : In the following questions, more than one of the answers given are correct. Select the correct answers and mark it according to the following codes:

Codes :

(a) 1, 2 and 3 are correct

(b) 1 and 2 are correct

(c) 2 and 4 are correct

(d) 1 and 3 are correct

Q.21 Which of the following statements are true ?

(1) Photographic plates are sensitive to ultraviolet rays.

(2) Photographic plates can be made sensitive to infrared rays.

(3) Infrared rays are emitted by hot objects.

(4) Infrared photon has more energy than the photon of visible light.

Q.22 Which of the following are electromagnetic waves ?

(1) Cosmic rays (2) Gamma rays

(3) X-rays (4) β– rays

Q.23 An electromagnetic wave of frequency $\nu = 3.0$ MHz passes from vacuum into a dielectric medium with permitivity $\varepsilon = 4.0$. Then the wrong statements are

(1) Wavelength is doubled and the frequency remains unchanged

(2) Wavelength is doubled and frequency becomes half

(3) Wavelength and frequency both remain unchanged

(4) Wavelength is halved and frequency remains unchanged

DIRECTIONS (Q.24-Q.26) : Read the passage given below and answer the questions that follows :

The electron density of a layer of ionosphere at a height 150 km from the earth surface is 9×10^{10} per m^3. For the sky wave transmission from this layer upto a range of 250 km, find

Q.24 The critical frequency of the layer

(a) 2.7×10^6 Hz (b) 2.7×10^5 Hz

(c) 4.7×10^6 Hz (d) 4.8×10^5Hz

Q.25 The maximum usuable frequency

(a) 3.17×10^8 Hz (b) 3.17×10^6 Hz

(c) 4.57×10^6 Hz (d) 4.57×10^6Hz

Q.26 The angle of incidence of this layer

(a) 34.5° (b) 25.2°

(c) 31.6° (d) 40°

DIRECTIONS (Q. 27-Q.29) : Each of these questions contains two statements: Statement-1 (Assertion) and Statement-2 (Reason). Each of these questions has four alternative choices, only one of which is the correct answer. You have to select the correct choice.

(a) Statement-1 is True, Statement-2 is True; Statement-2 is a correct explanation for Statement-1.

(b) Statement-1 is True, Statement-2 is True; Statement-2 is NOT a correct explanation for Statement-1.

(c) Statement -1 is False, Statement-2 is True.

(d) Statement -1 is True, Statement-2 is False.

RESPONSE GRID					
	19. ⓐⓑⓒⓓ	20. ⓐⓑⓒⓓ	21. ⓐⓑⓒⓓ	22. ⓐⓑⓒⓓ	23. ⓐⓑⓒⓓ
	24. ⓐⓑⓒⓓ	25. ⓐⓑⓒⓓ	26. ⓐⓑⓒⓓ		

Space for Rough Work

Q.27 Statement-1: The electromagnetic waves of shorter wavelength can travel longer distances on earth's surface than those of longer wavelengths.

Statement-2: Shorter the wavelength, the larger is the velocity of wave propagation.

Q.28 Statement-1: Ultraviolet radiation are of higher frequency waves and are dangerous to human being.

Statement-2: Ultraviolet radiation are absorbed by the atmosphere.

Q.29 Statement-1: Radio waves can be polarised.

Statement-2: Sound waves in air are longitudinal in nature.

RESPONSE GRID	27. ⓐⓑⓒⓓ	28. ⓐⓑⓒⓓ	29. ⓐⓑⓒⓓ

DAILY PRACTICE PROBLEM SHEET 48 - PHYSICS			
Total Questions	29	Total Marks	116
Attempted		Correct	
Incorrect		Net Score	
Cut-off Score	30	Qualifying Score	48
Success Gap = Net Score – Qualifying Score			
Net Score = (Correct × 4) – (Incorrect × 1)			

Space for Rough Work

DPP - Daily Practice Problems

Name : 　　　　　　　　　　　　Date :

Start Time : 　　　　　　　　　　End Time :

PHYSICS 49

SYLLABUS : RAY OPTICS-1 (Reflection on plane mirrors and curved mirrors)

Max. Marks : 120 　　　　　　　　　　**Time : 60 min.**

GENERAL INSTRUCTIONS

- The Daily Practice Problem Sheet contains 30 MCQ's. For each question only one option is correct. Darken the correct circle/ bubble in the Response Grid provided on each page.
- You have to evaluate your Response Grids yourself with the help of solution booklet.
- Each correct answer will get you 4 marks and 1 mark shall be deduced for each incorrect answer. No mark will be given/ deducted if no bubble is filled. Keep a timer in front of you and stop immediately at the end of 60 min.
- The sheet follows a particular syllabus. Do not attempt the sheet before you have completed your preparation for that syllabus. Refer syllabus sheet in the starting of the book for the syllabus of all the DPP sheets.
- After completing the sheet check your answers with the solution booklet and complete the Result Grid. Finally spend time to analyse your performance and revise the areas which emerge out as weak in your evaluation.

DIRECTIONS (Q.1-Q.22) : There are 22 multiple choice questions. Each question has 4 choices (a), (b), (c) and (d), out of which **ONLY ONE** choice is correct.

Q.1 Find the number of images formed by two mutually perpendicular mirrors –

(a) 3 　(b) 4 　(c) 1 　(d) 2

Q.2 The angle θ between two plane mirrors producing five images of a given object is given by.

(a) $30° \le \theta \le 72°$ 　(b) $45° \le \theta \le 72°$

(c) $60° \le \theta \le 72°$ 　(d) $15° \le \theta \le 72°$

Q.3 Two mirrors are inclined at an angle of 50°. Then what is the number of images formed for an object placed in between the mirrors ?

(a) 3 　(b) 6 　(c) 1 　(d) 7

Q.4 Two plane mirrors are inclined at an angle θ. A ray of light is incident on one mirror at an angle of incidence i. The ray is reflected from this mirror, falls on the second mirror from where it is reflected parallel to the first mirror. What is the value of i, the angle of incidence in terms θ ?

(a) $2\theta - 90°$ 　(b) $4\theta - 90°$

(c) $\theta - 90°$ 　(d) $3\theta - 90°$

Q.5 A girl stands at a distance 30 cm from the mirror. She is able to see her erect image but of 1/5 height of actual height. The mirror will be :

(a) plane mirror 　(b) concave mirror

(c) convex mirror 　(d) plane convex mirror

RESPONSE GRID	1. ⓐⓑⓒⓓ	2. ⓐⓑⓒⓓ	3. ⓐⓑⓒⓓ	4. ⓐⓑⓒⓓ	5. ⓐⓑⓒⓓ

Space for Rough Work

Q.6 An object is placed at a distance of 50 cm from a convex mirror. A plane mirror is placed in front of the convex mirror in such a way that it convers half of the convex mirror. If the distance between object and plane mirror is 30 cm then there is no parallax between the images formed by two mirrors , the radius of curvature of convex mirror will be :

(a) 50 cm (b) 25 cm (c) 12.5 cm (d) 100cm

Q.7 Two plane mirrors are inclined at an angle of 30°. Then the first four images of an object O placed between the two mirrors are correctly represented by

(a) 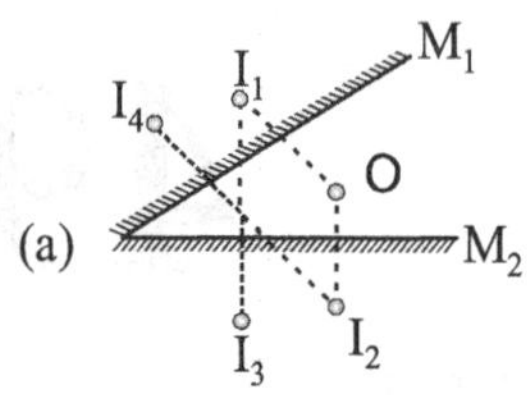

(b) 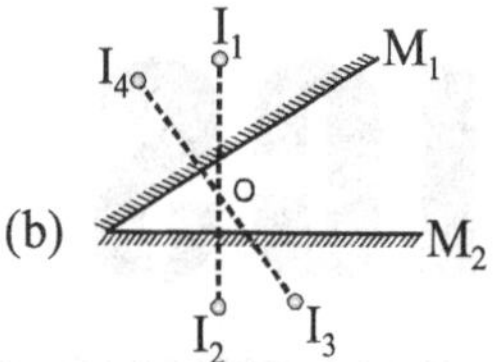

(c) 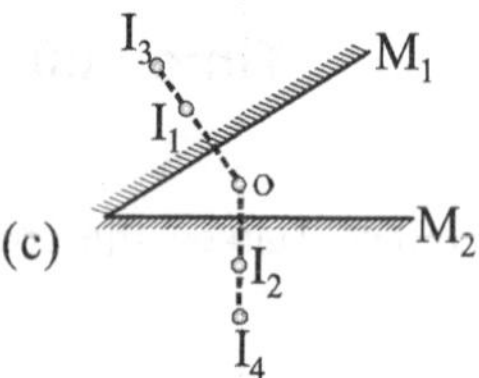

(d) 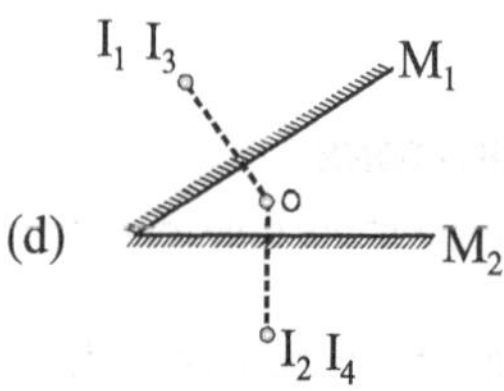

Q.8 The plane of a mirror makes an angle of 30° with horizontal. If a vertical ray is incident on a mirror, then what is the angle between mirror and reflected ray ?

(a) 60° (b) 90° (c) 45° (d) 30°

Q.9 Two plane mirrors are placed at an angle α so that a ray parallel to one mirror gets reflected parallel to the second mirror after two consecutive reflections. The value of α will be

(a) 30° (b) 60° (c) 75° (d) 90°

Q.10 A 0.2 cm high object is placed 15 cm from a concave mirror of focal length 5 cm. Find position and size of the image.

(a) 7.5 cm, 0.1 cm. (b) 7.5 cm, 0.4 cm.
(c) 10.0 cm, 0.5 cm. (d) 7.5 cm, 0.4 cm.

Q.11 A 0.5 cm high object is placed at 30 cm from a convex mirror whose focal length is 20 cm. Find the position and size of the image.

(a) 12 cm, 0.2 cm (b) 18 cm, 0.2 cm
(c) 6 cm, 0.5 cm (d) 5 cm, 0.1 cm

Q.12 There is a convex mirror of radius 50 cm. The image of a point at a distance 50 cm from the pole of mirror on its axis will be formed at :

O —— 50 cm —— 50 cm —— C

(a) infinity
(b) pole
(c) focus
(d) 16.67 cm behind the mirror

Q.13 A particle is moving at a constant speed v_0 from a large distance towards a concave mirror of radius R along its principle axis. Find the speed of the image formed by the mirror as a function of the distance u of the particles from the mirror.

(a) $\left(\frac{R}{2u-R}\right)^2 .v_0$ (b) $\left(\frac{R}{2u+R}\right)^2 .v_0$

(c) $\left(\frac{2R}{2u-R}\right)^2 .v_0$ (d) None of these

Q.14 A short linear object of length b lies along the axis of a concave mirror of focal length f at a distance u from the pole of the mirror. Find the approximate size of the image.

(a) $b\left(\frac{f}{u-f}\right)^2$ (b) $b\left(\frac{f}{u+f}\right)^2$

(c) $b\left(\frac{2f}{u-f}\right)^2$ (d) None of these

Q.15 The relation between the linear magnification m, the object distance u and the focal length f for a spherical mirror is

(a) $m=\frac{f-u}{f}$ (b) $m=\frac{f}{f-u}$

(c) $m=\frac{f+u}{f}$ (d) $m=\frac{f}{f+u}$

RESPONSE GRID					
	6. ⓐⓑⓒⓓ	7. ⓐⓑⓒⓓ	8. ⓐⓑⓒⓓ	9. ⓐⓑⓒⓓ	10. ⓐⓑⓒⓓ
	11. ⓐⓑⓒⓓ	12. ⓐⓑⓒⓓ	13. ⓐⓑⓒⓓ	14. ⓐⓑⓒⓓ	15. ⓐⓑⓒⓓ

Space for Rough Work

Q.16 An object of length 1 cm is placed at a distance of 15 cm from a concave mirror of focal length 10 cm. The nature and size of the image are

(a) real, inverted, 1.0 cm (b) real, inverted, 2.0 cm
(c) virtual, erect, 0.5 cm (d) virtual, erect, 1.0 cm

Q.17 In an experiment to determine the focal length (f) of a concave mirror by the u - v method, a student places the object pin A on the principal axis at a distance x from the pole P. The student looks at the pin and its inverted image from a distance keeping his/her eye in line with PA. When the student shifts his/her eye towards left, the image appears to the right of the object pin. Then,

(a) $x < f$ (b) $f<x<2f$ (c) $x = 2f$ (d) $x > 2f$

Q.18 Two plane mirrors are inclined to each other at some angle. A ray of light incident at 30° on one, after reflection from the other retraces its path. The angle between the mirrors is

(a) 30° (b) 45° (c) 60° (d) 90°

Q.19 A convex mirror is used to form the image of an object. Which of the following statements is wrong ?

(a) The image lies between the pole and the focus
(b) The image is diminished in size
(c) The image is erect
(d) The image is real

Q.20 A point source of light B is placed at a distance L in front of the centre of a mirror of width 'd' hung vertically on a wall. A man walks in front of the mirror along a line parallel to the mirror at a distance $2L$ from it as shown in fig. The greatest distance over which he can see the image of the light source in the mirror is

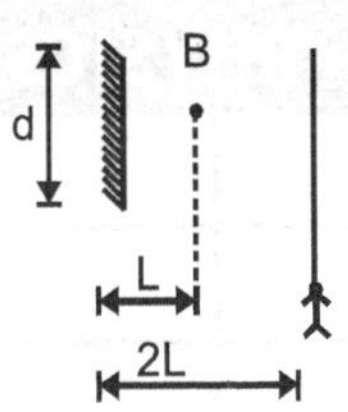

(a) $d/2$ (b) d (c) $2d$ (d) $3d$

Q.21 A concave mirror of focal length f_0(in magnitude) produces a real image n time the size of the object. What is the distance of the object from the mirror?

(a) $\frac{-(n+1)}{n}f_0$ (b) $\frac{(n+1)}{n}f_0$ (c) $\frac{(n-1)}{n}f_0$ (d) $\frac{-(n^2+1)}{n}f_0$

Q.22 The focal length of a concave mirror is 30 cm. Find the position of the object in front of the mirror, so that the image is three times the size of the object.

(a) 20 cm (only) (b) 40 cm (only)
(c) 30 cm (only) (d) 20 cm or 40 cm

DIRECTIONS (Q.23-Q.25) : In the following questions, more than one of the answers given are correct. Select the correct answers and mark it according to the following codes:

Codes :

(a) 1, 2 and 3 are correct **(b)** 1 and 2 are correct
(c) 2 and 4 are correct **(d)** 1 and 3 are correct

Q.23 A plane mirror reflecting a ray of incident light is rotated through an angle θ about an axis through the point of incidence in the plane of the mirror perpendicular to the plane of incidence, then

(1) The reflected ray rotates through an angle 2θ
(2) The incident ray is fixed
(3) The reflected ray does not rotate
(4) The reflected ray rotates through an angle θ

Q.24 The light reflected by a plane mirror will not form a real image

(1) If the rays incident on the mirror are diverging
(2) Under no circumstances
(3) If the object is real
(4) If the rays incident on the mirror are converging

RESPONSE GRID					
	16. ⓐⓑⓒⓓ	17. ⓐⓑⓒⓓ	18. ⓐⓑⓒⓓ	19. ⓐⓑⓒⓓ	20. ⓐⓑⓒⓓ
	21. ⓐⓑⓒⓓ	22. ⓐⓑⓒⓓ	23. ⓐⓑⓒⓓ	24. ⓐⓑⓒⓓ	

Space for Rough Work

Q.25 Which of the following form(s) a virtual and erect image for all positions of the object?

(1) Convex lens (2) Concave lens

(3) Convex mirror (4) Concave mirror

DIRECTIONS (Q.26-Q.27) : Read the passage given below and answer the questions that follows :

A plane mirror (M_1) and a concave mirror (M_2) of focal length 10 cm are arranged as shown in figure. An object is kept at origin. Answer the following questions. (Consider image formed by single reflection in all cases)

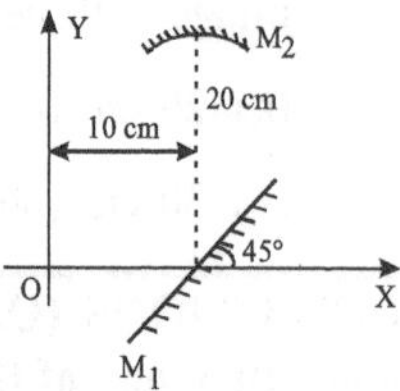

Q.26 The co-ordinates of image formed by plane mirror are

(a) (–20 cm, 0) (b) (10 cm, – 60 cm)

(c) (10 cm, –10 cm) (d) (10 cm, 10 cm)

Q.27 The co-ordinates of image formed by concave mirror are

(a) (10 cm, – 40 cm) (b) (10 cm, – 60 cm)

(c) (10 cm, 8 cm) (d) None of these

DIRECTIONS (Qs. 28-Q.30) : Each of these questions contains two statements: Statement-1 (Assertion) and Statement-2 (Reason). Each of these questions has four alternative choices, only one of which is the correct answer. You have to select the correct choice.

(a) Statement-1 is True, Statement-2 is True; Statement-2 is a correct explanation for Statement-1.

(b) Statement-1 is True, Statement-2 is True; Statement-2 is NOT a correct explanation for Statement-1.

(c) Statement -1 is False, Statement-2 is True.

(d) Statement -1 is True, Statement-2 is False.

Q.28 Statement-1: The mirrors used in search lights are parabolic and not concave spherical.

Statement-2: In a concave spherical mirror the image formed is always virtual.

Q.29 Statement-1: When an object is placed between two plane parallel mirors, then all the images found are of different intensity.

Statement-2: In case of plane parallel mirrors, only two images are possible.

Q.30 Statcment-1: Thc sizc of thc mirror docsn't affect the nature of the image.

Statement-2: Small mirror always forms a virtual image.

RESPONSE GRID	25. ⓐⓑⓒⓓ 26. ⓐⓑⓒⓓ 27. ⓐⓑⓒⓓ 28. ⓐⓑⓒⓓ 29. ⓐⓑⓒⓓ 30. ⓐⓑⓒⓓ

DAILY PRACTICE PROBLEM SHEET 49 - PHYSICS			
Total Questions	30	Total Marks	120
Attempted		Correct	
Incorrect		Net Score	
Cut-off Score	30	Qualifying Score	50
Success Gap = Net Score – Qualifying Score			
Net Score = (Correct × 4) – (Incorrect × 1)			

Space for Rough Work

DPP - Daily Practice Problems

Name :

Date :

Start Time :

End Time :

PHYSICS 50

SYLLABUS : RAY OPTICS - II (Refraction on plane surface, total internal reflection, prism)

Max. Marks : 108 **Time : 60 min.**

GENERAL INSTRUCTIONS

- The Daily Practice Problem Sheet contains 27 MCQ's. For each question only one option is correct. Darken the correct circle/ bubble in the Response Grid provided on each page.
- You have to evaluate your Response Grids yourself with the help of solution booklet.
- Each correct answer will get you 4 marks and 1 mark shall be deduced for each incorrect answer. No mark will be given/ deducted if no bubble is filled. Keep a timer in front of you and stop immediately at the end of 60 min.
- The sheet follows a particular syllabus. Do not attempt the sheet before you have completed your preparation for that syllabus. Refer syllabus sheet in the starting of the book for the syllabus of all the DPP sheets.
- After completing the sheet check your answers with the solution booklet and complete the Result Grid. Finally spend time to analyse your performance and revise the areas which emerge out as weak in your evaluation.

DIRECTIONS (Q.1-Q.21) : There are 21 multiple choice questions. Each question has 4 choices (a), (b), (c) and (d), out of which **ONLY ONE** choice is correct.

Q.1 A thin prism of angle $A = 6°$ produces a deviation $\delta = 3°$. Find the refractive index of the material of prism.

(a) 1.5 (b) 1.0
(c) 2.5 (d) 0.5

Q.2 A ray of light is incident at an angle of 60° on one face of a prism which has an angle of 30°. The ray emerging out of the prism makes an angle of 30° with the incident ray. Calculate the refractive index of the material of the prism

(a) 1 (b) $\sqrt{2}$
(c) $\sqrt{3}$ (d) 2

Q.3 Light of wavelength 6000Å enters from air into water (a medium of refractive index 4/3). Find the speed and wavelength $[c = 3 \times 10^8 \text{ m/s}]$

(a) 2.25×10^8 m/s, 4500Å (b) 1.25×10^8 m/s, 2500Å
(c) 3.15×10^8 m/s, 3500Å (d) 3.45×10^8 m/s, 5500Å

Q.4 A ray of light is incident on a transparent glass-slab of refractive index 1.5. If the reflected and refracted rays are mutually perpendicular, what is the angle of incidence ?

(a) 30° (b) $\sin^{-1}\frac{2}{3}$
(c) $\tan^{-1}\frac{2}{3}$ (d) $\tan^{-1}\frac{3}{2}$

RESPONSE GRID	1. ⓐⓑⓒⓓ	2. ⓐⓑⓒⓓ	3. ⓐⓑⓒⓓ	4. ⓐⓑⓒⓓ

Space for Rough Work

Q.5 An optical fibre consists of core of μ_1 surrounded by a cladding of $\mu_2 < \mu_1$. A beam of light enters from air at an angle α with axis of fibre. The highest α for which ray can be travelled through fibre is

(a) $\cos^{-1}\sqrt{\mu_2^2-\mu_1^2}$

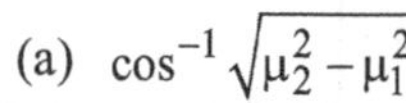

(b) $\sin^{-1}\sqrt{\mu_1^2-\mu_2^2}$

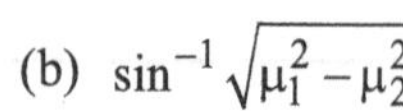

(c) $\tan^{-1}\sqrt{\mu_1^2-\mu_2^2}$

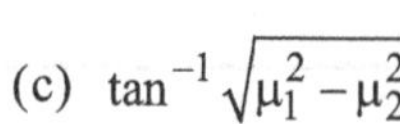

(d) $\sec^{-1}\sqrt{\mu_1^2-\mu_2^2}$

Q.6 A glass plate 4 mm thick is viewed from the above through a microscope. The microscope must be lowered 2.58 mm as the operator shifts from viewing the top surface to viewing the bottom surface through the glass. What is the index of refraction of the glass ?
(a) 1.61 (b) 1.55 (c) 3.24 (d) 1.21

Q.7 A vertical microscope is focussed on a point at the bottom of an empty tank. Water (μ = 4/3) is then poured into the tank. The height of the water column is 4cm. Another lighter liquid, which does not mix with water and which has refractive index 3/2 is then poured over the water.The height of liquid column is 2cm. What is the vertical distance through which the microscope must be moved to bring the object in focus again ?
(a) 2.61 m (b) 1.55 m (c) 3.12 m (d) 1.67 m

Q.8 Light from a sodium lamp (λ_0 = 589 nm) passes through a tank of glycerin (refractive index 1.47) 20m long in a time t_1. If it takes a time t_2 to traverse the same tank when filled with carbon disulphide (index 1.63), then the difference $t_2 - t_1$ is
(a) 6.67×10^{-8} sec (b) 1.09×10^{-7} sec
(c) 2.07×10^{-7} sec (d) 1.07×10^{-8} sec

Q.9 A light beam is travelling from Region I to Region IV (Refer Figure). The refractive index in Regions I, II, III and IV are $n_0, \frac{n_0}{2}, \frac{n_0}{6}$ and $\frac{n_0}{8}$, respectively. The angle of incidence θ for which the beam just misses entering Region IV is

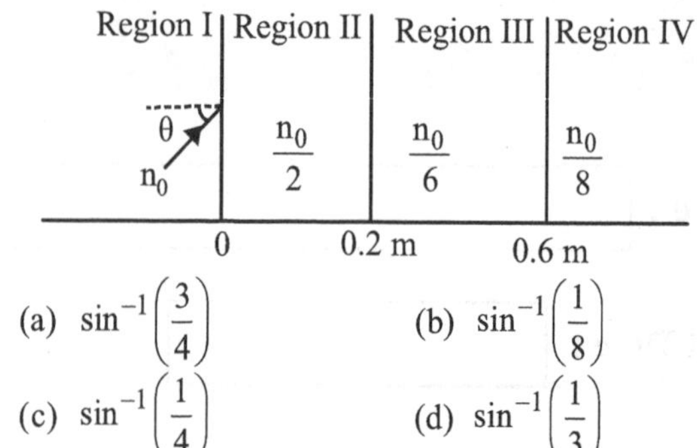

(a) $\sin^{-1}\left(\frac{3}{4}\right)$ (b) $\sin^{-1}\left(\frac{1}{8}\right)$
(c) $\sin^{-1}\left(\frac{1}{4}\right)$ (d) $\sin^{-1}\left(\frac{1}{3}\right)$

Q.10 The refractive index of the material of a prism is $\sqrt{2}$ and its prism angle is 30°. One of its refracting faces is polished. The incident beam of light will return back for the angle of incidence
(a) 60° (b) 45° (c) 30° (d) 0°

Q.11 A ray of light incident on a prism surface at an angle of 50° in the minimum deviation position. If the angle of prism is 60° then the values of δ_m and μ will be respectively – (sin 50° = 0.766)
(a) 40° and 1.532 (b) 60° and 1.532
(c) 90° and 1.532 (d) 0° and 1.532

Q.12 A glass prism of refractive index 1.5 and angle of prism 6° is put in contact with another prism of refractive index 1.6 when a ray of light is made incident on this combination normally then it emerges out undeviated. The angle of second prism will be –
(a) 6° (b) 5° (c) 4° (d) 3°

Q.13 A crown glass prism of angle 5° is to be combined with a flint glass prism in such a way that the mean ray passes undeviated. Find the angle of the flint glass prism needed and the angular dispersion produced by the combination when white light goes through it. Refractive indices for red, yellow and violet light are 1.514, 1.517 and 1.523 respectively for crown glass and 1.613, 1.620 and 1.632 for flint glass.
(a) 4.2°, 0.0348° (b) 4.2°, 0.0138°
(c) 1.2°, 0.0348° (d) 4.4°, 0.0218°

Q.14 Calculate the dispersive power for crown glass from the (given data : μ_v= 1.5230, μ_r =1.5145)
(a) 0.0163 (b) 0.0183 (c) 0.0142 (d) 0.0112

RESPONSE GRID					
	5. ⓐⓑⓒⓓ	6. ⓐⓑⓒⓓ	7. ⓐⓑⓒⓓ	8. ⓐⓑⓒⓓ	9. ⓐⓑⓒⓓ
	10. ⓐⓑⓒⓓ	11. ⓐⓑⓒⓓ	12. ⓐⓑⓒⓓ	13. ⓐⓑⓒⓓ	14. ⓐⓑⓒⓓ

Space for Rough Work

Q.15 A prism of dispersive power 0.021 and refractive index 1.53 form an achromatic combination with prism of angle 4.2° and dispersive power 0.045 having refractive index 1.65. Find the resultant deviation.
(a) 1.12° (b) 2.16° (c) 3.12° (d) 4.18°

Q.16 A ray of light fall normally on a refracting face of a prism of refractive index 1.5. Find the angle of the prism if the ray just fails to emerge from the prism.
(a) 55° (b) 22° (c) 12° (d) 42°

Q.17 The refractive indices of material of a prism for blue and red colours are 1.532 and 1.514 respectively. Calculate angular dispersion produced by the prism if angle of prism is 8°.
(a) 0.144° (b) 0.122° (c) 0.133° (d) 0.111°

Q.18 A ball is dropped from a height of 20 m above the surface of water in a lake. The refractive index of water is 4/3. A fish inside the lake, in the line of fall of the ball, is looking at the ball. At an instant, when the ball is 12.8 m above the water surface, the fish sees the speed of ball as [g = 10 m/s^2]
(a) 9 m/s (b) 12 m/s (c) 16 m/s (d) 21.33 m/s

Q.19 The dispersive powers of crown and flint glasses are 0.03 and 0.05 respectively. The refractive indices for yellow light for these glasses are 1.517 and 1.621 respectively. It is desired to form an achromatic combination of prisms of crown and flint glasses which can produce a deviation of 1° in the yellow ray. The refracting angle of the flint glass prism is
(a) 2.4° (b) 1.4° (c) 3.4° (d) 5.2°

Q.20 A glass prism (μ = 1.5) is dipped in water (μ = 4/3) as shown in figure. A light ray is incident normally on the surface AB. It reaches the surface BC after T.I.R

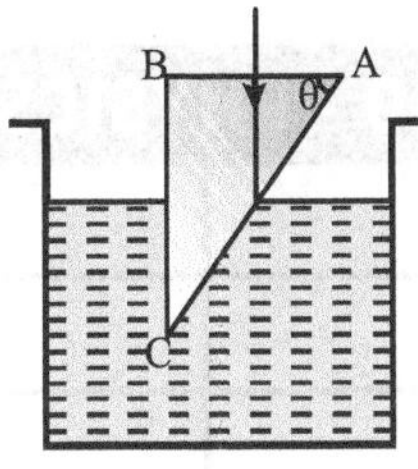

(a) $\sin\theta \geq 8/9$ (b) $2/3 < \sin\theta < 8/9$
(c) $\sin\theta \leq 2/3$ (d) It is not possible

Q.21 A prism having an apex angle 4° and refraction index 1.5 is located in front of a vertical plane mirror as shown in figure. Through what total angle is the ray deviated after reflection from the mirror

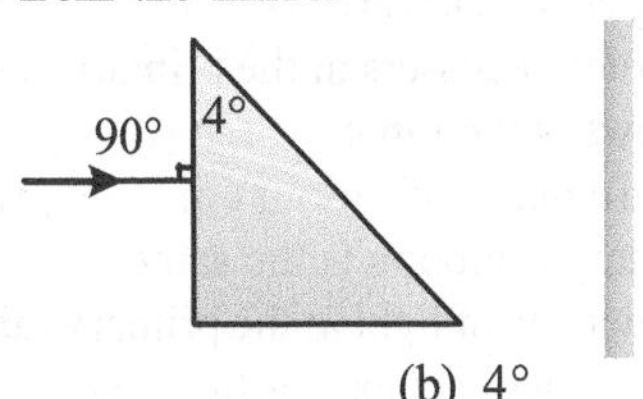

(a) 176° (b) 4°
(c) 178° (d) 2°

DIRECTIONS (Q.22-Q.24) : In the following questions, more than one of the answers given are correct. Select the correct answers and mark it according to the following codes:

Codes :
(a) 1, 2 and 3 are correct **(b)** 1 and 2 are correct
(c) 2 and 4 are correct **(d)** 1 and 3 are correct

Q.22 A ray of monochromatic light is incident on the plane surface of separation between two media *x* and *y* with angle of incidence *i* in the medium *x* and angle of refraction *r* in the medium *y*. The graph shows the relation between sin *i* and sin *r*.

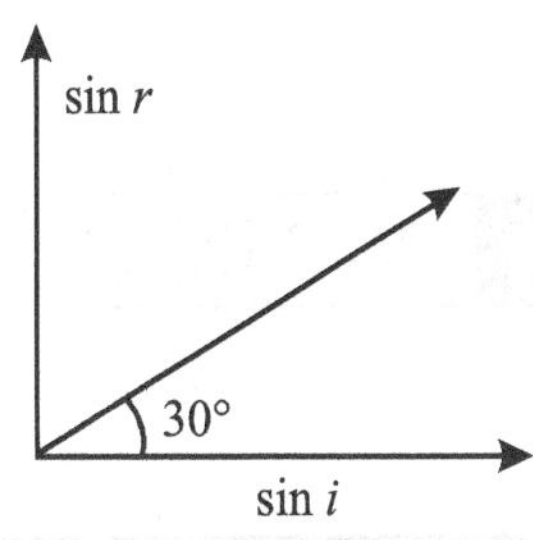

(1) The speed of light in the medium *y* is $\sqrt{3}$ times than in medium *x*
(2) The speed of light in the medium *y* is $\frac{1}{\sqrt{3}}$ times than in medium *x*.
(3) The total internal reflection can take place when the incidence is in *x*.
(4) The total internal reflection can take place when the incidence is in *y*

RESPONSE GRID					
	15. ⓐⓑⓒⓓ	16. ⓐⓑⓒⓓ	17. ⓐⓑⓒⓓ	18. ⓐⓑⓒⓓ	19. ⓐⓑⓒⓓ
	20. ⓐⓑⓒⓓ	21. ⓐⓑⓒⓓ	22. ⓐⓑⓒⓓ		

Space for Rough Work

Q.23 Dispersive power does not depend upon
(1) The shape of prism (2) Angle of prism
(3) Height of the prism (4) Material of prism

Q.24 The wrong statements are
(1) The order of colours in the primary and the secondary rainbows is the same
(2) The intensity of colours in the primary and the secondary rainbows is the same
(3) The intensity of light in the primary rainbow is greater and the order of colours is the same than the secondary rainbow
(4) The intensity of light for different colours in primary rainbow is greater and the order of colours is reverse as that in the secondary rainbow

DIRECTIONS (Qs. 25-Q.27) : Each of these questions contains two statements: Statement-1 (Assertion) and Statement-2 (Reason). Each of these questions has four alternative choices, only one of which is the correct answer. You have to select the correct choice.

(a) Statement-1 is True, Statement-2 is True; Statement-2 is a correct explanation for Statement-1.
(b) Statement-1 is True, Statement-2 is True; Statement-2 is NOT a correct explanation for Statement-1.
(c) Statement -1 is False, Statement-2 is True.
(d) Statement -1 is True, Statement-2 is False.

Q.25 **Statement-1:** There is no dispersion of light refracted through a rectangular glass slab.

Statement-2: Dispersion of light is the phenomenon of splitting of a beam of white light into its constituent colours.

Q.26 **Statement-1:** Dispersion of light occurs because velocity of light in a material depends upon its colour.

Statement-2: The dispersive power depends only upon the material of the prism, not upon the refracting angle of the prism.

Q.27 **Statement-1:** If a plane glass slab is placed on the letters of different colours all the letters appear to be raised up to the same height.

Statement-2: Different colours have different wavelengths.

RESPONSE GRID	23. ⓐⓑⓒⓓ	24. ⓐⓑⓒⓓ	25. ⓐⓑⓒⓓ	26. ⓐⓑⓒⓓ	27. ⓐⓑⓒⓓ

DAILY PRACTICE PROBLEM SHEET 50 - PHYSICS			
Total Questions	27	Total Marks	108
Attempted		Correct	
Incorrect		Net Score	
Cut-off Score	28	Qualifying Score	46
Success Gap = Net Score – Qualifying Score			
Net Score = (Correct × 4) – (Incorrect × 1)			

Space for Rough Work

DPP - Daily Practice Problems

Name : 　　　　　　　　Date :

Start Time : 　　　　　　　　End Time :

PHYSICS 51

SYLLABUS : RAY OPTICS - 3 (Refraction on curved surface lens, Optical instrument)

Max. Marks : 120 　　　　　　　　**Time : 60 min.**

GENERAL INSTRUCTIONS

- The Daily Practice Problem Sheet contains 30 MCQ's. For each question only one option is correct. Darken the correct circle/ bubble in the Response Grid provided on each page.
- You have to evaluate your Response Grids yourself with the help of solution booklet.
- Each correct answer will get you 4 marks and 1 mark shall be deduced for each incorrect answer. No mark will be given/ deducted if no bubble is filled. Keep a timer in front of you and stop immediately at the end of 60 min.
- The sheet follows a particular syllabus. Do not attempt the sheet before you have completed your preparation for that syllabus. Refer syllabus sheet in the starting of the book for the syllabus of all the DPP sheets.
- After completing the sheet check your answers with the solution booklet and complete the Result Grid. Finally spend time to analyse your performance and revise the areas which emerge out as weak in your evaluation.

DIRECTIONS (Q.1-Q.21) : There are 21 multiple choice questions. Each question has 4 choices (a), (b), (c) and (d), out of which **ONLY ONE** choice is correct.

Q.1 A small point object is placed at O, at a distance of 0.60 metre in air from a convex spherical surface of refractive index 1.5. If the radius of the curvature is 25 cm, then what is the position of the image on the principal axis ?

(a) 4.5 m　　(b) 2.5 m
(c) 1.5 m　　(d) 5.5 m

Q.2 The radius of a glass ball is 5 cm. There is an air bubble at 1cm from the centre of the ball and refractive index of glass is 1.5. The position of image viewed from surface near the bubble is.

(a) 3.63 cm　　(b) 4.63 cm
(c) 2.12 cm　　(d) 5.12 cm

Q.3 In case of thin lens of focal length f an object is placed at a distance x_1 from first focus and its image is formed at a distance x_2 from the second focus, find $x_1 x_2$

(a) f　　(b) f^3
(c) f^2　　(d) 1/f

Q.4 What is the refractive index of material of a plano-convex lens , if the radius of curvature of the convex surface is 10 cm and focal length of the lens is 30 cm ?

(a) 1/3　　(b) 4/3
(c) 2/3　　(d) 1/4

Q.5 A convex lens of focal length 10.0 cm is placed in contact with a convex lens of 15.0 cm focal length. What is the focal length of the combination ?

(a) 6 cm　　(b) 12 cm
(c) 8 cm　　(d) 4 cm

RESPONSE GRID	1. ⓐⓑⓒⓓ	2. ⓐⓑⓒⓓ	3. ⓐⓑⓒⓓ	4. ⓐⓑⓒⓓ	5. ⓐⓑⓒⓓ

Space for Rough Work

Q.6 A convex lens of focal length 20 cm is placed in contact with a diverging lens of unknown focal length. The lens combination acts as a converging lens and has a focal length of 30 cm. What is the focal length of the diverging lens ?

(a) – 10 cm (b) – 30 cm (c) – 60 cm (d) – 90 cm

Q.7 A pin is placed 10 cm in front of a convex lens of focal length 20 cm, made of material having refractive index 1.5. The surface of the lens farther away from the pin is silvered and has a radius of curvature 22 cm. Determine the position of the final image.

(a) 11 cm in front (b) 21 cm in front
(c) 15 cm in front (d) 31 cm in front

Q.8 An image is formed on the screen by a convex lens. When upper half part of lens is covered with black paper, then :

(a) half image is formed
(b) full image is formed
(c) intensity of image will be enhanced
(d) None of these

Q.9 A convex lens is made out of a substance of 1.2 refractive index. The two surfaces of lens are convex. If this lens is placed in water whose refractive index is 1.33, it will behave as :

(a) convergent lens (b) divergent lens
(c) plane glass plate (d) like a prism

Q.10 An equiconvex lens has a power of 5 diopter. If it is made of glass of refractive index 1.5 then the radius of the curvature of each surface will be

(a) 20 cm (b) 10 cm (c) 5 cm (d) zero

Q.11 A convex lens when placed in the first position forms a real image of an object on a fixed screen. The distance between the object and the screen is 75 cm. On displacing the lens from first position by 25 cm to the second position, again a real image is formed on the screen. Then the focal length of the lens is

D = 75 cm
x = 25 cm

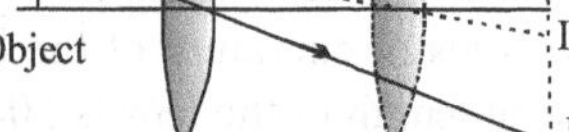
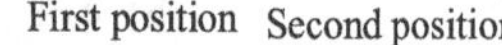

Object I_2 I_1
First position Second position

(a) 25.0 cm (b) 16.7 cm (c) 50.3 cm (d) 33.3 cm

Q.12 A lens is placed between a source of light and a wall. It forms images of area A_1 and A_2 on the wall for its two different positions. The area of the source of light is

(a) $\dfrac{A_1+A_2}{2}$ (b) $\left[\dfrac{1}{A_1}+\dfrac{1}{A_2}\right]^{-1}$

(c) $\sqrt{A_1A_2}$ (d) $\left[\dfrac{\sqrt{A_1}+\sqrt{A_2}}{2}\right]^2$

Q.13 A convex lens of power 4D is kept in contact with a concave lens of power 3D, the effective power of combination will be

(a) 7D (b) 4D/3
(c) 1D (d) 3D/4

Q.14 The power of a plano-convex lens is P. If this lens is cut longitudinally along its principal axis into two equal parts and then they are joined as given in the figure. The power of combination will be :

(a) P
(b) 2P
(c) P/2
(d) zero

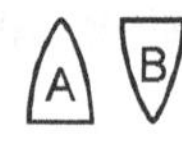

Q.15 The plane surface of a planoconvex lens is silvered. If radius of curved surface is R and refractive index is μ, then the system behaves like a concave mirror whose radius will be

(a) $\dfrac{R}{\mu}$ (b) $R\mu$

(c) $\dfrac{R}{\mu-1}$ (d) $R(\mu-1)$

Q.16 A slide projector lens has a focal length 10 cm. It throws an image of a 2cm × 2 cm slide on a screen 5 m from the lens. Find the size of the picture on the screen.

(a) $(98\times 98)\ cm^2$ (b) $(88\times 88)\ cm^2$
(c) $(64\times 64)\ cm^2$ (d) $(78\times 78)\ cm^2$

Q.17 If the focal length of a magnifier is 5 cm calculate the power of the lens.

(a) 20D (b) 10D
(c) 5D (d) 15D

RESPONSE GRID				
6. ⓐⓑⓒⓓ	7. ⓐⓑⓒⓓ	8. ⓐⓑⓒⓓ	9. ⓐⓑⓒⓓ	10. ⓐⓑⓒⓓ
11. ⓐⓑⓒⓓ	12. ⓐⓑⓒⓓ	13. ⓐⓑⓒⓓ	14. ⓐⓑⓒⓓ	15. ⓐⓑⓒⓓ
16. ⓐⓑⓒⓓ	17. ⓐⓑⓒⓓ			

Space for Rough Work

Q.18 In the above question, find the magnifying power of the lens for relaxed and strained eye.
(a) 2×, 3× (b) 5×, 6×
(c) 4×, 2× (d) 1×, 2×

Q.19 A 35 mm film is to be projected on a 20 m wide screen situated at a distance of 40 m from the film-projector. Calculate the focal length of projection lens.
(a) 70 mm (b) 35 mm
(c) 40 mm (d) 20 mm

Q.20 In a compound microscope the objective and the eye- piece have focal lengths of 0.95 cm and 5 cm respectively, and are kept at a distance of 20 cm. The last image is formed at a distance of 25 cm from the eye- piece. Calculate the total magnification.
(a) 94 (b) 84
(c) 75 (d) 88

Q.21 A Galilean telescope consists of an objective of focal length 12 cm and eye- piece of focal length 4 cm. What should be the separation of the two lenses when the virtual image of a distant object is formed at a distance of 24 cm from the eye- piece?
(a) 7.2 cm (b) 8.2 cm.
(c) 12.4 cm. (d) 2.8cm.

DIRECTIONS (Q.22-Q.24) : In the following questions, more than one of the answers given are correct. Select the correct answers and mark it according to the following codes:

Codes :
(a) 1, 2 and 3 are correct **(b)** 1 and 2 are correct
(c) 2 and 4 are correct **(d)** 1 and 3 are correct

Q.22 An astronomical telescope has an angular magnification of magnitude 5 for distant objects. The separation between the objective and eye- piece is 36 cm and the final image is formed at infinity.
(1) the focal length of objective is 30 cm
(2) the focal length of objective is 25 cm
(3) the focal length of eye piece is 6 cm
(4) the focal length of eye piece is 12 cm

Q.23 Resolving power of a microscope doesn't depend upon
(1) Velocity of light used
(2) Frequency of light used
(3) Focal length of objective
(4) Wavelength of light used

Q.24 The light gathering power of a camera lens doesn't depend on
(1) Ratio of focal length and diameter
(2) Product of focal length and diameter
(3) Wavelength of light used
(4) Its diameter

DIRECTIONS (Q.25-Q.27) : Read the passage given below and answer the questions that follows :

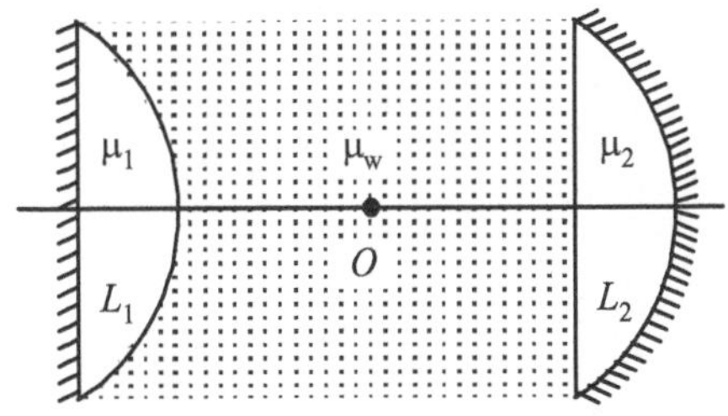

A cylindrical tube filled with water (μ_w = 4/3) is closed at its both ends by two silvered plano convex lenses as shown in the figure. Refractive index of lenses L_1 and L_2 are 2.0 and 1.5 while their radii of curvature are 5 cm and 9 cm respectively. A point object is placed somewhere at a point O on the axis of cylindrical tube. It is found that the object and image coincide each other.

Q.25 The position of object w.r.t lens L_1 is
(a) 8 cm (b) 10 cm
(c) 12 cm (d) 14 cm

Q.26 The position of object w.r.t lens L_2 is
(a) 8 cm (b) 10 cm
(c) 12 cm (d) 14 cm

Q.27 The length of the cylindrical tube is
(a) 16 cm (b) 18 cm
(c) 20 cm (d) 22 cm

RESPONSE GRID					
	18. ⓐⓑⓒⓓ	19. ⓐⓑⓒⓓ	20. ⓐⓑⓒⓓ	21. ⓐⓑⓒⓓ	22. ⓐⓑⓒⓓ
	23. ⓐⓑⓒⓓ	24. ⓐⓑⓒⓓ	25. ⓐⓑⓒⓓ	26. ⓐⓑⓒⓓ	27. ⓐⓑⓒⓓ

Space for Rough Work

DIRECTIONS (Q. 28-Q.30) : Each of these questions contains two statements: Statement-1 (Assertion) and Statement-2 (Reason). Each of these questions has four alternative choices, only one of which is the correct answer. You have to select the correct choice.

(a) Statement-1 is True, Statement-2 is True; Statement-2 is a correct explanation for Statement-1.

(b) Statement-1 is True, Statement-2 is True; Statement-2 is NOT a correct explanation for Statement-1.

(c) Statement -1 is False, Statement-2 is True.

(d) Statement -1 is True, Statement-2 is False.

Q.28 Statement-1: A double convex lens ($\mu = 1.5$) has focal length 10 cm. When the lens is immersed in water ($\mu = 4/3$) its focal length becomes 40 cm.

Statement-2: $\frac{1}{f} = \frac{\mu_\ell - \mu_m}{\mu_m}\left(\frac{1}{R_1} - \frac{1}{R_2}\right)$

Q.29 Statement-1: The focal length of lens changes when red light is replaced by blue light.
Statement-2: The focal length of lens does not depend on colour of light used.

Q.30 Statement-1: By increasing the diameter of the objective of telescope, we can increase its range.
Statement-2: The range of a telescope tells us how far away a star of some standard brightness can be spotted by telescope.

RESPONSE GRID	28. ⓐⓑⓒⓓ	29. ⓐⓑⓒⓓ	30. ⓐⓑⓒⓓ

DAILY PRACTICE PROBLEM SHEET 51 - PHYSICS			
Total Questions	30	Total Marks	120
Attempted		Correct	
Incorrect		Net Score	
Cut-off Score	30	Qualifying Score	48
Success Gap = Net Score – Qualifying Score			
Net Score = (Correct × 4) – (Incorrect × 1)			

Space for Rough Work

DPP - Daily Practice Problems

Name :

Date :

Start Time :

End Time :

PHYSICS 52

SYLLABUS : WAVE OPTICS - I (Interference of Light)

Max. Marks : 116 **Time : 60 min.**

GENERAL INSTRUCTIONS

- The Daily Practice Problem Sheet contains 29 MCQ's. For each question only one option is correct. Darken the correct circle/ bubble in the Response Grid provided on each page.
- You have to evaluate your Response Grids yourself with the help of solution booklet.
- Each correct answer will get you 4 marks and 1 mark shall be deduced for each incorrect answer. No mark will be given/ deducted if no bubble is filled. Keep a timer in front of you and stop immediately at the end of 60 min.
- The sheet follows a particular syllabus. Do not attempt the sheet before you have completed your preparation for that syllabus. Refer syllabus sheet in the starting of the book for the syllabus of all the DPP sheets.
- After completing the sheet check your answers with the solution booklet and complete the Result Grid. Finally spend time to analyse your performance and revise the areas which emerge out as weak in your evaluation.

DIRECTIONS (Q.1-Q.20) : There are 20 multiple choice questions. Each question has 4 choices (a), (b), (c) and (d), out of which **ONLY ONE** choice is correct.

Q.1 The intensity ratio of two waves is 9 : 1. These waves produce the event of interference. The ratio of maximum to minimum intensity will be

(a) 1 : 9 (b) 9 : 1 (c) 1 : 4 (d) 4 : 1

Q.2 The equation of two light waves are $y_1 = 6\cos \omega t$, $y_2 = 8\cos(\omega t + \phi)$. The ratio of maximum to minimum intensities produced by the superposition of these waves will be

(a) 49 : 1 (b) 1 : 49

(c) 1 : 7 (d) 7 : 1

Q.3 In a Young's double slit experiment, the separation between the slits is 0.10 mm, the wavelength of light used is 600 nm and the interference pattern is observed on a screen 1.0 m away. Find the separation between the successive bright fringes.

(a) 6.6 mm (b) 6.0 mm (c) 6 m (d) 6 cm.

Q.4 In Young's double slit experiment the two slits are illuminated by light of wavelength 5890 Å and the angular separation between the fringes obtained on the screen is 0.2°. If the whole apparatus is immersed in water then the angular fringe width will be, if the refractive index of water is 4/3?

(a) 0.30° (b) 0.15° (c) 15° (d) 30°

Q.5 The intensities of two light sources are I and 9I respectively. If the phase difference between the waves emitted by them is π then the resultant intensity at the point of observation will be –

(a) 3I (b) 4I (c) 10I (d) 82I

RESPONSE GRID	1. ⓐⓑⓒⓓ	2. ⓐⓑⓒⓓ	3. ⓐⓑⓒⓓ	4. ⓐⓑⓒⓓ	5. ⓐⓑⓒⓓ

Space for Rough Work

Q.6 In Fresnel's biprism experiment the width of 10 fringes is 2cm which are formed at a distance of 2 meter from the slit. If the wavelength of light is 5100Å then the distance between two coherent sources will be
(a) 5.1×10^{-4} m (b) 5.1×10^{4} cm
(c) 5.1×10^{-4} mm (d) 10.1×10^{-4}cm

Q.7 Two coherent sources of intensity ratio 1 : 4 produce an interference pattern. The fringe visibility will be -
(a) 1 (b) 0.8 (c) 0.4 (d) 0.6

Q.8 When a mica sheet (μ = 1.6) of thickness 7 microns is placed in the path of one of interfering beams in the biprism experiment then the central fringe gets shifted at the position of seventh bright fringe. The wavelength of light used will be –
(a) 4000Å (b) 5000Å (c) 6000 Å (d) 7000Å

Q.9 In Young's double slit experiment, the distance between two slits is made three times then the fringe width will become –
(a) 9 times (b) 1/9 times (c) 3 times (d) 1/3 times

Q.10 In the given diagram, CP represents a wavefront and AO & BP, the corresponding two rays. Find the condition on θ for constructive interference at P between the ray BP and reflected ray OP

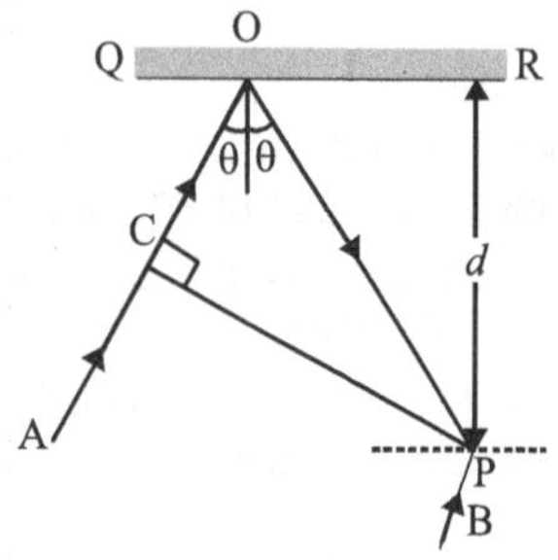

(a) $\cos\theta = 3\lambda/2d$ (b) $\cos\theta = \lambda/4d$
(c) $\sec\theta - \cos\theta = \lambda/d$ (d) $\sec\theta - \cos\lambda = 4\lambda/d$

Q.11 In Young's double slit experiment 10th order maximum is obtained at the point of observation in the interference pattern for λ = 7000 Å. If the source is replaced by another one of wavelength 5000 Å then the order of maximum at the same point will be–
(a) 12 th (b) 14 th
(c) 16 th (d) 18 th

Q.12 In Young's double slit experiment, white light is used. The separation between the slits is b. The screen is at a distance d (d > > b) from the slits. Some wavelengths are missing exactly in front of one slit. One of these wavelengths is
(a) $\lambda=\frac{b^2}{6d}$ (b) $\lambda=\frac{2b^2}{d}$ (c) $\lambda=\frac{b^2}{3d}$ (d) $\lambda=\frac{2b^2}{3d}$

Q.13 In Fresnel's biprism experiment distance of mth bright fringe from zeroth order fringe will be –
(a) $(2m-1)\frac{\lambda D}{2d}$ (b) $\frac{mD\lambda}{d}$
(c) $\frac{md}{\lambda D}$ (d) $(2m+1)\frac{\lambda D}{2d}$

Q.14 Consider interference between waves from two sources of Intensites I & 4I. Find intensities at points where the phase difference is $\frac{\pi}{2}$.
(a) I (b) 5 I (c) 4 I (d) 3 I

Q.15 The width of one of the two slits in a Young's double slit experiment is double of the other slit. Assuming that the amplitude of the light coming from a slit is proportional to slit-width. Find the ratio of the maximum to the minimum intensity in the interference pattern.
(a) 34 : 1 (b) 9 : 1 (c) 4 : 1 (d) 16 : 1

Q.16 The intensity of the light coming from one of the slits in a young's double slit experiment is double the intensity from the other slit. Find the ratio of the maximum intensity to the minimum intensity in the interference fringe pattern observed.
(a) 9 : 1 (b) 34 : 1 (c) 4 : 1 (d) 16 : 1

Q.17 Two waves originating from source S_1 and S_2 having zero phase difference and common wavelength λ will show completely destructive interference at a point P if $(S_1 P - S_2 P)$ is-
(a) 5λ (b) $3\lambda/4$
(c) 2λ (d) $11\lambda/2$

RESPONSE GRID					
	6. ⓐⓑⓒⓓ	7. ⓐⓑⓒⓓ	8. ⓐⓑⓒⓓ	9. ⓐⓑⓒⓓ	10. ⓐⓑⓒⓓ
	11. ⓐⓑⓒⓓ	12. ⓐⓑⓒⓓ	13. ⓐⓑⓒⓓ	14. ⓐⓑⓒⓓ	15. ⓐⓑⓒⓓ
	16. ⓐⓑⓒⓓ	17. ⓐⓑⓒⓓ			

Space for Rough Work

Q.18 In an interference pattern, at a point we observe the 16th order maximum for $\lambda_1 = 6000Å$. What order will be visible here if the source is replaced by light of wavelength ? $\lambda_2 = 4800$ Å.
(a) 40 (b) 20
(c) 10 (d) 80

Q.19 In Young's experiment the wavelength of red light is 7.5×10^{-5} cm. and that of blue light 5.0×10^{-5} cm. The value of n for which $(n + 1)^{th}$ blue bright band coincides with n^{th} red bright band is-
(a) 8 (b) 4
(c) 2 (d) 1

Q.20 In Young's double slit experiment, carried out with light of wavelength $\lambda = 5000Å$, the distance between the slits is 0.2mm and the screen is at 200 cm from the slits. The central maximum is at x = 0. The third maximum will be at x equal to.
(a) 1.67 cm (b) 1.5 cm
(c) 0.5 cm (d) 5.0 cm

DIRECTIONS (Q.21-Q.23) : In the following questions, more than one of the answers given are correct. Select the correct answers and mark it according to the following codes:

Codes :
(a) 1, 2 and 3 are correct **(b)** 1 and 2 are correct
(c) 2 and 4 are correct **(d)** 1 and 3 are correct

Q.21 The Young's double slit experiment, the ratio of intensities of bright and dark fringes is 9. This means that
(1) The intensities of individual sources are 5 and 4 units respectively
(2) The intensities of individual sources are 4 and 1 units respectively
(3) The ratio of the their amplitudes is 3
(4) The ratio of their amplitude is 2

Q.22 In an experiment similar to Young's experiment, interference is observed using waves associated with electrons. The electrons are being produced in an electron gun. In order to decrease the fringe width
(1) electron gun voltage be increased.
(2) the slits be moved away from each other.
(3) the screen be moved closer to interfering slits.
(4) electron gun voltage be decreased.

Q.23 Interference fringes were produced in Young's double slit experiment using light of wave length 5000 Å. When a film of material 2.5×10^{-3} cm thick was placed over one of the slits, the fringe pettern shifted by a distance equal to 20 fringe width. The refractive index of the material of the film cannot be
(1) 1.25 (2) 1.33
(3) 1.5 (4) 1.4

DIRECTIONS (Q.24-Q.26) : Read the passage given below and answer the questions that follows :

In a Young's double slit experiment a monochromatic light whose wavelength is λ strikes on the slits, separated by distance *d,* as shown in the figure. Refractive index of the medium between slits and screen varies with time *t* as $n = n_0 + kt$. Here n_0 and k are positive constants. Position of any point P on screen is measure by its *y*-coordinate as shown.

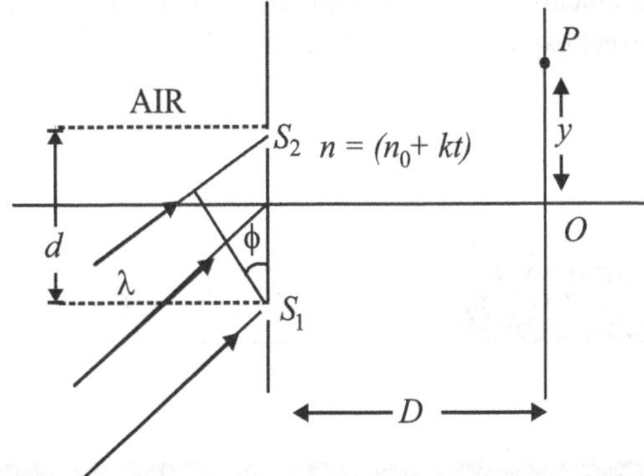

Q.24 The y co-ordinate of central maxima at any time *t* is
(a) $\frac{D\sin\phi}{n_0 + kt}$ (b) $\frac{D\cos\phi}{n_0 + kt}$
(c) $\frac{D\sin\phi}{(n_0 + kt)^2}$ (d) $\frac{D\cos\phi}{(n_0 + kt)^2}$

RESPONSE GRID	18. ⓐⓑⓒⓓ	19. ⓐⓑⓒⓓ	20. ⓐⓑⓒⓓ	21. ⓐⓑⓒⓓ	22. ⓐⓑⓒⓓ
	23. ⓐⓑⓒⓓ	24. ⓐⓑⓒⓓ			

Space for Rough Work

Q.25 The velocity of central maxima at any time t as a function of time t is

(a) $\frac{-2kD\sin\phi}{(n_0+kt)^2}$ (b) $\frac{-kD\sin\phi}{(n_0+kt)^2}$

(c) $\frac{-2kD\sin\phi}{(n_0+kt)}$ (d) $\frac{-kD\sin\phi}{(n_0+kt)}$

Q.26 If a glass plate of small thickness b is placed in front of S_1. How should its refractive index vary with time so that central maxima is formed at O.

(a) $n_0+kt+\frac{2d\sin\phi}{b}$ (b) $n_0+kt-\frac{2d\sin\phi}{b}$

(c) $n_0+kt-\frac{d\sin\phi}{b}$ (d) $n_0+kt+\frac{d\sin\phi}{b}$

DIRECTIONS (Q. 27-Q.29) : Each of these questions contains two statements: Statement-1 (Assertion) and Statement-2 (Reason). Each of these questions has four alternative choices, only one of which is the correct answer. You have to select the correct choice.

(a) Statement-1 is True, Statement-2 is True; Statement-2 is a correct explanation for Statement-1.

(b) Statement-1 is True, Statement-2 is True; Statement-2 is NOT a correct explanation for Statement-1.

(c) Statement -1 is False, Statement-2 is True.

(d) Statement -1 is True, Statement-2 is False.

Q.27 Statement-1 : No interference pattern is detected when two coherent sources are infinitely close to each other.

Statement-2 : The fringe width is directly proportional to the distance between the two slits.

Q.28 Statement-1 : In Young's experiment, the fringe width for dark fringes is same as that for white fringes.

Statement-2 : In Young's double slit experiment performed with a source of white light, only black and bright fringes are observed.

Q.29 Statement-1 : In Young's double slit experiment, the fringes become indistinct if one of the slits is covered with cellophane paper.

Statement-2 : The cellophane paper decreases the wavelength of light.

RESPONSE GRID	25. ⓐⓑⓒⓓ	26. ⓐⓑⓒⓓ	27. ⓐⓑⓒⓓ	28. ⓐⓑⓒⓓ	29. ⓐⓑⓒⓓ

DAILY PRACTICE PROBLEM SHEET 52 - PHYSICS			
Total Questions	29	Total Marks	116
Attempted		Correct	
Incorrect		Net Score	
Cut-off Score	28	Qualifying Score	46
Success Gap = Net Score – Qualifying Score			
Net Score = (Correct × 4) – (Incorrect × 1)			

Space for Rough Work

DPP - Daily Practice Problems

Name :

Date :

Start Time :

End Time :

PHYSICS 53

SYLLABUS : WAVE OPTICS - II (Diffraction and polarisation of light)

Max. Marks : 120 **Time : 60 min.**

GENERAL INSTRUCTIONS

- The Daily Practice Problem Sheet contains 30 MCQ's. For each question only one option is correct. Darken the correct circle/ bubble in the Response Grid provided on each page.
- You have to evaluate your Response Grids yourself with the help of solution booklet.
- Each correct answer will get you 4 marks and 1 mark shall be deduced for each incorrect answer. No mark will be given/ deducted if no bubble is filled. Keep a timer in front of you and stop immediately at the end of 60 min.
- The sheet follows a particular syllabus. Do not attempt the sheet before you have completed your preparation for that syllabus. Refer syllabus sheet in the starting of the book for the syllabus of all the DPP sheets.
- After completing the sheet check your answers with the solution booklet and complete the Result Grid. Finally spend time to analyse your performance and revise the areas which emerge out as weak in your evaluation.

DIRECTIONS (Q.1-Q.22) : There are 22 multiple choice questions. Each question has 4 choices (a), (b), (c) and (d), out of which **ONLY ONE** choice is correct.

Q.1 The first diffraction minima due to a single slit diffraction is at $\theta = 30°$ for a light of wavelength 5000Å. The width of the slit is-

(a) 5×10^{-5} cm (b) 1.0×10^{-4} cm
(c) 2.5×10^{-5} cm (d) 1.25×10^{-5} cm

Q.2 Two spectral line of sodium D_1 & D_2 have wavelengths of approximately 5890Å and 5896Å. A sodium lamp sends incident plane wave on to a slit of width 2 micrometre. A screen is located at 2m from the slit. Find the spacing between the first maxima of two sodium lines as measured on the screen.

(a) 10^{-4} m (b) 9×10^{-4} m
(c) 9×10^{4} m (d) None

Q.3 Width of slit is 0.3mm. Fraunhoffer diffraction is observed in focus plane of lense of a lense of focal length 1 m. If third minima is at 5 mm distance from central maxima, then wavelength of light is-

(a) 7000Å (b) 6500Å (c) 6000Å (d) 5000Å

Q.4 When a wave of wavelength 0.2 cm is made incident normally on a slit of width 0.004m, then the semi-angular width of central maximum of diffraction pattern will be-

(a) 60° (b) 30° (c) 90° (d) 0°

Q.5 A parallel beam of monochromatic light is incident on a narrow rectangular slit of width 1mm. When the diffraction pattern is seen on a screen placed at a distance of 2m. the width of principal maxima is found to be 2.5 mm. The wave length of light is-

(a) 6250 nm (b) 6200 nm
(c) 5890 nm (d) 6000 nm

RESPONSE GRID 1. ⓐⓑⓒⓓ 2. ⓐⓑⓒⓓ 3. ⓐⓑⓒⓓ 4. ⓐⓑⓒⓓ 5. ⓐⓑⓒⓓ

Space for Rough Work

Q.6 Light of wavelength 6328Å is incident normally on slit having a width of 0.2 mm. The width of the central maximum measured from minimum to minimum of diffraction pattern on a screen 9.0 meters away will be about -

(a) 0.36° (b) 0.18°
(c) 0.72° (d) 0.09°

Q.7 A screen is placed 2m away from the single narrow slit. Calculate the slit width if the first minimum lies 5mm on either side of the central maximum. Incident plane waves have a wavelenght of 5000Å.

(a) 2×10^{-4} m (b) 2×10^{-3} cm
(c) 2×10^{-4} cm (d) None

Q.8 Red light of wavelength 6500Å from a distant source falls on a slit 0.5 mm wide. What is the distance between two dark bands on each side of central bright band of diffraction pattern observed on a screen placed 1.8 m from the slit.

(a) 4.68×10^{-3} cm (b) 4.68×10^{-3} mm
(c) 4.68×10^{-3} nm (d) 4.68×10^{-3} m

Q.9 Fraunhoffer diffraction pattern is observed at a distance of 2m on screen, when a plane-wavefront of 6000Å is incident perpendicularly on 0.2 mm wide slit.Width of central maxima is:

(a) 10 mm (b) 6 mm
(c) 12 mm (d) None of these

Q.10 A diffraction pattern is produced by a single slit of width 0.5mm with the help of a convex lens of focal length 40cm. If the wavelength of light used is 5896Å. then the distance of first dark fringe from the axis will be-

(a) 0.047 cm (b) 0.047 m
(c) 0.047 mm (d) 47 cm

Q.11 What should be the size of the aperture of the objective of telescope which can just resolve the two stars of angular width of 10^{-3} degree by light of wavelength 5000Å?

(a) 3.5 cm (b) 3.5 mm
(c) 3.5 m (d) 3.5 km

Q.12 Image of sun formed due to reflection at air water interface is found to be very highly polarised. Refractive index of water being $\mu = 4/3$, find the angle of sun above the horizon.

(a) 36.9° (b) 26.9°
(c) 16.9° (d) 46.9°

Q.13 When light of a certain wavelength is incident on a plane surface of a material at a glancing angle 30°, the reflected light is found to be completely plane polarised. Determine refractive index of given material –

(a) $\sqrt{3}$ (b) $\sqrt{2}$
(c) $1/\sqrt{2}$ (d) 2

Q.14 Two polaroids are oriented with their planes perpendicular to incident light and transmission axis making an angle of 30° with each other. What fraction of incident unpolarised light is transmitted ?

(a) 57.5 % (b) 17.5 %
(c) 27.5 % (d) 37.5 %

Q.15 Unpolarised light of intensity 32 Wm^{-2} passes through three polarisers such that the transmission axis of the last polariser is crossed with the first. If the intensity of the emerging light is 3 Wm^{-2}. At what angle will the transmitted intensity be maximum ?

(a) 45° (b) 15°
(c) 35° (d) 75°

Q.16 V_0 and V_E represent the velocities, μ_0 and μ_E the refractive indices of ordinary and extraordinary rays for a doubly refracting crystal. Then

(a) $V_0 \geq V_E$, $\mu_0 \leq \mu_E$ if the crystal is calcite
(b) $V_0 \leq V_E$, $\mu_0 \leq \mu_E$ if the crystal is quartz
(c) $V_0 \leq V_E$, $\mu_0 \geq \mu_E$ if the crystal is calcite
(d) $V_0 \geq V_E$, $\mu_0 \geq \mu_E$ if the crystal is quartz

RESPONSE GRID					
	6. ⓐⓑⓒⓓ	7. ⓐⓑⓒⓓ	8. ⓐⓑⓒⓓ	9. ⓐⓑⓒⓓ	10. ⓐⓑⓒⓓ
	11. ⓐⓑⓒⓓ	12. ⓐⓑⓒⓓ	13. ⓐⓑⓒⓓ	14. ⓐⓑⓒⓓ	15. ⓐⓑⓒⓓ
	16. ⓐⓑⓒⓓ				

Space for Rough Work

Q.17 A ray of light is incident on the surface of a glass plate at an angle of incidence equal to Brewster's angle ϕ. If μ represents the refractive index of glass with respect to air. then the angle between reflected and refracted rays is

(a) $90° + \phi$ (b) $\sin^{-1}(\mu \cos\phi)$
(c) $90°$ (d) $90° - \sin^{-1}(\cos\phi/\mu)$

Q.18 A light has amplitude A and angle between analyser and polariser is 60°. Light transmitted by analyser has amplitude

(a) $A\sqrt{2}$ (b) $A/\sqrt{2}$ (c) $\sqrt{3}A/2$ (d) A/2

Q.19 A slit of size 0.15 cm is placed at 2.1 m from a screen. On illuminating it by a light of wavelength 5×10^{-5} cm, the width of central maxima will be

(a) 70 mm (b) 0.14 mm (c) 1.4 mm (d) 0.14 cm

Q.20 What will be the angle of diffraction for the first minimum due to Fraunhoffer diffraction with sources of light of wave lenght 550 nm and slit width 0.55 mm ?

(a) 0.001 rad (b) 0.01 rad
(c) 1 rad (d) 0.1 rad

Q.21 In Fresnel diffraction, if the distance between the disc and the screen is decreased, the intensity of central bright spot will

(a) increase (b) decrease
(c) remain constant (d) none of these

22. When an unpolarized light of intensity I_0 is incident on a polarizing sheet, the intensity of the light which does not get transmitted is

(a) $\frac{1}{4}I_0$ (b) $\frac{1}{2}I_0$ (c) I_0 (d) zero

DIRECTIONS (Q.23-Q.25) : In the following questions, more than one of the answers given are correct. Select the correct answers and mark it according to the following codes:

Codes :

(a) 1, 2 and 3 are correct **(b)** 1 and 2 are correct
(c) 2 and 4 are correct **(d)** 1 and 3 are correct

Q.23 Plane polarised light is passed through a polaroid. On viewing through the polaroid we find that when the polariod is given one complete rotation about the direction of the light, which of the following is not observed ?

(1) The intensity of light gradually decreases to zero and remains at zero
(2) The intensity of light gradually increases to a maximum and remains at maximum
(3) There is no change in intensity
(4) The intensity of light is twice maximum and twice zero

Q.24 Out of the following statements which are correct ?

(1) Nicol's prism works on the principle of double refraction and total internal reflection
(2) Nicol's prism can be used to produce and analyse polarised light
(3) Calcite and Quartz are both doubly refracting crystals
(4) When unpolarised light passes through a Nicol's prism, the emergent light is elliptically polarised

Q.25 Which statements are incorrect for a zone plate and a lens?

(1) Zone plate has one focus whereas lens has multiple focii
(2) Both zone plate and lens have multi focii
(3) Zone plate has one focus whereas a lens has infinite
(4) Zone plate has multi focii whereas lens has one

RESPONSE GRID					
	17. ⓐⓑⓒⓓ	18. ⓐⓑⓒⓓ	19. ⓐⓑⓒⓓ	20. ⓐⓑⓒⓓ	21. ⓐⓑⓒⓓ
	22. ⓐⓑⓒⓓ	23. ⓐⓑⓒⓓ	24. ⓐⓑⓒⓓ	25. ⓐⓑⓒⓓ	

Space for Rough Work

DIRECTIONS (Q.26-Q.27) : Read the passage given below and answer the questions that follows :

Angular width of central maximum in the Fraunhoffer-diffraction pattern of a slit is measured. The slit is illuminated by light of wavelength 6000 Å. When the slit is illuminated by light of another wavelength, the angular width decreases by 30%.

Q.26 The wavelength of the light is

(a) 4200Å (b) 3500Å (c) 5000 Å (d) 5200Å

Q.27 The same decrease in the angular width of central maximum is obtained when the original apparatus is immersed in a liquid. Find refractive index of the liquid.

(a) 1.23 (b) 1.43 (c) 2.2 (d) 2.43

DIRECTIONS (Q. 28-Q.30) : Each of these questions contains two statements: Statement-1 (Assertion) and Statement-2 (Reason). Each of these questions has four alternative choices, only one of which is the correct answer. You have to select the correct choice.

(a) Statement-1 is True, Statement-2 is True; Statement-2 is a correct explanation for Statement-1.

(b) Statement-1 is True, Statement-2 is True; Statement-2 is NOT a correct explanation for Statement-1.

(c) Statement-1 is False, Statement-2 is True.

(d) Statement-1 is True, Statement-2 is False.

Q.28 Statement-1 : The unpolarised light and polarised light can be distinguished from each other by using polaroid.

Statement-2 : A polaroid is capable of producing plane polarised beams of light.

Q.29 Statement-1 : Nicol prism is used to produce and analyse plane polarised light.

Statement-2 : Nicol prism reduces the intensity of light to zero.

Q.30 Statement-1 : The cloud in sky generally appear to be whitish.

Statement-2 : Diffraction due to clouds is efficient in equal measure at all wavelengths.

RESPONSE GRID	26. ⓐⓑⓒⓓ	27. ⓐⓑⓒⓓ	28. ⓐⓑⓒⓓ	29. ⓐⓑⓒⓓ	30. ⓐⓑⓒⓓ

DAILY PRACTICE PROBLEM SHEET 53 - PHYSICS			
Total Questions	30	Total Marks	120
Attempted		Correct	
Incorrect		Net Score	
Cut-off Score	26	Qualifying Score	46
Success Gap = Net Score – Qualifying Score			
Net Score = (Correct × 4) – (Incorrect × 1)			

Space for Rough Work

DPP - Daily Practice Problems

Name :

Date :

Start Time :

End Time :

PHYSICS 54

SYLLABUS : DUAL NATURE OF MATTER & RADIATION (Matter Waves, Photon, Photoelectric effect, X-ray)

Max. Marks : 120 **Time : 60 min.**

GENERAL INSTRUCTIONS

- The Daily Practice Problem Sheet contains 30 MCQ's. For each question only one option is correct. Darken the correct circle/ bubble in the Response Grid provided on each page.
- You have to evaluate your Response Grids yourself with the help of solution booklet.
- Each correct answer will get you 4 marks and 1 mark shall be deduced for each incorrect answer. No mark will be given/ deducted if no bubble is filled. Keep a timer in front of you and stop immediately at the end of 60 min.
- The sheet follows a particular syllabus. Do not attempt the sheet before you have completed your preparation for that syllabus. Refer syllabus sheet in the starting of the book for the syllabus of all the DPP sheets.
- After completing the sheet check your answers with the solution booklet and complete the Result Grid. Finally spend time to analyse your performance and revise the areas which emerge out as weak in your evaluation.

DIRECTIONS (Q.1-Q.21) : There are 21 multiple choice questions. Each question has 4 choices (a), (b), (c) and (d), out of which **ONLY ONE** choice is correct.

Q.1 Energy of a α-particle, having de broglie wavelength of 0.004 Å is approximately.

(a) 1275 eV (b) 1200 KeV

(c) 1200 MeV (d) 1200 GeV

Q.2 Velocity of a proton is c/20. Associated de-Broglie wavelength is (Take $h = 6.626 \times 10^{-34}$ J–s)

(a) 2.64×10^{-24} mm (b) 2.64×10^{-24} cm

(c) 2.64×10^{-14} Å (d) 2.64×10^{-14} m

Q.3 One electron & one proton is accelerated by equal potential. Ratio of their de-Broglie wavelengths is-

(a) $\sqrt{\frac{m_p}{m_e}}$ (b) $\frac{m_e}{m_p}$ (c) $\frac{m_p}{m_e}$ (d) 1

Q.4 de-Broglie wavelength of an electron is 10 Å then velocity will be-

(a) 7.2×10^7 m/s (b) 7.2×10^6 m/s

(c) 7.2×10^5 m/s (d) 7.2×10^4 m/s

Q.5 One electron & one proton have equal energies then ratio of associated de-Broglie wavelength will be-

(a) $1 : (1836)^2$ (b) $\sqrt{1836} : 1$

(c) 1836 : 1 (d) $(1836)^2 : 1$

RESPONSE GRID	1. ⓐⓑⓒⓓ	2. ⓐⓑⓒⓓ	3. ⓐⓑⓒⓓ	4. ⓐⓑⓒⓓ	5. ⓐⓑⓒⓓ

Space for Rough Work

Q.6 The ratio of wavelength of deutron & proton accelerated by an equal potential is

(a) $\frac{1}{\sqrt{2}}$ (b) $\sqrt{\frac{2}{1}}$

(c) $\frac{1}{2}$ (d) $\frac{2}{1}$

Q.7 In photoelectric effect if intensity of light is doubled then maximum kinetic energy of photoelectrons will become

(a) Double (b) Half

(c) Four time (d) No change

Q.8 Quantum nature of light is explained by which of the following phenomenon?

(a) Huygen wave theory

(b) Photoelectric effect

(c) Maxwell electromagnetic theory

(d) de- Broglie theory

Q.9 From rest an electron is accelerated between two such points which has potential 20 & 40 volts respectively. Associated de-Broglie wavelength of electron is-

(a) 0.75 Å (b) 7.5 Å

(c) 2.75 Å (d) 2.75 m

Q.10 An electron microscope uses 40 keV electrons. Find its resolving limit on the assumption that it is equal to the wavelength of the electron-

(a) 0.61 Å (b) 0.6 Å

(c) 0.06 Å (d) 0.061 Å

Q.11 A hydrogen atom moving at a speed v absorbs a photon of wavelength 122 nm and stops. Find the value of v. (Mass of hydrogen atom = 1.67×10^{-27} kg)

(a) 3.5 m/s (b) 32.5 m/s

(c) 3.05 m/s (d) 3.25 m/s

Q.12 The de-Broglie wavelength of an electron is 0.2 Å. Calculate the potential difference (approximate) required to retard it to rest-

(a) 3.76×10^{-3} V (b) 3.76×10^{3} V

(c) 3.76×10^{3} eV (d) 376.5 V

Q.13 A photon and an electron have equal energy E. $\lambda_{photon}/\lambda_{electron}$ is proportional to

(a) $\sqrt{E}$ (b) $\frac{1}{\sqrt{E}}$

(c) $\frac{1}{E}$ (d) Does not depend upon E.

Q.14 In a photoemissive cell with exciting wavelength λ, the fastest electron has speed v. If the exciting wavelength is changed to 3λ/4, the speed of the fastest emitted electron will be

(a) $v\,(3/4)^{1/2}$ (b) $v\,(4/3)^{1/2}$

(c) Less than $v\,(4/3)^{1/2}$ (d) Greater than $v\,(4/3)^{1/2}$

Q.15 Which of the following figure repesents variation of particle momentum and the associated de-Broglie wavelength?

(a) 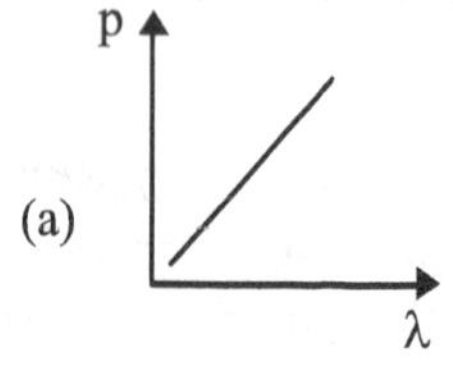

(b) 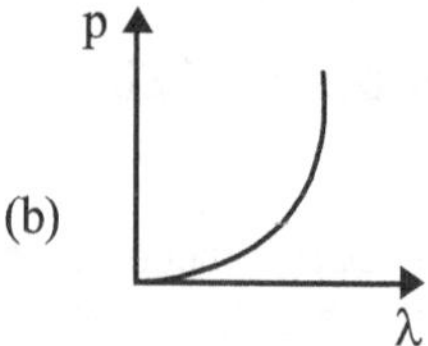

(c) 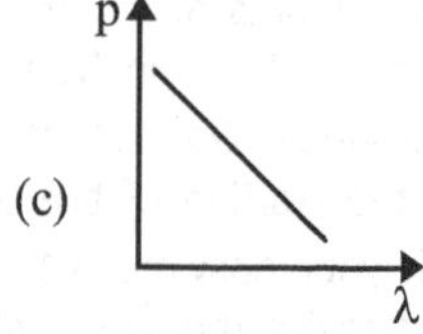

(d) 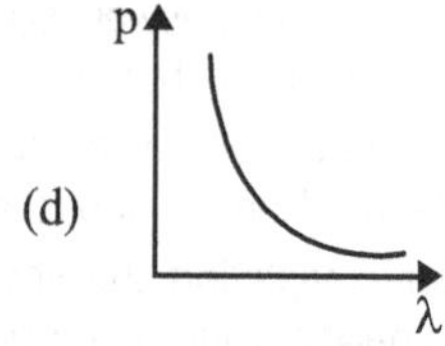

Q.16 The work function for the surface of aluminium is 4.2 eV. What will be the wavelength of that incident light for which the stopping potential will be zero.

($h \approx 6.6 \times 10^{-34}$ J–s $e \approx 1.6 \times 10^{-19}$ C)

(a) 2496 Å (b) 2946×10^{-7} m

(c) 2649 Å (d) 2946 Å

RESPONSE GRID					
	6. ⓐⓑⓒⓓ	7. ⓐⓑⓒⓓ	8. ⓐⓑⓒⓓ	9. ⓐⓑⓒⓓ	10. ⓐⓑⓒⓓ
	11. ⓐⓑⓒⓓ	12. ⓐⓑⓒⓓ	13. ⓐⓑⓒⓓ	14. ⓐⓑⓒⓓ	15. ⓐⓑⓒⓓ
	16. ⓐⓑⓒⓓ				

Space for Rough Work

Q.17 Slope of $V_0 - \nu$ curve is-
(where V_0 = Stopping potential and ν = frequency)

(a) e (b) $\frac{h}{e}$ (c) ϕ_0 (d) h

Q.18 A radio station is transmitting waves of wavelength 300 m. If diffracting power of transmitter is 10 kW, then numbers of photons diffracted per second is-

(a) 1.5×10^{35} (b) 1.5×10^{31}
(c) 1.5×10^{29} (d) 1.5×10^{33}

Q.19 Light of wavelength 3320 Å is incident on metal surface (work function = 1.07 eV). To stop emission of photo electron, retarding potential required to be
(Take $hc \approx 12420$ eV – Å)

(a) 3.74 V (b) 2.67 V (c) 1.07 V (d) 4.81 V

Q.20 The figure shows the variation of photocurrent with anode potential for a photo-sensitive surface for three different radiations. Let I_a, I_b and I_c be the intensities and f_a, f_b and f_c be the frequencies for the curves a, b and c respectively. Then

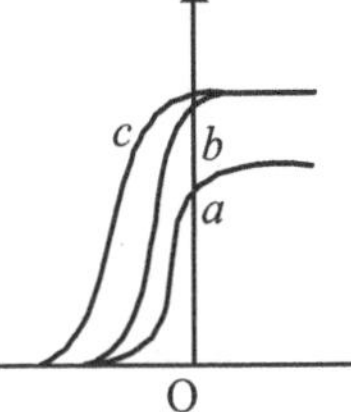

(a) $f_a = f_b$ and $I_a \neq I_b$
(b) $f_a = f_c$ and $I_a = I_c$
(c) $f_a = f_b$ and $I_a = I_b$
(d) $f_a = f_b$ and $I_a = I_c$

Q.21 An electromagnetic radiation of frequency 3×10^{15} cycles per second falls on a photo electric surface whose work function is 4.0 eV. Find out the maximum velocity of the photo electrons emitted by the surface-

(a) 13.4×10^{-19} m/s (b) 19.8×10^{-19}m/s
(c) 1.73×10^{6} m/s (d) None

DIRECTIONS (Q.22-Q.24) : In the following questions, more than one of the answers given are correct. Select the correct answers and mark it according to the following codes:

Codes :
(a) 1, 2 and 3 are correct **(b)** 1 and 2 are correct
(c) 2 and 4 are correct **(d)** 1 and 3 are correct

Q.22 Ultraviolet light of wavelength 280 nm is used in an experiment on photo electric effect with lithium (ϕ = 2.5 eV) cathode.

(1) The maximum kinetic energy is 1 .9 eV
(2) The stopping potential is1.9 V
(3) The maximum kinetic energy is 4.4 V
(4) The stopping potential is 4.4 eV

Q.23 The separation between Bragg's planes in a crystal is 10 Å. Then the wavelength of those X-rays which can be diffracted by this crystal is-

(1) 5 Å (2) 10 Å
(3) 20 Å (4) 25 Å

Q.24 Electrons are accelerated in television tubes through potential difference of about 10 KV.

(1) The lowest wavelength of the emitted X-rays is 12.4Å
(2) The lowest wavelength of the emitted X-rays is 1.24Å
(3) The highest frequency of the emitted X-rays is 2.4×10^{8} Hz
(4) The highest frequency of the emitted X-rays is 2.4×10^{18} Hz

DIRECTIONS (Q.25-Q.27) : Read the passage given below and answer the questions that follows :

A physicist wishes to eject electrons by shining light on a metal surface. The light source emits light of wavelength of 450 nm. The table lists the only available metals and their work functions.

Metal	W_0 (eV)
Barium	2.5
Lithium	2.3
Tantalum	4.2
Tungsten	4.5

Q.25 Which metal(s) can be used to produce electrons by the photoelectric effect from given source of light ?

(a) Barium only
(b) Barium or lithium
(c) Lithium, tantalum or tungsten
(d) Tungsten or tantalum

RESPONSE GRID					
	17. ⓐⓑⓒⓓ	18. ⓐⓑⓒⓓ	19. ⓐⓑⓒⓓ	20. ⓐⓑⓒⓓ	21. ⓐⓑⓒⓓ
	22. ⓐⓑⓒⓓ	23. ⓐⓑⓒⓓ	24. ⓐⓑⓒⓓ	25. ⓐⓑⓒⓓ	

Space for Rough Work

Q.26 Which option correctly identifies the metal that will produce the most energetic electrons and their energies ?
(a) Lithium, 0.45 eV (b) Tungsten, 1.75 eV
(c) Lithium, 2.30 eV (d) Tungsten, 2.75 eV

Q.27 Suppose photoelectric experiment is done separately with these metals with light of wavelength 450 nm. The maximum magnitude of stopping potential amongst all the metals is-
(a) 2.75 volt (b) 4.5 volt
(c) 0.45 volt (d) 0.25 volt

DIRECTIONS (Q. 28-Q.30) : Each of these questions contains two statements: Statement-1 (Assertion) and Statement-2 (Reason). Each of these questions has four alternative choices, only one of which is the correct answer. You have to select the correct choice.

(a) Statement-1 is True, Statement-2 is True; Statement-2 is a correct explanation for Statement-1.
(b) Statement-1 is True, Statement-2 is True; Statement-2 is NOT a correct explanation for Statement-1.
(c) Statement -1 is False, Statement-2 is True.
(d) Statement -1 is True, Statement-2 is False.

Q.28 Statement -1 : Mass of moving photon varies directly as the wavelength.
Statement -2 : Energy of the particle = Mass × (Speed of light)^2

Q.29 Statement -1 : Photosensitivity of a metal is large if its work function is small.
Statement -2 : Work function = hf_0 where f_0 is the threshold frequency.

Q.30 Statement -1 : The de-Broglie wavelength of a molecule varies inversely as the square root of temperature.
Statement -2 : The root mean square velocity of the molecule is proportional to square root of absolute temperature.

RESPONSE GRID	26. ⓐⓑⓒⓓ	27. ⓐⓑⓒⓓ	28. ⓐⓑⓒⓓ	29. ⓐⓑⓒⓓ	30. ⓐⓑⓒⓓ

DAILY PRACTICE PROBLEM SHEET 54 - PHYSICS			
Total Questions	30	Total Marks	120
Attempted		Correct	
Incorrect		Net Score	
Cut-off Score	30	Qualifying Score	50
Success Gap = Net Score – Qualifying Score			
Net Score = (Correct × 4) – (Incorrect × 1)			

Space for Rough Work

DPP - Daily Practice Problems

Name :

Date :

Start Time :

End Time :

PHYSICS 55

SYLLABUS : Atoms

Max. Marks : 120 **Time : 60 min.**

GENERAL INSTRUCTIONS

- The Daily Practice Problem Sheet contains 30 MCQ's. For each question only one option is correct. Darken the correct circle/ bubble in the Response Grid provided on each page.
- You have to evaluate your Response Grids yourself with the help of solution booklet.
- Each correct answer will get you 4 marks and 1 mark shall be deduced for each incorrect answer. No mark will be given/ deducted if no bubble is filled. Keep a timer in front of you and stop immediately at the end of 60 min.
- The sheet follows a particular syllabus. Do not attempt the sheet before you have completed your preparation for that syllabus. Refer syllabus sheet in the starting of the book for the syllabus of all the DPP sheets.
- After completing the sheet check your answers with the solution booklet and complete the Result Grid. Finally spend time to analyse your performance and revise the areas which emerge out as weak in your evaluation.

DIRECTIONS (Q.1-Q.21) : There are 21 multiple choice questions. Each question has 4 choices (a), (b), (c) and (d), out of which **ONLY ONE** choice is correct.

Q.1 In nature there may not be an element for which the principal quantum number $n > 4$, then the total possible number of elements will be

(a) 60 (b) 32 (c) 4 (d) 64

Q.2 In the following atoms and molecule for the transition from $n = 2$ to $n = 1$, the spectral line of minimum wavelength will be produced by

(a) Hydrogen atom (b) Deuterium atom

(c) Uni-ionized helium (d) Di-ionized lithium

Q.3 The Lyman series of hydrogen sperctum lies in the region

(a) Infrared (b) Visible

(c) Ultraviolet (d) X - rays

Q.4 The energy levels of the hydrogen spectrum is shown in figure. There are some transitions. A,B,C,D and E. Transition A, B and C respectively represent

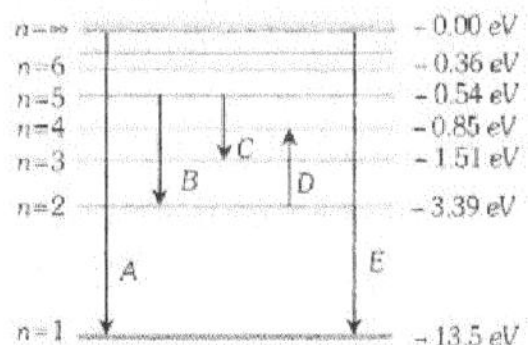

(a) First spectral line of Lyman series, third spectral line of Balmer series and the second spectral line of Paschen series.

(b) Ionization potential of hydrogen, second spectral line of Balmer series and third spectral line of Paschen series

(c) Series limit of Lyman series, third spectral line of Balmer series and second spectral line of Paschen series

(d) Series limit of Lyman series, second spectral line of Balmer series and third spectral line of Paschen series

RESPONSE GRID 1. (a)(b)(c)(d) 2. (a)(b)(c)(d) 3. (a)(b)(c)(d) 4. (a)(b)(c)(d)

Space for Rough Work

Q.5 Energy levels A, B, C of a certain atom corresponding to increasing values of *energy i.e.* $E_A < E_B < E_C$. If $\lambda_1, \lambda_2, \lambda_3$ are the wavelengths of radiations corresponding to the transitions C to B, B to A and C to A respectively, which of the following statements is correct?

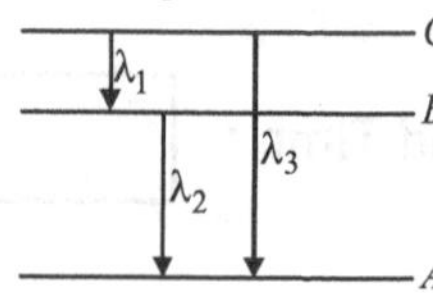

(a) $\lambda_3 = \lambda_1 + \lambda_2$ (b) $\lambda_3 = \frac{\lambda_1 \lambda_2}{\lambda_1 + \lambda_2}$

(c) $\lambda_1 + \lambda_2 + \lambda_3 = 0$ (d) $\lambda_3^2 = \lambda_1^2 + \lambda_2^2$

Q.6 If m is mass of electron, v its velocity, r the radius of stationary circular orbit around a nucleus with charge Ze, then from Bohr's first postulate, the kinetic energy $K = \frac{1}{2}mv^2$ of the electron in $C.G.S.$ system is equal to

(a) $\frac{1}{2}\frac{Ze^2}{r}$ (b) $\frac{1}{2}\frac{Ze^2}{r^2}$ (c) $\frac{Ze^2}{r}$ (d) $\frac{Ze}{r^2}$

Q.7 When a hydrogen atom is raised from the ground state to an excited state

(a) P. E. increases and K. E. decreases
(b) P. E. decreases and K. E. increases
(c) Both kinetic energy and potential energy increase
(d) Both K. E. and P. E. decrease

Q.8 The value of the kinetic energy divided by the total energy of an electron in a Bohr orbit is

(a) – 1 (b) 2
(c) 0.5 (d) None of these

Q.9 The ratio of the frequencies of the long wavelength limits of Lyman and Balmer series of hydrogen spectrum is

(a) 27 : 5 (b) 5 : 27
(c) 4 : 1 (d) 1 : 4

Q.10 Ratio of the wavelengths of first line of Lyman series and first line of Balmer series is

(a) 1 : 3 (b) 27 : 5 (c) 5 : 27 (d) 4 : 9

Q.11 According to Bohr's theory the moment of momentum of an electron revolving in second orbit of hydrogen atom will be

(a) $2\pi h$ (b) πh (c) $\frac{\pi}{h}$ (d) $\frac{h}{\pi}$

Q.12 In the Bohr model of a hydrogen atom, the centripetal force is furnished by the coulomb attraction between the proton and the electron. If a_0 is the radius of the ground state orbit, m is the mass, e is the charge on the electron and ε_0 is the vacuum permittivity, the speed of the electron is

(a) 0 (b) $\frac{e}{\sqrt{\varepsilon_0 a_0 m}}$

(c) $\frac{e}{\sqrt{4\pi\varepsilon_0 a_0 m}}$ (d) $\frac{\sqrt{4\pi\varepsilon_0 a_0 m}}{e}$

Q.13 Which of the following transitions in hydrogen atoms emit photons of highest frequency?

(a) $n = 1$ to $n = 2$ (b) $n = 2$ to $n = 6$
(c) $n = 6$ to $n = 2$ (d) $n = 2$ to $n = 1$

Q.14 As per Bohr model, the minimum energy (in eV) required to remove an electron from the ground state of doubly ionized Li atom $(Z = 3)$ is

(a) 1.51 (b) 13.6
(c) 40.8 (d) 122.4

Q.15 The third line of Balmer series of an ion equivalent to hydrogen atom has wavelength of 108.5 nm. The ground state energy of an electron of this ion will be

(a) 3.4 eV (b) 13.6 eV
(c) 54.4 eV (d) 122.4 eV

Q.16 The wavelength of radiation emitted is λ_0 when an electron jumps from the third to the second orbit of hydrogen atom. For the electron jump from the fourth to the second orbit of the hydrogen atom, the wavelength of radiation emitted will be

(a) $\frac{16}{25}\lambda_0$ (b) $\frac{20}{27}\lambda_0$ (c) $\frac{27}{20}\lambda_0$ (d) $\frac{25}{16}\lambda_0$

RESPONSE GRID					
	5. ⓐⓑⓒⓓ	6. ⓐⓑⓒⓓ	7. ⓐⓑⓒⓓ	8. ⓐⓑⓒⓓ	9. ⓐⓑⓒⓓ
	10. ⓐⓑⓒⓓ	11. ⓐⓑⓒⓓ	12. ⓐⓑⓒⓓ	13. ⓐⓑⓒⓓ	14. ⓐⓑⓒⓓ
	15. ⓐⓑⓒⓓ	16. ⓐⓑⓒⓓ			

Space for Rough Work

Q.17 The energy of electron in the nth orbit of hydrogen atom is expressed as $E_n = \frac{-13.6}{n^2} eV$. The shortest and longest wavelength of Lyman series will be

(a) 910 Å, 1213 Å (b) 5463 Å, 7858 Å
(c) 1315 Å, 1530 Å (d) None of these

Q.18 Consider a hydrogen like atom whose energy in n^{th} exicited state is given by $E_n = -\frac{13.6Z^2}{n^2}$ when this excited atom makes a transition from excited state to ground state, most energetic photons have energy $E_{max} = 52.224\ eV$ and least energetic photons have energy $E_{min} = 1.224\ eV$. The atomic number of atom is

(a) 2 (b) 5
(c) 4 (d) None of these

Q.19 In the Bohr model of the hydrogen atom, let R, v and E represent the radius of the orbit, the speed of electron and the total energy of the electron respectively. Which of the following quantity is proportional to the quantum number n

(a) R/E (b) E/v
(c) RE (d) vR

Q.20 An α – particle of 5 MeV energy strikes with a nucleus of uranium at stationary at an scattering angle of 180°. The nearest distance upto which α – particle reaches the nucleus will be closest to

(a) 1 Å (b) 10^{-10}cm
(c) 10^{-12}cm (d) 10^{-15}cm

Q.21 In a hypothetical Bohr hydrogen, the mass of the electron is doubled. The energy E_0 and the radius r_0 of the first orbit will be (a_0 is the Bohr radius)

(a) $E_0 = -27.2 eV; r_0 = a_0/2$
(b) $E_0 = -27.2 eV; r_0 = a_0$
(c) $E_0 = -13.6 eV; r_0 = a_0/2$
(d) $E_0 = -13.6 eV; r_0 = a_0$

DIRECTIONS (Q.22-Q.24) : In the following questions, more than one of the answers given are correct. Select the correct answers and mark it according to the following codes:

Codes :
(a) 1, 2 and 3 are correct **(b)** 1 and 2 are correct
(c) 2 and 4 are correct **(d)** 1 and 3 are correct

Q.22 The electron in a hydrogen atom makes a transition $n_1 \rightarrow n_2$, where n_1 and n_2 are the principal quantum numbers of two states. Assume the Bohr model to be valid. The time period of the electron in the initial state is eight times that in the final state. Then

(1) $n_1 = 4$ (2) $n_2 = 2$ (3) $n_2 = 5$ (4) $n_1 = 5$

Q.23 A free hydrogen atom in ground state is at rest. A neutron of kinetic energy K collides with the hydrogen atom. After collision hydrogen atom emits two photons in succession one of which has energy 2.55eV. Assume that the hydrogen atom and neutron has same mass.

(1) Minimum value of K is 25.5 eV
(2) Minimum value of K is 12.75 eV
(3) The other photon has energy 10.2eV
(4) The upper energy level is of excitation energy 12.5 eV

Q.24 Which of the series of hydrogen spectrum are not in the visible region?

(1) Lyman series (2) Paschen series
(3) Bracket series (4) Balmer series

DIRECTIONS (Q.25-Q.27) : Read the passage given below and answer the questions that follows :

A gas of identical hydrogen like atoms has some atoms in ground state and some atoms in a particular excited state and there are no atoms in any other energy level. The atoms of the gas make transition to a higher energy state by absorbing monochromatic light of wavelength 304Å. Subsequently, the atoms emit radiation of only six different photon energies. Some of emitted photons have wavelength 304 Å, some have wavelength more and some have less than 304Å (Take hc = 12420 eV–Å)

Q.25 Find the principal quantum number of the initially excited state.

(a) 1 (b) 2 (c) 3 (d) 4

RESPONSE GRID					
	17.ⓐⓑⓒⓓ	18.ⓐⓑⓒⓓ	19.ⓐⓑⓒⓓ	20.ⓐⓑⓒⓓ	21.ⓐⓑⓒⓓ
	22.ⓐⓑⓒⓓ	23.ⓐⓑⓒⓓ	24.ⓐⓑⓒⓓ	25.ⓐⓑⓒⓓ	

Space for Rough Work

Q.26 Identify the gas (Z = ?)
(a) 1 (b) 2 (c) 3 (d) 4

Q.27 Find the maximum and minimum energies of emitted photons (in eV)
(a) 20.4, 10.6 (b) 10.4, 3.6
(c) 40.8, 10.6 (d) None of these

DIRECTIONS (Q.28-Q.30) : Each of these questions contains two statements: Statement-1 (Assertion) and Statement-2 (Reason). Each of these questions has four alternative choices, only one of which is the correct answer. You have to select the correct choice.

(a) Statement-1 is True, Statement-2 is True; Statement-2 is a correct explanation for Statement-1.
(b) Statement-1 is True, Statement-2 is True; Statement-2 is NOT a correct explanation for Statement-1.
(c) Statement -1 is False, Statement-2 is True.
(d) Statement -1 is True, Statement-2 is False.

Q.28 **Statement-1 :** Bohr postulated that the electrons in stationary orbits around the nucleus do not radiate energy.
Statement-2 : According to classical physics all moving electrons radiate energy.

Q.29 **Statement-1 :** The force of repulsion between atomic nucleus and α-particle varies with distance according to inverse square law.
Statement-2 : Rutherford did α-particle scattering experiment.

Q.30 **Statement-1 :** Hydrogen atom consists of only one electron but its emission spectrum has many lines.
Statement-2 : Only Lyman series is found in the absorption spectrum of hydrogen atom whereas in the emission spectrum, all the series are found.

RESPONSE GRID	26. ⓐⓑⓒⓓ	27. ⓐⓑⓒⓓ	28. ⓐⓑⓒⓓ	29. ⓐⓑⓒⓓ	30. ⓐⓑⓒⓓ

DAILY PRACTICE PROBLEM SHEET 55 - PHYSICS			
Total Questions	30	Total Marks	120
Attempted		Correct	
Incorrect		Net Score	
Cut-off Score	30	Qualifying Score	50
Success Gap = Net Score – Qualifying Score			
Net Score = (Correct × 4) – (Incorrect × 1)			

Space for Rough Work

DPP - Daily Practice Problems

Name :   Date :

Start Time :   End Time :

PHYSICS 56

SYLLABUS : Nuclei

Max. Marks : 120   **Time : 60 min.**

GENERAL INSTRUCTIONS

- The Daily Practice Problem Sheet contains 30 MCQ's. For each question only one option is correct. Darken the correct circle/ bubble in the Response Grid provided on each page.
- You have to evaluate your Response Grids yourself with the help of solution booklet.
- Each correct answer will get you 4 marks and 1 mark shall be deduced for each incorrect answer. No mark will be given/ deducted if no bubble is filled. Keep a timer in front of you and stop immediately at the end of 60 min.
- The sheet follows a particular syllabus. Do not attempt the sheet before you have completed your preparation for that syllabus. Refer syllabus sheet in the starting of the book for the syllabus of all the DPP sheets.
- After completing the sheet check your answers with the solution booklet and complete the Result Grid. Finally spend time to analyse your performance and revise the areas which emerge out as weak in your evaluation.

DIRECTIONS (Q.1-Q.21) : There are 21 multiple choice questions. Each question has 4 choices (a), (b), (c) and (d), out of which **ONLY ONE** choice is correct.

Q.1 The energy released per fission of uranium 235 is about 200 MeV. A reactor using U-235 as fuel is producing 1000 kilowatt power. The number of U-235 nuclei undergoing fission per sec is, approximately-

(a) 10^6 (b) 2×10^8 (c) 3×10^{16} (d) 931

Q.2 Power output of ${}_{92}U^{235}$ reactor if it takes 30 days to use up 2kg of fuel, and if each fission gives 185 MeV of useable energy is-

(a) 5.846 kW (b) 58.46 MW
(c) .5846 kW (d) None

Q.3 How many electrons, protons and neutrons are there in a 6 gm of ${}_6C^{12}$.

(a) $6 \times 10^{23}, 6 \times 10^{23}, 6 \times 10^{23}$
(b) $36 \times 10^{23}, 36 \times 10^{23}, 36 \times 10^{23}$
(c) $12 \times 10^{23}, 12 \times 10^{23}, 12 \times 10^{23}$
(d) $18 \times 10^{23}, 18 \times 10^{23}, 18 \times 10^{23}$

Q.4 Nuclear radius of ${}_8O^{16}$ is 3×10^{-15} m. Find the density of nuclear matter.

(a) 7.5×10^{17} kg m^{-3} (b) 5.7×10^{17} kg m^{-3}
(c) 2.3×10^{17} kg m^{-3} (d) 1.66×10^{17} kg m^{-3}

Q.5 Consider the decay of radium-226 atom into an alpha particle and radon-222. Then, what is the mass defect of the reaction-

Mass of radium -226 atom = 226.0256 u
Mass of radon - 222 atom = 222.0715 u
Mass of helium - 4 atom = 4.0026 u

(a) 0.0053 u (b) 0.0083 u
(c) 0.083 u (d) None

RESPONSE GRID	1. ⓐⓑⓒⓓ	2. ⓐⓑⓒⓓ	3. ⓐⓑⓒⓓ	4. ⓐⓑⓒⓓ	5. ⓐⓑⓒⓓ

Space for Rough Work

Q.6 If mass equivalent to one mass of proton is completely converted into energy then determine the energy produced?

(a) 931.49 MeV (b) 731.49 MeV
(c) 911.49 MeV (d) 431.49 MeV

Q.7 If mass equivalent to one mass of electron is completely converted into energy then determine the energy liberated.

(a) 1.51 MeV (b) 0.51 MeV
(c) 3.12 MeV (d) 2.12 MeV

Q.8 If the mass defect in the formation of helium from hydrogen is 0.5%, then the energy obtained, in kWH, in forming helium from 1 kg of hydrogen will be-

(a) 1.25 (b) 125×10^4
(c) 1.25×10^8 (d) 1.25×10^6

Q.9 The half life of radioactive Radon is 3.8 days. The time at the end of which 1/20th of the Radon sample will remain undecayed is

(Given $\log_{10}e = 0.4343$)

(a) 3.8 days (b) 16.5 days
(c) 33 days (d) 76 days

Q.10 In the nuclear reaction,

${}_{92}U^{238} \rightarrow {}_{Z}Th^{A} + {}_{2}He^{4}$, the values of A and Z are-

(a) A = 234, Z = 94 (b) A = 234, Z = 90
(c) A = 238, Z = 94 (d) A = 238, Z = 90

Q.11 The mass of helium nucleus is less than that of its constituent particles by 0.03 a.m.u. The binding energy per nucleon of ${}_{2}He^{4}$ nucleus will be-

(a) 7 MeV (b) 14 MeV
(c) 3.5 MeV (d) 21 MeV

Q.12 If the binding energy of deuterium is 2.23 MeV, then the mass defect will be- (in a.m.u.)

(a) 0.0024 (b) – 0.0024
(c) – 0.0012 (d) 0.0012

Q.13 The ratio of the radii of the nuclei ${}^{27}_{13}Al$ and ${}_{52}Te^{125}$ is approximately -

(a) 6 : 10 (b) 13 : 52
(c) 40 : 177 (d) 14 : 73

Q.14 The radius of the ${}_{30}Zn^{64}$ nucleus is nearly (in fm)-

(a) 1.2 (b) 2.4
(c) 3.7 (d) 4.8

Q.15 How many electrons, protons, and neutrons are there in a nucleus of atomic number 11 and mass number 24?

(a) 11, 12, 13 (b) 11, 11, 13
(c) 12, 11, 13 (d) 11, 13, 12

Q.16 Energy of each photon obtained in the pair production process will be, if the mass of electron or positron is 1/2000 a.m.u-

(a) 0.213 MeV (b) 0.123 MeV
(c) 0.321 MeV (d) 0.465 MeV

Q.17 Deuterium is an isotope of hydrogen having a mass of 2.01470 amu. Find binding energy in MeV of this isotope

(a) 2.741 MeV (b) 2.174 MeV
(c) 1.741 MeV (d) 0.741 MeV

Q.18 The binding energy per nucleon for ${}_{3}Li^{7}$ will be, if the mass of ${}_{3}Li^{7}$ is 7.0163 a.m.u.

(a) 5.6 MeV (b) 39.25 MeV
(c) 1 MeV (d) zero

RESPONSE GRID					
	6. ⓐⓑⓒⓓ	7. ⓐⓑⓒⓓ	8. ⓐⓑⓒⓓ	9. ⓐⓑⓒⓓ	10. ⓐⓑⓒⓓ
	11. ⓐⓑⓒⓓ	12. ⓐⓑⓒⓓ	13. ⓐⓑⓒⓓ	14. ⓐⓑⓒⓓ	15. ⓐⓑⓒⓓ
	16. ⓐⓑⓒⓓ	17. ⓐⓑⓒⓓ	18. ⓐⓑⓒⓓ		

Space for Rough Work

Q.19 Sun radiates energy in all direction. The average energy recieved at earth is 1.4 kW/m^2. The average distance between the earth and the sun is 1.5×10^{11} m. If this energy is released by conservation of mass into energy, then the mass lost per day by the sun is approximately

(Use 1day = 86400 sec)

(a) 4.4×10^9 kg (b) 7.6×10^{14} kg
(c) 3.8×10^{12} kg (d) 3.8×10^{14} kg

Q.20 Fission of nuclei is possible because the binding energy per nucleon in them

(a) increases with mass number at high mass number
(b) decreases with mass number at high mass number
(c) increases with mass number at low mass numbers
(d) decreases with mass number at low mass numbers

Q.21 Half life of Bi^{210} is 5 days. If we start with 50,000 atoms of this isotope, the number of atoms left over after 10 days is

(a) 5,000 (b) 25,000
(c) 12,500 (d) 20,000

DIRECTIONS (Q.22-Q.24) : In the following questions, more than one of the answers given are correct. Select the correct answers and mark it according to the following codes:

Codes :

(a) 1, 2 and 3 are correct **(b)** 1 and 2 are correct
(c) 2 and 4 are correct **(d)** 1 and 3 are correct

Q.22 On disintegration of one atom of U^{235} the amount of energy obtained is 200 MeV. The power obtained in a reactor is 1000 KW. Then

(1) atoms disintegrated per second in reactor is 3.125×10^{16}
(2) atoms disintegrated per second in reactor is 3.125×10^{18}
(3) decay in mass per hour is 4×10^{-8} kg
(4) decay in mass per hour is 4×10^{-6} kg

Q.23 Which of the following are not examples of nuclear fusion?

(1) Formation of *Ba* and *Kr* from U^{235}
(2) Formation of *Pu* – 235 from U^{-235}
(3) Formation of water from hydrogen and oxygen
(4) Formation of *He* from *H*

Q.24 Which of the following are mode of radioactive decay?

(1) Positron emission (2) Electron capture
(3) Alpha decay (4) Fusion

DIRECTIONS (Q.25-Q.27) : Read the passage given below and answer the questions that follows :

In a living organism, the quantity of C^{14} is the same as in the atmosphere. But in organisms which are dead, no exchange takes place with the atmosphere and by measuring the decay rate of ^{14}C in the old bones or wood, the time taken for the activity to reduce to this level can be calculated. This gives the age of the wood or bone.

Given : $T_{1/2}$ for ^{14}C is 5370 years and the ratio of $^{14}C/^{12}C$ is 1.3×10^{-12}.

Q.25 The decay rate of ^{14}C in 1g of carbon in a living organism is

(a) 25 Bq (b) 2.5 Bq
(c) 0.25 Bq (d) 5 Bq

RESPONSE GRID					
	19. ⓐⓑⓒⓓ	20. ⓐⓑⓒⓓ	21. ⓐⓑⓒⓓ	22. ⓐⓑⓒⓓ	23. ⓐⓑⓒⓓ
	24. ⓐⓑⓒⓓ	25. ⓐⓑⓒⓓ			

Space for Rough Work

Q.26 If in an old sample of wood of 10g the decay rate is 30 decays per minute, the age of the wood is

(a) 50 years (b) 1000 years

(c) 13310 years (d) 15300 years

Q.27 The decay rate in another piece is found to be 0.30 Bq per gm then we can conclude

(a) the sample is very recent

(b) the observed decay is not that of ^{14}C alone

(c) there is a statistical error

(d) all of these

DIRECTIONS (Q. 28-Q.30) : Each of these questions contains two statements: Statement-1 (Assertion) and Statement-2 (Reason). Each of these questions has four alternative choices, only one of which is the correct answer. You have to select the correct choice.

(a) Statement-1 is True, Statement-2 is True; Statement-2 is a correct explanation for Statement-1.

(b) Statement-1 is True, Statement-2 is True; Statement-2 is NOT a correct explanation for Statement-1.

(c) Statement -1 is False, Statement-2 is True.

(d) Statement -1 is True, Statement-2 is False.

Q.28 **Statement-1 :** Amongst alpha, beta and gamma rays, γ-has maximum penetrating power.

Statement-2 : The alpha particle is heavier than beta and gamma rays.

Q.29 **Statement-1 :** The mass of β-particles when they are emitted is higher than the mass of electrons obtained by other means.

Statement-2 : β-particle and electron, both are similar particles.

Q.30 **Statement-1 :** Electron capture occurs more often than positron emission in heavy elements.

Statement-2 : Heavy elements exhibit radioactivity.

RESPONSE GRID	26. ⓐⓑⓒⓓ	27. ⓐⓑⓒⓓ	28. ⓐⓑⓒⓓ	29. ⓐⓑⓒⓓ	30. ⓐⓑⓒⓓ

DAILY PRACTICE PROBLEM SHEET 56 - PHYSICS			
Total Questions	30	Total Marks	120
Attempted		Correct	
Incorrect		Net Score	
Cut-off Score	28	Qualifying Score	48
Success Gap = Net Score – Qualifying Score			
Net Score = (Correct × 4) – (Incorrect × 1)			

Space for Rough Work

DPP - Daily Practice Problems

Name :

Date :

Start Time :

End Time :

PHYSICS 57

SYLLABUS : SEMICONDUCTOR ELECTRONICS - 1 (Semiconductors, LED, Photodiode, Zener diode)

Max. Marks : 120 **Time : 60 min.**

GENERAL INSTRUCTIONS

- The Daily Practice Problem Sheet contains 30 MCQ's. For each question only one option is correct. Darken the correct circle/ bubble in the Response Grid provided on each page.
- You have to evaluate your Response Grids yourself with the help of solution booklet.
- Each correct answer will get you 4 marks and 1 mark shall be deduced for each incorrect answer. No mark will be given/ deducted if no bubble is filled. Keep a timer in front of you and stop immediately at the end of 60 min.
- The sheet follows a particular syllabus. Do not attempt the sheet before you have completed your preparation for that syllabus. Refer syllabus sheet in the starting of the book for the syllabus of all the DPP sheets.
- After completing the sheet check your answers with the solution booklet and complete the Result Grid. Finally spend time to analyse your performance and revise the areas which emerge out as weak in your evaluation.

DIRECTIONS (Q.1-Q.21) : There are 21 multiple choice questions. Each question has 4 choices (a), (b), (c) and (d), out of which **ONLY ONE** choice is correct.

Q.1 When a semiconductor is heated, its resistance
(a) decreases (b) increases
(c) reamins unchanged (d) nothing is definite

Q.2 The energy band gap of *Si* is
(a) 0.70 *eV*
(b) 1.1 *eV*
(c) between 0.70 *eV* to 1.1*eV*
(d) 5 *eV*

Q.3 The forbidden energy band gap in conductors, semiconductors and insulators are EG_1, EG_2 and EG_3 respectively. The relation among them is
(a) $EG_1 = EG_2 = EG_3$ (b) $EG_1 < EG_2 < EG_3$
(c) $EG_1 > EG_2 > EG_3$ (d) $EG_1 < EG_2 > EG_3$

Q.4 Let n_h and n_e be the number of holes and conduction electrons respectively in a semiconductor. Then
(a) $n_h > n_e$ in an intrinsic semiconductor
(b) $n_h = n_e$ in an extrinsic semiconductor
(c) $n_h = n_e$ in an intrinsic semiconductor
(d) $n_e > n_h$ in an intrinsic semiconductor

Q.5 Which statement is correct?
(a) *N*-type germanium is negatively charged and *P*-type germanium is positively charged
(b) Both *N*-type and *P*-type germanium are neutral
(c) *N*-type germanium is positively charged and *P*-type germanium is negatively charged
(d) Both *N*-type and *P*-type germanium are negatively charged

RESPONSE GRID	1. ⓐⓑⓒⓓ	2. ⓐⓑⓒⓓ	3. ⓐⓑⓒⓓ	4. ⓐⓑⓒⓓ	5. ⓐⓑⓒⓓ

Space for Rough Work

Q.6 Wires P and Q have the same resistance at ordinary (room) temperature. When heated, resistance of P increases and that of Q decreases. We conclude that
(a) P and Q are conductors of different materials
(b) P is n-type semiconductor and Q is p-type semiconductor
(c) P is semiconductor and Q is conductor
(d) P is conductor and Q is semiconductor

Q.7 In extrinsic P and N-type, semiconductor materials, the ratio of the impurity atoms to the pure semiconductor atoms is about
(a) 1 (b) 10^{-1} (c) 10^{-4} (d) 10^{-7}

Q.8 At zero Kelvin a piece of germanium
(a) becomes semiconductor
(b) becomes good conductor
(c) becomes bad conductor
(d) has maximum conductivity

Q.9 Electronic configuration of germanium is 2, 8, 18 and 4, To make it extrinsic semiconductor small quantity of antimony is added
(a) The material obtained will be N-type germanium in which electrons and holes are equal in number
(b) The material obtained will be P-type germanium
(c) The material obtanied will be N-type germanium which has more electrons than holes at room temperature
(d) The material obtained will be N-type germanium which has less electrons than holes at room temperature

Q.10 The intrinsic semiconductor becomes an insulator at
(a) 0°C (b) –100°C
(c) 300 K (d) 0 K

Q.11 Energy bands in solids are a consequence of
(a) Ohm's Law
(b) Pauli's exclusion principle
(c) Bohr's theory
(d) Heisenberg's uncertainty principle

Q.12 The energy gap for diamond is nearly
(a) 1 ev (b) 2 ev (c) 4 ev (d) 6 ev

Q.13 The valence band and conduction band of a solid overlap at low temperature, the solid may be
(a) metal (b) semiconductor
(c) insulator (d) None of these

Q.14 Choose the correct statement
(a) When we heat a semiconductor its resistance increases
(b) When we heat a semiconductor its resistance decreases
(c) When we cool a semiconductor to 0 K then it becomes super conductor
(d) Resistance of a semiconductor is independent of temperature

Q.15 If n_e and v_d be the number of electrons and drift velocity in a semiconductor. When the temperature is increased
(a) n_e increases and v_d decreases
(b) n_e decreases and v_d increases
(c) Both n_e and v_d increases
(d) Both n_e and v_d decreases

Q.16 The reverse biasing in a PN junction diode
(a) decreases the potential barrier
(b) increases the potential barrier
(c) increases the number of minority charge carriers
(d) increases the number of majority charge carriers

Q.17 Two PN-junctions can be connected in series by three *different* methods as shown in the figure. If the potential difference in the junctions is the same, then the correct connections will be

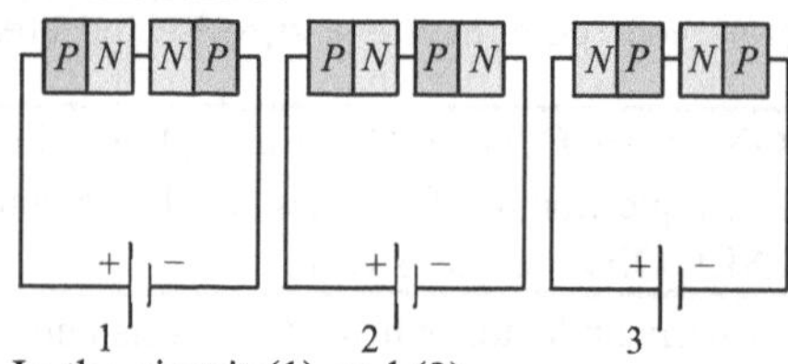

(a) In the circuit (1) and (2)
(b) In the circuit (2) and (3)
(c) In the circuit (1) and (3)
(d) Only in the circuit (1)

RESPONSE GRID					
	6. ⓐⓑⓒⓓ	7. ⓐⓑⓒⓓ	8. ⓐⓑⓒⓓ	9. ⓐⓑⓒⓓ	10. ⓐⓑⓒⓓ
	11. ⓐⓑⓒⓓ	12. ⓐⓑⓒⓓ	13. ⓐⓑⓒⓓ	14. ⓐⓑⓒⓓ	15. ⓐⓑⓒⓓ
	16. ⓐⓑⓒⓓ	17. ⓐⓑⓒⓓ			

Space for Rough Work

Q.18 The approximate ratio of resistances in the forward and reverse bias of the *PN*-junction diode is

(a) $10^2 : 1$ (b) $10^{-2} : 1$

(c) $1 : 10^{-4}$ (d) $1 : 10^4$

Q.19 The dominant mechanisms for motion of charge carriers in forward and reverse biased silicon *P-N* junctions are

(a) Drift in forward bias, diffusion in reverse bias

(b) Diffusion in forward bias, drift in reverse bias

(c) Diffusion in both forward and reverse bias

(d) Drift in both forward and reverse bias

Q.20 In a triclinic crystal system

(a) $a \neq b \neq c, \alpha \neq \beta \neq \gamma$

(b) $a = b = c, \alpha \neq \beta \neq \gamma$

(c) $a \neq b \neq c, \alpha \neq \beta = \gamma$

(d) $a = b = c, \alpha = \beta = \gamma$

Q.21 The correct cymbol for zener diode is

(a) 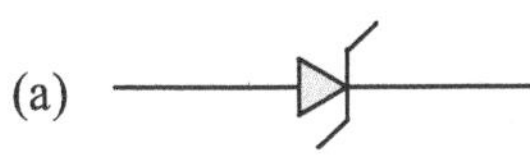

(b) 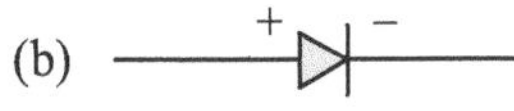

(c) 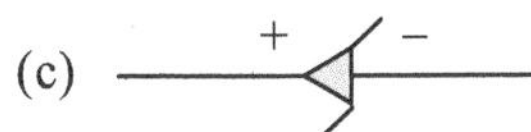

(d) 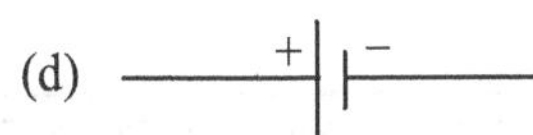

DIRECTIONS (Q.22-Q.24) : In the following questions, more than one of the answers given are correct. Select the correct answers and mark it according to the following codes:

Codes :

(a) 1, 2 and 3 are correct **(b)** 1 and 2 are correct

(c) 2 and 4 are correct **(d)** 1 and 3 are correct

Q.22 In the given figure, which of the diodes are forward biased?

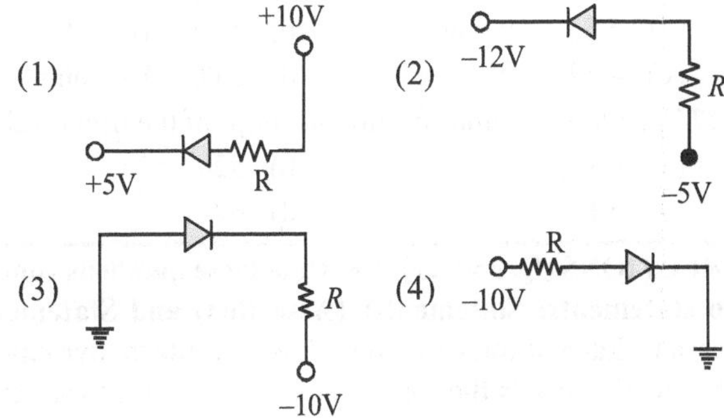

Q.23 Which of the following materials are crystalline?

(1) Copper (2) Sodium chloride

(3) Diamond (4) Wood

Q.24 A piece of copper and the other of germanium are cooled from the room temperature to 80 K, then which of the following would be wrong statements?

(1) Resistance of each increases

(2) Resistance of each decreases

(3) Resistance of copper increases while that of germanium decreases

(4) Resistance of copper decreases while that of germanium increases

DIRECTIONS (Q.25-Q.27) : Read the passage given below and answer the questions that follows :

A student performs an experiment for drawing the static characteristic curve of a triode valve in the laboratory. The following data were obtained from the linear portion of the curves:

Grid voltage V_g (volt)	–2.0	–3.5	–2.0
Plate voltage V_p (volt)	180	180	120
Plate current I_p (mA)	15	7	10

Q.25 Calculate the plate resistance r_p of the triode valve?

(a) 0.12×10^4 ohm (b) 1.2×10^4 ohm

(c) 1.3×10^4 ohm (d) 1.4×10^4 ohm

RESPONSE GRID					
	18. ⓐⓑⓒⓓ	19. ⓐⓑⓒⓓ	20. ⓐⓑⓒⓓ	21. ⓐⓑⓒⓓ	22. ⓐⓑⓒⓓ
	23. ⓐⓑⓒⓓ	24. ⓐⓑⓒⓓ	25. ⓐⓑⓒⓓ		

Space for Rough Work

Q.26 Calculate the mutual conductance g_m of the triode valve?

(a) 5.33×10^{-3} ohm^{-1} (b) 53.3×10^{-3} ohm^{-1}

(c) 4.32×10^{-3} ohm^{-1} (d) 5.00×10^{-3} ohm^{-1}

Q.27 Calculate the amplification factor μ, of the triode valve?

(a) 64 (b) 52

(c) 54 (d) 62

DIRECTIONS (Q. 28-Q.30) : Each of these questions contains two statements: Statement-1 (Assertion) and Statement-2 (Reason). Each of these questions has four alternative choices, only one of which is the correct answer. You have to select the correct choice.

(a) Statement-1 is True, Statement-2 is True; Statement-2 is a correct explanation for Statement-1.

(b) Statement-1 is True, Statement-2 is True; Statement-2 is NOT a correct explanation for Statement-1.

(c) Statement -1 is False, Statement-2 is True.

(d) Statement -1 is True, Statement-2 is False.

Q.28 Statement-1 : The number of electrons in a *P*-type silicon semiconductor is less than the number of electrons in a pure silicon semiconductor at room temperature.

Statement-2: It is due to law of mass action.

Q.29 Statement-1 : The resistivity of a semiconductor decreases with temperature.

Statement-2 : The atoms of a semiconductor vibrate with larger amplitude at higher temperature there by increasing its resistivity.

Q.30 Statement-1 : We can measure the potential barrier of a PN junction by putting a sensitive voltmeter across its terminals.

Statement-2 : The current through the PN junction is not same in forward and reversed bias.

RESPONSE GRID	26. ⓐⓑⓒⓓ	27. ⓐⓑⓒⓓ	28. ⓐⓑⓒⓓ	29. ⓐⓑⓒⓓ	30. ⓐⓑⓒⓓ

DAILY PRACTICE PROBLEM SHEET 57 - PHYSICS			
Total Questions	30	Total Marks	120
Attempted		Correct	
Incorrect		Net Score	
Cut-off Score	28	Qualifying Score	48
Success Gap = Net Score – Qualifying Score			
Net Score = (Correct × 4) – (Incorrect × 1)			

Space for Rough Work

DPP - Daily Practice Problems

Name :

Date :

Start Time :

End Time :

PHYSICS 58

SYLLABUS : SEMICONDUCTOR ELECTRONICS-2 (Junction transistor, transistor action, characteristics of a transistor, transistor as an amplifier, logic gates)

Max. Marks : 112 **Time : 60 min.**

GENERAL INSTRUCTIONS

- The Daily Practice Problem Sheet contains 28 MCQ's. For each question only one option is correct. Darken the correct circle/ bubble in the Response Grid provided on each page.
- You have to evaluate your Response Grids yourself with the help of solution booklet.
- Each correct answer will get you 4 marks and 1 mark shall be deduced for each incorrect answer. No mark will be given/ deducted if no bubble is filled. Keep a timer in front of you and stop immediately at the end of 60 min.
- The sheet follows a particular syllabus. Do not attempt the sheet before you have completed your preparation for that syllabus. Refer syllabus sheet in the starting of the book for the syllabus of all the DPP sheets.
- After completing the sheet check your answers with the solution booklet and complete the Result Grid. Finally spend time to analyse your performance and revise the areas which emerge out as weak in your evaluation.

DIRECTIONS (Q.1-Q.20) : There are 20 multiple choice questions. Each] question has 4 choices (a), (b), (c) and (d), out of which **ONLY ONE** choice is correct.

Q.1 A NPN transistor conducts when
(a) both collector and emitter are positive with respect to the base
(b) collector is positive and emitter is negative with respect to the base
(c) collector is positive and emitter is at same potential as the base
(d) both collector and emitter are negative with respect to the base

Q.2 In the case of constants α and β of a transistor
(a) $\alpha = \beta$ (b) $\beta < 1, \alpha > 1$
(c) $\alpha = \beta^2$ (d) $\beta > 1, \alpha < 1$

Q.3 In an NPN transistor 10^{10} electrons enter the emitter in 10^{-6} s and 2% electrons recombine with holes in base, then α and β respectively are
(a) $\alpha = 0.98, \beta = 49$ (b) $\alpha = 49, \beta = 0.98$
(c) $\alpha = 0.49, \beta = 98$ (d) $\alpha = 98, \beta = 0.49$

Q.4 If l_1, l_2, l_3 are the lengths of the emitter, base and collector of a transistor then
(a) $l_1 = l_2 = l_3$ (b) $l_3 < l_2 > l_1$
(c) $l_3 < l_1 < l_2$ (d) $l_3 > l_1 > l_2$

Q.5 In an NPN transistor circuit, the collector current is 10 mA. If 90% of the electrons emitted reach the collector, the emitter current (i_E) and base current (i_B) are given by
(a) $i_E = -1\text{mA}, i_B = 9\text{mA}$ (b) $i_E = 9\text{mA}, i_B = -1\text{mA}$
(c) $i_E = 1\text{mA}, i_B = 11\text{mA}$ (d) $i_E = 11\text{mA}, i_B = 1\text{mA}$

RESPONSE GRID 1. ⓐⓑⓒⓓ 2. ⓐⓑⓒⓓ 3. ⓐⓑⓒⓓ 4. ⓐⓑⓒⓓ 5. ⓐⓑⓒⓓ

Space for Rough Work

Q.6 The transfer ratio of a transistor is 50. The input resistance of the transistor when used in the common-emitter configuration is 1 kΩ. The peak value for an A.C input voltage of 0.01 V peak is

(a) 100 μA (b) 0.01 mA
(c) 0.25 mA (d) 500 μA

Q.7 For transistor, the current amplification factor is 0.8. The transistor is connected in common emitter configuration. The change in the collector current when the base current changes by 6 mA is

(a) 6 mA (b) 4.8 mA
(c) 24 mA (d) 8 mA

Q.8 In NPN transistor the collector current is 10 mA. If 90% of electrons emitted reach the collector, then

(a) emitter current will be 9 mA
(b) emitter current will be 11.1 mA
(c) base current will be 0.1 mA
(d) base current will be 0.01 mA

Q.9 In a transistor in CE configuration, the ratio of power gain to voltage gain is

(a) α (b) β/α
(c) $\beta\alpha$ (d) β

Q.10 The following truth table corresponds to the logic gate

A	0	0	1	1
B	0	1	0	1
X	0	1	1	1

(a) NAND (b) OR
(c) AND (d) XOR

Q.11 The truth table shown in figure is for

A	0	0	1	1
B	0	1	0	1
Y	1	0	0	1

(a) XOR (b) AND
(c) XNOR (d) OR

Q.12 For the given combination of gates, if the logic states of inputs A, B, C are as follows $A = B = C = 0$ and $A = B = 1$,
$C = 0$ then the logic states of output D are

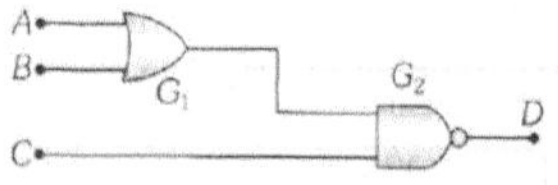

(a) 0, 0 (b) 0, 1 (c) 1, 0 (d) 1, 1

Q.13 Correct statement for 'NOR' gate is that, it gives

(a) high output when both the inputs are low
(b) low output when both the inputs are low
(c) high output when both the inputs are high
(d) None of these

Q.14 A gate has the following truth table

P	1	1	0	0
Q	1	0	1	0
R	1	0	0	0

The gate is

(a) NOR (b) OR (c) NAND (d) AND

Q.15 What will be the input of A and B for the Boolean expression $\overline{(A+B)}.\overline{(A.B)} = 1$

(a) 0, 0 (b) 0, 1 (c) 1, 0 (d) 1, 1

Q.16 To get an output 1 from the circuit shown in the figure, the input can be

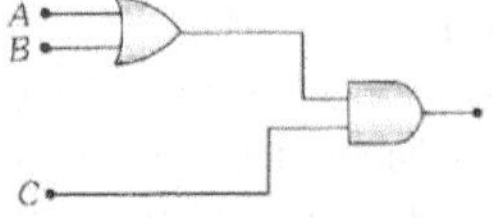

(a) $A = 0, B = 1, C = 0$ (b) $A = 1, B = 0, C = 0$
(c) $A = 1, B = 0, C = 1$ (d) $A = 1, B = 1, C = 0$

Q.17 The truth-table given below is for which gate?

A	0	0	1	1
B	0	1	0	1
C	1	1	1	0

(a) XOR (b) OR (c) AND (d) NAND

RESPONSE GRID					
	6. ⓐⓑⓒⓓ	7. ⓐⓑⓒⓓ	8. ⓐⓑⓒⓓ	9. ⓐⓑⓒⓓ	10. ⓐⓑⓒⓓ
	11. ⓐⓑⓒⓓ	12. ⓐⓑⓒⓓ	13. ⓐⓑⓒⓓ	14. ⓐⓑⓒⓓ	15. ⓐⓑⓒⓓ
	16. ⓐⓑⓒⓓ	17. ⓐⓑⓒⓓ			

Space for Rough Work

Q.18 The combination of gates shown below produces

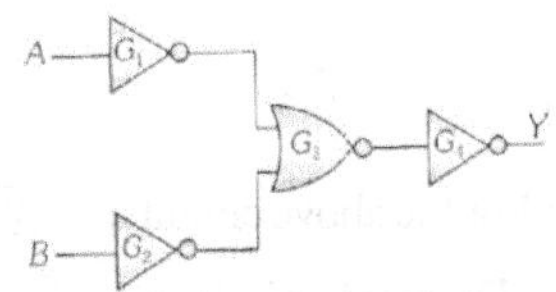

(a) AND gate (b) XOR gate
(c) NOR gate (d) NAND gate

Q.19 Figure gives a system of logic gates. From the study of truth table it can be found that to produce a high output (1) at R, we must have

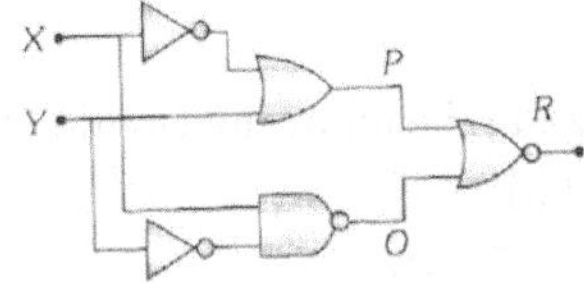

(a) X = 0, Y = 1 (b) X = 1, Y = 1
(c) X = 1, Y = 0 (d) X = 0, Y = 0

Q.20 In the case of a common emitter transistor as/an amplifier, the ratio I_c/I_e is 0.96, then the current gain (β) of the amplifier is

(a) 6 (b) 48
(c) 24 (d) 12

DIRECTIONS (Q.21-Q.23) : In the following questions, more than one of the answers given are correct. Select the correct answers and mark it according to the following codes:

Codes :

(a) 1, 2 and 3 are correct **(b)** 1 and 2 are correct
(c) 2 and 4 are correct **(d)** 1 and 3 are correct

Q.21 Which of the following are false?

(1) Common base transistor is commonly used because current gain is maximum
(2) Common collector is commonly used because current gain is maximum
(3) Common emitter is the least used transistor
(4) Common emitter is commonly used because current gain is maximum

Q.22 Given below are symbols for some logic gates. The XOR gate and NOR gate are respectively

(1) 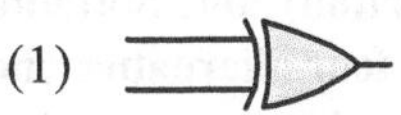 (2)

(3) (4)

Q.23 Given below are four logic gates symbol (figure). Those for OR, NOR and NAND are respectively

(1) A, B → y (2) A, B → y

(3) A, B → y (4) A, B → y

DIRECTIONS (Q.24-Q.25) : Read the passage given below and answer the questions that follows :

Doping changes the fermi energy of a semiconductor. Consider silicon, with a gap of 1.11 eV between the top of the valence bond and the bottom of the conduction band. At 300K the Fermi level of the pure material is nearly at the mid-point of the gap. Suppose that silicon is doped with donor atoms, each of which has a state 0.15 eV below the bottom of the silicon conduction band, and suppose further that doping raises the Fermi level to 0.11 eV below the bottom of that band.

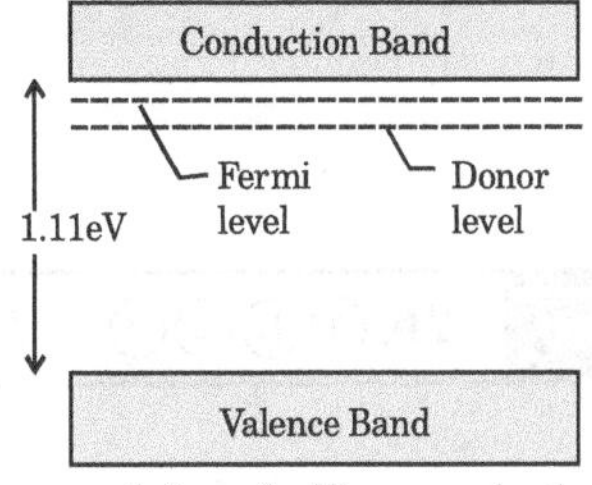

Q.24 For both pure and doped silicon, calculate the probability that a state at the bottom of the silicon conduction band is occupied? ($e^{4.524} = 70.38$)

(a) 5.20×10^{-2} (b) 1.40×10^{-2}
(c) 10.5×10^{-2} (d) 14×10^{-2}

Q.25 Calculate the probability that a donor state in the doped material is occupied? $e^{-1.547} = 0.212$

(a) 0.824 (b) 0.08
(c) 0.008 (d) 8.2

RESPONSE GRID					
	18. ⓐⓑⓒⓓ	19. ⓐⓑⓒⓓ	20. ⓐⓑⓒⓓ	21. ⓐⓑⓒⓓ	22. ⓐⓑⓒⓓ
	23. ⓐⓑⓒⓓ	24. ⓐⓑⓒⓓ	25. ⓐⓑⓒⓓ		

Space for Rough Work

DIRECTIONS (Q. 26-Q.28) : Each of these questions contains two statements: Statement-1 (Assertion) and Statement-2 (Reason). Each of these questions has four alternative choices, only one of which is the correct answer. You have to select the correct choice.

(a) Statement-1 is True, Statement-2 is True; Statement-2 is a correct explanation for Statement-1.
(b) Statement-1 is True, Statement-2 is True; Statement-2 is NOT a correct explanation for Statement-1.
(c) Statement -1 is False, Statement-2 is True.
(d) Statement -1 is True, Statement-2 is False.

Q.26 Statement -1 : The logic gate NOT cannot be built by using diode.

Statement -2 : The output voltage and the input voltage of the diode have 180° phase difference.

Q.27 Statement -1 : The following circuit represents 'OR' gate

Statement-2 : For the above circuit $Y = \overline{X} = \overline{\overline{A+B}} = A+B$

Q.28 Statement -1 : De-morgan's theorem $\overline{A+B} = \overline{A}.\overline{B}$ may be explained by the following circuit

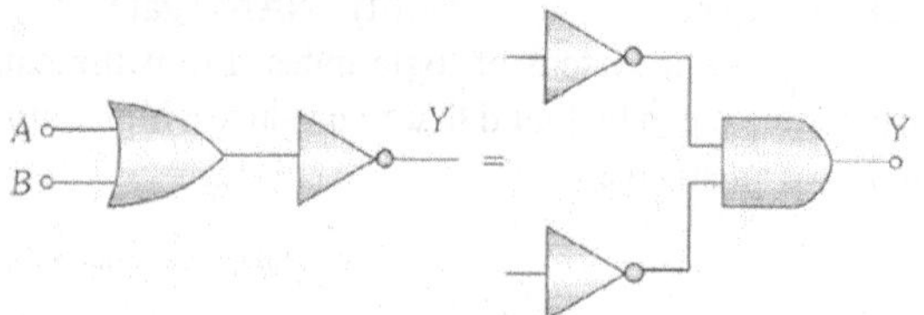

Statement -2 : In the following circuit, for output 1 inputs A,B,C are 1, 0, 1.

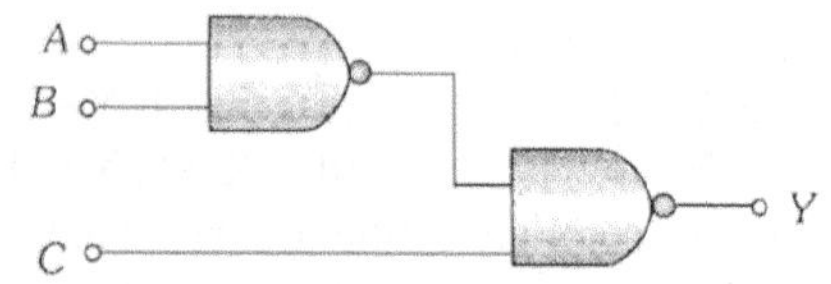

RESPONSE GRID	26. ⓐⓑⓒⓓ	27. ⓐⓑⓒⓓ	28. ⓐⓑⓒⓓ

DAILY PRACTICE PROBLEM SHEET 58 - PHYSICS			
Total Questions	28	Total Marks	112
Attempted		Correct	
Incorrect		Net Score	
Cut-off Score	26	Qualifying Score	46
Success Gap = Net Score – Qualifying Score			
Net Score = (Correct × 4) – (Incorrect × 1)			

Space for Rough Work

DPP - Daily Practice Problems

Name : 　　　　　　　　Date :

Start Time : 　　　　　　　　End Time :

PHYSICS 59

SYLLABUS : Communication Systems, Laser

Max. Marks : 120 　　　　　　　　**Time : 60 min.**

GENERAL INSTRUCTIONS

- The Daily Practice Problem Sheet contains 30 MCQ's. For each question only one option is correct. Darken the correct circle/ bubble in the Response Grid provided on each page.
- You have to evaluate your Response Grids yourself with the help of solution booklet.
- Each correct answer will get you 4 marks and 1 mark shall be deduced for each incorrect answer. No mark will be given/ deducted if no bubble is filled. Keep a timer in front of you and stop immediately at the end of 60 min.
- The sheet follows a particular syllabus. Do not attempt the sheet before you have completed your preparation for that syllabus. Refer syllabus sheet in the starting of the book for the syllabus of all the DPP sheets.
- After completing the sheet check your answers with the solution booklet and complete the Result Grid. Finally spend time to analyse your performance and revise the areas which emerge out as weak in your evaluation.

DIRECTIONS (Q.1-Q.21) : There are 21 multiple choice questions. Each question has 4 choices (a), (b), (c) and (d), out of which **ONLY ONE** choice is correct.

Q.1 In short wave communication which of the following frequencies will be reflected back by the ionospheric layer, having electron density 10^{11} per m^3.

(a) 2.84 MHz (b) 10.42 MHz
(c) 12.24 MHz (d) 18.1 MHz

Q.2 In an amplitude modulated wave for audio frequency of 500 cycle/second, the appropriate carrier frequency will be

(a) 50 cycles/sec (b) 100 cycles/sec
(c) 500 cycles/sec (d) 50, 000 cycles/sec

Q.3 Range of frequencies allotted for commercial FM radio broadcast is

(a) 88 to 108 MHz (b) 88 to 108 kHz
(c) 8 to 88 MHz (d) 88 to 108 GHz

Q.4 The process of superimposing signal frequency (*i.e.* audio wave) on the carrier wave is known as

(a) Transmission (b) Reception
(c) Modulation (d) Detection

Q.5 The characteristic impedance of a coaxial cable is of the order of

(a) 50Ω (b) 200Ω
(c) 270Ω (d) None of these

RESPONSE GRID	1. ⓐⓑⓒⓓ	2. ⓐⓑⓒⓓ	3. ⓐⓑⓒⓓ	4. ⓐⓑⓒⓓ	5. ⓐⓑⓒⓓ

Space for Rough Work

Q.6 If μ_1 and μ_2 are the refractive indices of the materials of core and cladding of an optical fibre, then the loss of light due to its leakage can be minimised by having

(a) $\mu_1 > \mu_2$ (b) $\mu_1 < \mu_2$

(c) $\mu_1 = \mu_2$ (d) None of these

Q.7 Maximum usable frequency (MUF) in *F*-region layer is *x*, when the critical frequency is 60 MHz and the angle of incidence is $70°$. Then x is (cos 70° = 0.34)

(a) 150 MHz (b) 170 MHz

(c) 175 MHz (d) 190 MHz

Q.8 A laser is a coherent source because it contains

(a) many wavelengths

(b) uncoordinated wave of a particular wavelength

(c) coordinated wave of many wavelengths

(d) coordinated waves of a particular wavelength

Q.9 A laser beam is used for carrying out surgery because it

(a) is highly monochromatic

(b) is highly coherent

(c) is highly directional

(d) can be sharply focussed

Q.10 Laser beams are used to measure long distances because

(a) they are monochromatic

(b) they are highly polarised

(c) they are coherent

(d) they have high degree of parallelism

Q.11 An oscillator is producing FM waves of frequency 2 kHz with a variation of 10 kHz. What is the modulating index?

(a) 0.20 (b) 5.0 (c) 0.67 (d) 1.5

Q.12 If f_a and f_f represent the carrier wave frequencies for amplitude and frequency modulations respectively, then

(a) $f_a > f_f$ (b) $f_a < f_f$

(c) $f_a = f_f$ (d) $f_a \geq f_f$

Q.13 An antenna is a device

(a) that converts electromagnetic energy into radio frequency signal

(b) that converts radio frequency signal into electromagnetic energy

(c) that converts guided electromagnetic waves into free space electromagnetic waves and vice-versa

(d) None of these

Q.14 While tuning in a certain broadcast station with a receiver, we are actually

(a) varying the local oscillator frequency

(b) varying the frequency of the radio signal to be picked up

(c) tuning the antenna

(d) None of these

Q.15 In a communication system, noise is most likely to affect the signal

(a) At the transmitter

(b) In the channel or in the transmission line

(c) In the information source

(d) At the receiver

Q.16 In an FM system a 7 kHz signal modulates 108 MHz carrier so that frequency deviation is 50 kHz. The carrier swing is

(a) 7.143 (b) 8

(c) 0.71 (d) 350

Q.17 The phenomenon by which light travels in an optical fibres is

(a) Reflection

(b) Refraction

(c) Total internal reflection

(d) Transmission

Q.18 In frequency modulation

(a) The amplitude of modulated wave varies as frequency of carrier wave

(b) The frequency of modulated wave varies as amplitude of modulating wave

(c) The amplitude of modulated wave varies as amplitude of carrier wave

(d) The frequency of modulated wave varies as frequency of modulating wave

RESPONSE GRID					
	6. ⓐⓑⓒⓓ	7. ⓐⓑⓒⓓ	8. ⓐⓑⓒⓓ	9. ⓐⓑⓒⓓ	10. ⓐⓑⓒⓓ
	11. ⓐⓑⓒⓓ	12. ⓐⓑⓒⓓ	13. ⓐⓑⓒⓓ	14. ⓐⓑⓒⓓ	15. ⓐⓑⓒⓓ
	16. ⓐⓑⓒⓓ	17. ⓐⓑⓒⓓ	18. ⓐⓑⓒⓓ		

Space for Rough Work

Q.19 Audio signal cannot be transmitted because
(a) The signal has more noise
(b) The signal cannot be amplified for distance communication
(c) The transmitting antenna length is very small to design
(d) The transmitting antenna length is very large and impracticable

Q.20 For sky wave propagation of a 10 MHz signal, what should be the minimum electron density in ionosphere
(a) $\sim 1.2\times10^{12}\,m^{-3}$
(b) $\sim 10^{6}\,m^{-3}$
(c) $\sim 10^{14}\,m^{-3}$
(d) $\sim 10^{22}\,m^{-3}$

Q.21 What should be the maximum acceptance angle at the aircore interface of an optical fibre if n_1 and n_2 are the refractive indices of the core and the cladding, respectively
(a) $\sin^{-1}(n_2/n_1)$ (b) $\sin^{-1}\sqrt{n_1^2-n_2^2}$
(c) $\left[\tan^{-1}\frac{n_2}{n_1}\right]$ (d) $\left[\tan^{-1}\frac{n_1}{n_2}\right]$

DIRECTIONS (Q.22-Q.24) : In the following questions, more than one of the answers given are correct. Select the correct answers and mark it according to the following codes:

Codes :
(a) 1, 2 and 3 are correct **(b)** 1 and 2 are correct
(c) 2 and 4 are correct **(d)** 1 and 3 are correct

Q.22 In which of the following remote sensing technique is used?
(1) Forest density
(2) Pollution
(3) Wetland mapping
(4) Medical treatment

Q.23 What type of modulation is not employed in India for radio transmission?
(1) A mixture of both frequency and pulse modulation.
(2) Pulse modulation
(3) Frequency modulation
(4) Amplitude modulation

Q.24 Which of the following are the characteristics of Laser beams?
(1) They are monochromatic
(2) They are coherent
(3) They have high degree of parallelism
(4) They are not monochromatic

DIRECTIONS (Q.25-Q.27) : Read the passage given below and answer the questions that follows :

The electron density of a layer of ionosphere at a height 150 km from the earth's surface is 9×10^9 per m^3. For the sky transmission from this layer up to a range of 250 km,

Q.25 The critical frequency of the layer is
(a) 2 Hz (b) 2.7 Hz
(c) 2.78 kHz (d) 2.7 MHz

Q.26 Maximum usable frequency is
(a) 3.17 Hz
(b) 3.17×10^6 HZ
(c) 3.17×10^3 Hz
(d) 3.17×10^{10}Hz

Q.27 Angle of incidence of this layer is
(a) 0° (b) $\sec^{-1}(1.51)$
(c) $\sec^{-1}(1.17)$ (d) 37°

RESPONSE GRID					
	19. ⓐⓑⓒⓓ	20. ⓐⓑⓒⓓ	21. ⓐⓑⓒⓓ	22. ⓐⓑⓒⓓ	23. ⓐⓑⓒⓓ
	24. ⓐⓑⓒⓓ	25. ⓐⓑⓒⓓ	26. ⓐⓑⓒⓓ	27. ⓐⓑⓒⓓ	

Space for Rough Work

DIRECTIONS (Qs. 28-Q.30) : Each of these questions contains two statements: Statement-1 (Assertion) and Statement-2 (Reason). Each of these questions has four alternative choices, only one of which is the correct answer. You have to select the correct choice.

(a) Statement-1 is True, Statement-2 is True; Statement-2 is a correct explanation for Statement-1.
(b) Statement-1 is True, Statement-2 is True; Statement-2 is NOT a correct explanation for Statement-1.
(c) Statement -1 is False, Statement-2 is True.
(d) Statement -1 is True, Statement-2 is False.

Q.28 Statement-1 : Television signals are not received through sky-wave propagation.

Statement-2 : The ionosphere reflects electromagnetic waves of frequencies greater than a certain critical frequency.

Q.29 Statement-1 : The electromagnetic waves of shorter wavelength can travel longer distances on earth's surface than those of longer wavelengths.

Statement-2 : Shorter the wavelength, the larger is the velocity of wave propagation.

Q.30 Statement-1 : A dish antenna is highly directional.

Statement-2 : This is because a dipole antenna is omni directional.

RESPONSE GRID	28. ⓐⓑⓒⓓ	29. ⓐⓑⓒⓓ	30. ⓐⓑⓒⓓ

DAILY PRACTICE PROBLEM SHEET 59 - PHYSICS			
Total Questions	30	Total Marks	120
Attempted		Correct	
Incorrect		Net Score	
Cut-off Score	30	Qualifying Score	50
Success Gap = Net Score – Qualifying Score			
Net Score = (Correct × 4) – (Incorrect × 1)			

Space for Rough Work

DPP - Daily Practice Problems

Name : 　　　　　　　　　　　　　　　　　　　Date :

Start Time : 　　　　　　　　　　　　　　　　End Time :

PHYSICS 60

SYLLABUS : Practical Physics - 2

Max. Marks : 112 　　　　　　　　　　　　　　　**Time : 60 min.**

GENERAL INSTRUCTIONS

- The Daily Practice Problem Sheet contains 28 MCQ's. For each question only one option is correct. Darken the correct circle/ bubble in the Response Grid provided on each page.
- You have to evaluate your Response Grids yourself with the help of solution booklet.
- Each correct answer will get you 4 marks and 1 mark shall be deduced for each incorrect answer. No mark will be given/ deducted if no bubble is filled. Keep a timer in front of you and stop immediately at the end of 60 min.
- The sheet follows a particular syllabus. Do not attempt the sheet before you have completed your preparation for that syllabus. Refer syllabus sheet in the starting of the book for the syllabus of all the DPP sheets.
- After completing the sheet check your answers with the solution booklet and complete the Result Grid. Finally spend time to analyse your performance and revise the areas which emerge out as weak in your evaluation.

DIRECTIONS (Q.1-Q.26) : There are 26 multiple choice questions. Each question has 4 choices (a), (b), (c) and (d), out of which **ONLY ONE** choice is correct.

Q.1 In making an Ohm's law circuit, which of the following connection is correct?
(a) Voltmeter in series and ammeter in parallel
(b) Voltmeter in parallel and ammeter in series
(c) Voltmeter and ammeter both are in parallel
(d) Voltmeter and ammeter both are in series

Q.2 To calculate an unknown resistance with the help of a meter bridge why is it advised to change the gap with the known and unknown resistance?
(a) To eliminate the resistance of the connecting wire and copper strip
(b) To include the resistance of the connecting wire and copper strip
(c) To balance the known and unknown resistance.
(d) To eliminate the resistance of the gap.

Q.3 Potential gradient of a potentiometer is equal to
(a) e.m.f per unit length
(b) potential drop per unit length
(c) current per unit length
(d) resistance per unit length

Q.4 The refractive index of the material of a prism does not depend on which of the following factor?
(a) Nature of the material
(b) Wavelength or colour of light
(c) Temperature
(d) Angle of the prism.

RESPONSE GRID	1. ⓐⓑⓒⓓ	2. ⓐⓑⓒⓓ	3. ⓐⓑⓒⓓ	4. ⓐⓑⓒⓓ

Space for Rough Work

Q.5 A meter-bridge is based on the principle of
(a) Wheatstone bridge
(b) Variation of resistance with temperature
(c) Galvanometer
(d) None of these

Q.6 A potentiometer works on the principle that
(a) when a current flows through a wire of uniform thickness and material, potential difference between its two points is directly proportional to the length of the wire between the two points
(b) when a current flows through a wire of uniform thickness and material, potential difference between its two points is inversely proportional to the length of the wire between the points
(c) when a current flows through a wire of uniform thickness and material, potential difference between its two points doesn't depend on the length of the wire between the points
(d) none of these

Q.7 Which of the following statement is wrong regarding a p–n junction diode?
(a) When the p-type section is connected to the positive terminal and the n-type section to the negative terminal of the battery the diode is called forward biased
(b) When the p-type section is connected to the negative terminal and the n-type section to the positive terminal of the battery the diode is said to be reverse biased
(c) When the diode is in reverse biased mode a forward current flows
(d) When the diode is in forward biased mode a forward current flows.

Q.8 A Zener diode operates on which of the following bias?
(a) Forward bias
(b) Reverse bias
(c) Both forward and reverse bias.
(d) No biasing is required for it.

Q.9 The transfer characteristics of a transistor means a plot of
(a) input voltage versus input current
(b) output voltage versus output current.
(c) output voltage versus input voltage
(d) input current versus output current.

Q.10 Current gain is maximum in which of the following configuration of a transistor ?
(a) common emitter configuration
(b) common base configuration
(c) common collector configuration
(d) equal in both common emitter and common base configuration

Q.11 Which of the following operations will not increase the sensitivity of a potentiometer?
(a) Increase in the number of wires of the potentiometer.
(b) Reducing the potential gradient.
(c) Increasing the current through the potentiometer
(d) Increasing the sensitivity of the galvanometer.

Q.12 Which two circuit components are connected in parallel in the following circuit diagram ?
(a) Rheostat and voltmeter
(b) Voltmeter and ammeter
(c) Voltmeter and resistor
(d) Ammeter and resistor

Q.13 A current of 4A produces a deflection of 30° in the galvanometer. The figure of merit is
(a) 6.5 A/rad (b) 7.6 A/rad
(c) 7.5 A/rad (d) 8.0 A/rad

Q.14 Two potentiometers A and B having 4 wires and 10 wires, each having 100 cm in length are used to compare e.m.f. of 2 cells. Which one will give a larger balancing length?
(a) Balancing length doesn't depend on the total length of the wire.
(b) Both A and B will give same balancing length
(c) Potentiometer B
(d) Potentiometer A

Q.15 An LED operates under which biasing condition?
(a) Forward bias
(b) Reverse bias
(c) Can operate both in forward and reverse bias
(d) No biasing is required.

Q.16 How are the currents flowing in the emitter, base and the collector related to each other?
(a) $I_c = I_b + I_e$ (b) $I_e = I_b + I_c$
(c) $I_b = I_e + I_c$ (d) $I_e = I_c - I_b$

RESPONSE GRID

5. ⓐⓑⓒⓓ 6. ⓐⓑⓒⓓ 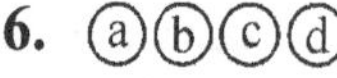 7. ⓐⓑⓒⓓ 8. ⓐⓑⓒⓓ 9. ⓐⓑⓒⓓ
10. ⓐⓑⓒⓓ 11. ⓐⓑⓒⓓ 12. ⓐⓑⓒⓓ 13. ⓐⓑⓒⓓ 14. ⓐⓑⓒⓓ
15. ⓐⓑⓒⓓ 16. ⓐⓑⓒⓓ

Space for Rough Work

Q.17 The potential gradient of a potentiometer can be increased by which of the following operation?
(a) By increasing the area of cross-section of the potentiometer wire.
(b) By decreasing the area of cross-section of the potentiometer wire.
(c) By decreasing the current through it.
(d) By using a wire of material of low specific resistance. when A is decreased, k will increase.

Q.18 Of the diodes shown in the following diagrams, which one is reverse biased ?

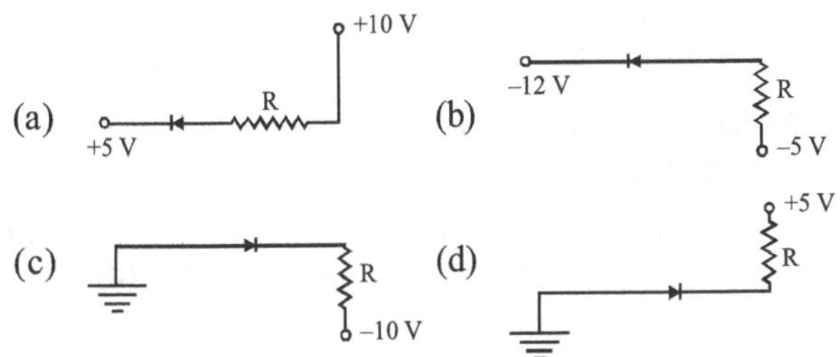

Q.19 To determine the equivalent resistance of two resistors when connected in series, a student arranged the circuit components as shown in the diagram. But he did not succeed to achieve the objective.

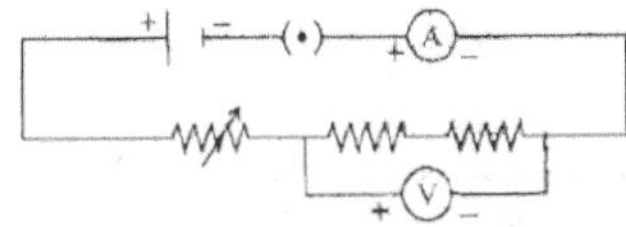

Which of the following mistakes has been committed by him in setting up the circuit ?
(a) Position of voltmeter is incorrect
(b) Position of ammeter is incorrect
(c) Terminals of voltmeter are wrongly connected
(d) Terminals of ammeter are wrongly connected

Q.20 In the circuit shown, voltmeter is ideal and its least count is 0.1 V. The least count of ammeter is 1 mA. Let reading of the voltmeter be 30.0 V and the reading of ammeter is 0.020 A. Calculate the value of resistance R within error limits.

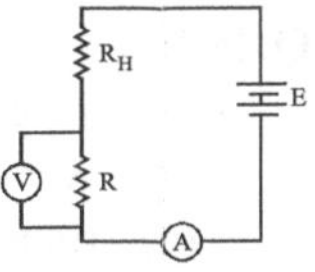

(a) (1.5 ± 0.05)kΩ
(b) (1.2 ± 0.05) k Ω
(c) (1.2 ± 0.08) kΩ
(d) (1.5 ± 0.08) kΩ

Q.21 In an experiment to measure the focal length of a concave mirror, it was found that for an object distance of 0.30 m, the image distance come out to be 0.60 m. Determine the focal length.
(a) (0.2 ± 0.01) m (b) (0.1 ± 0.01) m
(c) (0.2 ± 0.02) m (d) (0.1 ± 0.02) m

Q.22 In an experiment to determine an unknown resistance, a 100 cm long resistance wire is used. The unknown resistance is kept in the left gap and a known resistance is put into the right gap. The scale used to measure length has a least count 1 mm. The null point B is obtained at 40.0 cm from the left gap. Determine the percentage error in the computation of unknown resistance.
(a) 0.24% (b) 0.28% (c) 0.50% (d) 0.42%

Q.23 In an experiment to determine the focal length (f) of a concave mirror by the u - v method, a student places the object pin A on the principal axis at a distance x from the pole P. The student looks at the pin and its inverted image from a distance keeping his/her eye in line with PA. When the student shifts his/her eye towards left, the image appears to the right of the object pin. Then,
(a) $x < f$ (b) $f < x < 2f$
(c) $x = 2f$ (d) $x > 2f$

Q.24 For a convex spherical mirror, the graph of $\left(\frac{1}{v}\right)$ verses $\left(\frac{1}{u}\right)$ is

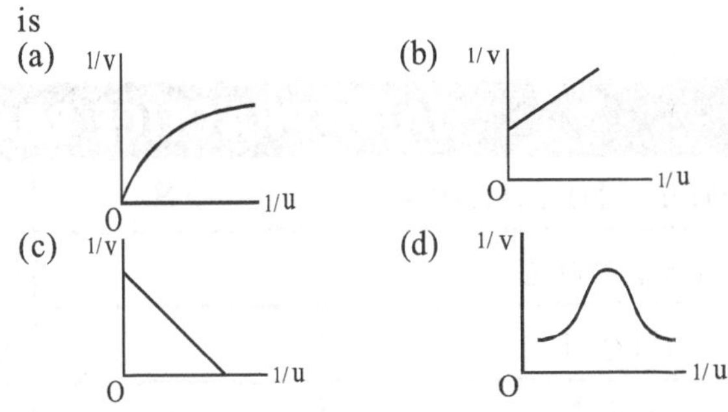

RESPONSE GRID	17. ⓐⓑⓒⓓ	18. ⓐⓑⓒⓓ	19. ⓐⓑⓒⓓ	20. ⓐⓑⓒⓓ	21. ⓐⓑⓒⓓ
	22. ⓐⓑⓒⓓ	23. ⓐⓑⓒⓓ	24. ⓐⓑⓒⓓ		

Space for Rough Work

Q.25 If the wire in the experiment to determine the resistivity of a material using meter bridge is replaced by copper or hollow wire the balance point i.e. null point shifts to

(a) right (b) left
(c) at same point (d) none of these

Q.26 Which device is used to measure the potential difference between two points of a conductor in the laboratory ?

(a) Voltameter (b) Ammeter
(c) Potentiometer (d) Galvanometer

DIRECTIONS (Q.27-Q.28) : Read the passage given below and answer the questions that follows :

Consider a block of conducting material of resistivity 'ρ' shown in the figure. Current 'I' enters at 'A' and leaves from 'D'. We apply superposition principle to find voltage 'ΔV' developed between 'B' and 'C'. The calculation is done in the following steps:

(i) Take current 'I' entering from 'A' and assume it to spread over a hemispherical surface in the block.

(ii) Calculate field E(r) at distance 'r' from A by using Ohm's law $E = \rho j$, where j is the current per unit area at 'r'.

(iii) From the 'r' dependence of E(r), obtain the potential V(r) at r.

(iv) Repeat (i), (ii) and (iii) for current 'I' leaving 'D' and superpose results for 'A' and 'D'.

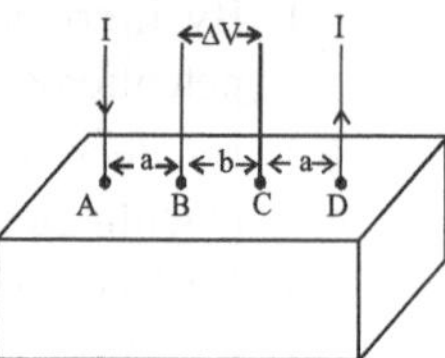

Q.27 ΔV measured between B and C is

(a) $\frac{\rho I}{\pi a} - \frac{\rho I}{\pi(a+b)}$ (b) $\frac{\rho I}{a} - \frac{\rho I}{(a+b)}$

(c) $\frac{\rho I}{2\pi a} - \frac{\rho I}{2\pi(a+b)}$ (d) $\frac{\rho I}{2\pi(a-b)}$

Q.28 For current entering at A, the electric field at a distance 'r' from A is

(a) $\frac{\rho I}{8\pi r^2}$ (b) $\frac{\rho I}{r^2}$ (c) $\frac{\rho I}{2\pi r^2}$ (d) $\frac{\rho I}{4\pi r^2}$

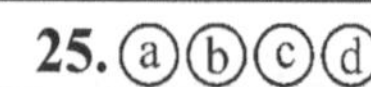

RESPONSE GRID	25. ⓐⓑⓒⓓ	26. ⓐⓑⓒⓓ	27. ⓐⓑⓒⓓ	28. ⓐⓑⓒⓓ

DAILY PRACTICE PROBLEM SHEET 60 - PHYSICS			
Total Questions	28	Total Marks	112
Attempted		Correct	
Incorrect		Net Score	
Cut-off Score	26	Qualifying Score	44
Success Gap = Net Score – Qualifying Score			
Net Score = (Correct × 4) – (Incorrect × 1)			

Space for Rough Work

1. (d) $f = \frac{1}{2\pi\sqrt{LC}}$

$\therefore \left(\frac{C}{L}\right)$ does not represent the dimension of frequency.

2. (d) $[n]$ = Number of particles crossing a unit area in unit time $= [L^{-2}T^{-1}]$
$[n_2] = [n_1]$ = number of particles per unit volume $= [L^{-3}]$
$[x_2] = [x_1]$ = positions

$\therefore \quad D = \frac{[n][x_2 - x_1]}{[n_2 - n_1]} = \frac{[L^{-2}T^{-1}]\times[L]}{[L^{-3}]} = [L^2T^{-1}]$

3. (d) $Y = \frac{X}{3Z^2} = \frac{M^{-1}L^{-2}T^4A^2}{[MT^{-2}A^{-1}]^2} = [M^{-3}L^{-2}T^8A^4]$

4. (a) In given equation, $\frac{\alpha Z}{k\theta}$ should be dimensionless

$\therefore \quad \alpha = \frac{k\theta}{Z} \Rightarrow [\alpha] = \frac{[ML^2T^{-2}K^{-1}\times K]}{[L]} = [MLT^{-2}]$

and $P = \frac{\alpha}{\beta} \Rightarrow [\beta] = \left[\frac{\alpha}{P}\right] = \frac{[MLT^{-2}]}{[ML^{-1}T^{-2}]} = [M^0L^2T^0]$

5. (c) $v = \frac{P}{2l}\left[\frac{F}{m}\right]^{1/2} \Rightarrow v^2 = \frac{P^2}{4l^2}\left[\frac{F}{m}\right] \therefore m \propto \frac{F}{l^2v^2}$

$\Rightarrow \quad [m] = \left[\frac{MLT^{-2}}{L^2T^{-2}}\right] = [ML^{-1}T^0]$

6. (d) By substituting the dimensions of mass $[M]$, length $[L]$ and coefficient of rigidity $[ML^{-1}T^{-2}]$ we get

$T = 2\pi\sqrt{\frac{M}{\eta L}}$ is the right formula for time period of oscillations.

7. (a) Time $\propto c^xG^yh^z \Rightarrow T = kc^xG^yh^z$
Putting the dimensions in the above relation

$\Rightarrow \quad [M^0L^0T^1] = [LT^{-1}]^x[M^{-1}L^3T^{-2}]^y[ML^2T^{-1}]^z$

$\Rightarrow \quad [M^0L^0T^1] = [M^{-y+z}L^{x+3y+2z}T^{-x-2y-z}]$

Comparing the powers of M, L and T
$-y + z = 0$ ....(i)
$x + 3y + 2z = 0$ ....(ii)
$-x - 2y - z = 1$ ....(iii)
On solving equations (i) and (ii) and (iii)

$x = \frac{-5}{2}, y = z = \frac{1}{2}$

Hence, dimension of time are $[G^{1/2}h^{1/2}c^{-5/2}]$.

8. (a) Let radius of gyration $[k] \propto [h]^x[c]^y[G]^z$
By substituting the dimension of

$[k] = [L]\ [h] = [ML^2T^{-1}], [c] = [LT^{-1}]$,

$[G] = [M^{-1}L^3T^{-2}]$

and by comparing the power of both sides we can get $x = 1/2, y = -3/2, z = 1/2$
So dimension of radius of gyration are $[h]^{1/2}[c]^{-3/2}[G]^{1/2}$

9. (b) Because magnitude is absolute.

10. (c) Stefan's law is $E = \sigma(T^4) \Rightarrow \sigma = \frac{E}{T^4}$

where, $E = \frac{\text{Energy}}{\text{Area} \times \text{Time}} = \frac{\text{Watt}}{\text{m}^2}$

$\sigma = \frac{\text{Watt} - \text{m}^{-2}}{\text{K}^4} = \text{Watt} - \text{m}^{-2}\text{K}^{-4}$

11. (d) ct^2 must have dimensions of L
$\Rightarrow$ c must have dimensions of $L/T^2 = LT^{-2}$

12. (b) $6 \times 10^{-5} = 60 \times 10^{-6} = 60$ microns

13. (d) $F = \frac{1}{4\pi\epsilon_0}\frac{q_1q_2}{r^2} \Rightarrow \epsilon_0 = \frac{1}{4\pi}\frac{q_1q_2}{Fr^2} = C^2m^{-2}N^{-1}$

14. (b) According to the defition.
15. (b) Pyrometer is used for measurement of temperature.
16. (d) $x = Ay + B\tan Cz$,
From the dimensional homogenity

$[x] = [Ay] = [B] \Rightarrow \left[\frac{x}{A}\right] = [y] = \left[\frac{B}{A}\right]$

$[Cz] = [M^0L^0T^0]$ = Dimensionless

x and B; C and z^{-1}; y and $\frac{B}{A}$ have the same dimension but x and A have the different dimensions.

17. (a) Let $T \propto S^xr^y\rho^z$
by substituting the dimension of

$[T] = [T], [S] = [MT^{-2}], [r] = [L], [\rho] = [ML^{-3}]$

and by comparing the power of both the sides $x = -1/2, y = 3/2, z = 1/2$

so $T \propto \sqrt{\rho r^3/S} \Rightarrow T = k\sqrt{\frac{\rho r^3}{S}}$

18. (c) $E = h\nu \Rightarrow [ML^2T^{-2}] = [h][T^{-1}] \Rightarrow [h] = [ML^2T^{-1}]$

19. (c) $\vec{P} = \vec{A} + \vec{B}$
$\vec{Q} = \vec{A} - \vec{B}$

$\vec{P}.\vec{Q} = 0 \Rightarrow (\vec{A}+\vec{B}).(\vec{A}-\vec{B}) = 0$

$\Rightarrow A^2 - B^2 = 0 \Rightarrow |\vec{A}| = |\vec{B}|$

$\because \vec{P} \perp \vec{Q}$

20. (a) By substituting dimension of each quantitity in R.H.S. of option (a) we get

$$\left[\frac{mg}{\eta r}\right] = \left[\frac{M \times LT^{-2}}{ML^{-1}T^{-1} \times L}\right] = [LT^{-1}].$$

This option gives the dimension of velocity.

21. (a) By principle of dimensional homogeneity

$$\left[\frac{a}{V^2}\right] = [P]$$

$\therefore \quad [a] = [P][V]^2 = [ML^{-1}T^{-2}] \times [L^6] = [ML^5T^{-2}]$

22. (c) $f = c\, m^x k^y$;

Spring constant k = force/length.

$[M^0L^0T^{-1}] = [M^x (MT^{-2})^y] = [M^{x+y} T^{-2y}]$

$\Rightarrow x + y = 0, -2y = -1 \text{ or } y = \frac{1}{2}$

Therefore, $x = -\frac{1}{2}$

23. (a) Try out the given alternatives.

When x = 1, y = –1, z = 1

$P^x S^y C^z = P^1 S^{-1} C^1 = \frac{PC}{S}$

$$= \frac{[ML^{-1}T^{-2}][LT^{-1}]}{[ML^2T^{-2}/L^2T]} = [M^0L^0T^0]$$

24. (c) Dimensions of angular momentum, $[L] = [ML^2T^{-1}]$

Dimensions of work, $[W] = [ML^2T^{-2}]$

Dimensions of torque, $[\tau] = [ML^2T^{-2}]$

Dimensions of energy, $[E] = [ML^2T^{-2}]$

Dimensions of Young's modulus,

$$[Y] = [ML^{-1}T^{-2}]$$

Dimensions of light year = [L]

Dimension of wavelength = [L]

25. (d), **26.** (b)

$$\sqrt{\frac{hG}{c^5}} = \sqrt{\frac{kgm^2s^{-1} \times m^3kg^{-1}s^{-2}}{m^5/s^5}} = \sqrt{s^2} = s$$

Putting the values of *h*, *G* and *c* in above relation Planck time = 1.3×10^{-43} s.

27. (a) $E = \frac{hc}{\lambda} = \frac{6.6\times10^{-34} \times 3\times10^8}{4\times10^{-7}} = 4.95\times10^{-19}$ J

= 3.0 eV

28. (a) Both statement -1 and statement -2 are correct and statement -1 follows from statement -2

29. (c) Let us write the dimension of various quantities on two sides of the given relation.

L.H.S. = *T* = [*T*],

R.H.S. = $2\pi\sqrt{g/l} = \sqrt{\frac{LT^{-2}}{L}} = [T^{-1}]$

($\therefore$ 2π has no dimension).

As dimension of L.H.S. is not equal to dimension of R.H.S. therefore according to principle of homogeneity the relation.

$T = 2\pi\sqrt{g/l}$.

30. (a) Unit of quantity (*L*/*R*) is Henry/ohm.

As Henry = ohm × sec,

hence unit of *L*/*R* is sec

i.e. [*L*/*R*] = [*T*].

Similarly, unit of product *CR* is farad × ohm

or $\frac{\text{Coulomb}}{\text{Volt}} \times \frac{\text{Volt}}{\text{Amp}}$

or $\frac{\text{Sec} \times \text{Amp}}{\text{Amp}}$ = second

i.e. [*CR*] = [*T*]

therefore, [*L*/*R*] and [*CR*] both have the same dimension.

DAILY PRACTICE PROBLEMS

PHYSICS SOLUTIONS 02

1. (d) Density $\rho = \frac{M}{V} = \frac{M}{\pi r^2 L}$

$\Rightarrow \frac{\Delta\rho}{\rho} = \frac{\Delta M}{M} + 2\frac{\Delta r}{r} + \frac{\Delta L}{L}$

$= \frac{0.003}{0.3} + 2\times\frac{0.005}{0.5} + \frac{0.06}{6}$

$= 0.01 + 0.02 + 0.01 = 0.04$

$\therefore$ Percentage error $= \frac{\Delta\rho}{\rho}\times 100 = 0.04\times 100 = 4\%$

2. (a) In division (or multiplication), the number of significant digits in the answer is equal to the number of significant digits which is the minimum in the given numbers.

3. (d) Percentage error in A

$= \left(2\times 1 + 3\times 3 + 1\times 2 + \frac{1}{2}\times 2\right)\% = 14\%$

4. (a) Percentage error in X = $a\alpha + b\beta + c\gamma$

5. (c) Errors in A and B will be added.

6. (c) Given, L = 2.331 cm
= 2.33 (correct upto two decimal places)
and B = 2.1 cm = 2.10 cm
$\therefore$ L + B = 2.33 + 2.10 = 4.43 cm. = 4.4 cm
Since minimum significant figure is 2.

7. (d) The number of significant figures in all of the given number is 4.

8. (c) $Y = \frac{4MgL}{\pi D^2 l}$ so maximum permissible error in Y

$= \frac{\Delta Y}{Y}\times 100 = \left(\frac{\Delta M}{M} + \frac{\Delta g}{g} + \frac{\Delta L}{L} + \frac{2\Delta D}{D} + \frac{\Delta l}{l}\right)\times 100$

$= \left(\frac{1}{300} + \frac{1}{981} + \frac{1}{2820} + 2\times\frac{1}{41} + \frac{1}{87}\right)\times 100$

$= 0.065\times 100 = 6.5\%$

9. (d) $a = \frac{b^{\alpha}c^{\beta}}{d^{\gamma}e^{\delta}}$

So, maximum error in a is given by

$\left(\frac{\Delta a}{a}\times 100\right)_{max} = \alpha.\frac{\Delta b}{b}\times 100 + \beta.\frac{\Delta c}{c}\times 100$

$+ \gamma.\frac{\Delta d}{d}\times 100 + \delta.\frac{\Delta e}{e}\times 100$

$= (\alpha b_1 + \beta c_1 + \gamma d_1 + \delta e_1)\%$

10. (c) $T = 2\pi\sqrt{\frac{l}{g}}$

$\Rightarrow T^2 = 4\pi^2\frac{l}{g} \Rightarrow g = \frac{4\pi^2 l}{T^2}$

Here % error in $l = \frac{1mm}{100cm}\times 100 = \frac{0.1}{100}\times 100 = 0.1\%$

and % error in $T = \frac{0.1}{2\times 100}\times 100 = 0.05\%$

$\therefore$ % error in g

= % error in l + 2(% error in T)
$= 0.1 + 2\times 0.05 = 0.2\%$

11. (c) Mean time period T = 2.00 sec
& Mean absolute error ΔT = 0.05 sec.
To express maximum estimate of error, the time period should be written as (2.00 ± 0.05) sec

12. (b) $V = \frac{4}{3}\pi r^3$

$\therefore$ % error in volume = 3 × % error in radius
= 3 × 1 = 3%

13. (a) Weight in air = (5.00 ± 0.05)N
Weight in water = $(4.00 \pm 0.05)N$
Loss of weight in water = $(1.00 \pm 0.1)N$

Now relative density = $\frac{\text{weight in air}}{\text{weight loss in water}}$

i.e. R.D = $\frac{5.00 \pm 0.05}{1.00 \pm 0.1}$

Now relative density with max permissible error

$= \frac{5.00}{1.00} \pm \left(\frac{0.05}{5.00} \pm \frac{0.1}{1.00}\right)\times 100 = 5.0 \pm (1+10)\%$

$= 5.0 \pm 11\%$

14. (b) $\therefore \left(\frac{\Delta R}{R}\times 100\right)_{max} = \frac{\Delta V}{V}\times 100 + \frac{\Delta l}{l}\times 100$

$= \frac{5}{100}\times 100 + \frac{0.2}{10}\times 100 = (5+2)\% = 7\%$

15. (c) Volume of cylinder $V = \pi r^2 \ell$
Percentage error in volume =

$\frac{\Delta V}{V}\times 100 = \frac{2\Delta r}{r}\times 100 + \frac{\Delta l}{l}\times 100$

$= \left(2\times\frac{0.01}{2.0}\times 100 + \frac{0.1}{5.0}\times 100\right) = (1+2)\% = 3\%$

$$\frac{\Delta V}{V}\times 100=\frac{2\Delta r}{r}\times 100+\frac{\Delta l}{l}\times 100$$

$$=\left(2\times\frac{0.01}{2.0}\times 100+\frac{0.1}{5.0}\times 100\right)=(1+2)\%=3\%$$

16. (b) $H=I^2Rt$

$$\frac{\Delta H}{H}\times 100=\left(\frac{2\Delta I}{I}+\frac{\Delta R}{R}+\frac{\Delta t}{t}\right)\times 100$$

$=(2\times 3+4+6)\%=16\%$

17. (c) Quantity C has maximum power. So it brings maximum error in *P*.

18. (d) Kinetic energy $E=\frac{1}{2}mv^2$

$$\therefore\ \frac{\Delta E}{E}\times 100=\frac{v'^2-v^2}{v^2}\times 100=[(1.5)^2-1]\times 100$$

$$\therefore\ \frac{\Delta E}{E}\times 100=125\%$$

19. (b) Required random error $=\frac{x}{4}$

20. (b) $\therefore E=\frac{1}{2}mv^2$

$\therefore$ % Error in K.E.

= % error in mass + 2 × % error in velocity

$=2+2\times 3=8\%$

21. (c)

22. (d) Since for 50.14 cm, significant number = 4 and for 0.00025, significant numbers = 2

23. (a) Since percentage increase in length = 2%

Hence, percentage increase in area of square sheet $=2\times 2\%=4\%$

24. (d) Here, $s=(13.8\pm 0.2)$ m

$t=(4.0\pm 0.3)$ s

velocity, $v=\frac{s}{t}=\frac{13.8}{4.0}=3.45\ ms^{-1}=3.5\ ms^{-1}$

(rounding off to two significant figures)

$$\frac{\Delta v}{v}=\pm\left(\frac{\Delta s}{s}+\frac{\Delta t}{t}\right)=\pm\left(\frac{0.2}{13.8}+\frac{0.3}{4.0}\right)=\pm\frac{(0.8+4.14)}{13.8\times 4.0}$$

$$\Rightarrow \frac{\Delta v}{v}=\pm\frac{4.94}{13.8\times 4.0}=\pm 0.0895$$

$\Delta v=\pm 0.0895\times v=\pm 0.0895\times 3.45=\pm 0.3087=\pm 0.31$

(rounding off to two significant fig.)

Hence, $v=(3.5\pm 0.31)\ ms^{-1}$

% age error in velocity $=\frac{\Delta v}{v}\times 100=\pm 0.0895\times 100=\pm 8.95$

$\%=\pm 9\%$

25. (a) Maximum percentage error in measurement of e, as given by Reyleigh's formula.

(Given error is measurement of radius is 0.1 cm)

$\Delta e=0.6\,\Delta R=0.6\times 0.1=0.06$ cm.

Percentage error is $\frac{\Delta e}{e}\times 100=\frac{0.06}{0.6\times 3}\times 100=3.33\%$

26. (b) Speed of sound at the room temperature.

$\ell_1=4.6$ cm, $\ell_2=14.0$ cm.,

$\lambda=2(\ell_2-\ell_1)=2\,(14.0-4.6)=18.8$ cm.

$v=f\lambda=2000\times\frac{18.8}{100}=376\ m/s$

27. (c) End correction obtained in the experiment.

$$e=\frac{\ell_2-3\ell_1}{2}=\frac{14.0-3\times 4.6}{2}=0.1\text{ cm.}$$

28. (d) Since zeros placed to the left of the number are never significant, but zeros placed to right of the number are significant.

29. (b) The last number is most accurate because it has greatest significant figure (3).

30. (a) As the distance of star increases, the parallex angle decreases, and great degree of accuracy is required for its measurement. Keeping in view the practical limitation in measuring the parallex angle, the maximum distance of a star we can measure is limited to 100 light year.

DAILY PRACTICE PROBLEMS

PHYSICS SOLUTIONS 03

1. (a) If t_1 and $2t_2$ are the time taken by particle to cover first and second half distance respectively.

$$t = \frac{x/2}{3} = \frac{x}{6} \quad ...(i)$$

$x_1 = 4.5t_2$ and $x_2 = 7.5t_2$

So, $x_1 + x_2 = \frac{x}{2} \Rightarrow 4.5t_2 + 7.5t_2 = \frac{x}{2}$

$$t_2 = \frac{x}{24} \quad ...(ii)$$

Total time $t = t_1 + 2t_2 = \frac{x}{6} + \frac{x}{12} = \frac{x}{4}$

So, average speed = 4 m/sec.

2. (c) $\frac{dv}{dt} = bt \Rightarrow dv = bt\,dt \Rightarrow v = \frac{bt^2}{2} + K_1$

At $t = 0, v = v_0 \Rightarrow K_1 = v_0$

We get $v = \frac{1}{2}bt^2 + v_0$

Again $\frac{dx}{dt} = \frac{1}{2}bt^2 + v_0 \Rightarrow x = \frac{1}{2}\frac{bt^2}{3} + v_0 t + K_2$

At $t = 0, x = 0 \Rightarrow K_2 = 0$

$\therefore \quad x = \frac{1}{6}bt^3 + v_0 t$

3. (d) $\frac{dv}{dt} = 6 - 3v \Rightarrow \frac{dv}{6-3v} = dt$

Integrating both sides, $\int \frac{dv}{(6-3v)} = \int dt$

$\Rightarrow \frac{\log_e(6-3v)}{-3} = t + K_1$

$\Rightarrow \log_e(6-3v) = -3t + K_2 \quad ....(i)$

At $t = 0, v = 0 \Rightarrow \log_e 6 = K_2$

Substituting the value of K_2 in equation (i)

$\log_e(6-3v) = -3t + \log_e 6$

$\Rightarrow \log_e\left(\frac{6-3v}{6}\right) = -3t \Rightarrow e^{-3t} = \frac{6-3v}{6}$

$\Rightarrow 6 - 3v = 6e^{-3t} \Rightarrow 3v = 6(1 - e^{-3t})$

$\Rightarrow v = 2(1 - e^{-3t})$

$\therefore v_{terminal} = 2$ m/s (when $t = \infty$)

Acceleration $a = \frac{dv}{dt} = \frac{d}{dt}\left[2(1-e^{-3t})\right] = 6e^{-3t}$

Initial acceleration $= 6$ m/s^2.

4. (a) The body starts from rest at $x = 0$ and then again comes to rest at $x = 1$. It means initially acceleration is positive and then negative.

So we can conclude that α can not remain positive for all t in the interval $0 \le t \le 1$ i.e. α must change sign during the motion.

5. (b) The area under acceleration time graph gives change in velocity. As acceleration is zero at the end of 11 sec. i.e. v_{max} = Area of ΔOAB

$$= \frac{1}{2} \times 11 \times 10 = 55 \text{ m/s}$$

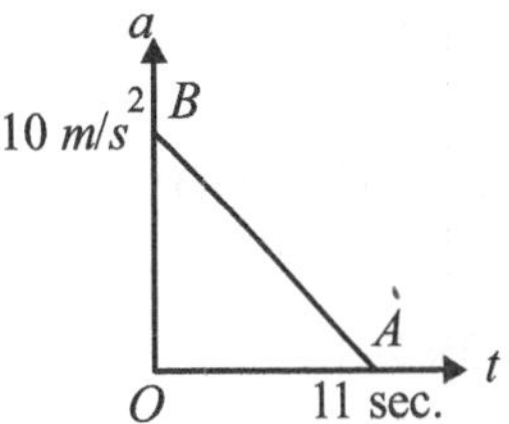

6. (d) Let the car accelerate at rate a for time t_1 then maximum velocity attained, $v = 0 + \alpha t_1 = \alpha t_1$

Now, the car decelerates at a rate β for time $(t - t_1)$ and finally comes to rest. Then,

$0 = v - \beta(t - t_1)$

$\Rightarrow 0 = \alpha t_1 - \beta t + \beta t_1$

$\Rightarrow t_1 = \frac{\beta}{\alpha + \beta}t$

$\Rightarrow v = \frac{\alpha\beta}{\alpha + \beta}t$

7. (c) $S_n = u + \frac{a}{2}(2n - 1)$

$= \frac{a}{2}(2n-1) \quad (\because u = 0)$

$S_{n+1} = \frac{a}{2}[2(n+1) - 1] = \frac{a}{2}(2n+1)$

$\therefore \frac{S_n}{S_{n+1}} = \frac{2n-1}{2n+1}$

8. (b) Distance = Area under v – t graph

$= A_1 + A_2 + A_3 + A_4$

$= \frac{1}{2} \times 1 \times 20 + (20 \times 1) + \frac{1}{2}(20 + 10) \times 1 + (10 \times 1)$

$= 10 + 20 + 15 + 10 = 55$ m

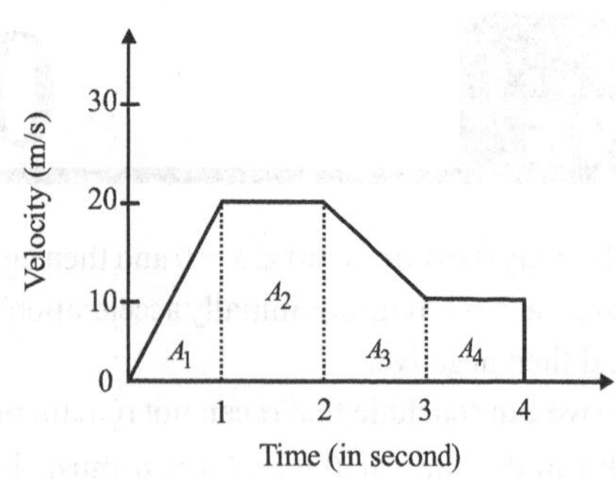

9. (c) As acc. is constant so from $s = ut + \frac{1}{2}at^2$ we have

$$x = \frac{1}{2}at^2 \quad [u = 0] \quad \text{....(i)}$$

Now if it travels a distance y in next t sec.
in 2t sec total distance travelled

$$x + y = \frac{1}{2}a(2t)^2 \quad \text{....(ii)} \quad (t + t = 2t)$$

Dividing eq^n. (ii) by eq^n (i), $\frac{x+y}{x} = 4$ or $y = 3x$

10. (b) $\frac{x(4)}{x(5)} = \frac{\frac{g}{2}(2\times4-1)}{\frac{g}{2}(2\times5-1)} = \frac{7}{9}$ $\quad [\because S_{n^{th}} = u + \frac{a}{2}(2n-1)]$

11. (c) Let body takes T sec to reach maximum height.
Then $v = u - gT$
$v = 0$, at highest point.

$$T = \frac{u}{g} \quad \text{..................(1)}$$

Velocity attained by body in $(T - t)$ sec
$v = u - g(T - t)$

$$= u - gT + gt = u - g\frac{u}{g} + gt$$

or $v = gt$ ....(2)

$\therefore$ Distance travelled in last t sec of its ascent

$$S = (gt)t - \frac{1}{2}gt^2 = \frac{1}{2}gt^2$$

12. (c) $v = \frac{dx}{dt} = \frac{d}{dt}(2t^2 + t + 5) = 4t + 1$ m/s

and $a = \frac{dv}{dt} = \frac{d}{dt}(4t + 1)$; $a = 4$ m/s^2

13. (a) $\frac{dv}{dt} = a_0\left(1 - \frac{t}{T}\right)$

$$\Rightarrow \int_0^v dv = \int_0^t a_0\left(1 - \frac{t}{T}\right)dt \Rightarrow v = a_0\left(t - \frac{t^2}{2T}\right)$$

$$\because \frac{dx}{dt} = v \text{ so, } \int dx = \int v\,dt \Rightarrow x = \int_0^t a_0\left(t - \frac{t^2}{2T}\right)dt$$

$$\Rightarrow x = a_0\left(\frac{t^2}{2} - \frac{t^3}{6T}\right) \quad \because a = 0 \Rightarrow t = T$$

$$\text{Average velocity} = \frac{\text{displacement}}{\text{time}}$$

$$= \frac{a_0\left(\frac{T^2}{2} - \frac{T^3}{6T}\right)}{T} = \frac{a_0T}{3}$$

14. (b) Let u_1, u_2, u_3 and u_4 be velocities at time $t = 0$, t_1, $(t_1 + t_2)$ and $(t_1 + t_2 + t_3)$ respectively and acceleration is a then

$$v_1 = \frac{u_1 + u_2}{2}, v_2 = \frac{u_2 + u_3}{2} \text{ and } v_3 = \frac{u_3 + u_4}{2}$$

Also $u_2 = u_1 + at_1$, $u_3 = u_1 + a(t_1 + t_2)$
and $u_4 = u_1 + a(t_1 + t_2 + t_3)$
By solving, we get

$$\frac{v_1 - v_2}{v_2 - v_3} = \frac{(t_1 + t_2)}{(t_2 + t_3)}$$

15. (c) $v_x = \frac{dx}{dt} = \frac{d}{dt}(3t^2 - 6t) = 6t - 6$. At $t = 1$, $v_x = 0$

$v_y = \frac{dy}{dt} = \frac{d}{dt}(t^2 - 2t) = 2t - 2$. At $t = 1$, $v_y = 0$

Hence $v = \sqrt{v_x^2 + v_y^2} = 0$

16. (c) Initial relative velocity $= v_1 - v_2$.
Final relative velocity $= 0$
From $v^2 = u^2 - 2as$
$\Rightarrow 0 = (v_1 - v_2)^2 - 2 \times a \times s$

$$\Rightarrow s = \frac{(v_1 - v_2)^2}{2a}$$

If the distance between two cars is 's' then collision will take place. To avoid collision $d > s$

$$\therefore d > \frac{(v_1 - v_2)^2}{2a}$$

where d = actual initial distance between two cars.

17. (c) If the body starts from rest and moves with constant acceleration then the ratio of distances in consecutive equal time interval $S_1 : S_2 : S_3 = 1 : 3 : 5$

18. (c) Instantaneous velocity $v = \frac{\Delta x}{\Delta t}$

By using the data from the table

$$v_1 = \frac{0-(-2)}{1} = 2\text{m/s}, \quad v_2 = \frac{6-0}{1} = 6\text{m/s},$$

$$v_3 = \frac{16-6}{1} = 10\text{m/s}$$

So, motion is non-uniform but accelerated.

19. (a) $\frac{1}{2}at^2 = vt \Rightarrow t = \frac{2v}{a}$

20. (b) $x = 4(t-2) + a(t-2)^2$

At $t = 0$, $x = -8 + 4a = 4a - 8$

$$v = \frac{dx}{dt} = 4 + 2a(t-2)$$

At $t = 0$, $v = 4 - 4a = 4(1-a)$

But acceleration, $a = \frac{d^2x}{dt^2} = 2a$

21. (d) $x = ae^{-\alpha t} + be^{\beta t}$

Velocity $v = \frac{dx}{dt} = \frac{d}{dt}(ae^{-\alpha t} + be^{\beta t})$

$= a.e^{-\alpha t}(-\alpha) + be^{\beta t}(\beta) = -a\alpha e^{-\alpha t} + b\beta e^{\beta t}$

Acceleration $= -a\alpha e^{-\alpha t}(-\alpha) + b\beta e^{\beta t}.\beta$

$= a\alpha^2 e^{-\alpha t} + b\beta^2 e^{\beta t}$

Acceleration is positive so velocity goes on increasing with time.

22. (d) (1) $a = 3\sin 4t$

$\Rightarrow \frac{dv}{dt} = 3\sin 4t$

$\Rightarrow \int dv = \int 3\sin 4t\, dt + c$

$\Rightarrow v = \frac{-3}{4}\cos 4t + c$

For initial velocity, $t = 0$

$v_0 = -\frac{3}{4} + C$

At particular value of $C = \frac{3}{4}$, $v_0 = 0$

Therefore, initial velocity may or may not be zero.

(2) Acceleration = 0

$\Rightarrow a = 3\sin 4t = 0 \Rightarrow \sin 4t = 0$

$\Rightarrow 4t = n\pi \quad \Rightarrow t = \frac{n\pi}{4}$

where n = 0, 1, 2, ...

Therefore, the acceleration of the particle becomes zero after each interval of $\frac{\pi}{4}$ second.

(3) As acceleration is sinusoidal function of time, so particle repeats its path periodically. Thus, the particle comes to its initial position after sometime (period of function).

(4) The particle moves in a straight line path as it performs S.H.M.

Since (1) & (3) are correct, hence correct answer is (d).

23. (c) For an inertial frame of reference, its acceleration should be zero. As reference frame attached to the earth i.e. a rotating or revolving frame is accelerating, therefore, it will be non-inertial.

Thus (2) & (4) are correct, so correct answer is (c).

24. (c) Average speed

$$= \frac{\text{Total distance}}{\text{Total time}} = \frac{2x}{\frac{x}{v_1} + \frac{x}{v_2}} = \frac{2v_1v_2}{v_1 + v_2}$$

Average velocity $= \frac{\text{Total displacement}}{\text{Total time}}$

∵ It comes back to its initial position

∴ Total displacement is zero.

Hence, average velocity is zero.

Sol. For Qs. 25-27. $a = \sin \pi t$

$\therefore \int dv = \int 2\sin \pi t\, dt$ or $v = -\frac{2}{\pi}\cos \pi t + C$

At $t = 0$, $v = 0$ $\therefore C = \frac{2}{\pi}$ or $v = \frac{2}{\pi}(1 - \cos \pi t)$

Velocity is always non-negative, hence particle always moves along positive x-direction.

∴ Distance from time $t = 0$ to $t = t$ is

$$S = \int_0^t \frac{2}{\pi}(1 - \cos \pi t)\, dt = \frac{2}{\pi}\left(t - \frac{1}{\pi}\sin \pi t\right)\Big|_0^t = \frac{2}{\pi}t - \frac{2}{\pi^2}\sin \pi t$$

Also displacement from time $t = 0$ to $t = \frac{2t}{\pi} - \frac{2}{\pi^2}\sin \pi t$

Distance from time $t = t$ to $t = 1s = \frac{2}{\pi}$ meters.

25. (a) **26.** (b) **27.** (b)

28. (d) Negative slope of position time graph represents that the body is moving towards the negative direction and if the slope of the graph decrease with time then it represents the decrease in speed i.e. retardation in motion.

29. (c) As per definition, acceleration is the rate of change of velocity,

i.e. $\vec{a} = \frac{d\vec{v}}{dt}$.

If velocity is constant

$\frac{d\vec{v}}{dt} = 0$, $\therefore \vec{a} = 0$

Therefore, if a body has constant velocity it cannot have non zero acceleration.

30. (d) The displacement is the shortest distance between initial and final position. When final position of a body coincides with its initial position, displacement is zero, but the distance travelled is not zero.

DAILY PRACTICE PROBLEMS

PHYSICS SOLUTIONS

04

1. **(c)** If a stone is dropped from height h

then $h=\frac{1}{2}gt^2$ .... (i)

If a stone is thrown upward with velocity u then

$h=-ut_1+\frac{1}{2}gt_1^2$ .... (ii)

If a stone is thrown downward with velocity u then

$h=ut_2+\frac{1}{2}gt_2^2$ .... (iii)

From (i), (ii) and (iii) we get

$-ut_1+\frac{1}{2}gt_1^2=\frac{1}{2}gt^2$ .... (iv)

$ut_2+\frac{1}{2}gt_2^2=\frac{1}{2}gt^2$ .... (v)

Dividing (iv) and (v) we get

$$\therefore \frac{-ut_1}{ut_2}=\frac{\frac{1}{2}g(t^2-t_1^2)}{\frac{1}{2}g(t^2-t_2^2)}$$

or $-\frac{t_1}{t_2}=\frac{t^2-t_1^2}{t^2-t_2^2}$

By solving $t=\sqrt{t_1t_2}$

2. **(c)** Since direction of v is opposite to the direction of g and h so from equation of motion

$$h=-vt+\frac{1}{2}gt^2$$

$$\Rightarrow gt^2-2vt-2h=0$$

$$\Rightarrow t=\frac{2v\pm\sqrt{4v^2+8gh}}{2g}$$

$$\Rightarrow t=\frac{v}{g}\left[1+\sqrt{1+\frac{2gh}{v^2}}\right]$$

3. **(c)** $h=ut+\frac{1}{2}gt^2 \Rightarrow 1=0\times t_1+\frac{1}{2}gt_1^2 \Rightarrow t_1=\sqrt{2/g}$

Velocity after travelling 1m distance

$v^2=u^2+2gh \Rightarrow v^2=(0)^2+2g\times 1 \Rightarrow v=\sqrt{2g}$

For second 1 meter distance

$1=\sqrt{2g}\times t_2+\frac{1}{2}gt_2^2 \Rightarrow gt_2^2+2\sqrt{2g}t_2-2=0$

$$t_2=\frac{-2\sqrt{2g}\pm\sqrt{8g+8g}}{2g}=\frac{-\sqrt{2}\pm 2}{\sqrt{g}}$$

Taking + ve sign $t_2=(2-\sqrt{2})/\sqrt{g}$

$$\therefore \frac{t_1}{t_2}=\frac{\sqrt{2/g}}{(2-\sqrt{2})/\sqrt{g}}=\frac{1}{\sqrt{2}-1}$$ and so on.

4. **(d)** Interval of ball throw = 2 sec.
If we want that minimum three (more than two) ball remain in air then time of flight of first ball must be greater than 4 sec.
T > 4 sec.

$\frac{2u}{g}>4\sec \Rightarrow u>19.6m/s$

for u = 19.6 First ball will just strike the ground(in sky)
Second ball will be at highest point (in sky)
Third ball will be at point of projection or at ground (not in sky)

5. **(a)** The distance covered by the ball during the last t seconds of its upward motion = Distance covered by it in first t seconds of its downward motion

From $h=ut+\frac{1}{2}gt^2$

$h=\frac{1}{2}gt^2$ [As u = 0 for it downward motion]

6. **(d)** In the positive region the velocity decreases linearly (during rise) and in the negative region velocity increases linearly (during fall) and the direction is opposite to each other during rise and fall, hence fall is shown in the negative region.

7. **(a)** For the given condition initial height h = d and velocity of the ball is zero. When the ball moves downward its velocity increases and it will be maximum when the ball hits the ground & just after the collision it becomes half and in opposite direction. As the ball moves upward its velocity again decreases and becomes zero at height d/2. This explanation match with graph (a).

8. **(c)** Acceleration of body along AB is g cos θ

Distance travelled in time t sec $=AB=\frac{1}{2}(g\cos\theta)t^2$

From ΔABC, AB = 2R cosθ; $2R\cos\theta=\frac{1}{2}g\cos\theta t^2$

$t^2=\frac{4R}{g}$ or $t=2\sqrt{\frac{R}{g}}$

9. **(b)** It has lesser initial upward velocity.

10. **(b)** At maximum height velocity v = 0

We know that $v = u + at$, hence

$0 = u - gT \Rightarrow u = gT$

When $v = \frac{u}{2}$, then

$\frac{u}{2} = u - gt \Rightarrow gt = \frac{u}{2} \Rightarrow gt = \frac{gT}{2} \Rightarrow t = \frac{T}{2}$

Hence at $t = \frac{T}{2}$, it acquires velocity $\frac{u}{2}$.

11. (b) $t = \sqrt{\frac{2h}{g}} \Rightarrow \frac{t_1}{t_2} = \sqrt{\frac{h_1}{h_2}}$

12. (c) Speed of the object at reaching the ground $v = \sqrt{2gh}$
If heights are equal then velocity will also be equal.

13. (b) $S_{3^{rd}} = 10 + \frac{10}{2}(2 \times 3 - 1) = 35m$

$S_{2^{nd}} = 10 + \frac{10}{2}(2 \times 2 - 1) = 25 \Rightarrow \frac{S_{3^{rd}}}{S_{2^{nd}}} = \frac{7}{5}$

14. (c) $S_n \propto (2n - 1)$. In equal time interval of 2 seconds
Ratio of distance = 1 : 3 : 5

15. (c) Net acceleration of a body when thrown upward
= acceleration of body – acceleration due to gravity
= a – g

16. (d) The initial velocity of aeroplane is horizontal, then the vertical component of velocity of packet will be zero.

So $t = \sqrt{\frac{2h}{g}}$

17. (b) $\text{Time} = \frac{\text{Total length}}{\text{Relative velocity}} = \frac{50 + 50}{10 + 15} = \frac{100}{25} = 4\,\text{sec}$

18. (d) Relative velocity
= 10 + 5 = 15 m/sec

$\therefore t = \frac{150}{15} = 10$ sec

19. (a) When two particles moves towards each other then
$v_1 + v_2 = 4$ .... (i)
When these particles moves in the same direction then
$v_1 - v_2 = 4$ .... (ii)
By solving $v_1 = 5$ and $v_2 = 1$ m/s

20. (b) $\overrightarrow{v_{ct}} = \overrightarrow{v_c} - \overrightarrow{v_t}$

$\overrightarrow{v_{ct}} = \overrightarrow{v_c} + (-\overrightarrow{v_t})$

Velocity of car w.r.t. train $(\overrightarrow{v_{ct}})$ is towards West – North

21. (a) As the trains are moving in the same direction. So the initial relative speed $(v_1 - v_2)$ and by applying retardation final relative speed becomes zero.

From $v = u - at \Rightarrow 0 = (v_1 - v_2) - at \Rightarrow t = \frac{v_1 - v_2}{a}$

22. (d) Let $\vec{v}_A$ and $\vec{v}_B$ be the respective velocities of the particles at A and B. The relative velocity of particle at A. w.r.t. to that at B is given by

$\vec{v}_A - \vec{v}_B = \vec{v}_A + (-\vec{v}_B)$

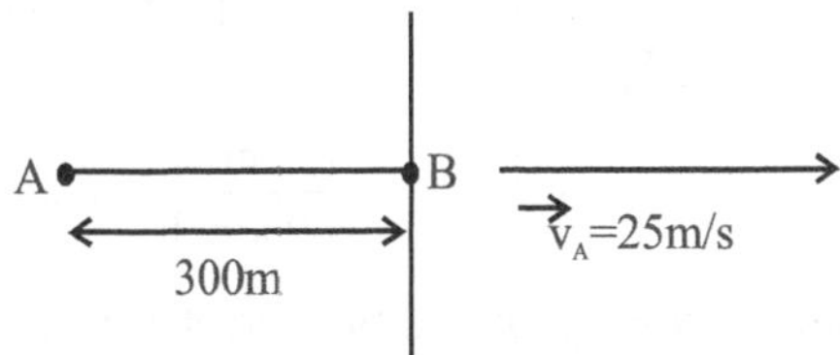

(see figure). From triangle law of velocities if $\overrightarrow{OP}$ and $\overrightarrow{PQ}$ represent $\vec{v}_A$ and $-\vec{v}_B$, then the required relative velocity $\vec{v}_R$ is given by $\overrightarrow{OQ}$.

$|\vec{v}_R| = \sqrt{25^2 + 20^2} = \sqrt{625 + 400} = 32.02$ m/s

If $\angle PQO = \theta$, then $\tan\theta = \frac{25}{20} \Rightarrow \theta = \tan^{-1}\left(\frac{5}{4}\right)$

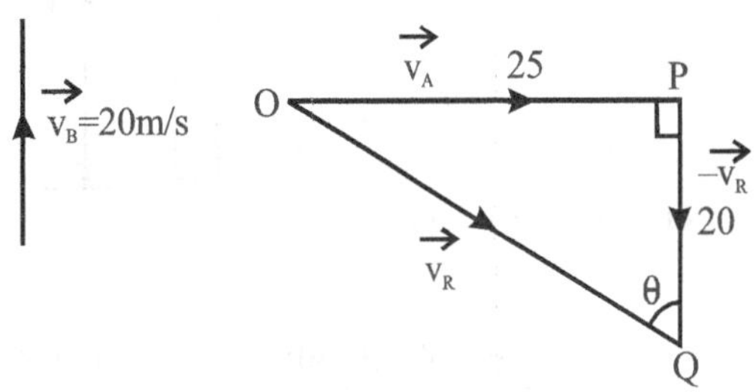

Thus, the particle at A, appears to approach B, in a direction making an angle of $\tan^{-1}(5/4)$ with its direction of motion.

Let us draw a line from A, as AC, such that $\angle BCA$ is equal to θ.

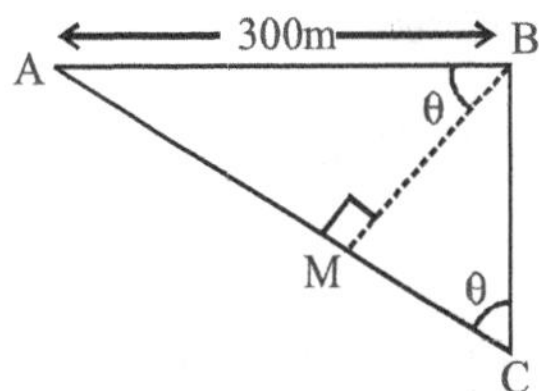

Thus, to B, A appears to move along AC. From B, draw a perpendicular to AC as BM.
BM is the shortest distance between them.

$\therefore \; BM = AB\cos\theta = 300 \times \frac{4}{\sqrt{41}} = 187.41$ m

Also, $AM = AB\sin\theta = 234.26$ m

$\therefore$ time taken to cover a distance

AB = 234.26 m with a velocity of 32.02 m/s

$= \frac{234.26}{32.02} = 7.32$ sec.

23. (d) Since the wind is blowing toward the east, the plane must head west of north as shown in figure. The velocity of the plane relative to the ground $\vec{v}_{pg}$ will be the sum of the velocity of the plane relative to the air $\vec{v}_{pa}$ and the velocity of the air relative to the ground $\vec{v}_{ag}$.

(i) 1. The velocity of the plane relative to the ground is given by equation :

$$\vec{v}_{pg} = \vec{v}_{pa} + \vec{v}_{ag}$$

2. The sine of the angle θ between the velocity of the plane and north equals the ratio of v_{ag} and v_{pa}.

$$\sin\theta = \frac{v_{ag}}{v_{pa}} = \frac{90 \text{ km/h}}{200 \text{ km/h}} = 0.45 \quad \therefore \theta = 26.74$$

(ii) Since v_{ag} and v_{pg} are perpendicular, we can use the Pythagorean theorem to find the magnitude of $\vec{v}_{pg}$.

$$v^2_{pg} = v^2_{ag} + v^2_{pg}$$

$$v_{pg} = \sqrt{v^2_{pa} - v^2_{ag}} = \sqrt{(200 \text{ km/h})^2 - (90 \text{ km/h})^2}$$

$= 179$ km/h.

24. (a) Using, $\vec{S} = \vec{u}T + \frac{1}{2}\vec{a}T^2$

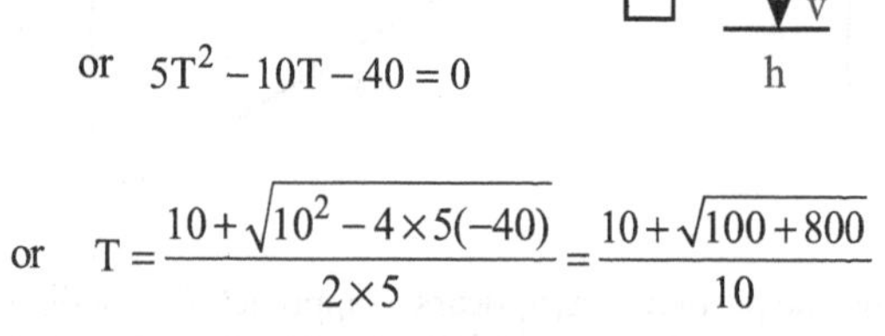

(i) $-40 = 10T - \frac{1}{2}gT^2$

or $-40 = 10T - 5T^2$

or $5T^2 - 10T - 40 = 0$

or $T = \frac{10 + \sqrt{10^2 - 4\times5(-40)}}{2\times5} = \frac{10+\sqrt{100+800}}{10}$

$= \frac{10+30}{10} = 4$ sec.

(ii) $t = \frac{2\times10}{g} = 2$ sec.

(iii) $v = 10 + g\times2 = 30$ m/s

25. (b) $V_{plane} = 100$ m/s

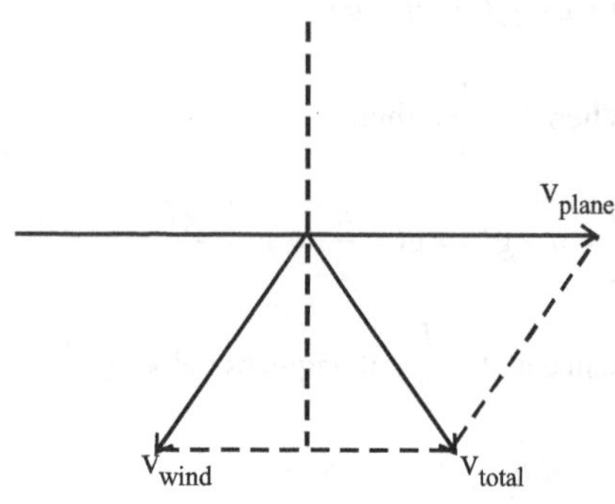

$$V_{total} = \sqrt{(20)^2 + (100)^2 + 2\times20\times100\times\cos135°}$$

$$= \sqrt{400 + 10000 + 2\times20\times100\times\left(-\frac{1}{\sqrt{2}}\right)}$$

$= 87$ m/s

26. (c) $\tan\phi = \frac{V_{wind}}{V_{plane}}$

$= \frac{20}{100}$

$\therefore \phi = \tan^{-1}\left(\frac{20}{100}\right)$

27. (d) $\cos\phi = \frac{V_{plane}}{V_{total}}$

$$\therefore V_{total} = \frac{V_{plane}}{\cos\phi} = \frac{100}{\cos\phi} \text{ m/s}$$

28. (a) If components of velocities of boat relative to river is same normal to river flow (as shown in figure) are same, both boats reach other bank simultaneously.

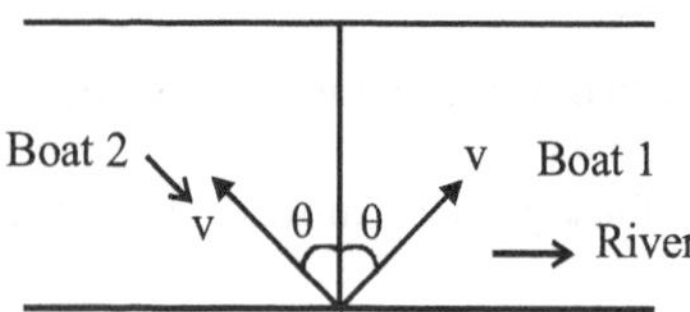

29. (a) Both statement - 1 & statement - 2 are correct and statement - 2 is correct explanation of statement - 2

30. (d) Statement - 1 is true but statement - 2 is false.

DAILY PRACTICE PROBLEMS

PHYSICS SOLUTIONS 05

1. **(d)** $\Delta v = 2v\sin\left(\frac{90°}{2}\right)$

$= 2v\sin 45° = 2v\times\frac{1}{\sqrt{2}} = \sqrt{2}\,v$

$= \sqrt{2}\times r\omega = \sqrt{2}\times 1\times\frac{2\pi}{60} = \frac{\sqrt{2}\,\pi}{30}$ cm/s

2. **(b)** $\Delta v = 2v\sin\left(\frac{\theta}{2}\right) = 2\times 5\times\sin 45° = \frac{10}{\sqrt{2}}$

$\therefore\ a = \frac{\Delta v}{\Delta t} = \frac{10/\sqrt{2}}{10} = \frac{1}{\sqrt{2}}$ m/s^2

3. **(c)** For motion of the particle from (0, 0) to (*a*, 0)

$\vec{F} = -K(0\hat{i} + a\hat{j}) \Rightarrow \vec{F} = -Ka\hat{j}$

Displacement $\vec{r} = (a\hat{i} + 0\hat{j}) - (0\hat{i} + 0\hat{j}) = a\hat{i}$

So work done from (0, 0) to (*a*, 0) is given by

$W = \vec{F}.\vec{r} = -Ka\hat{j}.a\hat{i} = 0$

For motion (*a*, 0) to (*a*, *a*)

$\vec{F} = -K(a\hat{i} + a\hat{j})$ and displacement

$\vec{r} = (a\hat{i} + a\hat{j}) - (a\hat{i} + 0\hat{j}) = a\hat{j}$

So work done from (*a*, 0) to (*a*, *a*)

$W = \vec{F}.\vec{r} = -K(a\hat{i} + a\hat{j}).a\hat{j} = -Ka^2$

So total work done $= -Ka^2$

4. **(d)**

As the metal sphere is in equilibrium under the effect of the three forces therefore

$\vec{T} + \vec{P} + \vec{W} = 0$

From the figure

$T\cos\theta = W$ .... (i)

$T\sin\theta = P$ .... (ii)

From equation (i) and (ii) we get

$P = W\tan\theta$ and $T^2 = P^2 + W^2$

5. **(b)** Time taken to cross the river along shortest possible path is given by

$t = \frac{d}{\sqrt{v^2 - u^2}}$

v = velocity of boat in still water
u = velocity of river water
d = width of river

$\therefore \frac{15}{60} = \frac{1}{\sqrt{5^2 - u^2}}$

$\Rightarrow u = 3$ km/h

6. **(d)** Here d = 320 m = $\frac{320}{1000}$ km

t = 4 min

$v = \frac{5}{3}u$

Putting values in $t = \frac{d}{\sqrt{v^2 - u^2}}$, u = 60 m/min

7. **(c)** $\frac{P}{\sin\theta_1} = \frac{Q}{\sin\theta_2} = \frac{R}{\sin 150°}$

$\Rightarrow \frac{1.93}{\sin\theta_1} = \frac{R}{\sin 150°}$

$\Rightarrow R = \frac{1.93\times\sin 150°}{\sin\theta_1} = \frac{1.93\times 0.5}{0.9659} = 1$

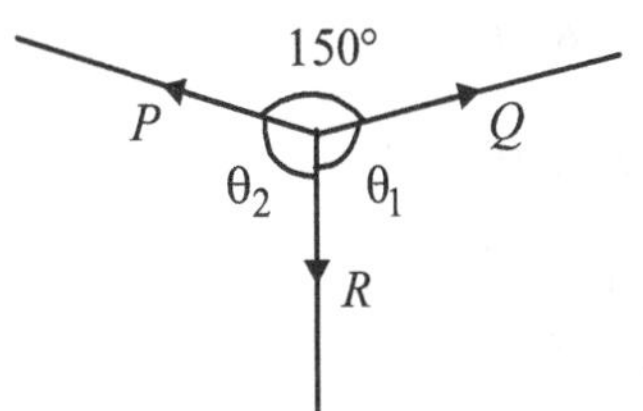

8. **(b)**

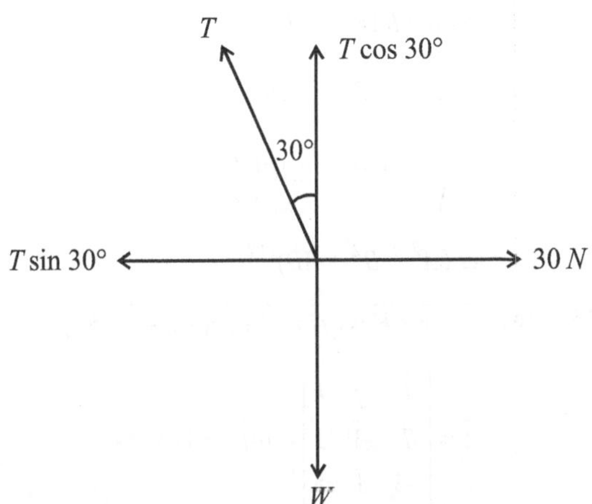

From the figure

$T\sin 30° = 30$ ...(i)

$T\cos 30° = W$ ...(ii)

By solving equation (i) and (ii) we get

$W = 30\sqrt{3}N$ and $T = 60\ N$

9. (c) Relative velocity $= (3\hat{i}+4\hat{j})-(-3\hat{i}-4\hat{j}) = 6\hat{i}+8\hat{j}$

10. (c)

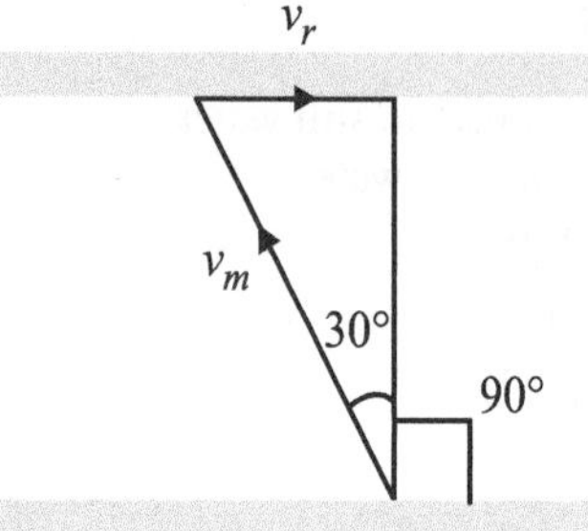

$\sin 30° = \frac{v_r}{v_m} = \frac{1}{2}$

$\Rightarrow v_r = \frac{v_m}{2} = \frac{0.5}{2} = 0.25$ m/s

11. (a) To cross the river in minimum time, the shift is given by $\frac{du}{v}$.

12. (d) Relative velocity = 10 + 5 = 15 m/s.

Time taken by the bird to cross the train $= \frac{120}{15} = 8$ sec

13. (d) $\vec{v} = \vec{\omega}\times\vec{r}$

$= \begin{vmatrix} \hat{i} & \hat{j} & \hat{k} \\ 3 & -4 & 1 \\ 5 & -6 & 6 \end{vmatrix} = -18\hat{i}-13\hat{j}+2\hat{k}$

14. (d) $|\vec{A}\times\vec{B}| = \sqrt{3}(\vec{A}.\vec{B})$

$AB\sin\theta = \sqrt{3}AB\cos\theta$

$\Rightarrow \tan\theta = \sqrt{3}$

$\Rightarrow \theta = 60°$

Now $|\vec{R}| = |\vec{A}+\vec{B}|$

$= \sqrt{A^2+B^2+2AB\cos\theta}$

$= \sqrt{A^2+B^2+2AB\left(\frac{1}{2}\right)}$

$= (A^2+B^2+AB)^{1/2}$

15. (a) $\vec{\tau} = \vec{r}\times\vec{F} = (7\hat{i}+3\hat{j}+\hat{k})(-3\hat{i}+\hat{j}+5\hat{k})$

$\vec{\tau} = \begin{vmatrix} \hat{i} & \hat{j} & \hat{k} \\ 7 & 3 & 1 \\ -3 & 1 & 5 \end{vmatrix} = 14\hat{i}-38\hat{j}+16\hat{k}$

16. (b) $|\vec{A}\times\vec{B}| = \vec{A}.\vec{B}$

$\Rightarrow AB\sin\theta = AB\cos\theta$

$\Rightarrow \tan\theta = 1$

$\therefore \theta = 45°$

17. (a) $\vec{P}.\vec{Q} = 0$

$\Rightarrow a^2 - 2a - 3 = 0$

$\Rightarrow a = 3$

18. (a) $\vec{S} = \vec{r_2} - \vec{r_1}$

$W = \vec{F}.\vec{S}$

$= (4\hat{i}+\hat{j}+3\hat{k}).(11\hat{i}+11\hat{j}+15\hat{k})$

$= (4\times 11 + 1\times 11 + 3\times 15) = 100$ J

19. (c) $\vec{A} = 3\hat{i}-2\hat{j}+\hat{k}$, $\vec{B} = \hat{i}-3\hat{j}+5\hat{k}$, $\vec{C} = 2\hat{i}+\hat{j}-4\hat{k}$

$|\vec{A}| = \sqrt{3^2+(-2)^2+1^2} = \sqrt{9+4+1} = \sqrt{14}$

$|\vec{B}| = \sqrt{1^2+(-3)^2+5^2} = \sqrt{1+9+25} = \sqrt{35}$

$|\vec{C}| = \sqrt{2^2+1^2+(-4)^2} = \sqrt{4+1+16} = \sqrt{21}$

As $B = \sqrt{A^2+C^2}$ therefore ABC will be right angled triangle.

20. (b) $\cos\theta = \frac{\vec{F_1}.\vec{F_2}}{|F_1||F_2|}$

$= \frac{(5\hat{i}+10\hat{j}-20\hat{k}).(10\hat{i}-5\hat{j}-15\hat{k})}{\sqrt{25+100+400}.\sqrt{100+25+225}}$

$= \frac{50-50+300}{\sqrt{525}.\sqrt{350}}$

$\Rightarrow \cos\theta = \frac{1}{\sqrt{2}}$

$\therefore \theta = 45°$

21. (a) $\vec{r} = \vec{a}+\vec{b}+\vec{c} = 4\hat{i}-\hat{j}-3\hat{i}+2\hat{j}-\hat{k} = \hat{i}+\hat{j}-\hat{k}$

$\hat{r} = \frac{\vec{r}}{|r|} = \frac{\hat{i}+\hat{j}-\hat{k}}{\sqrt{1^2+1^2+(-1)^2}} = \frac{\hat{i}+\hat{j}-\hat{k}}{\sqrt{3}}$

22. (a)

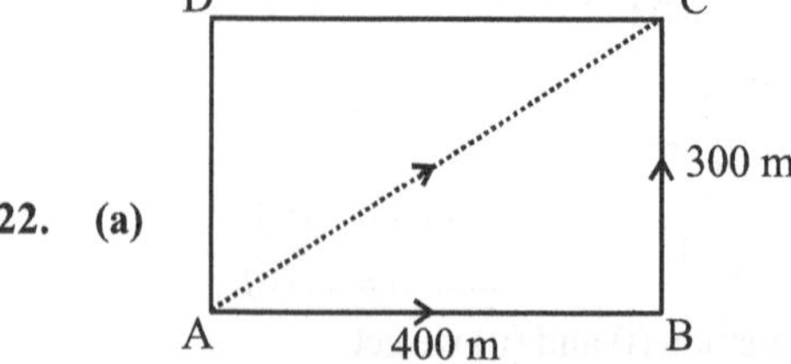

Displacement $\overrightarrow{AC} = \overrightarrow{AB} + \overrightarrow{BC}$

AC = $\sqrt{(AB)^2+(BC)^2} = \sqrt{(400)^2+(300)^2} = 500\,m$

Distance = AB+BC =400+300=700 m

23. **(a)** $\vec{A} = 3\hat{i}-2\hat{j}+\hat{k}, \vec{B} = \hat{i}-3\hat{j}+5\hat{k}, \vec{C} = 2\hat{i}-\hat{j}+4\hat{k}$

$|\vec{A}| = \sqrt{3^2+(-2)^2+1^2} = \sqrt{9+4+1} = \sqrt{14}$

$|\vec{B}| = \sqrt{1^2+(-3)^2+5^2} = \sqrt{1+9+25} = \sqrt{35}$

$|\vec{C}| = \sqrt{2^2+1^2+(-4)^2} = \sqrt{4+1+16} = \sqrt{21}$

As $B = \sqrt{A^2+C^2}$ therefore ABC will be right angled triangle.

24. **(a)** $\vec{A}\times\vec{B} = 0 \therefore \sin\theta = 0 \therefore \theta = 0°$

Two vectors will be parallel to each other.

25. (a), 26 (b), 27. (b)

$$\vec{A}\times\vec{B} = \begin{vmatrix} \hat{i} & \hat{j} & \hat{k} \\ 2 & 1 & 1 \\ 1 & 1 & 1 \end{vmatrix} = \hat{i}(1-1)-\hat{j}(2-1)+\hat{k}(2-1)$$

$= -\hat{j}+\hat{k}$

Unit vector perpendicular to $\vec{A}$ and $\vec{B}$ is $\left(\frac{-\hat{j}+\hat{k}}{\sqrt{2}}\right)$.

Any vector whose magnitude is k (constant) times $(2\hat{i}+\hat{j}+\hat{k})$ is parallel to $\vec{A}$ so, unit vector $\frac{2\hat{i}+\hat{j}+\hat{k}}{\sqrt{6}}$ is parallel to $\vec{A}$.

28. **(b)** $|\vec{A}+\vec{B}| = |\vec{A}-\vec{B}|$

$\Rightarrow \quad A^2+B^2+2AB\cos\theta = A^2+B^2+2AB\cos\theta$

Hence $\cos\theta = 0$ which gives $\theta = 90°$

Also vector addition is commutative.

Hence $\vec{A}+\vec{B} = \vec{B}+\vec{A}$

29. **(a)** Let $\vec{P}$ and $\vec{Q}$ are two vectors in opposite direction, then their sum $\vec{P}+(-\vec{Q}) = \vec{P}-\vec{Q}$

If $\vec{P} = \vec{Q}$ then sum equal to zero.

30. **(d)** The resultant of two vectors of unequal magnitude given by $R = \sqrt{A^2+B^2+2AB\cos\theta}$ cannot be zero for any value of θ.

DAILY PRACTICE PROBLEMS

PHYSICS SOLUTIONS

06

1. (b) Given, that $y = \sqrt{3}\,x - (1/2)x^2$ ....(1)
The above equation is similar to equation of trajectory of the projectiles
$$y = \tan\theta\, x - 1/2\,\frac{g}{u^2\cos^2\theta}\,x^2 \quad ....(2)$$
Comparing (1) & (2) we get
$\tan\theta = \sqrt{3} \Rightarrow \theta = 60°$
and $1/2 = (1/2)\dfrac{g}{u^2\cos^2\theta}$
$\Rightarrow u^2\cos^2\theta = g \Rightarrow u^2\cos^2 60° = 10$
$\Rightarrow u^2(1/4) = 10 \Rightarrow u^2 = 40 \Rightarrow u = 2\sqrt{10}$ m/s

2. (a) For angle of elevation of 60°, we have maximum height
$$h_1 = \frac{u^2\sin^2 60°}{2g} = \frac{3u^2}{8g}$$
For angle of elevation of 30°, we have maximum height
$$h_2 = \frac{u^2\sin^2 30°}{2g} = \frac{u^2}{8g};\ \frac{h_1}{h_2} = \frac{3}{1}$$

3. (b) Total time of flight = 2 × time taken to reach max. height
$\Rightarrow t_2 = 2t_1 \Rightarrow t_2/t_1 = 2/1$

4. (d) $v_x = dx/dt = 2ct,\ v_y = dy/dt = 2bt$
$\therefore v = \sqrt{v_x^2 + v_y^2} = 2t\sqrt{c^2 + b^2}$

5. (c) $v_y = dy/dt = 8 - 10\,t = 8$, when $t = 0$ (at the time of projection.)
$v_x = dx/dt = 6,\ v = \sqrt{v_x^2 + v_y^2} = \sqrt{8^2 + 6^2} = 10$ m/s

6. (a) Horizontal component of velocity
$v_x = u_x = u\cos\theta = 30 \times \cos 30° = 15\sqrt{3}$ m/s
Vertical component of the velocity
$v_y = u\sin\theta - gt = 30\sin 30° - 10 \times 1 = 5$ m/s
$v^2 = v_x{}^2 + v_y{}^2 = 700 \Rightarrow u = 10\sqrt{7}$ m/s

7. (a) Let u_x and u_y be the components of the velocity of the particle along the x- and y-directions. Then
$u_x = dx/dt = u_0$ and $u_y = dy/dt = \omega a\cos\omega t$
Integration : $x = u_0 t$ and $y = a\sin\omega t$
Eliminating t : $y = a\sin(\omega x/u_0)$
This is the equation of the trajectory
At $t = 3\pi/2\omega$, we have,
$x = u_0\,3\pi/2\omega$ and $y = a\sin 3\pi/2 = -a$
$\therefore$ The distance of the particle from the origin is
$$\sqrt{x^2 + y^2} = \sqrt{\left[\left(\frac{3\pi u_0}{2\omega}\right)^2 + a^2\right]}$$

8. (b) $T = \dfrac{2u\sin\theta}{g} \Rightarrow 2 = \dfrac{2u\sin\theta}{g} \Rightarrow u\sin\theta = g$
$$H = \frac{u^2\sin^2\theta}{2g} = \frac{g^2}{2g} = \frac{g}{2} = 5\text{ m}$$

9. (b) Let u_1 and u_2 be the initial velocities respectively. If h_1 and h_2 are the heights attained by them, then
$$h_1 = \frac{u_1^2}{2g} \text{ and } h_2 = \frac{u_2^2\sin^2\theta}{2g} \quad ...(1)$$
The times of ascent of balls are equal,
we have $t = u_1/g = u_2\sin\theta/g$
$\therefore u_1 = u_2\sin\theta$ ... (2)
From eq. (1) $\dfrac{h_1}{h_2} = \dfrac{u_1^2}{u_2^2\sin^2\theta}$ ... (3)
From (2) & (3), $\dfrac{h_1}{h_2} = \dfrac{1}{1}$

10. (a) $h_1 = \dfrac{u^2\sin^2\theta}{2g}$ and $h_2 = \dfrac{u^2\sin^2(90-\theta)}{2g}$
$\therefore h_1 + h_2 = u^2/2g\,(\sin^2\theta + \cos^2\theta)$
$= u^2/2g = \dfrac{98^2}{2\times 10} = 490$
$h_1 - h_2 = 50,\ \therefore h_1 = 270$ m and $h_2 = 220$ m

11. (c) Greatest height attained
$$h = \frac{u^2\sin^2\theta}{2g} \quad ... (1)$$
Horizontal range
$$R = \frac{u^2\sin 2\theta}{g} = \frac{2u^2\sin\theta\cos\theta}{g} \quad ... (2)$$
Given that R=2h
$$\Rightarrow \frac{2u^2\sin\theta\cos\theta}{g} = \frac{2u^2\sin^2\theta}{2g}$$
$\Rightarrow \tan\theta = 2$ ... (3)
Hence $\sin\theta = 2/\sqrt{5}$, $\cos\theta = 1/\sqrt{5}$,
$\therefore$ From (2) $R = 4u^2/5g$

12. (d) $R = \dfrac{u^2\sin 2\theta}{g} = (19.6)^2\sin 90°/10 = 39.2$ m
Man must run (67.4m – 39.2 m) = 28.2 m in the time taken by the ball to come to ground. Time taken by the ball.

$$t=\frac{2u\sin\theta}{g}=\frac{2\times 19.6\sin 45^\circ}{9.8}=\frac{4}{\sqrt{2}}=2.82 \text{ sec}$$

$$\text{Velocity of man}=\frac{28.2\text{m}}{2.82 \text{ sec}}=10 \text{ m/s}$$

13. (a) Referring to (fig.) let P be a point on the trajectory whose co-ordinates are (4, 4). As the ball strikes the ground at a distance 14 metre from the wall, the range is 4 + 14 = 18 metre. The equation of trajectory is

$$y=x\tan\theta-(1/2)g\frac{x^2}{u^2\cos^2\theta}$$

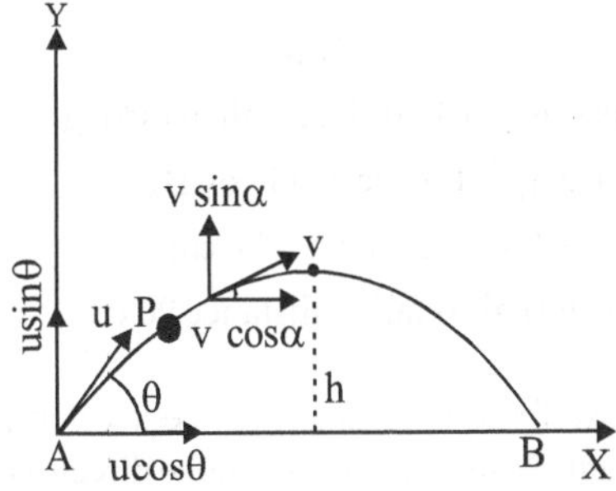

$$\text{or } y=x\tan\theta\left[1-\frac{gx}{2u^2\cos^2\theta.\tan\theta}\right]$$

$$\text{or } y=x\tan\theta\left[1-\frac{2u^2}{g}\frac{x}{\sin\theta\cos\theta}\right]$$

$$=x\tan\theta\left[1-\frac{x}{R}\right] \quad ...(1)$$

Here x = 4, y = 4 and R = 18

$$\therefore 4=4\tan\theta\left[1-\frac{4}{18}\right]=4\tan\theta\left(\frac{7}{9}\right)$$

or $\tan\theta=9/7$, $\sin\theta=9/\sqrt{130}$ and

$\cos\theta=7/\sqrt{130}$

Again $R=(2/g)\,u^2\sin\theta\cos\theta$

$=(2/9.8)\times u^2\times(9/\sqrt{130})\times(7/\sqrt{130})$

$$u^2=\frac{18\times 9.8\times\sqrt{130}\times\sqrt{130}}{2\times 9\times 7}=\frac{98\times 13}{7}=182,$$

$u=\sqrt{182}$ metre per second.

14. (b) The situation is shown in fig.

(a) Let the ball collide after t sec

From fig. OC = OB cos 37° = 500 cos 37°

$=500\times 0.8=400$ cm ...(1)

Horizontal velocity = 700 × cos 37°

$\therefore$ OC = 700 × cos 37° × t

$=700\times 0.8\times t=560t$ ... (2)

From eqs. (1) and (2) 560 t = 400

or t = (5/7) sec.

Now $h=(1/2)\,g\,t^2=(1/2)\times 1000\times(5/7)^2=255.1$ cm

(b) Let at the time of impact, v_x and v_y be the horizontal and vertical velocities respectively, then

$v_x=700\times\cos 37^\circ=700\times 0.8=560$ cm/s

and $v_y=-700\times\sin 37^\circ+1000\times(5/7)$

$=-700\times 0.6+(5000/7)=-420+714.3$

$=+294.3$ cm/sec (downward)

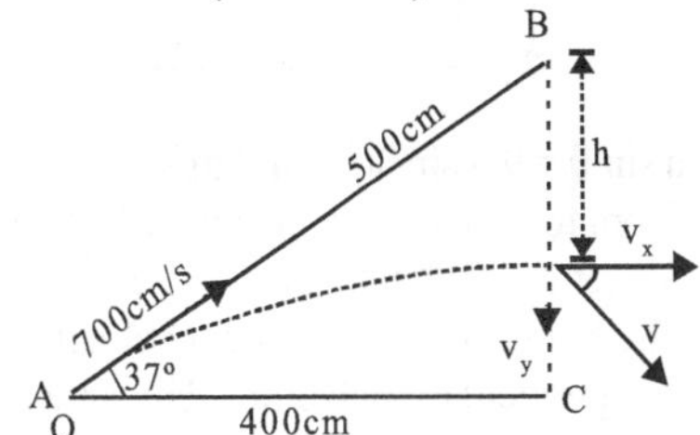

Velocity of the ball at the time of collision

$$v=\sqrt{(v_x^2+v_y^2)}$$

$$\therefore v=\sqrt{[(560)^2+(294.3)^2]}=632.6 \text{ cm/sec}$$

$$\text{Again } \tan\theta=\frac{v_y}{v_x}=\frac{294.3}{560}$$

$$\text{or } \theta=\tan^{-1}\left(\frac{294.3}{560}\right)=27^\circ 43'$$

15. (d) Initial velocity is constant let the ball touches the ground at an angle θ and velocity 3u

Hence $3u\cos\theta=u$ or $\cos\theta=1/3$ or $\sin\theta=\sqrt{8}/3$

The vertical component of velocity at the ground

$$=3u\sin\theta=\frac{3\sqrt{8}}{3}=\sqrt{8}\,u$$

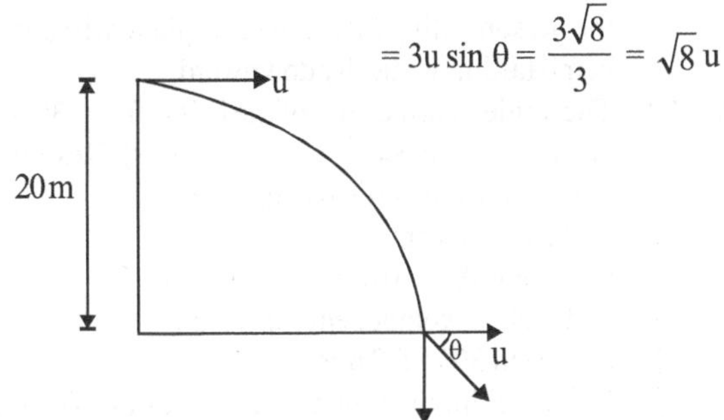

For a freely falling body it covers 20 m to acquire velocity $\sqrt{8}\,u$

$\therefore (\sqrt{8}\,u)^2-0=2\times 9.8\times 20$ or u = 7 m/s

16. (b) The horizontal range of the projectile on the ground

$$R=u\sqrt{\frac{2h}{g}}\Rightarrow R=\sqrt{2}\sqrt{\frac{2\times 10}{10}}=\sqrt{2}.\sqrt{2}=2\text{m}$$

17. (a) $R=ut\Rightarrow t=R/u=12/8$

Now $h=(1/2)\,gt^2=(1/2)\times 9.8\times(12/8)^2=11$ m

18. (b) The situation is shown in the adjoining figure.

The time taken by the body is equal to the time taken by the freely falling body from the height 29.4 m. Initial velocity of body

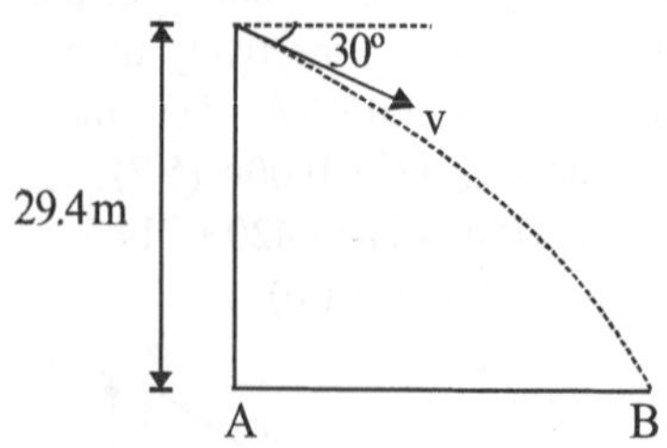

$u \sin\theta = 9.8 \sin 30^\circ = 4.9$ m/s

From the relation, $h = u \sin\theta\, t + (1/2)\, gt^2$,

we get $29.4 = 4.9\,t + (1/2) \times 9.8\, t^2 \Rightarrow t = 2$ sec

19. **(b)** The horizontal and vertical velocities of the bomb are independent to each other. The time taken by the bomb to hit the target can be calculated by its vertical motion. Let this time be t. Putting h = 490 m and g = 9.8 m/s^2 in the formula $h = 1/2\, gt^2$,

we have $490 = (1/2) \times 9.8 \times t^2$,

$$\therefore t = \sqrt{\frac{2 \times 490}{9.8}} = 10 \text{ sec}$$

The bomb will hit the target after 10 sec of its dropping. The horizontal velocity of the bomb is 60 km/hr which is constant. Hence the horizontal distance travelled by the bomb in 10 sec (horizontal velocity × time)

= 60 km/hr × 10 sec

= 60 km/hr × 10/(60 × 60) hr = 1/6 km

Hence the distance of aeroplane from the enemy post is 1/6 km = 1000/6 m = 500/3 meter.

The trajectory of the bomb as seen by an observer on the ground is parabola. Since the horizontal velocity of the bomb is the same as that of the aeroplane, the falling bomb will always remain below the aeroplane. Hence the person sitting inside the aeroplane will observe the bomb falling vertically downward.

20. **(a)** The angle of projection of the ball is θ_0 (= 30°) and the velocity of projection is u (= 10 m/s). Resolving u in horizontal and vertical components,

we have horizontal component,

$u_x = u \cos\theta_0 = 10 \cos 30^\circ = 8.65$ m/s

and vertical component (upward),

$u_y = u \sin 30^\circ = 5.0$ m/s

If the ball hit the ground after t sec of projection, then the horizontal range is $R = u_x \times t = 8.65\,t$ meter

$$\therefore t = \frac{R}{8.65} = \frac{17.3\text{m}}{8.65\text{m/s}} = 2.0 \text{ s}$$

If h be the height of the tower,

then $h = u'_y t + (1/2)\, g t^2$,

where u_y' is the vertical component (downward) of the velocity of the ball.

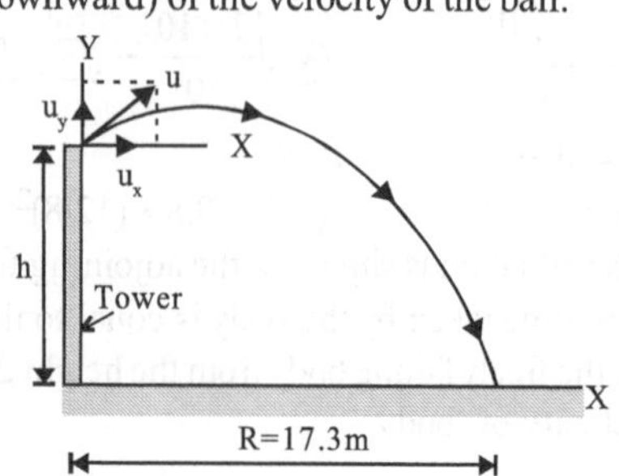

Here $u_y = -u_y' = -5.0$ m/s and t = 2.0 s

$\therefore h = (-5.0) \times 2.0 + 1/2 \times 10 \times (2.0)^2$

$= -10 + 20 = 10$ meter

21. **(b)** Let the ball B hits the ball A after t sec

The X-component of velocity of A is

$v_0 \cos 37^\circ = 700 \cos 37^\circ$

The X-compoment of position of B is 300 cos 37°

The collision will take place when the X-coordinate of A is the same as that of B.

As the collision takes place at a time t, hence

$700 \cos 37^\circ \times t = 300 \cos 37^\circ$

or t = (300/700) = (3/7) sec

In this time the ball B has fallen through a distance

$y = -1/2\ gt^2$ (Free fall of body B)

$= -1/2 \times 980 \times (3/7)^2 = -90$ cm

Hence the ball B falls a distance 90 cm

22. **(b)**

(1) Because force is constant hence acceleration will be constant. When force is in oblique direction with initial velocity, the resultant path is parabolic path.

(2) Total time of flight $= T = \dfrac{2u \sin\theta}{g}$,

Maximum height attained $H = \dfrac{u^2 \sin^2\theta}{2g}$

Now $\dfrac{H}{T} = \dfrac{u \sin\theta}{4}$

(3) Initially the height of the monkey = MB = y = x tan θ

Let the monkey drop to along line MA and the bullet reach along the parabolic path OA. If both reach at A simultaneously, the monkey is hit by the bullet.

$$AB = x \tan\theta - \frac{gx^2}{2u^2 \cos^2\theta},$$

$\therefore$ MA = MB – AB

$$MA = x \tan\theta - x \tan\theta + \frac{gx^2}{2u^2 \cos^2\theta}$$

$$= \frac{gx^2}{2u^2 \cos^2\theta} \qquad \text{....(i)}$$

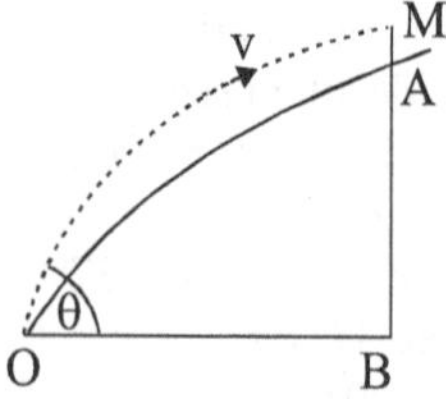

Time taken by the bullet to reach point A,

$$t = \frac{x}{u \cos\theta} \qquad \text{....(ii)}$$

Hence from (1), MA = (1/2) gt^2
The monkey drops through distance (1/2) gt^2 in the same time. So the monkey is hit by the bullet.

(4) The range $R = \frac{u^2 \sin 2\theta}{g}$

$\therefore$ Maximum range $R_{max} = d = \frac{u^2}{g}$ ....(iii)

Height $H = \frac{u^2 \sin^2 \theta}{2g}$

$\therefore$ Maximum height

$H_{max} = \frac{u^2}{2g}$ ....(iv)

From (iii) & (iv), $H_{max} = d/2$

23. (a) Range of projectile, $R = \frac{u^2 \sin 2\theta}{g}$

The range is same for two angle θ_1 and θ_2 provided $\theta_2 = 90° - \theta_1$

At an angle θ_1, range $R_1 = \frac{u^2 \sin 2\theta_1}{g}$

At an angle of projection θ_2,

Range $R_2 = \frac{u^2 \sin 2\theta_2}{g}$

$= \frac{u^2 \sin 2(90° - \theta_1)}{g} = \frac{u^2 \sin 2\theta_1}{g}$

$\Rightarrow R_1 = R_2$

$\therefore$ other angle $= 90° - \theta_1 = 90° - 15° = 75°$

24. (a) $t_1 = \frac{2u \sin\theta}{g}$

$t_2 = \frac{2u \sin(90° - \theta)}{g} = \frac{2u \cos\theta}{g}$

$t_1 t_2 = \frac{2}{g} \frac{u^2 \sin 2\theta}{g} = \frac{2}{g} . R$

where R is the range, Hence $t_1 t_2 \propto R$

25. (c) If the ball hits the nth step, the horizontal and vertical distances traversed are nb and nh respectively.

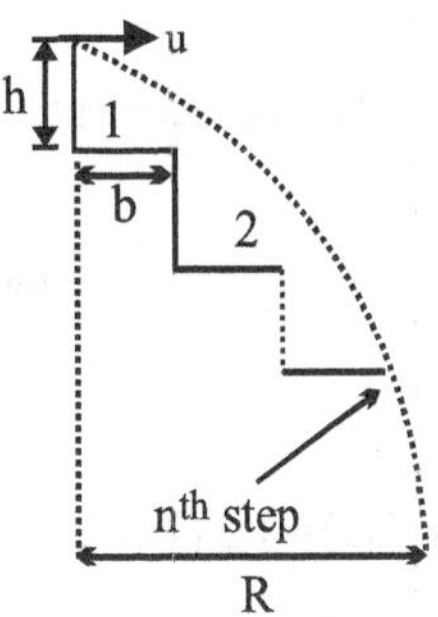

Let t be the time taken by the ball for these horizontal and vertical displacement. Then velocity along horizontal direction remains constant = u ; initial vertical velocity is zero

$\therefore$ nb = ut ....(1)

$nh = 0 + (1/2) gt^2$ ....(2)

From (1) & (2) we get

$nh = (1/2) g (nb/u)^2$

$\Rightarrow n = \frac{2hu^2}{gb^2}$ (eleminating t)

26. (a) $y = (1/2) gt^2$ (downward)

$\Rightarrow 1000 = (1/2) \times 10 \times t^2 \Rightarrow t = 14.15$ sec

$x = ut = \left(\frac{144 \times 10^3}{60 \times 60}\right) \times 14.15 = 571.43$ m

27. (b) Horizontal component of velocity
$= 720 \times 5/8 = 200$ m/s

Let t be the time taken for a freely falling body from 490. Then $y = (1/2) gt^2$

$\Rightarrow 490 = (1/2) \times 9.8 \times t^2 \Rightarrow t = 10$ second

Now horizontal distance = Velocity $\times$ time $= 200 \times 10 = 2000$m

Hence the bomb missed the target by 2000 m

28. (a) Since $W = \Delta K$ implies that the final speed will be same.

29. (a) The time of flight depends only on the vertical component of velocity which remains unchanged in collision with a vertical wall.

30. (a) In statement-2, if speed of both projectiles are same, horizontal ranges will be same. Hence statement-2 is correct explanation of statement-1.

DAILY PRACTICE PROBLEMS

PHYSICS SOLUTIONS

07

(1) **(b)** We have angular displacement

$$= \frac{\text{linear displacement}}{\text{radius of path}}$$

$$\Rightarrow \Delta\theta = \frac{\Delta S}{r}$$

Here, $\Delta S = n(2\pi r) = 1.5(2\pi \times 2 \times 10^{-2}) = 6\pi \times 10^{-2}$

$$\therefore \Delta\theta = \frac{6\pi \times 10^{-2}}{2 \times 10^{-2}} = 3\pi \text{ radian}$$

(2) **(a)** We have $\vec{\omega}_{av} = \frac{\text{Total angular displacement}}{\text{Total time}}$

For first one third part of circle,

angular displacement, $\theta_1 = \frac{S_1}{r} = \frac{2\pi r/3}{r}$

For second one third part of circle,

$$\theta_2 = \frac{2\pi r/3}{r} = \frac{2\pi}{3} \text{ rad}$$

Total angular displacement,
$\theta = \theta_1 + \theta_2 = 4\pi/3$ rad
Total time $= 2 + 1 = 3$ sec

$$\therefore \vec{\omega}_{av} = \frac{4\pi/3}{3} \text{ rad/s} = \frac{4\pi}{6} = \frac{2\pi}{3} \text{ rad/s}$$

(3) **(c)** Angular speed of hour hand,

$$\omega_1 = \frac{\Delta\theta}{\Delta t} = \frac{2\pi}{12 \times 60} \text{ rad/sec}$$

Angular speed of minute hand,

$$\omega_2 = \frac{2\pi}{60} \text{ rad/sec} \Rightarrow \frac{\omega_2}{\omega_1} = \frac{12}{1}$$

(4) **(d)** We have $\theta = \omega_0 t + \frac{1}{2}\alpha t^2 \Rightarrow \frac{d\theta}{dt} = \omega_0 + \alpha t$

This is angular velocity at time t.
Now angular velocity at t = 2 sec will be

$$\omega = \left(\frac{d\theta}{dt}\right)_{t=2\text{ sec}} = \omega_0 + 2\alpha = 1 + 2 \times 1.5 = 4 \text{ rad/sec}$$

(5) **(d)** The distance covered in completing the circle is
$2\pi r = 2\pi \times 10$ cm

The linear speed is $v = \frac{2\pi r}{t} = \frac{2\pi \times 10}{4} = 5\pi$ cm/s

The linear acceleration is,

$$a = \frac{v^2}{r} = \frac{(5\pi)^2}{10} = 2.5\,\pi^2 \text{ cm/s}^2$$

This acceleration is directed towards the centre of the circle

(6) **(d)** We know that

$$\text{Time period} = \frac{\text{Circumference}}{\text{Critical speed}} = \frac{2\pi r}{\sqrt{gr}}$$

$$= \frac{2 \times 22 \times 4}{7 \times \sqrt{10 \times 4}} = 4 \text{ sec}$$

(7) **(b)** Velocity $= \frac{\text{Circumference}}{\text{Time of revolution}} = \frac{2\pi r}{60}$

$$= \frac{2\pi \times 1}{60} = \frac{\pi}{30} \text{ cm/s}$$

Change in velocity $\Delta v = \sqrt{\left(\frac{\pi}{30}\right)^2 + \left(\frac{\pi}{30}\right)^2}$

$$= \frac{\pi}{30}\sqrt{2} \text{ cm/s}$$

(8) **(a)** Let the radius of the orbit be r and the number of revolutions per second be n. Then the velocity of electron is given by $v = 2\pi nr$,

$$\therefore \text{Acceleration } a = \frac{v^2}{r} = \frac{4\pi^2 r^2 n^2}{r} = 4\pi^2 r n^2$$

Substituting the given values, we have
$a = 4 \times (3.14)^2 \times (5.3 \times 10^{-11})(6.6 \times 10^{15})^2$
$= 9.1 \times 10^{22}$ m/s^2 towards the nucleus.

The centripetal force is
$F_C = ma = (9.1 \times 10^{-31})(9.1 \times 10^{22})$
$= 8.3 \times 10^{-8}$ N towards the nucleus.

(9) **(b)** Given that radius of horizontal loop
r = 1 km = 1000 m

Speed v = 900 km/h $= \frac{9000 \times 5}{18} = 250$ m/s

Centripetal acceleration $a_c = \frac{v^2}{r} = \frac{250 \times 250}{1000} = 62.5$ m/s^2

$$\therefore \frac{\text{Centripetal acceleration}}{\text{Gravitational acceleration}} = \frac{a_c}{g} = \frac{62.5}{9.8} = 6.38 : 1$$

(10) **(b)** We know that, $\tan\theta = \frac{v^2}{rg} = \frac{\left(18 \times \frac{5}{18}\right)^2}{100 \times 10}$

$$= \frac{1}{40} \Rightarrow \theta = \tan^{-1}\frac{1}{40}$$

(11) **(a)** The angular velocity is $\omega = \frac{v}{r}$

Hence, v = 10 m/s
r = 20 cm = 0.2 m $\therefore \omega = 50$ rad/s

(12) **(b)** Given that $\omega = 1.5t - 3t^2 + 2$

$$\alpha = \frac{d\omega}{dt} = 1.5 - 6t$$

When, $\alpha = 0$

$\Rightarrow 1.5 - 6t = 0$

$$\Rightarrow t = \frac{1.5}{6} = 0.25 \text{ sec}$$

(13) **(c)** Given $v = 1.5\,t^2 + 2t$

Linear acceleration $a = dv/dt = 3t + 2$

This is the linear acceleration at time t

Now angular acceleration at time t

$$\alpha = \frac{a}{r} \Rightarrow \alpha = \frac{3t + 2}{2 \times 10^{-2}}$$

Angular acceleration at $t = 2$ sec

$$(\alpha)_{\text{at } t = 2\text{sec}} = \frac{3 \times 2 + 2}{2 \times 10^{-2}} = \frac{8}{2} \times 10^2$$

$= 4 \times 10^2 = 400 \text{ rad/sec}^2$

(14) **(a)** Angular displacement after 4 sec is

$$\theta = \omega_0 t + \frac{1}{2}\alpha t^2 = \frac{1}{2}\alpha t^2 = \frac{1}{2} \times 4 \times 4^2 = 32 \text{ rad}$$

Angular velocity after 4 sec

$\omega = \omega_0 + \alpha t = 0 + 4 \times 4 = 16$ rad/sec

(15) **(a)** Given $\alpha = 3t - t^2$

$$\Rightarrow \frac{d\omega}{dt} = 3t - t^2 \Rightarrow d\omega = (3t - t^2)dt$$

$$\Rightarrow \omega = \frac{3t^2}{2} - \frac{t^3}{3} + c$$

At $t = 0, \omega = 0$

$$\therefore c = 0, \quad \omega = \frac{3t^2}{2} - \frac{t^3}{3}$$

Angular velocity at $t = 2$ sec, $(\omega)_{t = 2 \text{ sec}}$

$$= \frac{3}{2}(4) - \frac{8}{3} = \frac{10}{3} \text{ rad/sec}$$

Since there is no angular acceleration after 2 sec

$\therefore$ The angular velocity after 6 sec remains the same.

(16) **(d)** $\hat{i}x + \hat{y}j$, $x = r\cos\theta$,

$y = r\sin\theta$ where $\theta = \omega t$

$r = \hat{i}\,(r\cos\omega t) + \hat{j}\,(r\sin\omega t)$

$v = dr/dt = -\hat{i}\,(\omega r\sin\omega t) - \hat{j}\,(\omega r\cos\omega t)$

$a = d^2r/dt^2 = -\omega^2 r$

(17) **(c)** Let R be the normal reaction exerted by the road on the car. At the highest point, we have

$$\frac{mv^2}{(r+h)} = mg - R, \text{ R should not be negative.}$$

Therefore $v^2 \le (r + a)g = (8.9 + 1.1) \times 10$

or $v^2 \le 10 \times 10 v \le 10$ m/sec

$\therefore v_{max} - 10$ m/sec

(18) **(a)** Let $W = Mg$ be the weight of the car. Friction force $= 0.4$ W

$$\text{Centripetal force} = \frac{Mv^2}{r} = \frac{Wv^2}{gr}$$

$$0.4\,W = \frac{Wv^2}{gr}$$

$\Rightarrow v^2 = 0.4 \times g \times r = 0.4 \times 9.8 \times 30 = 117.6$

$\Rightarrow v = 10.84$ m/sec

(19) **(c)** Let v be the speed of earth's rotation.

We know that $W = mg$

$$\text{Hence } \frac{3}{5}W = mg - \frac{mv^2}{r}$$

$$\text{or } \frac{3}{5}mg = mg - \frac{mv^2}{r}$$

$$\therefore \frac{2}{3}mg = \frac{mv^2}{r} \text{ or } v^2 = \frac{2gr}{5}$$

$$\text{Now } v^2 = \frac{2 \times 9.8 \times (6400 \times 10^3)}{5}$$

Solving, we get $v = 5 \times 10^9$ m/sec,

$$\omega = \sqrt{\left(\frac{2g}{5r}\right)} = 7.8 \times 10^4 \text{ radian/sec.}$$

(20) **(a)** Let T be the tension produced in the stretched string. The centripetal force required for the mass m to move in a circle is provided by the tension T. The stretched length of the spring is r (radius of the circle). Now,

Elongation produced in the spring $= (r - \ell_0)$

Tension produced in the spring,

$T = k(r - \ell_0)$ .... (1)

Where k is the force constant

Linear velocity of the motion $v = 2\pi r n$

$$\therefore \text{Centripetal force} = \frac{mv^2}{r} = \frac{m(2\pi rn)^2}{r}$$

$= 4\pi^2 r n^2 m$ .... (2)

Equating equation. (1) and (2), we get

$k(r - \ell_0) = 4\pi^2 r n^2 m$ $(\therefore T = mv^2/r)$

$\Rightarrow kr - k\ell\ell_0 = 4\pi^2 r n^2 m$

$r(k - 4\pi^2 n^2 m) = k\ell_0$

$$\Rightarrow r = \frac{k\ell_0}{(k - 4\pi^2 n^2 m)} \quad ....(3)$$

Substituting the value of r in eqn. (1) we have

$$T = k\left[\frac{k\ell_0}{(k - 4\pi^2 n^2 m)} - \ell_0\right] \text{ or } T = \frac{4\pi^2 n^2 m\ell_0 k}{(k - 4\pi^2 n^2 m)}$$

(21) **(c)** Two types of acceleration are experienced by the car

(i) Radial acceleration due to circular path,

$$a_r = \frac{v^2}{r} = \frac{(30)^2}{500} = 1.8 \text{ m/s}^2$$

(ii) A tangential acceleration due to increase of tangential speed given by

$a_t = \Delta v/\Delta t = 2\ m/s^2$

Radial and tangential acceleration are perpendicular to each other.

Net acceleration of car

$$a = \sqrt{a_r^2 + a_t^2} = \sqrt{(1.8)^2 + (2)^2} = 2.7\ m/s^2$$

(22) **(a)** For A :

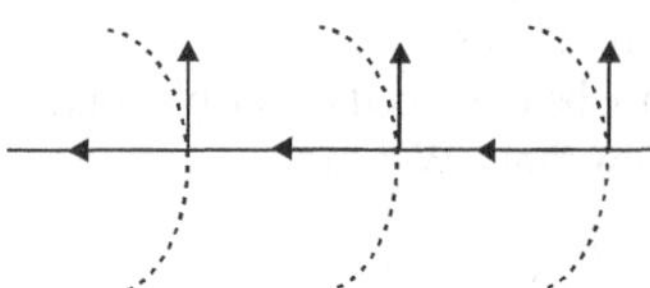

Required centripetal force $= \dfrac{mv_A^2}{3\ell}$

(net force towards centre $= T_1$)

This will provide required centripetal force

particle at A, $\therefore T_1 = \dfrac{mv_A^2}{3\ell}$

For B : Required centripetal force $= \dfrac{m(v_B^2)}{2\ell}$

Remember ω i.e. angular velocity, of all the particles is same

$$\therefore \omega = \frac{v_A}{3\ell} = \frac{v_B}{2\ell} = \frac{v_C}{\ell}$$

When a system of particles rotates about an axis, the angular velocity of all the particles will be same, but their linear velocity will be different, because of different distances from axis of rotation i.e. $v = r\omega$.

Thus for B, centripetal force $= \dfrac{2mv_A^2}{9\ell}$

Net force towards the centre $T_2 - T_1 = \dfrac{2mv_A^2}{9\ell}$

$$\Rightarrow T_2 = \frac{2mv_A^2}{9\ell} + T_1 = \frac{5mv_A^2}{9\ell} \text{ (Putting value of } T_1)$$

For C :

Centripetal force. $\dfrac{mv_C^2}{3\ell} = \dfrac{mv_A^2}{9\ell}$

Net force towards centre $= T_3 - T_2$

$$\therefore T_3 - T_2 = \frac{mv_A^2}{9\ell} \Rightarrow T_3 = \frac{mv_A^2}{9\ell} + T_2$$

$$T_3 = \frac{6mv_A^2}{9\ell} \text{ (on putting value of } T_2)$$

(23) **(b)** $N\cos\theta = \dfrac{mv^2}{r}$ and $N\sin\theta = mg$

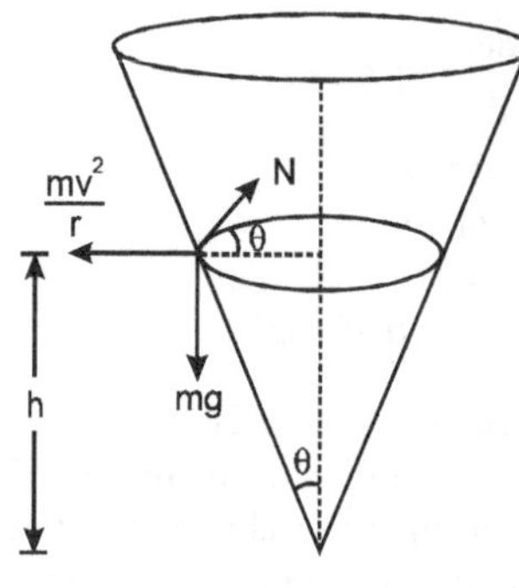

$$\Rightarrow \frac{N\sin\theta}{N\cos\theta} = \frac{mg}{mv^2/r} \Rightarrow \tan\theta = \frac{rg}{v^2}$$

But $\tan\theta = \dfrac{r}{h}$ $\therefore$ $\dfrac{r}{h} = \dfrac{rg}{v^2}$

$$\Rightarrow v = \sqrt{hg} = \sqrt{9.8 \times 9.8 \times 10^{-2}} = 0.98\ m/s$$

(24) **(d)** (1) Centripetal force is not a real force. It is only the requirement for circular motion. It is not a new kind of force. Any of the forces found in nature such as gravitational force, electric friction force, tension in string, reaction force may act as centripetal force.

(3) Work done by centripetal force is always zero.

(25) **(a)** We know, $a = \dfrac{v^2}{r}$

Hence $v = 10\ m/s, r = 5\ m \therefore a = \dfrac{(10)^2}{5} = 20\ m/s^2$

(26) **(a)** Given that the mass of the particle, m = 2 kg

Radius of circle = 3 m

Angular velocity = 60 rev/minute

$$= \frac{60 \times 2\pi}{60} \text{ rad/sec} = 2\pi \text{ rad/sec}$$

Because the angle described during 1 revolution is 2π radian

The linear velocity $v = r\omega = 2\pi \times 3\ m/s = 6\pi\ m/s$

The centripetal acceleration $= \dfrac{v^2}{r} = \dfrac{(6\pi)^2}{3}\ m/s^2$

$= 118.4\ m/s^2$

(27) **(a)** $F = \dfrac{mv^2}{r} = mr\omega^2$

Here m = 0.10 kg, r = 0.5 m

and $\omega = \dfrac{2\pi n}{t} = \dfrac{2 \times 3.14 \times 10}{31.4} = 2$ rad/s

$F = 0.10 \times 0.5 \times (2)^2 = 0.2$

(28) **(a)** In non-uniform circular motion acceleration vector makes some angle with radius hence it is not perpendicular to velocity vector.

(29) **(c)** If speed is increasing there is a tangential acceleration. Net acceleration is not pointing towards centre.

(30) **(b)** Both statements are true but statement-2 is not correct explanation for statement-1.

DAILY PRACTICE PROBLEMS

PHYSICS SOLUTIONS 08

(1) (a).

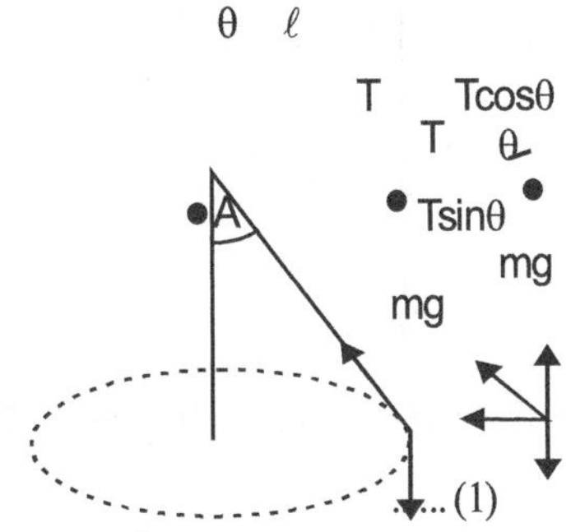

Form figure

$T\cos\theta = mg$ ...(1)

$T\sin\theta = \frac{mv^2}{r} = \frac{mv^2}{\ell\sin\theta}$ ......(2)

Form eq. (1), $T = \frac{mg}{\cos\theta}$

When the string is horizontal, θ must be 90° i.e., $\cos 90^\circ = 0$

$\therefore T = \frac{mg}{0} = \infty$

Thus the tension must be infinite which is impossible, so the string can not be in horizontal plane.

The maximum angle θ is given by the breaking tension of the string in the equation $T\cos\theta = m.g$

Here T (Maximum) = 8 N and m = 0.4 kg

$\therefore\ 8\cos\theta = 0.4 \times g = 0.4 \times 10 = 4$

$\cos\theta = (4/8) = \frac{1}{2}, \theta = 60^\circ$

The angle with horizontal = 90° – 60° = 30°

From equation (2), $8\sin 60^\circ = \frac{0.4 \times v^2}{4\sin 60^\circ}$

$v^2 = \frac{32\sin^2 60^\circ}{0.4} = 80\sin^2 60^\circ$

$\Rightarrow v = \sqrt{80}\ \sin 60^\circ = 7.7$ m/sec

(2) (a). Let m be the mass of the ball. When the ball comes down to B, its potential energy mgh which is converted into kinetic energy. Let v_B, be the velocity of the ball at B.

Then, $mgh = \frac{1}{2} m v_B^2$

The ball now rises to a point D, where its potential energy is mg(h – 2r). If v_D be the velocity of the ball at D, then,

$mg(h - 2r) = \frac{1}{2} m v_D^2$ ......(2)

Now to complete the circular path, it is necessary that the centrifugal force acting upward at point D, should be equal or greater than the force mg acting downward at point D should be equal or greater than the force mg acting downward. Therefore

$\frac{m v_D^2}{r} \geq mg$ or $v_D^2 \geq rg$

From equation (2) $v_D^2 = 2g(h - 2r)$,

$\therefore 2g(h - 2r) \geq rg \Rightarrow h \geq \frac{5}{2} r$

(3) (d). See fig, Here v = 360 km/hr = 100 m/sec

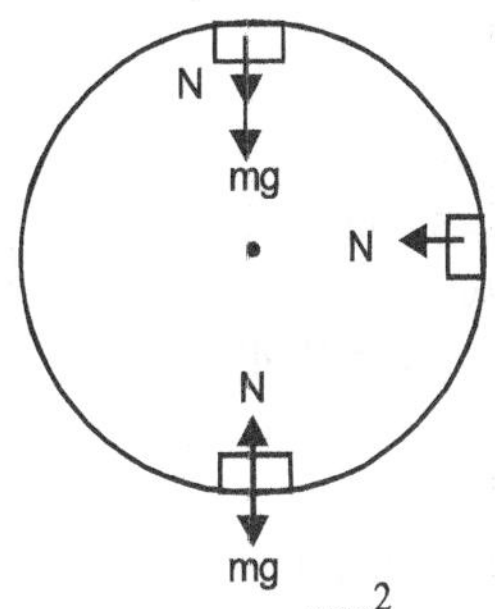

At lower point, $N - mg = \frac{mv^2}{R}$,

$N = \text{weight of the flyer} = mg + \frac{mv^2}{R}$

$N = 70 \times 10 + \frac{70 \times (10000)}{500} = 2100$ N

At upper point, $N + mg = \frac{mv^2}{R}$,

$N = \frac{mv^2}{R} - mg = 1400 - 700 = 700$ N

At middle point, $N = \frac{mv^2}{R} = 1400$ N

(4) (a). Given that $U(r) = 10r^3$

So the force F acting on the particle is given by,

$F = -\frac{\partial U}{\partial r} = -\frac{\partial}{\partial r}(10r^3) = -10 \times 3r^2 = -30r^2$

For circular motion of the particle,

$F = \frac{mv^2}{r} = 30r^2$

Substituting the given values, we have,

$\frac{3 \times v^2}{10} = 30 \times (10)^2$ or $v = 100$ m/s

The total energy in circular motion

$E = K.E. + P.E. = \frac{1}{2}mv^2 + U(r)$

$= \frac{1}{2} \times 3 \times (100)^2 + 10 + (10)^3 = 2.5 \times 10^4$ joule

Angular momentum
$= mvr = 3 \times 100 \times 10 = 3000$ kg–m²/sec

Also time period $T = \frac{2\pi r}{v} = \frac{2 \times \pi \times 10}{100} = \frac{\pi}{5}$ sec

(5) **(a).** Let T be the tension, θ the angle made by the string with the vertical through the point of suspension.

The time period $t = 2\pi\sqrt{\frac{h}{g}} = \frac{1}{\text{frequency}} = \pi/2$

Therefore $\omega = \sqrt{\frac{g}{h}} = 4 \Rightarrow \frac{h}{g} = \frac{1}{16}$

$\cos\theta = \frac{h}{\ell} = \frac{g}{16}$

$= 0.6125 \Rightarrow \theta = 52° 14'$

Linear velocity
$= (\ell \sin\theta)\omega = 1 \times \sin 52° 14' \times 4$
$= 3.16$ m/s

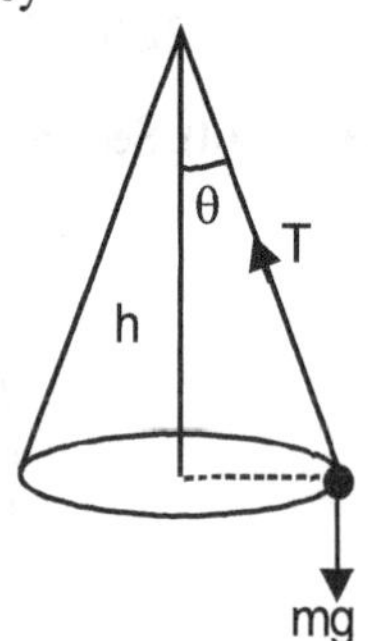

(6) **(d).** Centripetal acceleration, $a_c = \frac{v^2}{r} = k^2 rt^2$

$\therefore$ Variable velocity $v = \sqrt{k^2r^2t^2} = krt$

The force causing the velocity to varies

$F = m\frac{dv}{dt} = mkr$

The power delivered by the force is,
$P = Fv = mkr \times krt = mk^2r^2t$

(7) **(a).** We know centripetal acceleration

$a_c = \frac{(\text{tangential velocity})^2}{\text{radius}} = \frac{(200)^2}{100} = 400$ m/sec²

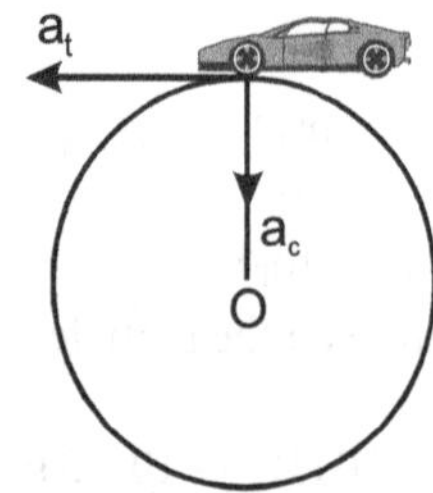

Tangential acceleration
$a_t = 100$ m/sec² (given)

$\therefore a_{net} = \sqrt{a_c^2 + a_t^2 + 2a_c a_t \cos 90°} = \sqrt{a_c^2 + a_t^2}$

$= \sqrt{(400)^2 + (100)^2} = 100\sqrt{17}$ m/s²

(8) **(b).** Suppose v be the velocity of particle at the lowest position B.
According to conservation of energy
(K.E. + P.E.) at A = (K.E. + P.E.) at B

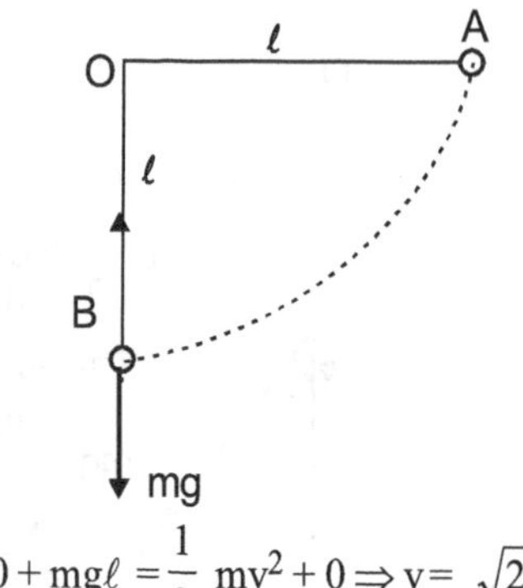

$\Rightarrow 0 + mg\ell = \frac{1}{2}mv^2 + 0 \Rightarrow v = \sqrt{2g\ell}$

(9) **(a).** Maximum tension $T = \frac{mv^2}{r} + mg$

$\therefore \frac{mv^2}{r} = T - mg$

or $\frac{mv^2}{r} = 163.6 - 4 \times 9.8 \Rightarrow v = 6$ m/s

(10) **(c).** The situation is shown in fig. Let v be the velocity of the bob at the lowest position. In this position the P.E. of bob is converted into K.E. hence -

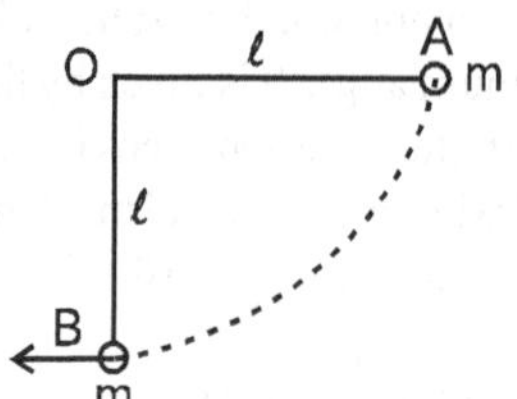

$mg\ell = \frac{1}{2}mv^2 \Rightarrow v^2 = 2g\ell$ ....(1)

If T be the tension in the string,

then $T - mg = \frac{mv^2}{\ell}$ ....(2)

From (1) & (2), T = 3 mg

(11) **(b).** The velocity of the swimmer w.r.t. water $\vec{v}_{SR} = 4.0$ km/h in the direction perpendicular to the river. The velocity of river w.r.t. the ground is $\vec{v}_{RG} = 3.0$ km/h along the length of river.

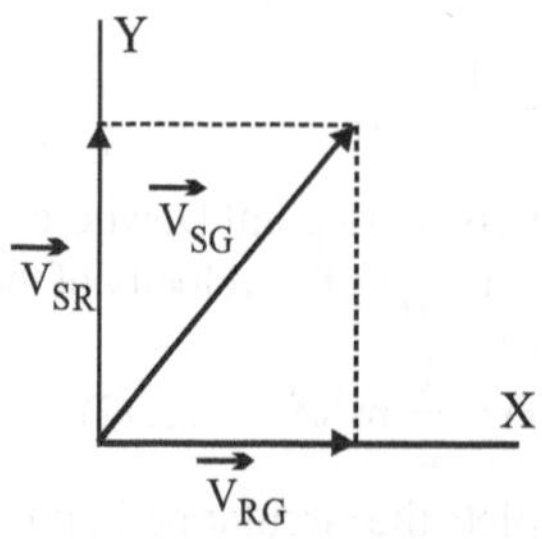

The velocity of the swimmer w.r.t. the ground is $\vec{v}_{SG}$ where

$$\vec{V}_{SG} = \vec{V}_{SR} + \vec{V}_{RG}$$

$$V_{SG} = \sqrt{V_{SR}^2 + V_{RG}^2} = \sqrt{4^2 + 3^2}$$

$$= \sqrt{16+9} = \sqrt{25} = 5 \text{ km/hr}$$

(12) **(b).** The minimum speed at highest point of a vertical circle is given by $v_c = \sqrt{rg} = \sqrt{20 \times 9.8} = 14$ m/s

(13) **(a).** The speed at highest point must be

$$v > \sqrt{gr}, v = r\omega = r\frac{2\pi}{T}$$

$$\therefore r\frac{2\pi}{T} > \sqrt{rg}$$

$$T < \frac{2\pi r}{\sqrt{rg}} < 2\pi\sqrt{\frac{r}{g}} < 2\pi\sqrt{\frac{0.5}{9.8}} < 1.4 \text{ sec}$$

Maximum period of revolution = 1.4 sec

(14) **(a).** Let the particles leaves the sphere at height h,

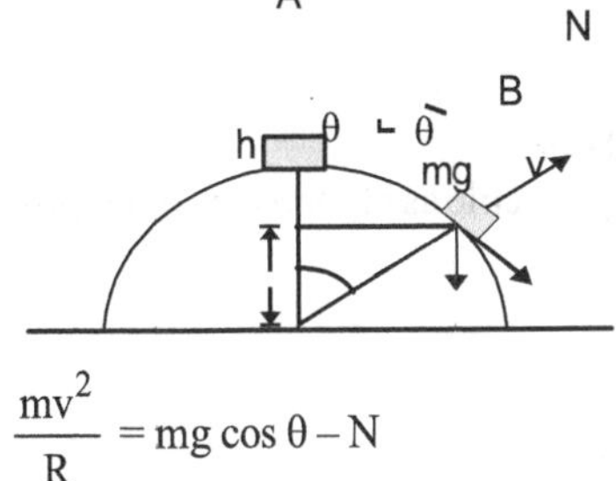

$$\frac{mv^2}{R} = mg\cos\theta - N$$

When the particle leaves the sphere i.e. N = 0

$$\frac{mv^2}{R} = mg\cos\theta$$

$$\Rightarrow v^2 = gR\cos\theta \qquad ....(1)$$

According to law of conservation of energy

(K.E. + P.E.) at A = (K.E. + P.E.) at B

$$\Rightarrow 0 + mgR = \frac{1}{2}mv^2 + mgh$$

$$\Rightarrow v^2 = 2g(R - h) \qquad ....(2)$$

From (1) & (2), $h = \frac{2}{3}R$

Also $\cos\theta = \frac{2}{3}$

(15) **(a).** Let the body will have the circular path at height h above the bottom of circle from figure

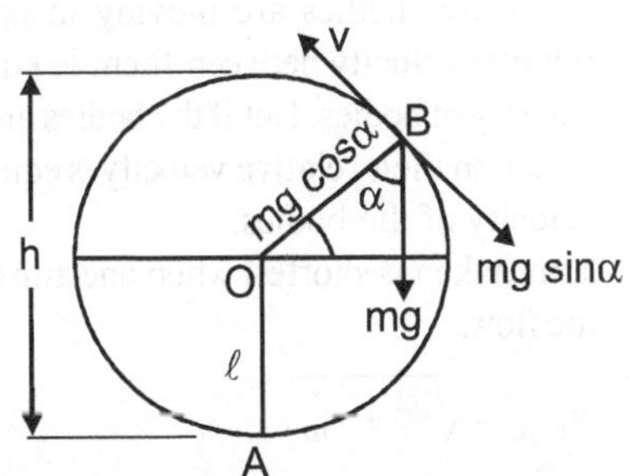

$$\frac{mv^2}{\ell} = T + mg\cos\alpha$$

On leaving the circular path

T = 0

$$\therefore \frac{mv^2}{\ell} = mg\cos\alpha$$

$$\Rightarrow v^2 = g\,\ell\cos\alpha \qquad ....(1)$$

According to law of conservation of energy

(K.E. + P.E.) at A = (K.E. + P.E.) at B

$$\Rightarrow 0 + 2mg\ell = \frac{1}{2}mv^2 + mgh$$

$$\Rightarrow v^2 = 2g(2\ell - h) \qquad ....(2)$$

From (1) & (2) $h = \frac{5}{3}\ell$

Also, $\cos\alpha = \frac{h - \ell}{\ell}$

(16) **(d)** $T\sin\theta = M\omega^2 R$

$T\sin\theta = M\omega^2 L\sin\theta$

From (i) and (ii)

$T = M\omega^2 L$

$= M4\pi^2 n^2 L$

$= M4\pi^2\left(\frac{2}{\pi}\right)^2 L$

$= 16$ ML

(17) **(a).** $v = 60$ km/hr $= \frac{50}{3}$ m/s

r = 0.1 km = 100m

$$\therefore \tan\theta = \frac{v^2}{rg} = 0.283$$

$$\therefore \theta = \tan^{-1}(0.283)$$

(18) **(c).** We know that $\tan\theta = \frac{v^2}{rg}$ ..... (1)

Let h be the relative raising of outer rail with respect to inner rail. Then

$$\tan\theta = \frac{h}{\ell} \qquad ......(2)$$

(ℓ = separation between rails)

From (1) & (2), $h = \frac{v^2}{rg} \times \ell$

Hence $v = 48$ km/hr $= \frac{120}{9}$ m/s, (r = 400 m, ℓ = 1m),

$$\therefore h = \frac{(120/9)^2 \times 1}{400 \times 9.8} = 0.045 \text{ m} = 4.5 \text{ cm}$$

(19) **(a).** The woman has two velocities simultaneously while running on the deck, one velocity is equal to the velocity of ship i.e. 12 m/s due east and other velocity is 5 m/s due north.

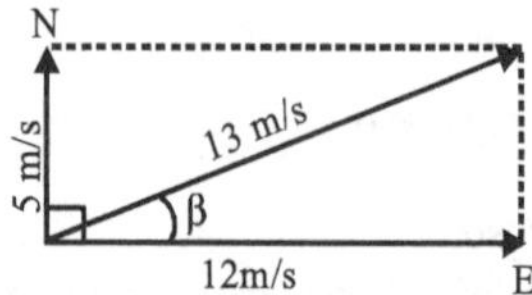

The resultant velocity of woman

$$= \sqrt{(12)^2 + (5)^2} = 13 \text{ m/s}$$

(20) **(c).** If we consider velocity of rain with respect to the man is V km/h.

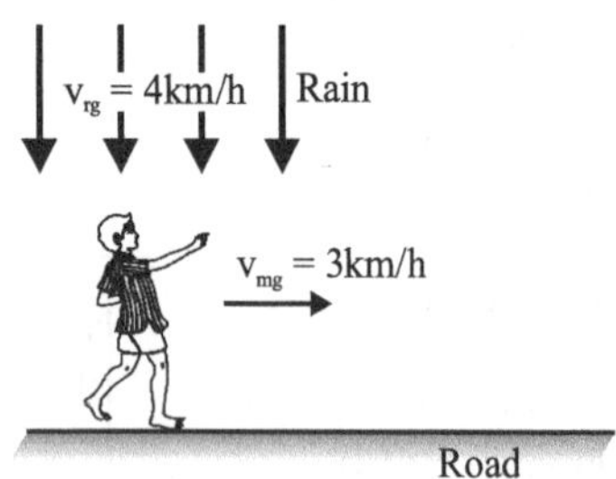

Relative velocity of man w.r.t. ground

$\vec{v}_{mg} = \vec{v}_m - \vec{v}_g$ ....(1)

Velocity of rain w.r.t. ground

$\vec{v}_{rg} = \vec{v}_r - \vec{v}_g$ ....(2)

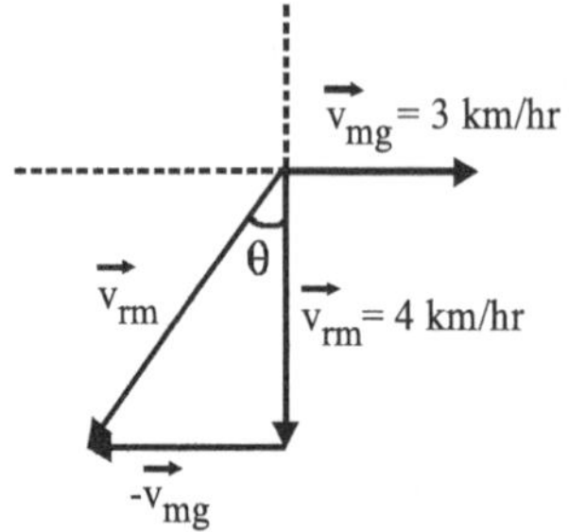

Velocity of rain w.r.t. man $\vec{v}_{rm} = \vec{v}_r - \vec{v}_m$

On subtracting eq^n. 1 from eq^n. 2

$\vec{v}_{rm} = \vec{v}_{rg} - \vec{v}_{mg}$

$$|v_{rm}| = \sqrt{v_{rg}^{\ 2} + v_{mg}^{\ 2}} = \sqrt{4^2 + 3^2} = 5 \text{ km/hr}$$

(21) **(d)** Since the maximum tension T_B in the string moving in the vertical circle is at the bottom and minimum tension T_T is at the top.

$$\therefore \quad T_B = \frac{mv_B^2}{L} + mg \text{ and } T_T = \frac{mv_T^2}{L} - mg$$

$$\therefore \quad \frac{T_B}{T_T} = \frac{\dfrac{mv_B^2}{L}}{\dfrac{mv_T^2}{L} - mg} = \frac{4}{1} \text{ or } \frac{v_B^2 + gL}{v_T^2 - gL} = \frac{4}{1}$$

or $\quad v_B^2 + gL = 4v_T^2 - 4gL$ but $v_B^2 = v_T^2 + 4gL$

$\therefore \quad v_T^2 + 4gL + gL = 4v_T^2 - 4gL \Rightarrow 3v_T^2 = 9gL$

$\therefore \quad v_T^2 = 3 \times g \times L = 3 \times 10 \times \dfrac{10}{3}$ or $v_T = 10 \text{m/sec}$

(22) **(d).** Use definition of relative velocity $\vec{V}_{PQ} = \vec{V}_P - \vec{V}_Q$

$\vec{V}_P = \text{const.}; \vec{V}_Q = \text{const.}$

$\therefore |\vec{V}_{PQ}| = |\vec{V}_{QP}| = \text{const.} \ ; |\vec{V}_P| > |\vec{V}_Q|$

$\therefore \vec{V}_{PQ} \to +\text{ve} ; \vec{V}_{QP} = -\text{ve}$ i.e. towards origin.

(23) **(c).** He can only reach the opposite point if he can cancel up the velocity of river by his component of velocity.

(24) **(a).** $v = R\omega$

$v_1 > v_2$

(25) (b), (26) (b), (27) (c).

The path of a projectile as observed by other projectile is a straight line.

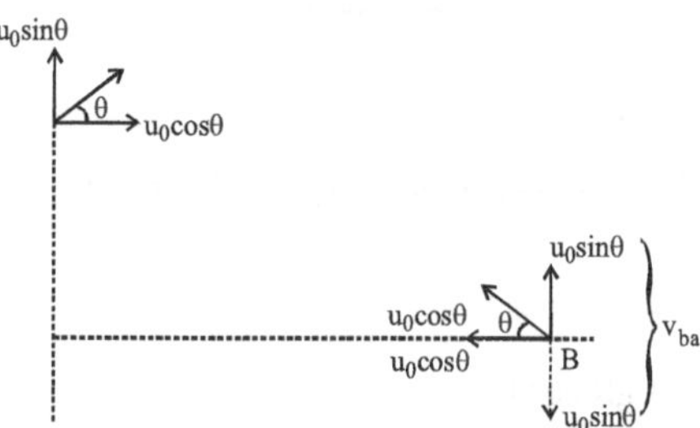

$v_A = u\cos\theta\,\hat{i} + (u\sin\theta - gt)\,\hat{j}. v_{AB} = (2u\cos\theta)\,\hat{i}$

$v_B = -u\cos\theta\,\hat{i} + (u\sin\theta - gt)\,\hat{j} \quad ; a_{BA} = g - g = 0$

The vertical component $u_0 \sin\theta$ will get cancelled.

The relative velocity will only be horizontal which is equal to $2u_0 \cos\theta$

Hence B will travel horiozontally towards left w.r.t A with constant speed $2u_0 \cos\theta$ and minimum distance will be h.

$$\frac{S_{rel}}{V_{rel}} = \frac{\ell}{2u_0 \cos\theta}$$

(28) **(a)** When two bodies are moving in opposite direction, relative velocity between them is equal to sum of the velocity of bodies. But if the bodies are moving in same direction their relative velocity is equal to difference in velocity of the bodies.

(29) **(b)** Time taken is shortest when one aims perpendicular to the flow.

(30) **(d)** $v_{r/m} = \sqrt{v_r^2 + v_m^2}$

DAILY PRACTICE PROBLEMS

PHYSICS SOLUTIONS 09

(1) **(a)** Force causing the acceleration = 400 – 200 = 200N
mass of the boy = 200/9.8

hence acceleration $= F/m = \frac{200}{200} \times 9.8 = 9.8\ m/s^2$

(2) **(a)** Acceleration $= \frac{\vec{F}}{m} = \frac{6\hat{i} + 8\hat{j}}{10}$ in the direction of force and displacement

$$\vec{S} = \vec{u}t + \frac{1}{2}\vec{a}t^2 = 0 + \frac{1}{2}\left(\frac{6\hat{i} + 8\hat{j}}{10}\right)100 = 30\hat{i} + 40\hat{j}$$

So the displacement is 50 m along $\tan^{-1}\frac{4}{3}$ with x-axis

(3) **(a)** From the law of conservation of momentum
$1000 \times 5 + 0 = (1000 + 60)\,v$

$\Rightarrow v = \frac{1000 \times 5}{1060} = 4.71\ m/s$

(4) **(b)** Weight of disc $= \frac{10}{1000}$ kg,
Let speed of the bullet = v
So rate of change of momentum of the bullets

$$= \frac{2 \times 10 \times 5}{1000}$$

v = applied force on the disc

Now $\frac{2 \times 10 \times 5}{1000} \times v = \frac{10 \times g}{1000}$

$\Rightarrow v = 0.98\ m/s^2 = 98\ cm/s^2$

(5) **(c)** Total mass = 80 + 40 = 120 kg
The rope cannot with stand this load so the fire man should slide down the rope with some acceleration
$\therefore$ The maximum tension = 100 × 9.8 N
m (g – a) = tension,
$120(9.8 - a) = 100 \times 9.8 \Rightarrow a = 1.63\ m/s^2$

(6) **(b)** Suppose the velocity of the body at the instant when it reaches the pile of sand be v. Then
$v^2 = 0 + 2(9.8) \times (5\ \text{metre}) = 98 (\because v^2 = u^2 + 2as)$

$$a = -\frac{98}{2 \times (0.05)} = -980\ m/sec^2$$

Now, retarding force
F = mass × acceleration = 0.02 kg × (– 980 m/sec²) = –19.6 N

(7) **(b)** Impulse = F . t = Area under F-t curve from 4 μs to 16 μs = Area under BCDFB
= Area of trapizium BCEF + area of ΔCDE

$$= \frac{1}{2}(200+800)(2\times10^{-6}) + \frac{1}{2}\times10\times10^{-6}\times800$$
$= 10 \times 10^{-4} + 40 \times 10^{-4}$ N-s $= 50 \times 10^{-4}$
$= 5.0 \times 10^{-3}$ N-s

(8) **(a)** (a) The elevator having an initial upward speed of 8 m/sec is brought to rest within a distance of 16 m
Hence, $0 = (8)^2 + 2a(16)\ (\because v^2 = u^2 + 2as)$,

$$a = -\frac{8 \times 8}{2 \times 16} = -2\ m/sec^2$$

Resultant upward force on elevator = T – mg. According to Newton's law.
T – mg = ma
or T = mg + ma = m (g + a) = 1000 (9.8 – 2) = 7800 N
(b) Let P be the upward force exerted on the man by the elevator floor. If m' be the mass of the man, then, weight of the man acting downward = m' g,
Upward force on the man = P – m'g
According to Newton's law. P–m' g = m' a or
P = m' (a + g) = (–2 + 9.8) = 624 N

(9) **(d)** As P and Q move down, the length ℓ decreases at the rate of U m/s

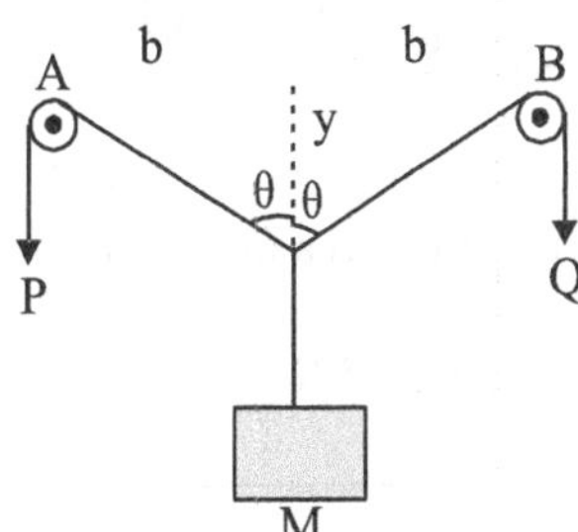

From figure, $\ell^2 = b^2 + y^2$
Differentiating with respect to time

$$2\ell\frac{d\ell}{dt} = 2y\frac{dy}{dt} \qquad (\because b \text{ is constant})$$

$$\therefore \frac{dy}{dt} = \frac{\ell}{y}\cdot\frac{d\ell}{dt} = \frac{1}{\cos\theta}\cdot\frac{d\ell}{dt} = \frac{U}{\cos\theta}$$

(10) **(a)** The engine, coach, coupling and resistance are, shown in figure.

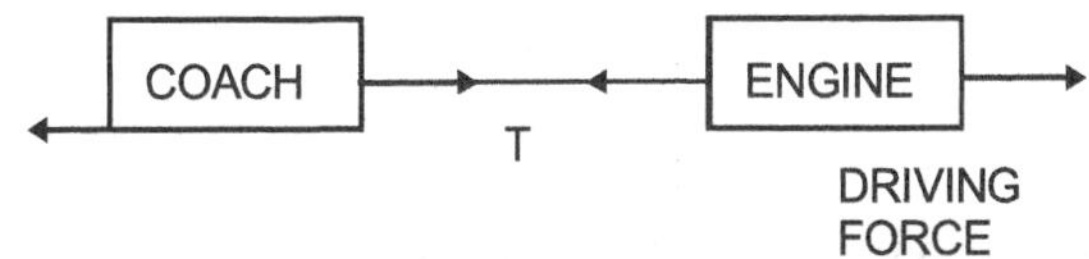

Driving force = 4500 N

Opposing force (Resistance) $= \frac{(5+4)10^4}{100} = 900$ N
Resultant force = 4500 – 900 = 3600 N
Mass of engine and coach = 9×10^4 kg
According to Newton's law, F = ma
$\therefore 3600 = 9 \times 10^4 a$
or $a = (3600)/(9 \times 10^4) = 0.04\ m/sec^2$
So acceleration of the train = $0.04\ m/sec^2$

Now considering the equilibrium of the coach only, we have $(T - R) = 4 \times 10^4 \times 0.04$ ($\because F = ma$)

or $T - \dfrac{4 \times 10^4}{100} = 4 \times 10^4 \times 0.04$,

$T = 4 \times 10^4 \times 0.04 + 4 \times 10^2 = 1600 + 400 = 2000\text{ N}$

(11) **(d)** Given that $\vec{F_1} = (8\hat{i} + 10\hat{j})$ and $\vec{F_2} = (4\hat{i} + 8\hat{j})$

Then the total force $\vec{F} = 12\hat{i} + 18\hat{j}$

So acceleration $\vec{a} = \dfrac{\vec{F}}{m} = \dfrac{12\hat{i} + 18\hat{j}}{6} = 2\hat{i} + 3\hat{j}\text{ m/sec}^2$

Net acceleration

$|\vec{a}| = \sqrt{2^2 + 3^2} = \sqrt{4+9} = \sqrt{13}\text{ m/sec}^2$

(12) **(c)** From the relation

$F = ma \Rightarrow a = \dfrac{F}{m} = \dfrac{1000}{1000} = 1\text{ m/s}^2$

As the force is brake force, acceleration is -1 m/s^2 using relation $v^2 = u^2 + 2as$, we obtain

$2as = u^2 \Rightarrow s = \dfrac{u^2}{2a} = \dfrac{\left(18 \times \dfrac{5}{18}\right)^2}{2} = 12.5\text{ m}$

(13) **(a)** The water jet striking the block at the rate of 1 kg/s at a speed of 5 m/s will exert a force on the block

$F = v\dfrac{dm}{dt} = 5 \times 1 = 5\text{ N}$

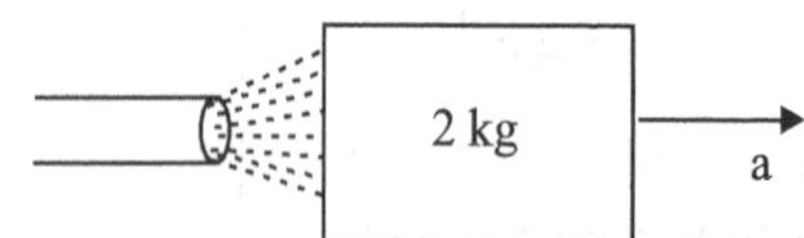

And under the action of this force of 5 N, the block of mass 2 kg will move with an acceleration given by,

$F = ma \Rightarrow a = F/m = 5/2 = 2.5\text{ m/s}^2$

(14) **(a)** Relative speed of the ball $= (v + u)$

Speed after rebouncing $= -(v + u)$

So, $F = m\dfrac{\Delta v}{\Delta t} = \dfrac{m[(v+u) - \{-(v+u)\}]}{t}$

$= \dfrac{2m(v+u)}{t}$

(15) **(b)** $F = \dfrac{dp}{dt} \Rightarrow F\,dt = dp = p_2 - p_1$

$\Rightarrow F \times 1 = mnv - 0$

$\Rightarrow F = mnv$

(Total mass of the bullets fired in 1 sec = mn)

(16) **(a)** The initial momentum $= 15 \times 10 = 150$ kgm/s and

$\text{Force} = \dfrac{\text{change in momentum}}{\text{time}} = \dfrac{0 - 150}{15} = -10\text{ N}$

A constant force of 10 N must be acting in opposite direction to the motion of body.

(17) **(a)** The change in momentum in the final direction is equal to the impulse $= \dfrac{2.50}{1000} \times 28 - \left(-\dfrac{250}{1000} \times 24\right) = 13\text{ Ns}$

and force $= \dfrac{\text{impulse}}{\text{time}} = \dfrac{13}{1/100} = 1300\text{ N}$

in the direction of the ball.

(18) **(b).** We know $\vec{F} = \dfrac{d\vec{p}}{dt} \Rightarrow \vec{F}dt = d\vec{p}$

$\Rightarrow 2 \times 2 = d\vec{p} \Rightarrow 4 = d\vec{p}$

Therefore change in momentum = 4 Ns

(19) **(a)** We know $\vec{F} = \dfrac{d\vec{p}}{dt}$

$\Rightarrow \vec{F}dt = d\vec{p} = \vec{p}_2 - \vec{p}_1 = m\vec{v}_2 - m\vec{v}_1$

$\Rightarrow 4\hat{j}.1 = 2.\vec{v}_2 - 2(2\hat{i})$

$\Rightarrow 2\vec{v}_2\ \ 4\hat{j}.1 = 2.\vec{v}_2 - 2(2\hat{i}) = 4\hat{j} + 4\hat{i}$

$\Rightarrow \vec{v}_2 = 2\hat{i} + 2\hat{j}$

$\Rightarrow |\vec{v}_2| = 2\sqrt{2}\text{ m/s}$

(20) **(c)** Initial momentum of the ball

$= \dfrac{150}{1000} \times 12 = 1.8\text{ kg.m/sec}$

Final momentum of the ball $= -\dfrac{150}{1000} \times 20 = -3.0\text{ kg m/sec}$

Change in momentum = 4.8 kg m/sec

Average force exerted = Impulse/ time $= \dfrac{4.8}{.01} = 480\text{ N}$

(21) **(b)** Initial momentum of the body $= mu = 20 \times 3 = 60$

and final momentum of the body $= -mu = -20 \times 3 = -60$

The change in momentum of body in initial direction

$= -60 - 60 = -120$

The change in momemtum imparted to the body in opposite direction = 120

$\therefore$ The impulse imparted to the body = 120 Ns

(22) **(a)** (1) Since the lift is moving down with an acceleration of 3 m/sec^2, then the inertial force F = ma, acts upwards on the body

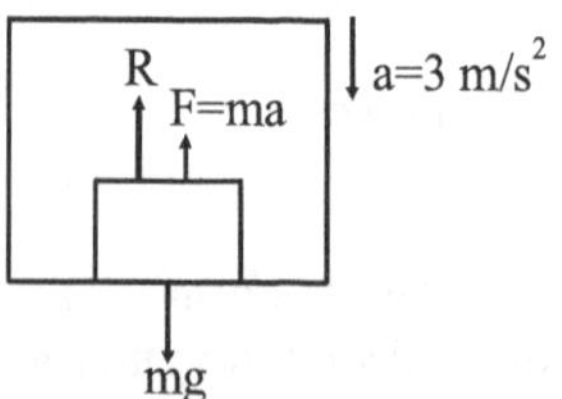

Now, $R + F = mg$

or $R = mg - F = mg - ma = m(g - a) = 60\,(9.8 - 3) = 408\text{ N}$

(2) When the lift is moving down with constant velocity $a = 0$ and hence, $R = mg = 60 \times 9.8 = 588\text{ N}$

(3) The lift is now moving down with a retardation of

3 m/sec^2.
The retardation is 3 m/sec^2 in the downward direction is equivalent to an acceleration of 3 m/sec^2 upwards.
Hence the direction of fictitious force is downwards.
Now, $R = mg + ma = m(g+a) = 60(12.8) = 768$ N

(23) **(b)** When the lift is moving up $m(g+a)$ = force

$$\text{The scale reading} = \frac{m(g+a)}{g} = \frac{10\left(g+\frac{g}{3}\right)}{g} = 13.3 \text{ kg}$$

When lift is moving down the scale reading

$$= \frac{m(g-a)}{g} = \frac{10\left(g-\frac{g}{3}\right)}{g} = 6.67 \text{ kg}$$

(24) **(a)**
(1) A reference frame in which Newton's first law is valid is called an inertial reference frame.
(2) Frame moving at constant velocity relative to a known inertial frame is also an inertial frame.
(3) Idealy, no inertial frame exists in the universe for practical purpose, a frame of reference may be considered as Inertial if its acceleration is negligible with respect to the acceleration of the object to be observed.
(4) To measure the acceleration of a falling apple, earth can be considered as an inertial frame.

(25) **(a)**
(i) In the case of constant velocity of lift, there is no reaction, therefore the apparent weight = actual weight. Hence the reading of machine is 50 kg wt.
(ii) In this case the acceleration is upward the reaction R = ma acts downward, therefore apparent weight is more than actual weight.
i.e. $W' = W + R = m(g+a)$
Hence, scale show a reading of

$$m(g+a)\text{ Newton} = \left(50 + \frac{50g}{a}\right) \text{ kg wt}$$

(26) **(a)** Tension = $m(g+a)$, when lift moving up, putting the values, we get
$175 = 25(9.8 + a) \Rightarrow a = 2.8$ m/s^2
[negative sign shows that lift is moving downward]

(27) **(b)** Apparent tension, $T = 2T_0$

$$\text{So, } T = 2T_0 = T_0\left(1 + \frac{a_0}{g}\right)$$

$$\text{or } 2 = 1 + \frac{a_0}{g} \Rightarrow a_0 = g = 9.8 \text{ m/s}^2$$

(28) **(b)** Cloth can be pulled out without dislodging the dishes from the table because of inertia. Therefore, statement- 1 is true.
This is Newton's third law and hence true. But statement 2 is not a correct explanation of statement 1.

(29) **(d)** According to Newton's second law

$\text{Acceleration} = \dfrac{\text{Force}}{\text{Mass}}$ i.e. if net external force on the body is zero then acceleration will be zero.

(30) **(c)** $F = \dfrac{dp}{dt}$ = Slope of momentum -time graph

i.e. Rate of change of meomentum = Slope of momentum-time graph = force.

DAILY PRACTICE PROBLEMS

PHYSICS SOLUTIONS 10

(1) (c) Force on the block
= Mass of the block × acceleration of the system

$$= M \times \frac{P}{M+m}$$

(2) (b) Mass of the rope $= 8 \times \frac{1}{2} = 4$ kg

Total mass $= 50 + 4 = 54$ kg

$\therefore \quad a = \frac{F}{m} = \frac{108}{54} = 2 \text{ m/s}^2$

Force utilised in pulling the rope $= 4 \times 2 = 8$ N
Force applied on mass $= 108 - 8 = 100$ N

(3) (b) Mass of the rope $= 15 \times 2 = 30$ kg

acceleration $= \frac{F}{m} = \frac{25}{30} = \frac{5}{6} \text{ m/s}^2$

At the point 7 m away from point of application the mass of first part of rope = 14 kg

$\therefore$ Force used in pulling 14 kg $= 14 \times \frac{5}{6} = 11.67$ N

The remaining force $= (25 - 11.67)$N $= 13.33$ N

(4) (b) The various forces acting are shown in fig.
The force of 100N has

(i) horizontal component of 100 cos 30° $= 50\sqrt{3}$ N
and (ii) A vertical component = 100 sin 30° = 50N

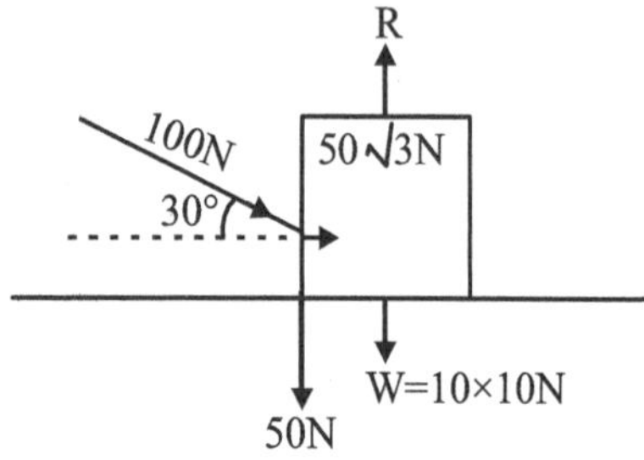

Since the block is always in contact with the table, the net vertical force
$R = mg + F\sin\theta = (10 \times 10 + 50)\text{ N} = 150\text{N}$
When the block moves along the table, work is done by the horizontal component of the force. Since the distance moves is 10 m, the work done is

$50\sqrt{3} \times 10 = 500\sqrt{3}$ Joule.

If v is the speed acquired by the block, the work done must be equal to the kinetic energy of the block. Therefore, we have

$$500\sqrt{3} = \frac{1}{2} \times 10 \times v^2 \Rightarrow v^2 = 100\sqrt{3} \Rightarrow v = 13.17 \text{ m/sec}$$

(5) (a) We have acceleration

$$a = \frac{F\cos\theta}{m} = \frac{50\sqrt{3}}{10} = 5\sqrt{3} \text{ m/sec}^2$$

The velocity after 2 sec, $v = u + at$

$\Rightarrow \quad v = 0 + 5\sqrt{3} \times 2 = 10\sqrt{3}$ m/sec

(6) (a) All the forces acting on the two blocks are shown in fig. As the blocks are rigid under the action of a force F, both will move together with same acceleration.

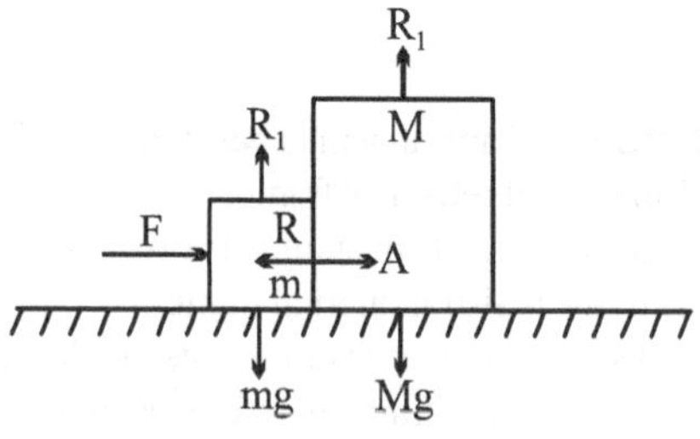

$a = F/(m+M) = 3/(1+2) = 1 \text{ m/s}^2$
Now as the mass of larger block is m and its acceleration a so force of contact i.e. action on it.

$$f = Ma = \frac{MF}{M+m} = \frac{2 \times 3}{2+1} = 2\text{N}$$

(7) (a) As the same force is applied to the combined mass, we have

$$\frac{1}{a} = \frac{1}{a_1} + \frac{1}{a_2} \quad \text{or} \quad a = \frac{a_1 a_2}{a_1 + a_2} = \frac{5 \times 15}{5+15} = 3.75 \text{ m/s}^2$$

(8) (a) As net force on the rod $= F_1 - F_2$ and its mass is M so acceleration of the rod will be
$a = (F_1 - F_2)/M$ ...(i)
Now considering the motion of part AB of the rod, which has mass (M/L)y,
Acceleration a given by

(i) Assuming that tension at B is T

$$F_1 - T = \frac{M}{L} y \times a \qquad \text{(from F = ma)}$$

$$\Rightarrow \quad F_1 - T = \frac{M}{L} y \frac{F_1 - F_2}{M} \qquad \text{(using eq. (1))}$$

$$\Rightarrow \quad T = F_1\left(1 - \frac{y}{L}\right) + F_2\left(\frac{y}{L}\right)$$

(9) (b) The net acceleration of the system is given by

$$\frac{1}{a} = \frac{1}{a_1} + \frac{1}{a_2}$$

$$\frac{1}{a} = \frac{1}{a_1} + \frac{1}{a_2} + \ldots\ldots \frac{1}{a_n} = 1 + 2 + 3 + \ldots\ldots\ldots + n$$

$$= \frac{n}{2}[2 + (n-1)1] = \frac{n}{2}[n+1] = \frac{2}{n(n+1)}$$

(10) (a) As the mass of the system is 6 + 4 + 2 = 12 kg and applied force is 60 N, the acceleration of the system

$$a=\frac{F}{m}=\frac{60}{12}=5\ m/s^2$$

Now at point A as tension in pulling the rope of mass 2kg and block Q of mass 4kg.

$T_A=(2+4)\times 5=30N$

Similarly for B and C,

$T_B=(1+4)\times 5=25N$

and $T_C=(0+4)\times 5=20N$

(11) (a) In case (a), the pulling force = 2mg – mg = mg and the mass is 2m + m = 3m

so acceleration a = mg/3m = g/3

While in case (b), the pulling force = 2mg – mg = mg but, the mass in motion = m + 0 = m

Acceleration, a = mg/m = g

(12) (c) It this problem as the pulling force is 2mg while opposing force is mg, so net force

F = 2mg – mg = mg,

and as the mass in motion = m + m + m = 3m

So the acceleration $=\frac{\text{force}}{\text{mass}}=\frac{mg}{3m}=\frac{g}{3}$

Now as A is accelerated up while B and C down. so tension T_1, is such that $mg < T_1 < 2mg$

Actually for the motion of A,

$$T_1=m(g+a)=m(g+g/3)=\frac{4}{3}mg$$

Now to calculate tension in the string BC we consider the downward motion of C,

i.e. $T_2=m(g-a)=m(g-g/3)=(2/3)\,mg$

(13) (a) As pulley Q is not fixed so if it moves a distance d the length of string between P and Q will changes by 2d (d from above and d from below) i.e. M will move 2d. This in turn implies that if a (→2d) is the acceleration of M, the acceleration of Q and so 2M will be of (a/2)

Now if we consider the motion of mass M, it is accelerated down so $T=M(g-a)$ ...(1)

And for the motion of Q,

$2T-T'=0\times(a/2)=0\Rightarrow T'=2T$ ...(2)

And for the motion of mass 2M,

$T'=2M(a/2)\Rightarrow T'=Ma$ ...(3)

From equation (2) and (3) $T=\frac{1}{2}Ma$, so eq. (1) reduces

$$\left(\frac{1}{2}\right)Ma=M(g-a)\Rightarrow a=\frac{2}{3}g$$

(14) (a) The tension is same in two segments

For B the equation is

$(40\times 9.8-T)=40a$ ...(1)

For C the equation is

$(T-50\times 9.8\times\frac{1}{2})=50a$ ...(2)

From equation (1) and (2) $a=1.63\ m/s^2$

distance of fall

$$S=\frac{1}{2}at^2=\frac{1}{2}\times 1.63\times 4^2=13.04\ m$$

(15) (b) The string is massless and inextensible the tension T is same. Let mass B move down the inclined plane.

For B the equation of motion $m_1 g\sin\theta-T=m_1 a$

$30\times 9.8\times\sin 53°-T=30a$

$\Rightarrow 235.2-T=30a$ ...(1)

and for A the equation of motion

$T-20\times 9.8\times\sin 37°=20a$

$T-117.6=20a$ ...(2)

From (1) & (2) T = 164.64 N

(16) (c)

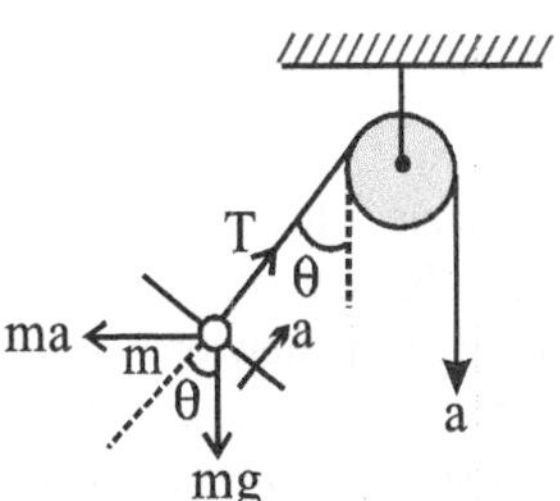

(Force diagram in the frame of the car)

Applying Newton's law perpendicular to string

$mg\sin\theta=ma\cos\theta$

$$\tan\theta=\frac{a}{g}$$

Applying Newton's law along string

$\Rightarrow T-m\sqrt{g^2+a^2}=ma$

or $T=m\sqrt{g^2+a^2}+ma$

(17) (a) As A moves up and B moves down with acceleration a for the motion of A,

$T-11g=11a$ ... (i)

for the motion of B,

$11.5g-T=11.5a$ ...(ii)

From (i) & (ii),

$$a=\frac{m_1-m_2}{m_1+m_2}$$

$$g=\frac{(11.5-11)9.8}{11.5+11}=0.218\ m/sec^2$$

Assuming that the particles are initially at rest, their velocity at the end of 4 sec will be

$v=u+at=0+0.218\times 4=0.872\ m/s$

(18) (a) The height ascended by A in 4 sec

$$h=ut+\frac{1}{2}at^2=0+\frac{1}{2}(0.218)4^2=1.744\ m$$

This is also the height descended by B in that time.

(19) (c) At the end of 4 sec the string is cut. Now A and B are no longer connected bodies but become free ones, falling under gravity.

Velocity of A, when the string was cut

= 0.872 m/s upwards.

Acceleration a = – g (acting downwards),

displacement from this position in the subsequent 2 sec

$h = ut + \frac{1}{2}at^2 = (0.872) \times 2 + \frac{1}{2}(-9.8)2^2$

$= 1.744 - 4.9 \times 4 = -17.856m$

A descends down by a distance of 17.856 m from the position it occupied at the end of 4 sec from its start. B has a free fall. Its position is given by

So the acceleration of mass M is (2/3)g

while tension in the string PQ will be

$T = M(g - (2/3)g) = (1/3)Mg$

The force exerted by clamp on the pulley

$= \sqrt{T^2 + T^2} = \frac{\sqrt{2}}{3}Mg$

(20) (a) Here the system behaves as a rigid system, therefore every part of the system will move with same acceleration. Thus applying newton's law

$mg - T = ma$ .... (i)

$2T - mg = ma$ .... (ii)

Doubling the first equation and adding

$mg = 3ma$ or acceleration $a = \frac{1}{3}g$

(21) (c) Tension in the string

$T = m(g - a) = m\left(g - \frac{g}{3}\right)$

$T = \frac{2}{3}mg$

(22) (a)

(1) Inertia $\propto$ mass

(2) 1 Newton = 10^5 dyne

(3) Thrust on rocket $\vec{F} = \frac{\Delta M}{\Delta t}\vec{v} - M\vec{g}$

(4) Apparent weight of a body in the lift accelerated up is $W = m(g + a)$.

(23) (b)

(1) If $a_1, a_2, \ldots a_n$ be the accelerations produced in n different bodies on applying the same force, the acceleration produced in their combination due to the same force will be $\frac{1}{a} = \frac{1}{a_1} + \frac{1}{a_2} + \ldots\ldots \frac{1}{a_n}$

(2) Newton's I[st] and IIIrd law can be derived from second law therefore IInd law is the most fundamental law out of the three law.

(24) (a)

(1) For equilibrium of a body under the action of concurrent forces $\vec{F_1} + \vec{F_2} + \vec{F_3} + \ldots\vec{F_n} = 0$

(2) If the downward acceleration of the lift is $a = g$, then the body will enjoy weightlessness.

(3) If the downward acceleration of the body is $a > g$, then the body will rise up to the ceiling of lift

(25) (d), (26) (d), (27) (a).

Initial elongation $= 2R\cos 30° = \sqrt{3}R$

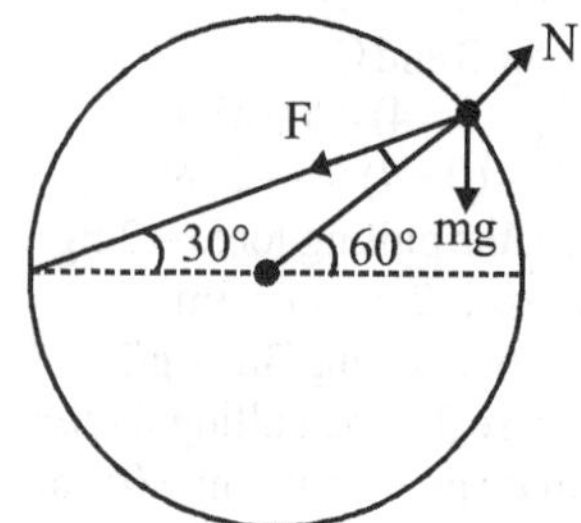

Extension in the spring is

$x = AB - R = 2R\cos 30° - R = (\sqrt{3} - 1)R$

$\therefore$ Spring force

$F = kx = \frac{(\sqrt{3}-1)\,mg}{R}(\sqrt{3}-1)R = 2mg$

FBD of bead is

$N = F(mg\cos 30°) = (2mg + mg)\frac{\sqrt{3}}{2} = \frac{3\sqrt{3}}{2}mg$

Tangential force $F_1 = F\sin 30° - mg\sin 30°$

$= (2mg - mg)\sin 30° = \frac{mg}{2}$

$\therefore$ tangential acceleration = g/2

28. (d) Here the acceleration of both will be same, but their masses are different. Hence, the net force acting on each of them will not be same.

29. (c) The FBD of block A in Figure is

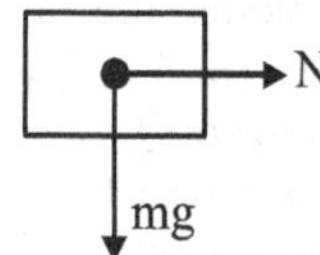

The force exerted by B on A is N (normal reaction). The force acting on A are N (horizontal) and mg (weight downwards). Hence statement I is false.

30. (d) $T - m_1g = m_1a$ – .... (1)

$m_2g - T = m_2a$ – .... (2)

Solving (1) and (2), $T_1 = \left(\frac{2m_1m_2}{m_1 + m_2}\right)g$ $a = \left(\frac{m_2 - m_1}{m_1 + m_2}\right)g$

DAILY PRACTICE PROBLEMS

PHYSICS SOLUTIONS 11

(1) **(a)** Let the contact force on the block by the surface be F which makes an angle θ with the vertical. The component of F perpendicular to the contact surface is the normal force N and the component F parallel to the surface is the friction f. As the surface is horizontal, N is vertically upward. For vertical equilibrium

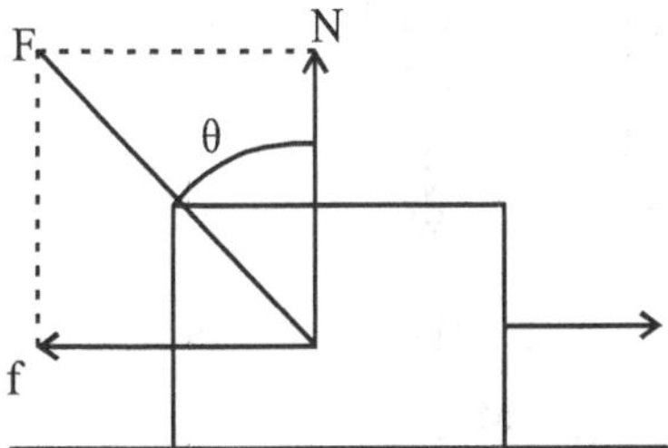

N = Mg = (0.400) (10) = 4.0 N
The frictional force is f = 3.0 N

$$\tan\theta = \frac{f}{N} = \frac{3}{4} \Rightarrow \theta = \tan^{-1}(3/4) = 37^{\circ}$$

(2) **(c)** The magnitude of the contact force is

$$F = \sqrt{N^2 + f^2} = \sqrt{(4)^2 + (3)^2} = 5.0\,N$$

(3) **(c)** The forces on the block are

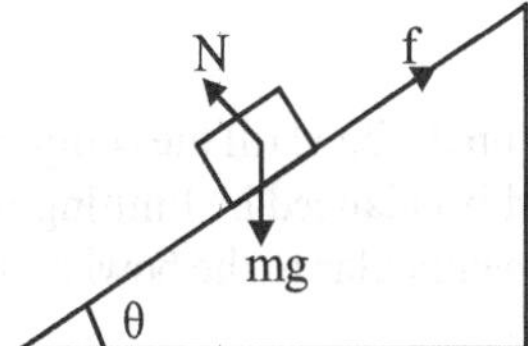

(i) the weight mg downward by the earth
(ii) the normal contact force N by the incline, and
(iii) the friction f parallel to the incline up the plane, by the incline.
As the block is at rest, these forces should add up to zero. Also since θ is the maximum angle to prevent slipping, this is a case of limiting equilibrium and so

$$f = \mu_s N$$

Taking component perpendicular to the Incline,
$N - mg\cos\theta = 0 \Rightarrow N = mg\cos\theta$ .... (1)
Taking component parallel to the incline
$f - mg\sin\theta = 0 \Rightarrow f = mg\sin\theta$ .... (2)
$\therefore\ \mu_s N = mg\sin\theta$
Dividing (2) by (1) $\mu_s = \tan\theta$
$\theta = \tan^{-1}\mu_s = \tan^{-1}(0.3)$

(4) **(a)** When the maximum force F is applied, both the blocks move together towards right. The only horizontal force on the upper block of mass m is that due to the friction by the lower block of mass M. Hence this force on m should be towards right. The force of friction on M by m should be towards left by Newton's third law. As we are talking of the minimum possible force F that can be applied, the friction is limiting and hence
f = μN, where N is normal force.
in the vertical direction, there is no acceleration
$\therefore$ N = mg
in the horizontal direction,

let the acceleration be a, then

$\mu N = ma$

$\mu mg = ma$

$a = \mu g$

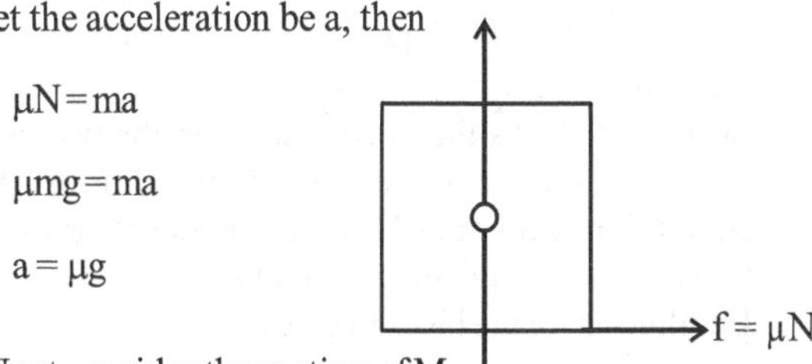

Next consider the motion of M
The equation of motion is
$F = \mu N = Ma$
$F - \mu mg = M\mu g$
$F = \mu g(M + m)$

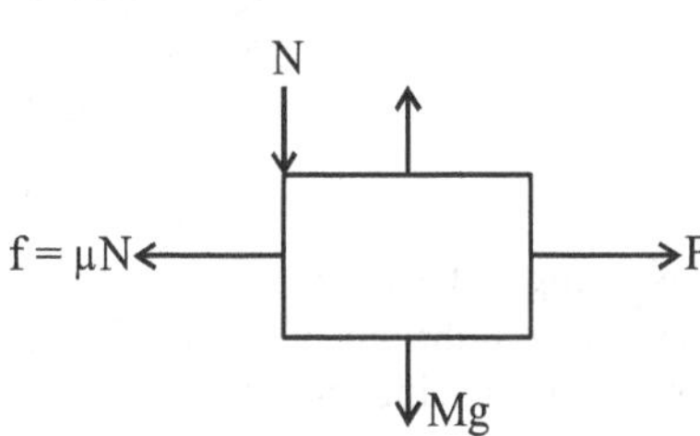

(5) **(b)** When A moves with B the force opposing the motion is the only force of friction between B and S the horizontal and velocity of the system is constant

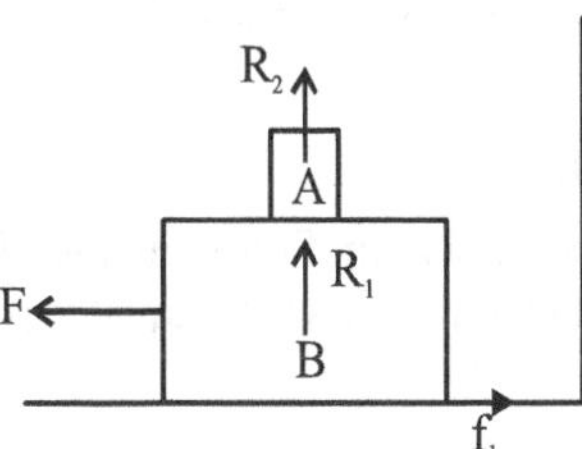

$F = f_1 = \mu R_1 = 0.25\,(4 + 8) = 3N$

(6) **(d)** When A is held stationary the friction opposing the motion is between A and B and B and S. So

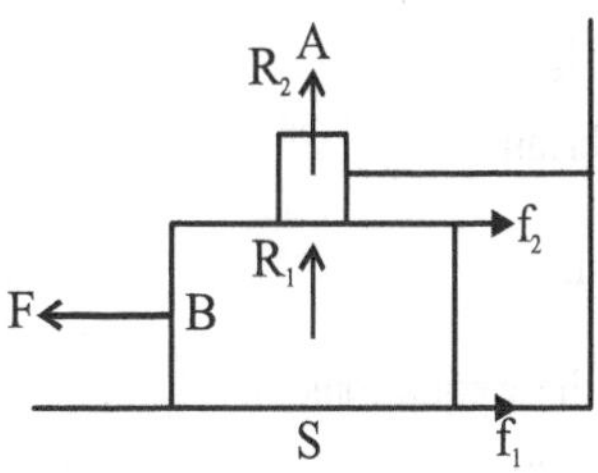

$F = \mu R_1 + \mu R_2 = 3 + 0.25\,(4)$
$F = 3 + 1 = 4\,N$

(7) **(d)** In this situation for dynamic equilibrium of B

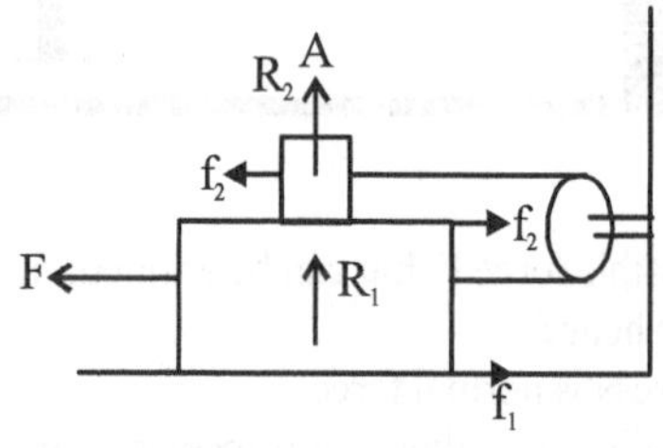

$F = \mu R_1 + \mu R_2 + T$ ..... (1)

While for the uniform motion of A

$T = \mu R_2$ .... (2)

Substituting T from Equation (2) in (1) we get

$F = \mu R_1 + 2\mu R_2 = 3 + 2 \times 1 = 5N$

(8) **(a)** Figure shows the forces acting on the two blocks. As we are looking for the maximum value of M/m, the equilibrium is limiting. Hence the frictional forces are equal to μ times corresponding normal force.

Equilibrium of the block m gives

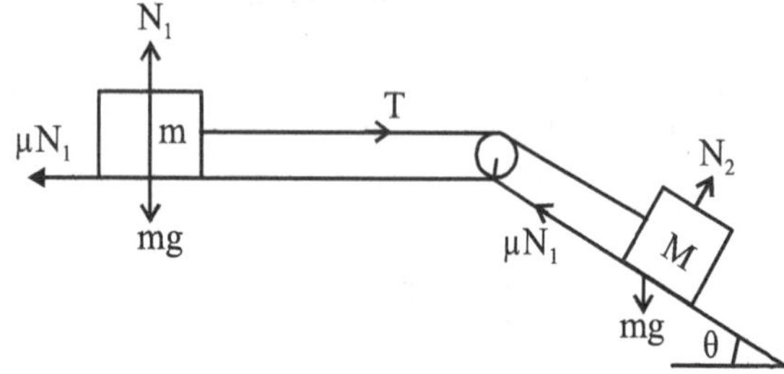

$T = \mu N_1$ and $N_1 = mg \Rightarrow T = \mu mg$ .... (1)

Next consider the equilibrium of the block M. Taking components parallel to the incline

$T + \mu N_2 = Mg \sin \theta$

Taking components normal to the Incline

$N_2 = Mg \cos \theta$

These give $T = Mg (\sin \theta - \mu \cos \theta)$ ...... (2)

From (1) and (2) $\mu mg = Mg (\sin \theta - \mu \cos \theta)$

$$\frac{M}{m} = \frac{\mu}{\sin \theta - \mu \cos \theta}$$

(9) **(a)** The situation is shown in figure in the limiting equilibriums the frictional force f will be equal to μN.

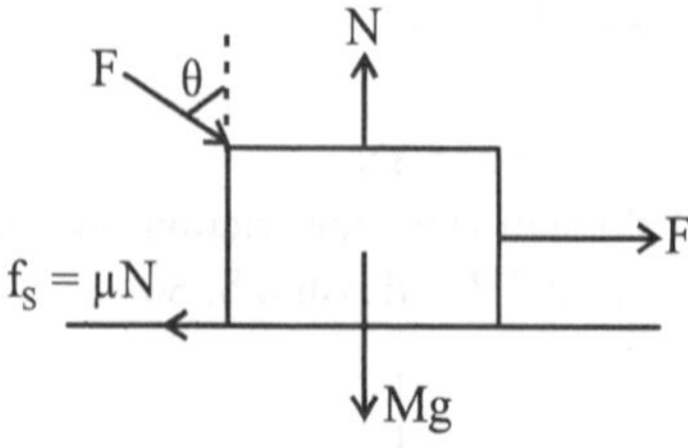

For horizontal equilibrium

$F \sin \theta = \mu N$

For vertical equilibrium

$F \cos \theta + mg = N$

Eliminating N from these equations

$F \sin \theta = \mu F \cos \theta + \mu mg$

$$F = \frac{\mu}{(\sin \theta - \mu \cos \theta)}$$

(10) **(c)** If y is the maximum length of chain which can be hang out side the table without sliding, then for equilibrium of the chain, the weight of hanging part must be balanced by force of friction from the portion on the table

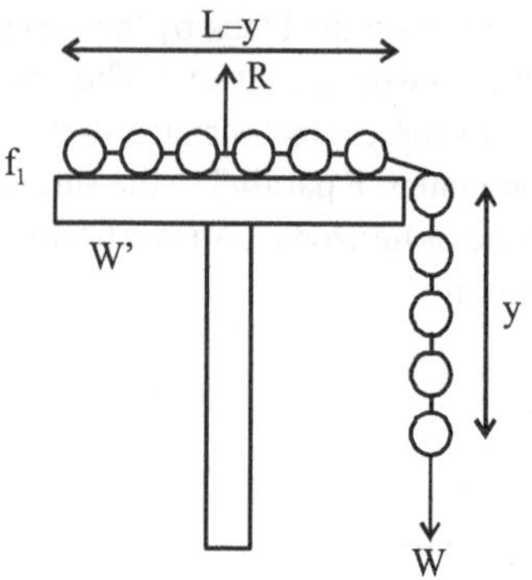

$W = f_L$ .... (1)

But from figure $W = \frac{M}{L} yg$ and

$$R = W' = \frac{M}{L}(L - y)g$$

So that $f_L = \mu R = \frac{\mu M}{L}(L - y)g$

Substituting these values of W and f_L in equation (1) we get

$$\frac{M}{L} yg = \mu \frac{M}{L}(L - y)g$$

(11) **(d)** The insect will crawl up the bowl till the component of its weight along the bowl is balanced by limiting friction so, resolving weight perpendicular to the bowl and along the bowl we get

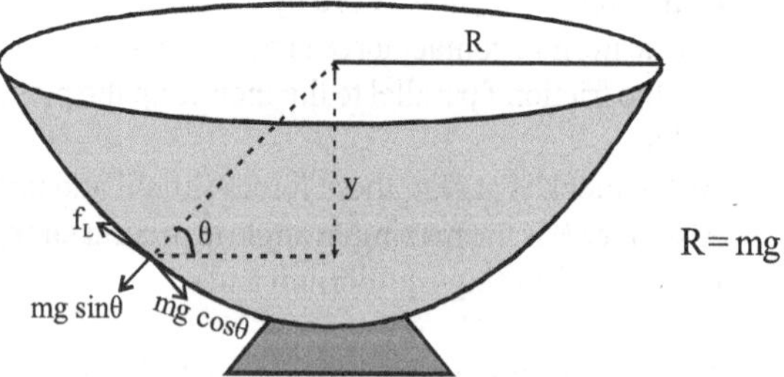

$R = mg \sin \theta$

$f_L = mg \cos \theta$

$$\tan\theta = \frac{R}{f_L} = \frac{R_1}{\mu R_1} = \frac{1}{\mu} \; ; \; \frac{y}{\sqrt{R^2 - y^2}} = \frac{1}{\mu}$$

$$\mu^2 y^2 = R^2 - y^2 \; ; \; y = \frac{R_1}{\sqrt{\mu^2 + 1}}$$

$$\text{So, } h = R - y = R - \frac{R}{\sqrt{\mu^2 + 1}} = R\left[1 - \frac{1}{\sqrt{(\mu^2 + 1)}}\right]$$

(12) (a)

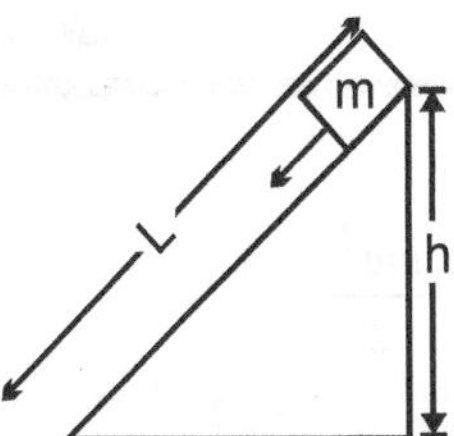

Loss in P.E. in reaching the bottom = mgh and gain in K.E.

reaching the bottom = $\frac{1}{2}mv^2$

where v is velocity gained by the body in reaching the bottom

$\therefore$ Net loss in energy = $mgh - \frac{1}{2}mv^2$

work done against friction = FL

$\therefore mgh - \frac{1}{2}mv^2 = FL; v = \sqrt{\frac{2}{m}(mgh - FL)}$

(13) (a) Let R be the normal reaction on the block exerted by the floor. The limiting (maximum) force of static friction is
$f_s = \mu_s R = \mu_s mg$
$= 0.4 \times 2kg \times 9.8\ ms^{-2} = 7.84\ N$
The applied force F is 2.5 N, that is less than the limiting frictional force. Hence under the force F, the block does not move. So long the block does not move, the (adjustable) frictional force is always equal to the applied force. Thus the frictional force is 2.5 N.

(14) (b) When the block does not slip on the table surface, it performs simple harmonic motion along with the table.
$x = a \sin \omega t$
The instantaneous acceleration of the block is

$$\frac{d^2x}{dt^2} = -\omega^2 a \sin \omega t$$

The maximum acceleration is $\left|\frac{d^2x}{dt^2}\right|_{max} = \omega^2 a$

The maximum force on the block is $f_{max} = m\omega^2 a$
where m is its mass. The frictional force on the block is μmg. since the block is at rest with respect to the table, we have $m\omega^2 a = \mu mg$
$(2\pi f)^2 a = \mu g$

$$\Rightarrow a = \frac{\mu g}{4\pi^2 f^2} = \frac{0.72 \times 10}{4 \times (3.14)^2 \times 3^2} = 0.02\ m$$

(15) (c) Stopping distance is independent on mass.

(16) (a) (i) coefficient of static friction is always greater than the coefficient of kinetic friction
(ii) limiting friction is always greater than the kinetic friction
(iii) limiting friction is never less than static friction

(17) (d) The system can not remain in equilibrium

(18) (a) (i) In the force applied v/s friction graph : The graph is a straight line of slope 45° for small F and a straight line parallel to the F-axis for large F.
(ii) There is small kink on the graph

(19) (a) (i) force of friction between two bodies may be equal to zero
(ii) bodies may be rough

(20) (b) It is easier to pull a body than to push, because the friction force is more in pushing than that in pulling

(21) (a) $ma = \mu mg$
$a = \mu g$

(22) (a)
(1) Kinetic friction is lesser than limiting friction.
(2) In rolling the surfaces at contact do not rub each other.
(3) If a body is at rest and no pulling force is acting on it, force of friction on it is zero.

(23) (a)
(1) Force of friction is partically independent of microscopic area of surface in contact and relative velocity between them. (if it is not high)
(2) Normally with increase in smoothness friction decreases. But if the surface area are made too smooth by polishing and cleaning the bonding force of adhesion will increase and so the friction will increase resulting in 'Cold welding'
(3) Friction is a non conservative force, i.e. work done against friction is path dependent.

(24) (c)
(2) Friction may opposes the motion
(4) If the applied force is increased the force of static friction also increases upto limiting friction.

(25) (a), (26) (a), (27) (a).
$F_{max} = kx + \mu mg$
$F_{min} = kx - \mu mg$
$\therefore F_{max} + F_{min} = 2\mu mg$
or $2 = 2\mu 10$
$\therefore \mu = 0.1$
$F_{max} + F_{min} = 2kx$ ....(1)
From graph, $F_{max} + F_{min} = 5$ and $x = 0.1$
Putting in eq. (1)
$t = 2k(0.1); k = 25\ N/m$
When $x = 0.03$
$kx = 25 \times 0.03 = 0.75\ N$, which is less than μmg
$= 0.1 \times 10 = 1N$
$\therefore$ The block will be at rest, without applying force F.

(28) (b) It is easier to pull a heavy object than to push it on a level ground. Statement-1 is true. This is because the normal reaction in the case of pulling is less as compared by pushing. ($f = \mu N$). Therefore the functional force is small in case of pulling.
Statement-2 is true but is not the correct explanation of statement-1.

(29) (c) W = (force) × (displacement of point of application)

(30) (d) Statement – 2 is false because friction force may be more than applied force when body is retarding and external force is acting on body.

DAILY PRACTICE PROBLEMS

PHYSICS SOLUTIONS

12

(1) (a) Here $\vec{F} = -\hat{i} + 2\hat{j} + 3\hat{k}$ &

$\vec{d} = (0-0)\,\hat{i} + (0-0)\,\hat{j} + (4-0)\,\hat{k} = 4\,\hat{k}$

$\therefore$ W (Work done) $= \vec{F}.\vec{d} = (-\hat{i} + 2\hat{j} + 3\hat{k})\,.\,4\hat{k} = 12$ J

(2) (a) The minimum force with a body is to be pulled up along the inclined plane is mg (sin θ + μ cos θ)

Work done, $W = \vec{F}.\vec{d}$

$= Fd\cos\theta = mg\,(\sin\theta + \mu\cos\theta) \times d$

$= 5 \times 9.8\,(\sin 60^o + 0.2\cos 60^o) \times 2 = 98.08$ J

(3) (d) $W = \int_0^5 F\,dx = \int_0^5 (7 - 2x + 3x^2)\,dx$

$$= [7x]_0^5 - \left[\frac{2x^2}{2}\right]_0^5 + \left[\frac{3x^3}{3}\right]_0^5 = 135 \text{ Joule}$$

(4) (d) Given that, power = Fv = P = constant

or $m\dfrac{dv}{dt}v = P$ $\quad$ [as $F = ma = \dfrac{mdv}{dt}$]

or $\int v\,dv = \int \dfrac{P}{m}dt \Rightarrow \dfrac{v^2}{2} = \dfrac{P}{m}t + C_1$

Now as initially, the body is at rest

i.e v = 0 at t = 0 so, $C_1 = 0$

$\therefore\ v = \sqrt{\dfrac{2Pt}{m}}$

(5) (b) By definition $v = \dfrac{ds}{dt}$ $\quad$ or $\dfrac{ds}{dt} = \left(\dfrac{2Pt}{m}\right)^{1/2}$

$\Rightarrow \int ds = \int\left(\dfrac{2Pt}{m}\right)^{1/2} dt \Rightarrow s = \left(\dfrac{2P}{m}\right)^{1/2}\dfrac{2}{3}t^{3/2} + C_2$

Now as t = 0, s = 0, so $C_2 = 0$

$s = \left(\dfrac{8P}{9m}\right)^{1/2} t^{3/2}$

(6) (c) The force acting on the particle $= \dfrac{mdv}{dt}$

Power of the force $= \left(\dfrac{mdv}{dt}\right) v = k$ (constant)

$\Rightarrow m\dfrac{v^2}{2} = kt + c$ $\quad$ .....(1)

At t = 0, v = u $\quad \therefore\ c = \dfrac{mu^2}{2}$

Now from (1), $m\dfrac{v^2}{2} = kt + \dfrac{mu^2}{2}$

$\Rightarrow \dfrac{1}{2}m(v^2 - u^2) = kt$ $\quad$ .....(2)

Again $\dfrac{mdv}{dt}v = k$

$\Rightarrow m.v\dfrac{dv}{dx}v = k \Rightarrow mv^2\,dv = kdx$

Intergrating,

$\dfrac{1}{3}m(v^3 - u^3) = kx$ $\quad$ .....(3)

From (2) and (3),

$t = \dfrac{3}{2}\left(\dfrac{v^2 - u^2}{v^3 - u^3}\right)(x)$

(7) (b) Mass of the chain hanging $= 4 \times 3 = 12$ kg

Shift in center of gravity $= 4/2 = 2$m

Work done, $W = mgh = 12 \times 9.8 \times 2 = 235.2$ J

(8) (b) Mass of 2 litre, water = 2 kg

Total mass to be lifted = 2 + 0.5 = 2.5 kg

Work done, $W = mgh = 2.5 \times 9.8 \times 6 = 147$ J

(9) (b) The following two forces are acting on the body

(i) Weight mg is acting vertically downward

(ii) The push of the air is acting upward.

As the body is accelerating downward, the resultant force is (mg – F)

Workdone by the resultant force to fall through a vertical distance of 20 m = (mg – F) × 20 joule

Gain in the kinetic energy $= \dfrac{1}{2}mv^2$

Now the workdone by the resultant force is equal to the change in kinetic energy i.e.

$(mg - F)\,20 = \dfrac{1}{2}mv^2$ (From work-energy theorem)

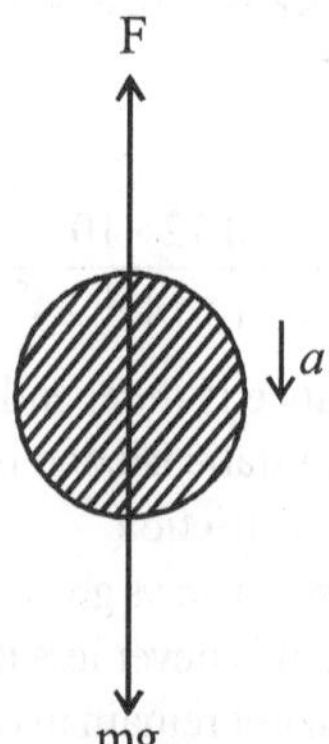

or $(50-F)\,20=\frac{1}{2}\times 5\times (10)^2$

or $50-F=12.5$ or $F=50-12.5$

$\therefore F=37.5\,N$

Work done by the force $=-37.5\times 20=-750$ joule

(The negative sign is used because the push of the air is upwards while the displacement is downwards.)

(10) **(a)**

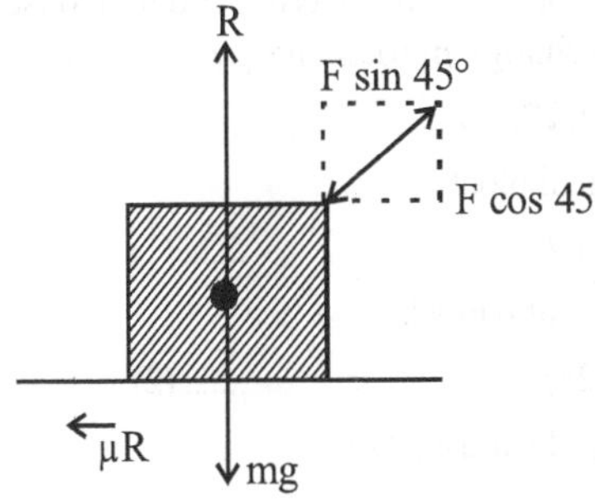

The different forces acting on the block are shown in fig. Now we have

$R+F\sin 45°=mg$ ....(1)

$F\cos 45°=\mu R$ ....(2)

From equation (1) and (2)

$$\therefore\ F=\frac{\mu\, mg}{\cos 45°+\mu\sin 45°}$$

Substituting the given values, we have

$$F=\frac{0.20\times(5\times 9.78)}{(0.707)+(0.20\times 0.707)}=11.55\,N$$

The block is pulled through a horizontal distance $r=20$ metre

Hence, the work done

$W=F\cos 45°\times r=(11.55\times 0.707)\times 20=163.32$ Joule

(11) **(c)**

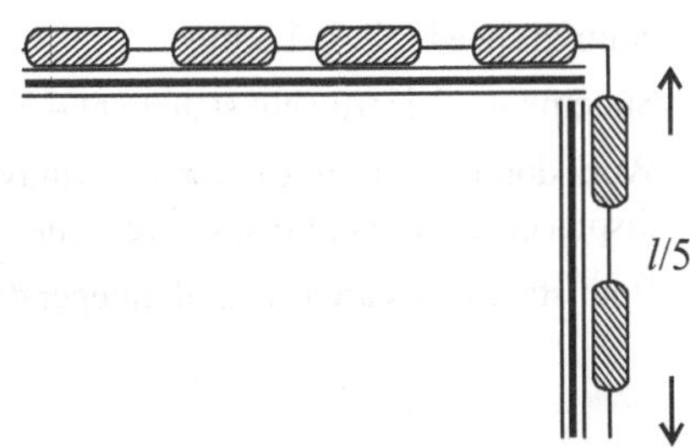

Mass of the hanging part of the chain = (m/5) The weight mg/5 acts at the centre of gravity of the hanging chain, i.e., at a distance = $\ell/10$ below the surface of a table.

The gain in potential energy in pulling the hanging part on the table.

$$U=\frac{mg}{5}\times\frac{\ell}{10}=\frac{mg\ell}{10}$$

$\therefore$ Work done $=U=mg\,\ell/50$

(12) **(a)** At maximum speed all the power is used to overcome the resistance to motion. Hence if the maximum speed is v, then $50000=1000\times v$ or $v=50$ m/s

At 25 m/s, let the pull of the engine be P, then the power

or $P=\frac{50,000}{25}=2000\,N$

Now resultant force $=2000-1000=1000$ N

Applying Newton's law ; $F=ma$, we have

$1000=1000\,a$ or $a=1.0\ m/s^2$

(13) **(a)** 1 mole i.e.235 gm of uranium contains 6×10^{23} atoms, so 2 kg i.e. 2×10^3 gm of uranium will contain

$$=\frac{2\times 10^3\times 6\times 10^{23}}{235}\text{ atoms}=5.106\times 10^{24}\text{ atoms}$$

Now as in each fission only one uranium atom is consumed i.e. Energy yield per uranium atom

$=185\text{ MeV}=185\times 1.6\times 10^{-13}\,J=2.96\times 10^{-11}\,J$

So Energy produced by 2 kg uranium

$=(\text{No. of atoms})\times(\text{energy/atom})$

$=5.106\times 10^{24}\times 2.96\times 10^{-11}=1.514\times 10^{-14}\,J$

As 2 kg uranium is consumed in 30 days i.e. 1.51×10^{-14} J of energy is produced in the reactor in 30 days i.e. 2.592×10^6 sec

So, power output of reactor

$$=\frac{E}{t}=\frac{1.514\times 10^{14}\,J}{2.592\times 10^6\,S}=58.4\text{ MW}$$

(14) **(c)** When the vehicle of mass m is moving with velocity v, the kinetic energy of the where $K=\frac{1}{2}mv^2$ and if S is the stopping distance, work done by the friction

$W=FS\cos\theta=\mu MgS\cos 180°=-\mu MgS$

So by Work-Energy theorem,

$W=\Delta K=K_f-k_i$

$$\Rightarrow -\mu MgS=0-\frac{1}{2}Mv^2\Rightarrow S=\frac{v^2}{2\mu g}$$

(15) **(a)** As $T=(2\pi/\omega)$,

so $\omega=2\pi/(3.15\times 10^7)=1.99\times 10^{-7}$ rad/s

Now $v=r\omega=1.5\times 10^{11}\times 1.99\times 10^{-7}\approx 3\times 10^4$ m/s

Now by work - energy theorem ,

$$W=K_f-K_i=0-\frac{1}{2}mv^2$$

$$=-\frac{1}{2}\times 6\times 10^{24}(3\times 10^4)^2=-2.7\times 10^{33}\,J$$

Negative sign means force is opposite to the motion.

(16) **(b)** As the particle is moving in a circle, so

$\frac{mv^2}{r}=\frac{k}{r^2}$ Now K.E $=\frac{1}{2}mv^2=\frac{k}{2r}$

Now as $F=-\frac{dU}{dr}$

$$\Rightarrow \text{P.E, } U=-\int_{\infty}^{r} F\,dr=\int_{\infty}^{r}+\left(\frac{k}{r^2}\right) dr=-\frac{k}{r}$$

So total energy $=U+K.E=-\frac{k}{r}+\frac{k}{2r}=-\frac{k}{2r}$

Negative energy means that particle is in bound state .

(17) **(c)** Let the mass of the person is m

Work done, W = P.E at height h above the earth surface
$=(M+m)\,gh$

or $4900=(M+10)\,9.8\times 10$ or $M=40$ kg

(18) **(b)** As the rod is kept in vertical position the shift in the centre of gravity is equal to the half the length = $\ell/2$

Work done $W=mgh=mg\frac{\ell}{2}=20\times 9.8\times\frac{4}{2}=392$ J

(19) **(a)** We know that the increase in the potential energy

$$\Delta U=GmM\left[\frac{1}{R}-\frac{1}{R'}\right]$$

According to question $R'=R+R=2R$

$$\Delta U=GMm\left[\frac{1}{R}-\frac{1}{2R}\right]=\frac{GMm}{2R}$$

(20) **(c)** In first case, $W_1=\frac{1}{2}m(v_1)^2+mgh$

$$=\frac{1}{2}m(12)^2+m\times 10\times 12$$

$=72\,m+120\,m=192\,m$

and in second case, $W_2=mgh=120$ m

The percentage of energy saved

$$=\frac{192m-120m}{192m}\times 100=38\%$$

(21) **(c)** Given that, $U(x)=\frac{a}{x^{12}}-\frac{b}{x^6}$

We know $F=-\frac{du}{dx}=(-12)\,a\,x^{-13}-(-6b)\,x^{-7}=0$

or $\frac{6b}{x^7}=\frac{12a}{x^{13}}$

or $x^6=12a/6b=2a/b$ or $x=\left(\frac{2a}{b}\right)^{1/6}$

(22) **(a)** $W=0$

(23) **(b)**

(1) There will be an increase in potential energy of the system if work is done upon the system by a conservative force.

(2) The work done by the external forces on a system equals the change in total energy

(24) **(a)**

(1) The work done by all forces equal to change in kinetic energy

(2) The work done by conservative forces equal to change in potential energy

(3) The work done by external and nonconservative forces equal to change in total energy

(25) (b), (26) (b), 27. (c)

For vertical block

$mg=kx+2T$ .... (1)

For horizontal block

$T=k(2x)$ .... (2)

From eq. (1) and eq. (2)

$$x=\frac{mg}{5k}=0.2\text{m}$$

∴ Extension of vertical spring = 0.2m

Extension of horizontal spring $=2x=0.4$m

From conservation of energy

$$mgx=\frac{1}{2}kx^2+\frac{1}{2}k(2x)^2+\frac{1}{2}mv^2+\frac{1}{2}m(2v)^2$$

$$mgx=\frac{3}{2}kx^2+\frac{3}{2}mv^2$$

$$\frac{7}{10}mgx=\frac{3}{2}mv^2$$

$$v=\sqrt{\frac{7}{15}gx}$$

Required speed = 2v = 1.9 m/s

(28) **(d)** Statement – 1 is true but statement – 2 is false.

(29) **(a)** Work done by action reaction force may be zero only if displacement of both bodies are same.

(30) **(b)** Both statements are true and independent.

DAILY PRACTICE PROBLEMS

PHYSICS SOLUTIONS 13

(1) (a) Let m_1 and m_2 be the masses of bullet and the rifleman and v_1 and v_2 their respective velocities after the first shot. Initially the rifleman and bullet are at rest, therefore initial momentum of system = 0.
As external force is zero, momentum of system is constant i.e. initial momentum = final momentum

$$= m_1v_1 + m_2v_2$$

$$\text{or } v_2 = \frac{m_1v_1}{m_2} = -\frac{(10\times10^{-3}\text{kg})(800\text{m/s})}{100\text{kg}} = -0.08 \text{ m/s}$$

Velocity acquired after 10 shots

$$= 10\,v_2 = 10\times(-0.08) = -0.8 \text{ m/s}$$

i.e, the velocity of rifle man is 0.8 m/s in a direction opposite to that of bullet.

(2) (c) Let the mass of block and bullet be M and m respectively If v is the velocity of bullet and V is the velocity of block with bullet embedded in it,
Now according to conservation of momentum,

$$mv = (M+m)V$$

$$(10\times10^{-3})(300) = (290\times10^{-3} + 10\times10^{-3})V \text{ or } V = 10 \text{ m/s}$$

The kinetic energy just after impact is $\frac{1}{2}(M+m)V^2$, which is lost due to work done on it by the force of friction F. Since force of friction F = μ (M+m)g and the work done is given by Fd, we have

$$\frac{1}{2}(M+m)V^2 = \mu(M+m)gd$$

$$\text{or } \mu = \frac{1}{2}\frac{V^2}{gd} = \frac{1}{2}\times\frac{10^2}{(10)(15)} = \frac{1}{3}$$

(3) (a) Let the initial velocity of the bullet of mass m = 20 g = 0.020 kg be u and v the velocity with which each plate moves.
The initial momentum of system (bullet + plate) = mu

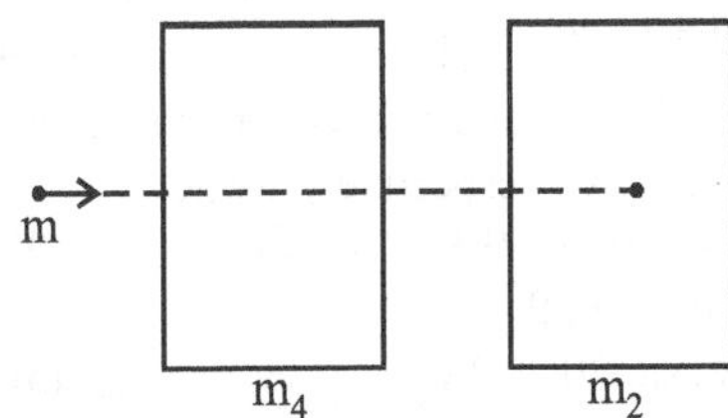

Final momentum of system = $m_1v + (m_2 + m)v$
(Since bullet remains in 2nd plate)
$\therefore$ According to principle of conservation of momentum
i.e. $mu = m_1v + (m_2 + m)v$,
i.e. $0.02u = 4v$

$$\text{or } u = \frac{4}{.02} = 200\,v \quad \text{.........(1)}$$

Let v_1 be the velocity of the bullet as it comes out of plate m_1. Applying conservation of linear momentum for the collision of bullet with plate m_2.
i.e. $mv_1 = (m_2 + m)v$

$$0.02\,v_1 = (2.98 + 0.02)\,v$$

$$\text{i.e. } v_1 = \frac{3}{.02}v = 150\,v \quad \text{.........(2)}$$

Required percentage loss in initial velocity of bullet

$$\frac{u - v_1}{u}\times100\% = \frac{200v - 150v}{200v}\times100 = 25\%$$

(4) (a) Part (I) - The horizontal component of the momentum of the bullet is equal to the momentum of the block with the bullet

$$mu\cos\alpha = (M+m)V \quad \text{..........(1)}$$

Where V is the velocity of the block plus bullet embedded in it.
Part (II) - As the block can move as a pendulum, the block rises till its kinetic energy is converted into potential energy. So, if the block rises upto a height h,

$$\frac{1}{2}(M+m)V^2 = (M+m)gh \quad \text{..........(2)}$$

From (1) & (2)

$$h = \left(\frac{m}{M+m}\right)^2\cdot\frac{u^2\cos^2\alpha}{2g}$$

$$= \left(\frac{20\times10^{-3}}{2}\right)^2.(200)^2\cdot\frac{\cos^2 30^\circ}{(2)(10)} = 0.15\text{ m}$$

(5) (c) Initial velocity of bullet, $u_1 = 500$ m/s
Let v_1 and v_2 be the speeds of bullet and block after collision

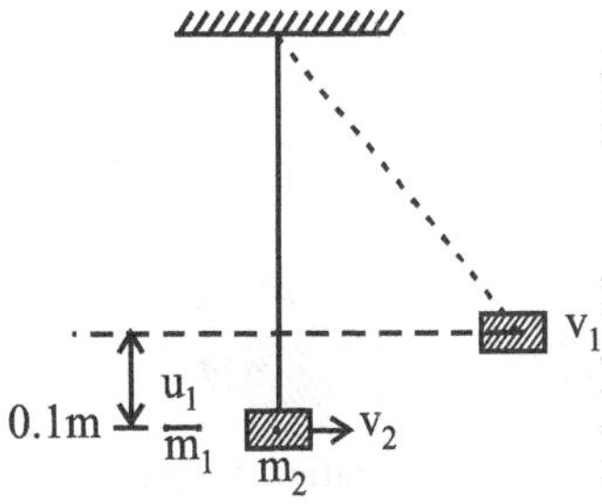

respectively then, $\frac{1}{2}mv_2^2 = mgh$

$$\Rightarrow v_2 = \sqrt{2gh} = \sqrt{2\times9.8\times0.1} = 1.4\text{ m/s}$$

According to principle of conservation of linear momentum, We have

$$m_1u_1 + 0 = m_1v_1 + m_2v_2$$

$$\text{or } 0.01\times500 = 0.01\,v_1 + 2\times1.4 \Rightarrow v_1 = 220\text{ m/s}$$

(6) **(a)** The rate of change of momentum is equal to force

$$F=\frac{dp}{dt}=v\frac{dm}{dt} \text{ (Here v is constant)}$$

Here $v=4\times10^3$ m/s & $\frac{dm}{dt}=50\times10^{-3}$ kg/s

$\therefore F=4\times10^3\times50\times10^{-3}=200$ N

(7) **(a)** Given that, Initial velocity = u

Final velocity $=\frac{u}{4}$

So by conservation of momentum, we have

$$1\times u+0=1\times\frac{u}{4}+m\times v_2 \Rightarrow mv_2=\frac{3u}{4} \quad ....(1)$$

and by conservation of energy, we have

$$\frac{1}{2}\times1\times u^2+0=\frac{1}{2}\times1\left(\frac{u}{4}\right)^2+\frac{1}{2}mv_2^2$$

$$\text{or } v_2^2=\frac{15}{16}u^2 \quad ....(2)$$

From equation (1) and (3),

$$\frac{(mv_2)^2}{mv_2^2}=\frac{(9/16)u^2}{(15/16)u^2} \text{ or } m=0.6\text{kg}$$

(8) **(a)** Initial momentum of the balls

$=m\times9+m\times0=9m$ ....(1)

where m is the mass of each ball.

Let after collision their velocities are v_1 and v_2 respectively.

Final momentum of the balls after collision along the same line $=mv_1\cos30°+mv_2\cos30°$

$$=\frac{mv_1\sqrt{3}}{2}+\frac{mv_2\sqrt{3}}{2} \quad ....(2)$$

According to law of conservation of momentum

$$9m=\frac{mv_1\sqrt{3}}{2}+\frac{mv_2\sqrt{3}}{2}$$

$$\frac{9\times2}{\sqrt{3}}=v_1+v_2 \quad ....(3)$$

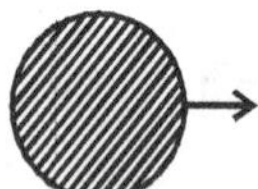

Stationary ball

(a) Before collision

The initial momentum of the balls along perpendicular direction = 0.

Final momentum of balls along the perpendicular direction

$$=mv_1\sin30°-mv_2\sin30°=\frac{m}{2}(v_1-v_2)$$

Again by the law of conservation of momentum

$(m/2)(v_1-v_2)=0$

$\therefore (v_1-v_2)=0$ ......(4)

Solving equations (3) and (4), we have

$v_1=3\sqrt{3}$ m/s and $v_2=3\sqrt{3}$ m/s

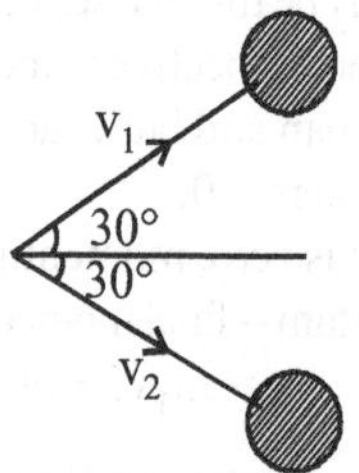

According to law of conservation of energy

Energy before collision = Energy after collision

$$\frac{1}{2}mu_1^2+\frac{1}{2}mu_2^2=\frac{1}{2}mv_1^2+\frac{1}{2}mv_2^2$$

$$\frac{1}{2}m(9)^2+0=\frac{1}{2}m(3\sqrt{3})^2+\frac{1}{2}m(3\sqrt{3})^2$$

$$\frac{81\text{ m}}{2}=\frac{54\text{ m}}{2}$$

L.H.S. # R.H.S.

i.e., energy is not conserved in this collision or this is a case of inelastic collision.

(9) **(a)** The situation is shown in fig.

Let v_1 and v_2 be the velocities of two pieces after explosion.

Applying the law of conservation of energy, we have

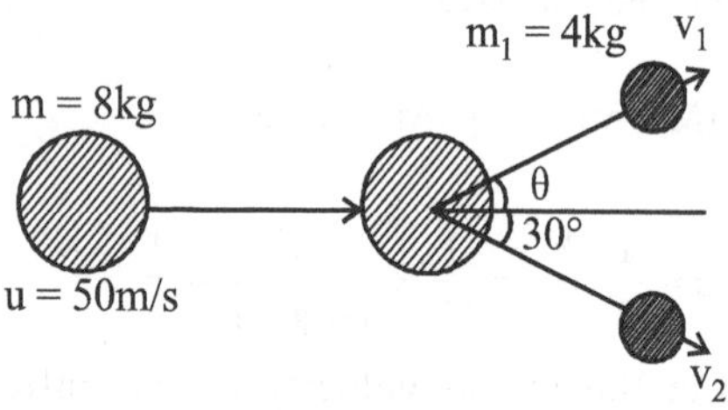

$$\frac{1}{2}(8)(50)^2+15000=\frac{1}{2}(D)v_1^2+\frac{1}{2}(D)v_2^2$$

or $25000=2(v_1^2+v_2^2)$ ......(1)

Applying the law of conservation of momentum along x-axis and y-axis respectively, we get

$8(50)=4v_1\cos\theta+v_2\cos30°$ ......(2)

and $0=4v_1\sin\theta$

$=4v_2\sin30°=2v_2$ ......(3)

$$\text{or } \sin\theta=\frac{v_2}{2v_1} \quad ......(4)$$

From eq. (2)

$100=v_1\cos\theta+v_2\cos30°$

(10) **(a)** Let m be the mass of the rocket and v_r the relative velocity of the gas ejecting from the rocket. Suppose the fuel is burnt at a rate (dm/dt) to provide the rocket an acceleration a.

Then $a = \frac{v_r}{m}\left(\frac{dm}{dt}\right) - g$ ......(1)

Here $v_r = 250$ m/s, m = 500 kg, g = 10 m/s^2 and a = 20 m/s^2

Now from (1) $\frac{dm}{dt} = \frac{m}{v_r}(a+g)$

$= \frac{500}{250}(20+10) = 60$ kg/s

(11) **(a)** Let m_1 and m_2 be the masses of electron and hydrogen atom respectively. If u_1 and v_1 be the initial and final velocities of electron, then initial kinetic energy of electron

$K_i = \left(\frac{1}{2}\right) m u_1^2$

Final kinetic energy of electron $K_f = \left(\frac{1}{2}\right) m v_1^2$

Fractional decrease in K.E.,

$\frac{K_i - K_f}{K_i} = 1 - \frac{v_1^2}{u_1^2}$ ....(1)

For such a collision, we have

$v_1 = \left(\frac{m_1 - m_2}{m_1 + m_2}\right) u_1$

$\therefore \frac{v_1}{u_1} = \left(\frac{m_1 - m_2}{m_1 + m_2}\right)$ ....(2)

From eqs. (1) and (2) we have

$\frac{K_i - K_f}{K_i} = 1 - \left(\frac{m_1 - m_2}{m_1 + m_2}\right)^2 = \frac{4m_1 m_2}{(m_1 + m_2)^2}$

or $\frac{K_i - K_f}{K_i} = \frac{4(m_2/m_1)}{(1+m_2/m_1)^2} = \frac{4 \times 1850}{(1+1850)^2}$

$= 0.00217 = 0.217\%$

(12) **(c)**

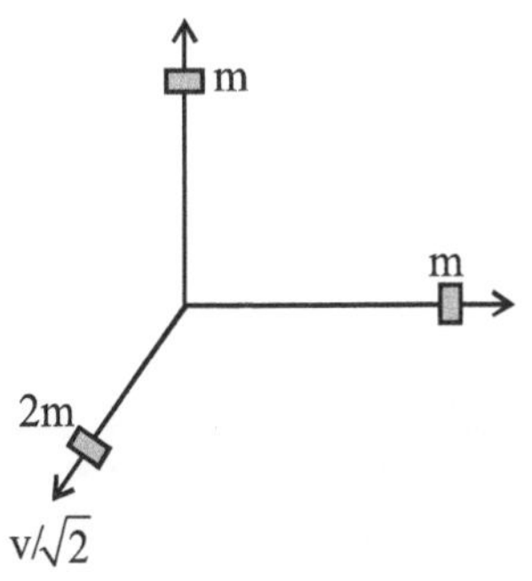

Now the total energy released in the explosion

$= \frac{1}{2}mv^2 + \frac{1}{2}mv^2 + \frac{1}{2}2m\left(\frac{v}{\sqrt{2}}\right)^2 = \frac{3}{2}mv^2$

(13) **(a)** Let the speed of the body before explosion be u. After explosion, if the two parts move with velocities u_1 and u_2 in the same direction, then according to conservation of momentum,

$\alpha M u_1 + (1-\alpha) M u_2 = Mu$

The kinetic energy T liberated during explosion is given

by $T = \frac{1}{2}\alpha M u_1^2 + \frac{1}{2}(1-\alpha) M u_2^2 - \frac{1}{2} Mu^2$

$= \frac{1}{2}\alpha M u_1^2 + \frac{1}{2}(1-\alpha) M u_2^2 - \frac{1}{2M}[\alpha M u_1 + (1-\alpha) M u_2]^2$

$= \frac{1}{2} M \alpha (1-\alpha)[u_1^2 + u_2^2 - 2u_1 u_2]$

$(u_1 - u_2)^2 = \frac{2T}{\alpha(1-\alpha)M}$

$\Rightarrow (u_1 - u_2) = \sqrt{\frac{2T}{\alpha(1-\alpha)M}}$

(14) **(a)** The situation is shown in fig.

Let A and B be two pieces of equal mass (1/5 kg) which fly off perpendicular to a each other with equal velocity (30 m/sec)

Momentum of A or B = (1/5 × 30)

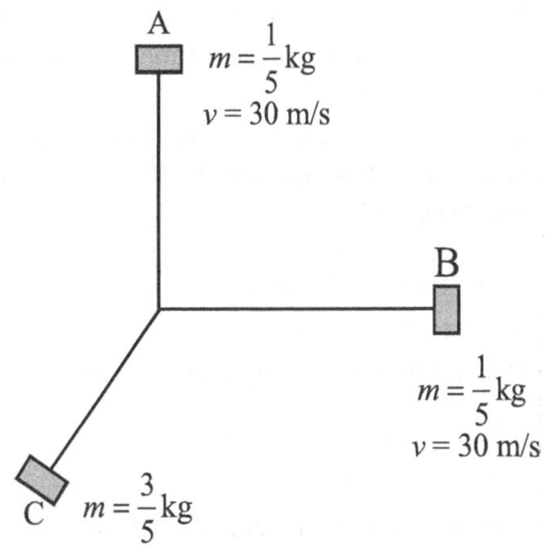

∴ Resultant momentum

$= \sqrt{\{(1/5) \times 30\}^2 + \{(1/5) \times 30\}^2} = 6\sqrt{2}$ kg m/sec

along the bisector of ∠ AOB

$(3/5) \times v = 6\sqrt{2} \Rightarrow v = 10\sqrt{2}$ m/sec

(15) **(c)** The situation is shown in fig.

Equating the total initial and final momentum along each axis, we get

$mv_1 + 0 = (M+m) v' \cos\theta$ ....(A)

$0 + Mv_2 = (M+m) v' \sin\theta$ ....(B)

Squaring and adding eq. (A) and (B), we get

$(mv_1)^2 + (Mv_2)^2 - (M+m)^2 v'^2$ (C)

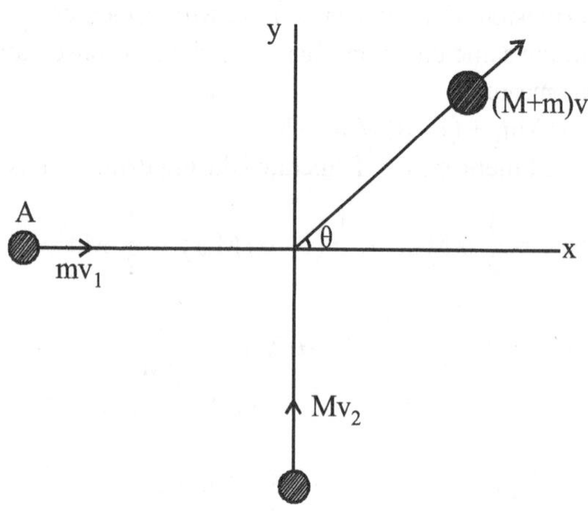

The final momentum

$P = (M+m)v' = \sqrt{[(mv_1)^2 + (Mv_2)^2]}$

[form eqn. (3)]

Dividing eqn. (2) by eqn. (1), we have

$\tan\theta = \frac{Mv_2}{mv_1}$ or $\theta = \tan^{-1}\left(\frac{Mv_2}{mv_1}\right)$

(16) **(a)** Let the angle of reflection be θ' and the magnitude of velocity after collision be v'. As there is no force parallel to the wall, the component of velocity parallel to the surface remains unchanged.

Therefore, $v' \sin\theta' = v \sin\theta$ ......(1)

As the coefficient of restitution is e, for perpendicular component of velocity

Velocity of separation = e x velocity of approach

$-(v' \cos\theta' - 0) = -e(v\cos\theta - 0)$ ......(2)

From (1) and (2)

$$v' = v\sqrt{\sin^2\theta + e^2\cos^2\theta}$$

and $\tan\theta' = \tan\theta/e$

(17) **(a)** The fraction of energy lost is given by,

$$\frac{\Delta E}{E} = \frac{mg(h-h')}{mgh} = \frac{h-h'}{h}$$

given that, h = 2 meter and h' = 1.5 meter

$$\therefore \frac{\Delta E}{E} = \frac{2-1.5}{2} = \frac{1}{4}$$

(18) **(a)** A bullet is fired from the gun. The gun recoils, the kinetic energy of the recoil shall be less than the kinetic energy of the bullet.

(19) **(a)** Conservation of linear momentum is equivalent to Newton's second law of motion

(20) **(a)** In an inelastic collision momentum is conserved but kinetic energy is not.

(21) **(a)** Inelastic collision is the collision of electron and positron to an inhilate each other.

(22) **(a)** Total kinetic energy is not conserved in inelastic collisions but momentum is conserved

(23) **(a)** (1) when $m_1 = m_2$ and m_2 is stationary, there is maximum transfer of kinetic energy in head an collision

(2) when $m_1 = m_2$ and m_2 is stationary, there is maximum transfer of momentum in head on collision

(3) when $m_1 >> m_2$ and m_2 is stationary, after head on collision m_2 moves with twice the velocity of m_1.

(24) **(a)** Momentum remains conserved

(25) **(a)** Speed of particle after the collision

$$= \sqrt{\left(\frac{15}{43} \times \sqrt{3}\right)^2 + 25} = 5.036 \text{ m/s}$$

(26) **(b)** Speed of the sphere just after collision $= \frac{30}{43}$ m/s

(27) **(a)** Angular speed of sphere is zero as impulse due to collision passes through centre of sphere.

(28) **(c)** When e = 0, velocity of separation along common normal zero, but there may be relative velocity along common tangent.

(29) **(c)** Statement – 1 is false but statement – 2 is true.

(30) **(d)** Momentum remains constant before, during and after the collision but *KE* does not remain constant during the collision as the energy gets converted into elastic potential energy due to deformation.

DAILY PRACTICE PROBLEMS

PHYSICS SOLUTIONS 14

1. (c) $m_1 = 1, m_2 = 35.5_1, \vec{r}_1 = 0, \vec{r}_2 = 1.27\hat{i}$

$$\vec{r} = \frac{m_1\vec{r}_1 \times m_2\vec{r}_2}{m_1+m_2} \Rightarrow \vec{r} = \frac{35.5\times1.27}{1+35.5}\hat{i}$$

$$\vec{r} = \frac{35.5}{36.5}\times1.27\hat{i} = 1.24\hat{i}$$

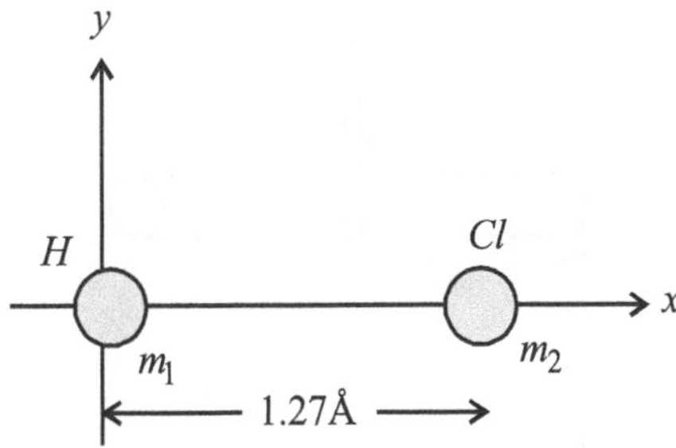

2. (d) $\vec{v}_{cm} = \frac{m_1\vec{v}_1 + m_2\vec{v}_2}{m_1+m_2}$

$$= \frac{2\times3+3\times2}{2+3} = \frac{12}{5} = 2.4 \text{ m/s}$$

3. (c) $m_1 = 12, m_2 = 16$

$\vec{r}_1 = 0\hat{i} + 0\hat{j}, r_2 = 1.1\hat{i} + 0\hat{j}$

$$\vec{r}_1 = \frac{m_1\vec{r}_1 + m_2\vec{r}_2}{m_1+m_2}$$

$$\vec{r}_1 = \frac{16\times1.1}{28}\hat{i} = 0.63\hat{i}$$

i.e. 0.63Å from carbon atom.

4. (a) $\vec{v}_{cm} = \frac{m_1\vec{v}_1 + m_2\vec{v}_2 + m_3\vec{v}_3}{m_1+m_2+m_3}$

$$= \frac{20\times10\hat{i} + 30\times10\hat{j} + 50\times10\hat{k}}{100}$$

$\therefore\ v_{cm} = 2\hat{i} + 3\hat{j} + 5\hat{k}$

5. (c) We can assume that three particles of equal mass m are placed at the corners of triangle.

$\vec{r}_1 = 0\hat{i} + 0\hat{j}, \vec{r}_2 = b\hat{i} + 0\hat{j}$

and $\vec{r}_3 = 0\hat{i} + h\hat{j}$

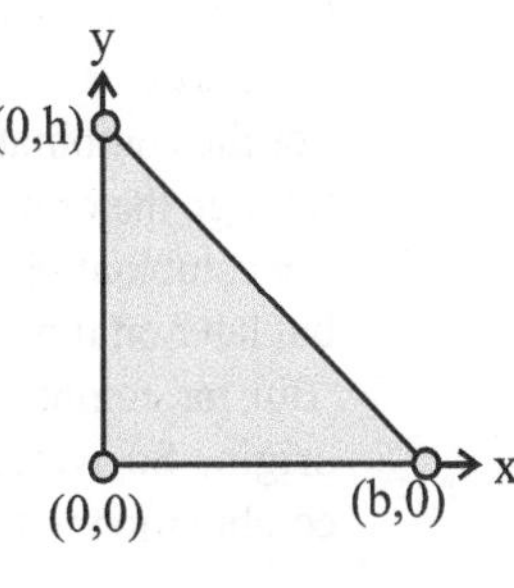

$$\therefore\ \vec{r}_{cm} = \frac{m_1\vec{r}_1 + m_2\vec{r}_2 + m_3\vec{r}_3}{m_1+m_2+m_3}$$

$$= \frac{b}{3}\hat{i} + \frac{h}{3}\hat{j}$$

i.e. coordinates of centre of mass is $\left(\frac{b}{3}, \frac{h}{3}\right)$

6. (b) $\vec{v}_{cm} = \frac{m_1\vec{v}_1 + m_2\vec{v}_2}{m_1+m_2}$

$$= \frac{2\times2+4\times10}{2+4} = 7.3 \text{ m/s}$$

7. (b) Let $m_1 = m, m_2 = 2m, m_3 = 3m, m_4 = 4m$

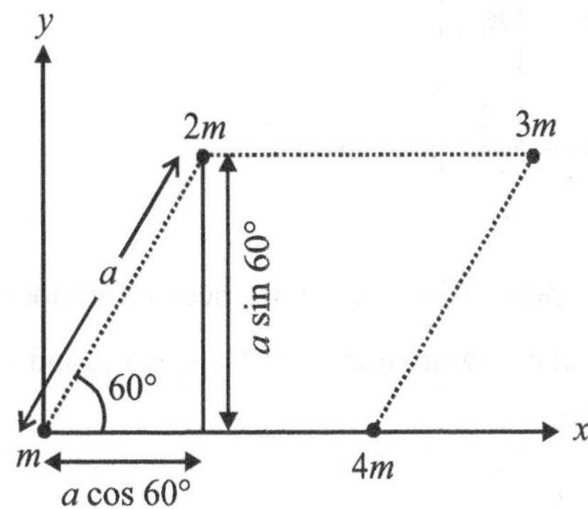

$\vec{r}_1 = 0\hat{i} + 0\hat{j}$

$\vec{r}_2 = a\cos60\hat{i} + a\sin60\hat{j} = \frac{a}{2}i + \frac{a\sqrt{3}}{2}\hat{j}$

$\vec{r}_3 = (a + a\cos60)\hat{i} + a\sin60\hat{j} = \frac{3}{2}a\hat{i} + \frac{a\sqrt{3}}{2}\hat{j}$

$\vec{r}_4 = a\hat{i} + 0\hat{j}$

by substituting above value in the following formula

$$\vec{r} = \frac{m_1\vec{r}_1 + m_2\vec{r}_2 + m_3\vec{r}_3 + m_4\vec{r}_4}{m_1+m_2+m_3+m_4} = 0.95ai + \frac{\sqrt{3}}{4}a\hat{j}$$

So the location of centre of mass $\left[0.95a, \frac{\sqrt{3}}{4}a\right]$

8. (d)

9. (d) $m_1 = 2\text{kg}, m_2 = 4\text{kg}, \vec{v}_1 = 2m/s, \vec{v}_2 = -10m/s$

$$\vec{v}_{cm} = \frac{m_1\vec{v}_1 + m_2\vec{v}_2}{m_1+m_2}$$

$$= \frac{2\times20 - 4\times10}{2+4} = 0m/s$$

10. (a) As initially both the particles were at rest therefore velocity of centre of mass was zero and there is no external force on the system so speed of centre of mass remains constant i.e. it should be equal to zero.

11. (a) For translatory motion the force should be applied on the centre of mass of the body, so we have to calculate the location of centre of mass of 'T' shaped object.
Let mass of rod AB is *m* so the mass of rod CD will be 2m.
Let y_1 is the centre of mass of rod AB and y_2 is the centre of mass of rod CD. We can consider that whole mass of the rod is placed at their respective centre of mass i.e., mass m is placed at y_1 and mass 2 m is placed at y_2.

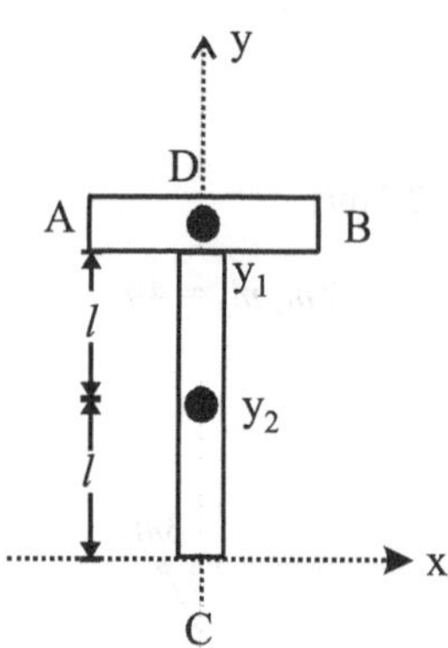

Taking point 'C' at the origin, position vector of point y_1 and y_2 can be written as $\vec{r_1} = 2l\,\hat{j}, \vec{r_2} = l\,\hat{j}$, and $m_1 = m$ and $m_2 = 2m$
Position vector of centre of mass of the system

$$\vec{r}_{cm} = \frac{m_1\vec{r_1} + m_2\vec{r_2}}{m_1 + m_2} = \frac{m2l\,\hat{j} + 2ml\,\hat{j}}{m + 2m}$$

$$= \frac{4ml\,\hat{j}}{3m} = \frac{4}{3}l\,\hat{j}$$

Hence the distance of centre of mass from $C = \frac{4}{3}l$

12. (a) Initial acceleration is zero of the system. So it will always remain zero because there is no external force on the system.

13. (b) According to figure let A is the origin and co-ordinates of centre of mass be (x, y) then,

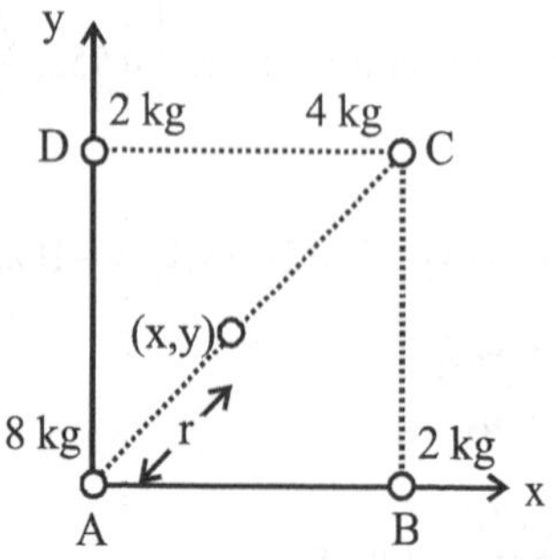

$$x = \frac{m_1x_1 + m_2x_2 + m_3x_3 + m_4x_4}{m_1 + m_2 + m_3 + m_4}$$

$$= \frac{0 + 2\times\frac{80}{\sqrt{2}} + 4\times\frac{80}{\sqrt{2}} + 0}{16} = \frac{30}{\sqrt{2}}$$

Similarly $y = \frac{30}{\sqrt{2}}$ so, $r = \sqrt{x^2 + y^2} = 30\,\text{cm}$

14. (b) Linear density of the rod varies with distance

$\frac{dm}{dx} = \lambda$ *(Given)* $\therefore dm = \lambda dx$

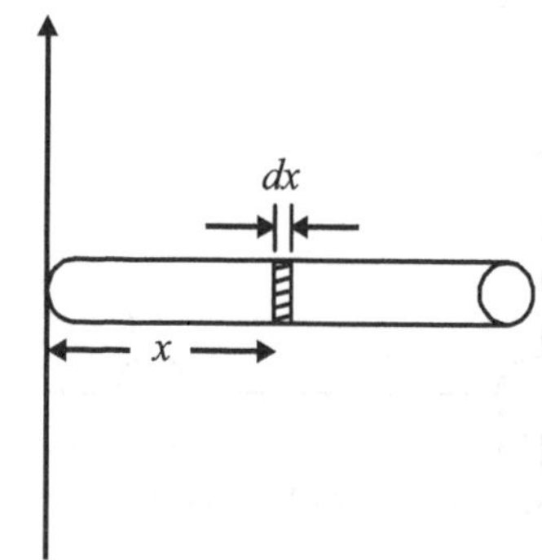

Position of centre of mass

$$x_{cm} = \frac{\int dm \times x}{\int dm}$$

$$= \frac{\int_0^3 (\lambda\,dx)\times x}{\int_0^3 \lambda\,dx}$$

$$= \frac{\int_0^3 (2+x)\times x dx}{\int_0^3 (2+x)dx} = \frac{\left[x^2 + \frac{x^3}{3}\right]_0^3}{\left[2x + \frac{x^2}{2}\right]_0^3}$$

$$= \frac{9+9}{6+\frac{9}{2}} = \frac{36}{21} = \frac{12}{7}m.$$

15. (c) Centre of mass lies always on the line that joins the two particles.
For the combination *cd* and *ab* this line does not pass through the origin.
For combination *bd*, initially it pass through the origin but later on it moves toward negative *x*-axis.
But for combination *ac* it will always pass through origin. So we can say that centre of mass of this combination will remain at origin.

16. (b) $x_{cm} = \frac{1\times 0+1\times PQ+1\times PR}{1+1+1} = \frac{PQ+PR}{3}$

and $y_{cm} = 0$

17. (a)

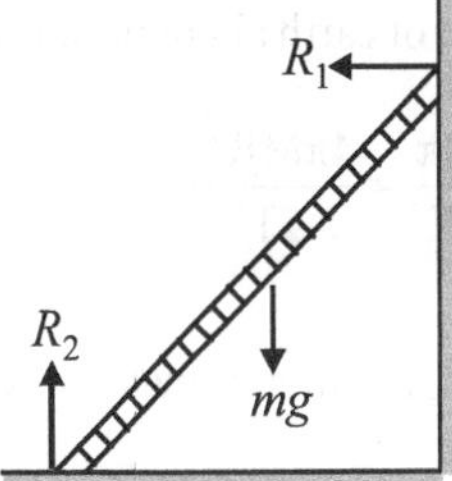

Due to net force in downward direction and towards left centre of mass will follow the path as shown in figure.

18. (a) Initially both the particles were at rest so $v_{cm} = 0$. As external force on the system is zero therefore velocity of centre of mass remains unaffected.

19. (a) $m_1\vec{r}_1 + m_2\vec{r}_2 = 0$

$\Rightarrow \frac{m}{4}15\hat{j} + \frac{3m}{4}\vec{r}_2 = 0$

$\Rightarrow \vec{r}_2 = -5\hat{j}$

i.e. larger fragment is at $y = -5$ cm.

20. (b) Centre of mass is closer to massive part of the body therefore the bottom piece of bat have larger mass.

21. (b) Initial position of centre of mass

$$r_{cm} = \frac{m_1x_1 + m_2x_2}{m_1+m_2} \quad ...(i)$$

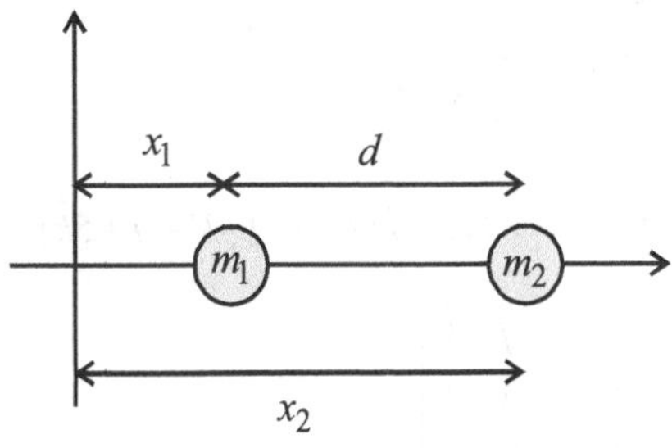

If the particle of mass m_1 is pushed towards the centre of mass of the system through distance *d* and to keep the centre of mass at the original position let second particle displaced through distance *d'* away from the centre of mass.

$$\text{Now } r_{cm} = \frac{m_1(x_1+d)+m_2(x_2+d')}{m_1+m_2} \quad ...(ii)$$

Equating (i) and (ii)

$$\frac{m_1x_1 ++m_2x_2}{m_1+m_2}$$

$$= \frac{m_1(x_1+d)+m_2(x_2+d')}{m_1+m_2}$$

By solving $d' = -\frac{m_1}{m_2}d$

Negative sign shows that particle m_2 should be displaced towards the centre of mass of the system.

22. (a) We know $m_1r_1 = m_2r_2 \Rightarrow m \times r = \text{constant} \therefore r \propto \frac{1}{m}$

23. (a) Depends on the distribution of mass in the body.

24. (a) $m_1r_1 = m_2r_2 \Rightarrow \frac{r_1}{r_2} = \frac{m_2}{m_1} \therefore r \propto \frac{1}{m}$

25. (b) $a_{cm} = \frac{m_1a_1 + m_2a_2}{m_1+m_2} = \frac{m\times 0 + m\times 3g}{m_1+m_2} = \frac{3g}{2}$

26. (b)

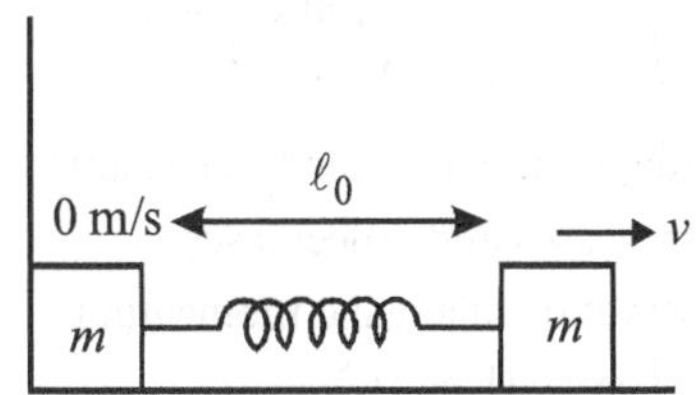

By *COE*, $\frac{1}{2}k\left(\frac{3mg}{k}\right)^2 = \frac{1}{2}mv^2$

$$v = \sqrt{\frac{9mg^2}{k}} = 3g\sqrt{\frac{m}{k}}$$

$$v_{cm} = \frac{m\times 0 + mv}{m+m} = \frac{v}{2} = \frac{3g}{2}\sqrt{\frac{m}{k}}$$

27. (a) By *COE* in CM-frame, $\frac{1}{2}\mu v_{ref}^2 = \frac{1}{2}kx^2$

$$\frac{1}{2}\frac{m}{2}\left(3g\sqrt{\frac{m}{k}}\right)^2 = \frac{1}{2}kx^2$$

$$\frac{9}{2}g^2\frac{m^2}{k} = kx^2 \;;\quad x = \frac{3mg}{\sqrt{2}k}$$

28. (b) Statement-1 is True, Statement-2 is True; Statement-2 is NOT a correct explanation for Statement-1.

29. (a) Initially the electron and proton were at rest so their centre of mass will be at rest. When they move towards each other under mutual attraction then velocity of centre of mass remains unaffected because external force on the system is zero.

30. (d) The centre of mass of a system of particles depends only on the masses of particles and the position of the particles relative to one another. The location of reference frame will not affect the location of centre of mass.

DAILY PRACTICE PROBLEMS

PHYSICS SOLUTIONS

15

1. (a) Initial angular momentum of ring. $L = I\omega = Mr^2\omega$
Final angular momentum of ring and four particles system $Mr^2\omega = \left(Mr^2 + 4mr^2\right)\omega' = \frac{M\omega}{M+4m}$

2. (b) The angular momontum of a system of particles is con served when no external torque acts on the system.

3. (c) Rotational kinetic energy $E\frac{L^2}{2I}$ $\therefore L = \sqrt{2EI}$

$$\Rightarrow \frac{L_A}{L_B} = \sqrt{\frac{E_A}{E_B} \times \frac{I_A}{I_B}} = \sqrt{100 \times \frac{1}{4}} = 5$$

4. (c) Angular momentum $L = I\omega$ constant
$\therefore$ I increases and ω decreases

5. (c) Conservation of angular momentum
$I_1\omega_1 + I_2\omega_2 = (I_1\omega_1 + I_2)_\omega$

Angular velocity of system $\omega = \frac{I_1\omega_1 + I_2\omega_2}{I_1 + I_2}$

$\therefore$ Rotational kinetic energy$= \frac{1}{2}(I_1 + I_2)\omega^2$

$$= \frac{1}{2}(I_1 + I_2)\left(\frac{I_1\omega_1 + I_2\omega_2}{I_1 + I_2}\right)^2 = \frac{(I_1\omega_1 + I_2\omega_2)^2}{2(I_1 + I_2)}$$

6. (d) Kinetic energy $E = \frac{1}{2}L\omega = \frac{1}{2}L \times 2\pi n$

$$\therefore E \propto L \times n \Rightarrow \frac{L_2}{L_1} = \frac{E_2}{E_1} \times \frac{n_1}{n_2}$$

$$\frac{L_2}{L_1} = \left[\frac{E_1/2}{E_1}\right] \times \left[\frac{n_1}{2n_1}\right] \Rightarrow L_2 = \frac{L_2}{4} = \frac{L}{4}$$

7. (b) $E = \frac{L^2}{2I}$. If boy stretches hgis arm then moment of inertia increases and accordingly kinetic energy of the system decreases because L = constant and $E \propto \frac{1}{I}$

8. (c)According to conservation of angular momentum

$$\therefore I_1\omega_1 = I_2\omega_2 \Rightarrow I_1\omega = (I_1 + I_2)\omega_2 \Rightarrow \omega_2 = \frac{I_1\omega}{I_1 + I_2}$$

9. (a) $L = \sqrt{2EI} = \sqrt{2 \times 10 \times 8 \times 10^{-7}} = 4 \times 10^{-3}\,kg\,m^2/s$

10. (d) Angular momentum, of earth about its axis of rotation,

$$L = I\omega = \frac{2}{5}MR^2 \times \frac{2\pi}{T} = \frac{4\pi MR^2}{5T}$$

11. (d) Angular momentum, $L = mvr = m\omega r^2 = m \times \frac{2\pi}{T} \times r^2$

$$= \frac{2 \times 3.14 \times 6 \times 10^{24} \times \left(1.5 \times 10^{11}\right)^2}{3.14 \times 10^7} = 2.7 \times 10^{40}\,kg - m^2/s$$

12. (c) $\omega = 2\pi n = \frac{2\pi \times 1800}{60} = 60\pi$ rad/s

$P = \tau \times \omega$

$$\Rightarrow \tau = \frac{P}{\omega} = \frac{100 \times 10^3}{60\pi} = 531\,N - m$$

13. (a) $\vec{\tau} = \frac{d\vec{L}}{dt} = \frac{L_2 - L_1}{\Delta t} = \frac{4A_0 - A_0}{4} = \frac{3A_0}{4}$

14. (c) $\alpha = \frac{2\pi(n_2 - n_1)}{t} = \frac{2\pi\left(0 - \frac{60}{60}\right)}{60}$

$$= \frac{-2\pi}{60} = \frac{-\pi}{30}\ \text{rad/sec}^2$$

$\therefore \tau = I\alpha$

$$= \frac{2 \times \pi}{30} = \frac{\pi}{15}N - m$$

15. (a) $\vec{\tau} = \vec{r} \times \vec{f} = (7\hat{i} + 3\hat{j} + \hat{k}) \times (-3\hat{i} + \hat{j} + 5\hat{k})$

$$\vec{\tau} = \begin{vmatrix} \hat{i} & \hat{j} & \hat{k} \\ 7 & 3 & 1 \\ -3 & 1 & 5 \end{vmatrix}$$

$$= \hat{i}(15 - 1) - \hat{j}(35 + 3) + \hat{k}(7 + 9)$$

$$= 14\hat{i} - 38\hat{j} + 16\hat{k}$$

16. (a) $\alpha = \frac{\tau}{I} = \frac{1000}{200} = 5\,rad/sec^2$

From $\omega = \omega_0 + \alpha t = 0 + 5 \times 3 = 15$ rad/s

17. (a) $\alpha = \frac{\tau}{I} = \frac{30}{2} = 15\ \text{rad/s}^2$

$$\because \theta = \omega_0 t + \frac{1}{2}\alpha t^2$$

$= 0 + \frac{1}{2} \times (15) \times (10)^2$

$= 750$ rad

18. (d) As the block remains stationary therefore
For translatory equilibrium

$\sum F_x = 0 \therefore F = N$

and $\sum F_y = 0 \therefore f = mg$

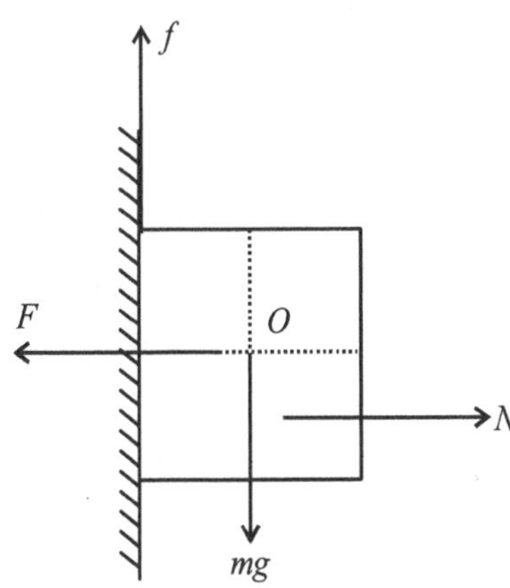

For rotational equilibrium $\sum \tau = 0$

By taking the torque of different forces about point 0

$\overrightarrow{\tau_F} + \overrightarrow{\tau_f} + \overrightarrow{\tau_N} + \overrightarrow{\tau_{mg}} = 0$

As F and mg passing through point O

$\therefore \overrightarrow{\tau_f} + \overrightarrow{\tau_N} = 0$

As $\tau_f \neq 0 \therefore \tau_N \neq 0$

and torque by friction and normal reaction will be in opposite direction.

19. (c) The velocity of the top point of the wheel is twice that of centre of mass and the speed of centre of mass is same for both the wheels (Angular speeds are different).

20. (d) $\alpha = \frac{2\pi(n_2 - n_1)}{t} = \frac{2\pi\left(\frac{4500-1200}{60}\right)}{10}$ rad/s^2

$= \frac{2\pi \frac{3300}{60}}{10} \times \frac{360}{2\pi} \frac{\text{degree}}{s^2}$

$\alpha = 1980$ degree/s^2

21. (b) $\theta = \omega_0 t + \frac{1}{2}\alpha t^2$

$\Rightarrow \theta = 100$ rad

$\therefore$ Number of revolution $= \frac{100}{2\pi} = 16$ (approx.)

22. (a) As mechanical contact is not made, total angular momentum remains constant.

$\therefore I\omega_0 =$ constant

Differentiating both sides,

$\Delta(I\omega_0) = 0$

$\Rightarrow I\,\Delta\omega_0 + \omega_0 \Delta I = 0$

$\Rightarrow \frac{\Delta\omega}{\omega} + \frac{\Delta I}{I} = 0 \Rightarrow \frac{\Delta I}{I} = -\frac{\Delta\omega_0}{\omega_0}$

Also, $\frac{\Delta\omega_0}{\omega_0} = -\frac{\Delta I}{I}$

$= -\frac{2\Delta R}{R} \left(\because \frac{\Delta I}{I} = \frac{2\Delta R}{R}\right) = -2\,\alpha\,\Delta T$

23. **(a)** $E = \frac{L^2}{2I} = K$ (given) $\therefore K \propto \frac{1}{I}$ (If L = constant)

When child stretches his arms the moment of inertia of system get doubled so kinetic energy will becomes half i.e. K/2.

24. **(c).** Angular impulse = change in angular momentum : Frt = $L \Rightarrow L_1 < L_2$

$K = \frac{L^2}{2I} \Rightarrow K_1 = K_2$

25. (b); 26. (a); 27. (c)

Drawing the F.B. D of the plank and the cylinder.

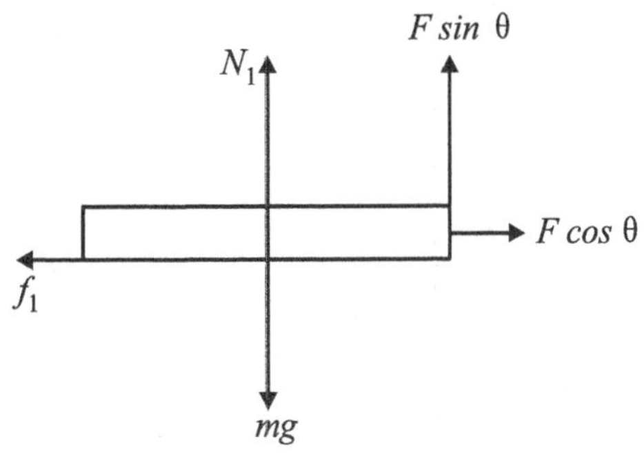

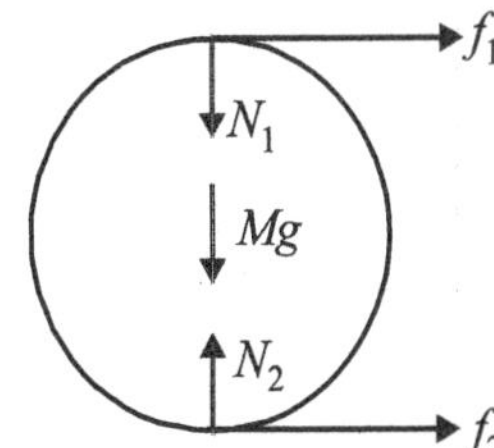

Equations of motion are

$F\cos\theta - f_1 = ma$ ....(1)

$F\sin\theta + N_1 = mg$ ....(2)

$f_1 + f_2 = MA$ .....(3)

$f_1 R - f_2 R = I\alpha$ .....(4)

$A = R\alpha$ .....(5)

$a = \frac{4F\cos\theta}{3M + 8m} = \frac{4 \times 55 \times \frac{1}{2}}{[(3\times 1) + (8 \times 1)]} = 10$ m/s^2

$$f_1 = \frac{3MF\cos\theta}{3M+8m} = \frac{3\times1\times55\times\frac{1}{2}}{3\times1+8\times1} = 7.5\ N$$

and $$f_2 = \frac{MF\cos\theta}{3M+8m} = \frac{1\times55\times\frac{1}{2}}{3\times1+8\times1} = 2.5\ N$$

28. **(b)** $\vec{\tau} = \frac{d\vec{L}}{dt}$ and $L = I\omega$

29. **(b)** $\tau = rF\sin\theta$. If $\theta = 90°$ then $\tau_{max} = rF$
Unit of torque is N-m.

30. **(d)** Torque = Force × perpendicular distance of the line of action of force from the axis of rotation (d).
Hence for a given applied force, torque or true tendency of rotation will be high for large value of *d*. If distance *d* is smaller, then greater force is required to cause the same torque, hence it is harder to open or shut down the door by applying a force near the hinge.

DAILY PRACTICE PROBLEMS

PHYSICS SOLUTIONS

16

(1) (b) As the mass of disc is negligible therefore only moment of inertia of five particles will be considered.

$I = \sum mr^2 = 5\ mr^2 = 5 \times 2 \times (0.1)^2 = 0.1$ kg-m^2

(2) (a) $I = \frac{1}{2}MR^2 = \frac{1}{2}\times\left(\pi R^2 t \times p\right)\times R^2$

$\Rightarrow I \propto R^4$ (As t and p are same)

$\therefore \frac{I_1}{I_2} = \left(\frac{R_1}{R_2}\right)^4 = \left(\frac{0.2}{0.6}\right)^4 = \frac{1}{81}$

(3) (a) $I = \frac{1}{2}MR^2 = \frac{1}{2}\times 0.5\times(0.1)^2 = 2.5\times10^{-3}\,kg-m^2$

(4) (a) $I = \frac{\tau}{\alpha} = \frac{31.4}{4\pi} = 2.5kg\,m^2$

(5) (d) Let the mass of loop P (radius = r) = m
So the mass of loop Q (radius = nr) = nm

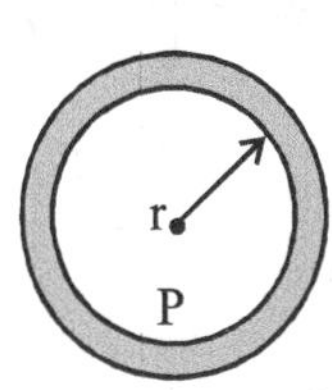

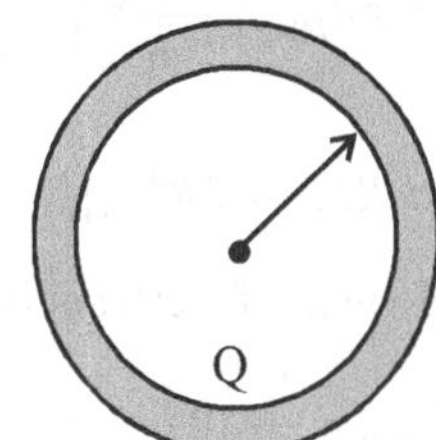

Moment of inertia of loop P, $I_P = mr^2$
Moment of inertia of loop Q. $I_Q = nm\,(nr)^2 = n^3\,mr^2$

$\therefore \frac{I_Q}{I_P} = n^3 = 8 \Rightarrow n = 2$

(6) (c) Moment of inertia of sphere about its tangent

$\frac{7}{5}MR^2 = MK^2 \Rightarrow K = \sqrt{\frac{7}{5}}R$

(7) (a) Moment of inertia of system about point P

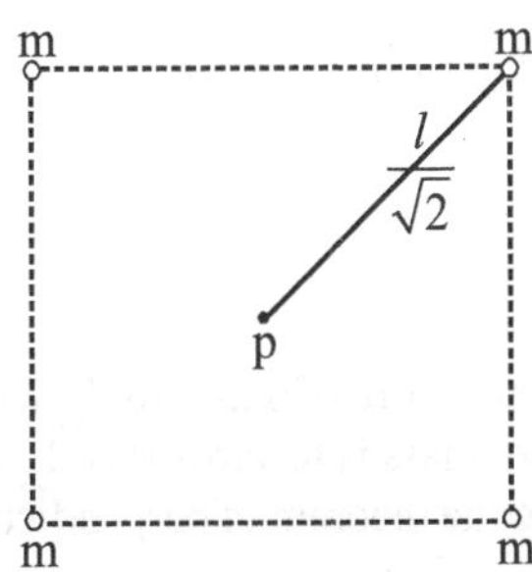

$= 4m\left(\frac{l}{\sqrt{2}}\right)^2 = 2ml^2$ and $4mK^2 = 2ml^2$

$\therefore K = \frac{l}{\sqrt{2}}$

(8) (b) $\frac{1}{2}MR^2 = MK^2 \Rightarrow K = \frac{R}{\sqrt{2}} = \frac{2.5}{\sqrt{2}} = 1.76\,\text{cm}$

(9) (c) $I = 2MR^2 = 2 \times 3 \times (1)^2 = 6$ gm-cm^2

(10) (a) $I = \frac{5}{4}Mr^2$

(11) (a)

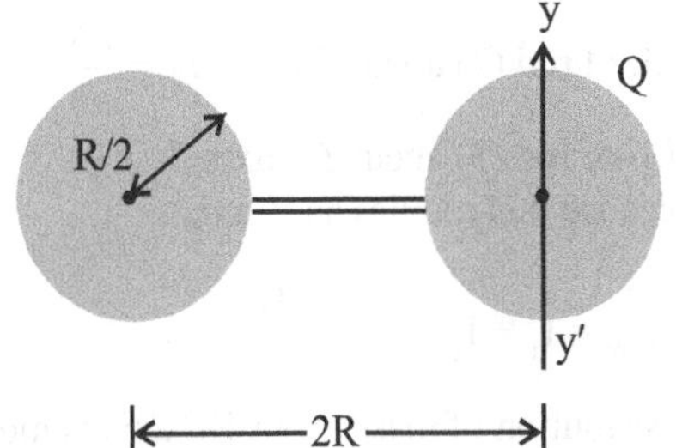

Moment of inertia of the system about yy′
$I_{yy'}$ = Moment of inertia of sphere P about yy′ + Moment of inertia of sphere Q about yy′
Moment of inertia of sphere P about yy′

$= \frac{2}{5}M\left(\frac{R}{2}\right)^2 + M(x)^2$ [Parallel axis theorem]

$= = \frac{2}{5}M\left(\frac{R}{2}\right)^2 + M(2R)^2 = \frac{MR^2}{10} + 4MR^2$

Moment of inertia of sphere Q about yy′ is $\frac{2}{5}M\left(\frac{R}{2}\right)^2$

Now $I_{yy'} = \frac{MR^2}{10} + 4MR^2 + \frac{2}{5}M\left(\frac{R}{2}\right)^2 = \frac{21}{5}MR^2$

(12) (a) M.I. of system about the axis which passing through m_1

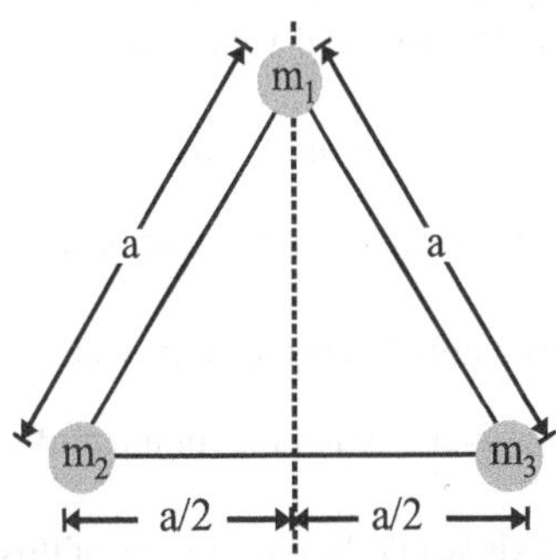

$I_{system} = m_1\,(0)^2 + m_2\left(\frac{a}{2}\right)^2 + m_3\left(\frac{a}{2}\right)^2$

$I_{system} = (m_2 + m_3)\frac{a^2}{4}$

(13) (a) M.I. of rod (1) about Z – axis $I_1 = \frac{Ml^2}{3}$

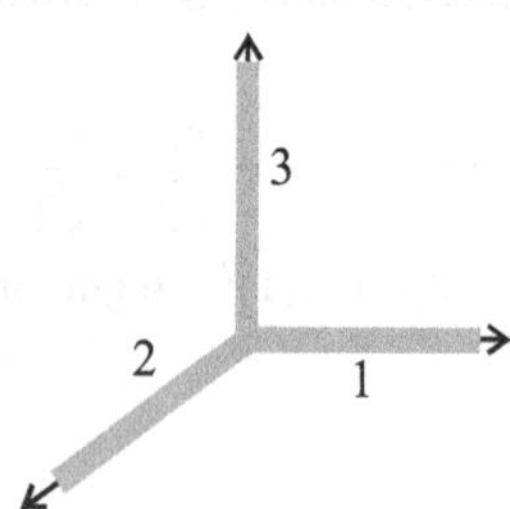

M.I. of rod (2) about Z-axis, $I_2 = \frac{Ml^2}{3}$

M.I. of rod (3) about Z – axis, $I_3 = 0$
Because this rod lies on Z-axis

$\therefore \quad I_{system} = I_1 + I_2 + I_3 = \frac{2Ml^2}{3}$

(14) (c) Distribution of mass about BC axis is more than that about AB axis i.e. radius of gyration about BC axis is more than that about AB axis.
i.e. $K_{BC} > K_{AB} \therefore I_{BC} > I_{AB} > I_{CA}$

(15) (a) $I = \frac{Ml^2}{12} = \frac{0.12 \times 1^2}{12} = 0.01\, kg - m^2$

(16) (c)

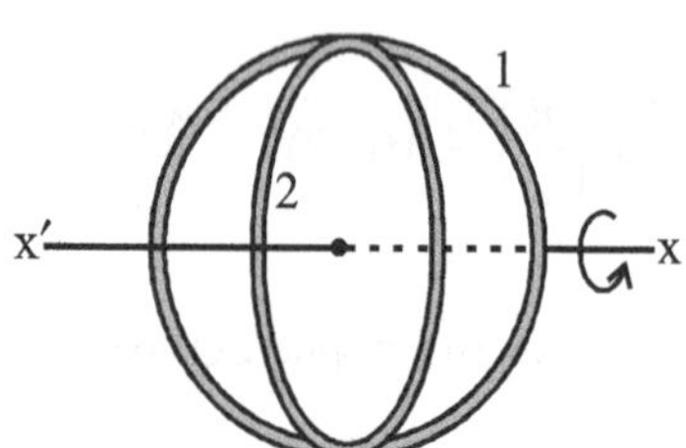

I_1 = M.I. of ring about its diameter = $\frac{1}{2} mR^2$

I_2 = M.I. of ring about the axis normal to plane and passing through centre = mR^2
Two rings are placed according to figure. Then

$I_{xx'} = I_1 + I_2 = \frac{1}{2} mR^2 + mR^2 = \frac{3}{2} mR^2$

(17) (a) Mass of the centre disc would be 4M and its moment of inertia about the given axis would be $\frac{1}{2}(4M)R^2$.
For the given section the moment of inertia about the same axis would be one quarter of this i.e. $\frac{1}{2} MR^2$.

(18) (d) Mass per unit length of the wire = ρ
Mass of L length, $M = \rho L$
and since the wire of length L is bent in a or of circular loop therefore $2\pi R = L \Rightarrow R = \frac{L}{2\pi}$

Moment of inertia of loop about given axis $= \frac{3}{2} MR^2$

$= \frac{3}{2}\rho L\left(\frac{L}{2\pi}\right)^2 = \frac{3\rho L^3}{8\pi^2}$

(19) (b) M.I. of disc $= \frac{1}{2} MR^2 = \frac{1}{2} M\left(\frac{M}{\pi t\rho}\right) = \frac{1}{2}\frac{M^2}{\pi t\rho}$

$\left(\text{As } \rho = \frac{M}{\pi R^2 t} \text{ Therefore } R^2 = \frac{M}{\pi t\rho}\right)$

If mass and thickness are same then, $I \propto \frac{1}{\rho}$

$\therefore \frac{I_1}{I_2} = \frac{\rho_2}{\rho_1} = \frac{3}{1}.$

(20) (c) According to problem disc is melted and recasted into a solid sphere so their volume will be same.

$V_{Disc} = V_{Sphere} \Rightarrow \pi R_{Disc}^2 t = \frac{4}{3}\pi R_{Sphere}^3$

$\Rightarrow \pi R_{Disc}^3\left(\frac{R_{Disc}}{6}\right) = \frac{4}{3}\pi R_{Sphere}^3 \quad \left[t = \frac{R_{Disc}}{6}, \text{given}\right]$

$\Rightarrow R_{Disc}^3 = 8R_{Sphere}^3 \Rightarrow R_{Sphere} = \frac{R_{Disc}}{2}$

Moment of inertia of disc

$I_{Disc} = \frac{1}{2} MR_{Disc}^2 = I \text{ (given)}$

$\therefore M(R_{Disc})^2 = 2I$

Moment of inertia of sphere $I_{sphere} = \frac{2}{5} MR_{Sphere}^2$

$= \frac{2}{5} M\left(\frac{R_{Disc}}{2}\right)^2 = \frac{M}{10}(R_{Discs})^2 = \frac{2I}{10} = \frac{1}{5}$

(21) (d) Moment of inertia of system about YY'
$I = I_1 + I_2 + I_3$

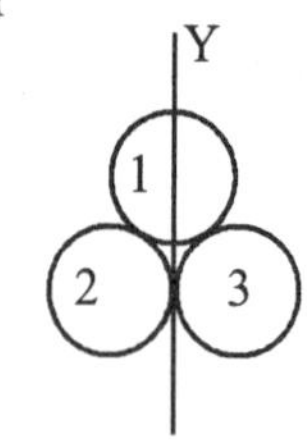

$= \frac{1}{2} MR^2 + \frac{3}{2} MR^2 + \frac{3}{2} MR^2$

$= \frac{7}{2} MR^2$

(22) (d) As C is the centre of mass, so, I_C will be minimum.
Also more mass is towards B so $I_A > I_B$.

(23) (a) Applying the theorem of perpendicular axis,
$I = I_1 + I_2 = I_3 + I_4$
Because of symmetry, we have $I_1 = I_2$ and $I_3 = I_4$ Hence
$I = 2I_1 = 2I_2 = 2I_3 = 2I_4$ or $I_1 = I_2 = I_3 = I_4$
i.e. sum of two moment of inertia of square plate about any axis in a plane (Passing through centre) should be equal to moment of inertia about the axis passing through the centre and perpendicular to the plane of the plate.

(24) (d) Moment of inertia depends on all the three factors given in (1), (2) & (3).

(25) (d) $I = 4\left[\frac{2}{5}MR^2 + M(R\sqrt{2})^2\right]$

$= 4MR^2\left[\frac{2}{5} + 2\right]$

$= \frac{4MR^2 \times 12}{5} = \frac{48MR^2}{5}.$

(26) (b) Let a be the acceleration of centre of mass

$Mg - T = 0$ ... (i)

$F.x = T.2x$ ... (ii)

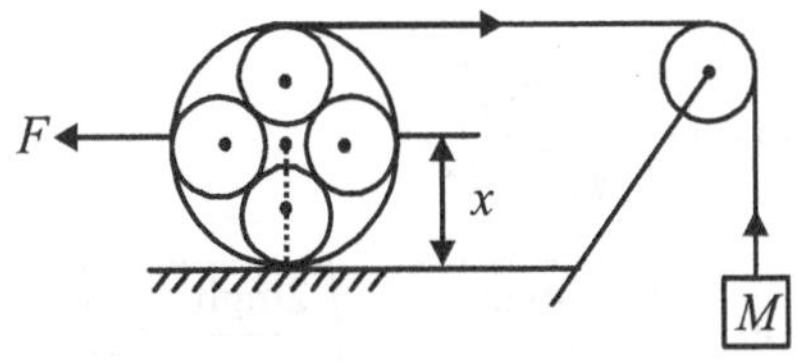

(27) (c) remain the same

(28) (c) Radius of gyration of a body is not a constant quantity. Its value changes with the change in location of the axis of rotation. Radius of gyration of a body about a given axis is given as $K = \sqrt{\frac{r_1^2 + r_2^2 + ... + r_n^2}{n}}$

(29) (c) The moment of inertia of a particle about an axis of rotation is given by the product of the mass of the particle and the square of the perpendicular distance of the particle from the axis of rotation. For different axis, distance would be different, therefore moment of inertia of a particle changes with the change in axis of rotation.

(30) (a) When earth shrinks, it angular momentum remains constant. i.e. $L = I\omega = \frac{2}{5}mR^2 \times \frac{2\pi}{T}$ = constant.

$\therefore T \propto I \propto R^2$. It means if size of the earth changes then its moment of inertia changes.

In the problem radius becomes half so time period (Length of the day) will becomes $\frac{1}{4}$ of the present value

i.e. $\frac{24}{4} = 6\,\text{hr}.$

DAILY PRACTICE PROBLEMS

PHYSICS SOLUTIONS

17

1. (a) Since disc is rolling (without slipping) about point O. Hence

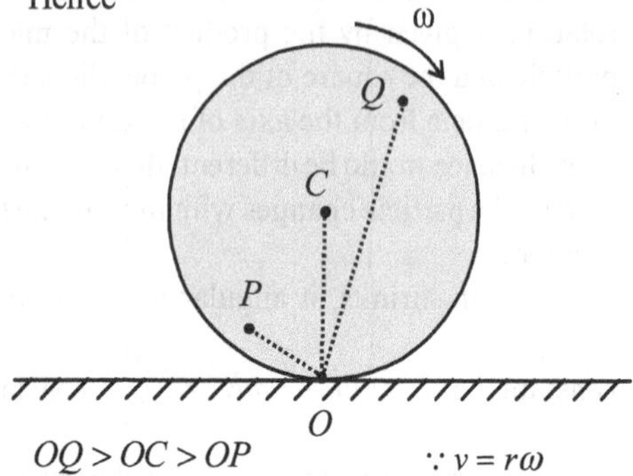

$OQ > OC > OP$ $\because v = r\omega$

$\therefore v_Q > v_C > v_P$

2. (d) Applying the theorem of perpendicular axis,

$I = I_1 + I_2 = I_3 + I_4$

Because of symmetry, we have

$I_1 = I_2$ and $I_3 = I_4$

Hence $I = 2I_1 = 2I_2 = 2I_3 = 2I_4$

or $I_1 = I_2 = I_3 = I_4$

i.e. sum of two moment of inertia of square plate about any axis in a plane (Passing through centre) should be equal to moment of inertia about the axis passing through the centre and perpendicular to the plane of the plate.

3. (a) By the conservation of energy

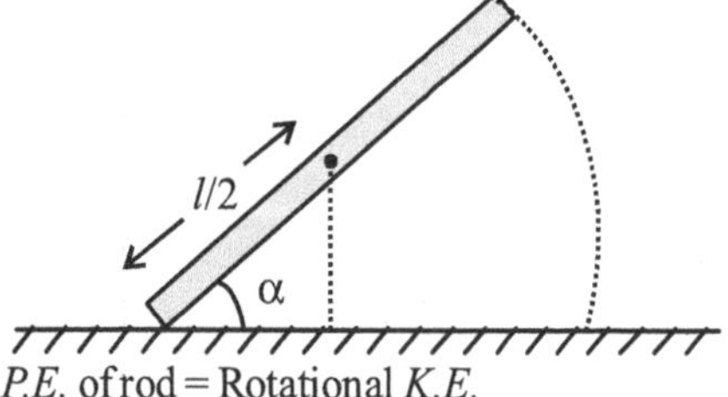

P.E. of rod = Rotational K.E.

$$mg\frac{1}{2}\sin\alpha = \frac{1}{2}I\omega^2$$

$$\Rightarrow mg\frac{1}{2}\sin\alpha = \frac{1}{2}\frac{ml^2}{3}\omega^2 \Rightarrow \omega = \sqrt{\frac{3g\sin\alpha}{l}}$$

But in the problem length of the rod $2L$ is given

$$\therefore \omega = \sqrt{\frac{3g\sin\alpha}{2L}}$$

4. (c) Graph should be parabola symmetric to I- axis, but it should not pass from origin because there is a constant value I_{cm} is present for $x = 0$.

5. (b) $v = \sqrt{\dfrac{2gh}{1+\dfrac{K^2}{R^2}}} = \sqrt{\dfrac{2gh}{1+\dfrac{1}{2}}} = \sqrt{\dfrac{4}{3}gh}$

6. (d) $a = \dfrac{g\sin\theta}{1+\dfrac{K^2}{R^2}} = \dfrac{g\sin\theta}{1+\dfrac{2}{5}} = \dfrac{g/2}{7/5} = \dfrac{5g}{14}$

As $\theta = 30^\circ$ and $\dfrac{K^2}{R^2} = \dfrac{2}{5}$

7. (b) We know $v = \sqrt{\dfrac{2gh}{1+\dfrac{k^2}{r^2}}}$

$$\therefore \omega = \frac{v}{r} = \sqrt{\frac{2gh}{r^2+k^2}}$$

$$\Rightarrow \omega = \sqrt{\frac{2mgh}{mr^2+mk^2}} = \sqrt{\frac{2mgh}{mr^2+I}} = \sqrt{\frac{2mgh}{I+mr^2}}$$

8. (a) Because its M.I. (or value of $\dfrac{K^2}{R^2}$) is minimum for sphere.

9. (b) As body is moving on a frictionless surface. Its mechanical energy is conserved. When body climbes up the inclined plane it keeps on rotating with same angular speed, as no friction force is present to provide retarding torque so

$$\frac{1}{2}I\omega^2 + \frac{1}{2}mv^2 \geq \frac{1}{2}I\omega^2 + mgh \Rightarrow v \geq \sqrt{2gh}$$

10. (a) $\dfrac{1}{2}MR^2 = I \Rightarrow MR^2 = 2I$

Moment of inertia of disc about a tangent in a plane

$$= \frac{5}{4}MR^2 = \frac{5}{4}(2I) = \frac{5}{2}I$$

11. (d) Moment of inertia of system about YY'

$I = I_1 + I_2 + I_3$

$$= \frac{1}{2}MR^2 + \frac{3}{2}MR^2 + \frac{3}{2}MR^2 = \frac{7}{2}MR^2$$

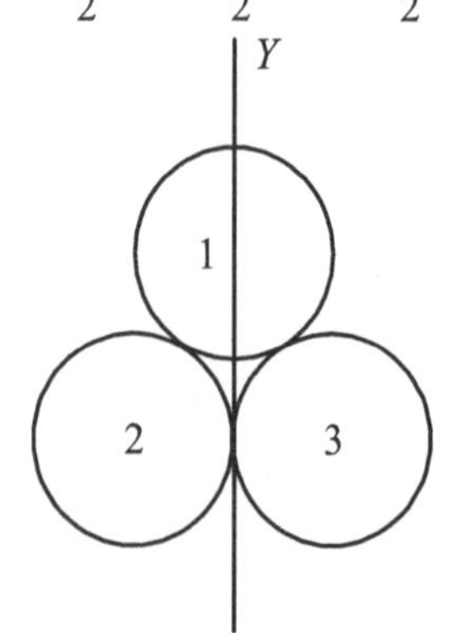

12. (b) $\frac{I_{Ring}}{I_{Disc}} = \frac{MR^2}{\frac{1}{2}MR^2} = 2:1$

13. (a)

14. (b) It follows from the theorem of parallel axes.

15. (a)

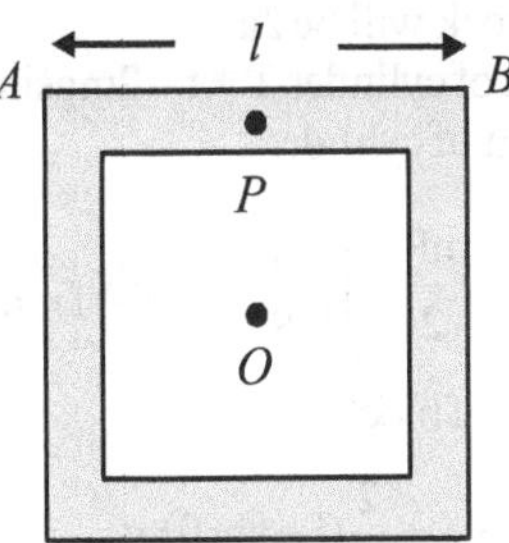

Moment of inertia of Rod AB about point P and perpendicular to the plane $= \frac{Ml^2}{12}$

M.I. of rod AB about point 'O'

$$= \frac{Ml^2}{12} + M\left(\frac{l}{2}\right)^2 = \frac{Ml^2}{3}$$

(By using parallel axis theorem)

but the system consists of four rods of similar type so by the symmetry

$$I_{system} = 4\left(\frac{Ml^2}{3}\right)$$

16. (a)

17. (d)

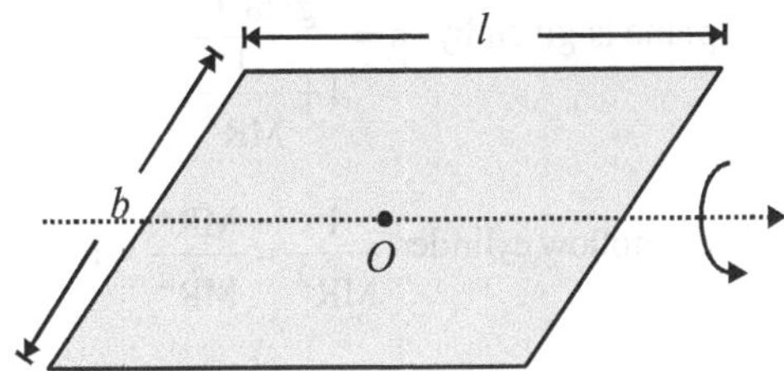

M.I. of plate about O and parallel to length $= \frac{Mb^2}{12}$

18. (d) $I_z = I_x + I_y$

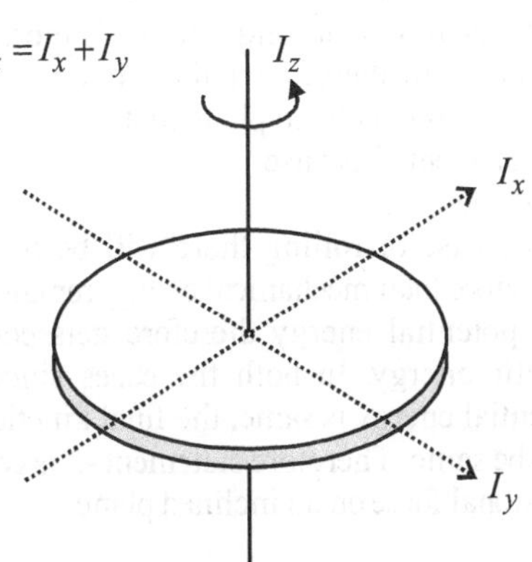

$$200 = I_D + I_D = 2I_d$$

$$\therefore I_D = 100 \ gm \times cm^2$$

19. (a) M.I. of complete disc about 'O' point

$$I_{Total} - \frac{1}{2}(9M)R^2$$

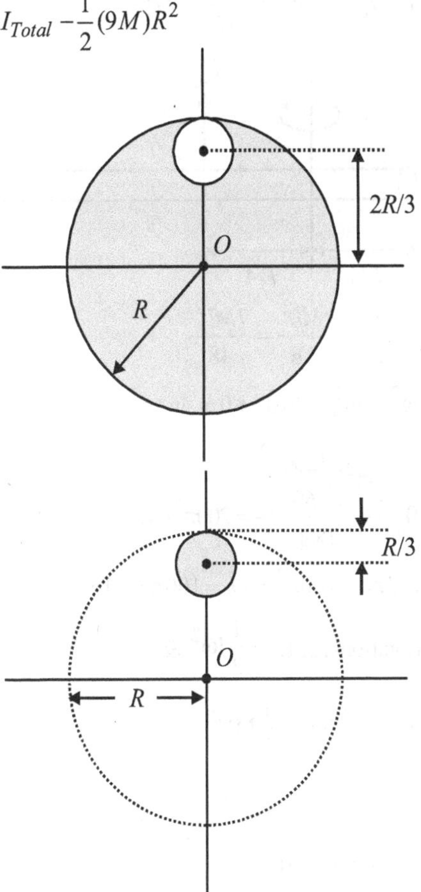

Radius of removed disc $= \frac{R}{3}$

$\therefore$ Mass of removed disc $= \frac{9M}{9} = M$

[As $M = \pi R^2 t \therefore M \propto R^2$]

M.I. of removed disc about its own axis

$$= \frac{1}{2}M\left(\frac{R}{3}\right)^2 = \frac{MR^2}{18}$$

Moment of inertia of removed disc about 'O'

$$I_{removed\ disc} = I_{cm} + mx^2 = \frac{MR^2}{18} + M\left(\frac{2R}{3}\right)^2 = \frac{MR^2}{2}$$

M. I. of complete disc can also be written as

$$I_{Total} = I_{Removed\ disc} + I_{Remaining\ disc}$$

$$I_{Total} = \frac{MR^2}{2} + I_{Remaining\ disc} \quad \text{..... (ii)}$$

Equating (i) and (ii) we get

$$\frac{MR^2}{2} + I_{Remaining\ disc} = \frac{9MR^2}{2}$$

$$\therefore I_{Remaining\ disc} = \frac{9MR^2}{2} - \frac{MR^2}{2} = \frac{8MR^2}{2} = 4MR^2$$

20. (d) $I = I_{cm} + Mx^2 = \frac{ML^2}{12} + M\left(\frac{L}{4}\right)^2$

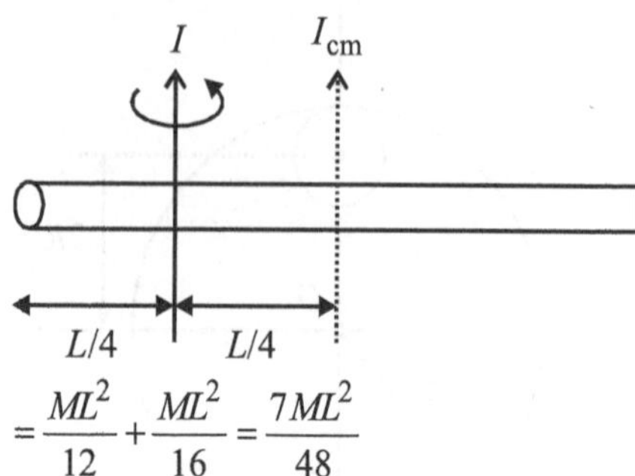

$= \frac{ML^2}{12} + \frac{ML^2}{16} = \frac{7ML^2}{48}$

21. (c) $\omega^2 = \omega_0^2 - 2\alpha\theta \Rightarrow 0 = 4\pi^2 n^2 - 2\alpha\theta$

$$\theta = \frac{4\pi^2\left(\frac{1200}{60}\right)^2}{2\times 4} = 200\pi^2 rad$$

$\therefore 2\pi n - 200\pi^2 \Rightarrow n = 100\pi = 314$ revolution

22. (b) Rotational K.E. $= \frac{1}{2}I\omega^2$ &

$$T.E. = \frac{1}{2}I\omega^2 + \frac{1}{2}MV^2$$

$$= \frac{1}{2}I\omega^2 + \frac{1}{2}MR^2\omega^2 = \frac{1}{2}\omega^2(I + MR^2)$$

For ring $I = MR^2$

$$\therefore T.E. = \frac{1}{2}\omega^2(MR^2 + MR^2) = \frac{1}{2}\omega^2 \times 2MR^2$$

Rotational K.E. $= \frac{1}{2}MR^2\omega^2$

$$\therefore \alpha = \frac{\frac{1}{2}MR^2\omega^2}{\frac{1}{2}\omega^2 \times 2MR^2} = \frac{1}{2}$$

For a solid sphere $I = \frac{2}{5}MR^2$

$$\therefore T.E. = \frac{1}{2}\omega^2\left(\frac{2}{5}MR^2 + MR^2\right) = \frac{1}{2}\omega^2 MR^2 \times \frac{7}{5}$$

Rotational K.E. $= \frac{1}{2} \times \frac{2}{5}MR^2\omega^2$

$$\beta = \frac{\frac{1}{2}\times\frac{2}{5}MR^2\omega^2}{\frac{1}{2}\omega^2 MR^2 \times \frac{7}{5}} = \frac{2}{7}$$

23. (a) Time of descent $\propto \frac{K^2}{R^2}$. Time of descent depends upon the value of radius of gyration (K) or moment of inertia (I). Actually radius of gyration is a measure of moment of inertia of the body.

24. (a) $K_T = K_R \Rightarrow \frac{1}{2}mv^2 = \frac{1}{2}mv^2\left(\frac{K^2}{R^2}\right) \Rightarrow \therefore \frac{K^2}{R^2} = 1$

This value of K^2/R^2 match with hollow cylinder.

25. (b) **26. (c)** **27. (d)**

(i) Let acceleration of centre of mass of cylinder be a then acceleration of block will be 2a.
For linear motion of cylinder $T + f - 2mg\sin\theta = 2m(a)$
For rolling motion of cylinder

$$(T - f)R = I\alpha = \left(\frac{2mR^2}{2}\right)\left(\frac{a}{R}\right) \Rightarrow T - f = ma$$

For linear motion of block

$$mg - T = m(2a) \Rightarrow a = \frac{2}{7}(1 - \sin\theta)g$$

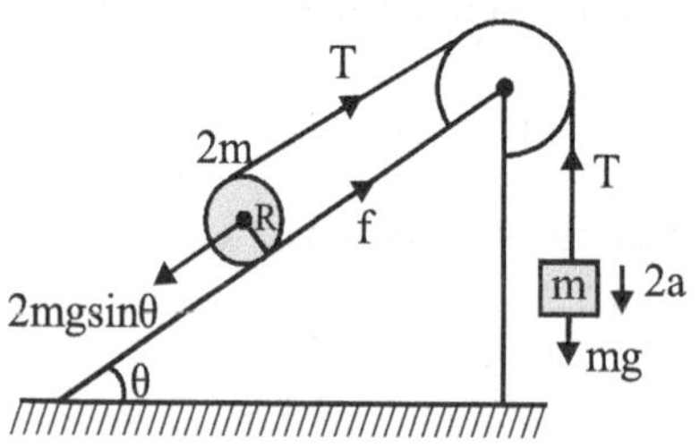

(ii) $T = mg - 2m\left(\frac{2}{7}g(1 - \sin\theta)\right) = \left(\frac{3 + 4\sin\theta}{7}\right)mg$

(iii) $F = T - ma = \left(\frac{1 + 6\sin\theta}{7}\right)mg$

28. (c) The acceleration of a body rolling down an inclined plane is given by $a = \frac{g\sin\theta}{1 + \frac{I}{MR^2}}$

For hollow cylinder $\frac{I}{MR^2} = \frac{MR^2}{MR^2} = 1$

For solid cylinder $\frac{I}{MR^2} = \frac{\frac{1}{2}MR^2}{MR^2} = \frac{1}{2}$

$\Rightarrow$ Acceleration of solid cylinder is more than hollow cylinder and therefore solid cylinder will reach the bottom of the inclined plane first.

$\therefore$ Statement -1 is false

• Statement - 2

In the case of rolling there will be no heat losses. Therefore total mechanical energy remains conserved. The potential energy therefore gets converted into kinetic energy. In both the cases since the initial potential energy is same, the final kinetic energy will also be same. Therefore statement -2 is correct.

29. (b) Frictional force on an inclined plane

$= \frac{1}{3}g\sin\alpha$ (for a disc).

30. (c) The moment of inertia about both the given axes shall be same if they are parallel. Hence statement–1 is false.

DAILY PRACTICE PROBLEMS — PHYSICS SOLUTIONS 18

(1) (c) If r is the distance between m and (M – m), the gravitational force will be -

$$F = G\frac{m(M-m)}{r^2} = \frac{G}{r^2}(mM - m^2)$$

The force will be maximum if, $\frac{dF}{dm} = 0$

i.e, $\frac{d}{dm}\left[\frac{G}{r^2}(mM - m^2)\right] = 0$

or $\frac{m}{M} = \frac{1}{2}$ (as M and r are constants)

(2) (c) $m_g = \sqrt{3}$ kg, $v = \frac{c}{2}$

$$m = \frac{m_0}{\sqrt{1-(v^2/c^2)}} = \frac{\sqrt{3}}{\sqrt{1-\frac{c^2}{4xc^2}}} = \frac{\sqrt{3}}{\frac{\sqrt{3}}{2}} \text{ kg}$$

(3) (a) $g' = g - R_e\omega^2$ (at equator $\lambda = 0$)
If a body is weightless,
$g' = 0$, $g - R_e\omega^2 = 0$

$$\Rightarrow \omega = \sqrt{\frac{g}{R}} = \sqrt{\frac{10}{6400 \times 10^3}} = 1.25 \times 10^{-3} \text{ rad/sec.}$$

(4) (b) The apparent weight of person on the equator (latitude $\lambda = 0$) is given by
$W' = W - m R_e\omega^2$,

$$W' = \frac{3}{5}W = \frac{3}{5}mg$$

$$\frac{3}{5}mg = mg - mR\omega^2 \text{ or } mR\omega^2 = mg - \frac{3}{5}mg$$

$$\omega = \sqrt{\frac{2}{5}\frac{g}{R}} = \sqrt{\frac{2}{5} \times \frac{9.8}{6400 \times 10^3}} \text{ rad/ sec}$$

$= 7.826 \times 10^{-4}$ rad/sec

(5) (c) According to question,

$g' = \frac{G \times 4M_p}{R_p^2}$ on the planet and $g = \frac{GM_e}{R_e^2}$ on the earth

$\because R_p = R_e$ and $M_p = M_e$

Now, $\frac{g'}{g} = 4 \Rightarrow g' = 4g = 40 \text{ m/sec}^2$

Energy needed to lift 2 kg mass through 2m distance $= mg'h = 2 \times 40 \times 2 = 160$ J

(6) (d) $V_g = V_{g1} + V_{g2} = -\frac{Gm_1}{r_1} - \frac{Gm_2}{r_2}$

$$= -6.67 \times 10^{-11}\left[\frac{10^2}{0.5} + \frac{10^3}{0.5}\right] = -1.47 \times 10^{-7} \text{ Joule/kg}$$

(7) (c) The P.E. of the object on the surface of earth is

$$U_1 = -\frac{GMm}{R}$$

The P.E. of object at a height R, $U_2 = -\frac{GMm}{(R+R)}$

The gain in P E is $U_2 - U_1 = \frac{GMm}{2R} = \frac{1}{2}mgR$

$\left[\because g = \frac{GM}{R^2} \text{ on surface of earth}\right]$

(8) (c) Resultant force on particle '1'
$F_r = \sqrt{2}\,F + F'$

$$\text{or } F_r = \sqrt{2}\frac{Gm^2}{2r^2} + \frac{Gm^2}{4r^2} = \frac{mv^2}{r}$$

$$\text{or } v = \sqrt{\frac{Gm}{r}\left(\frac{2\sqrt{2}+1}{4}\right)}$$

(9) (b) The resultant gravitational force on each particle provides it the necessary centripetal force

$$\therefore \frac{mv^2}{r} = \sqrt{F^2 + F^2 + 2F^2\cos 60^\circ} = \sqrt{3}F,$$

$$\text{But } r = \frac{\sqrt{3}}{2}\ell \times \frac{2}{3} = \frac{\ell}{\sqrt{3}}$$

$$\therefore v = \sqrt{\frac{GM}{\ell}}$$

(10) (b) The acceleration due to gravity on the surface of the earth, in terms of mass M_e and radius R_e of earth, is

given by $g = \frac{GM_e}{R_e^2}$

if M_m be the mass of the moon, R_m its radius, then the acceleration due to gravity on the surface of the moon

will be given by $g' = \frac{GM_m}{R_m^2}$

Dividing eq. (ii) by eq. (i), we get

$$\frac{g'}{g} = \frac{M_m}{M_e}\left(\frac{R_e}{R_m}\right)^2 = \frac{1}{80} \times \left(\frac{4}{1}\right)^2 = \frac{1}{5}$$

$\therefore g' = g/5$.

(11) (b) The value of g at the height h from the surface of earth

$$g' = g\left(1 - \frac{2h}{R}\right)$$

The value of g at depth x below the surface of earth

$$g' = g\left(1 - \frac{x}{R}\right)$$

These two are given equal, hence $\left(1-\frac{2h}{R}\right)=\left(1-\frac{x}{R}\right)$

On solving, we get $x = 2h$

(12) (a) If g be the acceleration due to gravity at the surface of the earth, then its value at a height h above the earth's surface will be -

$$g' = \frac{g}{\left(1+\frac{h}{R_e}\right)^2} \qquad \text{Here } \frac{g'}{g}=\frac{1}{9}$$

$$\therefore \frac{1}{9}=\frac{1}{\left(1+\frac{h}{R_e}\right)^2} \text{ or } 1+\frac{h}{R_e}=3$$

or $h = 2R_e = 2 \times 6400 = 12800$ km.

(13) (c) Consider the case of a body of mass m placed on the earth's surface (mass of the earth M and radius R). If g is acceleration due to gravity, then

$$mg = G\frac{M_e m}{R^2} \text{ or } g=\frac{GM_e}{R^2}$$

where G is universal constant of gravitation.

Now when the radius is reduced by 1%, i.e., radius becomes 0.99 R, let acceleration due to gravity be g',

$$\text{then } g'=\frac{GM_e}{(0.99R)^2}$$

From equation (A) and (B), we get

$$\frac{g'}{g}=\frac{R^2}{(0.99R)^2}=\frac{1}{(0.99)^2}$$

$$\therefore g' = g \times \left(\frac{1}{0.99}\right)^2 \text{ or } g' > g$$

Thus , the value of g is increased.

(14) (a) Force of gravity at surface of earth,

$F_1 = Gm\,M/R^2$ .... (1)

Force of gravity at height H is

$F_2 = Gm\,M\,(R+H)^2$ .... (2)

Dividing (A) by (B) and Rearranging

$$H = R\left(\sqrt{\frac{F_1}{F_2}}-1\right)=350 \text{ km where } (F_2 = .9F_1)$$

(15) (a) The extension in the length of spring is

$$x=\frac{mg}{k}=\frac{GMm}{r^2k},$$

$$\therefore x \propto \frac{1}{r^2}, \quad \therefore \frac{x_2}{x_1}=\frac{R^2}{(R+h)^2}$$

$$\text{or } x_2 = 1 \times \left(\frac{6400}{7200}\right)^2 = 0.79 \text{ cm}.$$

(16) (d) In the position of solar eclipse, net force on earth

$F_E = F_M + F_S$

In the position of lunar eclipse, net force on earth

$F'_E = F_S - F_M$

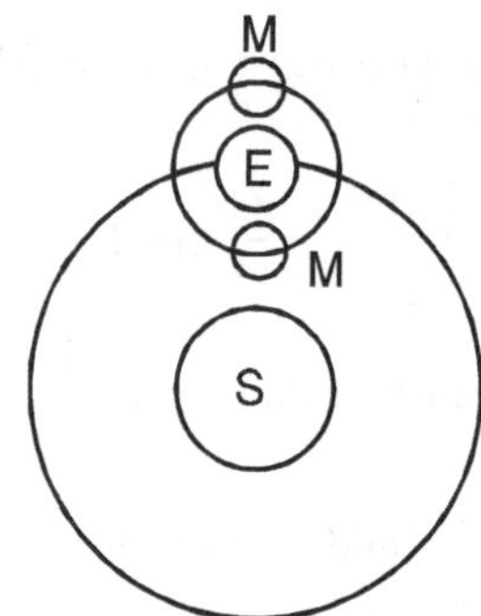

$\therefore$ Change in acceleration of earth,

$$\Delta f = \frac{2GM}{R^2}=\frac{2 \times 6.67 \times 10^{-11} \times 7.36 \times 10^{22}}{3.82^2 \times 10^{16}}$$

$= 6.73 \times 10^{-5}$ m/s^2

(17) (c) Let M_e be the mass of the earth. The work required

$$W = GM_e m\left[\frac{1}{R_e}-\frac{1}{R_e+h}\right]$$

$$=\frac{GM_e mh}{R_e(R_e+h)}=\frac{gR_e^2 mh}{R_e(R_e+h)} \quad [\therefore GM_e = gR_e^2]$$

$$=\frac{mgh}{\left(1+\frac{h}{R_e}\right)}$$

(18) (a) The P.E of the mass at d/2 due to the earth and moon is

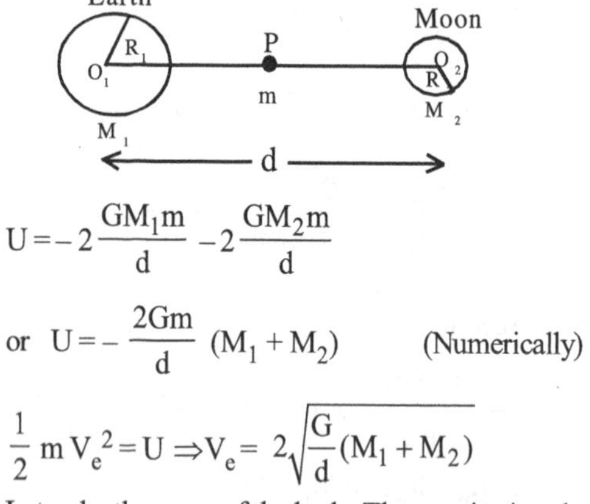

$$U = -2\frac{GM_1 m}{d} - 2\frac{GM_2 m}{d}$$

$$\text{or } U = -\frac{2Gm}{d}(M_1+M_2) \qquad \text{(Numerically)}$$

$$\frac{1}{2}mV_e^2 = U \Rightarrow V_e = 2\sqrt{\frac{G}{d}(M_1+M_2)}$$

(19) (d) Let m be the mass of the body. The gravitational potential energy of the body at the surface of the earth is

$$U=-\frac{GM_e m}{R_e}$$

The potential energy at a height 10 R_e above the surface of the earth will be

$$U' = -\frac{GM_e m}{(R_e + 10R_e)}$$

∴ Increase in potential energy

$$U' - U = -\frac{GM_e m}{11R_e} + \left(\frac{GM_e m}{R_e}\right) = \frac{10}{11}\frac{GM_e m}{R_e}$$

This increase will be obtained from the initial kinetic energy given to the body. Hence if the body be thrown with a v velocity then

$$\frac{1}{2}mv^2 = \frac{10}{11}\frac{GM_e m}{R_e} \Rightarrow v = \sqrt{\frac{20Gm_e}{11R_e}}$$

Substituting the given values, we get

$$v = \sqrt{\left(\frac{20 \times (6.67 \times 10^{-11}) \times (6 \times 10^{24})}{11 \times (6.4 \times 10^6)}\right)}$$

$= 1.07 \times 10^4$ m/s.

(20) (b) $F = \frac{Gm(M-m)}{r^2}$

For maximum force $\frac{dF}{dm} = 0$

$$\Rightarrow \frac{d}{dm}\left(\frac{GmM}{r^2} - \frac{Gm^2}{r^2}\right) = 0$$

$$\Rightarrow M - 2m = 0 \Rightarrow \frac{m}{M} = \frac{1}{2}$$

(21) (b) $g' = g - \omega^2 R \cos^2 \lambda$

For weightlessness at equator $\lambda = 0$ and $g' = 0$

$$\therefore 0 = g - \omega^2 R \Rightarrow \omega = \sqrt{\frac{g}{R}} = \frac{1}{800}\frac{\text{rad}}{\text{s}}$$

(22) (a) k represents gravitational constant which depends only on the system of units.

(23) (a) All statements except (4) are wrong.

(24) (a) Value of g decreases when we go from poles to equator.

(25) (b)

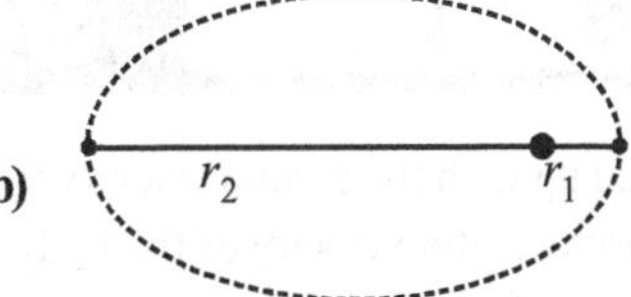

Gravitational PE at perihelion $= -GMm/r_1$ as r_1 is minimum Therefore, PE is minimum.

(26) (c) Total energy = constant.

(27) (c) As Pluto moves away, displacement has component opposite to air force, hence work done is –negative.

(28) (b) For two electron $\frac{F_g}{F_e} = 10^{-43}$ i.e. gravitational force is negligible in comparison to electrostatic force of attraction.

(29) (c) The universal gravitational constant G is totally different from g.

$$G = \frac{FR^2}{Mm}$$

The constant G is scalar and posses the dimensions $\left[M^{-1}L^3T^{-2}\right]$.

$$g = \frac{GM}{R^2}$$

g is a vector and has got the dimensions $\left[M^0LT^{-2}\right]$.

It is not a universal constant.

(30) (a) As the rotation of earth takes place about polar axis therefore body placed at poles will not feel any centrifugal force and its weight or acceleration due to gravity remains unaffected.

DAILY PRACTICE PROBLEMS

PHYSICS SOLUTIONS 19

(1) (a) A body projected up with the escape velocity v_e will go to infinity. Therefore, the velocity of the body falling on the earth from infinity will be v_e. Now, the escape velocity on the earth is

$$v_e = \sqrt{gR_e} = \sqrt{2\times(9.8 m/s^2)\times(6400\times10^3 m)}$$

$= 1.2 \times 10\ 10^4$ m/s $= 11.2$ km/s.

The kinetic energy acquired by the body is

$$K = \frac{1}{2}m v_e^2 = \frac{1}{2} \times 100\text{ kg} \times (11.2 \times 10^3\text{ m/s})^2$$

$= 6.27 \times 10^9$ J.

(2) (d) We know that $\frac{GMm}{r^2} = m\omega^2 r$ or $\frac{GM}{r^2} = \omega^2 r$.

$$\therefore r^3 = \frac{GM}{\omega^2}$$

where ω is the angular velocity of the satellite

In the present case, $\omega = 2\omega_0$,

where ω_0 is the angular velocity of the earth.

$\therefore\ \omega = 2 \times 7.3 \times 10^{-5}$ rad/ sec.

$G = 6.673 \times 10^{-11}$ n-m^2/kg^2

and $M = 6.00 \times 10^{24}$kg.

Substituting these values in equation (A), we get

$$r^3 = \frac{(6.673 \times 10^{-11})(6.00 \times 10^{24})}{(2 \times 7.3 \times 10^{-5})^2}$$

Solving we get $r = 2.66 \times 10^7$m.

(3) (a)

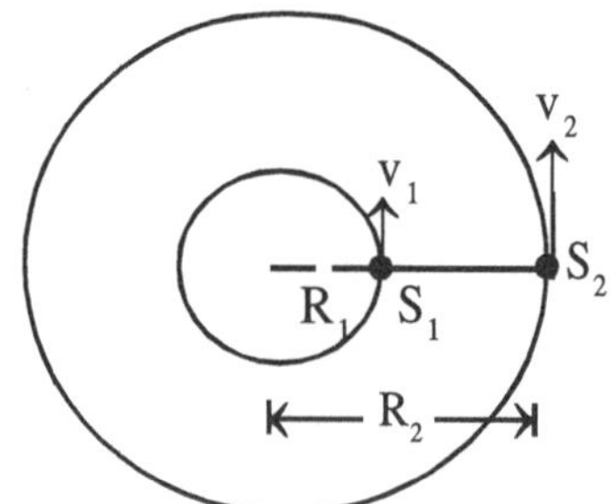

From Kepler's Law, $T^2 \propto r^3$

$$\therefore \left(\frac{T_1}{T_2}\right)^2 = \left(\frac{r_1}{r_2}\right)^3 \Rightarrow \left(\frac{1}{8}\right)^2 = \left(\frac{10^4}{r_2}\right)$$

$\Rightarrow r_2 = 4\times 10^4$ km

$$v = \omega r = \frac{2\pi r}{T}$$

$$\therefore |v_2 - v_1| = 2\pi\left(\frac{r_1}{T_1} - \frac{r_2}{T_2}\right) = \pi \times 10^4\text{ km/hr}$$

(4) (a) When S_2 is closest to S_1, the speed of S_2 relative to S_1 is $v_2 - v_1 = \pi \times 10^4$ km/hr. The angular speed of S_2 as observed from S_1 (when closest distance between them is $r_2 - r_1 = 3 \times 10^4$ km)

$$\omega = \frac{v_2 - v_1}{r_2 - r_1} = -\frac{\pi \times 10^4}{3 \times 10^4} = -\frac{\pi}{3}\text{ rad/hr},$$

$$|\omega| = \frac{\pi}{3}\text{ rad/hr}$$

(5) (c) Period of revolution of earth around sun

$$T_e^2 = \frac{4\pi^2 R_e^2}{GM_s}$$

Period of revolutions of moon around earth

$$T_n^2 = \frac{4\pi^2 R_m^2}{GM_e}$$

$$\therefore \left(\frac{T_e}{T_m}\right)^2 = \left(\frac{M_e}{M_s}\right)\left(\frac{R_e}{R_m}\right)^3$$

$$\therefore \frac{M_s}{M_e} = \left(\frac{T_m}{T_e}\right)^2\left(\frac{R_e}{R_m}\right)^3 = \frac{(393)^3}{13^2} = 3.56 \times 10^5$$

(6) (a) According to law of conservation of angular momentum, mvr = constant

$\Rightarrow$ vr = constant

$v_{max} . r_{min} = v_{min} . r_{max}$

$$\Rightarrow \frac{V_B}{V_A} = \frac{v_{max}}{v_{min}} = \frac{r_{max}}{r_{min}} = x$$

(7) (a) Angular momentum of satellite, J = mvr. But,

$$\frac{GMm}{r^2} = \frac{mv^2}{r} \Rightarrow v = \sqrt{\frac{GM}{r}}$$

$\therefore\ J = m\sqrt{GMr}$

(8) (a) The orbital velocity of space ship, $v_0 = \sqrt{\frac{GM}{r}}$

If space, ship is very near to earth's surface,

r = Radius of earth = R $\therefore v_0 = \sqrt{\frac{GM}{R}}$

$= \sqrt{Rg} = \sqrt{6.4 \times 10^6 \times 9.8}$

$= 7.9195 \times 10^3$ m/sec $= 7.195$ km/sec

The escape velocity of space-ship

$v_e = \sqrt{2Rg} = 7.9195\sqrt{2} = 11.2$ km/sec

Additional velocity required = 11.2– 7.9195=3.2805 km/ sec.

(9) (b) The escape velocity $v_e = \sqrt{2gR}$

Now, $(V_e)_{moon} = \sqrt{2gR}$

$(V_e)_{earth} = \sqrt{2 \times 6g \times 10R}$,

So $\frac{(V_e)_{earth}}{(V_e)_{moon}} = 8$

(10) (b) Escape velocity $= \sqrt{\frac{2GM}{R}} = \sqrt{2gR}$

$$\therefore \frac{V_p}{V_e} = \sqrt{\frac{g_p}{g_e} \times \frac{R_e}{R_p}} = \sqrt{10 \times 1} = \sqrt{10}$$

$V_p = \sqrt{10}\ V_e$

(11) (a) We know that $T^2 \propto a^3$

Given that $(12\,T)^2 \propto a_1^{\,3}$ and $T^2 \propto a_2^{\,3}$

$$\therefore \frac{a_1^{\,3}}{a_2^{\,3}} = \frac{(12T)^2}{T^2} = 144$$

or $\frac{a_1}{a_2} = (144)^{1/3} = 5.242$

Hence the jupiter's distance is 5.242 times that of the earth from the sun.

(12) (b) We know that $T^2 \propto a^3 \Rightarrow T \propto (a)^{3/2}$

$$\therefore \frac{T_{mars}}{T_{earth}} = \left(\frac{a_{mars}}{a_{earth}}\right)^{\frac{3}{2}} = (1.524)^{3/2} = 1.88$$

As the earth revolves round the sun in one year and hence, $T_{earth} = 1$ year.

$\therefore T_{mars} = T_{earth} \times 1.88 = 1 \times 1.88 = 1.88$ earth-year.

(13) (d) $\frac{T_{mercury}}{T_{mars}} = \left(\frac{a_{mercury}}{a_{mars}}\right)^{3/2} = \left(\frac{0.387}{1.524}\right)^{3/2}$

$$\therefore T_{mars} = T_{mercury} \times \left(\frac{1.524}{0.387}\right)^{3/2}$$

$= (0.241\text{years}) \times (7.8) = 1.9$ years.

(14) (a) $\frac{T^2}{r^3} = \frac{\left(\frac{2\pi r}{v_0}\right)^2}{r^3} = \frac{(2\pi r)^2}{r^3}\frac{1}{GM}r = \frac{4\pi^2}{GM}$

$$[\therefore \frac{mv_0^{\,2}}{r} = \frac{GMm}{r^2},\ v_0^2 = \frac{GM}{r}]$$

Slope of $T^2 - r^3$ curve $= \tan\theta = \frac{T^2}{r^3} = \frac{4\pi^2}{GM}$

(15) (c) Total energy of the particle at P

$$E = E_{kP} + U = \frac{1}{2}mv_e^2 - \frac{GM_1m}{d/2} - \frac{GM_2m}{d/2}$$

$$= \frac{1}{2}mv_e^2 - \frac{2Gm}{d}(M_1 + M_2)$$

At infinite distance from M_1 and M_2, the total energy of the particle is zero.

$$\therefore \frac{1}{2}mv_e^2 = \frac{2Gm}{d}(M_1 + M_2),$$

$$\therefore v_e = \sqrt{\frac{4G}{d}(M_1 + M_2)}$$

(16) (d) $v = \sqrt{\frac{GM}{r}} = \sqrt{\frac{gR^2}{r}} = \sqrt{\frac{9.8 \times 6.4^2 \times 10^{12}}{8 \times 10^6}}$

$= 7.08$ km/sec.

(17) (b) From conservation of energy,

The energy at height h = Total energy at earth's surface

$$0 - \frac{GMm}{R+h} = \frac{1}{2}mv^2 - \frac{GMm}{R},$$

$$\frac{1}{2}mv^2 = \frac{GMm}{R} - \frac{GMm}{R+h} = \frac{GMm}{R} - \frac{GMm}{2R}$$

$$\Rightarrow v = \sqrt{\frac{GM}{R}} = \sqrt{\frac{R^2g}{R}} = \sqrt{Rg}$$

$= \sqrt{6400 \times 10^3 \times 9.8} = 7.919 \times 10^3$ m/s

$= 7.919$ km/sec

(18) (a) If a body is projected from the surface of earth with a velocity v and reaches a height h, then using law of conservation of energy, $\frac{1}{2}mv^2 = \frac{mgh}{1 + h/R}$.

Given $v = Kv_e = K\sqrt{2gR}$ and $h = r - R$

Hence, $\frac{1}{2}mK^2\,2gR = \frac{mg(r-R)}{1 + \frac{r-R}{R}}$ or $r = \frac{R}{1-K^2}$

(19) (a) Orbital speed,

$v_0 = \sqrt{gR_e} = \sqrt{9.8 \times (6.4 \times 10^6)}$

$= 7.2 \times 10^3$ m/s $= 7.2$ km/s.

Period of revolution,

$T = 2\pi\sqrt{R/g}$

$= 2 \times 3.14\sqrt{(6.4 \times 10^6)/9.8} = 5075$ s $= 84.6$ minutes.

(20) (d) If the period of revolution of a satellite about the earth be T, then

$$T^2 = \frac{4\pi^2 (R_e + h)^3}{GM_e}$$

where h is the height of the satellite from earth's surface.

$$\therefore M_e = \frac{4\pi^2 (R_e + h)^3}{GT^2}$$

The satellite is revolving just above the earth, hence h is negligible compared to R_e.

$$\therefore M_e = \frac{4\pi^2 R_e^{\,3}}{GT^2}$$

But $M_e = \frac{4}{3}\pi R_e^{\,3}\rho$ where ρ is the density of the earth.

Thus $\frac{4}{3}\pi R_e^{\,3}\rho = \frac{4\pi^2 R_e^{\,3}}{GT^2}$

$$\therefore \rho T^2 = \frac{3\pi}{G}.$$

which is universal constant. To determine its value,

$$\rho T^2 = \frac{3\pi}{G} = \frac{3 \times 3.14}{6.67 \times 10^{-11}\ m^3/kg\text{-}s^2}$$

(21) (a) $\frac{E_{KQ}}{E_{KP}} = \frac{v_Q^2}{v_P^2}$.

Linear velocity of earth,

$$V_e = \frac{2\pi R_e}{T_e} = \frac{6.28 \times 6.4 \times 10^6}{24 \times 3600} = 463\ m/s$$

Orbital velocity, $V_0 = \sqrt{R_e g} = 7.9 \times 10^3\ m/s$

According to question,

$V_P = V_0 + V_e = 7900 - 463 = 7437\ m/s$

$V_Q = V_0 + V_e = 7900 + 463 = 8363\ m/s$

$$\therefore \frac{E_{KQ}}{E_{KP}} = \left(\frac{8363}{7437}\right)^2$$

(22) (b) $v_e = \sqrt{\frac{2GM}{R}}$ i.e. escape velocity depends upon the mass and radius of the planet.

(23) (a) $v_e = \sqrt{\frac{2GM}{R}} = R\sqrt{\frac{8}{3}\pi G\rho}$

If mean density is constant then $v_e \propto R$

$$\frac{v_e}{v_p} = \frac{R_e}{R_p} = \frac{1}{2} \Rightarrow v_e = \frac{v_p}{2}$$

(24) (a) $v_0 = \sqrt{\frac{GM}{r}}$

(25) (d) $T^2 \propto R^3$

$$T_e^2 = KR_e^3; T_m^2 = kR_m^3; T^2 = kR^3$$

$$R = \frac{R_e + R_m}{2}$$

$$\Rightarrow T^2 = k\left[\frac{T_e^{2/3}}{k^{1/3}} + \frac{T_m^{2/3}}{k^{1/3}} \times \frac{1}{2}\right]^3$$

$$\Rightarrow T = \left[\frac{T_e^{2/3} + T_m^{2/3}}{2}\right]^{3/2}$$

(26) (a) $E_e = -\frac{GM_s M_e}{2R_e} = -\frac{GM_s M}{2R}$

$$= \frac{2R_e E_e}{M_e} \times \frac{M}{2\left(\frac{R_e + R_m}{2}\right)}$$

$$= \frac{2M}{M_e}\left(\frac{R_e}{R_e + R_m}\right)E_e$$

(27) (c) Areal velocity of the artificial planet around the sun will be more than that of earth.

(28) (a) $v_0 = R_e\sqrt{\frac{g}{R_e + h}}$

For satellite revolving very near to earth $R_e + h = R_e$

As $(h << R)$

$$v_0 = \sqrt{R_e g} \simeq \sqrt{64 \times 10^5 \times 10} = 8 \times 10^3\ m/s = 8\ kms^{-1}$$

Which is independent of height of a satellite.

(29) (d) Due to resistance force of atmosphere, the satellite revolving around the earth losses kinetic energy. Therefore in a particular orbit the gravitational attraction of earth on satellite becomes greater than that required for circular orbit there. Therefore satellite moves down to a lower orbit. In the lower orbit as the potential energy $(U = -GMm/r)$ becomes more negative, Hence kinetic energy $(E_k = GMm/2r)$ increases, and hence speed of satellite increases.

(30) (a) Because gravitational force is always attractive in nature and every body is bound by this gravitational force of attraction of earth.

DAILY PRACTICE PROBLEMS — PHYSICS SOLUTIONS 20

1. **(a).** $Y = \frac{MgL}{\pi r^2 \Delta\ell}$
but $Mg/\pi r^2 = 20 \times 10^8$ & $\Delta\ell = L$ then
$Y = 20 \times 10^8 \text{ N/m}^2$

2. **(b).** $F = Y\alpha\Delta tA$
$A = 2 \times 10^{-6} \text{ m}^2, Y = 2 \times 11 \text{ N/m}^2$
$\alpha = 1.1 \times 10^{-5}, t = 50 - 30 = 20°C$
$F = 2 \times 10^{11} \times 1.1 \times 10^{-5} \times 20 \times 2 \times 10^{-6} = 88N.$

3. **(d).** Work done on the wire
$$W = \frac{1}{2} F \times \ell = \frac{1}{2} \times \text{stress x volume x strain}$$
$$W = \frac{1}{2} \times Y \times \text{strain}^2 \times \text{volume}$$
$$W = \frac{1}{2} \times Y \times \frac{\Delta\ell^2}{L^2} \times AL = \frac{YA\Delta L^2}{2L}$$
$$W = \frac{2 \times 10^{11} \times 10^{-6} \times 10^{-6}}{2 \times 1} = 0.1 \text{ J}$$

4. **(d).** $W = \frac{1}{2} \times \text{load} \times \text{elongation}$
$$W = \frac{1}{2} \times 5.4 \times 10^6 \times 3$$
$W = 8.1 \times 10^6$ ergs

5. **(d).** By Hook's law
$$Y = \frac{F/A}{\ell/L} = \frac{FL}{A\ell}$$
$$Y = \frac{16 \times 1}{(4 \times 10^{-8})(0.2 \times 10^{-2})} = 2 \times 10^{11} \text{ N/m}^2$$

6. **(a).** $B = -\frac{\Delta PV}{\Delta V}$
Given, $\Delta P = hdg = 200 \text{ x } 10^3 \text{ x } 10$
$\Delta P = 2 \times 10^5 \text{ N/m}^2$
$$\frac{\Delta V}{V} = \frac{0.1}{100} = 10^{-3}$$
$$\therefore B = \frac{2 \times 10^6}{10^{-3}} = 2 \times 10^9 \text{ N/m}^2$$

7. **(b).** $Y = \frac{\text{stress}}{\text{strain}} = \frac{F/A}{\ell/L} = \frac{FL}{A\ell}$
$$\therefore \ell = \frac{FL}{AY} = \frac{20 \times 9.8 \times 4}{\pi \times (10^{-3})^2 \times 1.96 \times 10^{11}}$$
$= 1.27 \times 10^{-3} \text{ m} = 1.27 \text{ mm}$

8. **(a).** Limiting stress $= 4.0 \times 10^8 \text{ N/m}^2$
$$\frac{F}{A} = \frac{400}{A} = 4.0 \times 10^8$$
or $A = 10^{-6} \text{ m}^2$
$$\therefore D = \left(\frac{4A}{\pi}\right)^{1/2} = \left(\frac{4 \times 10^{-6}}{\pi}\right)^{1/2}$$
$= 1.13 \times 10^{-3} \text{ m} = 1.13 \text{ mm}$

9. **(c).** Stress $= \frac{F}{A} = \frac{4.8 \times 10^3 \text{N}}{1.2 \times 10^{-4} \text{m}^2} = 4.0 \times 10^7 \text{ N/m}^2$

10. **(c).** Strain $= \frac{\Delta\ell}{\ell} = \frac{1 \times 10^{-3}}{2} = 5 \times 10^{-4}$, longitudinal

11. **(c).** $F = YA\alpha\Delta t$
$= 2 \times 10^{11} \times 3 \times 10^{-6} \times 10^{-5} \times 20$
$F = 120$ N.

12. **(c).** Compressibility
$$\chi = \frac{1}{K} = -\frac{\Delta V}{V\Delta p} = 5 \times 10^{-10}$$
$\therefore$ Fractional decrease in volume
$$= -\frac{\Delta V}{V} = \chi\Delta p = 5 \times 10^{-10} \times 15 \times 10^6$$
$= 7.5 \times 10^{-3}$

13. **(c).** Increase in length on heating $\Delta\ell = \alpha L \Delta T$
To annul this increase if pressure applied is p then
$$p = Y\frac{\Delta\ell}{L} = Y\alpha\Delta T$$
$= 2 \times 10^{11} \times 1.1 \times 10^{-5} \times 100 = 2.2 \times 10^8 \text{ N/m}^2$

14. **(c).** $y = 2\eta(1 + \sigma)$
$y = 2.4 \times \eta$
$2.4\eta = 2\eta(1 + \sigma)$
$(1 + \sigma) = 1.2$
$\sigma = 0.2$

15. **(c).** Stress $= F/A = 10/(2 \times 10^{-6}) = 5 \times 10^6 \text{ N/m}^2$
$$\text{Strain} = \frac{\text{Stress}}{Y} = \frac{5 \times 10^6}{2 \times 10^{11}} = 2.5 \times 10^{-5}$$
$\ell = L \times \text{strain} = 1 \times 2.5 \times 10^{-5}$
$\ell = 2.5 \times 10^{-5}$ m

16. **(b).** $\frac{\Delta\ell}{L} = \frac{Mg}{AY} = \frac{1000 \times 980 \times 100}{10^{12} \times 0.01}$
$\Delta\ell = 0.0098$ cm.

17. **(a).** Volume = Mass/density
Area of cross-section = volume/length

$$=\frac{\text{mass}}{\text{density}\times\text{length}}=\frac{15.6\times10^{-3}}{7800\times2.5}=8\times10^{-7}\,m^2$$

$$Y=\frac{F\ell}{A\Delta L}=\frac{8\times9.8\times2.5}{(8\times10^{-7})\times1.25\times10^{-3}}$$

$Y=1.96\times10^{11}\,N/m^2$

18. **(c).** Potential energy per unit volume = u

$$=\frac{1}{2}\times\text{stress}\times\text{strain}$$

But $Y=\frac{\text{stress}}{\text{strain}}$

$\therefore$ stress = Y x strain = Y x S

$\therefore$ Potential energy per unit volume = u

$$=\frac{1}{2}\times(YS)S=\frac{1}{2}YS^2$$

19. **(d).** $\frac{\ell_1}{\ell_2}=\frac{L_1 r_2^2}{L_2 r_1^2}$

$L_1=L, L_2=2L, r_1=2R., r_2=R$

$$\therefore\ \frac{\ell_1}{\ell_2}=\frac{L}{2L}\cdot\frac{R^2}{4R^2}=\frac{1}{8}$$

20. **(c).** stress $=\frac{\text{Force}}{\text{Area}}=\frac{F}{\pi r^2}$

$\therefore$ stress $S\propto\frac{1}{r^2}$

$$\therefore\ \left(\frac{S_1}{S_2}\right)=\left(\frac{r_2}{r_1}\right)^2$$

Given $\frac{r_1}{r_2}=\frac{2}{1}\ \therefore\ \frac{S_1}{S_2}=\frac{1}{4}$

21. **(c)** $F=YA\alpha\Delta\theta\ \therefore F\propto A$

22. **(a)** For twisting, angle of shear $\phi\propto\frac{1}{L}$

i.e. if L is more then ϕ will be small.

23. **(a)** $Y=2\eta(1+\sigma)$

24. **(a)** $Y=2\eta(1+\sigma)\Rightarrow\sigma=\frac{0.5Y-\eta}{\eta}$

25. **(a)** Tensile stress $=\frac{F\cos\theta}{a/\cos\theta}=\frac{F\cos^2\theta}{a}$

26. **(a)** Tensile stress is maximum when $\cos^2\theta$ is maximum, i.e., $\theta=0°$

27. **(b)** Shearing stress $=\frac{F\sin\theta}{a/\cos\theta}=\frac{F\sin\theta\cos\theta}{a}$

$$=\frac{F\sin2\theta}{2a}$$

28. **(a)** Elasticity is a measure of tendency of the body to regain its original configuration. As steel is deformed less than rubber therefore steel is more elastic than rubber.

29. **(a)** Bulk modulus of elasticity measures how good the body is to regain its original volume on being compressed. Therefore, it represents incompressibility of the material.

$K=\frac{-PV}{\Delta V}$ where P is increase in pressure, ΔV is change in volume.

30. **(a)** A bridge during its use undergoes alternating strains for a large number of times each day, depending upon the movement of vehicles on it when a bridge is used for long time, it losses its elastic strength. Due to which the amount of strain in the bridge for a given stress will become large and ultimately, the bridge may collapse. This may not happen, if the bridges are declared unsafe after long use.

DAILY PRACTICE PROBLEMS

PHYSICS SOLUTIONS 21

1. (a) Force required to separate the plates

$$F = \frac{2TA}{t} = \frac{2\times 70\times 10^{-3}\times 10^{-2}}{0.05\times 10^{-3}} = 28N$$

2. (c)

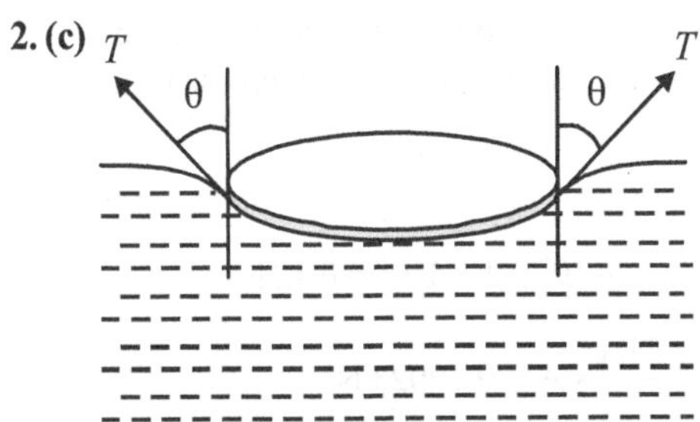

Weight of metal disc = total upward force
= upthrust force + force due to surface tension
= weight of displaced water + $T\cos\theta\,(2\pi r)$
$= W + 2\pi r T\cos\theta$

3. (d) $W = T\times 8\pi(r_2^2 - r_1^2) = T\times 8\pi\left(\frac{D^2}{4} - \frac{d^2}{4}\right)$

$= 2\pi(D^2 - d^2)T$

4. (b) Increment in area of soap film $= A_2 - A_1$

$= 2\times[(10\times 0.6) - (10\times 0.5)]\times 10^{-4} = 2\times 10^{-4}m^2$

Work done $= T\times\Delta A$

$= 7.2\times 10^{-2}\times 2\times 10^{-4} = 1.44\times 10^{-5}J$

5. (a)

6. (c) Excess pressure inside soap bubble is inversely proportional to the radius of bubble *i.e.* $\Delta P \propto \frac{1}{r}$

This means that bubbles *A* and *C* posses greater pressure inside it than *B*. So the air will move from *A* and *C* towards *B*.

7. (b) $r = \frac{r_1 r_2}{r_1 - r_2} = \frac{5\times 4}{5-4} = 20\ cm$

8. (c) The radius of resultant bubble is given by $R^2 = r_1^2 + r_2^2$.

9. (b)

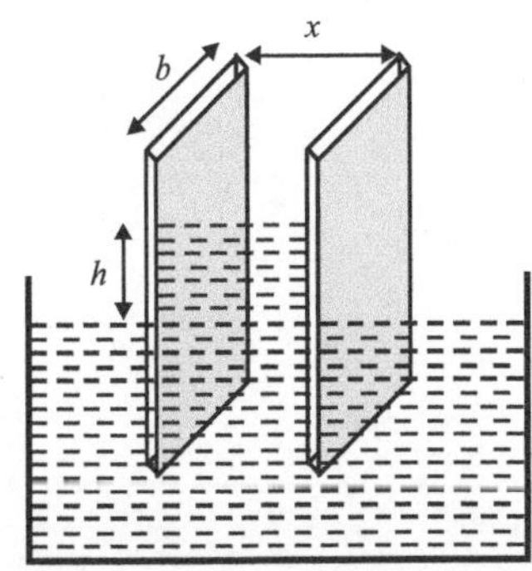

Let the width of each plate is *b* and due to surface tension liquid will rise upto height *h* then upward force due to Surface tension

$= 2Tb\cos\theta$ ....(i)

Weight of the liquid rises in between the plates

$= Vdg = (bxh)dg$ ....(ii)

Equating (i) and (ii) we get, $2T\cos\theta = bxhdg$

$\therefore h = \frac{2T\cos\theta}{xdg}$

10. (b) Mass of liquid in capillary tube

$M = \pi R^2 H\times\rho \therefore M \propto R^2\times\left(\frac{1}{R}\right)$ (As $H\propto 1/R$)

$\therefore\ M\propto R$. If radius becomes double then mass will becomes twice.

11. (d) In the satellite, the weight of the liquid column is zero. So the liquid will rise up to the top of the tube.

12. (b) $T_c = T_0(1 - \propto t)$

i.e. surface tension decreases with increase in temperature.

13. (a)

14. (d) Tension in spring T = upthrust - weight of sphere

$= V\sigma g - V\rho g = V\eta\rho g - V\rho g$ (As $\sigma = \eta\rho$)

$= (\eta - 1)V\rho g = (\eta - 1)\,mg.$

15. (b)

16. (c)

17. (c) A stream lined body has less resistance due to air.

18. (a) Weight of cylinder = upthrust due to both liquids

$$V\times D\times g = \left(\frac{A}{5}\times\frac{3}{4}L\right)\times d\times g + \left(\frac{A}{5}\times\frac{L}{4}\right)\times 2d\times g$$

$$\Rightarrow\left(\frac{A}{5}\times L\right)\times D\times g = \frac{A\times L\times d\times g}{4}$$

$$\Rightarrow\frac{D}{5} = \frac{d}{4}\therefore D = \frac{5}{4}d$$

19. (b) Velocity of efflux when the hole is at depth $h, v = \sqrt{2gh}$

Rate of flow of water from square hole

$Q_1 = a_1 v_1 = L^2\sqrt{2gy}$

Rate of flow of water from circular hole

$Q_2 = a_2 v_2 = \pi R^2\sqrt{2g(4y)}$

According to problem $Q_1 = Q_2$

$$\Rightarrow L^2\sqrt{2gy} = \pi R^2\sqrt{2g(4y)} \Rightarrow R = \frac{L}{\sqrt{2\pi}}$$

20. **(a)** Let A = cross-section of tank

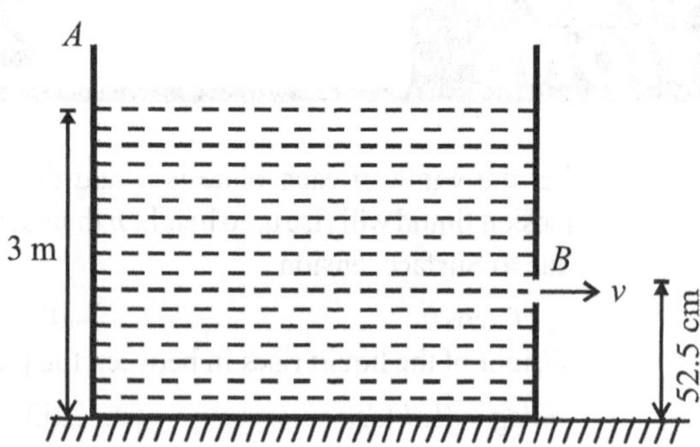

a = cross-section hole
V = velocity with which level decreases
v = velocity of efflux

From equation of continuity $av = AV \Rightarrow V = \frac{av}{A}$

By using Bernoulli's theorem for energy per unit volume
Energy per unit volume at point A
= Energy per unit volume at point B

$$P + \rho gh + \frac{1}{2}\rho V^2 = P + 0 + \frac{1}{2}\rho v^2$$

$$\Rightarrow v^2 = \frac{2gh}{1-\left(\frac{a}{A}\right)^2} = \frac{2\times 10\times(3-0.525)}{1-(0.1)^2} = 50(m/\sec)^2$$

21. **(c)** If the liquid is incompressible then mass of liquid entering through left end, should be equal to mass of liquid coming out from the right end.

$\therefore\ M = m_1 + m_2 \Rightarrow Av_1 = Av_2 + 1.5A.v$

$\Rightarrow A\times 3 = A\times 1.5 + 1.5A.v \Rightarrow v = 1\ m/s$

22. **(b)** $T = \frac{F}{2\pi r} = \frac{6.28\times 10^{-4}}{2\times 3.14\times 2\times 10^{-3}} = 5\times 10^{-2}\ N/m$

23. **(d)** At critical temperature $\left(T_c = 370^\circ C = 643\ K\right)$, the surface tension of water is zero.

24. **(d)**

25. **(b)**

$P_{Bottom} > P_{Surface}$. So bubble rises upward.

At constant temperature $V \propto \frac{1}{P}$ (Boyle's law)

Since as the bubble rises upward, pressure decreases, then from above law volume of bubble will increase i.e. its size increases.

26. **(a)**

27. **(d).**

$F = P_{atm} \times Area = 10^5 \times 1 \times 10^{-6} = 0.1\ N$

$F = \left(P_{atm} + \frac{2T}{r}\right) \times A = 0.10023\ N$

28. **(a)** Since the excess pressure due to surface tension is inversely proportional to its radius, it follows that smaller the bubble, greater is the excess pressure. Thus when the larger and the smaller bubbles are put in communication, air starts passing from the smaller into the large bubble because excess pressure inside the former is greater than inside the latter. As a result, the smaller bubble shrinks and the larger one swells.

29. **(b)** Statement-1 is True, Statement-2 is True; Statement-2 is NOT a correct explanation for Statement-1.

30. **(a)** In a stream line flow of a liquid, according to equation of continuity,

av = constant

Where a is the area of cross-section and v is the velocity of liquid flow. When water flowing in a broader pipe enters a narrow pipe, the area of cross-section of water decreases therefore the velocity of water increases.

DAILY PRACTICE PROBLEMS — PHYSICS SOLUTIONS 22

1. (c) Due to volume expansion of both mercury and flask, the change in volume of mercury relative to flask is given by $\Delta V = V_0[\gamma_L - \gamma_g]\Delta\theta = V[\gamma_m - 3\alpha_g]\Delta\theta$
$= 50[180 \times 10^{-6} - 3 \times 9 \times 10^{-6}](38 - 18)$
$= 0.153$ cc

2. (a) $\gamma_{real} = \gamma_{app.} + \gamma_{vessel}$
So $(\gamma_{app.} + \gamma_{vessel})_{glass} = (\gamma_{app.} + \gamma_{vessel})_{steel}$
$\Rightarrow 153 \times 10^{-6} + (\gamma_{vessel})_{glass} = (144 \times 10^{-6} + \gamma_{vessel})_{steel}$
Further, $(\gamma_{vessel})_{steel} = 3\alpha = 3 \times (12 \times 10^{-6}) = 36 \times 10^{-6}/°C$
$\Rightarrow 153 \times 10^{-6} + (\gamma_{vessel})_{glass} = 144 \times 10^{-6} + 36 \times 10^{-6}$
$\Rightarrow (\gamma_{vessel})_{glass} = 3\alpha = 27 \times 10^{-6}/°C$
$\Rightarrow \alpha = 9 \times 10^{-6}/°C$

3. (c) Initial diameter of tyre = (100 – 6) mm = 994 mm,
So, initial radius of tyre $R = \frac{994}{2} = 497$mm
and change in diameter ΔD = 6 mm, so
$\Delta R = \frac{6}{2} = 3$ mm
After increasing temperature by $\Delta\theta$, tyre will fit onto wheel
Increment in the length (circumference) of the iron tyre
$\Delta L = L \times \alpha \times \Delta\theta = L \times \frac{\gamma}{3} \times \Delta\theta$ [As $\alpha \times \frac{\gamma}{3}$]
$2\pi\Delta R = 2\pi R\left(\frac{\gamma}{3}\right)\Delta\theta$
$\Delta\theta \Rightarrow \frac{3}{\gamma}\frac{\Delta R}{R} = \frac{3 \times 3}{3.6 \times 10^{-5} \times 497}$
$\Rightarrow \Delta\theta = 500°C$

4. (b) $\Delta L = L_0\alpha\Delta\theta$
Rod A : $0.075 = 20 \times \alpha_A \times 100$
$\Rightarrow \alpha_A = \frac{75}{2} \times 10^{-6}/°C$
Rod B : $0.045 = 20 \times \alpha_B \times 100$
$\Rightarrow \alpha_B = \frac{45}{2} \times 10^{-6}/°C$
For composite rod: x cm of A and $(20 - x)$ cm of B we have

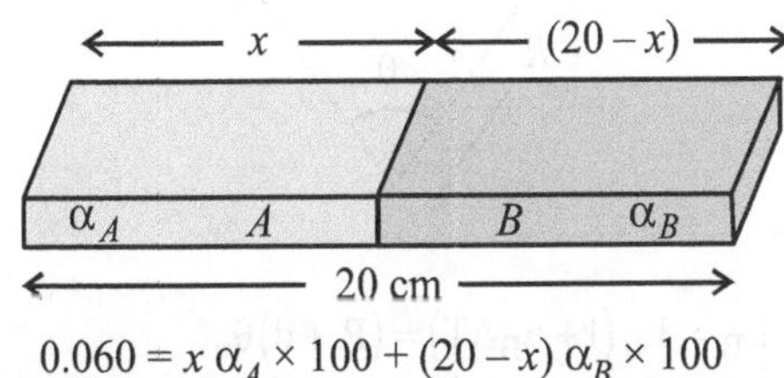

$0.060 = x\,\alpha_A \times 100 + (20 - x)\,\alpha_B \times 100$
$= x\left[\frac{75}{2} \times 10^{-6} \times 100 + (20 - x) \times \frac{45}{2} \times 10^{-6} \times 100\right]$
On solving we get $x = 10$ cm.

5. (b) Due to volume expansion of both liquid and vessel, the change in volume of liquid relative to container is given by $\Delta V = V_0[\gamma_L - \gamma_g]\Delta\theta$
Given $V_0 = 1000\ cc$, $\alpha_g = 0.1 \times 10^{-4}/°C$
$\therefore \gamma_g = 3\alpha_g = 3 \times 0.1 \times 10^{-4}/°C = 0.3 \times 10^{-4}/°C$
$\therefore \Delta V = 1000[1.82 \times 10^{-4} - 0.3 \times 10^{-4}] \times 100 = 15.2cc$

6. (b) $\gamma_r = \gamma_a + \gamma_v$; where γ_r = coefficient of real expansion,
γ_a = coefficient of apparent expansion and
γ_v = coefficient of expansion of vessel.
For copper $\gamma_r = C + 3\alpha_{Cu} = C + 3A$
For silver $\gamma_r = S + 3\alpha_{Ag}$
$\Rightarrow C + 3A = S + 3\alpha_{Ag} \Rightarrow \alpha_{Ag} = \frac{C - S + 3A}{3}$

7. (d) $V = V_0(1 + \gamma\Delta\theta) \Rightarrow$ Change in volume
$V - V_0 = \Delta V = A.\Delta l = V_0\gamma\Delta\theta$
$\Rightarrow \Delta l = \frac{V_0.\Delta\theta}{A} = \frac{10^{-6} \times 18 \times 10^{-5} \times (100 - 0)}{0.004 \times 10^{-4}}$
$= 45 \times 10^{-3}$m = 4.5 cm

8. (b) Loss of weight at 27°C is
$= 46 - 30 = 16 = V_1 \times 1.24\,\rho_1 \times g$ ...(i)
Loss of weight at 42°C is
$= 46 - 30.5 = 15.5 = V_2 \times 1.2\,\rho_1 \times g$ ...(ii)
Now dividing (i) by (ii), we get
$\frac{16}{15.5} = \frac{V_1}{V_2} \times \frac{1.24}{1.2}$
But $\frac{V_2}{V_1} = 1 + 3\alpha(t_2 - t_1) = \frac{15.5 \times 1.24}{16 \times 1.2} = 1.001042$
$\Rightarrow 3\alpha\,(42° - 27°) = 0.001042$
$\Rightarrow \alpha = 2.316 \times 10^{-5}/°C$

9. (b) Heat lost in t sec = mL or heat lost per sec = $\frac{mL}{t}$. This must be the heat supplied for keeping the substance in molten state per sec.
$\therefore \frac{mL}{t} = P$ or $L = \frac{Pt}{m}$

10. (b) Initially ice will absorb heat to raise it's temperature to 0°C then it's melting takes place
If m_1 = Initial mass of ice, m_1' = Mass of ice that melts and m_W = Initial mass of water

By Law of mixture

Heat gained by ice = Heat lost by water $\Rightarrow m_1 \times (20) + m_1' \times L = m_W c_W[20]$

$\Rightarrow 2 \times 0.5\ (20) + m_1' \times 80 = 5 \times 1 \times 20$

$\Rightarrow m_1' = 1$ kg

So final mass of water = Initial mass of water + Mass of ice that melts = 5 + 1 = 6 kg.

11. (a) If mass of the bullet is m gm,

then total heat required for bullet to just melt down

$Q_1 = mc\Delta\theta + mL = m \times 0.03\ (327 - 27) + m \times 6$

$= 15m$ cal $= (15m \times 4.2)$J

Now when bullet is stopped by the obstacle, the loss in its mechanical energy $= \frac{1}{2}(m \times 10^{-3})v^2 J$

(As m gm $= m \times 10^{-3}$kg)

As 25% of this energy is absorbed by the obstacle,

$$Q_2 = \frac{75}{100} \times \frac{1}{2} mv^2 \times 10^{-3} = \frac{3}{8} mv^2 \times 10^{-3} J$$

Now the bullet will melt if $Q_2 \geq Q_1$

i.e. $\frac{3}{8} mv^2 \times 10^{-3} \geq 15m \times 4.2 \Rightarrow v_{\min} = 410$ m/s

12. (c) Heat gain = heat lost

$$C_A(16-12) = C_B(19-16) \Rightarrow \frac{C_A}{C_B} = \frac{3}{4}$$

and $C_B(23-19) = C_C(28-23) \Rightarrow \frac{C_B}{C_C} = \frac{5}{4}$

$$\Rightarrow \frac{C_A}{C_C} = \frac{15}{16} \quad ....(i)$$

If θ is the temperature when A and C are mixed then,

$$C_A(\theta - 12) = C_C(28 - \theta) \Rightarrow \frac{C_A}{C_C} = \frac{28-\theta}{\theta - 12} \quad ...(ii)$$

On solving equation (i) and (ii) $\theta = 20.2$°C

13. (a) Same amount of heat is supplied to copper and water so $m_c c_c \Delta\theta_c = m_W c_W \Delta\theta_W$

$$\Rightarrow \Delta\theta_W = \frac{m_c c_c (\Delta\theta)_c}{m_W c_W} = \frac{50 \times 10^{-3} \times 420 \times 10}{10 \times 10^{-3} \times 4200} = 5°C$$

14. (b) Heat lost by hot water = Heat gained by cold water in beaker + Heat absorbed by beaker

$\Rightarrow 440\ (92 - \theta) = 200 \times (\theta - 20) + 20 \times (\theta - 20)$

$\Rightarrow \theta = 68$°C

15. (a)

16. (b) Firstly the temperature of bullet rises up to melting point, then it melts. Hence according to $W = JQ$.

$$\Rightarrow \frac{1}{2} mv^2 = J.[m.c.\Delta\theta + mL] = J[mS(475 - 25) + mL]$$

$$\Rightarrow mS(475 - 25) + mL = \frac{mv^2}{2J}$$

17. (b) Suppose m kg of ice melts then by using

W (Joules) $= H$ (Joules)

$\Rightarrow Mgh = mL$

$\Rightarrow 3.5 \times 10 \times 2000 = m \times 3.5 \times 10^5$

$\Rightarrow m = 0.2$ kg = 200 gm

18. (d) Coefficient of volume expansion

$$\gamma = \frac{\Delta\rho}{\rho.\Delta T} = \frac{(\rho_1 - \rho_2)}{\rho.(\Delta\theta)} = \frac{(10 - 9.7)}{10 \times (100 - 0)} = 3 \times 10^{-4}$$

Hence, coefficient of linear expansion

$$\alpha = \frac{\gamma}{3} = 10^{-4}/°C$$

19. (b) As we know

$\gamma_{real} = \gamma_{app.} + \gamma_{vessel}$

$\Rightarrow \gamma_{app.} = \gamma_{glycerine} - \gamma_{glass}$

$= 0.000597 - 0.000027$

$= 0.00057$/°C

20. (a) $t = \frac{(P_t - P_0)}{(P_{100} - P_0)} \times 100°C$

$$= \frac{(60-50)}{(90-50)} \times 100 = 25°C$$

21. (c) Since specific heat = 0.6 kcal/gm × °C
= 0.6 cal/gm × °C

From graph it is clear that in a minute, the temperature is raised from 0°C to 50°C.

$\Rightarrow$ Heat required for a minute

= 50 × 0.6 × 50 = 1500 cal.

Also from graph, boiling point of wax is 200°C.

22. (b) The horizontal parts of the curve, where the system absorbs heat at constant temperature must depict changes of state. Here the latent heats are proportional to lengths of the horizontal parts. In the sloping parts, specific heat capacity is inversely proportional to the slopes.

23. (d) Let L_0 be the initial length of each strip before heating. Length after heating will be

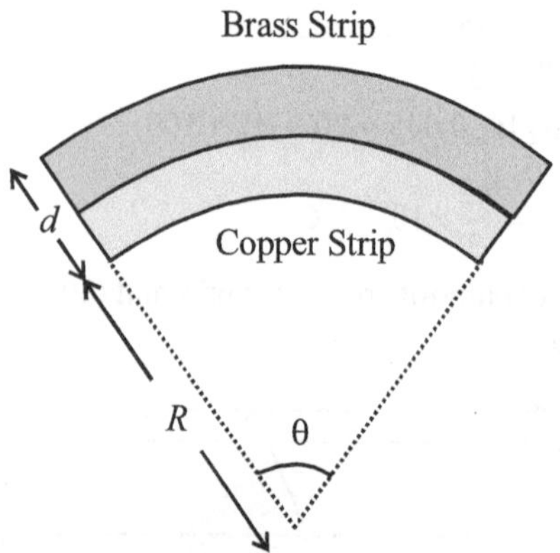

$$L_B = L_0(1 + \alpha_B \Delta T) = (R + d)\theta$$

$$L_C = L_0(1+\alpha_C\Delta T) = R\theta$$

$$\Rightarrow \frac{R+d}{R} = \frac{1+\alpha_B\Delta T}{1+\alpha_C\Delta T}$$

$$\Rightarrow 1+\frac{d}{R} = 1+(\alpha_B-\alpha_C)\Delta T$$

$$\Rightarrow R = \frac{d}{(\alpha_B-\alpha_C)\Delta T}$$

$$\Rightarrow R \propto \frac{1}{\Delta T} \text{ and } R \propto \frac{1}{(\alpha_B-\alpha_C)}$$

24. **(a)** A bimetallic strip on being heated bends in the form of an arc with more expandable metal (A) outside (as shown) correct.

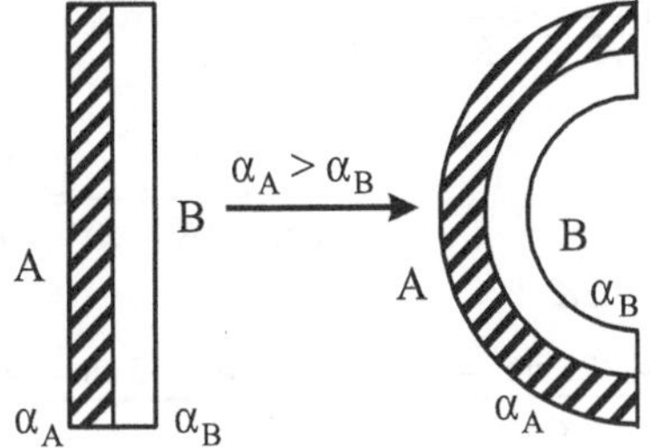

25. **(a)** **26.** **(c)** **27.** **(c)**

$\rho_1 v_1 A_1 = \rho_2 v_2 A_2$

$m = 1500\text{ kg/m}^3 \times 0.1\text{ m/s} \times 4\text{ (cm)}^2$

$ms\Delta T = 10000$

$1500 \times 0.1 \times 4 \times 10^{-4} \times 1500 \times \Delta T = 10000$

$$\Delta T = \frac{10000}{90} = \frac{1000}{9}\,°C$$

$$\rho_2 = \frac{\rho_1}{(1+\gamma\Delta T)} = \frac{1500}{\left(1+1\times10^{-3}\times\frac{1000}{9}\right)} = 1350\text{ kg/m}^3$$

$\rho_2 v_2 A_2 = \rho_1 v_1 A_1$

$\Rightarrow 1350 \times v_2 = 1500 \times 0.1$

$v_2 = 1/9\text{ m/s}$

$\therefore$ Volume rate of flow at the end of tube

$$= A_2 v_2 = 4\times10^{-4}\times\frac{1}{9}$$

$$= \frac{4}{9}\times10^{-4}\text{ m}^3 = \frac{40}{9}\times10^{-5}\text{ m}^3$$

Volume rate of flow at the entrance $= A_1 v_1$

$= 0.1 \times 4 \times 10^{-4} = 4 \times 10^{-5}\text{m}^3$

Hence, difference of volume rate of flow at the two ends

$$= \left(\frac{40}{9}-4\right)\times10^{-5} = \frac{4}{9}\times10^{-5}\text{ m}^3$$

28. **(d)** Celsius scale was the first temperature scale and Fahrenheit is the smallest unit measuring temperature.

29. **(a)** Linear expansion for brass (19×10^{-4}) > linear expansion for steel (11×10^{-4}). On cooling the disk shrinks to a greater extent than the hole and hence it will get loose.

30. **(b)** The latent heat of fusion of ice is amount of heat required to convert unit mass of ice at 0°C into water at 0°C. For fusion of ice

$L = 80$ cal/gm = 80000 cal/gm = 8000 × 4.2 j/kg

= 336000 J/kg.

DAILY PRACTICE PROBLEMS

PHYSICS SOLUTIONS

23

1. (a) $\frac{dQ}{dt} = \frac{KA\Delta\theta}{\ell}$

For both rods K, A and $\Delta\theta$ are same

$\Rightarrow \quad \frac{dQ}{dt} \propto \frac{1}{\ell}$

So, $\frac{(dQ/dt)_{semi\,circular}}{(dQ/dt)_{straight}} = \frac{\ell_{straight}}{\ell_{semi\,circular}} = \frac{2r}{\pi r} = \frac{2}{\pi}$.

2. (b) Suppose thickness of each wall is x then

$\left(\frac{Q}{t}\right)_{combination} = \left(\frac{Q}{t}\right)_A$

$\Rightarrow \quad \frac{K_s A(\theta_1 - \theta_2)}{2x} = \frac{2KA(\theta_1 - \theta)}{x}$

$\because \quad K_s = \frac{2 \times 2K \times K}{(2K + K)} = \frac{4}{3}K$

and $(\theta_1 - \theta_2) = 36°$

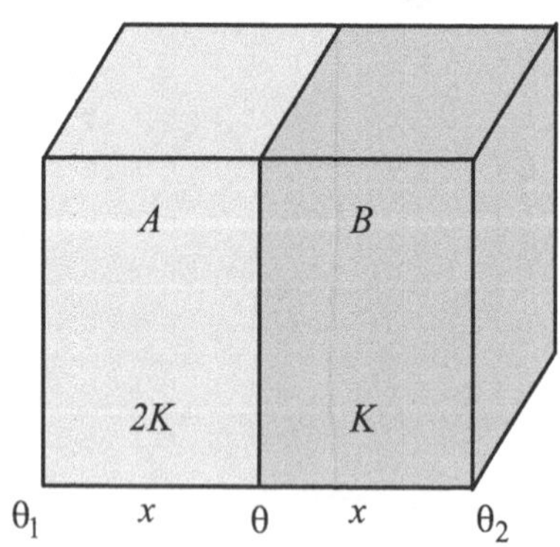

$\Rightarrow \quad \frac{\frac{4}{3}KA \times 36}{2x} = \frac{2KA(\theta_1 - \theta)}{x}$

Hence temperature difference across wall A is $(\theta_1 - \theta) = 12°$ C

3. (a) Heat developed by the heater

$H = \frac{V^2}{R}.\frac{t}{J} = \frac{(200)^2 \times t}{20 \times 4.2}$

Heat conducted by the glass

$H = \frac{0.2 \times 1 \times (20 - \theta)t}{0.002}$

Hence $\frac{(200)^2 \times t}{20 \times 4.2} = \frac{0.2 \times (20 - \theta)t}{0.002}$

$\Rightarrow \quad \theta = 15.24°$ C

4. (b) If suppose $K_{Ni} = K \Rightarrow K_{Al} = 3K$ and $K_{Cu} = 6K$
Since all metal bars are connected in series.

So, $\left(\frac{Q}{t}\right)_{Combination} = \left(\frac{Q}{t}\right)_{Cu} = \left(\frac{Q}{t}\right)_{Al} = \left(\frac{Q}{t}\right)_{Ni}$

and $\frac{3}{K_{eq}} = \frac{1}{K_{Cu}} + \frac{1}{K_{Al}} + \frac{1}{K_{Ni}}$

$= \frac{1}{6K} + \frac{1}{3K} + \frac{1}{K} = \frac{9}{6K}$

$\Rightarrow \quad K_{eq} = 2K$

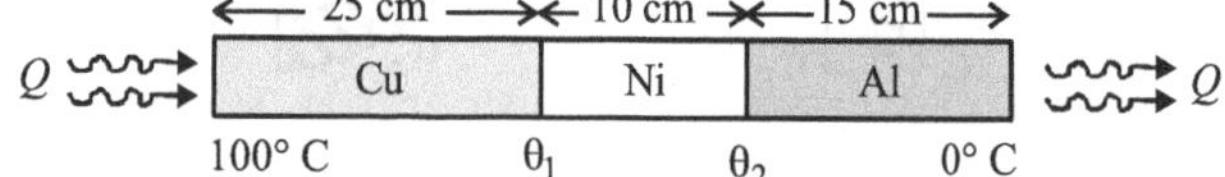

Hence, if $\left(\frac{Q}{t}\right)_{Combination} = \left(\frac{Q}{t}\right)_{Cu}$

$\Rightarrow \quad \frac{K_{eq}A(100 - 0)}{\ell_{Combination}} = \frac{K_{Cu}A(100 - \theta_1)}{\ell_{Cu}}$

$\Rightarrow \quad \frac{2KA(100 - 0)}{(25 + 10 + 15)} = \frac{6KA(100 - \theta_1)}{25}$

$\Rightarrow \quad \theta_1 = 83.33°$ C

Similar if $\left(\frac{Q}{t}\right)_{Combination} = \left(\frac{Q}{t}\right)_{Al}$

$\Rightarrow \quad \frac{2KA(100 - 0)}{50} = \frac{3KA(\theta_2 - 0)}{15}$

$\Rightarrow \quad \theta_2 = 20°$ C

5. (b) Let the temperature of junction be θ then according to following figure.

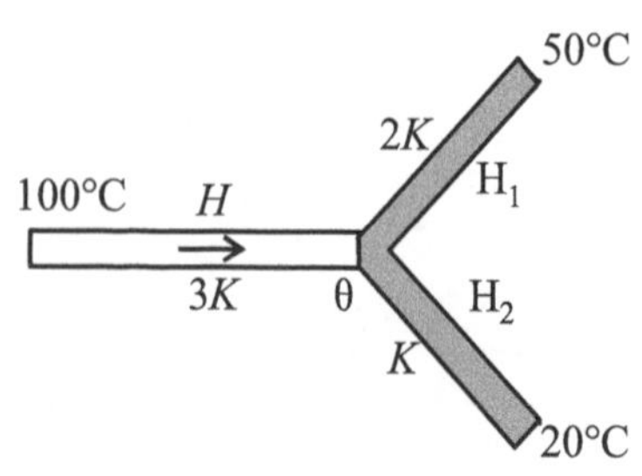

$$H = H_1 + H_2$$

$\Rightarrow$

$$\frac{3K \times A \times (100-\theta)}{\ell} = \frac{2KA(\theta-50)}{\ell} + \frac{KA(\theta-20)}{\ell}$$

$\Rightarrow$ $300 - 3\theta = 3\theta - 120$

$\Rightarrow$ $\theta = 70°\ C$

6. (d) Wein's displacement law is

$$\lambda_m T = b$$

$$\Rightarrow \quad \lambda_m = \frac{b}{T}$$

$$= \frac{2.88 \times 10^6}{2880} = 1000 \text{ nm}$$

Energy distribution with wavelength will be as follows

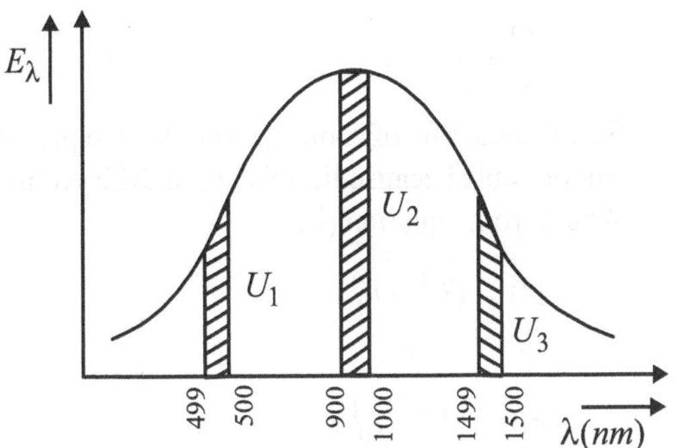

From the graph it is clear that

$$U_2 > U_1$$

7. (a) According to Newton law of cooling

$$\frac{\theta_1 - \theta_2}{t} = K\left[\frac{\theta_1 + \theta_2}{2} - \theta_0\right]$$

80° C —(1) 5 min→ 64°C

—(2) 10 min→ 52°C

—(3) 15 min→ $\theta = ?$

For first process :

$$\frac{(80-64)}{5} = K\left[\frac{80+64}{2} - \theta_0\right] \quad .... \text{(i)}$$

For second process :

$$\frac{(80-52)}{10} = K\left[\frac{80+52}{2} - \theta_0\right] \quad .... \text{(ii)}$$

For third process :

$$\frac{(80-\theta)}{15} = K\left[\frac{80+\theta}{2} - \theta_0\right] \quad .... \text{(iii)}$$

On solving equation (i) and (ii) we get $K = \frac{1}{15}$ and θ_0 = 24° C. Putting these values in equation (iii) we get θ = 42.7° C.

8. (c) $t = \dfrac{Q\ell}{KA(\theta_1 - \theta_2)}$

$$= \frac{mL\ell}{KA(\theta_1 - \theta_2)} = \frac{V\rho L\ell}{KA(\theta_1 - \theta_2)}$$

$$= \frac{5 \times A \times 0.92 \times 80 \times \frac{5+10}{2}}{0.004 \times A \times 10 \times 3600}$$

= 19.1 hours.

9. (c) $\dfrac{Q}{t} = \dfrac{KA\Delta\theta}{\ell}$

$$\Rightarrow \frac{mL}{t} = \frac{K(\pi r^2)\Delta\theta}{\ell}$$

$\Rightarrow$ Rate of melting of ice $\left(\dfrac{m}{t}\right) \propto \dfrac{Kr^2}{\ell}$

Since for second rod K becomes $\frac{1}{4}th$, r becomes double and length becomes half, so rate of melting will be twice *i.e.* $\left(\frac{m}{t}\right)_2 = 2\left(\frac{m}{t}\right)_1 = 2 \times 0.1 = 0.2$ gm/sec.

10. (d) $\dfrac{dQ}{dt} = \dfrac{KA}{\ell} d\theta$

$$= \frac{0.01 \times 1}{0.05} \times 30 = 6 \text{ J/sec}$$

Heat transferred in one day (86400 sec)

Q = 6 × 86400 = 518400 J

Now $Q = mL$

$$\Rightarrow \quad m = \frac{Q}{L} = \frac{518400}{334 \times 10^3} = 1.552 \text{ kg} = 1552 \text{ g.}$$

11. (b) $\dfrac{dT}{dt} = \dfrac{\sigma A}{mcJ}(T^4 - T_0^4)$

[In the given problem fall in temperature of body dT = (200 – 100) = 100 K, temp. of surrounding T_0 = 0 K, Initial temperature of body T = 200 K].

$$\frac{100}{dt} = \frac{\sigma 4\pi r^2}{\frac{4}{3}\pi r^3 \rho cJ}(200^4 - 0^4)$$

$$\Rightarrow \quad dt = \frac{r\rho cJ}{48\sigma} \times 10^{-6} s$$

$$= \frac{r\rho c}{\sigma} \times \frac{4.2}{48} \times 10^{-6}$$

$$= \frac{7}{80} \frac{r\rho c}{\sigma} \mu s \simeq \frac{7}{72} \frac{r\rho c}{\sigma} \mu s \qquad [\text{As J} = 4.2]$$

12. (c) Suppose temperature difference between A and B is $100°$ C and $\theta_A > \theta_B$

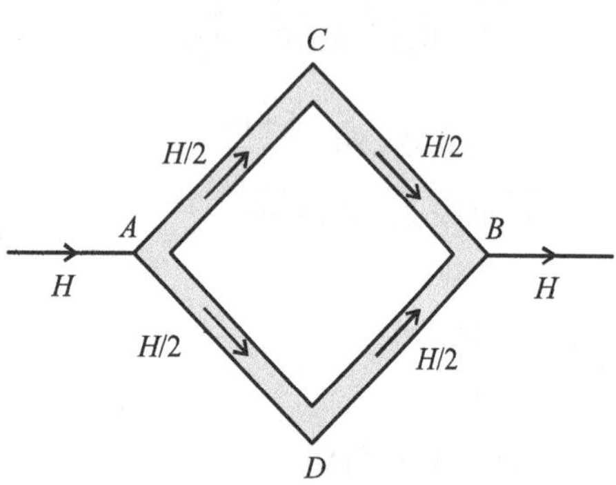

Heat current will flow from A to B via path ACB and ADB. Since all the rod are identical so

$$(\Delta\theta)_{AC} = (\Delta\theta)_{AD}$$

(Because heat current $H = \frac{\Delta\theta}{R}$; here R = same for all)

$$\Rightarrow \quad \theta_A - \theta_C = \theta_A - \theta_D$$

$$\Rightarrow \quad \theta_C = \theta_D$$

i.e. temperature difference between C and D is zero.

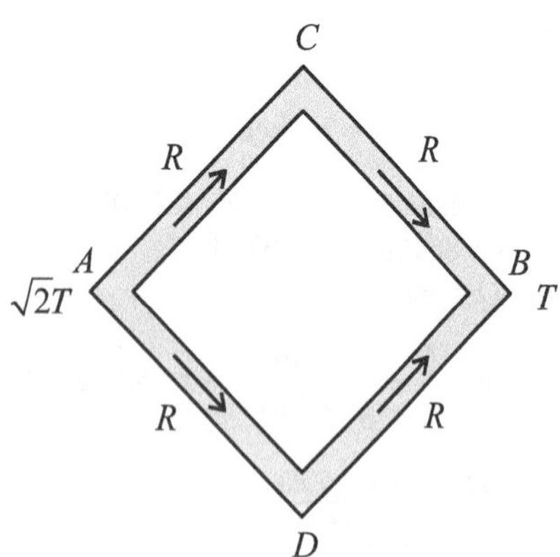

13. (a) Initially at $t = 0$

Rate of cooling $(R) \propto$ Fall in temperature of body $(\theta - \theta_0)$

$$\Rightarrow \quad \frac{R_1}{R_2} = \frac{\theta_1 - \theta_0}{\theta_2 - \theta_0}$$

$$= \frac{100 - 40}{80 - 40} = \frac{3}{2}$$

14. (c) $\frac{60 - 50}{10} = K\left(\frac{60 + 50}{2} - 25\right)$ .... (i)

$\frac{50 - \theta}{10} = K\left(\frac{50 + \theta}{2} - 25\right)$ .... (ii)

On dividing, we get

$$\frac{10}{50 - \theta} = \frac{60}{\theta}$$

$$\Rightarrow \quad \theta = 42.85° \text{ C}$$

15. (d) $\frac{d\theta}{dt} = \frac{\sigma A}{mc}(T^4 - T_0^4)$. If the liquids put in exactly similar calorimeters and identical surrounding then we can consider T_0 and A constant then

$$\frac{d\theta}{dt} \propto \frac{(T^4 - T_0^4)}{mc}$$

If we consider that equal masses of liquid (m) are taken at the same temperature then

$$\frac{d\theta}{dt} \propto \frac{1}{c}$$

So for same rate of cooling c should be equal which is not possible because liquids are of different nature. Again from equation (i),

$$\frac{d\theta}{dt} \propto \frac{(T^4 - T_0^4)}{mc}$$

$$\Rightarrow \quad \frac{d\theta}{dt} \propto \frac{(T^4 - T_0^4)}{V\rho c}$$

Now if we consider that equal volume of liquid (V) are taken at the same temperature then

$$\frac{d\theta}{dt} \propto \frac{1}{\rho c}.$$

So for same rate of cooling multiplication of $\rho \times c$ for two liquids of different nature can be possible. So, option (d) may be correct.

16. (d) For cooking utensils, low specific heat is preferred for it's material as it should need less heat to raise it's temperature and it should have high conductivity, because, it should transfer heat quickly.

17. (b) $\left(\frac{Q}{t}\right)_1 = \frac{K_1 A_1(\theta_1 - \theta_2)}{\ell}$

and $\left(\frac{Q}{t}\right)_2 = \frac{K_2 A_2(\theta_1 - \theta_2)}{\ell}$

Given, $\left(\frac{Q}{t}\right)_1 = \left(\frac{Q}{t}\right)_2$

$$\Rightarrow \quad K_1 A_1 = K_2 A_2$$

18. (a) Convection may be stopped

19. (d) Heated fluid becomes less dense than the cold fluid above it

20. (c) According to Kirchoff's law, the ratio of emissive power to absorptive power is same for all bodies is equal to the emissive power of a perfectly black body *i.e.*,

$$\left(\frac{e}{a}\right)_{body} = E_{Black\,body} \text{ for a particular wave length}$$

$$\left(\frac{e_\lambda}{a_\lambda}\right)_{body} = (E_\lambda)_{Black\,body}$$

$$\Rightarrow \qquad e_\lambda = a_\lambda E_\lambda$$

21. (a) As for a black body rate of absorption of heat is more. Hence thermometer *A* shows faster rise in temperature but finally both will acquire the atmospheric temperature.

22. (b)

According to Stefan's law

$$E = e A\sigma T^4 \Rightarrow E_1 = e_1 A\sigma T_1^4 \text{ and } E_2 = e_2 A\sigma T_2^4$$

$$\because E_1 = E_2 \;\therefore e_1 T_1^4 = e_2 T_2^4$$

$$\Rightarrow T_2 = \left(\frac{e_1}{e_2} T_1^4\right)^{\frac{1}{4}} = \left(\frac{1}{81} \times (5802)^4\right)^{\frac{1}{4}}$$

$$\Rightarrow T_B = 1934 \text{ K}$$

And, from Wein's law $\lambda_A \times T_A = \lambda_B \times T_B$

$$\Rightarrow \frac{\lambda_A}{\lambda_B} = \frac{T_B}{T_A} \Rightarrow \frac{\lambda_B - \lambda_A}{\lambda_B} = \frac{T_A - T_B}{T_A}$$

$$\Rightarrow \frac{1}{\lambda_B} = \frac{5802 - 1934}{5802} = \frac{3968}{5802} \Rightarrow \lambda_B = 1.5\,\mu\text{m}$$

23. (a) According to Newton's law of cooling.

24. (a) In forced convection rate of loss of heat

$$\frac{Q}{T} \propto A(T - T_0)$$

25(a), 26(c), 27(c)

Let c be the specific heat of turpentine

Mass of the solid, $M = 100$g

Mass of turpentine $m = 200$g

Water equivalent of calorimeter, $W = 4g$

Initial temperature of calorimeter, $T_1 = 15°C$

Temperature of ball, $T_2 = 100°C$

Final temperature of the liquid, $T = 23°C$

Specific heat of solid, $c_2 = 0.092$ cal/g°C

Heat gained by turpentine and calorimeter is

$$mc(T - T_1) + W(T - T_1) = 200c(23 - 15) + 4(23 - 15)$$
$$= (200c + 4)\,8$$

Heat lost by the ball is

$$Mc_2(T_2 - T) = 100(0.092)(100 - 23)$$
$$= 708.4 \text{ cal.}$$

According to the principle of calorimetry

Heat gained = Heat lost

$\therefore \quad (200c + 4)\,8 = 708.4$

$1600c + 32 = 708.4$

$$\text{or } c = \frac{708.4 - 32}{1600} = 0.42 \text{ cal/g°C}$$

28. (d) Equivalent thermal conductivity of two equally thick plates in series combination is given by

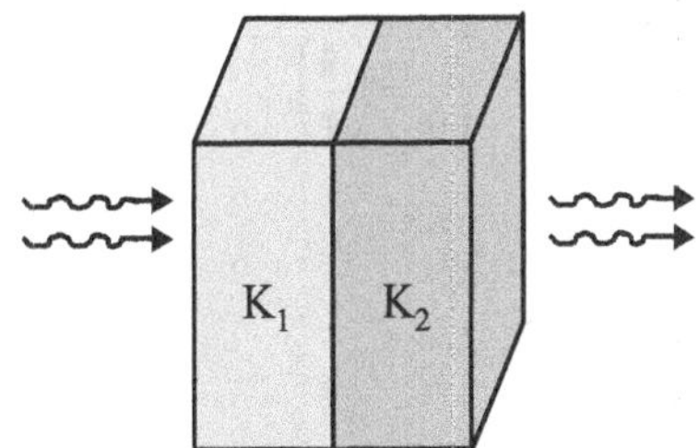

$$\frac{2}{K} = \frac{1}{K_1} + \frac{1}{K_2}$$

If $K_1 < K_2$

then $\qquad K_1 < K < K_2$

29. (b) Both statement-1 and statement-2 are true but statement-2 is not correctly explaining the statement-2.

30. (d) According to Wein's displacement law the

$$\lambda_m \propto \frac{1}{T}$$

Hence statement-1 is true but statement-2 is false.

DAILY PRACTICE PROBLEMS

PHYSICS SOLUTIONS

24

1. (b) Differentiate $PV = \text{constant}$ *w.r.t.* V

$$\Rightarrow P\Delta V + V\Delta P = 0 \Rightarrow \frac{\Delta P}{P} = \frac{\Delta V}{V}$$

2. (a) In isothermal compression, there is always an increase of heat which must flow out the gas.

$$\Delta Q = \Delta U + \Delta W \Rightarrow \Delta Q = \Delta W (\because \Delta U = 0)$$

$$\Rightarrow \Delta Q = -1.5 \times 10^4 J = \frac{1.5 \times 10^4}{4.18} cal = -3.6 \times 10^3 \text{ cal}$$

3. (c) $\Delta Q = \Delta U + \Delta W$

$$\Rightarrow \Delta U = \Delta Q - \Delta W = 2240 - 168 = 2072 J.$$

4. (b) $W_{iso} = \mu RT \log_e \frac{V_2}{V_1} = 1 \times 8.31 \times 300 \log_e \frac{20}{10} = 1728 J$

5. (d) For adiabatic process $\frac{T^\gamma}{P^\gamma - 1}$ = constant

$$\Rightarrow \frac{T_2}{T_1} = \left(\frac{P_1}{P_2}\right)^{\frac{1-\gamma}{\gamma}} \Rightarrow \frac{T_2}{300} = \left(\frac{4}{1}\right)^{\frac{(1-1.4)}{1.4}}$$

$$\Rightarrow T_2 = 300(4)^{\frac{0.4}{1.4}}$$

6. (d) PV^γ = constant

$$\Rightarrow \frac{P_2}{P_1} = \left(\frac{V_1}{V_2}\right)^\gamma \Rightarrow P_2 = (8)^{5/3} P_1 = 32 P_1$$

7. (b) $\frac{T_2}{T_1} = \left(\frac{V_1}{V_2}\right)^{\gamma - 1}$

$$\Rightarrow T_2 = 300\left(\frac{27}{8}\right)^{\frac{5}{3}-1} = 300\left(\frac{27}{8}\right)^{\frac{2}{3}}$$

$$= 300\left\{\left(\frac{27}{8}\right)^{1/3}\right\}^2 = 800\left(\frac{3}{2}\right)^2 = 675K$$

$$\Rightarrow \Delta T = 675 - 300 = 375 K$$

8. (b) In adiabatic change Q = constant $\Rightarrow \Delta Q = 0$

So $\Delta W = -\Delta U (\because \Delta Q = \Delta U + \Delta W)$

9. (d) $TV^{\gamma - 1}$ = constant

$$\Rightarrow T_2 = T_1\left(\frac{V_1}{V_2}\right)^{\gamma - 1} = 927^\circ C$$

10. (d) $W = \frac{R(T_i - T_f)}{\gamma - 1} \Rightarrow 6R = \frac{R(T - T_f)}{\left(\frac{5}{3} - 1\right)} \Rightarrow T_f = (T - 4)K$

11. (c) PV^γ constant : Differentiating both sides

$$P\gamma V^{\gamma - 1} dV + V^\gamma dP = 0 \Rightarrow \frac{dP}{P} = -\gamma \frac{dV}{V}$$

12. (c) V $\propto$ T at constant pressure

$$\Rightarrow \frac{V_1}{V_2} = \frac{T_1}{T_2} \Rightarrow V_2 = \frac{V_1 T_2}{T_1} = \frac{300 \times 280}{300} = 280 \text{ ml}$$

13. (b) In thermodynamic process, work done is equal to the area covered by the PV curve with volume axis.
Hence, according to graph shown
$W_{adiabatic} < W_{isothermal} < W_{isobaric}$

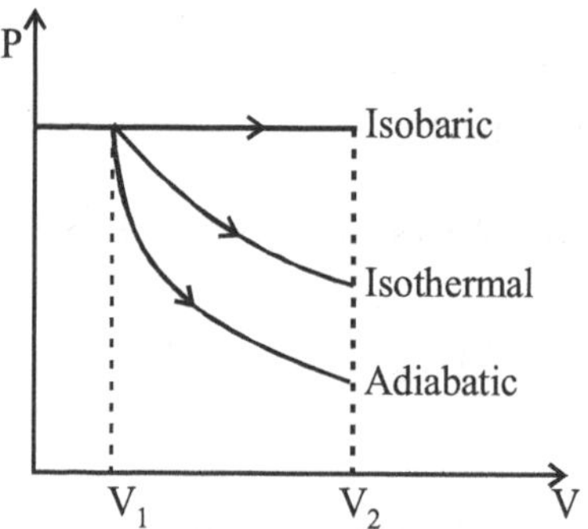

14. (d) $W = P\Delta V = 2.4 \times 10^4 \times 1 \times 10^5 = 24J$

15. (c) For isobaric process $\frac{V_2}{V_1} = \frac{T_2}{T_1} \Rightarrow V_2 = V \times \frac{274}{273}$

Increase $= \frac{274V}{273} - V = \frac{V}{273}$

16. (c) $W = P\Delta V = nR\Delta T = 0.1 \times 2 \times 300 = 60$ cal

17. (d) Fraction of supplied energy which increases the internal energy is given by

$$f = \frac{\Delta U}{(\Delta Q)_P} = \frac{(\Delta Q)_V}{(\Delta Q)_P} = \frac{\mu C_V \Delta T}{\mu C_P \Delta T} = \frac{1}{\gamma}$$

For diatomic gas, $\gamma = \frac{7}{5} \Rightarrow f = \frac{5}{7}$

18. (a) In isothermal change, temperature remains constant, Hence $\Delta U = 0$.

Also from $\Delta Q = \Delta U + \Delta W \Rightarrow \Delta Q = \Delta W$

19. **(c)** From graph it is clear that $P_3 > P_1$.

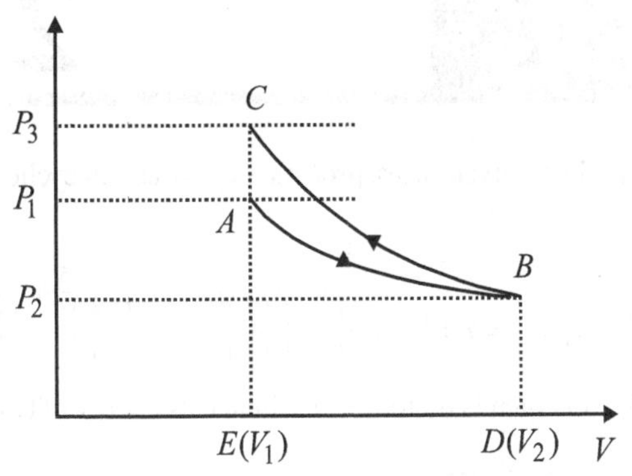

Since area under adiabatic process (*BCED*) is greater than that of isothermal process (*ABDE*). Therefore net work done

$W = W_i + (-W_A) \therefore W_A > W_i \Rightarrow W < 0$

20. **(b)** $PV^2 =$ constant represents adiabatic equation. So during the expansion of ideal gas internal energy of gas decreases and temperature falls.

21. **(a)** For adiabatic process

$$T_1 V_b^{\gamma-1} = \text{Constant}$$

For bc curve

$$T_1 V_b^{\gamma-1} = T_2 V_c^{\gamma-1}$$

or $$\frac{T_2}{T_1} = \left(\frac{V_b}{V_c}\right)^{\gamma-1} \quad \text{....(i)}$$

For ad curve

$$T_1 V_a^{\gamma-1} = T_2 V_d^{\gamma-1}$$

or $$\frac{T_2}{T_1} = \left(\frac{V_a}{V_d}\right)^{\gamma-1} \quad \text{.....(ii)}$$

From equation (i) and (ii)

$$\frac{V_b}{V_c} = \frac{V_a}{V_d}$$

22. **(d)** There is a decrease in volume during melting on an ice slab at 273K. Therefore, negative work is done by ice-water system on the atmosphere or positive work is done on the ice-water system by the atmosphere. Hence option (b) is correct. Secondly heat is absorbed during melting (i.e. ΔQ is positive) and as we have seen, work done by ice-water system is negative (ΔW is negative). Therefore, from first law of thermodynamics $\Delta U = \Delta Q - \Delta W$
Change in internal energy of ice-water system, ΔU will be positive or internal energy will increase.

23. **(a)** From graph it is clear that $P_3 > P_1$.
Since area under adiabatic process (BCED) is greater than that of isothermal process (ABDE). Therefore net work done

$W - W_1 + (-W_A) \because W_A > W_i \Rightarrow W < 0$

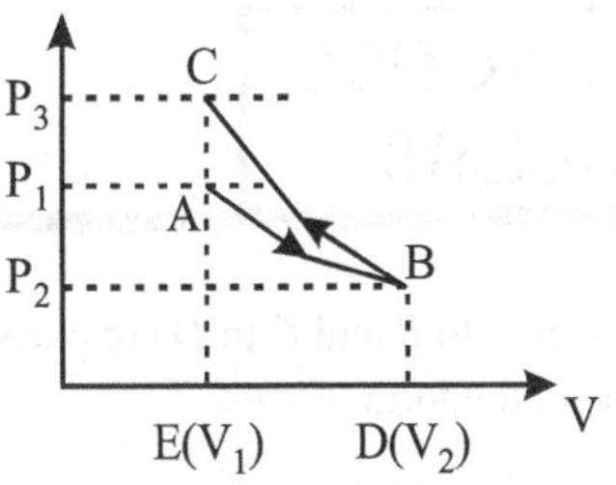

24. **(d)** Work done = Area of ABC with V-axis
$= P_0(2V_0 - V_0) + 0 = P_0V_0 = nRT_0 = RT_0$
Change in internal energy $= nC_V\Delta T$

$= 1 \times \frac{3}{2}R \times (4T_0 - T_0) = \frac{9}{2}RT_0$

Heat absorbed $= \frac{9}{2}RT_0 + RT_0 = \frac{11}{2}RT_0$

25. **(b)** AB is an isothermal process then

$P \times 2V = P_B \times 6V \Rightarrow P_B = \frac{P}{3}$

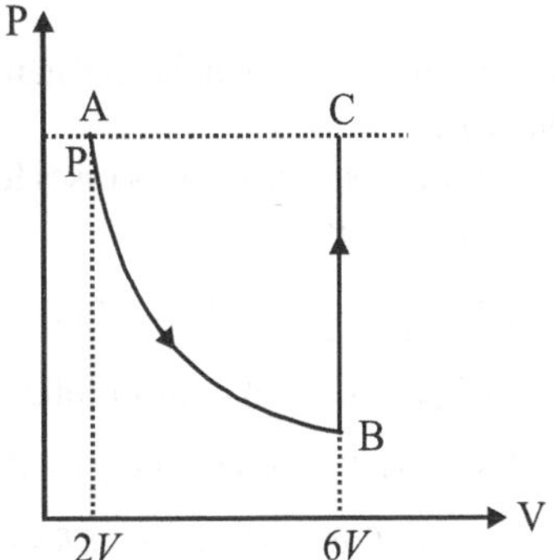

Now BC is an isochoric process then

$\frac{P_B}{T_B} = \frac{P_C}{T_C}$; $\frac{P}{3T_0} = \frac{P}{T_C}$; $T_C = 3T_0$

26. **(a)** Heat absorbed during BC is given by

$Q = nC_v\Delta T = n \times \frac{3R}{2}(T_C - T_B)$

$= n \times \frac{3R}{2}(2T_0) = 3nRT_0.$

27. **(b)** Heat capacity is given by

$C = \frac{1}{n}\frac{dQ}{dT}$; $C = \frac{1}{n}\frac{Q}{2T_0}$

28. **(c)** As isothermal processes are very slow and so the different isothermal curves have different slopes so they cannot intersect each other.

29. **(d)** Adiabatic compression is a rapid action and both the internal energy and the temperature increases.

30. **(a)** $c = \frac{Q}{m.\Delta\theta}$; a gas may be heated by putting pressure, so it can have values for 0 to ∞.
C_P and C_V are it's two principle specific heats, out of infinite possible values.
In adiabatic process $C = 0$ and in isothermal process $C = \infty$.

DAILY PRACTICE PROBLEMS

PHYSICS SOLUTIONS

25

1. (c) Processes A to B and C to D are parts of straight line graphs of the form $y = mx$

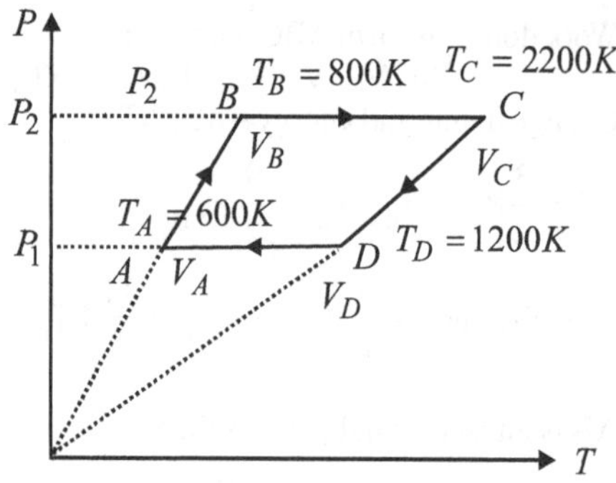

Also $P = \frac{\mu R}{V} T \; (\mu = 6)$

$\Rightarrow P \propto T$. So volume remains constant for the graphs AB and CD.

So no work is done during processes for A to B and C to D *i.e.*, $W_{AB} = W_{CD} = 0$ and

$W_{BC} = P_2(V_C - V_B) = \mu R(T_C - T_B)$

$= 6R(2200 - 800) = 6R \times 1400 J$

Also $W_{DA} = P_1(V_A - V_D) = \mu R(T_A - T_B)$

$= 6R(600 - 1200) = -6R \times 600\ J$

Hence work done in complete cycle

$W = W_{AB} + W_{BC} + W_{CD} + W_{DA}$

$= 0 + 6R \times 1400 + 0 - 6R \times 600$

$= 6R \times 900 = 6 \times 8.3 \times 800 = 40\ kJ$

2. (d) W= Area bonded by the indicator diagram with V-axis)

$= \frac{1}{2}(P_A + P_B)(V_B - V_A)$

3. (d) For path ab : $(\Delta U)_{ab} = 7000\ J$

By using $\Delta U = \mu C_V \Delta T$

$7000 = \mu \times \frac{5}{2} R \times 700 \Rightarrow \mu = 0.48$

For path ca :

$(\Delta Q)_{ca} = (\Delta U)_{ca} + (\Delta W)_{ca}$ ....(i)

$\therefore (\Delta U)_{ab} + (\Delta U)_{bc} + (\Delta U)_{ca} = 0$

$\therefore 7000 + 0 + (\Delta U)_{ca} = 0 \Rightarrow (\Delta U)_{ca} = -7000\ J$ .... (ii)

Also $(\Delta W)_{ca} = P_1(V_1 - V_2) = \mu R(T_1 - T_2)$

$= 0.48 \times 8.31 \times (300 - 1000) = -2792.16\ J$ ......(iii)

on solving equations (i), (ii) and (iii)

$(\Delta Q)_{ca} = -7000 - 2792.16 = -9792.16\ J = -9800 J$

4. (d) In all given cases, process is cyclic and in cyclic process $\Delta U = 0$.

5. (d) $T_1 V_1^{\gamma-1} = T_2 V_2^{\gamma-1} \Rightarrow \frac{T_1}{T_2} = \left(\frac{V_2}{V_1}\right)^{\gamma-1} = \left(\frac{L_2 A}{L_1 A}\right)^{\frac{5}{3}-1} = \left(\frac{L_2}{L_1}\right)^{\frac{2}{3}}$

6. (d) Oxygen is diatomic gas, hence its energy of two moles

$= 2 \times \frac{5}{2} RT = 5RT$

Argon is a monoatomic gas, hence its internal energy of 4 moles $= 4 \times \frac{3}{2} RT = 6RT$

Total internal energy $= (6+5)RT = 11RT$

7. (c) $\eta_A = \frac{T_1 - T_2}{T_1} = \frac{W_A}{Q_1} \Rightarrow \eta_B = \frac{T_2 - T_3}{T_2} = \frac{W_B}{Q_2}$

$\therefore \frac{Q_1}{Q_2} = \frac{T_1}{T_2} \times \frac{T_2 - T_3}{T_1 - T_2} = \frac{T_1}{T_2} \therefore W_A = W_B$

$\therefore T_2 = \frac{T_1 + T_3}{2} = \frac{800 + 300}{2} = 550K$

8. (d) Initially $\eta = \frac{T_1 - T_2}{T_1} \Rightarrow 0.5 = \frac{T_1 - (273+7)}{T_1}$

$\Rightarrow \frac{1}{2} = \frac{T_1 - 280}{T_1} \Rightarrow T_1 = 560K$

Finally $\eta_1' = \frac{T_1' - T_2}{T_1'}$

$\Rightarrow 0.7 = \frac{T_1' - (273+7)}{T_1'} \Rightarrow T_1' = 933K$

$\therefore$ increase in temperature $= 933 - 560 = 373K \approx 380K$

9. (d) In both cylinders A and B the gases are diatomic $(\gamma = 1.4)$. Piston A is free to move *i.e.* it is isobaric process. Piston B is fixed *i.e.* it is isochoric process. If same amount of heat ΔQ is given to both then

$(\Delta Q)_{isobaric} = (\Delta Q)_{isochoric}$

$\Rightarrow \mu C_P (\Delta T)_A = \mu C_v (\Delta T)_B$

$\Rightarrow (\Delta T)_B = \frac{C_P}{C_v}(\Delta T)_A = \gamma(\Delta T)_A = 1.4 \times 30 = 42K$

10. (b) In first case, $\eta_1 = \frac{T_1 - T_2}{T_1}$

In second case, $\eta_2 = \frac{2T_1 - 2T_2}{2T_1} = \frac{T_1 - T_2}{T_1} = \eta$

11. (b) $\eta = 1 - \frac{T_2}{T_1} \Rightarrow \frac{1}{2} = 1 - \frac{500}{T_1} \Rightarrow \frac{500}{T_1} = \frac{1}{2}$ ...(i)

$\frac{60}{100} = 1 - \frac{T_2'}{T_1} \Rightarrow \frac{T_2'}{T_1} = \frac{2}{5}$ ...(ii)

Dividing equation (i) by (ii),

$\frac{500}{T_2'} = \frac{5}{4} \Rightarrow T_2' = 400K$

12. (a) $\eta_{max} = 1 - \frac{T_2}{T_1} = 1 - \frac{300}{400} = \frac{1}{4} = 25\%$

So 26% efficiency is impossible.

13. (a) $\eta = 1 - \frac{T_2}{T_1} = 1 - \frac{(273+123)}{(273+27)} = 1 - \frac{150}{300} = \frac{1}{2} = 50\%$

14. (c) $\eta = 1 - \frac{T_2}{T_1} \Rightarrow \frac{25}{100} = 1 - \frac{300}{T_1} \Rightarrow \frac{1}{4} = 1 - \frac{300}{T_1}$

$T_1 = 400K = 127°C$

15. (b) $\Delta Q = \Delta U + \Delta W$

$\Rightarrow \Delta U = \Delta Q - \Delta W = Q - W$ (using proper sign)

16. (b)

17. (a) $J\Delta Q = \Delta U + \Delta W, \Delta U = J\Delta Q - \Delta W$

$\Delta U = 4.18 \times 300 - 600 = 654\ Joule$

18. (b) $\Delta Q = \Delta U + \Delta W$

$\Rightarrow \Delta W = \Delta Q - \Delta U = 110 - 40 = 70J$

19. (a)

20. (a) From FLOT

$\Rightarrow dU = dQ - dW \Rightarrow dU = dQ(<0)\ \ (\because dW = 0)$

$\Rightarrow dU < 0$ So, temperature will decrease.

21. (b) From FLOT $\Delta Q = \Delta U + \Delta W$

Work done at constant pressure

$(\Delta W)_P = (\Delta Q)_P - \Delta U$

$(\Delta Q)_P - (\Delta Q)_V$ (As we know $(\Delta Q)_V = \Delta U$)

Also $(\Delta Q)_P = mc_P\Delta T$ and $(\Delta Q)_V = mc_V\Delta T$

$\Rightarrow (\Delta W)_p = m(c_P - c_V)\Delta T$

$\Rightarrow (\Delta W)_P = 1 \times (3.4 \times 10^3 - 2.4 \times 10^3)10 = 10^4$ Cal.

22. (a) Slow isothermal expansion or compression of an ideal gas is reversible process, while the other given process are irreversible in nature.

23. (a) For a reversible process $\int \frac{dQ}{T} = 0$

24. (d)

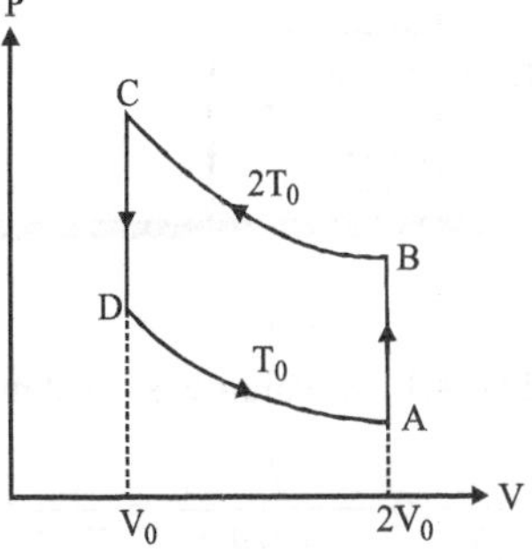

Here $W < 0 \Rightarrow Q < 0$ and $|W| = RT_0 \ln\left(\frac{V_0}{2V_0}\right) + 2RT_0$

$\ln\left(\frac{2V_0}{V_0}\right) = RT_0 \ln 2$

(25) (c) **(26) (b)** **(27) (b)**

(i)

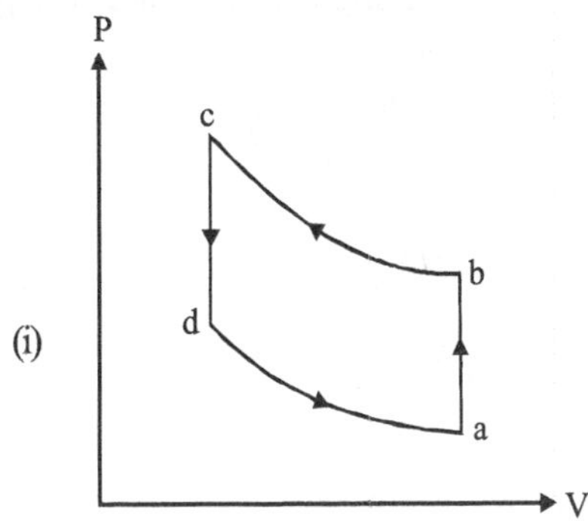

Anticlockwise cycle $\Rightarrow W < 0$

(ii) Process ab : $W_{ab} = 0, \Delta U_{ab} > 0 \Rightarrow Q_{ab} > 0$

Process bc : $W_{bc} < 0, \Delta U_{bc} = 0 \Rightarrow Q_{bc} < 0$

Process cd : $W_{cd} = 0, \Delta U_{cd} < 0 \Rightarrow Q_{cd} < 0$

Process da : $W_{da} > 0, \Delta U_{da} = 0 \Rightarrow Q_{da} > 0$

(iii) da and bc are isothermal process.

28. (a) Second law of thermodynamics can be explained with the help of example of refrigerator, as we know that refrigerator, the working substance extracts heat from colder body and rejcts a large amount of heat to a hotter body with the help of an external agency i.e., the electric supply of the refrigerator. No refrigerator can ever work without external supply of electric energy to it.

29. (d) When the door of refrigerator is kept open, heat rejected by the refrigerator to the room will be more than the heat taken by the refrigerator from the room (by an amount equal to work done by the compressor). Therefore, temperature of room will increase and so it will be warmed gradually. As according to second law of thermodynamics, heat cannot be transferred on its own, from a body at lower temperature to another at higher temperature.

30. (c) As there is no change in internal energy of the system during an isothermal change. Hence, the energy taken by the gas is utilised by doing work against external pressure. According to FLOT

$$\Delta Q = \Delta U + p\Delta V$$

Hence, $\Delta Q = \Delta U = p\Delta V$

Therefore, statement-2 is true and statement-1 is false.

DAILY PRACTICE PROBLEMS

PHYSICS SOLUTIONS 26

1. (a) Closed vessel *i.e.*, volume is constant

$$\Rightarrow \frac{P_1}{P_2}=\frac{T_1}{T_2}\Rightarrow \frac{P}{P+\left(\frac{0.4}{100}\right)P}=\frac{T}{T+1}\Rightarrow T=250K$$

2. (c) $V\propto T\Rightarrow \frac{V_1}{V_2}=\frac{T_1}{T_2}\Rightarrow \frac{V}{2V}=\frac{(273+27)}{T_2}=\frac{300}{T_2}$

$\Rightarrow T_2=600K=327^\circ C$

3. (c) At low pressure and high temperature real gases behaves like ideal gases.

4. (d) $PV=NkT\Rightarrow \frac{N_A}{N_B}=\frac{P_AV_A}{P_BV_B}\times\frac{T_B}{T_A}$

$$\Rightarrow \frac{N_A}{N_B}=\frac{P\times V\times(2T)}{2P\times\frac{V}{4}\times T}=\frac{4}{1}$$

5. (d) $PV=mrT$

Since $P, V, r\rightarrow$ remains same

Hence

$$m\propto\frac{1}{T}\Rightarrow \frac{m_1}{m_2}=\frac{T_2}{T_1}\Rightarrow \frac{13}{m_2}=\frac{(273+52)}{(273+27)}=\frac{325}{300}$$

$\Rightarrow m_2=12gm$

i.e., mass released $=13gm - 12gm = 1gm$

6. (c) $\mu_1=\frac{PV}{RT},\mu_2=\frac{PV}{RT}$

$$P'=\frac{(\mu_1+\mu_2)RT}{V}=\frac{2PV}{RT}\times\frac{RT}{V}=2P$$

7. (d) $v_{rms}=\sqrt{\frac{3kT}{m}}=v_{rms}\propto\frac{1}{\sqrt{m}}$

8. (a)

9. (a) $v_{rms}=\sqrt{\frac{3RT}{M}}\Rightarrow M=\frac{3RT}{v_{rms}^2}\therefore M=\frac{3\times8.3\times300}{(1920)^2}$

$=2\times10^{-3}kg=2gm\Rightarrow$ Gas is hydrogen.

10. (d) *r.m.s* velocity does not depend upon pressure.

11. (c) Average velocity of gas molecule is

$$v_{av}=\sqrt{\frac{8RT}{\pi M}}\Rightarrow v_{av}\propto\frac{1}{\sqrt{M}}$$

$$\Rightarrow \frac{<C_H>}{<C_{He}>}=\sqrt{\frac{M_{He}}{M_H}}=\sqrt{\frac{4}{1}}=2$$

$\Rightarrow <C_H>$

$=2<C_{He}>$

12. (b) $v_{rms}>v_{av}>v_{mp}$

13. (a) $$\gamma_{mix}=\frac{\frac{\mu_1\gamma_1}{\gamma_1-1}+\frac{\mu_2\gamma_2}{\gamma_2-1}}{\frac{\mu_1}{\gamma_1-1}+\frac{\mu_2}{\gamma_2-1}}=\frac{\frac{1\times\frac{5}{3}}{\left(\frac{5}{3}-1\right)}+\frac{1\times\frac{7}{5}}{\left(\frac{7}{5}-1\right)}}{\frac{1}{\left(\frac{5}{3}-1\right)}+\frac{1}{\left(\frac{7}{5}-1\right)}}=\frac{3}{2}=1.5$$

14. (c) We know that

$$C_P-C_V=\frac{R}{J}\Rightarrow J=\frac{R}{C_P-C_V}$$

$$C_P-C_V=1.98\frac{cal}{gm-mol-K},$$

$$R=8.32\frac{J}{gm-mol-K}$$

$$\therefore J=\frac{8.32}{1.98}=4.20\,J/cal$$

15. (b) $v_{rms}=\sqrt{\frac{3P}{\rho}}=\sqrt{\frac{3PV}{m}}\Rightarrow v_{rms}\propto\sqrt{\frac{P}{m}}$

$$\Rightarrow \frac{v_1}{v_2}=\sqrt{\frac{P_1}{P_2}\times\frac{m_2}{m_1}}$$

$$\Rightarrow \frac{v}{2v}=\sqrt{\frac{P_0}{P_2}\times\frac{m/2}{m}}\Rightarrow P_2=2P_0$$

16. (d) $P=\frac{2}{3}E$

17. (a) For one *gm* mole; average kinetic energy $=\frac{3}{2}RT$

18. (c) Average kinetic energy $\propto$ Temperature

$$\Rightarrow \frac{E_1}{E_2}=\frac{T_1}{T_2}\Rightarrow \frac{E}{2E}=\frac{T_1}{T_2}\Rightarrow T_2=2T_1$$

$T_2=2(273+20)=586K=313^\circ C$

19. (d) Kinetic energy per *gm* mole $E=\frac{f}{2}RT$

If nothing is said about gas then we should calculate

the translational kinetic energy *i.e.*,

$$E_{Trans} = \frac{3}{2}RT = \frac{3}{2}\times 8.31\times(273+0) = 3.4\times10^3 J$$

20. **(b)** $(\Delta Q)_P = \mu C_P \Delta T$

$$\Rightarrow 2\times C_P \times(35-30) \Rightarrow C_P = 7\frac{cal}{mole-K}$$

$$\because C_P - C_V = R$$

$$\Rightarrow C_V = C_P - R = 7-2 = 5\frac{cal}{mole-kelvin}$$

$$\therefore (\Delta Q)_V = \mu C_V \Delta T$$

$$= 2\times5\times(35-30) = 50\ cal$$

21. **(a)** Average kinetic energy per molecule per degree of freedom $= \frac{1}{2}kT$. Since both the gases are diatomic and at same temperature (300 *K*), both will have the same number of rotational degree of freedom *i.e.* two. Therefore, both the gases will have the same average rotational kinetic energy per molecule

$$\left(= 2\times\frac{1}{2}kT = kT\right).$$

Thus $\frac{E_1}{E_2} = \frac{1}{1}$

22. **(a)**

Coefficient of volume expansion at constant pressure is $\frac{1}{273}$ for all gases. The average transnational K.E. is same for molecule of all gases and for each molecules it is $\frac{3}{2}$kT

Mean free path $\lambda = \frac{kT}{\sqrt{2}\pi d^2 P}$ (as P decreases, λ increases)

23. **(b)** $v_{rms} = \sqrt{\frac{3RT}{M}}$, $v_P = \sqrt{\frac{2RT}{M}} = 0.816\ v_{rms}$

$$\bar{v} = \sqrt{\frac{8RT}{\pi M}} = 0.92\ v_{rms} \Rightarrow v_P < \bar{v} < v_{rms}$$

Further $E_{av} = \frac{1}{2}mv_{rms}^2 = \frac{1}{2}m\frac{3}{2}v_P^2 = \frac{3}{4}mv_P^2$

24. **(d)** According to problem mass of gases are equal so number of moles will not be equal i.e. $\mu_A \neq \mu_B$

From ideal gas equation $PV = \mu RT \Rightarrow \frac{P_A V_A}{\mu_A} = \frac{P_B V_B}{\mu_B}$

[As temperature of the container are equal]

From this relation it is clear that if $P_A = P_B$ then

$$\frac{V_A}{V_B} = \frac{\mu_A}{\mu_B} \neq 1 \text{ i.e. } V_A \neq V_B$$

Similarly if $V_A = V_B$ then $\frac{PA}{P_B} = \frac{\mu_A}{\mu_B} \neq 1$ i.e. $P_A \neq P_B$.

25. **(b)** $n_1 C_v (T - T_0) + n_2 C_v (T - 2T_0) = 0$

$$T = \frac{3}{2}T_0$$

$$P_f = \frac{P_i T_f}{T_i} = \frac{3}{2}P_0$$

26. **(c)** $\Delta Q = n_1 C_v (T_f - T_0)$

$$= \frac{P_0 V_0}{2RT_0}\times\frac{3}{2}R\times\left(\frac{3}{2}T_0 - T_0\right) = \frac{3}{8}P_0 V_0$$

27. **(c)** Let ΔV is change in volume in any compartment then

$$n_1 = \frac{P_0 V_0}{2RT_0} = \frac{P_f\left(\frac{V_0}{2} - \Delta V\right)}{RT_f}$$ and

$$n_2 = \frac{2P_0 V_0}{2RT_0} = \frac{P_f\left(\frac{V_0}{2} + \Delta V\right)}{RT_f} \Rightarrow \Delta V = 0$$

28. **(b)** Internal energy of an ideal gas does not depend upon volume of the gas, because there are no forces of attration/repulsion amongest the molecular of an ideal gas.
Also internal energy of an ideal gas depends on temperature.

29. **(b)** Helium is a monoatomic gas, while oxygen is diatomic. Therefore, the heat given to helium will be totally used up in increasing the translational kinetic energy of its molecules; whereas the heat given to oxygen will be used up in increasing the translational kinetic energy of the molecule and also in increasing the kinetic energy of rotation and vibration. Hence there will be a greater rise in the temperature of helium.

30. **(d)** Maxwell speed distribution graph is asymmetric graph, because it has a long "tail" that extends to infinity.
Also v_{rms} depends upon nature of the gas and it's temperature.

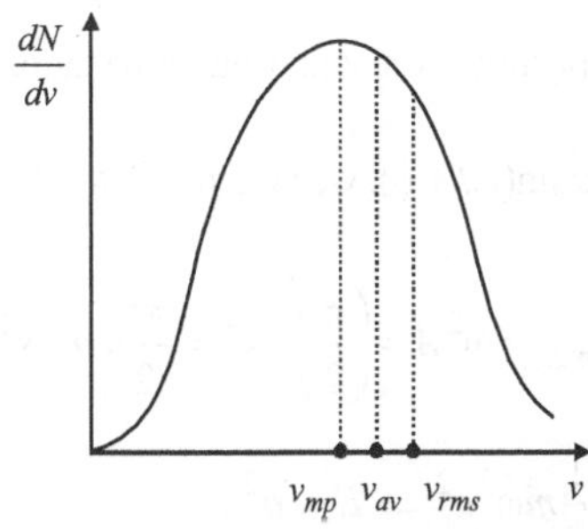

DAILY PRACTICE PROBLEMS

PHYSICS SOLUTIONS

27

1. (c) $a=10$

2. (d) $y=A\sin\omega t=\frac{A\sin 2\pi}{T}t\Rightarrow\frac{A}{2}=A\sin\frac{2\pi t}{T}\Rightarrow t=\frac{T}{12}$

3. (c) $y=a\sin\frac{2\pi}{T}t\Rightarrow\frac{a}{2}=a\sin\frac{2\pi t}{3}\Rightarrow\frac{1}{2}=\sin\frac{2\pi t}{3}$

$\Rightarrow\sin\frac{2\pi t}{3}=\sin\frac{\pi}{6}\Rightarrow\frac{2\pi t}{3}=\frac{\pi}{6}\Rightarrow t=\frac{1}{4}\sec$

4. (a) $x=a\sin\left(\omega t+\frac{\pi}{6}\right)$ and $x'=a\cos\omega t=a\sin\left(\omega t+\frac{\pi}{2}\right)$

$\therefore\Delta\phi=\left(\omega t+\frac{\pi}{6}\right)-\left(\omega t+\frac{\pi}{6}\right)=\frac{\pi}{3}$

5. (c) $v=\omega\sqrt{a^2-y^2}\Rightarrow 10=\omega\sqrt{a^2-(4)^2}$ and

$8=\omega\sqrt{a^2-(5)^2}$

On solving, $\omega=2\Rightarrow\omega=\frac{2\pi}{T}=2\Rightarrow T=\pi\sec$

6. (b) $v_{max}=a\omega=a\times\frac{2\pi}{T}=(50\times10^{-3})\times\frac{2\pi}{2}=0.15\,m/s$

7. (d) $v_{max}=a\omega$ and $A_{max}=a\omega^2$

$\Rightarrow\omega=\frac{A_{max}}{v_{max}}=\frac{4}{2}=2\,rad/sec$

8. (d) At mean position velocity is maximum

i.e., $v_{max}=\omega a\Rightarrow\omega=\frac{v_{max}}{a}=\frac{16}{4}=4$

$\therefore\upsilon=\omega\sqrt{a^2-y^2}\Rightarrow 8\sqrt{3}=4\sqrt{4^2-y^2}$

$\Rightarrow 192=16(16-y^2)\Rightarrow 12=16-y^2\Rightarrow y=2cm$

9. (a) Maximum acceleration $=a\omega^2=a\times4\pi^2n^2$

$=0.01\times4\times(\pi)^2\times(60)^2=144\pi^2 m/\sec$

10. (d) $A_{max}=a\omega^2\Rightarrow a=\frac{A_{max}}{\omega^2}=\frac{7.5}{(3.5)^2}=0.61\,m$

11. (b) Comparing given equation with standard equation,

$y=a\sin(\omega t+\phi)$, we get, $a=2cm,\omega=\frac{\pi}{2}$

$\therefore A_{max}=\omega^2A=\left(\frac{\pi}{2}\right)^2\times2=\frac{\pi^2}{2}cm/s^2$

12. (d) $E=\frac{1}{2}m\omega^2a^2\Rightarrow E\propto a^2$

13. (b) $\frac{U}{E}=\frac{\frac{1}{2}m\omega^2y^2}{\frac{1}{2}m\omega^2a^2}=\frac{y^2}{a^2}=\frac{\left(\frac{a}{2}\right)^2}{a}=\frac{1}{4}$

14. (c) In S.H.M., frequency of K.E. and P.E. $=2\times$ (Frequency of oscillating particle)

15. (c) Kinetic energy $K=\frac{1}{2}mv^2=\frac{1}{2}ma^2\omega^2\cos^2\omega t$

$=\frac{1}{2}m\omega^2a^2(1+\cos2\omega t)$

hence kinetic energy varies periodically with double the frequency of S.H.M. *i.e.* 2ω.

16. (a) At mean position, the kinetic energy is maximum.

Hence $\frac{1}{2}ma^2\omega^2=16$

On putting the values we get

$\omega=10\Rightarrow T=\frac{2\pi}{\omega}=\frac{\pi}{5}\sec$

17. (d) From the given equation, $\omega=2\pi n=4\pi\Rightarrow n=2Hz$

18. (a) Using $x=A\sin\omega t$

For $x=A/2$, $\sin\omega T_1=1/2\Rightarrow T_1=\frac{\pi}{6\omega}$

For $x=A$, $\sin\omega(T_1+T_2)=1\Rightarrow T_1+T_2=\frac{\pi}{2\omega}$

$\Rightarrow T_2=\frac{\pi}{2\omega}-T_1=\frac{\pi}{2\omega}-\frac{\pi}{6\omega}=\frac{\pi}{3\omega}$ *i.e.*, $T_1<T_2$

19. (a) Let the piston be displaced through distance x towards left, then volume decreases, pressure increases. If ΔP is increased in pressure and ΔV is decreased in volume, then considering the process to take place gradually (*i.e* isothermal)

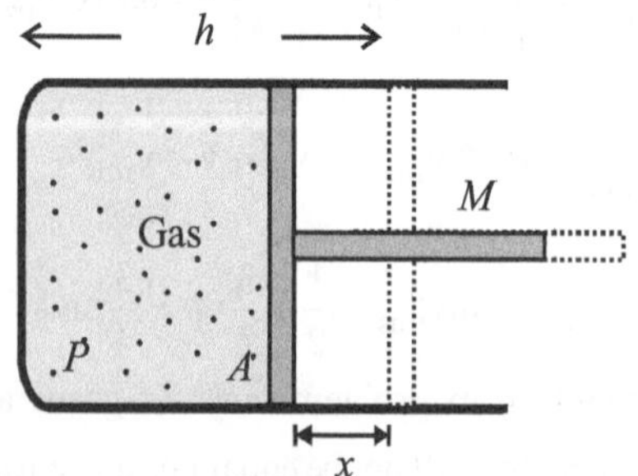

$P_1V_1=P_2V_2\Rightarrow PV=(P+\Delta P)(V-\Delta V)$

$\Rightarrow PV=PV+\Delta PV-P\Delta V-\Delta P\Delta V$

$\Rightarrow\Delta P.V-P.\Delta V=0$ (neglecting $\Delta P.\Delta V$)

$\Delta P(Ah)=P(Ax)\Rightarrow\Delta P=\frac{P.x}{h}$

This excess pressure is responsible for providing the restoring force (F) to the piston of mass M.

Hence $F = \Delta P.A = \dfrac{PAx}{h}$

Comparing it with $|F| = kx \Rightarrow k = M\omega^2 = \dfrac{PA}{h}$

$$\Rightarrow \omega = \sqrt{\frac{PA}{Mh}} \Rightarrow T = 2\pi\sqrt{\frac{Mh}{PA}}$$

20. (b) Time taken by particle to move from $x = 0$ (mean position) to $x = 4$ (extreme position)

$= \dfrac{T}{4} = \dfrac{1.2}{4} = 0.3\,s$

Let t be the time taken by the particle to move from $x = 0$ to $x = 2\,cm$

$$y = a\sin\omega t \Rightarrow 2 = 4\sin\frac{2\pi}{T}t \Rightarrow \frac{1}{2} = \sin\frac{2\pi}{1.2}t$$

$$\Rightarrow \frac{\pi}{6} = \frac{2\pi}{1.2}t \Rightarrow t = 0.1s.$$

Hence time to move from $x = 2$ to $x = 4$ will be equal to $0.3 - 0.1 = 0.2\,s$

Hence total time to move from $x = 2$ to $x = 4$ and back again $= 2 \times 0.2 = 0.4\,\text{sec}$

21. (b)

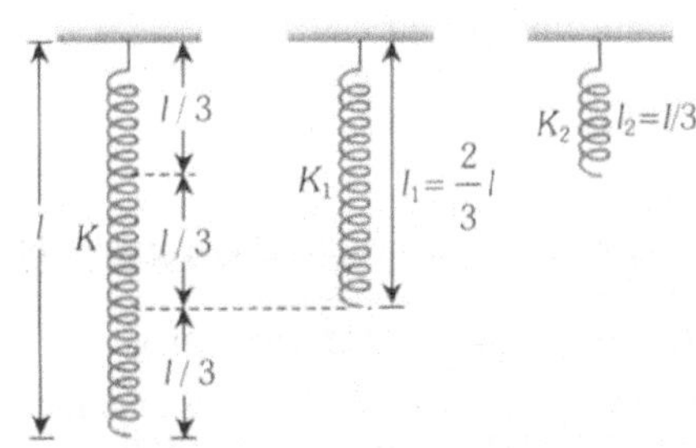

Force constant $(k) \propto \dfrac{1}{\text{Length of spring}}$

$$\Rightarrow \frac{K}{K_1} = \frac{l_1}{l} = \frac{\frac{2}{3}l}{l} \Rightarrow K_1 = \frac{3}{2}K.$$

22. (c) $y = Kt^2 \Rightarrow \dfrac{d^2y}{dt^2} = a_y = 2K = 2 \times 1 = 2\,m/s^2$ $(\because K = 1\,m/s^2)$

Now, $T_1 = 2\pi\sqrt{\dfrac{1}{g}}$ and $T_2 = 2\pi\sqrt{\dfrac{1}{(g + a_y)}}$

Dividing, $\dfrac{T_1}{T_2} = \sqrt{\dfrac{g + a_y}{g}} \Rightarrow \sqrt{\dfrac{6}{5}} \Rightarrow \dfrac{T_1^2}{T_2^2} = \dfrac{6}{5}$

23. (a) At x = 0, v = 0 and potential energy is minimum so particle will remain at rest.

24. (d) Let simple harmonic motions be represented by

$y_1 = a\sin\left(\omega t - \dfrac{\pi}{4}\right)$, $y_2 = a\sin\omega t$

and $y_3 = a\sin\left(\omega t + \dfrac{\pi}{4}\right)$

On superimposing, resultant SHM will be

$$y = a\left[\sin\left(\omega t - \frac{\pi}{4}\right) + \sin\omega t + \sin\left(\omega t + \frac{\pi}{4}\right)\right]$$

$$= a\left[2\sin\omega t\cos\frac{\pi}{4} + \sin\omega t\right] = a\,[\sqrt{2}\sin\omega t + \sin\omega t]$$

$$= a\,(1 + \sqrt{2})\sin\omega t$$

Resultant amplitude $= (1 + \sqrt{2})a$

Energy in SHM $\propto$ (Amplitude)^2

$$\therefore \frac{E_{Resultant}}{E_{Single}} = \left(\frac{A}{a}\right)^2 = (\sqrt{2} + 1)^2 = (3 + 2\sqrt{2})$$

$$\Rightarrow E_{resultant} = (3 + 2\sqrt{2})\,E_{single}$$

OR

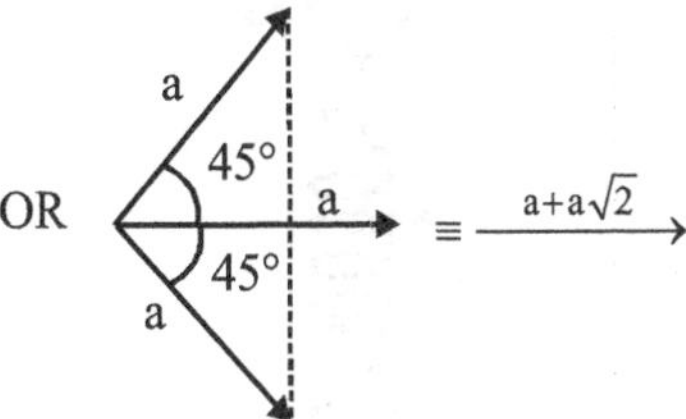

25. (a) Acceleration $\propto$ – displacement, and direction of acceleration is always directed towards the equilibrium position.

26. (d)

27. (b)

Compare given equation with $\dfrac{d^2x}{dt^2} + \omega^2 x = 0$; $\omega^2 = \dfrac{b}{a}$

$$\frac{a_{max}}{v_{max}} = \frac{\omega^2 A}{\omega A} = \omega = \sqrt{\frac{b}{a}}$$

At $t = 0$, $\phi = \pi/2$

$$x = A\sin(\omega t + \phi) = A\cos\sqrt{\frac{b}{a}}t$$

28. (b)

29. (b) $x = a\sin\omega t$ and $v = \dfrac{dx}{dt} = a\omega\cos\omega t$

It is clear that phase difference between 'x' and 'a' is $\dfrac{\pi}{2}$.

30. (a) The total energy of S.H.M. = Kinetic energy of particle + potential energy of particle.

The variation of total energy of the particle in SHM with time is shown in a graph.

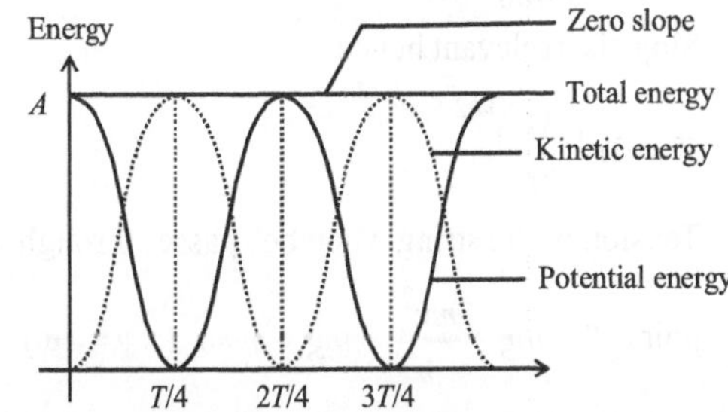

DAILY PRACTICE PROBLEMS

PHYSICS SOLUTIONS 28

1. **(b)** When the particle of mass m at O is pushed by y in the direction of A. The spring A will compressed by y while spring B and C will be stretched by $y' = y\cos 45^\circ$. So that the total restoring force on the mass m along OA.

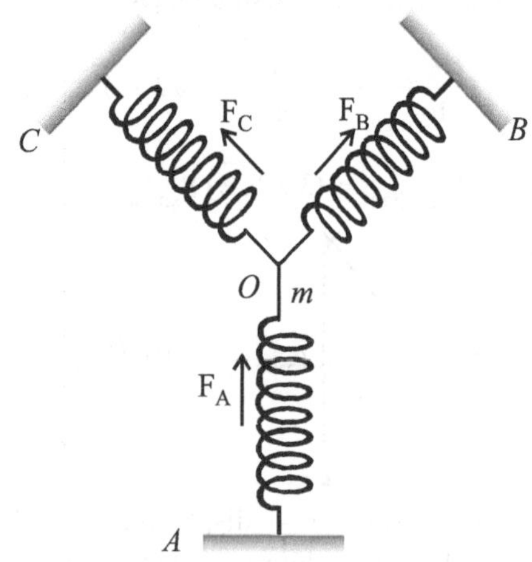

$$F_{net} = F_A + F_B\cos 45^\circ + F_C\cos 45^\circ$$

$$= ky + 2ky'45^\circ = ky + 2k(y\cos 45^\circ)\cos 45^\circ = 2ky$$

Also

$$F_{net} = k'y \Rightarrow k'y = 2ky \Rightarrow k' = 2k$$

$$T = 2\pi\sqrt{\frac{m}{k'}} = 2\pi\sqrt{\frac{m}{2k}}$$

2. **(b)** When mass 700 gm is removed, the left out mass (500 + 400) gm oscillates with a period of 3 sec

$$\therefore 3 = t = 2\pi\sqrt{\frac{(500+400)}{k}}$$ ......(i)

When 500 gm mass is also removed, the left out mass is 400 gm.

$$\therefore t' = 2\pi\sqrt{\frac{400}{k}} \quad ...(ii)$$

$$\Rightarrow \frac{3}{t'} = \sqrt{\frac{900}{400}} \Rightarrow t' = 2\,\text{sec}$$

3. **(a)** Slope is irrelevant hence

$$T = 2\pi\left(\frac{M}{2k}\right)^{1/2}$$

4. **(a)** Tension in the string when bob passes through lowest point $T = mg + \frac{mv^2}{r} = mg + mv\omega$ $(\because v = r\omega)$

Putting $v = \sqrt{2gh}$ and $\omega = \frac{2\pi}{T} = \frac{2\pi}{2} = \pi$

we get $T = m(g + \pi\sqrt{2gh})$

5. **(b)**

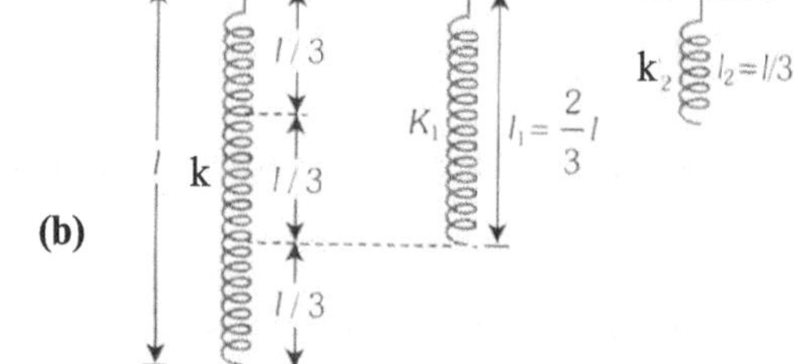

Force constant $(k) \propto \frac{1}{Length\ of\ spring}$

$$\Rightarrow \frac{k}{k_1} = \frac{l_1}{l} = \frac{\frac{2}{3}l}{l} \Rightarrow k_1 = \frac{3}{2}k$$

6. **(b)** Initially time period was

$$T = 2\pi\sqrt{\frac{l}{g}}$$

When train accelerates, the effective value of g becomes $\sqrt{(g^2 + a^2)}$ which is greater than g.

Hence, new time period, becomes less than the initial time period.

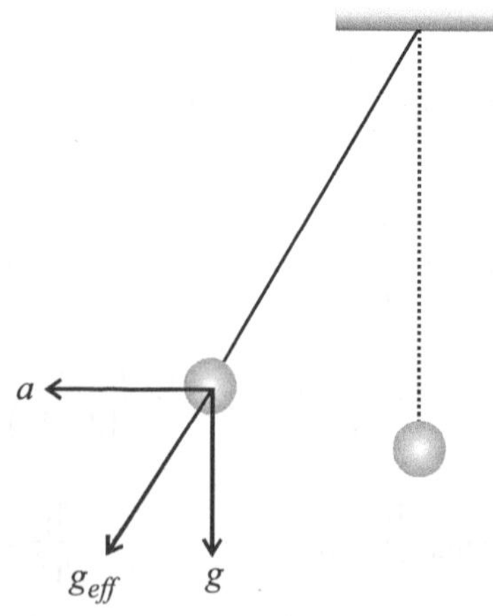

7. **(b)** In accelerated frame of reference, a fictitious force (pseudo force) ma acts on the bob of pendulum as shown in figure.

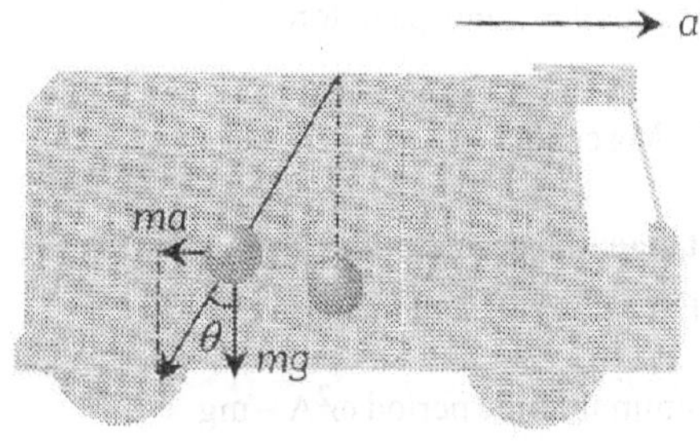

Hence $\tan\theta = \frac{ma}{mg} = \frac{a}{g}$

$\Rightarrow \theta = \tan^{-1}\left(\frac{a}{g}\right)$ in the backward direction.

8. (c) $T = 2\pi\sqrt{\frac{l}{g}}$ (Independent of mass)

9. (b) $T \propto \sqrt{l} \Rightarrow \frac{\Delta T}{T} = \frac{1}{2}\frac{\Delta l}{l} = \frac{1}{2}\times 1\% = 0.5\%$

10. (c) If suppose bob rises up to a height h as shown then after releasing potential energy at extreme position becomes kinetic *energy* of mean position

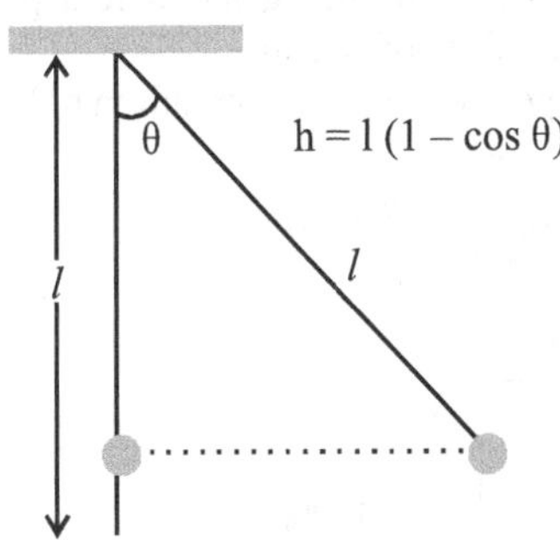

$\Rightarrow mgh = \frac{1}{2}mv_{\max}^2 \Rightarrow v_{\max} = \sqrt{2gh}$

Also, from figure $\cos\theta = \frac{l-h}{l}$

$\Rightarrow h = l(1-\cos\theta)$

So, $v_{\max} = \sqrt{2gl(1-\cos\theta)}$

11. (a) If initial length $l_1 = 100$ then $l_2 = 121$

By using $T = 2\pi\sqrt{\frac{l}{g}} \Rightarrow \frac{T_1}{T_2} = \sqrt{\frac{l_1}{l_2}}$

Hence, $\frac{T_1}{T_2} = \sqrt{\frac{100}{121}} \Rightarrow T_2 = 1.1T_1$

% increase $= \frac{T_2 - T_1}{T_1}\times 100 = 10\%$

12. (d) After standing centre of mass of the oscillating body will shift upward therefore effective length will decrease and by $T \propto \sqrt{l}$, time period will decrease.

13. (a) No momentum will be transferred because, at extreme position the velocity of bob is zero.

14. (c) The effective acceleration in a lift descending with acceleration $\frac{g}{3}$ is

$$g_{eff} = g - \frac{g}{3} = \frac{2g}{3}$$

$$\therefore T = 2\pi\sqrt{\left(\frac{L}{g_{eff}}\right)} = 2\pi\sqrt{\left(\frac{L}{2g/3}\right)} = 2\pi\sqrt{\left(\frac{3L}{2g}\right)}$$

15. (c) In series $k_{eq} = \frac{k_1k_2}{k_1+k_2}$ so time period

$$T = 2\pi\sqrt{\frac{m(k_1+k_2)}{k_1k_2}}$$

16. (c) Spring constant $(k) \propto \frac{1}{\text{Length of the spring } (l)}$

as length becomes half, k becomes twice i.e. $2k$

17. (b) Standard equation for given condition

$$x = a\cos\frac{2\pi}{T}t \Rightarrow x = -0.16\cos(\pi t)$$

[As $a = -0.16$ *meter*, $T = 2\,\text{sec}$]

18. (d) $t_1 = 2\pi\sqrt{\frac{m}{k_1}}$ and $t_2 = 2\pi\sqrt{\frac{m}{k_2}}$

Equivalent spring constant for shown combination is $k_1 + k_2$. So time period t is given by

$$t = 2\pi\sqrt{\frac{m}{k_1+k_2}}$$

By solving these equations we get

$$t^{-2} = t_1^{-2} + t_2^{-2}$$

19. (a) With mass m_2 alone, the extension of the spring l is given as

$$m_2g = kl \quad \text{........(i)}$$

With mass (m_1+m_2), the extension l' is given by

$$(m_1+m_2)g = k(l+\Delta l) \quad \text{.......(ii)}$$

The increase in extension is Δl which is the amplitude of vibration. Subtracting (i) from (ii), we get

$$m_1g = k\Delta l$$

or $$\Delta l = \frac{m_1g}{k}$$

20. (a) If $y_1 = a_1\sin\omega t$ and $a_2\sin(\omega t + \pi)$

$$\Rightarrow \frac{y_1}{a_1} + \frac{y_1}{a_2} \Rightarrow y_2 = \frac{a_2}{a_1}y_1$$

This is the equation of straight line.

21. **(c)** Energy of particle is maximum at resonant frequency *i.e.*, $\omega_2 = \omega_0$. For amplitude resonance (amplitude maximum) frequency of driver force

$$\omega = \sqrt{\omega_0^2 - b^2 2m^2} \Rightarrow \omega_1 \neq \omega_0$$

22. **(b)** $A = \frac{c}{a+b-c}$; when b = 0, a = c, amplitude $A \to \infty$. This corresponds to resonance.

23. **(b)** Let the velocity acquired by A and B be V, then

$$mv = mV + mV \Rightarrow V = \frac{v}{2}$$

Also $\frac{1}{2}mv^2 = \frac{1}{2}mV^2 + \frac{1}{2}mV^2 + \frac{1}{2}kx^2$

Where x is the maximum compression of the spring.

On solving the above equations, we get $x = v\left(\frac{m}{2k}\right)^{1/2}$

At maximum compression, kinetic energy of the

A – B system $= \frac{1}{2}mV^2 + \frac{1}{2}mV^2 = mV^2 = \frac{mv^2}{4}$

24. **(d)**

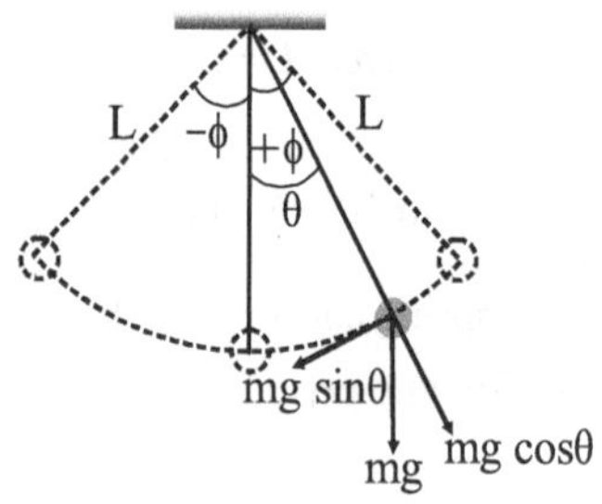

From following figure it is clear that

$$T - Mg\cos\theta = \text{Centripetal force}$$

$$\Rightarrow T - Mg\cos\theta = \frac{Mv^2}{L}$$

Also tangential acceleration $|a_T| = g\sin\theta$.

25. **(a)** Except (4) all statements are wrong.

26. (b) **27. (b).**

For minimum time period $\omega^2 A = mg$

$\frac{4\pi^2}{T^2}A = mg$, T = 0.2 sec,

At t = 0.05 sec.

$$y = A \sin \omega t = 1 \text{ sn } \frac{2\pi}{0.2} \times 0.05\text{cm.} = 1\text{cm.}$$

$$PE = mgy = 1 \times 10 \times \frac{1}{100} = 0.1 \text{ Joule}$$

28. **(c)** Statement -1 is False, Statement-2 is True.

29. **(a)** The time period of a oscillating spring is given by,

$$T = 2\pi\sqrt{\frac{m}{k}} \Rightarrow T \propto \frac{1}{\sqrt{k}}$$

Since the spring constant is large for hard spring, therefore hard spring has a less periodic time as compared to soft spring.

30. **(d)** Time period of simple pendulum of length *l* is,

$$T = 2\pi\sqrt{\frac{l}{g}} \Rightarrow T \propto \sqrt{l}$$

$$\Rightarrow \frac{\Delta T}{T} = \frac{1}{2}\frac{\Delta l}{l}$$

$$\therefore \frac{\Delta T}{T} = \frac{1}{2} \times 3 = 1.5\%$$

DAILY PRACTICE PROBLEMS

PHYSICS SOLUTIONS 29

1. (d) $v = n\lambda \Rightarrow \lambda = \frac{v}{n} = \frac{330}{256} = 1.29m$

2. (a) Time required for a point to move from maximum displacement to zero displacement is

$$t = \frac{T}{4} = \frac{1}{4n}$$

$$\Rightarrow n = \frac{1}{4t} = \frac{1}{4\times 0.170} = 1.47Hz$$

3. (b)

$$2d_1 + 2d_2 = v\times t_1 + v\times t_2 \Rightarrow 2(d_1+d_2) = v(t_1+t_2)$$

$$d_1 + d_2 = \frac{v(t_1+t_2)}{2} = \frac{340\times(1.5+3.5)}{2} = 850\ m$$

4. (c) At given temperature and pressure

$$v \propto \frac{1}{\sqrt{\rho}} \Rightarrow \frac{v_1}{v_2} = \sqrt{\frac{\rho_2}{\rho_1}} = \sqrt{\frac{4}{1}} = 2:1$$

5. (b) The distance between two points *i.e.* path difference between them

$$\Delta = \frac{\lambda}{2\pi}\times\phi = \frac{\lambda}{2\pi}\times\frac{\pi}{3} = \frac{\lambda}{6} = \frac{v}{6n} \qquad (\because v = n\lambda)$$

$$\Rightarrow \Delta = \frac{360}{6\times 500} = 0.12m = 12cm$$

6. (d) $y_1 = a\sin(\omega t - kx)$ and

$$y_2 = a\cos(\omega t - kx) = a\sin\left(\omega t - kx + \frac{\pi}{2}\right)$$

Hence phase difference between these two is $\frac{\pi}{2}$.

7. (c) $\frac{I_1}{I_2} = \frac{a_1^2}{a_2^2} = \left(\frac{0.06}{0.03}\right)^2 = \frac{4}{1}$

8. (d) Comparing the given equation with $y = a\sin(\omega t - kx)$,

We get $a = Y_0, \omega = 2\pi f, k = \frac{2\pi}{\lambda}$. Hence maximum particle velocity $(v_{\max})_{particle} = a\omega = Y_0\times 2\pi f$ and wave velocity

$$(v)_{wave} = \frac{\omega}{k} = \frac{2\pi f}{2\pi/\lambda} = f\lambda$$

$$\therefore (v_{\max})_{particle} = 4v_{wave} = Y_0\times 2\pi f = 4f\lambda \Rightarrow \lambda = \frac{\pi Y_0}{2}$$

9. (d) $y = f(x^2 - vt^2)$ doesn't follows the standard wave equation.

10. (a) $y_1 = a_1\sin\left(\omega t - \frac{2\pi x}{\lambda}\right)$ and

$$y_2 = a_2\cos\left(\omega t - \frac{2\pi x}{\lambda} + \phi\right) = a_2\sin\left(\omega t - \frac{2\pi x}{\lambda} + \phi + \frac{\pi}{2}\right)$$

So phase difference $= \phi + \frac{\pi}{2}$ and $\Delta = \frac{\lambda}{2\pi}\left(\phi + \frac{\pi}{2}\right)$

11. (d) On comparing the given equation with standard equation $y = a\sin\frac{2\pi}{\lambda}(vt - x)$. It is clear that wave speed $(v)_{wave} = v$ and maximum particle velocity

$(v_{\max})_{particle} = a\omega = y_0\times$ co-efficient of $t = y_0\times\frac{2\pi v}{\lambda}$

$$\therefore (v_{\max})_{particle} = 2(\omega)_{wave} \Rightarrow \frac{a\times 2\pi v}{\lambda} = 2v \Rightarrow \lambda = \pi y_0$$

12. (a) Compare the given equation with $y = a\sin(\omega t + kx)$

We get $\omega = 2\pi n = 100 \Rightarrow n = \frac{50}{\pi}Hz$

13. (c) A wave travelling in positive x-direction may be represented as $y = A\sin\frac{2\pi}{\lambda}(vt - x)$. On putting values

$$y = 0.2\sin\frac{2\pi}{60}(360t - x) \Rightarrow y = 0.2\sin 2\pi\left(6t - \frac{x}{60}\right)$$

14. (a) Comparing the given equation with $y = a\sin(\omega t - kx)$

We get $\omega = 3000\pi \Rightarrow n = \frac{\omega}{2\pi} = 1500Hz$

and $k = \frac{2\pi}{\lambda} = 12\pi \Rightarrow \lambda = \frac{1}{6}m$

So, $v = n\lambda \Rightarrow v = 1500\times\frac{1}{6} = 250\ m/s$

15. (b) Given, $y = 0.5\sin(20x - 400t)$

Comparing with $y = a\sin(\omega t - kx)$

Gives velocity of wave $v = \frac{\omega}{k} = \frac{400}{20} = 20\ m/s$.

16. (b) With path difference $\frac{\lambda}{2}$, waves are out of phase at the point of observation.

17. (b) $A^2 = a^2 = a^2 + a^2 + 2a^2\cos\theta \Rightarrow \cos\theta = -\frac{1}{2} \Rightarrow \theta = \frac{2\pi}{3}$

18. (c) For interference, two waves must have a constant phase relationship. Equation '1' and '3' and '2' and '4' have a constant phase relationship of $\frac{\pi}{2}$ out of two choices.

Only one S_2 emitting '2' and S_4 emitting '4' is given so only (c) option is correct.

19. (a) The resultant amplitude is given by

$$A_R = \sqrt{A^2 + A^2 + 2AA\cos\theta} = \sqrt{2A^2(1+\cos\theta)}$$

$$= 2A\cos\theta/2 \qquad (\because 1+\cos\theta = 2\cos^2\theta/2)$$

20. (d) $y = \frac{1}{\sqrt{a}}\sin\omega t \pm \frac{1}{\sqrt{b}}\sin\left(\omega t + \frac{\pi}{2}\right)$

Here phase difference $= \frac{\pi}{2}$

$\therefore$ The resultant amplitude

$$= \sqrt{\left(\frac{1}{\sqrt{a}}\right)^2 + \left(\frac{1}{\sqrt{b}}\right)^2} = \sqrt{\frac{1}{a} + \frac{1}{b}} = \sqrt{\frac{a+b}{ab}}$$

21. (a) In a wave equation, x and t must be related in the form $(x - vt)$

We rewrite the given equations $y = \frac{1}{1+(x-vt)^2}$

For $t = 0$, this becomes $y = \frac{1}{(1+x^2)}$, as given

For $t = 2$, this becomes

$$y = \frac{1}{[1+(x-2v)^2]} = \frac{1}{[1+(x-1)^2]}$$

$2v = 1$ or $v = 0$ m/s

22. (c)

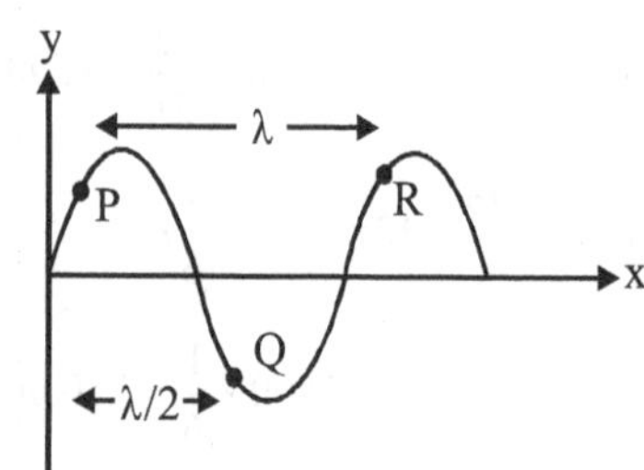

23. (b)

Standard wave equation which travel in negative x-direction is $y = A\sin(\omega t + kx + \phi_0)$

For the given wave $\omega = 2\pi n = 15\pi$, $k = \frac{2\pi}{\lambda} = 10\pi$

Now $v = \frac{\text{Coefficient of t}}{\text{Coefficient of x}} = \frac{\omega}{k} = \frac{15\pi}{10\pi} = 1.5\text{m/sec}$

and $\lambda = \frac{2\pi}{k} = \frac{2\pi}{10\pi} = 0.2$ m.

24. (a) $p_m = \frac{\omega BA}{v}$, $I = \frac{1}{2}\omega p_m A \Rightarrow I = \frac{p_m^2 v}{2B}$

25. (d), 26. (c), 27. (a).

(i) $I \propto \frac{1}{d^2}$, $d_{\text{initial}} = R$, $d_{\text{final}} = 3R$,

$\frac{I_{\text{initial}}}{I_{\text{final}}} = \frac{9}{1} \Rightarrow I_{\text{final}} = \frac{I_{\text{initial}}}{9}$

(ii) During the first half time, wavelength first increases as the component of velocity of source increases till it becomes equal to the velocity of source itself, then it decreases till it becomes zeros.

(iii) $t_1 = \frac{\pi/3}{\omega}, \frac{2\pi + \pi/3}{\omega}$......., $t_2 = \frac{5\pi/3}{\omega}, \frac{2\pi + 5\pi/3}{\omega}$

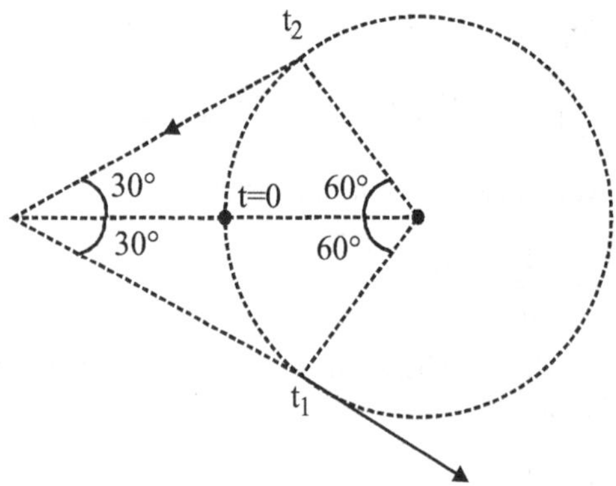

when $\omega = \frac{\pi}{3}$ rad/s, $t_1 = 1, 7, 13, ...$ $t_2 = 5, 11, 17,$

28. (c) The velocity of every oscillating particle of the medium is different of its different positions in one oscillation but the velocity of wave motion is always constant i.e., particle velocity vary with respect to time, while the wave velocity is independent of time.
Also for wave propagation medium must have the properties of elasticity and inertia.

29. (b) Velocity of wave $= \frac{\text{Distance travelled by wave } (\lambda)}{\text{Time period } (T)}$

Wavelength is also defined as the distance between two nearest points in phase.

30. (b) Transverse waves travel in the form of crests and through involving change in shape of the medium. As liquids and gases do not possess the elasticity of shape, therefore, transverse waves cannot be produced in liquid and gases. Also light wave is one example of transverse wave.

DAILY PRACTICE PROBLEMS — PHYSICS SOLUTIONS 30

1. (b) n_1 = Frequency of the police car horn observer heard by motorcyclist

n_1 = Frequency of the siren heard by motorcyclist.

v_2 = Speed of motor cyclist

$$n_1 = \frac{330-v}{330-22} \times 176; n_2 = \frac{330+v}{330} \times 165$$

$$\therefore n_1 - n_2 = 0 \Rightarrow v = 22\ m/s$$

2. (c) Frequency of first over tone of closed pipe = Frequency of first over tone of open pipe

$$\Rightarrow \frac{3v}{4L_1} = \frac{v}{L_2} \Rightarrow \frac{3}{4L_1}\sqrt{\frac{\gamma P}{\rho_1}} = \frac{1}{L_2}\sqrt{\frac{\gamma P}{\rho_2}}\left[\therefore v = \sqrt{\frac{\gamma P}{\rho}}\right]$$

$$\Rightarrow L_2 = \frac{4L_1}{3}\sqrt{\frac{\rho_1}{\rho_2}} = \frac{4L}{3}\sqrt{\frac{\rho_1}{\rho_2}}$$

3. (b) For observer note of B will not change due to zero relative motion.

Observed frequency of sound produced by A

$$= 660\frac{(330-30)}{330} = 600 Hz$$

$\therefore$ No of beats = 600 - 596 = 4

4. (c) Open pipe resonanace frequency $f_1 = \frac{2v}{2L}$

Closed pipe resonance frequency $f_2 = \frac{nv}{4L}$

$f_2 = \frac{n}{4} f_1$ (where n is odd and $f_2 > f_1$) $\therefore n = 5$

5. (b) $\frac{n(v - v_0)}{v - v_s}$

6. (a) Wave number $= \frac{1}{\lambda}$ but $= \frac{1}{\lambda'} == \frac{1}{\lambda}\left(\frac{v}{v-v_s}\right)$ and

$$v_s = \frac{v}{3}$$

$$\therefore (W.N.)' = (W.N.)\left(\frac{v}{v - v/3}\right) = 256 \times \frac{v}{2\upsilon/3}$$

$$= \frac{3}{2} \times 256 = 384$$

7. (d) Since there is no relative motion between observer and source, *therefore* there is no apparent change in frequency.

8. (c) $n' = n\left(\frac{v}{v-v_s}\right) = 1200 \times \left(\frac{350}{350-50}\right) = 1400\ cps$

9. (d) By using $n' = n\frac{v}{v-v_s} \Rightarrow \frac{n_1}{n} = \left(\frac{V}{V-S}\right)$

10. (a) $n' = n\frac{v-v_0}{v} = \left(\frac{330-33}{330}\right) \times 100 = 90\ Hz$

11. (d) The apparent frequency heard by the observer is given by

$$n' = \frac{v}{v-v_s} x = \frac{330}{330-33} \times 450 = \frac{330}{297} \times 450 = 500\ Hz$$

12. (b) Observer is moving ayay form siren 1 and towards the siren 2.

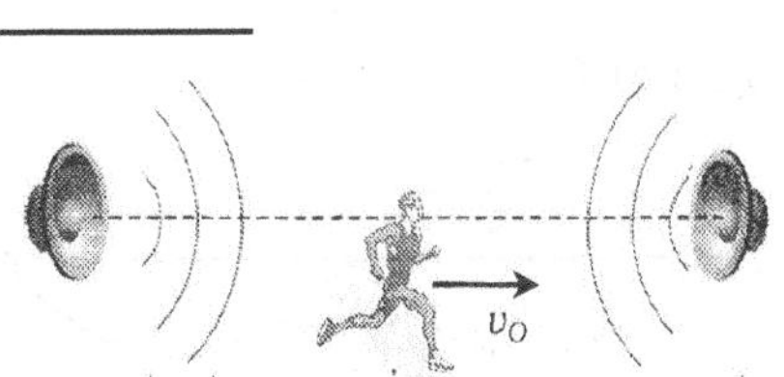

Hearing frequency of sound emitted by siren 1

$$n_1 = n\left(\frac{v-v_0}{v}\right) = 330\left(\frac{330-2}{330}\right) = 328 Hz$$

Hearing frequency of sound emitted by siren 2

$$n_2 = n\left(\frac{v-v_0}{v}\right) = 330\left(\frac{330+2}{330}\right) = 332 Hz$$

Hence, beat frequency $= n_2 - n_1 = 332 - 328 = 4$

13. (b) At point A, source is moving away from observer so apparent frequency $n_1 < n$ (actual frequency). At point B source is coming towards observer so apparent frequency $n_2 > n$ and point C source is moving perpendicular to observer so $n_3 = n$

Hence $n_2 > n_3 > n_1$

14. (c) According the concept of sound image

$$n' = \frac{v + v_{person}}{v - v_{person}}.272 = \frac{345+5}{345-5} \times 272 = 280 Hz$$

Δn = Number of beats = 280 – 272 = 8 Hz

15. (a) The observer will hear two sound, one directly from source and other from reflected image of sound

Hence number of beats heard per second

$$= \left(\frac{v}{v-v_s}\right)n - \left(\frac{v}{v+v_s}\right)n$$

$$= \frac{2n\upsilon\upsilon_s}{\upsilon^2-\upsilon_s^2} = \frac{2\times256\times330\times5}{335\times325} = 7.8Hz$$

16. (a) In closed pipe only odd harmonics are present.

17. (a) Maximum pressure at closed end will be atmospheric pressure adding with acoustic wave pressure

So $P_{\max} = P_A + P_0$ and $P_{\min} = P_A - P_0$

Thus $\frac{P_{\max}}{P_{\min}} = \frac{P_A+P_0}{P_A-P_0}$

18. (b) $n_{closed} = \frac{1}{2}(n_{open}) = \frac{1}{2}\times320 = 160Hz$

19. (a) $\lambda = 1.21\text{Å}$

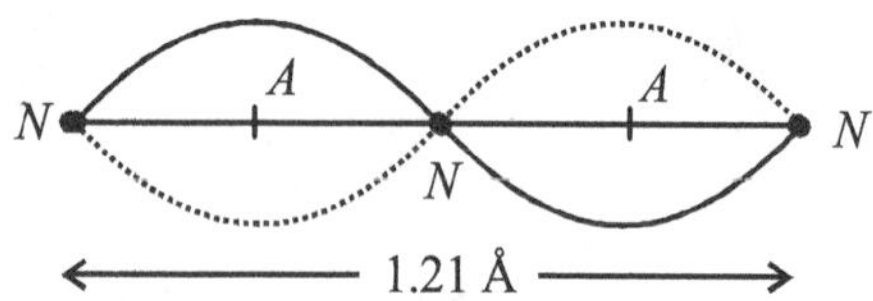

20. (a) $n \propto \frac{1}{l} \Rightarrow \frac{l_2}{l_1} = \frac{n_1}{n_2} \Rightarrow l_2 = l_1\left(\frac{n_1}{n_2}\right) = 50\times\frac{270}{1000} = 13.5cm$

21. (c) Loudness depends upon intensity while pitch depends upon frequency.

22. (d) Using $\lambda = 2(l_2 - l_1) \Rightarrow v = 2n(l_2 - l_1)$

$\Rightarrow 2\times215(63.2-30.7) = 33280\text{cm/s}$

Actual speed of sound $v_0 = 332\text{m/s} = 33200\text{cm/s}$

Hence error = 33280 – 33200 = 80cm / s

23. (c)

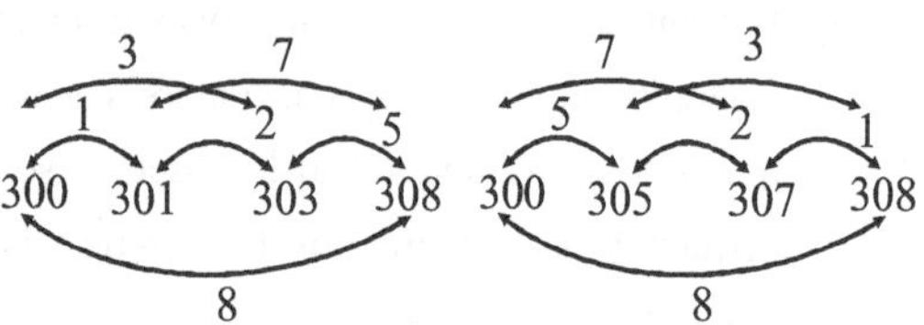

24. (a) Doppler shift doesn't depend upon the distance of listner from the source.

25. (b) Since the edges are clamped, displacement of the edges $u(x, y) = 0$ for line -

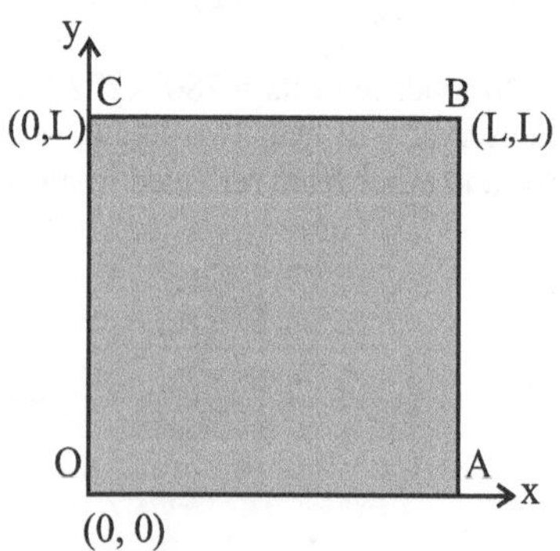

OA i.e $y = 0, 0 \le x \le L$

AB i.e. $y = L, 0 \le y \le L$

BC i.e. $y = L, 0 \le x \le L$

OC i.e. $x = 0, 0 \le y \le L$

The above conditions are satisfied only in alternatives (*b*) and (*c*).

Note that $u(x,y) = 0$ for all four values *e.g.* in alternative (*d*), $u(x,y) = 0$ for $y = 0, y = L$ But it is not zero for $x = 0$ or $x = L$. Similarly in option (a) $u(x,y) = 0$ at $x = L, y = L$ but it is not zero for $x = 0$ or $y = 0$, while in options (b) and (c), $u(x,y) = 0$ for $x = 0, y = 0, x = L$ and $y = L$

26. (b), 27. (d).

For fundamental force, $\frac{\lambda}{2} = s \Rightarrow \lambda = 2s$

Velocity of waves is, $v = \sqrt{\frac{Y}{\rho}}$ where Y is Young's modulus of quartz and ρ is its density.

From $f_0 = \frac{v}{\lambda} = \frac{2.87\times10^4}{s} \Rightarrow \sqrt{\frac{Y}{\rho}}\times\frac{1}{2s} = \frac{2.87\times10^4}{s}$

$\Rightarrow Y = 8.76\times10^{12}\ \text{N/m}^2$

For 3rd harmonic, $f = 3f_0 = 1.2\times10^6\ \text{Hz} \Rightarrow \frac{3\times2.87\times10^4}{s}$

$= 1.2\times10^6 \Rightarrow s = 0.07175$ cm.

28. (d) As emission of light from atom is a random and rapid phenomenon. The phase at a point due to two independent light source will change rapidly and randomly. Therefore, instead of beats, we shall get uniform intensity. However, if light sources are LASER beams of nearly equal frequencies, it may possible to observe the phenomenon of beats in light.

29. (d) The person will hear the loud sound at nodes than at antinodes. We know that at anti-nodes the displacement is maximum and pressure change is minimum while at nodes the displacement is zero and pressure change is maximum. The sound is heared due to variation of pressure.

Also in stationary waves particles in two different segment vibrates in opposite phase.

30. (a) Stationary wave

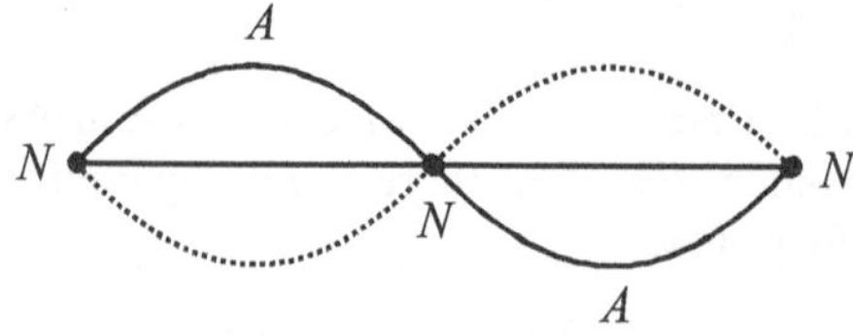

A node is a place of zero amplitude and an antinode is a place of maximum amplitude.

DAILY PRACTICE PROBLEMS

PHYSICS SOLUTIONS 31

1. **(b)**
20 division of vernier scale = 8 div. of main scale

$$\Rightarrow 1 \text{ V.S.D.} = \left(\frac{8}{20}\right) \text{M.S.D.} = \left(\frac{2}{5}\right) \text{M.S.D.}$$

Least count

$$= 1 \text{ M.S.D} - 1 \text{ V.S.D.} = 1 \text{ M.S.D.} - \left(\frac{2}{5}\right) \text{ M.S.D}$$

$$= \left(1-\frac{2}{5}\right) \text{M.S.D.} = \left(\frac{3}{5}\right) \text{M.S.D.} = \frac{3}{5} \times 0.1 \text{ cm.} = 0.06 \text{ cm.}$$

$$(\because 1 \text{ M.S.D.} = \frac{1}{10} \text{ cm.} = 0.1 \text{ cm.})$$

Directly we can use

$$\text{L.C.} = M - V = \left(\frac{b-a}{b}\right) M$$

$$= \left(\frac{20-8}{20}\right)\left(\frac{1}{10}\right) \text{cm.} = \frac{3}{50} \text{cm.} = 0.06 \text{ cm.}$$

2. **(c)** Within elastic limit it obeys Hooke's Law i.e., stress $\propto$ strain.

3. **(c)** Least count $= \frac{1}{N} \times \frac{1}{10} \text{cm} = \frac{1}{10N}$

4. **(b)** 5th division of vernier scale coincides with a main scale division. L.C. $= \frac{1}{10} = 0.1\text{mm}$

$\therefore$ Zero error $= -5 \times 0.1 = -0.5$ mm
This error is to be subtracted from the reading taken for measurement. Also, zero correction $= +0.5$ mm.

5. **(b)** If Y = Young's modulus of wire, M = mass of wire, g = acceleration due to gravity, x = extension in the wire, A = Area of cross-section of the wire, ℓ = length of the wire.

$$Y = \frac{Mgx}{A\ell} \Rightarrow \frac{\Delta Y}{Y} = \frac{\Delta M}{M} + \frac{\Delta x}{x} + \frac{\Delta A}{A} + \frac{\Delta \ell}{\ell}$$

$$\Rightarrow \frac{\Delta Y}{Y} = \frac{0.01}{3.00} + \frac{0.01}{0.87} + \frac{2 \times 0.001}{0.041} + \frac{0.001}{2.820} = 0.065$$

or $\frac{\Delta Y}{Y} \times 100 = \pm 6.5\%$

6. **(b)** The instrument has negative zero error.

7. **(c)** If A, B and C be the points corresponding to the impressions made by the legs of a spherometer then

$$c = \frac{a+b+c}{3}$$

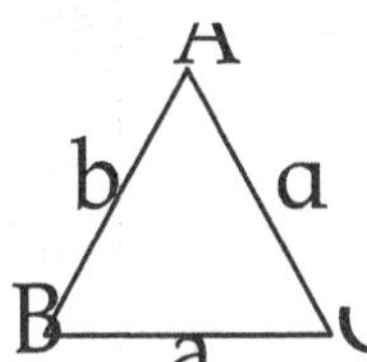

If h is the depression or elevation then the radius of curvature is given by $r = \frac{c^2}{6h} + \frac{h}{2}$

8. **(b)** $\text{L.C.} = \frac{\text{Pitch}}{\text{No. of circular divisions}}$

9. **(b)** The specific heat of a solid is determined by the method of mixture.

10. **(a)**

11. **(d)** Least count of screw gauge $= \frac{0.5}{50} \text{mm} = 0.01\text{mm}$

$\therefore$ Reading = [Main scale reading + circular scale reading × L.C] – (zero error)

$= [3 + 35 \times 0.01] - (-0.03) = 3.38$ mm

12. **(d)** 30 Divisions of vernier scale coincide with 29 divisions of main scales

Therefore 1 V.S.D $= \frac{29}{30}$ MSD

Least count = 1 MSD – 1VSD

$$= 1 \text{ MSD} - \frac{29}{30} \text{ MSD}$$

$$= \frac{1}{30} \text{MSD}$$

$$= \frac{1}{30} \times 0.5° = 1 \text{ minute.}$$

13. **(c)** Least count $= \frac{0.5}{50} = 0.01\text{mm}$

Zero error $= 5 \times L.C$
$= 5 \times 0.01$ mm
$= 0.05$ mm

Diameter of ball = [Reading on main scale] + [Reading on circular scale × $L.C$] – Zero error
$= 0.5 \times 2 + 25 \times 0.01 - 0.05$
$= 1.20$ mm

14. **(d)** $\frac{\Delta g}{g} = \frac{\Delta \ell}{\ell} + 2\frac{\Delta T}{T}$

$\Delta\ell$ and ΔT are least and number of readings are maximum in option (d), therefore the measurement of g is most accurate with data used in this option.

15. (b) We know that $Y = \dfrac{mg}{\pi\dfrac{D^2}{4}} \times \dfrac{L}{\ell}$

$$\Rightarrow \quad Y = \frac{4mgL}{\pi D^2 \ell} = \frac{4 \times 1 \times 9.8 \times 2}{\pi\left(0.4 \times 10^{-3}\right)^2 \times \left(0.8 \times 10^{-3}\right)}$$

$$= 2.0 \times 10^{11}\ \text{N/m}^2$$

Now $\dfrac{\Delta Y}{Y} = \dfrac{2\Delta D}{D} + \dfrac{\Delta \ell}{\ell}$

[$\because$ the value of m, g and L are exact]

$$= 2 \times \frac{0.01}{0.4} + \frac{0.05}{0.8} = 2 \times 0.025 + 0.0625$$

$$= 0.05 + 0.0625 = 0.1125$$

$$\Rightarrow \quad \Delta Y = 2 \times 10^{11} \times 0.1125 = 0.225 \times 10^{11}$$

$$= 0.2 \times 10^{11}\ \text{N/m}^2$$

16. (b) The time period of a simple pendulum is given by

$$T = 2\pi\sqrt{\frac{\ell}{g}} \therefore T^2 = 4\pi^2 \frac{\ell}{g} \Rightarrow g = 4\pi^2 \frac{\ell}{T^2}$$

$$\Rightarrow \quad \frac{\Delta g}{g} \times 100 = \frac{\Delta \ell}{\ell} \times 100 + 2\frac{\Delta T}{T} \times 100$$

Case (i)

$\Delta \ell = 0.1$ cm, $\ell = 64$cm, $\Delta T = 0.1$s, $T = 128$s

$$\therefore \frac{\Delta g}{g} \times 100 = 0.3125$$

Case (ii)

$\Delta \ell = 0.1$ cm, $\ell = 64$cm, $\Delta T = 0.1$s, $T = 64$s

$$\therefore \frac{\Delta g}{g} \times 100 = 0.46875$$

Case (iii)

$\Delta \ell = 0.1$ cm, $\ell = 20$cm, $\Delta T = 0.1$s, $T = 36$s

$$\therefore \frac{\Delta g}{g} \times 100 = 1.055$$

Clearly, the value of $\dfrac{\Delta g}{g} \times 100$ will be least in case (i)

17. (c) Terminal velocity, $v_T = \dfrac{2r^2(d_1 - d_2)g}{9\eta}$

$$\frac{v_{T_2}}{0.2} = \frac{(10.5 - 1.5)}{(19.5 - 1.5)} \Rightarrow v_{T_2} = 0.2 \times \frac{9}{18}$$

$$\therefore v_{T_2} = 0.1\ m/s$$

$$x\left\{-By\frac{d^2y}{dx^2} - B\left(\frac{dy}{dx}\right)^2\right\} + By\frac{dy}{dx} = 0$$

$$\Rightarrow \quad xy\frac{d^2y}{dx^2} + x\left(\frac{dy}{dx}\right)^2 - y\frac{dy}{dx} = 0$$

18. (a) The condition for terminal speed (v_t) is
Weight = Buoyant force + Viscous force

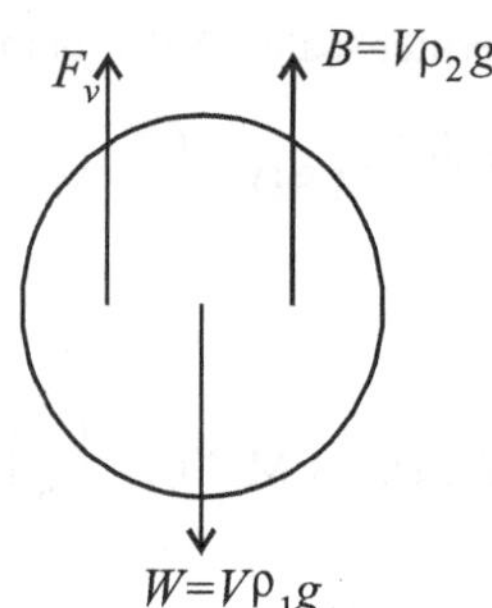

$$\therefore V\rho_1 g = V\rho_2 g + kv_t^2 \qquad \therefore v_t = \sqrt{\frac{Vg(\rho_1 - \rho_1)}{k}}$$

19. (d) From the figure it is clear that liquid 1 floats on liquid 2. The lighter liquid floats over heavier liquid. Therefore we can conclude that $\rho_1 < \rho_2$
Also $\rho_3 < \rho_2$ otherwise the ball would have sink to the bottom of the jar.
Also $\rho_3 > \rho_1$ otherwise the ball would have floated in liquid 1. From the above discussion we conclude that $\rho_1 < \rho_3 < \rho_2$.

20. (c) In case of water, the meniscus shape is concave upwards. Also according to ascent formula

$$h = \frac{2T\cos\theta}{r\rho g}$$

The surface tension (T) of soap solution is less than water. Therefore rise of soap solution in the capillary tube is less as compared to water. As in the case of water, the meniscus shape of soap solution is also concave upwards.

21. (c)

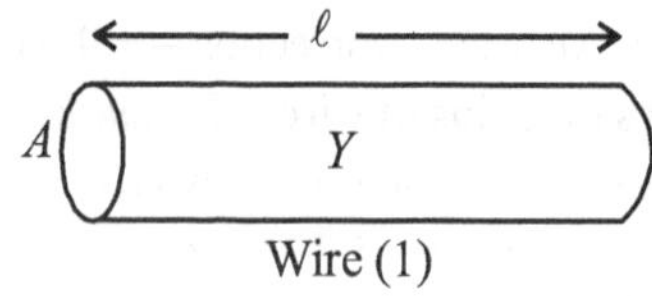

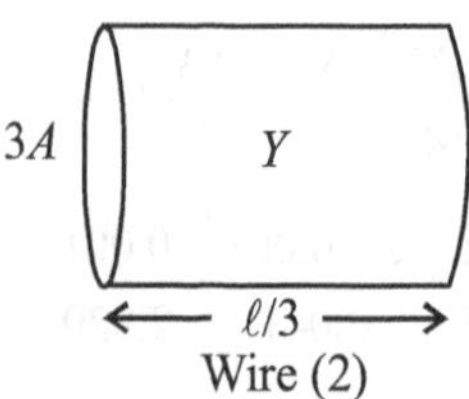

As shown in the figure, the wires will have the same Young's modulus (same material) and the length of the wire of area of cross-section $3A$ will be $\ell/3$ (same volume as wire 1).

For wire 1,

$$Y = \frac{F/A}{\Delta x/\ell} \quad ...(i)$$

For wire 2 ,

$$Y = \frac{F'/3A}{\Delta x/(\ell/3)} \quad ...(ii)$$

From (i) and (ii) , $\frac{F}{A} \times \frac{\ell}{\Delta x} = \frac{F'}{3A} \times \frac{\ell}{3\Delta x} \Rightarrow F' = 9F$

22. (b) Lower the vernier constant, more accurate measurement is possible by it.

23. (b) Effective length = MC = MN + NC = $\ell + \frac{d}{2}$

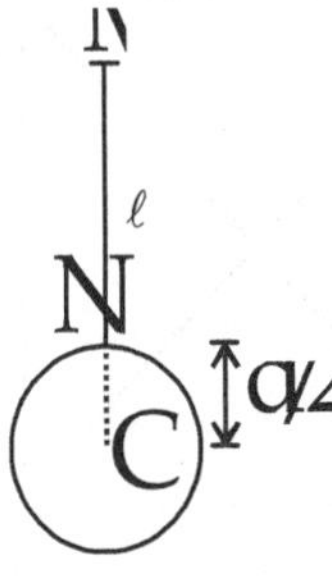

24. (c) Here, original length (L) = y,
Extension (ℓ) = x, Force applied (F) = p
Area of cross-section (A) = q

Now, Young's modulus (Y) = $\frac{FL}{AL}$

$$\Rightarrow Y = \frac{yp}{xq}$$

25. (a) Screw gauge is used to measure the diameter (d) of the wire so that the area of cross-section is calculated by the formula

$$A = \frac{\pi d^2}{4}$$

26. (b) Both the statements (1) & (2)are precautions to be taken during the experiment.

27. (a) The liquid cools faster first and slowly later on when its temperature gets close to surrounding temperature.

28. (a) Maximum percentage error in measurement of e, as given by Reyleigh's formula.
(Given error is measurement of radius is 0.1 cm)
$\Delta e = 0.6\ \Delta R = 0.6 \times 0.1 = 0.06$ cm.
Percentage error is

$$\frac{\Delta e}{e} \times 100 = \frac{0.06}{0.6 \times 3} \times 100 = 3.33\%$$

29. (b) Speed of sound at the room temperature.
$\ell_1 = 4.6$ cm, $\ell_2 = 14.0$ cm.,
$\lambda = 2(\ell_2 - \ell_1) = 2(14.0 - 4.6) = 18.8$ cm.

$$v = f\lambda = 2000 \times \frac{18.8}{100} = 376 \text{ m/s}$$

30. (c) End correction obtained in the experiment.

$$e = \frac{\ell_2 - 3\ell_1}{2} = \frac{14.0 - 3 \times 4.6}{2} = 0.1 \text{ cm.}$$

DAILY PRACTICE PROBLEMS

PHYSICS SOLUTIONS

32

1. **(d)** $Q_1 + Q_2 = Q$ .....(i)

and $F = k\dfrac{Q_1Q_2}{r^2}$ ..... (ii)

From (i) and (ii)

$$F = \frac{kQ_1(Q-Q_1)}{r^2}$$

For F to be maximum

$$\frac{dF}{dQ_1} = 0 \Rightarrow Q_1 = Q_2 = \frac{Q}{2}$$

2. **(a)** The position of the balls in the satellite will become as shown below

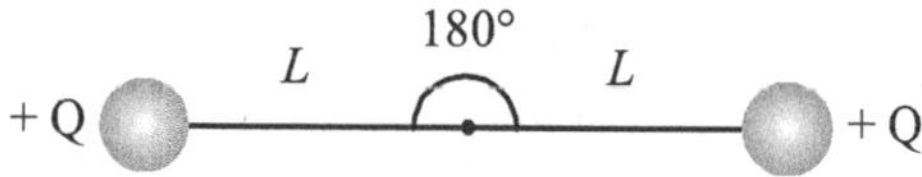

Thus angle $\theta = 180°$

and force $= \dfrac{1}{4\pi\epsilon_0}\cdot\dfrac{Q^2}{(2L)^2}$

3. **(b)** F_A = Force on C due to charge placed at A

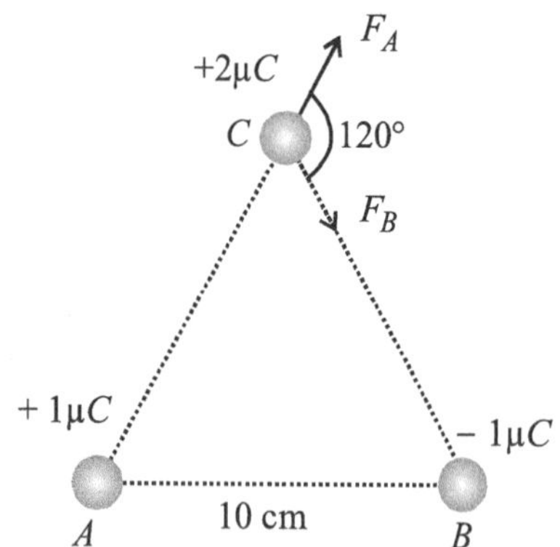

$$= 9\times10^9 \times \frac{10^{-6}\times2\times10^{-6}}{(10\times10^{-2})^2} = 1.8N$$

F_B = force on C due to charge placed at B

$$= 9\times10^9 \times \frac{10^{-6}\times2\times10^{-6}}{(0.1)^2} = 1.8N$$

Net force on C

$$F_{net} = \sqrt{(F_A)^2 + (F_B)^2 + 2F_AF_B\cos120°} = 1.8N$$

4. **(c)** $\vec{F} = -k\dfrac{e^2}{r^2}\hat{r} = -k.\dfrac{e^2}{r^3}\vec{r}$ $\left(\therefore \hat{r} = \dfrac{\vec{r}}{r}\right)$

5. **(c)** After following the guidelines mentioned above

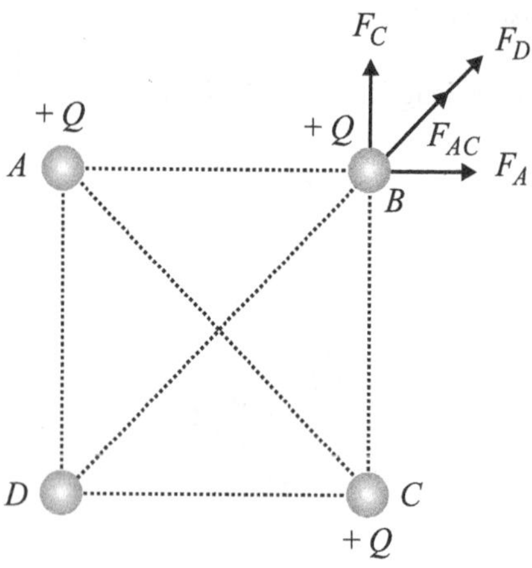

$$F_{net} = F_{AC} + F_D = \sqrt{F_A^2 + F_C^2} + F_D$$

Since $F_A = F_C = \dfrac{kq^2}{a^2}$ and $F_D = \dfrac{kq^2}{(a\sqrt{2})^2}$

$$F_{net} = \frac{\sqrt{2}kq^2}{a^2} + \frac{kq^2}{2a^2} = \frac{kq^2}{a^2}\left(\sqrt{2}+\frac{1}{2}\right) = \frac{q^2}{4\pi\epsilon_0 a^2}\left(\frac{1+2\sqrt{2}}{2}\right)$$

6. **(a)**

$$F = \frac{1}{4\pi\varepsilon_0}\frac{(+7\times10^{-6}(-5\times10^{-6})}{r^2} = -\frac{1}{4\pi\varepsilon_0}\frac{35\times10^{-12}}{r^2}N$$

$$F' = \frac{1}{4\pi\varepsilon_0}\frac{(+5\times10^{-6}(-7\times10^{-6})}{r^2} = -\frac{1}{4\pi\varepsilon_0}\frac{35\times10^{-12}}{r^2}N$$

7. **(c)** Electric field outside of the sphere $E_{out} = \dfrac{kQ}{r^2}$ ...(i)

Electric field inside the dielectric shphere $E_{in} = \dfrac{kQx}{R^3}$ ...(ii)

From (i) and (ii), $E_{in} = E_{out} \times \dfrac{r^2x}{R}$

$\Rightarrow$ At 3 cm, $E = 100 \times \dfrac{3(20)^2}{10^3} = 120$ V/m

8. **(c)** Electric lines force due to negative charge are radially inward.

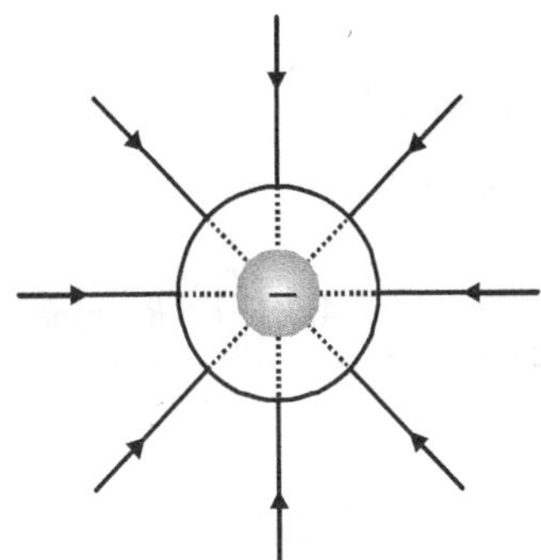

9. **(a)** In non-uniform electric field, intensity is more, where the lines are more denser.

10. **(b)** According to the question,

$$eE = mg \Rightarrow E = \frac{mg}{e}$$

11. **(b)** Because E points along the tangent to the lines of force. If initial velocity is zero, then due to the force, it always moves in the direction of E. Hence will always move on some lines of force.

12. (b) The field produced by charge - $3Q$ at A, this is E as mentioned in the example.

$\therefore E = \frac{3Q}{x^2}$ (along AB directed towards negative charge)

A B
Q ——— –3Q
←—x—→

Now field at location of $-3Q$ *i.e.* field at B due to charge Q will be $E' = \frac{Q}{x^2} = \frac{E}{3}$ (along AB directed away from positive charge)

13. **(c)** The electric field is due to all charges present whether inside or outside the given surface.

14. **(b)**

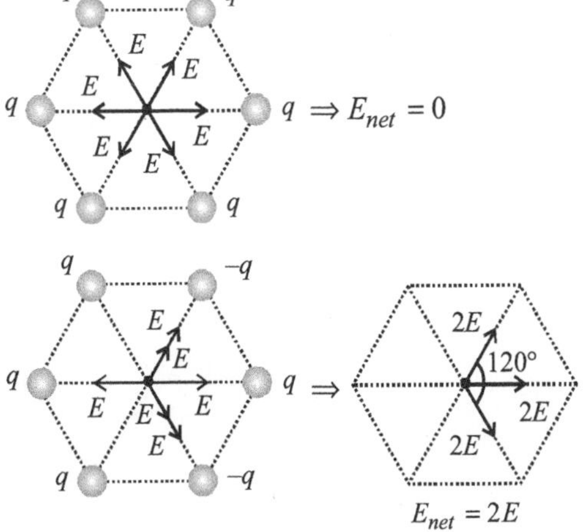

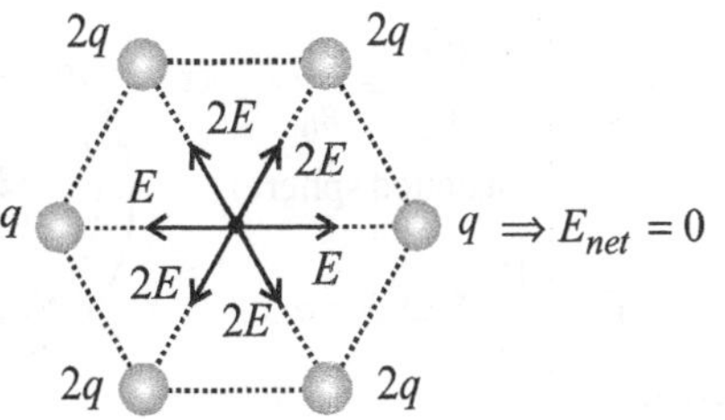

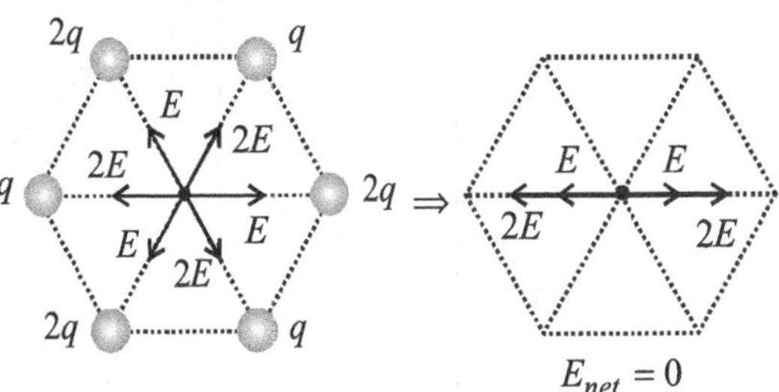

15. (b)

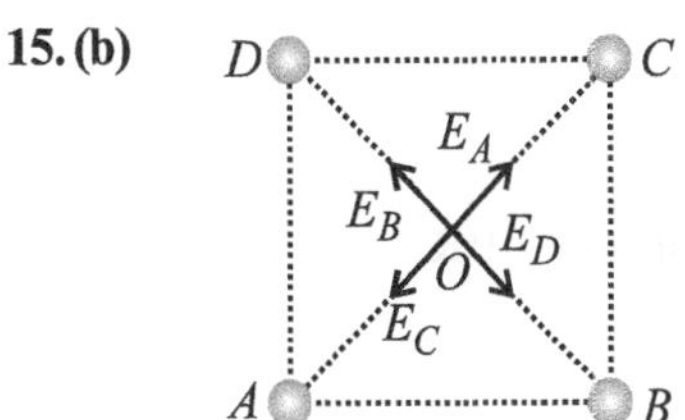

⇒

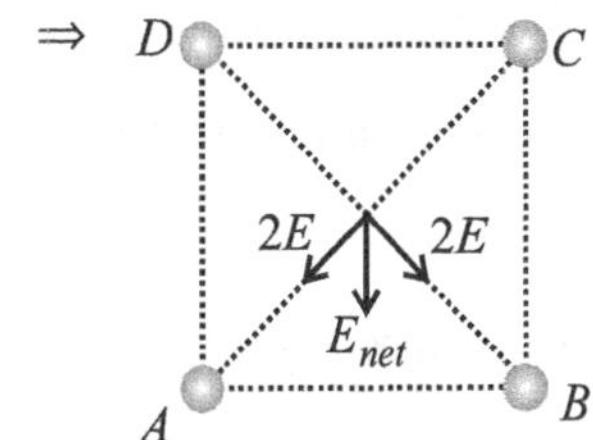

$$E_A = E, E_B = 2E, E_C = 3E, E_D = 4E$$

16. (b) $\vec{E} = -\frac{\sigma}{2\varepsilon_0}\hat{k} - \frac{2\sigma}{2\varepsilon_0}\hat{k} - \frac{\sigma}{2\varepsilon_0}\hat{k} = -\frac{2\sigma}{\varepsilon_0}\hat{k}$

17. (b)

18. **(c)** According to Gauss law $\oint E.ds = \frac{q_1}{\varepsilon_0}$

$\oint ds = 2\pi rl$; (E is constant)

$\therefore$ E. $2\pi rl = \frac{q_1}{\varepsilon_0} \Rightarrow E = \frac{q}{2\pi\varepsilon_0 r}$ *i.e.* $E \propto \frac{1}{r}$

19. **(c)** Let sphere has uniform chare density $\rho\left(\frac{3Q}{4\pi R^3}\right)$ and E is the electric field at distance x from the centre of the sphere.

Applying Gauss law,

$$E.4\pi x^2 = \frac{q}{\varepsilon_0} = \frac{\rho V'}{\varepsilon_0} = \frac{\rho}{\varepsilon_0} \times \frac{4}{3}\pi x^3$$

(V = Volume of dotted sphere)

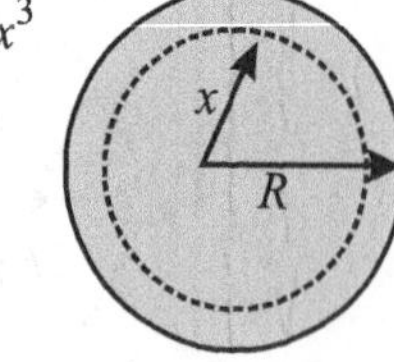

$\therefore \quad E = \frac{\rho}{3\varepsilon_0} x \Rightarrow E \propto x$

20. **(b)** $T \sin\theta = \theta E$
and $T \cos\theta = mg$

$\Rightarrow \quad \tan\theta = \frac{qE}{mg}$

$= \frac{q}{mg}\left(\frac{\sigma}{2\varepsilon_0}\right)$

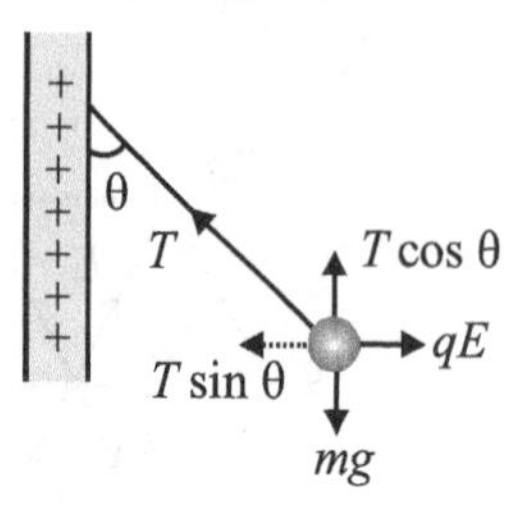

$\Rightarrow \quad \sigma \propto \tan\theta.$

21. **(d)** Next flux through the cube $\phi_{net} = \frac{Q}{\varepsilon_0}$; so flux through one face $\phi_{face} = \frac{q}{6\varepsilon_0}$

22. **(d).** For A : Power consumed $P = I^2R$

But $\frac{q}{\varepsilon_0} = \phi$,

so $q = \alpha\, \varepsilon_0 t^2 \Rightarrow I = \frac{dq}{dt} = 2\alpha\, \varepsilon_0 t \Rightarrow P = 4\alpha^2\, \varepsilon_0^2 R t^2$

For B : Assuming initial charge in reservoir be q_0 then electric flux through a closed

Spherical surface around S_2 will be $\phi_{S_2} = \frac{q_0 - \alpha\varepsilon_0 t^2}{\varepsilon_0}$

For C : $\frac{d\phi_{S_2}}{dt} = -2\alpha t$

23. **(c).** The time period will change only when the additional electrostatic force has a component along the direction of the displacement, which is always perpendicular to the string.

24. ()

25. **(b)** Net charge inside the sphere $= \int_{sphere} \rho\, dV$

Due to spherical symmetry, we get

$$Q = \int_0^R 4\pi r^2 \rho(r) dr = 4\pi A \int_0^R r^2 (R - r)\, dr$$

$$= 4\pi A\left(\frac{R^4}{3} - \frac{R^4}{4}\right)$$

$\therefore \quad A = \frac{3Q}{\pi R^4}$

26. **(a)** According to Gauss law

$$\oint_s \vec{E}.\vec{dS} = 4\pi r^2 E(r)$$

$$= \frac{\int_0^r 4\pi r^2 \rho(r)\, dr}{\epsilon_0}$$

$$\Rightarrow \quad 4\pi r^2 E(r) = \frac{4\pi A \int_0^r r^2 (R - r) dr}{\epsilon_0}$$

$$= \frac{4\pi A}{\epsilon_0}\left(\frac{r^3 R}{3} - \frac{r^4}{4}\right)$$

Hence, $E(r) = \frac{A}{\epsilon_0}\left(\frac{rR}{3} - \frac{r^2}{4}\right)$, for $0 < r < R$

But $A = \frac{3Q}{\pi R^4}$

$\therefore$ We get, $E(r) = \frac{3Q}{\epsilon_0 R^2}\left[\frac{1}{3}\left(\frac{r}{R}\right) - \frac{1}{4}\left(\frac{r}{R}\right)^2\right]$

27. **(b)** The electric field outside the sphere is given by :

$E(r) = \frac{kQ}{r^2}$, for $r \geq R$

28. **(c)** If electric lines of forces cross each other, then the electric field at the point of intersection will have two direction simultaneously which is not possible physically.

29. **(c)** Electric field at the nearby point will be resultant of existing field and field due to the charge brought. It may increase or decrease if the charge is positive or negative depending on the position of the point with respect to the charge brought.

30. ()

DAILY PRACTICE PROBLEMS

PHYSICS SOLUTIONS 33

1. (c) *ABCDE* is an equipotential surface, on equipotential surface no work is done in shifting a charge from one place to another.

2. (c) Potential at centre O of the square

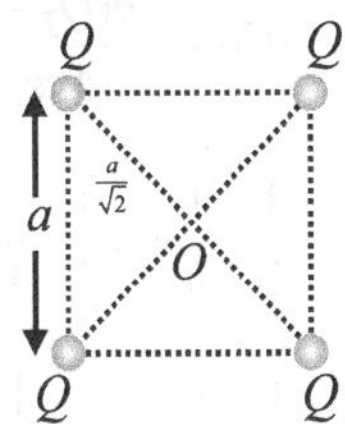

$$V_0 = \left(\frac{Q}{4\pi\varepsilon_0(a/\sqrt{2})}\right)$$

Work done in shifting (– Q) charge from centre to infinity

$$W = -Q(V_\infty - V_0) = QV_0$$

$$= Q.\frac{4\sqrt{2}Q}{4\pi\varepsilon_0 a} = \frac{\sqrt{2}Q^2}{\pi\varepsilon_0 a}$$

3. (b) Using

$$v = \sqrt{\frac{2QV}{M}} \Rightarrow v \propto \sqrt{Q} \Rightarrow \frac{v_A}{v_B} = \sqrt{\frac{Q_A}{Q_B}} = \sqrt{\frac{q}{4q}} = \frac{1}{2}$$

4. (a) Work done in moving a charge from P to L, P to M and P to N is zero while it is $q\,(V_P - V_k) > 0$ for motion from P to k.

5. (a) $KE = q(V_1 - V_2) = 2 \times (70 - 50)\Omega = 40\ eV$

6. (a) The electric potential $V(x, y, z) = 4x^2\ volt$

Now $\vec{E} = -\left(\hat{i}\frac{\partial V}{\partial x} + \hat{j}\frac{\partial V}{\partial y} + \hat{k}\frac{\partial V}{\partial z}\right)$

Now $\frac{\partial V}{\partial x} = 8x, \frac{\partial V}{\partial y} = 0$ and $\frac{\partial V}{\partial z} = 0$

Hence $\vec{E} = -8\hat{i}$, so at point $(1m, 0, 2m)$

$\vec{E} = -8x\hat{i}\ volt/meter$ or 8 along negative X - axis.

7. (b) Electric fields due to electrons on same line passing through centre cancel each other while electric potential due to each electron is negative at centre C. Therefore,

at centre $\vec{E} = 0, V \pm 0$

8. (a) By using $W = Q(\vec{E}.\Delta\vec{r})$

$$\Rightarrow W = Q[(e_1\hat{i} + e_2\hat{j} + e_3\hat{k}).(a\hat{i} + b\hat{j})] = Q(e_1 a + e_2 b)$$

9. (b) Potential at A= Potential due to (+q) charge + Potential due to (– q) charge

$$= \frac{1}{4\pi\varepsilon_0}.\frac{q}{\sqrt{a^2 + b^2}} + \frac{1}{4\pi\varepsilon_0}\frac{(-q)}{\sqrt{a^2 + b^2}} = 0$$

10. (c) Point P will lie near the charge which is smaller in magnitude *i.e.* $-6\mu C$. Hence potential at P

P -6μC 12μC

⟵ x ⟶⟵ 20 cm ⟶

$$V = \frac{1}{4\pi\varepsilon_0}\frac{(-6 \times 10^{-6})}{x} + \frac{1}{4\pi\varepsilon_0}\frac{(12 \times 10^{-6})}{(0.2 + x)} = 0 \Rightarrow x = 0.2m$$

11. (a) Work done $W = q^{-6}(V_A - V_B)$; where

$q = 3 \times 10^{-6}$ coulomb where

$$V_A = 10^{10}\left[\frac{(-5 \times 10^{-6})}{15 \times 10^{-2}} + \frac{2 \times 10^{-6}}{5 \times 10^{-2}}\right] = \frac{1}{15} \times 10^6\ volt$$

and $V_B = 10^{10}\left[\frac{(2 \times 10^{-6})}{15 \times 10^{-2}} - \frac{5 \times 10^{-6}}{5 \times 10^{-2}}\right]$

$$= -\frac{13}{15} \times 10^6\ volt$$

$$\therefore W = 3 \times 10^{-6}\left[\frac{1}{15} \times 10^6 - \left(\frac{13}{15} \times 10^6\right)\right]$$

$= 2.8$ J

12. (c)

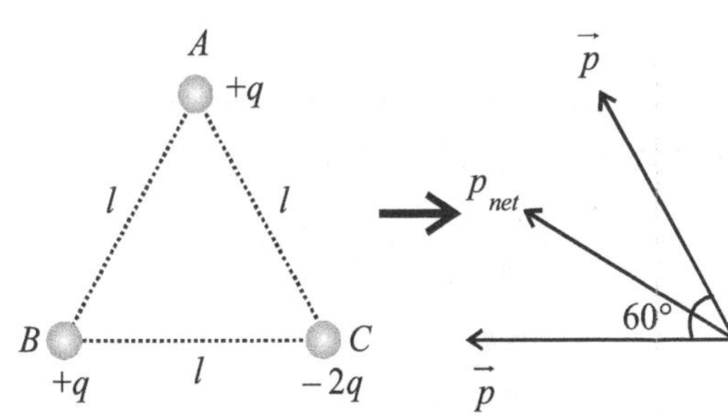

$$P_{net} = \sqrt{p^2 + p^2 + 2pp\cos 60^\circ} = \sqrt{3}p = \sqrt{3}\,ql (\because p = ql)$$

13. (d) According to figure, potential at A and C are both equal to kQ. Hence work done in moving $-q$ charge from A to $C = -q(V_A - V_C) = 0$

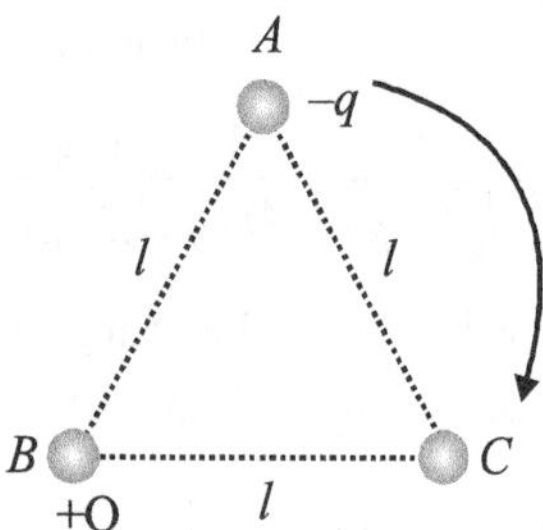

14. (c) $V = k \times \frac{Q}{r} = 9 \times 10^9 \times \frac{(+1.6 \times 10^{-19})}{0.53 \times 10^{-10}} = 27.2V$

15. (c) Potential will be zero at two points

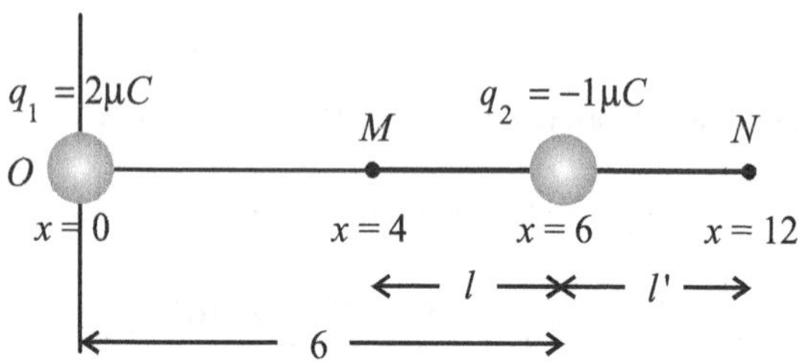

At internal point (M):

$$\frac{1}{4\pi\varepsilon_0} \times \left[\frac{2 \times 10^{-6}}{(6-l)} + \frac{(-1 \times 10^{-6})}{l}\right] = 0$$

$\Rightarrow l = 2$

So distance of M from origin;

$x = 6 - 2 = 4$

At exterior point (N):

$$\frac{1}{4\pi\varepsilon_0} \times \left[\frac{2 \times 10^{-6}}{(6-l')} + \frac{(-1 \times 10^{-6})}{l'}\right] = 0$$

$\Rightarrow l' = 6$

So distance of N from origin, $x = 6 + 6 = 12$

16. (a) $V = V_{AB} + V_{BC} + V_{CD}$

$$= \frac{k.5Q_0}{R} + \frac{k.(-2Q_0)}{R} + \frac{k.(3Q_0)}{R}$$

$$= \frac{6kQ_0}{R}$$

$$= \frac{3Q_0}{2\pi \epsilon_0 R}$$

17. (a) $V = 9 \times 10^9 . \frac{p}{r^2}$

$$= 9 \times 10^9 \times \frac{(1.6 \times 10^{-19}) \times 1.28 \times 10^{-10}}{(12 \times 10^{-10})^2} = 0.13V$$

18. (b) $W = q(V_{02} - V_{01})$

where $V_{01} = \frac{Q_1}{4\pi\varepsilon_0 R} + \frac{Q_2}{4\pi\varepsilon_0 R\sqrt{2}}$

and $V_{02} = \frac{Q_2}{4\pi\varepsilon_0 R} + \frac{Q_1}{4\pi\varepsilon_0 R\sqrt{2}}$

$$\Rightarrow W = q(V_{02} - V_{01}) = \frac{q(Q_2 - Q_1)}{4\pi\varepsilon_0 R}\frac{(\sqrt{2}-1)}{\sqrt{2}}$$

19. (d) $V = \frac{q}{4\pi\varepsilon_0 x_0}\left[1 + \frac{1}{3} + \frac{1}{5} + ...\right] - \frac{q}{4\pi\varepsilon_0 x_0}\left[\frac{1}{2} + \frac{1}{4} + \frac{1}{6} + ...\right]$

$$= \frac{q}{4\pi\varepsilon_0 x_0}\left[1 - \frac{1}{2} + \frac{1}{3} - \frac{1}{4} + ...\right] = \frac{q}{4\pi\varepsilon_0 x_0}\log_e 2$$

20. (b) Potential decreases in the direction of electric field. Dotted lines are equipotential surfaces

$\therefore V_A = V_C$ and $V_A > V_B$

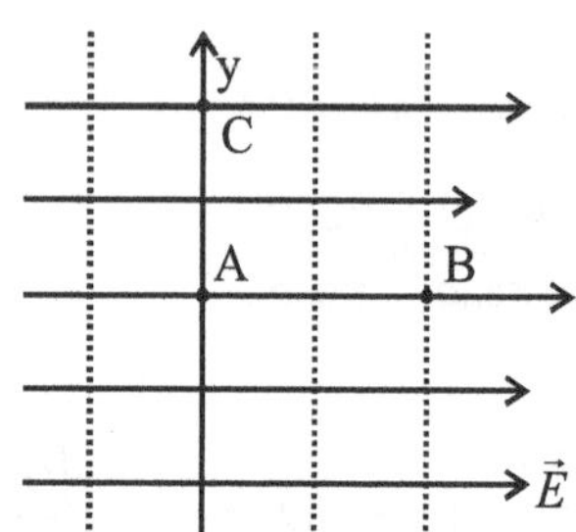

21. (d) $E_{equatorial} = \frac{kp}{r^3}$ i.e. $E \propto p$ and $E \propto r^{-3}$

22. (a) Suppose neutral point N lies at a distance x from dipole of moment p or at a distance x^2 from dipole of 64 p.

At N |E.F. due to dipole①| = |E.F. due to dipole②|

$$\Rightarrow \frac{1}{4\pi\varepsilon_0}.\frac{2p}{x^3} = \frac{1}{4\pi\varepsilon_0}.\frac{2(64p)}{(25-x)^3}$$

$$\Rightarrow \frac{1}{x^3} = \frac{64}{(25-x)^3} \Rightarrow x = 5 \text{ cm.}$$

23. (a). $BC = 2R\sin\left(\frac{120}{2}\right) = \sqrt{3}R$

Electric field at O $= \frac{1}{4\pi\varepsilon_0 R}\left(\frac{2q/3}{R^2}\right) = \frac{q}{6\pi\varepsilon_0 R^2}$

along negative X-axis.

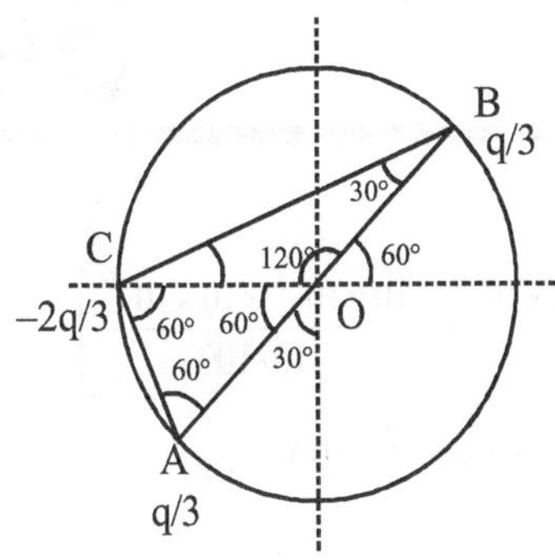

The potential energy of the system is non zero
Force between B & C

$$=\left|\frac{1}{4\pi\varepsilon_0}\frac{(q/3)(-2q/3)}{(\sqrt{3}R)^2}\right|=\frac{q^2}{54\pi\varepsilon_0 R^2}$$

$$\text{Potential at O}=\frac{1}{4\pi\varepsilon_0}\left(\frac{q}{3}+\frac{q}{3}-\frac{2q}{3}\right)=0$$

24. (d) The given graph is of charged conducting sphere of radius R_0. The whole charge q distributes on the surface of sphere

25 (b), 26 (b), 27 c

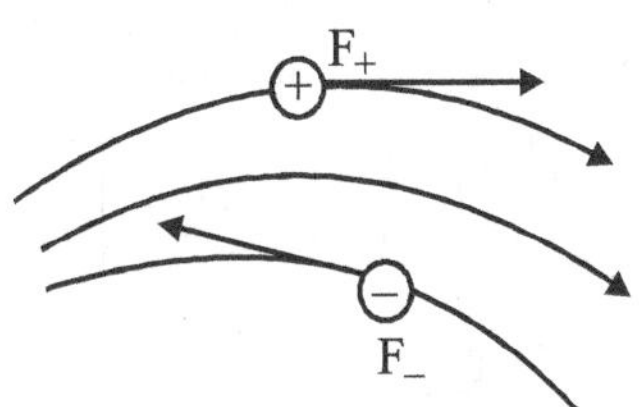

$(F_+ > F_- \text{ as } E_+ > E_-)$

Net force F :

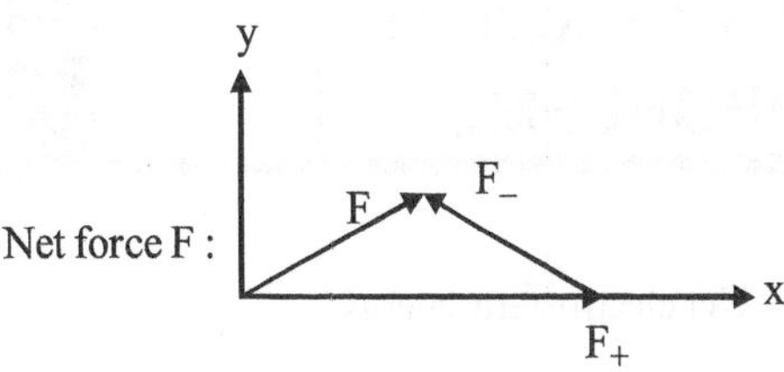

Net torque immediately after it is released ⇒ clockwise
A body cannot exert force on itself.

28. **(d)** When the bird perches on a single high power line, no current passes through its body because its body is at equipotential surface i.e., there is no potential difference. While when man touches the same line, standing bare foot on ground the electrical circuit is completed through the ground. The hands of man are at high potential and his feet's are at low potential. Hence large amount of current flows through the body of the man and person therefore gets a fatal shock.

29. **(a)** Electron has negative charge, in electric field negative charge moves from lower potential to higher potential.

30. **(b)** Potential is constant on the surface of a sphere so it behaves as an equipotential surface. Free charges (electrons) are available in conductor. The two statements are independent.

DAILY PRACTICE PROBLEMS

PHYSICS SOLUTIONS

34

1. **(b)** Net electrostatic energy

$$U = \frac{kQq}{a} + \frac{kq^2}{a} + \frac{kQq}{a\sqrt{2}} = 0$$

$$\Rightarrow \frac{kq}{a}\left(Q + q + \frac{Q}{\sqrt{2}}\right) = 0 \Rightarrow Q = -\frac{2q}{2+\sqrt{2}}$$

2. **(c)** Electric field is perpendicular to the equipotential surface and is zero every where inside the metal.

3. **(c)** $\Rightarrow \frac{kq}{l}(Q+q+Q) = 0 \Rightarrow Q = -\frac{q}{2}$ net potential energy

$$U_{net} = 3\times\frac{1}{4\pi\varepsilon_0}.\frac{q^2}{l}$$

4. **(d)** Length of the diagonal of a cube having each side b is $\sqrt{3}b$. So distance of centre of cube from each vertex is $\frac{\sqrt{3}b}{2}$.

Hence potential energy of the given system of charge is

$$U = 8\times\left\{\frac{1}{4\pi\varepsilon_0}.\frac{(-q)(q)}{\sqrt{3}b/2}\right\} = \frac{-4q^2}{\sqrt{3}\pi\varepsilon_0 b}$$

5. **(a)** Change in potential energy $(\Delta U) = U_f - U_i$

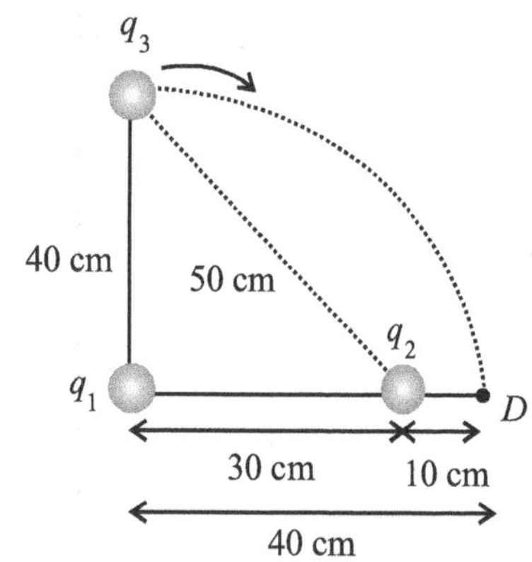

$$\Rightarrow \Delta U = \frac{1}{4\pi\varepsilon_0}\left[\left(\frac{q_1q_3}{0.4} + \frac{q_2q_3}{0.1}\right) - \left(\frac{q_1q_3}{0.4} + \frac{q_2q_3}{0.5}\right)\right]$$

$$\Rightarrow \Delta U = \frac{1}{4\pi\varepsilon_0}[8q_2q_3] = \frac{q_3}{4\pi\varepsilon_0}(8q_2)$$

$$\therefore k = 8q_2$$

6. **(c)** For pair of charge $U = \frac{1}{4\pi\varepsilon_0}.\frac{q_1q_2}{r}$

$$U_{System} = \frac{1}{4\pi\varepsilon_0}\left[\frac{10\times10^{-6}10\times10^{-6}}{10/100} + \frac{10\times10^{-6}\times10\times10^{-6}}{10/100} + \frac{10\times10^{-6}\times10\times10^{-6}}{10/100}\right]$$

$$= 3\times9\times10^9\times\frac{100\times10^{-12}\times100}{10} = 27J$$

7. **(c)** $U_{System} = \frac{1}{4\pi\varepsilon_0}\frac{(q)(-2q)}{a} + \frac{1}{4\pi\varepsilon_0}\frac{(-2q)(q)}{a} + \frac{1}{4\pi\varepsilon_0}\frac{(q)(q)}{2a}$

$$U_{System} = -\frac{7q^2}{8\pi\varepsilon_0 a}$$

8. **(c)** In the given condition angle between $\vec{p}$ and $\vec{E}$ is zero. Hence potential energy $U = -pE\cos 0 = -pE = \text{min}$. Also in uniform electric field $F_{net} = 0$

9. **(b)** $U = \frac{1}{4\pi\epsilon_0}\frac{(-e)(-e)}{r^2}$ As r decreases then U increases and sign of U is '+ve' so, U increases.

10. **(c)** $U = 2kq^2\left[-\frac{1}{a} + \frac{1}{2a} - \frac{1}{3a} + \frac{1}{4a} + ...\right]$

$$= -\frac{2q^2}{4\pi\varepsilon_0 a}\left[1 - \frac{1}{2} + \frac{1}{3} - \frac{1}{4} + ...\right]$$

$$= -\frac{2q^2\log_e 2}{4\pi\varepsilon_0 a}$$

11. **(b)** The initial energy of the system

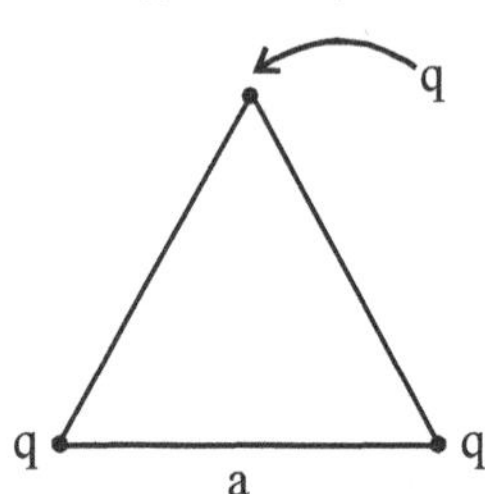

$$U_i = \frac{1}{4\pi\varepsilon_0}\frac{q^2}{a} = U$$

The final energy of the system

$$U_f = \frac{1}{4\pi\varepsilon_0}\left[\frac{q^2}{a} + \frac{q^2}{a} + \frac{q^2}{a}\right] = 3U$$

Thus work done, $W = U_f - U_i = 3U - U = 2U$

12. **(d)** $U = \frac{kq_1q_2}{r}$

13. **(c)** As potential at A and B is same, $V_A = V_B = \frac{kQ}{d}$. So, work done in both cases will be same.

14. **(b)** $U = \varepsilon\frac{kq_1q_2}{r}$. There will be 6 pairs, 4 on a side of square and 2 as diagonal.

15. **(c)** Apply conservation of mechanical energy between point a and b : (K.E. + P.E.)$_a$ = (K.E. + P.E.)$_a$

$$\Rightarrow 0+\frac{k(3\times10^{-9})q_0}{0.01}-\frac{k(3\times10^{-9})q_0}{0.02}$$

$$=\frac{1}{2}mv^2+\frac{k(3\times10^{-9})q_0}{0.02}-\frac{k(3\times10^{-9})q_0}{0.01}$$

Put the values we get : v = $12\sqrt{15}$ = 46 m/s

16. **(b)** $U=-\frac{kqQ}{r}-\frac{kqQ}{r}+\frac{kq^2}{2r}=0\Rightarrow Q/q=1/4$

17. **(b)** Find potential at A and C due to charge at B, then required work done is W = q ($V_A - V_C$)

18. **(d)** It depends whether both charges are of same or opposite sign.

19. **(a)** Because work is to be done by an external agent in moving a positive charge from low potential to high potential and this work gets stored in the form of potential energy of the system. Hence, it increases.

20. **(a)** $U=kq^2\left(-\frac{3}{d}+\frac{3}{\sqrt{2}d}-\frac{1}{\sqrt{3}d}\right)$

$$+kq^2\left(-\frac{2}{d}+\frac{3}{\sqrt{2}d}-\frac{1}{\sqrt{3}d}\right)+kq^2\left(-\frac{2}{d}+\frac{2}{\sqrt{2}d}-\frac{1}{\sqrt{3}d}\right)$$

$$+kq^2\left(-\frac{1}{d}+\frac{2}{\sqrt{2}d}-\frac{1}{\sqrt{3}d}\right)+kq^2\left(-\frac{2}{d}+\frac{1}{\sqrt{2}d}\right)$$

$$+kq^2\left(-\frac{1}{d}+\frac{2}{\sqrt{2}d}\right)+kq^2\left(-\frac{1}{d}\right)$$

$$U=kq^2\left(-\frac{12}{d}+\frac{12}{\sqrt{2}d}-\frac{4}{\sqrt{3}d}\right)$$

$$=-\frac{12kq^2}{d}\left(1-\frac{1}{\sqrt{2}}+\frac{1}{3\sqrt{3}}\right)$$

21. **(c)** $V=\frac{kQ}{r}\Rightarrow u=\frac{1}{2}\varepsilon_0E^2=\frac{1}{2}\frac{\varepsilon_0k^2Q^2}{r^4}$

$V^4\propto u$

22.

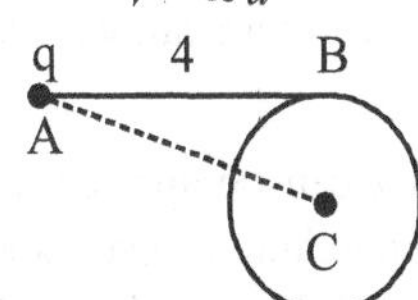

AC = 5m, $V=\frac{kq}{AC}=\frac{9\times10^9\times1\times10^{-6}}{5}$

$= 1.8\times10^3 = 1.8$ kV

$V_B = (V_B)_{\text{due to q}} + (V_B)_i$, where $(V_B)_i$ = Potential at B due to induced charge

$\therefore 1.8\times10^3=\frac{kq}{AB}+(V_B)_i$

$\Rightarrow 1.8\times10^3 = 2.25\times10^3 + (V_B)_i$

$\Rightarrow (V_B)_i = -0.45$ kV

23. **(b)** Force = eE

Work done = force × distance

Force and distance are in opposite direction, so work is negative.

W = – eE × d

Here, distance increases so, potential energy increases.

24. **(d)** Under electrostatic condition, all points lying on the conductor are in same potential. Therefore, potential at A = potential at B.

From Gauss's theorem, total flux through the surface of the cavity will be q/ε_0.

Note : Instead of an elliptical cavity, if it would and been a spherical cavity then options (a) and (b) were also correct.

25. **(b)** $q_1+q_2=0$

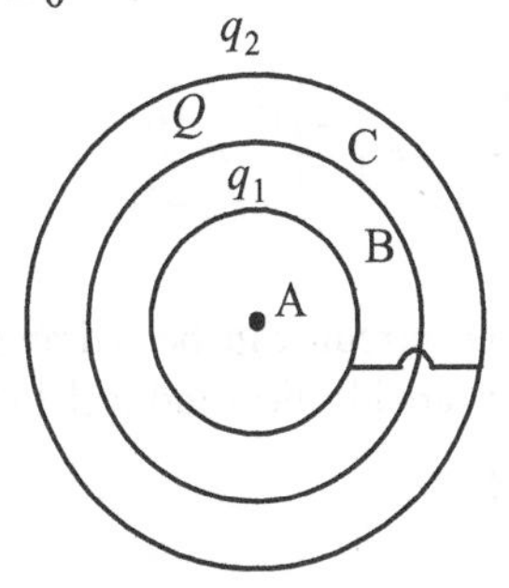

$$V_A=\frac{kq_1}{R}+\frac{kQ}{2R}+\frac{kq_2}{4R}$$

$$V_C=\frac{kq_1}{4R}+\frac{kQ}{4R}+\frac{kq_2}{4R}$$

$V_A = V_C$

$\therefore q_1 = -Q/3$ and $q_2 = Q/3$

26. **(b)** $V_A=k\left[\frac{-Q}{3R}+\frac{Q}{2R}+\frac{Q}{12R}\right]=\frac{Q}{16\pi\varepsilon_0R}$

27. **(c)** $V_B=k\left[\frac{-Q}{6R}+\frac{Q}{2R}+\frac{Q}{12R}\right]=\frac{5Q}{48\pi\varepsilon_0R}$

28. **(a)** Inside electric field is zero but not outside.

29. **(c)** Earth also has some surface charge density due to which it produces electric field in the surrounding space.

30. **(d)** Net potential at centre

$$+\frac{kq}{a/\sqrt{2}}-\frac{kq}{a/\sqrt{2}}+\frac{kq}{a/\sqrt{2}}-\frac{kq}{a/\sqrt{2}}=0$$

and field is zero due to symmetry.

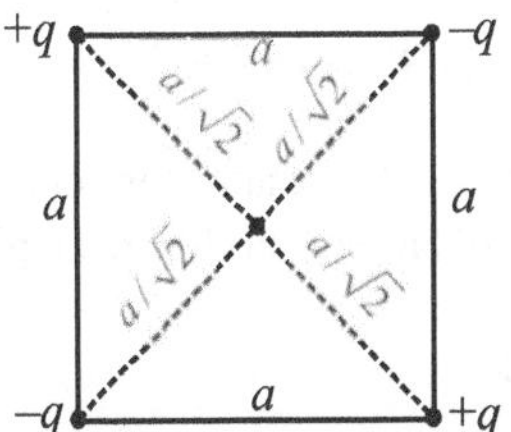

If electric potential at a point is zero then the magnitude of electric field at that point is not necessarily to be zero.

DAILY PRACTICE PROBLEMS

PHYSICS SOLUTIONS

35

1. (a) By using

$V = V_0 e^{-t/CR} \Rightarrow 40 = 50e^{-1/CR} \Rightarrow e^{-1/CR} = 4/5$

Potential difference after 2 sec

$$V' = V_0 e^{-2/CR} = 50(e^{-1/CR})^2 = 50\left(\frac{4}{5}\right)^2 = 32V$$

Fraction of energy after 1 sec

$$= \frac{\frac{1}{2}C(V_f)^2}{\frac{1}{2}C(V_i)^2} = \left(\frac{40}{50}\right)^2 = \frac{16}{25}$$

2. (c) The given circuit can be redrawn as follows. All capacitors are identical and each having capacitance

$C = \frac{\varepsilon_0 A}{d}$

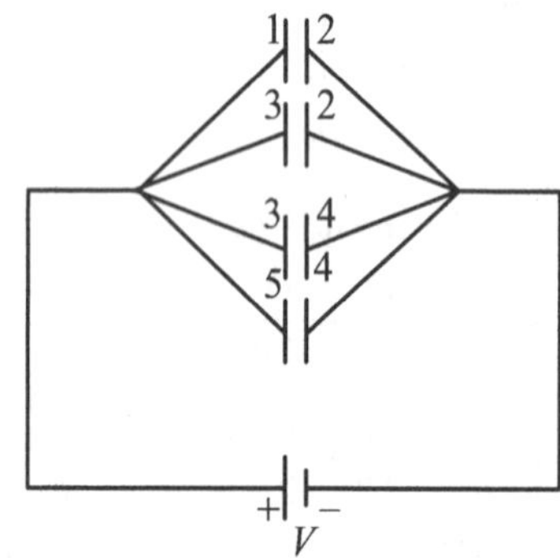

| Charge on each capacitor | = | Charge on each plate |

$= \frac{\varepsilon_0 A}{d} V$

Plate 1 is connected with positive terminal of battery so charge on it will be $+\frac{\varepsilon_0 A}{d}.V$

Plate 4 comes twice and it is connected with negative terminal of battery, so charge on plate 4 will be

$-\frac{2\varepsilon_0 A}{d} V$

3. (c) Initially potential difference across both the capacitor is same hence energy of the system is

$$U_1 = \frac{1}{2}CV^2 + \frac{1}{2}CV^2 = CV^2 \quad .....(i)$$

In the second case when key K is opened and dielectric medium is filled between the plates, capacitance of both the capacitors becomes $3C$, while potential difference across A is V and potential difference across B is $\frac{V}{3}$ hence energy of the system now is

$$U_2 = \frac{1}{2}(3C)V^2 + \frac{1}{2}(3C)\left(\frac{V}{3}\right)^2 = \frac{10}{6}CV^2 \quad ......(ii)$$

So, $\frac{U_1}{U_2} = \frac{3}{5}$

4. (c) Plane conducting surfaces facing each other must have equal and opposite charge densities. Here as the plate areas are equal, $Q_2 = -Q_3$.

The charge on a capacitor means the charge on the inner surface of the positive plate (here it is Q_2)

Potential difference between the plates

$$= \frac{\text{charge}}{\text{capacitance}} = \frac{Q_2}{C} = \frac{2Q_2}{2C}$$

$$= \frac{Q_2 - (-Q_2)}{2C} = \frac{Q_2 - Q_3}{2C}$$

5. (b) While drawing the dielectric plate outside, the capacitance decreases till the entire plate comes out and then becomes constant. So, V increases and then becomes constant.

6. (b) Given circuit can be reduced as follows

A —‖— 3C —‖— 3C — B

(C = capacitance of each capacitor)

The capacitor $3C, 3C$ shown in figure can with stand maximum 200 V.

∴ So maximum voltage that can be applied across A and B equally shared. Hence maximum voltage applied cross A and B be equally shared. Hence Maz. voltage applied across A and B will be $(200 + 200) = 400$ *volt.*

7. (c) Common potential

$$V' = \frac{C_1 V + C_2 \times 0}{C_1 + C_2} = \frac{C_1}{C_1 + C_2}.V$$

8. (b) $V = \frac{C_1V_1 - C_2V_2}{C_1 + C_2} = \frac{6 \times 12 - 3 \times 12}{3 + 6} = 4$ volt

9. (d) $C_1 = \frac{K_1 \varepsilon_0 \frac{A}{2}}{\left(\frac{d}{2}\right)} = \frac{K_1 \varepsilon_0 A}{d}$

$$C_2 = \frac{K_2\varepsilon_0 \frac{A}{2}}{\left(\frac{d}{2}\right)} = \frac{K_2\varepsilon_0 A}{d}$$

and $C_3 = \frac{K_3\varepsilon_0 A}{2d} = \frac{K_3\varepsilon_0 A}{2d}$

Now, $C_{eq} = C_3 + \frac{C_1C_2}{C_1+C_2} = \left(\frac{K_3}{2} + \frac{K_1K_2}{K_1+K_2}\right)\cdot\frac{\varepsilon_0 A}{d}$

10. (b) Given circuit is a balanced Wheatstone bridge.

11. (a)

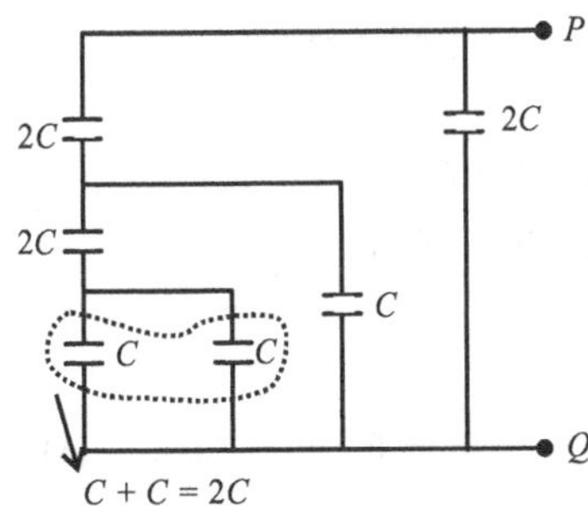

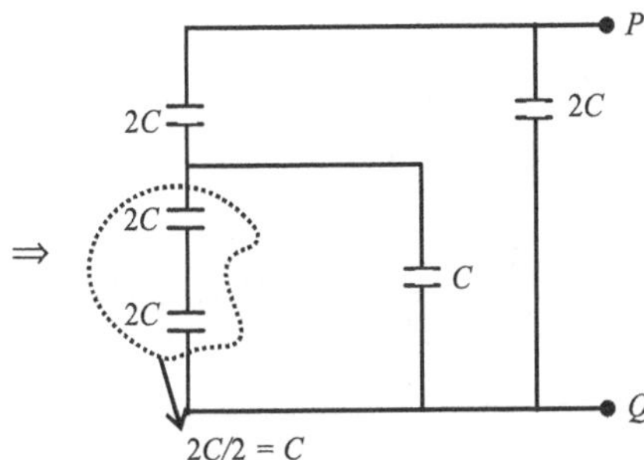

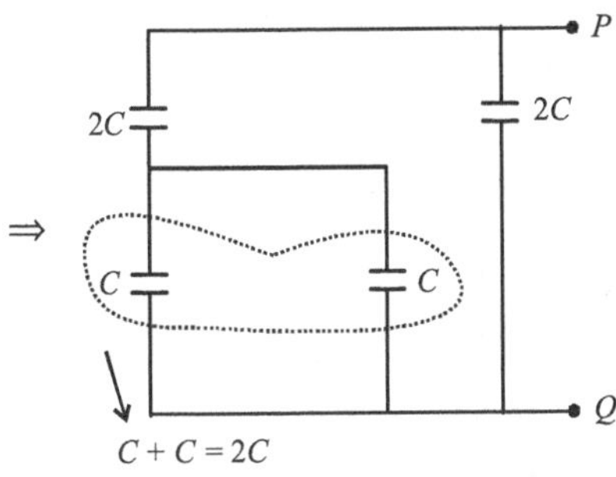

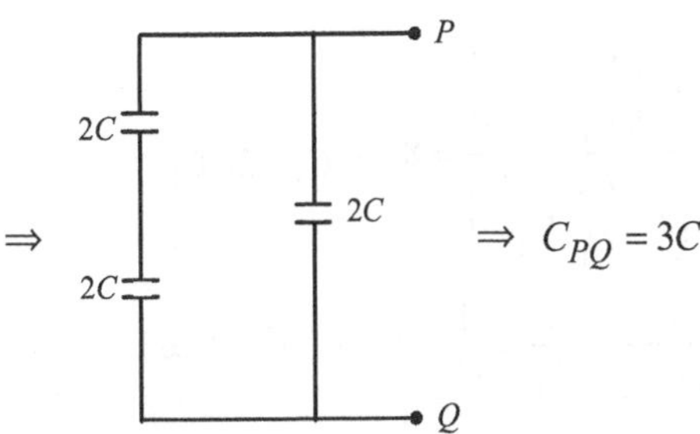

$\Rightarrow C_{PQ} = 3C$

12. (d) The two capacitors formed by the slabs may assumed to be in series combination.

13. (d) In the given system, no current will flow through the branch CD so it can be removed

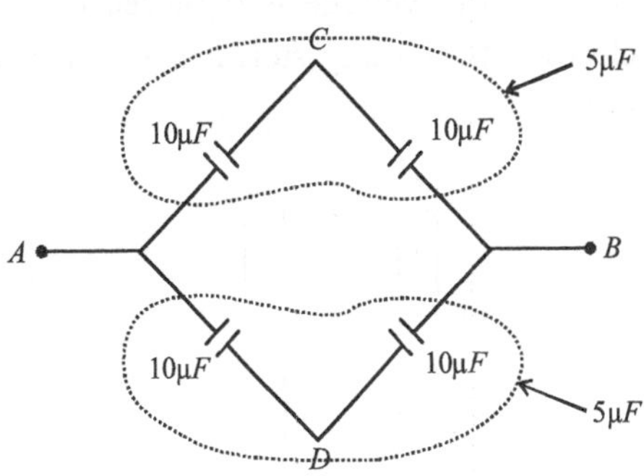

Effective capacitance of the system $= 5 + 5 = 10\mu F$

14. (c) Volume of 8 small drops = Volume of big drop

$$8 \times \frac{4}{3}\pi r^3 = \frac{4}{3}\pi R^3 \Rightarrow R = 2r$$

As capacity is proportional to r, hence capacity becomes 2 times.

15. (c) Potential difference between the plates

$V = V_{air} + V_{medium}$

$$= \frac{\sigma}{\varepsilon_0} \times (d-t) + \frac{\sigma}{K\varepsilon_0} \times t$$

$$\Rightarrow V = \frac{\sigma}{\varepsilon_0}\left(d - t + \frac{t}{K}\right)$$

$$= \frac{Q}{A\varepsilon_0}\left(d - t + \frac{t}{K}\right)$$

Hence capacitance

$$C = \frac{Q}{V} = \frac{Q}{\frac{Q}{A\varepsilon_0}\left(d - t + \frac{t}{K}\right)}$$

$$= \frac{\varepsilon_0 A}{\left(d - t + \frac{t}{K}\right)} = \frac{\varepsilon_0 A}{d - t\left(1 - \frac{1}{K}\right)}$$

16. (b) $C = \frac{\varepsilon_0 A}{d} = 1pF$ and $C' = \frac{K\varepsilon_0 A}{2d} = 2pF \therefore K = 4$

17. (a) Potential difference across the condenser

$$V = V_1 + V_2 = E_1t_1 + E_2t_2 = \frac{\sigma}{K_1\varepsilon_0}t_1 + \frac{\sigma}{K_2\varepsilon_0}t_2$$

$$\Rightarrow V = \frac{\sigma}{\varepsilon_0}\left(\frac{t_1}{K_1} + \frac{t_2}{K_2}\right) = \frac{Q}{A\varepsilon_0}\left(\frac{t_1}{K_1} + \frac{t_2}{K_2}\right)$$

18. (d) When the battery is disconnected, the charge will remain same in any case.

Capacitance of a parallel plate capacitor is given by

$$C = \frac{\varepsilon_0 A}{d}$$

When d is increased, capacitance will decreases and because the charge remains the same, so according to $q = CV$, the voltage will increase. Hence the electrostatics energy stored in the capacitor will increase.

19. (c)

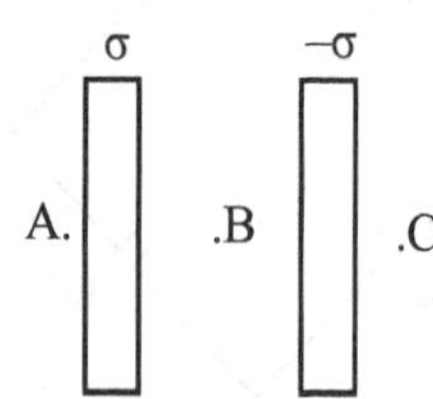

$E_A = \frac{\sigma}{2\varepsilon_0} - \frac{\sigma}{2\varepsilon_0} = 0$

$E_C = 0,$

$E_B = \frac{\sigma}{2\varepsilon_0} + \frac{\sigma}{2\varepsilon_0} = \frac{\sigma}{\varepsilon_0}.$

20. (b) $E = \frac{V}{d} = \frac{V_0/k}{d} = \frac{V_0}{kd}$

21. (c) $C_{eq} = \frac{(3+3)\times(1+1)}{(3+3)+(1+1)} + 1 = \left(\frac{6\times 2}{6+2}\right) + 1 = \frac{5}{2}\mu F$

$\therefore \quad Q = C \times V = \frac{5}{2} \times 100 = 250\ \mu C$

Change in 6 μF branch – $VC = \left(\frac{6\times 2}{6+2}\right)100 = 150\ \mu C$

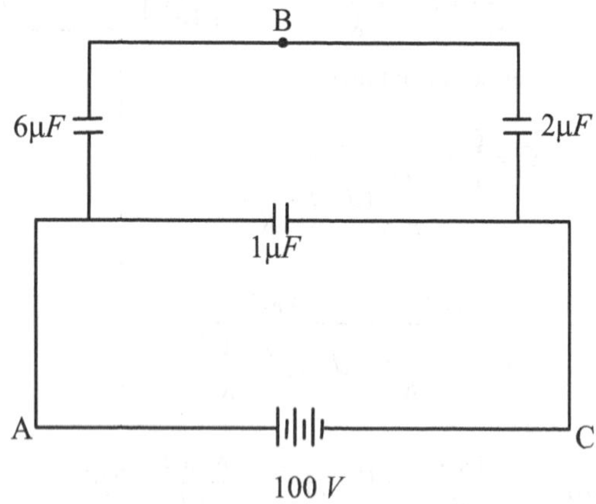

$V_{AB} = \frac{150}{6} = 25\ V$ and $V_{BC} = 100 - V_{AB} = 75\ V$

22. (c) Capacitance will be increased when a dielectric is introduced in the capacitor but potential difference will remain the same because battery is still connected. So according to q = CV, charge will increase i.e. $Q > Q_0$ and

$U = \frac{1}{2}QV_0, U_0 = \frac{1}{2}Q_0V_0 \Rightarrow Q > Q_0$ so $U > U_0$

23. (a) Electric field between the plates of a parallel plate capacitor $E = \frac{\sigma}{\varepsilon} = \frac{Q}{A\varepsilon_0}$ *i.e* $E \propto d^0$

24. (a) Capacitance of parallel plate condenser $= \frac{\varepsilon_0 A}{d}$

25. (a)

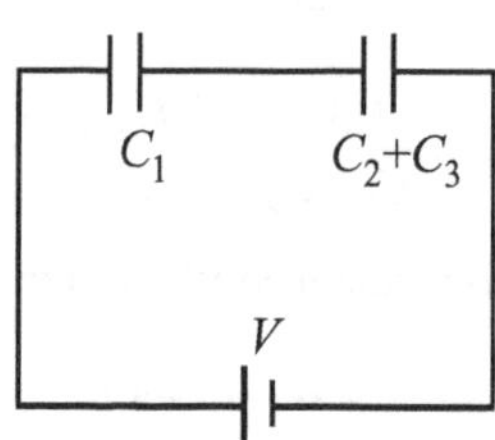

$V_{C_1} = \frac{V(C_2+C_3)}{C_1+(C_2+C_3)}$

Initially $C_3 = 0$

So $V_{C_1} = \frac{VC_2}{C_1+C_2} = 6$ .... (1)

Now, at $V_{C_1} = 10,\ C_3 = \infty$

$\Rightarrow 10 = \frac{V(C_2+C_3)}{C_1+(C_2+C_3)}$

$\Rightarrow 10 = \frac{V}{\left(\frac{C_1}{C_2+C_3}+1\right)} \Rightarrow 10 = V$ .... (2)

Eq. (1) and (2),

$\frac{10}{\left(\frac{C_1}{C_2}+1\right)} = 6 \Rightarrow 5 = 3\left(\frac{C_1}{C_2}+1\right)$

$\Rightarrow \frac{C_1}{C_2} = \frac{5}{3} - 1 = \frac{2}{3}$

26. (b) Now, $V_{C_1} = \frac{10(C_2+C_3)}{(C_1+C_2+C_3)} = 8$

$\Rightarrow \frac{10\left(\frac{C_2}{C_1}+\frac{C_3}{C_1}\right)}{\left(1+\frac{C_2}{C_1}+\frac{C_3}{C_1}\right)} = 8 \Rightarrow C_3 = 2.5C_1$

27. (c) $\frac{1}{C_3+C_2} + \frac{1}{C_1} \approx \frac{1}{C_1}$ (where $C_3 \to \infty$)

$\therefore$ Total energy = energy in C_1

$\therefore$ Required ratio = 1

28. **(b)** The electric field due to one charged plate at the location of the other is $E = \frac{\sigma}{2\varepsilon_0}$ and the force per unit area is $F = \sigma E = \frac{\sigma^2}{2\varepsilon_0}$.

29. **(d)** A charged capacitor, after removing the battery, does not discharge itself. If this capacitor is touched by someone, he may feel shock due to large charge still present on the capacitor. Hence it should be handled cautiously otherwise this may cause a severe shock.

30. **(b)** By the formula capacitance of a capacitor

$$C_1 = \varepsilon_0 \times \frac{KA}{d} \propto \frac{K}{d}$$

Hence, $\frac{C_1}{C_2} = \frac{K_1}{d_1} \times \frac{d_2}{K_2} = \frac{K_1}{K_2} \times \frac{d/2}{3K} = \frac{1}{6}$

or $C_2 = 6C_1$

Again for capacity of a capacitor $C = \frac{Q}{V}$

Therefore, capacity of a capacitor does not depend upon the nature of the material of the capacitor.

DAILY PRACTICE PROBLEMS

PHYSICS SOLUTIONS

36

1. **(b).** The current in 1Ω resistance is 3A. The current in 3Ω resistance is

$$I_1 = \frac{R_2}{R_1 + R_2} I = \frac{6}{3+6} \times 3 = 2A.$$

Therefore the ratio is $\frac{2}{3}$.

2. **(d).** $R = \frac{(R_1 + R_2)}{2} + \frac{1}{2}[(R_1 + R_2)^2 + 4R_3$
$(R_1 + R_2)]^{1/2}$ .....(a)
$R_1 = 1\Omega, R_2 = 0. R_3 = 2\Omega.$ ......(b)
From eqs. (a) and (b)

$$R = \frac{1}{2} + \frac{1}{2}[1 + 4 \times 2 \times 1]^{1/2} = \frac{1}{2}[1 + 3] = 2\Omega.$$

3. **(b).** Since Q is connected in parallel the net resistance becomes R/2, so the current I = 2V/R, double the value.

4. **(b).** Since there is no current in edcb part, the p.d. across be should be 2V. Let current in 500 Ω is I, then same current flows through X (think). Therefore, for loop abefa,
$12 = I(500) + IX$
or $12 = I(500) + 2$ ($\because$ IX = 2 volt)
Thus I = (1/50) A or from IX = 2,
$X = 2 \times 50 = 100\ \Omega.$

5. **(b).** Let ℓ_1 be the initial length of the wire. Then the new length will be

$$\ell_2 = \frac{110}{100}\ell_1 = \frac{11}{10}\ell_1$$

Since, the volume remains constant

$$A_1\ell_1 = A_2\ell_2 \text{ or } A_1/A_2 = \ell_2/\ell_1 = \frac{11}{10}$$

(where A_1 and A_2 are initial and final area of cross–section of the wire).
If R_1 and R_2 are the initial and final resistances, then

$$\frac{R_1}{R_2} = \frac{\ell_1 A_2}{\ell_2 A_1} = \frac{10}{11} \times \frac{10}{11} = \left(\frac{10}{11}\right)^2$$

$$\text{or } \frac{R_2}{R_1} = \left(\frac{11}{10}\right)^2$$

Now, percentage change in resistance is

$$\frac{\Delta R}{R_1} \times 100 = \left(\frac{R_2 - R_1}{R_1}\right) \times 100$$

$$= \left[\left(\frac{11}{10}\right)^2 - 1\right] \times 100 = 21\%$$

6. **(a).** The circuit is equivalent to Fig. It is a balanced wheatstone bridge between abcd, and then in parallel (2R) resistances. Thus ignoring resistance between bd arm. The circuit is equivalent to three (2R) resistances in parallel (abc, adc, aRRc).

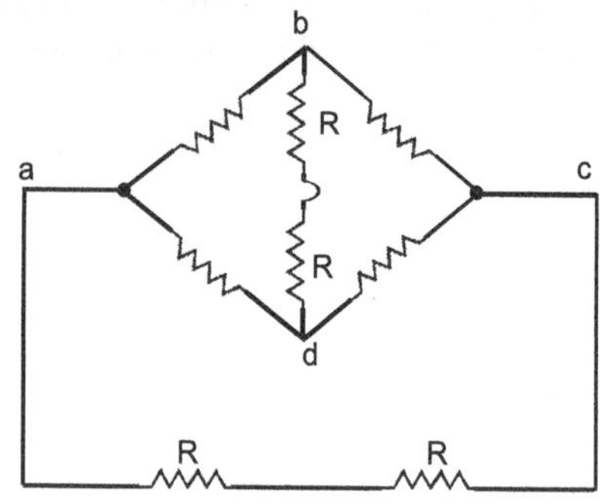

i.e. $\frac{1}{R_{eq}} = \frac{1}{2R} + \frac{1}{2R} + \frac{1}{2R} = \frac{3}{2R}$

$\Rightarrow R_{eq} = \frac{2}{3}R$

7. **(c).**

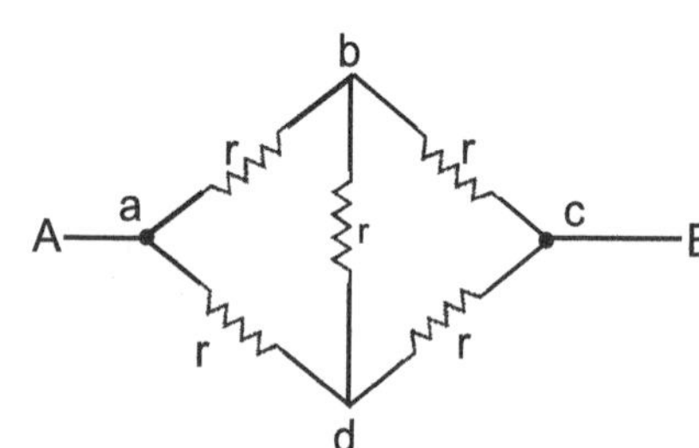

Imagine, A being pulled on the left side, then abcd becomes a balanced wheatstone bridge Fig. The arm bd can be ignored. Then resistance between A, B becomes = r.

i.e. $\frac{1}{R_{eq}} = \frac{1}{2r} + \frac{1}{2r} = \frac{1}{r} \Rightarrow R_{eq} = r$

8. **(d)** $R = 91 \times 10^2 \approx 9.1\ k\Omega.$

9. **(d)** Spacific resistance doesn't depend upon length and area.

10. **(b).** The diagram can be redrawn as shown in fig.

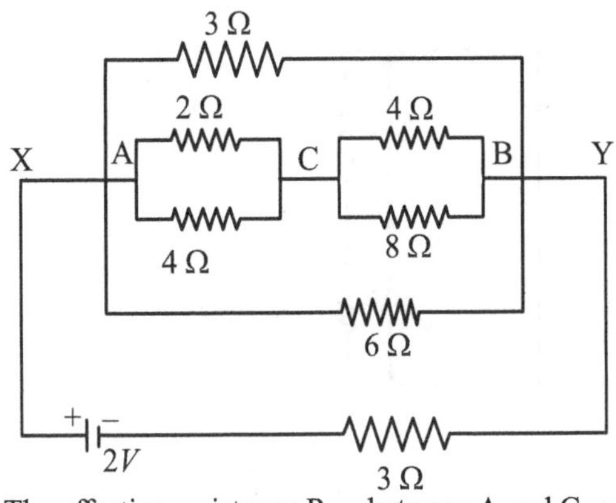

The effective resistance R_{AC} between A and C

$$\frac{1}{R_{AC}}=\frac{1}{2}+\frac{1}{4}=\frac{3}{4}\quad \therefore R_{AC}=\frac{3}{4}\text{ ohm}$$

The effective resistance R_{CB} between C and B

$$R_{CB}=\frac{1}{4}+\frac{1}{8}=\frac{3}{8}\quad \therefore R_{CB}=\frac{8}{3}\text{ ohm.}$$

Now, $R_{ACB}=R_{AC}+R_{CB}=\frac{4}{3}+\frac{8}{3}=4\text{ohm}.$

Corresponding to points X and Y, the resistances 3 ohm, 4 ohm and 6 ohm are in parallel, hence effective resistance R_{XY} is

$$\frac{1}{R_{XY}}=\frac{1}{3}+\frac{1}{4}+\frac{1}{6}=\frac{4+3+2}{12}=\frac{9}{12}$$

$$\therefore R_{XY}=\frac{12}{9}=\frac{4}{3}\text{ ohm.}$$

Total resistance R of the circuit $=\frac{4}{3}+\frac{2}{3}=2\Omega.$

Current in the circuit $=\frac{2}{2}=1\text{ A}$

Power dissipated in the circuit $=i^2R=1\times 2=2$ watts

Potential difference between X and

$$Y=i\times R_{XY}=1\times\frac{4}{3}=\frac{4}{3}\text{V.}$$

$\therefore$ Potential difference across 3 ohm resistor $=\frac{4}{3}$ V.

Current in 3 ohm resistor $=\frac{4/3}{3}=\frac{4}{9}=0.44$ amp.

11. **(c).** Requivalent $=\frac{(30+30)30}{(30+30)+30}=\frac{60\times 30}{90}=20\,\Omega$

$$\therefore\quad i=\frac{V}{R}=\frac{2}{20}=\frac{1}{10}\textit{ ampere}$$

12. **(c).** $R=2+2+\frac{2\times R}{2+R}\Rightarrow 2R+R^2=8+4R+2R$

$$\Rightarrow\quad R^2-4R-8=0\Rightarrow R=\frac{4\pm\sqrt{16+32}}{2}=2\pm 2\sqrt{3}$$

R cannot be negative, hence R = $2\pm 2\sqrt{3}=5.46\,\Omega$

13. **(c).** $P=\frac{V^2}{R}$. If resistance of heater coil is R, then resistance of parallel combination of two halves will be $\frac{R}{4}$

So $\quad\frac{P_1}{P_2}=\frac{P_2}{P_1}=\frac{R/4}{R}=\frac{1}{4}$

14. **(c).** Total kWh consumed $=\frac{60\times 8\times 301}{1000}=14.4$

Hence cost = 14.4 × 1.25 = ₹18

15. **(d).** Since all bulbs are identical they have the same resistances. The current I flowing through 1 branches at A. So current in 2 and 3, as well as in 4 will be less than I. The current through 5 is also I. Thus 1 and 5 glow equally brightly and more than 2, 3 or 4.

16. **(b).** Let R_1 and R_2 be the resistances of the coils, V the supply voltage, Q the heat required to boil the water.

Heat produced by first coil of resistance R_1 in time t_1

$$(=6\text{ min})=Q=\frac{V^2t_1}{JR_1}=\frac{V^2\times 6\times 60}{4.2R_1}\text{ cal}\quad\text{......(a)}$$

Heat produced in second coil of resistance R_2 in time t_2 (= 8 min)

$$=Q=\frac{V^2t_1}{JR_2}=\frac{V^2\times 6\times 60}{4.2R_2}\quad\text{.....(b)}$$

Equating (a) and (b), we get

$$\frac{6}{R_2}=\frac{8}{R_2}\text{ i.e. }\frac{R_2}{R_1}=\frac{8}{6}=\frac{4}{3}$$

$$\text{or } R_2=\frac{4}{3}R_1\quad\text{.....(c)}$$

(i) When the two heating coils are in series, the effective resistance is

$$R'=R_1+R_2=R_1+\frac{4}{3}.R_1=\frac{7}{3}R_1.$$

with two coils in series, let the kettle take t' time to boil.

$$Q=\frac{V^2t'}{JR'}=\frac{V^2t'}{4.2\times\left(\frac{7}{3}R_1\right)}\quad\text{.....(d)}$$

Comparing (a) and (d), we get $\dfrac{t'}{(7/3)} = 6 \times 60$

or $t' = \dfrac{7}{3} \times 6 \times 60 \text{ sec} = 14 \text{ min}.$

(ii) When the two heating coils are in parallel, the effective resistance is,

$$R'' = \frac{R_1 R_2}{R_1 + R_2} = \frac{R_1\left(\frac{4}{3}R_1\right)}{R_1 + \left(\frac{4}{3}R_1\right)} = \frac{4}{7}R_1$$

In parallel arrangement of heating coils, let t" be the time taken by kettle to boil, so

$$Q = \frac{V^2 t''}{JR''} = \frac{V^2 t''}{4.2 \times \left(\frac{4}{7}R_1\right)} \qquad .....(5)$$

Comparing (a) and (5), we get

$$\frac{t''}{(4/7)} = 6 \times 60 \text{ or } t'' = \frac{4}{7} \times 6 \times 60 \text{ sec} = 3.43 \text{ min.}$$

17. (c).

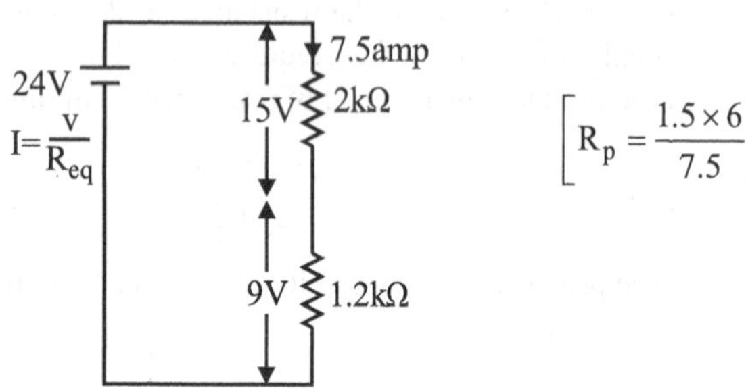

$$\left[R_p = \frac{1.5 \times 6}{7.5}\right]$$

$$I = \frac{240}{32} \Rightarrow \frac{60}{8} = 7.5\text{mA}$$

(1) Currrent I is 7.5mA

(2) Voltage drop across R_L is 9 volt

(3) $\dfrac{P_1}{P_2} = \dfrac{v_1^2}{R_1} \dfrac{R_2}{v_2^2} \Rightarrow \dfrac{225 \times 6}{2 \times 81} = 16.66$

(4) After intercharging the two resistor R_1 and R_2

$$I = \frac{V}{R_{eq}} = \frac{24}{(48)} \times 7 = 3.5\text{mA}$$

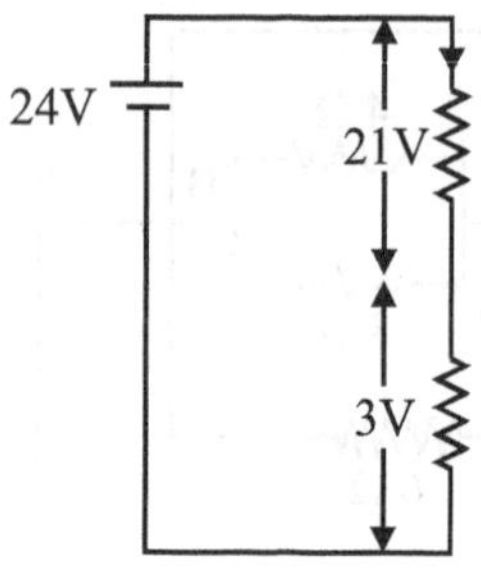

$$\frac{P_1}{P_2} = \frac{v_1^2}{R_L} \frac{R_L}{v_2^2} \Rightarrow \left(\frac{v_1}{v_2}\right)^2 = \left(\frac{9}{3}\right)^2 = 9$$

Sol. (18-20).

$$I = \frac{12V}{(1+5)\Omega} = 2A$$

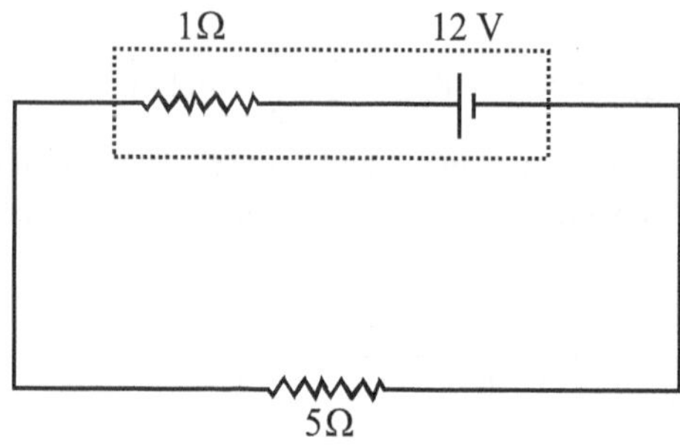

⇒ Rate of chemical energy conversion $= EI = 12 \times 2 = 24$ W
and P (in battery) $= I^2 r = 4$ W
Also, P (in resistor) $= I^2 r = 20$ W

18. (a) **19. (c)** **20. (a)**

21. (d) Resistivity of a semiconductor decreases with the temperature. The atoms of a semiconductor vibrate with larger amplitudes at higher temperatures thereby increasing it's conductivity not resistivity.

22. (d) It is quite clear that in a battery circuit, the point of lowest potential is the negative terminal of the battery and the current flows from higher potential to lower potential.

23. (b) The temperature co-efficient of resistance for metal is positive and that for semiconductor is negative.
In metals free electrons (negative charge) are charge carriers while in p-type semiconductors, holes (positive charge) are majority charge carriers.

DAILY PRACTICE PROBLEMS — PHYSICS SOLUTIONS 37

(1) (d). Let R is resistance of the voltmeter. The effective resistance across points A, B is

$$r = \frac{60 \times R}{60 + R} \quad ....(1)$$

The current in the circuit is $I = 12/(50 + r)$
The p.d. across AB points is $V = Ir$

$$\text{or } 6 = \frac{12}{50+r} \times r \quad \text{or} \quad 50 + r = 2r$$

$$\text{or } r = 50\,\Omega \quad ....(2)$$

using it in (1),

$$\text{we get } 50 = \frac{60}{60+R}$$

$$300 + 5R = 6R$$

or $R = 300\,\Omega$

(2) (c). $S = \frac{G}{n-1} = \frac{G}{100-1} = \frac{G}{99}\Omega$

(3) (b). $S = \frac{i_g}{i - i_g} G = \frac{10}{100-10} \times 99 = 11\Omega$

(4) (c). $R = 20 + \frac{80 \times 80}{80 + 80} = 60\Omega$

$$i = \frac{V}{R} = \frac{2}{60} = \frac{1}{30} \text{ amp.}$$

$$\therefore V = iR' = \frac{1}{30} \times 40 = 1.33 \text{ volt.}$$

(5) (a). According to Kirchhoff's first law
At junection A, $i_{AB} = 2 + 2 = 4A$
At junection B, $i_{AB} = i_{BC} - 1 = 3A$

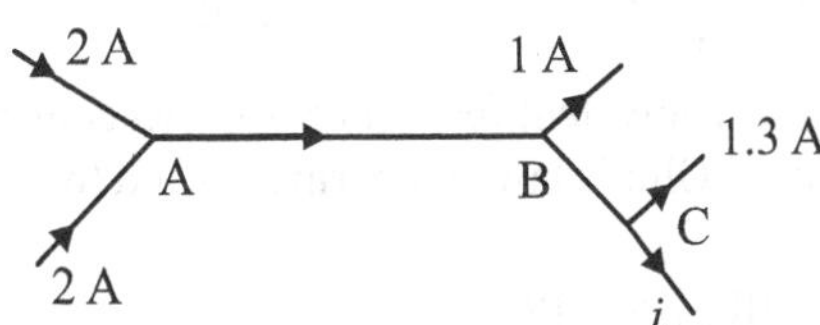

At junection C, $i = i_{BC} - 1.3 = 3 - 1.3 = 17$ amp

(6) (b). The current required for a full-scale deflection of the galvanometer is
$i = 4.0 \times 10^{-4} \times 25 = 10^{-2}$ A
Let a resistance R Ω is to be connected in series

Then by the ohm's law, we have $i = \frac{V}{G+R}$

Here $G = 50\,\Omega$, $V = 2.5$ V and $i = 10^{-2}$ A

$$\therefore G + R = \frac{V}{i} = \frac{2.5}{10^{-2}} = 250$$

$$\Rightarrow R = 250 - G = 250 - 50 = 200\Omega.$$

(7) (a). $\therefore i = \frac{V}{R} = \frac{25}{1000}$ A

Let R' be the required resistance to be connected in series with voltmeter.

$$\text{So } i = \frac{V'}{R + R'}$$

Here $V' = 250$, $R = 1000\,\Omega$ and $i = \frac{25}{1000}$ A

$$\therefore \frac{25}{1000} = \frac{250}{1000 + R'} \Rightarrow R' = 9000\,\Omega.$$

(8) (d). The potential difference between A and B in the absence of voltmeter = 2 volt.

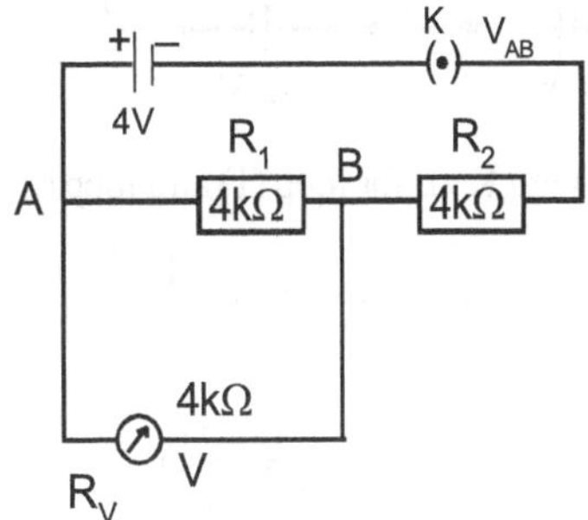

Current flowing in the circuit

$$I = \frac{E}{R_2 + \frac{R_1 R_V}{R_1 + R_V}} = \frac{E}{R_2 + R'}$$

$$I = \frac{4}{4 + \frac{4 \times 4}{4+4}} = \frac{2}{3} \text{ ampere}$$

Potential difference measured by voltmeter

$$V'_{AB} = IR' = \frac{2}{3} \times 2 = \frac{4}{3}$$

Error in the reading of voltmeter

$$= V_{AB} - V'_{AB} = 2 - \frac{4}{3} = \frac{2}{3} \text{ volt}$$

The error in voltmeter reading for 2 volt p.d. = $\frac{2}{3}$ volt

The error in voltmeter reading for 1 volt p.d.

$$= \frac{2}{3} \times \frac{1}{2} = \frac{1}{3} \text{ volt}$$

the error in voltmeter reading for 100 volt p.d.

$$= \frac{100}{3} = 33.3\% \text{ volt}$$

(9) (d). $E = V + Ir = IR + Ir$

$\Rightarrow E = 0.25 \times 10 + 0.25 \times r$

In second stage

$\Rightarrow E = 0.5 \times 4 + 0.5r$

Subtracting eq. (b) from eq. (a)

$2.5 + 0.25r - 2.0 - 0.5r = 0$

$0.5 = 0.25r$

$$r = \frac{0.5}{0.25} = 2\Omega.$$

(10) (d) Suppose current through different paths of the circuit is as follows.

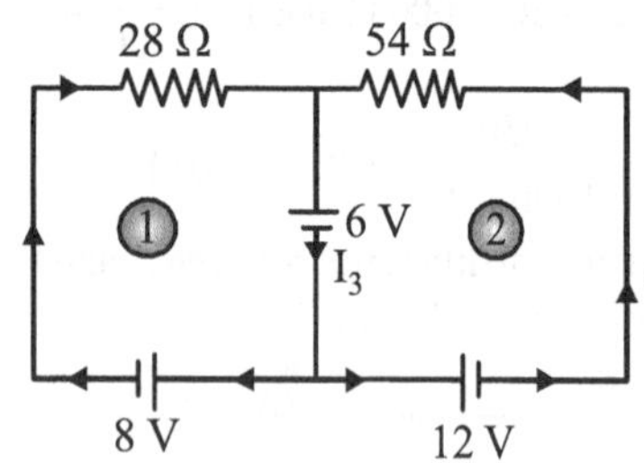

After applying KVL for loop (1) and loop (2)

We get $28i_1 = -6 - 8 \Rightarrow \quad i_1 = -\frac{1}{2}A$

and $\quad 54i_2 = -6 - 12 \Rightarrow i_2 = -\frac{1}{3}A$

Hence $i_3 = i_1 + i_2 = -\frac{5}{6}A$

(11) (d) $V_{AB} = 4 = \dfrac{5X + 2 \times 10}{X + 10} \Rightarrow = 20\,\Omega$

12. (b) The circuit can be simplified as follows

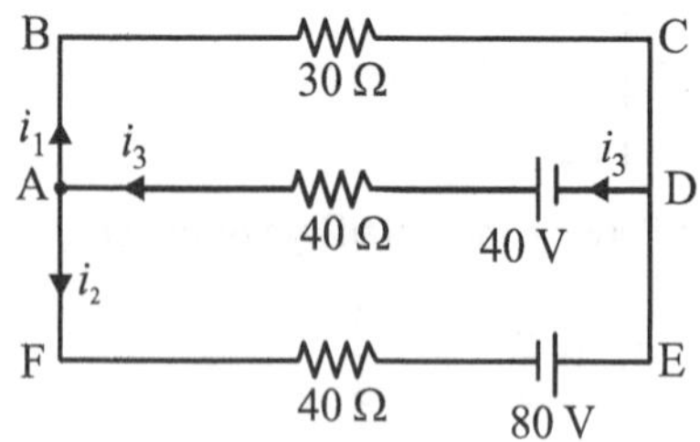

Applying KCL at junction A

$i_3 = i_1 + i_2$ ...(*i*)

Applying Kirchoff's voltage law for the loop ABCDA

$-30i_1 - 40i_3 + 40 = 0$

$\Rightarrow \quad -30i_1 - 40(i_1 + i_2) + 40 = 0$

$\Rightarrow \quad 7i + 4i_2 = 0$ ...(*ii*)

Applying Kirchoff's voltage law for the loop ADEFA.

$-40i_2 - 40i_3 + 80 + 4 = 0$

$\Rightarrow \quad -40i_2 - 40(i_1 + i_2) = -120$

$\Rightarrow \quad i_2 + 2i_2 = 3$ ...(*iii*)

On solving equation (*ii*) and (*iii*) $i_1 = -0.4$ A.

13. (b) Cells area joined in parallel when internal resistance is higher then a external resistance. ($R << r$)

$$i = \frac{E}{R + \frac{r}{n}}$$

14. (b). Current in the ammeter $I = \dfrac{E}{R' + r\left[1 + \dfrac{R'}{R}\right]}$

On increasing the value of R, the denominator will decrease and consequently the value of I will increase.

15. (a)

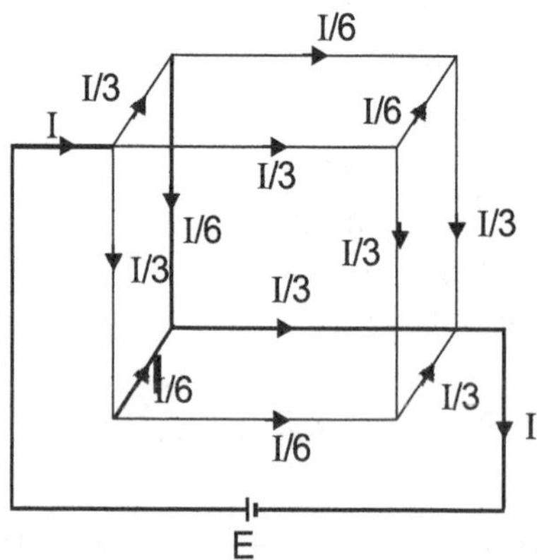

Let ABCDEFGH be skeleton cube formed of twelve equal wires each of resistance R. Let a battery of e.m.f. E be connected across A and G. Let the total current entering at the corner A and leaving the diagonally opposite corner G be I. By symmetry the distribution of currents in wires of cube, according to Kirchoff's I[st] law is shown in fig. ApplyingKirchoff's IInd law to mesh ADCGEA, we get

$$-\frac{1}{3}R - \frac{1}{6}R - \frac{1}{3}R + E = 0$$

or $E = \frac{5}{6}IR$ ......(a)

If R_{AB} is equivalent resistance between comers A and B, then from Ohm's law comparing (a) and (b), we get

$$IR_{AB} = \frac{5}{6}IR$$

16. (d).

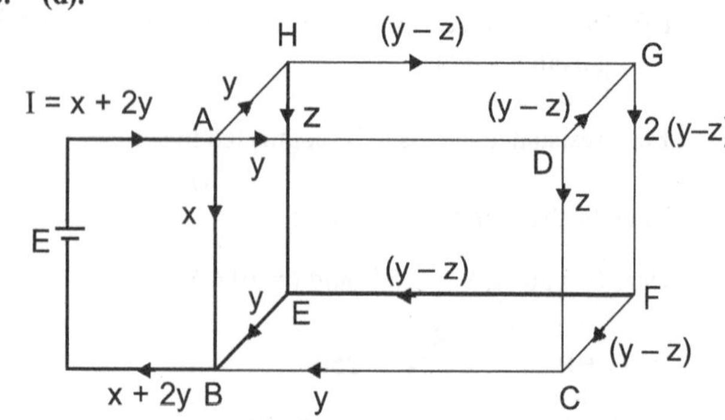

Let I = x + 2y current enter at point A, when a battery of e.m.f. E and no internal resistance is connected across edge AB. The edges AD and AH are symmetrically connected to A, therefore they will carry equal currents. The distribution of currents according to Kirchoff's I[st] law is shown in fig.

If R_{AB} is equivalent resistance, then from Ohm's law,

$E = R_{AB} I = R_{AB}(x+2y)$ ....(a)

and from Kirchoff's law applied to mesh containing AB and cell E is

$Rx = E$ .....(b)

(since R is resistance of each wire)

Applying Kirchoff's II law to mesh AHEB

$yR + zR + yR - xR = 0$ or $x - 2y - z = 0$ .....(c)

Applying Kirchoff's II law to mesh DGFC

$(y-z)R + 2(y-z)R - zR = 0$

or $4(y-z) - z = 0$ or $4y = 5z$ .....(d)

i.e. $z = (4/5)y$

.....(E)

Substituting this value in (c), we get

$$x - 2y - \frac{4}{5}y = 0$$

or $\frac{14}{5}y = x$ i.e. $y = \frac{5}{14}x$

Substituting value of y in (a), we get

$$E = R_{AB}\left(x + \frac{10}{14}x\right)$$

$$E = R_{AB}\frac{24}{14}x = R.x$$

$$R_{AB} = \frac{24}{14}R \therefore R_{AB} = \frac{7}{12}R.$$

17. (a).

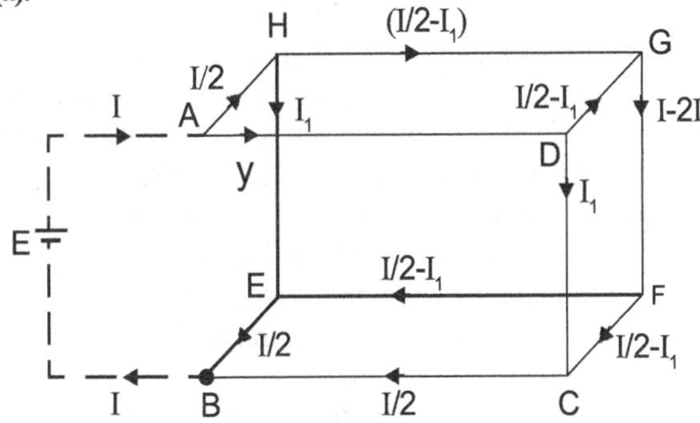

Let a battery of e.m.f. E is applied between points A and B. Let a current I, enter through point A.

If R_{AB} is equivalent resistance between points A and B, then from Ohm's law

$R_{AB} I = E$

The distribution of currents, keeping in mind symmetry condition, is shown in fig.

Let R (= 2Ω) be the resistance of each wire.

Applying Kirchoff's II law to mesh DGFC, we get

$$\left(\frac{1}{2} - I_1\right)R + (I - 2I_1)$$

$$R + \left(\frac{1}{2} - I_1\right)R - I_1 R = 0$$

$$\text{or } 2\left(\frac{1}{2} - I_1\right) + (I - 2I_1) - I_1 = 0$$

or $2I - 5I_1 = 0$ or $I_1 = \frac{2}{5}I$ ...... (b)

Applying Kirchoff's IInd law to external circuit AHEBE', we get

$$\frac{1}{2}R + I_1 R + \frac{1}{2}R = E$$

$$IR + \frac{2}{5}IR = E' \quad \text{[Using (b)]}$$

or $\frac{7}{5}IR = E$ ......(c)

Comparing (a) and (c), wet get

$$R_{AB} I = \frac{7}{5}IR \text{ i.e. } R_{AB} = \frac{7}{5}R = \frac{7}{5} \times 2 = 2.8\Omega$$

18. (c). In the first case $I = E/(r+R)$ and in the second case $I' = E/(r + R/2) = 2E/(2r + R)$

Using $E = I(r+R)$, we get

$$I' = I\left(\frac{2r+2R}{2r+R}\right) = I\left(1 + \frac{R}{2r+R}\right)$$

Thus the term in bracket is greater than 1 but less than 2. Thus $2I > I' > I$

19. (b). Let R be the combined resistance of galvanometer and an unknown resistance and r the internal resistance of each battery. When the batteries, each of e.m.f. E are connected in series, the net e.m.f. = 2E and net internal resistance = 2r

$\therefore$ Current $i_1 = \frac{2E}{R+2r}$ or $1.0 = \frac{2 \times 15}{R+2r}$

$\therefore R + 2r = 3.0.$ ...(i)

When the batteries are connected in parallel, the e.m.f. remains E and net internal resistance becomes r/2. therefore

$$\text{Current } i_2 = \frac{E}{R + \frac{r}{2}} = \frac{2E}{2R + r}$$

$\therefore 2R + r = \frac{2E}{i_2} = \frac{2 \times 15}{0.6} = 5.0$ ...(i)

Solving (i) and (ii), we get $r = 1/3\ \Omega$.

20. **(a).** The circuit with current distribution is shown in fig.

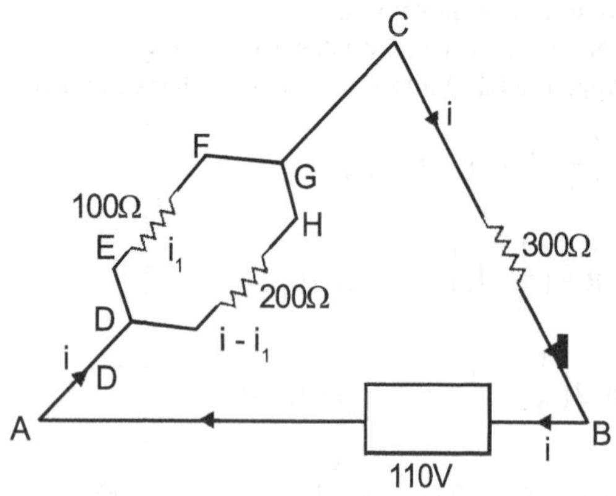

Applying Kirchoff's second law to the loop DEFGHID, we have $i_1 \times 100 - (i - i_1) \times 200 = 0$

$300\,i_1 - 200\,i = 0$ ......(1)

Now applying Kirchoff's second law to loop ADIHGCBA, we have. $(i - i_1)\,200 + i \times 300 = 110$

$500i - 200\,i_1 = 110$ .....(2)

Solving eqs. (1) and (2), we get

$i = \frac{3}{10}$ amp and $i_1 = \frac{1}{5}$ amp.

Current in 100 ohm resistance $i_1 = \frac{1}{5}$ amp.

Current in 200 ohm resistance $i - i_1 = \frac{1}{10}$

Current in 300 ohm resistance $i = \frac{3}{10}$ amp.

Potential difference between A and C

= Potential difference across 100 ohm resistance

or potential difference across 200 ohm resistance

$\therefore V_A - V_C =$ current × resistance

$= i_1 \times 100 = \frac{1}{5} \times 100 = 20$ volt.

Potential difference between C and B is given by

$V_C - V_B = i \times 300 = \frac{3}{10} \times 300 = 90$ volt.

21. **(a).** After full charging, the steady current in the condenser is zero, hence no current will flow in 4Ω resistance.

$$I = \frac{E}{R + R'} = \frac{6}{28 + \left(\frac{2 \times 3}{2 + 3}\right)} = \frac{6}{28 + 12} = 1.5\,A$$

Let current flowing in 2Ω resistance is I_1

$\therefore$ 2Ω and 3Ω resistance are connected in parallel

$\therefore 2I_1 = (1.5 - I_1) \times 3$

$5I_1 = 4.5$

$I_1 = 0.9$ amp.

24. **(d).**

I_g (G) R $I - I_g$

For Ammeter $I_g G = (I - I_g) R$

$50 \times 10^{-6} \times 100 = 5 \times 10^{-3} \times (R) \Rightarrow R \approx 1\Omega$

For voltmeter $I_g (R + G) = V$

$50\ \mu A\ (R + G) = 10V \Rightarrow R + G = 200\ k\Omega \Rightarrow R \approx 200 k\Omega$

25. **(a)** Potential at $A = 6V$

$V_A - V_C = 4$

$\Rightarrow V_C = 2V$

26. **(d)** $\frac{V_{AD}}{V_{AB}} = \frac{V_{AC}}{V_{AB}} = \frac{AD}{AB} = \frac{4}{6} = \frac{2}{3}$;

$AD = \frac{200}{3}$ cm.

27. **(a)** D is balance point, hence no current

28. **(a)** Voltameter measures current indirectly in terms of mass of ions deposited and electrochemical equivalent of the substance $\left(I = \frac{m}{Zt}\right)$. Since value of m and Z are measured to 3rd decimal place and 5th decimal place respectively. The relative error in the emasurement of current by voltmeter will be very small as compared to that when measured by ammeter directly.

29. **(c)** The e.m.f. of a dry cell is dependent upon the electrode potential of cathode and anode which in turn is dependent upon the reaction involved as well as concentration of the electrolyte. It has nothing to do with size of the cell.

So, statement-1 is false and statement-2 is true.

30. **(d)** $V = E - ir = 4 - 2 \times 2 = 0$, During charging $V > E$.

DAILY PRACTICE PROBLEMS

PHYSICS SOLUTIONS

38

1. (a) Potential gradient $x = \frac{e}{(R+R_h+r)}.\frac{R}{L}$

$\Rightarrow \frac{0.2\times10^{-3}}{10^{-2}} = \frac{2}{(R+490+0)}\times\frac{R}{1} \Rightarrow R = 4.9\Omega.$

2. (c) Let S be larger and R be smaller resistance connected in two gaps of meter bridge.

$\therefore S = \left(\frac{100-l}{l}\right)R = \frac{100-20}{20}R = 4R$ ......(i)

When 15Ω resistance is added to resistance R, then

$S = \left(\frac{100-40}{40}\right)(R+15) = \frac{6}{4}(R+15)$ ....(ii)

From equations (i) and (ii) $R = 9\Omega$

3. (b) $r = R\left(\frac{l_1}{l_2}-1\right) = 2\left(\frac{240}{120}-1\right) = 2\Omega$

4. (a) Potential difference per unit length

$= \frac{V}{L} = \frac{2}{4} = 0.5V/m$

5. (d) $E = \frac{e}{(R+R_h+r)}.\frac{R}{L}\times l \Rightarrow 0.4 = \frac{5}{(5+45+0)}\times\frac{5}{10}\times l$

$\Rightarrow l = 8m$

6. (d) Current through $2\Omega = 1.4\left\{\frac{(25+5)}{(10+2)+(25+5)}\right\} = 1A$

7. (c) Post office box is based on the principle of Wheatstone's bridge.

8. (b) Using $r = R\left(\frac{l_1}{l_2}-1\right) = 2\left(\frac{150}{100}-1\right) = 1\Omega$

9. (a) Since the given bridge is balanced, hence there will be no current through 9Ω resistance. This resistance has no effect and must be ignored in the calculations.

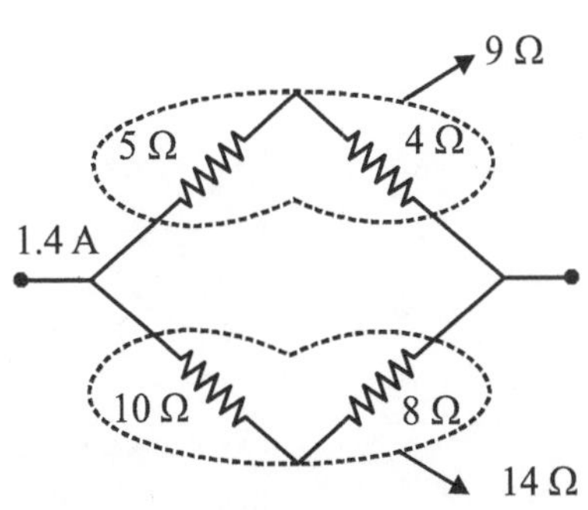

$R_{AB} = \frac{9\times18}{27} = 6\Omega$

10. (c) Potential gradient $(x) = \frac{i\rho}{A} = \frac{0.1\times10^{-7}}{10^{-6}} = 10^{-2}V/m$

11. (d) $S = \left(\frac{100-l}{l}\right).R$

Initially, $30 = \left(\frac{100-l}{l}\right)\times10 \Rightarrow l = 25cm$

Finally, $10 = \left(\frac{100-l}{l}\right)\times30 \Rightarrow l = 75cm$

So, shift $= 50cm$.

12. (d) $\frac{E_1}{E_2} = \frac{l_1+l_2}{l_1-l_2} = \frac{(6+2)}{(6-2)} = \frac{2}{1}$

13. (c) $\frac{E_1}{E_2} = \frac{l_1+l_2}{l_1-l_2} = \frac{58+29}{58-29} = \frac{3}{1}$

14. (d)

$E = \frac{e}{(R+R_h+r)}\frac{R}{L}\times l = \frac{2}{(10+40+0)}\times\frac{10}{1}\times0.4 = 0.16V$

15. (c)

X, 6Ω, G, 4Ω, 6Ω, A, C, B, 5V

Resistance of the part AC

$R_{AC} = 0.1\times40 = 4\Omega$ and $R_{CB} = 0.1\times60 = 6\Omega$

In balanced condition $\frac{X}{6} = \frac{4}{6} \Rightarrow X = 4\Omega$

Equivalent resistance $R_{eq} = 5\Omega$

so current drawn from battery $i = \frac{5}{5} = 1A$.

16. (a) $r = \left(\frac{l_1-l_2}{l_2}\right)\times R' \Rightarrow r = \left(\frac{55-50}{50}\right)\times10 = 1\Omega$

17. **(c)** Potential gradient

$$x = \frac{V}{L} = \frac{e}{(R+R_h+r)}\frac{R}{L}$$

$$\Rightarrow 2.2\times10^{-3} = \frac{2.2}{(10+R_h)}\times1 \Rightarrow R' = 990\Omega$$

18. **(a)** $E = xl = i\rho l \Rightarrow i = \frac{E}{\rho l} = \frac{2.4\times10^{-3}}{1.2\times5} = 4\times10^{-4}A$

19. **(b)** Give circuit is a balanced Wheatstone bridge circuit, hence it can be redrawn as follows

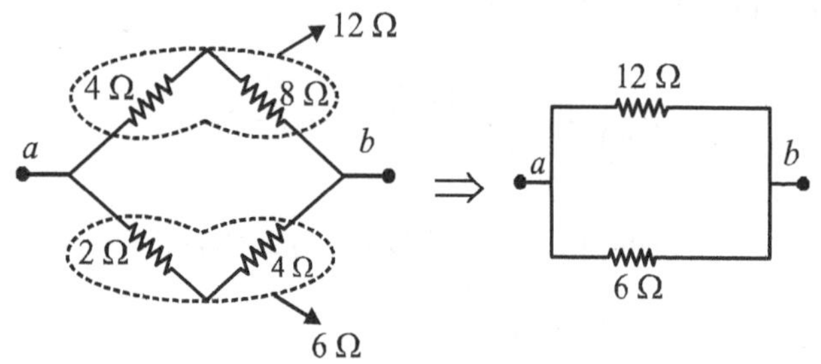

$$R^{AB} = \frac{12\times6}{(12+6)} = 4\Omega.$$

20. **(a)** Balancing length is independent of the cross sectional area of the wire.

21. **(a)** In meter bridge experiment, it is assumed that the resistance of the L shaped plate is negligible, but actually it is not so. The error created due to this is called, end error. To remove this the resistance box and the unknown resisance must be interchanged and then the mean reading must be taken.

22. **(a)** Ammeter is always connected in series with circuit.

23. **(a)** In balanced Wheastone bridge, the arms of galvanometer and cell can be interchanged without affecting the balance of the bridge.

24. **(d)**

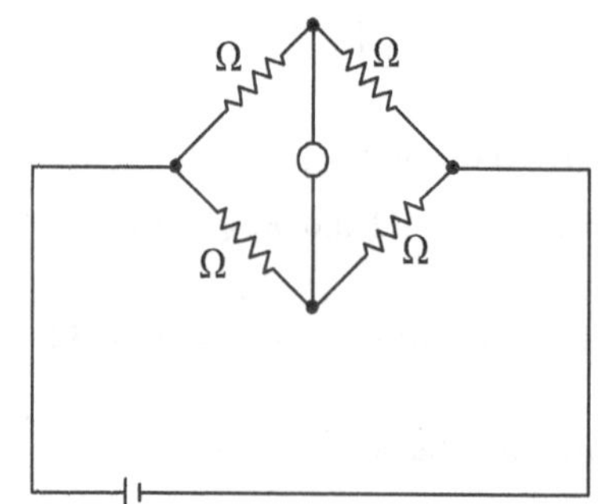

If P is slightly icnreased, potential of C will decrease.

Hence current will from A to C.

If Q is slightly increased, potential of C will increase.

Hence current will flow from C to A.

25-27

We have

$$R_s = \frac{V}{I_{fs}} - R_c = \frac{10.0V}{0.00100A} - 20.0\Omega = 9980\Omega$$

At full-scale deflection, $V_{ab} = 10.0V$, voltage across the meter is 0.0200 V, voltage across R_s is 9.98 V, and current through the voltmeter is 0.00100 A. In this case most of the voltage appears across the series resistor.

The equivalent meter resistance is $R_{eq} = 20.0\,\Omega + 9980\,\Omega = 10{,}000\Omega$. Such a meter is described as a "1,000 ohms-per-volt meter" referring to the ratio of resistance to full-scale deflection. In normal operation the current through the circuit element being measured is much greater than 0.00100 A, and the resistance between points a and b in the circuit is much less than 10,000Ω. So the voltmeter draws off only a small fraction of the current and disturbs, only slightly the circuit being measured.

25. **(d),** **26.** **(c),** **27.** **(b)**

28. **(d)** The resistance of the galvanometer is fixed. In meter bridge experiments, to protect the galvanometer from a high current, high resistance is connected to the galvanometer in order to protect it from damage.

29. **(a)** Sensitivity $\propto \frac{1}{\text{Potential gradiant}} \propto$ (Length of wire)

30. **(a)** If either the e.m.f. of the driver cell or potential difference across the whole potentiometer wire is lesser than the e.m.f. of the experimental cell, then balance point will not obtained.

DAILY PRACTICE PROBLEMS

PHYSICS SOLUTIONS

39

(1) (a) We know magnetic field due to a long straight current carrying wire

$$B = \frac{\mu_0 i}{2\pi r} = \frac{4\pi \times 10^{-7} \times 3}{2\pi \times 50 \times 10^{-3}}$$

(Note that $\mu_0 = 4\pi \times 10^{-7}$ in SI system)

$= 1.20 \times 10^{-5}$ Tesla = 0.12 G.

[As 1 Gauss = 10^{-4} Tesla]

(2) (c) The magnetic induction produced due to a current carrying arc at its centre of curvature is

$$B = \frac{\mu_0 i\alpha}{4\pi r} \quad \text{.......... (a)}$$

(subtending angle α at the centre of curvature)

$$\Rightarrow \quad B = \frac{\mu_0 i\pi}{4\pi r} \times \frac{\pi}{4} = \frac{\mu_0 i}{16r}$$

(3) (a)

R, O, Q, P, r, 90°, S, T, i, i, y, x, z

$\vec{B}_O = \vec{B}_{QRS} + \vec{B}_{ST}$.

$$\vec{B}_{PQ} = \text{zero}, \vec{B}_{QRS} = \frac{3}{4} \times \frac{\mu_0 i}{2r}\hat{k}, \vec{B}_{ST} = \frac{\mu_0 i}{4\pi r}\hat{k}$$

$$\Rightarrow \quad \vec{B}_O = \frac{\mu_0 i}{4\pi r}\hat{k} + \frac{3}{4}\frac{3\mu_0 i}{2r}\hat{k} = \frac{\mu_0 i}{4\pi r}\left[\frac{3\pi}{2} + 1\right]\hat{k}$$

(4) (a). $\vec{B}_O = \vec{B}_{PSR} + \vec{B}_{PQR}$ .... (a)

$$\vec{B}_{PSR} = \frac{\mu_0 i}{4\pi}\left[\frac{2\pi - 2\phi}{r}\right] = \frac{\mu_0 i}{2\pi r}[\pi - \phi](-\hat{k}) \quad \text{.... (b)}$$

$$\vec{B}_{PQR} = \frac{\mu_0 i}{4\pi} \cdot \frac{2\sin\phi}{OQ}(-\hat{k}) = \frac{\mu_0 i}{4\pi} \cdot \frac{2\sin\phi}{r\cos\phi} = \frac{\mu_0 i}{2\pi r}\tan\phi\,(-\hat{k}) \quad \text{.... (c)}$$

From eqs. (a), (b) and (c)

$$\vec{B} = \frac{\mu_0 i}{2\pi r}[\pi - \phi](-\hat{k}) + \frac{\mu_0 i}{2\pi r}\tan\phi\,(-\hat{k})$$

$$= \frac{\mu_0 i}{2\pi r}[\pi - \phi + \tan\phi](-\hat{k})$$

(5) (a). The rotating rod is a current-loop whose radius a = 0.6m. The magnetic field due to this current-loop at a point on its axis at a distance x from its centre is given by

$$B = \frac{\mu_0 i a^2}{2(a^2 + x^2)^{3/2}} \quad \text{...(i)}$$

Let T be the period of rotation of the rod. Then

$$i = \frac{q}{T} = \frac{q\omega}{2\pi} = \frac{1 \text{ colulomb} \times 10^4 \pi / \text{sec}}{2\pi}$$

$= 5 \times 10^3$ amp.

Now, a = 0.6 m, x = 0.8 m and $\mu_0 = 4\pi \times 10^{-7}$ V-s/A-m. Substituting these values in eq. (i) we get

$$B = \frac{(4\pi \times 10^{-7} V - s/A - m)(5 \times 10^3 A)(0.6m)^2}{2(0.36 + 0.64)^{3/2} m^3}$$

$= 0.36\pi \times 10^{-3} = 1.13 \times 10^{-3}$ tesla

In the second case the current remains the same because the rotating charge and the angular frequency are the same. However, the radius of the loop becomes half (a = 0.3 m) and the distance x is now 0.4 m.

$$\therefore \quad B = \frac{\mu_0 i a^2}{2(a^2 + x^2)^{3/2}}$$

$$= \frac{(4\pi \times 10^{-7} V - s/A - m)(5 \times 10^3 A)(0.3m)^2}{2(0.09 + 0.16)^{3/2} m^3}$$

$$= \frac{4\pi \times 10^{-7} \times 5 \times 10^3 \times 0.09}{2 \times 0.25 \times 0.5} \text{ tesla}$$

$= 0.72 \times 10^{-3}\pi = 2.26 \times 10^{-3}$ tesla.

(6) (a) The magnetic field at the centre of a current carrying coil having n turns is given by

$$B = \frac{\mu_0 ni}{2r} \text{ N/A.m}$$

where i, is the current in the coil and r is the radius of the coil.

Here i = 0.1 A, n = 1000 and r = 0.1 m.

$$\therefore \quad B = \frac{(4\pi \times 10^{-7}) \times 1000 \times 0.1}{2 \times 0.1} = 6.28 \times 10^{-4} \text{ N/A.m}$$

(7) (a). The two coils are perpendicular to each other. Coil 1 produces field along X axis and coil 2 produces field along Y axis. Thus the resultant field will be-

$$B = \sqrt{B_1^2 + B_2^2} \text{ making an angle}$$

$$\theta = \tan^{-1}\left(\frac{B_2}{B_1}\right) \text{ with x axis}$$

$$\text{As } B_1 = B_2 = \frac{\mu_0 NI}{2a}$$

$$\Rightarrow B = \sqrt{2} = \left(\frac{\mu_0 NI}{2a}\right) = \frac{\mu_0 NI}{\sqrt{2}a} \text{ and } \theta = 45^\circ.$$

(8) (d) Applying ampere's law at P, Q and R respectively, we find that there is no current enclosed by the circle of P. So magnetic induction at P is zero while that at Q and R is non- zero.

(9) (a). For a current carrying coil

$B = \frac{\mu_0 i}{2R}$ at centre and force on a current carrying conductor ie

$$F = i\ell B \Rightarrow F = \frac{\mu_0 i^2 \ell}{2R}$$

$$\Rightarrow [MLT^{-2}] = \frac{[\mu_0][A^2][L]}{[L]}$$

$$\Rightarrow [\mu_0] = [MLT^{-2}A^{-2}]$$

(10) (c) By Biat Savart Law,

$$\delta B = \frac{\mu_0}{4\pi}\frac{i\delta\ell \sin\theta}{r^2}$$

When $\theta = 90°$, then $\sin 90° = 1$ = maximum

$$\therefore \delta B = \frac{\mu_0 i \delta\ell}{4\pi r^2} = \text{maximum}$$

(11) (a) The magnitude of the magnetic field at the centroid O of the triangle due to a side PQ (say) is

$$\frac{\mu_0}{4\pi}\frac{i}{r}(\sin\phi_1 + \sin\phi_2)$$

Where r is the perpendicular distance of PQ from O, and ϕ_1, ϕ_2 the angles as shown. The field is perpendicular to the plane of paper and is directed into plane of paper. Since the magnetic field due to each of the three sides is the same in magnitude and direction, the magnitude of the resultant field at O is

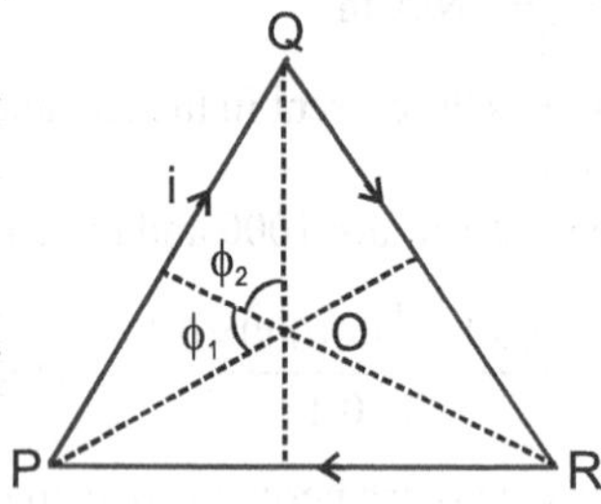

$$B = 3\frac{\mu_0}{4\pi}\frac{i}{r}(\sin\phi_1 + \sin\phi_2)$$

Here i = 1 ampere, $\phi_1 = \phi_2 = 60°$

and $r = \frac{l}{2}\cot 60° = \frac{l}{2}\times\frac{1}{\sqrt{3}}$

and ℓ is the side of the triangle (= 4.5×10^{-2} meter).

$$\therefore B = \frac{3\times10^{-7}\times1.0}{\left(\frac{1}{2}\times4.5\times10^{-2}\right)\times\left(\frac{1}{\sqrt{3}}\right)}\left(\frac{\sqrt{3}}{2}+\frac{\sqrt{3}}{2}\right)$$

$$= \frac{3\times10^{-7}\times2\times3}{4.5\times10^{-2}} = 4.0\times10^{-5}\ \text{weber/m}^2.$$

(12) (a). The magnetic field inside (near centre) a current carrying solenoid having n turns per unit length is given by
$B = \mu_0 ni$ newton/(ampere-meter),
where i (ampere) is the current in the solenoid and
$\mu_0 = 4\pi\times10^{-7}$ newton/ampere^2.
Here n = 500/0.40 = 1250 per meter, i = 1.0 amp.
$\therefore B = (4\times3.14\times10^{-7})\times 1250\times1.0$
$= 15.7\times10^{-4}$ newton/(ampere-meter) = 15.7 gauss.

(13) (b) We know, $B_{end} = \frac{\mu_0 ni}{2}$

Here $n = \frac{500}{0.2} = 2500$/metre,

$$\therefore i = \frac{2B_{end}}{\mu_0 n} = \frac{2\times8.71\times10^{-6}}{4\pi\times10^{-7}\times2500}$$

$$= \frac{17.42\times10^{-3}}{\pi} = \frac{0.01742}{\pi}\text{amp amp.}$$

(14) (b) $\frac{B_{centre}}{B_{axis}} = \left(1\times\frac{x^2}{R^2}\right)^{3/2}$, also Baxis $= \frac{1}{8}B_{centre}$

$$\Rightarrow \frac{8}{1} = \left(1\times\frac{x^2}{R^2}\right)^{3/2} \Rightarrow 2 = \left(1\times\frac{x^2}{R^2}\right)^{1/2}$$

$$\Rightarrow 4 = 1 + \frac{x^2}{R^2} \Rightarrow 3 = \frac{x^2}{R^2} \Rightarrow x^2 = 3R^2 \Rightarrow 3R^2$$

$\Rightarrow x = \sqrt{3}R$

(15) (a) $B_0 = \mu_0\frac{Ni}{2\pi R}$

$$= \frac{4\pi\times10^{-7}\times500\times0.5}{2\pi\times0.1} = 5\times10^{-4}\ \text{tesla}$$

(16) (c) At P : $B_{net} = \sqrt{B_1^2 + B_2^2}$

$$= \sqrt{\left(\frac{\mu_0}{4\pi}\frac{2i_1}{a}\right)^2 + \left(\frac{\mu_0}{4\pi}\frac{2i_2}{a}\right)^2}$$

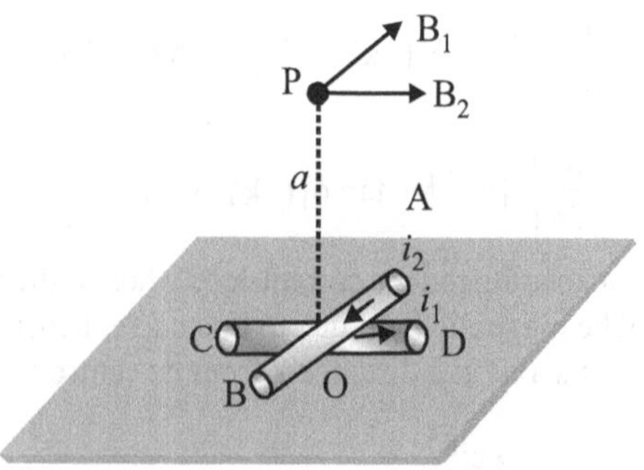

$$= \frac{\mu_0}{2\pi a}\left(i_1^2 + i_2^2\right)^{1/2}$$

(17) (d)

(18) (b) Current distribution in the network is as shown. Now, consider the pair of wires AB and GH. As current in these wires produce equal but opposite magnetic fields at centre O of the cube, resultant field due to the pair is zero.

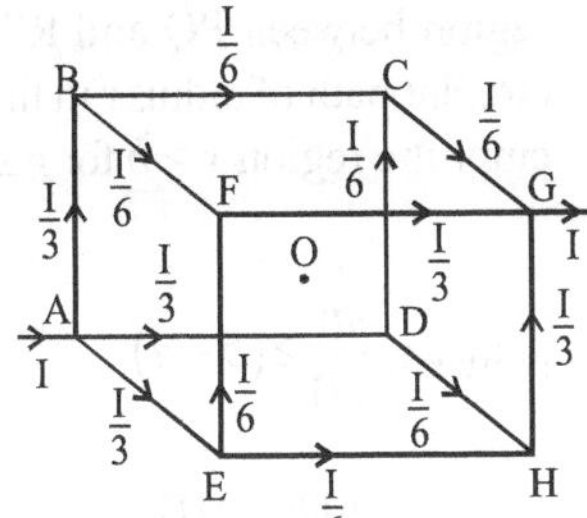

We can see five such more pairs namely :

(i) AE, CG

(ii) AD, FI

(iiii) BC, EH

(iv) EF, DC

(v) BF, OH

Magnetic field due to each of these pairs is zero.

Therefore, resultant magnetic field at centre O is zero.

(19) (a) Magnetic field inside a solid cylinder of current is

$$B_{inside} = \frac{\mu_0 ir}{2\pi R^2}$$

$$\Rightarrow \quad B_0 = \frac{\mu_0 i \frac{R}{2}}{2\pi R^2} \quad \text{(as per given information)}$$

$$\Rightarrow \quad i = \frac{4B_0 \pi R}{\mu_0}$$

Magnetic field outside a solid cylinder of current is

$$B_{outside} = \frac{\mu_0 i}{2\pi r}$$

$$\Rightarrow \quad B_{outside} \text{ at a distance } 2R = \frac{\mu_0\left(\frac{4B_0\pi R}{\mu_0}\right)}{2\pi(2R)} = B_0$$

(20) (d) As per sense of transversal,

$i_{crossing} = I_1 - I_2 - I_3$

By Ampere's law, $\oint \vec{B}.d\vec{\ell} = \mu_0\, i_{crossing}$

$$\Rightarrow \oint \vec{B}.d\vec{\ell} = \mu_0 (I_2 - I_1 - I_3)$$

(21) (a) $\ell = (2\pi r)\, n$ or $\quad n = \frac{\ell}{2\pi r}$

$$B = \frac{\mu_0 ni}{2r} = \frac{\mu_0 i\ell}{4\pi r^2}$$

or $B = \frac{4\pi \times 10^{-7} \times 6.28 \times 1}{2 \times 2 \times \pi \times (0.10)^2} = 6.28 \times 10^{-5}$ Tesla.

(22) (b). The arrangement is shown in fig.

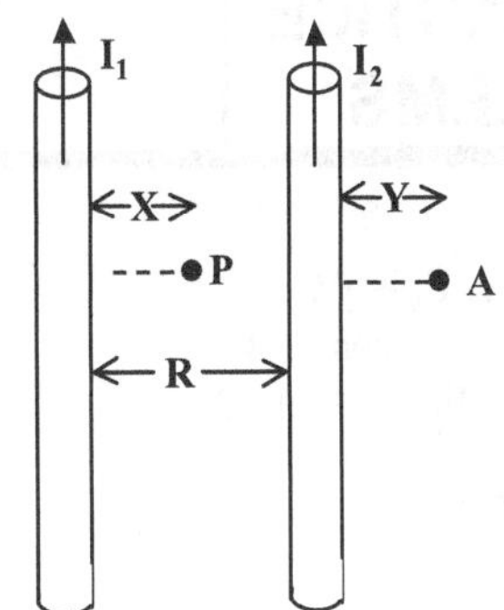

The magnetic field at a point P in between the two wires is

$\vec{B} = \vec{B}_1 + \vec{B}_2$. The field B_1 (due to current I_1) points down ward while B_2 (due to current I_2) points upwards. Thus field at point P is-

$$B = \frac{\mu_0}{2\pi}\left[\frac{I_1}{x} - \frac{I_2}{R-x}\right] \text{ in to the plane of paper.}$$

At $x = R/2$,

$$B = \frac{\mu_0 (I_1 - I_2)}{\pi R} \text{ into the plane of paper, } (\text{if } I_1 > I_2)$$

or $\quad B = \frac{\mu_0 (I_2 - I_1)}{\pi R}$ out of the plane of paper (if $I_2 > I_1$)

(23) (d) (i) Fields due to both coils are in the same direction

$$\Rightarrow B = \frac{\mu_0 N_1 I_1}{2R_1} + \frac{\mu_0 N_2 I_2}{2R_2}$$

If $I_1 = I_2 = I,\ N_1 = N_2 = N$,

$$B = \frac{\mu_0 NI(R_1 + R_2)}{2R_1 R_2}$$

(ii) Fields due to the two coils are in opposite direction,

$$\Rightarrow B = \frac{\mu_0 N_1 I_1}{2R_1} - \frac{\mu_0 N_2 I_2}{2R_2}$$

If $I_1 = I_2 = I,\ N_1 = N_2 = N$,

$$B = \frac{\mu_0 NI(R_2 - R_1)}{2R_1 R_2}$$

(24) **(a).** For circular coil $B_1 = \frac{\mu_0 I}{2r}$

Circumference of the coil $= 2\pi r = L$.

Thus $B_1 = \pi\mu_0 I/L = 3.14\, \mu_0\, I/L$

For square loop $B_2 = 2\sqrt{2}\, \mu_0\, I/L = 3.60\, \mu_0\, I/L$

Thus $B_1 < B_2$.

25. **(d)** Since $\vec{M} \parallel \vec{B}$ $\therefore$ Torque $= \vec{M} \times \vec{B}$ is zero.

26. **(d)** The field must be in $+\hat{k}$ direction.

27. **(a)**

28. **(b)** The statements are independently correct.

29. **(d)** $\vec{\tau} = \vec{m} \times \vec{B} \Rightarrow \vec{\tau} = 0$ for $\theta = 0°, 180°$.

30. **(b)**

DAILY PRACTICE PROBLEMS

PHYSICS SOLUTIONS 40

(1) (b) $E_{kp} = eV, \therefore E_k = qV,$

$\therefore \quad E_k \propto q, \therefore V = \text{constant}$

$E_{kp} : E_{kd} : E_{ka} :: 1 : 1 : 2.$

(2) (c) $E_K = \dfrac{q^2 r^2 B^2}{2m}$

$\therefore \quad E_k \propto \dfrac{q^2}{m} = \dfrac{q_a^2}{m_p^2} \times \dfrac{m_p}{E_{k_p}}$

$= \dfrac{4}{1} \times \dfrac{1}{4} = E_{K_a} = 8\text{eV}.$

(3) (a) $B = \dfrac{KVe}{r^2} = \dfrac{10^{-7} \times 3 \times 1.6^6 \times 10^{-19}}{(2 \times 10^{-10})^2} = 1.2 \text{ Tesla}.$

(4) (a) $F = qvB \sin\theta$

$= 2 \times 1.6 \times 10^{-19} \times 10^5 \times 0.8 \times \left(\dfrac{1}{2}\right)$

$= 1.28 \times 10^{-14}$ N $\quad$ [$\because$ charge on α particle = 2e]

(5) (b) The direction of $\vec{F}$ is along $(\vec{V} \times \vec{B})$ which is towards the right. Thus the beam deflects to your right side.

(6) (b) The particle is moving clockwise which shows that force on the particle is opposite to given by right hand palm rule of fleming left hand rule. These two laws are used for positive charge. Here since laws are disobeyed, we can say that charge is negative.

(7) (b) The point lies at the circumference hence it will come back after a time period T

$T = \dfrac{2\pi m}{qB}$

(8) (a) The magnetic force on a current carrying wire of length L, placed in a magnetic field B at an angleθ with the field is given by

$F = i\,\ell\, B\sin\theta.$

Here $B = 5.0 \times 10^{-4}$ N/A.m. $i = 2.0$ A,

$\ell = 50$ cm $= 0.50$ m,

$\theta = 60^\circ$

$F = 2.0 \times 0.50 \times (5.0 \times 10^{-4}) \times \sin 60^\circ$

$= 4.33 \times 10^{-4}$ N

According to the flemings left - hand rule, this force will act perpendicular to both the wire and the magnetic field.

(9) **(a, c)** $r \mu \dfrac{\sqrt{m}}{q} \Rightarrow r_H : r_{He} : r_o = \dfrac{\sqrt{1}}{1} : \dfrac{\sqrt{4}}{1} : \dfrac{\sqrt{16}}{2} = 1 : 2 : 2$

Radius is smallest for H^+, so it is deflected most.

(10) (b) In the figure, the z-axis points out of the paper, and the magnetic field is directed into the paper, existing in the region between PQ and RS. The particle moves in a circular path of radius r in the magnetic field. It can just enter the region $x > b$ for $r \geq (b - a)$

Now $r = \dfrac{mv}{qB} \geq (b-a)$

or $\quad v \geq \dfrac{q(b-a)B}{m} \Rightarrow v_{min} = \dfrac{q(b-a)B}{m}$

(11) (a) From figure it is clear that

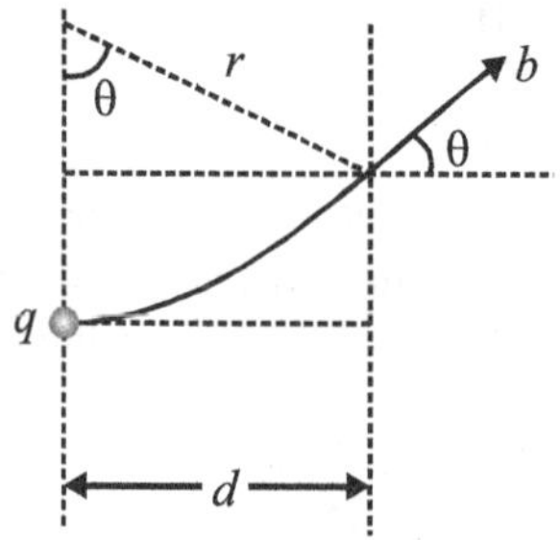

$\sin\theta = \dfrac{d}{r}$ also $r = \dfrac{p}{qB}$

$\therefore \quad \sin\theta = \dfrac{Bqd}{p}$

(12) (a) For on wire Q due to wire P is

$F_P = 10^{-7} \times \dfrac{2 \times 30 \times 10}{0.1} \times 0.1 = 6 \times 10^{-5}$ N (Towards left)

Force on wire Q due to wire R is

$F_R = 10^{-7} \times \dfrac{2 \times 20 \times 10}{0.02} \times 0.1 = 20 \times 10^{-5}$ (Towards right)

Hence $F_{net} = F_R - F_P = 14 \times 10^{-5}$N $= 14 \times 10^{-4}$ N (Towards right)

(13) (d) $\vec{F} = q(\vec{v} \times \vec{B})$

$$\vec{v} \times \vec{B} = \begin{vmatrix} \hat{i} & \hat{j} & \hat{k} \\ 3 & 2 & 0 \\ 5 \times 10^5 & 0 & 0 \end{vmatrix} = \hat{k}\,(-10 \times 10^5) = (-\hat{k}\ 10^6)$$

$q = 2e = 2 \times 1.6 \text{ x } 10^{-19} = 3.2 \times 10^{-19}$ Coulomb

$\vec{F} = 3.2 \times 10^{-19}\,(-\hat{k} \times 10^6)$

$\Rightarrow \quad \vec{F} = -3.2 \times 10^{-13}\,\hat{k}.$

$\therefore \quad |F| = 3.2 \times 10^{-13}$ Coulomb.

(14) (b)

$\therefore \quad F = q(\vec{v} \times \vec{B}) = 2evB \sin 90^\circ$

or $F = 2evB$

(15) (a) Force on side BC AND AD are equal but opposite so their net will be zero.

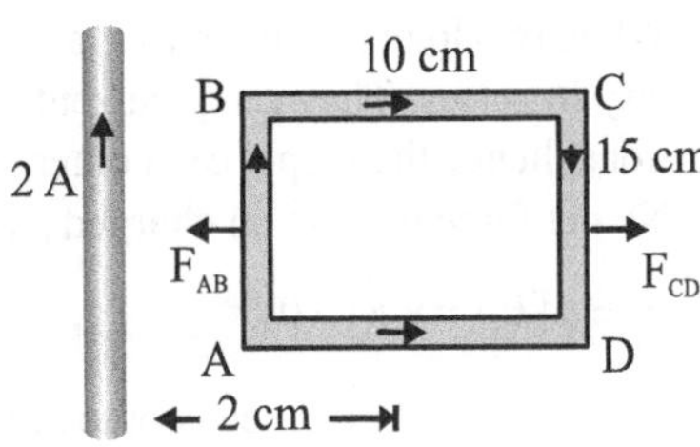

But $F_{AB} = 10^{-7} \times \frac{2\times2\times1}{2\times10^{-2}} \times 15\times10^{-2} = 3 \times 10^{-6}$ N

and $F_{CD} = 10^{-7} \times \frac{2\times2\times1}{(2\times10^{-2})} \times 15\times10^{-2} = 0.5 \times 10^{-6}$ N

$\Rightarrow F_{net} = F_{AB} - F_{CD} = 2.5 \times 10^{-6}$ N
$= 25 \times 10^{-7}$ N, towards the wire.

(16) (b) In order to make a proton circulate the earth along the equator, the minimum magnetic field induction $\vec{B}$ should be horizontal nad perpendicular to equator. The magnetic force provides the necessary centripetal force.

i.e. $qv B = \frac{mv^2}{r}$ or $B = \frac{mv}{qr}$

Here $m = 1.7 \times 10^{-27}$ kg, $v = 1.0 \times 10^7$ m/s
$q = e = 1.6 \times 10^{-19}$ coulomb, $r = 6.37 \times 10^6$ m

$$B = \frac{1.7\times10^{-27}\times1.0\times10^7}{1.6\times10^{-19}\times6.37\times10^6} = 1.67 \times 10^{-8} \text{ weber/m}^2.$$

(17) (c) We have $F = qvB = \frac{mv^2}{r}$ or $v = \frac{qBr}{m}$

$$= \frac{3.2\times10^{-19}\times1.2\times0.45}{6.8\times10^{-27}} = 2.6 \times 10^7 \text{ m/s}.$$

The frequency of rotation $n = \frac{v}{2\pi r}$

$$= \frac{2.6\times10^7}{2\times3.14\times0.45} = 9.2 \times 10^6 \text{ sec}^{-1}.$$

Kinetic energy of α-particle,

$$E_K = \frac{1}{2} \times 6.8 \times 10^{-27} \times (2.6 \times 10^7)^2$$

$= 2.3 \times 10^{-12}$ joule.

$$= \frac{2.3\times10^{-12}}{1.6\times10^{-19}} \text{ eVolt} = 14 \times 10^6 \text{ eV} = 14 \text{ MeVolt}.$$

If V is accelerating potential of α-particle, then Kinetic energy = qV
14×10^6 eVolt = 2eV (since charge on α-particle = 2e)

$$\therefore V = \frac{14\times10^6}{2} = 7 \times 10^6 \text{ Volt}.$$

(18) (a) If electron beam passes undeflected in simultaneous electric and magnetic fields $\vec{E}$ and $\vec{B}$ velocity of beam $\vec{v}$ much be mutually perpendicular and the required speed v is given by-

$$v = \frac{E}{B} = \frac{1\times10^4}{2\times10^{-3}} = 5 \times 10^6 \text{ m/s}.$$

If electric field is removed, the electron traverses a circular path of radius r given by $\frac{mv^2}{r} = evB$ or $r = \frac{mv}{eB}$.

Here $m = 9.1 \times 10^{-31}$ kg, $v = 5 \times 10^6$ m/s.
$e = 1.6 \times 10^{-19}$ coul and $B = 2 \times 10^{-3}$ weber/m^2

$$\therefore r = \frac{(9.1\times10^{-31})(5\times10^6)}{(1.6\times10^{-19})(2\times10^{-3})}$$

$= 1.43 \times 10^{-2}$ m = 1.43 cm.

(19) (a) For L length or wire, to balance,
$F_{magnetic} = mg \Rightarrow ILB = mg$,
Therefore $B = mg/IL = (m/L)g/I$

$$= \frac{45\times10^{-3}\times9.8}{30} = 1.47 \times 10^{-2} \text{ tesla}.$$

= 147 Gauss.

(20) **(b)** According to Fleming's left hand rule, magnetic force on electrons will be downward.

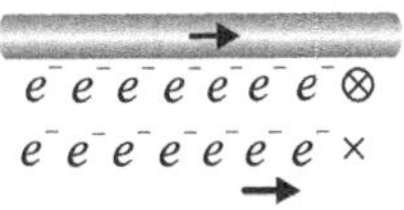

(21) (b) $\therefore F = mg = Bi\ell$
or $1 \times 9.8 = 0.98 \times i \times 1, \Rightarrow i = 10$A.

(22) (b) When currents flow in two long, parallel wires in the same direction, the wires exert a force of attraction on each other. The magnitude of this force acting per meter length of the wires is given by

$$F = \frac{\mu_0}{2\pi}\frac{i_1 i_2}{R} = 2 \times 10^{-7} \frac{i_1 i_2}{R} \text{ N/m}.$$

Here $i_1 = 10$ A, $i_2 = 15$ A, R = 30 cm = 0.3 m

$$\therefore F = 2 \times 10^{-7} \frac{10\times15}{0.3} = 1 \times 10^{-4} \text{ N/m}.$$

$\therefore$ Force on 5m length of the wire
$= 5 \times (1 \times 10^{-4}) = (5 \times 10^{-4}) = 5 \times 10^{-4}$ N (attraction).

(23) (d) The electron will pass undeviated if the electric force and magnetic force are equal and opposite. Thus
E.e. = Bev or B = E/v but E = V/d

Therefore, $B = \frac{V}{v.d.} = \frac{600}{3\times10^{-3}\times2\times10^6}$

$\therefore B = 0.1$ Wb/m^2.
The direction of field is perpendicular to the plane of paper vertically downward.

(24) (b) The component of velocity of the beam of protons, parallel to the field direction
$= v\cos\theta = 4 \times 10^5 \times \cos 60° = 2 \times 10^5$ m/sec.
and the component of velocity of the proton beam at right angle to the direction of field
$= v\sin\theta = 4 \times 10^5 \times \sin 60° = 2\sqrt{3} \times 10^5$ m/sec.
therefore,the radius of circular path = $(mv\sin\theta/Be)$

or $r = \dfrac{1.7\times10^{-27}\times2\sqrt{3}\times10^5}{0.3\times1.6\times10^{-19}} = 12.26 \times 10^{-3}$ metre

or $r = 1.226 \times 10^{-2}$ metre.
Pitch of the Helix = $v\cos\theta \times (2\pi m/Be)$

$\therefore$ Pitch $= \dfrac{2\times10^5\times2\times3.14\times1.7\times10^{-27}}{0.3\times1.6\times10^{-19}}$

$= 44.5 \times 10^{-3}$ m $= 4.45 \times 10^{-2}$ m.

(25) **(a), (26) (a), (27) (c).**

$\vec{F} = q(\vec{v}\times\vec{B}) = q(x^2 - y^2)\hat{k}$

(28) (c) When two long parallel wires, are connected to a battery in series. They carry currents in opposite directions, hence they repel each other.

(29) (c) No net force will act on charged particle if

$\vec{F} = q[\vec{E} + \vec{v}\times\vec{B}] = 0$

$\Rightarrow \vec{E} = -\vec{v}\times\vec{B} \Rightarrow v$ need not to be perpendicular to B

(30) (c) In this case we can not be sure about the absence of the magnetic field because if the electron moving parallel to the direction of magnetic field, the angle between velocity and applied magnetic field is zero (F = 0). Then also electron passes without deflection.

Also $F = evB\sin\theta \Rightarrow F \propto B$.

DAILY PRACTICE PROBLEMS

PHYSICS SOLUTIONS 41

1. (b) $M = NiA = 20 \times \frac{22}{7}(4 \times 10^{-2})^2 \times 3 = 0.3A - m^2$
2. (a) The magnetic moment of current carrying loop
$M = niA = ni(\pi r^2)$
Hence the work done in rotating it through 180°
$W = MB(1 - \cos\theta) = 2MB = 2(ni\pi r^2)B$
$= 2 \times (50 \times 2 \times 3.14 \times 16 \times 10^{-4}) \times 0.1 = 0.1\text{J}$
3. (b) $\theta = \frac{NiAB}{C} \Rightarrow \theta \propto N$ (Number of turns)
4. (d) $\tau = MB\sin\theta \Rightarrow \tau_{max} = NiAB$, $(\theta = 90°)$
5. (c) In equilibrium angle between $\vec{M}$ and $\vec{B}$ is zero. It happens, when plane of the coil is perpendicular to $\vec{B}$

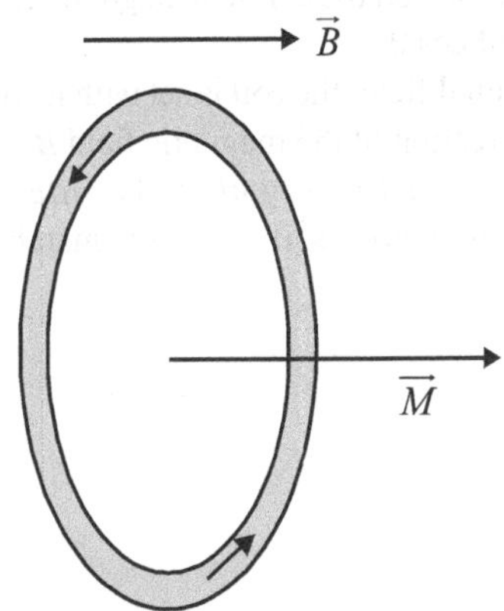

6. (a) $\tau = NBiA = 100 \times 0.2 \times 2 \times (0.08 \times 0.1) = 0.32\ \text{N} \times \text{m}$
7. (c) $\tau = NBiA = 100 \times 0.5 \times 1 \times 400 \times 10^{-4} = 2\ \text{N–m}$
8. (a) $\tau = NiAB\sin\theta = 0 \quad (\because \theta = 0°)$
9. (c) $M = NiA \Rightarrow M \propto A \Rightarrow M \propto r^2$ (As $l = 2\pi r \Rightarrow l \propto r$)
$\Rightarrow M \propto l^2$
10. (a)
11. (c) $\tau_{max} = NiAB = 1 \times i \times (\pi r^2) \times B$
$\left(2\pi r = L, \Rightarrow r = \frac{L}{2\pi}\right)$
$\tau_{max} = \pi i \left(\frac{L}{2\pi}\right)^2 B = \frac{L^2 iB}{4\pi}$
12. (b)
13. (b) The magnetic field at the centre of the solenoid is
$B = \mu_0 ni = (4\pi \times 10^{-7}) \times (500/0.4) \times 3$ N/A.m.
The torque acting on a current-carrying coil having N turns (say), placed perpendicular to the axis at the centre of the solenoid is-
$\tau = BiNA = (4\pi \times 10^{-7}) \times (500/0.4) \times 3 \times 0.4 \times 10 \times \pi(0.01)^2$
$= 6\pi^2 \times 10^{-7} = 5.92 \times 10^{-6}$ N.m.
14. (d) The equivalent magnetic moment is
$M = iA = ef(\pi r^2)$
As $f = \frac{v}{2\pi r}$
$\therefore M = \frac{ev}{2\pi r}\pi r^2 = \frac{evr}{2}$
15. (b) $i = \frac{C\theta}{NAB} \Rightarrow i \propto \theta$
16. (d) Initially for circular coil $L = 2\pi r$ and $M = i \times \pi r^2$
$= i \times \pi\left(\frac{L}{2\pi}\right)^2 = \frac{iL^2}{4\pi}$ ...(i)
Finally for square coil
$M' = i \times \left(\frac{L}{4}\right)^2 = \frac{iL^2}{16}$ ...(ii)

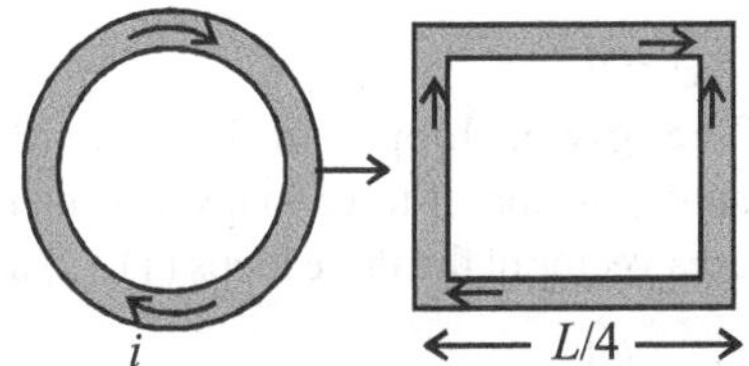

Solving equation (i) and (ii) $M' = \frac{\pi M}{4}$
17. (b) $M = iA = i \times \pi R^2$
also $i = \frac{Q\omega}{2\pi} \Rightarrow M = \frac{1}{2}Q\omega R^2$
18. (a) $\tau = NBiA\sin\theta$ so the graph between τ and θ is a sinusoidal graph.
19. (d) Initial magnetic moment = m1 = iL2

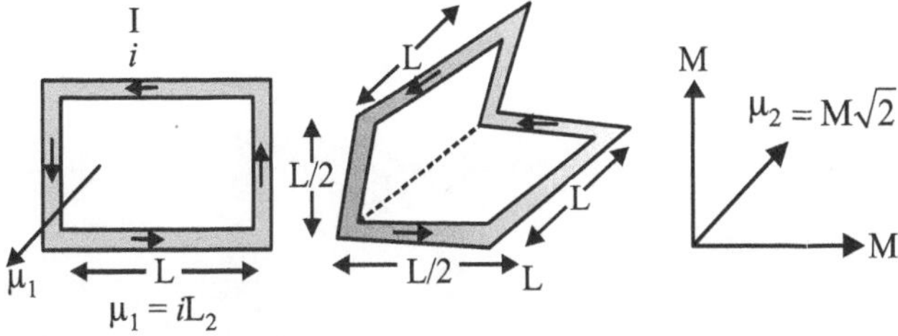

After folding the loop, M = magnetic moment due to each part $= i\,i\left(\frac{L}{2}\right) \times L = \frac{iL^2}{2} = \frac{\mu_1}{2}$

$$\Rightarrow \quad \mu_2 = M\sqrt{2} = \frac{\mu_1}{2}\times\sqrt{2} = \frac{\mu_1}{\sqrt{2}}$$

20. **(a)**

21. **(d)**

22. **(a)** Couple of force on loop S will be maximum because for same perimeter the area of loop will be maximum and magnetic moment of loop $= i \times A$. So, it will also be maximum for loop S.

23. **(b)** Sensitivity $S = \frac{\theta}{i} = \frac{nBA}{C}$

24. **(b)** $\tau = mB\sin\theta$ is zero for $\theta = 0°, 180°$.

25-27

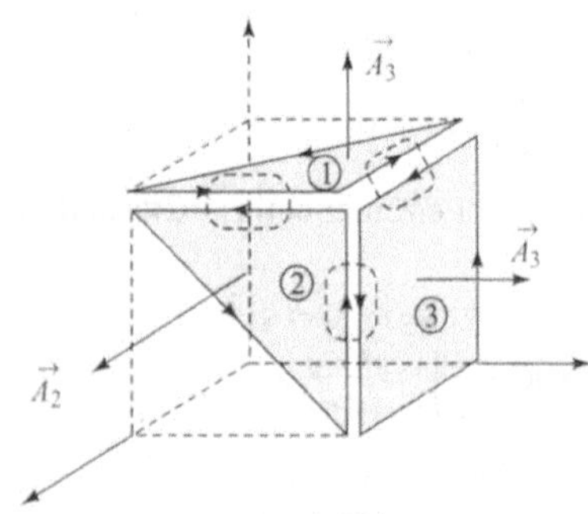

25. **(a)** **26. (b)** **27. (b)**

(a) The net force on a current carrying loop of any arbitrary shape in a uniform magnetic field is zero.

$\vec{F}_{net} = 0$

(b) The given loop can be considered to be a superposition of three loops as shown in figure. The area vector of the three loops (1), (2) and (3) are

$$\vec{A}_1 = \left(\frac{1}{2}\times 10\times 10\times 10^{-4}\right)\hat{j}\,m^2$$

$$\vec{A}_2 = \left(\frac{1}{2}\times 10\times 10\times 10^{-4}\right)\hat{k}\,m^2$$

$$\vec{A}_3 = \left(\frac{1}{2}\times 10\times 10\times 10^{-4}\right)\hat{i}\,m^2$$

Magnetic moment vector,

$$\vec{\mu} = i\vec{A} = 10(0.01\hat{i} + 0.005\hat{j} + 0.005\hat{k})Am^2$$

$$= (0.1\hat{i} + 0.05\hat{j} + 0.05\hat{k})Am^2$$

c. Torque,

$$\vec{\tau} = (0.1\hat{i} + 0.05\hat{j} + 0.05\hat{k})\times(2\hat{i} - 3\hat{j} + \hat{k})$$

$$= \begin{vmatrix} \hat{i} & \hat{j} & \hat{k} \\ 0.1 & 0.05 & 0.05 \\ 2 & -3 & 1 \end{vmatrix} = -0.1\hat{i} - 0.4\hat{k}\,Nm$$

28. **(a)** Due to metallic frame the deflection is only due to current in a coil and magnetic field, not due to vibration in the strings. If string start oscillating, presence of metallic frame in the field make these oscillations damped.

29. **(b)** The torque on the coil in a magnetic field is given by $\tau = nIBA\cos\theta$

For radial field, the coil is set with its plane parallel to the direction of the magnetic field B, then $\theta = 0°$ and $\cos\theta = 1 \Rightarrow$ Torque $= nIBA\,(1) = nIBA$ (maximum).

30. **(c)** Loop will not oscillate if in unstable equilibrium position.

DAILY PRACTICE PROBLEMS

PHYSICS SOLUTIONS 42

1. (a). Force between magnetic poles in air is given by

$$F = \frac{\mu_0}{4\pi} \times \frac{m_1 m_2}{r^2}$$

Given that $m_1 = 50$ Am, $m_2 = 100$Am, $r = 10$ cm $= 0.1$ m and μ_0 = permeability of air $= 4\pi \times 10^{-7}$ Hm^{-1}.

$$\therefore F = \frac{4\pi \times 10^{-7}}{4\pi} . \frac{50 \times 100}{0.1 \times 0.1} = 50 \times 10^{-3} \text{ N}$$

2. (a) Strength of a magnetic field due to a pole of strength m is given by

$$H = \frac{1}{4\pi} . \frac{m}{r^2}$$

Given that $m = 40$ Am, $r = 20$ cm $= 20 \times 10^{-2}$ m.

$$\therefore \quad H = \frac{1}{4\pi} \times \frac{40}{(20 \times 10^{-2})^2} = 79.57 \text{ Am}^{-1}$$

Now, magnetic induction at the same point :
$B = \mu_0 H = 4\pi \times 10^{-7} \times 79.57 = 10^{-4}$ wb/m^2

3. (c) Couple acting on a bar magnet of dipole moment M when placed in a magnetic field, is given by

$$\tau = MB \sin \theta$$

where θ is the angle made by the axis of magnet with the direction of field.
Given that $m = 5$ Am, $2\ell = 0.2$ m, $\theta = 30°$ and $B = 15$ Wbm^{-2}

$\therefore \quad \tau = MB \sin\theta = (m \times 2\ell) B \sin\theta$

$$= 5 \times 0.2 \times 15 \times \frac{1}{2} = 7.5 \text{ Nm.}$$

4. (a) $F = mB = \frac{\mu_0}{4\pi}\frac{2m'\ell}{x^3} m$

$$= \frac{10^{-7} \times 2 \times 200 \times 0.05 \times 100}{8 \times 10^{-3}}$$

$= 2.5 \times 10^{-2}$ N

5. (d). $W = MB(\cos\theta_1 - \cos\theta_2)$

$$\therefore W_1 = MB(\cos 0° - \cos 60°) = \frac{MB}{2}$$

$$W_2 = MB(\cos 30° - \cos 90°) = \frac{\sqrt{3}\, MB}{2}$$

$\therefore W_2 = \sqrt{3}\, W_1$.

6. (b). Loss in P.E. = gain in K.E.
$\therefore E_k = U_i - U_j = -MB\cos 90° - (-MB\cos 0°)$
$= 4 \times 25 \times 10^{-6} = 10^{-4}$ J

7. (a). $\tau = MB\sin\theta = m\ell B\sin\theta$
$= 10^{-3} \times 0.1 \times 4\pi \times 10^{3} \times 0.5$
$= 2\pi \times 10^{-7}$ N-m

8. (c)

9. (d) Here, $d = 10$ cm $= 0.1$ m,
$H = 0.4$ gauss $= 0.4 \times 10^{-4}$ T, $M = ?$
Neutral points in this case, lie on axial line of magnet, such that

$$\frac{\mu_0}{4\pi}\frac{2M}{d^3} = H$$

$$\therefore 10^{-7} \times \frac{2M}{(0.1)^3} = 0.4 \times 10^{-4}$$

$M = 0.2$ A m^2

10. (d). $\tau = F \times r = MB \sin 30$

$$F = \frac{\ell \times mB\sin 30}{r}$$

$$= \frac{25 \times 10^{-2} \times 24 \times 0.25}{12 \times 10^{-2}} \times \frac{1}{2} = 6.25 \text{ N}$$

11. (b). $B = \frac{\mu_0}{4\pi} \times \frac{2M}{x^3}$ = constant

$$\therefore x_2 = x_1 \left(\frac{M_2}{M_1}\right)^{1/3} = 20 \times \left(\frac{1}{2 \times 4}\right)^{1/3} = 10 \text{ cm}$$

12. (d). $B = \frac{\mu_0}{4\pi} = \frac{2M}{(\ell^2 + x^2)^{3/2}} = \frac{\mu_0}{4\pi}\frac{2M}{x^3}$

$$\therefore \frac{B_1}{B_2} = \left(\frac{x_2}{x_1}\right)^3 = 8 : 1 \text{ approximately.}$$

13. (c). According to tangent law
$B_A = B_B \tan\theta$

$$\text{or } \frac{\mu_0}{4\pi}\frac{2M}{d_1^3} = \frac{\mu_0}{4\pi}\frac{M}{d_1^3}\tan\theta$$

$$\therefore \frac{d_1}{d_2} = (2\cot\theta)^{1/3}$$

14. (a). $T = 2\pi\sqrt{\frac{I}{MB}} = 2\pi\sqrt{\frac{m\ell^2}{12 \times m_p \ell B}} = 4$ sec

$$T' = 2\pi\sqrt{\frac{\frac{m}{2}\ell^2}{12 \times \frac{m_p}{2}\ell B}} = 4 \text{ sec}$$

15. **(c).** $T = 2\pi\sqrt{\dfrac{I}{MB}}$ or $T \propto \dfrac{1}{\sqrt{M}}$

$$\frac{T_1}{T_2} = \sqrt{\frac{3M-2M}{3M+2M}} \text{ or } T_2 = 5\sqrt{5}\text{ s}$$

16. **(b).** $T = 2\pi\sqrt{\dfrac{I}{MB}}$

$$= 2\pi\sqrt{\frac{m\ell^2/12}{m_p\ell B}} \text{ or } T \propto m\ell$$

$$\frac{T'}{T} = \left(\frac{\frac{m}{n}\frac{\ell}{n}}{m\ell}\right)^{1/2} \text{ or } T' = \frac{T}{n}$$

17. **(a).** The volume of the bar magnet is

$$V = \frac{\text{mass}}{\text{density}}$$

$$= \frac{6.6 \times 10^{-3}\text{kg}}{7.9 \times 10^3\text{kg/m}^3} = 8.3 \times 10^{-7}\text{ m}^3.$$

The intensity of magnetization is

$$I = \frac{M}{V} = \frac{2.5\text{ A}-\text{m}^2}{8.3 \times 10^{-7}\text{ m}^2}$$

$$= 3.0 \times 10^6\text{ A/m}$$

18. **(d).** The compass box will be on the axial line of the magnet,

Hence, $\dfrac{1}{4\pi}\cdot\dfrac{2M}{r^3} = \dfrac{1}{4\pi}\cdot\dfrac{2\times 2\ell m}{r^3} = H\tan\theta$

Given that H = horizontal component of the earth's magnetic field = 30 Am^{-1}, $\theta = 45°$,

$r = 20\text{ cm} = 0.02\text{ m}$,

$M = 2\,\ell\,m = 4 \times 10^{-2}\text{ m}$

Hence, $\dfrac{2\times 4\times 10^{-2}\text{m}}{4\pi\,(0.2)^3} = 30 \times \tan 45° = 30 \times 1;$

$$\therefore m = \frac{30\times 4\pi\times(0.2)^3}{2\times 4\times 10^{-2}} = 37.7\text{ Am}$$

19. **(c).** As the magnet is placed with its south pole pointing south, hence the neutral point lies on the equatorial line. At the neutral point, the magnetic field B due to the magnet becomes equal and opposite to horizontal component of earth's magnetic field i.e., B_H.
Hence, if M be magnetic dipole moment of the magnet of length 2ℓ and r the distance of the neutral point from its centre, then

$$B = \frac{\mu_0}{4\pi}\frac{M}{(r^2+\ell^2)^{3/2}} = B_H$$

Given that $\mu_0 = 4\pi \times 10^{-1}\text{ T mA}^{-1}$,
$M = 13.4\text{ Am}^2$,
$r = 15\text{ cm} = 0.15\text{ m}$ and $\ell = 5.0\text{ cm} = 0.05\text{ m}$

$$\therefore B_H = 10^{-7} \times \frac{1.34}{[(0.15)^2+(0.5)^2]^{3/2}}$$

$$= 10^{-7} \times \frac{1.34}{0.025\sqrt{0.025}} = 0.34 \times 10^{-4}\text{ T}$$

20. **(a).** As the magnet is placed with its north pole pointing south, the neutral points are obtained on the axial line. At the neutral points the magnetic field B due to the magnet becomes equal and opposite to the horizontal component of earth's magnetic field i.e., B_H.
Hence, if M be the magnetic dipole moment of the magnet of length 2ℓ and r the distance of neutral point from the centre of the magnet, then we have

$$B = \frac{\mu_0}{4\pi}\cdot\frac{2Mr}{(r^2-\ell^2)^2} = B_H$$

Given that $\mu_0 = 4\pi \times 10^{-7}\text{ TmA}^{-1}$,
$r = 40\text{ cm} = 0.40\text{ m}$, $\ell = 15\text{ cm} = 0.15\text{ m}$ and
$H_H = 0.34\text{ Gauss} = 0.34 \times 10^{-4}\text{ T}$

$$\therefore M = \frac{4\pi}{\mu_0}\cdot\frac{B_H(r^2-\ell^2)^2}{2r}$$

$$= 10^{-7}\times\frac{(0.34\times 10^{-4})-[(0.40)^2-(0.15)^2]^2}{2\times 0.40}$$

$= 8.0\text{ Am}^2$

The pole strength of the magnet is,

$$m = \frac{M}{2\ell} = \frac{8.0}{0.30} = 26.7\text{ Am}$$

21. **(a).** The situation is shown in figure. The horizontal component of earth's magnetic field at the location of the cable (angle of dip $\theta = 0$ is)

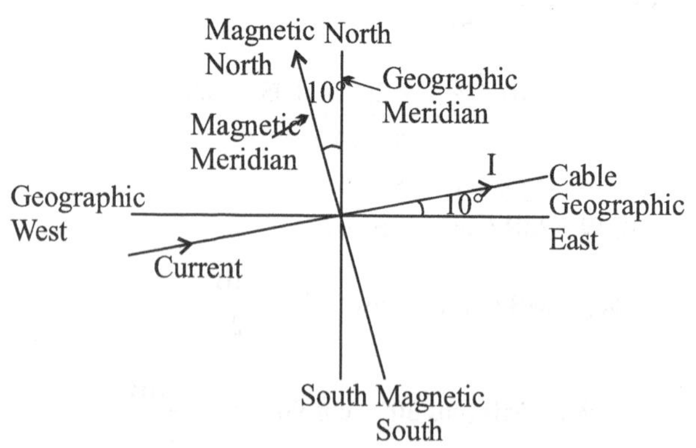

$B_H = B\cos\theta = B\cos 0 = B = 0.33\text{ Gauss}$
$= 0.33 \times 10^{-4}\text{ Tesla}$

B_H is directed horizontally in the magnetic meridian. The magnetic field produced by the cable at a distance of R meter is given by

$$B = \frac{\mu_0 I}{2\pi R}$$

$$= 2 \times 10^{-7} \times \frac{2.5}{R} = \frac{5.0 \times 10^{-7}}{R} T$$

According to right-hand-palm rule no, 1, the field B is directed horizontally along B_H at a point below the cable, and opposite to B_H at a point above the cable. Therefore, neutral points will be obtained above the cable. At these points, will be equal and opposite to B_H. Thus

$$\frac{5.0 \times 10^{-7}}{R} = B_H = 0.33 \times 10^{-4}.$$

$$\text{or } R = \frac{5.0 \times 10^{-7}}{0.33 \times 10^{-4}} = 15 \times 10^{-3} \text{ m} = 1.5 \text{ cm}$$

Thus, the line of neutral points lies above and parallel to the cable at a distance of 1.5 cm from it.

22. **(a)** The magnetic lines of force are in the form of closed curves whereas electric lines of force are open curves.

23. **(a)** Inside a magnet, magnetic lines of force move from south pole to north pole.

24. **(b)** Given that pole strength, $m = 5.25 \times 10^{-2}$ JT^{-1}, $\theta = 45°$ and $B = 0.42$ G $= 0.42 \times 10^{-4}$ T.

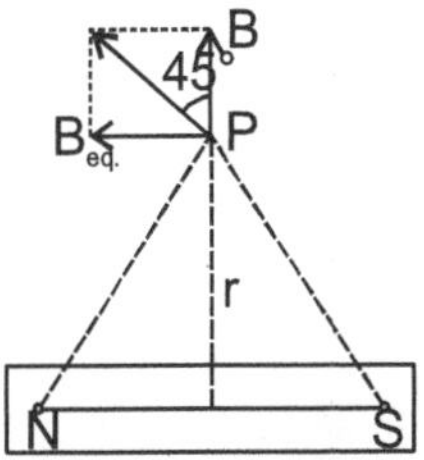

Figure shows a point P on the normal bisector of a short bar magnet lying at a distance r from the centre O of the magnet. It is given that resultant magnetic field at P makes an angle of 45° w.r.t. earth's field B.

From the figure, it is clear that

$B_{eq} = B \tan 45° = B.$

Now, for a short bar magnet in this position

$$B_{eq} = B = \frac{\mu_0}{4\pi} \cdot \frac{M}{r^3}$$

$$\text{or } r^3 = \frac{\mu_0}{4\pi} \cdot \frac{M}{B}$$

$$= 10^{-7} \times \frac{5.25 \times 10^{-2}}{0.42 \times 10^{-4}} = 125 \times 10^{-7}.$$

$\therefore r = 5 \times 10^{-2}$ m $= 5$ cm.

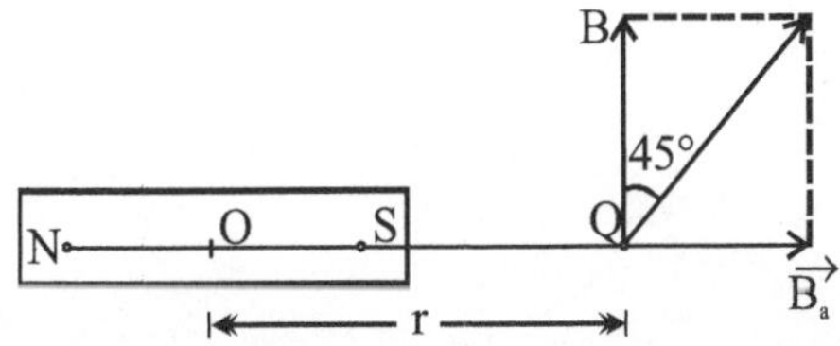

Figure shows a point Q on the axis of a short bar magnet lying at a distance r from the centre of the magnet. It is given that resultant magnetic field at Q is inclined at an angle of 45° w.r.t. earth's field B.

From the figure, it is clear that

$B_a = B \tan 45° = B$

Now for a short bar magnet in this position,

$$B_a = B = \frac{\mu_0}{4\pi} \cdot \frac{2M}{r^3}$$

$$\text{or } r^3 = \frac{\mu_0}{4\pi} \cdot \frac{2M}{B} = 2 \times 125 \times 10^{-6}$$

$\therefore r = (2 \times 125 \times 10^{-6})^{1/3} = 6.3$ cm

25-27

Given that earth's magnetic field, B = 0.39 G and angle of dip, $\theta = 35°$

Horizontal and vertical components of earth's magnetic field B at the location of the cable are

$B_H = B \cos \theta = 0.39 \cos 35°$

$= 0.39 \times 0.82 = 0.32$ Gauss and

$B_V = B \sin \theta = 0.39 \sin 35°$

$= 0.39 \times 0.57 = 0.22$ Gauss

The magnetic field produced by four current carrying straight cable wires at a distance R is

$$B' = \frac{\mu_0 I}{2\pi R} \times 4 = 2 \times 10^{-7} \times \frac{1.0}{0.04} \times 4$$

$= 0.2 \times 10^{-4}$ T $= 0.2$ Gauss

Resultant magnetic field below the cable

According to right - hand, palm rule no. 1, the direction of B′ below the cable will be opposite to that of B_H. Therefore, at a point 4 cm below the cable, resultant horizontal component of earth's magnetic field

$R_H = B_H - B' = 0.32 - 0.2 = 0.12$ Gauss.

Resultant vertical component of earth's magnetic field

$R_V = B_V = 0.22$ Gauss (unchanged)

$\therefore$ Resultant magnetic field below the cable is

$$R = \sqrt{[R^2{}_H + R^2{}_V]}$$

$$= \sqrt{[(0.12)^2 + (0.22)^2]}$$

$= 0.25$ Gauss

The angle that R makes with the horizontal is given by

$$\theta = \tan^{-1} \frac{R_V}{R_H}$$

$$= \tan^{-1} \frac{0.22}{0.12} = \tan^{-1}(1.8) \cong 62°$$

Resultant magnetic field above the cable

Again, according to right - hand - palm no. 1, the direction of B′ at a point above the cable is the same as that of B_H.

Therefore, at a point 4 cm below the cable, the horizontal component of resultant magnetic field will be

$R_H - B_H + B' = 0.32 \mid 0.20 = 0.52$ Gauss

Vertical component of resultant magnetic field will be $R_V = B_V - 0.22$ Gauss (unchanged)
Hence, magnitude of resultant magnetic field below the cable

$$R = \sqrt{[R^2_H + R^2_V]}$$

$$= \sqrt{[(0.52)^2 + (0.52)^2]}$$

$$= 0.56 \text{ Gauss}$$

The angle that R makes with the horizontal is given by

$$\theta = \tan^{-1}\left(\frac{R_V}{R_H}\right) = \tan^{-1}\left(\frac{0.22}{0.52}\right)$$

$$= \tan^{-1}(0.43) \cong 23°$$

25. (a) **26. (a)** **27. (b)**

28. (a)

29. (d) The earth has only vertical component of its magnetic field at the magnetic poles. Since compass needle is only free to rotate in horizontal plane. At north pole the vertical component of earth's field will exert torque on the magnetic needle so as to aligh it along its direction. As the compass needle can not rotate in vertical plane, it will rest horizontally, when placed ont he magnetic pole of the earth.

30. (c) It is quite clear that magnetic poles always exists in pairs. Since, one can imagine magnetic field configuration with three poles. When north poles or south poles of two magnets are glued together. They provide a three pole field configuration. It is also known that a bar magnet does not exert a torque on itself due to own its field.

DAILY PRACTICE PROBLEMS — PHYSICS SOLUTIONS 43

1. (a) Susceptibility of ferromagnetic substance is greater than 1.
2. (c) Susceptibility of diamagnetic substance is negative and it does not change with temperature.
3. (c) $\mu_d = 0$ and $\mu_p \neq 0$
4. (d) $\mu_r = \dfrac{B}{B_0} = 4$
5. (b) Dimagnetic
6. (b) Paramagnetic
7. (d) A super conductor exhibits perfect diamagnetism.
8. (a) Soft iron is highly ferromagnetic.
9. (a) Diamagnetism is the universal property of all substances.
10. (c) Diamagnetic substances are repelled by magnetic field.
11. (b) Coercive force
12. (b) Because, diamagnetic substance, moves from stronger magnetic field to weaker field.
13. (d) $\mu_r > 1, \chi > 0$
14. (b) Hysteresis curve for a given material estimates hysteresis loss.
15. (a) Neon atom is diamagnetic, hence it's net magnetic moment is zero.
16. (b) On heating, different domains have net magnetisation in them which are randomly distributes. Thus the net magnetisation of the substance due to various domains decreases to minimum.
17. (b) Repelled due to induction of similar poles.
18. (d) From the characteristic of $B-H$ curve.
19. (c) Diamagnetic substances are feebly repelled by magnets.
20. (d) Net magnetic induction $B = B_0 + B_m = \mu_0 H + \mu_0 M$
21. (d) $\chi_m = (\mu_r - 1) \Rightarrow \chi_m = (5500 - 1) = 5499$
22. (c) $\because \mu_r = \dfrac{\mu}{\mu_0} = \dfrac{B}{H\mu_0}$

    or $\mu_r = \dfrac{8\pi}{2\times10^3 \times 4\pi\times10^{-7}} = 10^4$
23. (a) $W_H = VAft = \dfrac{m}{d}Aft$

    or $W_H = \dfrac{0.6}{7.8\times10^3} \times 0.722 \times 50$
    $= 277.7 \times 10^{-5}$ Joule
24. (c)
25. (b) Given that : $H = 1600\ Am^{-1}$, $\phi = 2.4 \times 10^{-5}$ Wb, $A = 0.2\ cm^2 = 0.2 \times 10^{-4}\ m^2$.
    B = magnetic flux per unit cross - sectional are

    $= \dfrac{\phi}{A} = \dfrac{2.4\times10^{-5}}{0.2\times10^{-4}} = 1.2\ Wbm^{-2}$

    Magnetic permeability :

    $\mu = \dfrac{B}{H} = \dfrac{1.2\ Wbm^{-2} \text{ or } T}{1600\ Am^{-1}} = 7.5 \times 10^{-4}\ T\,A^{-1}\,m$

    As $\mu = \mu_0(1 + x_m)$

    $\therefore x_m = \dfrac{\mu}{\mu_0} - 1 = \dfrac{7.5\times10^{-4}}{4\times3.14\times10^{-7}}$
    $= 597.1 - 1 = 596.1.$
26. (a) The energy lost per unit volume of a substance in a complete cycle of magnetisation is equal to the area of the hysteresis loop.
27. (a) Statement (4) is the only true statement among the given choices.
28. (b) The susceptibility of ferromagnetic substance decreases with the rise of temperature in a complicated manner. After Curie's point the susceptibility of ferromagnetic substance varies inversely with its absolute temperature. Ferromagnetic substance obey's Curies law only above its Curie point.
29. (d) A paramagnetic sample display greater magnetisation when cooled, this is because at lower temperature, the tendency to disrupt the alignment of dipoles (due to magnetising field) decreases on account of reduced random thermal motion.
30. (d) The permeability of a ferromagnetic material dependent on magnetic field, $\vec{B} = K_m \vec{B}_0$, where B_0 is applied field. The total magnetic field $\vec{B}$ inside a ferromagnet may be 10^3 or 10^4 times the applied field B_0 The permeability K_m of a ferromagnetic material is not constant, neither the field $\vec{B}$ nor the magnetization $\vec{M}$ increases linearly with $\vec{B}$ even at small value of B_0. From the hysteresis curve, magnetic permeability is greater for lower field.

DAILY PRACTICE PROBLEMS

PHYSICS SOLUTIONS

44

(1) (a) $\phi = \vec{B}.\vec{A}$

$= (0.02\hat{i}).(30\hat{i}+16\hat{j}+23\hat{k}) \times 10^{-4}$

$= 0.6 \times 10^{-4}$ Wb $= 60\mu$ Wb

(2) (c) The induced emf

$$E = -d\phi/dt = -\frac{d}{dt}(3t^2+2t+3) \times 10^{-3}$$

(because given flux is in mWb).

Thus $E = (-6t-2) \times 10^{-3}$

At t = 2 sec,

$E = (-6 \times 2 - 2) \times 10^{-3} = -14$ mV.

(3) (a) The direction of current in the solenoid is clockwise. On displacing it towards the loop a current in the loop will be induced in opposite sense so as to oppose its approach. Therefore the direction of induced current as observed by the observer will be anticlockwise.

(4) (b) When north pole of the magnet is moved away, then south pole is induced on the face of the loop in front of the magnet i.e. as seen from the magnet side, a clockwise induced current flows in the loop. This makes free electrons to move in opposite direction, to plate a. Thus excess positive charge appear on plate b.

(5) **(d)** If electron is moving from left to right, the flux linked with the loop (which is into the page) will first increase and then decrease as the eletron passes by. So the induced current in the loop will be first anticlockwise and will change direction as the electron passes by.

(6) (c) $E = -\frac{d\phi}{dt}$

or $d\phi = -Edt = (0 - \phi)$

or $\phi = 4 \times 10^{-3} \times 0.1$

$= 4 \times 10^{-4}$ weber

(7) (d) $e = \frac{d\phi}{dt} = -\frac{d}{dt}[10t^2 + 5t + 1] \times 10^{-3}$

$= -[10 \times 10^{-3}(2t) + 5 \times 10^{-3}]$

at t = 5 second

$e = -[10 \times 10^{-2} + 5 \times 10^{-3}] = [0.1 + 0.005]$

$|e| = 0.105$V

(8) (a) For each spoke, the induced emf between the centre O and the rim will be the same

$$e = \frac{1}{2}B\omega L^2 = B\pi L^2 f (\because \omega = 2\pi f)$$

Further for all spokes, centre O will be positive while rim will be negative. Thus all emf's are in parallel giving total emf $e = B\pi L^2 f$

independent of the number of the spokes.

Substituting the values

$e = 4 \times 10^{-5} \times 3.14 \times (.5)^2 \times 2 = 6.28 \times 10^{-5}$ volt

(9) **(d)** Rate of decrease of area of the semicircular ring

$$-\frac{dA}{dt} = (2R)V$$

According to Faraday's law of induction, induced emf

$$e = -\frac{d\phi}{dt} = -B\frac{dA}{dt} = -B(2RV)$$

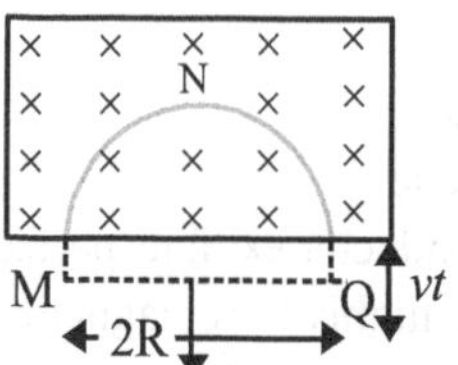

The induced current in the ring must generate magnetic field in the upward direction. Thus Q is at higher potential.

(10) (b) $E = B\ell v$

$$= 4 \times 10^{-4} \times 50 \times \frac{360 \times 1000}{60 \times 60} = 2\text{ V}$$

(11) (c) The flux through the area is

$\phi = BA\cos 57° = 42 \times 10^{-6} \times 2.5 \times 0.545$

$= 57 \times 10^{-6}$ Wb.

(12) (a) The magnetic flux linked with the loop at any instant of time t is given by

$\phi = BAN\cos\omega t$

or $\phi = 10Ba^2\cos\omega t$

Here N = 10, $A = a^2$

(13) (b) According to Lenz's Law

(14) (c) $e = Bv\ell = 0.4 \times 10^{-4} \times \frac{300 \times 10^3}{60 \times 60} \times 3 = 10^{-2}$ V

(15) (b) Magnetic flux passing through the disc is $\phi = BA$

$$= 0.01\ \frac{\text{weber}}{\text{meter}^2} \times 3.14 \times (15 \times 10^{-2}\text{ meter})^2$$

$= 7.065 \times 10^{-4}$ weber.

The line joining the centre and the circumference of the disc cuts 7.065 x 10^{-4} weber flux in one round. So, the rate of cutting flux (i.e. induced emf)

= flux × number of revolutions per second

$$= 7.065 \times 10^{-4} \times \frac{100}{60 \times 3} = 3.9 \times 10^{-4}\text{ volt.}$$

(16) (c) The magnetic flux passing through each turn of a coil of area A, perpendicular to a magnetic field B is given by

$\phi_1 = BA$.

The magnetic flux through it on rotating it through 180° will be

$\phi_2 = -BA$.(- sign is put because now the flux lines enters the coils through the outer face)

∴ change in magnetic flux

$\Delta \phi = \phi_1 - \phi_2 = -BA - (BA) = -2BA.$

Suppose this change takes in time Δt. According to Faraday's law, the emf induced in the coil is given by

$$e = -N\frac{\Delta\phi}{\Delta t} = \frac{2NBA}{\Delta t},$$

where N is number of turns in the coil. The current in the coil will be

$$i = \frac{e}{R} = \frac{1}{R}\frac{2NBA}{\Delta t}$$

where R is the resistance of the circuit. The current persists only during the change of flux i.e. for the time interval Δt second. So, the charge passed through the circuit is

$$q = i \times \Delta t = \frac{2NBA}{R}.$$

Here N = 500, B = 0.2 weber/meter^2,

A = 4.0 cm^2 = 4.0×10^{-4} meter^2 and R = 50 ohm.

$$\therefore \; q = \frac{2 \times 500 \times 0.2 \times 4.0 \times 10^{-4}}{50}$$

$= 1.6 \times 10^{-3}$ coulomb.

(17) (a) T

(18) (c) The induced emf between centre and rim of the rotating disc is

$$E = \frac{1}{2}B\omega R^2 = \frac{1}{2} \times 0.1 \times 2\pi \times 10 \times (0.1)^2$$

$= 10\pi \times 10^{-3}$ volt,

(19) (a) The induced emf

$E = B\ell v = 0.2 \times 10^{-4} \times 1 \times 180 \times 1000/3600$

$= 0.2 \times 18/3600 = 1 \times 10^{-3}$. V = 1mV

(20) (d) The induced emf is obtained by considering a strip on the disc fig. Then, the linear speed of a small element dr at a distance r from the centre is $= \omega r$. The induced emf across the ends of the small element is-

$de = B(dr)v = B\,\omega r\,dr$

Thus the induced emf across the inner and outer sides of the disc is

$$e = \int_a^b B\omega r\,dr = \frac{1}{2}B\omega(b^2 - a^2)$$

(21) (c) The induced emf $e = -(\vec{v} \times \vec{B}).\vec{\ell}$

For the part PX, $\vec{v} \perp \vec{B}$, and the angle between $(\vec{v} \times \vec{B})$ direction (the dotted line in figure and $\vec{\ell}$ is $(90 - \theta)$.

Thus

$e_P - e_X = vB\ell \cos(90 - \theta/2) = vB\ell \sin(\theta/2)$

Similarly $e_y - e_p = vB\ell \sin(\theta/2)$

Therefore induced emf

between X and Y is $e_{yx} = 2\,B\,v\,\ell \sin(\theta/2)$

(22) (a) Given, area = 10×20 cm^2 = 200×10^{-4} m^2

B = 0.5 T, N = 60, $\omega = 2\pi \times 1800/60$

$$\because e = -\frac{d(N\phi)}{dt}$$

$$= -N\frac{d}{dt}(BA\cos\omega t)$$

$= NBA\omega \sin \omega t$

$\therefore e_{max} = NAB\omega$

$= 60 \times 2 \times 10^{-2} \times 0.5 \times 2\pi \times 1800/60 = 113$ volt.

Hence, induced emf can be 111 V, 112 V and 113.04 V

(23) (b) The change is flux linked with the coil on rotating it through 180° is

$= nAB - (-nAB) = 2nAB$

$$\therefore \text{induced e.m.f.} = -\frac{d\phi}{dt}$$

$$= 2nAB/dt \text{ (numerically)} = \frac{2 \times 1 \times 0.1}{0.01} = 20\text{ V}$$

The coil is closed and has a resistance of 2.0 Ω.

Therefore i = 20/2 = 10A.

(24) (d) Initial magnetic flux $\phi_1 = 5.5 \times 10^{-4}$ weber.

Final magnetic flux $\phi_2 = 5 \times 10^{-5}$ weber.

∴ change in flux

$\Delta\phi = \phi_2 - \phi_1 = (5 \times 10^{-5}) - (5.5 \times 10^{-4}) = -50 \times 10^{-5}$ weber.

Time interval for this change, $\Delta t = 0.1$ sec.

∴ induced emf in the coil

$$e = -N\frac{\Delta\phi}{\Delta t} = -1000 \times \frac{(-50 \times 10^{-5})}{0.1} = 5\text{ volt.}$$

Resistance of the coil, R = 10 ohm. Hence induced current in the coil is

$$i = \frac{e}{R} = \frac{5\text{ volt}}{10\text{ ohm}} = 0.5\text{ ampere.}$$

(25) **(a), (26) (c), (27) (b)**

(a) The current of the battery at any instant, I = E/R.

The magnetic force due to this current

$$F_B = IBL = \frac{EB\ell}{R}$$

This magnetic force will accelerate the rod from its position of rest. The motional e.m.f. developed in the rod id B/v.

The induced current,

$$I_{induced} = \frac{B\ell v}{R}$$

The magnetic force due to the induced current,

$$F_{induced} = I_{induced} = \frac{B^2\ell^2 v}{R}$$

From Fleming's left hand rule, foce F_B is to the right and $F_{induced}$ is to the left.

Net force on the rod = $F_B - F_{induced}$.

From Newton's law,

$$F_B - F_{induced} = m\frac{dv}{dt}$$

$$\frac{EB\ell}{R} - \frac{B^2\ell^2 v}{R} = m\frac{dv}{dt}$$

On separating variables and integrating speed from v_0 to v and time from 0 to t, we have

$$\frac{dv}{E - Bv\ell} = \frac{B\ell}{mR}dt$$

$$\int_0^v \frac{dv}{E - Bv\ell} = \frac{B\ell}{mR}\int_0^t dt$$

$$-\ln\left(\frac{E - Bv\ell}{E}\right) = \frac{B^2\ell^2}{mR}t$$

$$\frac{E - Bv\ell}{E} = e^{-\frac{B^2\ell^2}{mR}t}$$

$$v = \frac{E}{B\ell}(1 - e^{-t/\tau})$$

where $t = \frac{mR}{(B\ell)^2}$

(b) The rod will attain a terminal velocity at $t \to \infty$, i.e., when $e^{-t/t} = 0$, the velocity is independent of time.

$$v_T = \frac{E}{B\ell}$$

(d) The induced current $I_{induced} = B\ell v/R$. When the rod has attained terminal speed.

$$I_{induced} = \frac{B\ell}{R} \times \left(\frac{E}{B\ell}\right) = E/R$$

The current of battery and the induced current are of same magnitude, hence net current through the circuit is zero.

(28) (c) Since both the loops are identical (same area and number of turns) and moving with a same speed in same magnetic field. Therefore same emf is induced in both the coils. But the induced current will be more in the copper loop as its resistance will be lesser as compared to that of the aluminium loop.

(29) (c) As the aircraft flies, magnetic flux changes through its wings due to the vertical component of the earth's magnetic field. Due to this, induced emf is produced across the wings of the aircraft. Therefore, the wings of the aircraft will not be at the same potential.

(30) (c) Lenz's Law is based on conservation of energy and induced emf opposes the cause of it *i.e.*, change in magnetic flux.

DAILY PRACTICE PROBLEMS

PHYSICS SOLUTIONS 45

1. **(a).** $e = -L\frac{dI}{dt} = -10\times10^{-3}\frac{1.0}{0.01} = -1\text{Volt.}$

$\therefore |e| = 1$ volt.

2. **(a).** $e = -L\frac{dI}{dt} = -L\frac{d}{dt}[3t^2+2t]$

$= -L[6t+2] = -10\times10^{-3}[6t+2]$

$(e)_{at\,t=2} = -10\times10^{-3}(6\times2+2)$

$= -10\times10^{-3}(14) = -0.14\text{ Volt}$

$|e| = 0.14$ volt.

3. **(a).** The mutual inductance between two coils depends on their degree of flux linkage, i.e., the fraction of flux linked with one coil which is also linked to the other coil. Here, the two coils in arrangement (a) are placed with their planes parallel. This will allow maximum flux linkage.

4. **(b).** $e_2 = -M\frac{dI_1}{dt} = -\frac{0-10}{3\times10^{-3}} = 3\times10^4\text{ volt} = 30\text{ kV.}$

5. **(d).** L_2 and L_3 are in parallel. Thus their combination gives

$$L' = \frac{L_2L_3}{L_2+L_3} = 0.25\text{ H}$$

The L' and L_1 are in series, thus the equivalent inductance is $L = L_1 + L' = 0.75 + 0.25 = 1\text{H}$

6. **(a).** We use $e = -L\Delta I/\Delta t$ to determine the value of induced emf. (i) $\Delta I = (7-0) = 7\text{A}$,

$\Delta t = (2-0)\text{ ms} = 2\text{ms}$

Thus $e = -4.6\times\frac{7}{2\times10^{-3}} = -16.1\times10^3$ volt

(ii) $\Delta I = 5-7 = -2\text{A}$,

$dt = (5-2)\text{ ms} = 3\text{ ms}$

Thus $e = -4.6\times\frac{(-2)}{3\times10^{-3}} = 3.07\times10^3\text{ V}$

7. **(b).** $I = I_0(1-e^{-t/\tau})$

Where $I = \frac{1}{2}I_0$ and $\tau = L/R$

Thus $\frac{1}{2}I_0 = I_0(1-e^{-t/\tau})$

or $\frac{1}{2} = e^{-t/\tau}$ or $2 = e^{+t/\tau}$ or $\log 2 = t/\tau$

Thus $t = \tau\log_e 2 = \frac{50\times10^{-3}}{0.025}\times0.693 = 1.385$

8. **(a).** $V_S = I_SZ_S \Rightarrow 22 = I_S\times220$

$\therefore I_S = 0.1\text{A}$

$$\frac{V_S}{V_P} = \frac{I_P}{I_S}$$

$$\frac{22}{220} = \frac{I_P}{0.1} \Rightarrow I_P = 0.01\text{ A.}$$

9. **(c).** $V_p = 220\text{V}, I_p = 5\text{S}, V_s = 2200\text{V}$

$P_s = \frac{P_p}{2}, I_s = ?$

$\because V_sI_s = \frac{V_pI_p}{2}$

After putting the given value you will find

$I_s = 0.25\text{ A.}$

10. **(d).** During decary of current

$$i = i_0e^{-\frac{Rt}{L}} = \frac{E}{R}e^{\frac{Rt}{L}} = \frac{100}{100}e^{\frac{100\times10^{-3}}{100\times10^{-3}}} = \frac{1}{e}\text{A.}$$

11. **(c).** In a generator e.m.f. is induced according as Lenz's rule.

The minus sign indicates that the direction of the induced e.m.f. is such as to oppose the change in current.

12. **(a).** 'Immediately' after pressing the switch S, the current in the coil L, due to its self-induction will be zero, that is $i_2 = 0$.

The current will only be found in the resistance R_1 and this will be the total current in the circuit.

$$\therefore i = i_1 = \frac{E}{R_2} = \frac{10\text{ volt}}{5.0\text{ volt}} = 2.0\text{ ampere.}$$

13. **(a).** (i) In series the same current i will be induced in both the inductors and the total magnetic-flux linked with them will be equal to the sum of the fluxes linked with them individualy, that is,

$\Phi = L_1i + L_2i.$

If the equivalent inductance be L, then $\Phi = Li$.

$\therefore Li = L_1i + L_2i$ or $L = L_1 + L_2$.

(ii) In parallel, let the induced currents in the two coils be i_1 and i_2. Then the total induced current is

$$i = i_1 + i_2 \quad \therefore \frac{di}{dt} = \frac{di_1}{dt} + \frac{di_2}{dt}$$

In parallel, the induced e.m.f. across each coil will be the same.

Hence $e = -L_1\frac{di_1}{dt} = -L_2\frac{di_2}{dt}$.

If the equivalent inductance be L, then $e = -L\frac{di}{dt}$.

$$\therefore \frac{e}{L} = -\frac{di}{dt} = -\left(\frac{di_1}{dt} + \frac{di_2}{dt}\right) = \frac{e}{L_1} + \frac{e}{L_2}$$

or $\frac{1}{L}=\frac{1}{L_1}+\frac{1}{L_2}$ or $L=\frac{L_1L_2}{L_1+L_2}$.

14. (d). If we try to find field of the small coil and then calculate flux through long solenoid, the problem becomes very difficult. So we use the following fact about mutual inductance.

$$M_{21}=M_{12},\ \frac{\phi_2}{I_1}=\frac{\phi_1}{I_2}$$

Thus if I current flows in long solenoid, then flux ϕ through small coil is the same as the flux ϕ_2 that is obtained when I current flows through the small coil. Therefore,

$\phi_2=\phi_1=$ (Field at small coil) × (area) × (turns)

$$=\left(\mu_0\frac{N_2}{\ell_2}I\right)(AN_1)=\frac{\mu_0N_1N_2AI}{\ell_2}$$

15. (a). The induced e.m.f. is

$$e=-M\frac{\Delta i}{\Delta t} \quad \text{or} \quad M=-\frac{e}{\Delta i/\Delta t}$$

Here e = 1500 volt.

$$\therefore M=-\frac{1500}{(0-3.0)/0.001}=\frac{1500\times0.001}{3.0}=0.5 \text{ henry.}$$

16. (c). $|E_S|=M\left|\frac{dI_p}{dt}\right|=M\left|\frac{d(I_0\sin\omega t)}{dt}\right|$

$=MI_0\omega|\cos\omega t|$

$\Rightarrow$ Crest value $=MI_0\omega$

$=1.5\times1\times2\pi\times50$

$=471$ V

17. (a) $U=\frac{1}{2}Li^2=\frac{1}{2}\times100\times10^{-3}\times(10)^2=5J$

18. (c) Transformation ratio $k=\frac{V_s}{V_p}\Rightarrow\frac{5}{3}=\frac{V_s}{60}\Rightarrow V_s=100$ V

19. (b) $\frac{i_p}{i_s}=\frac{N_s}{N_p}\Rightarrow\frac{i_p}{4}=\frac{1}{100}\Rightarrow i_p=0.04$ A

20. (a) $\eta=\frac{\text{Output}}{\text{Input}}\Rightarrow\frac{80}{100}=\frac{20\times120}{1000\times i_l}$

$$\Rightarrow i_l=\frac{20\times120\times100}{1000\times80}=3A.$$

21. ()
22. (b)
23. (a)
24. (b)
25. (a) 26. (a) 27. (b)

Consider a strip at a distance x from the wire of thickness dx. Magnetic flux associated with this strip

$$\Delta\phi=Badx=\frac{\mu_0 Ia}{2\pi x}dx$$

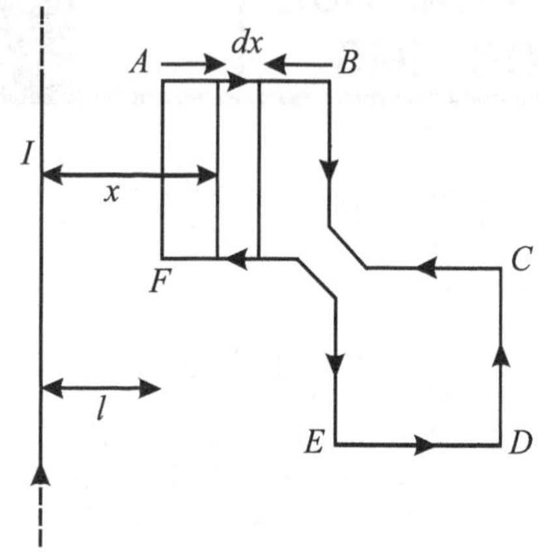

$$\phi=\frac{\mu_0Ia}{2\pi}\left[\int_l^{a+l}\frac{dx}{x}+\int_{a+l}^{2a+l}\frac{dx}{x}\right]=\frac{\mu_0Ia}{2\pi}\ln\left(\frac{2a+l}{l}\right)$$

$$M=\frac{\phi}{I_{(primary)}}\Rightarrow M=\frac{\mu_0a}{2\pi}\ln\left(\frac{2a+l}{l}\right)$$

$$e=-M\frac{dI}{dt}$$

$$e=-MI_0=-\frac{\mu_0I_0a}{2\pi}\ln\left(\frac{2a+l}{l}\right)$$

Heat produced $=\frac{e^2}{R}t=\frac{\left[\frac{\mu_0}{2\pi}\ln\left(\frac{2a+l}{l}\right)I_0\right]^2 at}{8\lambda}$

28. (a) Hysteresis loss in the core of transformer is directly proportional to the hysteresis loop area of the core material. Since soft iron has narrow hysteresis loop area, that is why soft iron core is used in the transformer.

29. (d) Efficiency of electric motor is maximum when the back emf set up in the armature is half the value of the applied battery emf.

30. (a) Transformer works on ac only, ac changes in magnitude as well as in direction.

DAILY PRACTICE PROBLEMS — PHYSICS SOLUTIONS 46

1. (b) The coil has inductance L besides the resistance R. Hence for ac it's impedance resistance $\sqrt{R^2+X_L^2}$ will be larger than it's impedance R for dc.

2. (b) $i_{r.m.s.}=\frac{i_0}{\sqrt{2}}=\frac{4}{\sqrt{2}}=2\sqrt{2}\ ampere$

3. (d) The current takes $\frac{T}{4}$ sec to reach the peak value.

In the given question $\frac{2\pi}{T}=200\pi \Rightarrow T=\frac{1}{100}$ sec

$\therefore$ Time to reach the peak value $=\frac{1}{400}$ sec

4. (b) 50 c/s or Hz

5. (b) $E=E_0\cos\omega t=E_0\cos\frac{2\pi t}{T}$

$=10\cos\frac{2\pi\times 50\times 1}{600}=10\cos\frac{\pi}{6}=5\sqrt{3}\ volt.$

6. (c) $i_{rms}=\sqrt{\frac{i_1^2+i_2^2}{2}}=\frac{1}{\sqrt{2}}(i_1^2+i_2^2)^{1/2}$

7. (c) Hot wire ammeter reads rms value of current. Hence its peak value $=i_{rms}\times\sqrt{2}=14.14\ amp$

8. (b) $V_{rms}=\frac{V_0}{\sqrt{2}}=\frac{120}{1.414}=84.8\ V$

9. (d) Peak value to $r.m.s.$ value means, current becomes $\frac{1}{\sqrt{2}}$ times.

If peak is at $t=0$, current is of the form

$i=i_0\cos 100\pi t \Rightarrow \frac{1}{\sqrt{2}}\times i_0=i_0\cos 100\pi t$

$\Rightarrow \cos\frac{\pi}{4}=\cos 100\pi t \Rightarrow t=\frac{1}{400}\sec=2.5\times 10^{-3}$ sec.

10. (b) $Z=\sqrt{R^2+X_L^2}, X_L=\omega L$ and $\omega=2\pi f$

$\therefore Z=\sqrt{R^2+4\pi^2 f^2 L^2}$

11. (b) The applied voltage is given by $V=\sqrt{V_R^2+V_L^2}$

$V=\sqrt{(200)^2+(150)^2}=250\ volt$

12. (b)

$$i=\frac{V}{\sqrt{R^2+\omega^2L^2}}=\frac{120}{\sqrt{100+4\pi^2\times 60^2\times 20^2}}=0.016A$$

13. (a) The voltage across a L-R combination is given by

$V^2=V_R^2+V_L^2$

$V_L=\sqrt{V^2-V_R^2}=\sqrt{400-144}=\sqrt{256}=16\ volt.$

14. (b) Reading of ammeter,

$$i_{rms}=\frac{V_{rms}}{X_c}=\frac{V_0\omega C}{\sqrt{2}}$$

$$=\frac{200\sqrt{2}\times 100\times(1\times 10^{-6})}{\sqrt{2}}$$

$=2\times 10^{-2}\ A=20\ mA$

15. (a)

Let the applied voltage be V, volt.
Here, $V_R=12V, V_C=5V$

$V=\sqrt{V_R^2+V_C^2}=\sqrt{(12)^2+(5)^2}=13V.$

16. (c) $Z=X_L=2\pi\times 60\times 0.7$

$\therefore i=\frac{120}{Z}=\frac{120}{2\pi\times 60\times 0.7}=0.455\ ampere$

17. (d) Current will be max at first time when $100\pi t+\pi/3=\pi/2 \Rightarrow 100\pi t=\pi/6 \Rightarrow t=1/600\ s.$

18. (d) $Z=\sqrt{R^2+X^2}=\sqrt{R^2+(2\pi fL)^2}$

$$=\sqrt{(30)^2+\left(2\pi\times 50\times\frac{0.4}{\pi}\right)^2}=\sqrt{900+1600}=50\Omega$$

$i=\frac{V}{Z}=\frac{200}{50}=4$ ampere

19. (d) In purely inductive circuit voltage leads the current by $90°$.

20. (a) Current through the bulb $i=\frac{P}{V}=\frac{60}{10}=6A$

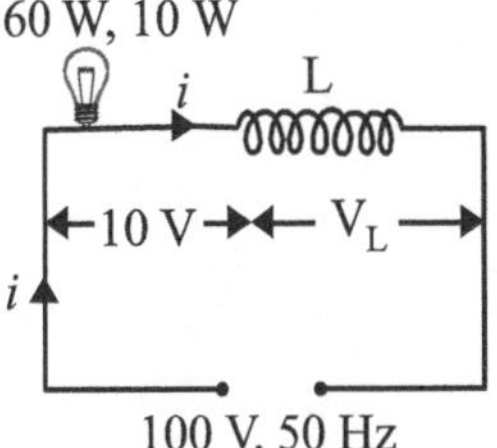

$V=\sqrt{V_R^2+V_L^2}$

$(100)^2=(10)^2+V_L^2 \Rightarrow V_L=99.5$ Volt

Also $V_L=iX_L=i\times(2\pi\nu L)$

$\Rightarrow$ $99.5=6\times2\times3.14\times50\times L=\Rightarrow L=0.052$ H.

21. **(a)** Phase angle $\tan\phi=\frac{\omega L}{R}=\frac{2\pi\times200}{300}\times\frac{1}{\pi}=\frac{4}{3}$

$\therefore\ \phi=\tan^{-1}\frac{4}{3}$

22. **(a)** The root mean square voltage is effective voltage.

23. **(c)** E = 141 sin (628t),

$E_{rms}=\frac{E_0}{\sqrt{2}}=\frac{141}{1.41}=100V$ and $2\pi f=628$

$\Rightarrow f=100Hz$

24. **(d)** Time taken by the current to reach the maximum value

$t=\frac{T}{4}=\frac{1}{4\nu}=\frac{1}{4\times50}=5\times10^{-3}$ sec

and $i_0=i_{rms}\sqrt{2}=10\sqrt{2}=14.14$ amp

25. **(a)** As in case of ac,

$V=V_0\sin(\omega t-\phi)$

The peak value $V_0=220\sqrt{2}=311$V

and as in case of ac,

$V_{rms}=\frac{V_0}{\sqrt{2}};V_{rms}=220$ V

26. **(b)** In case of ac,

$V_{av}=\frac{2}{\pi}V_0=\frac{2}{\pi}\times311=\frac{622}{\pi}V$

27. **(a)** As $\omega=2\pi f, 2\pi f=314$ i.e., $f=\frac{314}{2\times\pi}=50$Hz

28. **(d)** When ac flows through an inductor current lags behind the emf., by phase of $\pi/2$, inductive reactance, $X_L=\omega L=\pi.2f.L$, so when frequency increases correspondingly inductive reactance also increases.

29. **(c)** Like direct current, an alternating current also produces magnetic field. But the magnitude and direction of the field goes on changing continuously with time.

30. **(b)** We can use a capacitor of suitable capacitance as a choke coil, because average power consumed per cycle in an ideal capacitor is zero. Therefore, like a choke coil, a condenser can reduce ac without power dissipation.

DAILY PRACTICE PROBLEMS

PHYSICS SOLUTIONS

47

1. (a) For resonant frequency to remain same LC should be const. LC = const

$$\Rightarrow LC = L' \times 2C \Rightarrow L' = \frac{L}{2}$$

2. (b) At resonance, LCR circuit behaves as purely resistive circuit, for purely resistive circuit power factor = 1

3. (a) If the current is wattless than power is zero. Hence phase difference $\phi = 90°$

4. (c) $V_L = 46\ volts,\ V_C = 40\ volts,\ V_R = 8\ volts$

E.M.F. of source $V = \sqrt{8^2 + (46-40)^2} = 10\ volts$

5. (c) Resonant frequency $= \frac{1}{2\pi\sqrt{LC}}$ does not depend on resistance.

6. (a) At resonance LCR series circuit behaves as pure resistive circuit. For resistive circuit $\phi = 0°$

7. (b) $\cos\phi = \frac{R}{Z} = \frac{R}{\sqrt{R^2 + \omega^2 L^2}}$

$$= \frac{12}{\sqrt{(12)^2 + 4\times\pi^2\times(60)^2\times(0.1)^2}} \Rightarrow \cos\phi = 0.30$$

8. (a) $f = \frac{1}{2\pi\sqrt{LC}} \Rightarrow f \propto \frac{1}{\sqrt{C}}$

9. (b) In non resonant circuits

impedance $Z = \frac{1}{\sqrt{\frac{1}{R^2} + \left(\omega C - \frac{1}{\omega L}\right)^2}}$, with rise in frequency Z decreases *i.e.* current increases so circuit behaves as capacitive circuit.

10. (c) $\cos\phi = \frac{R}{Z} = \frac{10}{20} = \frac{1}{2} \Rightarrow \phi = 60°$

11. (d)

12. (a) $\tan\phi = \frac{X_C - X_L}{R} \Rightarrow \tan 45° = \frac{\frac{1}{2\pi fC} - 2\pi fL}{R}$

$$\Rightarrow C = \frac{1}{2\pi f(2\pi fL + R)}$$

13. (b) Resonance frequency

$$\omega = \frac{1}{\sqrt{LC}} = \frac{1}{\sqrt{8\times10^{-3}\times20\times10^{-6}}} = 2500\ rad/sec$$

Resonance current $= \frac{V}{R} = \frac{220}{44} = 5A$

14. (c) $P = V_{r.m.s.} \times i_{r.m.s.} \times \cos\phi = \frac{100}{\sqrt{2}} \times \frac{10^{-3}}{\sqrt{2}} \times \cos\frac{\pi}{3}$

$$= \frac{10^2\times10^{-3}}{2} \times \frac{1}{2} = \frac{10^{-1}}{4} = 0.025\ watt$$

15. (b) $R = X_L = 2X_C$

$Z = \sqrt{R^2 + (X_L - X_C)^2}$

$= \sqrt{(2X_C)^2 + (2X_C - X_C)^2}$

$= \sqrt{4X_C^2 + X_C^2}$

$= \sqrt{5}X_C = \frac{\sqrt{5}R}{2}$

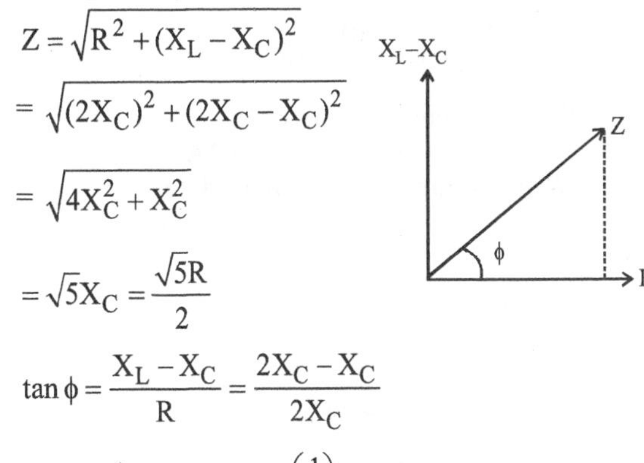

$\tan\phi = \frac{X_L - X_C}{R} = \frac{2X_C - X_C}{2X_C}$

$\tan\phi = \frac{1}{2}$; $\phi = \tan^{-1}\left(\frac{1}{2}\right)$

16. (d) Phase angle $\phi = 90°$, so power $P = V_{rms} I_{rms} \cos\phi = 0$

17. (b) $P = \frac{V_{rms}^2}{R} = \frac{(30)^2}{10} = 90\ W$

18. (c) At $A : X_C > X_L$

At $B : X_C = X_L$

At $C : X_C < X_L$

19. (d) The instantaneous values of emf and current in inductive circuit are given by $E = E_0 \sin at$ and $i = i_0 \sin\left(\omega - \frac{\pi}{2}\right)$ respectively.

So, $P_{inst} = Ei = E_0 \sin\omega t \times i_0 \sin\left(\omega t - \frac{\pi}{2}\right)$

$= E_0 i_0 \sin\omega t \cos\omega t$

$= \frac{1}{2} E_0 i_0 \sin 2\omega t \quad (\sin 2\omega t = 2\sin\omega t\cos\omega t)$

Hence, angular frequency of instantaneous power is 2ω.

20. **(d)** The voltage V_L and V_C are equal and opposite so voltmeter reading will be zero.

Also $R = 30\Omega, X_L = X_C = 25\Omega$

$$\text{So } i = \frac{V}{\sqrt{R^2 + (X_L - X_C)^2}} = \frac{V}{R} = \frac{240}{30} = 8A$$

21. **(d)** Since quality factor, $Q = \frac{1}{R}\sqrt{\frac{L}{C}}$

22. **(d)**

23. **(a)**

24. **(c)** Reactance $Z = \sqrt{(X_L - X_C)^2 + R^2}$

$= \sqrt{(80-50)^2 + 40^2}$

$= 50$ W

Power factor $= \cos\phi = \frac{R}{Z} = \frac{40}{50} = 0.8$

$I_{rms} = \frac{V_{rms}}{Z} = \frac{200V}{50\Omega} = 4$ A

Power current $= I_{rms}.\cos\phi = 4 \times 0.8$
$= 3.2$ A

Wattless current $= I_{rms}.\sin\phi = 4 \times 0.6$
$= 2.4$ A

Sol. **25-27**

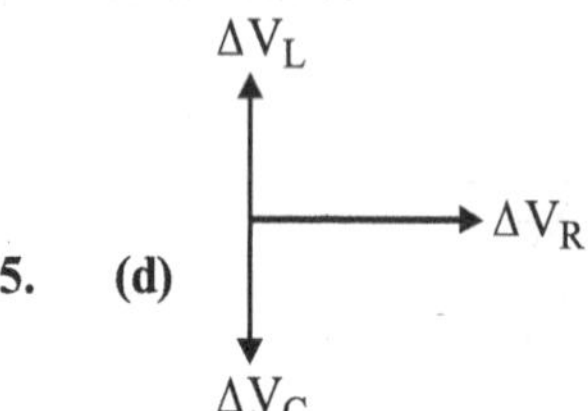

25. **(d)**

$(\Delta V_L + \Delta V_C)_{max} = \Delta V_C - \Delta V_L = 7.4 - 2.6 = 4.8$ volt

26. **(c)** $E_m = \sqrt{(\Delta V_R)^2_{max} + (\Delta V_C - \Delta V_L)^2_{max}}$

$= \sqrt{(8.8)^2 + (4.8)^2} = 10$ volt

27. **(a)** If f↑ then $(\Delta V_L)_{max}$ ↑

28. **(a)** Capacitive reactance $XC = \frac{1}{\omega C}$. When capacitance (C) increase, the capacitive reactance decreases. Due to decrease in its values, the current in the circuit will increase $\left(I = \frac{E}{\sqrt{R^2 + X_C^2}}\right)$ and hence brightness of source (or electric lamp) will also increase.

29. **(b)** The phase angle for the *LCR* circuit is given by

$$\tan\phi = \frac{X_L - X_C}{R} = \frac{\omega L - \frac{1}{\omega C}}{R}$$

where X_L, X_C are inductive reactance and capacitive reactance respectively when $X_L > X_C$ then $\tan\phi$ is positive i.e. ϕ is positive (between 0 and $\pi/2$). Hence emf leads the current.

30. **(a)** If resistor is used in controlling ac supply, electrical energy will be wasted in the form of heat energy across the resistance wire. However, ac supply can be controlled with choke without any wastage of energy. This is because, power factor ($\cos\phi$) for resistance is unity and is zero for an inductance. $[P = EI\cos\phi]$.

DAILY PRACTICE PROBLEMS

PHYSICS SOLUTIONS

48

1. (d) $\mu_0 = 4\pi \times 10_{-7},\ \varepsilon_0 = 8.85 \times 10^{-12}\ \dfrac{N-m^2}{C^2}$

$$\text{so } c = \frac{1}{\sqrt{\mu_0 \varepsilon_0}} = 3\times10^8\ \frac{\text{meter}}{\text{sec}}.$$

2. (b) Wavelength of visible spectrum is 3900 Å – 7800 Å.
3. (a) $\lambda_{\gamma-rays} < \lambda_{x\text{-}rays} < \lambda_{\alpha-rays} < \lambda_{\beta-rays}$.
4. (c) Electric field between the plates of the capacitor is given by

$$E = \frac{\sigma}{\epsilon_0} \text{ or } \frac{q}{A\,\epsilon_0}$$

Flux through the area considered

$$\phi = \frac{q}{A\,\epsilon_0} \times \frac{A}{4} = \frac{q}{4\,\epsilon_0}$$

Displacement corrent $i_d = \epsilon_0 \dfrac{d\phi_E}{dt}$

$$= \epsilon_0 \times \frac{d}{dt}\left(\frac{q}{4\,\epsilon_0}\right) = \frac{i}{4}$$

5. (a) Electric field between the plates is

$$E = \frac{Q}{\epsilon_0 A}$$

$$\therefore \phi_E = E.A \text{ or } \frac{Q}{\epsilon_0 A} \times A$$

$$\therefore i_d = \epsilon_0 \frac{d\phi_E}{dt} \text{ or } \epsilon_0 \frac{d}{dt}\left(\frac{Q}{\epsilon_0}\right)$$

$$\therefore i_d = \frac{dQ}{dt} = i \text{ (charging current)}$$

Hence $i_d = 1A$

6. (c) $E \times B$
7. (b) Since the direction of propogation of EM wave is given by $E \times B$ $\quad \therefore \left(\hat{j} \times \hat{i} = -\hat{k}\right)$
8. (d) Speed of E.M. wave $= \dfrac{1}{\sqrt{\mu_0 \epsilon_0 \mu_r \epsilon_r}}$ in medium hence it will travel with different speed in different medium.
9. (b) $I = \dfrac{1}{2} \epsilon_0 c E_0^{\ 2}$

$$E_0 = \sqrt{\frac{2I}{\epsilon_0 c}} \text{ or } \sqrt{\frac{2\times500\times10^9\times36\pi}{\pi\times3\times10^8}}$$

$$E_0 = 2\sqrt{3} \times 10^2 \text{N/C}$$

10. (b) $I = \dfrac{P}{A}$ or $\dfrac{2}{4\pi\times4} = \dfrac{1}{8\pi}$ W/m^2

$$I = \frac{1}{2} \epsilon_0 E_0^{\ 2} c$$

$$E_0 = \sqrt{\frac{2I}{\epsilon_0 c}} \text{ or } \sqrt{\frac{2\times1\times36\pi\times10^9}{8\pi\times3\times10^8}} = \sqrt{30} \text{ N/C}$$

11. (a) $U_B = \dfrac{B_0^2}{4\mu_0}$

Also $\dfrac{E_0}{B_0} = c \quad \therefore B_0 = \dfrac{E_0}{c}$

Hence $B_0 = \dfrac{E_0}{\dfrac{1}{\sqrt{\mu_0 \epsilon_0}}}$

$$\therefore U_B = \frac{E_0^2 \mu_0 \epsilon_0}{4\mu_0} \text{ or } \frac{100\times8.84\times10^{-12}}{4}$$

$\therefore U_B = 2.21 \times 10^{-10}$ J/m^3

12. (a) The speed of electromagnetic waves and in a medium is given by

$$v = \frac{1}{\sqrt{(\mu\varepsilon)}}$$

Where μ and ε are absolute permeability and absolute permittivity of the medium.

We know that, $\mu = \mu_0\mu_r$ and $\varepsilon = \varepsilon_0\varepsilon_r$. Hence

$$v = \frac{1}{\sqrt{(\mu_0\mu_r.\varepsilon_0\varepsilon_r)}} = \frac{1}{\sqrt{(\mu_0\varepsilon_0)}} \times \frac{1}{\sqrt{(\mu_r\varepsilon_r)}}$$

$$\text{or } v = \frac{c}{\sqrt{(\mu_r\varepsilon_r)}} \text{ or } \varepsilon_r = \frac{c^2}{v^2(\mu_r)}$$

$$\therefore \varepsilon_r = \frac{(3\times10^8)^2}{(2\times10^8)^2\times1} = 2.25$$

13. (a) Given $B_y = 2\times10^{-7} \sin(0.5\times10^3 x + 1.5\times10^{11} t)$

Comparing it with a standard equation for a progresive wave travelling along the negative direction of x-axis is

$$y = r \sin \frac{2\pi}{\lambda}(x + vt) = r \sin\left(\frac{2\pi x}{\lambda} + \frac{2\pi vt}{\lambda}\right)$$

$$= r\sin\left(\frac{2\pi x}{\lambda} + 2\pi vt\right)$$

$2\pi v = 1.5 \times 10^{11}$

$$v = \frac{1.5\times10^{11}}{2\pi} = 23.9 \times 10^9 \text{ Hz} = 23.9 \text{ Hz}$$

14. (c) The given equation
$E_y = 0.5 \cos[2\pi \times 10^8 (t - x/c)]$ ..... (1)
indicates that the electromagnetic waves are propagating along the positive direction of X-axis.
The standard equation of electromagnetic wave is given by $E_y = E_0 \cos\omega(t - x/c)$ ..... (2)
Comparing the given eq. (1) with the standard eq. (2), we get $\omega = 2\pi \times 10^8$
or $2\pi v = 2\pi \times 10^8$
$\therefore v = 10^8$ per second

$$\text{Now, } \lambda = \frac{c}{v} = \frac{3\times10^8}{10^8} = 3 \text{ m}$$

15. (a) The maximum value of magnetic field (B_0) is given by

$$B_0 = \frac{E_0}{c} = \frac{E_0}{c} = 10^{-6} \text{ tesla}$$

The magnetic field will be along Z-axis
The maximum magnetic force on the electron is
$F_b = |q(\mathbf{v} \times \mathbf{B})| = q v B_0$
$= (1.6 \times 10^{-19}) \times (2.0 \times 10^7) \times (10^{-6})$
$= 3.2 \times 10^{-18}$ N

16. (c) β-rays are beams of fast electrons.

17. (b)

18. (d) $v = \dfrac{c}{\sqrt{\mu_r \varepsilon_r}} = \dfrac{3\times10^8}{\sqrt{1.3\times2.14}} = 1.8\times10^8 \text{m/sec}$

19. (c) $\lambda_m > \lambda_v > \lambda_x$

20. (d) Direction of wave propagation is given by $\vec{E}\times\vec{B}$.

21. (c) Wave impedance $Z = \sqrt{\dfrac{\mu_r}{\varepsilon_r}} \times \sqrt{\dfrac{\mu_0}{\varepsilon_0}}$

$$= \sqrt{\frac{50}{2}} \times 376.6 = 1883\Omega$$

22. (a)

23. (a) β– rays are beams of fast electrons.

24. (a) Refractive index $= \sqrt{\dfrac{\mu\varepsilon}{\mu_0\varepsilon_0}}$

$\Rightarrow$ Then refractive index $= \sqrt{\dfrac{\varepsilon}{\varepsilon_0}} = 2$

$\therefore$ Speed and wavelength of wave becomes half and frequency remain unchanged.

For 25 - 27
$h = 150$ km $= 150 \times 10^3$ m
$N_m = 9 \times 10^{10}$ per m^3
$D = 250$ km $= 250 \times 10^3$ m

25. (a) Critical frequency of layer

$$f_c = 9\sqrt{N_m} = 9\times\sqrt{9\times10^{10}} = 2.7\times10^6 \text{ Hz.}$$

26. (b) Maximum usuable frequency

$$f = f_c\sqrt{1+\frac{D^2}{4h^2}} = 2.7\times10^6 \times \sqrt{1+\left(\frac{250\times10^3}{4\times150\times10^3}\right)^2}$$

$= 3.17 \times 10^6$ Hz.

27. (c) If angle of incidence at this layer is ϕ_i, from second law of $f = f_c \sec \phi_i$.

$$\sec\phi_i = \frac{f}{f_c} = \frac{3.17\times10^6}{2.7\times10^6} = 1.174$$

$$\phi_i = \sec^{-1}(1.174) = 31.6°$$

28. (d) The electromagnetic waves of shorter wavelength do not suffer much diffraction from the obstacles of earth's atmosphere so they can travel long distance.

29. (b) The wavelength of these waves ranges between 300 Å and 4000 Å that is smaller wavelength and higher frequency. They are absorbed by atmosphere and convert oxygen into ozone. They cause skin diseases and they are harmful to eye and cause permanent blindness.

30. (b) Radio waves can be polarised because they are transverse in nature. Sound waves in air are longitudinal in nature.

DAILY PRACTICE PROBLEMS

PHYSICS SOLUTIONS

49

(1) (a) Here, $n = \frac{360}{\theta} = \frac{360}{90} = 4$

$\therefore$ n is an even number.

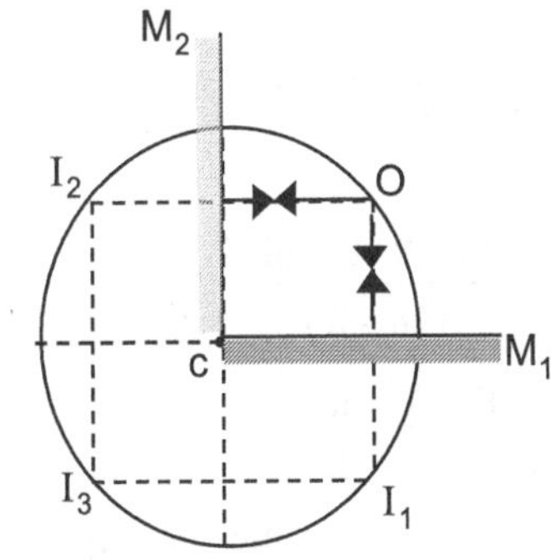

Thus, number of images formed = n – 1 = 3. All these three images lie on a circle with centre at C (The point of intersection of mirrors M_1 and M_2) and whose radius is equal to the distance between C and object.

(2) (c) Here, $[n] = 5 \Rightarrow n - 1 \le 5 \le n$

$\therefore \frac{360}{\theta} - 1 \le 5 \le \frac{360}{\theta}$

or, $\theta \ge \frac{360}{6}$ or $\theta \le \frac{360}{5}$

$\therefore 60° \le \theta \le 72°$

(3) (d) For the given $\theta = 50°$,

$n = \frac{360}{\theta} = \frac{360}{50} = 7.2$

The integer value of (7.2) is 7.
Thus number of images formed is 7.

(4) (a) The situation is illustrated in figure.

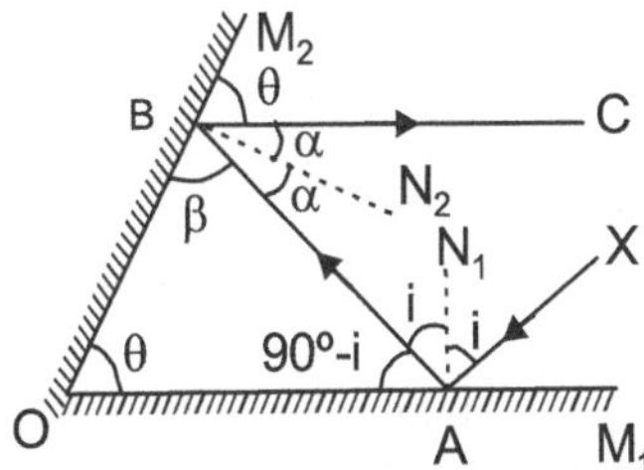

XA is the incident ray. BC is the final reflected ray. It is given that BC is parallel to mirror M_1. Look at the assignment of the angles carefully. Now N_2 is normal to mirror M_2. Therefore $\beta = \theta$

Then from Δ OAB

$\theta + \beta + 90° - i = 180°$

or $\theta + \theta + 90° - i = 180°$ or $i = 2\theta - 90°$

Thus if the angle of incidence is $i = 2\theta - 90°$, then the final reflected ray will be parallel to the first mirror.

(5) (c) Small and erect image is formed only by convex mirror. Plane mirror from images equal to object and concave mirror form images bigger than object.

(6) (b) The image will be formed by the plane mirror at a 30 cm behind it, while the image by convex mirror will be formed at 10 cm behind the convex mirror.

Since for convex mirror u = – 50 cm

v = 10 cm

$\frac{1}{f} = \frac{1}{-50} + \frac{1}{10} = \frac{-1+5}{50} = \frac{4}{50}$

$f = \frac{50}{4} = 12.5$ cm

Therefore the radius of curvature of convex mirror is r = 2 f = 25 cm

(7) (a) The image of object O from mirrror M_1 is I_1 and the image of I_1 (the vitual object) from mirror M_2 is I_3. The image of object O from mirror M_2 is I_2 and the image of I_2 (the virtual object) from mirror M_1 is I_4. Notice that this interpretation, according to ray diagram rules, is valid only for Fig. (A). All others are inconsistent.

(8) (a) Angle between incident ray and mirror = 90° – 30° = 60°

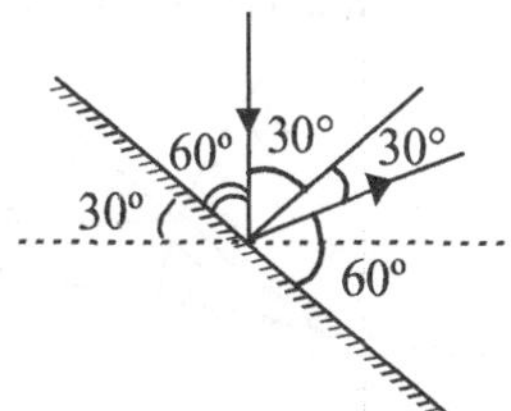

By law of reflection $\angle i = \angle r$

So angle of reflection $\angle r = 30°$.

Hence angle between mirror and reflected ray = 60°

(9) (b) As shown in figure, ray AB goes to mirror M_1, gets reflected and travels along BC and then gets reflected by M_2 and goes in CD direction. If the angle between M_1 and M_2 be α, then

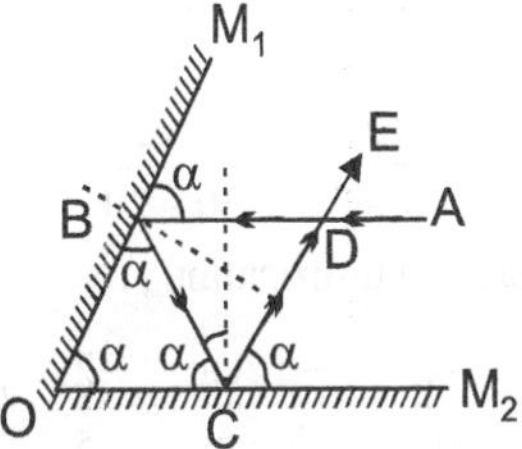

In Δ OBC, $\angle$OBC and $\angle$OCB are equal to α

$\therefore 3\alpha = 180°$

$\alpha = 60°$

(10) (a) Here, Object distance, u = – 15cm
focal length, f = – 5 cm
Object height, $h_0 = 0.2$ cm

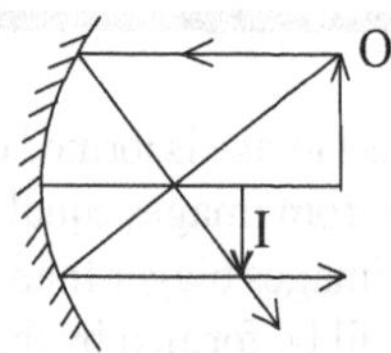

We know, mirror formula,

$$\frac{1}{v}+\frac{1}{u}=\frac{1}{f}$$

$$\Rightarrow V = \frac{uf}{u-f}$$

$$= \frac{(-15)(-5)}{-15+5} = -7.5 \text{ cm}$$

Again, magnification,

$$m = -\frac{v}{u} = -\frac{-7.5}{-15} = -\frac{1}{2}$$

Now, $|m| = \frac{h_i}{h_o}$

$$\Rightarrow h_i = |m| h_o = \frac{0.2}{2} \text{ cm} = 0.1 \text{ cm}$$

Thus, the image is formed at 7.5 cm from the pole of the mirror and its size is 0.1 cm.

(11) (a)

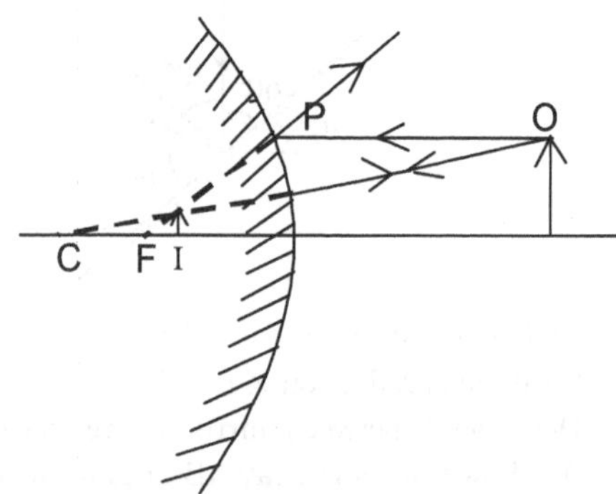

Here, Object distance, u = – 30 cm
Focal length, f = + 20 cm

We know, mirror formula, $\frac{1}{v}+\frac{1}{u}=\frac{1}{f}$

$$\Rightarrow v = \frac{uf}{u-f} = \frac{-30\times 20}{-30-20} = +12 \text{ cm}$$

Again, magnification,

$$m = \frac{-v}{u} = \frac{-12\text{cm}}{30\text{cm}} = \frac{2}{5}$$

Now, Image height = m × object height

$$= \frac{2}{5}\times 0.5 \text{ cm} = 0.2 \text{ cm}$$

Thus the image is formed behind the mirror at a distance of 12 cm from the pole. Image height is 0.2 cm.

(12) (d) u = –50 cm, f = 25cm

$$\frac{1}{25} = -\frac{1}{50}+\frac{1}{v};$$

$$\frac{1}{v} = \frac{1}{25}+\frac{1}{50} = \frac{2+1}{50} = \frac{3}{50}$$

$$v = \frac{50}{3} = 16.6 \text{ cm.}$$

(13) (a) We know, $\frac{1}{v}+\frac{1}{u}=\frac{2}{R}$

$$\Rightarrow v = \frac{uR}{2u-R}$$

$$\frac{dv}{dt} = \frac{(2u-R).R - uR.2}{(2u-R)^2}.\frac{du}{dt}$$

$$= -\left(\frac{R}{2u-R}\right)^2.\frac{du}{dt}$$

$\therefore$ Speed of image $\left|\frac{dv}{dt}\right| = \left(\frac{R}{2u-R}\right)^2.$

$$\left|\frac{du}{dt}\right| = \left(\frac{R}{2u-R}\right)^2 V_0$$

(14) (a) For shrot linear object du and dv represent size of object and image respectively.

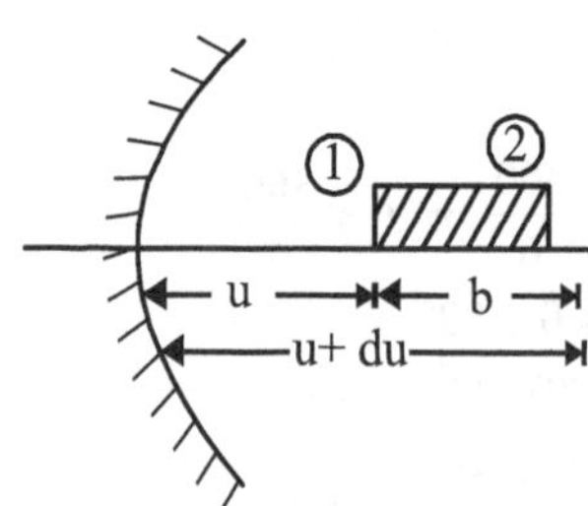

We know, $\frac{1}{v}+\frac{1}{u}=\frac{1}{f}$

$$\Rightarrow dv = -\frac{v^2}{u^2}.du$$

$$\Rightarrow |db| = \frac{v^2}{u^2}|du|$$

$$= \left(\frac{f}{f-u}\right)^2.b$$

15. (b) $\therefore m = -\frac{v}{u}$. Also $\frac{1}{f} = \frac{1}{v}+\frac{1}{u} \Rightarrow \frac{u}{f} = \frac{u}{v}+1$

$$\Rightarrow -\frac{u}{v} = 1-\frac{u}{f} \Rightarrow \frac{-v}{u} = \frac{f}{f-u}$$

so $m = \frac{f}{f-u}$

16. (b) Given u = – 15 cm, f = – 10 cm, O = 1 cm

$$\frac{1}{v}+\frac{1}{u}=\frac{1}{f},\ \frac{1}{v}=\frac{1}{f}-\frac{1}{u}=\frac{1}{-10}-\frac{1}{-15}$$

∴ v = – 30 cm

$$\frac{I}{O}=-\frac{v}{u}=-\frac{-30}{-15}=-2$$

I = – 2 × 1 = – 2 cm

Image is inverted and on the same side (real) of size 2 cm.

17. (b) As shown in the figure, when the object (*O*) is placed between *F* and *C*, the image (*I*) is formed beyond *C*. It is in this condition that when the student shifts his eyes towards left, the image appears to the right of the object pin.

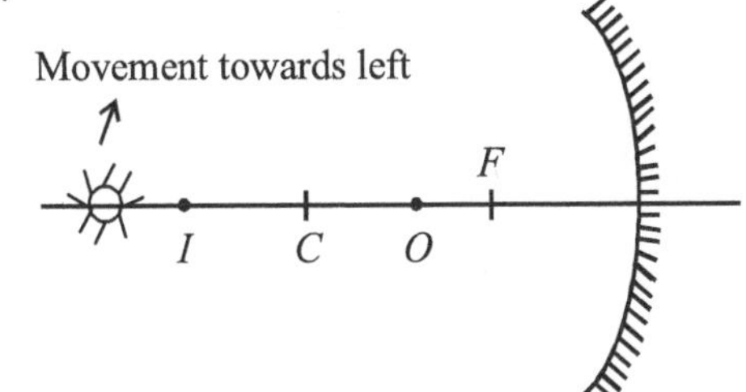

18. (a)

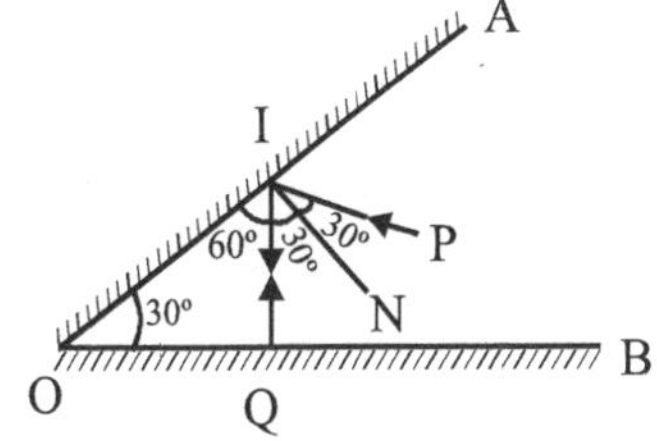

∠i = ∠r = 30°

∴ ∠ OIQ = 60°

∴ IOQ = 90° – 60° = 30°

19. (d) The image formed by a convex mirror is always virtual.

20. (d) From the ray diagram.

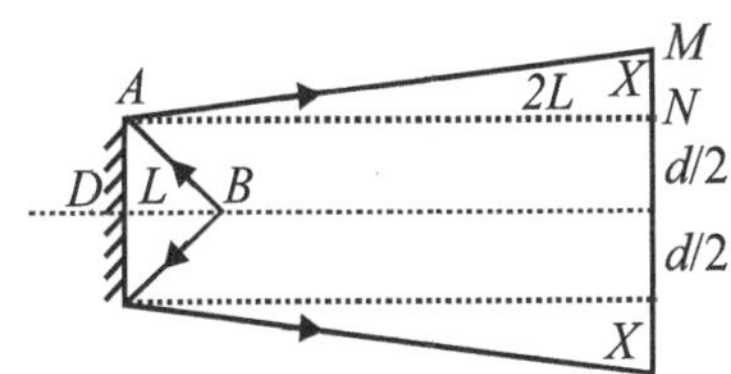

In Δ*ANM* and Δ*ADB*

∠*ADB* = ∠*ANM* = 90°

∠*MAN* = ∠*BAN* (laws of reflection)

Also ∠*BAN* = ∠*ABD*

⇒ ∠*MAN* = ∠*ABD*

∴ Δ*ANM* is similar to Δ*ADB*

$$\therefore\ \frac{x}{2L}=\frac{d/2}{L}\ \text{or}\ x=d$$

So, required distance = *d* + *d* + *d* = 3*d*.

21. (a) $m=-n\ ;\ m=\frac{f}{f-u}$

$$-n=\frac{-f}{-f-u}\ \Rightarrow nf+nu=-f$$

$$nu=-f-nf\quad\Rightarrow u=\frac{-(n+1)}{n}f$$

22. (d) Here image can be real or virtual. If the image is real f = –30, u = ?, m = –3

$$m=\frac{f}{f-u}\ \Rightarrow -3=\frac{-30}{-30-u}\ ;\ u=-40\text{ cm.}$$

If the image is virtual

$$m=\frac{f}{f-u}\ \Rightarrow\ 3=\frac{-30}{-30-u}\Rightarrow u=-20\text{ cm.}$$

23. (b) By keeping the incident ray is fixed, if plane mirror rotates through an angle θ reflected ray rotates through an angle 2θ

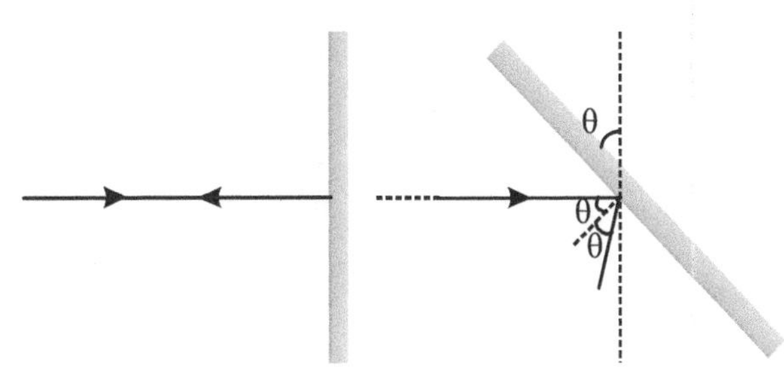

24. (d)

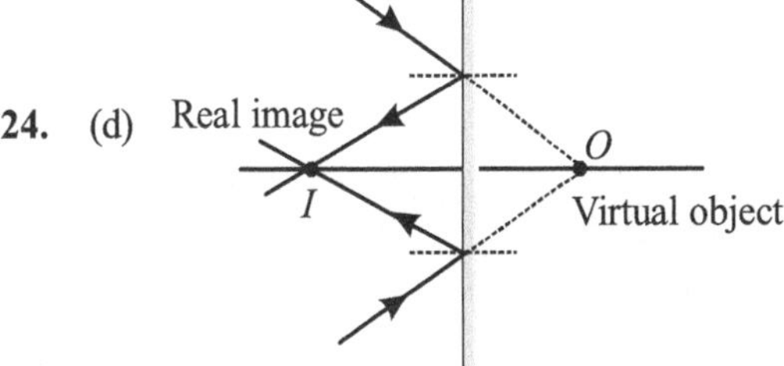

25. (b,c) Convex mirror and concave lens form, virtual image for all positions of object.

(26) (a); (27) (d)

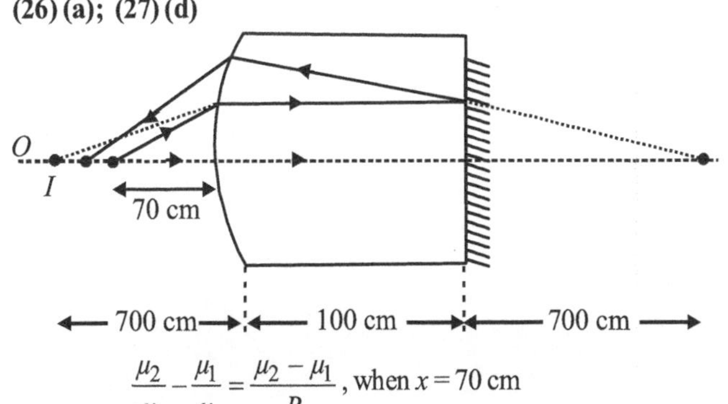

$$\frac{\mu_2}{v}-\frac{\mu_1}{u}=\frac{\mu_2-\mu_1}{R},\ \text{when } x=70\text{ cm}$$

$$\frac{1.5}{v}-\frac{1.2}{-70}=\frac{1.5-1.2}{20}$$

$$\Rightarrow v=\frac{20\times70\times1.5}{-1.2\times20+0.3\times70}$$

$\Rightarrow v = -700$ cm

$$\frac{\mu_2}{v} - \frac{\mu_1}{u} = \frac{\mu_2 - \mu_1}{R}$$

$$\frac{1.2}{v} - \frac{1.5}{900} = \frac{1.2 - 1.5}{-20}$$

$$\Rightarrow v = \frac{900 \times 200 \times 1.2}{1.5 \times 200 - 900 \times 3}$$

$\Rightarrow v = -90$ cm

Similarly, for x = 80 cm

$v = 80$ cm

and for $x = 90$ cm

$v = 70$ cm

28. (d)

29. (d) When an object is placed between two plane parallel mirrors, then infinite number of images are formed. Images are formed due to multiple reflections. At each reflection, a part of light energy is absorbed. Therefore, distant images get fainter.

30. (d) The size of the mirror does not affect the nature of the image except that a bigger mirror forms a brighter image.

DAILY PRACTICE PROBLEMS

PHYSICS SOLUTIONS

50

(1) (a) We know that $\delta = A(\mu - 1)$

or $\mu = 1 + \frac{\delta}{A}$

Here $A = 6°, \delta = 3°$, therefore

$\mu = 1 + \frac{3}{6} = 1.5$

(2) (c) According to given problem

$A = 30°, i_1 = 60°$ and $\delta = 30°$ and as in a prism

$\delta = (i_1 + i_2) - A,\ 30° = (60 + i_2) - 30$ i.e., $i_2 = 0$

So the emergent ray is perpendicular to the face from which it emerges.

Now as $i_2 = 0, r_2 = 0$

But as $r_1 + r_2 = A,\ r_1 = A = 30°$

So at first face

$1 \times \sin 60° = \mu \sin 30°$ i.e., $\mu = \sqrt{3}$

(3) (a) As are know, $\mu = \frac{c}{v} \Rightarrow v = \frac{c}{\mu} = \frac{3}{4} \times 3 \times 10^8$

$= 2.25 \times 10^8$ m/s

As, $c = \nu\lambda_0$ and $v = \nu\lambda$

$\Rightarrow \lambda/\lambda_0 = \frac{v}{c} = \frac{1}{\mu}$

i.e., $\lambda = \lambda_0/\mu = \frac{3}{4} \times 6000Å = 4500Å$

(4) (d) Herer $+ 90° + r' = 180°$

$\Rightarrow r' = 90° - r$

or, $r' = (90° - i)$ as $\angle i = \angle r$

Now, according to Snell's law :

$\sin i = \mu \sin r' = \mu \sin(90° - i)$

or, $\tan i = \mu$

or, $i = \tan^{-1}\mu = \tan^{-1}(1.5)$

(5) (b) Here the requirement is that $i > c$

$\Rightarrow \sin i > \sin c \Rightarrow \sin i > \frac{\mu_2}{\mu_1}$

From Snell's law $\mu_1 = \frac{\sin\alpha}{\sin r}$

Also in ΔOBA

$r + i = 90° \Rightarrow r = (90 - i)$

Hence from equation (ii)

$\sin\alpha = \mu_1 \sin(90 - i)$

$\Rightarrow \cos i = \frac{\sin\alpha}{\mu_1}$

$\sin i = \sqrt{1-\cos^2 i} = \sqrt{1-\left(\frac{\sin\alpha}{\mu_1}\right)^2}$ ...(iii)

From equation (i) and (ii)

$\Rightarrow \sin^2\alpha < \left(\mu_1^2 - \mu_2^2\right)\ \sin\alpha < \sqrt{\mu_1^2 - \mu_2^2}$

$\alpha_{max} = \sin^{-1}\sqrt{\mu_1^2 - \mu_2^2}$

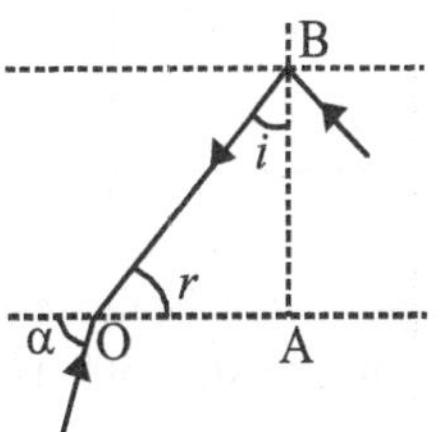

(6) (b) From the information given, it is clear that the apparent depth is 2.58 mm and the real depth is 4mm. Therefore, the refractive index will be

$\mu = \frac{R}{A} = \frac{4}{2.58} = 1.55$

(7) (d) The apparent shift of the bottom point upwards will be

$x = x_1 + x_2$

$= t_1\left(1 - \frac{1}{\mu_1}\right) + t_2\left(1 - \frac{1}{\mu_2}\right)$

$= 4\left(1 - \frac{1}{(4/3)}\right) + 2\left(1 - \frac{1}{(3/2)}\right)$

$= 4\left(1 - \frac{3}{4}\right) + 2\left(1 - \frac{2}{3}\right) = 1.67$ cm.

(8) (d) Since $v = \frac{C}{n}$

The time taken are

$t_2 = \frac{20\,(1.63)}{C},\ t_1 = \frac{20\,(1.47)}{C}$

Therefore , the difference is

$t_2 - t_1 = \frac{20(1.63 - 1.47)}{C} = \frac{20 \times 0.16}{3 \times 10^8} = 1.07 \times 10^{-8}$ sec.

(9) (b) As the beam just suffers TIR at interface of region III and IV

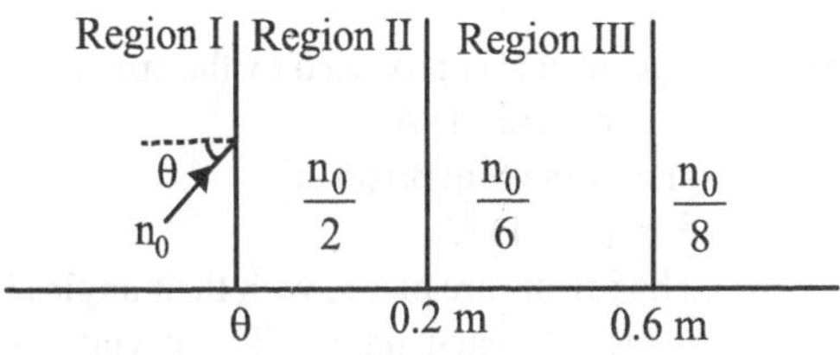

$n_0 \sin\theta = \frac{n_0}{2}\sin\theta_1 = \frac{n_0}{6}\sin\theta_2 = \frac{n_0}{8}\sin 90°,$

$\sin\theta\frac{1}{8} \Rightarrow \theta = \sin^{-1}\frac{1}{8}$

(10) (b) The ray of light returns back from the polished face AC.

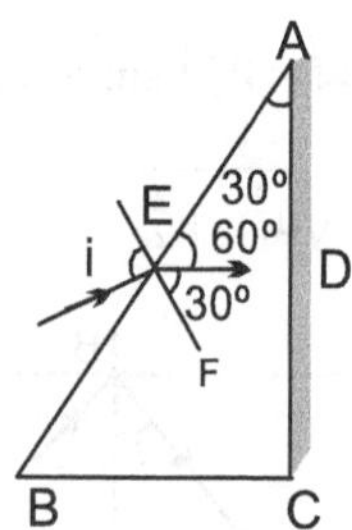

$\therefore \angle ADE = 90°$. From the figure it is clear that the angle of refraction at face AB is 30°. Hence from Snell's law

$$\mu = \frac{\sin i}{\sin r}$$

As $\mu = \sqrt{2}$ and $r = 30° \therefore \sqrt{2} = \frac{\sin i}{\sin 30°}$

or $\sin i = \frac{\sqrt{2}}{2} = \frac{1}{\sqrt{2}} = \sin 45°$

$\therefore i = 45°$

(11) (a) $A = r_1 + r_2 = 60$ ....(1)

In minimum deviation position

$r_1 = r_2$ ....(2)

From eqs. (1) and (2)

$A = 2r_1 = 60°$

$\therefore r_1 = 30°$

$$\therefore n = \frac{\sin i}{\sin r} = \frac{\sin 50°}{\sin 30°} = 1.532$$

$$n = \frac{\sin\frac{A+\delta_m}{2}}{\sin\frac{A}{2}} \text{ or } 1.532 = \frac{\sin\frac{60+\delta_m}{2}}{\sin 30°}$$

$$\sin\frac{60+\delta_m}{2} = \frac{1.532}{2} = 0.766$$

$\therefore \delta_m = 40°$

(12) (b) $\delta_1 = \delta_2$

$(\mu_1 - 1)A_1 = (\mu_2 - 1)A_2$

$(1.5 - 1)\,6 = (1.6 - 1)A_2$

$A_2 = 5°$

(13) (a) The deviation produced by the crown prism is

$\delta = (\mu - 1)A$

and by the flint prism is

$\delta' = (\mu' - 1)A'$

The prisms are placed with their angles inverted with respect to each other . The deviations are also in opposite directions. Thus, the net deviation is

$D = \delta - \delta' = (\mu - 1)A - (\mu' - 1)A'$ ....(1)

If the net deviation for the mean ray is zero,

$(\mu - 1)A = (\mu' - 1)A'$

or, $A' = \frac{(\mu - 1)}{(\mu' - 1)}A = \frac{1.517 - 1}{1.620 - 1} \times 5° = 4.2°$.

The angular dispersion produced by the crown prism is

$\delta_v - \delta_r = (\mu_v - \mu_r)A$

and that by the flint prism is

$\delta'_v - \delta'_r = (\mu'_v - \mu'_r)A'$

The net angular dispersion is ,

$\delta = (\mu_v - \mu_r)A - (\mu'_v - \mu'_r)A'$

$= (1.523 - 1.514) \times 5° - (1.632 - 1.613) \times 4.2°$

$= -0.0348°$.

The angular dispersion has magnitude 0.0348°.

(14) (a) $\mu_v = 1.5230, \mu_r = 1.5145, \omega = ?$

Mean refractive index,

$$\mu = \frac{\mu_v + \mu_r}{2} = \frac{1.5230 + 1.5145}{2}$$

$\mu = 1.5187$

$$\omega = \frac{\mu_v - \mu_r}{\mu - 1}$$

$$= \frac{1.5230 - 1.5145}{1.5187 - 1} = \frac{0.0085}{0.5187} = 0.0163$$

(15) (c) Here, $\omega = 0.021$; $\mu = 1.53$; $\omega' = 0.045$; $\mu' = 1.65$; $A' = 4.2°$

For no dispersion, $\omega\delta + \omega'\delta' = 0$

or $\omega A(\mu - 1) + \omega' A'(\mu - 1) = 0$

$$\text{or } A = -\frac{\omega' A'(\mu' - 1)}{\omega(\mu - 1)}$$

$$= -\frac{0.045 \times 4.2 \times (1.65 - 1)}{0.021 \times (1.53 - 1)} = -11.04°$$

Net deviation,

$\delta + \delta' = A(\mu - 1) + A'(\mu' - 1)$

$= -11.04\,(1.53 - 1) + 4.2\,(1.65 - 1)$

$= -11.04 \times 0.53 + 4.2 \times 0.65$

$= -5.85 + 2.73 = 3.12°$

(16) (d) At first face of the prism as $i_1 = 0$,

$\sin 0 = 1.5 \sin r_1$ i.e.,, $r_1 = 0$

And as for a prism

$r_1 + r_2 = A$ so $r_2 = A$ ......(1)

But at second face, as the ray just fails to emerge

i.e., $r_2 = \theta_C$ ......(2)

So from Eqn,.(1) and (2)

$A = r_2 = \theta_C$

But as $\theta_C = \sin^{-1}\left[\frac{1}{\mu}\right] = \sin^{-1}\left[\frac{2}{3}\right] = 42°$

So $A = 42°$

(17) (a) Here, $\mu_b = 1.532$ and $\mu_r = 1.514$ $A = 8°$.

Angular dispersion

$= (\mu_b - \mu_r)A = (1.532 - 1.514) \times 8$

$= 0.018 \times 8 = 0.144°$.

(18) (c) Velocity of the ball when it reaches at the height of 12.8 m. above the surface is $v = \sqrt{2\times10\times7.2} = 12\ m/s$

Height of the ball from surface as seen by fish

$$h' = \mu_h \Rightarrow \frac{dh'}{dt} = \mu\frac{dh}{dt}$$

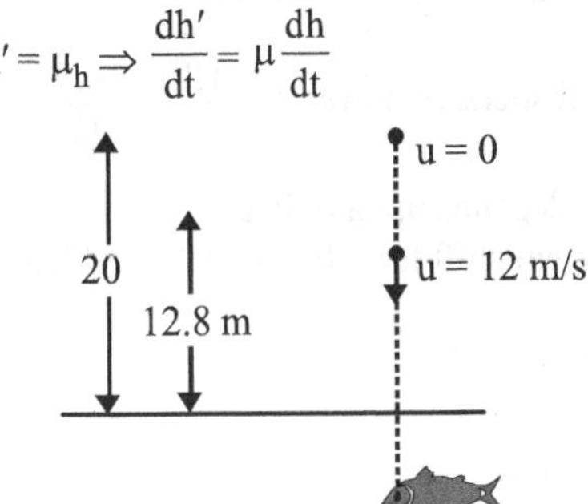

$$\Rightarrow v' = \mu v = \frac{4}{3}\times12 = 16\ m/s.$$

(19) (a) Suppose, the angle of the crown prism needed is A and that of the flint prism is A′. We have

$$\omega = \frac{\mu_v - \mu_r}{\mu - 1} \text{ or, } \mu_v - \mu_r = (\mu - 1)\,\omega$$

The angular dispersion produced by the crown prism is

$(\mu_v - \mu_r)A = (\mu - 1)\,\omega A$

Similarly, the angular dispersion produced by the flint prism is $(\mu' - 1)\,\omega' A'$

For achromatic combination, the net dispersion should be zero. Thus,

$(\mu - 1)\,\omega A = (\mu' - 1)\,\omega' A'$

or, $\frac{A'}{A} = \frac{(\mu-1)\,\omega}{(\mu'-1)\,\omega'} = \frac{0.517\times0.03}{0.621\times0.05} = 0.50$ ......(1)

The deviation in the yellow ray produced by the crown prism is $\delta = (\mu - 1)A$ and by the flint prism is $\delta' = (\mu' - 1)A'$. The net deviation produced by the combination is

$\delta - \delta' = (\mu - 1)A - (\mu' - 1)A'$

or $1° = 0.517\,A - 0.621\,A'$ ......(2)

Solving (1) and (2), A = 4.8° and A′ = 2.4°. Thus, the crown prism should have its refracting angle 4.8° and that of the flint prism should be 2.4°.

(20) (a) For TIR at AC

$\theta > C$

$\Rightarrow \sin\theta \geq \sin C$

$\Rightarrow \sin\theta \geq \frac{1}{{}_w\mu_g}$

$\Rightarrow \sin\theta \geq \frac{\mu_w}{\mu_g}$

$\Rightarrow \sin\theta \geq \frac{8}{9}$

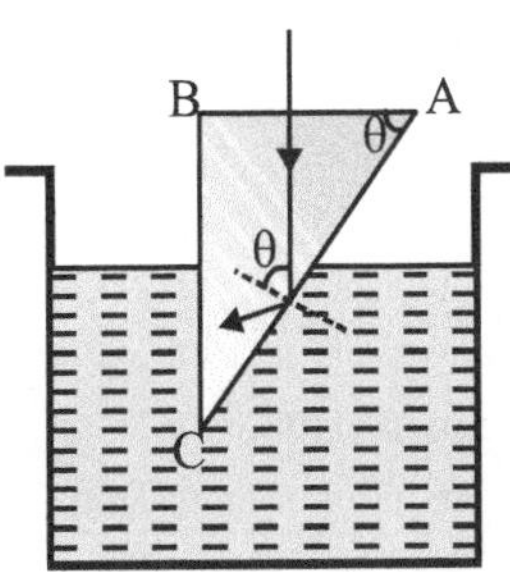

(21) (c) $\delta_{Prism} = (\mu - 1)A = (1.5 - 1)4° = 2°$

$\therefore\ \delta_{Total} = \delta_{Prism} + \delta_{Mirror}$

$= (\mu - 1)A + (180 - 2i) = 2° + (180 - 2\times2) = 178°$

(22) **(a)** From the following figure

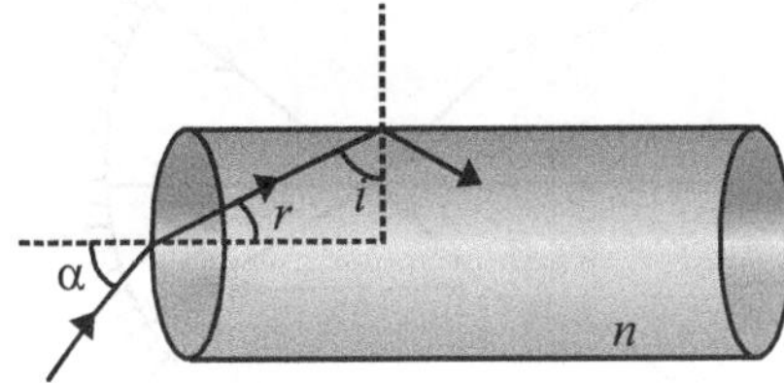

$r + i = 90° \Rightarrow i = 90° - r$

For ray not to emerge from curved surface $i > C$

$\Rightarrow\ \sin i > \sin C \Rightarrow \sin(90° - r) > \sin C \Rightarrow \cos r > \sin C$

$\Rightarrow\ \sqrt{1-\sin^2 r} > \frac{1}{n}$ $\quad\left\{\therefore\ \sin C = \frac{1}{n}\right\}$

$\Rightarrow 1 - \frac{\sin^2\alpha}{n^2} > \frac{1}{n^2} \Rightarrow 1 > \frac{1}{n^2}\left(1 + \sin^2\alpha\right)$

$\Rightarrow\ n^2 > 1 + \sin^2 a \Rightarrow n > \sqrt{2}$ $\quad\{\sin i \to 1\}$

$\Rightarrow$ Least value $= \sqrt{2}$

(23) (a) ω depends only on nature of material.

(24) (a)

(25) **(c);**

26. (b), 27 (a)

The normal shift produced by a glass slab is,

$$\Delta x = \left(1 - \frac{1}{\mu}\right)t = \left(1 - \frac{2}{3}\right)(6) = 2cm$$

i.e., for the mirror, the object is placed at a distance $(32 - \Delta x) = 30$ cm from it.

Applying mirror formula,

$$\frac{1}{v} + \frac{1}{u} = \frac{1}{f}$$

$$\frac{1}{v} + \frac{1}{30} = -\frac{1}{10}$$

or, $v = -15$ cm

(a) When x = 5 cm: The light falls on the slab on its return journey as shown. But the slab will again shift it by a distance.

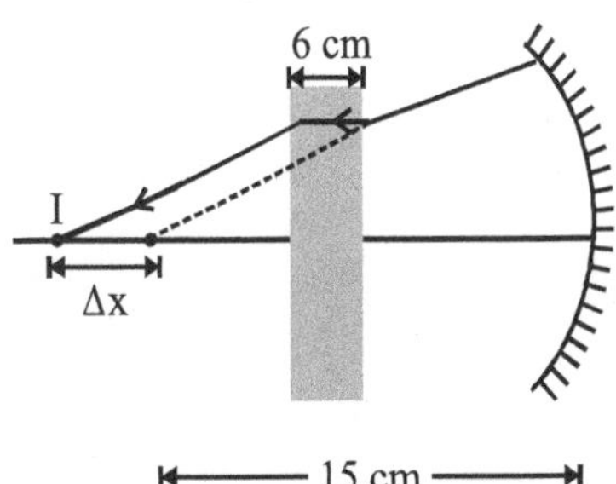

Δx = 2 cm. Hence, the final real image is formed at a distance (15 + 2) = 17 cm from the mirror.

(b) When x = 20 cm: This time also the final image is at a distance of 17 cm from the mirror but it is virtual as shown.

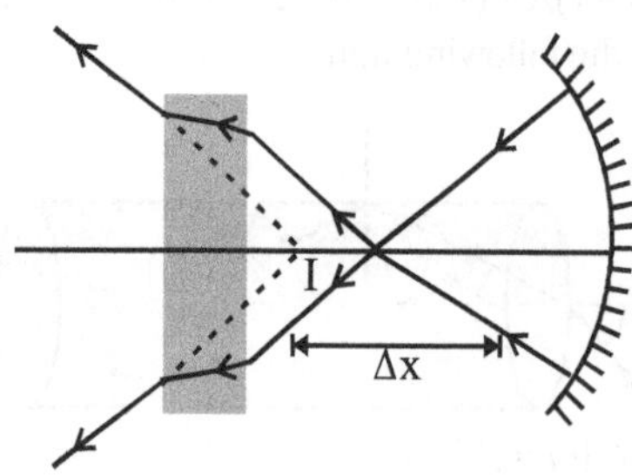

(28) (b) As rays of all colours emerge in the same direction (of incidence of white light), hence there is no dispersion, but only lateral displacement in a glass slab.

(29) (b) The velocity of light in a material medium depends upon its colour (wavelength). If a ray of white light is incident on a prism, then on emerging the different colours are deviated through different angles.

Also dispersive power $\omega = \frac{(\mu_V - \mu_R)}{(\mu_Y - 1)}$

i.e. ω depends upon only μ

(30) (c) Apparent shift for different coloured letter is

$$d = h\left(1 - \frac{1}{\mu}\right)$$

$\Rightarrow \lambda_R > \lambda_V$ so $\mu_R < \mu_V$

Hence $d_R < d_V$ i. e. red coloured letter raised least.

DAILY PRACTICE PROBLEMS

PHYSICS SOLUTIONS

51

1. **(a).** According to sign convention, it is given that $u = -0.6$ m, $R = 0.25$ m, $\mu_1 = 1$ (air), $\mu_2 = 1.5$

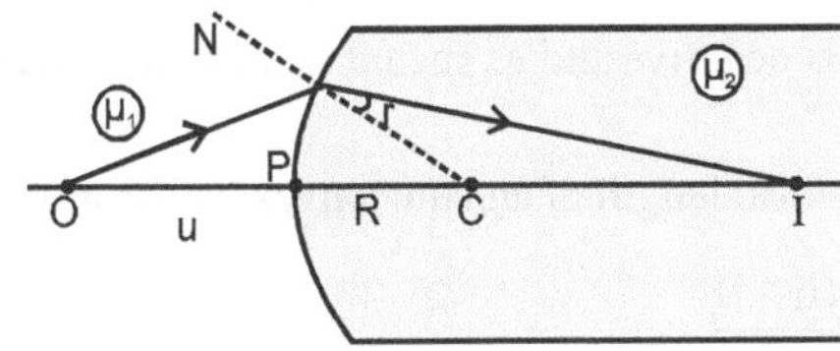

Therefore, using

$$-\frac{\mu_1}{u}+\frac{\mu_2}{v}=\frac{\mu_2-\mu_1}{R}, \text{ we get}$$

$$\frac{1.5}{v}=\frac{1}{(-0.6)}+\frac{1.5-1}{0.25}=-\frac{1}{0.6}+\frac{0.5}{0.25}$$

$$=-\frac{5}{3}+2=\frac{1}{3}$$

$\Rightarrow v = 4.5$ m

The image is formed on the other side of the object (i.e. inside the refracting surface).

2. **(a).** On viewing from the closer surface A (near to object) :

The final image is formed at I.

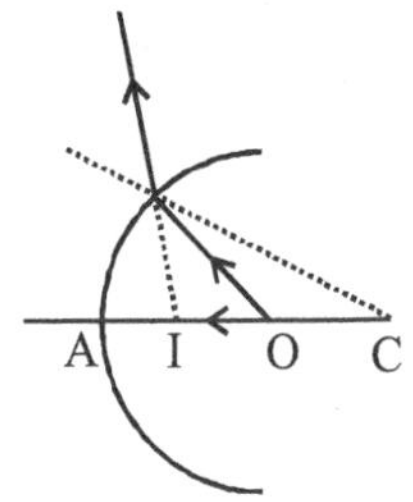

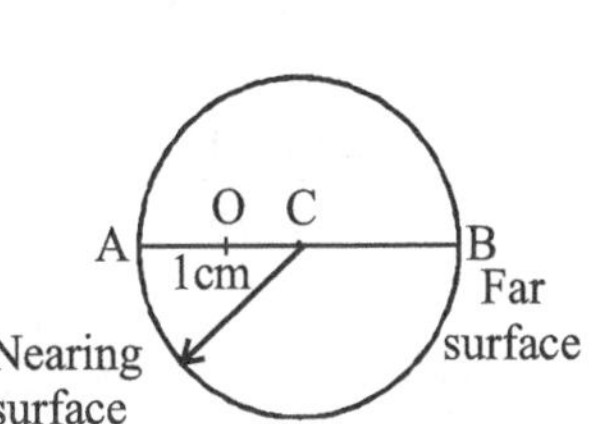

From sign convention

$u = OA = -4$ cm, $v = ?$

$R = AC = -5$cm

$$\mu=\frac{\mu_2}{\mu_1}=\frac{2}{3}$$

$$\frac{\mu}{v}-\frac{1}{u}=\frac{\mu-1}{R} \Rightarrow \frac{2/3}{v}+\frac{1}{4}=\frac{2/3-1}{-5}$$

$$\Rightarrow \frac{2}{3v}=\frac{1}{15}-\frac{1}{4}=\frac{4-15}{60}=\frac{-11}{60}$$

$$v=\frac{2}{3}\times\frac{-60}{11}=-\frac{40}{11}=-3.63 \text{ cm}$$

$\therefore$ AI = 3.63 cm

3. **(c).**

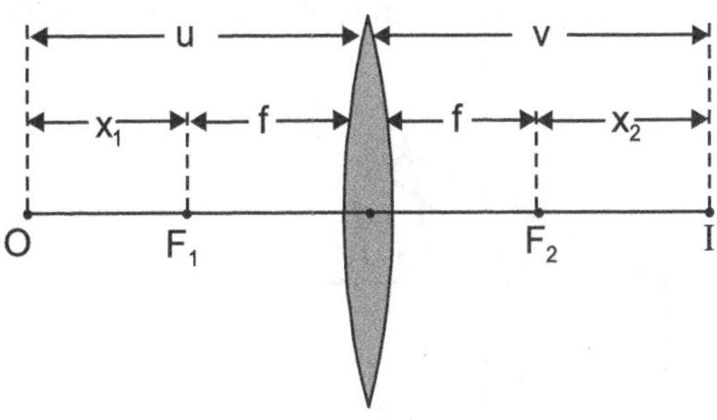

As in case of thin lens the distance of either foci from the optical centre is f,

$|u| = (f + x_1)$ and $|v| = (f + x_2)$

Substituting thses values of u and v in lens formula with proper sign

$$\frac{1}{(f+x_2)}-\frac{1}{-(f+x_1)}=\frac{1}{f}$$

or $$\frac{x_1+x_2+2f}{(f+x_1)(f+x_2)}=\frac{1}{f}$$

i.e., $fx_1 + fx_2 + 2f^2 = f^2 + fx_1 + fx_2 + x_1 x_2$

or, $x_1 x_2 = f^2$

4. **(b).** According to Lens-maker's formula :

$$\frac{1}{f}=(\mu-1)\left[\frac{1}{R_1}-\frac{1}{R_2}\right] \text{ with } \mu=\frac{\mu_L}{\mu_M}$$

Here f = 30 cm and R_1 = 10 cm and $R_2 = \infty$

So, $$\frac{1}{30}=(\mu-1)\left[\frac{1}{10}-\frac{1}{\infty}\right]$$

$3\mu - 3 = 1$ or, $\mu = (4/3)$

5. **(a).** For combination of lenses

$$\frac{1}{f}=\frac{1}{f_1}+\frac{1}{f_2}=\frac{1}{10}+\frac{1}{15}=\frac{25}{150}=\frac{1}{6}$$

Therefore, f = 6 cm.

6. **(c).** Let f_2 is the focal length of the diverging lens. Then ,

$$\frac{1}{f}=\frac{1}{f_1}+\frac{1}{f_2}$$

It is given that $f_1 = +20$ cm, $f = 30$ cm

$$\frac{1}{30}=\frac{1}{20}+\frac{1}{f_2}$$

or $\frac{1}{f_2}=\frac{1}{30}-\frac{1}{20}=\frac{2-3}{60}=-\frac{1}{60}$

Thus $f_2 = -60$ cm

7. **(a).** As radius of curvature of silvered surface is 22 cm, so,

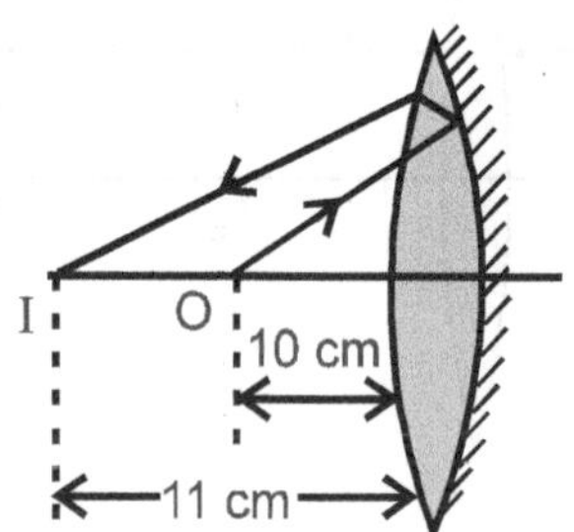

$$f_M=\frac{R}{2}=\frac{-22}{2}=-11 \text{ cm}=-0.11 \text{ m}$$

and hence, $P_M=-\frac{1}{f_M}=\frac{1}{-0.11}=\frac{1}{0.11}D$

Further as the focal length of lens is 20 cm, i.e. 0.20 m, its power will be given by :

$$P_L=\frac{1}{f_L}=\frac{1}{0.20}D$$

Now as in image formation, light after passing through the lens will be reflected back by the curved mirror the lens again $P = P_L + P_M + P_L = 2P_L + P_M$

i.e. $P=\frac{2}{0.20}+\frac{1}{0.11}=\frac{210}{11}D$

So the focal length of equivalent mirror

$$F=-\frac{1}{P}=-\frac{11}{210}\text{m}=-\frac{110}{21}\text{cm}$$

i.e. the silvered lens behaves as a concave mirror of focal length (110/21) cm. So for object at a distance 10 cm in front of it,

$$\frac{1}{v}+\frac{1}{-10}=-\frac{21}{110} \text{ i.e. } v=-11 \text{ cm}$$

i.e. image will be 11 cm in front of the silvered lens

8. **(b).** On covering the lens half by a black paper will reduce the intensity of image and not the part of image. So full image is formed.

9. **(b).** The focal length of lens in water is given by

$$f_\ell=\frac{{}_a\mu_g-1}{\frac{{}_a\mu_g}{{}_a\mu_\ell}-1}f_a=\frac{1.2-1}{\frac{1.2}{1.33}-1}f_a$$

$$f_\ell=-\frac{0.2\times1.33}{0.13}f_a$$

Hence f is negative and as such it behaves as a divergent lens.

10. **(a).** The focal length of an equiconvex lens is given by

$$\frac{1}{f}=\frac{2(\mu-1)}{R}$$

It is given that $\frac{1}{f}=+5$ and $\mu=1.5$

Therefore, $5=\frac{2(1.5-1)}{R}$

or $R=\frac{1}{5}$ metre $=20$ cm

11. **(b).** The question is based on the conventional method of measurement of focal length by displacement method. According to this method where D is the distance between object and the image, and x is the displacement given to the object.

From the data x = 25 cm and D = 75 cm .

Thus

$$f=\frac{(75)^2-(25)^2}{4\times75}=\frac{(75-25)\,(75+25)}{4\times75}$$

$$=\frac{50\times100}{4\times75}=\frac{50}{3}=16.7 \text{ cm}$$

12. **(c).** $m_1=\frac{A_1}{O}$ and $m_2\,\frac{A_2}{O}$

$$\Rightarrow m_1m_2=\frac{A_1A_2}{O^2}$$

Also it can be proved that $m_1m_2 = 1$

So $O=\sqrt{A_1A_2}$

13. **(c).** Effective power $P = P_1 + P_2 = 4 - 3 = 1D$

14. **(d).** One part of combination will behave as converging lens and the other as diverging lens of same focal length. As such total power will be zero.

15. **(c).** Let the image of an object at O is formed at the same point as shown in figure. The distance of O from the plane surface is x. The rays suffer refraction at first surface (curved) as they reach lens. After wards become parallel and gets reflected form plane surface and so retrace the path and image is formed at O itself.

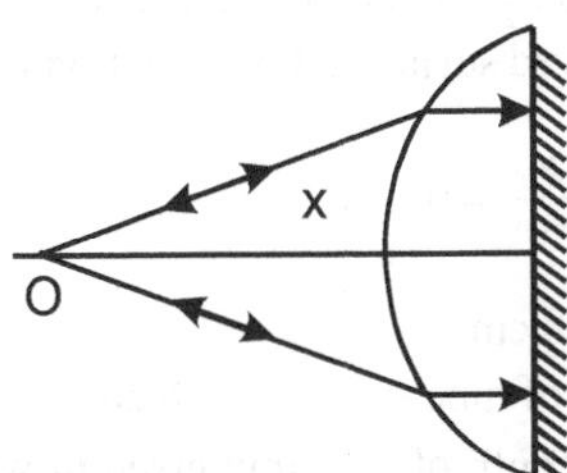

$$\frac{\mu}{v}-\frac{1}{u}=\frac{\mu-1}{R}$$

$u=-x, v=\infty$

$$\frac{\mu}{\infty}+\frac{1}{x}=\frac{\mu-1}{R}$$

$$x=\frac{R}{\mu-1}$$

As such O behaves as equivalent to centre of curvature of equivalent concave mirror.

$\therefore$ Radius $= x = \dfrac{R}{\mu-1}$

16. **(a).** As here f = 10 cm and v = 5m = 500 cm

So from lens formula $\frac{1}{v}-\frac{1}{u}=\frac{1}{f}$, we have

$$\frac{1}{500}-\frac{1}{u}=\frac{1}{10} \quad \text{i.e.,} \quad u=-\left[\frac{500}{49}\right]\text{cm}$$

So that $m=\dfrac{v}{u}=\dfrac{500}{-500/49}=-49$

Here negative sign means the image is inverted with respect to object. Now as here object is (2 cm × 2cm) so the size of picture on the screen

$A_i = (2 \times 49 \text{ cm}) \times (2 \times 49\text{cm}) = (98 \times 98)\text{ cm}^2$

17. **(a).** As power of a lens is reciprocal of focal length in m,

$$P=\frac{1}{5\times10^{-2}\text{m}}=\frac{1}{0.05}\text{ diopter} = 20\text{ D}$$

18. **(b).** For relaxed eye, MP is minimum and will be

$$MP=\frac{D}{f}=\frac{25}{5}=5\times$$

While for strained eye, MP is maximum and will be

$$MP=1+\frac{D}{f}=1+5=6\times$$

19. **(a).** As in case of projector, $m=\dfrac{I}{O}=\dfrac{v}{u}$

So $-\dfrac{20\times100\text{ cm}}{3.5\text{ cm}}=\dfrac{40\times100}{u}$

i.e., u = – 7 cm

i.e., film is at a distance of 7 cm in front of projection lens.

And from lens formula $\frac{1}{v}-\frac{1}{u}=\frac{1}{f}$, here we have

$$\frac{1}{4000}-\frac{1}{-7}=\frac{1}{f} \text{ or } f \cong 7\text{ cm} = 70\text{ mm}$$

[as (1/4000) << (1/7)]

i.e., focal length of projection lens is 70 mm.

20. **(a).** As final image is at 25 cm in front of eye piece

$$\frac{1}{-25}-\frac{1}{u_e}=\frac{1}{5} \text{ i.e., } u_e=-\frac{25}{6}$$

And so, $m_e=\dfrac{v_e}{u_e}=\dfrac{-25}{(-25/6)}=6$ ...(1)

Now for objective,

$v = L - u_e = 20 - (25/6) = (95/6)$

So if object is at a distance u from the objective,

$$\frac{6}{95}-\frac{1}{u}=\frac{1}{0.95}\text{ i.e., } u=-\frac{95}{94}\text{ cm}$$

i.e. object is at a distance (95/94) cm in front of field lens.

Also, $m=\dfrac{v}{u}=\dfrac{(95/6)}{(-95/94)}=-\left[\dfrac{94}{6}\right]$ ...(2)

So total magnification,

$$M = m \times m_e = -\left[\frac{94}{6}\right]\times(6) = -94$$

i.e., final image is inverted, virtual and 94 times that of object.

21. **(a).** As object is distant, i.e., u = – ∞ , so

$$\frac{1}{v}-\frac{1}{-\infty}=\frac{1}{f_0}\text{ i.e. } v=f_0=12\text{ cm}$$

i.e. objective will form the image I_M at its focus which is at a distance of 12 cm from O. Now as eye- piece of focal length – 4 cm forms image I at a distance of 24 cm from it,

$$\frac{1}{-24}-\frac{1}{u_e}=\frac{1}{-4} \Rightarrow u_e=\frac{24}{5}=4.8\text{ cm.}$$

i.e, the distance of I_M from eye lens EA is 4.8 cm. So the length of tube L = OA – EA = 12 – 4.8 = 7.2 cm.

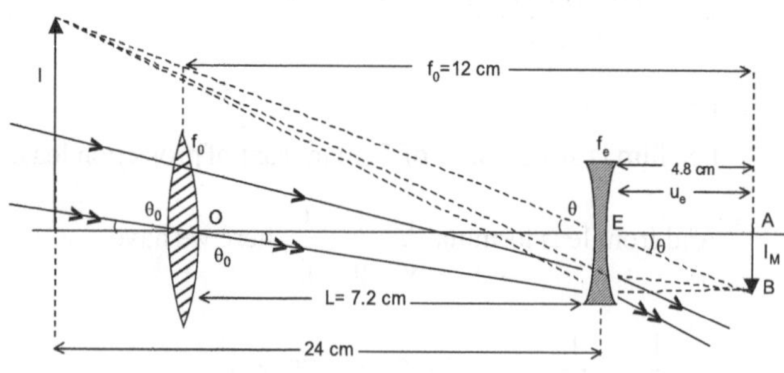

22. **(d)** In case of astronomical telescope if object and final image both are at infinity.
$MP = -(f_0/f_e)$ and $L = f_0 + f_e$
So here $-(f_0/f_e) = -5$ and $f_0 + f_e = 36$
Solving these for f_0 and f_e, we get
$f_0 = 30$ cm and $f_e = 6$ cm

23. **(a)** Resolving power of microscope $\propto \frac{1}{\lambda}$

24. **(a)** Light gathering power $\propto$ Area of lens aperture or d^2

25. **(b)** For lens L_1, ray must move parallel to the axis after refraction

$$\frac{\mu_1}{\infty} + \frac{\mu_\omega}{x} = \frac{\mu_1 - \mu_\omega}{R_1} \Rightarrow x = 10cm$$

26. **(a)** For lens L_2, image must form at centre of curvature of the curved surface after refraction through plane part

$$\frac{\mu_2}{-R_2} + \frac{\mu_\omega}{x'} = 0$$

$\Rightarrow x' = 8$ cm

27. **(b)** Length of tube $= x + x' = 18$ cm

28. **(a)** Focal length of lens immersed in water is four times the focal length of lens in air. It means
$f_w = 4f_a = 4 \times 10 = 40$ cm

29. **(d)** Focal length of the lens depends upon its refractive index as $\frac{1}{f} \propto (\mu - 1)$.

Since $\mu_b > \mu_r$ so $f_b < f_r$
Therefore, the focal length of a lens decreases when red light is replaced by blue light.

30. **(b)** The light gathering power (or brightness) of a telescope $\propto$ (diameter)^2. So by increasing the objective diameter even far off stars may produce images of optimum brightness.

DAILY PRACTICE PROBLEMS

PHYSICS SOLUTIONS 52

1. (d) $\frac{I_1}{I_2}=\frac{9}{1}$

$$\frac{I_{max}}{I_{min}}=\left[\frac{\sqrt{\frac{I_1}{I_2}}+1}{\sqrt{\frac{I_1}{I_2}}-1}\right]^2=\left[\frac{\sqrt{9}+1}{\sqrt{9}-1}\right]^2$$

$$\frac{I_{max}}{I_{min}}=\frac{4^2}{2^2}=\frac{4}{1}$$

2. (a) $a_1=6$ units, $a_2=8$ units

$$\frac{I_{max}}{I_{min}}=\frac{\left[\frac{a_1}{a_2}+1\right]^2}{\left[\frac{a_1}{a_2}-1\right]^2}=\frac{\left[\frac{6}{8}+1\right]^2}{\left[\frac{6}{8}-1\right]^2}$$

$$\frac{I_{max}}{I_{min}}=\frac{49}{1}$$

3. (b) The separation between the sucessive bright fringes is-

$$\beta=\frac{D\lambda}{d}=\frac{1\times600\times10^{-9}}{0.1\times10^{-3}}$$

$\beta=6.0$ mm.

4. (b) $\omega_a=\lambda/d$

$$\therefore\ \omega_a\propto\lambda\ \Rightarrow\frac{(\omega_a)_{water}}{\omega_a}=\frac{\lambda_{water}}{\lambda}$$

$$\Rightarrow\frac{(\omega_a)_{water}}{\omega_a}=\frac{\lambda}{\mu_{water}\lambda}$$

$$\Rightarrow(\omega_a)_{water}=0.15°$$

5. (b) $I'=I_1+I_2+2\sqrt{I_1I_2}\cos\phi$

$I_1=I,\ I_2=9I,\ \phi=\pi$

$$I'=I+9I+2\sqrt{9I^2}\cos\pi=10I-6I=4I$$

6. (a) $d=\frac{D\lambda}{\beta}$ .... (1)

According to quesion.

$\lambda=5100\times10^{-10}$m

$\beta=\frac{2}{10}\times10^{-2}$ m .... (2)

D = 2m, d = ?.

From eqs. (1) and (2)

$$d=\frac{2\times51\times10^{-8}}{2\times10^{-3}}=5.1\times10^{-4}\ \text{m}$$

7. (b) $V=\frac{I_{max}-I_{min}}{I_{max}+I_{min}}=\frac{\frac{I_{max}}{I_{min}}-1}{\frac{I_{max}}{I_{min}}+1}$ .... (1)

$$\frac{I_{max}}{I_{min}}=\left(\frac{\sqrt{\frac{I_1}{I_2}}+1}{\sqrt{\frac{I_1}{I_2}}-1}\right)^2 \quad .... (2)$$

According to question

$\frac{I_1}{I_2}=\frac{1}{4}$ .... (3)

From eqs. (2) and (3)

$$\frac{I_{max}}{I_{min}}=\left(\frac{\sqrt{\frac{1}{4}}+1}{\sqrt{\frac{1}{4}}-1}\right)^2=\frac{\frac{9}{4}}{\frac{1}{4}}=9 \quad .... (4)$$

From eqs. (1) and (4)

$$V=\frac{[9-1]}{[9+1]}=\frac{8}{10}=0.8$$

8. (c) $\lambda=\frac{(\mu-1)t}{n}=\frac{(1.6-1)\times7\times10^{-6}}{7}$

$\lambda=6\times10^{-7}$ meter $\Rightarrow\lambda=6000$ Å

9. (d) $\beta\propto\frac{1}{d}$ $\therefore$ d On increasing d three times β will become 1/3 times.

10. (b) $\therefore$ PR = d $\Rightarrow$ PO = d secθ and CO = PO cos 2θ = d sec θ cos 2θ is

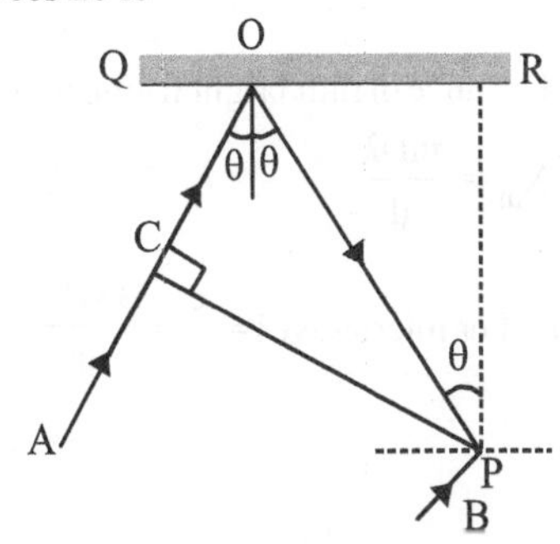

Path difference between the two rays
$\Delta = CO + PO = (d \sec\theta + d \sec\theta \cos 2\theta)$
Phase difference between the two rays is
$\phi = \pi$ (One is reflected, while another is direct)
Therefore condition for constructive interference should be

$$\Delta = \frac{\lambda}{2}, \frac{3\lambda}{2} ....$$

or $\quad d \sec\theta(1 + \cos 2\theta) = \frac{\lambda}{2}$

or $\quad \frac{d}{\cos\theta}\left(2\cos^2\theta\right) = \frac{\lambda}{2} \Rightarrow \cos\theta = \frac{\lambda}{4d}$

11. (b) $n_1\lambda_1 = n_2\lambda_2$
$10 \times 7000 = n_2 \times 5000$
$n_2 = 14$

12. (a, c) Path difference between the rays reaching infront of slit S_1 is.
$S_1P - S_2P = (b^2 + d^2)^{1/2} - d$
For distructive interference at P

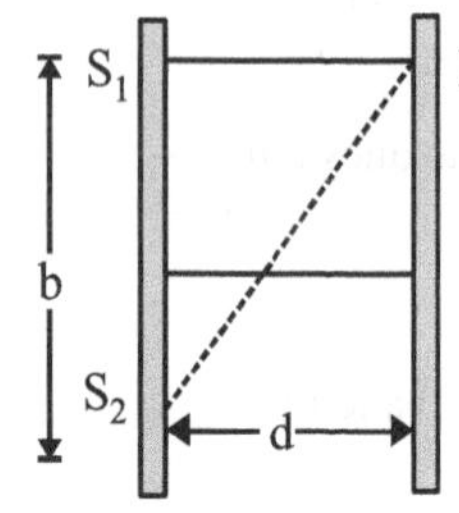

$$S_1P - S_2P = \frac{(2n-1)\lambda}{2}$$

i.e., $(b_2 + d_2)^{1/2} - d = \frac{(2n-1)\lambda}{2}$

$$\Rightarrow \quad d\left(1 + \frac{b^2}{d^2}\right)^{1/2} - d = \frac{(2n-1)\lambda}{2}$$

$$\Rightarrow \quad d\left(1 + \frac{b^2}{2d^2} + ...\right) - d = \frac{(2n-1)\lambda}{2}$$

(Binomial Expansion)

$$\Rightarrow \quad \frac{b}{2d} = \frac{(2n-1)\lambda}{2} \Rightarrow = \frac{b^2}{(2n-1)d}$$

For $n = 1, 2$ ...., $\lambda = \frac{b^2}{d}, \frac{b^2}{3d}$

13. (b) Distance of mth bright fringe from central fringes is

$$X_m = \frac{mD\lambda}{d}$$

14. (a, b) For microwave $l = \frac{c}{f} = \frac{3\times10^8}{10^6} = 300$ m

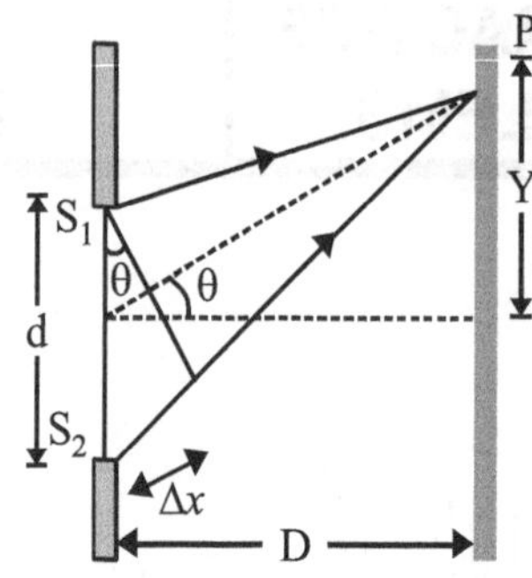

As Dx = d sin q

Phase difference f= $\frac{2\pi}{\lambda}$ (Path difference)

$$= \frac{2\pi}{\lambda}(d\sin\theta) = \frac{2\pi}{300}\,(150 \sin q) = p \sin q$$

$I_R = I_1 + I_2 + 2\sqrt{I_1 I_2}\cos\theta$
Here $I_1 = I_2$ and $\phi = \pi \sin\theta$

$$\therefore \quad I_R = 2I_1\,[(\pi\sin\theta)] = 4I_1\cos^2\left(\frac{\pi\sin\theta}{2}\right)$$

I_R will be maximum when $\cos^2\left(\frac{\pi\sin\theta}{2}\right) = 1$

$$\therefore \quad (I_R)_{max} = 4I_1 = I_o$$

Hence $I = I_o \cos^2\left(\frac{\pi\sin\theta}{2}\right)$

If $\theta = 0$, then $I = I_o \cos\theta = Io$
If $\theta = 30°$, then $I = I_o \cos^2(\pi/4) = I_o/2$
If $\theta = 90°$, then $I = I_o \cos^2(\pi/2) = 0$

15. (b) $I = R^2 = a_1^2 + a_2^2 + 2a_1 a_2 \cos\delta$

$$= I + 4I + 4I\cos\frac{\pi}{2} = 5I$$

16. (b) Suppose the amplitude of the light wave coming from the narrower slit is A and that coming from the wider slit is 2A. The maximum intensity occurs at a place where constructive interference takes place. Then the resultant amplitude is the sum of the individual amplitudes.
Thus,
$A_{max} = 2A + A = 3A$
The minimum intensity occurs at a place where destructive interference takes place. The resultant amplitude is then difference of the individual amplitudes.
Thus, $A_{min} = 2A - A = A$.

$$\therefore \quad \frac{I_{max}}{I_{min}} = \frac{(A_{max})^2}{(A_{min})^2} = \frac{(3A)^2}{(A)^2} = 9$$

17. (b) $\frac{I_1}{I_2}=\frac{2}{1}$

$\frac{a_1}{a_2}=\sqrt{\frac{I_1}{I_2}}=\frac{\sqrt{2}}{1}$

At the point of constructive interference, the resultant amplitude becomes $(a_1+a_2)=\sqrt{2}+1$ at the point of destructive interference, the resultant amplitude is

$(a_1-a_2)=\sqrt{2}-1$

$\therefore \frac{I_{max}}{I_{min}}=\frac{(a_1+a_2)^2}{(a_1-a_2)^2}=\frac{(\sqrt{2}+1)^2}{(\sqrt{2}-1)^2}=34$

18. (d) For destructive interference :

Path difference= $S_1P-S_2P=(2n-1)\lambda/2$.

For $n=1$, $S_1P-S_2P=(2\times1-1)\lambda/2=\lambda/2$.

$n=2$, $S_1P-S_2P=(2\times2-1)\lambda/2=3\lambda/2$.

$n=3$, $S_1P-S_2P=(2\times3-1)\lambda/2=5\lambda/2$.

$n=4$, $S_1P-S_2P=(2\times4-1)\lambda/2=7\lambda/2$.

$n=5$, $S_1P-S_2P=(2\times5-1)\lambda/2=9\lambda/2$.

$n=6$, $S_1P-S_2P=(2\times6-1)\lambda/2=11\lambda/2$.

So, destructive pattern is possible only for path difference = $11\lambda/2$.

19. (b) The distance of a bright fringe from zero order fringe is given by-

$X_n=\frac{n\lambda D}{d}$

D & d is constant

$n_1\lambda_1=n_2\lambda_2$

$n_1=16, \lambda_1=6000\,A°, \lambda_2=4800\,Å$

$n_2=\frac{n_1\lambda_1}{\lambda_2}=\frac{16\times6000}{4800}=20$

20. (c) $n_1\lambda_1=n_2\lambda_2$ for bright fringe

$n(7.5\times10^{-5})=(n+1)(5\times10^{-5})$

$n=\frac{5.0\times10^{-5}}{2.5\times10^{-5}}=2.$

21. (b) $X_n=\frac{n\lambda D}{d}$ or $X_3=\frac{3\lambda D}{d}$

$X_3=\frac{3\times(5000\times10^{-8}\,cm)\times200\,cm}{0.02\,cm}=1.5\,cm.$

22. (b, d) $\frac{I_{max}}{I_{max}}=9\Rightarrow\left(\frac{a_1+a_2}{a_1-a_2}\right)^2=9\Rightarrow\frac{a_1+a_2}{a_1-a_2}=3$

$\Rightarrow\frac{a_1}{a_2}=\frac{3+1}{3-1}\Rightarrow\frac{a_1}{a_2}=5$ There $I_1:I_2=4:1$

23. (a) Fringe width $\beta=\frac{\lambda D}{d}$

According to de Broglie,

Wavelength $\lambda=\frac{h}{p}=\frac{h}{\sqrt{2meV}}$

As V increases, λ decreases, β decreases.

Also $\beta\propto\frac{1}{d}$ and $\beta\propto D$.

24. (a) $n=\frac{(\mu-1)tD}{d}$

but $\beta=\frac{\lambda D}{d}\Rightarrow\frac{D}{d}=\frac{\beta}{\lambda}$

$n=(\mu-1)t\beta/\lambda$

$20\beta=(\mu-1)2.5\times10^{-3}\{\beta/5000\times10^{-8}\}$

$\mu-1=\frac{20\times5000\times10^{-8}}{2.5\times10^{-3}}=0.4$

$\mu=1.4$.

25. (a) $S_1P-S_2P=\frac{dy}{D}$

For central maxima,

$\Delta x=(n_0+kt)\frac{dy}{D}-d\sin\phi=0$

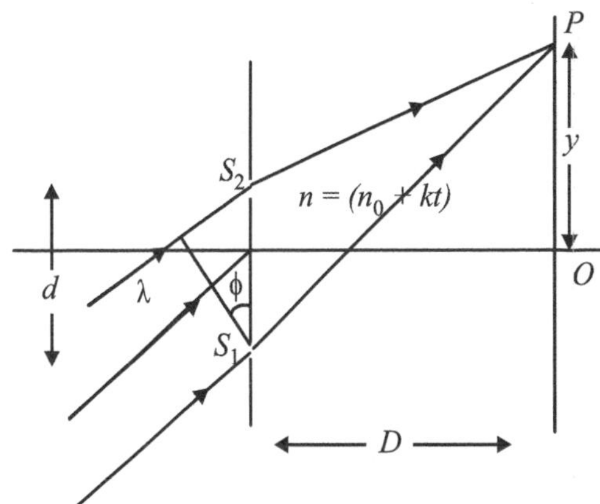

$\therefore y=\frac{D\sin\phi}{n_0+kt}$ (y-coordinates of central maximum).

26. (b) $\frac{dy}{dt}=\frac{-kD\sin\phi}{(n_0+kt)^2}$ = velocity of central maximum.

27. (d) For central maxima to be formed at O

$n'\left(\frac{n}{n'}-1\right)b=d\sin\phi$

Here $n' = n_0 + kt$, n = refractive index of plate.

$$n = n_0 + kt + \frac{d \sin \phi}{b}$$

28. (d) When d is negligibly small, fringe width β which is proportional to 1/d may become too large. Even a single fringe may occupy the whole screen. Hence the pattern cannot be detected.

29. (d) In Young's experiments fringe width for dark and white fringes are same while in Young's double slit experiment when a white light as a source is used, the central fringe is white around which few coloured fringes are observed on either side.

30. (d) When one of slits is covered with cellophane paper, the intensity of light emerging from the slit is decreased (because this medium is translucent). Now the two interfering beam have different intensities or amplitudes. Hence intensity at minima will not be zero and fringes will become indistinct.

DAILY PRACTICE PROBLEMS

PHYSICS SOLUTIONS

53

1. **(b).** The distance of first diffraction minimum from the central principal maximum $x = \lambda D/d$

$$\therefore \sin\theta = \frac{x}{D} = \frac{\lambda}{d} \Rightarrow d = \frac{\lambda}{\sin\theta}$$

$$\Rightarrow d = \frac{5000\times10^{-8}}{\sin 30^\circ} = 2\times5\times10^{-5}$$

$\Rightarrow d = 1.0\times10^{-4}$ cm.

2. **(b).** Here, $\lambda_1 = 5890Å = 5890\times10^{-10}$ m
$\lambda_2 = 5896$ Å $= 5896\times10^{-10}$ m
$a = 2\mu m = 2\times10^{-6}$ m, D = 2m

For first maxima, $\sin\theta = \frac{3\lambda_1}{2a} = \frac{x_1}{D}$

$$\Rightarrow x_1 = \frac{3\lambda_1 D}{2a} \text{ and } x_2 = \frac{3\lambda_2 D}{2a}$$

$\therefore$ spacing between the first maxima of two sodium lines

$$= x_2 - x_1 = \frac{3D}{2a}(\lambda_2 - \lambda_1)$$

$$= \frac{3\times2(5896-5890)\times10^{-10}}{2\times2\times10^{-6}} = 9\times10^{-4}\text{ m.}$$

3. **(d).** $\frac{ax}{f} = n\lambda$

$$\lambda = \frac{ax}{n.f} = \frac{0.3\times10^{-3}\times5\times10^{-3}}{3\times1}$$

$\lambda = 5\times10^{-7}$ m
$\lambda = 5000Å$

4. **(b).** $\theta = \sin^{-1}\left(\frac{\lambda}{a}\right)$ .....(1)

According to question
$\lambda = 2\times10^{-3}$ m
$a = 4\times10^{-3}$ m .....(2)
From equation (1) and (2)

$$\theta = \sin^{-1}\left(\frac{1}{2}\right)$$

$\theta = 30^\circ$

5. **(a).** Here the width of principal maxima is 2.5 mm, therefore its half width is

$$\frac{\beta}{2} = \frac{2.5}{2} = 1.25\times10^{-3}\text{m}$$

Diffraction angle $\theta = \frac{\beta/2}{D} = \frac{12.5\times10^{-3}}{2}$

$\therefore a\theta = \lambda$

$$\theta = \lambda/a = \frac{12.5\times10^{-3}}{2}$$

$$\lambda = \frac{12.5\times10^{-3}}{2}\times a = \frac{12.5\times10^{-3}\times10^{-3}}{2}$$

$\lambda = 6.25\times10^{-6}\text{m} = 6250\text{mm}$

6. **(a).** Slit width = a = 0.2mm,

$$\beta = \frac{2\lambda d}{a}$$

Angular width $W_\theta = \frac{\beta}{D} = \frac{2\lambda}{a}$

$$\theta = \frac{2\times6328}{0.2} = 0.36^\circ$$

7. **(a).** Here distance of the screen from the slit,
D = 2m, a = ?, x = 5 mm
$= 5\times10^{-3}$ m, $\lambda = 5000Å$
$= 5000\times10^{-10}$ m
For the first minima, $\sin\theta = \lambda/a = x/D$,

$$a = \frac{D\lambda}{x} = \frac{2\times5000\times10^{-10}}{5\times10^{-3}} = 2\times10^{-4}\text{ m}$$

8. **(d).** Here, $\lambda = 6500Å = 6.5\times10^{-7}$ m, a = 0.5 mm = 5×10^{-4} m,
D = 1.8 m
Angular separation of two dark bands on each side of central bright band $2\theta = 2\lambda/a$
Actual distance between them,
$2x = 2\lambda/a \times D$

$$2x = \frac{2\times6.5\times10^{-7}\times18}{5\times10^{-4}}$$

$2x = 4.68\times10^{-3}$ m

9. **(c).** Width of central maxima $= \frac{2f\lambda}{a}$

$$= \frac{2\times2\times6000\times10^{-10}}{0.2\times10^{-3}} = 12\text{ mm}$$

10. **(a).** $\theta = \frac{\lambda}{a}$ .....(a)

$$\theta = \frac{x}{f}$$ .....(b)

From eqs. (a) and (b)

$$\frac{\lambda}{a} = \frac{x}{f}$$

$$x = \frac{f\lambda}{a}$$ .....(c)

According to question x = ?, f = 40 cm

$\lambda = 5896 \times 10^{-8}$ cm

$a = 0.5 \times 10^{-1}$ cm .....(d)

From eqs. (c) and (d)

$$x = \frac{40 \times 5896 \times 10^{-8}}{5 \times 10^{-2}} 96 = 0.047\text{cm}$$

11. (a). $d\theta = \frac{1.22\lambda}{a}$ or $a = \frac{1.22\lambda}{d\theta}$

According to question

$$d\theta = 10^{-3} \text{ degree} = \frac{10^{-3} \times \pi}{180} \text{ Radian,}$$

$\lambda = 5 \times 10^{-5}$

$$a = \frac{1.22 \times 5 \times 10^{-5} \times 180}{10^{-3} \times 3.14}$$

a = 3.5 cm

12. (a). Since the reflected light is very highly polarised, it implies that incident light falls at polarising angle of incidence θ_P. From Brewster's law,

$\mu = \tan\theta_p$

$\therefore \theta_P = \tan^{-1}(\mu) = \tan^{-1}(4/3) = 53.1°$

Since θ_P is the angle which the rays from sun make with the normal to the interface, angle with the interface will be 90° – 53.1° = 36.9°.

13. (a). Angle of incident light with the surface is 30°. Hence angle of incidence = 90° – 30° = 60°. Since reflected light is completely polarised, therefore, incidence takes place at polarising angle of incidence θ_p.

$\therefore \theta_p = 60°$

Using Brewster's law

$\mu = \tan\theta_p = \tan 60°$

$\therefore \mu = \sqrt{3}$

14. (d). If unpolarised light is passed through a polariod P_1, its intensity will become half.

So $I_1 = \frac{1}{2} I_0$ with vibrations parallel to the axis of P_1.

Now this light will pass through second polaroid P_2 whose axis is inclined at ana angle of 30° to the axis of P_1 and hence, vibrations of I_1. So in accordance with Malus law, the intensity of light emerging from P_2 will be

$$I_2 = I_1 \cos^2 30° = \left(\frac{1}{2}I_0\right)\left(\frac{\sqrt{3}}{2}\right)^2 = \frac{3}{8}I_0$$

$$\frac{I_2}{I_0} = \frac{3}{8} = 37.5\ \%$$

15. (a). If θ is the angle between the transmission axes of first polaroid P_1 and second P_2 while ϕ between the transmission axes of second polaroids P_2 and P_3, then according to given problem,

$\theta + \phi = 90°$ or $\phi = (90° - \theta)$ ....(1)

Now, if I_0 is the intensity of unpolarised light incident on polaroid P_1, the intensity of light transmitted through it,

$$I_1 = \frac{1}{2}I_0 = I_0 = \frac{1}{2}(32) = 16\frac{W}{m^2} \quad ....(2)$$

Now as angle between transmission axes of polaroids P_1 and P_2 is θ, in accordance with Malus law, intensity of light transmitted through P_2 will be

$I_2 = I_1 \cos^2\theta = 16 \cos^2\theta$ [from Eq. (2)] ....(3)

And as angle between transmission axes of P_2 and P_3 is ϕ, light transmitted through P_3 will be

$I_3 = I_2\cos^2\phi = 16 \cos^2\theta \cos^2\phi$ [from Eq.(3)]

Above equation in the light of (1) becomes,

$I_3 = 16 \cos^2\theta \cos^2(90° - \theta) = 4(\sin 2\theta)^2$ ....(4)

According to given problem, $I_3 = 3\ W/m^2$

So, $4(\sin 2\theta)^2 = 3$ i.e., $\sin 2\theta = (\sqrt{3}/2)$

or $2\theta = 60°$ i.e. $\theta = 30°$

Further in accordance with Eq. (4), I_3 will be max. when $\sin 2\theta$ = max., i.e.,

$\sin 2\theta = 1$ or $2\theta = 90°$, i.e., $\theta = 45°$

16. (c) In double refraction light rays always splits into two rays (O–ray & E–ray). O–ray has same velocity in all direction but E– ray has different velocity in different direction.

For calcite $\mu_e < \mu_0 \Rightarrow v_e > v_0$

For quartz $\mu_e > \mu_0 \Rightarrow v_0 > v_e$

17. (c) At polarizing angle, the reflected and refracted rays are mutually perpendicular to each other.

18. (d) The amplitude will be A cos60° = A/2

19. (c) Width of central maxima $= \frac{2\lambda D}{d}$

$$= \frac{2 \times 2.1 \times 5 \times 10^{-7}}{0.15 \times 10^{-2}} = 1.4 \times 10^{-3}\ m = 1.4\ mm$$

20. (a) Using $d \sin\theta = n\lambda$, for n = 1

$$\sin\theta = \frac{\lambda}{d} = \frac{550 \times 10^{-9}}{0.55 \times 10^{-3}} = 10^{-3} = 0.001 \text{ rad}$$

21. (b) $A = n\pi\, d\lambda \Rightarrow nd = \frac{A}{\pi\lambda} = \text{constant} \Rightarrow n \propto \frac{1}{d}$

(n = number of blocked HPZ) on decreasing d, n increases, hence intensity decreases.

22. (b) Intensity of polarized light $= \frac{I_0}{2}$

$\Rightarrow$ Intensity of untransmitted light $= I_0 - \frac{I_0}{2} = \frac{I_0}{2}$

23. (a)

24. (a) It magnitude of light vector varies periodically during it's rotation, the tip of vector traces an ellipse and light is said to be elliptically polarised. This is not in nicol prism.

25. (a) Multiple focii of zone plate given by $f_p = \frac{r_n^2}{(2p-1)\lambda}$,

where p = 1, 2, 3 ....

26. (a) Angular width is the angle subtended by the distance between first minima on either side at the centre of the slit. It is given by $\phi = 2\,\theta$, where θ is the angle of diffraction.

For first diffraction minimum, $a \sin\theta = 1\,\lambda$

or $\sin\theta = \lambda/a$ or $\theta = \frac{\lambda}{a}$

$\therefore$ Angular width $\phi = 2\theta = \frac{2\lambda}{a}$ i.e. $\phi \propto \lambda$

$$\frac{\phi_1}{\phi_2} = \frac{\lambda_1}{\lambda_2};\ \therefore\ \lambda_2 = \lambda_1\frac{\phi_2}{\phi_1} = 6000 \times \frac{70}{100} = 4200\,\text{Å}$$

27. (b) On immersing in liquid, a wavelength $\lambda = 6000$ Å must be behaving as $\lambda' = 4200$ Å to get the same decrease in angular width. Therefore, refractive index of medium

$$\mu = \frac{\lambda}{\lambda'} = \frac{6000}{4200} = 1.43.$$

28. (a) When a polaroid is rotated in the path of unpolarised light, the intensity of light transmitted from polaroid remains undiminished (because unpolarised light contains waves vibrating in all possible planes with rotated in path of plane polarised light, its intensity will vary from maximum (when the vibrations of the plane polarised light are parallel to the axis of the polaroid) to minimum (when the direction of the vibrations becomes perpendicular to the axis of the crystal). Thus using polaroid we can easily verify that whether the light is polarised or not.

29. (d) The nicol prism is made of calcite crystal. When light is passed through calcite crystal, it breaks up into two rays

(i) the ordinary ray which has its electric vector perpendicular to the principal section of the crystal and

(ii) the extra ordinary ray which has its electric vector parallel to the principal section. The nicol prism is made in such a way that it eliminates one of the two rays by total internal reflection, thus produces plane polarised light. It is generally found that the ordinary ray is eliminated and only the extra ordinary ray is transmitted through the prism. The nicol prism consists of two calcite crystal cut at - 68° with its principal axis joined by a glue called Canada balsam.

30. (d) The clouds consists of dust particles and water droplets. Their size is very large as compared to the wavelength of the incident light from the sun. So there is very little scattering of light. Hence the light which we receive through the clouds has all the colours of light. As a result of this, we receive almost white light. Therefore, the cloud are generally white.

DAILY PRACTICE PROBLEMS

PHYSICS SOLUTIONS 54

1. (a). $\lambda_\alpha = \frac{0.101}{\sqrt{V}}$ Å, $\sqrt{V} = \frac{0.101}{0.004}$

$\sqrt{V} = 25.25\,V, V = 637.5\,V$

$E_\alpha = q_\alpha \times V_\alpha \approx 1275\,eV$

2. (d). $\lambda = \frac{h}{mv}$

$\because v = \frac{c}{20} = \frac{3\times10^8}{20} = 1.5 \times 10^7$ m/sec

$h = 6.626 \times 10^{-34}$ J-s, $m = 1.67 \times 10^{-27}$ kg

$\therefore \lambda = \frac{6.626\times10^{-34}}{1.67\times10^{-27}\times1.5\times10^7}$

$\Rightarrow \lambda = 2.64 \times 10^{-14}$ m

3. (a). $\because \lambda \propto \frac{1}{\sqrt{m}} \Rightarrow \lambda_e \propto \frac{1}{\sqrt{m_e}}, \lambda_p \propto \frac{1}{\sqrt{m_p}}$

$\therefore \frac{\lambda_e}{\lambda_p} = \sqrt{\frac{m_p}{m_e}}$

4. (c). $\lambda = \frac{h}{mv} \Rightarrow v = \frac{h}{m\lambda}$,

$v = \frac{6.6\times10^{-34}}{9.1\times10^{-31}\times10\times10^{-10}} = 7.2 \times 10^5$ m/s

5. (b). $\lambda \propto \frac{1}{\sqrt{m}}$,

$\frac{\lambda_e}{\lambda_p} = \sqrt{\frac{m_p}{m_e}} = \sqrt{\frac{1836}{1}}$

6. (a). $\lambda_p = \frac{h}{\sqrt{2m_pe_pV}}$

$\Rightarrow \lambda_d = \frac{h}{\sqrt{2m_de_dV}}$

$\therefore \frac{\lambda_d}{\lambda_p} = \sqrt{\frac{m_pe_p}{m_de_d}} \quad \because m_d = 2m_p,$

$e_d = e_p \Rightarrow \frac{\lambda_d}{\lambda_p} = \sqrt{\frac{m_pe_p}{2m_pe_p}} = \frac{1}{\sqrt{2}}$

7. (d). K_{max} of photoelectrons doesn't depends upon intensity of incident light.

8. (b).

$\lambda = \frac{6.62\times10^{-34}}{\sqrt{2\times9.1\times10^{-31}\times13.6\times1.6\times10^{-19}}}$

$\Rightarrow \lambda = 3.3 \times 10^{-10}$ m = 3.3 Å

9. (c). $\lambda = \frac{12.27}{\sqrt{V}}$ Å

$\Rightarrow V = 40 - 20 = 20$ Volt

$\Rightarrow \lambda = \frac{12.27}{\sqrt{20}}$ Å = 2.75 Å

10. (d). Wavelength of electrons is $\lambda = \sqrt{\frac{150}{V}}$ Å

Now, electrons have energy of 40 KeV, therefore they are accelerated through a potential difference of 40×10^3 volt.

$\lambda = \sqrt{\frac{150}{40\times10^3}} = 0.061$ Å

$\therefore$ Resolving limit of electron microscope = 0.061 Å

11. (d). The linear momentum of the photon

$= \frac{h}{\lambda} = \frac{6.63\times10^{-34}}{122\times10^{-9}} = 5.43 \times 10^{-27}\ \frac{kg-m}{s}$

$\because p = mv \Rightarrow v = \frac{p}{m}$

$\Rightarrow v = \frac{5.43\times10^{-27}}{1.67\times10^{-27}} = 3.25$ m/s

12. (b). $V = \frac{150}{\lambda_e^2}$ volt, to determine the p.d. through which it was accelerated to achieve the given de-broglie wavelength. Then the same p.d. will retard it to rest. Thus,

$V = \frac{150}{0.2\times0.2}$ volt, V = 3765 Volt = 3.76 kV

13. (b). $\lambda_{photon} = \frac{hc}{E}$ and $\lambda_{proton} = \frac{h}{\sqrt{2me}}$

$\Rightarrow \frac{\lambda_{photon}}{\lambda_{electron}} = c\sqrt{\frac{2m}{E}} \Rightarrow \frac{\lambda_{photon}}{\lambda_{electron}} \propto \frac{1}{\sqrt{E}}$

14. **(d).** $hv - W_0 = \frac{1}{2}mv_{max}^2 \Rightarrow \frac{hc}{\lambda} - \frac{hc}{\lambda_0} = \frac{1}{2}mv_{max}^2$

$$\Rightarrow hc\left(\frac{\lambda_0 - \lambda}{\lambda\lambda_0}\right) = \frac{1}{2}mv_{max}^2 \Rightarrow v_{max} = \sqrt{\frac{2hv}{m}\left(\frac{\lambda_0 - \lambda}{\lambda\lambda_0}\right)}$$

When wavelength is λ and velocity is v, then

$$v = \sqrt{\frac{2hv}{m}\left(\frac{\lambda_0 - \lambda}{\lambda\lambda_0}\right)} \quad ....(i)$$

When wavelength is $\frac{3\lambda}{4}$ and velocity is 'v' then

$$v' = \sqrt{\frac{2hc}{m}\left[\frac{\lambda_0 - (3\lambda/4)}{(3\lambda/4)\times\lambda_0}\right]} \quad ...(ii)$$

Divide equation (ii) by (i), we get

$$\frac{v'}{v} = \sqrt{\frac{[\lambda_0 - (3\lambda/4)]}{\frac{3}{4}\lambda\lambda_0} \times \frac{\lambda\lambda_0}{\lambda_0 - \lambda}}$$

$$v' = v\left(\frac{4}{3}\right)^{1/2}\sqrt{\frac{[\lambda_0 - (3\lambda - 4)]}{\lambda_0\lambda}}$$

i.e $\quad v' > v\left(\frac{4}{3}\right)^{1/2}$

15. **(d).** De-Broglie wavelength $\lambda = \frac{h}{p}$

$$\Rightarrow \quad \lambda \propto \frac{1}{p}$$

i.e. graph will be a rectangular hyperbola.

16. **(d).** If the incident light be of threshold wavelength (λ_0), then the stopping potential shall be zero. Thus

$$\lambda_0 = \frac{hc}{\phi}, \ \lambda_0 = \frac{6.6\times10^{-34}\times3\times10^8}{4.2\times1.6\times10^{-19}}$$

$\lambda_0 = 2.946 \times 10^{-7}$ m $= 2946$ Å

17. **(b).** Relation between $V_0 - v$., $V_0 = \frac{hv}{e} - \frac{hv_0}{e}$

Put it in the form of $y = mx - c$,

here $V_0 = y$, $v = x$, $\frac{hv_0}{e} = c$

$$\therefore y = \left(\frac{h}{e}\right)x - c$$

$$\therefore m = \frac{h}{e}$$

18. **(b).** $P = 10 \times 10^3$ watt, $n = ?$, $\lambda = 300$ m

$$P = \frac{nhc}{\lambda t}$$

$$10^4 = \frac{6.62\times10^{-34}\times3\times10^8\times n}{300\times1}$$

$$n = \frac{300\times10^4}{6.62\times10^{-34}\times3\times10^8} = 1.5\times10^{31}$$

19. **(b).** $V_0 = \frac{hc}{e\lambda} - \frac{\phi_0}{e} = 3.74 - 1.07 = 2.67$ V

20. **(a).** The stopping potential for curves a and b is same.

$\therefore \quad f_a = f_b$

Also saturation current is proportional to intensity

$\therefore \quad I_a < I_b$

21. **(c).** $hv = hv_0 + E_k$

$6.6 \times 10^{-34} \times 3 \times 10^{15} = 4 \times 1.6 \times 10^{-19} + E_k$

$19.8 \times 10^{-19} - 6.4 \times 10^{-19} = E_k$

$E_k = 13.4 \times 10^{-19}$ J

$$\Rightarrow \frac{1}{2}mv^2_{max} = 13.4\times10^{-19}$$

$$v_{max} = \sqrt{\frac{2\times13.4\times10^{-19}}{m}}$$

$$= \sqrt{\frac{2\times13.4\times10^{-19}}{9\times10^{-31}}} = 1.73\times10^6 \text{ m/s}$$

22. **(b).** The maximum kinetic energy is

$$K_{max} = \frac{hc}{\lambda} - \phi = \frac{1242 \text{ eV} - \text{nm}}{280 \quad \text{nm}} - 2.5 \text{ eV}$$

$= 4.4$ eV $- 2.5$ eV $= 1.9$ eV

Stopping potential V is given by $eV = K_{max}$

$$V = \frac{K_{max}}{e} = \frac{1.9}{e}\text{ eV} = 1.9 \text{ V}$$

23. **(a).** $\because 2d\sin\phi = n\lambda$

$$\lambda_{max} = \frac{(2d\sin\phi)_{max}}{n_{min}} = \frac{2d\sin90^\circ}{1} = 2\times10 \text{ Å}$$

$\lambda_{max} = 20$Å

$\therefore$ Possible wavelengths are 5Å, 10Å and 20Å.

24. **(c).** $\lambda_{min} = \frac{12400}{10000}$ Å $= 1.24$ Å

$$v_{max} = \frac{c}{\lambda_{min}} = \frac{3\times10^8}{1.24\times10^{-10}} = 2.4\times10^{18} \text{ Hz.}$$

25. **(b)** $\Delta E = \frac{12400}{4500Å}$

$\Delta = 2.75$ eV

For photoelectric effect, $\Delta E > W_0$ (work function).

26. **(a)** $\Delta E = W_0 + E$; $(E_k) = \Delta E - W_0$
For maximum value of (E_k), W_0 should be minimum
W_0 for lithium = 2.3 eV
$\therefore$ $(E_k) = 2.75 - 2.3 = 0.45$ eV

27. **(c)** The maximum magnitude of stopping potential will be for metal of least work function.
$\therefore$ required stopping potential is

$$V_s = \frac{hv - \phi_0}{e} = 0.45 \text{ volt.}$$

28. (c) Mass of moving photon $m = \frac{hv}{c^2} = \frac{h}{c\lambda}$ and $E = mc^2$

29. **(c)** Less work function means less energy is required for ejecting out the electrons.

30. **(a)** de-Broglie wavelength associated with gas molecules varies as $\lambda \propto \frac{1}{\sqrt{T}}$

DAILY PRACTICE PROBLEMS

PHYSICS SOLUTIONS

55

1. **(a)** For $n=1$, maximum number of states $= 2n^2 = 2$ and for $n = 2, 3, 4$ maximum number of states would be 8, 18, 32 respectively, Hence number of possible elements $= 2+8+18+32 = 60$

2. **(d)** $\frac{1}{\lambda} = RZ^2\left(\frac{1}{1^2}-\frac{1}{2^2}\right)$

   For di-ionised lithium the value of Z is maximum

3. **(c)** Lyman series lies in the UV region.

4. **(c)** Transition A ($n = \infty$ to 1): Series limit of Lyman series

   Transition B ($n = 5$ to $n = 2$) Third spectral line of Balmer series

   Transition C ($n = 5$ to $n = 3$): Second spectral line of Paschen series

5. **(b)** Let the energy in A, B and C state be E_A, E_B and E_C then from the figure

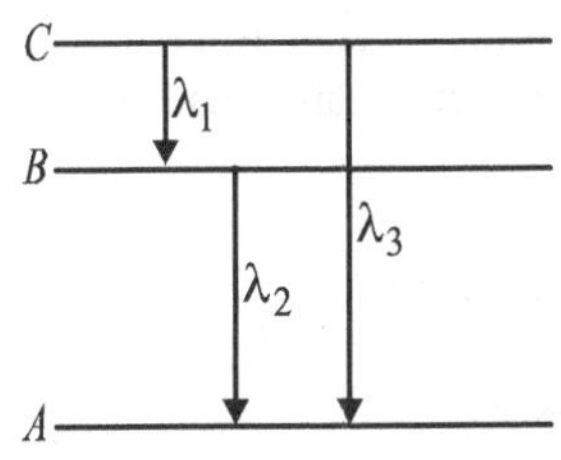

$$(E_C - E_B) + (E_B - E_A) = (E_C - E_A)$$

or $$\frac{hc}{\lambda_1} + \frac{hc}{\lambda_2} = \frac{hc}{\lambda_3}$$

$$\Rightarrow \lambda_3 = \frac{\lambda_1 \lambda_2}{\lambda_1 + \lambda_2}$$

6. **(a)** In the revolution of electron, coulomb force provides the necessary centripetal force

$$\Rightarrow \frac{Ze^2}{r^2} = \frac{mv^2}{r} \Rightarrow mv^2 = \frac{Ze^2}{r}$$

$$\therefore \text{K.E.} = \frac{1}{2}mv^2 = \frac{Ze^2}{2r}$$

7. **(a)** P.E. $\propto -\frac{1}{r}$ and K.E. $\propto \frac{1}{r}$

   As r increases so K.E. decreases but P.E. increases.

8. **(a)** K.E. $= -$(T.E.)

9. **(a)** For Lyman series

$$\nu_{Lyman} = \frac{c}{\lambda_{\max}} = Rc\left[\frac{1}{(1)^2} - \frac{1}{(2)^2}\right] = \frac{3RC}{4}$$

For Balmer series

$$\nu_{Balmer} = \frac{c}{\lambda_{\max}} = Rc\left[\frac{1}{(2)^2} - \frac{1}{(3)^2}\right] = \frac{5RC}{36}$$

$$\therefore \frac{\nu_{Lyman}}{\nu_{Balmer}} = \frac{27}{5}$$

10 **(c)** $\frac{1}{\lambda} = R\left(\frac{1}{n_1^2} - \frac{1}{n_2^2}\right)$

   For first line of Lymen series $n_1 = 1$ and $n_2 = 2$

   For first line of Balmer series $n_2 = 2$ and $n_2 = 3$

   So, $\frac{\lambda_{Lymen}}{\lambda_{Balmer}} = \frac{5}{27}$

11. **(d)** Angular momentum $L = n\left(\frac{h}{2\pi}\right)$

    For this case $n = 2$, hence $L = 2 \times \frac{h}{2\pi} = \frac{h}{\pi}$

12. **(c)** $\frac{mv^2}{a_0} = \frac{1}{4\pi\varepsilon_0}\frac{e^2}{a_0^2} \Rightarrow v = \frac{e}{\sqrt{4\pi\varepsilon_0 a_0 m}}$

13. **(d)** We have to find the frequency of emitted photons. For emission of photons the transition must take place from a higher energy level to a lower energy level which are given only in options (c) and (d).
    Frequency is given by

$$h\nu = -13.6\left(\frac{1}{n_2^2} - \frac{1}{n_1^2}\right)$$

For transition from $n = 6$ to $n = 2$,

$$\nu_1 = \frac{-13.6}{h}\left(\frac{1}{6^2} - \frac{1}{2^2}\right) = \frac{2}{9} \times \left(\frac{13.6}{h}\right)$$

For transition from n = 2 to n = 1,

$$\nu_2 = \frac{-13.6}{h}\left(\frac{1}{2^2} - \frac{1}{1^2}\right) = \frac{3}{4} \times \left(\frac{13.6}{h}\right).$$

$$\therefore \nu_2 > \nu_1$$

14. (d) $E = -Z^2 \times 13.6\, eV = -9 \times 13.6\, eV = -122.4\, eV$

So ionisation energy $= +122.4\, eV$

15. (c) For third line of Balmer series $n_1 = 2, n_2 = 5$

$$\therefore \frac{1}{\lambda} = RZ^2\left[\frac{1}{n_1^2} - \frac{1}{n_2^2}\right] \text{ gives } Z^2 = \frac{n_1^2 n_2^2}{(n_2^2 - n_1^2)\lambda R}$$

On putting values Z = 2

From $E = -\frac{13.6Z^2}{n^2} = \frac{-13.6(2)^2}{(1)^2} = -54.4\, eV$

16. (b) $\frac{1}{\lambda} = R\left[\frac{1}{n_1^2} - \frac{1}{n_2^2}\right] \Rightarrow \frac{1}{\lambda_{3\to2}} = R\left[\frac{1}{(2)^2} - \frac{1}{(3)^2}\right] = \frac{5R}{36}$

and $\frac{1}{\lambda_{4\to2}} = R\left[\frac{1}{(2)^2} - \frac{1}{(4)^2}\right] = \frac{3R}{16}$

$$\therefore \frac{\lambda_{4\to2}}{\lambda_{3\to2}} = \frac{20}{27} \Rightarrow \lambda_{4\to2} = \frac{20}{27}\lambda_0$$

17. (a) $\frac{1}{\lambda_{max}} = R\left[\frac{1}{(1)^2} - \frac{1}{(2)^2}\right] \Rightarrow \lambda_{max} = \frac{4}{3R} \approx 1213 Å$

and $\frac{1}{\lambda_{min}} = R\left[\frac{1}{(1)^2} - \frac{1}{\infty}\right] \Rightarrow \lambda_{min} = \frac{1}{R} \approx 910 Å$

18. (a) Maximum energy is liberated for transition $E_n \to 1$ and minimum energy for $E_n \to E_{n-1}$

Hence $\frac{E_1}{n^2} - E_1 = 52.224\, eV$ .....(i)

and $\frac{E_1}{n^2} - \frac{E_1}{(n-1)^2} = 1.224\, eV$ ......(ii)

Solving equations (i) and (ii) we get

$E_1 = -54.4\, eV$ and $n = 5$

Now $E_1 = -\frac{13.6Z^2}{1^2} = -54.4\, eV$. Hence $Z = 2$

19. (d) Radius $R = \frac{\varepsilon_0 n^2 h^2}{\pi n Z e^2}$

Velocity $v = \frac{Ze^2}{2\varepsilon_0 nh}$ and energy $E = -\frac{mZ^2e^4}{8\varepsilon_0^2 n^2 h^2}$

Now, it is clear from above expressions $R.v \propto n$

20. (c) At closest distance of approach

Kinetic energy = Potential energy

$$\Rightarrow 5 \times 10^6 \times 1.6 \times 10^{-19} = \frac{1}{4\pi\varepsilon_0} \times \frac{(Ze)(2e)}{r}$$

For uranium Z = 92, so $r = 5.3 \times 10^{-12}\, cm$

21. (a) Here radius of electron orbit $r \propto 1/m$ and energy $E \propto m$, where m is the mass of the electron.

Hence energy of hypothetical atom

$E_0 = 2 \times (-13.6\, eV) = -27.2\, eV$ and radius $r_0 = \frac{a_0}{2}$

22. (a) Time period, $T_n = \frac{2\pi r_n}{v_n}$ (in n[th] state)

i.e. $T_n \propto \frac{r_n}{v_n}$ But $r_n \propto n^2$ and $v_n \propto \frac{1}{n}$

Therefore, $T_n \propto n^3$.

Given $T_{n_1} = 8T_{n_1}$, Hence $n_1 = 2n_2$. $\Rightarrow n_1$ is even

23. (d) $2.55 eV = E_4 - E_2$.

Therefore other photon will have energy

$= E_2 - E_1 = 10.2$ eV.

Energy given to H-atom excitation $= E_2 - E_1 = 12.75$ eV.

Consider perfectly inelastic collision for other answer.

24. (a) Balmer series lies in the visible region.

25. (b), 26. (d), 27. (a)

Since 6 different types of photons are emitted implies ${}^4C_2$

i.e. highest excitation state is n = 4

Since emission energies are equal, lesser and greater so initial state

$$e = \frac{12420}{\lambda} = 13.6Z^2\left[\frac{1}{4} - \frac{1}{16}\right]$$

$\Rightarrow Z^2 = 16 \Rightarrow Z = 4$

$$E_{4\to1} = 13.6\,(16)\left|\frac{1}{1} - \frac{1}{16}\right| = 20.4 eV$$

$$E_{4\to3} = 13.6\,(16)\left|\frac{1}{9} - \frac{1}{16}\right| = 10.6\ eV$$

28. (b) Bohr postulated that electrons in stationary orbits around the nucleus do not radiate.

This is the one of Bohr's postulate. According to this the moving electrons radiate only when they go from one orbit to the next lower orbit.

29. (b) Rutherford confirmed the repulsive force on α-particle due to nucleus varies with distance according to inverse square law and that the positive charges are concentrated at the centre and not distributed throughout the atom.

30. (b) When the atom gets appropriate energy from outside, then this electron rises to some higher energy level. Now it can return either directly to the lower energy level or come to the lowest energy level after passing through other lower energy lends, hence all possible transitions take place in the source and many lines are seen in the spectrum.

DAILY PRACTICE PROBLEMS

PHYSICS SOLUTIONS

56

1. **(c).** The energy produced per second is

$$= 1000 \times 10^3 \text{ J} = \frac{10^6}{1.6\times10^{-19}} \text{ eV } 6.25 \times 10^{24} \text{ eV}$$

The number of fissions should be, thus

$$\text{number} = \frac{6.25\times10^{24}}{200\times10^6} = 3.125 \times 10^{16}$$

2. **(b).** No. of atoms in 2kg $_{92}U^{235} = \frac{2}{235} \times N_A$

$$= \frac{2}{235} \times (6.02 \times 10^{26}) = 5.12 \times 10^{24}$$

$$\text{Fission rate} = \frac{5.12\times10^{24}}{30\times24\times60\times60} = 1.975 \times 10^{18} \text{ per sec}$$

Usable energy per fission = 185 MeV

$\therefore$ Power output

$= (185 \times 10^6)(1.975 \times 10^{18})(1.6 \times 10^{-19})$ watt

$= 58.4 \times 10^6$ watt = 58.46 MW

3. **(d).** $\therefore$ 6 gm of $_6C^{12}$ contains atoms $= \frac{6\times10^{23}}{2}$ and each atom of $_6C^{12}$ contains electron, protons and neutrons = 6, 6, 6

$\therefore$ No. of electron, protons and neutron in 6 gm of $_6C^{12} = 18 \times 10^{23}, 18 \times 10^{23}, 18 \times 10^{23}$

4. **(c).** Use ρ = Mass/volume

$$= \frac{1.66\times10^{-27}\times16}{(4/3)\pi(3\times10^{-15})} = 2.35 \times 10^{17} \text{ kg m}^{-3}$$

5. **(a).** Mass defect Δm = M (Ra 226) – M(Rn 222) – M (α)
= 226.0256 – 222.0175 – 4.00026 = 0.0053 u.

6. **(a).** $E = mc^2 = (1.66 \times 10^{-27}) (3 \times 10^8)^2$ J
$= 1.49 \times 10^{-10}$ J

$$= \frac{1.49\times10^{-10}}{1.6\times10^{-13}} \text{ MeV} = 931.49 \text{ MeV}$$

7. **(b).** $E = mc^2$
$= (9.1 \times 10^{-31}) (3 \times 10^8)^2$ J = 0.51 MeV

8. **(c).** $\Delta E = \Delta mc^2$

$$\Delta m = \frac{0.5}{100} \text{kg} = 0.005 \text{ kg}$$

$c = 3 \times 10^8$ m/s

$\Delta E = 0.005 \times (3 \times 10^8)^2$

$\Delta E = 4.5 \times 10^{14}$ J or watt-sec

$$\Delta E = \frac{4.5\times10^{14}}{60\times60} = 1.25 \times 10^{11} \text{ watt hour}$$

$\Delta E = 1.25 \times 10^8$ kWH

9. **(b).** By the forumula $N = N_0 e^{-\lambda t}$

Given $\frac{N}{N_0} = \frac{1}{20}$ and $\lambda = \frac{0.6931}{3.8} \Rightarrow 20 = e^{\frac{0.631\times t}{3.8}}$

Taking log of both sides

$$\text{or } \log 20 = \frac{0.6931\times t}{3.8}\log_{10} e$$

$$\text{or } 1.3010 = \frac{0.6931\times t\times0.4343}{3.8} \Rightarrow t = 16.5 \text{ days}$$

10. **(b).** A = 238 - 4 = 234, Z = 92 - 2 = 90

11. **(a).** Δm = 0.03 a.m.u., A = 4

$$\Rightarrow \Delta E = \frac{\Delta m\times931}{A}$$

$$\Rightarrow \Delta E = \frac{0.03\times931}{4} = 7 \text{ MeV}$$

12. **(a).** $\because \Delta E = \Delta m \times 931$ MeV

$$\Rightarrow \Delta m = \frac{\Delta E}{931} = \frac{2.23}{931} = 0.0024 \text{ a.m.u.}$$

13. **(a).** $\frac{R_{Al}}{R_{Te}} = \frac{(27)^{1/3}}{(125)^{1/3}} = \frac{3}{5} = \frac{6}{10}$

14. **(d).** $R = R_0 A^{1/3} = 1.2 \times 10^{-15} \times (64)^{1/3}$
$= 1.2 \times 10^{-15} \times 4 = 4.8$ fm

15. **(b).** Number of protrons in nucleus = atomic number = 11
Number of electrons = number of protons = 11.
Number of neutrons = mass number A – atomic number Z
N = 24 – 11 = 13

16. **(d).** $\because$ equivalent mass of each photon = 1/2000 amu
$\because$ 1 amu = 931 MeV

$$\therefore \text{Energy of each photon} = \frac{931}{2000} = 0.465 \text{ MeV}$$

17. **(c).** Deuterium, the isotope of hydrogen consits of one proton and neutron. Therefore mass of nuclear constituents of deuterium = mass of proton + mass of neutron
= 1.00759 + 1.00898 = 2.01657 amu.
mass of nucleus of deuterium = 2.01470 amu.
Mass defect = 2.01657 – 2.01470 = 0.00187 amu.
Binding energy = ΔE = 0.00187 × 931 MeV = 1.741 MeV.

18. **(a).** $E = \frac{\Delta E}{A} = \frac{\Delta m\times931}{A}$ MeV

$\Delta m = (3m_p + 4m_n)$ – mass of Li^7

$\Delta m = (3 \times 1.00759 + 4 \times 1.00898) - 7.01653$

Δm = 0.04216 a.m.u.

$$\Delta E = \frac{0.04216\times931}{7} = \frac{39.25}{7} = 5.6 \text{ MeV}$$

19. **(d).** The sun radiates energy in all directions in a sphere. At a distance R, the energy received per unit area per second is 1.4 KJ (given). Therefore the energy released in area $4\pi R^2$ per sec is $= 1400 \times 4\pi R^2$ Joule the energy released per day $= 1400 \times 4\pi R^2 \times 86400$J

where $R = 1.5 \times 10^{11}$ m, Thus

$\Delta E = 1400 \times 4 \times 3.14 \times (1.5 \times 10^{11})^2 \times 86400$

The equivalent mass is

$\Delta m = \Delta E/c^2$

$$\Delta m = \frac{1400 \times 4 \times 3.14 \times (1.5 \times 10^{11})^2 \times 86400}{9 \times 10^{16}}$$

$\Delta m = 3.8 \times 10^{14}$ kg

20. **(b)**

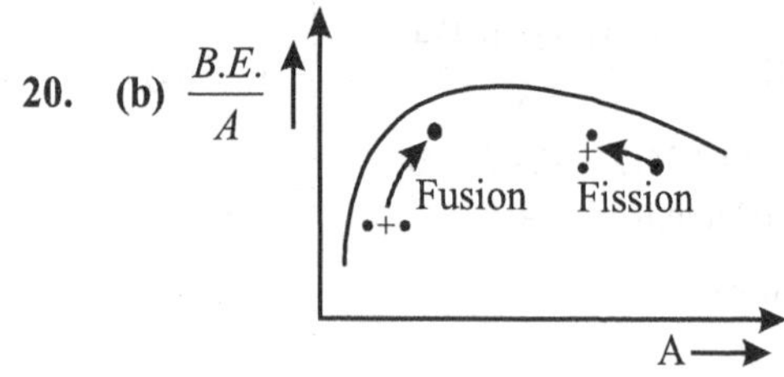

21. **(c)** $N_t = N_0\left(\frac{1}{2}\right)^{t/T} = 50000\left(\frac{1}{2}\right)^{10/5} = 12500$

22. **(d).** Power received from the reactor,

$P = 1000$ KW $= 1000 \times 1000$ W $= 10^6$ J/s

$$P = \frac{10^6}{1.6 \times 10^{-19}} \text{ eV/sec.}$$

$P = 6.25 \times 10^{18}$ MeV/sec

$\therefore$ number of atoms disintegrated per sec

$$= \frac{6.25 \times 10^{18}}{200} = 3.125 \times 10^{16}$$

Energy released per hour $= 10^6 \times 60 \times 60$ Joule

Mass decay per hour $= \Delta m = \dfrac{\Delta E}{c^2}$

$$\Rightarrow \Delta m = \frac{10^6 \times 60 \times 60}{(3 \times 10^8)^2}$$

$\Rightarrow \Delta m = 4 \times 10^{-8}$ kg

23. **(a)**

24. **(a)** In fusion two lighter nuclei combines, it is not the radioactive decay.

25. **(c)** The number of ^{12}C atoms in 1g of carbon,

$$N = \frac{N_A}{12} \times m \Rightarrow N = \frac{6.022 \times 10^{23}}{12} \times 1$$

$= 5.02 \times 10^{22}$ atoms.

The ratio of $^{14}C/^{12}C$ atoms $= 1.3 \times 10^{-12}$ (Given)

$\therefore$ Number of ^{14}C atoms $= 5.02 \times 10^{22} \times 1.3 \times 10^{-12}$

$= 6.5 \times 10^{10}$

$\therefore$ Rate of decay $R_0 = \lambda N_0 = \dfrac{0.693}{T_{1/2}} N_0$

$$\therefore R_0 = \frac{0.693 \times 6.5 \times 10^{10}}{5730 \times 365 \times 24 \times 3600}$$

$= 0.25$ Bq $= 0.25$ (decays/s)

26. **(c)** For 10g sample, number of decays = 0.5 per second.
i.e. $R = 0.05$ and $R_0 = 0.25$ for each gram of ^{14}C

$$\frac{R}{R_0} = e^{-\lambda t} \Rightarrow t = \frac{1}{\lambda}\frac{\ln(R_0/R)}{1} = \frac{\ln(R_0/R)}{(0.693/T_{1/2})}$$

$$\Rightarrow t = \frac{5730 \text{ years}}{0.693} \times \ln\left(\frac{0.25}{0.05}\right) = 13310 \text{ years}$$

27. **(d)** If there are no other radioactive ingredients, the sample is very recent. But the error of measurement must be high unless the statistical error itself is large. In any case, for an old sample, the activity will not be higher than that of a recent one.

28. **(d)** The penetrating power is maximum in case of gamma rays because gamma rays are an electromagnetic radiation of very small wavelength.

29. **(b)** β-particles, being emitted with very high velocity (up to 0.99 c). So, according to Einstein's theory of relatively, the mass of a β-particle is much higher compared to is its rest mass (m_0). The velocity of electrons obtained by other means is very small compared to c (Velocity of light). So its mass remains nearly m_0. But β-particle and electron both are similar particles.

30. **(b)** Electron capture occurs more often than positron emission in heavy elements. This is because if positron emission is energetically allowed, electron capture is necessarily allowed, but the reverse is not true *i.e.* when electron caputre is energetically allowed, positron emission is not necessarily allowed.

DAILY PRACTICE PROBLEMS

PHYSICS SOLUTIONS

57

1. (a) With temperature rise conductivity of semiconductors increases.
2. (b)
3. (b) In insulators, the forbidden energy gap is very large, in case of semiconductor it is moderate and in conductors the energy gap is zero.
4. (c) In intrinsic semiconductors, the creation or liberation of one free electron by the thermal energy creates one hole. Thus in intrinsic semiconductors $n_e = n_h$
5. (b) Both P-type and N-type semiconductors are neutral because neutral atoms are added during doping.
6. (d) Conductor has positive temperature coefficient of resistance but semiconductor has negative temperature coefficient of resistance.
7. (d)
8. (c) At zero Kelvin, there is no thermal agitation and therefore no electrons from valence band are able to shift to conduction band.
9. (c) Antimony is a fifth group impurity and is therefore a donor of electrons.
10. (d) At 0K temperature semiconductor behaves as an insulator, because at very low temperature electrons cannot jump from the valence band to conduction band.
11. (b) Formation of energy bands in solids are due to Pauli's exclusion principle.
12. ()
13. (a) In conductors valence band and conduction band may overlaps.
14. (b) With rise in temperature, conductivity of semiconductor increases while resistance decreases.
15. (a) Because $v_d = \dfrac{i}{(n_e)eA}$
16. (b) In reverse biasing, width of depletion layer increases.
17. (b) Because in case (1) N is connected with N. This is not a series combination of transistor.
18. (d) Resistance in forward biasing $R_{fr} \approx 10\Omega$ and resistance in reverse biasing

$$R_{Rw} \approx 10^5\,\Omega \Rightarrow \frac{R_{fr}}{R_{Rw}} = \frac{1}{10^4}$$

19. (b) In forward biasing the diffusion current increases and drift current remains constant so not current is due to the diffusion.
In reverse biasing diffusion becomes more difficult so net current (very small) is due to the drift.
20. (a) In a triclinic crystal $a \neq b \neq c$ and $\alpha \neq \beta \neq \gamma \neq 90°$
21. (a)
22. (a) In figure (1), (2) and (3). *P*-crystals are more positive as compared to *N*-crystals.
23. (a) Wood is non-crystalline and others are crystalleine.
24. (a) Resistance of conductors (*Cu*) decreases with decrease in temperature while that of semi-conductors (*Ge*) increases with decrease in temperature.
25. (b) $r_p = \dfrac{\Delta V_p}{\Delta I_p} = \dfrac{(180-120)}{(15-10)\times 10^{-3}} = 1.2\times 10^4\ ohm$
26. (a) $g_m = \dfrac{\Delta I_p}{\Delta V_g} = \dfrac{(15-7)\times 10^{-3}}{(-2.0)-(-3.5)} = 5.33\times 10^{-3}\ \text{ohm}^{-1}$
27. (a) $\mu = r_p \times g_m = (1.2\times 10^4)\times(5.33\times 10^{-3}) = 64$
28. (a) According to law of mass action, $n_i^2 = n_e n_h$. In intrinsic semiconductors $n_i = n_e = n_h$ and for *P*-type semiconductor n_e would be less than n_i, since n_h is necessarily more than n_i.
29. (d) Resistivity of semiconductors decreases with temperature. The atoms of a semiconductor vibrate with larger amplitudes at higher temperatures there by increasing its conductivity not resistivity.
30. (c) We cannot measure the potential barrier of a PN-junction by connecting a sensitive voltmeter across its terminals because in the depletion region, there are no free electrons and holes and in the absence of forward biasing, PN- junction offers infinite resistance.

DAILY PRACTICE PROBLEMS

PHYSICS SOLUTIONS

58

1. (b)

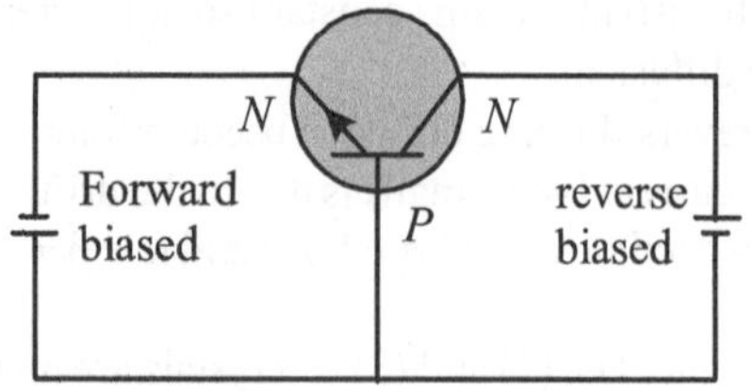

N N P
Forward biased
reverse biased

2. (d) α is the ratio of collector current and *emitter* current while β is the ratio of collector current and base current.

3. (a) Emitter current $I_e = \frac{Ne}{t} = \frac{10^{10} \times 1.6 \times 10^{-19}}{10^{-6}} = 1.6\text{ mA}$

Base current $I_b = \frac{2}{100} \times 1.6 = 0.032\text{ mA}$

But, $I_e = I_c + I_b$

$\therefore I_c = I_e - I_b = 1.6 - 0.032 = 1.568\text{ mA}$

$\therefore \alpha = \frac{I_c}{I_e} = \frac{1.568}{1.6} = 0.98$ and $\beta = \frac{I_c}{I_b} = \frac{1.568}{0.032} = 49$

4. (d)

5. (d) $i_C = \frac{90}{100} \times i_E \Rightarrow 10 = 0.9 \times i_E \Rightarrow I_E = 11mA$

Also $i_E = i_B + i_C \Rightarrow i_B = 11 - 10 = 1mA$

6. (d) $\beta = 50, R = 1000\Omega, V_i = 0.01V$

$\beta = \frac{i_c}{i_b}$ and $i_b = \frac{V_i}{R_i} = \frac{0.01}{10^3} = 10^{-5} A$

Hence $i_c = 50 \times 10^{-5} A = 500\ \mu A$

7. (c) $\alpha = 0.8 \Rightarrow \beta = \frac{0.8}{(1-0.8)} = 4$

Also $\beta = \frac{\Delta i_c}{\Delta i_b} \Rightarrow \Delta i_c = \beta \times \Delta i_b = 4 \times 6 = 24mA$

8. (b) $i_e = i_b + i_c \Rightarrow i_c = i_e - i_b$

9. (b)

10. (d) For CE configuration voltage gain $= \beta \times R_L / R_i$

Power gain $= \beta^2 \times R_L / R_i \Rightarrow \frac{Power\ gain}{Voltage\ gain} = \beta$

11. (b) For 'OR' gate $X = A + B$

i.e. $0+0=0,\ 0+1=1, 1+0=1, 1+1=1$

12. (a)

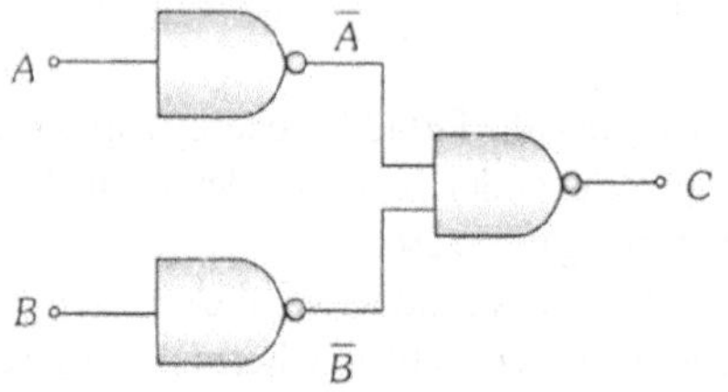

A B $\bar{A}$ $\bar{B}$ C

$C = \overline{\bar{A}.\bar{B}} = \bar{\bar{A}} + \bar{\bar{B}} = A + B$ (De morgan's theorem)

Hence output C is equivalent to OR gate.

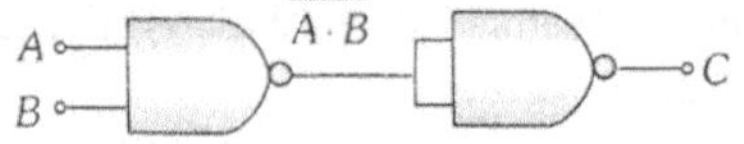

A B $\overline{A \cdot B}$ C

$C = \overline{\overline{AB}.\overline{AB}} = \overline{\overline{A.B}} + \overline{\overline{A.B}} = AB + AB = AB$

In this case output C is equivalent to AND gate.

13. (c) For 'XNOR' gate $Y = \bar{A}\bar{B} + AB$

i.e. $\bar{0}.\bar{0} + 0.0 = 1.1 + 0.0 = 1 + 0 = 1$

$\bar{0}.\bar{1} + 0.1 = 1.0 + 0.1 = 0 + 0 = 0$

$\bar{1}.\bar{0} + 1.0 = 0.1 + 1.0 = 0 + 0 = 0$

$\bar{1}.\bar{1} + 1.1 = 0.0 + 1.1 = 0 + 1 = 1$

14. (d) The output D for the given combination

$D = \overline{(A+B).C} = \overline{(A+B)} + \bar{C}$

If $A = B = C = 0$ then

$D = \overline{(0+0)} + \bar{0} = \bar{0} + \bar{0} = 1 + 1 = 1$

If $A = B = 1, C = 0$ then

$D = \overline{(1+1)} + \bar{0} = \bar{1} + \bar{0} = 0 + 1 = 1$

15. (a) The Boolean expression for 'NOR' gate is $Y = \overline{A+B}$

i.e. if $A = B = 0$ (Low), $Y = \overline{0+0} = \bar{0} = 1$ (High)

16. (d) The Boolean expression for 'AND' gate is $R = P.Q$

$\Rightarrow 1.1 = 1, 1.0 = 0, 0.1 = 0, 0.0 = 0$

17. (a) The given Boolean expression can be written as

$Y = \overline{(A+B)}.\overline{(A.B)} = (\bar{A}.\bar{B}).(\bar{A} + \bar{B}) = (\bar{A}\bar{A}).\bar{B} + \bar{A}(\bar{B}.\bar{B})$

$= \bar{A}.\bar{B} + \bar{A}\bar{B} = \bar{A}\bar{B}$

A	B	Y
0	0	1
1	0	0
0	1	0
1	1	0

18. (c) The Boolean expression for the given combination is output $Y = (A + B).C$

The truth table is

A	B	C	$Y = (A+B).C$
0	0	0	0
1	0	0	0
0	1	0	0
0	0	1	0
1	1	0	0
0	1	1	1
1	0	1	1
1	1	1	1

Hence A = 1, B = 0, C = 1

19. (d) For 'NAND' gate $C = \overline{A.B}$

i.e. $\overline{0.0} = \overline{0} = 1,\ \overline{0.1} = \overline{0} = 1$

$\overline{1.0} = \overline{0} = 1, \overline{1.1} = \overline{1} = 0$

20. (d)

Hence option (d) is true.

21. (c)

True Table

X	Y	$\overline{X}$	$\overline{Y}$	$P = \overline{X} + Y$	$Q = \overline{X.\overline{Y}}$	$R = \overline{P+Q}$
0	1	1	0	1	1	0
1	1	0	0	1	1	0
1	0	0	1	0	0	1
0	0	1	1	1	1	0

Hence X = 1, Y = 0 gives output R = 1

22. (c) $\frac{I_c}{I_e} = 0.96$

$\Rightarrow\ I_c = 0.96 I_e$

But $I_e = I_c + I_b = 0.96 I_e + I_b$

$\Rightarrow\ I_b = 0.04 I_e$

$\therefore$ Current gain, $\beta = \frac{I_c}{I_b} = \frac{0.96 I_e}{0.04 I_e} = 24$

23. (a)

24. (b)

25. (a)

26. (b) The probability that a state with energy E is occupied is given by

$P(E) = \frac{1}{e^{(E-E_F)/kT} + 1}$, where E_F is the Fermi energy,

T is the temperature on the Kelvin scale, and k is the Boltzmann constant. If energies are measured from the top of the valence band, then the energy associated with a state at the bottom of the conduction band is E = 1.11 eV. Furthermore, $kT = (8.62 \times 10^{-5}$ eV/K) (300K) = 0.02586 eV. For pure silicon, E_F = 0.555 eV and $(E - E_F)/kT$ = (0.555eV) / (0.02586eV) = 21.46. Thus,

$$P(E) = \frac{1}{e^{21.46} + 1} = 4.79 \times 10^{-10}$$

For the doped semi-conductor,

$(E - E_F) / kT$ = (0.11 eV) / (0.02586 eV) = 4.254

and $P(E) = \frac{1}{e^{4.254} + 1} = 1.40 \times 10^{-2}$.

27. (a) The energy of the donor state, relative to the top of the valence bond, is 1.11 eV – 0.15 eV = 0.96 eV. The Fermi energy is 1.11 eV – 0.11 eV = 1.00 eV. Hence,

$(E - E_F) / kT = (0.96\text{eV} - 1.00\text{eV}) / (0.02586\text{eV}) = -1.547$

and $P(E) = \frac{1}{e^{-1.547} + 1} = 0.824$

28. (d) In diode the output is in same phase with the input therefore it cannot be used to built NOT gate.

29. (a) This is Boolean expression for 'OR' gate.

30. (d) Statement -1 is true but statement -2 is false.

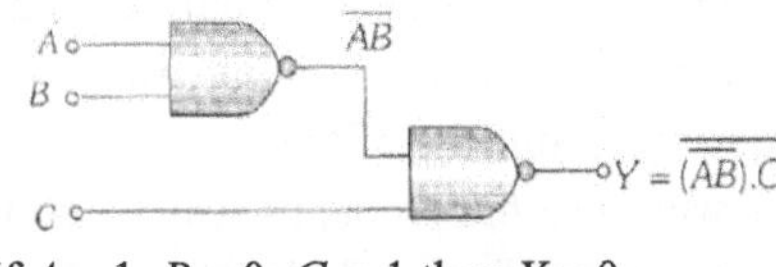

If $A = 1, B = 0, C = 1$ then $Y = 0$

DAILY PRACTICE PROBLEMS

PHYSICS SOLUTIONS

59

1. (a) By using $f_c \approx 9(N_{max})^{1/2} \Rightarrow f_c = 2.84 MHz$
2. (d) Carrier frequency > audio frequency
3. (a) A maximum frequency deviation of 75 kHz is permitted for commercial FM broadcast stations in the 88 to 108 MHz VHF band.
4. (c) Carrier + signal → modulation.
5. (c) 270Ω
6. (a) $\mu_1 > \mu_2$
7. (c) $MUF = \frac{f_c}{\cos\theta} = \frac{60}{\cos 70^\circ} = 175 MHz$
8. (d) coordinated waves of a particular wavelength
9. (d) Surgery needs sharply focused beam of light and laser can be sharply focused.
10. (d) Laser beams are perfectly parallel. So that they are very narrow and can travel a long distance without spreading. This is the feature of laser while they are monochromatic and coherent, these are characteristics only.
11. (b) The formula for modulating index is given by

$$m_f = \frac{\delta}{v_m} = \frac{\text{Frequency variation}}{\text{Modulating frequency}} = \frac{10\times10^3}{2\times10^3} = 5$$

12. (b) $f_a < f_f$
13. (c) An antenna is a metallic structure used to radiate or receive EM waves.
14. (a) Varying the local oscillator frequency
15. (b) In the channel or in the transmission line
16. (a) Carrier swing $= \frac{\text{Frequency deviation}}{\text{Modulating frequency}} = \frac{50}{7} = 7.143$
17. (c) In optical fibre, light travels inside it, due to total internal reflection.
18. (b) The process of changing the frequency of a carrier wave (modulated wave) in accordance with the audio frequency signal (modulating wave) is known as frequency modulation (FM).
19. (d) Following are the problems which are faced while transmitting audio signals directly.
    (i) These signals are relatively of short range.
    (ii) If *every* body started transmitting these low frequency signals directly, mutual interference will render all of them ineffective.
    (iii) Size of antenna required for their efficient radiation would be larger *i.e.* about 75 *km*.
20. (a) The critical frequency of a sky wave for relection from a layer of atmosphere is given by $f_c = 9(N_{max})1^{1/2}$

$$\Rightarrow 10\times10^6 = 9(N_{max})1^{1/2}$$

$$\Rightarrow N_{max} = \left(\frac{10\times10^6}{9}\right)^2 \approx 1.2\times10^{12} m^{-3}$$

21. (b) Core of acceptance angle $\theta = \sin^{-1}\sqrt{n_1^2 - n_2^2}$
22 (a) Remote sensing is the technique to collect information about an object in respect of its size, colour, nature, location, temperature etc without physically touching it. There are some areas or location which are inaccessible. So to explore sensing is used. Remote sensing is done through a satellite.
23. (b)
24. (a) Laser beams are perfectly parallel. They are monochromatic and coherent. These are characteristics only.
25. (d) $f_c = 9\sqrt{N_m} = 9\times\sqrt{9\times10^{10}}$

$$= 2.7\times10^6 \text{ Hz} = 2.7 \text{ MHz}$$

26. (b) $f = f_c\sqrt{1+\frac{D^2}{4h^2}} = 2.7\times10^6 \times \sqrt{1+\frac{(250\times10^3)^2}{4\times(150\times10^3)^2}}$

$$= 3.17 \times 10^6 Hz$$

27. (c) $f = f_c = \sec\phi_i$

$$Sec\phi_i = \frac{f}{f_c} = \frac{3.17\times10^6}{2.7\times10^6} = 1.174$$

$$\phi_i = \sec^{-1}(1.174) = 31.6^\circ.$$

28. (d) TV signals (frequency greater than 30 MHz) cannot be propagated through sky wave propagation.
    Above critical frequency, an electromagnetic wave penetrates the ionosphere and is not reflected by it.
29. (d) The electromagnetic waves of shorter wavelength do not suffer much diffraction from the obstacles of earth's atmosphere so they can travel long distance.
    Also, shorter the wavelength, shorter is the velocity of wave propagation.
30. (b) A dish antenna is a directional antenna because it can transmit or signals in all direction.

DAILY PRACTICE PROBLEMS — PHYSICS SOLUTIONS 60

1. (b) Voltmeter measures voltage across two points so it is connected in parallel and ammeter measures current so it is connected in series.

2. (a) $\frac{P}{Q} = \frac{R}{S}$ so by changing the gap resistance of copper strip gets cancelled.

3. (b) Potential gradient = Potential drop/unit length

$$V = IR = I\frac{\rho\ell}{A}$$

$$\therefore \frac{V}{\ell} = \frac{I\rho}{A}]$$

4. (d) Refractive index is the property of the material, hence it does not depend on angle of the prism.

5. (a) A meter-bridge is a device which is based on the principle of Wheatstone bridge.

6. (a) A potentiometer is device which is used to compare e.m.f.'s of two cells as well as to determine the internal resistance of a cell. It is based on the principle that when a current flows through a wire of uniform thickness and material, potential difference between its two points is directly proportional to the length of the wire between the two points.

7. (c) When a p-n junction diode is connected in reverse biased mode, a reverse current flows.

8. (b) A Zener diode is a heavily doped p-n junction diode which operates on reverse bias beyond breakdown voltage.

9. (c) In a transfer characteristics V_i is plotted along x-axis and V_0 along y-axis.

10. (a) Current gain in CE configuration is $\beta = \frac{\Delta I_C}{\Delta I_B}$.

$I_c >> I_B$, hence it is maximum.

11. (c) $V = IR = I\frac{\rho\ell}{A} \therefore \frac{V}{\ell} = k = \frac{I\rho}{A}$,

increase in I, will increase k, so it will decrease sensitivity.

12. (c)

13. (c) According to the figure the voltmeter and the resistor are connected in parallel.

14. (b) Here I = 4A

$$\theta = 30° = \left(\frac{30 \times \pi}{180}\right)^c = \frac{\pi^c}{6}$$

$$\text{Now, } k = \frac{I}{\theta} = \frac{4}{\left(\frac{\pi}{6}\right)} = \frac{4 \times 6 \times 7}{22} = \frac{2 \times 6 \times 7}{11}$$

$$= \frac{84}{11} = 7.6 \text{ A/rad}$$

15. (c) In reverse-bias mode, a reverse current flows. Therefore, (c) represents the form.

16. (c) Larger the length, lesser will be the potential gradient, so more balancing length will be required.

17. (a) LED is a p-n junction diode which always operates on forward bias.

18. (b) Emitter current is the sum of base and collector current by Kirchhoff's 1st law.

19. (b) $V = IR = \frac{I\rho\ell}{A}$ $\therefore$ potential gradient = k

when A is decreased, k will increase.

20. (d) Positive terminal is at lower potential (0V) and negative terminal is at higher potential + 5V.

21. (d)

22. (d) V = 30.0, I = 0.020 A, $R = \frac{V}{I} = \frac{30.0}{0.020} = 1.50 \text{ k}\Omega$

Error : As $R = \frac{V}{I}$ $\therefore \frac{\Delta R}{R} = \frac{\Delta V}{V} + \frac{\Delta I}{I}$

$$\Rightarrow \Delta R = R\left(\frac{\Delta V}{V} + \frac{\Delta I}{I}\right)$$

$$= 1.50 \times 10^3 \left(\frac{0.1}{30.0} + \frac{0.001}{0.020}\right) = 0.080 \text{ k}\Omega$$

23. (a) u = – 0.30 m, v = – 0.60m

By mirror formula,

$$\frac{1}{v} + \frac{1}{u} = \frac{1}{f}$$

$$\Rightarrow \frac{1}{f} = \frac{-1}{0.30} - \frac{1}{0.60} \Rightarrow f = \frac{-3.0}{0.60} \Rightarrow f = 0.20\text{m}$$

$$\Rightarrow \frac{1}{f} = \frac{1}{v} + \frac{1}{u} \Rightarrow \frac{-df}{f^2} = \frac{-dv}{v^2} - \frac{-du}{u^2}$$

$$\Rightarrow df = (0.20)^2 \left[\frac{0.01}{(0.60)^2} + \frac{0.01}{(0.30)^2}\right]$$

$\Rightarrow df = 0.0055 \approx 0.01\text{m}$

$\Rightarrow$ Focal length $f = (0.20 \pm 0.01)$ m

24. (d) As shown in the figure.

$$\frac{P}{Q} = \frac{\ell}{100 - \ell}, \quad P \propto \frac{\ell}{100 - \ell}$$

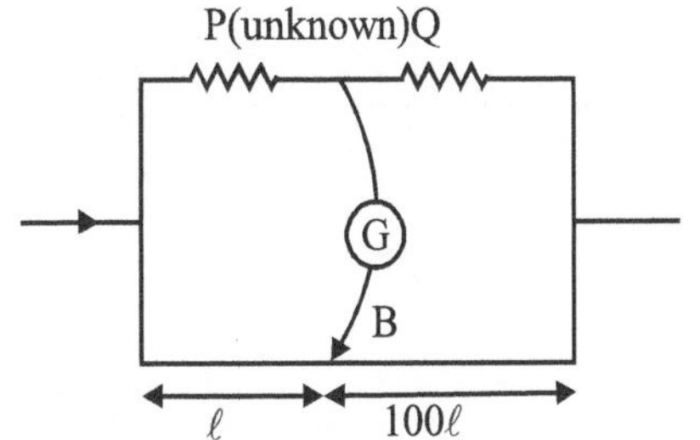

$$\frac{\Delta P}{P}=\frac{\Delta \ell}{\ell}+\frac{\Delta(100-\ell)}{100-\ell}$$

$$=\frac{\Delta \ell}{\ell}+\frac{\Delta \ell}{100-\ell}$$

$$=\frac{0.1}{40.0}+\frac{0.1}{60.0} \Rightarrow \frac{\Delta P}{P}\times 100=0.42\%$$

25. (b) As shown in the figure, when the object (O) is placed between F and C, the image (I) is formed beyond C. It is in this condition that movement towards left

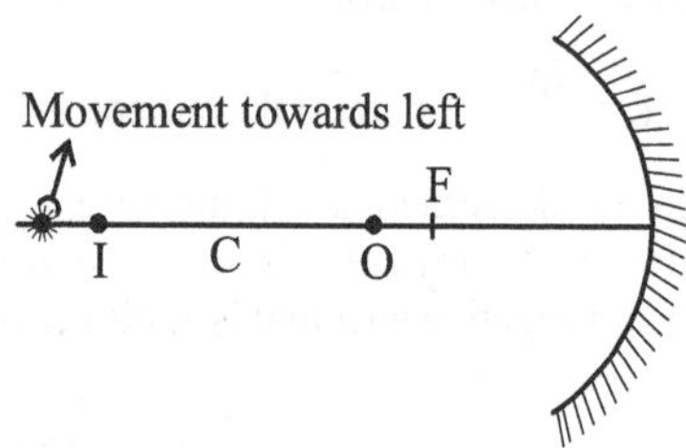

when the student shifts his eye towards left, the image appears to the right of the object pin.

26. (c) For a spherical mirror, the graph plotted between (1/u) and (1/v) will be a straight line with a negative slope of (– 1) and position intercept (1/f) on the (1/v) axis

$$\frac{1}{v}=-\frac{1}{u}+\frac{1}{f}$$

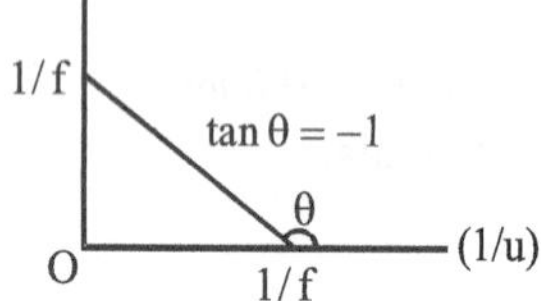

27. (c) because balance point depends upon the value of unknown and known resistance only.

28. (c) Potentiometer is used to measure the potential difference between the two points of a wire.

29. (a) Let j be the current density.

Then $j\times 2\pi r^2=I \Rightarrow j=\frac{I}{2\pi r^2}$

$$\therefore\ E=\rho j=\frac{\rho I}{2\pi r^2}$$

Now, $\Delta V'_{BC}=-\int_{a+b}^{a}\vec{E}.\vec{dr}=-\int_{a+b}^{a}\frac{\rho I}{2\pi r^2}dr$

$$=-\frac{\rho I}{2\pi}\left[-\frac{1}{r}\right]_{a+b}^{a}=\frac{\rho I}{2\pi a}-\frac{\rho I}{2\pi(a+b)}$$

On applying superposition as mentioned we get

$$\Delta V_{BC}=2\times\Delta V'_{BC}=\frac{\rho I}{\pi a}-\frac{\rho I}{\pi(a+b)}$$

30. (c) $E=\frac{\rho I}{2\pi r^2}$

CHAPTER-WISE DPP SHEETS WITH SOLUTIONS

INDEX/CHAPTERS

DPP - Daily Practice Problems

Chapter-wise Sheets

Date : ______ **Start Time :** ______ **End Time :** ______

PHYSICS

CP01

SYLLABUS : Physical World, Units & Measurements

Max. Marks : 180 **Marking Scheme :** (+4) for correct & (–1) for incorrect answer **Time : 60 min.**

INSTRUCTIONS : This Daily Practice Problem Sheet contains 45 MCQs. For each question only one option is correct. Darken the correct circle/ bubble in the Response Grid provided on each page.

1. The density of material in CGS system of units is $4 g/cm^3$. In a system of units in which unit of length is 10 cm and unit of mass is 100 g, the value of density of material will be
   (a) 0.4 unit (b) 40 unit
   (c) 400 unit (d) 0.04 unit
2. The time period of a body under S.H.M. is represented by: $T = P^a D^b S^c$ where P is pressure, D is density and S is surface tension, then values of a, b and c are
   (a) $-\frac{3}{2}, \frac{1}{2}, 1$ (b) $-1, -2, 3$
   (c) $\frac{1}{2}, -\frac{3}{2}, -\frac{1}{2}$ (d) $1, 2, \frac{1}{3}$
3. The respective number of significant figures for the numbers 23.023, 0.0003 and 2.1×10^{-3} are
   (a) 5, 1, 2 (b) 5, 1, 5 (c) 5, 5, 2 (d) 4, 4, 2
4. Young's modulus of a material has the same unit as that of
   (a) pressure (b) strain
   (c) compressibility (d) force
5. Of the following quantities, which one has dimensions different from the remaining three?
   (a) Energy per unit volume
   (b) Force per unit area
   (c) Product of voltage and charge per unit volume
   (d) Angular momentum
6. The pressure on a square plate is measured by measuring the force on the plate and length of the sides of the plate by using the formula $P = \frac{F}{\ell^2}$. If the maximum errors in the measurement of force and length are 4% and 2% respectively, then the maximum error in the measurement of pressure is
   (a) 1% (b) 2% (c) 8% (d) 10%

RESPONSE GRID	1. ⓐⓑⓒⓓ	2. ⓐⓑⓒⓓ	3. ⓐⓑⓒⓓ	4. ⓐⓑⓒⓓ	5. ⓐⓑⓒⓓ
	6. ⓐⓑⓒⓓ				

Space for Rough Work

7. The siemen is the SI unit of
(a) resistivity (b) resistance
(c) conductivity (d) conductance

8. An object is moving through the liquid. The viscous damping force acting on it is proportional to the velocity. Then dimensions of constant of proportionality are
(a) $[ML^{-1}T^{-1}]$ (b) $[MLT^{-1}]$
(c) $[M^0LT^{-1}]$ (d) $[ML^0T^{-1}]$

9. The least count of a stop watch is 0.2 second. The time of 20 oscillations of a pendulum is measured to be 25 second. The percentage error in the measurement of time will be
(a) 8% (b) 1.8% (c) 0.8% (d) 0.1%

10. Weber is the unit of
(a) magnetic susceptibility
(b) intensity of magnetisation
(c) magnetic flux
(d) magnetic permeability

11. The physical quantity which has the dimensional formula $[M^1T^{-3}]$ is
(a) surface tension (b) solar constant
(c) density (d) compressibility

12. The dimensions of Wien's constant are
(a) $[ML^0 T K]$ (b) $[M^0 LT^0 K]$
(c) $[M^0 L^0 T K]$ (d) $[MLTK]$

13. If the capacitance of a nanocapacitor is measured in terms of a unit '*u*' made by combining the electric charge '*e*', Bohr radius 'a_0', Planck's constant 'h' and speed of light '*c*' then
(a) $u = \frac{e^2h}{a_0}$ (b) $u = \frac{hc}{e^2a_0}$
(c) $u = \frac{e^2c}{ha_0}$ (d) $u = \frac{e^2a_0}{hc}$

14. The dimensions of $\frac{1}{\epsilon_0}\frac{e^2}{hc}$ are
(a) $M^{-1} L^{-3} T^4 A^2$ (b) $ML^3 T^{-4} A^{-2}$
(c) $M^0 L^0 T^0 A^0$ (d) $M^{-1} L^{-3} T^2 A$

15. The density of a cube is measured by measuring its mass and length of its sides. If the maximum error in the measurement of mass and length are 4% and 3% respectively, the maximum error in the measurement of density will be
(a) 7% (b) 9% (c) 12% (d) 13%

16. Which is different from others by units ?
(a) Phase difference (b) Mechanical equivalent
(c) Loudness of sound (d) Poisson's ratio

17. A quantity *X* is given by $\varepsilon_0 L\frac{\Delta V}{\Delta t}$ where ϵ_0 is the permittivity of the free space, *L* is a length, D*V* is a potential difference and D*t* is a time interval. The dimensional formula for *X* is the same as that of
(a) resistance (b) charge
(c) voltage (d) current

18. If the error in the measurement of the volume of sphere is 6%, then the error in the measurement of its surface area will be
(a) 2% (b) 3% (c) 4% (d) 7.5%

19. If velocity (V), force (F) and energy (E) are taken as fundamental units, then dimensional formula for mass will be
(a) $V^{-2}F^0E$ (b) V^0FE^2 (c) $VF^{-2}E^0$ (d) $V^{-2}F^0E$

20. Multiply 107.88 by 0.610 and express the result with correct number of significant figures.
(a) 65.8068 (b) 65.807 (c) 65.81 (d) 65.8

21. Which of the following is a dimensional constant?
(a) Refractive index (b) Poissons ratio
(c) Strain (d) Gravitational constant

22. If E, m, J and G represent energy, mass, angular momentum and gravitational constant respectively, then the dimensional formula of EJ^2/m^5G^2 is same as that of the
(a) angle (b) length (c) mass (d) time

23. The refractive index of water measured by the relation $m = \frac{\text{real depth}}{\text{apparent depth}}$ is found to have values of 1.34, 1.38, 1.32 and 1.36; the mean value of refractive index with percentage error is
(a) $1.35 \pm 1.48\%$ (b) $1.35 \pm 0\%$
(c) $1.36 \pm 6\%$ (d) $1.36 \pm 0\%$

RESPONSE GRID					
	7. ⓐⓑⓒⓓ	8. ⓐⓑⓒⓓ	9. ⓐⓑⓒⓓ	10. ⓐⓑⓒⓓ	11. ⓐⓑⓒⓓ
	12. ⓐⓑⓒⓓ	13. ⓐⓑⓒⓓ	14. ⓐⓑⓒⓓ	15. ⓐⓑⓒⓓ	16. ⓐⓑⓒⓓ
	17. ⓐⓑⓒⓓ	18. ⓐⓑⓒⓓ	19. ⓐⓑⓒⓓ	20. ⓐⓑⓒⓓ	21. ⓐⓑⓒⓓ
	22. ⓐⓑⓒⓓ	23. ⓐⓑⓒⓓ			

Space for Rough Work

24. If e is the charge, V the potential difference, T the temperature, then the units of $\frac{eV}{T}$ are the same as that of

(a) Planck's constant (b) Stefan's constant
(c) Boltzmann's constant (d) gravitational constant

25. The dimensions of mobility are

(a) $M^{-2}T^2A$ (b) $M^{-1}T^2A$
(c) $M^{-2}T^3A$ (d) $M^{-1}T^3A$

26. Two quantities A and B have different dimensions which mathematical operation given below is physically meaningful?

(a) A/B (b) A+B (c) A−B (d) A=B

27. The velocity of water waves (v) may depend on their wavelength l, the density of water r and the acceleration due to gravity, g. The method of dimensions gives the relation between these quantities is

(a) v (b) $v^2 \propto g\lambda$
(c) $v^2 \propto g\lambda^2$ (d) $v^2 \propto g^{-1}\lambda^2$

28. The physical quantities not having same dimensions are

(a) torque and work
(b) momentum and Planck's constant
(c) stress and Young's modulus
(d) speed and $(m_0e_0)^{-1/2}$

29. A physical quantity of the dimensions of length that can be formed out of c, G and $\frac{e^2}{4\pi\varepsilon_0}$ is [c is velocity of light, G is universal constant of gravitation and e is charge]

(a) $c^2\left[G\frac{e^2}{4\pi\varepsilon_0}\right]^{1/2}$ (b) $\frac{1}{c^2}\left[\frac{e^2}{G4\pi\varepsilon_0}\right]^{1/2}$

(c) $\frac{1}{c}G\frac{e^2}{4\pi\varepsilon_0}$ (d) $\frac{1}{c^2}\left[G\frac{e^2}{4\pi\varepsilon_0}\right]^{1/2}$

30. The unit of impulse is the same as that of

(a) energy (b) power
(c) momentum (d) velocity

31. If Q denote the charge on the plate of a capacitor of capacitance C then the dimensional formula for $\frac{Q^2}{C}$ is

(a) $[L^2M^2T]$ (b) $[LMT^2]$
(c) $[L^2MT^{-2}]$ (d) $[L^2M^2T^2]$

32. The mass of the liquid flowing per second per unit area of cross-section of the tube is proportional to (pressure difference across the ends)^n and (average velocity of the liquid)^m. Which of the following relations between m and n is correct?

(a) m = n (b) m = −n (c) $m^2 = n$ (d) $m = -n^2$

33. The Richardson equation is given by $I = AT^2e^{-B/kT}$. The dimensional formula for AB^2 is same as that for

(a) IT^2 (b) kT (c) Ik^2 (d) Ik^2/T

34. Turpentine oil is flowing through a capillary tube of length ℓ and radius r. The pressure difference between the two ends of the tube is p. The viscosity of oil is given by :

$\eta = \frac{p(r^2 - x^2)}{4v\ell}$. Here v is velocity of oil at a distance x from the axis of the tube. From this relation, the dimensional formula of η is

(a) $[ML^{-1}T^{-1}]$ (b) $[MLT^{-1}]$

(c) $[ML^2T^{-2}]$ (d) $[M^0L^0T^0]$

35. Given that $y = A\sin\left[\left(\frac{2\pi}{\lambda}(ct - x)\right)\right]$, where y and x are measured in metre. Which of the following statements is true?

(a) The unit of λ is same as that of x and A
(b) The unit of λ is same as that of x but not of A
(c) The unit of c is same as that of $\frac{2\pi}{\lambda}$
(d) The unit of (ct − x) is same as that of $\frac{2\pi}{\lambda}$

36. If L = 2.331 cm, B = 2.1 cm, then L + B =

(a) 4.431 cm (b) 4.43 cm (c) 4.4 cm (d) 4 cm

37. In the relation $x = \cos(\omega t + kx)$, the dimension(s) of ω is/are

(a) $[M^0LT]$ (b) $[M^0L^{-1}T^0]$
(c) $[M^0L^0T^{-1}]$ (d) $[M^0LT^{-1}]$

RESPONSE GRID					
	24. ⓐⓑⓒⓓ	25. ⓐⓑⓒⓓ	26. ⓐⓑⓒⓓ	27. ⓐⓑⓒⓓ	28. ⓐⓑⓒⓓ
	29. ⓐⓑⓒⓓ	30. ⓐⓑⓒⓓ	31. ⓐⓑⓒⓓ	32. ⓐⓑⓒⓓ	33. ⓐⓑⓒⓓ
	34. ⓐⓑⓒⓓ	35. ⓐⓑⓒⓓ	36. ⓐⓑⓒⓓ	37. ⓐⓑⓒⓓ	

Space for Rough Work

38. In a vernier callipers, ten smallest divisions of the vernier scale are equal to nine smallest division on the main scale. If the smallest division on the main scale is half millimeter, then the vernier constant is

(a) 0.5 mm (b) 0.1 mm (c) 0.05 mm (d) 0.005 mm

39. Which two of the following five physical parameters have the same dimensions?

(A) Energy density (B) Refractive index
(C) Dielectric constant (D) Young's modulus
(E) Magnetic field

(a) (B) and (D) (b) (C) and (E)
(c) (A) and (D) (d) (A) and (E)

40. In the eqn. $\left(P+\frac{a}{V^2}\right)(V-b) = \text{constant}$, the unit of a is

(a) dyne cm^5 (b) dyne cm^4
(c) dyne/cm^3 (d) dyne cm^2

41. The dimensions of Reynold's constant are

(a) $[M^0L^0T^0]$ (b) $[ML^{-1}T^{-1}]$
(c) $[ML^{-1}T^{-2}]$ (d) $[ML^{-2}T^{-2}]$

42. Which of the following do not have the same dimensional formula as the velocity?

Given that m_0 = permeability of free space, e_0 = permittivity of free space, n = frequency, l = wavelength, P = pressure, r = density, w = angular frequency, k = wave number,

(a) $1/\sqrt{\mu_0 \varepsilon_0}$ (b) $n l$ (c) $\sqrt{P/\rho}$ (d) ωk

43. Unit of magnetic moment is

(a) ampere–metre^2 (b) ampere–metre
(c) weber–metre^2 (d) weber/metre

44. An experiment is performed to obtain the value of acceleration due to gravity g by using a simple pendulum of length L. In this experiment time for 100 oscillations is measured by using a watch of 1 second least count and the value is 90.0 seconds. The length L is measured by using a meter scale of least count 1 mm and the value is 20.0 cm. The error in the determination of g would be:

(a) 1.7% (b) 2.7% (c) 4.4% (d) 2.27%

45. The dimensional formula for magnetic flux is

(a) $[ML^2T^{-2}A^{-1}]$ (b) $[ML^3T^{-2}A^{-2}]$
(c) $[M^0L^{-2}T^2A^{-2}]$ (d) $[ML^2T^{-1}A^2]$

RESPONSE GRID	
	38. ⓐⓑⓒⓓ 39. ⓐⓑⓒⓓ 40. ⓐⓑⓒⓓ 41. ⓐⓑⓒⓓ 42. ⓐⓑⓒⓓ
	43. ⓐⓑⓒⓓ 44. ⓐⓑⓒⓓ 45. ⓐⓑⓒⓓ

DAILY PRACTICE PROBLEM DPP CHAPTERWISE CP01 - PHYSICS			
Total Questions	45	Total Marks	180
Attempted		Correct	
Incorrect		Net Score	
Cut-off Score	50	Qualifying Score	70
Success Gap = Net Score – Qualifying Score			
Net Score = (Correct × 4) – (Incorrect × 1)			

Space for Rough Work

DPP - Daily Practice Problems

Chapter-wise Sheets

Date : ______ **Start Time :** ______ **End Time :** ______

PHYSICS (CP02)

SYLLABUS : Motion in a Straight Line

Max. Marks : 180 **Marking Scheme :** (+4) for correct & (–1) for incorrect answer **Time : 60 min.**

INSTRUCTIONS : This Daily Practice Problem Sheet contains 45 MCQs. For each question only one option is correct. Darken the correct circle/ bubble in the Response Grid provided on each page.

1. A particle starts moving rectilinearly at time t = 0 such that its velocity v changes with time t according to the equation $v = t^2 - t$ where t is in seconds and v is in m/s. Find the time interval for which the particle retards.

(a) $\frac{1}{2} < t < 1$ (b) $\frac{1}{2} > t > 1$

(c) $\frac{1}{4} < t < 1$ (d) $\frac{1}{2} < t < \frac{3}{4}$

2. The co-ordinates of a moving particle at any time 't' are given by $x = \alpha t^3$ and $y = \beta t^3$. The speed of the particle at time 't' is given by

(a) $3t\sqrt{\alpha^2 + \beta^2}$ (b) $3t^2\sqrt{\alpha^2 + \beta^2}$

(c) $t^2\sqrt{\alpha^2 + \beta^2}$ (d) $\sqrt{\alpha^2 + \beta^2}$

3. If a car covers $2/5^{th}$ of the total distance with v_1 speed and $3/5^{th}$ distance with v_2 then average speed is

(a) $\frac{1}{2}\sqrt{v_1 v_2}$ (b) $\frac{v_1 + v_2}{2}$ (c) $\frac{2v_1 v_2}{v_1 + v_2}$ (d) $\frac{5v_1 v_2}{3v_1 + 2v_2}$

4. Choose the correct statements from the following.

(a) The magnitude of instantaneous velocity of a particle is equal to its instantaneous speed

(b) The magnitude of the average velocity in an interval is equal to its average speed in that interval.

(c) It is possible to have a situation in which the speed of the particle is never zero but the average speed in an interval is zero.

(d) It is possible to have a situation in which the speed of particle is zero but the average speed is not zero.

5. A particle located at $x = 0$ at time $t = 0$, starts moving along with the positive x-direction with a velocity 'v' that varies as $v = \alpha\sqrt{x}$. The displacement of the particle varies with time as

(a) t^2 (b) t (c) $t^{1/2}$ (d) t^3

6. Figure here gives the speed-time graph for a body. The displacement travelled between t = 1.0 second and t = 7.0 second is nearest to

(a) 1.5m

(b) 2m

(c) 3m

(d) 4m

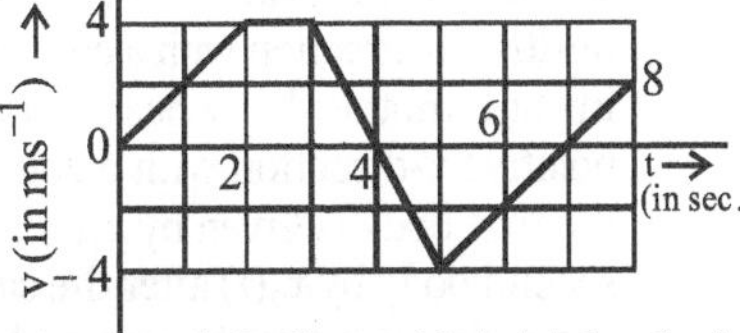

7. A particle is moving in a straight line with initial velocity and uniform acceleration a. If the sum of the distance travelled in t^{th} and $(t + 1)^{th}$ seconds is 100 cm, then its velocity after t seconds, in cm/s, is

(a) 80 (b) 50 (c) 20 (d) 30

RESPONSE GRID					
	1. ⓐⓑⓒⓓ	2. ⓐⓑⓒⓓ	3. ⓐⓑⓒⓓ	4. ⓐⓑⓒⓓ	5. ⓐⓑⓒⓓ
	6. ⓐⓑⓒⓓ	7. ⓐⓑⓒⓓ			

Space for Rough Work

8. A thief is running away on a straight road on a jeep moving with a speed of 9 m/s. A police man chases him on a motor cycle moving at a speed of 10 m/s. If the instantaneous separation of jeep from the motor cycle is 100 m, how long will it take for the police man to catch the thief?
(a) 1 second (b) 19 second
(c) 90 second (d) 100 second

9. The displacement x of a particle varies with time according to the relation $x = \frac{a}{b}(1 - e^{-bt})$. Then select the false alternative.
(a) At $t = \frac{1}{b}$, the displacement of the particle is nearly $\frac{2}{3}\left(\frac{a}{b}\right)$
(b) The velocity and acceleration of the particle at t = 0 are a and –ab respectively
(c) The particle cannot go beyond $x = \frac{a}{b}$
(d) The particle will not come back to its starting point at $t \to \infty$

10. A metro train starts from rest and in five seconds achieves a speed 108 km/h. After that it moves with constant velocity and comes to rest after travelling 45m with uniform retardation. If total distance travelled is 395 m, find total time of travelling.
(a) 12.2 s (b) 15.3 s (c) 9 s (d) 17.2 s

11. The deceleration experienced by a moving motor boat after its engine is cut off, is given by $dv/dt = -kv^3$ where k is a constant. If v_0 is the magnitude of the velocity at cut-off, the magnitude of the velocity at a time t after the cut-off is
(a) $\frac{v_0}{\sqrt{(2v_0^2 kt + 1)}}$ (b) $v_0 e^{-kt}$
(c) $v_0 / 2$ (d) v_0

12. The velocity of a particle is $v = v_0 + gt + ft^2$. If its position is $x = 0$ at $t = 0$, then its displacement after unit time ($t = 1$) is
(a) $v_0 + g/2 + f$ (b) $v_0 + 2g + 3f$
(c) $v_0 + g/2 + f/3$ (d) $v_0 + g + f$

13. A man is 45 m behind the bus when the bus starts accelerating from rest with acceleration 2.5 m/s^2. With what minimum velocity should the man start running to catch the bus?
(a) 12 m/s (b) 14 m/s (c) 15 m/s (d) 16 m/s

14. A body is at rest at $x = 0$. At $t = 0$, it starts moving in the positive x-direction with a constant acceleration. At the same instant another body passes through $x = 0$ moving in the positive x-direction with a constant speed. The position of the first body is given by $x_1(t)$ after time 't'; and that of the second body by $x_2(t)$ after the same time interval. Which of the following graphs correctly describes $(x_1 - x_2)$ as a function of time 't'?

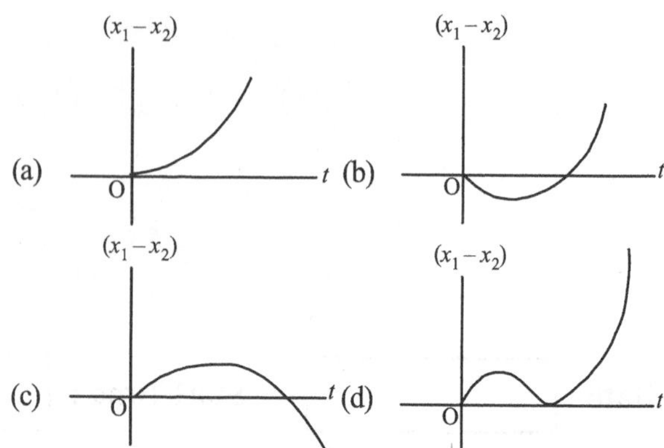

15. From the top of a building 40 m tall, a boy projects a stone vertically upwards with an initial velocity 10 m/s such that it eventually falls to the ground. After how long will the stone strike the ground ? Take g = 10 m/s^2.
(a) 1 s (b) 2 s (c) 3 s (d) 4 s

16. Two bodies begin to fall freely from the same height but the second falls T second after the first. The time (after which the first body begins to fall) when the distance between the bodies equals L is
(a) $\frac{1}{2}T$ (b) $\frac{T}{2} + \frac{L}{gT}$ (c) $\frac{L}{gT}$ (d) $T + \frac{2L}{gT}$

17. Let A, B, C, D be points on a vertical line such that AB = BC = CD. If a body is released from position A, the times of descent through AB, BC and CD are in the ratio.
(a) $1 : \sqrt{3} - \sqrt{2} : \sqrt{3} + \sqrt{2}$ (b) $1 : \sqrt{2} - 1 : \sqrt{3} - \sqrt{2}$
(c) $1 : \sqrt{2} - 1 : \sqrt{3}$ (d) $1 : \sqrt{2} : \sqrt{3} - 1$

18. The water drops fall at regular intervals from a tap 5 m above the ground. The third drop is leaving the tap at an instant when the first drop touches the ground. How far above the ground is the second drop at that instant? (Take g = 10 m/s^2)
(a) 1.25m (b) 2.50m (c) 3.75m (d) 5.00m

19. The displacement 'x' (in meter) of a particle of mass 'm' (in kg) moving in one dimension under the action of a force, is related to time 't' (in sec) by $t = \sqrt{x} + 3$. The displacement of the particle when its velocity is zero, will be
(a) 2m (b) 4m (c) zero (d) 6m

20. A body moving with a uniform acceleration crosses a distance of 65 m in the 5 th second and 105 m in 9th second. How far will it go in 20 s?
(a) 2040m (b) 240m (c) 2400m (d) 2004m

21. An automobile travelling with a speed of 60 km/h, can brake to stop within a distance of 20m. If the car is going twice as fast i.e., 120 km/h, the stopping distance will be
(a) 60m (b) 40m (c) 20m (d) 80m

RESPONSE GRID					
	8. ⓐⓑⓒⓓ	9. ⓐⓑⓒⓓ	10. ⓐⓑⓒⓓ	11. ⓐⓑⓒⓓ	12. ⓐⓑⓒⓓ
	13. ⓐⓑⓒⓓ	14. ⓐⓑⓒⓓ	15. ⓐⓑⓒⓓ	16. ⓐⓑⓒⓓ	17. ⓐⓑⓒⓓ
	18. ⓐⓑⓒⓓ	19. ⓐⓑⓒⓓ	20. ⓐⓑⓒⓓ	21. ⓐⓑⓒⓓ	

Space for Rough Work

22. A particle accelerates from rest at a constant rate for some time and attains a velocity of 8 m/sec. Afterwards it decelerates with the constant rate and comes to rest. If the total time taken is 4 sec, the distance travelled is

(a) 32m (b) 16m
(c) 4m (d) None of the above

23. The equation represented by the graph below is :

(a) $y = \frac{1}{2} gt$

(b) $y = \frac{-1}{2} gt$

(c) $y = \frac{1}{2} gt^2$

(d) $y = \frac{-1}{2} gt^2$

24. A particle moves a distance x in time t according to equation $x = (t + 5)^{-1}$. The acceleration of particle is proportional to:

(a) $(\text{velocity})^{3/2}$ (b) $(\text{distance})^2$
(c) $(\text{distance})^{-2}$ (d) $(\text{velocity})^{2/3}$

25. A particle when thrown, moves such that it passes from same height at 2 and 10 seconds, then this height h is :

(a) 5g (b) g (c) 8g (d) 10g

26. The distance through which a body falls in the nth second is h. The distance through which it falls in the next second is

(a) h (b) $h + \frac{g}{2}$ (c) $h - g$ (d) $h + g$

27. A stone thrown upward with a speed u from the top of the tower reaches the ground with a velocity 3u. The height of the tower is

(a) $3u^2/g$ (b) $4u^2/g$ (c) $6u^2/g$ (d) $9u^2/g$

28. A particle moves along a straight line OX. At a time t (in seconds) the distance x (in metres) of the particle from O is given by $x = 40 + 12t - t^3$. How long would the particle travel before coming to rest?

(a) 40m (b) 56m (c) 16m (d) 24m

29. The graph shown in figure shows the velocity v versus time t for a body.
Which of the graphs represents the corresponding acceleration versus time graphs?

(a) 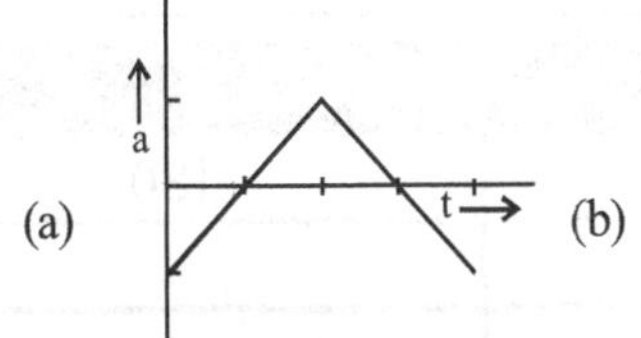

(b) 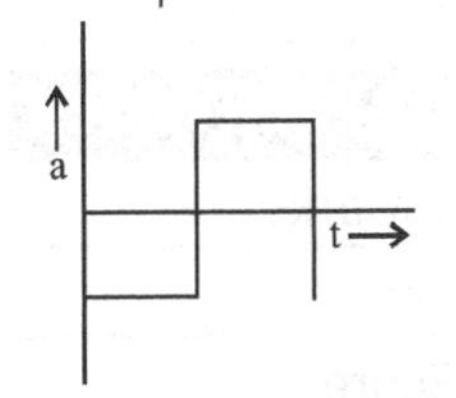

(c) (d)

30. A particle moving along x-axis has acceleration f, at time t, given by $f = f_0\left(1 - \frac{t}{T}\right)$, where f_0 and T are constants. The particle at t = 0 has zero velocity. In the time interval between $t = 0$ and the instant when f = 0, the particle's velocity (v_x) is

(a) $\frac{1}{2} f_0 T^2$ (b) $f_0 T^2$ (c) $\frac{1}{2} f_0 T$ (d) $f_0 T$

31. A body is thrown vertically up with a velocity u. It passes three points A, B and C in its upward journey with velocities $\frac{u}{2}, \frac{u}{3}$ and $\frac{u}{4}$ respectively. The ratio of AB and BC is

(a) 20 : 7 (b) 2 (c) 10 : 7 (d) 1

32. A boat takes 2 hours to travel 8 km and back in still water lake. With water velocity of 4 km h^{-1}, the time taken for going upstream of 8 km and coming back is

(a) 160 minutes (b) 80 minutes
(c) 100 minutes (d) 120 minutes

33. A body starts from rest and travels a distance x with uniform acceleration, then it travels a distance 2x with uniform speed, finally it travels a distance 3x with uniform retardation and comes to rest. If the complete motion of the particle is along a straight line, then the ratio of its average velocity to maximum velocity is

(a) 2/5 (b) 3/5 (c) 4/5 (d) 6/7

34. A man of 50 kg mass is standing in a gravity free space at a height of 10 m above the floor. He throws a stone of 0.5 kg mass downwards with a speed 2 m/s. When the stone reaches the floor, the distance of the man above the floor will be:

(a) 9.9m (b) 10.1m (c) 10m (d) 20m

35. A boy moving with a velocity of 20 km h^{-1} along a straight line joining two stationary objects. According to him both objects

(a) move in the same direction with the same speed of 20 km h^{-1}
(b) move in different direction with the same speed of 20 km h^{-1}
(c) move towards him
(d) remain stationary

RESPONSE GRID					
	22. ⓐⓑⓒⓓ	23. ⓐⓑⓒⓓ	24. ⓐⓑⓒⓓ	25. ⓐⓑⓒⓓ	26. ⓐⓑⓒⓓ
	27. ⓐⓑⓒⓓ	28. ⓐⓑⓒⓓ	29. ⓐⓑⓒⓓ	30. ⓐⓑⓒⓓ	31. ⓐⓑⓒⓓ
	32. ⓐⓑⓒⓓ	33. ⓐⓑⓒⓓ	34. ⓐⓑⓒⓓ	35. ⓐⓑⓒⓓ	

Space for Rough Work

36. A rubber ball is dropped from a height of 5 metre on a plane where the acceleration due to gravity is same as that onto the surface of the earth. On bouncing, it rises to a height of 1.8 m. On bouncing, the ball loses its velocity by a factor of

(a) $\frac{3}{5}$ (b) $\frac{9}{25}$ (c) $\frac{2}{5}$ (d) $\frac{16}{25}$

37. A stone falls freely from rest from a height h and it travels a distance $\frac{9h}{25}$ in the last second. The value of h is

(a) 145 m (b) 100 m (c) 122.5m (d) 200 m

38. Which one of the following equations represents the motion of a body with finite constant acceleration ? In these equations, y denotes the displacement of the body at time t and a, b and c are constants of motion.

(a) $y = at$ (b) $y = at + bt^2$

(b) $y = at + bt^2 + ct^3$ (d) $y = \frac{a}{t} + bt$

39. A particle travels half the distance with a velocity of 6 ms^{-1}. The remaining half distance is covered with a velocity of 4 ms^{-1} for half the time and with a velocity of 8 ms^{-1} for the rest of the half time. What is the velocity of the particle averaged over the whole time of motion ?

(a) 9 ms^{-1} (b) 6 ms^{-1} (c) 5.35 ms^{-1} (d) 5 ms^{-1}

40. A bullet is fired with a speed of 1000 m/sec in order to penetrate a target situated at 100 m away. If $g = 10$ m/s^2, the gun should be aimed

(a) directly towards the target
(b) 5 cm above the target
(c) 10 cm above the target
(d) 15 cm above the target

41. A body covers 26, 28, 30, 32 meters in 10th, 11th, 12th and 13th seconds respectively. The body starts

(a) from rest and moves with uniform velocity
(b) from rest and moves with uniform acceleration
(c) with an initial velocity and moves with uniform acceleration
(d) with an initial velocity and moves with uniform velocity

42. A particle is moving with uniform acceleration along a straight line. The average velocity of the particle from P to Q is 8ms^{-1} and that Q to S is 12ms^{-1}. If QS = PQ, then the average velocity from P to S is

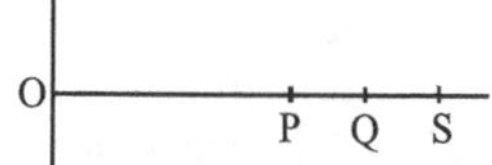

(a) 9.6 ms^{-1} (b) 12.87 ms^{-1}

(c) 64 ms^{-1} (d) 327 ms^{-1}

43. The variation of velocity of a particle with time moving along a straight line is illustrated in the figure. The distance travelled by the particle in four seconds is

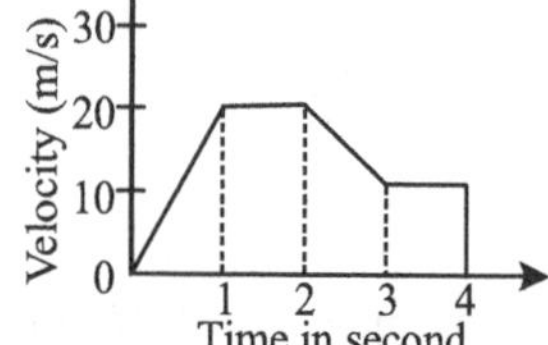

(a) 60m

(b) 55m

(c) 25m

(d) 30m

44. A stone falls freely under gravity. It covers distances h_1, h_2 and h_3 in the first 5 seconds, the next 5 seconds and the next 5 seconds respectively. The relation between h_1, h_2 and h_3 is

(a) $h_1 = \frac{h_2}{3} = \frac{h_3}{5}$ (b) $h_2 = 3h_1$ and $h_3 = 3h_2$

(c) $h_1 = h_2 = h_3$ (d) $h_1 = 2h_2 = 3h_3$

45. A car, starting from rest, accelerates at the rate f through a distance S, then continues at constant speed for time t and then decelerates at the rate $\frac{f}{2}$ to come to rest. If the total distance traversed is 15 S, then

(a) $S = \frac{1}{6}ft^2$ (b) $S = ft$

(c) $S = \frac{1}{4}ft^2$ (d) $S = \frac{1}{72}ft^2$

RESPONSE GRID					
	36. ⓐⓑⓒⓓ	37. ⓐⓑⓒⓓ	38. ⓐⓑⓒⓓ	39. ⓐⓑⓒⓓ	40. ⓐⓑⓒⓓ
	41. ⓐⓑⓒⓓ	42. ⓐⓑⓒⓓ	43. ⓐⓑⓒⓓ	44. ⓐⓑⓒⓓ	45. ⓐⓑⓒⓓ

DAILY PRACTICE PROBLEM DPP CHAPTERWISE CP02 - PHYSICS			
Total Questions	45	Total Marks	180
Attempted		Correct	
Incorrect		Net Score	
Cut-off Score	50	Qualifying Score	70
Success Gap = Net Score – Qualifying Score			
Net Score = (Correct × 4) – (Incorrect × 1)			

Space for Rough Work

DPP - Daily Practice Problems

Chapter-wise Sheets

Date : ______ **Start Time :** ______ **End Time :** ______

PHYSICS (CP03)

SYLLABUS : Motion in a Plane

Max. Marks : 180 **Marking Scheme :** (+4) for correct & (–1) for incorrect answer **Time : 60 min.**

INSTRUCTIONS : This Daily Practice Problem Sheet contains 45 MCQs. For each question only one option is correct. Darken the correct circle/ bubble in the Response Grid provided on each page.

1. A projectile is given an initial velocity of $(\hat{i}+2\hat{j})$ m/s, where $\hat{i}$ is along the ground and $\hat{j}$ is along the vertical. If g = 10 m/s^2, the equation of its trajectory is :
   (a) $y = x - 5x^2$ (b) $y = 2x - 5x^2$
   (c) $4y = 2x - 5x^2$ (d) $4y = 2x - 25x^2$

2. An aircraft moving with a speed of 250 m/s is at a height of 6000 m, just overhead of an anti aircraft–gun. If the muzzle velocity is 500 m/s, the firing angle θ should be:
   (a) 30°
   (b) 45°
   (c) 60°
   (d) 75°

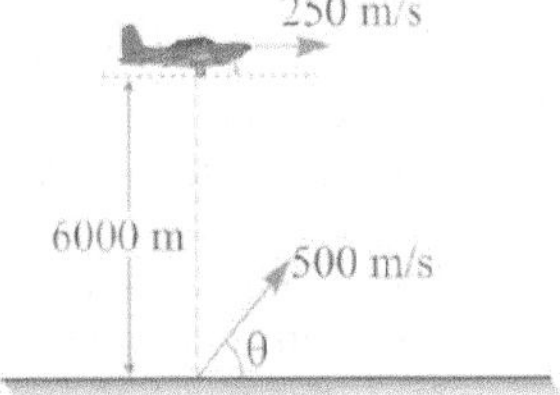

3. Two racing cars of masses m_1 and m_2 are moving in circles of radii r_1 and r_2 respectively. Their speeds are such that each makes a complete circle in the same duration of time t. The ratio of the angular speed of the first to the second car is
   (a) $m_1 : m_2$ (b) $r_1 : r_2$
   (c) 1 : 1 (d) $m_1r_1 : m_2r_2$

4. A boy playing on the roof of a 10 m high building throws a ball with a speed of 10m/s at an angle of 30° with the horizontal. How far from the throwing point will the ball be at the height of 10 m from the ground ?
   $[g = 10\text{m/s}^2,\ \sin 30^o = \frac{1}{2},\ \cos 30^o = \frac{\sqrt{3}}{2}]$
   (a) 5.20m (b) 4.33m (c) 2.60m (d) 8.66m

5. A bomber plane moves horizontally with a speed of 500 m/s and a bomb released from it, strikes the ground in 10 sec. Angle at which it strikes the ground wil be (g = 10 m/s^2)
   (a) $\tan^{-1}\left(\frac{1}{5}\right)$ (b) $\tan\left(\frac{1}{5}\right)$
   (c) $\tan^{-1}(1)$ (d) $\tan^{-1}(5)$

6. Two particles start simultaneously from the same point and move along two straight lines, one with uniform velocity v and other with a uniform acceleration a. If α is the angle between the lines of motion of two particles then the least value of relative velocity will be at time given by
   (a) $\frac{v}{a}\sin\alpha$ (b) $\frac{v}{a}\cos\alpha$ (c) $\frac{v}{a}\tan\alpha$ (d) $\frac{v}{a}\cot\alpha$

7. Initial velocity with which a body is projected is 10 m/sec and angle of projection is 60°. Find the range R
   (a) $\frac{15\sqrt{3}\text{m}}{2}$
   (b) $\frac{40}{3}$m
   (c) $5\sqrt{3}$m
   (d) $\frac{20}{3}$m

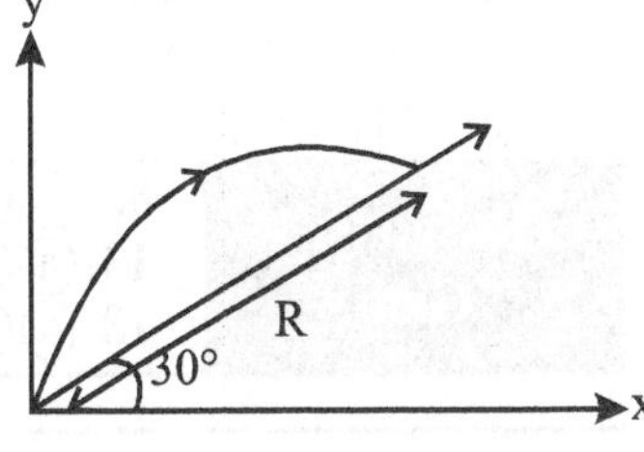

RESPONSE GRID					
	1. ⓐⓑⓒⓓ	2. ⓐⓑⓒⓓ	3. ⓐⓑⓒⓓ	4. ⓐⓑⓒⓓ	5. ⓐⓑⓒⓓ
	6. ⓐⓑⓒⓓ	7. ⓐⓑⓒⓓ			

Space for Rough Work

8. The position vectors of points A, B, C and D are $A = 3\hat{i}+4\hat{j}+5\hat{k}, B = 4\hat{i}+5\hat{j}+6\hat{k}, C = 7\hat{i}+9\hat{j}+3\hat{k}$ and $D = 4\hat{i}+6\hat{j}$ then the displacement vectors $\overrightarrow{AB}$ and $\overrightarrow{CD}$ are
(a) perpendicular (b) parallel
(c) antiparallel (d) inclined at an angle of 60°

9. A person swims in a river aiming to reach exactly on the opposite point on the bank of a river. His speed of swimming is 0.5 m/s at an angle of 120° with the direction of flow of water. The speed of water is
(a) 1.0 m/s (b) 0.5 m/s (c) 0.25 m/s (d) 0.43 m/s

10. A projectile thrown with velocity v making angle θ with vertical gains maximum height H in the time for which the projectile remains in air, the time period is
(a) $\sqrt{H\cos\theta/g}$ (b) $\sqrt{2H\cos\theta/g}$
(c) $\sqrt{4H/g}$ (d) $\sqrt{8H/g}$

11. A ball is thrown from a point with a speed 'v_0' at an elevation angle of θ. From the same point and at the same instant, a person starts running with a constant speed $\frac{'v_0'}{2}$ to catch the ball. Will the person be able to catch the ball? If yes, what should be the angle of projection θ ?
(a) No, 0° (b) Yes, 30° (c) Yes, 60° (d) Yes, 45°

12. If vectors $\vec{A} = \cos\omega t\hat{i} + \sin\omega t\hat{j}$ and $\vec{B} = \cos\frac{\omega t}{2}\hat{i} + \sin\frac{\omega t}{2}\hat{j}$ are functions of time, then the value of t at which they are orthogonal to each other is :
(a) $t=\frac{\pi}{2\omega}$ (b) $t=\frac{\pi}{\omega}$ (c) $t=0$ (d) $t=\frac{\pi}{4\omega}$

13. A bus is moving on a straight road towards north with a uniform speed of 50 km/hour turns through 90°. If the speed remains unchanged after turning, the increase in the velocity of bus in the turning process is
(a) 70.7 km/hour along south-west direction
(b) 70.7 km/hour along north-west direction.
(c) 50 km/hour along west
(d) zero

14. The velocity of projection of oblique projectile is $(6\hat{i}+8\hat{j})\,ms^{-1}$. The horizontal range of the projectile is
(a) 4.9 m (b) 9.6 m (c) 19.6 m (d) 14 m

15. A point P moves in counter-clockwise direction on a circular path as shown in the figure. The movement of 'P' is such that it sweeps out a length $s = t^3 + 5$, where s is in metres and t is in seconds. The radius of the path is 20 m. The acceleration of 'P' when $t = 2$ s is nearly

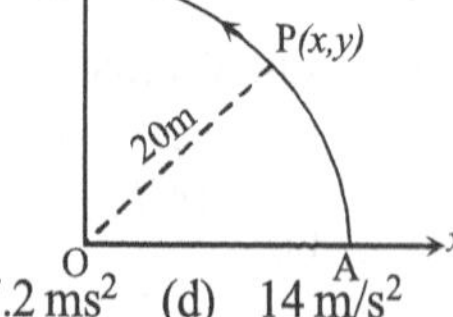

(a) 13 m/s² (b) 12 m/s² (c) 7.2 ms² (d) 14 m/s²

16. The resultant of two vectors $\vec{A}$ and $\vec{B}$ is perpendicular to the vector $\vec{A}$ and its magnitude is equal to half the magnitude of vector $\vec{B}$. The angle between $\vec{A}$ and $\vec{B}$ is
(a) 120° (b) 150° (c) 135° (d) 180°

17. A man running along a straight road with uniform velocity $\vec{u} = u\hat{i}$ feels that the rain is falling vertically down along $-\hat{j}$. If he doubles his speed, he finds that the rain is coming at an angle θ with the vertical. The velocity of the rain with respect to the ground is
(a) $ui - uj$ (b) $ui - \frac{u}{\tan\theta}\hat{j}$
(c) $2u\hat{i} + u\cot\theta\hat{j}$ (d) $ui + u\sin\theta\hat{j}$

18. Two projectiles A and B thrown with speeds in the ratio $1:\sqrt{2}$ acquired the same heights. If A is thrown at an angle of 45° with the horizontal, the angle of projection of B will be
(a) 0° (b) 60° (c) 30° (d) 45°

19. A projectile can have the same range 'R' for two angles of projection. If 'T_1' and 'T_2' be time of flights in the two cases, then the product of the two time of flights is directly proportional to
(a) R (b) $\frac{1}{R}$ (c) $\frac{1}{R^2}$ (d) R^2

20. A man standing on the roof of a house of height h throws one particle vertically downwards and another particle horizontally with the same velocity u. The ratio of their velocities when they reach the earth's surface will be
(a) $\sqrt{2gh+u^2}:u$ (b) $1:2$
(c) $1:1$ (d) $\sqrt{2gh+u^2}:\sqrt{2gh}$

21. If a unit vector is represented by $0.5\hat{i}+0.8\hat{j}+c\hat{k}$, the value of c is
(a) 1 (b) $\sqrt{0.11}$ (c) $\sqrt{0.01}$ (d) 0.39

22. An aeroplane is flying at a constant horizontal velocity of 600 km/hr at an elevation of 6 km towards a point directly above the target on the earth's surface. At an appropriate time, the pilot releases a ball so that it strikes the target at the earth. The ball will appear to be falling
(a) on a parabolic path as seen by pilot in the plane
(b) vertically along a straight path as seen by an observer on the ground near the target
(c) on a parabolic path as seen by an observer on the ground near the target
(d) on a zig-zag path as seen by pilot in the plane

RESPONSE GRID					
	8. ⓐⓑⓒⓓ	9. ⓐⓑⓒⓓ	10. ⓐⓑⓒⓓ	11. ⓐⓑⓒⓓ	12. ⓐⓑⓒⓓ
	13. ⓐⓑⓒⓓ	14. ⓐⓑⓒⓓ	15. ⓐⓑⓒⓓ	16. ⓐⓑⓒⓓ	17. ⓐⓑⓒⓓ
	18. ⓐⓑⓒⓓ	19. ⓐⓑⓒⓓ	20. ⓐⓑⓒⓓ	21. ⓐⓑⓒⓓ	22. ⓐⓑⓒⓓ

Space for Rough Work

23. A particle is projected with a velocity v such that its range on the horizontal plane is twice the greatest height attained by it. The range of the projectile is (where g is acceleration due to gravity)

(a) $\frac{4v^2}{5g}$ (b) $\frac{4g}{5v^2}$ (c) $\frac{v^2}{g}$ (d) $\frac{4v^2}{\sqrt{5}g}$

24. Two stones are projected from the same point with same speed making angles $(45° + \theta)$ and $(45° - \theta)$ with the horizontal respectively. If $\theta \le 45°$, then the horizontal ranges of the two stones are in the ratio of

(a) 1 : 1 (b) 1 : 2 (c) 1 : 3 (d) 1 : 4

25. Three forces acting on a body are shown in the figure. To have the resultant force only along the y-direction, the magnitude of the minimum additional force needed is:

(a) 0.5 N

(b) 1.5 N

(c) $\frac{\sqrt{3}}{4}$ N

(d) $\sqrt{3}$N

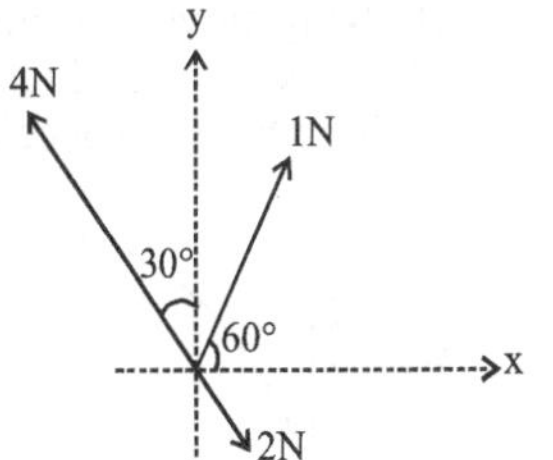

26. A particle moves in x-y plane under the action of force $\vec{F}$ and $\vec{p}$ at a given time t $p_x = 2\cos\theta$, $p_y = 2\sin\theta$. Then the angle θ between $\vec{F}$ and $\vec{p}$ at a given time t is :

(a) $\theta = 30°$ (b) $\theta = 180°$ (c) $\theta = 0°$ (d) $\theta = 90°$

27. A person sitting in the rear end of the compartment throws a ball towards the front end. The ball follows a parabolic path. The train is moving with velocity of 20 m/s. A person standing outside on the ground also observes the ball. How will the maximum heights (y_m) attained and the ranges (R) seen by the thrower and the outside observer compare with each other?

(a) Same y_m different R (b) Same y_m and R
(c) Different y_m same R (d) Different y_m and R

28. A car moves on a circular road. It describes equal angles about the centre in equal intervals of time. Which of the following statement about the velocity of the car is true ?

(a) Magnitude of velocity is not constant
(b) Both magnitude and direction of velocity change
(c) Velocity is directed towards the centre of the circle
(d) Magnitude of velocity is constant but direction changes

29. Three particles A, B and C are thrown from the top of a tower with the same speed. A is thrown up, B is thrown down and C is horizontally. They hit the ground with speeds v_A, v_B and v_C respectively then,

(a) $v_A = v_B = v_C$ (b) $v_A = v_B > v_C$
(c) $v_B > v_C > v_A$ (d) $v_A > v_B = v_C$

30. A particle is moving such that its position coordinate (x, y) are

(2m, 3m) at time t = 0
(6m, 7m) at time t = 2 s and
(13m, 14m) at time t = 5s.

Average velocity vector $(\vec{V}_{av})$ from t = 0 to t = 5s is :

(a) $\frac{1}{5}(13\hat{i} + 14\hat{j})$ (b) $\frac{7}{3}(\hat{i} + \hat{j})$
(c) $2(\hat{i} + \hat{j})$ (d) $\frac{11}{5}(\hat{i} + \hat{j})$

31. A particle moves so that its position vector is given by $\vec{r} = \cos\omega t\hat{x} + \sin\omega t\hat{y}$. Where ω is a constant. Which of the following is true ?

(a) Velocity and acceleration both are perpendicular to $\vec{r}$
(b) Velocity and acceleration both are parallel to $\vec{r}$
(c) Velocity is perpendicular to $\vec{r}$ and acceleration is directed towards the origin
(d) Velocity is perpendicular to $\vec{r}$ and acceleration is directed away from the origin

32. Two boys are standing at the ends A and B of a ground where $AB = a$. The boy at B starts running in a direction perpendicular to AB with velocity v_1. The boy at A starts running simultaneously with velocity v and catches the other boy in a time t, where t is

(a) $a/\sqrt{v^2 + v_1^2}$ (b) $a/(v + v_1)$
(c) $a/(v - v_1)$ (d) $\sqrt{a^2/(v^2 - v_1^2)}$

33. A projectile is fired at an angle of 45° with the horizontal. Elevation angle of the projectile at its highest point as seen from the point of projection is

(a) 60° (b) $\tan^{-1}\left(\frac{1}{2}\right)$ (c) $\tan^{-1}\left(\frac{\sqrt{3}}{2}\right)$ (d) 45°

34. The position vector of a particle $\vec{R}$ as a function of time is given by $\vec{R} = 4\sin(2\pi t)\hat{i} + 4\cos(2\pi t)\hat{j}$

where R is in meter, t in seconds and $\hat{i}$ and $\hat{j}$ denote unit vectors along x-and y-directions, respectively.
Which one of the following statements is wrong for the motion of particle?

(a) Magnitude of acceleration vector is $\frac{v^2}{R}$, where v is the velocity of particle
(b) Magnitude of the velocity of particle is 8 meter/second
(c) Path of the particle is a circle of radius 4 meter.
(d) Acceleration vector is along - $\vec{R}$

35. The vectors $\vec{A}$ and $\vec{B}$ are such that $|\vec{A} + \vec{B}| = |\vec{A} - \vec{B}|$
The angle between the two vectors is

(a) 60° (b) 75° (c) 45° (d) 90°

RESPONSE GRID					
	23. ⓐⓑⓒⓓ	24. ⓐⓑⓒⓓ	25. ⓐⓑⓒⓓ	26. ⓐⓑⓒⓓ	27. ⓐⓑⓒⓓ
	28. ⓐⓑⓒⓓ	29. ⓐⓑⓒⓓ	30. ⓐⓑⓒⓓ	31. ⓐⓑⓒⓓ	32. ⓐⓑⓒⓓ
	33. ⓐⓑⓒⓓ	34. ⓐⓑⓒⓓ	35. ⓐⓑⓒⓓ		

Space for Rough Work

36. The velocity of projection of oblique projectile is $(6\hat{i}+8\hat{j})\,\text{m s}^{-1}$. The horizontal range of the projectile is
(a) 4.9m (b) 9.6m (c) 19.6m (d) 14m

37. An artillary piece which consistently shoots its shells with the same muzzle speed has a maximum range R. To hit a target which is R/2 from the gun and on the same level, the elevation angle of the gun should be
(a) 15° (b) 45° (c) 30° (d) 60°

38. A car runs at a constant speed on a circular track of radius 100 m, taking 62.8 seconds in every circular loop. The average velocity and average speed for each circular loop respectively, is
(a) 0, 10 m/s (b) 10 m/s, 10 m/s
(c) 10 m/s, 0 (d) 0, 0

39. A vector of magnitude b is rotated through angle θ. What is the change in magnitude of the vector?
(a) $2b\sin\frac{\theta}{2}$ (b) $2b\cos\frac{\theta}{2}$ (c) $2b\sin\theta$ (d) $2b\cos\theta$

40. A stone projected with a velocity u at an angle θ with the horizontal reaches maximum height H_1. When it is projected with velocity u at an angle $\left(\frac{\pi}{2}-\theta\right)$ with the horizontal, it reaches maximum height H_2. The relation between the horizontal range R of the projectile, heights H_1 and H_2 is
(a) $R=4\sqrt{H_1H_2}$ (b) $R=4(H_1-H_2)$
(c) $R=4(H_1+H_2)$ (d) $R=\frac{H_1^2}{H_2^2}$

41. The vector sum of two forces is perpendicular to their vector differences. In that case, the forces
(a) cannot be predicted
(b) are equal to each other
(c) are equal to each other in magnitude
(d) are not equal to each other in magnitude

42. A particle crossing the origin of co-ordinates at time t = 0, moves in the xy-plane with a constant acceleration a in the y-direction. If its equation of motion is $y = bx^2$ (b is a constant), its velocity component in the x-direction is
(a) $\sqrt{\frac{2b}{a}}$ (b) $\sqrt{\frac{a}{2b}}$ (c) $\sqrt{\frac{a}{b}}$ (d) $\sqrt{\frac{b}{a}}$

43. A vector $\vec{A}$ is rotated by a small angle Δθ radian (Δθ << 1) to get a new vector $\vec{B}$ In that case $|\vec{B}-\vec{A}|$ is :
(a) $|\vec{A}|\Delta\theta$ (b) $|\vec{B}|\Delta\theta-|\vec{A}|$
(c) $|\vec{A}|\left(1-\frac{\Delta\theta^2}{2}\right)$ (d) 0

44. If a body moving in circular path maintains constant speed of 10 ms^{-1}, then which of the following correctly describes relation between acceleration and radius ?

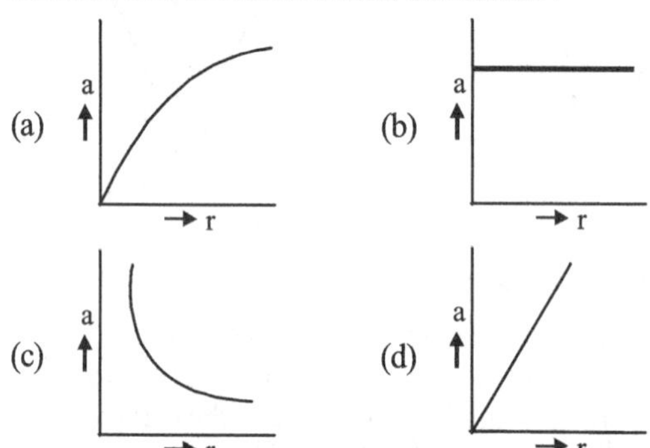

45. The position of a projectile launched from the origin at $t = 0$ is given by $\vec{r}=(40\hat{i}+50\hat{j})$m at $t = 2$s. If the projectile was launched at an angle θ from the horizontal, then θ is (take g = 10 ms^{-2})
(a) $\tan^{-1}\frac{2}{3}$ (b) $\tan^{-1}\frac{3}{2}$ (c) $\tan^{-1}\frac{7}{4}$ (d) $\tan^{-1}\frac{4}{5}$

RESPONSE GRID					
	36. ⓐⓑⓒⓓ	37. ⓐⓑⓒⓓ	38. ⓐⓑⓒⓓ	39. ⓐⓑⓒⓓ	40. ⓐⓑⓒⓓ
	41. ⓐⓑⓒⓓ	42. ⓐⓑⓒⓓ	43. ⓐⓑⓒⓓ	44. ⓐⓑⓒⓓ	45. ⓐⓑⓒⓓ

DAILY PRACTICE PROBLEM DPP CHAPTERWISE CP03 - PHYSICS			
Total Questions	45	Total Marks	180
Attempted		Correct	
Incorrect		Net Score	
Cut-off Score	50	Qualifying Score	70
Success Gap = Net Score – Qualifying Score			
Net Score = (Correct × 4) – (Incorrect × 1)			

Space for Rough Work

DPP - Daily Practice Problems

Chapter-wise Sheets

Date : ______ **Start Time :** ______ **End Time :** ______

PHYSICS

CP04

SYLLABUS : Laws of Motion

Max. Marks : 180 **Marking Scheme :** (+4) for correct & (–1) for incorrect answer **Time : 60 min.**

INSTRUCTIONS : This Daily Practice Problem Sheet contains 45 MCQs. For each question only one option is correct. Darken the correct circle/ bubble in the Response Grid provided on each page.

1. A player stops a football weighing 0.5 kg which comes flying towards him with a velocity of 10m/s. If the impact lasts for 1/50th sec. and the ball bounces back with a velocity of 15 m/s, then the average force involved is
(a) 250 N (b) 1250 N (c) 500 N (d) 625 N

2. For the given situation as shown in the figure, the value of θ to keep the system in equilibrium will be

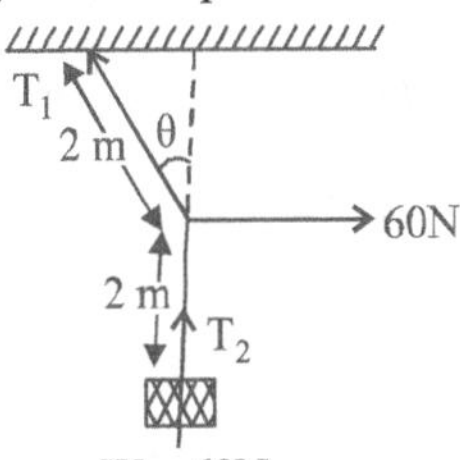

(a) 30° (b) 45° (c) 0° (d) 90°

3. A 5000 kg rocket is set for vertical firing. The exhaust speed is 800 m/s. To give an initial upward acceleration of 20 m/s^2, the amount of gas ejected per second to supply the needed thrust will be (Take g = 10 m/s^2)
(a) 127.5 kg/s (b) 137.5 kg/s
(c) 155.5 kg/s (d) 187.5 kg/s

4. Which one of the following statements is correct?
(a) If there were no friction, work need to be done to move a body up an inclined plane is zero.
(b) If there were no friction, moving vehicles could not be stopped even by locking the brakes.
(c) As the angle of inclination is increased, the normal reaction on the body placed on it increases.
(d) A duster weighing 0.5 kg is pressed against a vertical board with force of 11 N. If the coefficient of friction is 0.5, the work done in rubbing it upward through a distance of 10 cm is 0.55 J.

5. A stone is dropped from a height *h*. It hits the ground with a certain momentum *P*. If the same stone is dropped from a height 100% more than the previous height, the momentum when it hits the ground will change by :
(a) 68% (b) 41% (c) 200% (d) 100%

6. A 3 kg ball strikes a heavy rigid wall with a speed of 10 m/s at an angle of 60°. It gets reflected with the same speed and angle as shown here. If the ball is in contact with the wall for 0.20s, what is the average force exerted on the ball by the wall?
(a) 150N (b) zero
(c) $150\sqrt{3}$N (d) 300N

RESPONSE GRID					
	1. ⓐⓑⓒⓓ	2. ⓐⓑⓒⓓ	3. ⓐⓑⓒⓓ	4. ⓐⓑⓒⓓ	5. ⓐⓑⓒⓓ
	6. ⓐⓑⓒⓓ				

Space for Rough Work

7. The upper half of an inclined plane of inclination θ is perfectly smooth while lower half is rough. A block starting from rest at the top of the plane will again come to rest at the bottom, if the coefficient of friction between the block and lower half of the plane is given by

(a) $\mu = \frac{2}{\tan\theta}$ (b) $\mu = 2\tan\theta$

(c) $\mu = \tan\theta$ (d) $\mu = \frac{1}{\tan\theta}$

8. A block of mass m is in contact with the cart C as shown in the figure.

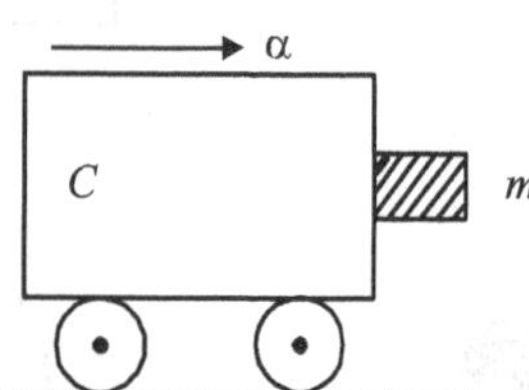

The coefficient of static friction between the block and the cart is μ. The acceleration α of the cart that will prevent the block from falling satisfies:

(a) $\alpha > \frac{mg}{\mu}$ (b) $\alpha > \frac{g}{\mu m}$ (c) $\alpha \geq \frac{g}{\mu}$ (d) $\alpha < \frac{g}{\mu}$

9. A bridge is in the from of a semi-circle of radius 40m. The greatest speed with which a motor cycle can cross the bridge without leaving the ground at the highest point is ($g = 10\ m\ s^{-2}$) (frictional force is negligibly small)

(a) $40\ m\ s^{-1}$ (b) $20\ m\ s^{-1}$
(c) $30\ m\ s^{-1}$ (d) $15\ m\ s^{-1}$

10. An explosion blows a rock into three parts. Two parts go off at right angles to each other. These two are, 1 kg first part moving with a velocity of $12\ ms^{-1}$ and 2 kg second part moving with a velocity of $8\ ms^{-1}$. If the third part flies off with a velocity of $4\ ms^{-1}$, its mass would be

(a) 5 kg (b) 7 kg (c) 17 kg (d) 3 kg

11. A monkey is decending from the branch of a tree with constant acceleration. If the breaking strength is 75% of the weight of the monkey, the minimum acceleration with which monkey can slide down without breaking the branch is

(a) g (b) $\frac{3g}{4}$ (c) $\frac{g}{4}$ (d) $\frac{g}{2}$

12. A car having a mass of 1000 kg is moving at a speed of 30 metres/sec. Brakes are applied to bring the car to rest. If the frictional force between the tyres and the road surface is 5000 newtons, the car will come to rest in

(a) 5 seconds (b) 10 seconds
(c) 12 seconds (d) 6 seconds

13. A spring is compressed between two toy carts of mass m_1 and m_2. When the toy carts are released, the springs exert equal and opposite average forces for the same time on each toy cart. If v_1 and v_2 are the velocities of the toy carts and there is no friction between the toy carts and the ground, then :

(a) $v_1/v_2 = m_1/m_2$ (b) $v_1/v_2 = m_2/m_1$
(c) $v_1/v_2 = -m_2/m_1$ (d) $v_1/v_2 = -m_1/m_2$

14. A plate of mass M is placed on a horizontal frictionless surface (see figure), and a body of mass m is placed on this plate. The coefficient of dynamic friction between this body and the plate is μ. If a force 2μ mg is applied to the body of mass m along the horizontal, the acceleration of the plate will be

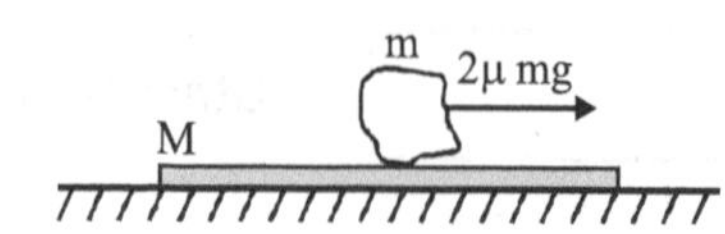

(a) $\frac{\mu m}{M}g$ (b) $\frac{\mu m}{(M+m)}g$ (c) $\frac{2\mu m}{M}g$ (d) $\frac{2\mu m}{(M+m)}g$

15. The rate of mass of the gas emitted from rear of a rocket is initially 0.1 kg/sec. If the speed of the gas relative to the rocket is 50 m/sec and mass of the rocket is 2 kg, then the acceleration of the rocket in m/sec^2 is

(a) 5 (b) 5.2 (c) 2.5 (d) 25

16. A plank with a box on it at one end is gradually raised about the other end. As the angle of inclination with the horizontal reaches 30° the box starts to slip and slides 4.0 m down the plank in 4.0s. The coefficients of static and kinetic friction between the box and the plank will be, respectively :

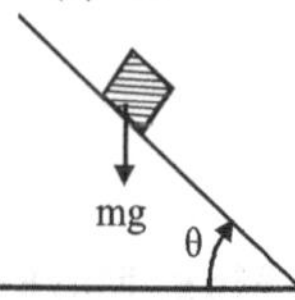

(a) 0.6 and 0.5 (b) 0.5 and 0.6
(c) 0.4 and 0.3 (d) 0.6 and 0.6

17. Four blocks of same mass connected by cords are pulled by a force F on a smooth horizontal surface, as shown in fig. The tensions T_1, T_2 and T_3 will be

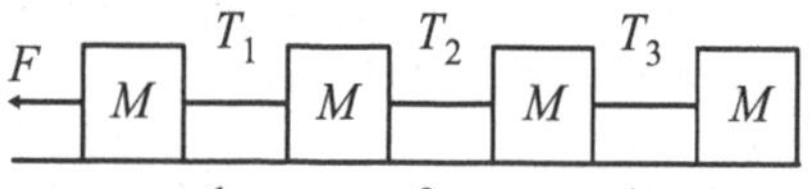

(a) $T_1 = \frac{1}{4}F,\ T_2 = \frac{3}{2}F,\ T_3 = \frac{1}{4}F$

(b) $T_1 = \frac{1}{4}F,\ T_2 = \frac{1}{2}F,\ T_3 = \frac{1}{2}F$

(c) $T_1 = \frac{3}{4}F,\ T_2 = \frac{1}{2}F,\ T_3 = \frac{1}{4}F$

(d) $T_1 = \frac{3}{4}F,\ T_2 = \frac{1}{2}F,\ T_3 = \frac{1}{2}F$

18. A body of mass M is kept on a rough horizontal surface (friction coefficient μ). A person is trying to pull the body by applying a horizontal force but the body is not moving. The force by the surface on the body is F, then

(a) F = Mg (b) F = μMg

(c) $Mg \leq F \leq Mg\sqrt{1+\mu^2}$ (d) $Mg \geq F \geq Mg\sqrt{1+\mu^2}$

RESPONSE GRID					
	7. ⓐⓑⓒⓓ	8. ⓐⓑⓒⓓ	9. ⓐⓑⓒⓓ	10. ⓐⓑⓒⓓ	11. ⓐⓑⓒⓓ
	12. ⓐⓑⓒⓓ	13. ⓐⓑⓒⓓ	14. ⓐⓑⓒⓓ	15. ⓐⓑⓒⓓ	16. ⓐⓑⓒⓓ
	17. ⓐⓑⓒⓓ	18. ⓐⓑⓒⓓ			

Space for Rough Work

19. Which one of the following motions on a smooth plane surface does not involve force?
(a) Accelerated motion in a straight line
(b) Retarded motion in a straight line
(c) Motion with constant momentum along a straight line
(d) Motion along a straight line with varying velocity

20. A block A of mass m_1 rests on a horizontal table. A light string connected to it passes over a frictionless pulley at the edge of table and from its other end another block B of mass m_2 is suspended. The coefficient of kinetic friction between the block and the table is μ_k. When the block A is sliding on the table, the tension in the string is
(a) $\frac{(m_2 - \mu_k m_1)g}{(m_1 + m_2)}$ (b) $\frac{m_1 m_2(1+\mu_k)g}{(m_1 + m_2)}$
(c) $\frac{m_1 m_2(1-\mu_k)g}{(m_1 + m_2)}$ (d) $\frac{(m_2 + \mu_k m_1)g}{(m_1 + m_2)}$

21. The upper half of an inclined plane with inclination f is perfectly smooth while the lower half is rough. A body starting from rest at the top will again come to rest at the bottom if the coefficient of friction for the lower half is given by
(a) $2 \cos \phi$ (b) $2 \sin \phi$ (c) $\tan \phi$ (d) $2 \tan \phi$

22. A particle describes a horizontal circle in a conical funnel whose inner surface is smooth with speed of 0.5 m/s. What is the height of the plane of circle from vertex of the funnel?
(a) 0.25 cm (b) 2 cm (c) 4 cm (d) 2.5 cm

23. You are on a frictionless horizontal plane. How can you get off if no horizontal force is exerted by pushing against the surface?
(a) By jumping
(b) By spitting or sneezing
(c) by rolling your body on the surface
(d) By running on the plane

24. The coefficient of static and dynamic friction between a body and the surface are 0.75 and 0.5 respectively. A force is applied to the body to make it just slide with a constant acceleration which is
(a) $\frac{g}{4}$ (b) $\frac{g}{2}$ (c) $\frac{3g}{2}$ (d) g

25. In the system shown in figure, the pulley is smooth and massless, the string has a total mass 5g, and the two suspended blocks have masses 25 g and 15 g. The system is released from state $\ell = 0$ and is studied upto stage $\ell' = 0$. During the process, the acceleration of block A will be
(a) constant at $\frac{g}{9}$
(b) constant at $\frac{g}{4}$
(c) increasing by factor of 3
(d) increasing by factor of 2

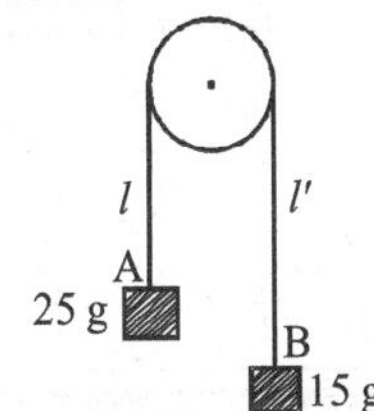

26. The minimum force required to start pushing a body up rough (frictional coefficient μ) inclined plane is F_1 while the minimum force needed to prevent it from sliding down is F_2. If the inclined plane makes an angle θ from the horizontal such that $\tan \theta = 2\mu$ then the ratio $\frac{F_1}{F_2}$ is
(a) 1 (b) 2 (c) 3 (d) 4

27. Two blocks are connected over a massless pulley as shown in fig. The mass of block A is 10 kg and the coefficient of kinetic friction is 0.2. Block A slides down the incline at constant speed. The mass of block B in kg is
(a) 3.5 (b) 3.3 (c) 3.0 (d) 2.5

28. Tension in the cable supporting an elevator, is equal to the weight of the elevator. From this, we can conclude that the elevator is going up or down with a
(a) uniform velocity (b) uniform acceleration
(c) variable acceleration (d) either (b) or (c)

29. A particle tied to a string describes a vertical circular motion of radius r continually. If it has a velocity $\sqrt{3\,gr}$ at the highest point, then the ratio of the respective tensions in the string holding it at the highest and lowest points is
(a) 4 : 3 (b) 5 : 4 (c) 1 : 4 (d) 3 : 2

30. It is difficult to move a cycle with brakes on because
(a) rolling friction opposes motion on road
(b) sliding friction opposes motion on road
(c) rolling friction is more than sliding friction
(d) sliding friction is more than rolling friction

31. A plumb line is suspended from a celling of a car moving with horizontal acceleration of a. What will be the angle of inclination with vertical?
(a) $\tan^{-1}(a/g)$ (b) $\tan^{-1}(g/a)$
(c) $\cos^{-1}(a/g)$ (d) $\cos^{-1}(g/a)$

32. A cart of mass M has a block of mass m attached to it as shown in fig. The coefficient of friction between the block and the cart is μ. What is the minimum acceleration of the cart so that the block m does not fall?

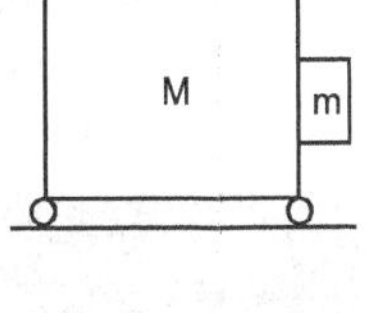

(a) μg (b) g/μ
(c) μ/g (d) $M\mu g/m$

RESPONSE GRID					
	19. ⓐⓑⓒⓓ	20. ⓐⓑⓒⓓ	21. ⓐⓑⓒⓓ	22. ⓐⓑⓒⓓ	23. ⓐⓑⓒⓓ
	24. ⓐⓑⓒⓓ	25. ⓐⓑⓒⓓ	26. ⓐⓑⓒⓓ	27. ⓐⓑⓒⓓ	28. ⓐⓑⓒⓓ
	29. ⓐⓑⓒⓓ	30. ⓐⓑⓒⓓ	31. ⓐⓑⓒⓓ	32. ⓐⓑⓒⓓ	

Space for Rough Work

33. What is the maximum value of the force F such that the block shown in the arrangement, does not move?

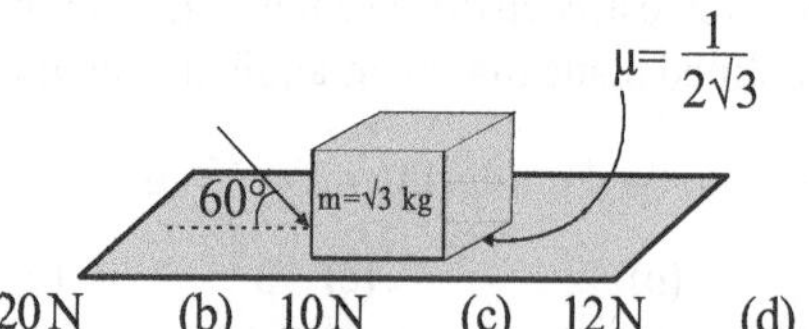

(a) 20 N (b) 10 N (c) 12 N (d) 15 N

34. A block has been placed on an inclined plane with the slope angle θ, block slides down the plane at constant speed. The coefficient of kinetic friction is equal to

(a) sin θ (b) cos θ (c) g (d) tan θ

35. A block of mass *m* is connected to another block of mass *M* by a spring (massless) of spring constant *k*. The block are kept on a smooth horizontal plane. Initially the blocks are at rest and the spring is unstretched. Then a constant force *F* starts acting on the block of mass *M* to pull it. Find the force of the block of mass *m*.

(a) $\frac{MF}{(m+M)}$ (b) $\frac{mF}{M}$ (c) $\frac{(M+m)F}{m}$ (d) $\frac{mF}{(m+M)}$

36. A block of mass m is placed on a surface with a vertical cross section given by $y = \frac{x^3}{6}$. If the coefficient of friction is 0.5, the maximum height above the ground at which the block can be placed without slipping is:

(a) $\frac{1}{6}$m (b) $\frac{2}{3}$m (c) $\frac{1}{3}$m (d) $\frac{1}{2}$m

37. A ball of mass 10 g moving perpendicular to the plane of the wall strikes it and rebounds in the same line with the same velocity. If the impulse experienced by the wall is 0.54 Ns, the velocity of the ball is

(a) 27 ms^{-1} (b) 3.7 ms^{-1} (c) 54 ms^{-1} (d) 37 ms^{-1}

38. A block is kept on a inclined plane of inclination θ of length ℓ. The velocity of particle at the bottom of inclined is (the coefficient of friction is μ)

(a) $[2g\ell(\mu\cos\theta - \sin\theta)]^{1/2}$ (b) $\sqrt{2g\ell(\sin\theta - \mu\cos\theta)}$

(c) $\sqrt{2g\ell(\sin\theta + \mu\cos\theta)}$ (d) $\sqrt{2g\ell(\cos\theta + \mu\sin\theta)}$

39. A 100 g iron ball having velocity 10 m/s collides with a wall at an angle 30° and rebounds with the same angle. If the period of contact between the ball and wall is 0.1 second, then the force experienced by the wall is

(a) 10 N (b) 100 N (c) 1.0 N (d) 0.1 N

40. A bullet is fired from a gun. The force on the bullet is given by $F = 600 - 2 \times 10^5\, t$ where, F is in newton and t in second. The force on the bullet becomes zero as soon as it leaves the barrel. What is the average impulse imparted to the bullet?

(a) 1.8 N-s (b) zero (c) 9 N-s (d) 0.9 N-s

41. Two stones of masses m and 2 m are whirled in horizontal circles, the heavier one in radius $\frac{r}{2}$ and the lighter one in radius r. The tangential speed of lighter stone is n times that of the value of heavier stone when they experience same centripetal forces. The value of n is :

(a) 3 (b) 4 (c) 1 (d) 2

42. A 0.1 kg block suspended from a massless string is moved first vertically up with an acceleration of $5ms^{-2}$ and then moved vertically down with an acceleration of $5ms^{-2}$. If T_1 and T_2 are the respective tensions in the two cases, then

(a) $T_2 > T_1$
(b) $T_1 - T_2 = 1$ N, if $g = 10ms^{-2}$
(c) $T_1 - T_2 = 1$ kg f
(d) $T_1 - T_2 = 9.8$ N, if $g = 9.8\ ms^{-2}$

43. Three forces start acting simultaneously on a particle moving with velocity, $\vec{v}$. These forces are represented in magnitude and direction by the three sides of a triangle ABC. The particle will now move with velocity

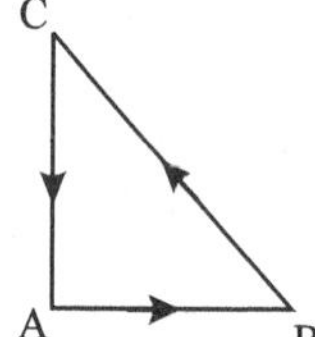

(a) less than $\vec{v}$
(b) greater than $\vec{v}$
(c) |v| in the direction of the largest force BC
(d) $\vec{v}$, remaining unchanged

44. If in a stationary lift, a man is standing with a bucket full of water, having a hole at its bottom. The rate of flow of water through this hole is R_0. If the lift starts to move up and down with same acceleration and then the rates of flow of water are R_u and R_d, then

(a) $R_0 > R_u > R_d$ (b) $R_u > R_0 > R_d$
(c) $R_d > R_0 > R_u$ (d) $R_u > R_d > R_0$

45. A stationary body of mass 3 kg explodes into three equal pieces. Two of the pieces fly off in two mutually perpendicular directions, one with a velocity of $3\hat{i}\ ms^{-1}$ and the other with a velocity of $4\hat{j}\ ms^{-1}$. If the explosion occurs in 10^{-4} s, the average force acting on the third piece in newton is

(a) $(3\hat{i} + 4\hat{j}) \times 10^{-4}$ (b) $(3\hat{i} - 4\hat{j}) \times 10^{-4}$
(c) $(3\hat{i} - 4\hat{j}) \times 10^{4}$ (d) $-(3\hat{i} + 4\hat{j}) \times 10^{4}$

RESPONSE GRID					
	33. ⓐⓑⓒⓓ	34. ⓐⓑⓒⓓ	35. ⓐⓑⓒⓓ	36. ⓐⓑⓒⓓ	
	37. ⓐⓑⓒⓓ	38. ⓐⓑⓒⓓ	39. ⓐⓑⓒⓓ	40. ⓐⓑⓒⓓ	41. ⓐⓑⓒⓓ
	42. ⓐⓑⓒⓓ	43. ⓐⓑⓒⓓ	44. ⓐⓑⓒⓓ	45. ⓐⓑⓒⓓ	

DAILY PRACTICE PROBLEM DPP CHAPTERWISE CP04 - PHYSICS			
Total Questions	45	Total Marks	180
Attempted		Correct	
Incorrect		Net Score	
Cut-off Score	45	Qualifying Score	60
Success Gap = Net Score – Qualifying Score			
Net Score = (Correct × 4) – (Incorrect × 1)			

Space for Rough Work

DPP - Daily Practice Problems

Chapter-wise Sheets

Date : ______ **Start Time :** ______ **End Time :** ______

PHYSICS

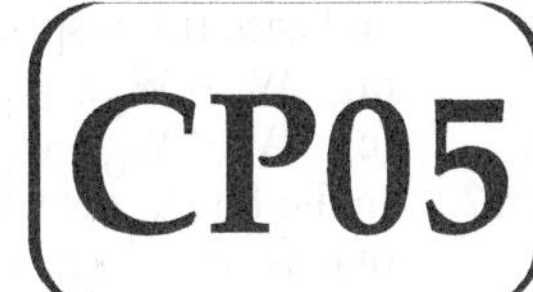

SYLLABUS : Work, Energy and Power

Max. Marks : 180 **Marking Scheme :** (+4) for correct & (–1) for incorrect answer **Time : 60 min.**

INSTRUCTIONS : This Daily Practice Problem Sheet contains 45 MCQs. For each question only one option is correct. Darken the correct circle/ bubble in the Response Grid provided on each page.

1. A spring of spring constant 5×10^3 N/m is stretched initially by 5cm from the unstretched position. Then the work required to stretch it further by another 5 cm is
   (a) 12.50 Nm (b) 18.75 Nm
   (c) 25.00 Nm (d) 6.25 Nm
2. A particle of mass 10 g moves along a circle of radius 6.4 cm with a constant tangential acceleration. What is the magnitude of this acceleration if the kinetic energy of the particle becomes equal to 8×10^{-4} J by the end of the second revolution after the beginning of the motion ?
   (a) 0.1 m/s^2 (b) 0.15 m/s^2 (c) 0.18 m/s^2 (d) 0.2 m/s^2
3. A body is moved along a straight line by a machine delivering a constant power. The distance moved by the body in time '*t*' is proportional to
   (a) $t^{3/4}$ (b) $t^{3/2}$ (c) $t^{1/4}$ (d) $t^{1/2}$
4. A ball is thrown vertically downwards from a height of 20 m with an initial velocity v_0. It collides with the ground and loses 50% of its energy in collision and rebounds to the same height. The initial velocity v_0 is : (Take g = 10 ms^{-2})
   (a) 20 ms^{-1} (b) 28 ms^{-1}
   (c) 10 ms^{-1} (d) 14 ms^{-1}
5. A cord is used to lower vertically a block of mass M, a distance d at a constant downward acceleration of g/4. The work done by the cord on the block is
   (a) $Mg\frac{d}{4}$ (b) $3Mg\frac{d}{4}$ (c) $-3Mg\frac{d}{4}$ (d) Mg d
6. A rubber ball is dropped from a height of 5m on a plane, where the acceleration due to gravity is not shown. On bouncing it rises to 1.8 m. The ball loses its velocity on bouncing by a factor of
   (a) $\frac{16}{25}$ (b) $\frac{2}{5}$ (c) $\frac{3}{5}$ (d) $\frac{9}{25}$
7. A ball of mass m moving with a constant velocity strikes against a ball of same mass at rest. If e = coefficient of restitution, then what will be the ratio of velocity of two balls after collision?
   (a) $\frac{1-e}{1+e}$ (b) $\frac{e-1}{e+1}$ (c) $\frac{1+e}{1-e}$ (d) $\frac{2+e}{e-1}$
8. A particle of mass m is driven by a machine that delivers a constant power of k watts. If the particle starts from rest the force on the particle at time t is
   (a) $\sqrt{mk}\, t^{-1/2}$ (b) $\sqrt{2mk}\, t^{-1/2}$
   (c) $\frac{1}{2}\sqrt{mk}\, t^{-1/2}$ (d) $\sqrt{\frac{mk}{2}}\, t^{-1/2}$

RESPONSE GRID					
	1. ⓐⓑⓒⓓ	2. ⓐⓑⓒⓓ	3. ⓐⓑⓒⓓ	4. ⓐⓑⓒⓓ	5. ⓐⓑⓒⓓ
	6. ⓐⓑⓒⓓ	7. ⓐⓑⓒⓓ	8. ⓐⓑⓒⓓ		

Space for Rough Work

9. A body of mass 2 kg moving under a force has relation between displacement x and time t as $x = \frac{t^3}{3}$ where x is in metre and t is in sec. The work done by the body in first two second will be
(a) 1.6 joule (b) 16 joule
(c) 160 joule (d) 1600 joule

10. A sphere of mass 8m collides elastically (in one dimension) with a block of mass 2m. If the initial energy of sphere is E. What is the final energy of sphere?
(a) 0.8 E (b) 0.36 E
(c) 0.08 E (d) 0.64 E

11. Two similar springs P and Q have spring constants K_P and K_Q, such that $K_P > K_Q$. They are stretched, first by the same amount (case a,) then by the same force (case b). The work done by the springs W_P and W_Q are related as, in case (a) and case (b), respectively
(a) $W_P = W_Q$; $W_P = W_Q$ (b) $W_P > W_Q$; $W_Q > W_P$
(c) $W_P < W_Q$; $W_Q < W_P$ (d) $W_P = W_Q$; $W_P > W_Q$

12. In the figure, the variation of potential energy of a particle of mass m = 2 kg is represented w.r.t. its x-coordinate. The particle moves under the effect of this conservative force along the x-axis.

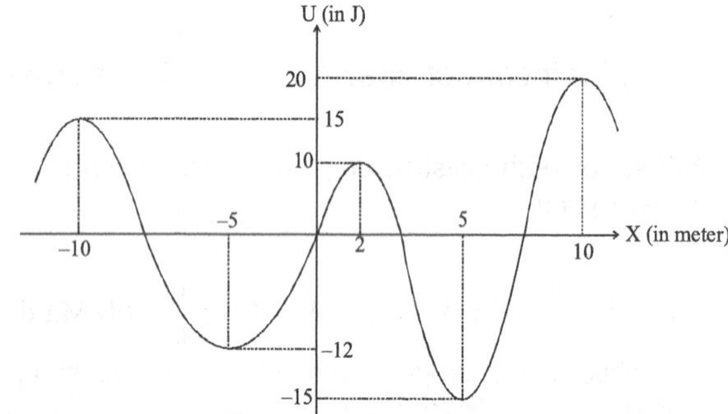

If the particle is released at the origin then
(a) it will move towards positive x-axis
(b) it will move towards negative x-axis
(c) it will remain stationary at the origin
(d) its subsequent motion cannot be decided due to lack of information

13. The potential energy of a certain spring when stretched through distance S is 10 joule. The amount of work done (in joule) that must be done on this spring to stretch it through an additional distance s, will be
(a) 20 (b) 10 (c) 30 (d) 40

14. A force applied by an engine of a train of mass 2.05×10^6 kg changes its velocity from 5m/s to 25 m/s in 5 minutes. The power of the engine is
(a) 1.025 MW (b) 2.05 MW
(c) 5 MW (d) 6 MW

15. The relationship between the force F and position x of a body is as shown in figure. The work done in displacing the body form x = 1 m to x = 5 m will be

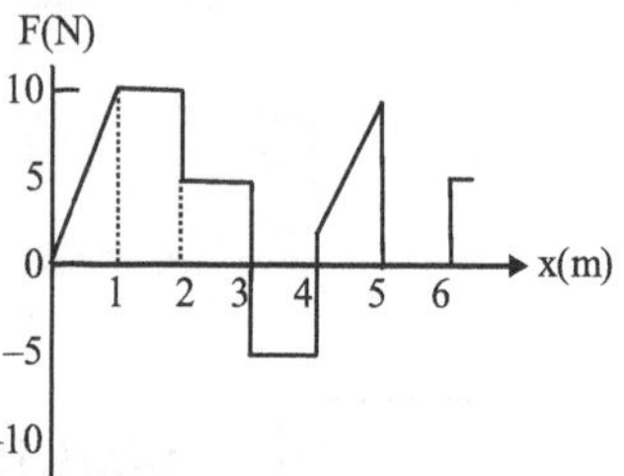

(a) 30 J (b) 15 J (c) 25 J (d) 20 J

16. A body is allowed to fall freely under gravity from a height of 10m. If it looses 25% of its energy due to impact with the ground, then the maximum height it rises after one impact is
(a) 2.5m (b) 5.0m (c) 7.5m (d) 8.2m

17. A block C of mass m is moving with velocity v_0 and collides elastically with block A of mass m and connected to another block B of mass 2m through spring constant k. What is k if x_0 is compression of spring when velocity of A and B is same?

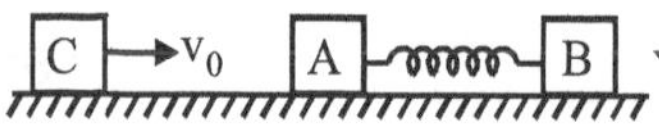

(a) $\frac{mv_0^2}{x_0^2}$ (b) $\frac{mv_0^2}{2x_0^2}$

(c) $\frac{3}{2}\frac{mv_0^2}{x_0^2}$ (d) $\frac{2}{3}\frac{mv_0^2}{x_0^2}$

18. Two springs of force constants 300 N/m (Spring A) and 400 N/m (Spring B) are joined together in series. The combination is compressed by 8.75 cm. The ratio of energy stored in A and B is $\frac{E_A}{E_B}$. Then $\frac{E_A}{E_B}$ is equal to :
(a) $\frac{4}{3}$ (b) $\frac{16}{9}$ (c) $\frac{3}{4}$ (d) $\frac{9}{16}$

19. A body of mass 1 kg begins to move under the action of a time dependent force $\vec{F} = (2t\hat{i} + 3t^2\hat{j})$ N, where $\hat{i}$ and $\hat{j}$ are unit vectors alogn x and y axis. What power will be developed by the force at the time t?
(a) $(2t^2 + 3t^3)$W (b) $(2t^2 + 4t^4)$W
(c) $(2t^3 + 3t^4)$ W (d) $(2t^3 + 3t^5)$W

20. A bullet of mass 20 g and moving with 600 m/s collides with a block of mass 4 kg hanging with the string. What is the velocity of bullet when it comes out of block, if block rises to height 0.2 m after collision?
(a) 200 m/s (b) 150 m/s (c) 400 m/s (d) 300 m/s

RESPONSE GRID					
	9. ⓐⓑⓒⓓ	10. ⓐⓑⓒⓓ	11. ⓐⓑⓒⓓ	12. ⓐⓑⓒⓓ	13. ⓐⓑⓒⓓ
	14. ⓐⓑⓒⓓ	15. ⓐⓑⓒⓓ	16. ⓐⓑⓒⓓ	17. ⓐⓑⓒⓓ	18. ⓐⓑⓒⓓ
	19. ⓐⓑⓒⓓ	20. ⓐⓑⓒⓓ			

Space for Rough Work

21. A body of mass m kg is ascending on a smooth inclined plane of inclination $\theta\left(\sin\theta = \frac{1}{x}\right)$ with constant acceleration of a m/s^2. The final velocity of the body is v m/s. The work done by the body during this motion is
(Initial velocity of the body = 0)

(a) $\frac{1}{2}mv^2(g+xa)$ (b) $\frac{mv^2}{2}\left(\frac{g}{2}+a\right)$

(c) $\frac{2mv^2x}{a}(a+gx)$ (d) $\frac{mv^2}{2ax}(g+xa)$

22. A glass marble dropped from a certain height above the horizontal surface reaches the surface in time t and then continues to bounce up and down. The time in which the marble finally comes to rest is

(a) $e^n t$ (b) $e^2 t$ (c) $t\left[\frac{1+e}{1-e}\right]$ (d) $t\left[\frac{1-e}{1+e}\right]$

23. The potential energy of a 1 kg particle free to move along the x-axis is given by $V(x) = \left(\frac{x^4}{4} - \frac{x^2}{2}\right)$ J.
The total mechanical energy of the particle is 2 J. Then, the maximum speed (in m/s) is

(a) $\frac{3}{\sqrt{2}}$ (b) $\sqrt{2}$ (c) $\frac{1}{\sqrt{2}}$ (d) 2

24. Water falls from a height of 60 m at the rate of 15 kg/s to operate a turbine. The losses due to frictional force are 10% of energy. How much power is generated by the turbine?(g = 10 m/s^2)
(a) 8.1 kW (b) 10.2 kW (c) 12.3 kW (d) 7.0 kW

24. A car of mass m starts from rest and accelerates so that the instantaneous power delivered to the car has a constant magnitude P_0. The instantaneous velocity of this car is proportional to :

(a) t^2P_0 (b) $t^{1/2}$ (c) $t^{-1/2}$ (d) $\frac{t}{\sqrt{m}}$

25. When a 1.0kg mass hangs attached to a spring of length 50 cm, the spring stretches by 2 cm. The mass is pulled down until the length of the spring becomes 60 cm. What is the amount of elastic energy stored in the spring in this condition. if g = 10 m/s^2.
(a) 1.5 joule (b) 2.0 joule (c) 2.5 joule (d) 3.0 joule

26. A block of mass m rests on a rough horizontal surface (Coefficient of friction is μ). When a bullet of mass m/2 strikes horizontally, and get embedded in it, the block moves a distance d before coming to rest. The initial velocity of the bullet is $k\sqrt{2\mu gd}$, then the value of k is

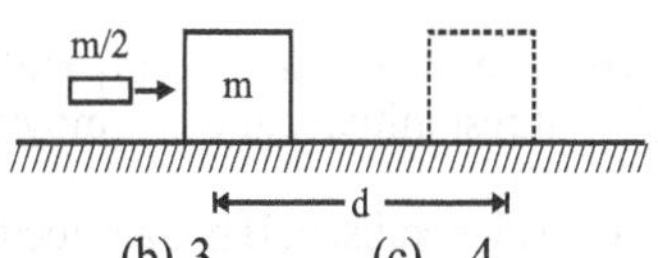

(a) 2 (b) 3 (c) 4 (d) 5

27. A force acts on a 30 gm particle in such a way that the position of the particle as a function of time is given by $x = 3t - 4t^2 + t^3$, where x is in metres and t is in seconds. The work done during the first 4 seconds is
(a) 576mJ (b) 450mJ (c) 490mJ (d) 530mJ

28. A particle of mass m_1 moving with velocity v strikes with a mass m_2 at rest, then the condition for maximum transfer of kinetic energy is
(a) $m_1 >> m_2$ (b) $m_2 >> m_2$ (c) $m_1 = m_2$ (d) $m_1 = 2m_2$

29. A mass *m* is moving with velocity v collides inelastically with a bob of simple pendulum of mass m and gets embedded into it. The total height to which the masses will rise after collision is

(a) $\frac{v^2}{8g}$ (b) $\frac{v^2}{4g}$ (c) $\frac{v^2}{2g}$ (d) $\frac{2v^2}{g}$

30. A 10 H.P. motor pumps out water from a well of depth 20 m and fills a water tank of volume 22380 litres at a height of 10 m from the ground. The running time of the motor to fill the empty water tank is (g = 10ms^{-2})
(a) 5 minutes (b) 10 minutes
(c) 15 minutes (d) 20 minutes

31. A particle of mass m_1 is moving with a velocity v_1 and another particle of mass m_2 is moving with a velocity v_2. Both of them have the same momentum but their different kinetic energies are E_1 and E_2 respectively. If $m_1 > m_2$ then

(a) $E_1 = E_2$ (b) $E_1 < E_2$ (c) $\frac{E_1}{E_2} = \frac{m_1}{m_2}$ (d) $E_1 > E_2$

32. A block of mass 10 kg, moving in x direction with a constant speed of 10 ms^{-1}, is subject to a retarding force F = 0.1 × J m during its travel from x = 20 m to 30 m. Its final KE will be :
(a) 450 J (b) 275 J (c) 250 J (d) 475 J

33. Identify the false statement from the following
(a) Work-energy theorem is not independent of Newton's second law.
(b) Work-energy theorem holds in all inertial frames.
(c) Work done by friction over a closed path is zero.
(d) No potential energy can be associated with friction.

34. A one-ton car moves with a constant velocity of 15 ms^{-1} on a rough horizontal road. The total resistance to the motion of the car is 12% of the weight of the car. The power required to keep the car moving with the same constant velocity of 15ms^{-1} is [Take g = 10 ms^{-2}]
(a) 9 kW (b) 18 kW (c) 24 kW (d) 36 kW

35. A ball is released from the top of a tower. The ratio of work done by force of gravity in first, second and third second of the motion of the ball is
(a) 1 : 2 : 3 (b) 1 : 4 : 9 (c) 1 : 3 : 5 (d) 1 : 5 : 3

RESPONSE GRID					
	21. ⓐⓑⓒⓓ	22. ⓐⓑⓒⓓ	23. ⓐⓑⓒⓓ	24. ⓐⓑⓒⓓ	25. ⓐⓑⓒⓓ
	26. ⓐⓑⓒⓓ	27. ⓐⓑⓒⓓ	28. ⓐⓑⓒⓓ	29. ⓐⓑⓒⓓ	30. ⓐⓑⓒⓓ
	31. ⓐⓑⓒⓓ	32. ⓐⓑⓒⓓ	33. ⓐⓑⓒⓓ	34. ⓐⓑⓒⓓ	35. ⓐⓑⓒⓓ

Space for Rough Work

36. Two spheres A and B of masses m_1 and m_2 respectively collide. A is at rest initially and B is moving with velocity v along x-axis. After collision B has a velocity $\frac{v}{2}$ in a direction perpendicular to the original direction. The mass A moves after collision in the direction.
(a) Same as that of B
(b) Opposite to that of B
(c) $\theta = \tan^{-1}(1/2)$ to the x-axis
(d) $\theta = \tan^{-1}(-1/2)$ to the x-axis

37. A 2 kg block slides on a horizontal floor with a speed of 4m/s. It strikes a uncompressed spring, and compresses it till the block is motionless. The kinetic friction force is 15N and spring constant is 10,000 N/m. The spring compresses by
(a) 8.5 cm (b) 5.5 cm (c) 2.5 cm (d) 11.0 cm

38. An engine pumps water through a hose pipe. Water passes through the pipe and leaves it with a velocity of 2 m/s. The mass per unit length of water in the pipe is 100 kg/m. What is the power of the engine?
(a) 400 W (b) 200 W (c) 100 W (d) 800 W

39. A uniform chain of length 2 m is kept on a table such that a length of 60 cm hangs freely from the edge of the table. The total mass of the chain is 4 kg. What is the work done in pulling the entire chain on the table ?
(a) 12 J (b) 3.6 J (c) 7.2 J (d) 1200 J

40. A mass 'm' moves with a velocity 'v' and collides inelastically with another identical mass. After collision the 1st mass moves with velocity $\frac{v}{\sqrt{3}}$ in a direction perpendicular to the initial direction of motion. Find the speed of the 2nd mass after collision.

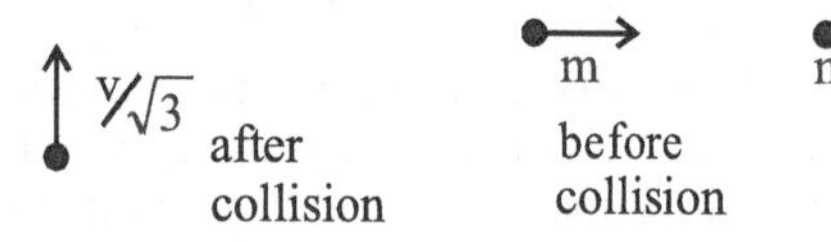

(a) $\sqrt{3}v$ (b) v (c) $\frac{v}{\sqrt{3}}$ (d) $\frac{2}{\sqrt{3}}v$

41. A spherical ball of mass 20 kg is stationary at the top of a hill of height 100 m. It rolls down a smooth surface to the ground, then climbs up another hill of height 30 m and finally rolls down to a horizontal base at a height of 20 m above the ground. The velocity attained by the ball is
(a) 20 m/s (b) 40 m/s (c) $10\sqrt{30}$ m/s (d) 10 m/s

42. A block of mass M is kept on a platform which is accelerated upward with a constant acceleration 'a' during the time interval T. The work done by normal reaction between the block and platform is

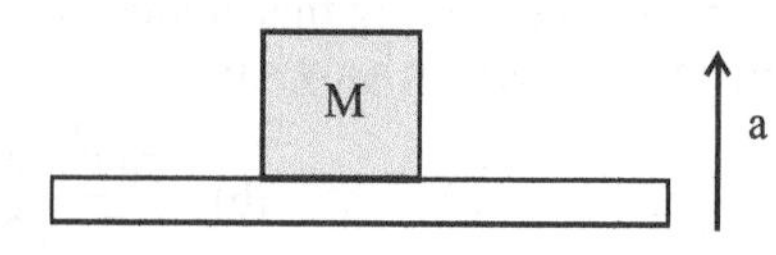

(a) $-\frac{MgaT^2}{2}$ (b) $\frac{1}{2}M(g+a)aT^2$
(c) $\frac{1}{2}Ma^2T$ (d) Zero

43. A spring lies along an x axis attached to a wall at one end and a block at the other end. The block rests on a frictionless surface at x = 0. A force of constant magnitude F is applied to the block that begins to compress the spring, until the block comes to a maximum displacement x_{max}.

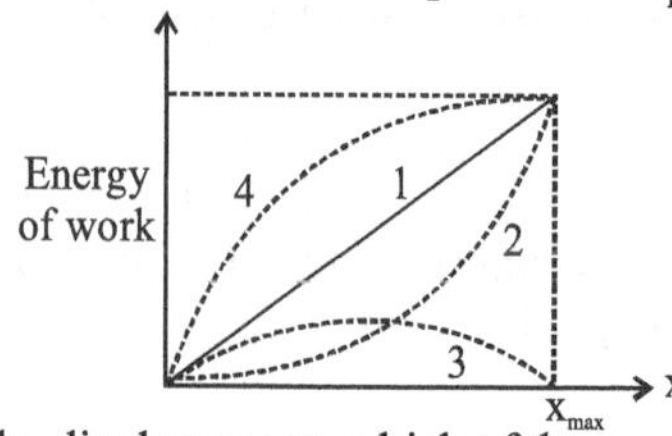

During the displacement, which of the curves shown in the graph best represents the kinetic energy of the block ?
(a) 1 (b) 2 (c) 3 (d) 4

44. The K.E. acquired by a mass m in travelling a certain distance d, starting form rest, under the action of a constant force is directly proportional to
(a) m (b) $\sqrt{m}$
(c) $\frac{1}{\sqrt{m}}$ (d) independent of m

45. A vertical spring with force constant k is fixed on a table. A ball of mass m at a height h above the free upper end of the spring falls vertically on the spring so that the spring is compressed by a distance d. The net work done in the process is
(a) $mg(h+d)-\frac{1}{2}kd^2$ (b) $mg(h-d)-\frac{1}{2}kd^2$
(c) $mg(h-d)+\frac{1}{2}kd^2$ (d) $mg(h+d)+\frac{1}{2}kd^2$

RESPONSE GRID					
	36. ⓐⓑⓒⓓ	37. ⓐⓑⓒⓓ	38. ⓐⓑⓒⓓ	39. ⓐⓑⓒⓓ	40. ⓐⓑⓒⓓ
	41. ⓐⓑⓒⓓ	42. ⓐⓑⓒⓓ	43. ⓐⓑⓒⓓ	44. ⓐⓑⓒⓓ	45. ⓐⓑⓒⓓ

DAILY PRACTICE PROBLEM DPP CHAPTERWISE CP05 - PHYSICS			
Total Questions	45	Total Marks	180
Attempted		Correct	
Incorrect		Net Score	
Cut-off Score	50	Qualifying Score	70
Success Gap = Net Score – Qualifying Score			
Net Score = (Correct × 4) – (Incorrect × 1)			

Space for Rough Work

DPP - Daily Practice Problems

Chapter-wise Sheets

Date : ______ **Start Time :** ______ **End Time :** ______

PHYSICS (CP06)

SYLLABUS : System of Particles and Rotational Motion

Max. Marks : 180 **Marking Scheme :** (+4) for correct & (–1) for incorrect answer **Time : 60 min.**

INSTRUCTIONS : This Daily Practice Problem Sheet contains 45 MCQs. For each question only one option is correct. Darken the correct circle/ bubble in the Response Grid provided on each page.

1. From a solid sphere of mass M and radius R, a cube of maximum possible volume is cut. Moment of inertia of cube about an axis passing through its center and perpendicular to one of its faces is :

   (a) $\frac{4MR^2}{9\sqrt{3}\pi}$ (b) $\frac{4MR^2}{3\sqrt{3}\pi}$ (c) $\frac{MR^2}{32\sqrt{2}\pi}$ (d) $\frac{MR^2}{16\sqrt{2}\pi}$

2. A hollow sphere is held suspended. Sand is now poured into it in stages.
   The centre of mass of the sphere with the sand
   (a) rises continuously
   (b) remains unchanged in the process
   (c) first rises and then falls to the original position
   (d) first falls and then rises to the original position

3. A body A of mass M while falling vertically downwards under gravity breaks into two parts; a body B of mass $\frac{1}{3}M$ and a body C of mass $\frac{2}{3}M$. The centre of mass of bodies B and C taken together shifts compared to that of body A towards
   (a) does not shift
   (b) depends on height of breaking
   (c) body B (d) body C

4. From a uniform wire, two circular loops are made (i) P of radius r and (ii) Q of radius nr. If the moment of inertia of Q about an axis passing through its centre and perpendicular to its plane is 8 times that of P about a similar axis, the value of n is (diameter of the wire is very much smaller than r or nr)
   (a) 8 (b) 6 (c) 4 (d) 2

5. A billiard ball of mass m and radius r, when hit in a horizontal direction by a cue at a height h above its centre, acquired a linear velocity v_0. The angular velocity ω_0 acquired by the ball is

   (a) $\frac{5v_0r^2}{2h}$ (b) $\frac{2v_0r^2}{5h}$ (c) $\frac{2v_0h}{5r^2}$ (d) $\frac{5v_0h}{2r^2}$

6. Three bricks each of length L and mass M are arranged as shown from the wall. The distance of the centre of mass of the system from the wall is

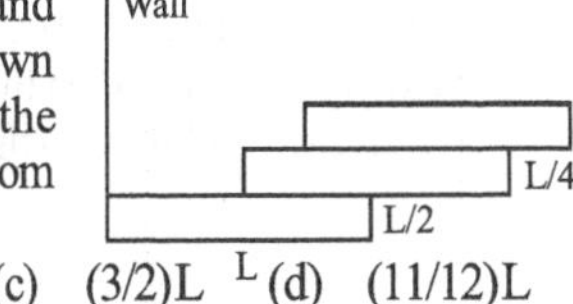

   (a) L/4 (b) L/2 (c) (3/2)L (d) (11/12)L

7. Four point masses, each of value m, are placed at the corners of a square $ABCD$ of side ℓ. The moment of inertia of this system about an axis passing through A and parallel to BD is
   (a) $2m\ell^2$ (b) $\sqrt{3}m\ell^2$ (c) $3m\ell^2$ (d) $m\ell^2$

8. A loop of radius r and mass m rotating with an angular velocity ω_0 is placed on a rough horizontal surface. The initial velocity of the centre of the hoop is zero. What will be the velocity of the centre of the hoop when it ceases to slip?

   (a) $\frac{r\omega_0}{4}$ (b) $\frac{r\omega_0}{3}$ (c) $\frac{r\omega_0}{2}$ (d) $r\omega_0$

RESPONSE GRID					
	1. ⓐⓑⓒⓓ	2. ⓐⓑⓒⓓ	3. ⓐⓑⓒⓓ	4. ⓐⓑⓒⓓ	5. ⓐⓑⓒⓓ
	6. ⓐⓑⓒⓓ	7. ⓐⓑⓒⓓ	8. ⓐⓑⓒⓓ		

Space for Rough Work

9. Two masses m_1 and m_2 are connected by a massless spring of spring constant k and unstretched length ℓ. The masses are placed on a frictionless straight channel, which are consider our x-axis. They are initially at $x = 0$ and $x = \ell$ respectively. At $t = 0$, a velocity v_0 is suddenly imparted to the first particle. At a later time t, the centre of mass of the two masses is at :

(a) $x = \dfrac{m_2\ell}{m_1 + m_2}$

(b) $x = \dfrac{m_1\ell}{m_1 + m_2} + \dfrac{m_2 v_0 t}{m_1 + m_2}$

(c) $x = \dfrac{m_2\ell}{m_1 + m_1} + \dfrac{m_2 v_0 t}{m_1 + m_2}$ (d) $x = \dfrac{m_2\ell}{m_1 + m_2} + \dfrac{m_1 v_0 t}{m_1 + m_2}$

10. A body of mass 1.5 kg rotating about an axis with angular velocity of 0.3 rad s^{-1} has the angular momentum of 1.8 kg m^2s^{-1}. The radius of gyration of the body about an axis is

(a) 2m (b) 1.2m (c) 0.2m (d) 1.6m

11. If $\vec{F}$ is the force acting on a particle having position vector $\vec{r}$ and $\vec{\tau}$ be the torque of this force about the origin, then:

(a) $\vec{r}\cdot\vec{\tau} > 0$ and $\vec{F}\cdot\vec{\tau} < 0$

(b) $\vec{r}\cdot\vec{\tau} = 0$ and $\vec{F}\cdot\vec{\tau} = 0$

(c) $\vec{r}\cdot\vec{\tau} = 0$ and $\vec{F}\cdot\vec{\tau} \neq 0$

(d) $\vec{r}\cdot\vec{\tau} \neq 0$ and $\vec{F}\cdot\vec{\tau} = 0$

12. A thin uniform rod of length l and mass m is swinging freely about a horizontal axis passing through its end. Its maximum angular speed is ω. Its centre of mass rises to a maximum height of

(a) $\dfrac{1}{6}\dfrac{l\omega}{g}$ (b) $\dfrac{1}{2}\dfrac{l^2\omega^2}{g}$ (c) $\dfrac{1}{6}\dfrac{l^2\omega^2}{g}$ (d) $\dfrac{1}{3}\dfrac{l^2\omega^2}{g}$

13. A wheel is rolling straight on ground without slipping. If the axis of the wheel has speed v, the instantenous velocity of a point P on the rim, defined by angle θ, relative to the ground will be

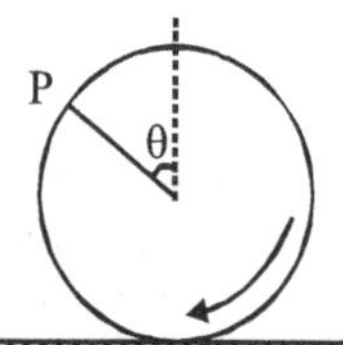

(a) $v\cos\left(\dfrac{1}{2}\theta\right)$ (b) $2v\cos\left(\dfrac{1}{2}\theta\right)$

(c) $v(1+\sin\theta)$ (d) $v(1+\cos\theta)$

14. A solid sphere having mass m and radius r rolls down an inclined plane. Then its kinetic energy is

(a) $\dfrac{5}{7}$ rotational and $\dfrac{2}{7}$ translational

(b) $\dfrac{2}{7}$ rotational and $\dfrac{5}{7}$ translational

(c) $\dfrac{2}{5}$ rotational and $\dfrac{3}{5}$ translational

(d) $\dfrac{1}{2}$ rotational and $\dfrac{1}{2}$ translational

15. A ring of mass M and radius R is rotating about its axis with angular velocity ω. Two identical bodies each of mass m are now gently attached at the two ends of a diameter of the ring. Because of this, the kinetic energy loss will be :

(a) $\dfrac{m(M+2m)}{M}\omega^2R^2$ (b) $\dfrac{Mm}{(M+m)}\omega^2R^2$

(c) $\dfrac{Mm}{(M+2m)}\omega^2R^2$ (d) $\dfrac{(M+m)M}{(M+2m)}\omega^2R^2$

16. Acertain bicycle can go up a gentle incline with constant speed when the frictional force of ground pushing the rear wheel is $F_2 = 4$ N. With what force F_1 must the chain pull on the sprocket wheel if R_1=5 cm and R_2 = 30 cm?

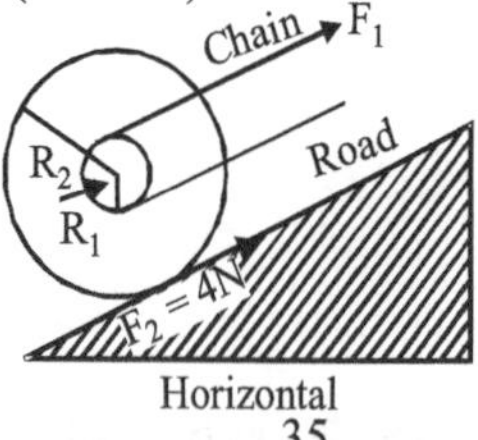

(a) 4N (b) 24N (c) 140N (d) $\dfrac{35}{4}$N

17. A wooden cube is placed on a rough horizontal table, a force is applied to the cube. Gradually the force is increased. Whether the cube slides before toppling or topples before sliding is independent of:

(a) the position of point of application of the force
(b) the length of the edge of the cube
(c) mass of the cube
(d) Coefficient of friction between the cube and the table

18. From a circular ring of mass M and radius R, an arc corresponding to a 90° sector is removed. The moment of inertia of the ramaining part of the ring about an axis passing through the centre of the ring and perpendicular to the plane of the ring is k times MR^2. Then the value of k is

(a) 3/4 (b) 7/8 (c) 1/4 (d) 1

19. A mass m moves in a circle on a smooth horizontal plane with velocity v_0 at a radius R_0. The mass is attached to string which passes through a smooth hole in the plane as shown.

The tension in the string is increased gradually and finally m moves in a circle of radius $\dfrac{R_0}{2}$. The final value of the kinetic energy is

(a) $\dfrac{1}{4}mv_0^2$ (b) $2mv_0^2$ (c) $\dfrac{1}{2}mv_0^2$ (d) mv_0^2

20. A rod PQ of length L revolves in a horizontal plane about the axis YY′. The angular velocity of the rod is ω. If A is the area of cross-section of the rod and ρ be its density, its rotational kinetic energy is

RESPONSE GRID

9. ⓐⓑⓒⓓ 10. ⓐⓑⓒⓓ 11. ⓐⓑⓒⓓ 12. ⓐⓑⓒⓓ 13. ⓐⓑⓒⓓ
14. ⓐⓑⓒⓓ 15. ⓐⓑⓒⓓ 16. ⓐⓑⓒⓓ 17. ⓐⓑⓒⓓ 18. ⓐⓑⓒⓓ
19. ⓐⓑⓒⓓ

Space for Rough Work

(a) $\frac{1}{3}AL^3\rho\omega^2$ (b) $\frac{1}{2}AL^3\rho\omega^2$

(c) $\frac{1}{24}AL^3\rho\omega^2$ (d) $\frac{1}{18}AL^3\rho\omega^2$

21. A solid sphere of mass 2 kg rolls on a smooth horizontal surface at 10 m/s. It then rolls up a smooth inclined plane of inclination 30° with the horizontal. The height attained by the sphere before it stops is

(a) 700 cm (b) 701 cm (c) 7.1 m (d) 70 m

22. A hollow smooth uniform sphere *A* of mass m rolls without sliding on a smooth horizontal surface. It collides head on elastically with another stationary smooth solid sphere *B* of the same mass m and same radius. The ratio of kinetic energy of *B* to that of *A* just after the collision is

(a) 1 : 1
(b) 2 : 3
(c) 3 : 2
(d) 4 : 3

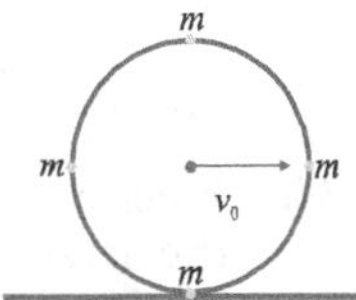

23. Two discs of same thickness but of different radii are made of two different materials such that their masses are same. The densities of the materials are in the ratio of 1 : 3. The moments of inertia of these discs about the respective axes passing through their centres and perpendicular to their planes will be in the ratio of

(a) 1 : 3 (b) 3 : 1 (c) 1 : 9 (d) 9 : 1

24. A pulley fixed to the ceiling carries a string with blocks of mass m and 3 m attached to its ends. The masses of string and pulley are negligible. When the system is released, its centre of mass moves with what acceleration ?

(a) 0 (b) – g/4 (c) g/2 (d) – g/2

25. A ring of mass m and radius *R* has four particles each of mass m attached to the ring as shown in figure. The centre of ring has a speed v_0. The kinetic energy of the system is

(a) mv_0^2 (b) $3mv_0^2$ (c) $5mv_0^2$ (d) $6mv_0^2$

26. Consider a uniform square plate of side '*a*' and mass '*M*'. The moment of inertia of this plate about an axis perpendicular to its plane and passing through one of its corners is

(a) $\frac{5}{6}Ma^2$ (b) $\frac{1}{12}Ma^2$ (c) $\frac{7}{12}Ma^2$ (d) $\frac{2}{3}Ma^2$

27. A dancer is standing on a stool rotating about the vertical axis passing through its centre. She pulls her arms towards the body reducing her moment of inertia by a factor of n. The new angular speed of turn table is proportional to

(a) n (b) n^{-1} (c) n^0 (d) n^2

28. A uniform square plate has a small piece Q of an irregular shape removed and glued to the centre of the plate leaving a hole behind. Then the moment of inertia about the z-axis

(a) increases
(b) decreases
(c) remains same
(d) changed in unpredicted manner.

29. A circular turn table has a block of ice placed at its centre. The system rotates with an angular speed ω about an axis passing through the centre of the table. If the ice melts on its own without any evaporation, the speed of rotation of the system

(a) becomes zero
(b) remains constant at the same value ω
(c) increases to a value greater than ω
(d) decreases to a value less than ω

30. Seven identical coins are rigidly arranged on a flat table in the pattern shown below so that each coin touches it neighbors. Each coin is a thin disc of mass *m* and radius *r*. The moment of inertia of the system of seven coins about an axis that passes through point *P* and perpendicular to the plane of the coin is :

(a) $\frac{55}{2}mr^2$ (b) $\frac{127}{2}mr^2$ (c) $\frac{111}{2}mr^2$ (d) $55\,mr^2$

31. In a two-particle system with particle masses m_1 and m_2, the first particle is pushed towards the centre of mass through a distance d, the distance through which second particle must be moved to keep the centre of mass at the same position is

(a) $\frac{m_2 d}{m_1}$ (b) d (c) $\frac{m_1 d}{(m_1+m_2)}$ (d) $\frac{m_1 d}{m_2}$

32. A uniform bar of mass M and length L is horizontally suspended from the ceiling by two vertical light cables as shown. Cable A is connected 1/4th distance from the left end of the bar. Cable B is attached at the far right end of the bar. What is the tension in cable A?

(a) 1/4 Mg (b) 1/3 Mg (c) 2/3 Mg (d) 3/4 Mg

33. A couple produces

(a) purely linear motion
(b) purely rotational motion
(c) linear and rotational motion
(d) no motion

34. Point masses 1, 2, 3 and 4 kg are lying at the point (0, 0, 0), (2, 0, 0), (0, 3, 0) and (–2, –2, 0) respectively. The moment of inertia of this system about x-axis will be

(a) 43 kgm^2 (b) 34 kgm^2 (c) 27 kgm^2 (d) 72 kgm^2

RESPONSE GRID					
	20. ⓐⓑⓒⓓ	21. ⓐⓑⓒⓓ	22. ⓐⓑⓒⓓ	23. ⓐⓑⓒⓓ	24. ⓐⓑⓒⓓ
	25. ⓐⓑⓒⓓ	26. ⓐⓑⓒⓓ	27. ⓐⓑⓒⓓ	28. ⓐⓑⓒⓓ	29. ⓐⓑⓒⓓ
	30. ⓐⓑⓒⓓ	31. ⓐⓑⓒⓓ	32. ⓐⓑⓒⓓ	33. ⓐⓑⓒⓓ	34. ⓐⓑⓒⓓ

Space for Rough Work

35. A solid sphere of mass M and radius R is pulled horizontally on a sufficiently rough surface as shown in the figure.

Choose the correct alternative.

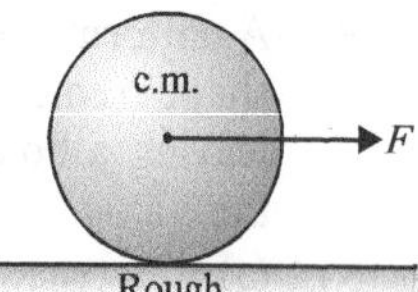

(a) The acceleration of the centre of mass is F/M

(b) The acceleration of the centre of mass is $\frac{2}{3}\frac{F}{M}$

(c) The friction force on the sphere acts forward

(d) The magnitude of the friction force is F/3

36. The moment of inertia of a body about a given axis is $1.2\,kg\,m^2$. Initially, the body is at rest. In order to produce a rotational kinetic energy of 1500 joule, an angular acceleration of 25 radian/sec^2 must be applied about that axis for a duration of

(a) 4 sec (b) 2 sec (c) 8 sec (d) 10 sec

37. A gymnast takes turns with her arms and legs stretched. When she pulls her arms and legs in

(a) the angular velocity decreases

(b) the moment of inertia decreases

(c) the angular velocity stays constant

(d) the angular momentum increases

38. An equilateral triangle *ABC* formed from a uniform wire has two small identical beads initially located at *A*. The triangle is set rotating about the vertical axis *AO*. Then the beads are released from rest simultaneously and allowed to slide down, one along *AB* and the other along *AC* as shown. Neglecting frictional effects, the quantities that are conserved as the beads slide down, are

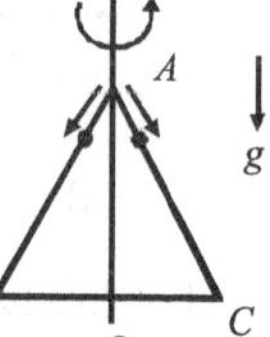

(a) angular velocity and total energy (kinetic and potential)

(b) total angular momentum and total energy

(c) angular velocity and moment of inertia about the axis of rotation

(d) total angular momentum and moment of inertia about the axis of rotation

39. The moment of inertia of a uniform semicircular wire of mass m and radius r, about an axis passing through its centre of mass and perpendicular to its plane is $mr^2\left(1-\frac{k}{\pi^2}\right)$. Find the value of k.

(a) 2 (b) 3 (c) 4 (d) 5

40. Initial angular velocity of a circular disc of mass *M* is ω_1. Then two small spheres of mass *m* are attached gently to diametrically opposite points on the edge of the disc. What is the final angular velocity of the disc?

(a) $\left(\frac{M+m}{M}\right)\omega_1$ (b) $\left(\frac{M+m}{m}\right)\omega_1$

(c) $\left(\frac{M}{M+4m}\right)\omega_1$ (d) $\left(\frac{M}{M+2m}\right)\omega_1$.

41. Two identical discs of mass m and radius r are arranged as shown in the figure. If α is the angular acceleration of the lower disc and a_{cm} is acceleration of centre of mass of the lower disc, then relation between a_{cm}, α and r is

(a) $a_{cm} = \alpha/r$ (b) $a_{cm} = 2\alpha r$

(c) $a_{cm} = \alpha r$ (d) None of these

42. Five masses are placed in a plane as shown in figure. The coordinates of the centre of mass are nearest to

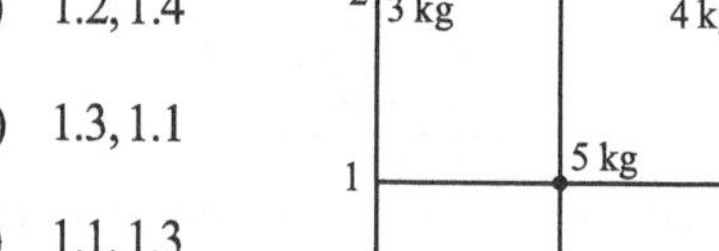

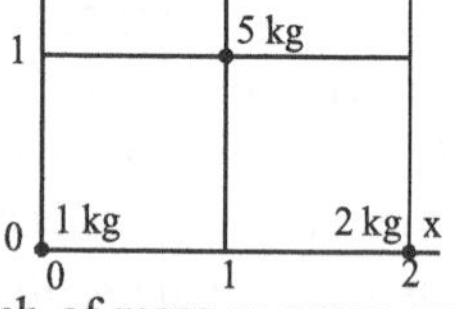

(a) 1.2, 1.4

(b) 1.3, 1.1

(c) 1.1, 1.3

(d) 1.0, 1.0

43. Three particles, each of mass *m* gram, are situated at the vertices of an equilateral triangle ABC of side ℓ cm (as shown in the figure). The moment of inertia of the system about a line AX perpendicular to AB and in the plane of ABC, in gram-cm^2 units will be

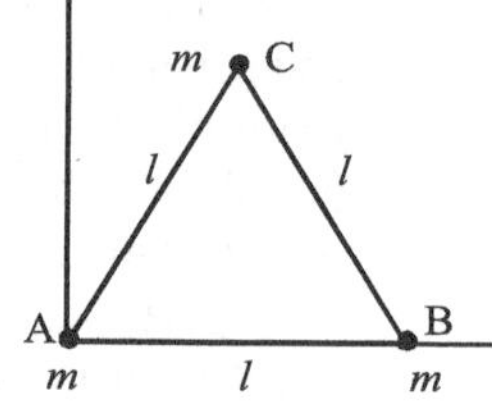

(a) $\frac{3}{2}m\ell^2$

(b) $\frac{3}{4}m\ell^2$

(c) $2\,m\ell^2$

(d) $\frac{5}{4}m\ell^2$

44. When a ceiling fan is switched on, it makes 10 rotations in the first 3 seconds. Assuming a uniform angular acceleration, how many rotation it will make in the next 3 seconds?

(a) 10 (b) 20 (c) 30 (d) 40

45. A solid sphere spinning about a horizontal axis with an angular velocity ω is placed on a horizontal surface. Subsequently it rolls without slipping with an angular velocity of :

(a) $\frac{2\omega}{5}$ (b) $\frac{7\omega}{5}$ (c) $\frac{2\omega}{7}$ (d) ω

RESPONSE GRID					
	35. ⓐⓑⓒⓓ	36. ⓐⓑⓒⓓ	37. ⓐⓑⓒⓓ	38. ⓐⓑⓒⓓ	39. ⓐⓑⓒⓓ
	40. ⓐⓑⓒⓓ	41. ⓐⓑⓒⓓ	42. ⓐⓑⓒⓓ	43. ⓐⓑⓒⓓ	44. ⓐⓑⓒⓓ
	45. ⓐⓑⓒⓓ				

DAILY PRACTICE PROBLEM DPP CHAPTERWISE CP06 - PHYSICS

Total Questions	45	Total Marks	180
Attempted		Correct	
Incorrect		Net Score	
Cut-off Score	45	Qualifying Score	60
Success Gap = Net Score – Qualifying Score			
Net Score = (Correct × 4) – (Incorrect × 1)			

Space for Rough Work

DPP - Daily Practice Problems

Chapter-wise Sheets

Date : ______ **Start Time :** ______ **End Time :** ______

PHYSICS

CP07

SYLLABUS : Gravitation

Max. Marks : 180 **Marking Scheme :** (+4) for correct & (–1) for incorrect answer **Time : 60 min.**

INSTRUCTIONS : This Daily Practice Problem Sheet contains 45 MCQs. For each question only one option is correct. Darken the correct circle/ bubble in the Response Grid provided on each page.

1. The radius of a planet is $1/4^{th}$ of R_e and its acc. due to gravity is 2g. What would be the value of escape velocity on the planet, if escape velocity on earth is v_e.

(a) $\frac{v_e}{\sqrt{2}}$ (b) $v_e\sqrt{2}$ (c) $2\,v_e$ (d) $\frac{v_e}{2}$

2. A projectile is fired vertically from the Earth with a velocity kv_e where v_e is the escape velocity and k is a constant less than unity. The maximum height to which projectile rises, as measured from the centre of Earth, is

(a) $\frac{R}{k}$ (b) $\frac{R}{k-1}$ (c) $\frac{R}{1-k^2}$ (d) $\frac{R}{1+k^2}$

3. A solid sphere of uniform density and radius R applies a gravitational force of attraction equal to F_1 on a particle placed at A, distance 2R from the centre of the sphere.

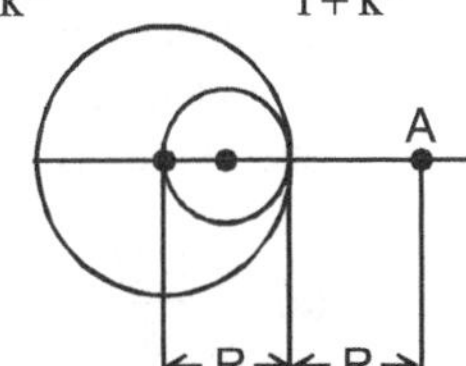

A spherical cavity of radius R/2 is now made in the sphere as shown in the figure. The sphere with cavity now applies a gravitational force F_2 on the same particle placed at A. The ratio F_2/F_1 will be

(a) 1/2 (b) 3 (c) 7 (d) 1/9

4. A geostationary satellite is orbiting the earth at a height of 5R above that surface of the earth, R being the radius of the earth. The time period of another satellite in hours at a height of 2R from the surface of the earth is :

(a) 5 (b) 10 (c) $6\sqrt{2}$ (d) $\frac{6}{\sqrt{2}}$

5. A satellite of mass m is orbiting around the earth in a circular orbit with a velocity v. What will be its total energy?

(a) $(3/4)\,mv^2$ (b) $(1/2)\,mv^2$
(c) mv^2 (d) $-(1/2)m\,v^2$

RESPONSE GRID	1. ⓐⓑⓒⓓ	2. ⓐⓑⓒⓓ	3. ⓐⓑⓒⓓ	4. ⓐⓑⓒⓓ	5. ⓐⓑⓒⓓ

Space for Rough Work

6. The gravitational force of attraction between a uniform sphere of mass M and a uniform rod of length l and mass m oriented as shown is

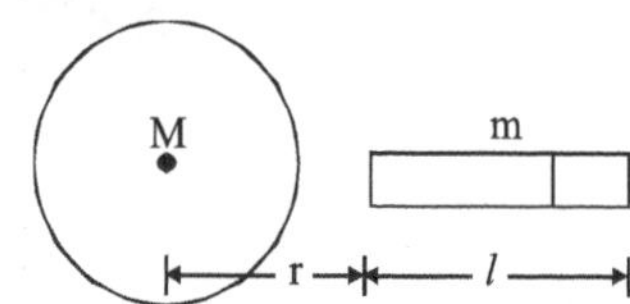

(a) $\frac{GMm}{r(r+l)}$ (b) $\frac{GM}{r^2}$ (c) Mmr^2+l (d) $(r^2+l)mM$

7. If the gravitational force between two objects were proportional to $1/R$ (and not as $1/R^2$) where R is separation between them, then a particle in circular orbit under such a force would have its orbital speed v proportional to

(a) $1/R^2$ (b) R^0 (c) R^1 (d) $1/R$

8. A satellite of mass m revolves around the earth of radius R at a height 'x' from its surface. If g is the acceleration due to gravity on the surface of the earth, the orbital speed of the satellite is

(a) $\frac{gR^2}{R+x}$ (b) $\frac{gR}{R-x}$ (c) gx (d) $\left(\frac{gR^2}{R+x}\right)^{1/2}$

9. A body is projected up with a velocity equal to 3/4th of the escape velocity from the surface of the earth. The height it reaches from the centre of the earth is (Radius of the earth = R)

(a) $\frac{10R}{9}$ (b) $\frac{16R}{7}$ (c) $\frac{9R}{8}$ (d) $\frac{10R}{3}$

10. A Planet is revolving around the sun.

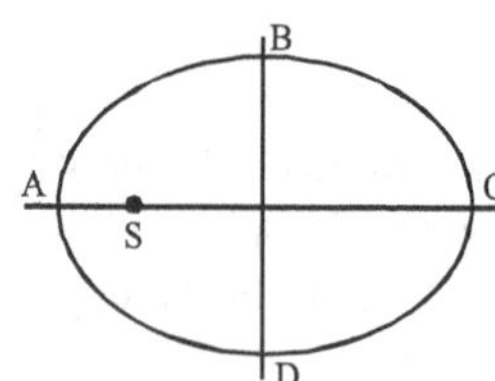

Which of the following is correct option?

(a) The time taken in travelling DAB is less than that for BCD

(b) The time taken in travelling DAB is greater than that for BCD

(c) The time taken in travelling CDA is less than that for ABC

(d) The time taken in travelling CDA is greater than that for ABC

11. The acceleration due to gravity on the planet A is 9 times the acceleration due to gravity on planet B. A man jumps to a height of 2m on the surface of A. What is the height of jump by the same person on the planet B?

(a) $\frac{2}{3}$m (b) $\frac{2}{9}$m (c) 18m (d) 6m

12. If suddenly the gravitational force of attraction between the earth and a satellite revolving around it becomes zero, then the satellite will

(a) continue to move in its orbit with same speed
(b) move tangentially to the original orbit with same speed
(c) become stationary in its orbit
(d) move towards the earth

13. Mass M is divided into two parts xM and (1 – x)M. For a given separation, the value of x for which the gravitational attraction between the two pieces becomes maximum is

(a) $\frac{1}{2}$ (b) $\frac{3}{5}$ (c) 1 (d) 2

14. The potential energy of a satellite, having mass m and rotating at a height of 6.4×10^6 m from the earth surface, is

(a) $-mgR_e$ (b) $-0.67\,mgR_e$
(c) $-0.5\,mgR_e$ (d) $-0.33\,mgR_e$

15. If the radius of the earth were to shrink by 1%, with its mass remaining the same, the acceleration due to gravity on the earth's surface would

(a) decrease by 1% (b) decrease by 2%
(c) increase by 1% (d) increase by 2%

16. Suppose the law of gravitational attraction suddenly changes and becomes an inverse cube law i.e. $F \propto \frac{1}{r^3}$, but still remaining a central force. Then

(a) Kepler's law of area still holds
(b) Kepler's law of period still holds
(c) Kepler's law of area and period still holds
(d) neither the law of area nor the law of period still holds

17. Four equal masses (each of mass M) are placed at the corners of a square of side a. The escape velocity of a body from the centre O of the square is

(a) $4\sqrt{\frac{2GM}{a}}$ (b) $\sqrt{\frac{8\sqrt{2}GM}{a}}$ (c) $\frac{4GM}{a}$ (d) $\sqrt{\frac{4\sqrt{2}GM}{a}}$

18. If the gravitational force had varied as $r^{-5/2}$ instead of r^{-2}; the potential energy of a particle at a distance 'r' from the centre of the earth would be directly proportional to

(a) r^{-1} (b) r^{-2} (c) $r^{-3/2}$ (d) $r^{-5/2}$

19. A particle of mass 'm' is kept at rest at a height 3R from the surface of earth, where 'R' is radius of earth and 'M' is mass of earth. The minimum speed with which it should be projected, so that it does not return back, is (g is acceleration due to gravity on the surface of earth)

(a) $\left(\frac{GM}{R}\right)^{\frac{1}{2}}$ (b) $\left(\frac{GM}{2R}\right)^{\frac{1}{2}}$ (c) $\left(\frac{gR}{4}\right)^{\frac{1}{2}}$ (d) $\left(\frac{2g}{4}\right)^{\frac{1}{2}}$

RESPONSE GRID					
	6. ⓐⓑⓒⓓ	7. ⓐⓑⓒⓓ	8. ⓐⓑⓒⓓ	9. ⓐⓑⓒⓓ	10. ⓐⓑⓒⓓ
	11. ⓐⓑⓒⓓ	12. ⓐⓑⓒⓓ	13. ⓐⓑⓒⓓ	14. ⓐⓑⓒⓓ	15. ⓐⓑⓒⓓ
	16. ⓐⓑⓒⓓ	17. ⓐⓑⓒⓓ	18. ⓐⓑⓒⓓ	19. ⓐⓑⓒⓓ	

Space for Rough Work

20. The ratio between the values of acceleration due to gravity at a height 1 km above and at a depth of 1 km below the Earth's surface is (radius of Earth is R)

(a) $\frac{R-2}{R-1}$ (b) $\frac{R}{R-1}$ (c) $\frac{R-2}{R}$ (d) 1

21. The weight of an object in the coal mine, sea level and at the top of the mountain, are respectively W_1, W_2 and W_3 then

(a) $W_1 < W_2 > W_3$ (b) $W_1 = W_2 = W_3$
(c) $W_1 < W_2 < W_3$ (d) $W_1 > W_2 > W_3$

22. The period of moon's rotation around the earth is nearly 29 days. If moon's mass were 2 fold its present value and all other things remain unchanged, the period of moon's rotation would be nearly

(a) $29\sqrt{2}$ days (b) $29/\sqrt{2}$ days
(c) 29×2 days (d) 29 days

23. The mean radius of earth is R, its angular speed on its own axis is ω and the acceleration due to gravity at earth's surface is g. What will be the radius of the orbit of a geostationary satellite ?

(a) $(R^2g/\omega^2)^{1/3}$ (b) $(Rg/\omega^2)^{1/3}$
(c) $(R^2\omega^2/g)^{1/3}$ (d) $(R^2g/\omega)^{1/3}$

24. In order to make the effective acceleration due to gravity equal to zero at the equator, the angular velocity of rotation of the earth about its axis should be ($g = 10\ ms^{-2}$ and radius of earth is 64000 km)

(a) Zero (b) $\frac{1}{800}$ rad sec $^{-1}$
(c) $\frac{1}{80}$ rad sec^{-1} (d) $\frac{1}{8}$ rad sec $^{-1}$

25. A body weighs 72 N on the surface of the earth. What is the gravitational force on it due to earth at a height equal to half the radius of the earth from the surface?

(a) 32 N (b) 28 N (c) 16 N (d) 72 N

26. A body weighs W newton at the surface of the earth. Its weight at a height equal to half the radius of the earth, will be

(a) $\frac{W}{2}$ (b) $\frac{2W}{3}$ (c) $\frac{4W}{9}$ (d) $\frac{8W}{27}$

27. A shell of mass M and radius R has a point mass m placed at a distance r from its centre. The graph of gravitational potential energy U(r) vs distance r will be

(a)

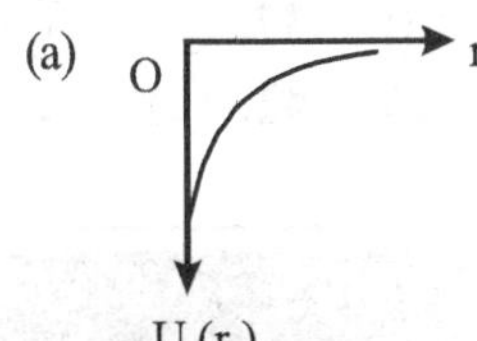

(b)

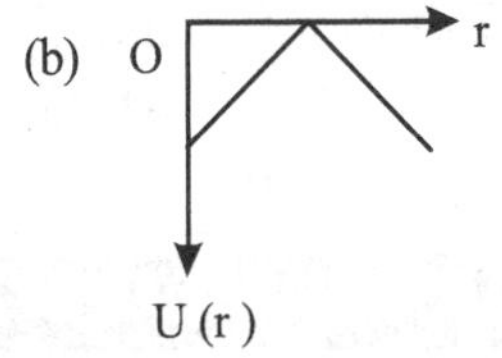

(c) O, r, $-\frac{GMm}{R}$, U (r)

(d) O, r, U (r)

28. The largest and the shortest distance of the earth from the sun are r_1 and r_2. Its distance from the sun when it is at perpendicular to the major-axis of the orbit drawn from the sun

(a) $(r_1 + r_2)/4$ (b) $(r_1 + r_2)/(r_1 - r_2)$
(c) $2r_1 r_2/(r_1 + r_2)$ (d) $(r_1 + r_2)/3$

29. A planet is moving in an elliptical orbit around the sun. If T, V, E and L stand respectively for its kinetic energy, gravitational potential energy, total energy and magnitude of angular momentum about the centre of force, then which of the following is correct ?

(a) T is conserved
(b) V is always positive
(c) E is always negative
(d) L is conserved but direction of vector L changes continuously

30. The earth is assumed to be sphere of radius R. A platform is arranged at a height R from the surface of Earth. The escape velocity of a body from this platform is kv, where v is its escape velocity from the surface of the earth. The value of k is

(a) $\frac{1}{\sqrt{2}}$ (b) $\frac{1}{3}$ (c) $\frac{1}{2}$ (d) $\sqrt{2}$

31. A solid sphere of mass M and radius R is surrounded by a spherical shell of same mass M and radius 2R as shown. A small particle of mass m is released from rest from a height h [<< R] above the shell. There is a hole in the shell.

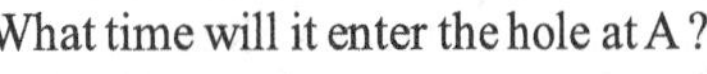

What time will it enter the hole at A ?

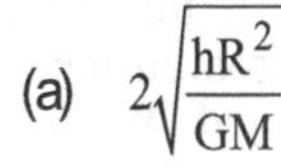

(a) $2\sqrt{\frac{hR^2}{GM}}$ (b) $\sqrt{\frac{2hR^2}{GM}}$
(c) $\sqrt{\frac{hR^2}{GM}}$ (d) $\sqrt{\frac{3hR^2}{GM}}$

32. A body starts from rest from a point distance R_0 from the centre of the earth. The velocity acquired by the body when it reaches the surface of the earth will be (R represents radius of the earth).

(a) $2GM\left(\frac{1}{R} - \frac{1}{R_0}\right)$ (b) $\sqrt{2GM\left(\frac{1}{R_0} - \frac{1}{R}\right)}$
(c) $GM\left(\frac{1}{R} - \frac{1}{R_0}\right)$ (d) $2GM\sqrt{\left(\frac{1}{R} - \frac{1}{R_0}\right)}$

RESPONSE GRID					
	20. ⓐⓑⓒⓓ	21. ⓐⓑⓒⓓ	22. ⓐⓑⓒⓓ	23. ⓐⓑⓒⓓ	24. ⓐⓑⓒⓓ
	25. ⓐⓑⓒⓓ	26. ⓐⓑⓒⓓ	27. ⓐⓑⓒⓓ	28. ⓐⓑⓒⓓ	29. ⓐⓑⓒⓓ
	30. ⓐⓑⓒⓓ	31. ⓐⓑⓒⓓ	32. ⓐⓑⓒⓓ		

Space for Rough Work

33. A satellite of mass M is moving in a circle of radius R under a centripetal force given by $(-k/R^2)$, where k is a constant. Then

(a) The kinetic energy of the particle is $\frac{k}{12}R$

(b) The total energy of the particle is $\left(-\frac{k}{2R}\right)$

(c) The kinetic energy of the particle is $\left(-\frac{k}{R}\right)$

(d) The potential energy of the particle is $\left(\frac{k}{2R}\right)$

34. The change in the value of 'g' at a height 'h' above the surface of the earth is the same as at a depth 'd' below the surface of earth. When both 'd' and 'h' are much smaller than the radius of earth, then which one of the following is correct?

(a) $d=\frac{3h}{2}$ (b) $d=\frac{h}{2}$ (c) $d=h$ (d) $d=2h$

35. Two identical geostationary satellites are moving with equal speeds in the same orbit but their sense of rotation brings them on a collision course. The debris will

(a) fall down
(b) move up
(c) begin to move from east to west in the same orbit
(d) begin to move from west to east in the same orbit

36. A diametrical tunnel is dug across the Earth. A ball is dropped into the tunnel from one side. The velocity of the ball when it reaches the centre of the Earth is .... (Given : gravitational potential at the centre of Earth = $-\frac{3}{2}\frac{GM}{R}$)

(a) $\sqrt{R}$ (b) $\sqrt{gR}$ (c) $\sqrt{2.5gR}$ (d) $\sqrt{7.1gR}$

37. A satellite revolves around the earth of radius R in a circular orbit of radius 3R. The percentage increase in energy required to lift it to an orbit of radius 5R is

(a) 10% (b) 20% (c) 30% (d) 40%

38. A (nonrotating) star collapses onto itself from an initial radius R_i with its mass remaining unchanged. Which curve in figure best gives the gravitational acceleration a_g on the surface of the star as a function of the radius of the star during the collapse

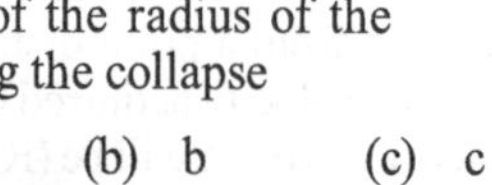

(a) a (b) b (c) c (d) d

39. If the earth is treated as a sphere of radius R and mass M; its angular momentum about the axis of its rotation with period T, is

(a) $\frac{\pi MR^3}{T}$ (b) $\frac{MR^2\pi}{T}$ (c) $\frac{2\pi MR^2}{5T}$ (d) $\frac{4\pi MR^2}{5T}$

40. A satellite is launched into a circular orbit of radius R around the earth. A second satellite is launched into an orbit of radius 1.01 R. The period of second satellite is larger than the first one by approximately

(a) 0.5% (b) 1.0% (c) 1.5% (d) 3.0%

41. A uniform spherical shell gradually shrinks maintaining its shape. The gravitational potential at the centre

(a) increases (b) decreases
(c) remains constant (d) cannot say

42. The depth *d* at which the value of acceleration due to gravity becomes $\frac{1}{n}$ times the value at the surface of the earth, is [R = radius of the earth]

(a) $\frac{R}{n}$ (b) $R\left(\frac{n-1}{n}\right)$ (c) $\frac{R}{n^2}$ (d) $R\left(\frac{n}{n+1}\right)$

43. Radius of moon is 1/4 times that of earth and mass is 1/81 times that of earth. The point at which gravitational field due to earth becomes equal and opposite to that of moon, is (Distance between centres of earth and moon is 60R, where R is radius of earth)

(a) 5.75 R from centre of moon
(b) 16 R from surface of moon
(c) 53 R from centre of earth
(d) 54 R from centre of earth

44. If earth is supposed to be a sphere of radius R, if g_{30} is value of acceleration due to gravity at lattitude of 30° and g at the equator, the value of $g - g_{30}$ is

(a) $\frac{1}{4}\omega^2R$ (b) $\frac{3}{4}\omega^2R$ (c) ω^2R (d) $\frac{1}{2}\omega^2R$

45. What is the minimum energy required to launch a satellite of mass m from the surface of a planet of mass M and radius R in a circular orbit at an altitude of 2R?

(a) $\frac{5GmM}{6R}$ (b) $\frac{2GmM}{3R}$ (c) $\frac{GmM}{2R}$ (d) $\frac{GmM}{2R}$

RESPONSE GRID					
	33. ⓐⓑⓒⓓ	34. ⓐⓑⓒⓓ	35. ⓐⓑⓒⓓ	36. ⓐⓑⓒⓓ	37. ⓐⓑⓒⓓ
	38. ⓐⓑⓒⓓ	39. ⓐⓑⓒⓓ	40. ⓐⓑⓒⓓ	41. ⓐⓑⓒⓓ	42. ⓐⓑⓒⓓ
	43. ⓐⓑⓒⓓ	44. ⓐⓑⓒⓓ	45. ⓐⓑⓒⓓ		

DAILY PRACTICE PROBLEM DPP CHAPTERWISE CP07 - PHYSICS			
Total Questions	45	Total Marks	180
Attempted		Correct	
Incorrect		Net Score	
Cut-off Score	45	Qualifying Score	60
Success Gap = Net Score – Qualifying Score			
Net Score = (Correct × 4) – (Incorrect × 1)			

Space for Rough Work

DPP - Daily Practice Problems

Chapter-wise Sheets

Date : 　　 **Start Time :** 　　 **End Time :**

PHYSICS (CP08)

SYLLABUS : Mechanical Properties of Solids

Max. Marks : 180 **Marking Scheme :** (+4) for correct & (–1) for incorrect answer **Time : 60 min.**

INSTRUCTIONS : This Daily Practice Problem Sheet contains 45 MCQs. For each question only one option is correct. Darken the correct circle/ bubble in the Response Grid provided on each page.

1. Two wires A and B are of the same material. Their lengths are in the ratio 1 : 2 and the diameter are in the ratio 2 : 1. If they are pulled by the same force, then increase in length will be in the ratio
   (a) 2 : 1 (b) 1 : 4 (c) 1 : 8 (d) 8 : 1
2. The load versus elongation graphs for four wires of same length and made of the same material are shown in the figure. The thinnest wire is represented by the line
   (a) *OA*
   (b) *OC*
   (c) *OD*
   (d) *OB*

Load, D, C, B, A, O, Elongation

3. A spring of force constant 800 N/m has an extension of 5 cm. The work done in extending it from 5 cm to 15 cm is
   (a) 16 J (b) 8 J (c) 32 J (d) 24 J
4. A metal wire of length L_1 and area of cross-section A is attached to a rigid support. Another metal wire of length L_2 and of the same cross-sectional area is attached to the free end of the first wire. A body of mass M is then suspended from the free end of the second wire. If Y_1 and Y_2 are the Young's moduli of the wires respectively, the effective force constant of the system of two wires is
   (a) $\dfrac{(Y_1Y_2)A}{2(Y_1L_2+Y_2L_1)}$ (b) $\dfrac{(Y_1Y_2)A}{(L_1L_2)^{1/2}}$
   (c) $\dfrac{(Y_1Y_2)A}{Y_1L_2+Y_2L_1}$ (d) $\dfrac{(Y_1Y_2)^{1/2}A}{(L_2L_1)^{1/2}}$
5. The approximate depth of an ocean is 2700 m. The compressibility of water is 45.4×10^{-11} Pa^{-1} and density of water is 10^3 kg/m^3. What fractional compression of water will be obtained at the bottom of the ocean ?
   (a) 1.0×10^{-2} (b) 1.2×10^{-2}
   (c) 1.4×10^{-2} (d) 0.8×10^{-2}
6. The Young's modulus of steel is twice that of brass. Two wires of same length and of same area of cross section, one of steel and another of brass are suspended from the same roof. If we want the lower ends of the wires to be at the same level, then the weights added to the steel and brass wires must be in the ratio of :
   (a) 2 : 1 (b) 4 : 1 (c) 1 : 1 (d) 1 : 2
7. Choose the wrong statement.
   (a) The bulk modulus for solids is much larger than for liquids.
   (b) Gases are least compressible.
   (c) The incompressibility of the solids is due to the tight coupling between neighbouring atoms.
   (d) The reciprocal of the bulk modulus is called compressibility.

RESPONSE GRID	1. ⓐⓑⓒⓓ	2. ⓐⓑⓒⓓ	3. ⓐⓑⓒⓓ	4. ⓐⓑⓒⓓ	5. ⓐⓑⓒⓓ
	6. ⓐⓑⓒⓓ	7. ⓐⓑⓒⓓ			

Space for Rough Work

8. A copper wire of length 1.0 m and a steel wire of length 0.5 m having equal cross-sectional areas are joined end to end. The composite wire is stretched by a certain load which stretches the copper wire by 1 mm. If the Young's modulii of copper and steel are respectively 1.0×10^{11} Nm^{-2} and 2.0×10^{11} Nm^{-2}, the total extension of the composite wire is :

(a) 1.75 mm (b) 2.0 mm (c) 1.50 mm (d) 1.25 mm

9. A cube at temperature 0°C is compressed equally from all sides by an external pressure P. By what amount should its temperature be raised to bring it back to the size it had before the external pressure was applied. The bulk modulus of the material of the cube is B and the coefficient of linear expansion is a.

(a) P/Bα (b) P/3Bα (c) 3$\pi\alpha$/B (d) 3B/P

10. The diagram below shows the change in the length X of a thin uniform wire caused by the application of stress F at two different temperatures T_1 and T_2. The variation shown suggests that

(a) $T_1 > T_2$

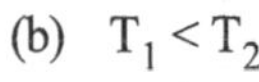

(b) $T_1 < T_2$

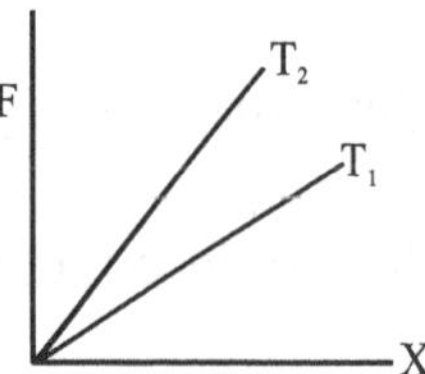

(c) $T_2 > T_1$

(d) $T_1 \geq T_2$

11. If the ratio of lengths, radii and Young's moduli of steel and brass wires in the figure are *a*, *b* and *c* respectively, then the corresponding ratio of increase in their lengths is :

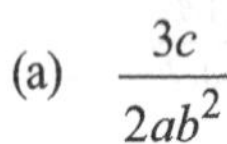

(a) $\frac{3c}{2ab^2}$

(b) $\frac{2a^2c}{b}$

(c) $\frac{3a}{2b^2c}$

(d) $\frac{2ac}{b^2}$

Steel
M
Brass
2M

12. The Young's modulus of brass and steel are respectively 10^{10} N/m^2. and 2×10^{10} N/m^2. A brass wire and a steel wire of the same length are extended by 1 mm under the same force, the radii of brass and steel wires are R_B and R_S respectively. Then

(a) $R_S = \sqrt{2}\, R_B$ (b) $R_S = R_B/\sqrt{2}$

(c) $R_S = 4R_B$ (d) $R_S = R_B/4$

13. Steel ruptures when a shear of 3.5×10^8 N m^{-2} is applied. The force needed to punch a 1 cm diameter hole in a steel sheet 0.3 cm thick is nearly:

(a) 1.4×10^4 N (b) 2.7×10^4 N

(c) 3.3×10^4 N (d) 1.1×10^4 N

14. A ball falling in a lake of depth 400 m has a decrease of 0.2% in its volume at the bottom. The bulk modulus of the material of the ball is (in N m^{-2})

(a) 9.8×10^9 (b) 9.8×10^{10}

(c) 1.96×10^{10} (d) 1.96×10^9

15. A circular tube of mean radius 8 cm and thickness 0.04 cm is melted up and recast into a solid rod of the same length. The ratio of the torsional rigidities of the circular tube and the solid rod is

(a) $\frac{(8.02)^4 - (7.98)^4}{(0.8)^4}$ (b) $\frac{(8.02)^2 - (7.98)^2}{(0.8)^2}$

(c) $\frac{(0.8)^2}{(8.02)^4 - (7.98)^4}$ (d) $\frac{(0.8)^2}{(8.02)^3 - (7.98)^2}$

16. Two wires are made of the same material and have the same volume. However wire 1 has cross-sectional area *A* and wire 2 has cross-sectional area 3*A*. If the length of wire 1 increases by Δx on applying force *F*, how much force is needed to stretch wire 2 by the same amount?

(a) 4*F* (b) 6*F* (c) 9*F* (d) *F*

17. In materials like aluminium and copper, the correct order of magnitude of various elastic modului is:

(a) Young's modulus < shear modulus < bulk modulus.

(b) Bulk modulus < shear modulus < Young's modulus

(c) Shear modulus < Young's modulus < bulk modulus.

(d) Bulk modulus < Young's modulus < shear modulus.

18. What per cent of length of wire increases by applying a stress of 1 kg weight/mm^2 on it?

($Y = 1 \times 10^{11}$ N/m^2 and 1 kg weight = 9.8 newton)

(a) 0.0067% (b) 0.0098%

(c) 0.0088% (d) 0.0078%

19. An elastic string of unstretched length *L* and force constant *k* is stretched by a small length *x*. It is further stretched by another small length *y*. The work done in the second stretching is :

(a) $\frac{1}{2}ky^2$ (b) $\frac{1}{2}k(x^2+y^2)$

(c) $\frac{1}{2}k(x+y)^2$ (d) $\frac{1}{2}ky(2x+y)$

20. Two, spring *P* and *Q* of force constants k_p and $k_Q\left(k_Q = \frac{k_p}{2}\right)$ are stretched by applying forces of equal magnitude. If the energy stored in *Q* is E, then the energy stored in *P* is

(a) E (b) 2E (c) E/2 (d) E/4

RESPONSE GRID					
	8. ⓐⓑⓒⓓ	9. ⓐⓑⓒⓓ	10. ⓐⓑⓒⓓ	11. ⓐⓑⓒⓓ	12. ⓐⓑⓒⓓ
	13. ⓐⓑⓒⓓ	14. ⓐⓑⓒⓓ	15. ⓐⓑⓒⓓ	16. ⓐⓑⓒⓓ	17. ⓐⓑⓒⓓ
	18. ⓐⓑⓒⓓ	19. ⓐⓑⓒⓓ	20. ⓐⓑⓒⓓ		

Space for Rough Work

21. The pressure that has to be applied to the ends of a steel wire of length 10 cm to keep its length constant when its temperature is raised by 100°C is:

(For steel Young's modulus is $2 \times 10^{11}\,Nm^{-2}$ and coefficient of thermal expansion is $1.1 \times 10^{-5}\,K^{-1}$)

(a) 2.2×10^8 Pa (b) 2.2×10^9 Pa

(c) 2.2×10^7 Pa (d) 2.2×10^6 Pa

22. A steel ring of radius r and cross sectional area A is fitted onto a wooden disc of radius R (R > r). If the Young's modulus of steel is Y, then the force with which the steel ring is expanded is

(a) AY(R/r) (b) AY(R – r)/r

(c) (Y/A)[(R – r)/r] (d) Yr/AR

23. Two wires A and B of same material and of equal length with the radii in the ratio 1 : 2 are subjected to identical loads. If the length of A increases by 8 mm, then the increase in length of B is

(a) 2 mm (b) 4 mm (c) 8 mm (d) 16 mm

24. A material has poisson's ratio 0.50. If a uniform rod of it suffers a longitudinal strain of 2×10^{-3}, then the percentage change in volume is

(a) 0.6 (b) 0.4 (c) 0.2 (d) Zero

25. The upper end of a wire of diameter 12mm and length 1m is clamped and its other end is twisted through an angle of 30°. The angle of shear is

(a) 18° (b) 0.18° (c) 36° (d) 0.36°

26. The pressure on an object of bulk modulus B undergoing hydraulic compression due to a stress exerted by surrounding fluid having volume strain $\left(\frac{\Delta V}{V}\right)^2$ is

(a) $B^2\left(\frac{\Delta V}{V}\right)$ (b) $B\left(\frac{\Delta V}{V}\right)^2$

(c) $\frac{1}{B}\left(\frac{\Delta V}{V}\right)$ (d) $B\left(\frac{\Delta V}{V}\right)$

27. A structural steel rod has a radius of 10 mm and length of 1.0 m. A 100 kN force stretches it along its length. Young's modulus of structural steel is $2 \times 10^{11}\,Nm^{-2}$. The percentage strain is about

(a) 0.16% (b) 0.32% (c) 0.08% (d) 0.24%

28. A beam of metal supported at the two edges is loaded at the centre. The depression at the centre is proportional to

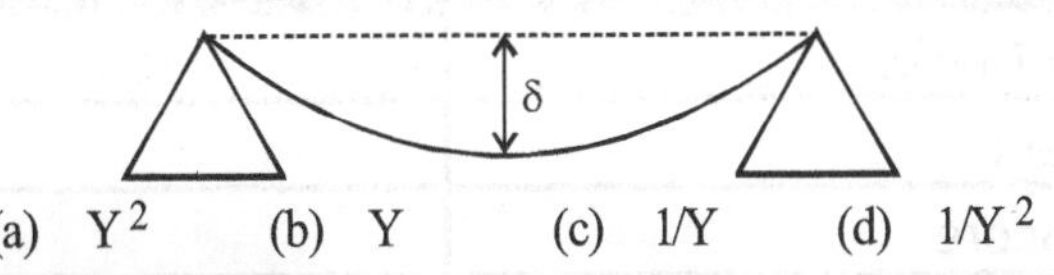

(a) Y^2 (b) Y (c) 1/Y (d) $1/Y^2$

29. When a 4 kg mass is hung vertically on a light spring that obeys Hooke's law, the spring stretches by 2 cms. The work required to be done by an external agent in stretching this spring by 5 cms will be ($g = 9.8\,m/sec^2$)

(a) 4.900 joule (b) 2.450 joule

(c) 0.495 joule (d) 0.245 joule

30. The length of a metal is ℓ_1 when the tension in it is T_1 and is ℓ_2 when the tension is T_2. The original length of the wire is

(a) $\frac{\ell_1 + \ell_2}{2}$ (b) $\frac{\ell_1 T_2 + \ell_2 T_1}{T_1 + T_2}$

(c) $\frac{\ell_1 T_2 - \ell_2 T_1}{T_2 - T_1}$ (d) $\sqrt{T_1 T_2 \ell_1 \ell_2}$

31. For the same cross-sectional area and for a given load, the ratio of depressions for the beam of a square cross-section and circular cross-section is

(a) $3 : \pi$ (b) $\pi : 3$ (c) $1 : \pi$ (d) $\pi : 1$

32. The bulk moduli of ethanol, mercury and water are given as 0.9, 25 and 2.2 respectively in units of $10^9\,Nm^{-2}$. For a given value of pressure, the fractional compression in volume is $\frac{\Delta V}{V}$. Which of the following statements about $\frac{\Delta V}{V}$ for these three liquids is correct ?

(a) Ethanol > Water > Mercury

(b) Water > Ethanol > Mercury

(c) Mercury > Ethanol > Water

(d) Ethanol > Mercury > Water

33. The graph given is a stress-strain curve for

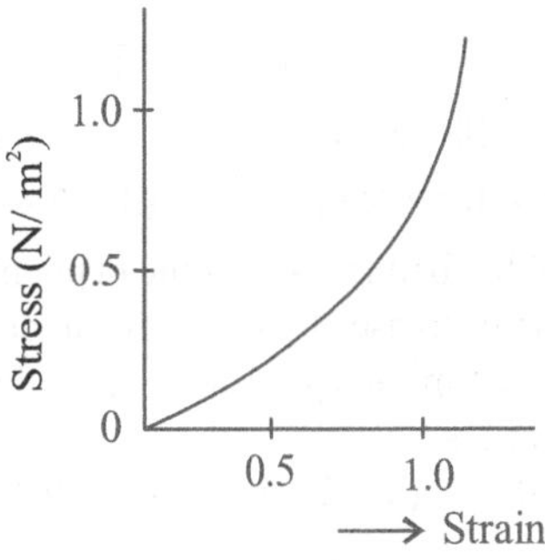

(a) elastic objects (b) plastics

(c) elastomers (d) None of these

34. A metal rod of Young's modulus $2 \times 10^{10}\,N\,m^{-2}$ undergoes an elastic strain of 0.06%. The energy per unit volume stored in $J\,m^{-3}$ is

(a) 3600 (b) 7200 (c) 10800 (d) 14400

35. Two wires of the same material and same length but diameters in the ratio 1 : 2 are stretched by the same force. The potential energy per unit volume of the two wires will be in the ratio

(a) 1 : 2 (b) 4 : 1 (c) 2 : 1 (d) 16 : 1

RESPONSE GRID					
	21. ⓐⓑⓒⓓ	22. ⓐⓑⓒⓓ	23. ⓐⓑⓒⓓ	24. ⓐⓑⓒⓓ	25. ⓐⓑⓒⓓ
	26. ⓐⓑⓒⓓ	27. ⓐⓑⓒⓓ	28. ⓐⓑⓒⓓ	29. ⓐⓑⓒⓓ	30. ⓐⓑⓒⓓ
	31. ⓐⓑⓒⓓ	32. ⓐⓑⓒⓓ	33. ⓐⓑⓒⓓ	34. ⓐⓑⓒⓓ	35. ⓐⓑⓒⓓ

Space for Rough Work

36. The length of an elastic string is a metre when the longitudinal tension is 4 N and b metre when the longitudinal tension is 5 N. The length of the string in metre when the longitudinal tension is 9 N is

(a) $a-b$ (b) $5b-4a$ (c) $2b-\frac{1}{4}a$ (d) $4a-3b$

37. A force of 10^3 newton, stretches the length of a hanging wire by 1 millimetre. The force required to stretch a wire of same material and length but having four times the diameter by 1 millimetre is

(a) 4×10^3 N (b) 16×10^3 N

(c) $\frac{1}{4}\times10^3$ N (d) $\frac{1}{16}\times10^3$ N

38. A steel wire of length l and cross sectional area A is stretched by 1 cm under a given load. When the same load is applied to another steel wire of double its length and half of its cross section area, the amount of stretching (extension) is

(a) 0.5 cm (b) 2 cm (c) 4 cm (d) 1.5 cm

39. The adjacent graph shows the extension (Δl) of a wire of length 1 m suspended from the top of a roof at one end with a load W connected to the other end. If the cross-sectional area of the wire is 10^{-6} m^2, calculate the Young's modulus of the material of the wire :

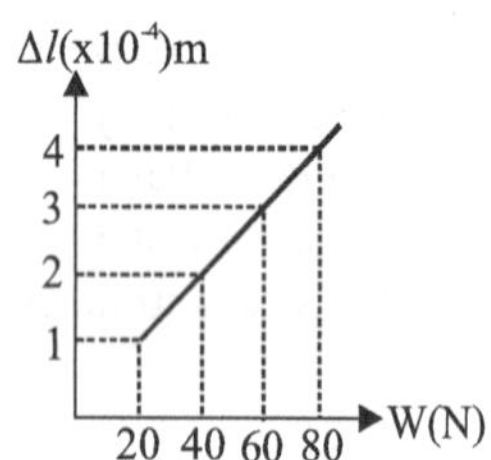

(a) 2×10^{11} N/m^2 (b) 2×10^{-11} N/m^2

(c) 3×10^{-12} N/m^2 (d) 2×10^{-13} N/m^2

40. If a rubber ball is taken at the depth of 200 m in a pool, its volume decreases by 0.1%. If the density of the water is 1×10^3 kg/m^3 and g = 10m/s^2, then the volume elasticity in N/m^2 will be

(a) 10^8 (b) 2×10^8 (c) 10^9 (d) 2×10^9

41. A ball is falling in a lake of depth 200 m creates a decrease 0.1 % in its volume at the bottom. The bulk modulus of the material of the ball will be

(a) 19.6×10^{-8} N/m^2 (b) 19.6×10^{10} N/m^2

(c) 19.6×10^{-10} N/m^2 (d) 19.6×10^{8} N/m^2

42. The diagram shows a force-extension graph for a rubber band. Consider the following statements :

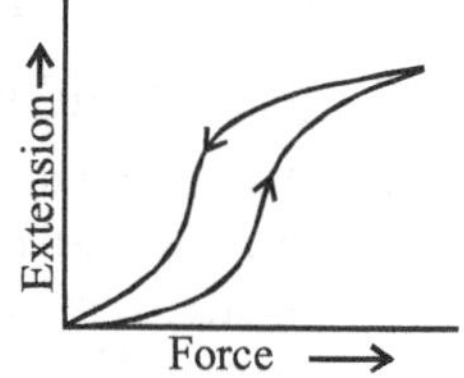

I. It will be easier to compress this rubber than expand it

II. Rubber does not return to its original length after it is stretched

III. The rubber band will get heated if it is stretched and released

Which of these can be deduced from the graph:

(a) III only (b) II and III (c) I and III (d) I only

43. The Poisson's ratio of a material is 0.5. If a force is applied to a wire of this material, there is a decrease in the cross-sectional area by 4%. The percentage increase in the length is:

(a) 1% (b) 2% (c) 2.5% (d) 4%

44. Copper of fixed volume 'V; is drawn into wire of length 'l'. When this wire is subjected to a constant force 'F', the extension produced in the wire is 'Δl'. Which of the following graphs is a straight line?

(a) Δl versus $\frac{1}{l}$ (b) Δl versus l^2

(c) Δl versus $\frac{1}{l^2}$ (d) Δl versus l

45. When a 4 kg mass is hung vertically on a light spring that obeys Hooke's law, the spring stretches by 2 cms. The work required to be done by an external agent in stretching this spring by 5 cm will be (g = 9.8 m/sec^2)

(a) 4.900 joule (b) 2.450 joule

(c) 0.495 joule (d) 0.245 joule

RESPONSE GRID					
	36. ⓐⓑⓒⓓ	37. ⓐⓑⓒⓓ	38. ⓐⓑⓒⓓ	39. ⓐⓑⓒⓓ	40. ⓐⓑⓒⓓ
	41. ⓐⓑⓒⓓ	42. ⓐⓑⓒⓓ	43. ⓐⓑⓒⓓ	44. ⓐⓑⓒⓓ	45. ⓐⓑⓒⓓ

DAILY PRACTICE PROBLEM DPP CHAPTERWISE CP08 - PHYSICS			
Total Questions	45	Total Marks	180
Attempted		Correct	
Incorrect		Net Score	
Cut-off Score	50	Qualifying Score	70
Success Gap = Net Score – Qualifying Score			
Net Score = (Correct × 4) – (Incorrect × 1)			

Space for Rough Work

DPP - Daily Practice Problems

Chapter-wise Sheets

Date :   **Start Time :**   **End Time :**

PHYSICS (CP09)

SYLLABUS : Mechanical Properties of Fluids

Max. Marks : 180   **Marking Scheme :** (+4) for correct & (–1) for incorrect answer   **Time : 60 min.**

INSTRUCTIONS : This Daily Practice Problem Sheet contains 45 MCQs. For each question only one option is correct. Darken the correct circle/ bubble in the Response Grid provided on each page.

1. The density of water at the surface of ocean is ρ. If the bulk modulus of water is B, what is the density of ocean water at a depth where the pressure is nP_0, where P_0 is the atmospheric pressure ?

(a) $\frac{\rho B}{B-(n-1)P_0}$ (b) $\frac{\rho B}{B+(n-1)P_0}$

(c) $\frac{\rho B}{B-nP_0}$ (d) $\frac{\rho B}{B+nP_0}$

2. A ball of radius r and density ρ falls freely under gravity through a distance h before entering water. Velocity of ball does not change even on entering water. If viscosity of water is η the value of h is given by

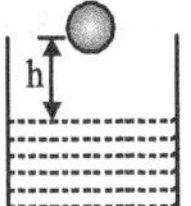

(a) $\frac{2}{9}r^2\left(\frac{1-\rho}{\eta}\right)g$ (b) $\frac{2}{81}r^2\left(\frac{\rho-1}{\eta}\right)g$

(c) $\frac{2}{81}r^4\left(\frac{\rho-1}{\eta}\right)^2 g$ (d) $\frac{2}{9}r^4\left(\frac{\rho-1}{\eta}\right)^2 g$

3. Two parallel glass plates are dipped partly in the liquid of density 'd' keeping them vertical . If the distance between the plates is 'x', surface tension for liquids is T and angle of contact is θ, then rise of liquid between the plates due to capillary will be

(a) $\frac{T\cos\theta}{xd}$ (b) $\frac{2T\cos\theta}{xdg}$ (c) $\frac{2T}{xdg\cos\theta}$ (d) $\frac{T\cos\theta}{xdg}$

4. A liquid is allowed to flow into a tube of truncated cone shape. Identify the correct statement from the following

(a) The speed is high at the wider end and high at the narrow end

(b) The speed is low at the wider end and high at the narrow end

(c) The speed is same at both ends in a streamline flow

(d) The liquid flows with uniform velocity in the tube

5. A wide vessel with a small hole at the bottom is filled with water (density ρ_1, height h_1) and kerosene (density ρ_2, height h_2). Neglecting viscosity effects, the speed with which water flows out is :

(a) $[2g(h_1+h_2)]^{1/2}$ (b) $[2g(h_1\rho_1+h_2\rho_2)]^{1/2}$

(c) $[2g(h_1+h_2(\rho_2/\rho_1))]^{1/2}$ (d) $[2g\,(h_1+h_2(\rho_1/\rho_2))]^{1/2}$

RESPONSE GRID	1. ⓐⓑⓒⓓ	2. ⓐⓑⓒⓓ	3. ⓐⓑⓒⓓ	4. ⓐⓑⓒⓓ	5. ⓐⓑⓒⓓ

Space for Rough Work

6. A capillary tube of radius r is immersed vertically in a liquid such that liquid rises in it to height h (less than the length of the tube). Mass of liquid in the capillary tube is m. If radius of the capillary tube is increased by 50%, then mass of liquid that will rise in the tube, is

(a) $\frac{2}{3}$m (b) $\frac{4}{9}$m (c) $\frac{3}{2}$m (d) $\frac{9}{4}$m

7. A lead shot of 1 mm diameter falls through a long column of glycerine. The variation of its velocity v with distance covered is represented by

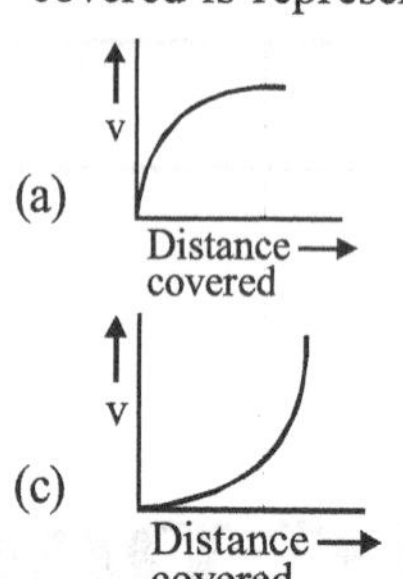

(a) (b) (c) (d)

8. Two mercury drops (each of radius '*r*') merge to form bigger drop. The surface energy of the bigger drop, if *T* is the surface tension, is :

(a) $4\pi r^2 T$ (b) $2\pi r^2 T$

(c) $2^{8/3}\pi r^2 T$ (d) $2^{5/3}\pi r^2 T$

9. Wax is coated on the inner wall of a capillary tube and the tube is then dipped in water. Then, compared to the unwaxed capillary, the angle of contact θ and the height *h* upto which water rises change. These changes are :

(a) θ increases and *h* also increases
(b) θ decreases and *h* also decreases
(c) θ increases and *h* decreases
(d) θ decreases and *h* increases

10. A rain drop of radius 0.3 mm has a terminal velocity in air = 1 m/s. The viscosity of air is 8×10^{-5} poise. The viscous force on it is

(a) 45.2×10^{-4} dyne (b) 101.73×10^{-5} dyne
(c) 16.95×10^{-4} dyne (d) 16.95×10^{-5} dyne

11. A water tank of height 10m, completely filled with water is placed on a level ground. It has two holes one at 3 m and the other at 7 m from its base. The water ejecting from

(a) both the holes will fall at the same spot
(b) upper hole will fall farther than that from the lower hole
(c) upper hole will fall closer than that from the lower hole
(d) more information is required

12. Two capillary of length L and 2L and of radius R and 2R are connected in series. The net rate of flow of fluid through them will be (given rate to the flow through single capillary, $X = \frac{\pi P R^4}{8\eta L}$)

(a) $\frac{8}{9}$X (b) $\frac{9}{8}$X (c) $\frac{5}{7}$X (d) $\frac{7}{5}$X

13. A candle of diameter d is floating on a liquid in a cylindrical container of diameter D (D >> d) as shown in figure. If it is burning at the rate of 2 cm/hour then the top of the candle will

(a) remain at the same height
(b) fall at the rate of 1 cm/hour
(c) fall at the rate of 2 cm/hour
(d) go up at the rate of 1 cm/hour

14. An isolated and charged spherical soap bubble has a radius r and the pressure inside is atmospheric. T is the surface tension of soap solution. If charge on drop is $X\,\pi r\sqrt{2rT\varepsilon_0}$ then find the value of X.

(a) 8 (b) 9 (c) 7 (d) 2

15. A thread is tied slightly loose to a wire frame as in figure and the frame is dipped into a soap solution and taken out. The frame is completely covered with the film. When the portion A is punctured with a pin, the thread

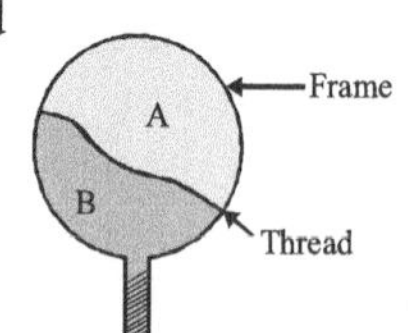

(a) becomes concave towards A
(b) becomes convex towards A
(c) remains in the initial position
(d) either (a) or (b) depending on the size of A w.r.t. B

16. Which of the following expressions represents the excess of pressure inside the soap bubble?

(a) $P_i - P_o = \frac{s}{r}$ (b) $P_i - P_o = \frac{2s}{r}$

(c) $P_i - P_o = \frac{2s}{r} + h\rho g$ (d) $P_i - P_o = \frac{4s}{r}$

17. A spherical solid ball of volume *V* is made of a material of density ρ_1. It is falling through a liquid of density ρ_1 ($\rho_2 < \rho_1$). Assume that the liquid applies a viscous force on the ball that is proportional to the square of its speed *v*, i.e., $F_{viscous} = -kv^2$ ($k > 0$). The terminal speed of the ball is

(a) $\sqrt{\frac{Vg(\rho_1 - \rho_2)}{k}}$ (b) $\frac{Vg\rho_1}{k}$

(c) $\sqrt{\frac{Vg\rho_1}{k}}$ (d) $\frac{Vg(\rho_1 - \rho_2)}{k}$

18. Select the correct statements from the following.

(a) Bunsen burner and sprayers work on Bernoulli's principle
(b) Blood flow in arteries is explained by Bernoulli's principle
(c) A siphon works on account of atmospheric pressure.
(d) All are correct

RESPONSE GRID					
	6. ⓐⓑⓒⓓ	7. ⓐⓑⓒⓓ	8. ⓐⓑⓒⓓ	9. ⓐⓑⓒⓓ	10. ⓐⓑⓒⓓ
	11. ⓐⓑⓒⓓ	12. ⓐⓑⓒⓓ	13. ⓐⓑⓒⓓ	14. ⓐⓑⓒⓓ	15. ⓐⓑⓒⓓ
	16. ⓐⓑⓒⓓ	17. ⓐⓑⓒⓓ	18. ⓐⓑⓒⓓ		

Space for Rough Work

19. The wetability of a surface by a liquid depends primarily on
(a) surface tension
(b) density
(c) angle of contact between the surface and the liquid
(d) viscosity

20. The relative velocity of two parallel layers of water is 8 cm/sec. If the perpendicular distance between the layers is 0.1 cm, then velocity gradient will be
(a) 80/sec (b) 60 /sec (c) 50/sec (d) 40/sec

21. Choose the correct statement
(a) Terminal velocities of rain drops are proportional to square of their radii
(b) Water proof agents decrease the angle of contact between water and fibres
(c) Detergents increase the surface tension of water
(d) Hydraulic machines work on the principle of Torricelli's law

22. When a ball is released from rest in a very long column of viscous liquid, its downward acceleration is 'a' (just after release). Its acceleration when it has acquired two third of the maximum velocity is a/X. Find the value of X.
(a) 2 (b) 3 (c) 4 (d) 5

23. A ring is cut from a platinum tube 8.5 cm internal and 8.7 cm external diameter. It is supported horizontally from the pan of a balance, so that it comes in contact with the water in a glass vessel. If an extra 3.97. If is required to pull it away from water, the surface tension of water is
(a) 72 dyne cm^{-1} (b) 70.80 dyne cm^{-1}
(c) 63.35 dyne cm^{-1} (d) 60 dyne cm^{-1}

24. Which of the following graph represents the variation of surface tension with temperature over small temperature ranges for water?

(a) 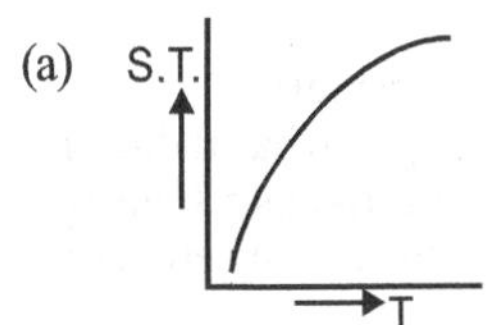

(b) 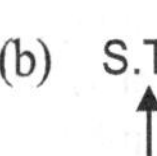

(c)
S.T.
T

(d) 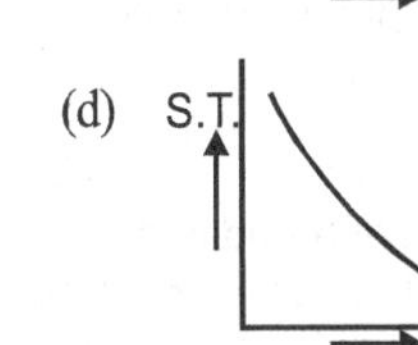

25. When a large air bubble rises from the bottom of a lake to the surface, its radius doubles. If the atmospheric pressure is equal to that of a column of water of height H, then depth of the lake is
(a) H (b) 2H (c) 7H (d) 8H

26. What is the velocity v of a metallic ball of radius r falling in a tank of liquid at the instant when its acceleration is one -half that of a freely falling body ? (The densities of metal and of liquid are ρ and σ respectively, and the viscosity of the liquid is η).

(a) $\frac{r^2 g}{9\eta}(\rho - 2\sigma)$ (b) $\frac{r^2 g}{9\eta}(2\rho - \sigma)$

(c) $\frac{r^2 g}{9\eta}(\rho - \sigma)$ (d) $\frac{2r^2 g}{9\eta}(\rho - \sigma)$

27. Two pieces of metals are suspended from the arms of a balance and are found to be in equilibrium when kept immersed in water. The mass of one piece is 32 g and its density 8 g cm^{-3}. The density of the other is 5 g per cm^3. Then the mass of the other is
(a) 28 g (b) 35 g (c) 21 g (d) 33.6 g

28. A block of material of specific gravity 0.4 is held submerged at a depth of 1m in a vessel filled with water. The vessel is accelerated upwards with acceleration of $a_o = g/5$. If the block is released at $t = 0$, neglecting viscous effects, it will reach the water surface at t equal to ($g = 10$ m/s²) :
(a) 0.60 s (b) 0.33 s (c) 3.3 s (d) 1.2 s

29. Figure shows a capillary rise H. If the air is blown through the horizontal tube in the direction as shown then rise in capillary tube will be
(a) =H
(b) >H
(c) <H
(d) zero

H

30. A small spherical ball falling through a viscous medium of negligible density has terminal velocity v. Another ball of the same mass but of radius twice that of the earlier falling through the same viscous medium will have terminal velocity
(a) v (b) v/4 (c) v/2 (d) 2v

31. Two non-mixing liquids of densities ρ and nρ (n > 1) are put in a container. The height of each liquid is h. A solid cylinder of length L and density d is put in this container. The cylinder floats with its axis vertical and length pL(p < 1) in the denser liquid. The density d is equal to :
(a) $\{1+(n+1)p\}\rho$ (b) $\{2+(n+1)p\}\rho$
(c) $\{2+(n-1)p\}\rho$ (d) $\{1+(n-1)p\}\rho$

32. A thin liquid film formed between a U-shaped wire and a light slider supports a weight of 1.5×10^{-2} N (see figure). The length of the slider is 30 cm and its weight negligible. The surface tension of the liquid film is

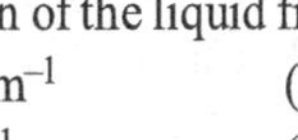

(a) 0.0125 Nm^{-1} (b) 0.1 Nm^{-1}
(c) 0.05 Nm^{-1} (d) 0.025 Nm^{-1}

RESPONSE GRID					
	19. ⓐⓑⓒⓓ	20. ⓐⓑⓒⓓ	21. ⓐⓑⓒⓓ	22. ⓐⓑⓒⓓ	23. ⓐⓑⓒⓓ
	24. ⓐⓑⓒⓓ	25. ⓐⓑⓒⓓ	26. ⓐⓑⓒⓓ	27. ⓐⓑⓒⓓ	28. ⓐⓑⓒⓓ
	29. ⓐⓑⓒⓓ	30. ⓐⓑⓒⓓ	31. ⓐⓑⓒⓓ	32. ⓐⓑⓒⓓ	

Space for Rough Work

33. Two liquids of densities d_1 and d_2 are flowing in identical capillary tubes uder the same pressure difference. If t_1 and t_2 are time taken for the flow of equal quantities (mass) of liquids, then the ratio of coefficient of viscosity of liquids must be

(a) $\frac{d_1 t_1}{d_2 t_2}$ (b) $\frac{t_1}{t_2}$ (c) $\frac{d_2\, t_2}{d_1\, t_1}$ (d) $\sqrt{\frac{d_1 t_1}{d_2 t_2}}$

34. Let T_1 be surface tension between solid and air, T_2 be the surface tension between solid and liquid and T be the surface tension between liquid and air. Then in equilibrium, for a drop of liquid on a clean glass plate, the correct relation is (θ is angle of contact)

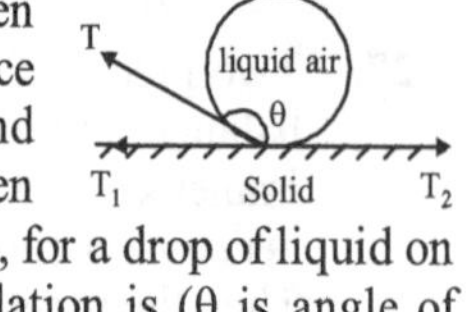

(a) $\cos\theta = \frac{T}{T_1 + T_2}$ (b) $\cos\theta = \frac{T}{T_1 - T_2}$

(c) $\cos\theta = \frac{T_1 + T_2}{T}$ (d) $\cos\theta = \frac{T_1 - T_2}{T}$

35. A uniform rod of density ρ is placed in a wide tank containing a liquid of density $\rho_0(\rho_0 > \rho)$. The depth of liquid in the tank is half the length of the rod. The rod is in equilibrium, with its lower end resting on the bottom of the tank. In this position the rod makes an angle θ with the horizontal

(a) $\sin\theta = \frac{1}{2}\sqrt{\rho_0/\rho}$ (b) $\sin\theta = \frac{1}{2}.\frac{\rho_0}{\rho}$

(c) $\sin\theta = \sqrt{\rho/\rho_0}$ (d) $\sin\theta = \rho_0/\rho$

36. A spherical ball of iron of radius 2 mm is falling through a column of glycerine. If densities of glycerine and iron are respectively 1.3×10^3 kg/m^3 and 8×10^3 kg/m^3. η for glycerine = 0.83 Nm^{-2} sec, then the terminal velocity is

(a) 0.7 m/s (b) 0.07 m/s (c) 0.007 m/s (d) 0.0007 m/s

37. A water film is formed between two straight parallel wires of 10 cm length 0.5 cm apart. If the distance between wires is increased by 1 mm. What will be the work done? (surface tension of water = 72 dyne/cm)

(a) 36 erg (b) 288 erg (c) 144 erg (d) 72 erg

38. A waterproofing agent changes the angle of contact

(a) from obtuse to acute.

(b) from acute to obtuse.

(c) from obtuse to $\pi/2$.

(d) from acute to $\pi/2$.

39. A thin metal disc of radius r floats on water surface and bends the surface downwards along the perimeter making an angle θ with vertical edge of the disc. If the disc displaces a weight of water W and surface tension of water is T, then the weight of metal disc is:

(a) $2\pi r T + W$ (b) $2\pi r T\cos\theta - W$

(c) $2\pi r T\cos\theta + W$ (d) $W - 2\pi r T\cos\theta$

40. A tank has a small hole at its botom of area of cross-section a. Liquid is being poured in the tank at the rate V m^3/s, the maximum level of liquid in the container will be (Area of tank = A)

(a) $\frac{V}{gaA}$ (b) $\frac{V^2}{2gAa}$ (c) $\frac{V^2}{gAa}$ (d) $\frac{V}{2gaA}$

41. A jar is filled with two non-mixing liquids 1 and 2 having densities ρ_1 and, ρ_2 respectively. A solid ball, made of a material of density ρ_3, is dropped in the jar. It comes to equilibrium in the position shown in the figure. Which of the following is true for ρ_1, ρ_2 and ρ_3?

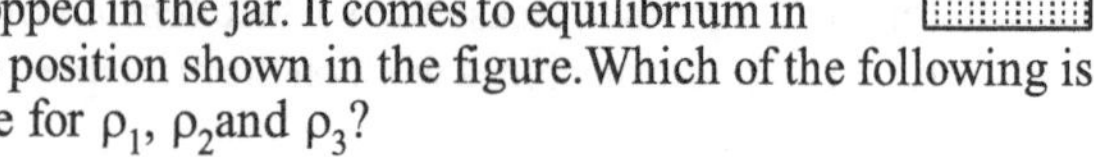

(a) $\rho_3 < \rho_1 < \rho_2$ (b) $\rho_1 > \rho_3 > \rho_2$

(c) $\rho_1 < \rho_2 < \rho_3$ (d) $\rho_1 < \rho_3 < \rho_2$

42. On heating water, bubbles being formed at the bottom of the vessel detach and rise. Take the bubbles to be spheres of radius R and making a circular contact of radius r with the bottom of the vessel. If r ≪ R and the surface tension of water is T, value of r just before bubbles detach is: (density of water is ρ_w)

(a) $R^2\sqrt{\frac{2\rho_w g}{3T}}$ (b) $R^2\sqrt{\frac{\rho_w g}{6T}}$ (c) $R^2\sqrt{\frac{\rho_w g}{T}}$ (d) $R^2\sqrt{\frac{3\rho_w g}{T}}$

43. The lift of an air plane is based on

(a) Torricelli's theorem

(b) Bernoulli's theorem

(c) Law of gravitation

(d) conservation of linear momentum

44. The cylindrical tube of a spray pump has radius, R, one end of which has n fine holes, each of radius r. If the speed of the liquid in the tube is V, the speed of the ejection of the liquid through the holes is :

(a) $\frac{VR^2}{nr^2}$ (b) $\frac{VR^2}{n^3 r^2}$ (c) $\frac{V^2 R}{nr}$ (d) $\frac{VR^2}{n^2 r^2}$

45. Drops of liquid of density ρ are floating half immersed in a liquid of density σ. If the surface tension of liquid is T, the radius of the drop will be

(a) $\sqrt{\frac{3T}{g(3\rho - \sigma)}}$ (b) $\sqrt{\frac{6T}{g(2\rho - \sigma)}}$ (c) $\sqrt{\frac{3T}{g(2\rho - \sigma)}}$ (d) $\sqrt{\frac{3T}{g(4\rho - 3\sigma)}}$

RESPONSE GRID					
	33. ⓐⓑⓒⓓ	34. ⓐⓑⓒⓓ	35. ⓐⓑⓒⓓ	36. ⓐⓑⓒⓓ	37. ⓐⓑⓒⓓ
	38. ⓐⓑⓒⓓ	39. ⓐⓑⓒⓓ	40. ⓐⓑⓒⓓ	41. ⓐⓑⓒⓓ	42. ⓐⓑⓒⓓ
	43. ⓐⓑⓒⓓ	44. ⓐⓑⓒⓓ	45. ⓐⓑⓒⓓ		

DAILY PRACTICE PROBLEM DPP CHAPTERWISE CP09 - PHYSICS

Total Questions	45	Total Marks	180
Attempted		Correct	
Incorrect		Net Score	
Cut-off Score	45	Qualifying Score	60
Success Gap = Net Score – Qualifying Score			
Net Score = (Correct × 4) – (Incorrect × 1)			

Space for Rough Work

DPP - Daily Practice Problems

Chapter-wise Sheets

Date :   **Start Time :**   **End Time :**

PHYSICS (CP10)

SYLLABUS : Thermal Properties of Matter

Max. Marks : 180   **Marking Scheme :** (+4) for correct & (–1) for incorrect answer   **Time : 60 min.**

INSTRUCTIONS : This Daily Practice Problem Sheet contains 45 MCQs. For each question only one option is correct. Darken the correct circle/ bubble in the Response Grid provided on each page.

1. The total radiant energy per unit area, normal to the direction of incidence, received at a distance R from the centre of a star of radius r, whose outer surface radiates as a black body at a temperature T K is given by: (σ is Stefan's constant)

(a) $\dfrac{\sigma r^2 T^4}{R^2}$ (b) $\dfrac{\sigma r^2 T^4}{4\pi r^2}$

(c) $\dfrac{\sigma r^4 T^4}{r^4}$ (d) $\dfrac{4\pi\sigma r^2 T^4}{R^2}$

2. Three rods of same dimensions are arranged as shown in figure, have thermal conductivities K_1, K_2 and K_3. The points P and Q are maintained at different temeratures for the heat to flow at the same rate along PRQ and PQ. Then which of the following option is correct?

(a) $K_3 = \frac{1}{2}(K_1 + K_2)$

(b) $K_3 = K_1 + K_2$

(c) $K_3 = \dfrac{K_1 K_2}{K_1 + K_2}$

(d) $K_3 = -2(K_1 + K_2)$

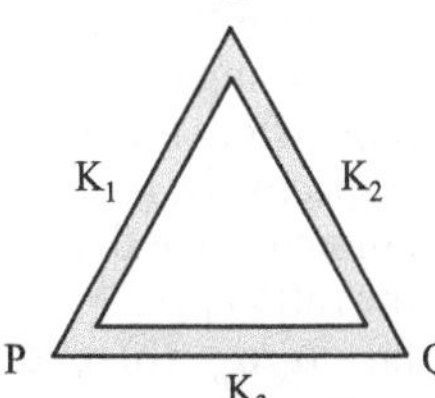

3. The sprinkling of water slightly reduces the temperature of a closed room because

(a) temperature of water is less than that of the room
(b) specific heat of water is high
(c) water has large latent heat of vaporisation
(d) water is a bad conductor of heat

4. The value of molar heat capacity at constant temperature is

(a) zero   (b) infinity
(c) unity   (d) 4.2

5. The specific heat capacity of a metal at low temperature (T) is given as

$$C_p(kJK^{-1}kg^{-1}) = 32\left(\frac{T}{400}\right)^3$$

A 100 gram vessel of this metal is to be cooled from 20°K to 4°K by a special refrigerator operating at room temperature (27°C). The amount of work required to cool the vessel is

(a) greater than 0.148 kJ
(b) between 0.148 kJ and 0.028 kJ
(c) less than 0.028 kJ
(d) equal to 0.002 kJ

RESPONSE GRID 1. ⓐⓑⓒⓓ 2. ⓐⓑⓒⓓ 3. ⓐⓑⓒⓓ 4. 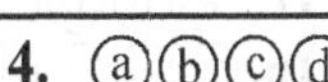 5. 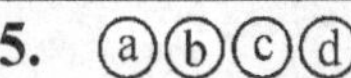

Space for Rough Work

6. The emissive power of a black body at T = 300K is 100 Watt/ m^2. Consider a body B of area A = $10m^2$ coefficient of reflectivity r = 0.3 and coefficient of transmission t = 0.5. Its temperature is 300K. Then which of the following is incorrect?
(a) The emissive power of B is 20 W/m^2
(b) The emissive power of B is 200 W/m^2
(c) The power emitted by B is 200 Watts
(d) The emissivity of B is 0.2

7. A solid cube and a solid sphere of the same material have equal surface area. Both are at the same temperture 120°C, then
(a) both the cube and the sphere cool down at the same rate
(b) the cube cools down faster than the sphere
(c) the sphere cools down faster than the cube
(d) whichever is having more mass will cool down faster

8. The density of water at 20°C is 998 kg/m^3 and at 40°C 992 kg/m^3. The coefficient of volume expansion of water is
(a) 10^{-4}/°C (b) 3×10^{-4}/°C
(c) 2×10^{-4}/°C (d) 6×10^{-4}/°C

9. A metallic rod ℓ cm long, A square cm in cross-section is heated through t°C. If Young's modulus of elasticity of the metal is E and the mean coefficient of linear expansion is α per degree celsius, then the compressional force required to prevent the rod from expanding along its length is
(a) $EA\alpha t$ (b) $EA\alpha t/(1+\alpha t)$
(c) $EA\alpha t/(1-\alpha t)$ (d) $E\ell\alpha t$

10. If liquefied oxygen at 1 atmospheric pressure is heated from 50 K to 300 K by supplying heat at constant rate. The graph of temperature vs time will be

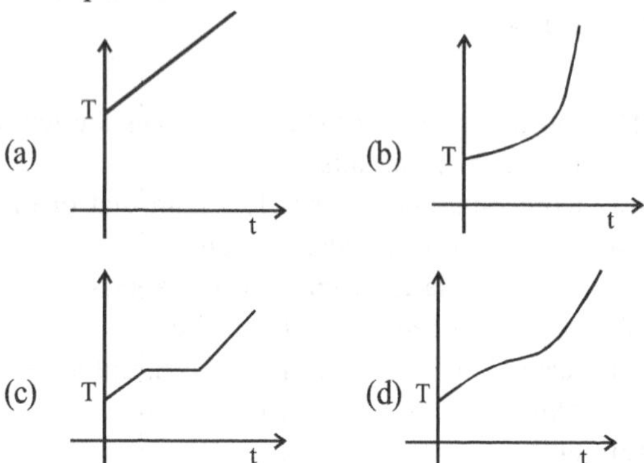

11. If a bar is made of copper whose coefficient of linear expansion is one and a half times that of iron, the ratio of force developed in the copper bar to the iron bar of identical lengths and cross-sections, when heated through the same temperature range (Young's modulus of copper may be taken to be equal to that of iron) is
(a) 3/2 (b) 2/3 (c) 9/4 (d) 4/9

12. A piece of ice falls from a height h so that it melts completely. Only one-quarter of the heat produced is absorbed by the ice and all energy of ice gets converted into heat during its fall. The value of h is :
[Latent heat of ice is 3.4×10^5 J/kg and g = 10 N/kg]
(a) 34 km (b) 544 km (c) 136 km (d) 68 km

13. A body of mass 5 kg falls from a height of 20 metres on the ground and it rebounds to a height of 0.2 m. If the loss in potential energy is used up by the body, then what will be the temperature rise?
(specific heat of material = 0.09 cal gm^{-1} °C^{-1})
(a) 0°C (b) 4°C (c) 8°C (d) None of these

14. Two straight metallic strips each of thickness t and length ℓ are rivetted together. Their coefficients of linear expansions are α_1 and α_2. If they are heated through temperature ΔT, the bimetallic strip will bend to form an arc of radius
(a) $t/\{(\alpha_1+\alpha_2)\Delta T\}$ (b) $t/\{(\alpha_2-\alpha_1)\Delta T\}$
(c) $t(\alpha_1-\alpha_2)\Delta T$ (d) $t(\alpha_2-\alpha_1)\Delta T$

15. The figure shows a system of two concentric spheres of radii r_1 and r_2 are kept at temperatures T_1 and T_2, respectively. The radial rate of flow of heat in a substance between the two concentric spheres is proportional to

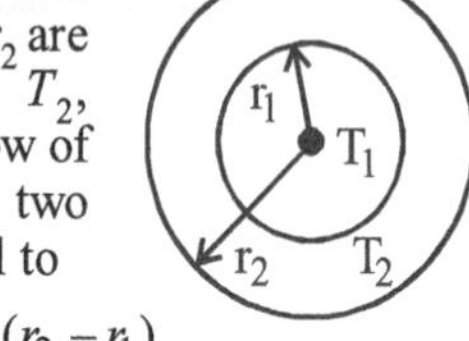

(a) $\ell n\left(\frac{r_2}{r_1}\right)$ (b) $\frac{(r_2-r_1)}{(r_1 r_2)}$
(c) (r_2-r_1) (d) $\frac{r_1 r_2}{(r_2-r_1)}$

16. A block of steel heated to 100°C is left in a room to cool. Which of the curves shown in fig., represents the correct behaviour?

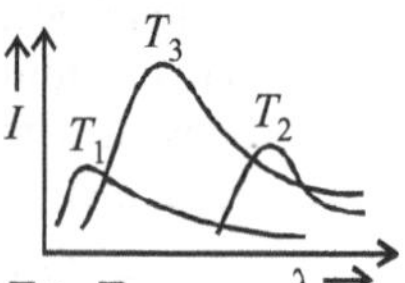

(a) A (b) B (c) C (d) None of these

17. Which of the following will expand the most for same rise in temperature?
(a) Aluminium (b) Glass
(c) Wood (d) All will expand same

18. The plots of intensity versus wavelength for three black bodies at temperatures T_1, T_2 and T_3 respectively are as shown. Their temperature are such that
(a) $T_1 > T_2 > T_3$ (b) $T_1 > T_3 > T_2$
(c) $T_2 > T_3 > T_1$ (d) $T_3 > T_2 > T_1$

19. When the temperature of a rod increases from t to t + Δt, its moment of inertia increases from I to I + ΔI. If α be the coefficient of linear expansion of the rod, then the value of $\frac{\Delta I}{I}$ is
(a) $2\alpha\Delta t$ (b) $\alpha\Delta T$ (c) $\frac{\alpha\Delta t}{2}$ (d) $\frac{\Delta t}{\alpha}$

20. Two rods, one of aluminum and the other made of steel, having initial length ℓ_1 and ℓ_2 are connected together to form a single rod of length $\ell_1+\ell_2$. The coefficients of linear expansion for aluminum and steel are α_a and α_s and respectively. If the length of each rod increases by the same amount when their temperature are raised by t°C, then find the ratio $\ell_1/(\ell_1+\ell_2)$
(a) α_s/α_a (b) α_a/α_s (c) $\alpha_s/(\alpha_a+\alpha_s)$ (d) $\alpha_a/(\alpha_a+\alpha_s)$

RESPONSE GRID					
	6. ⓐⓑⓒⓓ	7. ⓐⓑⓒⓓ	8. ⓐⓑⓒⓓ	9. ⓐⓑⓒⓓ	10. ⓐⓑⓒⓓ
	11. ⓐⓑⓒⓓ	12. ⓐⓑⓒⓓ	13. ⓐⓑⓒⓓ	14. ⓐⓑⓒⓓ	15. ⓐⓑⓒⓓ
	16. ⓐⓑⓒⓓ	17. ⓐⓑⓒⓓ	18. ⓐⓑⓒⓓ	19. ⓐⓑⓒⓓ	20. ⓐⓑⓒⓓ

Space for Rough Work

21. A polished metal plate with a rough black spot on it is heated to about 1400 K and quickly taken into a dark room. Which one of the following statements will be true?
(a) The spot will appear brighter than the plate
(b) The spot will appear darker than the plate
(c) The spot and plate will appear equally bright
(d) The spot and the plate will not be visible in the dark room

22. On observing light from three different stars P, Q and R, it was found that intensity of violet colour is maximum in the spectrum of P, the intensity of green colour is maximum in the spectrum of R and the intensity of red colour is maximum in the spectrum of Q. If T_P, T_Q and T_R are the respective absolute temperature of P, Q and R, then it can be concluded from the above observations that
(a) $T_P > T_R > T_Q$ (b) $T_P < T_R < T_Q$
(c) $T_P < T_Q < T_R$ (d) $T_P > T_Q > T_R$

23. A partition wall has two layers of different materials A and B in contact with each other. They have the same thickness but the thermal conductivity of layer A is twice that of layer B. At steady state the temperature difference across the layer B is 50 K, then the corresponding difference across the layer A is
(a) 50 K (b) 12.5 K (c) 25 K (d) 60 K

24. Which of the following statements is/are false about mode of heat transfer?
(a) In radiation, heat is transfered from one medium to another without affecting the intervening medium
(b) Radiation and convection are possible in vaccum while conduction requires material medium.
(c) Conduction is possible in solids while convection occurs in liquids and gases.
(d) All are correct

25. In a vertical U-tube containing a liquid, the two arms are maintained at different temperatures t_1 and t_2. The liquid columns in the two arms have heights l_1 and l_2 respectively. The coefficient of volume expansion of the liquid is equal to

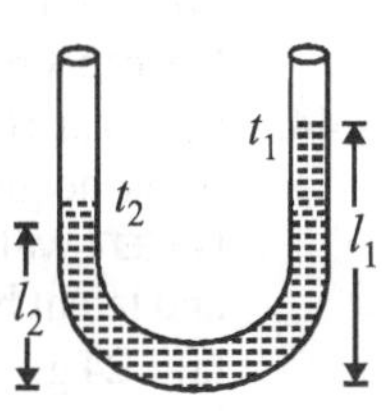

(a) $\frac{l_1 - l_2}{l_2 t_1 - l_1 t_2}$ (b) $\frac{l_1 - l_2}{l_1 t_1 - l_2 t_2}$ (c) $\frac{l_1 + l_2}{l_2 t_1 + l_1 t_2}$ (d) $\frac{l_1 + l_2}{l_1 t_1 + l_2 t_2}$

26. The top of an insulated cylindrical container is covered by a disc having emissivity 0.6 and conductivity 0.167 $WK^{-1}m^{-1}$ and thickness 1 cm. The temperature is maintained by circulating oil as shown in figure. Find the radiation loss to the surrounding in $Jm^{-2}s^{-1}$ if temperature of the upper surface of the disc is 27°C and temperature of the surrounding is 27°C.

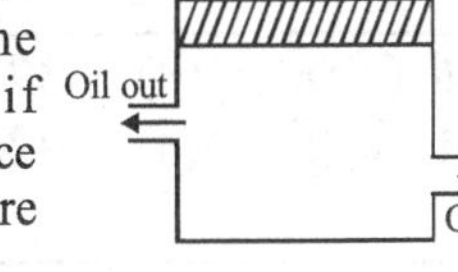

(a) 595 $Jm^{-2}s^{-1}$ (b) 545 $Jm^{-2}s^{-1}$
(c) 495 $Jm^{-2}s^{-1}$ (d) None of these

27. Wien's law is concerned with
(a) relation between emissivity and absorptivity of a radiating surface
(b) total radiation, emitted by a hot surface
(c) an expression for spectral distribution of energy of a radiation from any source
(d) a relation between the temperature of a black body and the wavelength at which there is maximum radiant energy per unit wavelength

28. If a piece of metal is heated to temperature θ and then allowed to cool in a room which is at temperature θ_0, the graph between the temperature T of the metal and time t will be closest to

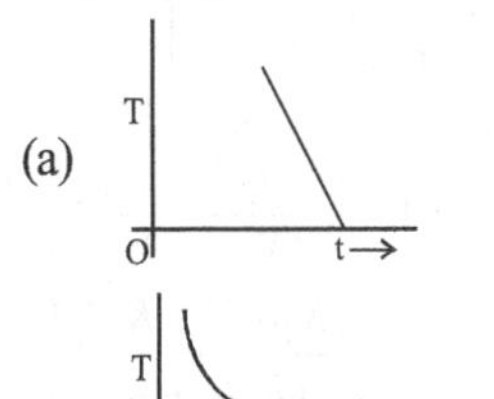
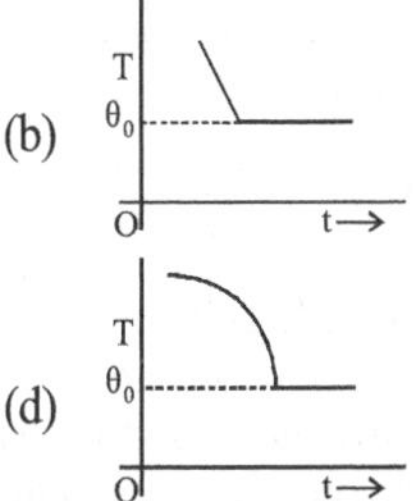

29. Two rods of same length and transfer a given amount of heat 12 second, when they are joined as shown in figure (i). But when they are joined as shwon in figure (ii), then they will transfer same heat in same conditions in

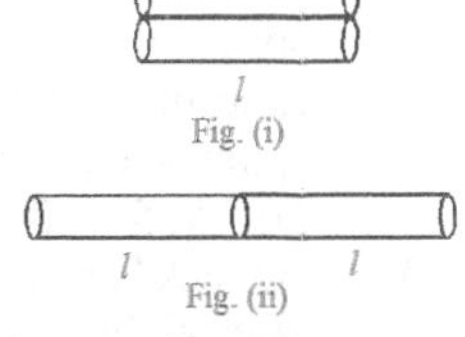

(a) 24 s (b) 13 s (c) 15 s (d) 48 s

30. Consider a compound slab consisting of two different materials having equal thicknesses and thermal conductivities *K* and 2*K*, respectively. The equivalent thermal conductivity of the slab is
(a) $\frac{4}{3}K$ (b) $\frac{2}{3}K$ (c) $\sqrt{3}\,K$ (d) $3K$

31. The coefficient of thermal conductivity of copper, mercury and glass are respectively K_c, K_m and K_g such that $K_c > K_m > K_g$. If the same quantity of heat is to flow per sec per unit area of each and corresponding temperature gradients are X_c, X_m and X_g then
(a) $X_c = X_m = X_g$ (b) $X_c > X_m > X_g$
(c) $X_c < X_m < X_g$ (d) $X_m < X_c < X_g$

32. The radiation energy density per unit wavelength at a temperature T has a maximum at a wavelength λ_0. At temperature 2T, it will have a maximum wavelength
(a) $4\lambda_0$ (b) $2\lambda_0$ (c) $\frac{\lambda_0}{2}$ (d) $\frac{\lambda_0}{4}$

33. Assuming the Sun to be a spherical body of radius *R* at a temperature of *TK*, evaluate the total radiant powerd incident of Earth at a distance *r* from the Sun
(a) $4\pi r_0^2 R^2 \sigma \frac{T^4}{r^2}$ (b) $\pi r_0^2 R^2 \sigma \frac{T^4}{r^2}$
(c) $r_0^2 R^2 \sigma \frac{T^4}{4\pi r^2}$ (d) $R^2 \sigma \frac{T^4}{r^2}$

RESPONSE GRID					
	21. ⓐⓑⓒⓓ	22. ⓐⓑⓒⓓ	23. ⓐⓑⓒⓓ	24. ⓐⓑⓒⓓ	25. ⓐⓑⓒⓓ
	26. ⓐⓑⓒⓓ	27. ⓐⓑⓒⓓ	28. ⓐⓑⓒⓓ	29. ⓐⓑⓒⓓ	30. ⓐⓑⓒⓓ
	31. ⓐⓑⓒⓓ	32. ⓐⓑⓒⓓ	33. ⓐⓑⓒⓓ		

Space for Rough Work

34. A metal ball immersed in alcohol weighs W_1 at 0ºC and W_2 at 59ºC. The coefficient of cubical expansion of the metal is less than that of alcohol. Assuming that the density of the metal is large compared to that of alcohol, it can be shown that

(a) $W_1 > W_2$ (b) $W_1 = W_2$
(c) $W_1 < W_2$ (d) $W_1 = (W_2/2)$

35. One end of a thermally insulated rod is kept at a temperature T_1 and the other at T_2. The rod is composed of two sections of length l_1 and l_2 and thermal conductivities K_1 and K_2 respectively. The temperature at the interface of the two sections is

(a) $\frac{(K_1l_1T_1 + K_2l_2T_2)}{(K_1l_1 + K_2l_2)}$ (b) $\frac{(K_2l_2T_1 + K_1l_1T_2)}{(K_1l_1 + K_2l_2)}$

(c) $\frac{(K_2l_1T_1 + K_1l_2T_2)}{(K_2l_1 + K_1l_2)}$ (d) $\frac{(K_1l_2T_1 + K_2l_1T_2)}{(K_1l_2 + K_2l_1)}$

36. Two spheres of different materials one with double the radius and one-fourth wall thickness of the other are filled with ice. If the time taken for complete melting of ice in the larger sphere is 25 minute and for smaller one is 16 minute, the ratio of thermal conductivities of the materials of larger spheres to that of smaller sphere is

(a) 4 : 5 (b) 5 : 4 (c) 25 : 8 (d) 8 : 25

37. A black body has maximum wavelength λ_m at temperature 2000 K. Its corresponding wavelength at temperature 3000 K will be

(a) $\frac{3}{2}\lambda_m$ (b) $\frac{2}{3}\lambda_m$ (c) $\frac{4}{9}\lambda_m$ (d) $\frac{9}{4}\lambda_m$

38. A solid material is supplied with heat at constant rate and the temperature of the material changes as shown. From the graph, the FALSE conclusion drawn is

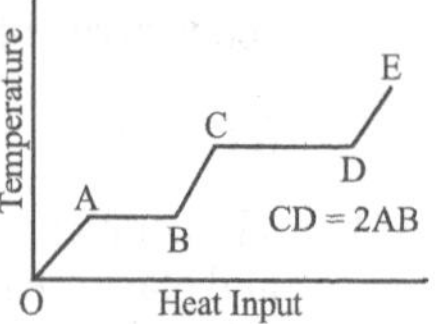

(a) AB and CD of the graph represent phase changes
(b) AB represents the change of state from solid to liquid
(c) latent heat of fusion is twice the latent heat of vaporization
(d) CD represents change of state from liquid to vapour

39. 10 gm of ice cubes at 0°C are released in a tumbler (water equivalent 55 g) at 40°C. Assuming that negligible heat is taken from the surroundings, the temperature of water in the tumbler becomes nearly (L = 80 cal/g)

(a) 31ºC (b) 22ºC (c) 19ºC (d) 15ºC

40. In a surrounding medium of temperature 10°C, a body takes 7 min for a fall of temperature from 60°C to 40°C. In what time the temperature of the body will fall from 40°C to 28°C?

(a) 7 min (b) 11 min (c) 14 min (d) 21 min

41. Two rods of same length and area of cross-section A_1 and A_2 have their ends at the same temperature. If K_1 and K_2 are their thermal conductivities, c_1 and c_2 are their specific heats and d_1 and d_2 are their densities, then the rate of flow of heat is the same in both the rods if

(a) $\frac{A_1}{A_2} = \frac{-k_1}{k_2}$ (b) $\frac{A_1}{A_2} = \frac{k_1 c_1 d_1}{k_2 c_2 d_2}$

(c) $\frac{A_1}{A_2} = \frac{k_2 c_1 d_1}{c_2 d_2 k_1}$ (d) $\frac{A_1}{A_2} = \frac{k_2}{k_1}$

42. A student takes 50gm wax (specific heat = 0.6 kcal/kg°C) and heats it till it boils. The graph between temperature and time is as follows. Heat supplied to the wax per minute and boiling point are respectively

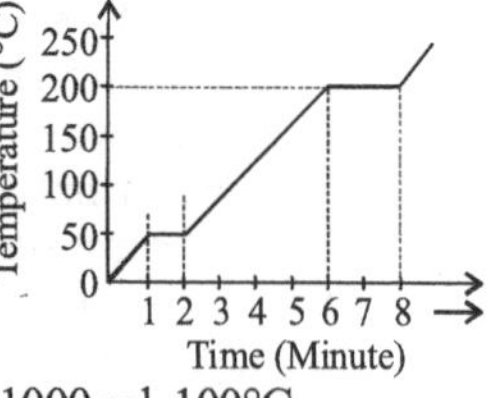

(a) 500 cal, 50°C (b) 1000 cal, 100°C
(c) 1500 cal, 200°C (d) 1000 cal, 200°C

43. Consider two identical iron spheres, one which lie on a thermally insulating plate, while the other hangs from an insulatory thread. Equal amount of heat is supplied to the two spheres, then

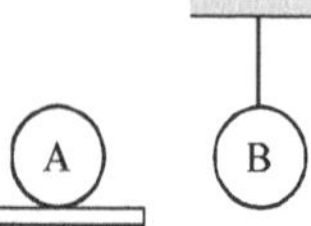

(a) temperature of A will be greater than B
(b) temperature of B will be greater than A
(c) their temperature will be equal
(d) can't be predicted

44. Steam at 100°C is passed into 20 g of water at 10°C. When water acquires a temperature of 80°C, the mass of water present will be: [Take specific heat of water = 1 cal g^{-1} $°C^{-1}$ and latent heat of steam = 540 cal g^{-1}]

(a) 24 g (b) 31.5 g (c) 42.5 g (d) 22.5 g

45. Two solid spheres, of radii R_1 and R_2 are made of the same material and have similar surfaces. The spheres are raised to the same temperature and then allowed to cool under identical conditions. Assuming spheres to be perfect conductors of heat, their initial rates of loss of heat are

(a) R_1^2 / R_2^2 (b) R_1 / R_2 (c) R_2 / R_1 (d) R_2^2 / R_1^2

RESPONSE GRID					
	34. ⓐⓑⓒⓓ	35. ⓐⓑⓒⓓ	36. ⓐⓑⓒⓓ	37. ⓐⓑⓒⓓ	38. ⓐⓑⓒⓓ
	39. ⓐⓑⓒⓓ	40. ⓐⓑⓒⓓ	41. ⓐⓑⓒⓓ	42. ⓐⓑⓒⓓ	43. ⓐⓑⓒⓓ
	44. ⓐⓑⓒⓓ	45. ⓐⓑⓒⓓ			

DAILY PRACTICE PROBLEM DPP CHAPTERWISE CP10 - PHYSICS			
Total Questions	45	Total Marks	180
Attempted		Correct	
Incorrect		Net Score	
Cut-off Score	45	Qualifying Score	60
Success Gap = Net Score – Qualifying Score			
Net Score = (Correct × 4) – (Incorrect × 1)			

Space for Rough Work

DPP - Daily Practice Problems

Chapter-wise Sheets

Date : ______ **Start Time :** ______ **End Time :** ______

PHYSICS

SYLLABUS : Thermodynamics

Max. Marks : 180 **Marking Scheme :** (+4) for correct & (–1) for incorrect answer **Time : 60 min.**

INSTRUCTIONS : This Daily Practice Problem Sheet contains 45 MCQs. For each question only one option is correct. Darken the correct circle/ bubble in the Response Grid provided on each page.

1. The relation between U, P and V for an ideal gas in an adiabatic process is given by relation $U = a + bP\,V$. Find the value of adiabatic exponent (γ) of this gas

   (a) $\frac{b+1}{b}$ (b) $\frac{b+1}{a}$ (c) $\frac{a+1}{b}$ (d) $\frac{a}{a+b}$

2. Carbon monoxide is carried around a closed cycle abc in which bc is an isothermal process as shown in the figure. The gas absorbs 7000 J of heat as its temperture increases from 300 K to 1000 K in going from a to b. The quantity of heat rejected by the gas during the process ca is

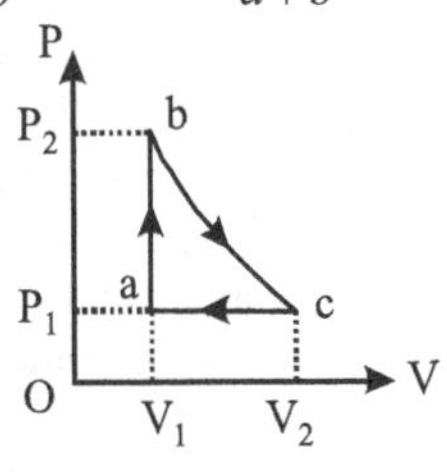

   (a) 4200 J (b) 5000 J (c) 9000 J (d) 9800 J

3. A Carnot engine, having an efficiency of $\eta = 1/10$ as heat engine, is used as a refrigerator. If the work done on the system is 10 J, the amount of energy absorbed from the reservoir at lower temperature is

   (a) 100 J (b) 99 J (c) 90 J (d) 1 J

4. In a thermodynamic process, fixed mass of a gas is changed in such a manner that the gas release 20 J of heat and 8 J of work was done on the gas. If the initial internal energy of the gas was 30 J. Then the final internal energy will be

   (a) 2 joule (b) 18 joule (c) 42 joule (d) 58 joule

5. A closed gas cylinder is divided into two parts by a piston held tight. The pressure and volume of gas in two parts respectively are (P, 5V) and (10P, V). If now the piston is left free and the system undergoes isothermal process, then the volumes of the gas in two parts respectively are

   (a) 2V,4V (b) 3V,3V (c) 5V,V (d) $\frac{10}{11}V, \frac{20}{11}V$

6. The efficiency of an ideal gas with adiabatic exponent 'γ' for the shown cyclic process would be

   (a) $\frac{(2\ln 2 - 1)}{\gamma/(\gamma-1)}$

   (b) $\frac{(1 - 2\ln 2)}{\gamma/(\gamma-1)}$

   (c) $\frac{(2\ln 2 + 1)}{\gamma/(\gamma-1)}$

   (d) $\frac{(2\ln 2 - 1)}{\gamma/(\gamma+1)}$

7. A mass of diatomic gas ($\gamma = 1.4$) at a pressure of 2 atmospheres is compressed adiabatically so that its temperature rises from 27°C to 927°C. The pressure of the gas in final state is

   (a) 28 atm (b) 68.7 atm (c) 256 atm (d) 8 atm

RESPONSE GRID					
	1. (a)(b)(c)(d)	2. (a)(b)(c)(d)	3. (a)(b)(c)(d)	4. (a)(b)(c)(d)	5. (a)(b)(c)(d)
	6. (a)(b)(c)(d)	7. (a)(b)(c)(d)			

Space for Rough Work

8. A diatomic ideal gas is used in a Carnot engine as the working substance. If during the adiabatic expansion part of the cycle the volume of the gas increases from V to 32 V, the efficiency of the engine is

(a) 0.5 (b) 0.75 (c) 0.99 (d) 0.25

9. The P-V diagram of a gas system undergoing cyclic process is shown here. The work done during isobaric compression is

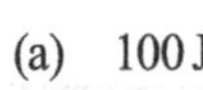

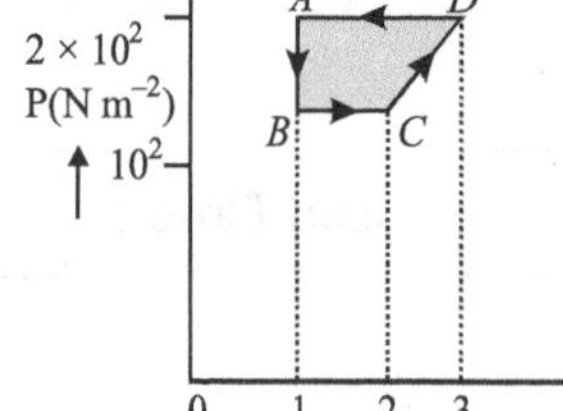

(a) 100 J

(b) 200 J

(c) 600 J

(d) 400 J

10. During an adiabatic process of an ideal gas, if P is proportional to $\frac{1}{V^{1.5}}$, then the ratio of specific heat capacities at constant pressure to that at constant volume for the gas is

(a) 1.5 (b) 0.25 (c) 0.75 (d) 0.4

11. The work of 146 kJ is performed in order to compress one kilo mole of gas adiabatically and in this process the temperature of the gas increases by 7°C. The gas is ($R = 8.3\ \text{J mol}^{-1}\text{K}^{-1}$)

(a) diatomic
(b) triatomic
(c) a mixture of monoatomic and diatomic
(d) monoatomic

12. Consider a spherical shell of radius R at temperature T. The black body radiation inside it can be considered as an ideal gas of photons with internal energy per unit volume $u = \frac{U}{V} \propto T^4$ and pressure $p = \frac{1}{3}\left(\frac{U}{V}\right)$. If the shell now undergoes an adiabatic expansion the relation between T and R is :

(a) $T \propto \frac{1}{R}$ (b) $T \propto \frac{1}{R^3}$ (c) $T \propto e^{-R}$ (d) $T \propto e^{-3R}$

13. The specific heat capacity of a metal at low temperature (T) is given as $C(\text{kJK}^{-1}\text{kg}^{-1}) = 32\left(\frac{T}{400}\right)^3$. A 100 g vessel of this metal is to be cooled from 20 K to 4 K by a special refrigerator operating at room temperature (27°C). The amount of work required to cool in vessel is

(a) equal to 0.002 kJ
(b) greater than 0.148 kJ
(c) between 0.148 kJ and 0.028 kJ
(d) less than 0.028 kJ

14. 5.6 litre of helium gas at STP is adiabatically compressed to 0.7 litre. Taking the initial temperature to be T_1, the work done in the process is

(a) $\frac{9}{8}RT_1$ (b) $\frac{3}{2}RT_1$ (c) $\frac{15}{8}RT_1$ (d) $\frac{9}{2}RT_1$

15. Four curves A, B, C and D are drawn in the figure for a given amount of a gas. The curves which represent adiabatic and isothermal changes are

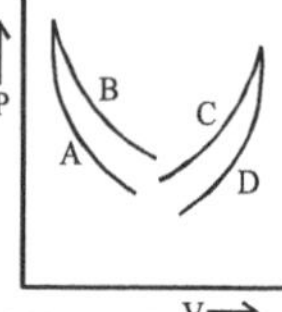

(a) C and D respectively
(b) D and C respectively
(c) A and B respectively
(d) B and A respectively

16. In an adiabatic process, the pressure is increased by $\frac{2}{3}\%$. If $\gamma = \frac{3}{2}$, then the volume decreases by nearly

(a) $\frac{4}{9}\%$ (b) $\frac{2}{3}\%$ (c) 1% (d) $\frac{9}{4}\%$

17. A reversible engine converts one-sixth of the heat input into work. When the temperature of the sink is reduced by 62°C, the efficiency of the engine is doubled. The temperatures of the source and sink are

(a) 99°C, 37°C (b) 80°C, 37°C
(c) 95°C, 37°C (d) 90°C, 37°C

18. A diatomic ideal gas is compressed adiabatically to $\frac{1}{32}$ of its initial volume. If the initial temperature of the gas is T_i (in Kelvin) and the final temperature is aT_i, the value of a is

(a) 8 (b) 4 (c) 3 (d) 5

19. When the state of a gas adiabatically changed from an equilibrium state A to another equilibrium state B an amount of work done on the stystem is 35 J. If the gas is taken from state A to B via process in which the net heat absorbed by the system is 12 cal, then the net work done by the system is (1 cal = 4.19 J)

(a) 13.2 J (b) 15.4 J (c) 12.6 J (d) 16.8 J

20. Calculate the work done when 1 mole of a perfect gas is compressed adiabatically. The initial pressure and volume of the gas are 10^5 N/m² and 6 litre respectively. The final volume of the gas is 2 litres. Molar specific heat of the gas at constant volume is 3R/2. [Given $(3)^{5/3} = 6.19$]

(a) –957 J (b) +957 J (c) – 805 J (d) + 805 J

21. A Carnot engine whose efficiency is 40%, receives heat at 500K. If the efficiency is to be 50%, the source temperature for the same exhaust temperature is

(a) 900 K (b) 600 K (c) 700 K (d) 800 K

22. 1 gm of water at a pressure of 1.01×10^5 Pa is converted into steam without any change of temperature. The volume of 1 g of steam is 1671 cc and the latent heat of evaporation is 540 cal. The change in internal energy due to evaporation of 1 gm of water is

(a) ≈ 167 cal (b) 500 cal (c) 540 cal (d) 581 cal

RESPONSE GRID					
	8. ⓐⓑⓒⓓ	9. ⓐⓑⓒⓓ	10. ⓐⓑⓒⓓ	11. ⓐⓑⓒⓓ	12. ⓐⓑⓒⓓ
	13. ⓐⓑⓒⓓ	14. ⓐⓑⓒⓓ	15. ⓐⓑⓒⓓ	16. ⓐⓑⓒⓓ	17. ⓐⓑⓒⓓ
	18. ⓐⓑⓒⓓ	19. ⓐⓑⓒⓓ	20. ⓐⓑⓒⓓ	21. ⓐⓑⓒⓓ	22. ⓐⓑⓒⓓ

Space for Rough Work

23. One mole of an ideal gas at temperature T was cooled isochorically till the gas pressure fell from P to $\frac{P}{n}$. Then, by an isobaric process, the gas was restored to the initial temperature. The net amount of heat absorbed by the gas in the process is

(a) nRT (b) $\frac{RT}{n}$

(c) $RT(1-n^{-1})$ (d) $RT(n-1)$

24. A Carnot engine, having an efficiency of $\eta = \frac{1}{10}$ as heat engine, is used as a refrigerator. If the work done on the system is 10 J, the amount of energy absorbed from the reservoir at lower temperature is

(a) 99 J (b) 90 J (c) 1 J (d) 100 J

25. The volume of an ideal gas is 1 litre and its pressure is equal to 72 cm of mercury column. The volume of gas is made 900 cm^3 by compressing it isothermally. The stress of the gas will be

(a) 8 cm of Hg (b) 7 cm of Hg

(c) 6 cm of Hg (d) 4 cm of Hg

26. An ideal gas is taken through the cycle $A \to B \to C \to A$, as shown in figure. If the net heat supplied to the gas in the cycle is 5 J, the work done by the gas in the process $C \to A$ is

(a) – 5 J

(b) – 10 J

(c) – 15 J

(d) – 20 J

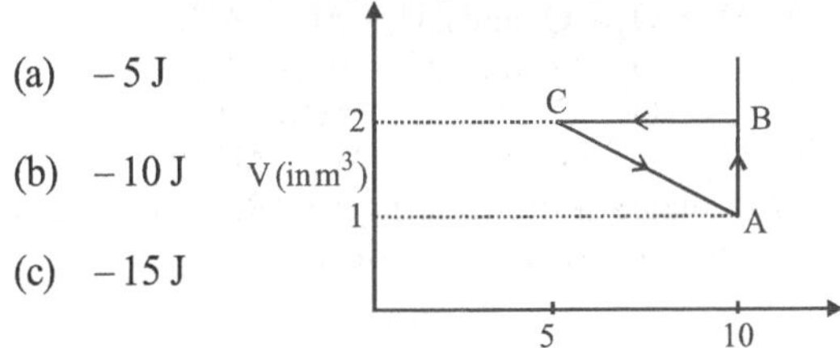

27. An ideal gas undergoing adiabatic change has the following pressure-temperature relationship

(a) $P^{\gamma-1}T^{\gamma}$ = constant (b) $P^{\gamma}T^{\gamma-1}$ = constant

(c) $P^{\gamma}T^{1-\gamma}$ = constant (d) $P^{1-\gamma}T^{\gamma}$ = constant

28. In a thermodynamic process, fixed mass of a gas is changed in such a manner that the gas release 20 J of heat and 8 J of work was done on the gas. If the initial internal energy of the gas was 30 J, the final internal energy will be

(a) 2 joule (b) 18 joule (c) 42 joule (d) 58 joule

29. The coefficient of performance of a refrigerator is 5. If the inside temperature of freezer is –20°C, then the temperature of the surroundings to which it rejects heat is

(a) 41°C (b) 11°C (c) 21°C (d) 31°C

30. Monatomic, diatomic and polyatomic ideal gases each undergo slow adiabatic expansions from the same initial volume and same initial pressure to the same final volume. The magnitude of the work done by the environment on the gas is

(a) the greatest for the polyatomic gas

(b) the greatest for the monatomic gas

(c) the greatest for the diatomic gas

(d) the question is irrelevant, there is no meaning of slow adiabatic expansion

31. The given p-v diagram represents the thermodynamic cycle of an engine, operating with an ideal monatomic gas. The amount of heat, extracted from the source in a single cycle is

(a) p_0v_0

(b) $\left(\frac{13}{2}\right)p_0v_0$

(c) $\left(\frac{11}{2}\right)p_0v_0$

(d) $4p_0v_0$

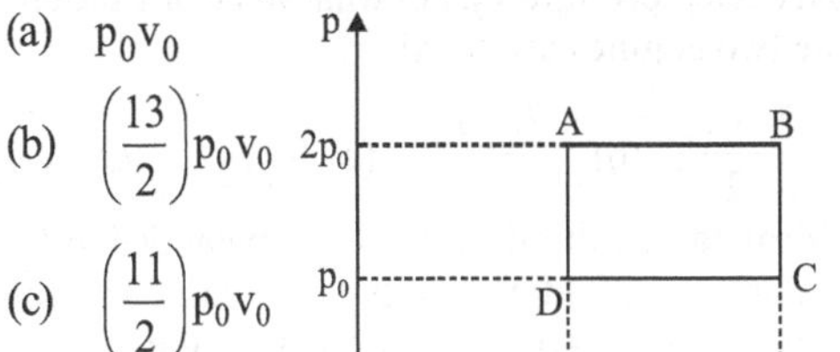

32. For an ideal gas graph is shown for three processes. Process 1, 2 and 3 are respectively.

(a) Isobaric, adiabatic, isochoric

(b) Adiabatic, isobaric, isochoric

(c) Isochoric, adiabatic, isobaric

(d) Isochoric, isobaric, adiabatic

33. During an adiabatic process an object does 100J of work and its temperature decreases by 5K. During another process it does 25J of work and its temperature decreases by 5K. Its heat capacity for 2nd process is

(a) 20 J/K (b) 24 J/K (c) 15 J/K (d) 100 J/K

34. A refrigerator works between 4°C and 30°C. It is required to remove 600 calories of heat every second in order to keep the temperature of the refrigerated space constant. The power required is: (Take 1 cal = 4.2 joule)

(a) 2.365 W (b) 23.65 W (c) 236.5 W (d) 2365 W

35. A perfect gas goes from a state A to another state B by absorbing 8×10^5 J of heat and doing 6.5×10^5 J of external work. It is now transferred between the same two states in another process in which it absorbs 10^5 J of heat. In the second process

(a) work done by gas is 10^5 J

(b) work done on gas is 10^5 J

(c) work done by gas is 0.5×10^5 J

(d) work done on the gas is 0.5×10^5 J

36. One mole of a diatomic ideal gas undergoes a cyclic process ABC as shown in figure. The process BC is adiabatic. The temperatures at A, B and C are 400 K, 800 K and 600 K respectively. Choose the correct statement:

(a) The change in internal energy in whole cyclic process is 250 R.

(b) The change in internal energy in the process CA is 700 R.

(c) The change in internal energy in the process AB is -350 R.

(d) The change in internal energy in the process BC is – 500 R.

RESPONSE GRID					
	23. ⓐⓑⓒⓓ	24. ⓐⓑⓒⓓ	25. ⓐⓑⓒⓓ	26. ⓐⓑⓒⓓ	27. ⓐⓑⓒⓓ
	28. ⓐⓑⓒⓓ	29. ⓐⓑⓒⓓ	30. ⓐⓑⓒⓓ	31. ⓐⓑⓒⓓ	32. ⓐⓑⓒⓓ
	33. ⓐⓑⓒⓓ	34. ⓐⓑⓒⓓ	35. ⓐⓑⓒⓓ	36. ⓐⓑⓒⓓ	

Space for Rough Work

37. Two Carnot engines A and B are operated in series. The engine A receives heat from the source at temperature T_1 and rejects the heat to the sink at temperature T. The second engine B receives the heat at temperature T and rejects to its sink at temperature T_2. For what value of T the efficiencies of the two engines are equal?

(a) $\frac{T_1+T_2}{2}$ (b) $\frac{T_1-T_2}{2}$ (c) T_1T_2 (d) $\sqrt{T_1T_2}$

38. An ideal gas is initially at P_1, V_1 is expanded to P_2, V_2 and then compressed adiabatically to the same volume V_1 and pressure P_3. If W is the net work done by the gas in complete process which of the following is true ?

(a) $W>0; P_3>P_1$ (b) $W<0; P_3>P_1$
(c) $W>0; P_3<P_1$ (d) $W<0; P_3<P_1$

39. Which of the following statements is correct for any thermodynamic system ?

(a) The change in entropy can never be zero
(b) Internal energy and entropy are state functions
(c) The internal energy changes in all processes
(d) The work done in an adiabatic process is always zero.

40. One mole of an ideal gas goes from an initial state A to final state B via two processes : It first undergoes isothermal expansion from volume V to 3V and then its volume is reduced from 3V to V at constant pressure. The correct P-V diagram representing the two processes is :

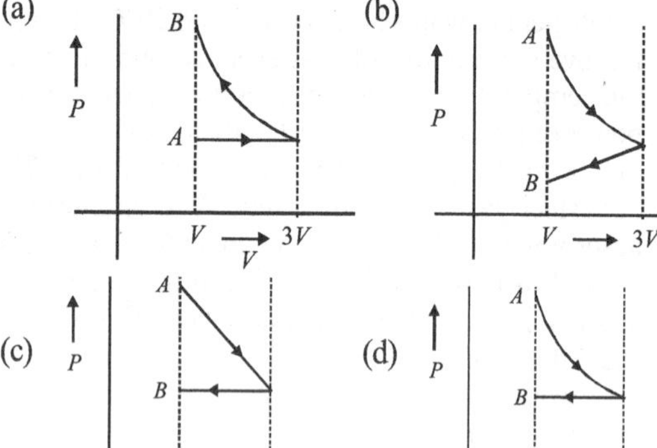

41. What will be the final pressure if an ideal gas in a cylinder is compressed adiabatically to $\frac{1}{3}$rd of its volume?

(a) Final pressure will be three times less than initial pressure.
(b) Final pressure will be three times more than initial pressure.
(c) Change in pressure will be more than three times the initial pressure.
(d) Change in pressure will be less than three times the initial pressure.

42. A gas is compressed isothermally to half its initial volume. The same gas is compressed separately through an adiabatic process until its volume is again reduced to half. Then :

(a) Compressing the gas isothermally will require more work to be done.
(b) Compressing the gas through adiabatic process will require more work to be done.
(c) Compressing the gas isothermally or adiabatically will require the same amount of work.
(d) Which of the case (whether compression through isothermal or through adiabatic process) requires more work will depend upon the atomicity of the gas.

43. An ideal gas goes from state A to state B via three different processes as indicated in the P-V diagram : If Q_1, Q_2, Q_3 indicate the heat a absorbed by the gas along the three processes and ΔU_1, ΔU_2, ΔU_3 indicate the change in internal energy along the three processes respectively, then

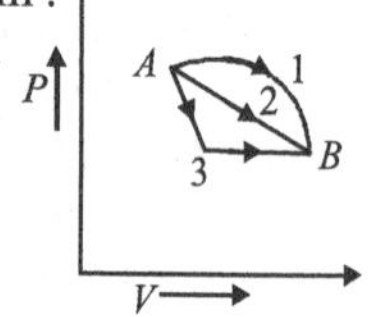

(a) $Q_1 > Q_2 > Q_3$ and $\Delta U_1 = \Delta U_2 = \Delta U_3$
(b) $Q_3 > Q_2 > Q_1$ and $\Delta U_1 = \Delta U_2 = \Delta U_3$
(c) $Q_1 = Q_2 = Q_3$ and $\Delta U_1 > \Delta U_2 > \Delta U_3$
(d) $Q_3 > Q_2 > Q_1$ and $\Delta U_1 > \Delta U_2 > \Delta U_3$

44. In *P-V* diagram shown in figure *ABC* is a semicircle. The work done in the process *ABC* is

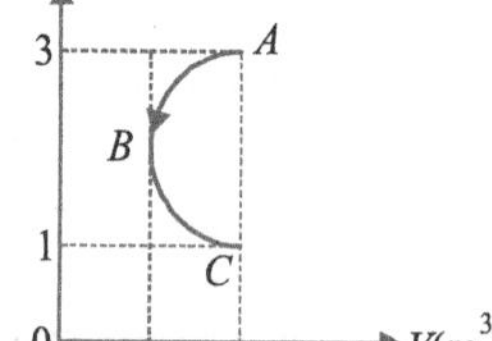

(a) 4 J
(b) $\frac{-\pi}{2}$ J
(c) $\frac{\pi}{2}$ J
(d) zero

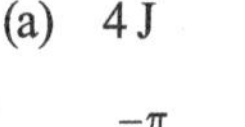

45. For an isothermal expansion of a perfect gas, the value of $\frac{\Delta P}{P}$ is equal to

(a) $-\gamma^{1/2}\frac{\Delta V}{V}$ (b) $-\frac{\Delta V}{V}$ (c) $-\gamma\frac{\Delta V}{V}$ (d) $-\gamma^2\frac{\Delta V}{V}$

RESPONSE GRID	37. ⓐⓑⓒⓓ	38. ⓐⓑⓒⓓ	39. ⓐⓑⓒⓓ	40. ⓐⓑⓒⓓ	41. ⓐⓑⓒⓓ
	42. ⓐⓑⓒⓓ	43. ⓐⓑⓒⓓ	44. ⓐⓑⓒⓓ	45. ⓐⓑⓒⓓ	

DAILY PRACTICE PROBLEM DPP CHAPTERWISE CP11 - PHYSICS			
Total Questions	45	Total Marks	180
Attempted		Correct	
Incorrect		Net Score	
Cut-off Score	45	Qualifying Score	60
Success Gap = Net Score – Qualifying Score			
Net Score = (Correct × 4) – (Incorrect × 1)			

Space for Rough Work

DPP - Daily Practice Problems

Chapter-wise Sheets

Date : | **Start Time :** | **End Time :**

PHYSICS

CP12

SYLLABUS : Kinetic Theory

Max. Marks : 180 **Marking Scheme :** (+4) for correct & (–1) for incorrect answer **Time : 60 min.**

INSTRUCTIONS : This Daily Practice Problem Sheet contains 45 MCQs. For each question only one option is correct. Darken the correct circle/ bubble in the Response Grid provided on each page.

1. 4.0 g of a gas occupies 22.4 litres at NTP. The specific heat capacity of the gas at constant volume is $5.0 JK^{-1}$. If the speed of any quantity x in this gas at NTP is $952 ms^{-1}$, then the heat capacity at constant pressure is (Take gas constant $R = 8.3 JK^{-1} mol^{-1}$)
   (a) $7.5 JK^{-1} mol^{-1}$ (b) $7.0 JK^{-1} mol^{-1}$
   (c) $8.5 JK^{-1} mol^{-1}$ (d) $8.0 JK^{-1} mol^{-1}$

2. A fixed mass of gas at constant pressure occupies a volume V. The gas undergoes a rise in temperature so that the root mean square velocity of its molecules is doubled. The new volume will be
   (a) V/2 (b) $V/\sqrt{2}$ (c) 2V (d) 4V

3. A gaseous mixture consists of 16 g of helium and 16 g of oxygen. The ratio $\frac{C_p}{C_v}$ of the mixture is
   (a) 1.62 (b) 1.59 (c) 1.54 (d) 1.4

4. Air is pumped into an automobile tube upto a pressure of 200 kPa in the morning when the air temperature is 22°C. During the day, temperature rises to 42°C and the tube expands by 2%. The pressure of the air in the tube at this temperature, will be approximately
   (a) 212 kPa (b) 209 kPa (c) 206 kPa (d) 200 kPa

5. The rms speed of the particles of fume of mass 5×10^{-17} kg executing Brownian motion in air at N.T.P. is ($k = 1.38 \times 10^{-23}$ J/K)
   (a) 1.5 m/s (b) 3.0 m/s (c) 1.5 cm/s (d) 3 cm/s

6. One mole of an ideal monoatomic gas requires 207 J heat to raise the temperature by 10 K when heated at constant pressure. If the same gas is heated at constant volume to raise the temperature by the same 10 K, the heat required is [Given the gas constant R = 8.3 J/ mol. K]
   (a) 198.7 J (b) 29 J (c) 215.3 J (d) 124 J

RESPONSE GRID	
	1. ⓐⓑⓒⓓ 2. ⓐⓑⓒⓓ 3. ⓐⓑⓒⓓ 4. ⓐⓑⓒⓓ 5. ⓐⓑⓒⓓ
	6. ⓐⓑⓒⓓ

Space for Rough Work

7. Figure shows the variation in temperature (ΔT) with the amount of heat supplied (Q) in an isobaric process corresponding to a monoatomic (M), diatomic (D) and a polyatomic (P) gas. The initial state of all the gases are the same and the scales for the two axes coincide. Ignoring vibrational degrees of freedom, the lines *a*, *b* and *c* respectively correspond to

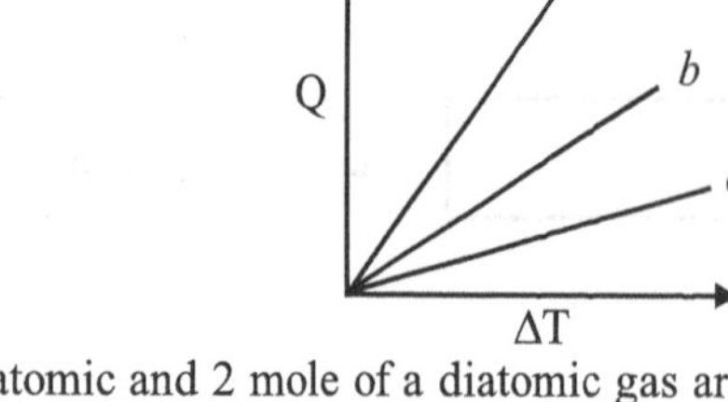

(a) P, M and D
(b) M, D and P
(c) P, D and M
(d) D, M and P

8. 1 mole of a monatomic and 2 mole of a diatomic gas are mixed. The resulting gas is taken through a process in which molar heat capacity was found 3R. Polytropic constant in the process is
(a) –1/5 (b) 1/5 (c) 2/5 (d) –2/5

9. The density of a gas is 6×10^{-2} kg/m^3 and the root mean square velocity of the gas molecules is 500 m/s. The pressure exerted by the gas on the walls of the vessel is
(a) 5×10^3 N/m^2 (b) 1.2×10^{-4} N/m^2
(c) 0.83×10^{-4} N/m^2 (d) 30 N/m^2

10. The absolute temperature of a gas is increases 3 times. The root mean square velocity of the moelcules will be
(a) 3 times (b) 9 times
(c) 1/3 times (d) $\sqrt{3}$ times

11 Consider an ideal gas confined in an isolated closed chamber. As the gas undergoes an adiabatic expansion, the average time of collision between molecules increases as V^q, where V is the volume of the gas. The value of q is : $\left(\gamma = \frac{C_p}{C_v}\right)$
(a) $\frac{\gamma+1}{2}$ (b) $\frac{\gamma-1}{2}$ (c) $\frac{3\gamma+5}{6}$ (d) $\frac{3\gamma-5}{6}$

12. One kg of a diatomic gas is at a pressure of 8×10^4 N/m^2. The density of the gas is 4kg/m^3. What is the energy of the gas due to its thermal motion?
(a) 5×10^4 J (b) 6×10^4 J
(c) 7×10^4 J (d) 3×10^4 J

13. A thermally insulated vessel contains an ideal gas of molecular mass *M* and ratio of specific heats γ. It is moving with speed *v* and it suddenly brought to rest. Assuming no heat is lost to the surroundings, its temperature increases by
(a) $\frac{(\gamma-1)}{2\gamma R} Mv^2 K$ (b) $\frac{\gamma M v^2}{2R} K$
(c) $\frac{(\gamma-1)}{2R} Mv^2 K$ (d) $\frac{(\gamma-1)}{2(\gamma+1)R} Mv^2 K$

14. Figure shows a parabolic graph between T and 1/V for a mixture of a gases undergoing an adiabatic process. What is the ratio of V_{rms} of molecules and speed of sound in mixture?

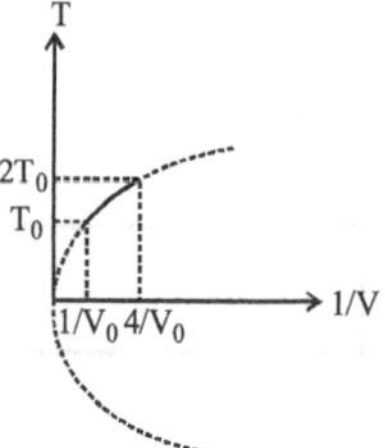

(a) $\sqrt{3/2}$
(b) $\sqrt{2}$
(c) $\sqrt{2/3}$
(d) $\sqrt{3}$

15. The work of 146 kJ is performed in order to compress one kilomole of gas adiabatically and in this process the temperature of the gas increases by 7°C. The gas is ($R = 8.3$ J mol^{-1} K^{-1})
(a) diatomic
(b) triatomic
(c) a mixture of monatomic and diatomic
(d) monatomic

16. At what temperature is root mean square velocity of gaseous hydrogen molecules equal to that of oxygen molecules at 47°C?
(a) 40K (b) 80K (c) –73K (d) 3K

17. The kinetic theory of gases states that the average squared velocity of molecules varies linearly with the mean molecular weight of the gas. If the root mean square (rms) velocity of oxygen molecules at a certain temperature is 0.5 km/sec. The rms velocity for hydrogen molecules at the same temperature will be :
(a) 2 km/sec (b) 4 km/sec (c) 8 km/sec (d) 16 km/sec

18. If 2 moles of an ideal monatomic gas at temperature T_0 is mixed with 4 moles of another ideal monatomic gas at temperature $2T_0$, then the temperature of the mixture is
(a) $\frac{5}{3}T_0$ (b) $\frac{3}{2}T_0$ (c) $\frac{4}{3}T_0$ (d) $\frac{5}{4}T_0$

19. From the following statements, concerning ideal gas at any given temperature *T*, select the incorrect one(s)
(a) The coefficient of volume expansion at constant pressure is same for all ideal gas
(b) The average translational kinetic energy per molecule of oxygen gas is 3 *KT* (*K* being Boltzmann constant)
(c) In a gaseous mixture, the average translational kinetic energy of the molecules of each component is same
(d) The mean free path of molecules increases with decrease in pressure

RESPONSE GRID					
	7. ⓐⓑⓒⓓ	8. ⓐⓑⓒⓓ	9. ⓐⓑⓒⓓ	10. ⓐⓑⓒⓓ	11. ⓐⓑⓒⓓ
	12. ⓐⓑⓒⓓ	13. ⓐⓑⓒⓓ	14. ⓐⓑⓒⓓ	15. ⓐⓑⓒⓓ	16. ⓐⓑⓒⓓ
	17. ⓐⓑⓒⓓ	18. ⓐⓑⓒⓓ	19. ⓐⓑⓒⓓ		

Space for Rough Work

20. The adjoining figure shows graph of pressure and volume of a gas at two tempertures T_1 and T_2. Which of the following inferences is correct?

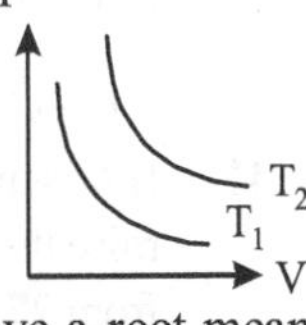

(a) $T_1 > T_2$
(b) $T_1 = T_2$
(c) $T_1 < T_2$
(d) None of these

21. The molecules of a given mass of gas have a root mean square velocity of 200m s^{-1} at 27°C and 1.0×10^5 N m^{-2} pressure. When the temperature is 127°C and the pressure $0.5 \times 10^5 Nm^{-2}$, the root mean square velocity in ms^{-1}, is

(a) $\frac{400}{\sqrt{3}}$ (b) $100\sqrt{2}$ (c) $\frac{100\sqrt{2}}{3}$ (d) $\frac{100}{3}$

22. A graph is plotted with PV/T on y-axis and mass of the gas along x-axis for different gases. The graph is
(a) a straight line parallel to x-axis for all the gases
(b) a straight line passing through origin with a slope having a constant value for all the gases
(c) a straight line passing through origin with a slope having different values for different gases
(d) a straight line parallel to y-axis for all the gases

23. At identical temperatures, the rms speed of hydrogen molecules is 4 times that for oxygen molecules. In a mixture of these in mass ratio $H_2 : O_2 = 1:8$, the rms speed of all molecules is n times the rms speed for O_2 molecules, where n is
(a) 3 (b) 4/3 (c) $(8/3)^{1/2}$ (d) $(11)^{1/2}$

24. Work done by a system under isothermal change from a volume V_1 to V_2 for a gases which obeys Vander Waal's equation $(V-\beta n)\left(P+\frac{\alpha n^2}{V}\right)=nRT$ is

(a) $nRT\log_e\left(\frac{V_2-n\beta}{V_1-n\beta}\right)+\alpha n^2\left(\frac{V_1-V_2}{V_1V_2}\right)$

(b) $nRT\log_{10}\left(\frac{V_2-n\beta}{V_1-n\beta}\right)+\alpha n^2\left(\frac{V_1-V_2}{V_1V_2}\right)$

(c) $nRT\log_e\left(\frac{V_2-n\beta}{V_1-n\beta}\right)+\beta n^2\left(\frac{V_1-V_2}{V_1V_2}\right)$

(d) $nRT\log_e\left(\frac{V_1-n\beta}{V_2-n\beta}\right)+\alpha n^2\left(\frac{V_1V_2}{V_1-V_2}\right)$

25. Two vessels separately contain two ideal gases A and B at the same temperature. The pressure of A being twice that of B. Under such conditions, the density of A is found to be 1.5 times the density of B. The ratio of molecular weight of A and B is :

(a) $\frac{3}{4}$ (b) 2 (c) $\frac{1}{2}$ (d) $\frac{2}{3}$

26. The temperature of the mixture of one mole of helium and one mole of hydrogen is increased from 0°C to 100°C at constant pressure. The amount of heat delivered will be
(a) 600 cal (b) 1200 cal (c) 1800 cal (d) 3600 cal

27. If the intermolecular forces vanish away, the volume occupied by the molecules contained in 4.5 g water at standard temperature and pressure will be
(a) 5.6 litre (b) 4.5 litre (c) 11.2 litre (d) 6.5 litre

28. A gas mixture consists of 2 moles of oxygen and 4 moles of Argon at temperature T. Neglecting all vibrational moles, the total internal energy of the system is
(a) 4 RT (b) 15 RT (c) 9 RT (d) 11RT

29. A vessel has 6g of hydrogen at pressure P and temperature 500K. A small hole is made in it so that hydrogen leaks out. How much hydrogen leaks out if the final pressure is P/2 and temperature falls to 300 K ?
(a) 2g (b) 3g (c) 4g (d) 1g

30. For a gas if ratio of specific heats at constant pressure and volume is γ then value of degrees of freedom is

(a) $\frac{3\gamma-1}{2\gamma-1}$ (b) $\frac{2}{\gamma-1}$ (c) $\frac{9}{2}(\gamma-1)$ (d) $\frac{25}{2}(\gamma-1)$

31. The given *P-V* curve is predicted by

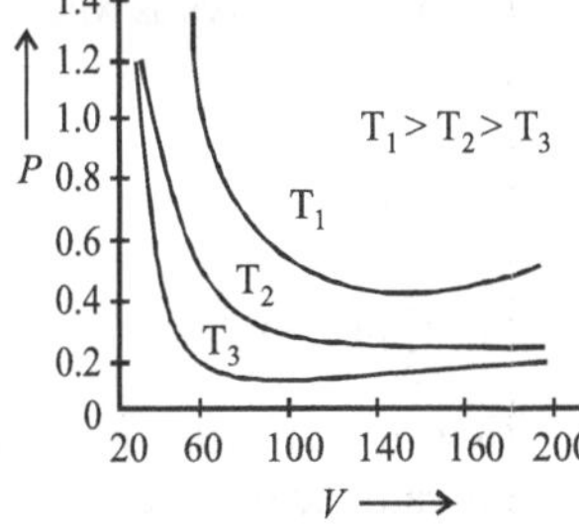

(a) Boyle's law
(b) Charle's law
(c) Avogadro's law
(d) Gaylussac's law

32. Three perfect gases at absolute temperatures T_1, T_2 and T_3 are mixed. The masses of molecules are m_1, m_2 and m_3 and the number of molecules are n_1, n_2 and n_3 respectively. Assuming no loss of energy, the final temperature of the mixture is :

(a) $\frac{n_1T_1+n_2T_2+n_3T_3}{n_1+n_2+n_3}$ (b) $\frac{n_1T_1^2+n_2T_2^2+n_3T_3^2}{n_1T_1+n_2T_2+n_3T_3}$

(c) $\frac{n_1^2T_1^2+n_2^2T_2^2+n_3^2T_3^2}{n_1T_1+n_2T_2+n_3T_3}$ (d) $\frac{(T_1+T_2+T_3)}{3}$

33. A gas is enclosed in a cube of side *l*. What will be the change in momentum of the molecule, if it suffers an elastic collision with the plane wall parallel to *yz*-plane and rebounds with the same velocity ?
[(V_x, V_y & V_z) initial velocities of the gas molecules]
(a) mv_x (b) zero (c) $-mv_x$ (d) $-2mv_x$

RESPONSE GRID					
	20. ⓐⓑⓒⓓ	21. ⓐⓑⓒⓓ	22. ⓐⓑⓒⓓ	23. ⓐⓑⓒⓓ	24. ⓐⓑⓒⓓ
	25. ⓐⓑⓒⓓ	26. ⓐⓑⓒⓓ	27. ⓐⓑⓒⓓ	28. ⓐⓑⓒⓓ	29. ⓐⓑⓒⓓ
	30. ⓐⓑⓒⓓ	31. ⓐⓑⓒⓓ	32. ⓐⓑⓒⓓ	33. ⓐⓑⓒⓓ	

Space for Rough Work

34. What will be the ratio of number of molecules of a monoatomic and a diatomic gas in a vessel, if the ratio of their partial pressures is 5 : 3 ?

(a) 5 : 1 (b) 3 : 1 (c) 5 : 3 (d) 3 : 5

35. The average transitional energy and the rms speed of molecules in a sample of oxygen gas at 300 K are 6.21×10^{-21} J and 484 m/s respectively. The corresponding values at 600 K are nearly (assuming ideal gas behaviour)

(a) 12.42×10^{-21} J, 968 m/s (b) 8.78×10^{-21} J, 684 m/s

(c) 6.21×10^{-21} J, 968 m/s (d) 12.42×10^{-21} J, 684 m/s

36. At 10° C the value of the density of a fixed mass of an ideal gas divided by its pressure is x. At 110°C this ratio is:

(a) x (b) $\frac{383}{283}x$ (c) $\frac{10}{110}x$ (d) $\frac{283}{383}x$

37. If the potential energy of a gas molecule is $U = M/r^6 - N/r^{12}$, M and N being positive constants. Then the potential energy at equilibrium must be

(a) zero (b) $M^2/4N$ (c) $N^2/4M$ (d) $MN^2/4$

38. Consider a gas with density ρ and $\bar{c}$ as the root mean square velocity of its molecules contained in a volume. If the system moves as whole with velocity v, then the pressure exerted by the gas is

(a) $\frac{1}{3}\rho\bar{c}^2$ (b) $\frac{1}{3}\rho(c+v)^2$

(c) $\frac{1}{3}\rho(\bar{c}-v)^2$ (d) $\frac{1}{3}\rho(c^{-2}-v)^2$

39. How is the mean free path (λ) in a gas related to the interatomic distance?

(a) λ is 10 times the interatomic distance

(b) λ is 100 times the interatomic distance

(c) λ is 1000 times the interatomic distance

(d) λ is $\frac{1}{10}$ times of the interatomic distance

40. Four molecules have speeds 2 km/sec, 3 km/sec, 4 km/sec and 5 km/sec. The root mean square speed of these molecules (in km/sec) is

(a) $\sqrt{54/4}$ (b) $\sqrt{54/2}$ (c) 3.5 (d) $3\sqrt{3}$

41. If R is universal gas constant, the amount of heat needed to raise the temperature of 2 moles of an ideal monoatomic gas from 273 K to 373 K, when no work is done, is

(a) $100R$ (b) $150R$ (c) $300R$ (d) $500R$

42. N molecules, each of mass m, of gas A and 2 N molecules, each of mass 2 m, of gas B are contained in the same vessel which is maintained at a temperature T. The mean square velocity of molecules of B type is denoted by V_2 and the mean square velocity of A type is denoted by V_1, then $\frac{V_1}{V_2}$ is

(a) 2 (b) 1 (c) 1/3 (d) 2/3

43. The root mean square value of the speed of the molecules in a fixed mass of an ideal gas is increased by increasing

(a) the volume while keeping the temperature constant

(b) the pressure while keeping the volume constant

(c) the temperature while keeping the pressure constant

(d) the pressure while keeping the temperature constant

44. The P-V diagram of a diatomic gas is a straight line passing through origin. The molar heat capacity of the gas in the process will be

(a) 4R (b) 2.5R (c) 3R (d) $\frac{4R}{3}$

45. For a gas, difference between two specific heats is 5000 J/mole°C. If the ratio of specific heats is 1.6, the two specific heats in J/mole-°C are

(a) $C_P = 1.33 \times 10^4, C_V = 2.66 \times 10^4$

(b) $C_P = 13.3 \times 10^4, C_V = 8.33 \times 10^3$

(c) $C_P = 1.33 \times 10^4, C_V = 8.33 \times 10^3$

(d) $C_P = 2.6 \times 10^4, C_V = 8.33 \times 10^4$

RESPONSE GRID					
	34. ⓐⓑⓒⓓ	35. ⓐⓑⓒⓓ	36. ⓐⓑⓒⓓ	37. ⓐⓑⓒⓓ	38. ⓐⓑⓒⓓ
	39. ⓐⓑⓒⓓ	40. ⓐⓑⓒⓓ	41. ⓐⓑⓒⓓ	42. ⓐⓑⓒⓓ	43. ⓐⓑⓒⓓ
	44. ⓐⓑⓒⓓ	45. ⓐⓑⓒⓓ			

DAILY PRACTICE PROBLEM DPP CHAPTERWISE CP12 - PHYSICS			
Total Questions	45	Total Marks	180
Attempted		Correct	
Incorrect		Net Score	
Cut-off Score	50	Qualifying Score	70
Success Gap = Net Score – Qualifying Score			
Net Score = (Correct × 4) – (Incorrect × 1)			

Space for Rough Work

DPP - Daily Practice Problems

Chapter-wise Sheets

Date :   **Start Time :**   **End Time :**

PHYSICS

CP13

SYLLABUS : Oscillations

Max. Marks : 180   **Marking Scheme :** (+4) for correct & (–1) for incorrect answer   **Time : 60 min.**

INSTRUCTIONS : This Daily Practice Problem Sheet contains 45 MCQs. For each question only one option is correct. Darken the correct circle/ bubble in the Response Grid provided on each page.

1. If x, v and a denote the displacement, the velocity and the acceleration of a particle executing simple harmonic motion of time period T, then, which of the following does not change with time?
   (a) aT/x (b) $aT + 2\pi v$
   (c) aT/v (d) $a^2T^2 + 4\pi^2v^2$

2. A mass is suspended separately by two different springs in successive order, then time periods is t_1 and t_2 respectively. It is connected by both springs as shown in fig. then time period is t_0. The correct relation is

   (a) 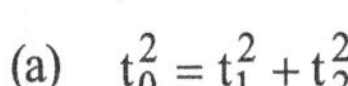 $t_0^2 = t_1^2 + t_2^2$
   (b) $t_0^{-2} = t_1^{-2} + t_2^{-2}$
   (c) $t_0^{-1} = t_1^{-1} + t_2^{-1}$
   (d) $t_0 = t_1 + t_2$

3. A rod of length ℓ is in motion such that its ends A and B are moving along x-axis and y-axis respectively. It is given that $\frac{d\theta}{dt} = 2$ rad/sec always. P is a fixed point on the rod. Let M be the projection of P on x-axis. For the time interval in which θ changes from 0 to $\frac{\pi}{2}$, the correct statement is
   (a) The acceleration of M is always directed towards right
   (b) M executes SHM
   (c) M moves with constant speed
   (d) M moves with constant acceleration

4. A particle of mass m executes simple harmonic motion with amplitude a and frequency ν. The average kinetic energy during its motion from the position of equilibrium to the end is
   (a) $2\pi^2 ma^2\nu^2$ (b) $\pi^2 ma^2\nu^2$
   (c) $\frac{1}{4}ma^2\nu^2$ (d) $4\pi^2 ma^2\nu^2$

5. A mass M attached to a spring oscillates with a period of 2s. If the mass is increased by 2 kg, then the period increases by 2s. Find the initial mass M assuming that Hooke's law is obeyed.
   (a) $\frac{2}{3}$kg   (b) $\frac{1}{3}$kg   (c) $\frac{1}{2}$kg   (d) 1 kg

RESPONSE GRID	1. ⓐⓑⓒⓓ	2. ⓐⓑⓒⓓ	3. ⓐⓑⓒⓓ	4. ⓐⓑⓒⓓ	5. ⓐⓑⓒⓓ

Space for Rough Work

6. The amplitude of a damped oscillator becomes $\left(\frac{1}{3}\right)^{rd}$ in 2 seconds. If its amplitude after 6 seconds is $\frac{1}{n}$ times the original amplitude, the value of n is
(a) 3^2 (b) 3^3 (c) $\sqrt[3]{3}$ (d) 2^3

7. Assume the earth to be perfect sphere of uniform density. If a body is dropped at one end of a tunnel dug along a diameter of the earth (remember that inside the tunnel the force on the body is – k times the displacement from the centre, k being a constant), it (body) will
(a) reach the earth's centre and stay there
(b) go through the tunnel and comes out at the other end
(c) oscillate simple harmonically in the tunnel
(d) stay somewhere between the earth's centre and one of the ends of tunnel.

8. A particle undergoes simple harmonic motion having time period T. The time taken in 3/8th oscillation is
(a) $\frac{3}{8}T$ (b) $\frac{5}{8}T$ (c) $\frac{5}{12}T$ (d) $\frac{7}{12}T$

9. A particle is executing simple harmonic motion with amplitude A. When the ratio of its kinetic energy to the potential energy is $\frac{1}{4}$, its displacement from its mean position is
(a) $\frac{2}{\sqrt{5}}A$ (b) $\frac{\sqrt{3}}{2}A$ (c) $\frac{3}{4}A$ (d) $\frac{1}{4}A$

10. The length of a simple pendulum executing simple harmonic motion is increased by 21%. The percentage increase in the time period of the pendulum of increased length is
(a) 11% (b) 21% (c) 42% (d) 10%

11. The time period of a mass suspended from a spring is T. If the spring is cut into four equal parts and the same mass is suspended from one of the parts, then the new time period will be
(a) $2T$ (b) $\frac{T}{4}$ (c) 2 (d) $\frac{T}{2}$

12. Two simple harmonic motions act on a particle. These harmonic motions are $x = A\cos(\omega t + \delta)$, $y = A\cos(\omega t + \alpha)$ when $\delta = \alpha + \frac{\pi}{2}$, the resulting motion is
(a) a circle and the actual motion is clockwise
(b) an ellipse and the actual motion is counterclockwise
(c) an elllipse and the actual motion is clockwise
(d) a circle and the actual motion is counter clockwise

13. A point mass oscillates along the x-axis according to the law $x = x_0\cos(\omega t - \pi/4)$. If the acceleration of the particle is written as $a = A\cos(\omega t - \delta)$, then
(a) $A = x_0\omega^2, \delta = 3\pi/4$ (b) $A = x_0, \delta = -\pi/4$
(c) $A = x_0\omega^2, \delta = \pi/4$ (d) $A = x_0\omega^2, \delta = -\pi/4$

14. A mass M is suspended from a spring of negligible mass. The spring is pulled a little and then released so that the mass executes SHM of time period T. If the mass is increased by m, the time period becomes $\frac{5T}{3}$. Then the ratio of $\frac{m}{M}$ is
(a) $\frac{3}{5}$ (b) $\frac{25}{9}$ (c) $\frac{16}{9}$ (d) $\frac{5}{3}$

15. A body oscillates with a simple harmonic motion having amplitude 0.05 m. At a certain instant of time, its displacement is 0.01 m and acceleration is 1.0 m/s^2. The period of oscillation is
(a) 0.1 s (b) 0.2 s (c) $\frac{\pi}{10}$s (d) $\frac{\pi}{5}$s

16. The particle executing simple harmonic motion has a kinetic energy $K_0\cos^2\omega t$. The maximum values of the potential energy and the total energy are respectively
(a) $K_0/2$ and K_0 (b) K_0 and $2K_0$
(c) K_0 and K_0 (d) 0 and $2K_0$

17. A simple pendulum attached to the ceiling of a stationary lift has a time period T. The distance y covered by the lift moving upwards varies with time t as $y = t^2$ where y is in metres and t in seconds. If g = 10 m/s^2, the time period of pendulum will be
(a) $\sqrt{\frac{4}{5}}T$ (b) $\sqrt{\frac{5}{6}}T$ (c) $\sqrt{\frac{5}{4}}T$ (d) $\sqrt{\frac{6}{5}}T$

18. A particle moves with simple harmonic motion in a straight line. In first τs, after starting from rest it travels a distance a, and in next τ s it travels 2a, in same direction, then:
(a) amplitude of motion is 3a
(b) time period of oscillations is 8τ
(c) amplitude of motion is 4a
(d) time period of oscillations is 6τ

19. Two simple harmonic motions are represented by the equations $y_1 = 0.1\sin\left(100\pi t + \frac{\pi}{3}\right)$ and $y_2 = 0.1\cos\pi t$. The phase difference of the velocity of particle 1 with respect to the velocity of particle 2 is
(a) $\frac{\pi}{3}$ (b) $\frac{-\pi}{6}$ (c) $\frac{\pi}{6}$ (d) $\frac{-\pi}{3}$

20. Masses M_A and M_B hanging from the ends of strings of lengths L_A and L_B are executing simple harmonic motions. If their frequencies are $f_A = 2f_B$, then
(a) $L_A = 2L_B$ and $M_A = M_B/2$
(b) $L_A = 4L_B$ regardless of masses
(c) $L_A = L_B/4$ regardless of masses
(d) $L_A = 2L_B$ and $M_A = 2M_B$

RESPONSE GRID					
	6. ⓐⓑⓒⓓ	7. ⓐⓑⓒⓓ	8. ⓐⓑⓒⓓ	9. ⓐⓑⓒⓓ	10. ⓐⓑⓒⓓ
	11. ⓐⓑⓒⓓ	12. ⓐⓑⓒⓓ	13. ⓐⓑⓒⓓ	14. ⓐⓑⓒⓓ	15. ⓐⓑⓒⓓ
	16. ⓐⓑⓒⓓ	17. ⓐⓑⓒⓓ	18. ⓐⓑⓒⓓ	19. ⓐⓑⓒⓓ	20. ⓐⓑⓒⓓ

Space for Rough Work

21. In damped oscillations, the amplitude of oscillations is reduced to one-third of its inital value a_0 at the end of 100 oscillations. When the oscillator completes 200 oscillations, its amplitude must be

(a) $a_0/2$ (b) $a_0/4$ (c) $a_0/6$ (d) $a_0/9$

22. The spring constant from the adjoining combination of springs is

(a) K
(b) 2K
(c) 4K
(d) 5K/2

23. A body executes simple harmonic motion. The potential energy (P.E), the kinetic energy (K.E) and total energy (T.E) are measured as a function of displacement x. Which of the following statements is true ?

(a) K.E. is maximum when $x = 0$
(b) T.E is zero when $x = 0$
(c) K.E is maximum when x is maximum
(d) P.E is maximum when $x = 0$

24. A simple harmonic wave having an amplitude a and time period T is represented by the equation $y = 5\sin \pi(t+4)m$. Then the value of amplitude (a) in (m) and time period (T) in second are

(a) a = 10, T = 2 (b) a = 5, T = 1
(c) a = 10, T = 1 (d) a = 5, T = 2

25. A particle moves such that its acceleration 'a' is given by a = – zx where x is the displacement from equilibrium position and z is constant. The period of oscillation is

(a) $2\pi/z$ (b) $2\pi/\sqrt{z}$ (c) $\sqrt{2\pi/z}$ (d) $2\sqrt{\pi/z}$

26. The displacement of an object attached to a spring and executing simple harmonic motion is given by $x = 2\times10^{-2}\cos \pi t$ metre. The time at which the maximum speed first occurs is

(a) 0.25 s (b) 0.5 s (c) 0.75 s (d) 0.125 s

27. A tunnel has been dug through the centre of the earth and a ball is released in it. It executes S.H.M. with time period

(a) 42 minutes (b) 1 day
(c) 1 hour (d) 84.6 minutes

28. The displacement equation of a particle is $x = 3\sin 2t + 4\cos 2t$. The amplitude and maximum velocity will be respectively

(a) 5, 10 (b) 3, 2 (c) 4, 2 (d) 3, 4

29. A body of mass 0.01 kg executes simple harmonic motion about x = 0 under the influence of a force as shown in figure. The time period of SHM is

(a) 1.05 s
(b) 0.52 s
(c) 0.25 s
(d) 0.03 s

30. Two oscillators are started simultaneously in same phase. After 50 oscillations of one, they get out of phase by π, that is half oscillation. The percentage difference of frequencies of the two oscillators is nearest to

(a) 2% (b) 1% (c) 0.5% (d) 0.25%

31. The length of a second's pendulum at the surface of earth is 1 m. The length of second's pendulum at the surface of moon where g is 1/6th that at earth's surface is

(a) 1/6 m (b) 6 m (c) 1/36 m (d) 36 m

32. A simple spring has length l and force constant K. It is cut into two springs of lengths l_1 and l_2 such that $l_1 = n\,l_2$ (n = an integer). The force constant of spring of length l_1 is

(a) K(1 + n) (b) (K/n)(1 + n)
(c) K (d) K/(n + 1)

33. The displacement of a particle from its mean position (in metre) is given by $y = 0.2\sin(10\pi t + 1.5\pi)\cos(10\pi t + 1.5\pi)$. The motion of particle is

(a) periodic but not SHM
(b) non-periodic
(c) simple harmonic motion with period 0.1 s
(d) simple harmonic motion with period 0.2s.

34. A point particle of mass 0.1 kg is executing S.H.M. of amplitude 0.1 m. When the particle passes through the mean position, its kinetic energy is 8×10^{-3} joule. Obtain the equation of motion of this particle, if the initial phase of oscillation is 45°.

(a) $y = 0.1\sin\left(\pm4t+\frac{\pi}{4}\right)$ (b) $y = 0.2\sin\left(\pm4t+\frac{\pi}{4}\right)$

(c) $y = 0.1\sin\left(\pm2t+\frac{\pi}{4}\right)$ (d) $y = 0.2\sin\left(\pm2t+\frac{\pi}{4}\right)$

35. For a simple pendulum, a graph is plotted between its kinetic energy (KE) and potential energy (PE) against its displacement d. Which one of the following represents these correctly? (*graphs are schematic and not drawn to scale*)

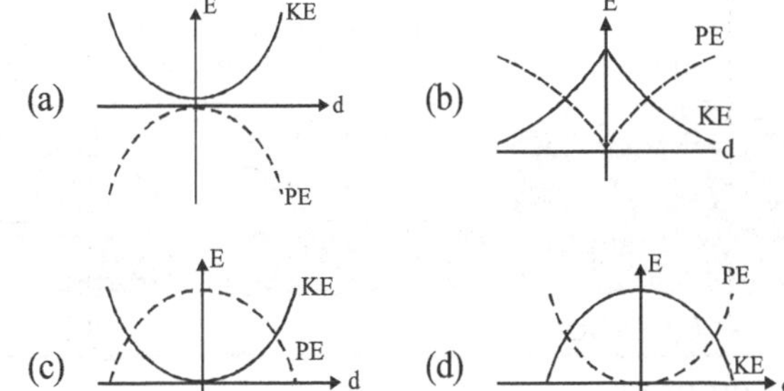

RESPONSE GRID

21. (a)(b)(c)(d) 22. (a)(b)(c)(d) 23. (a)(b)(c)(d) 24. (a)(b)(c)(d) 25. (a)(b)(c)(d)
26. (a)(b)(c)(d) 27. (a)(b)(c)(d) 28. (a)(b)(c)(d) 29. (a)(b)(c)(d) 30. (a)(b)(c)(d)
31. (a)(b)(c)(d) 32. (a)(b)(c)(d) 33. (a)(b)(c)(d) 34. (a)(b)(c)(d) 35. (a)(b)(c)(d)

Space for Rough Work

36. The equation of a simple harmonic wave is given by

$$y = 3\sin\frac{\pi}{2}(50t - x)$$

Where x and y are in meters and t is in seconds. The ratio of maximum particle velocity to the wave velocity is

(a) 2π (b) $\frac{3}{2}\pi$ (c) 3π (d) $\frac{2}{3}\pi$

37. If the mass shown in figure is slightly displaced and then let go, then the system shall oscillate with a time period of

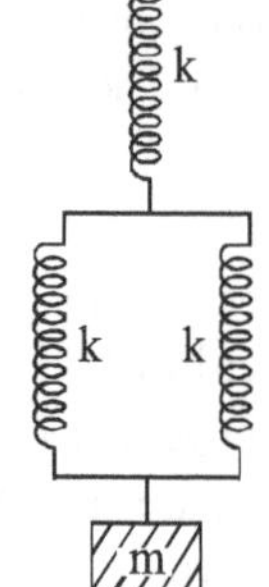

(a) 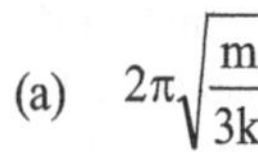 $2\pi\sqrt{\frac{m}{3k}}$

(b) $2\pi\sqrt{\frac{3m}{2k}}$

(c) $2\pi\sqrt{\frac{2m}{3k}}$

(d) $2\pi\sqrt{\frac{3k}{m}}$

38. A hollow sphere is filled with water. It is hung by a long thread. As the water flows out of a hole at the bottom, the period of oscillation will

(a) first increase and then decrease
(b) first decrease and then increase
(c) go on increasing
(d) go on decreasing

39. The figure shows a position time graph of a particle executing SHM. If the time period of SHM is 2 sec, then the equation of SHM is

(a) $x = 10\cos\pi t$

(b) $x = 5\sin\left(\pi t + \frac{\pi}{3}\right)$

(c) $x = 10\sin\left(\pi t + \frac{\pi}{3}\right)$

(d) $x = 10\sin\left(\pi t + \frac{\pi}{6}\right)$

40. A coin is placed on a horizontal platform which undergoes vertical simple harmonic motion of angular frequency ω. The amplitude of oscillation is gradually increased. The coin will leave contact with the platform for the first time

(a) at the mean position of the platform
(b) for an amplitude of $\frac{g}{\omega^2}$
(c) for an amplitude of $\frac{g^2}{\omega^2}$
(d) at the highest position of the platform

41. The bob of a simple pendulum executes simple harmonic motion in water with a period t, while the period of oscillation of the bob is t_0 in air. Neglecting frictional force of water and given that the density of the bob is $(4/3) \times 1000$ kg/m^3. The relationship between t and t_0 is

(a) $t = 2t_0$ (b) $t = t_0/2$ (c) $t = t_0$ (d) $t = 4t_0$

42. Starting from the origin a body oscillates simple harmonically with a period of 2 s. After what time will its kinetic energy be 75% of the total energy?

(a) $\frac{1}{6}$s (b) $\frac{1}{4}$s (c) $\frac{1}{3}$s (d) $\frac{1}{12}$s

43. A body executes simple harmonic motion under the action of a force F_1 with a time period $\frac{4}{5}$ s. If the force is changed to F_2, it executes S.H.M. with time period $\frac{3}{5}$ s. If both the forces F_1 and F_2 act simultaneously in the same direction on the body, its time period in second is

(a) $\frac{12}{25}$ (b) $\frac{7}{5}$ (c) $\frac{24}{25}$ (d) $\frac{5}{7}$

44. A block connected to a spring oscillates vertically. A damping force F_d, acts on the block by the surrounding medium. Given as $F_d = -bV$, b is a positive constant which depends on :

(a) viscosity of the medium
(b) size of the block
(c) shape of the block
(d) All of these

45. If a simple pendulum of length l has maximum angular displacement θ, then the maximum K.E. of bob of mass m is

(a) $\frac{1}{2}ml/g$ (b) $mg/2l$
(c) $mgl(1 - \cos\theta)$ (d) $mgl\sin\theta/2$

RESPONSE GRID					
	36. ⓐⓑⓒⓓ	37. ⓐⓑⓒⓓ	38. ⓐⓑⓒⓓ	39. ⓐⓑⓒⓓ	40. ⓐⓑⓒⓓ
	41. ⓐⓑⓒⓓ	42. ⓐⓑⓒⓓ	43. ⓐⓑⓒⓓ	44. ⓐⓑⓒⓓ	45. ⓐⓑⓒⓓ

DAILY PRACTICE PROBLEM DPP CHAPTERWISE CP13 - PHYSICS			
Total Questions	45	Total Marks	180
Attempted		Correct	
Incorrect		Net Score	
Cut-off Score	45	Qualifying Score	60
Success Gap = Net Score – Qualifying Score			
Net Score = (Correct × 4) – (Incorrect × 1)			

Space for Rough Work

DPP - Daily Practice Problems

Chapter-wise Sheets

Date : ______ **Start Time :** ______ **End Time :** ______

PHYSICS

CP14

SYLLABUS : Waves

Max. Marks : 180 **Marking Scheme :** (+4) for correct & (–1) for incorrect answer **Time : 60 min.**

INSTRUCTIONS : This Daily Practice Problem Sheet contains 45 MCQs. For each question only one option is correct. Darken the correct circle/ bubble in the Response Grid provided on each page.

1. Where should the two bridges be set in a 110cm long wire so that it is divided into three parts and the ratio of the frequencies are 3 : 2 : 1 ?
   (a) 20cm from one end and 60cm from other end
   (b) 30cm from one end and 70cm from other end
   (c) 10cm from one end and 50cm from other end
   (d) 50cm from one end and 40cm from other end
2. When a wave travel in a medium, the particle displacement is given by the equation $y = a \sin 2\pi (bt - cx)$ where a, b and c are constants. The maximum particle velocity will be twice the wave velocity if
   (a) $c = \frac{1}{\pi a}$ (b) $c = \pi a$ (c) $b = ac$ (d) $b = \frac{1}{ac}$
3. The wave described by $y = 0.25 \sin (10\pi x - 2\pi t)$, where x and y are in meters and t in seconds, is a wave travelling along the:
   (a) –ve x direction with frequency 1 Hz.
   (b) +ve x direction with frequency π Hz and wavelength $\lambda = 0.2$ m.
   (c) +ve x direction with frequency 1 Hz and wavelength $\lambda = 0.2$ m
   (d) –ve x direction with amplitude 0.25 m and wavelength $\lambda = 0.2$ m
4. The equation of a plane progressive wave is $y = 0.9 \sin 4\pi\left[t - \frac{x}{2}\right]$. When it is reflected at a rigid support, its amplitude becomes $\frac{2}{3}$ of its previous value. The equation of the reflected wave is
   (a) $y = 0.6 \sin 4\pi\left[t + \frac{x}{2}\right]$
   (b) $y = -0.6 \sin 4\pi\left[t + \frac{x}{2}\right]$
   (c) $y = -0.9 \sin 8\pi\left[t - \frac{x}{2}\right]$
   (d) $y = -0.6 \sin 4\pi\left[t + \frac{x}{2}\right]$
5. A person carrying a whistle emitting continuously a note of 272 Hz is running towards a reflecting surface with a speed of 18 km h^{-1}. The speed of sound in air is 345 m s^{-1}. The number of beats heard by him is
   (a) 4 (b) 6 (c) 8 (d) zero

RESPONSE GRID	1. ⓐⓑⓒⓓ	2. ⓐⓑⓒⓓ	3. ⓐⓑⓒⓓ	4. ⓐⓑⓒⓓ	5. ⓐⓑⓒⓓ

Space for Rough Work

6. A closed organ pipe (closed at one end) is excited to support the third overtone. It is found that air in the pipe has
(a) three nodes and three antinodes
(b) three nodes and four antinodes
(c) four nodes and three antinodes
(d) four nodes and four antinodes

7. A wave disturbance in a medium is described by $y(x,t) = 0.02\cos\left(50\pi t + \frac{\pi}{2}\right)\cos(10\pi x)$ where x and y are in metre and t is in second. Which of the following is correct?
(a) A node occurs at x = 0.15 m
(b) An antinode occurs at x = 0.3 m
(c) The speed wave is 5 ms^{-1}
(d) The wavelength is 0.3 m

8. In a resonance column, first and second resonance are obtained at depths 22.7 cm and 70.2 cm. The third resonance will be obtained at a depth
(a) 117.7 cm (b) 92.9 cm
(c) 115.5 cm (d) 113.5 cm

9. An engine approaches a hill with a constant speed. When it is at a distance of 0.9 km, it blows a whistle whose echo is heard by the driver after 5 seconds. If the speed of sound in air is 330 m/s, then the speed of the engine is :
(a) 32 m/s (b) 27.5 m/s (c) 60 m/s (d) 30 m/s

10. Two identical piano wires kept under the same tension T have a fundamental frequency of 600 Hz. The fractional increase in the tension of one of the wires which will lead to occurrence of 6 beats/s when both the wires oscillate together would be
(a) 0.02 (b) 0.03 (c) 0.04 (d) 0.01

11. Two sound sources emitting sound each of wavelength λ are fixed at a given distance apart. A listener moves with a velocity u along the line joining the two sources. The number of beats heard by him per second is
(a) $\frac{u}{2\lambda}$ (b) $\frac{2u}{\lambda}$ (c) $\frac{u}{\lambda}$ (d) $\frac{u}{3\lambda}$

12. An observer moves towards a stationary source of sound, with a velocity one-fifth of the velocity of sound. What is the percentage increase in the apparent frequency ?
(a) 0.5% (b) zero (c) 20% (d) 5%

13. Velocity of sound in air is 320 m s^{-1}. A pipe closed at one end has a length of 1 m. Neglecting end correction, the air column in the pipe cannot resonate with sound of frequency
(a) 80 Hz (b) 240 Hz (c) 320 Hz (d) 400 Hz

14. The driver of a car travelling with speed 30 m/sec towards a hill sounds a horn of frequency 600 Hz. If the velocity of sound in air is 330 m/s, the frequency of reflected sound as heard by driver is
(a) 555.5 Hz (b) 720 Hz (c) 500 Hz (d) 550 Hz

15. What will be the frequency of beats formed from the superposition of two harmonic waves shown below?

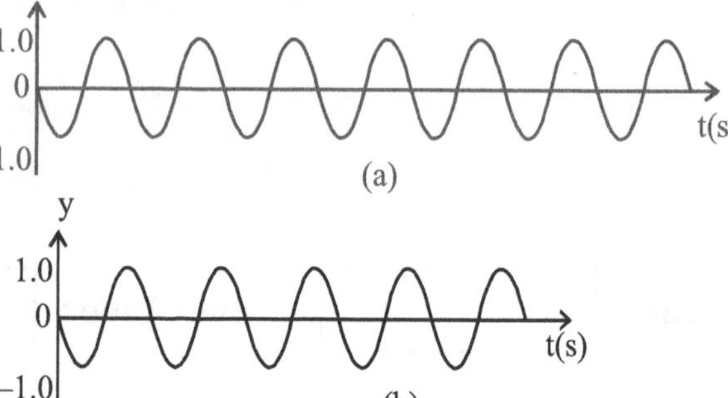

(a) 20 Hz (b) 11 Hz (c) 9 Hz (d) 2 Hz

16. What is the effect of increase in temperature on the frequency of sound produced by an organ pipe?
(a) increases (b) decreases
(c) no effect (d) erratic change

17. A cylinderical tube open at both ends, has a fundamental frequency f in air. The tube is dipped vertically in water so that half of it is in water. The fundamental frequency of air column is now
(a) f/2 (b) f (c) 3f/4 (d) 2f

18. The transverse displacement y (x, t) of a wave on a string is given by $y(x,t) = e^{-\left(ax^2+bt^2+2\sqrt{ab}xt\right)}$.
This represents a:
(a) wave moving in –x direction with speed $\sqrt{\frac{b}{a}}$
(b) standing wave of frequency $\sqrt{b}$
(c) standing wave of frequency $\frac{1}{\sqrt{b}}$
(d) wave moving in + x direction with speed $\sqrt{\frac{a}{b}}$

19. A longitudinal wave is represented by
$x = x_0 \sin 2\pi\left(nt - \frac{x}{\lambda}\right)$
The maximum particle velocity will be four times the wave velocity if
(a) $\lambda = \frac{\pi x_0}{4}$ (b) $\lambda = 2\pi x_0$
(c) $\lambda = \frac{\pi x_0}{2}$ (d) $\lambda = 4\pi x_0$

20. Two tones of frequencies n_1 and n_2 are sounded together. The beats can be heard distinctly when
(a) $10 < (n_1 - n_2) < 20$ (b) $5 < (n_1 - n_2) > 20$
(c) $5 < (n_1 - n_2) < 20$ (d) $0 < (n_1 - n_2) < 10$

RESPONSE GRID					
	6. ⓐⓑⓒⓓ	7. ⓐⓑⓒⓓ	8. ⓐⓑⓒⓓ	9. ⓐⓑⓒⓓ	10. ⓐⓑⓒⓓ
	11. ⓐⓑⓒⓓ	12. ⓐⓑⓒⓓ	13. ⓐⓑⓒⓓ	14. ⓐⓑⓒⓓ	15. ⓐⓑⓒⓓ
	16. ⓐⓑⓒⓓ	17. ⓐⓑⓒⓓ	18. ⓐⓑⓒⓓ	19. ⓐⓑⓒⓓ	20. ⓐⓑⓒⓓ

Space for Rough Work

21. A pipe of length 85 cm is closed from one end. Find the number of possible natural oscillations of air column in the pipe whose frequencies lie below 1250 Hz. The velocity of sound in air is 340 m/s.
(a) 12 (b) 8 (c) 6 (d) 4

22. A vehicle, with a horn of frequency n is moving with a velocity of 30 m/s in a direction perpendicular to the straight line joining the observer and the vehicle. The observer perceives the sound to have a frequency $n + n_1$. Then (if the sound velocity in air is 300 m/s)
(a) $n_1 = 10n$ (b) $n_1 = 0$
(c) $n_1 = 0.1n$ (d) $n_1 = -0.1n$

23. A source of sound gives 5 beats per second, when sounded with another source of frequency 100/sec. The second harmonic of the source, together with a source of frequency 205/sec gives 5 beats per second. What is the frequency of the source?
(a) 95 sec^{-1} (b) 100 sec^{-1}
(c) 105 sec^{-1} (d) 205 sec^{-1}

24. If we study the vibration of a pipe open at both ends, then which of the following statements is not true ?
(a) Odd harmonics of the fundamental frequency will be generated
(b) All harmonics of the fundamental frequency will be generated
(c) Pressure change will be maximum at both ends
(d) Antinode will be at open end

25. 41 forks are so arranged that each produces 5 beats per sec when sounded with its near fork. If the frequency of last fork is double the frequency of first fork, then the frequencies (in Hz) of the first and the last fork are respectively.
(a) 200, 400 (b) 205, 410
(c) 195, 390 (d) 100, 200

26. Two points are located at a distance of 10 m and 15 m from the source of oscillation. The period of oscillation is 0.05 sec and the velocity of the wave is 300 m/sec. What is the phase difference between the oscillations of two points?
(a) $\frac{\pi}{3}$ (b) $\frac{2\pi}{3}$ (c) π (d) $\frac{\pi}{6}$

27. A sound absorber attenuates the sound level by 20 dB. The intensity decreases by a factor of
(a) 100 (b) 1000 (c) 10000 (d) 10

28. A wave travelling along the x-axis is described by the equation $y(x, t) = 0.005 \cos(\alpha x - \beta t)$. If the wavelength and the time period of the wave are 0.08 m and 2.0s, respectively, then α and β in appropriate units are
(a) $\alpha = 25.00\pi, \beta = \pi$ (b) $\alpha = \frac{0.08}{\pi}, \beta = \frac{2.0}{\pi}$
(c) $\alpha = \frac{0.04}{\pi}, \beta = \frac{1.0}{\pi}$ (d) $\alpha = 12.50\pi, \beta = \frac{\pi}{2.0}$

29. The equation $Y = 0.02 \sin(500\pi t) \cos(4.5x)$ represents
(a) progressive wave of frequency 250 Hz along x-axis
(b) a stationary wave of wavelength 1.4 m
(c) a transverse progressive wave of amplitude 0.02 m
(d) progressive wave of speed of about 350 m s^{-1}

30. Which of the following statements is/are incorrect about waves ?
(a) Waves are patterns of disturbance which move without the actual physical transfer of flow of matter as a whole.
(b) Waves cannot transport energy.
(c) The pattern of disturbance in the form of waves carry information that propagate from one point to another.
(d) All our communications essentially depend on transmission of signals through waves.

31. An organ pipe P_1, closed at one end vibrating in its first harmonic and another pipe P_2, open at both ends vibrating in its third harmonic, are in resonance with a given tuning fork. The ratio of the lengths of P_1 and P_2 is :
(a) $\frac{8}{3}$ (b) $\frac{1}{6}$ (c) $\frac{1}{2}$ (d) $\frac{1}{3}$

32. Two vibrating tuning forks producing waves given by $y_1 = 27 \sin 600\pi t$ and $y_2 = 27 \sin 604\pi t$ are held near the ear of a person, how many beats will be heard in three seconds by him ?
(a) 4 (b) 2 (c) 6 (d) 12

33. A source of sound A emitting waves of frequency 1800 Hz is falling towards ground with a terminal speed v. The observer B on the ground directly beneath the source receives waves of frequency 2150 Hz. The source A receives waves, reflected from ground of frequency nearly: (Speed of sound = 343 m/s)
(a) 2150 Hz (b) 2500 Hz (c) 1800 Hz (d) 2400 Hz

34. Consider the three waves z_1, z_2 and z_3 as
$z_1 = A \sin(kx - \omega t)$
$z_2 = A \sin(kx + \omega t)$
$z_3 = A \sin(ky - \omega t)$
Which of the following represents a standing wave?
(a) $z_1 + z_2$ (b) $z_2 + z_3$
(c) $z_3 + z_1$ (d) $z_1 + z_2 + z_3$

35. A sonometer wire supports a 4 kg load and vibrates in fundamental mode with a tuning fork of frequency 416 Hz. The length of the wire between the bridges is now doubled. In order to maintain fundamental mode, the load should be changed to
(a) 1 kg (b) 2 kg (c) 4 kg (d) 16 kg

RESPONSE GRID				
21. ⓐⓑⓒⓓ	22. ⓐⓑⓒⓓ	23. ⓐⓑⓒⓓ	24. ⓐⓑⓒⓓ	25. ⓐⓑⓒⓓ
26. ⓐⓑⓒⓓ	27. ⓐⓑⓒⓓ	28. ⓐⓑⓒⓓ	29. ⓐⓑⓒⓓ	30. ⓐⓑⓒⓓ
31. ⓐⓑⓒⓓ	32. ⓐⓑⓒⓓ	33. ⓐⓑⓒⓓ	34. ⓐⓑⓒⓓ	35. ⓐⓑⓒⓓ

Space for Rough Work

36. The vibrations of a string of length 60 cm fixed at both the ends are represented by the equation $y = 2\sin\left(\frac{4\pi x}{15}\right)\cos$ $(96\pi t)$ where x and y are in cm. The maximum number of loops that can be formed in it is
(a) 4 (b) 16 (c) 5 (d) 15

37. If n_1, n_2 and n_3 are the fundamental frequencies of three segments into which a string is divided, then the original fundamental frequency n of the string is given by
(a) $n = n_1 + n_2 + n_3$
(b) $\frac{1}{n} = \frac{1}{n_1} + \frac{1}{n_2} + \frac{1}{n_3}$
(c) $\frac{1}{\sqrt{n}} = \frac{1}{\sqrt{n_1}} + \frac{1}{\sqrt{n_2}} + \frac{1}{\sqrt{n_3}}$
(d) $\sqrt{n} = \sqrt{n_1} + \sqrt{n_2} + \sqrt{n_3}$

38. An echo repeats two syllables. If the velocity of sound is 330 m/s, then the distance of the reflecting surface is
(a) 66.0m (b) 33.0m (c) 99.0m (d) 16.5m

39. What is the effect of humidity on sound waves when humidity increases?
(a) Speed of sound waves is more
(b) Speed of sound waves is less
(c) Speed of sound waves remains same
(d) Speed of sound waves becomes zero

40. If the ratio of maximum to minimum intensity in beats is 49, then the ratio of amplitudes of two progressive wave trains is
(a) 7 : 1 (b) 4 : 3 (c) 49 : 1 (d) 16 : 9

41. A whistle of frequency 1000 Hz is sounded on a car travelling towards a cliff with velocity of 18 ms^{-1} normal to the cliff. If velocity of sound (v) = 330 m s^{-1}, then the apparent frequency of the echo as heard by the car driver is nearly
(a) 1115 Hz (b) 115 Hz (c) 67 Hz (d) 47.2 Hz

42. The transverse wave represented by the equation $y = 4\sin\left(\frac{\pi}{6}\right)\sin(3x - 15t)$ has
(a) amplitude = 4
(b) wavelength $= 4\frac{\pi}{3}$
(c) speed of propagation = 5
(d) period $= \frac{\pi}{15}$

43. If the intensities of two interfering waves be I_1 and I_2, the contrast between maximum and minimum intensity is maximum, when
(a) $I_1 >> I_2$ (b) $I_1 << I_2$
(c) $I_1 = I_2$ (d) either I_1 or I_2 is zero

44. The fundamental frequency of a closed organ pipe of length 20 cm is equal to the second overtone of an organ pipe open at both the ends. The length of organ pipe open at both the ends is
(a) 100 cm (b) 120 cm (c) 140 cm (d) 80 cm

45. The equation of a travelling wave is $y = 60\cos(180t - 6x)$ where y is in μm, *t* in second and x in metres. The ratio of maximum particle velocity to velocity of wave propagation is
(a) 3.6×10^{-2} (b) 3.6×10^{-4}
(c) 3.6×10^{-6} (d) 3.6×10^{-11}

RESPONSE GRID					
	36. ⓐⓑⓒⓓ	37. ⓐⓑⓒⓓ	38. ⓐⓑⓒⓓ	39. ⓐⓑⓒⓓ	40. ⓐⓑⓒⓓ
	41. ⓐⓑⓒⓓ	42. ⓐⓑⓒⓓ	43. ⓐⓑⓒⓓ	44. ⓐⓑⓒⓓ	45. ⓐⓑⓒⓓ

DAILY PRACTICE PROBLEM DPP CHAPTERWISE CP14 - PHYSICS			
Total Questions	45	Total Marks	180
Attempted		Correct	
Incorrect		Net Score	
Cut-off Score	50	Qualifying Score	70
Success Gap = Net Score – Qualifying Score			
Net Score = (Correct × 4) – (Incorrect × 1)			

Space for Rough Work

DPP - Daily Practice Problems

Chapter-wise Sheets

Date : ______ **Start Time :** ______ **End Time :** ______

PHYSICS (CP15)

SYLLABUS : Electric Charges and Fields

Max. Marks : 180 **Marking Scheme :** (+4) for correct & (–1) for incorrect answer **Time : 60 min.**

INSTRUCTIONS : This Daily Practice Problem Sheet contains 45 MCQs. For each question only one option is correct. Darken the correct circle/ bubble in the Response Grid provided on each page.

1. The surface charge density of a thin charged disc of radius R is σ. The value of the electric field at the centre of the disc is $\frac{\sigma}{2\epsilon_0}$. With respect to the field at the centre, the electric field along the axis at a distance R from the centre of the disc reduces by
   (a) 70.7% (b) 29.3% (c) 9.7% (d) 14.6%

2. A solid conducting sphere of radius a has a net positive charge 2Q. A conducting spherical shell of inner radius b and outer radius c is concentric with the solid sphere and has a net charge – Q. The surface charge density on the inner and outer surfaces of the spherical shell will be respectively

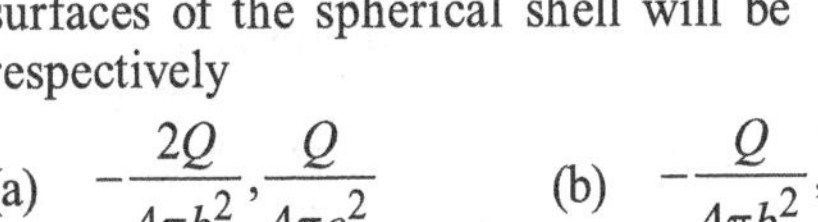

   (a) $-\frac{2Q}{4\pi b^2}, \frac{Q}{4\pi c^2}$ (b) $-\frac{Q}{4\pi b^2}, \frac{Q}{4\pi c^2}$
   (c) $0, \frac{Q}{4\pi c^2}$ (d) $\frac{Q}{4\pi c^2}, 0$

3. Two equally charged, identical metal spheres A and B repel each other with a force 'F'. The spheres are kept fixed with a distance 'r' between them. A third identical, but uncharged sphere C is brought in contact with A and then placed at the mid point of the line joining A and B. The magnitude of the net electric force on C is
   (a) F (b) $\frac{3F}{4}$ (c) $\frac{F}{2}$ (d) $\frac{F}{4}$

4. In the figure, the net electric flux through the area A is $\phi = \vec{E} \cdot \vec{A}$ when the system is in air. On immersing the system in water the net electric flux through the area
   (a) becomes zero
   (b) remains same
   (c) increases
   (d) decreases

5. ABC is an equilateral triangle. Charges +q are placed at each corner as shown in fig. The electric intensity at centre O will be

   (a) 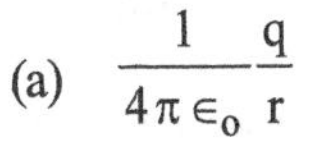 $\frac{1}{4\pi\epsilon_0}\frac{q}{r}$

   (b) 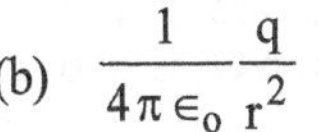 $\frac{1}{4\pi\epsilon_0}\frac{q}{r^2}$

   (c) 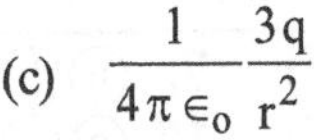 $\frac{1}{4\pi\epsilon_0}\frac{3q}{r^2}$

   (d) zero

RESPONSE GRID	1. ⓐⓑⓒⓓ	2. ⓐⓑⓒⓓ	3. ⓐⓑⓒⓓ	4. ⓐⓑⓒⓓ	5. ⓐⓑⓒⓓ

Space for Rough Work

6. An electric dipole is placed in a uniform electric field. The dipole will experience
(a) a force that will displace it in the direction of the field
(b) a force that will displace it in a direction opposite to the field.
(c) a torque which will rotate it without displacement
(d) a torque which will rotate it and a force that will displace it

7. An uniform electric field E exists along positive x-axis. The work done in moving a charge 0.5 C through a distance 2 m along a direction making an angle 60° with x-axis is 10 J. Then the magnitude of electric field is
(a) $5\,Vm^{-1}$ (b) $2\,Vm^{-1}$ (c) $\sqrt{5}\ Vm^{-1}$ (d) $20\,Vm^{-1}$

8. Which one of the following graphs represents the variation of electric field with distance r from the centre of a charged spherical conductor of radius R?

(a) 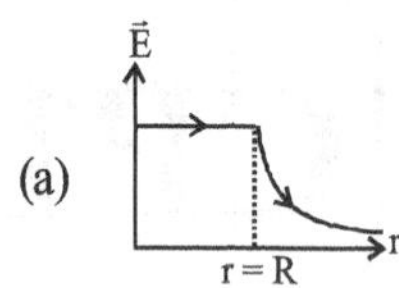

(b) 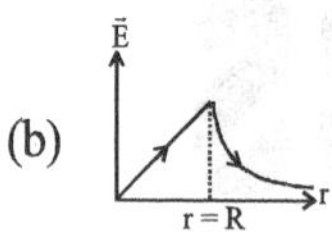

(c) 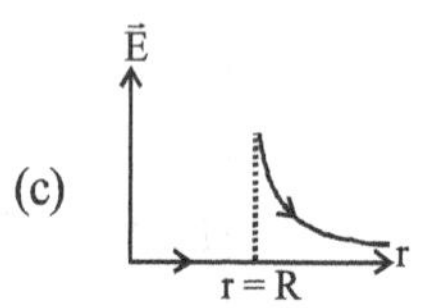

(d) 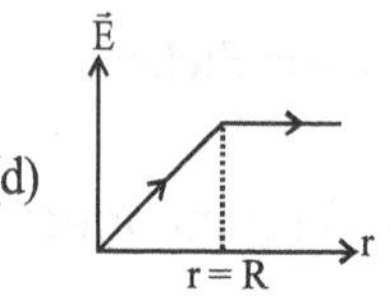

9. A hollow cylinder has a charge q coulomb within it. If ϕ is the electric flux in units of voltmeter associated with the curved surface B, the flux linked with the plane surface A in units of voltmeter will be

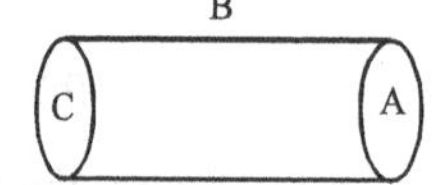

(a) $\frac{q}{2\varepsilon_0}$ (b) $\frac{\phi}{3}$
(c) $\frac{q}{\varepsilon_0}-\phi$ (d) $\frac{1}{2}\left(\frac{q}{\varepsilon_0}-\phi\right)$

10. If E_a be the electric field strength of a short dipole at a point on its axial line and E_e that on the equatorial line at the same distance, then
(a) $E_e = 2E_a$ (b) $E_a = 2E_e$
(c) $E_a = E_e$ (d) None of the above

11. Three positive charges of equal value q are placed at vertices of an equilateral triangle. The resulting lines of force should be sketched as in

12. Three point charges Q_1, Q_2, Q_3 in the order are placed equally spaced along a straight line. Q_2 and Q_3 are equal in magnitude but opposite in sign. If the net force on Q_3 is zero. The value of Q_1 is
(a) $Q_1 = 4(Q_3)$ (b) $Q = 2(Q_3)$
(c) $Q_1 = \sqrt{2}(Q_3)$ (d) $Q_1 = |Q_3|$

13. Electric charge is uniformly distributed along a long straight wire of radius 1 mm. The charge per cm length of the wire is Q coulomb. Another cylindrical surface of radius 50 cm and length 1 m symmetrically encloses the wire. The total electric flux passing through the cylindrical surface is
(a) $\frac{Q}{\varepsilon_0}$ (b) $\frac{100Q}{\varepsilon_0}$ (c) $\frac{10Q}{\pi\varepsilon_0}$ (d) $\frac{100Q}{\pi\varepsilon_0}$

14. A small sphere carrying a charge 'q' is hanging in between two parallel plates by a string of length L. Time period of pendulum is T_0. When parallel plates are charged, the time period changes to T. The ratio T/T_0 is equal to

+ + + + + + + + + + +
L
m
- - - - - - - - - - -

(a) $\left(\frac{g+\frac{qE}{m}}{g}\right)^{1/2}$ (b) $\left(\frac{g}{g+\frac{qE}{m}}\right)^{3/2}$
(c) $\left(\frac{g}{g+\frac{qE}{m}}\right)^{1/2}$ (d) $\left(\frac{g}{g+\frac{qE}{m}}\right)^{5/2}$

15. An electric dipole, consisting of two opposite charges of 2×10^{-6} C each separated by a distance 3 cm is placed in an electric field of 2×10^{5} N/C. Torque acting on the dipole is
(a) $12\times10^{-1}N-m$ (b) $12\times10^{-2}N-m$
(c) $12\times10^{-3}N-m$ (d) $12\times10^{-4}N-m$

16. The electric field in a certain region is acting radially outward and is given by E = Ar. A charge contained in a sphere of radius 'a' centred at the origin of the field, will be given by
(a) $A\,\varepsilon_0 a^2$ (b) $4\pi\varepsilon_0 Aa^3$ (c) $\varepsilon_0 Aa^3$ (d) $4\,\pi\varepsilon_0 Aa^2$

17. The spatial distribution of electric field due to charges (A, B) is shown in figure. Which one of the following statements is correct?

A B

(a) A is +ve and B –ve, |A| > |B|
(b) A is –ve and B +ve, |A| = |B|
(c) Both are +ve but A > B
(d) Both are –ve but A > B

18. Point charges + $4q$, $-q$ and $+4q$ are kept on the X-axis at points $x = 0$, $x = a$ and $x = 2a$ respectively.
(a) only $-q$ is in stable equilibrium
(b) none of the charges is in equilibrium
(c) all the charges are in unstable equilibrium
(d) all the charges are in stable equilibrium.

19. Figure shows some of the electric field lines corresponding to an electric field. The figure suggests that

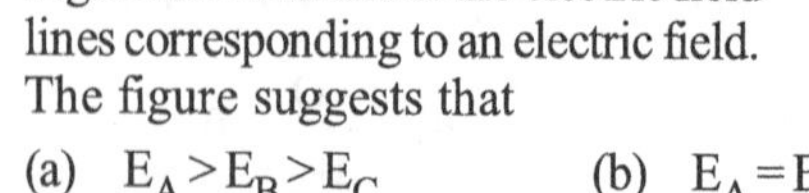

(a) $E_A > E_B > E_C$ (b) $E_A = E_B = E_C$
(c) $E_A = E_C > E_B$ (d) $E_A = E_C < E_B$

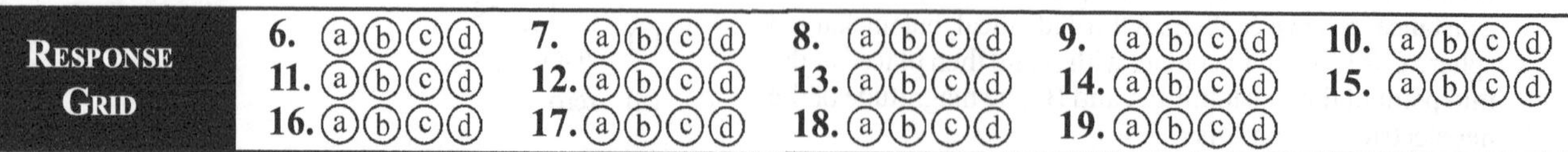

RESPONSE GRID
6. ⓐⓑⓒⓓ 7. ⓐⓑⓒⓓ 8. ⓐⓑⓒⓓ 9. ⓐⓑⓒⓓ 10. ⓐⓑⓒⓓ
11. ⓐⓑⓒⓓ 12. ⓐⓑⓒⓓ 13. ⓐⓑⓒⓓ 14. ⓐⓑⓒⓓ 15. ⓐⓑⓒⓓ
16. ⓐⓑⓒⓓ 17. ⓐⓑⓒⓓ 18. ⓐⓑⓒⓓ 19. ⓐⓑⓒⓓ

Space for Rough Work

20. For distance far away from centre of dipole the change in magnitude of electric field with change in distance from the centre of dipole is
(a) zero.
(b) same in equatorial plane as well as axis of dipole.
(c) more in case of equatorial plane of dipole as compared to axis of dipole.
(d) more in case of axis of dipole as compared to equatorial plane of dipole.

21. Two charge q and $-3q$ are placed fixed on x–axis separated by distance d. Where should a third charge $2q$ be placed such that it will not experience any force ?

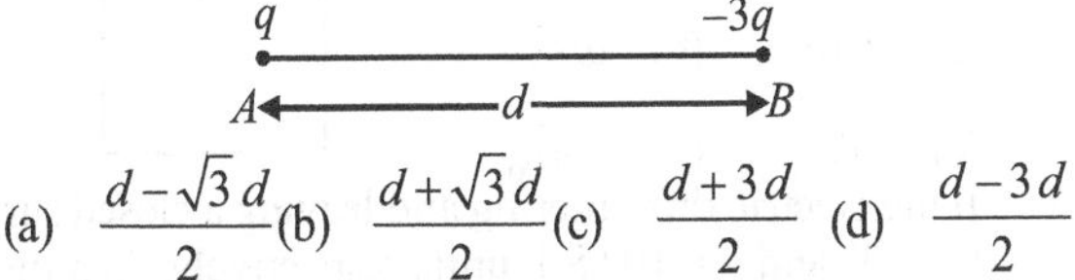

(a) $\frac{d-\sqrt{3}\,d}{2}$ (b) $\frac{d+\sqrt{3}\,d}{2}$ (c) $\frac{d+3d}{2}$ (d) $\frac{d-3d}{2}$

22. A charge Q is placed at each of the opposite corners of a square. A charge q is placed at each of the other two corners. If the net electrical force on Q is zero, then Q/q equals:
(a) -1 (b) 1 (c) $-\frac{1}{\sqrt{2}}$ (d) $-2\sqrt{2}$

23. Identify the wrong statement in the following. Coulomb's law correctly describes the electric force that
(a) binds the electrons of an atom to its nucleus
(b) binds the protons and neutrons in the nucleus of an atom
(c) binds atoms together to form molecules
(d) binds atoms and molecules together to form solids

24. An oil drop of radius r and density ρ is held stationary in a uniform vertically upwards electric field 'E'. If ρ_0 ($<\rho$) is the density of air and e is quanta of charge, then the drop has–
(a) $\frac{4\pi r^3(\rho-\rho_0)g}{3eE}$ excess electrons
(b) $\frac{4\pi r^2(\rho-\rho_0)g}{eE}$ excess electrons
(c) deficiency of $\frac{4\pi r^3(\rho-\rho_0)g}{3eE}$ electrons
(d) deficiency of $\frac{4\pi r^2(\rho-\rho_0)g}{eE}$ electrons

25. A square surface of side L meter in the plane of the paper is placed in a uniform electric field E (volt/m) acting along the same plane at an angle θ with the horizontal side of the square as shown in Figure. The electric flux linked to the surface, in units of volt. m, is

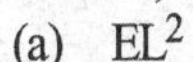
(a) EL^2 (b) $EL^2 \cos\theta$
(c) $EL^2 \sin\theta$ (d) zero

26. An electric dipole of moment $\vec{p}$ placed in a uniform electric field $\vec{E}$ has minimum potential energy when the angle between $\vec{P}$ and $\vec{E}$ is
(a) zero (b) $\frac{\pi}{2}$ (c) π (d) $\frac{3\pi}{2}$

27. Which of the following statements is incorrect?
(a) The charge q on a body is always given by q = ne, where n is any integer, positive or negative.
(b) By convention, the charge on an electron is taken to be negative.
(c) The fact that electric charge is always an integral multiple of e is termed as quantisation of charge.
(d) The quatisation of charge was experimentally demonstrated by Newton in 1912.

28. Two positive ions, each carrying a charge q, are separated by a distance d. If F is the force of repulsion between the ions, the number of electrons missing from each ion will be (e being the charge of an electron)

(a) $\frac{4\pi\varepsilon_0 Fd^2}{e^2}$ (b) $\sqrt{\frac{4\pi\varepsilon_0 Fe^2}{d^2}}$
(c) $\sqrt{\frac{4\pi\varepsilon_0 Fd^2}{e^2}}$ (d) $\frac{4\pi\varepsilon_0 Fd^2}{q^2}$

29. Two small similar metal spheres A and B having charges 4q and – 4q, when placed at a certain distance apart, exert an electric force F on each other. When another identical uncharged sphere C, first touched with A then with B and then removed to infinity, the force of interaction between A and B for the same separation will be
(a) F/2 (b) F/8 (c) F/16 (d) F/32

30. The electric field intensity just sufficient to balance the earth's gravitational attraction on an electron will be: (given mass and charge of an electron respectively are 9.1×10^{-31}kg and 1.6×10^{-19}C.)
(a) -5.6×10^{-11}N/C (b) -4.8×10^{-15}N/C
(c) -1.6×10^{-19}N/C (d) -3.2×10^{-19}N/C

31. An electric dipole is placed at an angle of 30° with an electric field of intensity 2×10^5 NC^{-1}, It experiences a torque of 4 Nm. Calculate the charge on the dipole if the dipole length is 2 cm.
(a) 8mC (b) 4mC (c) 8mC (d) 2mC

32. A particle of mass m and charge q is placed at rest in a uniform electric field E and then released. The kinetic energy attained by the particle after moving a distance y is
(a) qEy^2 (b) qE^2y (c) qEy (d) q^2Ey

33. There is an electric field E in x-direction. If the work done on moving a charge of 0.2 C through a distance of 2 m along a line making an angle 60° with x-axis is 4 J, then what is the value of E?
(a) 3 N/C (b) 4 N/C (c) 5 N/C (d) 20 N/C

RESPONSE GRID					
	20. ⓐⓑⓒⓓ	21. ⓐⓑⓒⓓ	22. ⓐⓑⓒⓓ	23. ⓐⓑⓒⓓ	24. ⓐⓑⓒⓓ
	25. ⓐⓑⓒⓓ	26. ⓐⓑⓒⓓ	27. ⓐⓑⓒⓓ	28. ⓐⓑⓒⓓ	29. ⓐⓑⓒⓓ
	30. ⓐⓑⓒⓓ	31. ⓐⓑⓒⓓ	32. ⓐⓑⓒⓓ	33. ⓐⓑⓒⓓ	

Space for Rough Work

34. A surface has the area vector $\vec{A} = (2\hat{i} + 3\hat{j})m^2$. The flux of an electric field through it if the field is $\vec{E} = 4\hat{i}\frac{V}{m}$:
(a) 8V-m (b) 12V-m (c) 20V-m (d) zero

35. There exists a non!-uniform electric field along x-axis as shown in the figure below. The field increases at a uniform rate along +ve x-axis. A dipole is placed inside the field as shown. Which one of the following is correct for the dipole?

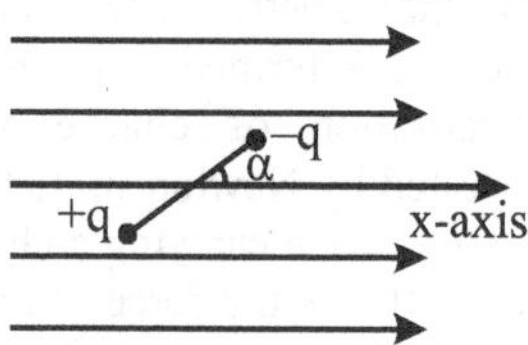

(a) Dipole moves along positive x-axis and undergoes a clockwise rotation
(b) Dipole moves along negative x-axis and undergoes a clockwise rotation
(c) Dipole moves along positive x-axis and undergoes a anticlockwise rotation
(d) Dipole moves along negative x-axis and undergoes a anticlockwise rotation

36. A square surface of side L metres is in the plane of the paper. A uniform electric field $\vec{E}$ (volt /m), also in the plane of the paper, is limited only to the lower half of the square surface (see figure). The electric flux in SI units associated with the surface is

(a) $EL^2/2$ (b) zero (c) EL^2 (d) $EL^2/(2\varepsilon_0)$

37. Among two discs A and B, first have radius 10 cm and charge 10^{-6} μC and second have radius 30 cm and charge 10^{-5} C. When they are touched, charge on both q_A and q_B respectively will, be
(a) q_A=2.75μC, q_B=3.15μC
(b) q_A=1.09μC, q_B=1.53μC
(c) q_A = q_B=5.5μC (d) None of these

38. The total electric flux emanating from a closed surface enclosing an α-particle (e-electronic charge) is
(a) $\frac{2e}{\varepsilon_0}$ (b) $\frac{e}{\varepsilon_0}$ (c) $e\varepsilon_0$ (d) $\frac{\varepsilon_0 e}{4}$

39. Which of the following is a wrong statement?
(a) The charge of an isolated system is conserved
(b) It is not possible to create or destroy charged particles
(c) It is possible to create or destroy charged particles
(d) It is not possible to create or destroy net charge

40. A charge q is placed at the centre of the open end of a cylindrical vessel. The flux of the electric field through the surface of the vessel is
(a) zero (b) q/ε_o
(c) $q/2\varepsilon_o$ (d) $2q/\varepsilon_o$

41. If the electric flux entering and leaving a closed surface are 6×10^6 and 9×10^6 S.I. units respectively, then the charge inside the surface of permittivity of free space ε_0 is
(a) $\varepsilon_0 \times 10^6$ (b) $-\varepsilon_0 \times 10^6$
(c) $-2\varepsilon_0 \times 10^6$ (d) $3\varepsilon_0 \times 10^6$

42. Two particle of equal mass m and charge q are placed at a distance of 16 cm. They do not experience any force. The value of $\frac{q}{m}$ is
(a) 1 (b) $\sqrt{\frac{\pi\varepsilon_0}{G}}$ (c) $\sqrt{\frac{G}{4\pi\varepsilon_0}}$ (d) $\sqrt{4\pi\varepsilon_0 G}$

43. A rod of length 2.4 *m* and radius 4.6 *mm* carries a negative charge of 4.2×10^{-7} C spread uniformly over it surface. The electric field near the mid–point of the rod, at a point on its surface is
(a) -8.6×10^5 N C^{-1} (b) 8.6×10^4 N C^{-1}
(c) -6.7×10^5 N C^{-1} (d) 6.7×10^4 N C^{-1}

44. A hollow insulated conduction sphere is given a positive charge of 10 μC. What will be the electric field at the centre of the sphere if its radius is 2 m?
(a) Zero (b) 5 μCm^{-2}
(c) 20 μCm^{-2} (d) 8 μCm^{-2}

45. A charge Q is enclosed by a Gaussian spherical surface of radius R. If the radius is doubled, then the outward electric flux will
(a) increase four times (b) be reduced to half
(c) remain the same (d) be doubled

RESPONSE GRID					
	34. ⓐⓑⓒⓓ	35. ⓐⓑⓒⓓ	36. ⓐⓑⓒⓓ	37. ⓐⓑⓒⓓ	38. ⓐⓑⓒⓓ
	39. ⓐⓑⓒⓓ	40. ⓐⓑⓒⓓ	41. ⓐⓑⓒⓓ	42. ⓐⓑⓒⓓ	43. ⓐⓑⓒⓓ
	44. ⓐⓑⓒⓓ	45. ⓐⓑⓒⓓ			

DAILY PRACTICE PROBLEM DPP CHAPTERWISE CP15 - PHYSICS			
Total Questions	45	Total Marks	180
Attempted		Correct	
Incorrect		Net Score	
Cut-off Score	50	Qualifying Score	70
Success Gap = Net Score – Qualifying Score			
Net Score = (Correct × 4) – (Incorrect × 1)			

Space for Rough Work

DPP - Daily Practice Problems

Chapter-wise Sheets

Date : ☐ **Start Time :** ☐ **End Time :** ☐

PHYSICS

CP16

SYLLABUS : Electrostatic Potential & Capacitance

Max. Marks : 180 **Marking Scheme :** (+4) for correct & (–1) for incorrect answer **Time : 60 min.**

INSTRUCTIONS : This Daily Practice Problem Sheet contains 45 MCQs. For each question only one option is correct. Darken the correct circle/ bubble in the Response Grid provided on each page.

1. If n drops, each charged to a potential V, coalesce to form a single drop. The potential of the big drop will be

   (a) $\frac{V}{n^{2/3}}$ (b) $\frac{V}{n^{1/3}}$ (c) $Vn^{1/3}$ (d) $Vn^{2/3}$

2. The capacitance of a parallel plate capacitor is C_a (Fig. a). A dielectric of dielectric constant K is inserted as shown in fig. (b) and (c). If C_b and C_c denote the capacitances in fig. (b) and (c), then

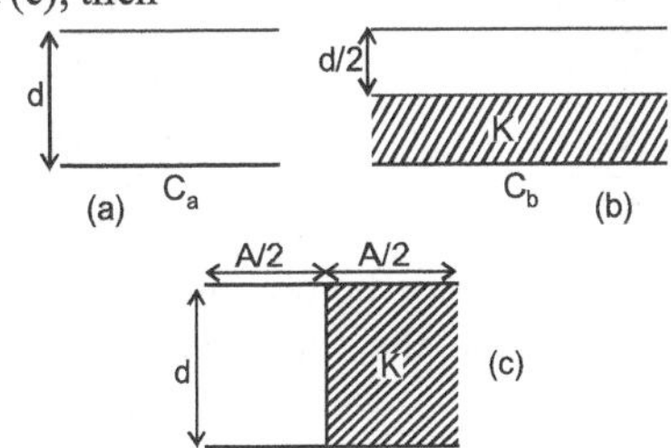

   (a) both $C_b, C_c > C_a$ (b) $C_c > C_a$ while $C_b > C_a$
   (c) both $C_b, C_c < C_a$ (d) $C_a = C_b = C_c$

3. The electric potential $V(x)$ in a region around the origin is given by $V(x) = 4x^2$ volts. The electric charge enclosed in a cube of 1 m side with its centre at the origin is (in coulomb)

   (a) $8\varepsilon_0$ (b) $-4\varepsilon_0$ (c) 0 (d) $-8\varepsilon_0$

4. A parallel plate condenser is immersed in an oil of dielectric constant 2. The field between the plates is

   (a) increased, proportional to 2
   (b) decreased, proportional to $\frac{1}{2}$
   (c) increased, proportional to – 2
   (d) decreased, proportional to $-\frac{1}{2}$

5. What is the effective capacitance between points X and Y?

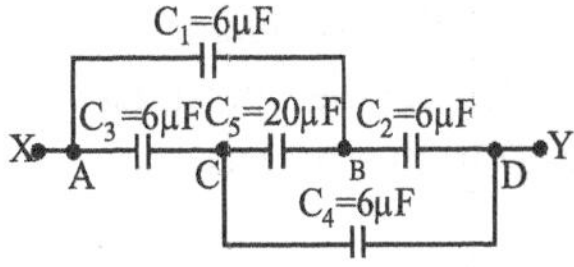

   (a) $24\,\mu F$ (b) $18\,\mu F$ (c) $12\,\mu F$ (d) $6\,\mu F$

6. Two identical particles each of mass m and having charges –q and +q are revolving in a circle of radius r under the influence of electric attraction. Kinetic energy of each particle is $\left(k = \frac{1}{4\pi\varepsilon_0}\right)$

   (a) $kq^2/4r$ (b) $kq^2/2r$ (c) $kq^2/8r$ (d) kq^2/r

RESPONSE GRID	1. ⓐⓑⓒⓓ	2. ⓐⓑⓒⓓ	3. ⓐⓑⓒⓓ	4. ⓐⓑⓒⓓ	5. ⓐⓑⓒⓓ
	6. ⓐⓑⓒⓓ				

Space for Rough Work

7. Four metallic plates each with a surface area of one side A, are placed at a distance d from each other. The two outer plates are connected to one point A and the two other inner plates to another point B as shown in the figure. Then the capacitance of the system is

(a) $\frac{\varepsilon_0 A}{d}$ (b) $\frac{2\varepsilon_0 A}{d}$ (c) $\frac{3\varepsilon_0 A}{d}$ (d) $\frac{4\varepsilon_0 A}{d}$

8. A parallel plate condenser with a dielectric of dielectric constant K between the plates has a capacity C and is charged to a potential V volt. The dielectric slab is slowly removed from between the plates and then reinserted. The net work done by the system in this process is

(a) zero (b) $\frac{1}{2}(K-1)\,CV^2$

(c) $\frac{CV^2(K-1)}{K}$ (d) $(K-1)\,CV^2$

9. If a slab of insulating material 4×10^{-5} m thick is introduced between the plates of a parallel plate capacitor, the distance between the plates has to be increased by 3.5×10^{-5} m to restore the capacity to original value. Then the dielectric constant of the material of slab is

(a) 8 (b) 6 (c) 12 (d) 10

10. A unit charge moves on an equipotential surface from a point A to point B, then

(a) $V_A - V_B = +ve$ (b) $V_A - V_B = 0$
(c) $V_A - V_B = -ve$ (d) it is stationary

11. Identify the false statement.

(a) Inside a charged or neutral conductor, electrostatic field is zero
(b) The electrostatic field at the surface of the charged conductor must be tangential to the surface at any point
(c) There is no net charge at any point inside the conductor
(d) Electrostatic potential is constant throughout the volume of the conductor

12. In a hollow spherical shell, potential (V) changes with respect to distance (s) from centre as

(a) 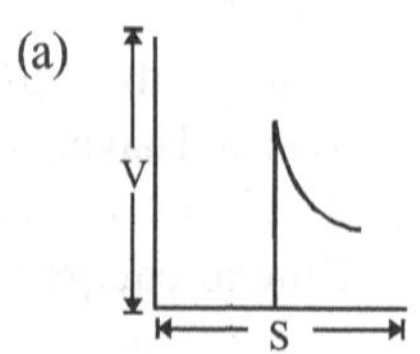

(b) 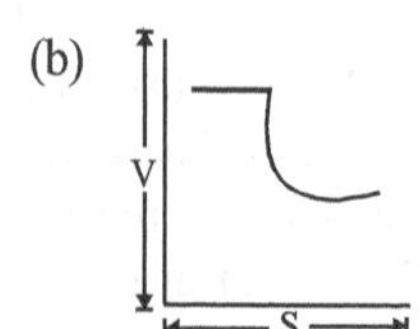

(c) 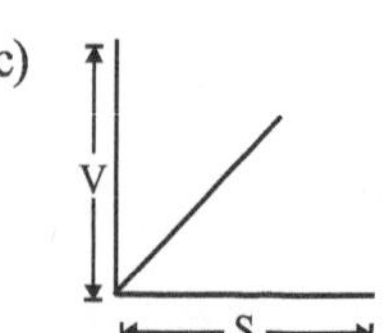

(d) 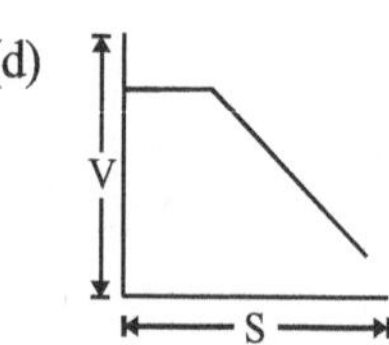

13. The 1000 small droplets of water each of radius r and charge Q, make a big drop of spherical shape. The potential of big drop is how many times the potential of one small droplet ?

(a) 1 (b) 10 (c) 100 (d) 1000

14. The work done in carrying a charge q once around a circle of radius r with a charge Q placed at the centre will be

(a) $Qq(4\pi\varepsilon_0 r^2)$ (b) $Qq/(4\pi\varepsilon_0 r)$
(c) zero (d) $Qq^2/(4\pi\varepsilon_0 r)$

15. A parallel plate condenser is filled with two dielectrics as shown. Area of each plate is A m^2 and the separation is t m. The dielectric constants are k_1 and k_2 respectively. Its capacitance in farad will be

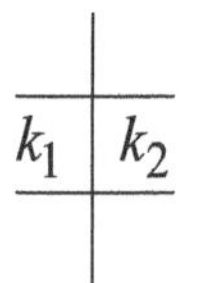

(a) $\frac{\varepsilon_0 A}{t}(k_1 + k_2)$ (b) $\frac{\varepsilon_0 A}{t}.\frac{k_1 + k_2}{2}$

(c) $\frac{2\varepsilon_0 A}{t}(k_1 + k_2)$ (d) $\frac{\varepsilon_0 A}{t}.\frac{k_1 - k_2}{2}$

16. Two metal pieces having a potential difference of 800 V are 0.02 m apart horizontally. A particle of mass 1.96×10^{-15} kg is suspended in equilibrium between the plates. If e is the elementary charge, then charge on the particle is

(a) 8 (b) 6 (c) 0.1 (d) 3

17. A one microfarad capacitor of a TV is subjected to 4000 V potential difference. The energy stored in capacitor is

(a) 8 J (b) 16 J
(c) 4×10^{-3} J (d) 2×10^{-3} J

18. An unchanged parallel plate capacitor filled with a dielectric constant K is connected to an air filled identical parallel capacitor charged to potential V_1. If the common potential is V_2, the value of K is

(a) $\frac{V_1 - V_2}{V_1}$ (b) $\frac{V_1}{V_1 - V_2}$

(c) $\frac{V_2}{V_1 - V_2}$ (d) $\frac{V_1 - V_2}{V_2}$

19. In the circuit given below, the charge in μC, on the capacitor having 5 μF is

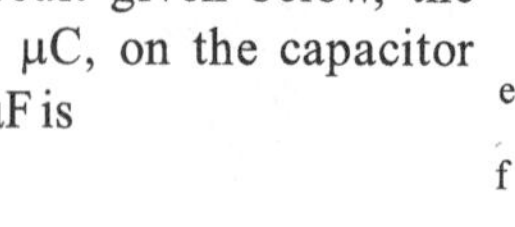

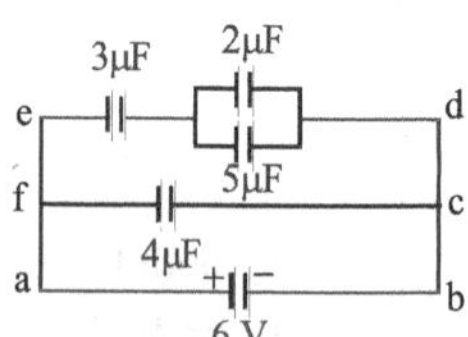

(a) 4.5
(b) 9
(c) 7
(d) 15

RESPONSE GRID				
7. ⓐⓑⓒⓓ	8. ⓐⓑⓒⓓ	9. ⓐⓑⓒⓓ	10. ⓐⓑⓒⓓ	11. ⓐⓑⓒⓓ
12. ⓐⓑⓒⓓ	13. ⓐⓑⓒⓓ	14. ⓐⓑⓒⓓ	15. ⓐⓑⓒⓓ	16. ⓐⓑⓒⓓ
17. ⓐⓑⓒⓓ	18. ⓐⓑⓒⓓ	19. ⓐⓑⓒⓓ		

Space for Rough Work

20. Two concentric, thin metallic spheres of radii R_1 and R_2 $(R_1 > R_2)$ bear charges Q_1 and Q_2 respectively. Then the potential at distance r between R_1 and R_2 will be

(a) $k\left(\frac{Q_1+Q_2}{r}\right)$ (b) $k\left(\frac{Q_1}{r}+\frac{Q_2}{R_2}\right)$

(c) $k\left(\frac{Q_2}{r}+\frac{Q_1}{R_1}\right)$ (d) $k\left(\frac{Q_1}{R_1}+\frac{Q_2}{R_2}\right)$

21. Charge Q on a capacitor varies with voltage V as shown in the figure, where Q is taken along the X-axis and V along the Y-axis. The area of triangle OAB represents

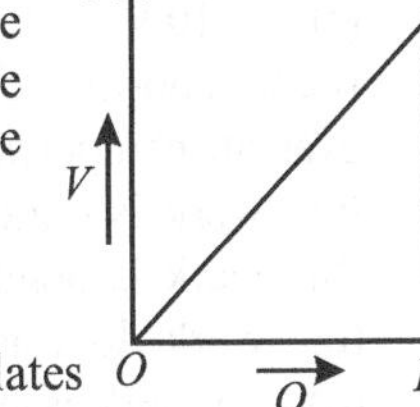

(a) capacitance
(b) capacitive reactance
(c) magnetic field between the plates
(d) energy stored in the capacitor

22. An alpha particle is accelerated through a potential difference of 10^6 volt. Its kinetic energy will be

(a) 1 MeV (b) 2 MeV (c) 4 MeV (d) 8 MeV

23. Four point charges $-Q$, $-q$, $2q$ and $2Q$ are placed, one at each corner of the square. The relation between Q and q for which the potential at the centre of the square is zero is :

(a) $Q=-q$ (b) $Q=-\frac{1}{q}$ (c) $Q=q$ (d) $Q=\frac{1}{q}$

24. A parallel plate capacitor having a separation between the plates d, plate area A and material with dielectric constant K has capacitance C_0. Now one-third of the material is replaced by another material with dielectric constant 2K, so that effectively there are two capacitors one with area 1/3A, dielectric constant 2K and another with area 2/3A and dielectric constant K. If the capacitance of this new capacitor is C then $\frac{C}{C_0}$ is

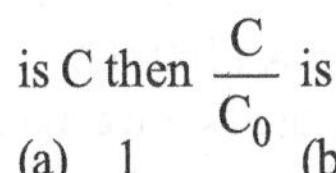

(a) 1 (b) 4/3 (c) 2/3 (d) 1/3

25. Two condensers, one of capacity C and other of capacity C/2 are connected to a V-volt battery, as shown. The work done in charging fully both the condensers is

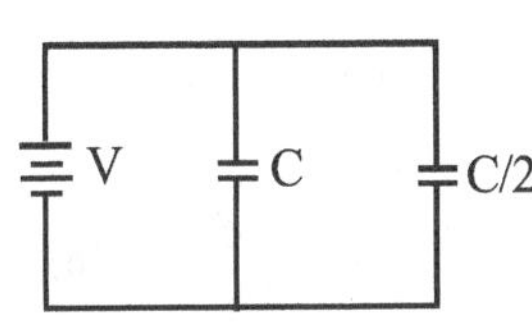

(a) $\frac{1}{4}CV^2$ (b) $\frac{3}{4}CV^2$ (c) $\frac{1}{2}CV^2$ (d) $2CV^2$.

26. A, B and C are three points in a uniform electric field. The electric potential is

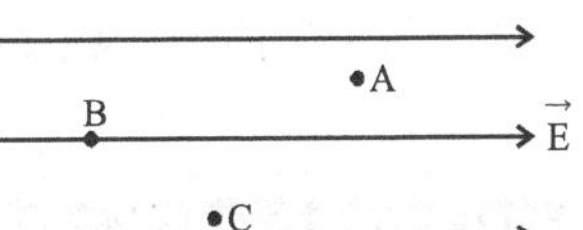

(a) maximum at B
(b) maximum at C
(c) same at all the three points A, B and C
(d) maximum at A

27. Three capacitors are connected in the arms of a triangle ABC as shown in figure 5 V is applied between A and B. The voltage between B and C is

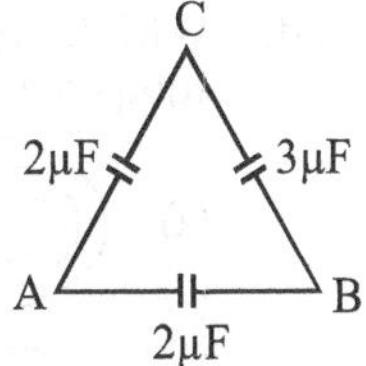

(a) 2V (b) 1V
(c) 3V (d) 1.5V

28. Two parallel metal plates having charges + Q and –Q face each other at a certain distance between them. If the plaves are now dipped in kerosene oil tank, the electric field between the plates will

(a) remain same (b) become zero
(c) increases (d) decrease

29. An air capacitor C connected to a battery of e.m.f. V acquires a charge q and energy E. The capacitor is disconnected from the battery and a dielectric slab is placed between the plates. Which of the following statements is correct ?

(a) V and q decrease but C and E increase
(b) V remains unchange, but q, E and C increase
(c) q remains unchanged, C increases, V and E decrease
(d) q and C increase but V and E decrease.

30. Choose the wrong statement about equipotential surfaces.

(a) It is a surface over which the potential is constant
(b) The electric field is parallel to the equipotential surface
(c) The electric field is perpendicular to the equipotential surface
(d) The electric field is in the direction of steepest decrease of potential

31. Two spherical conductors A and B of radii a and b (b>a) are placed concentrically in air. The two are connected by a copper wire as shown in figure. Then the equivalent capacitance of the system is

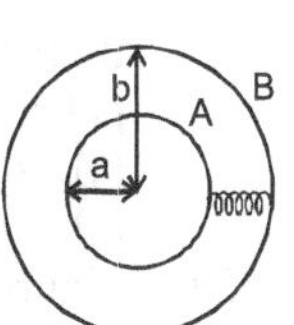

(a) $4\pi\varepsilon_0\frac{ab}{b-a}$ (b) $4\pi\varepsilon_0(a+b)$

(c) $4\pi\varepsilon_0 b$ (d) $4\pi\varepsilon_0 a$

32. A capacitor is charged to store an energy U. The charging battery is disconnected. An identical capacitor is now connected to the first capacitor in parallel. The energy in each of the capacitors is

(a) U/2 (b) 3U/2 (c) U (d) U/4

33. Equipotentials at a great distance from a collection of charges whose total sum is not zero are approximately

(a) spheres (b) planes
(c) paraboloids (d) ellipsoids

RESPONSE GRID					
	20. ⓐⓑⓒⓓ	21. ⓐⓑⓒⓓ	22. ⓐⓑⓒⓓ	23. ⓐⓑⓒⓓ	24. ⓐⓑⓒⓓ
	25. ⓐⓑⓒⓓ	26. ⓐⓑⓒⓓ	27. ⓐⓑⓒⓓ	28. ⓐⓑⓒⓓ	29. ⓐⓑⓒⓓ
	30. ⓐⓑⓒⓓ	31. ⓐⓑⓒⓓ	32. ⓐⓑⓒⓓ	33. ⓐⓑⓒⓓ	

34. Which of the following figure shows the correct equipotential surfaces of a system of two positive charges?

(a) 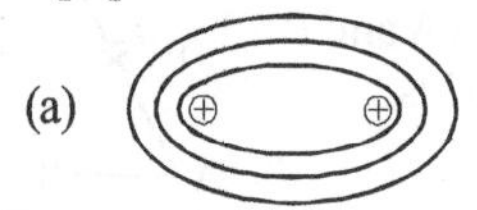 (b) 

(c)  (d)

35. Two identical metal plates are given positive charges Q_1 and Q_2 ($< Q_1$) respectively. If they are now brought close together to form a parallel plate capacitor with capacitance C, the potential difference between them is

(a) $\frac{Q_1+Q_2}{2C}$ (b) $\frac{Q_1+Q_2}{C}$ (c) $\frac{Q_1-Q_2}{C}$ (d) $\frac{Q_1-Q_2}{2C}$

36. The capacitance of the capacitor of plate areas A_1 and A_2 ($A_1 < A_2$) at a distance d, as shown in figure is

(a) 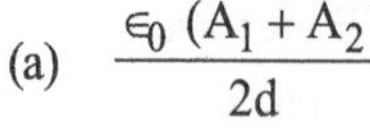 $\frac{\epsilon_0 (A_1+A_2)}{2d}$ (b) $\frac{\epsilon_0 A_2}{d}$

(c) $\frac{\epsilon_0 \sqrt{A_1 A_2}}{d}$ (d) $\frac{\epsilon_0 A_1}{d}$

37. In a given network the equivalent capacitance between A and B is [$C_1 = C_4 = 1\ \mu F$, $C_2 = C_3 = 2\mu F$]

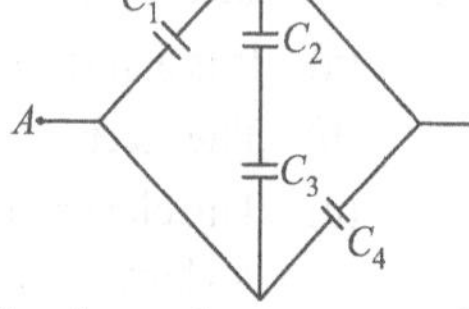

(a) $3\ \mu F$ (b) $6\ \mu F$

(c) $4.5\ \mu F$ (d) $2.5\ \mu F$

38. A parallel plate air capacitor is charged to a potential difference of V volts. After disconnecting the charging battery the distance between the plates of the capacitor is increased using an insulating handle. As a result the potential difference between the plates

(a) does not change (b) becomes zero

(c) increases (d) decreases

39. Figure shows three circular arcs, each of radius R and total charge as indicated. The net electric potential at the centre of curvature is

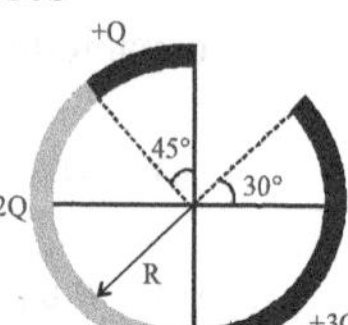

(a) $\frac{Q}{2\pi\varepsilon_0 R}$ (b) $\frac{Q}{4\pi\varepsilon_0 R}$

(c) $\frac{2Q}{\pi\varepsilon_0 R}$ (d) $\frac{Q}{\pi\varepsilon_0 R}$

40. An electric field $\vec{E} = (25\hat{i} + 30\hat{j}) NC^{-1}$ exists in a region of space. If the potential at the origin is taken to be zero then the potential at $x = 2\ m$, $y = 2\ m$ is :

(a) −110 V (b) −140 V (c) −120 V (d) −130 V

41. If a unit positive charge is taken from one point to another over an equipotential surface, then

(a) work is done on the charge

(b) work is done by the charge

(c) work done is constant

(d) no work is done

42. Three large plates A, B and C are placed parallel to each other and charges are given as shown. The charge that appears on the left surface of plate B is

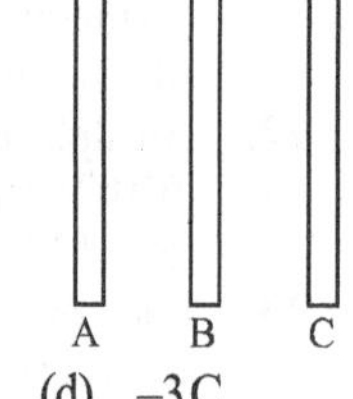

(a) 5C (b) 6C (c) 3C (d) −3C

43. Three charges 2 q, – q and – q are located at the vertices of an equilateral triangle. At the centre of the triangle

(a) the field is zero but potential is non-zero

(b) the field is non-zero, but potential is zero

(c) both field and potential are zero

(d) both field and potential are non-zero

44. If a charge – 150 nC is given to a concentric spherical shell and a charge +50 nC is placed at its centre then the charge on inner and outer surface of the shell is

(a) –50 nC, –100 nC (b) +50 nC, –200 nC

(c) –50 nC, –200 nC (d) 50 nC, 100 nC

45. Two capacitors of capacitances C_1 and C_2 are connected in parallel across a battery. If Q_1 and Q_2 respectively be the charges on the capacitors, then $\frac{Q_1}{Q_2}$ will be equal to

(a) $\frac{C_2}{C_1}$ (b) $\frac{C_1}{C_2}$ (c) $\frac{C_1^2}{C_2^2}$ (d) $\frac{C_2^2}{C_1^2}$

RESPONSE GRID					
	34. ⓐⓑⓒⓓ	35. ⓐⓑⓒⓓ	36. ⓐⓑⓒⓓ	37. ⓐⓑⓒⓓ	38. ⓐⓑⓒⓓ
	39. ⓐⓑⓒⓓ	40. ⓐⓑⓒⓓ	41. ⓐⓑⓒⓓ	42. ⓐⓑⓒⓓ	43. ⓐⓑⓒⓓ
	44. ⓐⓑⓒⓓ	45. ⓐⓑⓒⓓ			

DAILY PRACTICE PROBLEM DPP CHAPTERWISE CP16 - PHYSICS			
Total Questions	45	Total Marks	180
Attempted		Correct	
Incorrect		Net Score	
Cut-off Score	50	Qualifying Score	70
Success Gap = Net Score – Qualifying Score			
Net Score = (Correct × 4) – (Incorrect × 1)			

Space for Rough Work

DPP - Daily Practice Problems

Chapter-wise Sheets

Date : ☐ **Start Time :** ☐ **End Time :** ☐

PHYSICS

SYLLABUS : Current Electricity

Max. Marks : 180 **Marking Scheme :** (+4) for correct & (–1) for incorrect answer **Time : 60 min.**

INSTRUCTIONS : This Daily Practice Problem Sheet contains 45 MCQs. For each question only one option is correct. Darken the correct circle/ bubble in the Response Grid provided on each page.

1. When 5V potential difference is applied across a wire of length 0.1 m, the drift speed of electrons is $2.5 \times 10^{-4}\ ms^{-1}$. If the electron density in the wire is $8 \times 10^{28}\ m^{-3}$, the resistivity of the material is close to :

(a) $1.6 \times 10^{-6}\ \Omega m$ (b) $1.6 \times 10^{-5}\ \Omega m$
(c) $1.6 \times 10^{-8}\ \Omega m$ (d) $1.6 \times 10^{-7}\ \Omega m$

2. Variation of current passing through a conductor as the voltage applied across its ends is varied as shown in the adjoining diagram. If the resistance (R) is determined at the points A, B, C and D, we will find that

(a) $R_C = R_D$
(b) $R_B > R_A$
(c) $R_C > R_B$
(d) $R_A > R_B$

3. The length of a wire of a potentiometer is 100 cm, and the e.m.f. of its standard cell is E volt. It is employed to measure the e.m.f. of a battery whose internal resistance is 0.5Ω. If the balance point is obtained at $\ell = 30$ cm from the positive end, the e.m.f. of the battery is

(a) $\dfrac{30E}{100.5}$ (b) $\dfrac{30E}{(100-0.5)}$
(c) $\dfrac{30(E-0.5i)}{100}$ (d) $\dfrac{30E}{100}$

4. The masses of the three wires of copper are in the ratio of 1 : 3 : 5 and their lengths are in the ratio of 5 : 3 : 1. The ratio of their electrical resistance is

(a) 1 : 3 : 5 (b) 5 : 3 : 1
(c) 1 : 25 : 125 (d) 125 : 15 : 1

5. n equal resistors are first connected in series and then connected in parallel. What is the ratio of the maximum to the minimum resistance ?

(a) n (b) $1/n^2$ (c) n^2 (d) $1/n$

6. A battery is charged at a potential of 15V for 8 hours when the current flowing is 10A. The battery on discharge supplies a current of 5A for 15 hours. The mean terminal voltage during discharge is 14V. The "watt-hour" efficiency of the battery is

(a) 87.5% (b) 82.5% (c) 80% (d) 90%

RESPONSE GRID	1. ⓐⓑⓒⓓ 2. ⓐⓑⓒⓓ 3. ⓐⓑⓒⓓ 4. ⓐⓑⓒⓓ 5. ⓐⓑⓒⓓ 6. ⓐⓑⓒⓓ

Space for Rough Work

7. Shown in the figure below is a meter-bridge set up with null deflection in the galvanometer.

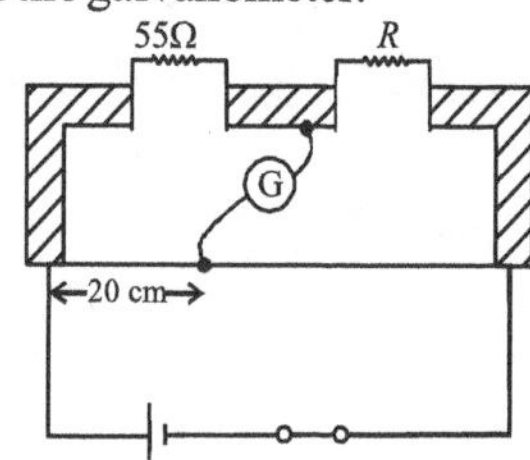

The value of the unknown resistor R is
(a) 13.75 Ω (b) 220 Ω (c) 110 Ω (d) 55 Ω

8. In the equation AB = C, A is the current density, C is the electric field, Then B is
(a) resistivity (b) conductivity
(c) potential difference (d) resistance

9. The Kirchhoff's first law $(\Sigma i = 0)$ and second law $(\Sigma iR = \Sigma E)$, are respectively based on
(a) conservation of charge, conservation of momentum
(b) conservation of energy, conservation of charge
(c) conservation of momentum, conservation of charge
(d) conservation of charge, conservation of energy

10. You are given a resistance coil and a battery. In which of the following cases the largest amount of heat generated ?
(a) When the coil is connected to the battery directly
(b) When the coil is divided into two equal parts and both the parts are connected to the battery in parallel
(c) When the coil is divided into four equal parts and all the four parts are connected to the battery in parallel
(d) When only half the coil is connected to the battery

11. The resistance of the coil of an ammeter is R. The shunt required to increase its range n-fold should have a resistance
(a) $\frac{R}{n}$ (b) $\frac{R}{n-1}$ (c) $\frac{R}{n+1}$ (d) nR

12. On increasing the temperature of a conductor, its resistance increases because the
(a) relaxation time increases
(b) mass of electron increases
(c) electron density decreases
(d) relaxation time decreases

13. An electric current is passed through a circuit containing two wires of the same material, connected in parallel. If the lengths and radii are in the ratio of $\frac{4}{3}$ and $\frac{2}{3}$, then the ratio of the current passing through the wires will be
(a) 8/9 (b) 1/3 (c) 3 (d) 2

14. In a meter bridge experiment null point is obtained at 20 cm. from one end of the wire when resistance X is balanced against another resistance Y. If $X < Y$, then where will be the new position of the null point from the same end, if one decides to balance a resistance of 4 X against Y
(a) 40 cm (b) 80 cm (c) 50 cm (d) 70 cm

15. In the circuit shown, the current through 8 ohm is same before and after connecting E. The value of E is
(a) 12 V
(b) 6 V
(c) 4 V
(d) 2 V

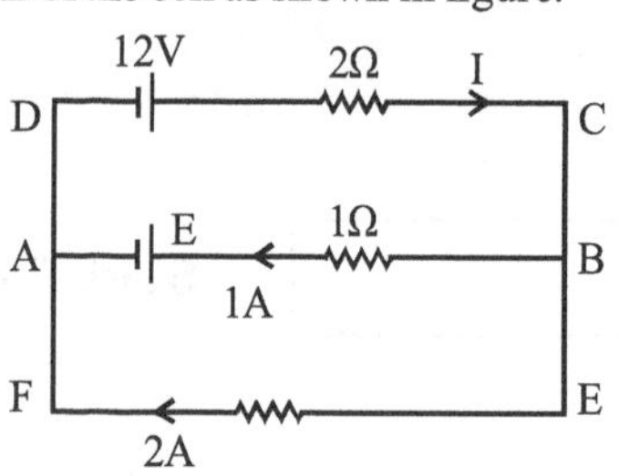

16. Find emf E of the cell as shown in figure.

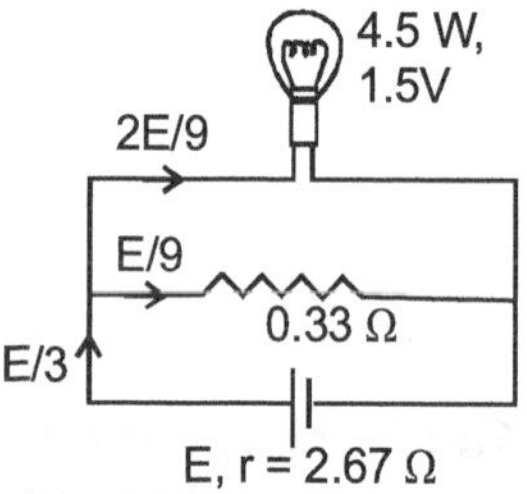

(a) 15V (b) 10V (c) 12V (d) 5V

17. A torch bulb rated as 4.5 W, 1.5 V is connected as shown in fig. The e.m.f. of the cell, needed to make the bulb glow at full intensity is

4.5 W,
1.5V
2E/9
E/9
0.33 Ω
E/3
E, r = 2.67 Ω

(a) 4.5 V (b) 1.5 V (c) 2.67 V (d) 13.5 V

18. In a given network, each resistance has value of 6Ω. The point X is connected to point A by a copper wire of negligible resistance and point Y is connected to point B by the same wire. The effective resistance between X and Y will be

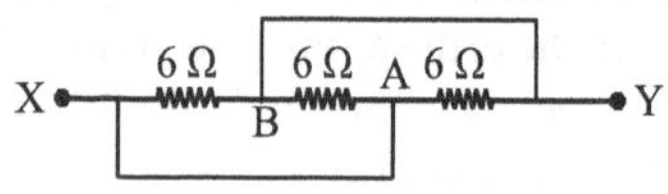

(a) 18Ω (b) 6Ω (c) 3Ω (d) 2Ω

19. If N, e, τ and m are representing electron density, charge, relaxation time and mass of an electron respectively, then the resistance of wire of length ℓ and cross-sectional area A is given by
(a) $\frac{2m\ell}{Ne^2A\tau}$ (b) $\frac{2m\tau A}{Ne^2\ell}$ (c) $\frac{Ne^2\tau A}{2m\ell}$ (d) $\frac{Ne^2A}{2m\tau\ell}$

20. Cell having an emf ε and internal resistance r is connected across a variable external resistance R. As the resistance R is increased, the plot of potential difference V across R is given by :

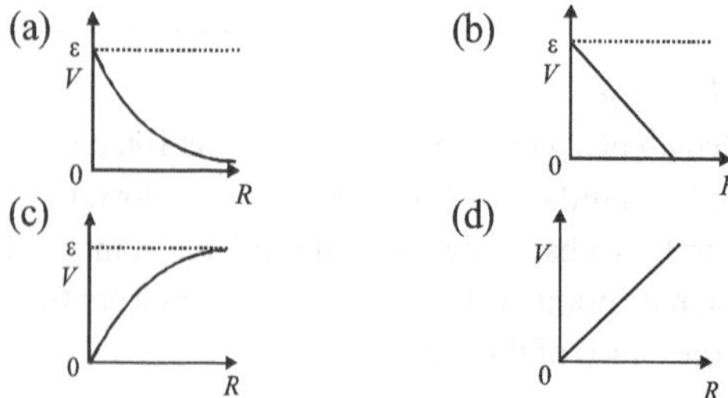

RESPONSE GRID				
7. ⓐⓑⓒⓓ	8. ⓐⓑⓒⓓ	9. ⓐⓑⓒⓓ	10. ⓐⓑⓒⓓ	11. ⓐⓑⓒⓓ
12. ⓐⓑⓒⓓ	13. ⓐⓑⓒⓓ	14. ⓐⓑⓒⓓ	15. ⓐⓑⓒⓓ	16. ⓐⓑⓒⓓ
17. ⓐⓑⓒⓓ	18. ⓐⓑⓒⓓ	19. ⓐⓑⓒⓓ	20. ⓐⓑⓒⓓ	

21. If voltage across a bulb rated 220 Volt-100 Watt drops by 2.5% of its rated value, the percentage of the rated value by which the power would decrease is :
(a) 20% (b) 2.5% (c) 5% (d) 10%

22. If specific resistance of a potentiometer wire is $10^{-7}\ \Omega$m, the current flow through it is 0.1 A and the cross-sectional area of wire is $10^{-6}\ m^2$ then potential gradient will be
(a) 10^{-2} volt/m (b) 10^{-4} volt/m
(c) 10^{-6} volt/m (d) 10^{-8} volt/m

23. Two resistances R_1 and R_2 are made of different materials. The temperature coefficient of the material of R_1 is α and that of material of R_2 is $-\beta$. The resistance of the series combination of R_1 and R_2 will not change with temperature if $\frac{R_1}{R_2}$ equal to
(a) $\frac{\alpha}{\beta}$ (b) $\frac{\alpha+\beta}{\alpha-\beta}$ (c) $\frac{\alpha^2+\beta^2}{2\alpha\beta}$ (d) $\frac{\beta}{\alpha}$

24. Five cells each of emf E and internal resistance r send the same amount of current through an external resistance R whether the cells are connected in parallel or in series. Then the ratio $\left(\frac{R}{r}\right)$ is
(a) 2 (b) $\frac{1}{2}$ (c) $\frac{1}{5}$ (d) 1

25. The length of a given cylindrical wire is increased by 100%. Due to the consequent decrease in diameter the change in the resistance of the wire will be
(a) 200% (b) 100% (c) 50% (d) 300%

26. Potentiometer wire of length 1 m is connected in series with 490Ω resistance and 2 V battery. If 0.2 mV/cm is the potential gradient, then resistance of the potentiometer wire is
(a) 4.9 Ω (b) 7.9 Ω (c) 5.9 Ω (d) 6.9 Ω

27. See the electric circuit shown in the figure. Which of the following equations is a correct equation for it?
(a) $\varepsilon_2 - i_2 r_2 - \varepsilon_1 - i_1 r_1 = 0$
(b) $-\varepsilon_2 - (i_1+i_2)R + i_2 r_2 = 0$
(c) $\varepsilon_1 - (i_1+i_2)R + i_1 r_1 = 0$
(d) $\varepsilon_1 - (i_1+i_2)R - i_1 r_1 = 0$

28. In a large building, there are 15 bulbs of 40 W, 5 bulbs of 100 W, 5 fans of 80 W and 1 heater of 1 kW. The voltage of electric mains is 220 V. The minimum capacity of the main fuse of the building will be:
(a) 8 A (b) 10 A (c) 12 A (d) 14 A

29. Two sources of equal emf are connected to an external resistance R. The internal resistance of the two sources are R_1 and R_2 ($R_1 > R_1$). If the potential difference across the source having internal resistance R_2 is zero, then
(a) $R = R_2 - R_1$
(b) $R = R_2 \times (R_1 + R_2)/(R_2 - R_1)$
(c) $R = R_1R_2/(R_2 - R_1)$
(d) $R = R_1R_2/(R_1 - R_2)$

30. The resistance of the series combination of two resistances is S. when they are joined in parallel the total resistance is P. If $S = nP$ then the minimum possible value of n is
(a) 2 (b) 3 (c) 4 (d) 1

31. If an ammeter is to be used in place of a voltmeter, then we must connect with the ammeter a
(a) low resistance in parallel
(b) high resistance in parallel
(c) high resistance in series
(d) low resistance in series.

32. A d.c. main supply of e.m.f. 220 V is connected across a storage battery of e.m.f. 200 V through a resistance of 1Ω. The battery terminals are connected to an external resistance 'R'. The minimum value of 'R', so that a current passes through the battery to charge it is:
(a) 7 Ω (b) 9 Ω (c) 11 Ω (d) Zero

33. In the given circuit diagram when the current reaches steady state in the circuit, the charge on the capacitor of capacitance C will be :
(a) $CE\frac{r_2}{(r+r_2)}$
(b) $CE\frac{r_1}{(r_1+r)}$
(c) 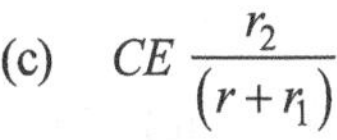 $CE\frac{r_2}{(r+r_1)}$
(d) $CE\frac{r_1}{(r_2+r)}$

34. Suppose the drift velocity v_d in a material varied with the applied electric field E as $v_d \propto \sqrt{E}$. Then V – I graph for a wire made of such a material is best given by :

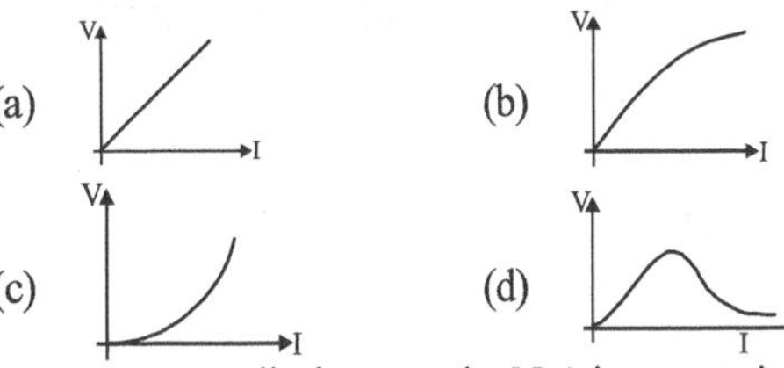

35. In a neon gas discharge tube Ne^+ ions moving through a cross-section of the tube each second to the right is 2.9×10^{18}, while 1.2×10^{18} electrons move towards left in the same time; the electronic charge being 1.6×10^{-19} C, the net electric current is
(a) 0.27 A to the right (b) 0.66 A to the right
(c) 0.66 A to the left (d) zero

36. Two rods are joined end to end, as shown. Both have a cross-sectional area of 0.01 cm^2. Each is 1 meter long. One rod is of copper with a resistivity of 1.7×10^{-6} ohm-centimeter, the other is of iron with a resistivity of 10^{-5} ohm-centimeter. How much voltage is required to produce a current of 1 ampere in the rods?
(a) 0.117V
(b) 0.00145V
(c) 0.0145V
(d) 1.7×10^{-6} V

RESPONSE GRID					
	21. ⓐⓑⓒⓓ	22. ⓐⓑⓒⓓ	23. ⓐⓑⓒⓓ	24. ⓐⓑⓒⓓ	25. ⓐⓑⓒⓓ
	26. ⓐⓑⓒⓓ	27. ⓐⓑⓒⓓ	28. ⓐⓑⓒⓓ	29. ⓐⓑⓒⓓ	30. ⓐⓑⓒⓓ
	31. ⓐⓑⓒⓓ	32. ⓐⓑⓒⓓ	33. ⓐⓑⓒⓓ	34. ⓐⓑⓒⓓ	35. ⓐⓑⓒⓓ
	36. ⓐⓑⓒⓓ				

Space for Rough Work

37. An energy source will supply a constant current into the load if its internal resistance is
(a) very large as compared to the load resistance
(b) equal to the resistance of the load
(c) non-zero but less than the resistance of the load
(d) zero

38. The resistance of a wire at room temperature 30°C is found to be 10 Ω. Now to increase the resistance by 10%, the temperature of the wire must be [The temperature coefficient of resistance of the material of the wire is 0.002 per °C]
(a) 36°C (b) 83°C
(c) 63°C (d) 33°C

39. If current flowing in a conductor changes by 1% then power consumed will change by
(a) 10% (b) 2% (c) 1% (d) 100%

40. In the circuit shown in figure, the 5Ω resistance develops 20.00 cal/s due to the current flowing through it. The heat developed in 2 Ω resistance (in cal/s) is

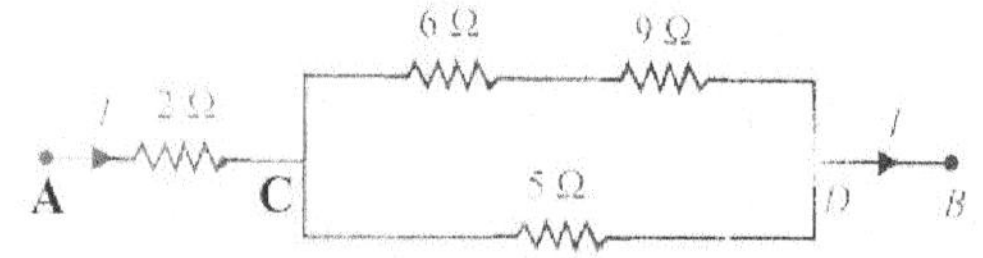

(a) 23.8 (b) 14.2 (c) 11.9 (d) 7.1

41. In a Wheatstone's bridge, three resistances P, Q and R connected in the three arms and the fourth arm is formed by two resistances S_1 and S_2 connected in parallel. The condition for the bridge to be balanced will be
(a) $\frac{P}{Q}=\frac{2R}{S_1+S_2}$ (b) $\frac{P}{Q}=\frac{R(S_1+S_2)}{S_1S_2}$
(c) $\frac{P}{Q}=\frac{R(S_1+S_2)}{2S_1S_2}$ (d) $\frac{P}{Q}=\frac{R}{S_1+S_2}$

42. The electric resistance of a certain wire of iron is R. If its length and radius are both doubled, then
(a) the resistance and the specific resistance, will both remain unchanged
(b) the resistance will be doubled and the specific resistance will be halved
(c) the resistance will be halved and the specific resistance will remain unchanged
(d) the resistance will be halved and the specific resistance will be doubled

43. A car battery has e.m.f. 12 volt and internal resistance 5×10^{-2} ohm. If it draws 60 amp current, the terminal voltage of the battery will be
(a) 15 volt (b) 3 volt (c) 5 volt (d) 9 volt

44. A conducting wire of cross-sectional area 1 cm^2 has 3×10^{23} charge carriers per m^3. If wire carries a current of 24 mA, then drift velocity of carriers is
(a) 5×10^{-2} m/s (b) 0.5 m/s
(c) 5×10^{-3} m/s (d) 5×10^{-6} m/s

45. In the series combination of n cells each cell having emf ε and internal resistance r. If three cells are wrongly connected, then total emf and internal resistance of this combination will be
(a) $n\varepsilon$, $(nr - 3r)$ (b) $(n\varepsilon - 2\varepsilon)$ nr
(c) $(n\varepsilon - 4\varepsilon)$, nr (d) $(n\varepsilon - 6\varepsilon)$, nr

RESPONSE GRID	37. ⓐⓑⓒⓓ	38. ⓐⓑⓒⓓ	39. ⓐⓑⓒⓓ	40. ⓐⓑⓒⓓ	41. ⓐⓑⓒⓓ
	42. ⓐⓑⓒⓓ	43. ⓐⓑⓒⓓ	44. ⓐⓑⓒⓓ	45. ⓐⓑⓒⓓ	

DAILY PRACTICE PROBLEM DPP CHAPTERWISE CP17 - PHYSICS

Total Questions	45	Total Marks	180
Attempted		Correct	
Incorrect		Net Score	
Cut-off Score	45	Qualifying Score	60
Success Gap = Net Score – Qualifying Score			
Net Score = (Correct × 4) – (Incorrect × 1)			

Space for Rough Work

DPP - Daily Practice Problems

Chapter-wise Sheets

Date : **Start Time :** **End Time :**

PHYSICS

CP18

SYLLABUS : Moving Charges and Magnetism

Max. Marks : 180 **Marking Scheme :** (+4) for correct & (–1) for incorrect answer **Time : 60 min.**

INSTRUCTIONS : This Daily Practice Problem Sheet contains 45 MCQs. For each question only one option is correct. Darken the correct circle/ bubble in the Response Grid provided on each page.

1. An insulating rod of length ℓ carries a charge q distributed uniformly on it. The rod is pivoted at its mid point and is rotated at a frequency f about a fixed axis perpendicular to rod and passing through the pivot. The magnetic moment of the rod system is $\frac{1}{2a}\pi q f \ell^2$. Find the value of a.

   (a) 6 (b) 4
   (c) 5 (d) 8

2. A portion of a conductive wire is bent in the form of a semicircle of radius r as shown below in fig. At the centre of semicircle, the magnetic induction will be

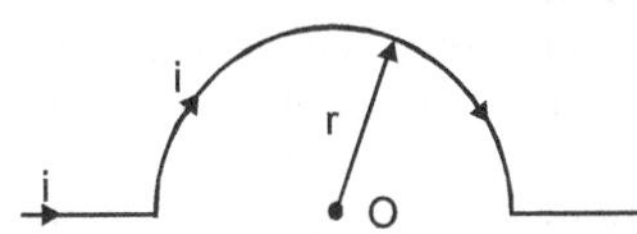

   (a) zero (b) infinite
   (c) $\frac{\mu_0}{4\pi}.\frac{2\pi i}{r}$ (d) $\frac{\mu_0}{4\pi}.\frac{\pi i}{r}$

3. A closely wound solenoid of 2000 turns and area of cross-section 1.5×10^{-4} m^2 carries a current of 2.0 *A*. It suspended through its centre and perpendicular to its length, allowing it to turn in a horizontal plane in a uniform magnetic field 5×10^{-2} tesla making an angle of 30° with the axis of the solenoid. The torque on the solenoid will be:

   (a) 3×10^{-2} N-m (b) 3×10^{-3} N-m
   (c) 1.5×10^{-3} N-m (d) 1.5×10^{-2} N-m

4. An alternating electric field, of frequency *v*, is applied across the dees (radius = *R*) of a cyclotron that is being used to accelerate protons (mass = *m*). The operating magnetic field (*B*) used in the cyclotron and the kinetic energy (*K*) of the proton beam, produced by it, are given by :

   (a) $B = \frac{m\nu}{e}$ and $K = 2m\pi^2\nu^2R^2$
   (b) $B = \frac{2\pi m\nu}{e}$ and $K = m^2\pi\nu R^2$
   (c) $B = \frac{2\pi m\nu}{e}$ and $K = 2m\pi^2\nu^2R^2$
   (d) $B = \frac{m\nu}{e}$ and $K = m^2\pi\nu R^2$

5. A galvanometer of 50 ohm resistance has 25 divisions. A current of 4×10^{-4} ampere gives a deflection of one per division. To convert this galvanometer into a voltmeter having a range of 25 volts, it should be connected with a resistance of

   (a) 2450 Ω in series (b) 2500 Ω in series.
   (c) 245 Ω in series. (d) 2550 Ω in series.

RESPONSE GRID	1. ⓐⓑⓒⓓ	2. ⓐⓑⓒⓓ	3. ⓐⓑⓒⓓ	4. ⓐⓑⓒⓓ	5. ⓐⓑⓒⓓ

Space for Rough Work

6. If we double the radius of a coil keeping the current through it unchanged, then the magnetic field at any point at a large distance from the centre becomes approximately
(a) double (b) three times
(c) four times (d) one-fourth

7. A particle of mass m, charge Q and kinetic energy T enters a transverse uniform magnetic field of induction $\vec{B}$. After 3 seconds, the kinetic energy of the particle will be:
(a) 3T (b) 2T
(c) T (d) 4T

8. A 10 eV electron is circulating in a plane at right angles to a uniform field at magnetic induction 10^{-4} Wb/m^2 (= 1.0 gauss). The orbital radius of the electron is
(a) 12 cm (b) 16 cm
(c) 11 cm (d) 18 cm

9. A uniform electric field and a uniform magnetic field exist in a region in the same direction. An electron is projected with velocity pointed in the same direction. The electron will
(a) turn to its right
(b) turn to its left
(c) keep moving in the same direction but its speed will increase
(d) keep moving in the same direction but its speed will decrease

10. Proton, deuteron and alpha particle of same kinetic energy are moving in circular trajectories in a constant magnetic field. The radii of proton, deuteron and alpha particle are respectively r_p, r_d and r_α. Which one of the following relation is correct?
(a) $r_\alpha = r_p = r_d$ (b) $r_\alpha = r_p < r_d$
(c) $r_\alpha > r_d > r_p$ (d) $r_\alpha = r_d > r_p$

11. A moving coil galvanometer has 150 equal divisions. Its current sensitivity is 10-divisions per milliampere and voltage sensitivity is 2 divisions per millivolt. In order that each division reads 1 volt, the resistance in ohms needed to be connected in series with the coil will be
(a) 10^5 (b) 10^3 (c) 9995 (d) 99995

12. A 2 μC charge moving around a circle with a frequency of 6.25×10^{12} Hz produces a magnetic field 6.28 tesla at the centre of the circle. The radius of the circle is
(a) 2.25 m (b) 0.25 m (c) 13.0 m (d) 1.25 m

13. A charged particle with charge q enters a region of constant, uniform and mutually orthogonal fields $\vec{E}$ and $\vec{B}$ with a velocity $\vec{v}$ perpendicular to both $\vec{E}$ and $\vec{B}$, and comes out without any change in magnitude or direction of $\vec{v}$. Then
(a) $\vec{v} = \vec{B} \times \vec{E} / E^2$ (b) $\vec{v} = \vec{E} \times \vec{B} / B^2$
(c) $\vec{v} = \vec{B} \times \vec{E} / B^2$ (d) $\vec{v} = \vec{E} \times \vec{B} / E^2$

14. A square current carrying loop is suspended in a uniform magnetic field acting in the plane of the loop. If the force on one arm of the loop is $\vec{F}$, the net force on the remaining three arms of the loop is
(a) $3\vec{F}$ (b) $-\vec{F}$
(c) $-3\vec{F}$ (d) $\vec{F}$

15. A straight section PQ of a circuit lies along the X-axis from $x = -\frac{a}{2}$ to $x = \frac{a}{2}$ and carries a steady current i. The magnetic field due to the section PQ at a point $X = +a$ will be
(a) proportional to a (b) proportional to a^2
(c) proportional to $1/a$ (d) zero

16. A and B are two conductors carrying a current i in the same direction. x and y are two electron beams moving in the same direction. Then

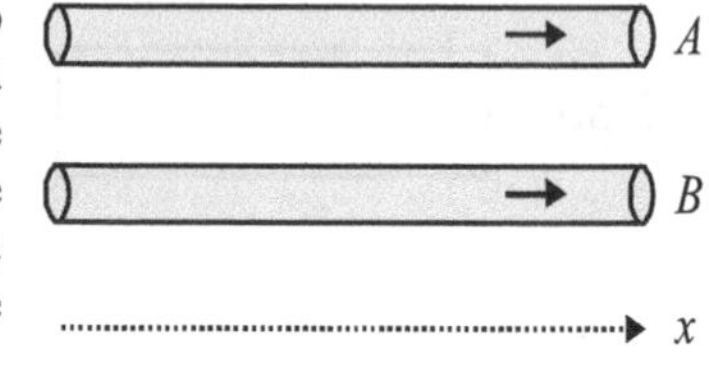

(a) there will be repulsion between A and B, attraction between x and y
(b) there will be attraction between A and B, repulsion between x and y
(c) there will be repulsion between A and B and also x and y
(d) there will be attraction between A and B and also x and y

17. A galvanometer of resistance, G is shunted by a resistance S ohm. To keep the main current in the circuit unchanged, the resistance to be put in series with the galvanometer is
(a) $\frac{S^2}{(S+G)}$ (b) $\frac{SG}{(S+G)}$ (c) $\frac{G^2}{(S+G)}$ (d) $\frac{G}{(S+G)}$

18. A current I flows in an infinitely long wire with cross section in the form of a semi-circular ring of radius R. The magnitude of the magnetic induction along its axis is:
(a) $\frac{\mu_0 I}{2\pi^2 R}$ (b) $\frac{\mu_0 I}{2\pi R}$ (c) $\frac{\mu_0 I}{4\pi R}$ (d) $\frac{\mu_0 I}{\pi^2 R}$

19. Two equal electric currents are flowing perpendicular to each other as shown in the figure. AB and CD are perpendicular to each other and symmetrically placed with respect to the current flow. Where do we expect the resultant magnetic field to be zero?

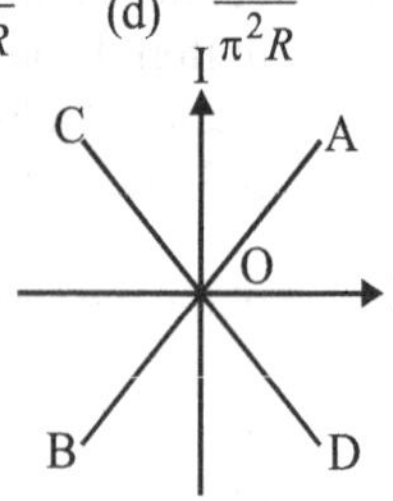

(a) On AB (b) On CD
(c) On both AB and CD (d) On both OD and BO

20. A closed loop PQRS carrying a current is placed in a uniform magnetic field.
If the magnetic forces on segments PS, SR, and RQ are F_1, F_2 and F_3 respectively and are in the plane of the paper and along the directions shown, the force on the segment QP is

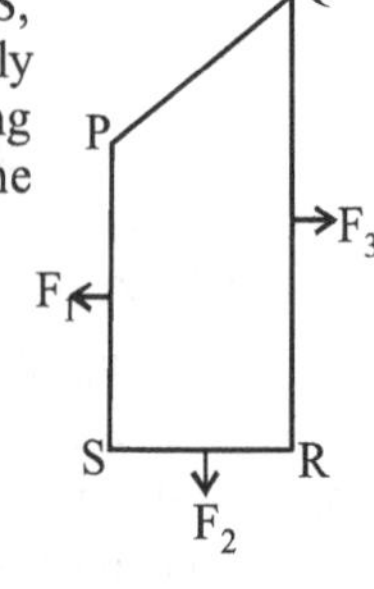

(a) $F_3 - F_1 - F_2$
(b) $\sqrt{(F_3 - F_1)^2 + F_2^2}$
(c) $\sqrt{(F_3 - F_1)^2 - F_2^2}$
(d) $F_3 - F_1 + F_2$

RESPONSE GRID					
	6. ⓐⓑⓒⓓ	7. ⓐⓑⓒⓓ	8. ⓐⓑⓒⓓ	9. ⓐⓑⓒⓓ	10. ⓐⓑⓒⓓ
	11. ⓐⓑⓒⓓ	12. ⓐⓑⓒⓓ	13. ⓐⓑⓒⓓ	14. ⓐⓑⓒⓓ	15. ⓐⓑⓒⓓ
	16. ⓐⓑⓒⓓ	17. ⓐⓑⓒⓓ	18. ⓐⓑⓒⓓ	19. ⓐⓑⓒⓓ	20. ⓐⓑⓒⓓ

Space for Rough Work

21. A long solenoid carrying a current produces a magnetic field B along its axis. If the current is double and the number of turns per cm is halved, the new value of the magnetic field is
(a) $4B$ (b) $B/2$ (c) B (d) $2B$

22. A particle of charge q and mass m moves in a circular orbit of radius r with angular speed ω. The ratio of the magnitude of its magnetic moment to that of its angular momentum depends on
(a) ω and q (b) ω, q and m
(c) q and m (d) ω and m

23. A current loop in a magnetic field
(a) can be in equilibrium in one orientation
(b) can be in equilibrium in two orientations, both the equilibrium states are unstable
(c) can be in equilibrium in two orientations, one stable while the other is unstable
(d) experiences a torque whether the field is uniform or non-uniform in all orientations

24. Two long parallel wires P and Q are held perpendicular to the plane of paper with distance of 5 m between them. If P and Q carry current of 2.5 amp. and 5 amp. respectively in the same direction, then the magnetic field at a point half-way between the wires is
(a) $\mu_0/17$ (b) $\sqrt{3}\mu_0/2\pi$
(c) $\mu_0/2\pi$ (d) $3\mu_0/2\pi$

25. A very long straight wire carries a current I. At the instant when a charge $+Q$ at point P has velocity $\vec{v}$, as shown, the force on the charge is

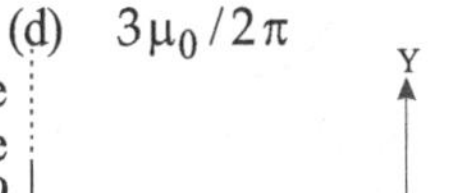

(a) along OY (b) opposite to OY
(c) along OX (d) opposite to OX

26. Two wires with currents 2 A and 1 A are enclosed in a circular loop. Another wire with current 3 A is situated outside the loop as shown. The $\oint \vec{B}.d\vec{l}$ around the loop is
(a) μ_0
(b) $3\mu_0$
(c) $6\mu_0$
(d) $2\mu_0$

27. If in a circular coil A of radius R, current I is flowing and in another coil B of radius $2R$ a current $2I$ is flowing, then the ratio of the magnetic fields B_A and B_B, produced by them will be
(a) 1 (b) 2 (c) 1/2 (d) 4

28. A charged particle moves through a magnetic field perpendicular to its direction. Then
(a) kinetic energy changes but the momentum is constant
(b) the momentum changes but the kinetic energy is constant
(c) both momentum and kinetic energy of the particle are not constant
(d) both momentum and kinetic energy of the particle are constant

29. The deflection in a galvanometer falls from 50 division to 20 when a 12 ohm shunt is applied. The galvanometer resistance is
(a) 18 ohm (b) 36 ohm (c) 24 ohm (d) 30 ohm

30. When a long wire carrying a steady current is bent into a circular coil of one turn, the magnetic induction at its centre is B. When the same wire carrying the same current is bent to form a circular coil of n turns of a smaller radius, the magnetic induction at the centre will be
(a) B/n (b) nB (c) B/n^2 (d) n^2B

31. The magnetic field due to a current carrying circular loop of radius 3 cm at a point on the axis at a distance of 4 cm from the centre is 54 μT. What will be its value at the centre of loop ?
(a) 125 μT (b) 150 μT
(c) 250 μT (d) 75 μT

32. A charge moving with velocity v in X-direction is subjected to a field of magnetic induction in negative X-direction. As a result, the charge will
(a) remain unaffected
(b) start moving in a circular path Y–Z plane
(c) retard along X-axis
(d) move along a helical path around X-axis

33. An electron travelling with a speed u along the positive x-axis enters into a region of magnetic field where $B = -B_0\hat{k}$ $(x > 0)$. It comes out of the region with speed v then

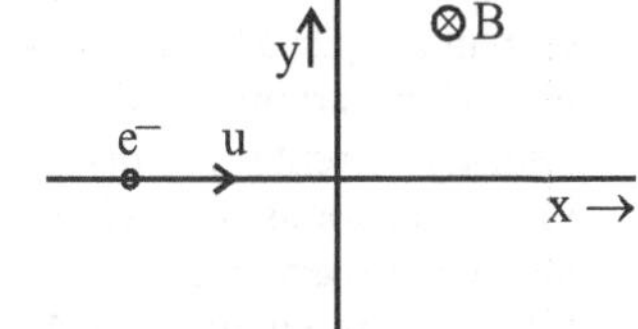

(a) $v = u$ at $y > 0$
(b) $v = u$ at $y < 0$
(c) $v > u$ at $y > 0$
(d) $v > u$ at $y < 0$

34. If an ammeter is to be used in place of a voltmeter, then we must connect with the ammeter a
(a) low resistance in parallel
(b) high resistance in parallel
(c) high resistance in series
(d) low resistance in series

35. An infinite straight conductor carrying current 2 I is split into a loop of radius r as shown in fig. The magnetic field at the centre of the coil is

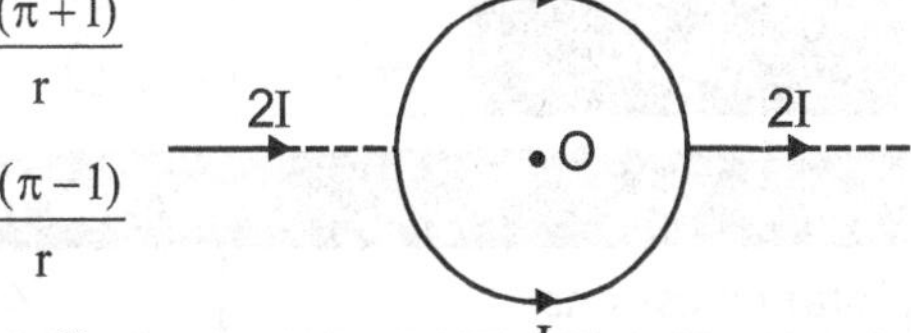

(a) $\frac{\mu_0}{4\pi}\frac{2(\pi+1)}{r}$
(b) $\frac{\mu_0}{4\pi}\frac{2(\pi-1)}{r}$
(c) $\frac{\mu_0}{4\pi}\frac{(\pi+1)}{r}$
(d) zero

RESPONSE GRID				
21. ⓐⓑⓒⓓ	22. ⓐⓑⓒⓓ	23. ⓐⓑⓒⓓ	24. ⓐⓑⓒⓓ	25. ⓐⓑⓒⓓ
26. ⓐⓑⓒⓓ	27. ⓐⓑⓒⓓ	28. ⓐⓑⓒⓓ	29. ⓐⓑⓒⓓ	30. ⓐⓑⓒⓓ
31. ⓐⓑⓒⓓ	32. ⓐⓑⓒⓓ	33. ⓐⓑⓒⓓ	34. ⓐⓑⓒⓓ	35. ⓐⓑⓒⓓ

Space for Rough Work

36. A parallel plate capacitor of area 60 cm^2 and separation 3 mm is charged initially to 90 μC. If the medium between the plate gets slightly conducting and the plate loses the charge initially at the rate of 2.5×10^{-8} C/s, then what is the magnetic field between the plates ?

(a) 2.5×10^{-8} T (b) 2.0×10^{-7} T
(c) 1.63×10^{-11} T (d) Zero

37. Four wires, each of length 2.0 m, are bent into four loops P, Q, R and S and then suspended in a uniform magnetic field. If the same current is passed in each, then the torque will be maximum on the loop

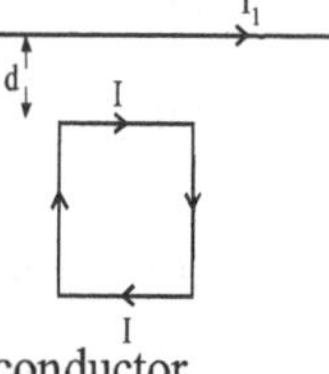

(a) P (b) Q (c) R (d) S

38. A certain region has an electric field $\vec{E} = (2\hat{i} - 3\hat{j})$ N/C and a uniform magnetic field $\vec{B} = (5\hat{i} + 3\hat{j} + 4\hat{k})$ T. The force experienced by a charge 1C moving with velocity $(\hat{i} + 2\hat{j})$ ms^{-1} is

(a) $(10\hat{i} - 7\hat{j} - 7\hat{k})$ (b) $(10\hat{i} + 7\hat{j} + 7\hat{k})$
(c) $(-10\hat{i} + 7\hat{j} + 7\hat{k})$ (d) $(10\hat{i} + 7\hat{j} - 7\hat{k})$

39. A galvanometer of resistance 100 Ω gives a full scale deflection for a current of 10^{-5} A. To convert it into a ammeter capable of measuring upto 1 A, we should connect a resistance of

(a) 1 Ω in parallel (b) 10^{-3} Ω in parallel
(c) 10^5 Ω in series (d) 100 Ω in series

40. A square loop, carrying a steady current I, is placed in a horizontal plane near a long straight conductor carrying a steady current I_1 at a distance d from the conductor as shown in figure. The loop will experience

(a) a net repulsive force away from the conductor
(b) a net torque acting upward perpendicular to the horizontal plane
(c) a net torque acting downward normal to the horizontal plane
(d) a net attractive force towards the conductor

41. Two coaxial solenoids of different radius carry current I in the same direction. $\vec{F_1}$ be the magnetic force on the inner solenoid due to the outer one and $\vec{F_2}$ be the magnetic force on the outer solenoid due to the inner one. Then :

(a) $\vec{F_1}$ is radially inwards and $\vec{F_2} = 0$
(b) $\vec{F_1}$ is radially outwards and $\vec{F_2} = 0$
(c) $\vec{F_1} = \vec{F_2} = 0$
(d) $\vec{F_1}$ is radially inwards and $\vec{F_2}$ is radially outwards

42. A beam of electrons is moving with constant velocity in a region having simultaneous perpendicular electric and magnetic fields of strength 20 Vm^{-1} and 0.5 T respectively at right angles to the direction of motion of the electrons. Then the velocity of electrons must be

(a) 8 m/s (b) 20 m/s (c) 40 m/s (d) $\frac{1}{40}$ m/s

43. The magnetic flux density B at a distance r from a long straight wire carrying a steady current varies with r as

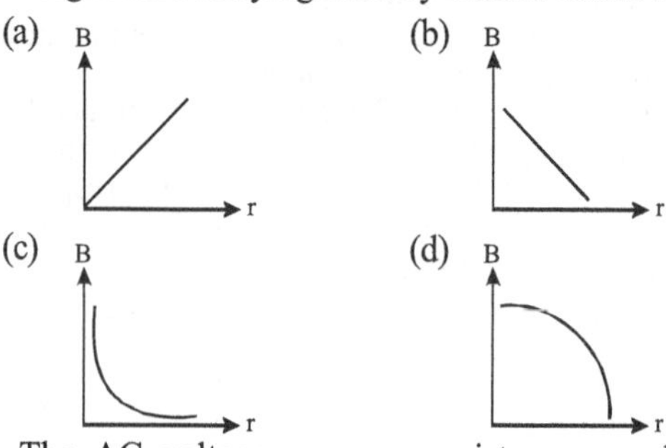

44. The AC voltage across a resistance can be measured using a :

(a) hot wire voltmeter
(b) moving coil galvanometer
(c) potential coil galvanometer
(d) moving magnet galvanometer

45. When a charged particle moving with velocity $\vec{v}$ is subjected to a magnetic field of induction $\vec{B}$, the force on it is non-zero. This implies that

(a) angle between $\vec{v}$ and $\vec{B}$ is necessarily 90°
(b) angle between $\vec{v}$ and $\vec{B}$ can have any value other than 90°
(c) angle between $\vec{v}$ and $\vec{B}$ can have any value other than zero and 180°
(d) angle between $\vec{v}$ and $\vec{B}$ is either zero or 180°

RESPONSE GRID					
	36. ⓐⓑⓒⓓ	37. ⓐⓑⓒⓓ	38. ⓐⓑⓒⓓ	39. ⓐⓑⓒⓓ	40. ⓐⓑⓒⓓ
	41. ⓐⓑⓒⓓ	42. ⓐⓑⓒⓓ	43. ⓐⓑⓒⓓ	44. ⓐⓑⓒⓓ	45. ⓐⓑⓒⓓ

DAILY PRACTICE PROBLEM DPP CHAPTERWISE CP18 - PHYSICS			
Total Questions	45	Total Marks	180
Attempted		Correct	
Incorrect		Net Score	
Cut-off Score	45	Qualifying Score	60
Success Gap = Net Score – Qualifying Score			
Net Score = (Correct × 4) – (Incorrect × 1)			

Space for Rough Work

DPP - Daily Practice Problems

Chapter-wise Sheets

Date : ____ **Start Time :** ____ **End Time :** ____

PHYSICS

CP19

SYLLABUS : Magnetism and Matter

Max. Marks : 180 **Marking Scheme :** (+4) for correct & (–1) for incorrect answer **Time : 60 min.**

INSTRUCTIONS : This Daily Practice Problem Sheet contains 45 MCQs. For each question only one option is correct. Darken the correct circle/ bubble in the Response Grid provided on each page.

1. Two identical magnetic dipoles of magnetic moments 1.0 A-m^2 each, placed at a separation of 2 m with their axis perpendicular to each other. The resultant magnetic field at point midway between the dipole is
   (a) 5×10^{-7} T (b) $\sqrt{5} \times 10^{-7}$ T
   (c) 10^{-7} T (d) 2×10^{-7} T
2. Two identical thin bar magnets each of length ℓ and pole strength m are placed at right angles to each other, with north pole of one touching south pole of the other, then the magnetic moment of the system is

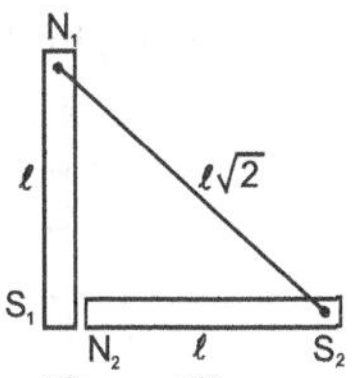

   (a) 1 mℓ (b) 2 mℓ (c) $\sqrt{2}$ mℓ (d) mℓ/2
3. The magnetic lines of force inside a bar magnet
   (a) are from north-pole to south-pole of the magnet
   (b) do not exist
   (c) depend upon the area of cross-section of the bar magnet
   (d) are from south-pole to north-pole of the magnet
4. Relative permittivity and permeability of a material ε_r and μ_r, respectively. Which of the following values of these quantities are allowed for a diamagnetic material?
   (a) $\varepsilon_r = 0.5, \mu_r = 1.5$ (b) $\varepsilon_r = 1.5, \mu_r = 0.5$
   (c) $\varepsilon_r = 0.5, \mu_r = 0.5$ (d) $\varepsilon_r = 1.5, \mu_r = 1.5$
5. If the period of oscillation of freely suspended bar magnet in earth's horizontal field H is 4 sec. When another magnet is brought near it, the period of oscillation is reduced to 2s. The magnetic field of second bar magnet is
   (a) 4H (b) 3H (c) 2H (d) $\sqrt{3}$ H
6. Three identical bars A, B and C are made of different magnetic materials. When kept in a uniform magnetic field, the field lines around them look as follows:

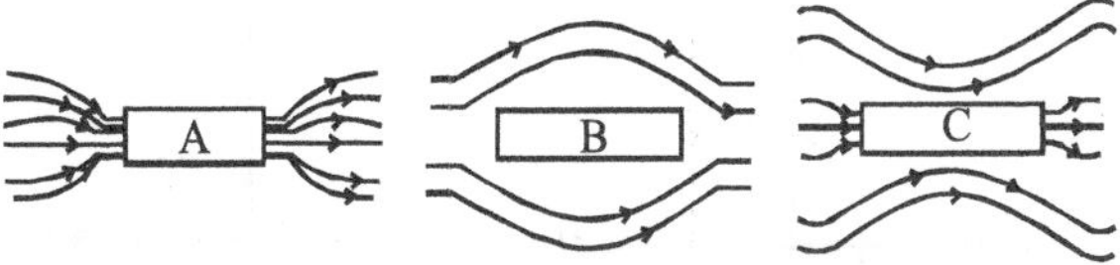

   Make the correspondence of these bars with their material being diamagnetic (D), ferromagnetic (F) and paramagnetic (P):
   (a) $A \leftrightarrow D, B \leftrightarrow P, C \leftrightarrow F$
   (b) $A \leftrightarrow F, B \leftrightarrow D, C \leftrightarrow P$
   (c) $A \leftrightarrow P, B \leftrightarrow F, C \leftrightarrow D$
   (d) $A \leftrightarrow F, B \leftrightarrow P, C \leftrightarrow D$

RESPONSE GRID	1. ⓐⓑⓒⓓ	2. ⓐⓑⓒⓓ	3. ⓐⓑⓒⓓ	4. ⓐⓑⓒⓓ	5. ⓐⓑⓒⓓ
	6. ⓐⓑⓒⓓ				

Space for Rough Work

7. Curie temperature is the temperature above which
(a) a ferromagnetic material becomes paramagnetic
(b) a paramagnetic material becomes diamagnetic
(c) a ferromagnetic material becomes diamagnetic
(d) a paramagnetic material becomes ferromagnetic

8. A watch glass containing some powdered substance is placed between the pole pieces of a magnet. Deep concavity is observed at the centre. The substance in the watch glass is
(a) iron (b) chromium (c) carbon (d) wood

9. A coil in the shape of an equilateral triangle of side l is suspended between the pole pieces of a permanent magnet such that $\vec{B}$ is in the plane of the coil. If due to a current i in the triangle a torque τ acts on it, the side l of the triangle is
(a) $\frac{2}{\sqrt{3}}\left(\frac{\tau}{B.i}\right)^{\frac{1}{2}}$ (b) $2\left(\frac{\tau}{\sqrt{3}B.i}\right)^{\frac{1}{2}}$
(c) $\frac{2}{\sqrt{3}}\left(\frac{\tau}{B.i}\right)$ (d) $\frac{1}{\sqrt{3}}\frac{\tau}{B.i}$

10. A compass needle whose magnetic moment is 60 Am^2, is directed towards geographical north at any place experiencing moment of force of 1.2×10^{-3} Nm. At that place the horizontal component of earth field is 40 micro W/m^2. What is thc valuc of dip angle at that place?
(a) 30° (b) 60° (c) 45° (d) 15°

11. The materials suitable for making electromagnets should have
(a) high retentivity and low coercivity
(b) low retentivity and low coercivity
(c) high retentivity and high coercivity
(d) low retentivity and high coercivity

12. The length of a magnet is large compared to its width and breadth. The time period of its oscillation in a vibration magnetometer is 2s. The magnet is cut along its length into three equal parts and these parts are then placed on each other with their like poles together. The time period of this combination will be
(a) $2\sqrt{3}$ s (b) $\frac{2}{3}$s (c) 2 s (d) $\frac{2}{\sqrt{3}}$s

13. Hysteresis loops for two magnetic materials A and B are given below :

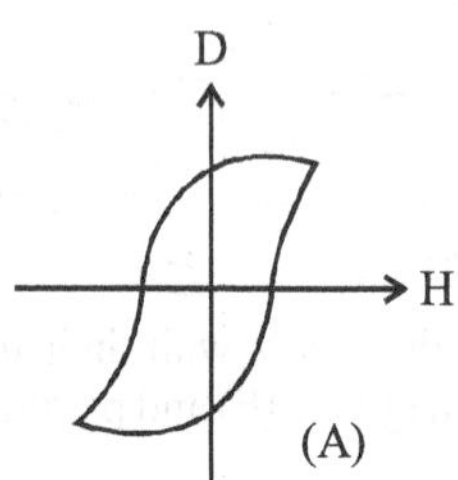

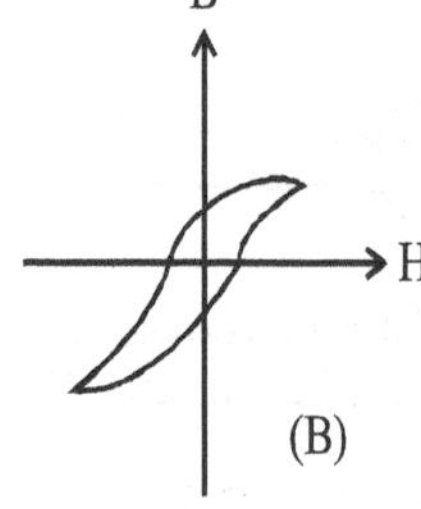

These materials are used to make magnets for elecric generators, transformer core and electromagnet core. Then it is proper to use :
(a) A for transformers and B for electric generators.
(b) B for electromagnets and transformers.
(c) A for electric generators and trasformers.
(d) A for electromagnets and B for electric generators.

14. Which of the following is responsible for the earth's magnetic field?
(a) Convective currents in earth's core.
(b) Diversive current in earth's core.
(c) Rotational motion of earth.
(d) Translational motion of earth.

15. In a vibration megnetometer, the time period of a bar magnet oscillating in horizontal component of earth's magnetic field is 2 sec. When a magnet is brought near and parallel to it, the time period reduces to 1 sec. The ratio H/F of the horizontal component H and the field F due to magnet will be
(a) 3 (b) 1/3 (c) $\sqrt{3}$ (d) $1/\sqrt{3}$

16. Let V and H be the vertical and horizontal components of earth's magnetic field at any point on earth. Near the north pole
(a) V>>H (b) V<<H (c) V=H (d) V=H=0

17. A thin circular wire carrying a current I has a magnetic moment M. The shape of the wire is changed to a square and it carries the same current. It will have a magnetic moment
(a) M (b) $\frac{4}{\pi^2}M$ (c) $\frac{4}{\pi}M$ (d) $\frac{\pi}{4}M$

18. A bar magnet of magnetic moment M is placed at right angles to a magnetic induction B. If a force F is experienced by each pole of the magnet, the length of the magnet will be
(a) F/MB (b) MB/F (c) BF/M (d) MF/B

19. If the susceptibility of dia, para and ferro magnetic materials are χ_d, χ_p, χ_f respectively, then
(a) $\chi_d < \chi_p < \chi_f$ (b) $\chi_d < \chi_f < \chi_p$
(c) $\chi_f < \chi_d < \chi_p$ (d) $\chi_f < \chi_p < \chi_d$

20. The basic magnetization curve for a ferromagnetic material is shown in figure. Then, the value of relative permeability is highest for the point

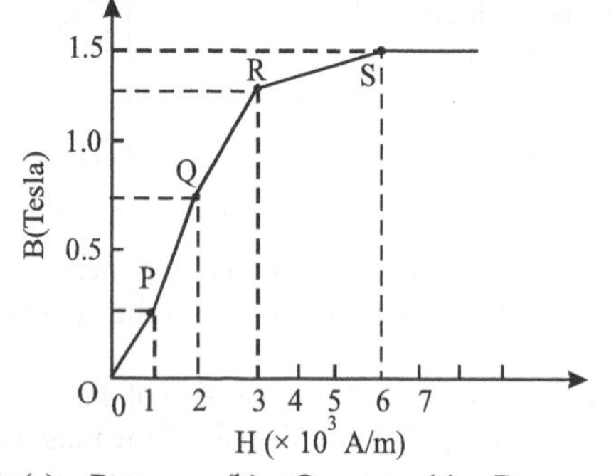

(a) P (b) Q (c) R (d) S

RESPONSE GRID					
	7. ⓐⓑⓒⓓ	8. ⓐⓑⓒⓓ	9. ⓐⓑⓒⓓ	10. ⓐⓑⓒⓓ	11. ⓐⓑⓒⓓ
	12. ⓐⓑⓒⓓ	13. ⓐⓑⓒⓓ	14. ⓐⓑⓒⓓ	15. ⓐⓑⓒⓓ	16. ⓐⓑⓒⓓ
	17. ⓐⓑⓒⓓ	18. ⓐⓑⓒⓓ	19. ⓐⓑⓒⓓ	20. ⓐⓑⓒⓓ	

Space for Rough Work

21. A magnetic needle suspended by a silk thread is vibrating in the earth's magnetic field. if the temperature of the needle is increased by 700°C, then
(a) time period decreases
(b) time period increases
(c) time period remains unchanged
(d) the needle stops vibrating

22. Torques τ_1 and τ_2 are required for a magnetic needle to remain perpendicular to the magnetic fields at two different places. The magnetic fields at those places are B_1 and B_2 respectively; then ratio $\frac{B_1}{B_2}$ is
(a) $\frac{\tau_2}{\tau_1}$ (b) $\frac{\tau_1}{\tau_2}$
(c) $\frac{\tau_1+\tau_2}{\tau_1-\tau_2}$ (d) $\frac{\tau_1-\tau_2}{\tau_1+\tau_2}$

23. A bar magnet has a length 8 cm. The magnetic field at a point at a distance 3 cm from the centre in the broad side-on position is found to be $4\times10^{-6}\,T$. The pole strength of the magnet is.
(a) 6×10^{-5} Am (b) 5×10^{-5} Am
(c) 2×10^{-4} Am (d) 3×10^{-4} Am

24. A vibration magnetometer consists of two identical bar magnets placed one over the other such that they are perpendicular and bisect each other. The time period of oscillation in a horizontal magnetic field is $2^{5/4}$ seconds. One of the magnets is removed and if the other magnet oscillates in the same field, then the time period in seconds is
(a) $2^{1/4}$ (b) $2^{1/2}$ (c) 2 (d) $2^{3/4}$

25. A magnetic needle is kept in a non-uniform magnetic field. It experiences
(a) neither a force nor a torque
(b) a torque but not a force
(c) a force but not a torque
(d) a force and a torque

26. The angle of dip at a place is 37° and the vertical component of the earth's magnetic field is 6×10^{-5}T. The earth's magnetic field at this place is (tan 37° = 3/4)
(a) 7×10^{-5} T (b) 6×10^{-5} T
(c) 5×10^{-5} T (d) 10^{-4} T

27. Needles N_1, N_2 and N_3 are made of a ferromagnetic, a paramagnetic and a diamagnetic substance respectively. A magnet when brought close to them will
(a) attract N_1 and N_2 strongly but repel N_3
(b) attract N_1 strongly, N_2 weakly and repel N_3 weakly
(c) attract N_1 strongly, but repel N_2 and N_3 weakly
(d) attract all three of them

28. The figure shows the various positions (labelled by subscripts) of small magnetised needles P and Q. The arrows show the direction of their magnetic moment. Which configuration corresponds to the lowest potential energy among all the configurations shown ?

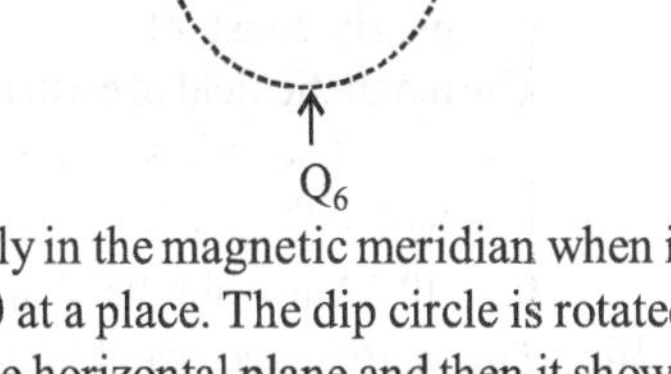

(a) PQ_3
(b) PQ_4
(c) PQ_5
(d) PQ_6

29. A dip needle lies initially in the magnetic meridian when it shows an angle of dip θ at a place. The dip circle is rotated through an angle x in the horizontal plane and then it shows an angle of dip θ′.
Then $\frac{\tan\theta'}{\tan\theta}$ is
(a) $\frac{1}{\cos x}$ (b) $\frac{1}{\sin x}$ (c) $\frac{1}{\tan x}$ (d) cos x

30. Two tangent galvanometers having coils of the same radius are connected in series. A current flowing in them produces deflections of 60° and 45° respectively. The ratio of the number of turns in the coils is
(a) 4/3 (b) $\frac{\sqrt{3}+1}{1}$ (c) $\frac{\sqrt{3}+1}{\sqrt{3}-1}$ (d) $\frac{\sqrt{3}}{1}$

31. Following figures show the arrangement of bar magnets in different configurations. Each magnet has magnet ic dipole moment $\vec{m}$. Which configuration has highest net magnetic dipole moment?

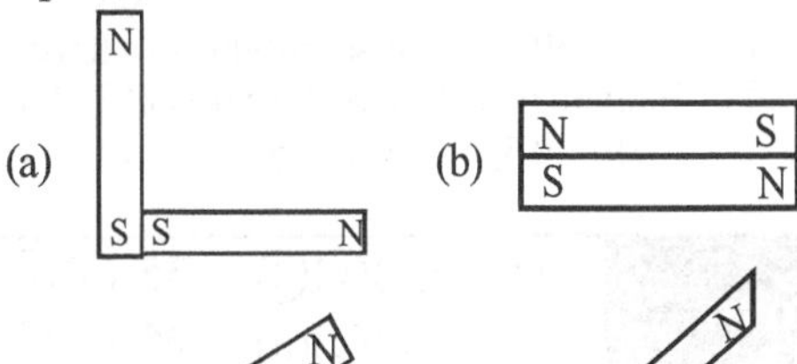

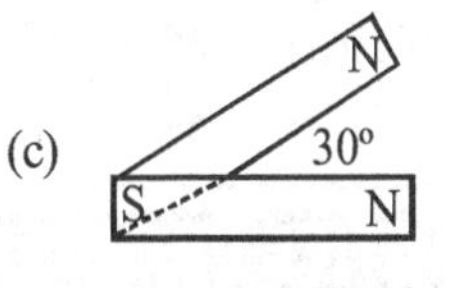

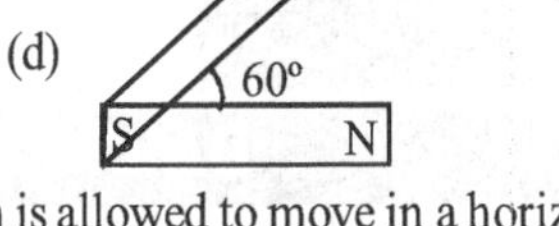

32. A compass needle which is allowed to move in a horizontal plane is taken to a geomagnetic pole. It :
(a) will become rigid showing no movement
(b) will stay in any position
(c) will stay in north-south direction only
(d) will stay in east-west direction only

33. If a magnetic dipole of moment *M* situated in the direction of a magnetic field *B* is rotated by 180°, then the amount of work done is
(a) MB (b) 2MB (c) $\frac{MB}{\sqrt{2}}$ (d) 0

RESPONSE GRID				
21. ⓐⓑⓒⓓ	22. ⓐⓑⓒⓓ	23. ⓐⓑⓒⓓ	24. ⓐⓑⓒⓓ	25. ⓐⓑⓒⓓ
26. ⓐⓑⓒⓓ	27. ⓐⓑⓒⓓ	28. ⓐⓑⓒⓓ	29. ⓐⓑⓒⓓ	30. ⓐⓑⓒⓓ
31. ⓐⓑⓒⓓ	32. ⓐⓑⓒⓓ	33. ⓐⓑⓒⓓ		

Space for Rough Work

34. A bar magnet is oscillating in the earth's magnetic field with a period T. What happens to its period of motion, if its mass is quadrupled
(a) motion remains simple harmonic with new period = T/2
(b) motion remains simple harmonic with new period = 2 T
(c) motion remains simple harmonic with new period = 4T
(d) motion remains simple harmonic and the period stays nearly constant

35. The magnetic field of earth at the equator is approximately 4×10^{-5} T. The radius of earth is 6.4×10^6 m. Then the dipole moment of the earth will be nearly of the order of:
(a) 10^{23} Am^2 (b) 10^{20} Am^2 (c) 10^{16} Am^2 (d) 10^{10} Am^2

36. The relative permeability of a medium is 0.075. What is its magnetic susceptibility?
(a) 0.925 (b) −0.925 (c) 1.075 (d) −1.075

37. A dip circle is so set that its needle moves freely in the magnetic meridian. In this position, the angle of dip is 40°. Now the dip circle is rotated so that the plane in which the needle moves makes an angle of 30° with the magnetic meridian. In this position, the needle will dip by an angle
(a) 40° (b) 30° (c) more than 40° (d) less than 40°

38. The mid points of two small magnetic dipoles of length d in end-on positions, are separated by a distance x, (x >> d). The force between them is proportional to x^{-n} where n is:

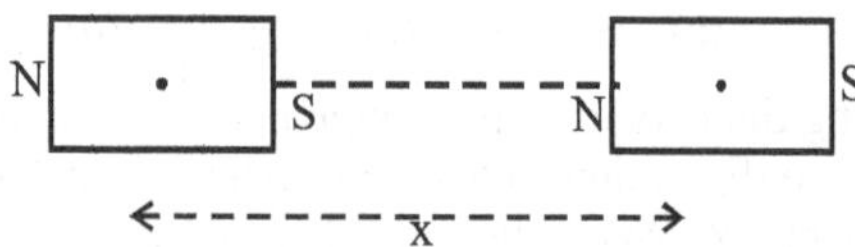

(a) 1 (b) 2 (c) 3 (d) 4

39. At a temperatur of 30°C, the susceptibility of a ferromagnetic material is found to be χ. Its susceptibility at 333°C is
(a) χ (b) 0.5χ (c) 2χ (d) 11.1χ

40. The susceptibility of annealed iron at saturation is 5500. Find the permeability of annealed iron at saturation.
(a) 6.9×10^{-3} (b) 5.1×10^{-2} (c) 5×10^2 (d) 3.2×10^{-5}

41. A short magnet oscillates in an oscillation magnetometer with a time period of 0.10s where the earth's horizontal magnetic field is 24 μT. A downward current of 18 A is established in a vertical wire placed 20 cm east of the magnet. Find the new time period.
(a) 0.076 s (b) 0.5 s (c) 0.1 s (d) 0.2 s

42. A permanent magnet in the shape of a thin cylinder of length 10 cm has magnetisation (M) = 10^6 A m^{-1}. Its magnetization current I_M is
(a) 10^5 A (b) 10^6 A (c) 10^7 A (d) 10^8 A

43. The earth's magnetic field lines resemble that of a dipole at the centre of the earth. If the magnetic moment of this dipole is close to 8×10^{22} Am^2, the value of earth's magnetic field near the equator is close to (radius of the earth = 6.4×10^6 m)
(a) 0.6 Gauss (b) 1.2 Gauss
(c) 1.8 Gauss (d) 0.32 Gauss

44. The coercivity of a small magnet where the ferromagnet gets demagnetized is 3×10^3 Am^{-1}. The current required to be passed in a solenoid of length 10 cm and number of turns 100, so that the magnet gets demagnetized when inside the solenoid, is:
(a) 30 mA (b) 60 mA (c) 3 A (d) 6 A

45. A thin bar magnet of length 2ℓ and breadth 2 b pole strength m and magnetic moment M is divided into four equal parts with length and breadth of each part being half of original magnet.
Then, the magnetic moment of each part is
(a) M/4 (b) M (c) M/2 (d) 2 M

RESPONSE GRID					
	34. ⓐⓑⓒⓓ	35. ⓐⓑⓒⓓ	36. ⓐⓑⓒⓓ	37. ⓐⓑⓒⓓ	38. ⓐⓑⓒⓓ
	39. ⓐⓑⓒⓓ	40. ⓐⓑⓒⓓ	41. ⓐⓑⓒⓓ	42. ⓐⓑⓒⓓ	43. ⓐⓑⓒⓓ
	44. ⓐⓑⓒⓓ	45. ⓐⓑⓒⓓ			

DAILY PRACTICE PROBLEM DPP CHAPTERWISE CP19 - PHYSICS			
Total Questions	45	Total Marks	180
Attempted		Correct	
Incorrect		Net Score	
Cut-off Score	50	Qualifying Score	70
Success Gap = Net Score – Qualifying Score			
Net Score = (Correct × 4) – (Incorrect × 1)			

Space for Rough Work

DPP - Daily Practice Problems

Chapter-wise Sheets

Date : | **Start Time :** | **End Time :**

PHYSICS

SYLLABUS : Electromagnetic Induction

Max. Marks : 180 **Marking Scheme :** (+4) for correct & (–1) for incorrect answer **Time : 60 min.**

INSTRUCTIONS : This Daily Practice Problem Sheet contains 45 MCQs. For each question only one option is correct. Darken the correct circle/ bubble in the Response Grid provided on each page.

1. A metal disc of radius 100 cm is rotated at a constant angular speed of 60 rad/s in a plane at right angles to an external field of magnetic induction 0.05 Wb/m^2. The emf induced between the centre and a point on the rim will be
(a) 3V (b) 1.5V (c) 6V (d) 9V

2. In a coil of resistance $100\,\Omega$, a current is induced by changing the magnetic flux through it as shown in the figure. The magnitude of change in flux through the coil is
(a) 250 Wb
(b) 275 Wb
(c) 200 Wb
(d) 225 Wb

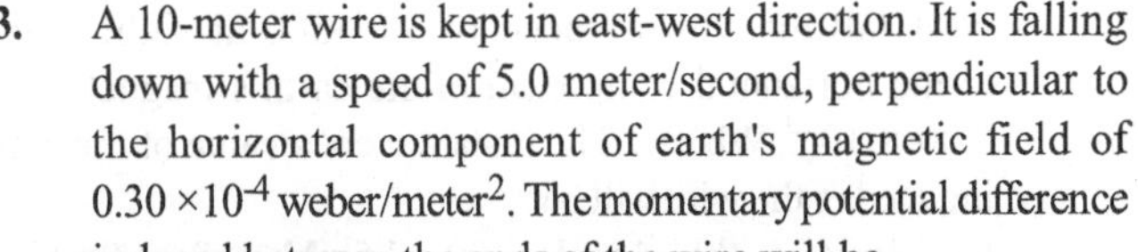

3. A 10-meter wire is kept in east-west direction. It is falling down with a speed of 5.0 meter/second, perpendicular to the horizontal component of earth's magnetic field of 0.30×10^{-4} weber/meter^2. The momentary potential difference induced between the ends of the wire will be
(a) 0.0015V (b) 0.015V
(c) 0.15V (d) 1.5V

4. The figure shows certain wire segments joined together to form a coplanar loop. The loop is placed in a perpendicular magnetic field in the direction going into the plane of the figure.

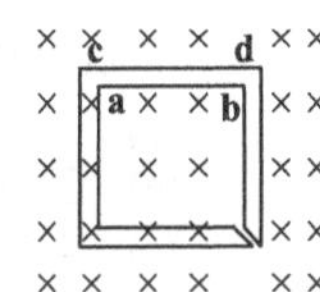

The magnitude of the field increases with time. I_1 and I_2 are the currents in the segments ab and cd. Then,
(a) $I_1 > I_2$ (b) $I_1 < I_2$
(c) I_1 is in the direction ba and I_2 is in the direction cd
(d) I_1 is in the direction ab and I_2 is in the direction dc

5. Two solenoids of equal number of turns have their lengths and the radii in the same ratio 1 : 2. The ratio of their self inductances will be
(a) 1 : 2 (b) 2 : 1 (c) 1 : 1 (d) 1 : 4

6. A metal conductor of length 1 m rotates vertically about one of its ends at angular velocity 5 radians per second. If the horizontal component of earth's magnetic field is 0.2×10^{-4}T, then the e.m.f. developed between the two ends of the conductor is
(a) 5mV (b) 50μV (c) 5μV (d) 50mV

7. Eddy currents do not produce
(a) heat (b) a loss of energy
(c) spark (d) damping of motion

RESPONSE GRID					
	1. ⓐⓑⓒⓓ	2. ⓐⓑⓒⓓ	3. ⓐⓑⓒⓓ	4. ⓐⓑⓒⓓ	5. ⓐⓑⓒⓓ
	6. ⓐⓑⓒⓓ	7. ⓐⓑⓒⓓ			

Space for Rough Work

8. A conducting square frame of side 'a' and a long staight wire carrying current I are located in the same plane as shown in the figure. The frame moves to the right with a constant velocity 'V'. The emf induced in the frame will be proportional to

(a) 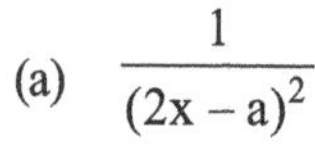
$\frac{1}{(2x-a)^2}$

(b) 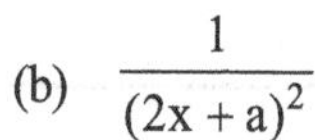
$\frac{1}{(2x+a)^2}$

(c) 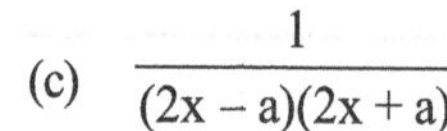
$\frac{1}{(2x-a)(2x+a)}$

(d) $\frac{1}{x^2}$

9. Which of the following figure correctly depicts the Lenz's law. The arrows show the movement of the labelled pole of a bar magnet into a closed circular loop and the arrows on the circle show the direction of the induced current

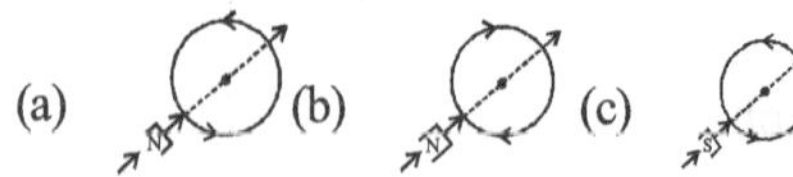

10. The magnetic flux (in weber) linked with a coil of resistance 10 Ω is varying with respect to time t as $\phi = 4t^2 + 2t + 1$. Then the current in the coil at time t = 1 second is

(a) 0.5 A (b) 2 A (c) 1.5 A (d) 1 A

11. Two coaxial solenoids are made by winding thin insulated wire over a pipe of cross-sectional area $A = 10\ cm^2$ and length = 20 cm. If one of the solenoid has 300 turns and the other 400 turns, their mutual inductance is $(\mu_0 = 4\pi \times 10^{-7}\ TmA^{-1})$

(a) $2.4\pi \times 10^{-5}$ H (b) $4.8\pi \times 10^{-4}$ H
(c) $4.8\pi \times 10^{-5}$ H (d) $2.4\pi \times 10^{-4}$ H

12. When the current changes from +2 A to –2A in 0.05 second, an e.m.f. of 8 V is induced in a coil. The coefficient of self - induction of the coil is

(a) 0.2 H (b) 0.4 H (c) 0.8 H (d) 0.1 H

13. A long solenoid has 500 turns. When a current of 2 ampere is passed through it, the resulting magnetic flux linked with each turn of the solenoid is 4×10^{-3} Wb. The self- inductance of the solenoid is

(a) 2.5 henry (b) 2.0 henry
(c) 1.0 henry (d) 40 henry

14. A metallic square loop ABCD is moving in its own plane with velocity v in a uniform magnetic field perpendicular to its plane as shown in the figure. An electric field is induced

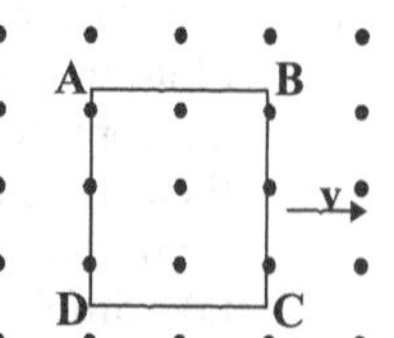

(a) in AD, but not in BC (b) in BC, but not in AD
(c) neither in AD nor in BC (d) in both AD and BC

15. In an AC generator, a coil with N turns, all of the same area A and total resistance R, rotates with frequency ω in a magnetic field B. The maximum value of emf generated in the coil is

(a) N.A.B.R.ω (b) N.A.B.
(c) N.A.B.R. (d) N.A.B.ω

16. In an inductor of self-inductance L = 2 mH, current changes with time according to relation $i = t^2e^{-t}$. At what time emf is zero?

(a) 4s (b) 3s (c) 2s (d) 1s

17. Choke coil works on the principle of

(a) transient current (b) self induction
(c) mutual induction (d) wattless current

18. A coil having n turns and resistance R Ω is connected with a galvanometer of resistance 4R Ω. This combination is moved in time t seconds from a magnetic field W_1 weber to W_2 weber. The induced current in the circuit is

(a) $-\frac{(W_1 - W_2)}{Rnt}$ (b) $-\frac{n(W_2 - W_1)}{5\ Rt}$

(c) $-\frac{(W_2 - W_1)}{5\ Rnt}$ (d) $-\frac{n(W_2 - W_1)}{Rt}$

19. A thin circular ring of area A is held perpendicular to a uniform magnetic field of induction B. A small cut is made in the ring and a galvanometer is connected across the ends such that the total resistance of the circuit is R. When the ring is suddenly squeezed to zero area, the charge flowing through the galvanometer is

(a) $\frac{BR}{A}$ (b) $\frac{AB}{R}$ (c) ABR (d) $\frac{B^2A}{R^2}$

20. A boat is moving due east in a region where the earth's magnetic field is $5.0 \times 10^{-5}\ NA^{-1}m^{-1}$ due north and horizontal. The boat carries a vertical aerial 2 m long. If the speed of the boat is 1.50 ms^{-1}, the magnitude of the induced emf in the wire of aerial is:

(a) 0.75 mV (b) 0.50 mV (c) 0.15 mV (d) 1 mV

21. In a coil of area 10 cm^2 and 10 turns with magnetic field directed perpendicular to the plane and is changing at the rate of 10^8 Gauss/second. The resistance of the coil is 20Ω. The current in the coil will be

(a) 0.5 A (b) 5 A (c) 50 A (d) 5×10^8 A

22. A horizontal straight wire 20 m long extending from east to west falling with a speed of 5.0 m/s, at right angles to the horizontal component of the earth's magnetic field $0.30 \times 10^{-4}\ Wb/m^2$. The instantaneous value of the e.m.f. induced in the wire will be

(a) 3mV (b) 4.5mV (c) 1.5mV (d) 6.0mV

RESPONSE GRID					
	8. ⓐⓑⓒⓓ	9. ⓐⓑⓒⓓ	10. ⓐⓑⓒⓓ	11. ⓐⓑⓒⓓ	12. ⓐⓑⓒⓓ
	13. ⓐⓑⓒⓓ	14. ⓐⓑⓒⓓ	15. ⓐⓑⓒⓓ	16. ⓐⓑⓒⓓ	17. ⓐⓑⓒⓓ
	18. ⓐⓑⓒⓓ	19. ⓐⓑⓒⓓ	20. ⓐⓑⓒⓓ	21. ⓐⓑⓒⓓ	22. ⓐⓑⓒⓓ

Space for Rough Work

23. The self inductance of a long solenoid cannot be increased by
(a) increasing its area of cross section
(b) increasing its length
(c) changing the medium with greater permeability
(d) increasing the current through it

24. A metallic rod of length 'ℓ' is tied to a string of length 2ℓ and made to rotate with angular speed ω on a horizontal table with one end of the string fixed. If there is a vertical magnetic field 'B' in the region, the e.m.f. induced across the ends of the rod is

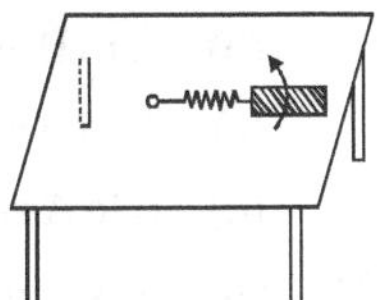

(a) $\frac{2B\omega\ell^2}{2}$ (b) $\frac{3B\omega\ell^2}{2}$ (c) $\frac{4B\omega\ell^2}{2}$ (d) $\frac{5B\omega\ell^2}{2}$

25. Lenz's law gives
(a) the magnitude of the induced e.m.f.
(b) the direction of the induced current
(c) both the magnitude and direction of the induced current
(d) the magnitude of the induced current

26. A metal ring is held horizontally and bar magnet is dropped through the ring with its length along the axis of the ring. The acceleration of the falling magnet
(a) is equal to g
(b) is less than g
(c) is more than g
(d) depends on the diameter of ring and length of magnet

27. The pointer of a dead-beat galvanometer gives a steady deflection because
(a) eddy currents are produced in the conducting frame over which the coil is wound.
(b) its magnet is very strong.
(c) its pointer is very light.
(d) its frame is made of ebonite.

28. A metal rod of length l cuts across a uniform magnetic field B with a velocity v. If the resistance of the circuit of which the rod forms a part is r, then the force required to move the rod is

(a) $\frac{B^2l^2v}{r}$ (b) $\frac{Blv}{r}$ (c) $\frac{B^2lv}{r}$ (d) $\frac{B^2l^2v^2}{r}$

29. In an A.C. generator, when the plane of the armature is perpendicular to the magnetic field
(a) both magnetic flux and emf are maximum
(b) both magnetic flux and emf are zero
(c) both magnetic flux and emf are half of their respective maximum values
(d) magnetic flux is maximum and emf is zero

30. A copper disc of radius 0.1 m rotated about its centre with 10 revolutions per second in a uniform magnetic field of 0.1 tesla with its plane perpendicular to the field. The e.m.f. induced across the radius of disc is

(a) $\frac{\pi}{10}$ volt (b) $\frac{2\pi}{10}$ volt
(c) $\pi \times 10^{-2}$ volt (d) $2\pi \times 10^{-2}$ volt

31. The mutual inductance of a pair of coils, each of N turns, is M henry. If a current of I ampere in one of the coils is brought to zero in t second, the emf induced per turn in the other coil, in volt, will be

(a) $\frac{MI}{t}$ (b) $\frac{NMI}{t}$ (c) $\frac{MN}{It}$ (d) $\frac{MI}{Nt}$

32. Consider the situation shown in figure. If the switch is closed and after some time it is opened again, the closed loop will show
(a) a clockwise current
(b) an anticlockwise current
(c) an anticlockwise current and then clockwise
(d) a clockwise current and then an anticlock wise current.

33. A magnet is moved towards a coil (i) quickly (ii) slowly, then the induced e.m.f. is
(a) larger in case (i)
(b) smaller in case (i)
(c) equal to both the cases
(d) larger or smaller depending upon the radius of the coil

34. A circular wire of radius r rotates about its own axis with angular speed ω in a magnetic field B perpendicular to its plane, then the induced e.m.f. is

(a) $\frac{1}{2}Br\omega^2$ (b) $Br\omega^2$ (c) $2Br\omega^2$ (d) zero

35. A conducting ring of radius 1 *m* kept in a uniform magnetic field *B* of 0.01 *T*, rotates uniformly with an angular velocity 100 rad s^{-1} with its axis of rotation perpendicular to *B*. The maximum induced emf in it is
(a) $1.5\pi V$ (b) πV (c) $2\pi V$ (d) $0.5\pi V$

36. A magnetic field of 2×10^{-2} T acts at right angles to a coil of area 100 cm^2, with 50 turns. The average e.m.f. induced in the coil is 0.1 V, when it is removed from the field in t sec. The value of t is
(a) 10 s (b) 0.1 s (c) 0.01 s (d) 1 s

37. The magnetic flux through a circuit of resistance R changes by an amount $\Delta\phi$ in a time Δt. Then the total quantity of electric charge Q that passes any point in the circuit during the time Δt is represented by

(a) $Q = R.\frac{\Delta\phi}{\Delta t}$ (b) $Q = \frac{1}{R}.\frac{\Delta\phi}{\Delta t}$

(c) $Q = \frac{\Delta\phi}{R}$ (d) $Q = \frac{\Delta\phi}{\Delta t}$

RESPONSE GRID					
	23. ⓐⓑⓒⓓ	24. ⓐⓑⓒⓓ	25. ⓐⓑⓒⓓ	26. ⓐⓑⓒⓓ	27. ⓐⓑⓒⓓ
	28. ⓐⓑⓒⓓ	29. ⓐⓑⓒⓓ	30. ⓐⓑⓒⓓ	31. ⓐⓑⓒⓓ	32. ⓐⓑⓒⓓ
	33. ⓐⓑⓒⓓ	34. ⓐⓑⓒⓓ	35. ⓐⓑⓒⓓ	36. ⓐⓑⓒⓓ	37. ⓐⓑⓒⓓ

Space for Rough Work

38. Fig shown below represents an area $A = 0.5\ m^2$ situated in a uniform magnetic field $B = 2.0\ weber/m^2$ and making an angle of 60° with respect to magnetic field.

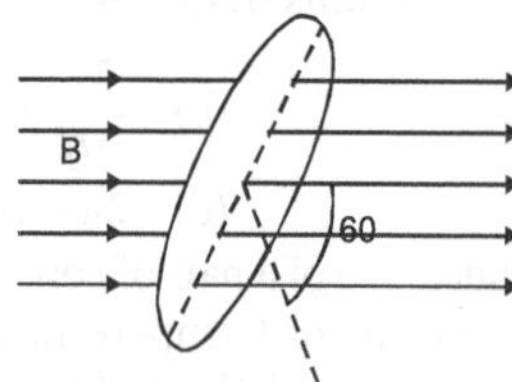

The value of the magnetic flux through the area would be equal to

(a) 2.0 weber (b) $\sqrt{3}$ weber

(c) $\sqrt{3}/2$ weber (d) 0.5 weber

39. As a result of change in the magnetic flux linked to the closed loop shown in the fig, an e.m.f. V volt is induced in the loop. The work done (joule) in taking a charge Q coulomb once along the loop is

(a) QV (b) 2QV (c) QV/2 (d) Zero

40. Two coils are placed close to each other. The mutual inductance of the pair of coils depends upon

(a) the rates at which currents are changing in the two coils
(b) relative position and orientation of the two coils
(c) the materials of the wires of the coils
(d) the currents in the the two coils

41. When current i passes through an inductor of self inductance L, energy stored in it is $1/2.\ L i^2$. This is stored in the

(a) current (b) voltage

(c) magnetic field (d) electric field

42. A conducting wire frame is placed in a magnetic field which is directed into the paper. The magnetic field is increasing at a constant rate. The directions of induced current in wires AB and CD are

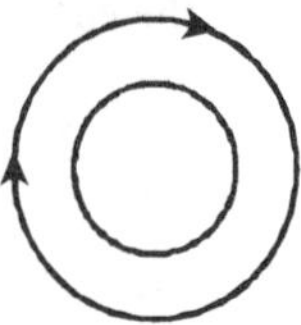

(a) B to A and D to C
(b) A to B and C to D
(c) A to B and D to C
(d) B to A and C to D

43. Two different wire loops are concentric and lie in the same plane. The current in the outer loop (I) is clockwise and increases with time. The induced current in the inner loop

(a) is clockwise
(b) is zero
(c) is counter clockwise
(d) has a direction that depends on the ratio of the loop radii.

44. When current in a coil changes from 5 A to 2 A in 0.1 s, average voltage of 50 V is produced. The self - inductance of the coil is :

(a) 6 H (b) 0.67 H (c) 3 H (d) 1.67 H

45. Two conducting circular loops of radii R_1 and R_2 are placed in the same plane with their centres coinciding. If $R_1 >> R_2$, the mutual inductance M between them will be directly proportional to

(a) R_1/R_2 (b) R_2/R_1 (c) R_1^2/R_2 (d) R_2^2/R_1

RESPONSE GRID					
	38. ⓐⓑⓒⓓ	39. ⓐⓑⓒⓓ	40. ⓐⓑⓒⓓ	41. ⓐⓑⓒⓓ	42. ⓐⓑⓒⓓ
	43. ⓐⓑⓒⓓ	44. ⓐⓑⓒⓓ	45. ⓐⓑⓒⓓ		

DAILY PRACTICE PROBLEM DPP CHAPTERWISE CP20 - PHYSICS			
Total Questions	45	Total Marks	180
Attempted		Correct	
Incorrect		Net Score	
Cut-off Score	50	Qualifying Score	70
Success Gap = Net Score – Qualifying Score			
Net Score = (Correct × 4) – (Incorrect × 1)			

Space for Rough Work

DPP - Daily Practice Problems

Chapter-wise Sheets

Date : ☐ **Start Time :** ☐ **End Time :** ☐

PHYSICS

CP21

SYLLABUS : Alternating Current

Max. Marks : 180 **Marking Scheme :** (+4) for correct & (–1) for incorrect answer **Time : 60 min.**

INSTRUCTIONS : This Daily Practice Problem Sheet contains 45 MCQs. For each question only one option is correct. Darken the correct circle/ bubble in the Response Grid provided on each page.

1. In a series resonant LCR circuit, the voltage across R is 100 volts and $R = 1\ k\Omega$ with $C = 2\mu F$. The resonant frequency ω is 200 rad/s. At resonance, the voltage across L is
   (a) 2.5×10^{-2} V (b) 40V
   (c) 250V (d) 4×10^{-3} V
2. An alternating voltage $V = V_0 \sin \omega t$ is applied across a circuit. As a result, a current $I = I_0 \sin(\omega t - \pi/2)$ flows in it. The power consumed per cycle is
   (a) zero (b) $0.5\ V_0I_0$
   (c) $0.707\ V_0I_0$ (d) $1.414\ V_0I_0$
3. For the circuit shown in the fig., the current through the inductor is 0.9 A while the current through the condenser is 0.4 A. Then

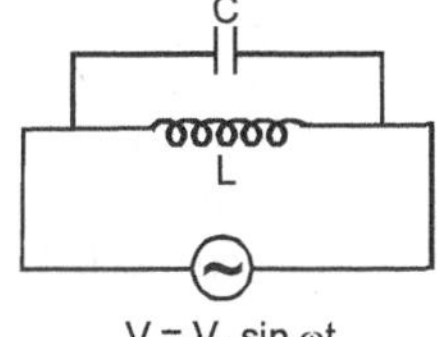

   (a) current drawn from generator I = 1.13 A
   (b) $\omega = 1/(1.5\ LC)$
   (c) I = 0.5 A
   (d) I = 0.6 A
4. A capacitor has capacity C and reactance X. If capacitance and frequency become double, then reactance will be
   (a) 4X (b) X/2 (c) X/4 (d) 2X
5. A coil of inductance 300 mH and resistance 2Ω is connected to a source of voltage 2V. The current reaches half of its steady state value in
   (a) 0.1 s (b) 0.05 s (c) 0.3 s (d) 0.15 s
6. In an A.C. circuit, a resistance of R ohm is connected in series with an inductance L. If phase angle between voltage and current be 45°, the value of inductive reactance will be
   (a) R/4 (b) R/2 (c) R (d) R/5
7. A bulb is rated at 100 V, 100 W, it can be treated as a resistor. Find out the inductance of an inductor (called choke coil) that should be connected in series with the bulb to operate the bulb at its rated power with the help of an ac source of 200 V and 50 Hz.
   (a) $\frac{\pi}{\sqrt{3}}$H (b) 100 H (c) $\frac{\sqrt{2}}{\pi}$H (d) $\frac{\sqrt{3}}{\pi}$H
8. An ac source of angular frequency ω is fed across a resistor r and a capacitor C in series. The current registered is I. If now the frequency of source is changed to $\omega/3$ (but maintaining the same voltage), the current in the circuit is found to be halved. The ratio of reactance to resistance at the original frequency ω is
   (a) $\sqrt{\frac{3}{5}}$ (b) $\sqrt{\frac{2}{5}}$ (c) $\sqrt{\frac{1}{5}}$ (d) $\sqrt{\frac{4}{5}}$

RESPONSE GRID					
	1. ⓐⓑⓒⓓ	2. ⓐⓑⓒⓓ	3. ⓐⓑⓒⓓ	4. ⓐⓑⓒⓓ	5. ⓐⓑⓒⓓ
	6. ⓐⓑⓒⓓ	7. ⓐⓑⓒⓓ	8. ⓐⓑⓒⓓ		

Space for Rough Work

9. Large transformers, when used for some time, become hot and are cooled by circulating oil. The heating of transformer is due to

(a) heating effect of current alone
(b) hysteresis loss alone
(c) both the hysteresis loss and heating effect of current
(d) none of the above

10. An inductor of inductance $L = 400$ mH and resistors of resistance $R_1 = 2\Omega$ and $R_2 = 2\Omega$ are connected to a battery of emf 12 V as shown in the figure. The internal resistance of the battery is negligible. The switch S is closed at $t = 0$. The potential drop across L as a function of time is

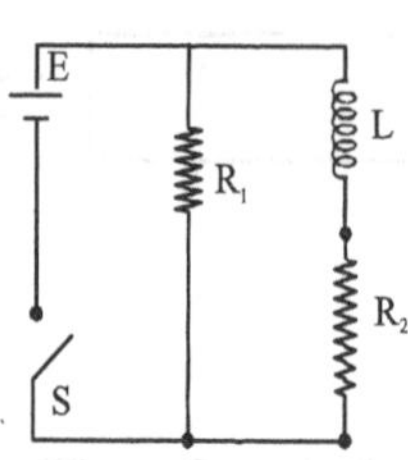

(a) $\frac{12}{t}e^{-3t}$V (b) $6(1-e^{-t/0.2})$V
(c) $12e^{-5t}$V (d) $6e^{-5t}$V

11. An ideal coil of 10H is connected in series with a resistance of 5Ω and a battery of 5V. 2second after the connection is made, the current flowing in ampere in the circuit is

(a) $(1-e^{-1})$ (b) $(1-e)$ (c) e (d) e^{-1}

12. In an A.C. circuit, the current flowing in inductance is $I = 5\sin(100t - \pi/2)$ amperes and the potential difference is $V = 200\sin(100t)$ volts. The power consumption is equal to

(a) 1000 watt (b) 40 watt
(c) 20 watt (d) Zero

13. In an oscillating LC circuit the maximum charge on the capacitor is Q. The charge on the capacitor when the energy is stored equally between the electric and magnetic field is

(a) $\frac{Q}{2}$ (b) $\frac{Q}{\sqrt{3}}$ (c) $\frac{Q}{\sqrt{2}}$ (d) Q

14. A fully charged capacitor C with initial charge q_0 is connected to a coil of self inductance L at $t = 0$. The time at which the energy is stored equally between the electric and the magnetic fields is:

(a) $\frac{\pi}{4}\sqrt{LC}$ (b) $2\pi\sqrt{LC}$ (c) $\sqrt{LC}$ (d) $\pi\sqrt{LC}$

15. For an LCR series circuit with an A.C. source of angular frequency ω

(a) circuit will be capacitive if $\omega > \frac{1}{\sqrt{LC}}$

(b) circuit will be inductive if $\omega = \frac{1}{\sqrt{LC}}$

(c) power factor of circuit will be unity if capacitive reactance equals inductive reactance

(d) current will be leading voltage if $\omega > \frac{1}{\sqrt{LC}}$

16. The r.m.s. value of potential difference V shown in the figure is

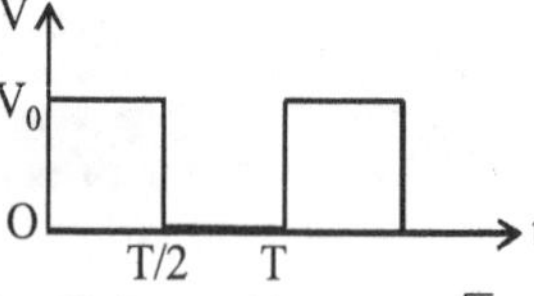

(a) V_0 (b) $V_0/\sqrt{2}$ (c) $V_0/2$ (d) $V_0/\sqrt{3}$

17. Which of the following statements is/are incorrect?

(a) If the resonance is less sharp, not only is the maximum current less, the circuit is close to resonance for a larger range Δω of frequencies and the tuning of the circuit will not be good.
(b) Less sharp the resonance less is the selectivity of the circuit or *vice–versa.*
(c) If quality factor is large, i.e., R is low or L is large, the circuit is more selective.
(d) Below resonance, voltage leads the current while above it, current leads the voltage.

18. A lamp consumes only 50% of peak power in an a.c. circuit. What is the phase difference between the applied voltage and the circuit current?

(a) $\frac{\pi}{6}$ (b) $\frac{\pi}{3}$ (c) $\frac{\pi}{4}$ (d) $\frac{\pi}{2}$

19. A step down transformer reduces 220 V to 110 V. The primary draws 5 ampere of current and secondary supplies 9 ampere. The efficiency of transformer is

(a) 20% (b) 44% (c) 90% (d) 100%

20. The voltage time (V-t) graph for triangular wave having peak value V_0 is as shown in figure. The rms value of V in time interval from t = 0 to T/4 is

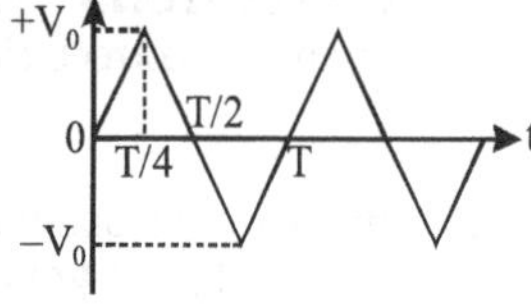

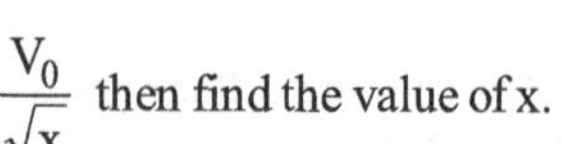
$\frac{V_0}{\sqrt{x}}$ then find the value of x.

(a) 5 (b) 4 (c) 7 (d) 3

21. The tuning circuit of a radio receiver has a resistance of 50 Ω, an inductor of 10 mH and a variable capacitor. A 1 MHz radio wave produces a potential difference of 0.1 mV. The values of the capacitor to produce resonance is (Take $\pi^2 = 10$)

(a) 2.5 pF (b) 5.0 pF (c) 25 pF (d) 50 pF

22. In an alternating current circuit in which an inductance and capacitance are joined in series, current is found to be maximum when the value of inductance is 0.5 henry and the value of capacitance is 8μF. The angular frequency of applied alternating voltage will be

(a) 5000 rad/sec (b) 4000 rad/sec
(c) 2×10^5 rad/sec (d) 500 rad/sec

23. A coil has resistance 30 ohm and inductive reactance 20 ohm at 50 Hz frequency. If an ac source, of 200 volt, 100 Hz, is connected across the coil, the current in the coil will be

(a) 4.0A (b) 8.0A (c) $\frac{20}{\sqrt{13}}$A (d) 2.0A

RESPONSE GRID					
	9. ⓐⓑⓒⓓ	10. ⓐⓑⓒⓓ	11. ⓐⓑⓒⓓ	12. ⓐⓑⓒⓓ	13. ⓐⓑⓒⓓ
	14. ⓐⓑⓒⓓ	15. ⓐⓑⓒⓓ	16. ⓐⓑⓒⓓ	17. ⓐⓑⓒⓓ	18. ⓐⓑⓒⓓ
	19. ⓐⓑⓒⓓ	20. ⓐⓑⓒⓓ	21. ⓐⓑⓒⓓ	22. ⓐⓑⓒⓓ	23. ⓐⓑⓒⓓ

Space for Rough Work

24. In the figure shown, three AC voltmeters are connected. At resonance

(a) $V_2 = 0$ (b) $V_1 = 0$

(c) $V_3 = 0$ (d) $V_1 = V_2 \neq 0$

25. A.C. power is transmitted from a power house at a high voltage as

(a) the rate of transmission is faster at high voltages

(b) it is more economical due to less power loss

(c) power cannot be transmitted at low voltages

(d) a precaution against theft of transmission lines

26. A transformer has an efficiency of 80%. It works at 4 kW and 100 V. If secondary voltage is 240 V, the current in primary coil is

(a) 0.4 A (b) 4 A (c) 10 A (d) 40 A

27. A 12 Ω resistor and a 0.21 henry inductor are connected in series to an a.c. source operating at 20 volt, 50 cycle. The phase angle between the current and source voltage is

(a) 30º (b) 40º (c) 80º (d) 90º

28. In LCR series circuit fed by a DC source, how does the amplitude of charge oscillations vary with time during discharge ?

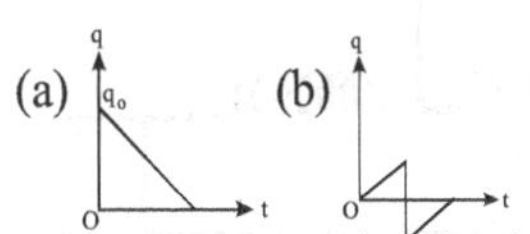

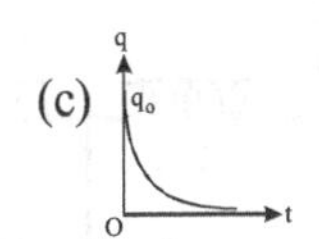

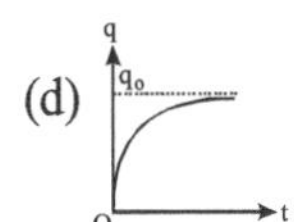

29. The primary and secondary coil of a transformer have 50 and 1500 turns respectively. If the magnetic flux ϕ linked with the primary coil is given by $\phi = \phi_0 + 4t$, where ϕ is in webers, t is time in seconds and ϕ_0 is a constant, the output voltage across the secondary coil is

(a) 120 volts (b) 220 volts

(c) 30 volts (d) 90 volts

30. The primary winding of a transformer has 100 turns and its secondary winding has 200 turns. The primary is connected to an A.C. supply of 120 V and the current flowing in it is 10 A. The voltage and the current in the secondary are

(a) 240 V, 5 A (b) 240 V, 10 A

(c) 60 V, 20 A (d) 120 V, 20 A

31. The resistance in the following circuit is increased at a particular instant. At this instant the value of resistance is 10Ω. The current in the circuit will be now

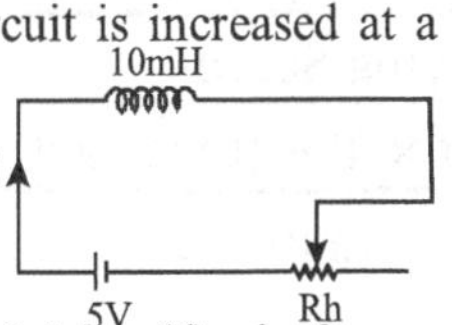

(a) i = 0.5 A (b) i > 0.5 A (c) i < 0.5 A (d) i = 0

32. The current in a *LR* circuit builds up to $\frac{3}{4}th$ of its steady state value in 4s. The time constant of this circuit is

(a) $\frac{1}{\ell n 2} s$ (b) $\frac{2}{\ell n 2} s$ (c) $\frac{3}{\ell n 2} s$ (d) $\frac{4}{\ell n 2} s$

33. An LCR series circuit is connected to a source of alternating current. At resonance, the applied voltage and the current flowing through the circuit will have a phase difference of

(a) π (b) $\frac{\pi}{2}$ (c) $\frac{\pi}{4}$ (d) 0

34. What is the value of inductance L for which the current is maximum in a series LCR circuit with $C = 10\ \mu F$ and $\omega = 1000 s^{-1}$?

(a) 1 mH

(b) cannot be calculated unless R is known

(c) 10 mH

(d) 100 mH

35. In the circuit of Fig, the bulb will become suddenly bright if

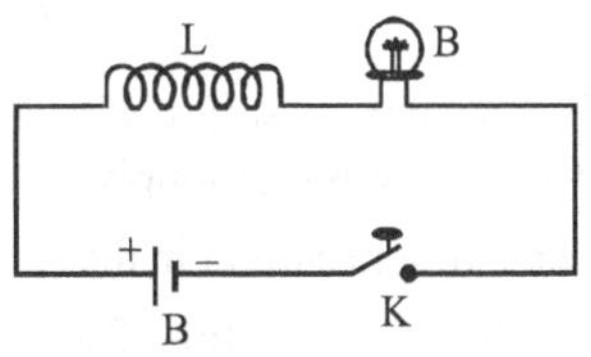

(a) contact is made or broken

(b) contact is made

(c) contact is broken

(d) won't become bright at all

36. The voltage of an ac source varies with time according to the equation $V = 100 \sin 100\pi t \cos 100\pi t$ where t is in seconds and V is in volt. Then

(a) the peak voltage of the source is 100 volt

(b) the peak voltage of the source is 50 volt

(c) the peak voltage of the source is $100/\sqrt{2}$ volt

(d) the frequency of the source is 50 Hz

37. The current (I) in the inductance is varying with time according to the plot shown in figure.

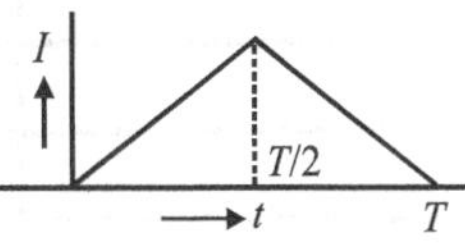

Which one of the following is the correct variation of voltage with time in the coil?

(a)
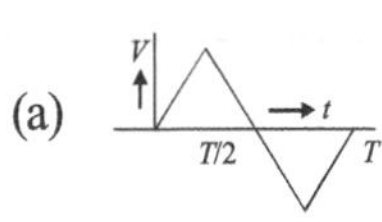

(b)
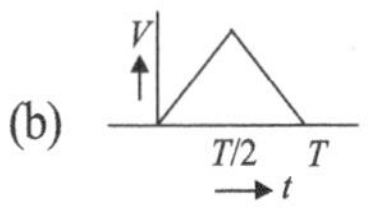

(c)
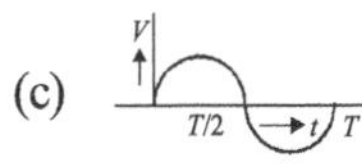

(d)
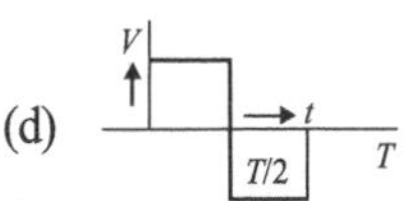

RESPONSE GRID					
	24. ⓐⓑⓒⓓ	25. ⓐⓑⓒⓓ	26. ⓐⓑⓒⓓ	27. ⓐⓑⓒⓓ	28. ⓐⓑⓒⓓ
	29. ⓐⓑⓒⓓ	30. ⓐⓑⓒⓓ	31. ⓐⓑⓒⓓ	32. ⓐⓑⓒⓓ	33. ⓐⓑⓒⓓ
	34. ⓐⓑⓒⓓ	35. ⓐⓑⓒⓓ	36. ⓐⓑⓒⓓ	37. ⓐⓑⓒⓓ	

Space for Rough Work

38. Using an A.C. voltmeter the potential difference in the electrical line in a house is read to be 234 volt. If the line frequency is known to be 50 cycles/second, the equation for the line voltage is

(a) $V = 165 \sin(100\pi t)$ (b) $V = 331 \sin(100\pi t)$
(c) $V = 220 \sin(100\pi t)$ (d) $V = 440 \sin(100\pi t)$

39. In the circuit shown, when the switch is closed, the capacitor charges with a time constant

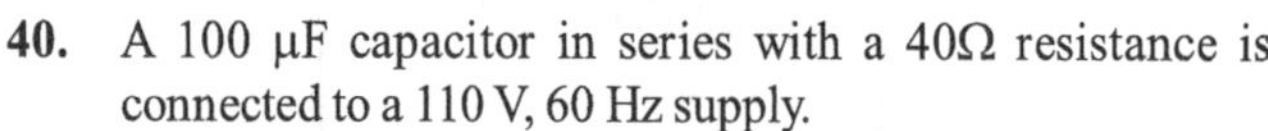

(a) RC
(b) 2RC
(c) $\frac{1}{2}$RC
(d) RC ln 2

40. A 100 μF capacitor in series with a 40Ω resistance is connected to a 110 V, 60 Hz supply.

What is the maximum current in the circuit?

(a) 3.24A (b) 4.25A (c) 2.25A (d) 5.20A

41. The core of any transformer is laminated so as to

(a) reduce the energy loss due to eddy currents
(b) make it light weight
(c) make it robust and strong
(d) increase the secondary voltage

42. An AC generator of 220 V having internal resistance r = 10Ω and external resistance R = 100Ω. What is the power developed in the external circuit?

(a) 484 W (b) 400 W (c) 441 W (d) 369 W

43. What is increased in step-down transformer?

(a) Voltage (b) Current
(c) Power (d) Current density

44. In the circuit shown below, the key K is closed at $t = 0$. The current through the battery is

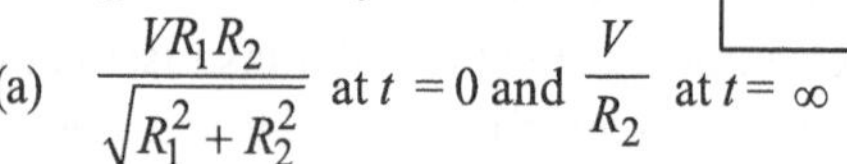

(a) $\frac{VR_1R_2}{\sqrt{R_1^2 + R_2^2}}$ at $t = 0$ and $\frac{V}{R_2}$ at $t = \infty$

(b) $\frac{V}{R_2}$ at $t = 0$ and $\frac{V(R_1 + R_2)}{R_1R_2}$ at $t = \infty$

(c) $\frac{V}{R_2}$ at $t = 0$ and $\frac{VR_1R_2}{\sqrt{R_1^2 + R_2^2}}$ at $t = \infty$

(d) $\frac{V(R_1 + R_2)}{R_1R_2}$ at $t = 0$ and $\frac{V}{R_2}$ at $t = \infty$

45. The inductance between A and D is

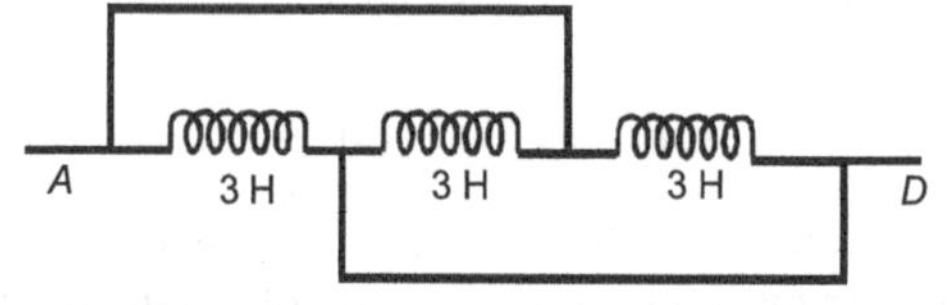

(a) 3.66H (b) 9H (c) 0.66H (d) 1H

RESPONSE GRID					
	38. ⓐⓑⓒⓓ	39. ⓐⓑⓒⓓ	40. ⓐⓑⓒⓓ	41. ⓐⓑⓒⓓ	42. ⓐⓑⓒⓓ
	43. ⓐⓑⓒⓓ	44. ⓐⓑⓒⓓ	45. ⓐⓑⓒⓓ		

DAILY PRACTICE PROBLEM DPP CHAPTERWISE CP21 - PHYSICS

Total Questions	45	Total Marks	180
Attempted		Correct	
Incorrect		Net Score	
Cut-off Score	50	Qualifying Score	70
Success Gap = Net Score – Qualifying Score			
Net Score = (Correct × 4) – (Incorrect × 1)			

Space for Rough Work

DPP - Daily Practice Problems

Chapter-wise Sheets

Date : ________ **Start Time :** ________ **End Time :** ________

PHYSICS (CP22)

SYLLABUS : Electromagnetic Waves

Max. Marks : 180 **Marking Scheme :** (+4) for correct & (–1) for incorrect answer **Time : 60 min.**

INSTRUCTIONS : This Daily Practice Problem Sheet contains 45 MCQs. For each question only one option is correct. Darken the correct circle/ bubble in the Response Grid provided on each page.

1. An electromagnetic wave in vacuum has the electric and magnetic field $\vec{E}$ and $\vec{B}$, which are always perpendicular to each other. The direction of polarization is given by $\vec{X}$ and that of wave propagation by $\vec{k}$. Then

   (a) $\vec{X} \parallel \vec{B}$ and $\vec{k} \parallel \vec{B}\times\vec{E}$ (b) $\vec{X} \parallel \vec{E}$ and $\vec{k} \parallel \vec{E}\times\vec{B}$

   (c) $\vec{X} \parallel \vec{B}$ and $\vec{k} \parallel \vec{E}\times\vec{B}$ (d) $\vec{X} \parallel \vec{E}$ and $\vec{k} \parallel \vec{B}\times\vec{E}$

2. The rms value of the electric field of the light coming from the Sun is 720 N/C. The average total energy density of the electromagnetic wave is

   (a) 4.58×10^{-6} J/m^3 (b) 6.37×10^{-9} J/m^3

   (c) 81.35×10^{-12} J/m^3 (d) 3.3×10^{-3} J/m^3

3. In order to establish an instantaneous displacemet current of 1 mA in the space between the plates of 2μF parallel plate capacitor, the potential difference need to apply is

   (a) 100 Vs^{-1} (b) 200 Vs^{-1} (c) 300 Vs^{-1} (d) 500 Vs^{-1}

4. During the propagation of electromagnetic waves in a medium:

   (a) Electric energy density is double of the magnetic energy density.

   (b) Electric energy density is half of the magnetic energy density.

   (c) Electric energy density is equal to the magnetic energy density.

   (d) Both electric and magnetic energy densities are zero.

5. An electromagnetic wave with frequency ω and wavelength λ travels in the + *y* direction. Its magnetic field is along + *x*-axis. The vector equation for the associated electric field (of amplitude E_0) is

   (a) $\vec{E} = -E_0 \cos\left(\omega t + \frac{2\pi}{\lambda} y\right)\hat{x}$

   (b) $\vec{E} = E_0 \cos\left(\omega t - \frac{2\pi}{\lambda} y\right)\hat{x}$

   (c) $\vec{E} = E_0 \cos\left(\omega t - \frac{2\pi}{\lambda} y\right)\hat{z}$

   (d) $\vec{E} = -E_0 \cos\left(\omega t + \frac{2\pi}{\lambda} y\right)\hat{z}$

RESPONSE GRID	1. ⓐⓑⓒⓓ	2. ⓐⓑⓒⓓ	3. ⓐⓑⓒⓓ	4. ⓐⓑⓒⓓ	5. ⓐⓑⓒⓓ

Space for Rough Work

6. An electromagnetic wave of frequency $\nu = 3.0$ MHz passes from vacuum into a dielectric medium with permittivity $\in = 4.0$. Then
   (a) wavelength is halved and frequency remains unchanged
   (b) wavelength is doubled and frequency becomes half
   (c) wavelength is doubled and the frequency remains unchanged
   (d) wavelength and frequency both remain unchanged.
7. The average electric field of electromagnetic waves in certain region of free space is 9×10^{-4} NC^{-1}. Then the average magnetic field in the same region is of the order of
   (a) 27×10^{-4} T (b) 3×10^{-12} T
   (c) $\left(\frac{1}{3}\right) \times 10^{-12}$ T (d) 3×10^{12} T
8. The electric field of an electromagnetic wave travelling through vaccum is given by the equation $E = E_0 \sin(kx - \omega t)$. The quantity that is independent of wavelength is
   (a) $k\omega$ (b) $\frac{k}{\omega}$ (c) $k^2\omega$ (d) ω
9. The electric and the magnetic field associated with an E.M. wave, propagating along the +z-axis, can be represented by
   (a) $\left[\vec{E} = E_0\hat{i}, \vec{B} = B_0\hat{j}\right]$ (b) $\left[\vec{E} = E_0\hat{k}, \vec{B} = B_0\hat{i}\right]$
   (c) $\left[\vec{E} = E_0\hat{j}, \vec{B} = B_0\hat{i}\right]$ (d) $\left[\vec{E} = E_0\hat{j}, \vec{B} = B_0\hat{k}\right]$
10. The energy of electromagnetic wave in vacuum is given by the relation
   (a) $\frac{E^2}{2\varepsilon_0} + \frac{B^2}{2\mu_0}$ (b) $\frac{1}{2}\varepsilon_0 E^2 + \frac{1}{2}\mu_0 B^2$
   (c) $\frac{E^2 + B^2}{c}$ (d) $\frac{1}{2}\varepsilon_0 E^2 + \frac{B^2}{2\mu_0}$
11. A plane electromagnetic wave is incident on a plane surface of area A, normally and is perfectly reflected. If energy E strikes the surface in time t then average pressure exerted on the surface is (c = speed of light)
   (a) zero (b) E/Atc (c) 2E/Atc (d) E/c
12. An electromagnetic wave travels along z-axis. Which of the following pairs of space and time varying fields would generate such a wave ?
   (a) E_x, B_y (b) E_y, B_x (c) E_z, B_x (d) E_y, B_z
13. The magnetic field in a travelling electromagnetic wave has a peak value of 20 nT. The peak value of electric field strength is :
   (a) 3 V/m (b) 6 V/m (c) 9 V/m (d) 12 V/m
14. Microwave oven acts on the principle of :
   (a) giving rotational energy to water molecules
   (b) giving translational energy to water molecules
   (c) giving vibrational energy to water molecules
   (d) transferring electrons from lower to higher energy levels in water molecule
15. Displacement current is
   (a) continuous when electric field is changing in the circuit
   (b) continuous when magnetic field is changing in the circuit
   (c) continuous in both types of fields
   (d) continuous through wires and resistance only
16. The electric field associated with an e.m. wave in vacuum is given by $\vec{E} = \hat{i}\, 40 \cos(kz - 6 \times 10^8 t)$, where E, z and t are in volt/m, meter and seconds respectively. The value of wave vector k is
   (a) $2\,m^{-1}$ (b) $0.5\,m^{-1}$ (c) $6\,m^{-1}$ (d) $3\,m^{-1}$
17. The charge on a parallel plate capacitor varies as $q = q_0 \cos 2\pi\upsilon t$. The plates are very large and close together (area = A, separation = d). The displacement current through the capacitor is
   (a) $q_0\, 2\pi\upsilon \sin\pi\upsilon t$ (b) $-q_0\, 2\pi\upsilon \sin 2\pi\upsilon t$
   (c) $q_0\, 2\pi \sin\pi\upsilon t$ (d) $q_0\, \pi\upsilon \sin 2\pi\upsilon t$
18. A radiation of energy 'E' falls normally on a perfectly reflecting surface. The momentum transferred to the surface is (C = Velocity of light)
   (a) $\frac{2E}{C}$ (b) $\frac{2E}{C^2}$ (c) $\frac{E}{C^2}$ (d) $\frac{E}{C}$
19. Match List - I (Electromagnetic wave type) with List - II (Its association/application) and select the correct option from the choices given below the lists:

	List 1		List 2
1.	Infrared waves	(i)	To treat muscular strain
2.	Radio waves	(ii)	For broadcasting
3.	X-rays	(iii)	To detect fracture of bones
4.	Ultraviolet rays	(iv)	Absorbed by the ozone layer of the atmosphere

	1	2	3	4
(a)	(iv)	(iii)	(ii)	(i)
(b)	(i)	(ii)	(iv)	(iii)
(c)	(iii)	(ii)	(i)	(iv)
(d)	(i)	(ii)	(iii)	(iv)

20. A plane electromagnetic wave travels in free space along X-direction. If the value of $\vec{B}$ (in tesla) at a particular point in space and time is $1.2 \times 10^{-8}\ \hat{k}$. The value of $\vec{E}$ (in Vm^{-1}) at that point is
   (a) $1.2\,\hat{j}$ (b) $3.6\,\hat{k}$ (c) $1.2\,\hat{k}$ (d) $3.6\,\hat{j}$

RESPONSE GRID				
6. ⓐⓑⓒⓓ	7. ⓐⓑⓒⓓ	8. ⓐⓑⓒⓓ	9. ⓐⓑⓒⓓ	10. ⓐⓑⓒⓓ
11. ⓐⓑⓒⓓ	12. ⓐⓑⓒⓓ	13. ⓐⓑⓒⓓ	14. ⓐⓑⓒⓓ	15. ⓐⓑⓒⓓ
16. ⓐⓑⓒⓓ	17. ⓐⓑⓒⓓ	18. ⓐⓑⓒⓓ	19. ⓐⓑⓒⓓ	20. ⓐⓑⓒⓓ

Space for Rough Work

21. If v_s, v_x and v_m are the speed of soft gamma rays, X-rays and microwaves respectively in vacuum, then

(a) $v_s > v_x > v_m$ (b) $v_s < v_x < v_m$

(c) $v_s > v_x < v_m$ (d) $v_s = v_x = v_m$

22. Photons of an electromagnetic radiation has an energy 11 keV each. To which region of electromagnetic spectrum does it belong ?

(a) X-ray region (b) Ultra violet region

(c) Infrared region (d) Visible region

23. A plane electromagnetic wave travels in free space along x-axis. At a particular point in space, the electric field along y-axis is 9.3 V m^{-1}. The magnetic induction (B) along z-axis is

(a) 3.1×10^{-8} T (b) 3×10^{-5} T

(c) 3×10^{-6} T (d) 9.3×10^{-6} T

24. The ratio of amplitude of magnetic field to the amplitude of electric field for an electromagnetic wave propagating in vacuum is equal to :

(a) the speed of light in vacuum

(b) reciprocal of speed of light in vacuum

(c) the ratio of magnetic permeability to the electric susceptibility of vacuum

(d) unity

25. A plane electromagnetic wave is incident on a material surface. If the wave delivers momentum p and energy E, then

(a) $p = 0, E = 0$ (b) $p \neq 0, E \neq 0$

(c) $p \neq 0, E = 0$ (d) $p = 0, E \neq 0$

26. Intensity of electromagnetic wave will be

(a) $I = c\mu_0 B_0^2 / 2$ (b) $I = c\varepsilon_0 B_0^2 / 2$

(c) $I = B_0^2 / c\mu_0$ (d) $I = E_0^2 / 2c\varepsilon_0$

27. The decreasing order of wavelength of infrared, microwave, ultraviolet and gamma rays is

(a) microwave, infrared, ultraviolet, gamma rays

(b) gamma rays, ultraviolet, infrared, micro-waves

(c) microwaves, gamma rays, infrared, ultraviolet

(d) infrared, microwave, ultraviolet, gamma rays

28. Which radiation in sunlight, causes heating effect ?

(a) Ultraviolet (b) Infrared

(c) Visible light (d) All of these

29. The speed of electromagnetic wave in vacuum depends upon the source of radiation. It

(a) increases as we move from γ-rays to radio waves

(b) decreases as we move from γ-rays to radio waves

(c) is same for all of them

(d) None of these

30. When an electromagnetic waves enter the ionised layer of ionosphere, the motion of electron cloud produces a space current and the electric field has its own capacitative displacement current, then

(a) the space current is in phase of displacement current

(b) the space current lags behind the displacement current by a phase 180°.

(c) the space current lags behind the displacement current by a phase 90°.

(d) the space current leads the displacement current by a phase 90°.

31. The displacement current is

(a) $\varepsilon_0 \, d\phi_E / dt$ (b) $\frac{\varepsilon_0}{R} d\phi_E / dt$

(c) $\varepsilon_0 E / R$ (d) $\varepsilon_0 q C / R$

32. Electromagnetic radiation of highest frequency is

(a) infrared radiations (b) visible radiation

(c) radio waves (d) γ-rays

33. A point source of electromagnetic radiation has an average power output of 1500 W. The maximum value of electric field at a distance of 3m from this sources in Vm^{-1} is

(a) 500 (b) 100 (c) $\frac{500}{3}$ (d) $\frac{250}{3}$

34. Frequency of a wave is 6×10^{15} Hz. The wave is

(a) radiowave (b) microwave

(c) x-ray (d) ultraviolet

35. The electromagnetic waves do not transport

(a) energy (b) charge

(c) momentum (d) information

36. Which of the following statement is false for the properties of electromagnetic waves?

(a) Both electric and magnetic field vectors attain the maxima and minima at the same place and same time.

(b) The energy in electromagnetic wave is divided equally between electric and magnetic vectors

(c) Both electric and magnetic field vectors are parallel to each other and perpendicular to the direction of propagation of wave

(d) These waves do not require any material medium for propagation.

RESPONSE GRID				
21. ⓐⓑⓒⓓ	22. ⓐⓑⓒⓓ	23. ⓐⓑⓒⓓ	24. ⓐⓑⓒⓓ	25. ⓐⓑⓒⓓ
26. ⓐⓑⓒⓓ	27. ⓐⓑⓒⓓ	28. ⓐⓑⓒⓓ	29. ⓐⓑⓒⓓ	30. ⓐⓑⓒⓓ
31. ⓐⓑⓒⓓ	32. ⓐⓑⓒⓓ	33. ⓐⓑⓒⓓ	34. ⓐⓑⓒⓓ	35. ⓐⓑⓒⓓ
36. ⓐⓑⓒⓓ				

Space for Rough Work

37. Which of the following electromagnetic waves has minimum frequency ?
(a) Microwaves (b) Audible waves
(c) Ultrasonic wave (d) Radiowaves

38. The wave impendance of free space is
(a) zero (b) 376.6 Ω (c) 33.66 Ω (d) 3.76 Ω

39. A plane electromagnetic wave in a non-magnetic dielectric medium is given by $\vec{E} = \vec{E}_0 (4\times10^{-7} x - 50t)$ with distance being in meter and time in seconds. The dielectric constant of the medium is :
(a) 2.4 (b) 5.8 (c) 8.2 (d) 4.8

40. We consider the radiation emitted by the human body. Which of the following statements is true?
(a) the radiation emitted lies in the ultraviolet region and hence is not visible.
(b) the radiation emitted is in the infra-red region.
(c) the radiation is emitted only during the day.
(d) the radiation is emitted during the summers and absorbed during the winters.

41. In a plane electromagnetic wave propagating in space has an electric field of amplitude 9×10^3 V/m, then the amplitude of the magnetic field is
(a) 2.7×10^{12} T (b) 9.0×10^{-3} T
(c) 3.0×10^{-4} T (d) 3.0×10^{-5} T

42. Out of the following options which one can be used to produce a propagating electromagnetic wave ?
(a) A charge moving at constant velocity
(b) A stationary charge
(c) A chargeless particle
(d) An accelerating charge

43. Radio waves of constant amplitude can be generated with
(a) rectifier (b) filter
(c) F.E.T. (d) oscillator

44. In an electromagnetic wave
(a) power is transmitted along the magnetic field
(b) power is transmitted along the electric field
(c) power is equally transferred along the electric and magnetic fields
(d) power is transmitted in a direction perpendicular to both the fields

45. If c is the speed of electromagnetic waves in vacuum, its speed in a medium of dielectric constant K and relative permeability μ_r is
(a) $v = \frac{1}{\sqrt{\mu_r K}}$ (b) $v = c\sqrt{\mu_r K}$
(c) $v = \frac{c}{\sqrt{\mu_r K}}$ (d) $v = \frac{K}{\sqrt{\mu_r C}}$

RESPONSE GRID					
	37. ⓐⓑⓒⓓ	38. ⓐⓑⓒⓓ	39. ⓐⓑⓒⓓ	40. ⓐⓑⓒⓓ	41. ⓐⓑⓒⓓ
	42. ⓐⓑⓒⓓ	43. ⓐⓑⓒⓓ	44. ⓐⓑⓒⓓ	45. ⓐⓑⓒⓓ	

DAILY PRACTICE PROBLEM DPP CHAPTERWISE CP22 - PHYSICS

Total Questions	45	Total Marks	180
Attempted		Correct	
Incorrect		Net Score	
Cut-off Score	50	Qualifying Score	70
Success Gap = Net Score – Qualifying Score			
Net Score = (Correct × 4) – (Incorrect × 1)			

Space for Rough Work

DPP - Daily Practice Problems

Chapter-wise Sheets

Date : ______ **Start Time :** ______ **End Time :** ______

PHYSICS (CP23)

SYLLABUS : Ray Optics and Optical Instruments

Max. Marks : 180 **Marking Scheme :** (+4) for correct & (–1) for incorrect answer **Time : 60 min.**

INSTRUCTIONS : This Daily Practice Problem Sheet contains 45 MCQs. For each question only one option is correct. Darken the correct circle/ bubble in the Response Grid provided on each page.

1. A double convex lens is made of glass which has its refractive index 1.45 for violet rays and 1.50 for red rays. If the focal length for violet ray is 20cm, the focal length for red ray will be
(a) 9 cm (b) 18 cm (c) 20 cm (d) 22 cm

2. If the refractive index of the material of a prism is $\cot\frac{A}{2}$ and the angle of prism is A, then angle of minimum deviation is
(a) $\pi - 2A$ (b) $\pi - A$ (c) $\frac{\pi}{2} - 2A$ (d) $\frac{\pi}{2} - A$

3. If two + 5 diopter lenses are mounted at some distance apart, the equivalent power will always be negative if the distance is
(a) greater than 40 cm (b) equal to 40 cm
(c) equal to 10 cm (d) less than 10 cm

4. Refraction of light from air to glass and from air to water are shown in figure (i) and figure (ii) below. The value of the angle θ in the case of refraction as shown in figure (iii) will be

(i) Glass / Air, 35°, 60° (ii) Air / Water, 60°, 41° (iii) Glass / Water, θ, 41°

(a) 30° (b) 35° (c) 60° (d) 41°

5. A fish looking up through the water sees the outside world contained in a circular horizon. If the refractive index of water is $\frac{4}{3}$ and the fish is 12 cm below the surface, the radius of this circle in cm is
(a) $36\sqrt{5}$ (b) $4\sqrt{5}$ (c) $36\sqrt{7}$ (d) $36/\sqrt{7}$

6. If f_V and f_R are the focal lengths of a convex lens for violet and red light respectively and F_V and F_R are the focal lengths of concave lens for violet and red light respectively, then we have
(a) $f_V < f_R$ and $F_V > F_R$ (b) $f_V < f_R$ and $F_V < F_R$
(c) $f_V > f_R$ and $F_V > F_R$ (d) $f_V > f_R$ and $F_V < F_R$

RESPONSE GRID	1. ⓐⓑⓒⓓ	2. ⓐⓑⓒⓓ	3. ⓐⓑⓒⓓ	4. ⓐⓑⓒⓓ	5. ⓐⓑⓒⓓ
	6. ⓐⓑⓒⓓ				

Space for Rough Work

7. Spherical aberration in a lens :
   (a) is minimum when most of the deviation is at first surface
   (b) is minimum when most of the deviation is at the second surface
   (c) is minimum when the total deviation is equally distributed over the two surfaces
   (d) does not depend on the above considerations
8. A rod of length 10 cm lies along the principal axis of a concave mirror of focal length 10 cm in such a way that its end closer to the pole is 20 cm away from the mirror. The length of the image is :
   (a) 10 cm (b) 15 cm (c) 2.5 cm (d) 5 cm
9. A telescope consists of two thin lenses of focal lengths, 0.3 m and 3 cm respectively. It is focused on moon which subtends an angle of 0.5° at the objective. Then the angle subtended at the eye by the final image will be
   (a) 5° (b) 0.25° (c) 0.5° (d) 0.35°
10. The layered lens as shown is made of two types of transparent materials-one indicated by horizontal lines and the other by vertical lines. The number of images formed of an object will be

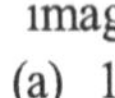

   (a) 1 (b) 2 (c) 3 (d) 6
11. A man's near point is 0.5 m and far point is 3 m. Power of spectacle lenses required for (i) reading purposes, (ii) seeing distant objects, respectively, are
   (a) –2 D and + 3 D (b) +2 D and –3 D
   (c) +2 D and –0.33 D (d) –2 D and + 0.33 D
12. A ray of light falls on a transparent glass slab of refractive index 1.62. If the reflected ray and the refracted ray are mutually perpendicular, the angle of incidence is
   (a) $\tan^{-1}(1.62)$ (b) $\tan^{-1}\left(\frac{1}{1.62}\right)$
   (c) $\tan^{-1}(1.33)$ (d) $\tan^{-1}\left(\frac{1}{1.33}\right)$
13. A telescope has an objective of focal length 100 cm and an eyepiece of focal length 5 cm. What is the magnifying power of the telescope when the final image is formed at the least distance of distinct vision ?
   (a) 20 (b) 24 (c) 28 (d) 32
14. Which light rays undergoes two internal reflection inside a raindrop, which of the rainbow is formed?
   (a) Primary rainbow (b) Secondary rainbow
   (c) Both (a) and (b) (d) Can't say
15. When a plane mirror is placed horizontally on a level ground at a distance of 60 *m* from the foot of a tower, the top of the tower and its image in the mirror subtend an angle of 90° at the eye. The height of the tower will be
   (a) 30m (b) 60m (c) 90m (d) 120m
16. A parallel beam of light is incident on the surface of a transparent hemisphere of radius *R* and refractive index 2.0 as shown in figure. The position of the image formed by refraction at the first surface is :
   (a) *R*/2
   (b) *R*
   (c) 2*R*
   (d) 3*R*

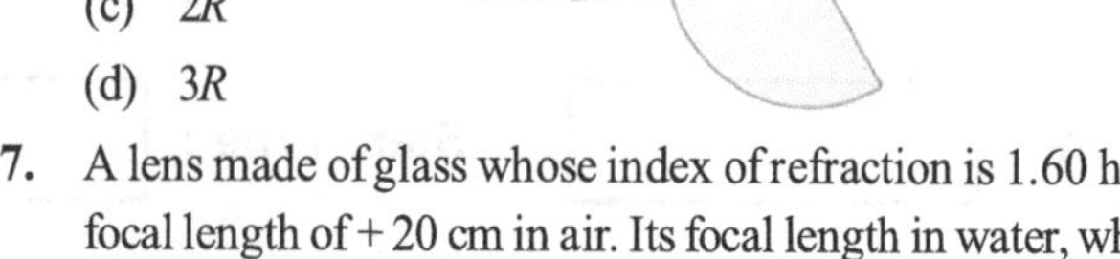

17. A lens made of glass whose index of refraction is 1.60 has a focal length of + 20 cm in air. Its focal length in water, whose refractive index is 1.33, will be
   (a) three times longer than in air
   (b) two times longer than in air
   (c) same as in air
   (d) None of these
18. A compound microscope has an eye piece of focal length 10 cm and an objective of focal length 4 cm. Calculate the magnification, if an object is kept at a distance of 5 cm from the objective so that final image is formed at the least distance vision (20 cm) :
   (a) 12 (b) 11 (c) 10 (d) 13
19. For a prism kept in air it is found that for an angle of incidence 60°, the angle of Prism A, angle of deviation δ and angle of emergence 'e' become equal. Then the refractive index of the prism is
   (a) 1.73 (b) 1.15 (c) 1.5 (d) 1.33
20. A person can see clearly only upto a distance of 30 cm. He wants to read a book placed at a distance of 50 cm from his eyes. What is the power of the lens of his spectacles ?
   (a) –1.0 D (b) –1.33 D (c) –1.67 D (d) –2.0 D
21. An object is placed at a distance of 40 cm in front of a concave mirror of focal length 20 cm. The image produced is
   (a) real, inverted and smaller in size
   (b) real, inverted and of same size
   (c) real and erect
   (d) virtual and inverted
22. A vessel of depth *x* is half filled with oil of refractive index μ_1 and the other half is filled with water of refractive index μ_2. The apparent depth of the vessel when viewed from above is
   (a) $\frac{x(\mu_1+\mu_2)}{2\mu_1\mu_2}$ (b) $\frac{x\mu_1\mu_2}{2(\mu_1+\mu_2)}$
   (c) $\frac{x\mu_1\mu_2}{(\mu_1+\mu_2)}$ (d) $\frac{2x(\mu_1+\mu_2)}{\mu_1\mu_2}$

RESPONSE GRID					
	7. ⓐⓑⓒⓓ	8. ⓐⓑⓒⓓ	9. ⓐⓑⓒⓓ	10. ⓐⓑⓒⓓ	11. ⓐⓑⓒⓓ
	12. ⓐⓑⓒⓓ	13. ⓐⓑⓒⓓ	14. ⓐⓑⓒⓓ	15. ⓐⓑⓒⓓ	16. ⓐⓑⓒⓓ
	17. ⓐⓑⓒⓓ	18. ⓐⓑⓒⓓ	19. ⓐⓑⓒⓓ	20. ⓐⓑⓒⓓ	21. ⓐⓑⓒⓓ
	22. ⓐⓑⓒⓓ				

Space for Rough Work

23. The following figure shows refraction of light at the interface of three media Correct the order of optical density (d) of the media is

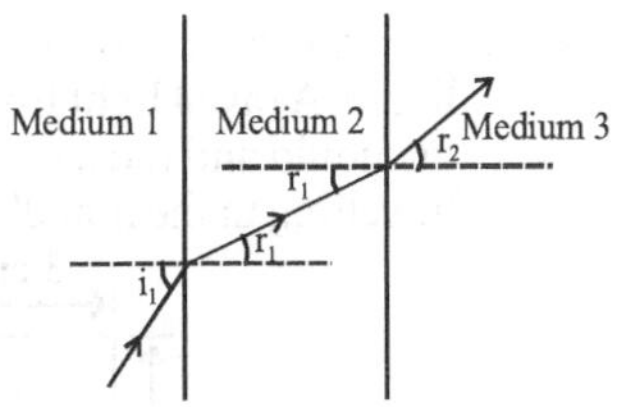

(a) $d_1 > d_2 > d_3$ (b) $d_2 > d_1 > d_3$
(c) $d_3 > d_3 > d_2$ (d) $d_2 > d_3 > d_1$

24. Light travels in two media A and B with speeds 1.8×10^8 m s^{-1} and 2.4×10^8 m s^{-1} respectively. Then the critical angle between them is

(a) $\sin^{-1}\left(\frac{2}{3}\right)$ (b) $\tan^{-1}\left(\frac{3}{4}\right)$
(c) $\tan^{-1}\left(\frac{2}{3}\right)$ (d) $\sin^{-1}\left(\frac{3}{4}\right)$

25. The refractive index of a glass is 1.520 for red light and 1.525 for blue light. Let D_1 and D_2 be angles of minimum deviation for red and blue light respectively in a prism of this glass. Then,
(a) $D_1 < D_2$ (b) $D_1 = D_2$
(c) D_1 can be less than or greater than D_2 depending upon the angle of prism
(d) $D_1 > D_2$

26. Which of the following is not due to total internal reflection?
(a) Working of optical fibre
(b) Difference between apparent and real depth of pond
(c) Mirage on hot summer days
(d) Brilliance of diamond

27. A body is located on a wall. Its image of equal size is to be obtained on a parallel wall with the help of a convex lens. The lens is placed at a distance 'd' ahead of second wall, then the required focal length will be
(a) only $\frac{d}{4}$
(b) only $\frac{d}{2}$
(c) more than $\frac{d}{4}$ but less than $\frac{d}{2}$
(d) less than $\frac{d}{4}$

28. A concave mirror forms the image of an object on a screen. If the lower half of the mirror is covered with an opaque card, the effect would be to make the
(a) image less bright.
(b) lower half of the image disappear.
(c) upper half of the image disappear.
(d) image blurred.

29. A ray of light passes through an equilateral prism such that the angle of incidence is equal to the angle of emergence and the latter is equal to $\frac{3}{4}$ th of angle of prism. The angle of deviation is
(a) 25° (b) 30° (c) 45° (d) 35°

30. The power of a biconvex lens is 10 dioptre and the radius of curvature of each surface is 10 cm. Then the refractive index of the material of the lens is
(a) $\frac{3}{2}$ (b) $\frac{4}{3}$ (c) $\frac{9}{8}$ (d) $\frac{5}{3}$

31. A microscope is focussed on a mark on a piece of paper and then a slab of glass of thickness 3 cm and refractive index 1.5 is placed over the mark. How should the microscope be moved to get the mark in focus again ?
(a) 4.5 cm downward (b) 1 cm downward
(c) 2 cm upward (d) 1 cm upward

32. What causes chromatic aberration?
(a) Marginal rays
(b) Central rays
(c) Difference in radii of curvature of its surfaces
(d) Variation of focal length of lens with colour

33. The graph between angle of deviation (δ) and angle of incidence (i) for a triangular prism is represented by

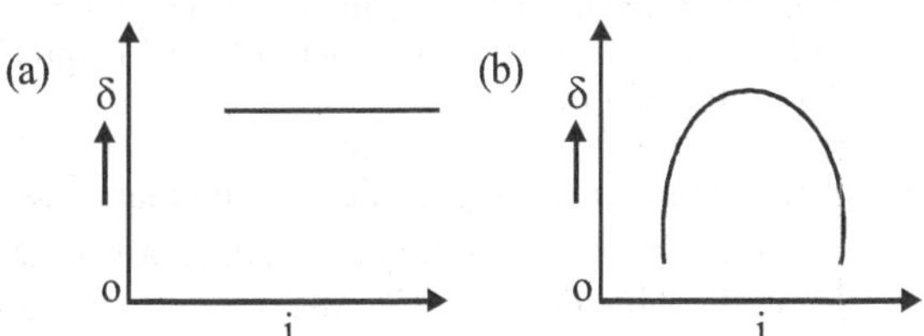

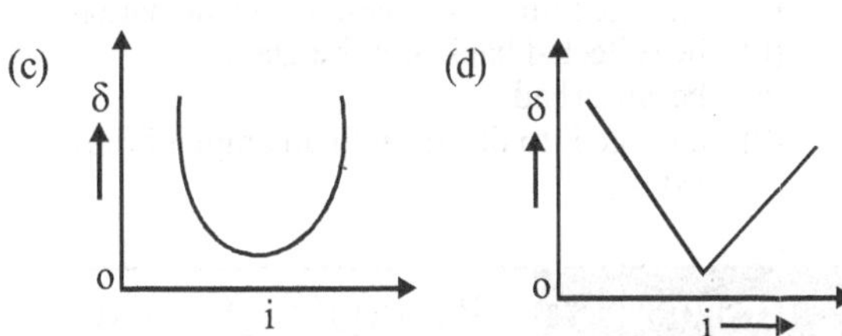

34. The ratio of thickness of plates of two transparent medium A and B is 6 : 4. If light takes equal time in passing through them, then refractive index of A with respect to B will be
(a) 1.33 (b) 1.75 (c) 1.4 (d) 1.5

35. A rectangular block of glass is placed on a mark made on the surface of the table and it is viewed from the vertical position of eye. If refractive index of glass be μ and its thickness d, then the mark will appear to be raised up by
(a) $\frac{(\mu+1)d}{\mu}$ (b) $\frac{(\mu-1)d}{\mu}$ (c) $\frac{(\mu+1)}{\mu d}$ (d) $\frac{(\mu-1)\mu}{d}$

RESPONSE GRID					
	23. ⓐⓑⓒⓓ	24. ⓐⓑⓒⓓ	25. ⓐⓑⓒⓓ	26. ⓐⓑⓒⓓ	27. ⓐⓑⓒⓓ
	28. ⓐⓑⓒⓓ	29. ⓐⓑⓒⓓ	30. ⓐⓑⓒⓓ	31. ⓐⓑⓒⓓ	32. ⓐⓑⓒⓓ
	33. ⓐⓑⓒⓓ	34. ⓐⓑⓒⓓ	35. ⓐⓑⓒⓓ		

Space for Rough Work

36. If a glass prism is dipped in water, its dispersive power
(a) increases
(b) decreases
(c) does not change
(d) may increase or decrease depending on whether the angle of the prism is less than or greater than 60°

37. A planoconcave lens is placed on a paper on which a flower is drawn. How far above its actual position does the flower appear to be?

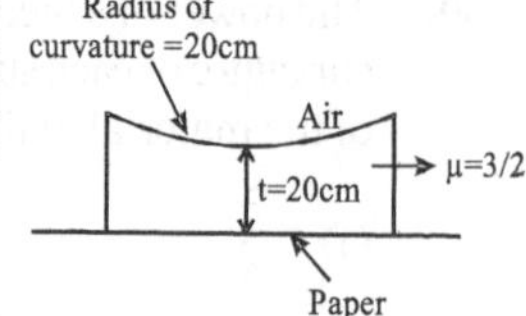

(a) 10 cm (b) 15 cm
(c) 50 cm (d) None of these

38. To get three images of a single object, one should have two plane mirrors at an angle of
(a) 60° (b) 90° (c) 120° (d) 30°

39. Light propagates with speed of $2.2\times10^8 m/s$ and $2.4\times10^8 m/s$ in the media P and Q respectively. The critical angle of incidence for light undergoing reflection from P and Q is

(a) $\sin^{-1}\left(\frac{1}{11}\right)$ (b) $\sin^{-1}\left(\frac{11}{12}\right)$

(c) $\sin^{-1}\left(\frac{5}{12}\right)$ (d) $\sin^{-1}\left(\frac{5}{11}\right)$

40. A thin convergent glass lens (μ_g = 1.5) has a power of + 5.0 D. When this lens is immersed in a liquid of refractive index μ, it acts as a divergent lens of focal length 100 cm. The value of μ must be
(a) 4/3 (b) 5/3 (c) 5/4 (d) 6/5

41. A ray of light travelling inside a rectangular glass block of refractive index $\sqrt{2}$ is incident on the glass-air surface at an angle of incidence of 45°. The refractive index of air is one. Under these conditions the ray will
(a) emerge into the air without any deviation
(b) be reflected back into the glass
(c) be absorbed
(d) emerge into the air with an angle of refraction equal to 90°

42. A small coin is resting on the bottom of a beaker filled with liquid. A ray of light from the coin travels upto the surface of the liquid and moves along its surface. How fast is the light travelling in the liquid?

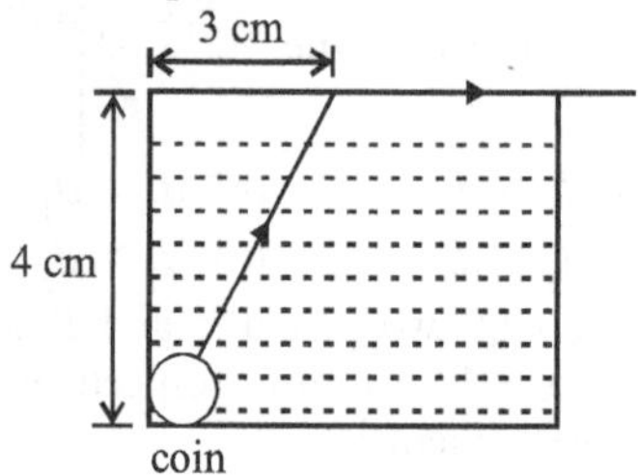

(a) 2.4×10^8 m/s (b) 3.0×10^8 m/s
(c) 1.2×10^8 m/s (d) 1.8×10^8 m/s

43. A ray PQ incident on the refracting face BA is refracted in the prism BAC as shown in the figure and emerges from the other refracting face AC as RS such that AQ = AR. If the angle of prism A = 60° and the refractive index of the material of prism is $\sqrt{3}$, then the angle of deviation of the ray is

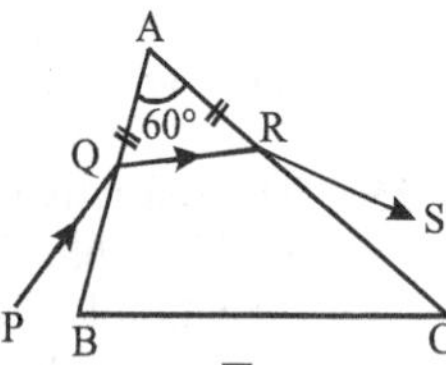

(a) 60° (b) 45°
(c) 30° (d) None of these

44. When a biconvex lens of glass having refractive index 1.47 is dipped in a liquid, it acts as a plane sheet of glass. This implies that the liquid must have refractive index.
(a) equal to that of glass
(b) less then one
(c) greater than that of glass
(d) less then that of glass

45. If a thin prism of glass is dipped in water then minimum deviation (with respect to air) of light produced by prism will be $\left({}_w\mu_g=\frac{3}{2},\ {}_a\mu_w=\frac{4}{3}\right)$

(a) $\frac{1}{5}$ (b) $\frac{1}{4}$ (c) $\frac{1}{2}$ (d) $\frac{1}{3}$

RESPONSE GRID					
	36. ⓐⓑⓒⓓ	37. ⓐⓑⓒⓓ	38. ⓐⓑⓒⓓ	39. ⓐⓑⓒⓓ	40. ⓐⓑⓒⓓ
	41. ⓐⓑⓒⓓ	42. ⓐⓑⓒⓓ	43. ⓐⓑⓒⓓ	44. ⓐⓑⓒⓓ	45. ⓐⓑⓒⓓ

DAILY PRACTICE PROBLEM DPP CHAPTERWISE CP23 - PHYSICS			
Total Questions	45	Total Marks	180
Attempted		Correct	
Incorrect		Net Score	
Cut-off Score	45	Qualifying Score	60
Success Gap = Net Score – Qualifying Score			
Net Score = (Correct × 4) – (Incorrect × 1)			

Space for Rough Work

DPP - Daily Practice Problems

Chapter-wise Sheets

Date : | **Start Time :** | **End Time :**

PHYSICS

SYLLABUS : Wave Optics

Max. Marks : 180 **Marking Scheme :** (+4) for correct & (–1) for incorrect answer **Time : 60 min.**

INSTRUCTIONS : This Daily Practice Problem Sheet contains 45 MCQs. For each question only one option is correct. Darken the correct circle/ bubble in the Response Grid provided on each page.

1. In young's double-slit experiment, the intensity of light at a point on the screen where the path difference is λ is I, λ being the wavelength of light used. The intensity at a point where the path difference is $\frac{\lambda}{4}$ will be
   (a) $\frac{I}{4}$ (b) $\frac{I}{2}$ (c) I (d) zero

2. A beam of light is incident on a glass slab ($\mu = 1.54$) in a direction as shown in the figure. The reflected light is analysed by a polaroid prism. On rotating the polaroid, ($\tan 57° = 1.54$)

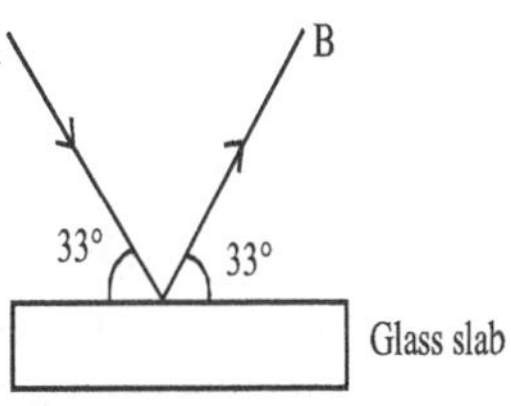

   (a) the intensity remains unchanged
   (b) the intensity is reduced to zero and remains at zero
   (c) the intensity gradually reduces to zero and then again increase
   (d) the intensity increases continuously

3. Two sources of light of wavelengths 2500 Å and 3500 Å are used in Young's double slit expt. simultaneously. Which orders of fringes of two wavelength patterns coincide?
   (a) 3rd order of 1st source and 5th of the 2nd
   (b) 7th order of 1st and 5th order of 2nd
   (c) 5th order of 1st and 3rd order of 2nd
   (d) 5th order of 1st and 7th order of 2nd

4. Figure shows behavior of a wavefront when it passes through a prism.

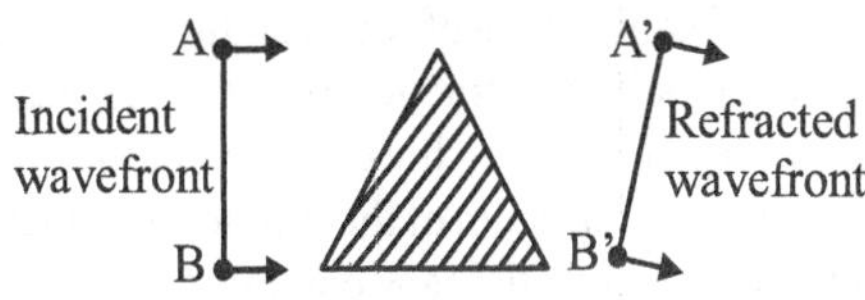

   Which of the following statements is/are correct ?
   (a) Lower portion of wavefront (*B*') is delayed resulting in a tilt.
   (b) Time taken by light to reach *A*' is equal to the time taken to reach *B*' from *B*.
   (c) Speed of wavefront is same everywhere.
   (d) A particle on wavefront *A*' *B*' is in phase with a particle on wavefront *AB*.

5. When the angle of incidence is 60° on the surface of a glass slab, it is found that the reflected ray is completely polarised. The velocity of light in glass is
   (a) $\sqrt{2} \times 10^8 ms^{-1}$ (b) $\sqrt{3} \times 10^8 ms^{-1}$
   (c) $2 \times 10^8 ms^{-1}$ (d) $3 \times 10^8 ms^{-1}$

RESPONSE GRID	1. ⓐⓑⓒⓓ	2. ⓐⓑⓒⓓ	3. ⓐⓑⓒⓓ	4. ⓐⓑⓒⓓ	5. ⓐⓑⓒⓓ

Space for Rough Work

6. Figure shows two coherent sources S_1 and S_2 vibrating in same phase. AB is an irregular wire lying at a far distance from the sources S_1 and S_2. Let $\frac{\lambda}{d} = 10^{-3}$ and $\angle BOA = 0.12°$. How many bright spots will be seen on the wire, including points A and B?
(a) 5
(b) 4
(c) 2
(d) 7

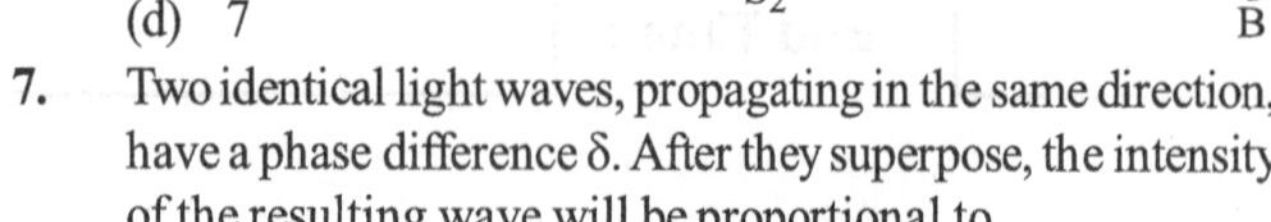

7. Two identical light waves, propagating in the same direction, have a phase difference δ. After they superpose, the intensity of the resulting wave will be proportional to
(a) $\cos\delta$ (b) $\cos(\delta/2)$
(c) $\cos^2(\delta/2)$ (d) $\cos^2\delta$

8. In YSDE, both slits are covered by transparent slab. Upper slit is covered by slab of R.I. 1.5 and thickness t and lower is covered by R.I. $\frac{4}{3}$ and thickness 2t, then central maxima

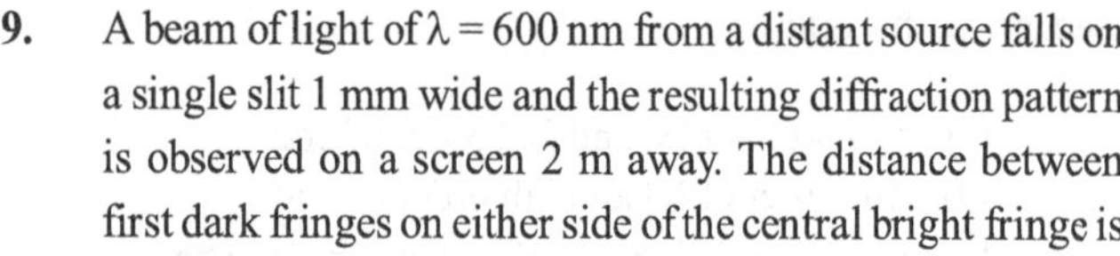

(a) shifts in +ve y-axis direction
(b) shifts in –ve y-axis direction
(c) remains at same position
(d) may shift in upward or downward depending upon wavelength of light

9. A beam of light of $\lambda = 600$ nm from a distant source falls on a single slit 1 mm wide and the resulting diffraction pattern is observed on a screen 2 m away. The distance between first dark fringes on either side of the central bright fringe is
(a) 1.2 cm (b) 1.2 mm
(c) 2.4 cm (d) 2.4 mm

10. A parallel beam of light of wavelength λ is incident normally on a narrow slit. A diffraction pattern is formed on a screen placed perpendicular to the direction of the incident beam. At the second minimum of the diffraction pattern, the phase difference between the rays coming from the two edges of slit is
(a) $\pi\lambda$ (b) 2π
(c) 3π (d) 4π

11. The diffraction effects in a microscopic specimen become important when the separation between two points is
(a) much greater than the wavelength of light used.
(b) much less than the wavelength of light used.
(c) comparable to the wavelength of light used.
(d) independent of the wavelength of light used.

12. On a rainy day, if there is an oil drop on tar road coloured rings are seen around this drop. This is due to
(a) total internal reflection of light
(b) polarisation
(c) diffraction pattern
(d) interference pattern produced due to oil film

13. In a Young's double slit experiment, the intensity at a point where the path difference $\frac{\lambda}{6}$ (λ – is wavelength of the light) is I. If I_0 denotes the maximum intensity, then $\frac{I}{I_0}$ is equal to
(a) $\frac{1}{2}$ (b) $\frac{\sqrt{3}}{2}$ (c) $\frac{1}{\sqrt{2}}$ (d) $\frac{3}{4}$

14. According to Huygens, medium through which light waves travel is
(a) vacuum only (b) luminiferous ether
(c) liquid only (d) solid only

15. If we observe the single slit Fraunhofer diffraction with wavelength λ and slit width b, the width of the central maxima is 2θ. On decreasing the slit width for the same λ
(a) θ increases
(b) θ remains unchanged
(c) θ decreases
(d) θ increases or decreases depending on the intensity of light

16. Aperture of the human eye is 2 mm. Assuming the mean wavelength of light to be 5000 Å, the angular resolution limit of the eye is nearly
(a) 2 minute (b) 1 minute
(c) 0.5 minute (d) 1.5 minute

17. Unpolarised light is incident on a dielectric of refractive index $\sqrt{3}$. What is the angle of incidence if the reflected beam is completely polarised?
(a) 30° (b) 45°
(c) 60° (d) 75°

18. The figure shows the interference pattern obtained in a double-slit experiment using light of wavelength 600nm. 1, 2, 3, 4 and 5 are marked on five fringes.

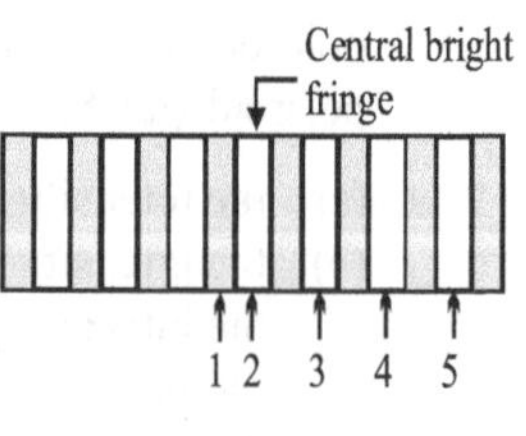

The third order bright fringe is
(a) 2 (b) 3 (c) 4 (d) 5

RESPONSE GRID

6. ⓐⓑⓒⓓ	7. ⓐⓑⓒⓓ	8. ⓐⓑⓒⓓ	9. ⓐⓑⓒⓓ	10. ⓐⓑⓒⓓ
11. ⓐⓑⓒⓓ	12. ⓐⓑⓒⓓ	13. ⓐⓑⓒⓓ	14. ⓐⓑⓒⓓ	15. ⓐⓑⓒⓓ
16. ⓐⓑⓒⓓ	17. ⓐⓑⓒⓓ	18. ⓐⓑⓒⓓ		

Space for Rough Work

19. Which of the following diagrams represent the variation of electric field vector with time for a circularly polarised light ?

(a) (b) (c) (d)

20. With a monochromatic light, the fringe-width obtained in a Young's double slit experiment is 0.133 cm. The whole set-up is immersed in water of refractive index 1.33, then the new fringe-width is

(a) 0.133 cm (b) 0.1 cm
(c) 1.33 cm (d) 0.2 cm

21. The condition for obtaining secondary maxima in the diffraction pattern due to single slit is

(a) $a\sin\theta = n\lambda$ (b) $a\sin\theta = (2n-1)\frac{\lambda}{2}$

(c) $a\sin\theta = (2n-1)\lambda$ (d) $a\sin\theta = \frac{n\lambda}{2}$

22. In double slit experiment, the angular width of the fringes is 0.20° for the sodium light $(\lambda = 5890\text{Å})$. In order to increase the angular width of the fringes by 10%, the necessary change in wavelength is

(a) zero (b) increased by 6479 Å
(c) decreased by 589 Å (d) increased by 589 Å

23. In Young's double slit experiment with sodium vapour lamp of wavelength 589 nm and the slits 0.589 mm apart, the half angular width of the central maximum is

(a) $\sin^{-1}(0.01)$ (b) $\sin^{-1}(0.0001)$
(c) $\sin^{-1}(0.001)$ (d) $\sin^{-1}(0.1)$

24. The adjacent figure shows Fraunhoffer's diffraction due to a single slit. If first minimum is obtained in the direction shown, then the path difference between rays 1 and 3 is

(a) 0
(b) $\lambda/4$
(c) $\lambda/2$
(d) λ

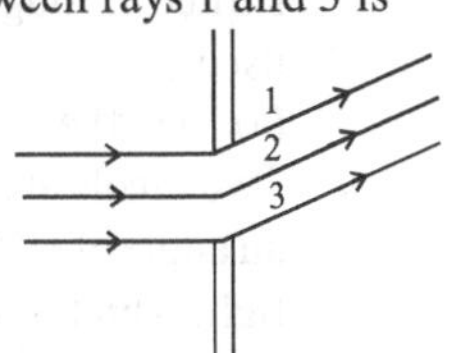

25. A YDSE is conducted in water (μ_1) as shown in figure. A glass plate of thickness t and refractive index μ_2 is placed in the path of S_2. The optical path difference at O is

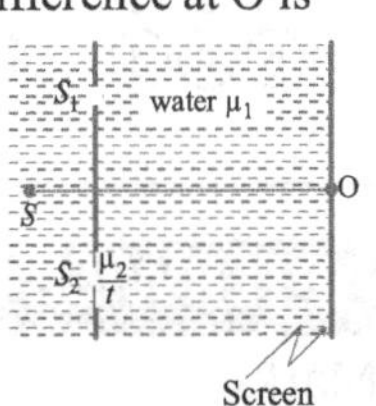

(a) $(\mu_2 - 1)t$
(b) $(\mu_1 - 1)t$
(c) 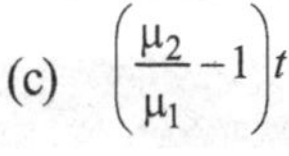 $\left(\frac{\mu_2}{\mu_1} - 1\right)t$
(d) $(\mu_2 - \mu_1)t$

26. In a Fresnel biprism experiment, the two positions of lens give separation between the slits as 16 cm and 9 cm respectively. What is the actual distance of separation?

(a) 12.5 cm (b) 12 cm (c) 13 cm (d) 14 cm

27. If two waves represented by $y_1 = 4\sin\omega t$ and $y_2 = \left(\omega t + \frac{\pi}{3}\right)$ interfere at a point, then the amplitude of the resulting wave will be about

(a) 7 (b) 6 (c) 5 (d) 3.5

28. In Young's double slit experiment, the separation between the slits is halved and the distance between the slits and screen is doubled. The fringe width will

(a) be halved (b) be doubled
(c) be quadrupled (d) remain unchanged

29. At the first minimum adjacent to the central maximum of a single-slit diffraction pattern, the phase difference between the Huygen's wavelet from the edge of the slit and the wavelet from the midpoint of the slit is :

(a) $\frac{\pi}{2}$ radian (b) π radian

(c) $\frac{\pi}{8}$ radian (d) $\frac{\pi}{4}$ radian

30. The central fringe of the interference pattern produced by light of wavelength 6000Å is found to shift to the position of 4th bright fringe after a glass plate of refractive index 1.5 is introduced in front of one of slits in Young's experiment. The thickness of the glass plate will be

(a) 4.8 μm (b) 8.23 μm
(c) 14.98 μm (d) 3.78 μm

31. Sodium light $(\lambda = 6\times10^{-7}\text{m})$ is used to produce interference pattern. The observed fringe width is 0.12 mm. The angle between two interfering wave trains, is

(a) 1×10^{-3} rad (b) 1×10^{-2} rad
(c) 5×10^{-3} rad (d) 5×10^{-2} rad

32. The Young's double slit experiment is performed with blue and with green light of wavelengths 4360Å and 5460Å respectively. If x is the distance of 4th maxima from the central one, then

(a) x (blue) = x (green) (b) x (blue) > x (green)

(c) x (blue) < x (green) (d) $\frac{x(\text{blue})}{x(\text{green})} = \frac{5460}{4360}$

33. If yellow light emitted by sodium lamp in Young's double slit experiment is replaced by a monochromatic blue light of the same intensity

(a) fringe width will decrease
(b) finge width will increase
(c) fringe width will remain unchanged
(d) fringes will become less intense

RESPONSE GRID					
	19. ⓐⓑⓒⓓ	20. ⓐⓑⓒⓓ	21. ⓐⓑⓒⓓ	22. ⓐⓑⓒⓓ	23. ⓐⓑⓒⓓ
	24. ⓐⓑⓒⓓ	25. ⓐⓑⓒⓓ	26. ⓐⓑⓒⓓ	27. ⓐⓑⓒⓓ	28. ⓐⓑⓒⓓ
	29. ⓐⓑⓒⓓ	30. ⓐⓑⓒⓓ	31. ⓐⓑⓒⓓ	32. ⓐⓑⓒⓓ	33. ⓐⓑⓒⓓ

Space for Rough Work

34. When unpolarised light is incident on a plane glass plate at Brewster's angle, then which of the following statements is correct?
(a) Reflected and refracted rays are completely polarised with their planes of polarization parallel to each other
(b) Reflected and refracted rays are completely polarised with their planes of polarization perpendicular to each other
(c) Reflected light is plane polarised but transmitted light is partially polarised
(d) Reflected light is partially polarised but refracted light is plane polarised

35. The maximum number of possible interference maxima for slit- separation equal to twice the wavelength in Young's double-slit experiment is
(a) infinite (b) five (c) three (d) zero

36. In the figure shown if a parallel beam of white light is incident on the plane of the slits then the distance of the nearest white spot on the screen from O is d/A. Find the value of A. (assume $d << D$, $\lambda << d$]

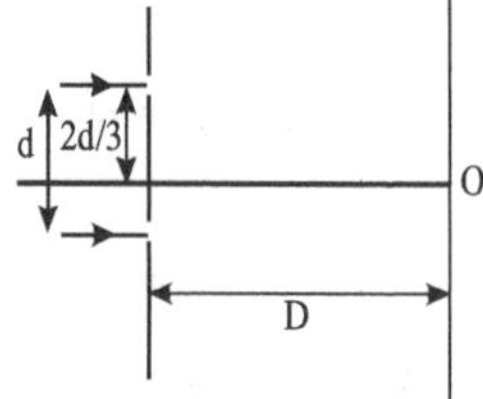

(a) 3 (b) 5 (c) 6 (d) 4

37. Two light waves superimposing at the mid-point of the screen are coming from coherent sources of light with phase difference 3π rad. Their amplitudes are 1 cm each. The resultant amplitude at the given point will be.
(a) 5 cm (b) 3 cm (c) 2 cm (d) zero

38. Spherical wavefronts, emanating from a point source, strike a plane reflecting surface. What will happen to these wave fronts, immediately after reflection?
(a) They will remain spherical with the same curvature, both in magnitude and sign.
(b) They will become plane wave fronts.
(c) They will remain spherical, with the same curvature, but sign of curvature reversed.
(d) They will remain spherical, but with different curvature, both in magnitude and sign.

39. Two coherent point sources S_1 and S_2 are separated by a small distance d as shown. The fringes obtained on the vertical screen will be :

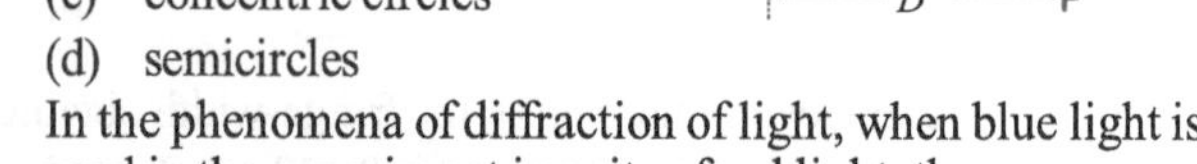

(a) points
(b) straight bands
(c) concentric circles
(d) semicircles

40. In the phenomena of diffraction of light, when blue light is used in the experiment in spite of red light, then
(a) fringes will become narrower
(b) fringes will become broader
(c) no change in fringe width
(d) None of these

41. On a hot summer night, the refractive index of air is smallest near the ground and increases with height from the ground. When a light beam is directed horizontally, the Huygens' principle leads us to conclude that as it travels, the light beam :
(a) bends downwards
(b) bends upwards
(c) becomes narrower
(d) goes horizontally without any deflection

42. If I_0 is the intensity of the principal maximum in the single slit diffraction pattern, then what will be its intensity when the slit width is doubled?
(a) $4I_0$ (b) $2I_0$ (c) $\frac{I_0}{2}$ (d) I_0

43. Conditions of diffraction is
(a) $\frac{a}{\lambda}=1$ (b) $\frac{a}{\lambda}>>1$ (c) $\frac{a}{\lambda}<<1$
(d) None of these

44. In Fresnel's biprism expt., a mica sheet of refractive index 1.5 and thickness 6×10^{-6} m is placed in the path of one of interfering beams as a result of which the central fringe gets shifted through 5 fringe widths. The wavelength of light used is
(a) 6000 Å (b) 8000 Å (c) 4000 Å (d) 2000 Å

45. Two nicols are oriented with their principal planes making an angle of 60°. Then the percentage of incident unpolarised light which passes through the system is
(a) 100 (b) 50 (c) 12.5 (d) 37.5

RESPONSE GRID					
	34. ⓐⓑⓒⓓ	35. ⓐⓑⓒⓓ	36. ⓐⓑⓒⓓ	37. ⓐⓑⓒⓓ	38. ⓐⓑⓒⓓ
	39. ⓐⓑⓒⓓ	40. ⓐⓑⓒⓓ	41. ⓐⓑⓒⓓ	42. ⓐⓑⓒⓓ	43. ⓐⓑⓒⓓ
	44. ⓐⓑⓒⓓ	45. ⓐⓑⓒⓓ			

DAILY PRACTICE PROBLEM DPP CHAPTERWISE CP24 - PHYSICS			
Total Questions	45	Total Marks	180
Attempted		Correct	
Incorrect		Net Score	
Cut-off Score	45	Qualifying Score	60
Success Gap = Net Score – Qualifying Score			
Net Score = (Correct × 4) – (Incorrect × 1)			

Space for Rough Work

DPP - Daily Practice Problems

Chapter-wise Sheets

Date : ______ **Start Time :** ______ **End Time :** ______

PHYSICS

CP25

SYLLABUS : Dual Nature of Radiation and Matter

Max. Marks : 180 **Marking Scheme :** (+4) for correct & (–1) for incorrect answer **Time : 60 min.**

INSTRUCTIONS : This Daily Practice Problem Sheet contains 45 MCQs. For each question only one option is correct. Darken the correct circle/ bubble in the Response Grid provided on each page.

1. A particle of mass 1 mg has the same wavelength as an electron moving with a velocity of $3\times10^6\ ms^{-1}$. The velocity of the particle is:

(a) $2.7\times10^{-18}\ ms^{-1}$ (b) $9\times10^{-2}\ ms^{-1}$

(c) $3\times10^{-31}\ ms^{-1}$ (d) $2.7\times10^{-21}\ ms^{-1}$

2. An electron of mass m and a photon have same energy E. The ratio of de-Broglie wavelengths associated with them is :

(a) $\frac{1}{c}\left(\frac{E}{2m}\right)^{\frac{1}{2}}$ (b) $\left(\frac{E}{2m}\right)^{\frac{1}{2}}$

(c) $c(2mE)^{\frac{1}{2}}$ (d) $\frac{1}{c}\left(\frac{2m}{E}\right)^{\frac{1}{2}}$

3. All electrons ejected from a surface by incident light of wavelength 200nm can be stopped before travelling 1m in the direction of uniform electric field of 4N/C. The work function of the surface is

(a) 4 eV (b) 6.2 eV (c) 2 eV (d) 2.2 eV

4. The maximum kinetic energy of the electrons hitting a target so as to produce X-ray of wavelength 1 Å is

(a) 1.24 keV (b) 12.4 keV

(c) 124 keV (d) None of these

5. An X-ray tube is operated at 15 kV. Calculate the upper limit of the speed of the electrons striking the target.

(a) 7.26×10^7 m/s (b) 7.62×10^9 m/s

(c) 7.62×10^7 cm/s (d) 7.26×10^9 m/s

6. A and B are two metals with threshold frequencies 1.8×10^{14} Hz and 2.2×10^{14} Hz. Two identical photons of energy 0.825 eV each are incident on them. Then photoelectrons are emitted in (Take $h = 6.6\times10^{-34}$ Js)

(a) B alone (b) A alone

(c) neither A nor B (d) both A and B.

7. If E_1, E_2, E_3 are the respective kinetic energies of an electron, an alpha-particle and a proton, each having the same de-Broglie wavelength, then

(a) $E_1 > E_3 > E_2$ (b) $E_2 > E_3 > E_1$

(c) $E_1 > E_2 > E_3$ (d) $E_1 = E_2 = E_3$

RESPONSE GRID	
	1. ⓐⓑⓒⓓ 2. ⓐⓑⓒⓓ 3. ⓐⓑⓒⓓ 4. ⓐⓑⓒⓓ 5. ⓐⓑⓒⓓ
	6. ⓐⓑⓒⓓ 7. ⓐⓑⓒⓓ

Space for Rough Work

8. Which of the following when falls on a metal will emit photoelectrons ?
(a) UV radiations (b) Infrared radiation
(c) Radio waves (d) Microwaves

9. The stopping potential (V_0) versus frequency (ν) plot of a substance is shown in figure, the threshold wavelength is

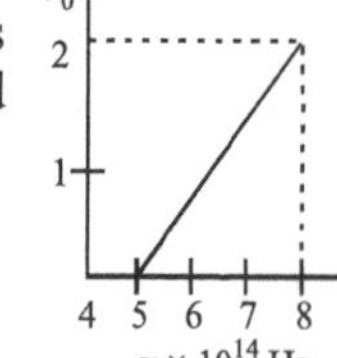

(a) 5×10^{14}m
(b) 6000 Å
(c) 5000 Å
(d) Cannot be estimated from given data

10. A material particle with a rest mass m_0 is moving with speed of light c. The de-Broglie wavelength associated is given by
(a) $\frac{h}{m_0 c}$ (b) $\frac{m_0 c}{h}$ (c) zero (d) ∞

11. A 200 W sodium street lamp emits yellow light of wavelength 0.6 μm. Assuming it to be 25% efficient in converting electrical energy to light, the number of photons of yellow light it emits per second is
(a) 1.5×10^{20} (b) 6×10^{18}
(c) 62×10^{20} (d) 3×10^{19}

12. A proton has kinetic energy E = 100 keV which is equal to that of a photon. The wavelength of photon is λ_2 and that of proton is λ_1. The ratio of λ_2/λ_1 is proportional to
(a) E^2 (b) $E^{1/2}$ (c) E^{-1} (d) $E^{-1/2}$

13. In photoelectric effect the work function of a metal is 3.5 eV. The emitted electrons can be stopped by applying a potential of –1.2 V. Then
(a) the energy of the incident photon is 4.7 eV
(b) the energy of the incident photon is 2.3 eV
(c) if higher frequency photon be used, the photoelectric current will rise
(d) when the energy of photon is 3.5 eV, the photoelectric current will be maximum

14. The threshold frequency for a metallic surface corresponds to an energy of 6.2 eV and the stopping potential for a radiation incident on this surface is 5 V. The incident radiation lies in
(a) ultra-violet region (b) infra-red region
(c) visible region (d) X-ray region

15. When photons of energy $h\nu$ fall on an aluminium plate (of work function E_0), photoelectrons of maximum kinetic energy K are ejected. If the frequency of the radiation is doubled, the maximum kinetic energy of the ejected photoelectrons will be
(a) 2K (b) K (c) $K + h\nu$ (d) $K + E_0$

16. Which metal will be suitable for a photoelectric cell using light of wavelength 4000Å. The work functions of sodium and copper are respectively 2.0 eV and 4.0 eV.
(a) Sodium (b) Copper
(c) Both (d) None of these

17. The maximum velocity of an electron emitted by light of wavelength λ incident on the surface of a metal of work-function ϕ is
(a) $\sqrt{\frac{2(hc + \lambda\phi)}{m\lambda}}$ (b) $\frac{2(hc + \lambda\phi)}{m\lambda}$
(c) $\sqrt{\frac{2(hc - \lambda\phi)}{m\lambda}}$ (d) $\sqrt{\frac{2(h\lambda - \phi)}{m}}$

18. If the kinetic energy of a free electron doubles, it's deBroglie wavelength changes by the factor
(a) 2 (b) $\frac{1}{2}$ (c) $\sqrt{2}$ (d) $\frac{1}{\sqrt{2}}$

19. Radiations of two photon's energy, twice and ten times the work function of metal are incident on the metal surface successsively. The ratio of maximum velocities of photoelectrons emitted in two cases is
(a) 1 : 2 (b) 1 : 3 (c) 1 : 4 (d) 1 : 1

20. The cathode of a photoelectric cell is changed such that the work function changes from W_1 to W_2 ($W_2 > W_1$). If the current before and after changes are I_1 and I_2, all other conditions remaining unchanged, then (assuming $h\nu > W_2$)
(a) $I_1 = I_2$ (b) $I_1 < I_2$
(c) $I_1 > I_2$ (d) $I_1 < I_2 < 2I_1$

21. Monochromatic radiation emitted when electron on hydrogen atom jumps from first excited to the ground state irradiates a photosensitive material. The stopping potential is measured to be 3.57 V. The threshold frequency of the materials is :
(a) 4×10^{15} Hz (b) 5×10^{15} Hz
(c) 1.6×10^{15} Hz (d) 2.5×10^{15} Hz

22. Photoelectric work function of a metal is 1eV. Light of wavelength $\lambda = 3000$ Å falls on it. The photo electrons come out with velocity
(a) 10 metres/sec (b) 10^2 metres/sec
(c) 10^4 metres/sec (d) 10^6 metres/sec

23. When the energy of the incident radiation is incredased by 20%, the kinetic energy of the photoelectrons emitted from a metal surface increased from 0.5 eV to 0.8 eV. The work function of the metal is :
(a) 0.65 eV (b) 1.0 eV (c) 1.3 eV (d) 1.5 eV

24. The maximum distance between interatomic lattice planes is 15 Å. The maximum wavelength of X-rays which are diffracted by this crystal will be
(a) 15 Å (b) 20 Å (c) 30 Å (d) 45 Å

RESPONSE GRID					
	8. ⓐⓑⓒⓓ	9. ⓐⓑⓒⓓ	10. ⓐⓑⓒⓓ	11. ⓐⓑⓒⓓ	12. ⓐⓑⓒⓓ
	13. ⓐⓑⓒⓓ	14. ⓐⓑⓒⓓ	15. ⓐⓑⓒⓓ	16. ⓐⓑⓒⓓ	17. ⓐⓑⓒⓓ
	18. ⓐⓑⓒⓓ	19. ⓐⓑⓒⓓ	20. ⓐⓑⓒⓓ	21. ⓐⓑⓒⓓ	22. ⓐⓑⓒⓓ
	23. ⓐⓑⓒⓓ	24. ⓐⓑⓒⓓ			

Space for Rough Work

25. In photoelectric effect, stopping potential for a light of frequency n_1 is V_1. If light is replaced by another having a frequency n_2 then its stopping potential will be

(a) $V_1 - \frac{h}{e}(n_2 - n_1)$ (b) $V_1 + \frac{h}{e}(n_2 + n_1)$

(c) $V_1 + \frac{h}{e}(n_2 - 2n_1)$ (d) $V_1 + \frac{h}{e}(n_2 - n_1)$

26. The maximum kinetic energy of the photoelectrons ejected from a photocathode when it is irradiated with light of wavelength 440nm is 1eV. If the threshold energy of the surface is 1.9eV, then which of the following statement is/are incorrect?

(a) The threshold frequency for photo sensitive metal is 4.6×10^{14}Hz
(b) The minimum wavelength of incident light required for photoemission is 6513 Å.
(c) The maximum wavelength of incident light required for photoemission is 6513 Å.
(d) The energy of incident photon is 2.9 eV.

27. The work functions of metals A and B are in the raio 1 : 2. If light of frequencies f and 2f are incident on the surfaces of A and B respectively, the ratio of the maximum kinetic energies of photoelectrons emitted is (f is greater than threshold frequency of A, 2f is greater than threshold frequency of B)

(a) 1 : 1 (b) 1 : 2 (c) 1 : 3 (d) 1 : 4

28. Which one of the following graphs represents the variation of maximum kinetic energy (E_K) of the emitted electrons with frequency υ in photoelectric effect correctly ?

(a)
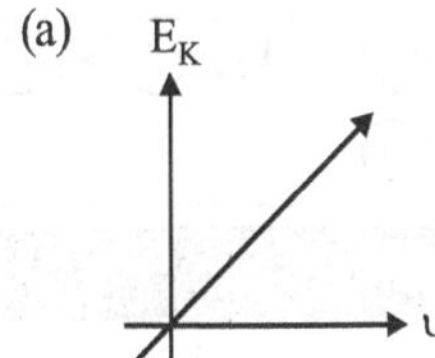

(b)
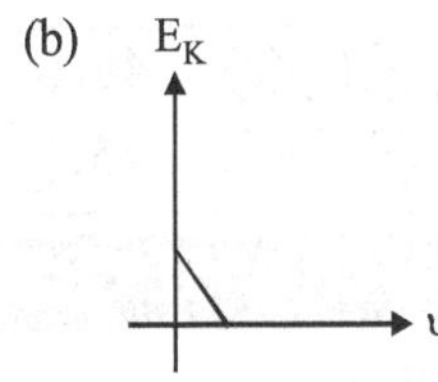

(c)
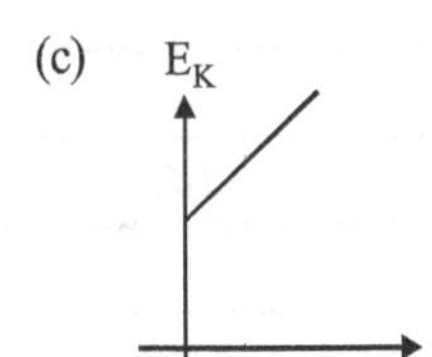

(d)
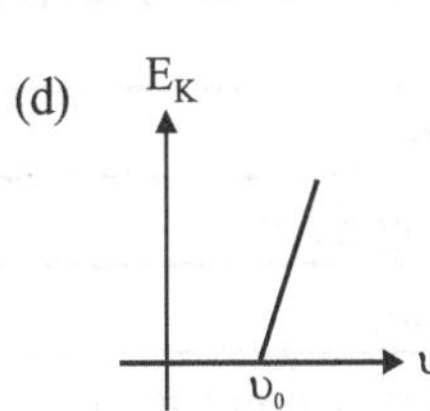

29. The potential difference that must be applied to stop the fastest photoelectrons emitted by a nickel surface, having work function 5.01 eV, when ultraviolet light of 200 nm falls on it, must be:

(a) 2.4V (b) −1.2 V (c) −2.4 V (d) 1.2V

30. X-rays are produced in X-ray tube operating at a given accelerating voltage. The wavelength of the continuous X-rays has values from

(a) 0 to ∞
(b) λ_{min} to ∞, where $\lambda_{min} > 0$
(c) 0 to λ_{max}, where $\lambda_{max} < \infty$
(d) λ_{min} to λ_{max}, where $0 < \lambda_{min} < \lambda_{max} < \infty$

31. Electrons used in an electron microscope are accelerated by a voltage of 25 kV. If the voltage is increased to 100kV then the de–Broglie wavelength associated with the electrons would

(a) increase by 2 times (b) decrease by 2 times
(c) decrease by 4 times (d) increase by 4 times

32. In the Davisson and Germer experiment, the velocity of electrons emitted from the electron gun can be increased by

(a) increasing the potential difference between the anode and filament
(b) increasing the filament current
(c) decreasing the filament current
(d) decreasing the potential difference between the anode and filament

33. Two radiations of photons energies 1 eV and 2.5 eV, successively illuminate a photosensitive metallic surface of work function 0.5 eV. The ratio of the maximum speeds of the emitted electrons is :

(a) 1 : 4 (b) 1 : 2 (c) 1 : 1 (d) 1 : 5

34. Photoelectric emission is observed from a metallic surface for frequencies v_1 and v_2 of the incident light rays ($v_1 > v_2$). If the maximum values of kinetic energy of the photoelectrons emitted in the two cases are in the ratio of 1 : k, then the threshold frequency of the metallic surface is

(a) $\frac{v_1 - v_2}{k-1}$ (b) $\frac{kv_1 - v_2}{k-1}$

(c) $\frac{kv_2 - v_1}{k-1}$ (d) $\frac{v_2 - v_1}{k}$

35. Which of the following is/are false regarding cathode rays?

(a) They produce heating effect
(b) They don't deflect in electric field
(c) They cast shadow
(d) They produce fluorescence

36. The ratio of the respective de Broglie wavelengths associated with electrons accelerated from rest with the voltages 100 V, 200 V and 300 V is

(a) 1 : 2 : 3 (b) 1 : 4 : 9 (c) $1 : \frac{1}{\sqrt{2}} : \frac{1}{\sqrt{3}}$ (d) $1 : \frac{1}{2} : \frac{1}{3}$

RESPONSE GRID					
	25. ⓐⓑⓒⓓ	26. ⓐⓑⓒⓓ	27. ⓐⓑⓒⓓ	28. ⓐⓑⓒⓓ	29. ⓐⓑⓒⓓ
	30. ⓐⓑⓒⓓ	31. ⓐⓑⓒⓓ	32. ⓐⓑⓒⓓ	33. ⓐⓑⓒⓓ	34. ⓐⓑⓒⓓ
	35. ⓐⓑⓒⓓ	36. ⓐⓑⓒⓓ			

Space for Rough Work

37. A 5 watt source emits monochromatic light of wavelength 5000 Å. When placed 0.5 m away, it liberates photoelectrons from a photosensitive metallic surface. When the source is moved to a distance of 1.0 m, the number of photoelectrons liberated will be reduced by a factor of

(a) 8 (b) 16 (c) 2 (d) 4

38. In the photoeletric effect, electrons are emitted

(a) at a rate that is proportional to the amplitude of the incident radiation
(b) with a maximum velocity proportional to the frequency of the incident radiation
(c) at a rate that is independent of the emitter
(d) only if the frequency of the incident radiations is above a certain threshold value

39. The threshold frequency for a photosensitive metal is 3.3×10^{14} Hz. If light of frequency 8.2×10^{14} Hz is incident on this metal, the cut-off voltage for the photoelectric emission is nearly

(a) 2V (b) 3V (c) 5V (d) 1V

40. In an experiment on photoelectric effect, a student plots stopping potential V_0 against reciprocal of the wavelength λ of the incident light for two different metals A and B. These are shown in the figure.

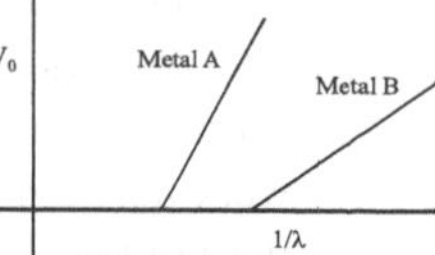

Looking at the graphs, you can most appropriately say that:

(a) Work function of metal B is greater than that of metal A
(b) For light of certain wavelength falling on both metal, maximum kinetic energy of electrons emitted from A will be greater than those emitted from B.
(c) Work function of metal A is greater than that of metal B
(d) Students data is not correct

41. White X-rays are called white due to the fact that

(a) they are electromagnetic radiations having nature same as that of white light.
(b) they are produced most abundantly in X ray tubes.
(c) they have a continuous wavelength range.
(d) they can be converted to visible light using coated screens and photographic plates are affected by them just like light.

42. The wavelength associated with an electron, accelerated through a potential difference of 100 V, is of the order of

(a) 1000 Å (b) 100 Å (c) 10.5 Å (d) 1.2 Å

43. Monochromatic light of frequency 6.0×10^{14} Hz is produced by a laser. The power emitted is 2×10^{-3} w. The number of photons emitted, on the average, by the sources per second is

(a) 5×10^{16} (b) 5×10^{17} (c) 5×10^{14} (d) 5×10^{15}

44. The de-Broglie wavelength of neutron in thermal equilibrium at temperature T is

(a) $\frac{30.8}{\sqrt{T}}$ Å (b) $\frac{3.08}{\sqrt{T}}$ Å (c) $\frac{0.308}{\sqrt{T}}$ Å (d) $\frac{0.0308}{\sqrt{T}}$ Å

45. Which of the following cannot be explained on the basis of photoelectric theory?

(a) Instantaneous emission of photoelectrons
(b) Existence of threshold frequency
(c) Sufficiently intense beam of radiation can emit photoelectrons
(d) Existence of stopping potential

RESPONSE GRID					
	37.ⓐⓑⓒⓓ	38.ⓐⓑⓒⓓ	39.ⓐⓑⓒⓓ	40.ⓐⓑⓒⓓ	41.ⓐⓑⓒⓓ
	42.ⓐⓑⓒⓓ	43.ⓐⓑⓒⓓ	44.ⓐⓑⓒⓓ	45.ⓐⓑⓒⓓ	

DAILY PRACTICE PROBLEM DPP CHAPTERWISE CP25 - PHYSICS			
Total Questions	45	Total Marks	180
Attempted		Correct	
Incorrect		Net Score	
Cut-off Score	45	Qualifying Score	60
Success Gap = Net Score – Qualifying Score			
Net Score = (Correct × 4) – (Incorrect × 1)			

Space for Rough Work

DPP - Daily Practice Problems

Chapter-wise Sheets

Date : ______ **Start Time :** ______ **End Time :** ______

PHYSICS

CP26

SYLLABUS : Atoms

Max. Marks : 180 **Marking Scheme : (+4) for correct & (–1) for incorrect answer** **Time : 60 min.**

INSTRUCTIONS : This Daily Practice Problem Sheet contains 45 MCQs. For each question only one option is correct. Darken the correct circle/ bubble in the Response Grid provided on each page.

1. The potential energy associated with an electron in the orbit
   (a) increases with the increases in radii of the orbit
   (b) decreases with the increase in the radii of the orbit
   (c) remains the same with the change in the radii of the orbit
   (d) None of these

2. The diagram shows the energy levels for an electron in a certain atom. Which transition shown represents the emission of a photon with the most energy?

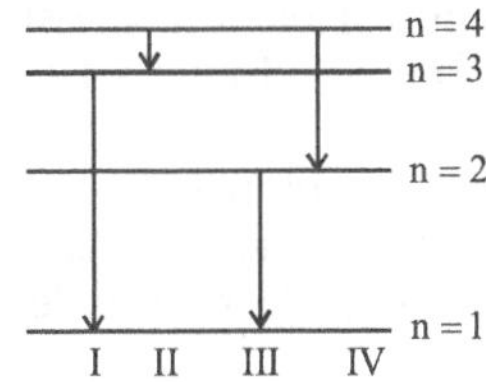

   (a) IV (b) III (c) II (d) I

3. Electrons in a certain energy level $n = n_1$, can emit 3 spectral lines. When they are in another energy level, $n = n_2$. They can emit 6 spectral lines. The orbital speed of the electrons in the two orbits are in the ratio of
   (a) 4 : 3 (b) 3 : 4 (c) 2 : 1 (d) 1 : 2

4. In Rutherford scattering experiment, the number of α-particles scattered at 60° is 5×10^6. The number of α-particles scattered at 120° will be
   (a) 15×10^6 (b) $\frac{3}{5} \times 10^6$
   (c) $\frac{5}{9} \times 10^6$ (d) None of these

5. In the Bohr model an electron moves in a circular orbit around the proton. Considering the orbiting electron to be a circular current loop, the magnetic moment of the hydrogen atom, when the electron is in n^{th} excited state, is :
   (a) $\left(\frac{e}{2m}\frac{n^2 h}{2\pi}\right)$ (b) $\left(\frac{e}{m}\right)\frac{nh}{2\pi}$
   (c) $\left(\frac{e}{2m}\right)\frac{nh}{2\pi}$ (d) $\left(\frac{e}{m}\right)\frac{n^2 h}{2\pi}$

RESPONSE GRID	1. ⓐⓑⓒⓓ	2. ⓐⓑⓒⓓ	3. ⓐⓑⓒⓓ	4. ⓐⓑⓒⓓ	5. ⓐⓑⓒⓓ

Space for Rough Work

6. A 12.5 eV electron beam is used to bombard gaseous hydrogen at room temperature. It will emit :
   (a) 2 lines in the Lyman series and 1 line in the Balmar series
   (b) 3 lines in the Lyman series
   (c) 1 line in the Lyman series and 2 lines in the Balmar series
   (d) 3 lines in the Balmer series

7. A Hydrogen atom and a Li^{++} ion are both in the second excited state. If ℓ_H and ℓ_{Li} are their respective electronic angular momenta, and E_H and E_{Li} their respective energies, then
   (a) $\ell_H > \ell_{Li}$ and $|E_H| > |E_{Li}|$ (b) $\ell_H = \ell_{Li}$ and $|E_H| < |E_{Li}|$
   (c) $\ell_H = \ell_{Li}$ and $|E_H| > |E_{Li}|$ (d) $\ell_H < \ell_{Li}$ and $|E_H| < |E_{Li}|$

8. The radius of hydrogen atom in its ground state is 5.3×10^{-11} m. After collision with an electron it is found to have a radius of 21.2×10^{-11} m. What is the principal quantum number n of the final state of the atom
   (a) $n=4$ (b) $n=2$ (c) $n=16$ (d) $n=3$

9. When hydrogen atom is in its first excited level, its radius is
   (a) four times its ground state radius
   (b) twice
   (c) same
   (d) half

10. Consider 3rd orbit of He^+ (Helium), using non-relativistic approach, the speed of electron in this orbit will be [given K $= 9 \times 10^9$ constant, Z = 2 and h (Plank's Constant) $= 6.6 \times 10^{-34}$ J s]
   (a) 1.46×10^6 m/s (b) 0.73×10^6 m/s
   (c) 3.0×10^8 m/s (d) 2.92×10^6 m/s

11. An electron in the hydrogen atom jumps from excited state n to the ground state. The wavelength so emitted illuminates a photosensitive material having work function 2.75 eV. If the stopping potential of the photoelectron is 10 V, the value of n is
   (a) 3 (b) 4 (c) 5 (d) 2

12. The electron in a hydrogen atom makes a transition from an excited state to the ground state. Which of the following statements is true?
   (a) Its kinetic energy increases and its potential energy decreases.
   (b) Its kinetic energy decreases, potential energy increases.
   (c) Its kinetic and its potential energy increases.
   (d) Its kinetic, potential energy decrease.

13. An energy of 24.6 eV is required to remove one of the electrons from a neutral helium atom. The energy in (eV) required to remove both the electrons from a neutral helium atom is
   (a) 38.2 (b) 49.2 (c) 51.8 (d) 79.0

14. One of the lines in the emission spectrum of Li^{2+} has the same wavelength as that of the 2nd line of Balmer series in hydrogen spectrum. The electronic transition corresponding to this line is n = 12 → n = x. Find the value of x.
   (a) 8 (b) 6 (c) 7 (c) 5

15. If the atom ${}_{100}Fm^{257}$ follows the Bohr model and the radius of ${}_{100}Fm^{257}$ is n times the Bohr radius, then find n.
   (a) 100 (b) 200 (c) 4 (d) 1/4

16. The energy of He^+ in the ground state is – 54.4 eV, then the energy of Li^{++} in the first excited state will be
   (a) -30.6 eV (b) 27.2 eV
   (c) -13.6 eV (d) -27.2 eV

17. If the angular momentum of an electron in an orbit is J then the K.E. of the electron in that orbit is
   (a) $\frac{J^2}{2mr^2}$ (b) $\frac{Jv}{r}$ (c) $\frac{J^2}{2m}$ (d) $\frac{J^2}{2\pi}$

18. Suppose an electron is attracted towards the origin by a force $\frac{k}{r}$ where 'k' is a constant and 'r' is the distance of the electron from the origin. By applying Bohr model to this system, the radius of the nth orbital of the electron is found to be 'r_n' and the kinetic energy of the electron to be 'T_n'. Then which of the following is true?
   (a) $T_n \propto \frac{1}{n^2}, r_n \propto n^2$ (b) T_n independent of n, $r_n \propto n$
   (c) $T_n \propto \frac{1}{n}, r_n \propto n$ (d) $T_n \propto \frac{1}{n}, r_n \propto n^2$

19. In Hydrogen spectrum, the wavelength of H_α line is 656 nm, whereas in the spectrum of a distant galaxy, H_α line wavelength is 706 nm. Estimated speed of the galaxy with respect to earth is
   (a) 2×10^8 m/s (b) 2×10^7m/s
   (c) 2×10^6 m/s (d) 2×10^5 m/s

20. In the hydrogen atom, an electron makes a transition from n = 2 to n = 1. The magnetic field produced by the circulating electron at the nucleus
   (a) decreases 16 times (b) increases 4 times
   (c) decreases 4 times (d) increases 32 times

21. What is the radius of iodine atom (At. no. 53, mass no. 126)
   (a) 2.5×10^{-11} m (b) 2.5×10^{-9} m
   (c) 7×10^{-9}m (d) 7×10^{-6}m

22. When an α-particle of mass 'm' moving with velocity 'v' bombards on a heavy nucleus of charge 'Ze', its distance of closest approach from the nucleus depends on m as :
   (a) $\frac{1}{m}$ (b) $\frac{1}{\sqrt{m}}$ (c) $\frac{1}{m^2}$ (d) m

RESPONSE GRID					
	6. ⓐⓑⓒⓓ	7. ⓐⓑⓒⓓ	8. ⓐⓑⓒⓓ	9. ⓐⓑⓒⓓ	10. ⓐⓑⓒⓓ
	11. ⓐⓑⓒⓓ	12. ⓐⓑⓒⓓ	13. ⓐⓑⓒⓓ	14. ⓐⓑⓒⓓ	15. ⓐⓑⓒⓓ
	16. ⓐⓑⓒⓓ	17. ⓐⓑⓒⓓ	18. ⓐⓑⓒⓓ	19. ⓐⓑⓒⓓ	20. ⓐⓑⓒⓓ
	21. ⓐⓑⓒⓓ	22. ⓐⓑⓒⓓ			

Space for Rough Work

23. The ionization energy of the electron in the hydrogen atom in its ground state is 13.6 eV. The atoms are excited to higher energy levels to emit radiations of 6 wavelengths. Maximum wavelength of emitted radiation corresponds to the transition between
(a) n = 3 to n = 1 states (b) n = 2 to n = 1 states
(c) n = 4 to n = 3 states (d) n = 3 to n = 2 states

24. The wavelengths involved in the spectrum of deuterium $\left({}_1^2D\right)$ are slightly different from that of hydrogen spectrum, because
(a) the size of the two nuclei are different
(b) the nuclear forces are different in the two cases
(c) the masses of the two nuclei are different
(d) the attraction between the electron and the nucleus is differernt in the two cases

25. An electron in hydrogen atom makes a transition $n_1 \to n_2$ where n_1 and n_2 are principal quantum numbers of the two states. Assuming Bohr's model to be valid the time period of the electron in the initial state is eight times that in the final state. The possible values of n_1 and n_2 are
(a) $n_1 = 4$ and $n_2 = 2$ (b) $n_1 = 6$ and $n_2 = 2$
(c) $n_1 = 8$ and $n_2 = 1$ (d) $n_1 = 8$ and $n_2 = 2$

26. Ina hydrogen like atom electronmake transition from an energy level with quantum number n to another with quantum number (n – 1). If n>>1, the frequency of radiation emitted is proportional to :
(a) $\frac{1}{n}$ (b) $\frac{1}{n^2}$ (c) $\frac{1}{n^{3/2}}$ (d) $\frac{1}{n^3}$

27. The spectrum obtained from a sodium vapour lamp is an example of
(a) band spectrum
(b) continuous spectrum
(c) emission spectrum
(d) absorption spectrum

28. Ionization potential of hydrogen atom is 13.6eV. Hydrogen atoms in the ground state are excited by monochromatic radiation of photon energy 12.1 eV. According to Bohr's theory, the spectral lines emitted by hydrogen will be
(a) three (b) four (c) one (d) two

29. The Bohr model of atoms
(a) predicts the same emission spectra for all types of atoms
(b) assumes that the angular momentum of electrons is quantised
(c) uses Einstein's photoelectric equation
(d) predicts continuous emission spectra for atoms

30. The largest wavelength in the ultraviolet region of the hydrogen spectrum is 122 nm. The smallest wavelength in the infrared region of the hydrogen spectrum (to the nearest integer) is
(a) 802 nm (b) 823 nm (c) 1882 nm (d) 1648 nm

31. A doubly ionised Li atom is excited from its ground state(n = 1) to n = 3 state. The wavelengths of the spectral lines are given by λ_{32}, λ_{31} and λ_{21}. The ratio $\lambda_{32}/\lambda_{31}$ and $\lambda_{21}/\lambda_{31}$ are, respectively
(a) 8.1, 0.67 (b) 8.1, 1.2
(c) 6.4, 1.2 (d) 6.4, 0.67

32. In Rutherford scattering experiment, what will be the correct angle for α-scattering for an impact parameter, $b = 0$?
(a) 90° (b) 270° (c) 0° (d) 180°

33. Consider 3rd orbit of He^+ (Helium), using non-relativistic approach, the speed of electron in this orbit will be [given K = 9 × 10^9 constant, Z = 2 and h (Plank's Constant) = 6.6×10^{-34} J s]
(a) 1.46×10^6 m/s (b) 0.73×10^6 m/s
(c) 3.0×10^8 m/s (d) 2.92×10^6 m/s

34. The ionization energy of hydrogen atom is 13.6 eV. Following Bohr's theory, the energy corresponding to a transition between 3rd and 4th orbit is
(a) 3.40 eV (b) 1.51 eV (c) 0.85 eV (d) 0.66 eV

35. The transition from the state $n = 3$ to $n = 1$ in a hydrogen like atom results in ultraviolet radiation. Infrared radiation will be obtained in the transition from :
(a) 2→1 (b) 3→2 (c) 4→2 (d) 4→3

36. Given the value of Rydberg constant is $10^7 m^{-1}$, the wave number of the last line of the Balmer series in hydrogen spectrum will be :
(a) $0.025 \times 10^4 m^{-1}$ (b) $0.5 \times 10^7 m^{-1}$
(c) $0.25 \times 10^7 m^{-1}$ (d) $2.5 \times 10^7 m^{-1}$

37. Which of the plots shown in the figure represents speed (v_n) of the electron in a hydrogen atom as a function of the principal quantum number (n)?

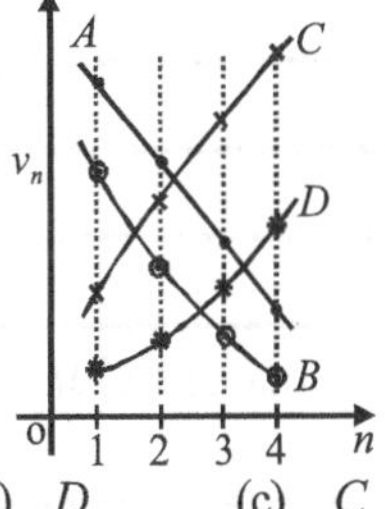

(a) B (b) D (c) C (d) A

RESPONSE GRID					
	23. ⓐⓑⓒⓓ	24. ⓐⓑⓒⓓ	25. ⓐⓑⓒⓓ	26. ⓐⓑⓒⓓ	27. ⓐⓑⓒⓓ
	28. ⓐⓑⓒⓓ	29. ⓐⓑⓒⓓ	30. ⓐⓑⓒⓓ	31. ⓐⓑⓒⓓ	32. ⓐⓑⓒⓓ
	33. ⓐⓑⓒⓓ	34. ⓐⓑⓒⓓ	35. ⓐⓑⓒⓓ	36. ⓐⓑⓒⓓ	37. ⓐⓑⓒⓓ

Space for Rough Work

38. The ionisation potential of H-atom is 13.6 V. When it is excited from ground state by monochromatic radiations of 970.6 Å, the number of emission lines will be (according to Bohr's theory)
(a) 10 (b) 8 (c) 6 (d) 4

39. The energy of hydrogen atom in nth orbit is E_n, then the energy in nth orbit of single ionised helium atom will be
(a) $4E_n$ (b) $E_n/4$ (c) $2E_n$ (d) $E_n/2$

40. In the Rutherford experiment, α-particles are scattered from a nucleus as shown. Out of the four paths, which path is not possible?

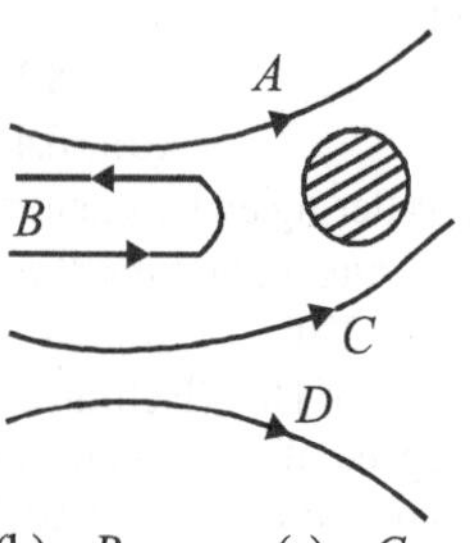

(a) D (b) B (c) C (d) A

41. An electron changes its position from orbit $n = 2$ to the orbit $n = 4$ of an atom. The wavelength of the emitted radiations is (R = Rydberg's constant)
(a) $\frac{16}{R}$ (b) $\frac{16}{3R}$ (c) $\frac{16}{5R}$ (d) $\frac{16}{7R}$

42. In a Rutherford scattering experiment when a projectile of charge Z_1 and mass M_1 approaches a target nucleus of charge Z_2 and mass M_2, the distance of closest approach is r_0. The energy of the projectile is
(a) directly proportional to $Z_1 Z_2$
(b) inversely proportional to Z_1
(c) directly proportional to mass M_1
(d) directly proportional to $M_1 \times M_2$

43. The wavelength of the first spectral line in the Balmer series of hydrogen atom is 6561 A°. The wavelength of the second spectral line in the Balmer series of singly-ionized helium atom is
(a) 1215 Å (b) 1640 Å (c) 2430 Å (d) 4687 Å

44. If υ_1 is the frequency of the series limit of Lyman series, υ_2 is the frequency of the first line of Lyman series and υ_3 is the frequency of the series limit of the Balmer series then
(a) $\upsilon_1 - \upsilon_2 = \upsilon_3$ (b) $\upsilon_1 = \upsilon_2 - \upsilon_3$
(c) $\frac{1}{\upsilon_2} = \frac{1}{\upsilon_1} + \frac{1}{\upsilon_3}$ (d) $\frac{1}{\upsilon_1} = \frac{1}{\upsilon_2} + \frac{1}{\upsilon_3}$

45. In a hypothetical Bohr hydrogen atom, the mass of the electron is doubled. The energy E'_0 and radius r'_0 of the first orbit will be (r_0 is the Bohr radius)
(a) –11.2 eV (b) –6.8 eV (c) –13.6 eV (d) –27.2 eV

RESPONSE GRID					
	38. ⓐⓑⓒⓓ	39. ⓐⓑⓒⓓ	40. ⓐⓑⓒⓓ	41. ⓐⓑⓒⓓ	42. ⓐⓑⓒⓓ
	43. ⓐⓑⓒⓓ	44. ⓐⓑⓒⓓ	45. ⓐⓑⓒⓓ		

DAILY PRACTICE PROBLEM DPP CHAPTERWISE CP26 - PHYSICS			
Total Questions	45	Total Marks	180
Attempted		Correct	
Incorrect		Net Score	
Cut-off Score	50	Qualifying Score	70
Success Gap = Net Score – Qualifying Score			
Net Score = (Correct × 4) – (Incorrect × 1)			

Space for Rough Work

DPP - Daily Practice Problems

Chapter-wise Sheets

Date : ______ **Start Time :** ______ **End Time :** ______

PHYSICS

SYLLABUS : Nuclei

Max. Marks : 180 **Marking Scheme :** (+4) for correct & (–1) for incorrect answer **Time : 60 min.**

INSTRUCTIONS : This Daily Practice Problem Sheet contains 45 MCQs. For each question only one option is correct. Darken the correct circle/ bubble in the Response Grid provided on each page.

1. The mass of a ${}_3^7Li$ nucleus is 0.042 u less than the sum of the masses of all its nucleons. The binding energy per nucleon of ${}_3^7Li$ nucleus is nearly
   (a) 46 MeV (b) 5.6 MeV
   (c) 3.9 MeV (d) 23 MeV
2. In the nuclear decay given below:
   $${}_Z^AX \longrightarrow {}_{Z+1}^{\ A}Y \longrightarrow {}_{Z-1}^{A-4}B^* \longrightarrow {}_{Z-1}^{A-4}B,$$
   the particles emitted in the sequence are
   (a) γ, β, α (b) β, γ, α
   (c) α, β, γ (d) β, α, γ
3. If the nuclear radius of ^{27}Al is 3.6 Fermi, the approximate nuclear radius of ^{64}Cu in Fermi is :
   (a) 2.4 (b) 1.2 (c) 4.8 (d) 3.6
4. Which of the following statements is true for nuclear forces?
   (a) they obey the inverse square law of distance
   (b) they obey the inverse third power law of distance
   (c) they are short range forces
   (d) they are equal in strength to electromagnetic forces.
5. A radioactive sample at any instant has its disintegration rate 5000 disintegrations per minute. After 5 minutes, the rate is 1250 disintegrations per minute. Then, the decay constant (per minute) is
   (a) 0.4 ln 2 (b) 0.2 ln 2 (c) 0.1 ln 2 (d) 0.8 ln 2
6. The radioactivity of a sample is R_1 at a time T_1 and R_2 at a time T_2. If the half-life of the specimen is T, the number of atoms that have disintegrated in the time $(T_1 - T_2)$ is proportional to
   (a) $(R_1T_1 - R_2T_2)$ (b) $(R_1 - R_2)$
   (c) $(R_1 - R_2)/T$ (d) $(R_1 - R_2)\,T$
7. In the reaction, ${}_1^2H + {}_1^3H \longrightarrow {}_2^4He + {}_0^1n$, if the binding energies of ${}_1^2H$, ${}_1^3H$ and ${}_2^4He$ are respectively, *a*, *b* and *c* (in MeV), then the energy (in MeV) released in this reaction is
   (a) $a+b+c$ (b) $a+b-c$
   (c) $c-a-b$ (d) $c+a-b$
8. If M (A; Z), M_p and M_n denote the masses of the nucleus ${}_Z^AX$, proton and neutron respectively in units of u (1u = 931.5 MeV/c^2) and BE represents its bonding energy in MeV, then
   (a) $M(A, Z) = ZM_p + (A - Z)M_n - BE/c^2$
   (b) $M(A, Z) = ZM_p + (A - Z)M_n + BE$
   (c) $M(A, Z) = ZM_p + (A - Z)M_n - BE$
   (d) $M(A, Z) = ZM_p + (A - Z)M_n + BE/c^2$

RESPONSE GRID					
	1. ⓐⓑⓒⓓ	2. ⓐⓑⓒⓓ	3. ⓐⓑⓒⓓ	4. ⓐⓑⓒⓓ	5. ⓐⓑⓒⓓ
	6. ⓐⓑⓒⓓ	7. ⓐⓑⓒⓓ	8. ⓐⓑⓒⓓ		

Space for Rough Work

9. How does the binding energy per nucleon vary with the increase in the number of nucleons?

(a) Increases continuously with mass number
(b) Decreases continuously with mass number
(c) First decreases and then increases with increase in mass number
(d) First increases and then decreases with increase in mass number

10. The energy spectrum of β-particles [Number N(E) as a function of β-energy E] emitted from a radioactive source is

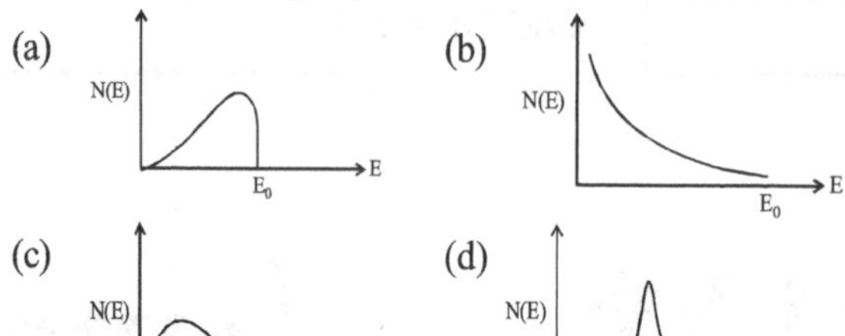

11. A radioactive nucleus undergoes a series of decay according to the scheme

$$A \xrightarrow{a} A_1 \xrightarrow{\beta} A_2 \xrightarrow{\alpha} A_3 \xrightarrow{\gamma} A_4$$

If the mass number and atomic number of 'A' are 180 and 72 respectively, then what are these numbers for A_4

(a) 172 and 69 (b) 174 and 70
(c) 176 and 69 (d) 176 and 70

12. The activity of a radioactive sample is measured as 9750 counts per minute at $t = 0$ and as 975 counts per minute at $t = 5$ minutes. The decay constant is approximately

(a) 0.922 per minute (b) 0.691 per minute
(c) 0.461 per minute (d) 0.230 per minute

13. Actinium 231, $^{231}AC_{89}$, emit in succession two β particles, four α-particles, one β and one α plus several γ rays. What is the resultant isotope?

(a) $^{221}Au_{79}$ (b) $^{211}Au_{79}$
(c) $^{221}Pb_{82}$ (d) $^{211}Pb_{82}$

14. Fusion reactions take place at high temperature because

(a) atoms are ionised at high temperature
(b) molecules break up at high temperature
(c) nuclei break up at high temperature
(d) kinetic energy is high enough to overcome repulsion between nuclei

15. If M_O is the mass of an oxygen isotope $_8O^{17}$, M_P and M_N are the masses of a proton and a neutron respectively, the nuclear binding energy of the isotope is

(a) $(M_O - 17M_N)c^2$ (b) $(M_O - 8M_P)c^2$
(c) $(M_O - 8M_P - 9M_N)c^2$ (d) M_Oc^2

16. Which of the following nuclear reactions is not possible?

(a) $^{12}_{6}C + ^{12}_{6}C \longrightarrow ^{20}_{10}Ne + ^{4}_{2}He$
(b) $^{9}_{4}Be + ^{1}_{1}H \longrightarrow ^{6}_{3}Li + ^{4}_{2}He$
(c) $^{11}_{5}Be + ^{1}_{1}H \longrightarrow ^{9}_{4}Be + ^{4}_{2}He$
(d) $^{7}_{3}Li + ^{4}_{2}He \longrightarrow ^{1}_{1}H + ^{10}_{4}B$

17. The ratio of half-life times of two elements A and B is $\frac{T_A}{T_B}$. The ratio of respective decay constant $\frac{\lambda_A}{\lambda_B}$, is

(a) T_B/T_A (b) T_A/T_B
(c) $\frac{T_A + T_B}{T_A}$ (d) $\frac{T_A - T_B}{T_A}$

18. Two radioactive materials X_1 and X_2 have decay constants 10λ and λ respectively. If initially they have the same number of nuclei, then the ratio of the number of nuclei of X_1 to that of X_2 will be 1/e after a time

(a) $1/10\lambda$ (b) $1/11\lambda$
(c) $11/10\lambda$ (d) $1/9\lambda$

19. In a radioactive material the activity at time t_1 is R_1 and at a later time t_2, it is R_2. If the decay constant of the material is λ, then

(a) $R_1 = R_2 e^{\lambda(t_1 - t_2)}$ (b) $R_1 = R_2 e^{(t_2/t_1)}$
(c) $R_1 = R_2$ (d) $R_1 = R_2 e^{-\lambda(t_1 - t_2)}$

20. The correct relation between t_{av} = average life and $t_{1/2}$ = half life for a radioactive nuclei.

(a) $t_{av} = t_{1/2}$ (b) $t_{av} = \frac{1}{2} t_{1/2}$
(c) $0.693\, t_{av} = t_{1/2}$ (d) $t_{av} = 0.693\, t_{1/2}$

21. If the nuclear force between two protons, two neutrons and between proton and neutron is denoted by F_{pp}, F_{nn} and F_{pn} respectively, then

(a) $F_{pp} \approx F_{nn} \approx F_{pn}$ (b) $F_{pp} \neq F_{nn}$ and $F_{pp} = F_{nn}$
(c) $F_{pp} = F_{nn} = F_{pn}$ (d) $F_{pp} \neq F_{nn} \neq F_{pn}$

22. Which one is correct about fission?

(a) Approx. 0.1% mass converts into energy
(b) Most of energy of fission is in the form of heat
(c) In a fission of U^{235} about 200 eV energy is released
(d) On an average, one neutron is released per fission of U^{235}

RESPONSE GRID					
	9. ⓐⓑⓒⓓ	10. ⓐⓑⓒⓓ	11. ⓐⓑⓒⓓ	12. ⓐⓑⓒⓓ	13. ⓐⓑⓒⓓ
	14. ⓐⓑⓒⓓ	15. ⓐⓑⓒⓓ	16. ⓐⓑⓒⓓ	17. ⓐⓑⓒⓓ	18. ⓐⓑⓒⓓ
	19. ⓐⓑⓒⓓ	20. ⓐⓑⓒⓓ	21. ⓐⓑⓒⓓ	22. ⓐⓑⓒⓓ	

Space for Rough Work

23. If 200 MeV energy is released in the fission of a single U^{235} nucleus, the number of fissions required per second to produce 1 kilowatt power shall be (Given 1 eV = 1.6×10^{-19} J)
(a) 3.125×10^{13} (b) 3.125×10^{14}
(c) 3.125×10^{15} (d) 3.125×10^{16}

24. In any fission process, the ratio of
$\frac{\text{mass of fission products}}{\text{mass of parent nucleus}}$ is
(a) equal to 1
(b) greater than 1
(c) less than 1
(d) depends on the mass of the parent nucleus

25. In an α-decay the kinetic energy of α-particle is 48 MeV and Q-value of the reaction is 50 MeV. The mass number of the mother nucleus is X. Find value of X/25.
(Assume that daughter nucleus is in ground state)
(a) 2 (b) 4 (c) 6 (d) 8

26. A sample of radioactive element has a mass of 10gm at an instant t=0. The approximate mass of this element in the sample after two mean lives is
(a) 6.30 gm (b) 1.35 gm
(c) 2.50 gm (d) 3.70 gm

27. Consider a radioactive material of half-life 1.0 minute. If one of the nuclei decays now, the next one will decay
(a) after 1 minute
(b) after $\frac{1}{\log_e 2}$ minute
(c) after $\frac{1}{N}$ minute, where N is the number of nuclei present at that moment
(d) after any time

28. The mass of α-particle is
(a) less than the sum of masses of two protons and two neutrons
(b) equal to mass of four protons
(c) equal to mass of four neutrons
(d) equal to sum of masses of two protons and two neutron

29. The decay constants of a radioactive substance for α and β emission are λ_α and λ_β respectively. If the substance emits α and β simultaneously, then the average half life of the material will be
(a) $\frac{2T_\alpha T_\beta}{T_\alpha + T_\beta}$ (b) $T_\alpha + T_\beta$
(c) $\frac{T_\alpha T_\beta}{T_\alpha + T_\beta}$ (d) $\frac{1}{2}\left(T_\alpha + T_\beta\right)$

30. If the end A of a wire is irradiated with α-rays and the other end B is irradiated with β-rays. Then
(a) a current will flow from A to B
(b) a current will flow from B to A
(c) there will be no current in the wire
(d) a current will flow from each end to the mid-point of the wire

31. A radioactive nucleus of mass M emits a photon of frequency ν and the nucleus recoils. The recoil energy will be
(a) $Mc^2 - h\nu$ (b) $h^2\nu^2 / 2Mc^2$
(c) zero (d) hν

32. Radioactive element decays to form a stable nuclide. The rate of decay of reactant is correctly depicted by

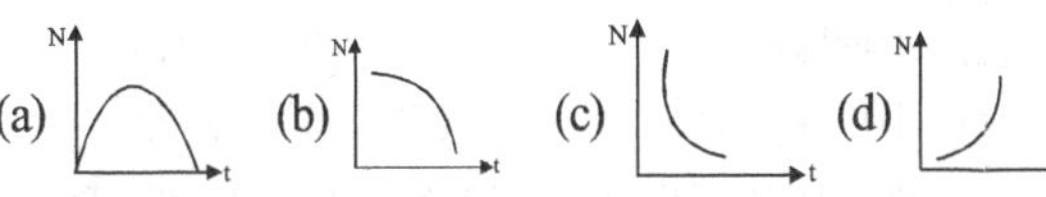

33. A nucleus of mass M + Δm is at rest and decays into two daughter nuclei of equal mass $\frac{M}{2}$ each. Speed of light is c. The speed of daughter nuclei is
(a) $c\frac{\Delta m}{M+\Delta m}$ (b) $c\sqrt{\frac{2\Delta m}{M}}$ (c) $c\sqrt{\frac{\Delta m}{M}}$ (d) $c\sqrt{\frac{\Delta m}{M+\Delta m}}$

34. Atomic weight of Boron is 10.81 and it has two isotopes ${}_5B^{10}$ and ${}_5B^{11}$. Then the ratio ${}_5B^{10} : {}_5B^{11}$ in nature would be
(a) 19 : 81 (b) 10 : 11 (c) 15 : 16 (d) 81 : 19

35. A nucleus ruptures into two nuclear parts, which have their velocity ratio equal to 2:1. What will be the ratio of their nuclear size (nuclear radius)?
(a) $2^{1/3} : 1$ (b) $1 : 2^{1/3}$ (c) $3^{1/2} : 1$ (d) $1 : 3^{1/2}$

36. A nucleus of uranium decays at rest into nuclei of thorium and helium. Then :
(a) the helium nucleus has less momentum than the thorium nucleus.
(b) the helium nucleus has more momentum than the thorium nucleus.
(c) the helium nucleus has less kinetic energy than the thorium nucleus.
(d) the helium nucleus has more kinetic energy than the thorium nucleus.

37. If radius of the ${}^{27}_{12}Al$ nucleus is taken to be R_{Al}, then the radius of ${}^{125}_{53}Te$ nucleus is nearly:
(a) $\frac{5}{3}R_{Al}$ (b) $\frac{3}{5}R_{Al}$ (c) $\left(\frac{13}{53}\right)^{1/3} R_{Al}$ (d) $\left(\frac{53}{13}\right)^{1/3} R_{Al}$

RESPONSE GRID					
	23. ⓐⓑⓒⓓ	24. ⓐⓑⓒⓓ	25. ⓐⓑⓒⓓ	26. ⓐⓑⓒⓓ	27. ⓐⓑⓒⓓ
	28. ⓐⓑⓒⓓ	28. ⓐⓑⓒⓓ	29. ⓐⓑⓒⓓ	30. ⓐⓑⓒⓓ	31. ⓐⓑⓒⓓ
	32. ⓐⓑⓒⓓ	33. ⓐⓑⓒⓓ	34. ⓐⓑⓒⓓ	35. ⓐⓑⓒⓓ	36. ⓐⓑⓒⓓ
	37. ⓐⓑⓒⓓ				

Space for Rough Work

38. M_n and M_p represent mass of neutron and proton respectively. If an element having atomic mass M has N-neutron and Z-proton, then the correct relation will be

(a) $M < [NM_n + ZM_p]$ (b) $M > [NM_n + ZM_p]$

(c) $M = [NM_n + ZM_p]$ (d) $M = N[M_n + M_p]$

39. After 300 days, the activity of a radioactive sample is 5000 dps (disintegrations per sec). The activity becomes 2500 dps after another 150 days. The initial activity of the sample in dps is

(a) 20,000 (b) 10,000

(c) 7,000 (d) 25,000

40. Order of magnitude of density of uranium nucleus is $(m_p = 1.67 \times 10^{-27}\,kg)$

(a) $10^{20}\,kg/m^3$ (b) $10^{17}\,kg/m^3$

(c) $10^{14}\,kg/m^3$ (d) $10^{11}\,kg/m^3$

41. The electrons cannot exist inside the nucleus because

(a) de-Broglie wavelength associated with electron in β-decay is much less than the size of nucleus

(b) de-Broglie wavelength associated with electron in β-decay is much greater than the size of nucleus

(c) de-Broglie wavelength associated with electron in β-decay is equal to the size of nucleus

(d) negative charge cannot exist in the nucleus

42. If the total binding energies of ${}^2_1H$, ${}^4_2He$, ${}^{56}_{26}Fe$ & ${}^{235}_{92}U$ nuclei are 2.22, 28.3, 492 and 1786 MeV respectively, identify the most stable nucleus of the following.

(a) ${}^{56}_{26}Fe$ (b) ${}^2_1H$

(c) ${}^{235}_{92}U$ (d) ${}^4_2He$

43. At a specific instant emission of radioactive compound is deflected in a magnetic field. The compound cannot emit

(a) electrons (b) protons

(c) He^{2+} (d) neutrons

44. A nuclear reaction is given by

${}_ZX^A \rightarrow {}_{Z+1}Y^A + {}_{-1}e^0 + \bar{\nu}$, represents

(a) fission (b) β-decay

(c) ∝-decay (d) fusion

45. Radioactive material 'A' has decay constant '8 λ' and material 'B' has decay constant 'λ'. Initially they have same number of nuclei. After what time, the ratio of number of nuclei of material 'B' to that 'A' will be $\frac{1}{e}$?

(a) $\frac{1}{7\lambda}$ (b) $\frac{1}{8\lambda}$ (c) $\frac{1}{9\lambda}$ (d) $\frac{1}{\lambda}$

RESPONSE GRID	38. ⓐⓑⓒⓓ	39. ⓐⓑⓒⓓ	40. ⓐⓑⓒⓓ	41. ⓐⓑⓒⓓ	42. ⓐⓑⓒⓓ
	43. ⓐⓑⓒⓓ	44. ⓐⓑⓒⓓ	45. ⓐⓑⓒⓓ		

DAILY PRACTICE PROBLEM DPP CHAPTERWISE CP27 - PHYSICS

Total Questions	45	Total Marks	180
Attempted		Correct	
Incorrect		Net Score	
Cut-off Score	50	Qualifying Score	70
Success Gap = Net Score – Qualifying Score			
Net Score = (Correct × 4) – (Incorrect × 1)			

Space for Rough Work

DPP - Daily Practice Problems

Chapter-wise Sheets

Date : ______ **Start Time :** ______ **End Time :** ______

PHYSICS

CP28

SYLLABUS : Semiconductor Electronics: Materials, Devices and Simple Circuits

Max. Marks : 180 **Marking Scheme :** (+4) for correct & (–1) for incorrect answer **Time : 60 min.**

INSTRUCTIONS : This Daily Practice Problem Sheet contains 45 MCQs. For each question only one option is correct. Darken the correct circle/ bubble in the Response Grid provided on each page.

1. A change of 8.0 mA in the emitter current bring a change of 7.9 mA in the collector current. The values of parameters α and β are respectively
(a) 0.99,90 (b) 0.96,79 (c) 0.97,99 (d) 0.99,79

2. A pure semiconductor has equal electron and hole concentration of 10^{16} m^{-3}. Doping by indium increases number of hole concentration n_h to 5×10^{22} m^{-3}. Then, the value of number of electron concentration n_e in the doped semiconductor is
(a) $10^6/m^3$ (b) $10^{22}/m^3$
(c) $2 \times 10^6/m^3$ (d) $2 \times 10^9/m^3$

3. For LED's to emit light in visible region of electromagnetic light, it should have energy band gap in the range of:
(a) 0.1 eV to 0.4 eV (b) 0.5 eV to 0.8 eV
(c) 0.9 eV to 1.6 eV (d) 1.7 eV to 3.0 eV

4. A common emitter amplifier has a voltage gain of 50, an input impedance of 100Ω and an output impedance of 200Ω. The power gain of the amplifier is
(a) 1000 (b) 1250 (c) 100 (d) 500

5. Which logic gate with inputs A and B performs the same operation as that performed by the following circuit?

(a) NAND gate (b) OR gate
(c) NOR gate (d) AND gate

6. In an unbiased p-n junction, holes diffuse from the p-region to n-region because of
(a) the potential difference across the p-n junction
(b) the attraction of free electrons of n-region
(c) the higher hole concentration in p-region than that in n-region
(d) the higher concentration of electrons in the n-region than that in the p-region

7. A silicon diode has a threshold voltage of 0.7 V. If an input voltage given by 2 sin(πt) is supplied to a half wave rectifier circuit using this diode, the rectified output has a peak value of
(a) 2V (b) 1.4V (c) 1.3V (d) 0.7V

RESPONSE GRID					
	1. ⓐⓑⓒⓓ	2. ⓐⓑⓒⓓ	3. ⓐⓑⓒⓓ	4. ⓐⓑⓒⓓ	5. ⓐⓑⓒⓓ
	6. ⓐⓑⓒⓓ	7. ⓐⓑⓒⓓ			

Space for Rough Work

8. The current gain for a transistor working as common-base amplifier is 0.96. If the emitter current is 7.2 mA, then the base current is
(a) 0.29 mA (b) 0.35 mA (c) 0.39 mA (d) 0.43 mA

9. In a npn transistor 10^{10} electrons enter the emitter in 10^{-6} s. 4% of the electrons are lost in the base. The current transfer ratio will be
(a) 0.98 (b) 0.97 (c) 0.96 (d) 0.94

10. Assuming that the silicon diode having resistance of 20 Ω, the current through the diode is (knee voltage 0.7 V)

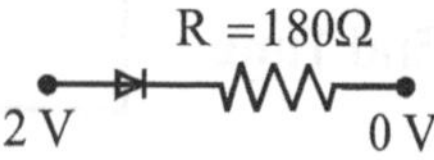

(a) 0 mA (b) 10 mA (c) 6.5 mA (d) 13.5 mA

11. Transfer characteristics [output voltage (V_0) vs input voltage (V_i)] for a base biased transistor in CE configuration is as shown in the figure. For using transistor as a switch, it is used

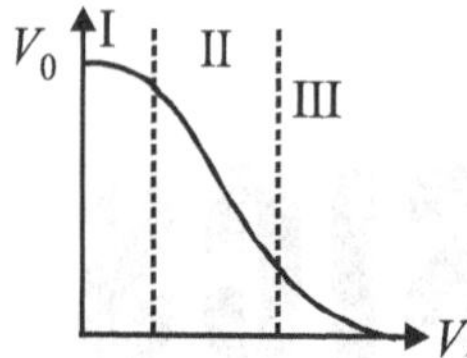

(a) in region III
(b) both in region (I) and (III)
(c) in region II
(d) in region (I)

12. A half-wave rectifier is being used to rectify an alternating voltage of frequency 50 Hz. The number of pulses of rectified current obtained in one second is
(a) 50 (b) 25 (c) 100 (d) 2000

13. A diode having potential difference 0.5 V across its junction which does not depend on current, is connected in series with resistance of 20Ω across source. If 0.1 A current passes through resistance then what is the voltage of the source?
(a) 1.5 V (b) 2.0 V (c) 2.5 V (d) 5 V

14. In common emitter amplifier, the current gain is 62. The collector resistance and input resistance are 5 kΩ an 500Ω respectively. If the input voltage is 0.01V, the output voltage is
(a) 0.62 V (b) 6.2 V (c) 62 V (d) 620 V

15. On doping germanium with donor atoms of density 10^{17} cm^{-3} its conductivity in mho/cm will be
[Given : $\mu_e = 3800$ cm^2/V–s and $n_i = 2.5 \times 10^{13}$ cm^{-13}]
(a) 30.4 (b) 60.8 (c) 91.2 (d) 121.6

16. The voltage gain of an amplifier with 9% negative feedback is 10. The voltage gain without feedback will be
(a) 90 (b) 10 (c) 1.25 (d) 100

17. A system of four gates is set up as shown. The 'truth table' corresponding to this system is :

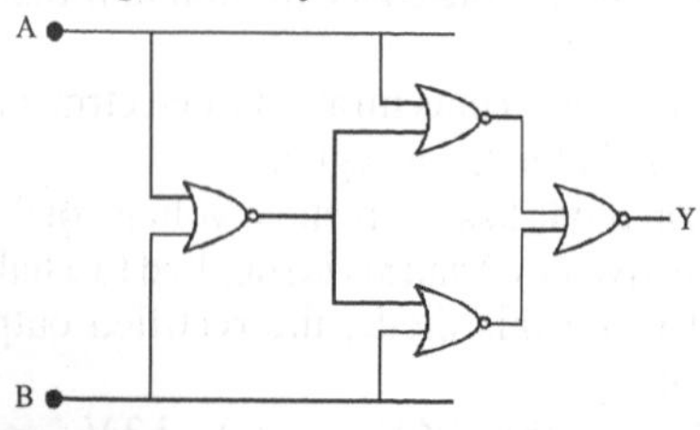

(a)

A	B	Y
0	0	1
0	1	0
1	0	0
1	1	1

(b)

A	B	Y
0	0	0
0	1	0
1	0	1
1	1	0

(c)

A	B	Y
0	0	1
0	1	0
1	0	1
1	1	0

(d)

A	B	Y
0	0	1
0	1	1
1	0	0
1	1	0

18. The intrinsic conductivity of germanium at 27° is 2.13 mho m^{-1} and mobilities of electrons and holes are 0.38 and 0.18 m^{2}V^{-1}s^{-1} respectively. The density of charge carriers is
(a) 2.37×10^{19} m^{-3} (b) 3.28×10^{19} m^{-3}
(c) 7.83×10^{19} m^{-3} (d) 8.47×10^{19} m^{-3}

19. The logic circuit shown below has the input waveforms '*A*' and '*B*' as shown. Pick out the correct output waveform

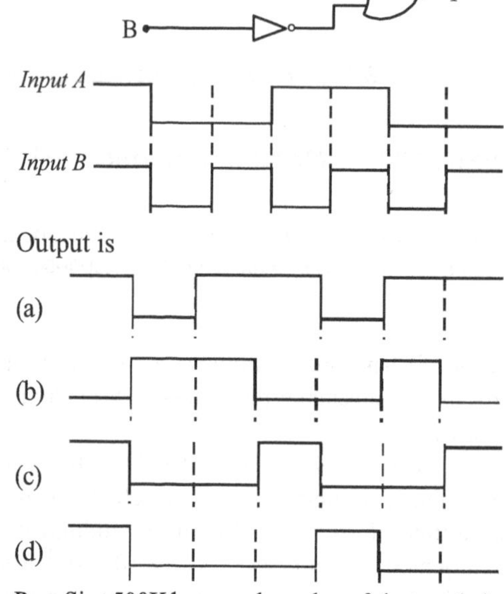

20. Pure Si at 500K has equal number of electron (n_e) and hole (n_h) concentrations of 1.5×10^{16} m^{-3}. Doping by indium increases n_h to 4.5×10^{22} m^{-3}. The doped semiconductor is of
(a) n–type with electron concentration $n_e = 5 \times 10^{22}$ m^{-3}
(b) p–type with electron concentration $n_e = 2.5 \times 10^{10}$ m^{-3}
(c) n–type with electron concentration $n_e = 2.5 \times 10^{23}$ m^{-3}
(d) p–type having electron concentration $n_e = 5 \times 10^{9}$ m^{-3}

21. Which of the following statements is incorrect?
(a) The resistance of intrinsic semiconductors decrease with increase of temperature
(b) Doping pure Si with trivalent impurities give *p*-type semiconductors
(c) The majority carriers in *n*-type semiconductors are holes
(d) A *p-n* junction can act as a semiconductor diode

RESPONSE GRID

8. ⓐⓑⓒⓓ 9. ⓐⓑⓒⓓ 10. ⓐⓑⓒⓓ 11. ⓐⓑⓒⓓ 12. ⓐⓑⓒⓓ
13. ⓐⓑⓒⓓ 14. ⓐⓑⓒⓓ 15. ⓐⓑⓒⓓ 16. ⓐⓑⓒⓓ 17. ⓐⓑⓒⓓ
18. ⓐⓑⓒⓓ 19. ⓐⓑⓒⓓ 20. ⓐⓑⓒⓓ 21. ⓐⓑⓒⓓ

Space for Rough Work

22. The relation between number of free electrons (n) in a semiconductor and temperature (T) is given by

(a) $n \propto T$ (b) $n \propto T^2$ (c) $n \propto \sqrt{T}$ (d) $n \propto T^{3/2}$

23. If a PN junction diode of depletion layer width W and barrier height V_0 is forward biased, then
(a) W increases, V_0 decreases
(b) W decreases, V_0 increases
(c) both W and V_0 increase
(d) both W and V_0 decrease

24. The circuit has two oppositively connected ideal diodes in parallel. The current flowing in the circuit is

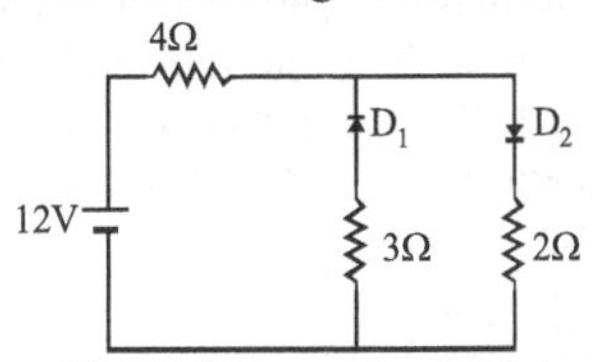

(a) 1.71 A (b) 2.00 A (c) 2.31 A (d) 1.33 A

25. For a transistor amplifier in common emitter configuration for load impedance of 1kΩ ($h_{fe} = 50$ and $h_{0e} = 25$) the current gain is
(a) −24.8 (b) −15.7 (c) −5.2 (d) −48.78

26. A PN-junction has a thickness of the order of
(a) 1 cm (b) 1 mm (c) 10^{-6} m (d) 10^{-12} cm

27. A working transistor with its three legs marked *P*, *Q* and *R* is tested using a multimeter. No conduction is found between *P* and *Q*. By connecting the common (negative) terminal of the multimeter to *R* and the other (positive) terminal to *P* or *Q*, some resistance is seen on the multimeter. Which of the following is true for the transistor?
(a) It is an npn transistor with *R* as base
(b) It is a pnp transistor with *R* as base
(c) It is a pnp transistor with *R* as emitter
(d) It is an npn transistor with *R* as collector

28. If in a p-n junction, a square input signal of 10 V is applied as shown, then the output across R_L will be

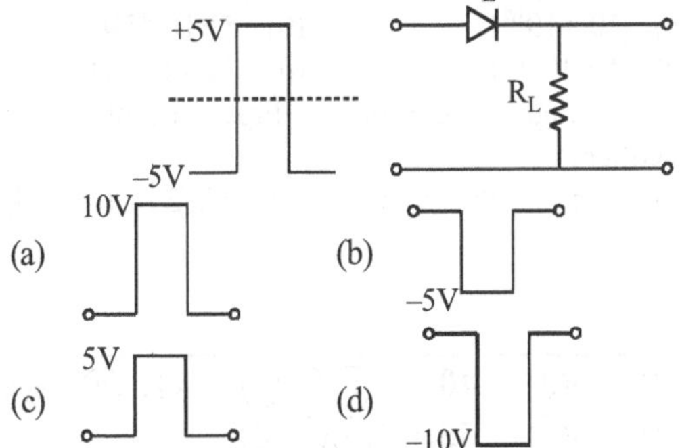

29. When *n*-type semiconductor is heated
(a) number of electrons increases while that of holes decreases
(b) number of holes increases while that of electrons decreases
(c) number of electrons and holes remain same
(d) number of electrons and holes increases equally.

30. The ratio of electron and hole currents in a semiconductor is 7/4 and the ratio of drift velocities of electrons and holes is 5/4, then the ratio of concentrations of electrons and holes will be
(a) 5/7 (b) 7/5 (c) 25/49 (d) 49/25

31. C and Si both have same lattice structure, having 4 bonding electrons in each. However, C is insulator whereas Si is intrinsic semiconductor. This is because :
(a) In case of C the valence band is not completely filled at absolute zero temperature.
(b) In case of C the conduction band is partly filled even at absolute zero temperature.
(c) The four bonding electrons in the case of C lie in the second orbit, whereas in the case of Si they lie in the third.
(d) The four bonding electrons in the case of C lie in the third orbit, whereas for Si they lie in the fourth orbit.

32. Which one of the following represents forward bias diode ?
(a) −4V —▷|— R —−3V
(b) −2V —▷|— R — +2V
(c) 3V —▷|— R — 5V
(d) 0V —▷|— R — −2V

33. An oscillator is nothing but an amplifer with
(a) positive feedback (b) negative feedback
(c) large gain (d) no feedback

34. The current gain in the common emitter mode of a transistor is 10. The input impedance is 20kΩ and load of resistance is 100kΩ. The power gain is
(a) 300 (b) 500 (c) 200 (d) 100

35. The input signal given to a CE amplifier having a voltage gain of 150 is $V_i = 2\cos\left(15t + \frac{\pi}{3}\right)$. The corresponding output signal will be :
(a) $75\cos\left(15t + \frac{2\pi}{3}\right)$ (b) $2\cos\left(15t + \frac{5\pi}{6}\right)$
(c) $300\cos\left(15t + \frac{4\pi}{3}\right)$ (d) $300\cos\left(15t + \frac{\pi}{3}\right)$

36. To use a transistor as an amplifier
(a) the emitter base junction is forward biased and the base collector junction is reverse biased
(b) no bias voltage is required
(c) both junctions are forward biased
(d) both junctions are reverse biased.

RESPONSE GRID					
	22. ⓐⓑⓒⓓ	23. ⓐⓑⓒⓓ	24. ⓐⓑⓒⓓ	25. ⓐⓑⓒⓓ	26. ⓐⓑⓒⓓ
	27. ⓐⓑⓒⓓ	28. ⓐⓑⓒⓓ	29. ⓐⓑⓒⓓ	30. ⓐⓑⓒⓓ	31. ⓐⓑⓒⓓ
	32. ⓐⓑⓒⓓ	33. ⓐⓑⓒⓓ	34. ⓐⓑⓒⓓ	35. ⓐⓑⓒⓓ	36. ⓐⓑⓒⓓ

Space for Rough Work

37. A piece of copper and another of germanium are cooled from room temperature to 77K. The resistance of
(a) copper increases and germanium decreases
(b) each of them decreases
(c) each of them increases
(d) copper decreases and germanium increases

38. A d.c. battery of V volt is connected to a series combination of a resistor R and an ideal diode D as shown in the figure below. The potential difference across R will be

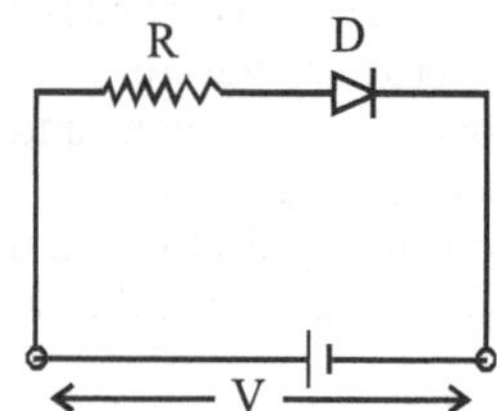

(a) 2V when diode is forward biased
(b) Zero when diode is forward biased
(c) 5V when diode is reverse biased
(d) 6V when diode is forward biased

39. The current gain for a transistor working as common-base amplifier is 0.96. If the emitter current is 7.2 mA, then the base current is
(a) 0.29 mA (b) 0.35 mA (c) 0.39 mA (d) 0.43 mA

40. In the circuit given below, *A* and *B* represent two inputs and *C* represents the output.

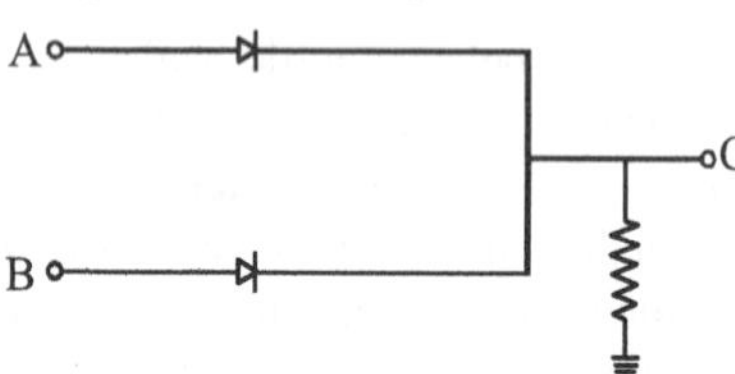

The circuit represents
(a) NOR gate (b) AND gate
(c) NAND gate (d) OR gate

41. The I-V characteristic of a P-N junction diode is shown below. The approximate dynamic resistance of the p-n junction when a forward bias voltage of 2 volt is applied is

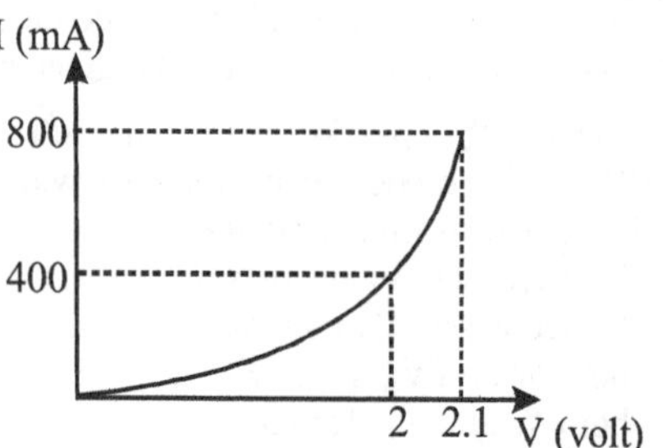

(a) 1 Ω (b) 0.25 Ω (c) 0.5 Ω (d) 5 Ω

42. The circuit diagram shows a logic combination with the states of outputs X, Y and Z given for inputs P, Q, R and S all at state 1. When inputs P and R change to state 0 with inputs Q and S still at 1, the states of outputs X, Y and Z change to

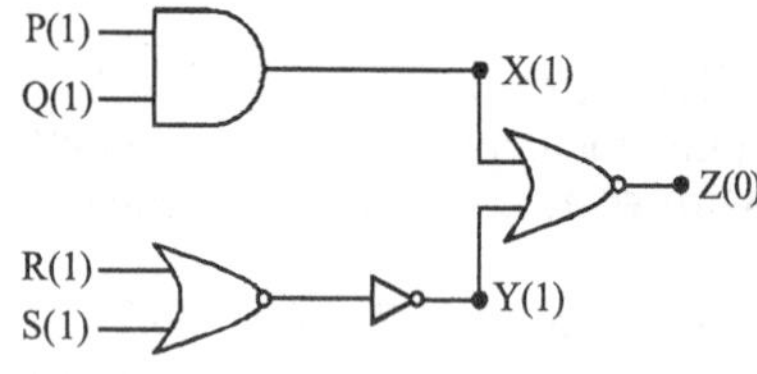

(a) 1, 0, 0 (b) 1, 1, 1 (c) 0, 1, 0 (d) 0, 0, 1

43. The following configuration of gate is equivalent to

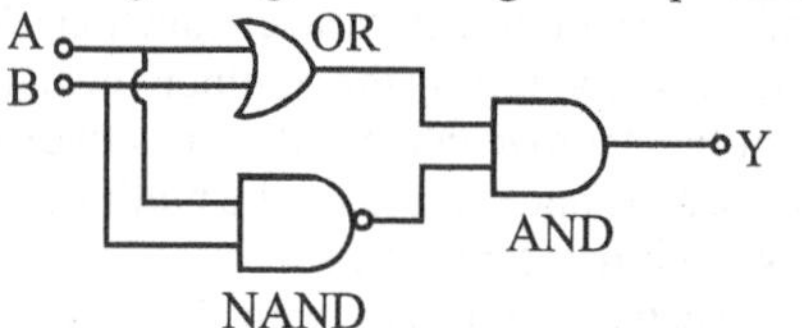

(a) NAND gate (b) XOR gate
(c) OR gate (d) NOR gate

44. A p-n photodiode is made of a material with a band gap of 2.0 eV. The minimum frequency of the radiation that can be absorbed by the material is nearly
(a) 10×10^{14} Hz (b) 5×10^{14} Hz
(c) 1×10^{14} Hz (d) 20×10^{14} Hz

45. The average value of output direct current in a full wave rectifier is
(a) I_0/π (b) $I_0/2$ (c) $\pi I_0/2$ (d) $2I_0/\pi$

RESPONSE GRID					
	37. ⓐⓑⓒⓓ	38. ⓐⓑⓒⓓ	39. ⓐⓑⓒⓓ	40. ⓐⓑⓒⓓ	41. ⓐⓑⓒⓓ
	42. ⓐⓑⓒⓓ	43. ⓐⓑⓒⓓ	44. ⓐⓑⓒⓓ	45. ⓐⓑⓒⓓ	

DAILY PRACTICE PROBLEM DPP CHAPTERWISE CP28 - PHYSICS			
Total Questions	45	Total Marks	180
Attempted		Correct	
Incorrect		Net Score	
Cut-off Score	50	Qualifying Score	70
Success Gap = Net Score – Qualifying Score			
Net Score = (Correct × 4) – (Incorrect × 1)			

Space for Rough Work

1. (b) In CGS system,

$$d = 4\frac{g}{cm^3}$$

The unit of mass is 100g and unit of length is 10 cm, so

$$\text{density} = \frac{4\left(\frac{100g}{100}\right)}{\left(\frac{10}{10}cm\right)^3}$$

$$= \frac{\left(\frac{4}{100}\right)}{\left(\frac{1}{10}\right)^3}\frac{(100g)}{(10cm)^3}$$

$$= \frac{4}{100}\times(10)^3\cdot\frac{100g}{(10cm)^3}$$

$= 40$ unit

2. (a) $T = P^a D^b S^c$

$M^0L^0T^1 = (ML^{-1}T^{-2})^a (ML^{-3})^b (MT^{-2})^c$

$= M^{a+b+c} L^{-a-3b} T^{-2a-2c}$

Applying principle of homogeneity

$a + b + c = 0;\ -a - 3b = 0;\ -2a - 2c = 1$

on solving, we get $a = -3/2$, $b = 1/2$, $c = 1$

3. (a) Number of significant figures in 23.023= 5

Number of significant figures in 0.0003 = 1

Number of significant figures in $2.1 \times 10^{-3} = 2$

4. (a) $Y = \frac{\text{Stress}}{\text{Strain}} = \frac{\text{Force / Area}}{\text{Dimensionless}} \Rightarrow Y = \text{Pressure}.$

5. (d) For angular momentum, the dimensional formula is $[ML^2T^{-1}]$. For other three, it is $[ML^2T^{-2}]$.

6. (c) $\frac{\Delta P}{P}\times 100 = \frac{\Delta F}{F}\times 100 + 2\frac{\Delta \ell}{\ell}\times 100 = 4\% + 2\times 2\%$

$= 8\%$

7. (d) Conductance,

$$G = \frac{1}{\text{resistance}} = \text{mho}(\Omega^{-1}) \text{ or siemen (S)}$$

8. (d) $F \propto v \Rightarrow F = kv \Rightarrow [k] = \left[\frac{F}{v}\right] = \left[\frac{MLT^{-2}}{LT^{-1}}\right] = [ML^0T^{-1}]$

9. (c) $\frac{0.2}{25}\times 100 = 0.8\,\%$

10. (c) Weber is the unit of magnetic flux in S.I. system.

1 Wb(S.I unit) = 10^8 maxwell

11. (b) Solar constant = energy/area/time

$$= \frac{ML^2T^{-2}}{L^2T} = [M^1T^{-3}].$$

12. (b) $b = \lambda_m T = LK = [M^0 L^1 T^0 K^1]$

13. (d) Let unit 'u' related with e, a_0, h and c as follows.

$$[u] = [e]^a [a_0]^b [h]^c [C]^d$$

Using dimensional method,

$[M^{-1}L^{-2}T^{+4}A^{+2}] = [A^1T^1]^a[L]^b[ML2T^{-1}]^c[LT^{-1}]^d$

$[M^{-1}L^{-2}T^{+4}A^{+2}] = [M^c L^{b+2c+d} T^{a-c-d} A^a]$

$a = 2,\ b = 1,\ c = -1,\ d = -1$

$$\therefore\ u = \frac{e^2 a_0}{hc}$$

14. (c) From $F = \frac{1}{4\pi\varepsilon_o}\frac{e^2}{r^2}$

$\therefore\ \frac{e^2}{\varepsilon_o} = 4\pi F r^2$ (dimensionally)

$$\frac{e^2}{\varepsilon_o hc} = \frac{4\pi Fr^2}{hc} = \frac{(MLT^{-2})L^2}{ML^2T^{-1}[LT^{-1}]} = [M^0L^0T^0A^0],$$

$\frac{e^2}{\varepsilon_o hc}$ is called fine structure constant & has value $\frac{1}{137}$.

15. (d) $\text{Density} = \frac{\text{Mass}}{\text{Volume}}$

$\rho = \frac{M}{L^3}\qquad \therefore\ \frac{\Delta\rho}{\rho} = \frac{\Delta M}{M} + 3\frac{\Delta L}{L}$

% error in density = % error in Mass

+ 3 (% error in length]

= 4 + 3(3) = 13%

16. (d) Poisson's ratio is a unitless quantity.

17. (d) Dimensionally $\varepsilon_0 L$ = Capacitance (c)

$$\therefore\ \varepsilon_0 L\frac{\Delta V}{\Delta t} = \frac{C\Delta V}{\Delta t} = \frac{q}{\Delta t} = I$$

18. (c) $\frac{\Delta V}{V} = 3\frac{\Delta r}{r}$ or $6\% = 3\frac{\Delta r}{r}$ or $\frac{\Delta r}{r} = 2\%$

Now surface area $s = 4\pi r^2$ or $\log s = \log 4\pi + 2\log r$

$$\therefore\ \frac{\Delta s}{s} = 2\frac{\Delta r}{r} = 2\times 2\% = 4\%.$$

19. (d) Let $(M) = V^a F^b E^c$

Putting the dimensions of V, F and E, we have

$(M) = (LT^{-1})^a \times (MLT^{-2})^b \times (ML^2T^{-2})^c$

or $M^1 = M^{b+c} L^{a+b+2c} T^{-a-2b-2c}$

Equating the powers of dimensions, we have

$b+c=1$
$a+b+2c=0; \quad -a-2b-2c=0$
which give $a=-2, b=0$ and $c=1$.
Therefore $(M)=(V^{-2}F^0E)$.

20. (d) Number of significant figures in multiplication is three, corresponding to the minimum number
$107.88\times 0.610=65.8068=65.8$

21. (d) A quantity which has dimensions and a constant value is called dimensional constant. Therefore, gravitational constant (G) is a dimensional constant.

22. (a) $\frac{[ML^2T^{-2}][ML^2T^{-1}]^2}{[M^5][M^{-1}L^3T^{-2}]^2}=[M^0L^0T^0]=\text{angle.}$

23. (a) The mean value of refractive index,

$$\mu=\frac{1.34+1.38+1.32+1.36}{4}=1.35$$

and

$$\Delta\mu=\frac{|(1.35-1.34)|+|(1.35-1.38)|+|(1.35-1.32)|+|(1.35-1.36)|}{4}$$

$$=\ 0.02$$

Thus $\frac{\Delta\mu}{\mu}\times 100=\frac{0.02}{1.35}\times 100=1.48$

24. (c) $\frac{eV}{T}=\frac{W}{T}=\frac{PV}{T}=R$

and $\frac{R}{N}=$ Boltzmann constant.

25. (b) Mobility $\mu=\frac{\text{drift velocity } V_d}{\text{electric field } E}=\frac{(ms^{-1})}{(Vm^{-1})}=\frac{m^2s^{-3}}{V}$

$$\left(\because \text{Volt}=V=\frac{\text{joule(J)}}{\text{coulomb(C)}}\right)$$

$$=\frac{m^2s^{-1}C}{J}=\frac{m^2s^{-1}As}{kgm^2s^{-2}}[\text{Coulomb}, c=As]$$

$$=kg^{-1}s^2A=M^{-1}T^2A$$

26. (a)

27. (b) $v=k\lambda^a\rho^b g^c$

$$[M^0LT^{-1}]=L^a(ML^{-3})^b(LT^{-2})^c$$
$$=M^bL^{a-3b+c}T^{-2c}$$

$\therefore\ b=0;\ a-3b+c=1$

$-2c=-1\Rightarrow c=1/2 \qquad \therefore\ a=\frac{1}{2}$

$v\propto\lambda^{1/2}\rho^0g^{1/2}$ or $v^2\propto\lambda g$

28. (b) $[\text{momentum}]=[M][L][T^{-1}]=[MLT^{-1}]$

Planck's constant $=\frac{E}{\nu}=\frac{[M][LT^{-1}]^2}{T^{-1}}=ML^2T^{-1}$

29. (d) Let dimensions of length is related as,

$$L=[c]^x[G]^y\left[\frac{e^2}{4\pi\varepsilon_0}\right]^z$$

$$\frac{e^2}{4\pi\varepsilon_0}=ML^3T^{-2}$$

$L=[LT^{-1}]^x[M^{-1}L^3T^{-2}]^y[ML^3T^{-2}]^z$

$[L]=[L^{x+3y+3z}M^{-y+z}T^{-x-2y-2z}]$

Comparing both sides

$-y+z=0\Rightarrow y=z$ ...(i)

$x+3y+3z=1$ ...(ii)

$-x-4z=0 \quad (\because y=z)$ ...(iii)

From (i), (ii) & (iii)

$z=y=\frac{1}{2},\ x=-2$

Hence, $L=c^{-2}\left[G\cdot\frac{e^2}{4\pi\varepsilon_0}\right]^{1/2}$

30. (c) Impulse = change in momentum

31. (c) We know that $\frac{Q^2}{2C}$ is energy of capacitor so it represent the dimension of energy $=[ML^2T^{-2}]$.

32. (b) Let $M=p^nv^m$

$$ML^{-2}T^{-1}=(ML^{-1}T^{-2})^n(LT^{-1})^m$$
$$=M^nL^{-n+m}T^{-2n-m}$$

$\therefore\ n=1;\ -n+m=-2$

$\therefore\ m=-2+n=-2+1=-1 \qquad \therefore\ m=-n$

33. (c) $I=AT^2e^{-B/kT}$
Dimensions of $A=I/T^2$; Dimensions of $B=kT$
($\because$ power of exponential is dimensionless)

$$AB^2=\frac{I}{T^2}(kT)^2=Ik^2$$

34. (a) $\eta=\frac{p(r^2-x^2)}{4vl}=\frac{[ML^{-1}T^{-2}][L^2]}{[LT^{-1}][L]}=[ML^{-1}T^{-1}]$

35. (a) The unit of λ, x and A are the same

36. (c) $L+B=2.331+2.1\cong 4.4$ cm
Since minimum significant figure is 2.

37. (c) Given, $x=\cos(\omega t+kx)$
$(\omega t+kx)$ is an angle and hence it is a dimension less quantity.

$[(\omega t+kx)]=[M^0L^0T^0]$

or $[\omega t]=[M^0L^0T^0]$

$$[\omega]=\frac{[M^0L^0T^0]}{[T]}=[M^0L^0T^{-1}]$$

38. (c) $10\,VD=9\,MD,\ 1\,VD=\frac{9}{10}MD$

Vernier constant = 1 MD – 1 VD

$$=\left(1-\frac{9}{10}\right)MD=\frac{1}{10}MD=\frac{1}{10}\times\frac{1}{2}=0.05\,mm$$

39. (c) $[\text{Energy density}] = \frac{[\text{Work done}]}{[\text{Volume}]}$

$$= \frac{ML^2T^{-2}}{L^3} = ML^{-1}T^{-2}$$

$$[\text{Young's Modulus}] = \left[\frac{F}{A} \times \frac{l}{\Delta l}\right]$$

$$= \frac{MLT^{-2}}{L^2} \cdot \frac{L}{L} = [ML^{-1}T^{-2}]$$

40. (b) As $\frac{a}{V^2} = P$

$$\therefore \; a = PV^2 = \frac{\text{dyne}}{\text{cm}^2}(\text{cm}^3)^2 = \text{dyne cm}^4$$

41. (a) Reyonld's constant is a pure number, hence it has no dimensions.

42. (d) $\omega k = \frac{1}{T} \times \frac{1}{L} = [L^{-1}T^{-1}]$
The dimensions of the quantities in a, b, c are of velocity $[LT^{-1}]$

43. (a) M = Pole strength × length
= amp – metre × metre = amp – metre^2

44. (b) According to the question.

$$t = (90 \pm 1) \text{ or, } \frac{\Delta t}{t} = \frac{1}{90}$$

$$l = (20 \pm 0.1) \text{ or, } \frac{\Delta l}{l} = \frac{0.1}{20}$$

$$\frac{\Delta g}{g}\% = ?$$

As we know,

$$t = 2\pi\sqrt{\frac{l}{g}}$$

$$\Rightarrow \quad g = \frac{4\pi^2 l}{t^2}$$

$$\text{or, } \quad \frac{\Delta g}{g} = \pm\left(\frac{\Delta l}{l} + 2\frac{\Delta t}{t}\right)$$

$$= \left(\frac{0.1}{20} + 2 \times \frac{1}{90}\right)$$

$$= 0.027$$

$$\therefore \quad \frac{\Delta g}{g}\% = 2.7\%$$

45. (a) Dimension of magnetic flux
= Dimension of voltage × Dimension of time
$= [ML^2T^{-3}A^{-1}]\,[T] = [ML^2T^{-2}A^{-1}]$

$$\because \text{Voltage} = \frac{\text{work}}{\text{charge}}$$

1. (a) Acceleration of the particle $a = 2t - 1$
The particle retards when acceleration is opposite to velocity.
$\Rightarrow a \cdot v < 0 \Rightarrow (2t-1)(t^2-t) < 0 \Rightarrow t(2t-1)(t-1) < 0$
Now t is always positive
$\therefore\ (2t-1)(t-1) < 0$
or $2t-1<0$ and $t-1>0 \Rightarrow t<\frac{1}{2}$ and $t>1$.
This is not possible
or $2t-1>0\ \&\ t-1<0 \Rightarrow 1/2<t<1$

2. (b) $x = \alpha t^3$ and $y = \beta t^3$

$$v_x = \frac{dx}{dt} = 3\alpha t^2 \text{ and } v_y = \frac{dy}{dt} = 3\beta t^2$$

$$\therefore\ v = \sqrt{v_x^2 + v_y^2} = \sqrt{9\alpha^2 t^4 + 9\beta^2 t^4}$$

$$= 3t^2\sqrt{\alpha^2 + \beta^2}$$

3. (d) Average speed $= \frac{\text{Total distance travelled}}{\text{Total time taken}}$

$$= \frac{x}{\frac{2x/5}{v_1} + \frac{3x/5}{v_2}} = \frac{5v_1v_2}{3v_1 + 2v_2}$$

4. (a) Instantaneous speed is the distance being covered by the particle per unit time at the given instant. It is equal to the magnitude of the instantaneous velocity at the given instant.

5. (a) $v = \alpha\sqrt{x}$, $\frac{dx}{dt} = \alpha\sqrt{x} \Rightarrow \frac{dx}{\sqrt{x}} = \alpha\, dt$

$$\int_0^x \frac{dx}{\sqrt{x}} = \alpha \int_0^t dt$$

$$\left[\frac{2\sqrt{x}}{1}\right]_0^x = \alpha[t]_0^t$$

$$\Rightarrow 2\sqrt{x} = \alpha t \Rightarrow x = \frac{\alpha^2}{4}t^2$$

6. (c) $\frac{1}{2}(1+4)\times 4 - \frac{1}{2}\times 1\times 2 - \frac{1}{2}\times 3\times 4 = 3\text{ m}$

7. (b) The distance travel in n^{th} second is
$S_n = u + ½(2n-1)a$ ....(1)
so distance travel in t^{th} & $(t+1)^{th}$ second are
$S_t = u + ½(2t-1)a$ ....(2)
$S_{t+1} = u + ½(2t+1)a$ ....(3)
As per question,
$S_t + S_{t+1} = 100 = 2(u + at)$ ....(4)
Now from first equation of motion the velocity, of particle after time t, if it moves with an accleration a is
$v = u + at$ ....(5)
where u is initial velocity
So from eq(4) and (5), we get v = 50 cm/sec.

8. (d) Relative speed of police with respect to thief
$= 10 - 9 = 1$ m/s
Instantaneous separation = 100 m

$$\text{Time} = \frac{\text{Distance}}{\text{Velocity}} = \frac{100}{1} = 100\text{sec.}$$

9. (d) $x = \frac{a}{b}(1 - e^{-b\times\frac{1}{b}}) = \frac{a}{b}(1-e^{-1}) = \frac{a}{b}(1-\frac{1}{e})$

$$= \frac{a}{b}\frac{(e-1)}{e} = \frac{a}{b}\frac{(2.718-1)}{2.718} = \frac{a}{b}\frac{(1.718)}{2.718} = 0.637\frac{a}{b} \approx \frac{2}{3}a/b$$

velocity $v = \frac{dx}{dt} = ae^{-bt}$, $v_0 = a$

accleration $a = \frac{dv}{dt} = -abe^{-bt}$ & $a_0 = -ab$

At $t = 0$, $x = \frac{a}{b}(1-1) = 0$ and

At $t = \frac{1}{b}$, $x = \frac{a}{b}(1-e^{-1}) = \frac{a}{b}(1-\frac{1}{e}) = \frac{2}{3}a/b$

At $t = \infty$, $x = \frac{a}{b}$

It cannot go beyond this, so point $x > \frac{a}{b}$ is not reached by the particle.

At $t = 0$, $x = 0$, at $t = \infty$, $x = \frac{a}{b}$, therefore the particle does not come back to its starting point at $t = \infty$.

10. (d) Ist part: $u = 0$, $t = 5s$, $v = 108$ km/hr $= 30$ m/s
$v = u + at \Rightarrow 30 = 0 + a\times 5 \Rightarrow a = 6\text{ m/s}^2$

$$s = ut + \frac{1}{2}at^2 = 0\times 5 + \frac{1}{2}\times 6\times 5^2 = 75\text{ m}$$

IIIrd part: $s = 45$m, $u = 30$m/s, $v = 0$

$$a = \frac{v^2 - u^2}{2s} = \frac{-30\times 30}{2\times 45} = -10\text{m/s}^2$$

$v = u + at \Rightarrow 0 = 30 - 10\times t \Rightarrow t = 3s$
IInd part :
$s = s_1 + s_2 + s_3$
$395 = 75 + s_2 + 45 \Rightarrow s_2 = 275$ m

$$t = \frac{275}{30} = 9.16 = 9.2s.$$

Total time taken = (5 + 9.2 + 3) sec = 17.2 sec

11. (a) $\frac{dv}{dt} = -kv^3$ or $\frac{dv}{v^3} = -k\ dt$

Integrating we get, $-\frac{1}{2v^2} = -kt + c$ ...(1)

At $t = 0$, $v = v_0$ $\therefore\ -\frac{1}{2v_0^2} = c$

Putting in (1)

$$-\frac{1}{2v^2} = -kt - \frac{1}{2v_0^2} \text{ or } \frac{1}{2v_0^2} - \frac{1}{2v^2} = -kt$$

$$\text{or } \left[\frac{1}{2v_0^2} + kt\right] = \frac{1}{2v^2} \text{ or } \left[1 + 2v_0^2\, kt\right] = \frac{v_0^2}{v^2}$$

$$\text{or } v^2 = \frac{v_0^2}{1 + 2v_0^2 kt} \quad \text{or} \quad v = \frac{v_0}{\sqrt{1 + 2v_0^2 kt}}$$

12. **(c)** We know that, $v = \frac{dx}{dt} \Rightarrow dx = v\,dt$

Integrating, $\int_0^x dx = \int_0^t v\,dt$

$$\text{or } x = \int_0^t (v_0 + gt + ft^2)\,dt$$

$$= \left[v_0 t + \frac{gt^2}{2} + \frac{ft^3}{3}\right]_0^t$$

$$\text{or, } x = v_0 t + \frac{gt^2}{2} + \frac{ft^3}{3}$$

At $t = 1$, $x = v_0 + \frac{g}{2} + \frac{f}{3}$.

13. **(c)** Let man will catch the bus after 't' sec. So he will cover distance ut.

Similarly, distance travelled by the bus will be $\frac{1}{2}at^2$

For the given condition

$ut = 45 + \frac{1}{2}at^2 = 45 + 1.25t^2$ [As $a = 2.5$ m/s^2]

$$\Rightarrow u = \frac{45}{t} + 1.25t$$

To find the minimum value of u $\frac{du}{dt} = 0$

so we get t = 6 sec then,

$$u = \frac{45}{6} + 1.25 \times 6 = 7.5 + 7.5 = 15 \text{ m/s}$$

14. **(b)** **For the body starting from rest**

$$x_1 = 0 + \frac{1}{2}at^2$$

$$\Rightarrow x_1 = \frac{1}{2}at^2$$

For the body moving with constant speed

$x_2 = vt$

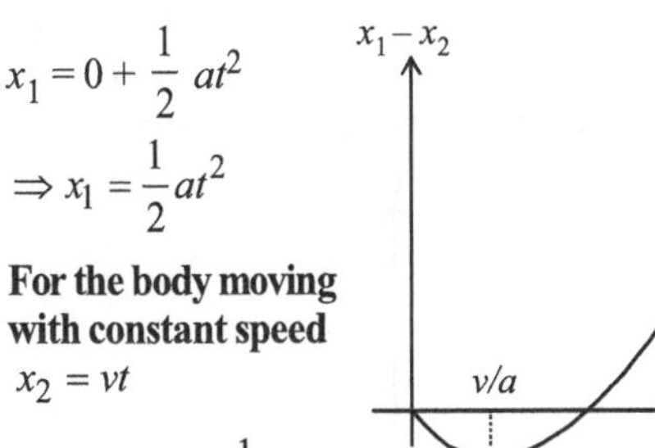

$$\therefore\ x_1 - x_2 = \frac{1}{2}at^2 - vt$$

at $t = 0$, $x_1 - x_2 = 0$

For $t < \frac{v}{a}$; the slope is negative

For $t = \frac{v}{a}$; the slope is zero

For $t > \frac{v}{a}$; the slope is positive

These characteristics are represented by graph (b).

15. **(d)** The stone reaches its maximum height after time t_1 given by

u = 10 m/s

40 m

$$t_1 = \frac{u}{g} \quad (\because v = u - gt)$$

$$= \frac{10}{10} = 1 \text{ sec}$$

Again it reaches to its initial position in 1 sec and falls with same initial speed of 10 m/s.

Let t_2 be the time taken to reach the ground, then

$$v_{ground} = u + gt_2$$

$$\text{But } v_{ground}^2 = u^2 + 2gh$$

$$= (10)^2 + 2 \times 10 \times 40 = 900$$

$$\Rightarrow v_{ground} = \sqrt{900} = 30 \text{ m/s}$$

$$\therefore\ t_2 = \frac{v_{ground} - u}{g} = \frac{30 - 10}{10} = 2 \text{ sec.}$$

$\therefore$ Total required time $= (1 + 1 + 2)$ sec $= 4$ sec

16. **(b)**
$$L = \frac{1}{2}gt^2 - \frac{1}{2}g(t - T)^2$$

$$\Rightarrow t = \frac{T}{2} + \frac{L}{gt}.$$

17. **(b)** $S = AB = \frac{1}{2}g{t_1}^2 \Rightarrow 2S = AC = \frac{1}{2}g\,(t_1 + t_2)^2$

and $3S = AD = \frac{1}{2}g\,(t_1 + t_2 + t_3)^2$

$$t_1 = \sqrt{\frac{2S}{g}}$$

$$t_1 + t_2 = \sqrt{\frac{4S}{g}},\ t_2 = \sqrt{\frac{4S}{g}} - \sqrt{\frac{2S}{g}}$$

$$t_1 + t_2 + t_3 = \sqrt{\frac{6S}{g}}$$

$$t_3 = \sqrt{\frac{6S}{g}} - \sqrt{\frac{4S}{g}}$$

A, B, C, D; S, S, S; 2S; 3S

$$t_1 : t_2 : t_3 :: 1 : (\sqrt{2} - 1) : (\sqrt{3} - \sqrt{2})$$

18. **(c)** Height of tap = 5m and (g) = 10 m/sec^2.

For the first drop,

$$5 = ut + \frac{1}{2}gt^2 = (0 \times t) + \frac{1}{2} \times 10t^2 = 5t^2 \text{ or } t^2 = 1 \text{ or } t = 1$$

It means that the third drop leaves after one second of the first drop. Or, each drop leaves after every 0.5 sec. Distance covered by the second drop in 0.5 sec

$= ut + \frac{1}{2}gt^2 = (0 \times 0.5) + \frac{1}{2} \times 10 = (0.5)^2 = 1.25\text{m}$.

Therefore, distance of the second drop above the ground $= 5 - 1.25 = 3.75$ m.

19. (c) $\because t = \sqrt{x} + 3$

$\Rightarrow \sqrt{x} = t - 3 \Rightarrow x = (t-3)^2$

$v = \frac{dx}{dt} = 2(t-3) = 0$

$\Rightarrow t = 3$

$\therefore x = (3-3)^2$

$\Rightarrow x = 0.$

20. (c) We have, $S_n = u + \frac{a}{2}(2n-1)$

or $65 = u + \frac{a}{2}(2 \times 5 - 1)$

or $65 = u + \frac{9}{2}a$ ..... (1)

Also, $105 = u + \frac{a}{2}(2 \times 9 - 1)$

or $105 = u + \frac{17}{2}a$ ..... (2)

Equation (2) – (1) gives,

$40 = \frac{17}{2}a - \frac{9}{2}a = 4a$ or $a = 10\text{ m/s}^2$.

Substitute this value in (1) we get,

$u = 65 - \frac{9}{2} \times 10 = 65 - 45 = 20\text{ m/s}$

$\therefore$ The distance travelled by the body in 20 s is,

$s = ut + \frac{1}{2}at^2 = 20 \times 20 + \frac{1}{2} \times 10 \times (20)^2$

$= 400 + 2000 = 2400$ m.

21. (d) Speed, $u = 60 \times \frac{5}{18}\text{m/s} = \frac{50}{3}\text{m/s}$

$d = 20\text{m}$, $u' = 120 \times \frac{5}{18} = \frac{100}{3}\text{m/s}$

Let declaration be a then $(0)^2 - u^2 = -2ad$

or $u^2 = 2ad$ ...(1)

and $(0)^2 - u'^2 = -2ad'$

or $u'^2 = 2ad'$ ...(2)

(2) divided by (1) gives,

$4 = \frac{d'}{d} \Rightarrow d' = 4 \times 20 = 80\text{m}$

22. (b) $8 = a\,t_1$ and $0 = 8 - a\,(4 - t_1)$

or $t_1 = \frac{8}{a}$ $\therefore 8 = a\left(4 - \frac{8}{a}\right)$

$8 = 4\,a - 8$ or $a = 4$ and $t_1 = 8/4 = 2$ sec

Now, $s_1 = 0 \times 2 + \frac{1}{2} \times 4(2)^2$ or $s_1 = 8\text{ m}$

$s_2 = 8 \times 2 - \frac{1}{2} \times 4 \times (2)^2$ or $s_2 = 8\text{ m}$

$\therefore s_1 + s_2 = 16$ m

23. (d)

24. (a) $x = \frac{1}{t+5}$

$\therefore v = \frac{dx}{dt} = \frac{-1}{(t+5)^2}$

$\therefore a = \frac{d^2x}{dt^2} = \frac{2}{(t+5)^3} = 2x^3$

Now $\frac{1}{(t+5)} \propto v^{\frac{1}{2}}$

$\therefore \frac{1}{(t+5)^3} \propto v^{\frac{3}{2}} \propto a$

25. (d)

B (v = 0)
4 sec 4 sec
(2 sec) A C
(t = 0) D

As the time taken from D to A = 2 sec.
and D → A → B → C = 10 sec (given).
As ball goes from B → C (u = 0, t = 4 sec)
$v_c = 0 + 4g$.

As it moves from C to D, $s = ut + \frac{1}{2}gt^2$

$s = 4g \times 2 + \frac{1}{2}g \times 4 = 10$ g.

26. (d) $y = \frac{1}{2}g(n+1)^2 - \frac{1}{2}gn^2$

$= \frac{g}{2}[(n+1)^2 - n^2] = \frac{g}{2}(2n+1)$ ......(i)

Also, $h = \frac{g}{2}(2n-1)$ ......(ii)

From (i) and (ii)

$y = h + g$

27. (b) The stone rises up till its vertical velocity is zero and again reached the top of the tower with a speed u (downward). The speed of the stone at the base is 3u.

u ↑ –

↓ +
v, g, h

Hence $(3u)^2 = (-u)^2 + 2gh$ or $h = \frac{4u^2}{g}$

28. **(b)** $x = 40 + 12t - t^3$

$$v = \frac{dx}{dt} = 12 - 3t^2$$

For $v = 0$; $t = \sqrt{\frac{12}{3}} = 2$ sec

So, after 2 seconds velocity becomes zero.
Value of x in 2 secs $= 40 + 12 \times 2 - 2^3$
$= 40 + 24 - 8 = 56$ m

29. **(b)** The slope of v-t graph is constant and velocity decreasing for first half. It is positive and constant over next half.

30. **(c)** Here, $f = f_0\left(1-\frac{t}{T}\right)$ or, $\frac{dv}{dt} = f_0\left(1-\frac{t}{T}\right)$

or, $dv = f_0\left(1-\frac{t}{T}\right)dt$

$$\therefore v = \int dv = \int\left[f_0\left(1-\frac{t}{T}\right)\right]dt$$

or, $v = f_0\left(t - \frac{t^2}{2T}\right) + C$

where C is the constant of integration.
At $t = 0$, $v = 0$.

$$\therefore 0 = f_0\left(0 - \frac{0}{2T}\right) + C \Rightarrow C = 0$$

$$\therefore v = f_0\left(t - \frac{t^2}{2T}\right)$$

If $f = 0$, then

$$0 = f_0\left(1-\frac{t}{T}\right) \Rightarrow t = T$$

Hence, particle's velocity in the time interval $t = 0$ and $t = T$ is given by

$$v_x = \int_{t=0}^{t=T} dv = \int_{t=0}^{T}\left[f_0\left(1-\frac{t}{T}\right)\right]dt$$

$$= f_0\left[\left(t - \frac{t^2}{2T}\right)\right]_0^T$$

$$= f_0\left(T - \frac{T^2}{2T}\right) = f_0\left(T - \frac{T}{2}\right)$$

$$= \frac{1}{2}f_0T.$$

31. **(a)** Using $v^2 = u^2 - 2gh$ i.e., $h = \frac{u^2 - v^2}{2g}$,

$$AB = \frac{\left(\frac{u}{2}\right)^2 - \left(\frac{u}{3}\right)^2}{2g}$$

and $BC = \dfrac{\left(\frac{u}{3}\right)^2 - \left(\frac{u}{4}\right)^2}{2g}$

C u/4↑
B u/3↑
A u/2↑
O u ↑

$$\therefore \frac{AB}{BC} = \frac{\left(\frac{u}{2}\right)^2 - \left(\frac{u}{3}\right)^2}{\left(\frac{u}{3}\right)^2 - \left(\frac{u}{4}\right)^2} = \frac{\left(\frac{1}{2}\right)^2 - \left(\frac{1}{3}\right)^2}{\left(\frac{1}{3}\right)^2 - \left(\frac{1}{4}\right)^2} = \frac{20}{7}$$

32. **(a)** Velocity of boat $= \frac{8+8}{2} = 8\,\text{km h}^{-1}$

Velocity of water $= 4\,\text{km h}^{-1}$

$$t = \frac{8}{8-4} + \frac{8}{8+4} = \frac{8}{3}\text{h} = 160 \text{ minutes}$$

33. **(b)** $v_{av} = \frac{x + 2x + 3x}{t_1 + t_2 + t_3}$

$$t_1 = \frac{2x}{v_{max}},\ t_2 = \frac{2x}{v_{max}},\ t_3 = \frac{6x}{v_{max}}$$

$$v_{av} = \frac{6x\ v_{max}}{10x}$$

$$\frac{v_{av}}{v_{max}} = \frac{3}{5}$$

34. **(b)** No external force is acting, therefore,
$50u + 0.5 \times 2 = 0$
where u is the velocity of man.

$$u = -\frac{1}{50}\text{ms}^{-1}$$

Negative sign of u shows that man moves upward.
Time taken by the stone to reach the ground

$$= \frac{10}{2} = 5\text{S}$$

50 *kg*
$2\,ms^{-1}$ 0.5 *kg*
10 *m*

Distance moved by the man

$$= 5 \times \frac{1}{50} = 0.1\text{m}$$

$\therefore$ when the stone reaches the floor, the distance of the man above floor = 10.1 *m*

35. **(a)** Use $\vec{v}_{AB} = \vec{v}_A - \vec{v}_B$.

36. **(c)** Downward motion

$v^2 - 0^2 = 2 \times 9.8 \times 5$

$\Rightarrow\ v=\sqrt{98}=9.9$

Also for upward motion

$0^2-u^2=2\times(-9.8)\times1.8$

$\Rightarrow\ u=\sqrt{3528}=5.94$

Fractional loss $=\dfrac{9.9-5.94}{9.9}=0.4$

37. (c) Distance travelled by the stone in the last second is

$\dfrac{9h}{25}=\dfrac{g}{2}(2t-1)\quad(\because\ u=0)$ ...(i)

Distance travelled by the stone in t s is

$h=\dfrac{1}{2}gt^2\quad(\text{using } s=ut+\dfrac{1}{2}at^2)$ ...(ii)

Divide (i) by (ii), we get

$\dfrac{9}{25}=\dfrac{(2t-1)}{t^2}$

$9t^2=50t-25,\ 9t^2-50t+25=0$

Solving, we get

$t=5s$ or $t=\dfrac{5}{9}s$

Substituting $t=5s$ in (ii), we get

$h=\dfrac{1}{2}\times9.8\times(5)^2=122.5\text{ m}$

38. (b) $y\propto t^2; v\propto t'; a\propto t^\circ$

39. (b) Average velocity for the second half of the distance is

$=\dfrac{v_1+v_2}{2}=\dfrac{4+8}{2}=6\,ms^{-1}$

Given that first half distance is covered with a velocity of $6\,ms^{-1}$. Therefore, the average velocity for the whole time of motion is $6\,ms^{-1}$

40. (b) Bullet will take $\dfrac{100}{1000}=0.1$ sec to reach target.

During this period vertical distance (downward) travelled by the bullet

$=\dfrac{1}{2}gt^2=\dfrac{1}{2}\times10\times(0.1)^2=0.05\text{m}=5\text{cm}$

So the gun should be aimed 5 cm above the target.

41. (c) The distance covered in n^{th} second is

$S_n=u+\dfrac{1}{2}(2n-1)a$

where u is initial velocity & a is acceleration

then $26=u+\dfrac{19a}{2}$ ....(1)

$28=u+\dfrac{21a}{2}$ ....(2)

$30=u+\dfrac{23a}{2}$ ....(3)

$32=u+\dfrac{25a}{2}$ ....(4)

From eqs. (1) and (2) we get u = 7m/sec, a=2m/sec^2

∴ The body starts with initial velocity u =7m/sec and moves with uniform acceleration a = 2m/sec^2

42. (a) $8=\dfrac{x}{t_1},\ 12=\dfrac{x}{t_2}$

$\bar{v}=\dfrac{2x}{t_1+t_2}=\dfrac{2x}{\dfrac{x}{8}+\dfrac{x}{12}}=\dfrac{2\times8\times12}{12+8}=9.6\text{ ms}^{-1}$

43. (b) Distance = Area under v – t graph $=A_1+A_2+A_3+A_4$

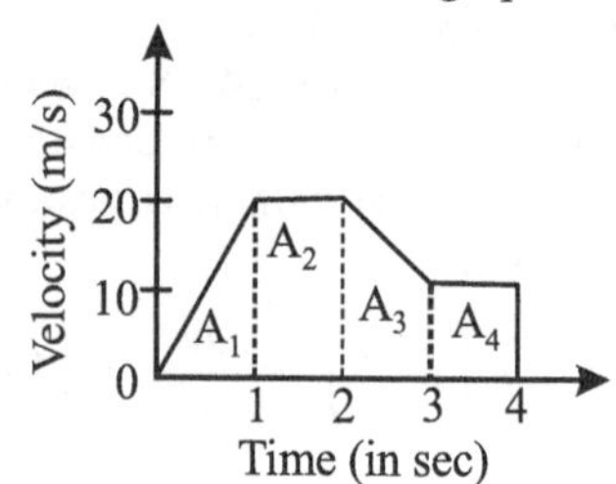

$=\dfrac{1}{2}\times1\times20+(20\times1)+\dfrac{1}{2}(20+10)\times1+(10\times1)$

$=10+20+15+10=55\text{ m}$

44. (a) $\because\ h=\dfrac{1}{2}gt^2$

$\therefore\ h_1=\dfrac{1}{2}g(5)^2=125$

$h_1+h_2=\dfrac{1}{2}g(10)^2=500$

$\Rightarrow\ h_2=375$

$h_1+h_2+h_3=\dfrac{1}{2}g(15)^2=1125$

$\Rightarrow\ h_3=625$

$h_2=3h_1,\ h_3=5h_1$

or $h_1=\dfrac{h_2}{3}=\dfrac{h_3}{5}$

45. (d) Distance from A to $B=S=\dfrac{1}{2}ft_1^2$

Distance from B to $C=(ft_1)t$

Distance from C to $D=\dfrac{u^2}{2a}=\dfrac{(ft_1)^2}{2(f/2)}=ft_1^2=2S$

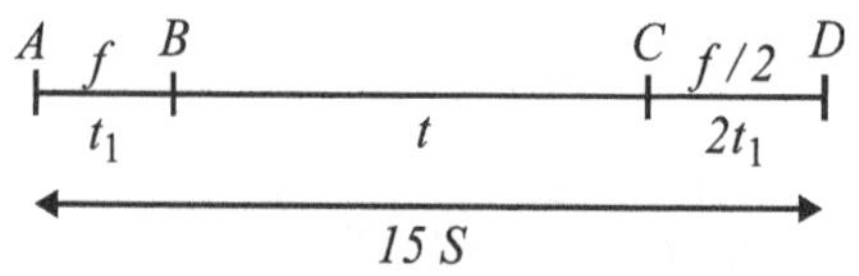

$\Rightarrow\ S+ft_1t+2S=15S$

$\Rightarrow\ ft_1t=12S$ .... (i)

$\dfrac{1}{2}ft_1^2=S$ .... (ii)

Dividing (i) by (ii), we get $t_1=\dfrac{t}{6}$

$\Rightarrow\ S=\dfrac{1}{2}f\left(\dfrac{t}{6}\right)^2=\dfrac{ft^2}{72}$

1. (b) $\vec{u}=\hat{i}+2\hat{j}=u_x\hat{i}+u_y\hat{j} \Rightarrow u\cos\theta=1,\ u\sin\theta=2$

$$y = x\tan\theta - \frac{1}{2}\frac{gx^2}{u_x^2}$$

$$\therefore\ y = 2x - \frac{1}{2}gx^2 = 2x - 5x^2$$

2. (c) $500\cos\theta = 250 \Rightarrow \cos\theta = \frac{1}{2}$

or $\theta = 60°$.

3. (c) As time periods are equal therefore ratio of angular speeds will be 1 : 1. $\left(\omega=\frac{2\pi}{T}\right)$.

4. (d)

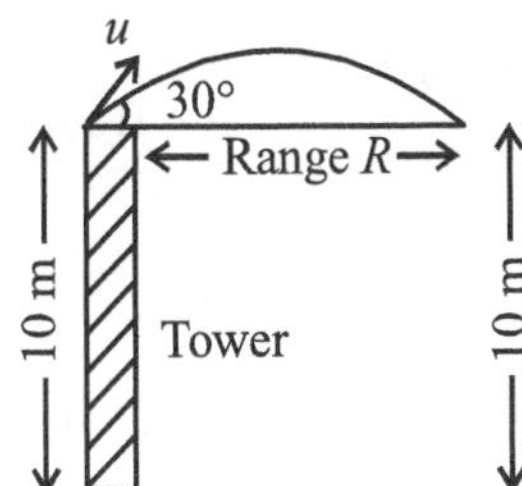

From the figure it is clear that range is required

$$R = \frac{u^2 \sin 2\theta}{g} = \frac{(10)^2 \sin(2\times 30°)}{10} = 5\sqrt{3} = 8.66 \text{ m}$$

5. (a) Horizontal component of velocity v_x = 500 m/s and vertical component of velocity while striking the ground.

$u_v = 0 + 10 \times 10 = 100$ m/s

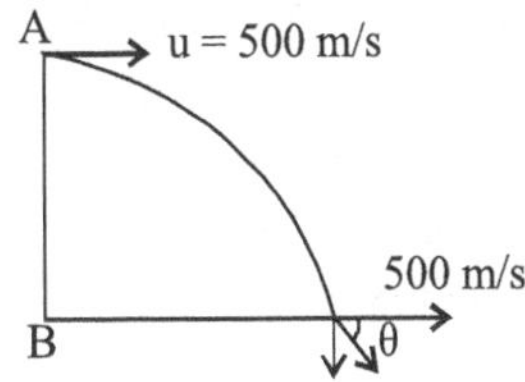

$\therefore$ Angle with which it strikes the ground

$$\theta = \tan^{-1}\left(\frac{u_v}{u_x}\right) = \tan^{-1}\left(\frac{100}{500}\right) = \tan^{-1}\left(\frac{1}{5}\right)$$

6. (b)

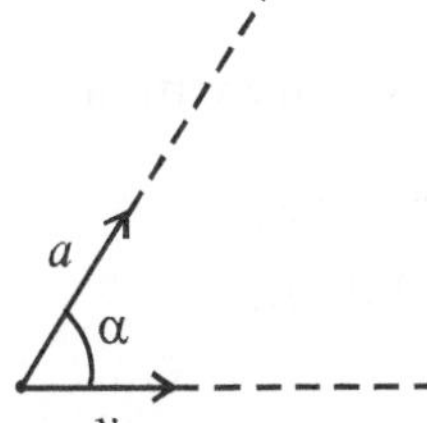

The velocity of first particle, $v_1 = v$

The velocity of second particle, $v_2 = at$

Relative velocity, $\vec{v}_{12} = \vec{v}_1 - \vec{v}_2$

or $v_{12}^2 = v^2 + (at)^2 - 2v(at\cos\alpha)$

For least value of relative velocity, $\frac{dv_{12}}{dt} = 0$

or $\frac{d}{dt}\left[v^2 + a^2t^2 - 2vat\cos\alpha\right] = 0$

or $0 + a^2 \times 2t - 2va\cos\alpha = 0$

or $t = \frac{v\cos\alpha}{a}$

7. (d) $t = \frac{2u\sin 30°}{g\cos 30°} = \frac{2(10)\,(1/2)}{10\,(\sqrt{3}/2)} = \frac{2}{\sqrt{3}}$ sec

$$R = 10\cos 30°\, t - \frac{1}{2}\, g\sin 30°\, t^2$$

$$= \frac{10\sqrt{3}}{2}\left(\frac{2}{\sqrt{3}}\right) - \frac{1}{2}(10)\left(\frac{1}{2}\right)\frac{4}{3} = 10 - \frac{10}{3} = \frac{20}{3}\text{ m}$$

8. (b) $\overline{AB} = (4\hat{i} + 5\hat{j} + 6\hat{k}) - (3\hat{i} + 4\hat{j} + 5\hat{k}) = \hat{i} + \hat{j} + \hat{k}$

$\overline{CD} = (4\hat{i} + 6\hat{j}) - (7\hat{i} + 9\hat{j} + 3\hat{k}) = 3\hat{i} - 3\hat{j} + 3\hat{k}$

$\overline{AB}$ and $\overline{CD}$ are parallel, because its cross-product is 0.

9. (c) Here v = 0.5 m/sec. u = ?

so $\sin\theta = \frac{u}{v} \Rightarrow \frac{u}{.5} = \frac{1}{2}$ or u = 0.25 ms^{-1}

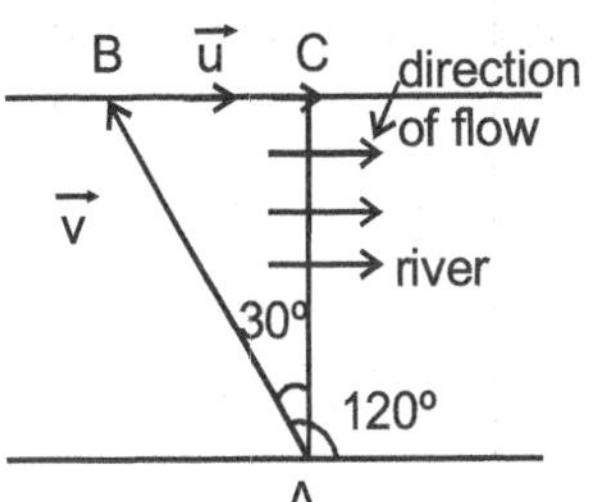

10. (d) Max. height = $H = \frac{v^2\sin^2(90-\theta)}{2g}$ .....(i)

Time of flight, $T = \frac{2v\sin(90-\theta)}{g}$ ...(ii)

From (i), $\frac{v\cos\theta}{g} = \sqrt{\frac{2H}{g}}$

From (ii), $T = 2\sqrt{\frac{2H}{g}} = \sqrt{\frac{8H}{g}}$

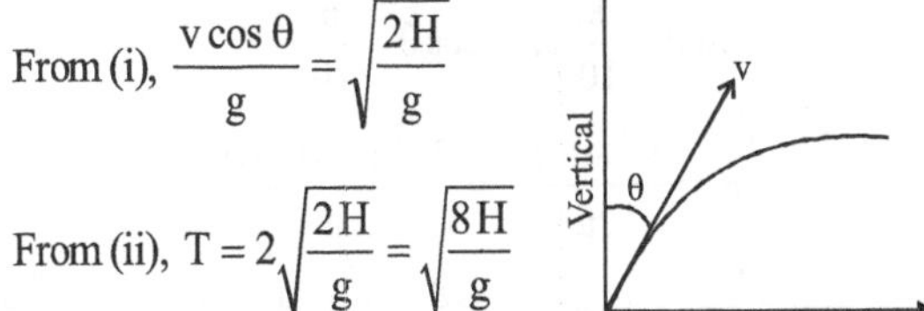

11. (c) Yes, the person can catch the ball when horizontal velocity is equal to the horizontal component of ball's velocity, the motion of ball will be only in vertical direction with respect to person for that,

$$\frac{v_o}{2} = v_o \cos\theta \text{ or } \theta = 60°$$

12. (b) Two vectors are

$\vec{A} = \cos\omega t\hat{i} + \sin\omega t\hat{j}$

$\vec{B} = \cos\frac{\omega t}{2}\hat{i} + \sin\frac{\omega t}{2}\hat{j}$

For two vectors $\vec{A}$ and $\vec{B}$ to be orthogonal A.B = 0

$\vec{A}.\vec{B} = 0 = \cos\omega t.\cos\frac{\omega t}{2} + \sin\omega t.\sin\frac{\omega t}{2}$

$= \cos\left(\omega t - \frac{\omega t}{2}\right) = \cos\left(\frac{\omega t}{2}\right)$

So, $\frac{\omega t}{2} = \frac{\pi}{2}$ $\quad\therefore\ t = \frac{\pi}{\omega}$

13. (a) $\vec{v_1} = 50\,\text{km h}^{-1}$ due North;

$\vec{v_2} = 50\,\text{km h}^{-1}$ due West. Angle between $\vec{v_1}$ and $\vec{v_2} = 90°$

$-\vec{v_1} = 50\ \text{km h}^{-1}$ due South

∴ Change in velocity

$= |\vec{v_2} - \vec{v_1}| = |\vec{v_2} + (-\vec{v_1})|$

$= \sqrt{v_2^2 + v_1^2} = \sqrt{50^2 + 50^2} = 70.7$ km/h

The direction of this change in velocity is in South-West.

14. (b) $\vec{v} = 6\hat{i} + 8\hat{j}$

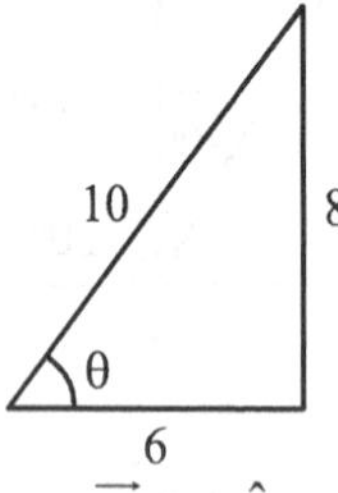

Comparing with $\vec{v} = v_x\hat{i} + v_y\hat{j}$, we get $v_x = 6\text{ms}^{-1}$ and $v_y = 8\ \text{ms}^{-1}$

Also, $v^2 = v_x^2 + v_y^2 = 36 + 64 = 100$ or $v = 10\ \text{ms}^{-1}$

$\sin\theta = \frac{8}{10}$ and $\cos\theta = \frac{6}{10}$

$R = \frac{v^2\sin 2\theta}{g} = \frac{2v^2\sin\theta\cos\theta}{g}$

$R = 2\times10\times10\times\frac{8}{10}\times\frac{6}{10}\times\frac{1}{10} = 9.6\text{m}$

15. (d) $s = t^3 + 5$

$\Rightarrow$ velocity, $v = \frac{ds}{dt} = 3t^2$

Tangential acceleration $a_t = \frac{dv}{dt} = 6t$

Radial acceleration $a_c = \frac{v^2}{R} = \frac{9t^4}{R}$

At $t = 2s$, $a_t = 6\times2 = 12\ \text{m/s}^2$

$a_c = \frac{9\times16}{20} = 7.2\ \text{m/s}^2$

∴ Resultant acceleration

$= \sqrt{a_t^2 + a_c^2} = \sqrt{(12)^2 + (7.2)^2} = \sqrt{144 + 51.84}$

$= \sqrt{195.84} = 14\ \text{m/s}^2$

16. (b) $\frac{B}{2} = \sqrt{A^2 + B^2 + 2AB\cos\theta}$ ...... (i)

$\therefore\ \tan 90° = \frac{B\sin\theta}{A + B\cos\theta} \Rightarrow A + B\cos\theta = 0$

$\therefore\ \cos\theta = -\frac{A}{B}$

Hence, from (i) $\frac{B^2}{A} = A^2 + B^2 - 2A^2 \Rightarrow A = \sqrt{3}\frac{B}{2}$

$\Rightarrow\ \cos\theta = -\frac{A}{B} = -\frac{\sqrt{3}}{2}$ $\quad\therefore\ \theta = 150°$

17. (b) Suppose velocity of rain

$\vec{v}_R = v_x\hat{i} - v_y\hat{j}$

and the velocity of the man

$\vec{v}_m = u\hat{i}$

∴ Velocity of rain relative to man

$\vec{v}_{Rm} = \vec{v}_R - \vec{v}_m = (v_x - u)\hat{i} - v_y\hat{j}$

According to given condition that rain appears to fall vertically, so $(v_x - u)$ must be zero.

∴ $v_x - u = 0$ or $v_x = u$

When he doubles his speed,

$\vec{v'}_m = 2u\hat{i}$

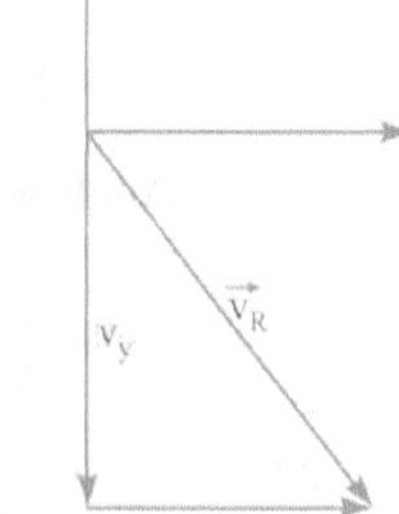

Now $\vec{v}_{Rm} = \vec{v}_R - \vec{v'}_m$

$= (v_x\hat{i} - v_y\hat{j}) - (2u\hat{i})$

$= (v_x - 2u)\hat{i} - v_y\hat{j}$

The $\vec{v}_{Rm}$ makes an angle θ with the vertical

$\tan\theta = \frac{x - \text{componend of } \vec{v}_{Rm}}{y - \text{componend of } \vec{v}_{Rm}}$

$= \frac{(v_x - 2u)}{-v_y}$

$= \frac{u - 2u}{-v_y}$

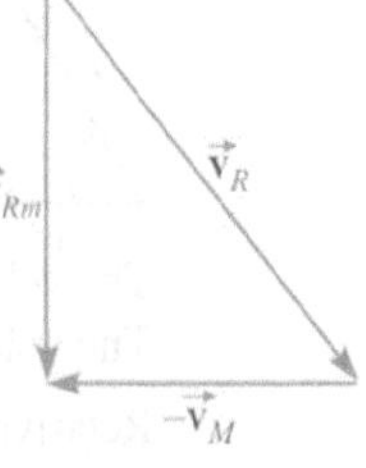

which gives

$$v_y = \frac{u}{\tan\theta}$$

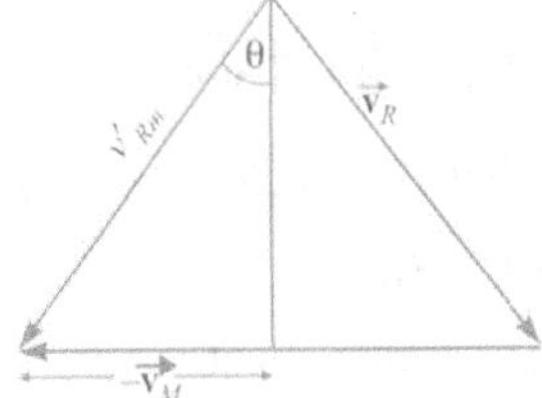

Thus the velocity of rain

$$\vec{v}_R = v_x\hat{i} - v_y\hat{i}$$

$$= u\hat{i} - \frac{u}{\tan\theta}\hat{j}.$$

18. (c) For projectile A

Maximum height, $H_A = \frac{u_A^2 \sin^2 45°}{2g}$

For projectile B

Maximum height, $H_B = \frac{u_B^2 \sin^2 \theta}{2g}$

As we know, $H_A = H_B$

$$\frac{u_A^2 \sin^2 45°}{2g} = \frac{u_B^2 \sin^2 \theta}{2g}$$

$$\frac{\sin^2\theta}{\sin^2 45°} = \frac{u_A^2}{u_B^2}$$

$$\sin^2\theta = \left(\frac{u_A}{u_B}\right)^2 \sin^2 45°$$

$$\sin^2\theta = \left(\frac{1}{\sqrt{2}}\right)^2\left(\frac{1}{\sqrt{2}}\right)^2 = \frac{1}{4}$$

$$\sin\theta = \frac{1}{2} \Rightarrow \theta = \sin^{-1}\left(\frac{1}{2}\right) = 30°$$

19. (a) The angle for which the ranges are same is complementary.

Let one angle be θ, then other is 90° – θ

$$T_1 = \frac{2u\sin\theta}{g}, T_2 = \frac{2u\cos\theta}{g}$$

$$T_1T_2 = \frac{4u^2\sin\theta\cos\theta}{g} = 2R \quad \left(\because R = \frac{u^2\sin^2\theta}{g}\right)$$

Hence it is proportional to R.

20. (c) When particle thrown in vertically downward direction with velocity u then final velocity at the ground level

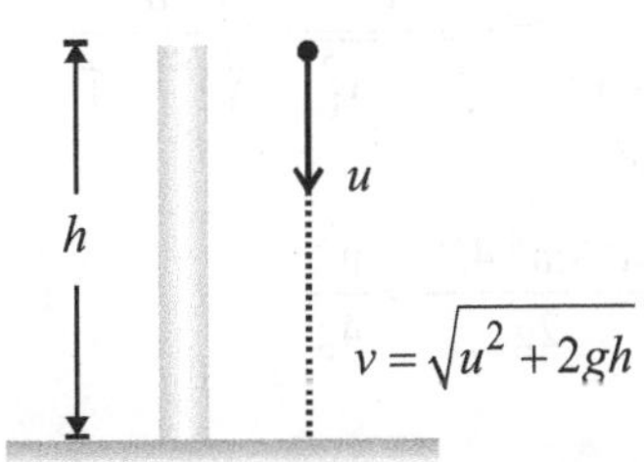

$$v^2 = u^2 + 2gh \; \therefore \; v = \sqrt{u^2 + 2gh}$$

Another particle is thrown horizontally with same velocity then velocity of particle at the surface of earth.

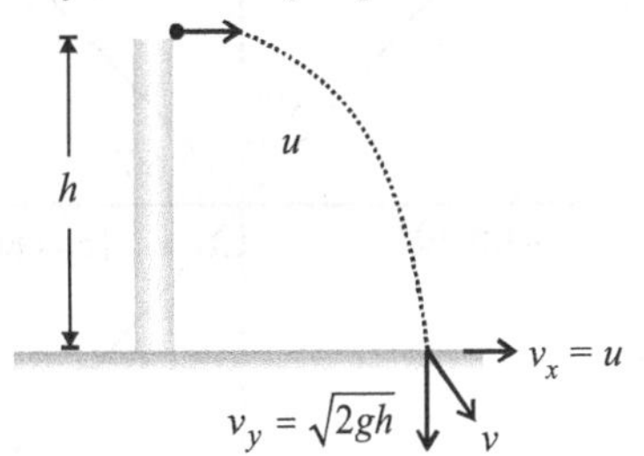

Horizontal component of velocity $v_x = u$

$\therefore$ Resultant velocity, $v = \sqrt{u^2 + 2gh}$

For both the particles, final velocities when they reach the earth's surface are equal.

21. (b) $\hat{r} = 0.5\hat{i} + 0.8\hat{j} + c\hat{k}$

$$|\hat{r}| = 1 = \sqrt{(0.5)^2 + (0.8)^2 + c^2}$$

$$(0.5)^2 + (0.8)^2 + c^2 = 1$$

$$c^2 = 0.11 \Rightarrow c = \sqrt{0.11}$$

22. (c) The pilot will see the ball falling in straight line because the reference frame is moving with the same horizontal velocity but the observer at rest will see the ball falling in parabolic path.

23. (a) R = 2H (given)

We know, $R = 4H\cot\theta \Rightarrow \cot\theta = \frac{1}{2}$

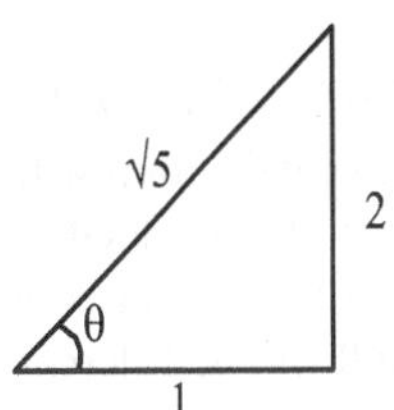

From triangle we can say that $\sin\theta = \frac{2}{\sqrt{5}}, \cos\theta = \frac{1}{\sqrt{5}}$

$\therefore$ Range of projectile $R = \frac{2v^2\sin\theta\cos\theta}{g}$

$$= \frac{2v^2}{g} \times \frac{2}{\sqrt{5}} \times \frac{1}{\sqrt{5}} = \frac{4v^2}{5g}$$

24. (a) Note that the given angles of projection add upto 90°. For complementary angles of projection (45° + α) and (45° – α) with same initial velocity u, range R is same.

$\theta_1 + \theta_2 = (45° + \alpha) + (45° - \alpha) = 90°$

So, the ratio of horizontal ranges is 1 : 1.

25. (a) The components of 1 N and 2N forces along + x axis = 1 cos 60° + 2 sin 30°

$$-1\times\frac{1}{2} + 2\times\frac{1}{2} - \frac{1}{2} + 1 - \frac{3}{2} = 1.5\text{N}$$

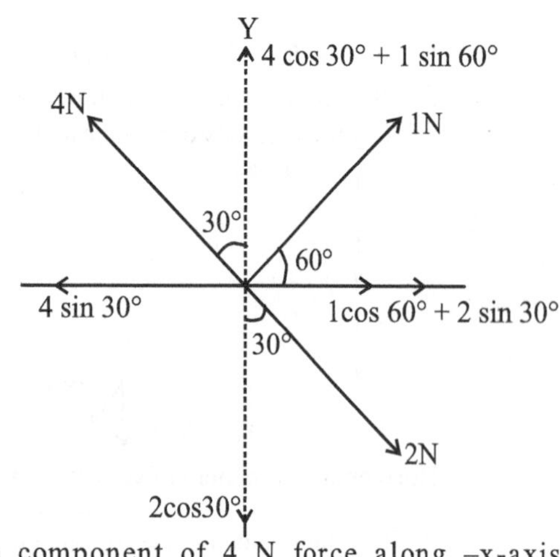

The component of 4 N force along –x-axis

$= 4\sin 30° = 4 \times \frac{1}{2} = 2N$.

Therefore, if a force of 0.5N is applied along + x-axis, the resultant force along x-axis will become zero and the resultant force will be obtained only along y-axis.

26. **(d)** $F_x = \frac{dp_x}{dt} = -2\sin\theta.$

Similarly, $F_y = \frac{dp_y}{dx} = 2\cos\theta.$

Angle θ between two vectors

$$\cos\theta = \frac{F_x p_x + F_y p_y}{|\vec{F}||\vec{p}|}$$

$$= \frac{(-2\sin\theta)(2\cos\theta) + (2\cos\theta)(2\sin\theta)}{|\vec{F}||\vec{p}|}$$

$\Rightarrow \cos\theta = 0 \Rightarrow \theta = 90°$

27. **(a)** The motion of the train will affect only the horizontal component of the velocity of the ball. Since, vertical component is same for both observers, the y_m will be same, but R will be different.

28. **(d)** As body covers equal angle in equal time intervals. Its angular velocity and hence magnitude of linear velocity is constant.

29. **(a)** **For A:** It goes up with velocity *u* will it reaches its maximum height (i.e. velocity becomes zero) and comes back to O and attains velocity *u*.

Using $v^2 = u^2 + 2as \Rightarrow v_A = \sqrt{u^2 + 2gh}$

For B, going down with velocity u

$\Rightarrow v_B = \sqrt{u^2 + 2gh}$

For C, horizontal velocity remains same, i.e. u. Vertical velocity $= \sqrt{0 + 2gh} = \sqrt{2gh}$

The resultant $v_C = \sqrt{v_x^2 + v_y^2} = \sqrt{u^2 + 2gh}$.

Hence $v_A = v_B = v_C$

30. **(d)** $\vec{v}_{av} = \frac{\Delta\vec{r}\text{ (displacement)}}{\Delta t\text{ (time taken)}}$

$$= \frac{(13-2)\hat{i} + (14-3)\hat{j}}{5-0} = \frac{11}{5}(\hat{i} + \hat{j})$$

31. **(c)** Position vector

$\vec{r} = \cos wt\ \hat{x} + \sin \omega t\ \hat{y}$

$\therefore$ Velocity, $\vec{v} = -\omega\sin\omega t\ \hat{x} + \omega\cos\omega t\ \hat{y}$

and acceleration,

$\vec{a} = -\omega^2\cos\omega t\ \hat{x} + \omega\sin\omega t\ \hat{y} = -\omega 2\ \vec{r}$

$\vec{r}\cdot\vec{v} = 0$ hence $\vec{r} \perp \vec{v}$ and

$\vec{a}$ is directed towards the origin.

32. **(d)**

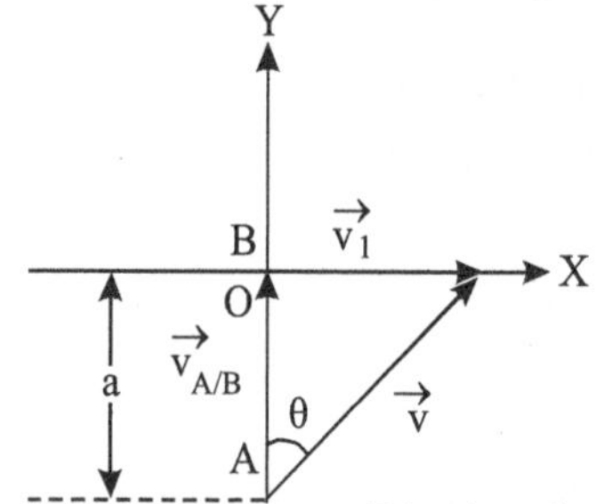

Velocity of A relative to *B* is given by

$\vec{v}_{A/B} = \vec{v}_A - \vec{v}_B = \vec{v} - \vec{v}_1$ .... (1)

By taking x-components of equation (1), we get

$0 = v\sin\theta - v_1 \Rightarrow \sin\theta = \frac{v_1}{v}$ .... (2)

By taking *Y*-components of equation (1), we get

$v_y = v\cos\theta$ .....(3)

Time taken by boy at *A* to catch the boy at *B* is given by

$$t = \frac{\text{Relative displacement along Y - axis}}{\text{Relative velocity along Y - axis}}$$

$$= \frac{a}{v\cos\theta} = \frac{a}{v.\sqrt{1-\sin^2\theta}} = \frac{a}{v.\sqrt{1-\left(\frac{v_1}{v}\right)^2}}$$

[From equation (1)]

$$= \frac{a}{v.\sqrt{\frac{v^2 - v_1^2}{v^2}}} = \frac{a}{\sqrt{v^2 - v_1^2}} = \sqrt{\frac{a^2}{v^2 - v_1^2}}$$

33. **(b)** $H = \frac{u^2\sin^2 45°}{2g} = \frac{u^2}{4g}$ ...(1)

$$R = \frac{u^2 \sin 90°}{g} = \frac{u^2}{g}$$

$$\therefore \frac{R}{2} = \frac{u^2}{2g} \qquad ...(2)$$

$$\therefore \tan\alpha = \frac{H}{R/2}$$

$$= \frac{\frac{u^2}{4g}}{\frac{u^2}{2g}} = \frac{1}{2} \qquad \therefore \alpha = \tan^{-1}\left(\frac{1}{2}\right)$$

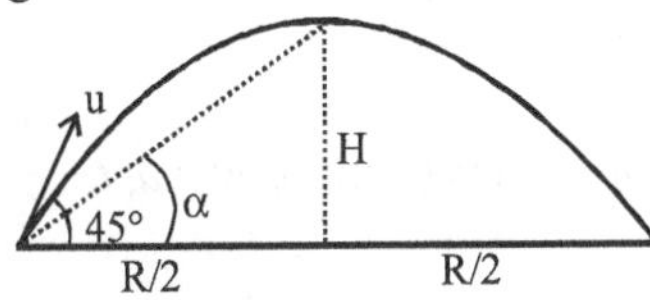

34. (b) Here, $x = 4\sin(2\pi t)$ ...(*i*)

$y = 4\cos(2\pi t)$ ...(*ii*)

Squaring and adding equation (*i*) and (*ii*)

$x^2 + y^2 = 4^2 \Rightarrow R = 4$

Motion of the particle is circular motion, acceleration vector is along $-\vec{R}$ and its magnitude $= \frac{v^2}{R}$

Velocity of particle, $v = \omega R = (2\pi)(4) = 8\pi$

35. (d) $|\vec{A}+\vec{B}|^2 = |\vec{A}-\vec{B}|^2$

$|\vec{A}+\vec{B}|^2 = |\vec{A}|^2 + |\vec{B}|^2 + 2\vec{A}.\vec{B} = A^2 + B^2 + 2AB\cos\theta$

$|\vec{A}-\vec{B}|^2 = |\vec{A}|^2 + |\vec{B}|^2 - 2\vec{A}.\vec{B}$

$= A^2 + B^2 - 2AB\cos\theta$

So, $A^2 + B^2 + 2AB\cos\theta$

$= A^2 + B^2 - 2AB\cos\theta$

$4AB\cos\theta = 0 \Rightarrow \cos\theta = 0$

$\therefore \theta = 90°$

So, angle between *A* & *B* is 90°.

36. (b) $\vec{v} = 6\hat{i} + 8\hat{j}$

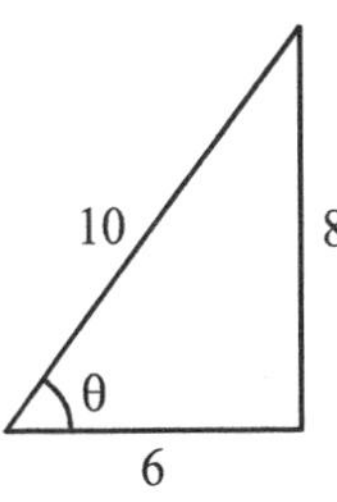

Comparing with $\vec{v} = v_x\hat{i} + v_y\hat{j}$, we get

$v_x = 6\,ms^{-1}$ and $v_y = 8\,ms^{-1}$

Also, $v^2 = {v_x}^2 + {v_y}^2 = 36 + 64 = 100$

or $v = 10\ ms^{-1}$

$\sin\theta = \frac{8}{10}$ and $\cos\theta = \frac{6}{10}$

$$R = \frac{v^2 \sin 2\theta}{g} = \frac{2v^2 \sin\theta\cos\theta}{g}$$

$$R = 2 \times 10 \times 10 \times \frac{8}{10} \times \frac{6}{10} \times \frac{1}{10} = 9.6\,m$$

37. (a) Range of a projectile is maximum when it is projected at an angle of 45° and is given by

$R_{max} = \frac{u^2}{g}$, where u is the velocity of projection

$\Rightarrow R = \frac{u^2}{g} \qquad \therefore u^2 = Rg \quad ...(i)$

Now, to hit a target at a distance (R/2) from the gun, we must have

$\frac{R}{2} = \frac{u^2 \sin 2\theta}{g}$, where θ is the angle of projection.

$\Rightarrow \frac{R}{2} = \frac{Rg\sin 2\theta}{g}$; from (i)

$\Rightarrow \sin 2\theta = \frac{1}{2} \Rightarrow \sin 2\theta = \sin 30°$

$\Rightarrow 2\theta = 30° \quad \therefore \theta = 15°$

38. (a) Distance covered in one circular loop $= 2\pi r$

$= 2 \times 3.14 \times 100 = 628\,m$

Speed $= \frac{628}{62.8} = 10\ m/sec$

Displacement in one circular loop = 0

Velocity $= \frac{0}{time} = 0$

39. (a) $\overrightarrow{PQ} + \overrightarrow{QR} = \overrightarrow{PR}$

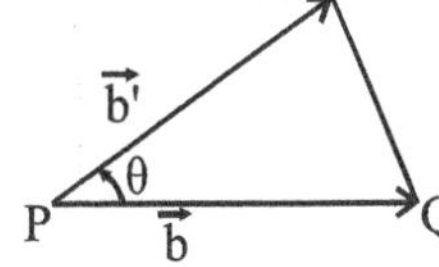

$\therefore \overrightarrow{QR} = \vec{b'} - \vec{b}$

Now $|\vec{b'} - \vec{b}|^2 = (\vec{b'} - \vec{b}).(\vec{b'} - \vec{b})$

$= b'^2 - 2bb'\cos\theta + b^2$

$= 2b^2(1 - \cos\theta) \qquad [\because b' = b]$

$\vec{b'} - \vec{b} = \sqrt{2}b\sqrt{1-\cos\theta}$

$= \sqrt{2}b\left(\sqrt{2}\sin\frac{\theta}{2}\right) = 2b\sin\frac{\theta}{2}$

40. (a) $H_1 = \frac{u^2\sin^2\theta}{2g}$

and $H_2 = \frac{u^2\sin^2(90° - \theta)}{2g} = \frac{u^2\cos^2\theta}{2g}$

$H_1H_2 = \frac{u^2\sin^2\theta}{2g} \times \frac{u^2\cos^2\theta}{2g} = \frac{(u^2\sin 2\theta)^2}{16g^2} = \frac{R^2}{16}$

$\therefore R = 4\sqrt{H_1H_2}$

41. (c) $\vec{P}$ = vector sum $= \vec{A} + \vec{B}$

$\vec{Q}$ = vector differences $= \vec{A} - \vec{B}$

Since $\vec{P}$ and $\vec{Q}$ are perpendicular

$\therefore \vec{P}.\vec{Q}=0$

$\Rightarrow (\vec{A}+\vec{B}).(\vec{A}-\vec{B})=0 \Rightarrow A^2=B^2 \Rightarrow |A|=|B|$

42. (b) $y=bx^2$

Differentiating w.r.t to t an both sides, we get

$$\frac{dy}{dx}=b2x\frac{dx}{dt}$$

$v_y=2bxv_x$

Again differentiating w.r.t to t on both sides we get

$$\frac{dv_y}{dt}=2bv_x\frac{dx}{dt}+2bx\frac{dv_x}{dt}=2bv_x^2+0$$

[$\frac{dv_x}{dt}$ = 0, because the particle has constant acceleration along y-direction]

Now, $\frac{dv_y}{dt}=a=2bv_x^2$;

$$v_x^2=\frac{a}{2b}$$

$$v_x=\sqrt{\frac{a}{2b}}$$

43. (a) Arc length = radius × angle

So, $|\vec{B}-\vec{A}|=|\vec{A}|\Delta\theta$

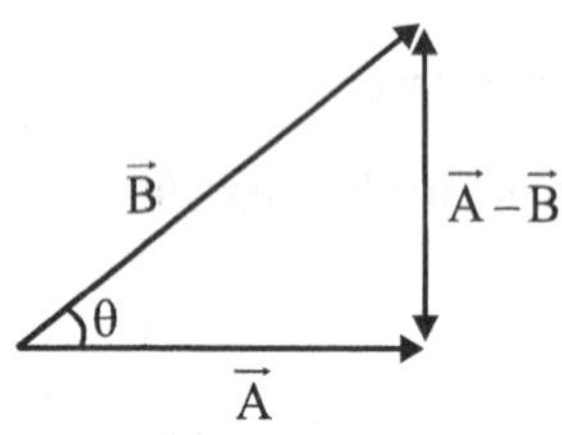

44. (c) Speed, V = constant (from question)

Centripetal acceleration,

$$a=\frac{V^2}{r}$$

ra = constant

Hence graph (c) correctly describes relation between acceleration and radius.

45. (c) From question,

Horizontal velocity (initial),

$$u_x=\frac{40}{2}=20\text{m/s}$$

Vertical velocity (initial), $50=u_y t+\frac{1}{2}gt^2$

$\Rightarrow \quad u_y\times 2+\frac{1}{2}(-10)\times 4$

or, $\quad 50=2u_y-20$

or, $\quad u_y=\frac{70}{2}=35\text{m/s}$

$\therefore \quad \tan\theta=\frac{u_y}{u_x}=\frac{35}{20}=\frac{7}{4}$

$\Rightarrow \quad$ Angle $\theta=\tan^{-1}\frac{7}{4}$

1. (d) Here m = 0.5 kg ; u = – 10 m/s;
t = 1/50 s ; v = + 15 ms^{-1}
Force = m (v– u)/t = 0.5 (10 + 15) × 50 = 625 N

2. (b)

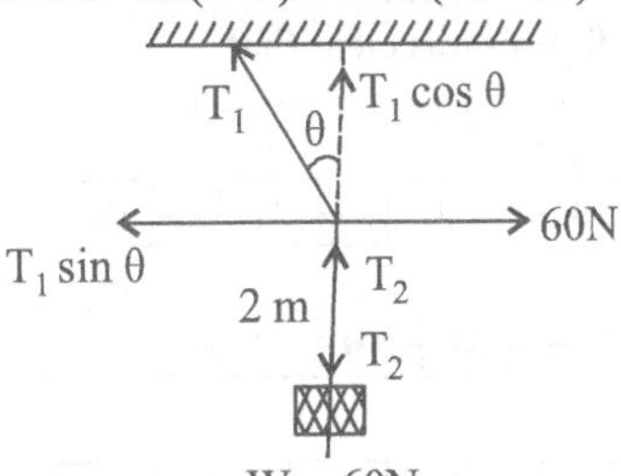

In eqbm $T_1 \cos\theta = T_2 = 60N$. ...(1)
$T_1 \sin\theta = 60$ N ...(2)
∴ $\tan\theta = 1$
$\theta = 45°$.

3. (d) Mass of rocket (m) = 5000 Kg
Exhaust speed (v) = 800 m/s
Acceleration of rocket (a) = 20 m/s^2
Gravitational acceleration (g) = 10 m/s^2
We know that upward force
F = m (g + a) = 5000 (10 +20)
= 5000 × 30 = 150000 N.
We also know that amount of gas ejected

$$\left(\frac{dm}{dt}\right) = \frac{F}{v} = \frac{150000}{800} = 187.5 \text{ kg/s}$$

4. (b) (i) If a body is moved up an inclined plane, then the work done against friction will be zero as there is no friction. But work must be done against gravity. So this statement is incorrect.
(ii) This statement is correct, because moving vehicles are stopped by air friction only.
(iii) The normal reaction acting on a body on an inclined plane is given by,
R = mg cosθ
Where θ is the angle of inclination.
As θ increases, cos θ decreases and hence R decreases. So this statement is incorrect.
(iv) The applied force needed to rub the duster upward,

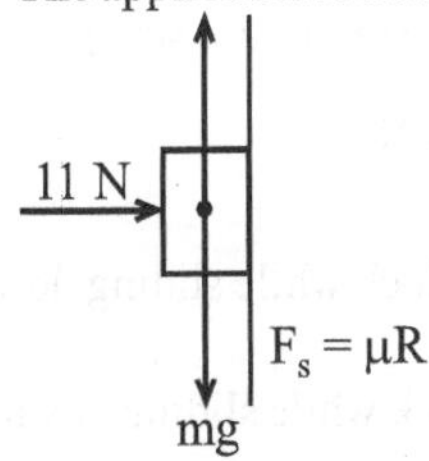

$F_{applied} = mg + \mu R = 0.5 \times 10 + 0.5 \times 11$
$= 5 + 5.5 = 10.5$ N
∴ The work done in rubbing it upward through a distance of 10 cm,
$W = F_{applied} \times d = 10.5 \times 0.10 = 1.05$ J
Hence this statement is incorrect.

5. (b) Momentum $P = mv = m\sqrt{2gh}$
($\because v^2 = u^2 + 2gh$; Here $u = 0$)
When stone hits the ground momentum
$P = m\sqrt{2gh}$
when same stone dropped from $2h$ (100% of initial) then momentum
$P' = m\sqrt{2g(2h)} = \sqrt{2}P$
Which is changed by 41% of initial.

6. (c) Change in momentum along the wall
$= mv \cos 60° - mv \cos 60° = 0$
Change in momentum perpendicular to the wall
$= mv \sin 60° - (-mv \sin 60°) = 2mv \sin 60°$

$$\therefore \text{Applied force} = \frac{\text{Change in momentum}}{\text{Time}}$$

$$= \frac{2\,mv \sin 60°}{0.20}$$

$$= \frac{2 \times 3 \times 10 \times \sqrt{3}}{2 \times 20} = 50 \times 3\sqrt{3}$$

$= 150\sqrt{3}$ newton

7. (b)

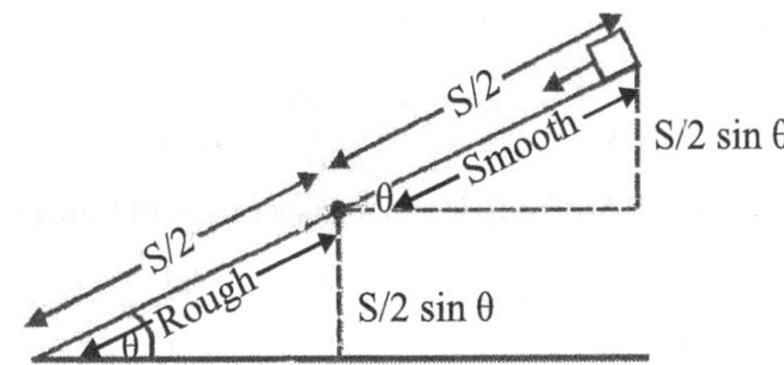

For upper half of inclined plane
$v^2 = u^2 + 2a\, S/2 = 2\,(g \sin\theta)\, S/2 = gS \sin\theta$
For lower half of inclined plane
$0 = u^2 + 2\,g\,(\sin\theta - \mu \cos\theta)\, S/2$
$\Rightarrow -gS \sin\theta = gS\,(\sin\theta - \mu \cos\theta)$
$\Rightarrow 2 \sin\theta = \mu \cos\theta$

$$\Rightarrow \mu = \frac{2 \sin\theta}{\cos\theta} = 2 \tan\theta$$

8. (c) Forces acting on the block are as shown in the fig. Normal reaction N is provided by the force $m\alpha$ due to acceleration α

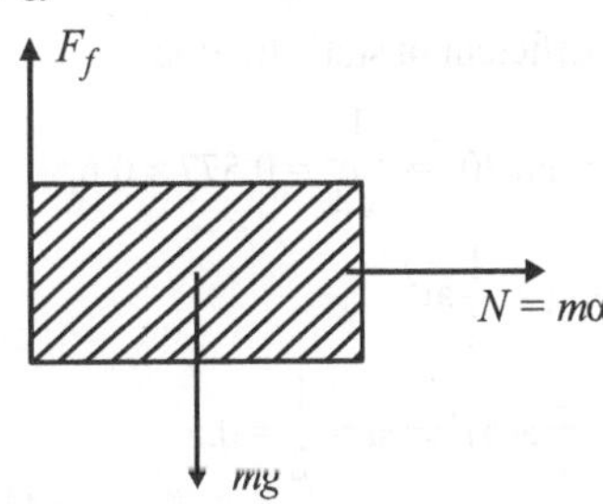

$\therefore$ N = m α

For the block not to fall, frictional force, $F_f \geq mg$

$\Rightarrow \mu N \geq mg$

$\Rightarrow \mu m\alpha \geq mg$

$\Rightarrow \alpha \geq g/\mu$

9. (b) $v = \sqrt{gr} = \sqrt{10 \times 40} = 20 \text{ m s}^{-1}$

10. (a)

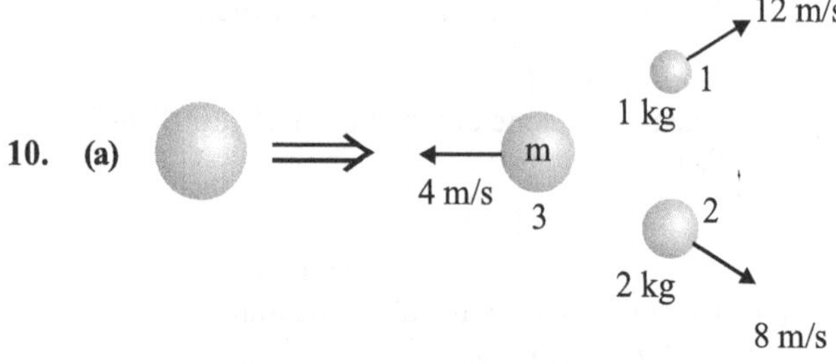

According to conservation of linear momentum

$P_3 = \sqrt{p_1^2 + p_2^2}$

$\Rightarrow m \times 4 = \sqrt{(1\times12)^2 + (2\times8)^2} = 20 \Rightarrow m = 5$ kg.

11. (c) Let T be the tension in the branch of a tree when monkey is descending with acceleration a. Then mg – T = ma; and T = 75% of weight of monkey

$= \left(\frac{75}{100}\right)mg = \left(\frac{1}{4}\right)mg$ or $a = \frac{g}{4}$.

12. (d) $v = u - at \Rightarrow t = \frac{u}{a}$ [As v = 0]

$t = \frac{u \times m}{F} = \frac{30 \times 1000}{5000} = 6\text{sec}$

13. (c) Applying law of conservation of linear momentum

$m_1v_1 + m_2v_2 = 0, \frac{m_1}{m_2} = -\frac{v_2}{v_1}$ or $\frac{v_1}{v_2} = -\frac{m_2}{m_1}$

14. (a) The frictional force acting on M is μmg

$\therefore$ Acceleration $= \frac{\mu mg}{M}$

15. (c) $\frac{dM}{dt} = 0.1$ kg/s, $v_{gas} = 50$ m/s,

Mass of the rocket = 2 kg. Mv = constant

$-v\frac{dM}{dt} + M\frac{dv}{dt} = 0$. $\therefore \frac{dv}{dt} = \frac{1}{M}v\frac{dM}{dt}$

$\Rightarrow$ Acceleration $= \frac{1}{2} \times 50 \times 0.1 = 2.5 \text{ m/s}^2$

16. (a) Coefficient of static friction,

$\mu_s = \tan 30° = \frac{1}{\sqrt{3}} = 0.577 \cong 0.6$

$S = ut + \frac{1}{2}at^2$

$4 = \frac{1}{2}a(4)^2 \Rightarrow a = \frac{1}{2} = 0.5$

[$\because$ s = 4m and t = 4s given]

$a = g\sin\theta - \mu_k(g)\cos\theta$

$\Rightarrow \mu_k = \frac{0.9}{\sqrt{3}} = 0.5$

17. (c) All blocks will move with the same aceleration Let it be a. Then

$F = 4Ma \Rightarrow a = \frac{F}{4M}$

From the figures it is clear that

$T_1 = 3\,Ma, T_2 = 2\,Ma$ and $T_3 = Ma$

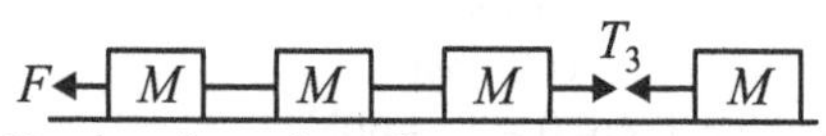

F ← M — M →T₂← M — M

F ← M — M — M →T₃← M

Putting the value of a, we get

$T_1 = \frac{3}{4}F,\ T_2 = \frac{F}{2}$ and $T_3 = \frac{F}{4}$

18. (c) Maximum force by surface when friction works

$F = \sqrt{f^2 + R^2} = \sqrt{(\mu R)^2 + R^2} = R\sqrt{\mu^2 + 1}$

Minimum force = R when there is no friction

Hence ranging from R to $R\sqrt{\mu^2 + 1}$ [where, R = mg]

19. (c) Motion with constant momentum along a straight line. According to Newton's second law rate of change of momentum is directly proportional to force applied.

20. (b) For the motion of both the blocks

$m_1a = T - \mu_k m_1 g$

$m_2 g - T = m_2 a$

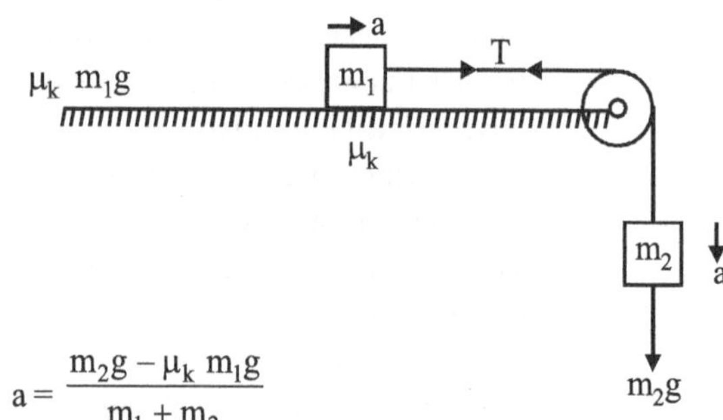

$a = \frac{m_2 g - \mu_k m_1 g}{m_1 + m_2}$

$m_2 g - T = (m_2)\left(\frac{m_2 g - \mu_k m_1 g}{m_1 + m_2}\right)$

solving we get tension in the string

$T = \frac{m_1 m_2 (1 + \mu_k)g}{m_1 + m_2}$

21. (d) Acceleration of block while sliding down upper half = g sin ϕ;

retardation of block while sliding down lower half = –$(g\sin\phi - \mu g\cos\phi)$

For the block to come to rest at the bottom, acceleration in I half = retardation in II half.

$g\sin\phi = -(g\sin\phi - \mu g\cos\phi)$

$\Rightarrow \mu = 2\tan\phi$

22. (d) The particle is moving in circular path

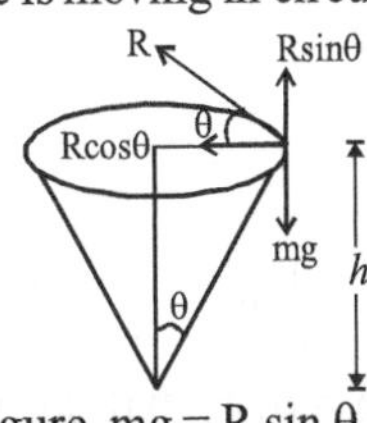

From the figure, $mg = R\sin\theta$ ... (i)

$\frac{mv^2}{r} = R\cos\theta$ ... (ii)

From equations (i) and (ii) we get

$\tan\theta = \frac{rg}{v^2}$ but $\tan\theta = \frac{r}{h}$

$\therefore h = \frac{v^2}{g} = \frac{(0.5)^2}{10} = 0.025\text{m} = 2.5\text{cm}$

23. (b) By spitting or sneezing we get a momentum in opposite direction which will help us in getting off the plane. In all other cases we will slip on ice as there is no friction.

24. (a) Force required to just move a body (F) = force due to static friction = $\mu_s mg$

When body moves with a constant acceleration (a) then

$F - f_k = ma$, where f_k is the force of kinetic friction = μ_k mg

$\therefore a = \frac{F - f_k}{m} = \frac{F - f_k}{m} = \frac{\mu_s mg - \mu_k mg}{m}$

$= (\mu_s - \mu_k) g = (0.75 - 0.5) g = \frac{g}{4}$.

25. (c) Considering the two masses and the rope a system, then

Initial net force $= [25 - (15 + 5)]g = 5g$

Final net force $= [(25 + 5) - 15]g = 15\text{ g}$

$\Rightarrow (\text{acceleration})_{\text{final}} = 3\ (\text{acceleration})_{\text{initial}}$

26. (c)

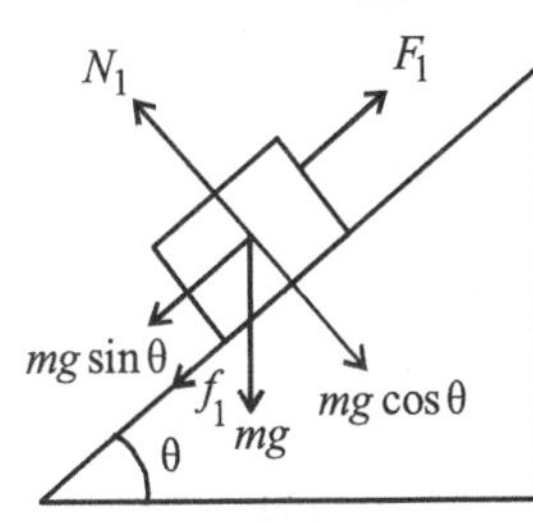

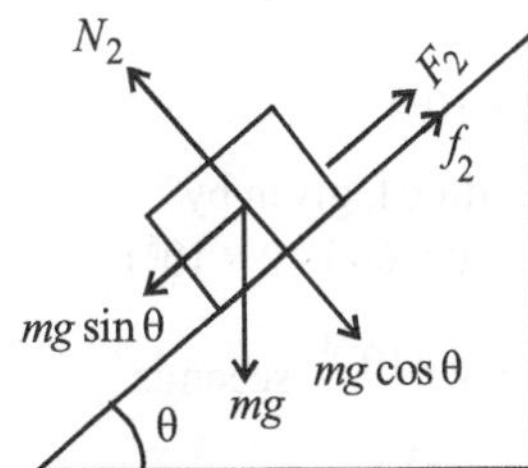

For the upward motion of the body

$mg\sin\theta + f_1 = F_1$

or, $F_1 = mg\sin\theta + \mu mg\cos\theta$

For the downward motion of the body,

$mg\sin\theta - f_2 = F_2$

or $F_2 = mg\sin\theta - \mu mg\cos\theta$

$\therefore \frac{F_1}{F_2} = \frac{\sin\theta + \mu\cos\theta}{\sin\theta - \mu\cos\theta}$

$\Rightarrow \frac{\tan\theta + \mu}{\tan\theta - \mu} = \frac{2\mu + \mu}{2\mu - \mu} = \frac{3\mu}{\mu} = 3$

27. (b) Considering the equilibrium of B

$-m_B g + T = m_B a$

Since the block A slides down with constant speed.

a = 0.

Therefore $T = m_B g$

Considering the equilibrium of A, we get

$10a = 10g\sin 30° - T - \mu N$

where $N = 10g\cos 30°$

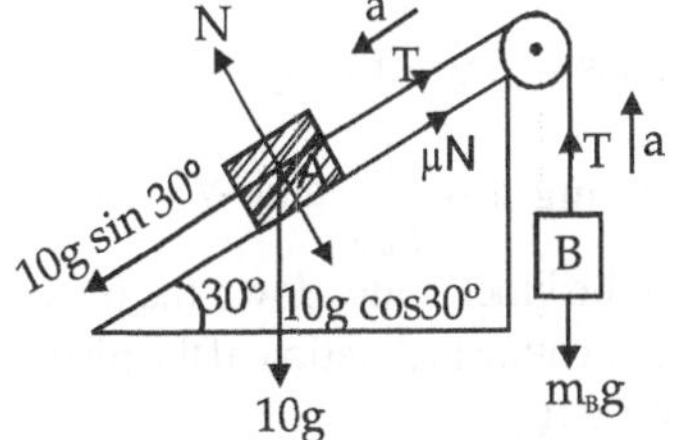

$\therefore 10a = \frac{10}{2}g - T - \mu \times 10g\cos 30°$

but a = 0, $T = m_B g$

$0 = 5g - m_B g - \frac{0.2\sqrt{3}}{2} \times 10 \times g$

$\Rightarrow m_B = 3.268 \approx 3.3\text{ kg}$

28. (a) When tension in the cable is equal to the weight of cable, the system is in equilibrium. It means the system is at rest or moving with uniform velocity.

29. (c) Tension at the highest point

$T_{top} = \frac{mv^2}{r} - mg = 2mg \quad (\because v_{top} = \sqrt{3gr})$

Tension at the lowest point

$T_{bottom} = 2mg + 6mg = 8mg$

$\therefore \frac{T_{top}}{T_{bottom}} = \frac{2mg}{8mg} = \frac{1}{4}$.

30. (d) When brakes are on, the wheels of the cycle will slide on the road instead of rolling there. It means the sliding friction will come into play instead of rolling friction. The value of sliding friction is more than that of rolling friction.

31. (a) When car moves towards right with acceleration a then due to pseudo force the plumb line will tilt in backward direction making an angle θ with vertical

From the figure

$\tan\theta = a/g$

$\therefore \theta = \tan^{-1}(a/g)$

32. (b) See fig.

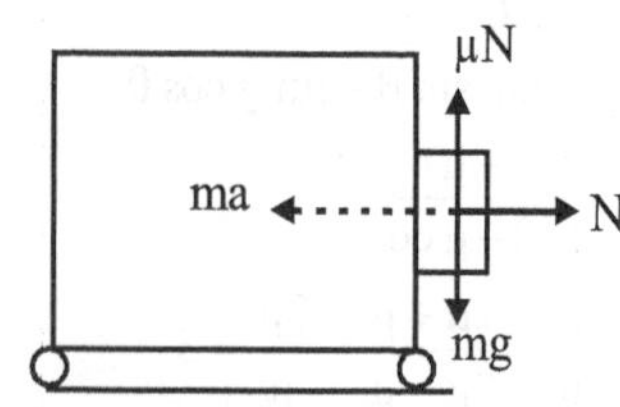

If a = acceleration of the cart, then N = ma

$\therefore$ μN = mg or μ ma = mg or a = g/μ

33. (a)

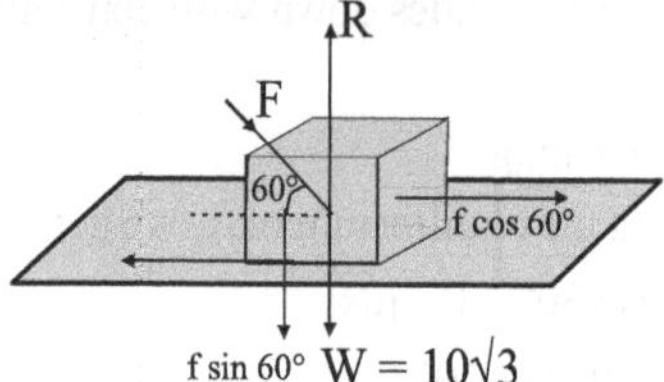

f = μR

$$F\cos 60° = \mu(W + F\sin 60°)$$

Substituing $\mu = \frac{1}{2\sqrt{3}}$ and $W = 10\sqrt{3}$ we get F = 20 N

34. (d) When the block slides down the plane with a constant speed, then the inclination of the plane is equal to angle of repose (θ).
Coeff. of friction = tan of the angle of repose = tan θ.

35. (d) Writing free body-diagrams for m & M,

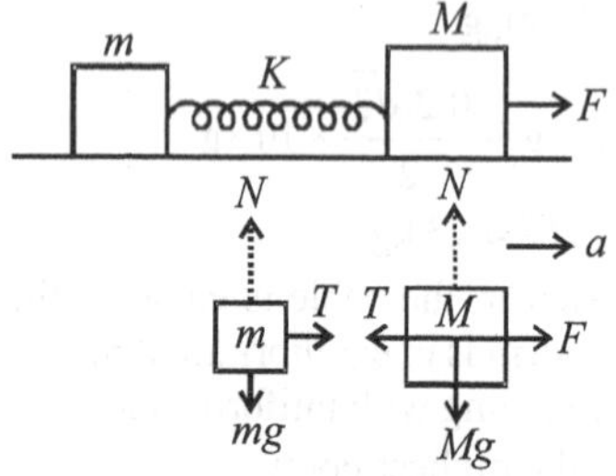

we get $T = ma$ and $F - T = Ma$
where T is force due to spring
$\Rightarrow F - ma = Ma$ or, $F = Ma + ma$

$$\therefore \quad a = \frac{F}{M+m}.$$

Now, force acting on the block of mass m is

$$ma = m\left(\frac{F}{M+m}\right) = \frac{mF}{m+M}.$$

36. (a) At limiting equilibrium, μ = tan θ

$$\tan\theta = \mu = \frac{dy}{dx} = \frac{x^2}{2} \text{ (from question)}$$

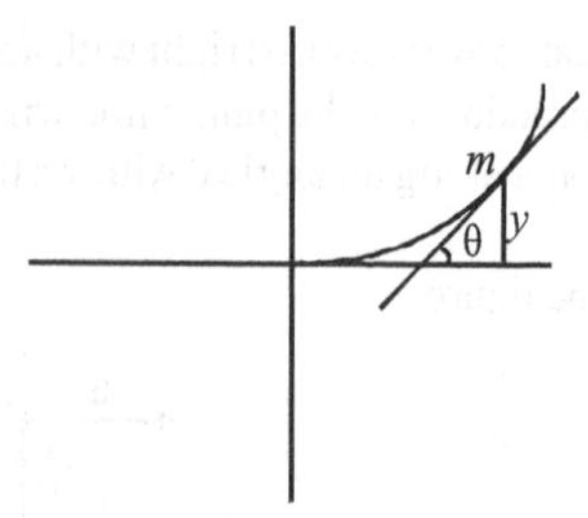

$\because$ Coefficient of friction μ = 0.5

$$\therefore \quad 0.5 = \frac{x^2}{2}$$

$\Rightarrow \quad x = \pm 1$

$$\text{Now, } y = \frac{x^3}{6} = \frac{1}{6}m$$

37. (a) As the ball, m = 10 g = 0.01 kg rebounds after striking the wall
$\therefore$ Change in momentum $= mv - (-mv) = 2mv$
Inpulse = Change in momentum = $2mv$

$$\therefore v = \frac{\text{Impulse}}{2m} = \frac{0.54 \text{ N s}}{2 \times 0.01 \text{ kg}} = 27 \text{ m s}^{-1}$$

38. (b) From the F.B.D.
N = mg cos θ
F = ma = mg sin θ − μN
$\Rightarrow$ a = g(sin θ − μ cos θ)

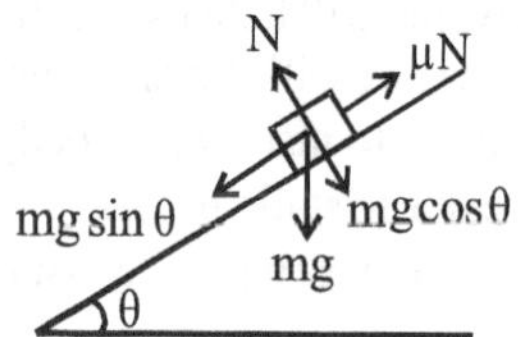

Now using, $v^2 - u^2 = 2as$

or, $v^2 = 2 \times g(\sin\theta - \mu\cos\theta)\ell$
(ℓ = length of incline)

or, $v = \sqrt{2g\ell(\sin\theta - \mu\cos\theta)}$

39. (a) During collision of ball with the wall horizontal momentum changes (vertical momentum remains constant)

$$\therefore \quad F = \frac{\text{Change in horizontal momentum}}{\text{Time of contact}}$$

$$= \frac{2P\cos\theta}{0.1} = \frac{2mv\cos\theta}{0.1}$$

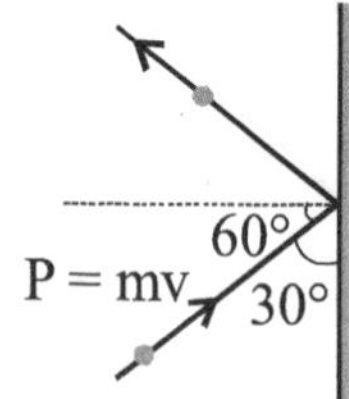

$$= \frac{2 \times 0.1 \times 10 \times \cos 60°}{0.1} = 10\text{N}$$

40. (d) Given F = 600 − (2 × 10^5 t)
The force is zero at time t, given by
$0 = 600 - 2 \times 10^5 t$

$$\Rightarrow \quad t = \frac{600}{2 \times 10^5} = 3 \times 10^{-3} \text{ seconds}$$

$$\therefore \text{Impulse} = \int_0^t F dt = \int_0^{3\times10^{-3}} (600 - 2 \times 10^5 t)\, dt$$

$$= \left[600t - \frac{2\times10^5 t^2}{2}\right]_0^{3\times10^{-3}}$$

$$= 600\times3\times10^{-3} - 10^5(3\times10^{-3})^2$$

$$= 1.8 - 0.9 = 0.9\text{Ns}$$

41. (d) According to question, two stones experience same centripetal force

i.e. $F_{C_1} = F_{C_2}$

or, $\frac{mv_1^2}{r} = \frac{2mv_2^2}{(r/2)}$ or, $V_1^2 = 4V_2^2$

So, $V_1 = 2V_2$ i.e., n = 2

42. (b) $T_1 = m(g+a) = 0.1(10+5) = 1.5\text{N}$

$T_2 = m(g-a) = 0.1(10-5) = 0.5\text{N}$

$\Rightarrow T_1 - T_2 = (1.5-0.5)\text{N} = 1\text{N}$

43. (d) As shown in the figure, the three forces are represented by the sides of a triangle taken in the same order. Therefore the resultant force is zero. $\vec{F}_{net} = m\vec{a}$. Therefore acceleration is also zero i.e., velocity remains unchanged.

44. (b) Rate of flow of water will depend on the net acceleration due to gravity.

When the lift is moving upward with acceleration a,

$g'_u = g + a$

When the lift is moving downward with acceleration on a, $g'_d = g - a$

$\therefore\ g'_u > g > g'_d \quad \therefore R_u > R_0 > R_d$

45. (d) According to law of conservation of momentum the third piece has momentum

$= 1\times -(3\hat{i} + 4\hat{j})\text{ kg ms}^{-1}$

Impulse = Average force × time

$\Rightarrow$ Average force $= \frac{\text{Impulse}}{\text{time}}$

$$= \frac{\text{Change in momentum}}{\text{time}} = \frac{-(3\hat{i}+4\hat{j})\text{kg ms}^{-1}}{10^{-4}\text{s}}$$

1. **(b)** $k = 5 \times 10^3$ N/m

$$W = \frac{1}{2}k\left(x_2^2 - x_1^2\right) = \frac{1}{2} \times 5 \times 10^3\left[(0.1)^2 - (0.05)^2\right]$$

$$= \frac{5000}{2} \times 0.15 \times 0.05 = 18.75 \text{ Nm}$$

2. **(a)** Given: Mass of particle, $M = 10g = \frac{10}{1000}$ kg

radius of circle R = 6.4 cm

Kinetic energy E of particle = 8×10^{-4}J

acceleration $a_t = ?$

$$\frac{1}{2}mv^2 = E$$

$$\Rightarrow \frac{1}{2}\left(\frac{10}{1000}\right)v^2 = 8 \times 10^{-4}$$

$$\Rightarrow v^2 = 16 \times 10^{-2}$$

$$\Rightarrow v = 4 \times 10^{-1} = 0.4 \text{ m/s}$$

Now, using

$v^2 = u^2 + 2a_t s$ $\qquad (s = 4\pi R)$

$$(0.4)^2 = 0^2 + 2a_t\left(4 \times \frac{22}{7} \times \frac{6.4}{100}\right)$$

$$\Rightarrow a_t = (0.4)^2 \times \frac{7 \times 100}{8 \times 22 \times 6.4} = 0.1 \text{ m/s}^2$$

3. **(b)** We know that $F \times v$ = Power

$\therefore F \times v = c$ where c = constant

$$m\frac{dv}{dt} \times v = c \qquad \left(\because F = ma = \frac{mdv}{dt}\right)$$

$$m\int_0^v vdv = c\int_0^t dt \qquad \Rightarrow \frac{1}{2}mv^2 = ct$$

$$v = \sqrt{\frac{2c}{m}} \times t^{1/2}$$

$$\frac{dx}{dt} = \sqrt{\frac{2c}{m}} \times t^{1/2} \quad \text{where } v = \frac{dx}{dt}$$

$$\int_0^x dx = \sqrt{\frac{2c}{m}} \times \int_0^t t^{1/2}dt$$

$$x = \sqrt{\frac{2c}{m}} \times \frac{2t^{3/2}}{3} \Rightarrow x \propto t^{3/2}$$

4. **(a)** When ball collides with the ground it loses its 50% of energy

$$\therefore \frac{KE_f}{KE_i} = \frac{1}{2} \Rightarrow \frac{\frac{1}{2}mV_f^2}{\frac{1}{2}mV_i^2} = \frac{1}{2}$$

$$\text{or } \frac{V_f}{V_i} = \frac{1}{\sqrt{2}}$$

$$\text{or, } \frac{\sqrt{2gh}}{\sqrt{v_0^2 + 2gh}} = \frac{1}{\sqrt{2}}$$

$$\text{or, } 4gh = v_0^2 + 2gh$$

$$\therefore v_0 = 20\text{ms}^{-1}$$

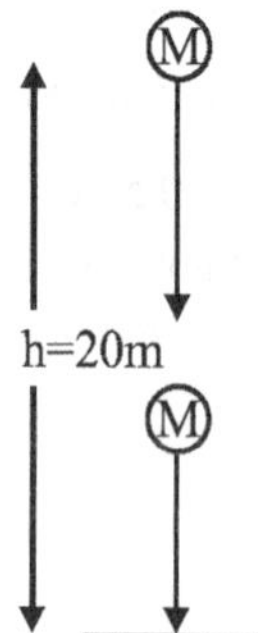

5. **(c)** As the cord is trying to hold the motion of the block, work done by the cord is negative.

$$W = -M(g-a)d = -M\left(g - \frac{g}{4}\right)d = \frac{-3Mgd}{4}$$

6. **(b)** According to principle of conservation of energy

Loss in potential energy = Gain in kinetic energy

$$\Rightarrow mgh = \frac{1}{2}mv^2 \Rightarrow v = \sqrt{2gh}$$

If h_1 and h_2 are initial and final heights, then

$$v_1 = \sqrt{2gh_1}, v_2 = \sqrt{2gh_2}$$

Loss in velocity

$$\Delta v = v_1 - v_2 = \sqrt{2gh_1} - \sqrt{2gh_2}$$

$\therefore$ Fractional loss in velocity

$$= \frac{\Delta v}{v_1} = \frac{\sqrt{2gh_1} - \sqrt{2gh_2}}{\sqrt{2gh_1}} = 1 - \sqrt{\frac{h_2}{h_1}}$$

$$= 1 - \sqrt{\frac{1.8}{5}} = 1 - \sqrt{0.36} = 1 - 0.6 = 0.4 = \frac{2}{5}$$

7. **(a)** As $u_2 = 0$ and $m_1 = m_2$, therefore from

$m_1 u_1 + m_2 u_2 = m_1 v_1 + m_2 v_2$ we get $u_1 = v_1 + v_2$

$$\text{Also, } e = \frac{v_2 - v_1}{u_1} = \frac{v_2 - v_1}{v_2 + v_1} = \frac{1 - v_1/v_2}{1 + v_1/v_2},$$

$$\text{which gives } \frac{v_1}{v_2} = \frac{1-e}{1+e}$$

8. (d) As we know power $P = \frac{dw}{dt}$

$\Rightarrow \quad w = Pt = \frac{1}{2} mv^2$

So, $v = \sqrt{\frac{2Pt}{m}}$

Hence, acceleration $a = \frac{dv}{dt} = \sqrt{\frac{2P}{m}} \cdot \frac{1}{2\sqrt{t}}$

Therefore, force on the particle at time 't'

$$= ma = \sqrt{\frac{2Km^2}{m}} \cdot \frac{1}{2\sqrt{t}} = \sqrt{\frac{Km}{2t}} = \sqrt{\frac{mK}{2}}\, t^{-1/2}$$

9. (b) $x = \frac{t^3}{3} \Rightarrow \frac{dx}{dt} = \frac{3t^2}{3} = t^2 \Rightarrow v = t^2$

when, $t = 2$ sec, $v = t^2 = (2)^2 = 4$ m/s

Work done = K.E. acquired $= \frac{1}{2} mv^2$

$$= \frac{1}{2} \times (2) \times (4)^2 = 16\,J$$

10. (b) For elastic collision in one dimension

$$v_1 = \frac{2m_2u_2}{m_1 + m_2} + \frac{(m_1 - m_2)u_1}{(m_1 + m_2)}$$

As mass 2m, is at rest, So $u_2 = 0$

$$\Rightarrow \quad v_1 = \frac{(8m - 2m)u}{8m + 2m} = \frac{3}{5}u$$

Final energy of sphere = $(K.E.)_f$

$$= \frac{1}{2}(8m)\left(\frac{3u}{5}\right)^2 = \frac{1}{2}(8m)u^2 \times \left(\frac{3}{5}\right)^2$$

$$= \frac{9}{25}E = 0.36\,E$$

11. (b) As we know work done in stretching spring

$$w = \frac{1}{2}kx^2$$

where k = spring constant

x = extension

Case (a) If extension (x) is same,

$$W = \frac{1}{2}K\,x^2$$

So, $W_P > W_Q \qquad (\because K_P > K_Q)$

Case (b) If spring force (F) is same $W = \frac{F^2}{2K}$

So, $W_Q > W_P$

12. (b) If the particle is released at the origin, it will try to go in the direction of force. Here $\frac{dU}{dx}$ is positive and hence force is negative, as a result it will move towards –ve x-axis.

13. (c) The potential energy of a spring is given by,

$$U = \frac{1}{2}kx^2 \Rightarrow 10\,J = \frac{1}{2}ks^2 \quad \text{.... (i)}$$

The potential energy stored when stretched through $(2s) = \frac{1}{2}k(2s^2) = \frac{1}{2}ks^2 \times 4$

Substituting from (i)

P.E. = 40 J.

But to increase 's' to '2s', the work done

= 40 – 10 = 30 J.

14. (b) $\text{Power} = \frac{\text{Work done}}{\text{Time}} = \frac{\frac{1}{2}m\left(v^2 - u^2\right)}{t}$

$$P = \frac{1}{2} \times \frac{2.05 \times 10^6 \times \left[(25)^2 - (5^2)\right]}{5 \times 60}$$

$P = 2.05 \times 10^6$ W = 2.05 MW

15. (b) Work done = Area under F-x graph

= area of rectangle ABCD + area of rectangle LCFE

+ area of rectangle GFIH + area of triangle IJK

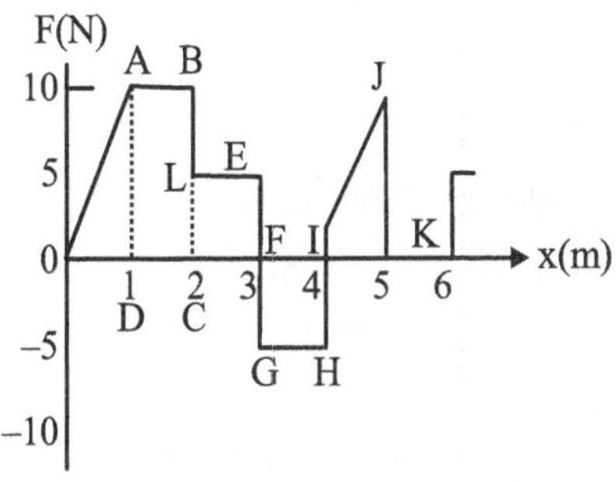

$$= (2–1)\times(10–0) + (3–2)(5–0)\ (4–3)(–5–0)$$

$$+ \frac{1}{2}(5-4)(10-0) = 15\,J$$

16. (c)

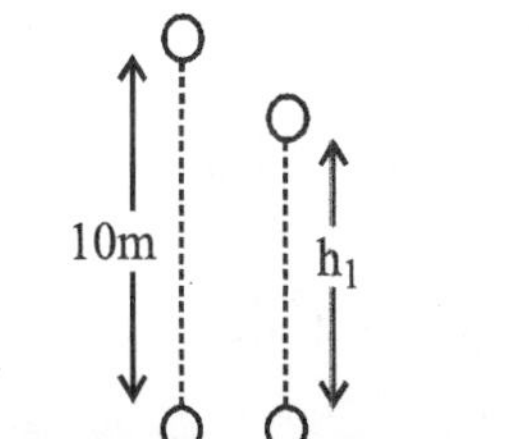

Just before impact, energy

E = mgh = 10mg .... (1)

Just after impact

$$E_1 = mgh - \frac{25}{100}mgh = 0.75\,mgh$$

Hence, $mgh_1 = E_1$ (from given figure)

$mgh_1 = 0.75\, mg(10)$

$h_1 = 7.5m$

17. (d) When C strikes A

$\frac{1}{2}mv_0^2 = \frac{1}{2}mv'^2 + \frac{1}{2}kx_0^2$ (v' = velocity of A)

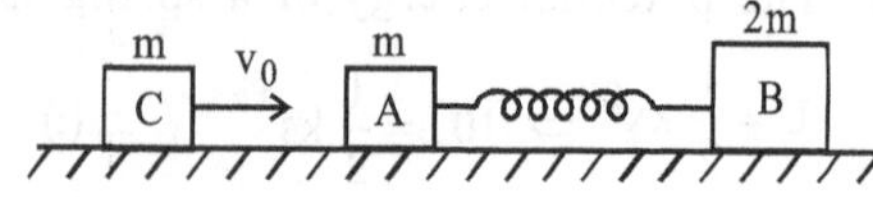

$kx_0^2 = m(v_0^2 - v'^2)$ .... (i)

$\frac{1}{2}2mv'^2 = \frac{1}{2}kx_0^2$

(When A and B Block attains K.E.)

$\therefore \frac{1}{2}kx_0^2 = mv'^2$ .... (ii)

From (i) and (ii),

$kx_0^2 = mv_0^2 - mv'^2 = mv_0^2 - \frac{k}{2}x_0^2$

$\Rightarrow kx_0^2 + \frac{k}{2}x_0^2 = mv_0^2$

$\frac{3}{2}kx_0^2 = mv_0^2 \therefore k = \frac{2}{3}m\frac{v_0^2}{x_0^2}$

18. (a) Given : $k_A = 300\,N/m,\ k_B = 400\,N/m$

Let the combination of springs is compressed by force F. Spring A is compressed by x. Therefore compression in spring B

$x_B = (8.75 - x)$ cm

$F = 300 \times x = 400(8.75 - x)$

Solving we get, x = 5 cm

$x_B = 8.75 - 5 = 3.75$cm

$\frac{E_A}{E_B} = \frac{\frac{1}{2}k_A(x_A)^2}{\frac{1}{2}k_B(x_B)^2} = \frac{300 \times (5)^2}{400 \times (3.75)^2} = \frac{4}{3}$

19. (d) Given force $\vec{F} = 2t\hat{i} + 3t^2\hat{j}$

According to Newton's second law of motion,

$m\frac{d\vec{v}}{dt} = 2t\hat{i} + 3t^2\hat{j}$ (m = 1 kg)

$\Rightarrow \int_0^{\vec{v}} d\vec{v} = \int_0^t (2t\hat{i} + 3t^2\hat{j})dt$

$\Rightarrow \vec{v} = t^2\hat{i} + t^3\hat{j}$

Power $P = \vec{F}\cdot\vec{v} = (2t\hat{i} + 3t^2\hat{j})\cdot(t^2\hat{i} + t^3\hat{j})$
$= (2t^3 + 3t^5)W$

20. (a) According to conservation of linear momentum,

$M_bV_b = M_{bl}V_{bl} + M_bV_b^1$ ....(i)

where v_b is velocity of bullet before collision v_b^1 velocity of bullet after collision and v_{bl} is the velocity of block.

K.E. of block = P.E. of block

$\frac{1}{2}M_{bl}V_{bl}^2 = M_{bl}gh$ (h = 0.2m)

Solving we get $V_{bl} = 2ms^{-1}$

Now from eq (i)

$20 \times 10^{-3} \times 600 = 4 \times 2 + 20 \times 10^{-3}\ V_b^1$

Solving we get $V_b^1 = 200$ m/s

21. (d) $\sin\theta = \frac{1}{x}$

From free body diagram of the body

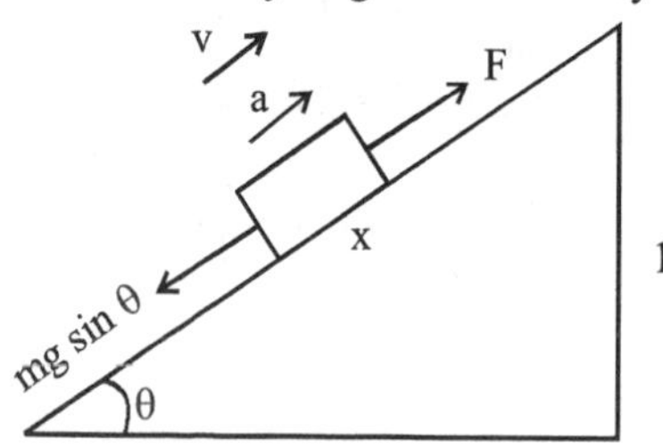

$F - mg\sin\theta = ma$

$F = m(g\sin\theta + a) = m\left(\frac{g}{x} + a\right)$ .... (1)

Displacement of the body till its velocity reaches v

$v^2 = 0 + 2as \Rightarrow s = \frac{v^2}{2a}$

Now, work done $= F\,s\cos 0° = \frac{m}{x}(g + ax) \times \frac{v^2}{2a}$

$= \frac{mv^2}{2ax}(g + ax)$

22. (c) $t_{AB} = \sqrt{\frac{2h}{g}}$

$t_{BC} + t_{CB} = 2\sqrt{\frac{2h_1}{g}}$

$= 2\sqrt{\frac{2e^2h}{g}} = 2e\sqrt{\frac{2h}{g}}$

$t_{BD} + t_{DB} = 2e^2\sqrt{\frac{2h}{g}}$

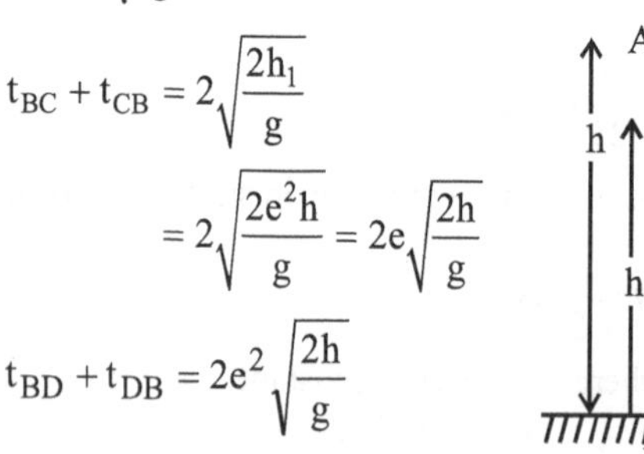

$\therefore$ Total time taken by the body in coming to rest

$= \sqrt{\frac{2h}{g}} + 2e\sqrt{\frac{2h}{g}} + 2e^2\sqrt{\frac{2h}{g}} + ...$

$= \sqrt{\frac{2h}{g}} + 2e\sqrt{\frac{2h}{g}}[1 + e + e^2 + ...]$

$$= \sqrt{\frac{2h}{g}} + 2e\sqrt{\frac{2h}{g}} \times \frac{1}{1-e} = \sqrt{\frac{2h}{g}}\left[\frac{1+e}{1-e}\right] = t\left(\frac{1+e}{1-e}\right)$$

23. (a) Velocity is maximum when K.E. is maximum

For minimum. P.E.,

$$\frac{dV}{dx} = 0 \Rightarrow x^3 - x = 0 \Rightarrow x = \pm 1$$

$$\Rightarrow \text{Min. P.E.} = \frac{1}{4} - \frac{1}{2} = -\frac{1}{4} J$$

$$K.E._{(max.)} + P.E._{(min.)} = 2 \text{ (Given)}$$

$$\therefore K.E._{(max.)} = 2 + \frac{1}{4} = \frac{9}{4}$$

$$K.E._{max.} = \frac{1}{2} mv^2_{max.}$$

$$\Rightarrow \frac{1}{2} \times 1 \times v^2_{max.} = \frac{9}{4} \Rightarrow v_{max.} = \frac{3}{\sqrt{2}}$$

24. (a) Given, $h = 60m$, $g = 10\ ms^{-2}$,

Rate of flow of water = 15 kg/s

∴ Power of the falling water

$= 15\ kgs^{-1} \times 10\ ms^{-2} \times 60\ m = 900$ watt.

Loss in energy due to friction

$$= 9000 \times \frac{10}{100} = 900 \text{ watt.}$$

∴ Power generated by the turbine

$= (9000 - 900)$ watt $= 8100$ watt $= 8.1$ kW

24. (b) Let initial velocity of the bullet be v.

By linear momentum conservation

$$\frac{m}{2} v = \left(\frac{m}{2} + m\right) v_1$$

(v_1 = combined velocity)

$$v_1 = \frac{v}{3} \quad \text{....(1)}$$

retardation = μg

$$0 = \left(\frac{v}{3}\right)^2 - 2\mu g d \Rightarrow v = 3\sqrt{2\mu g d}$$

25. (c) Force constant of a spring

$$k = \frac{F}{x} = \frac{mg}{x} = \frac{1 \times 10}{2 \times 10^{-2}} \Rightarrow k = 500 N/m$$

Increment in the length = 60 – 50 = 10 cm

$$U = \frac{1}{2} kx^2 = \frac{1}{2} 500 \left(10 \times 10^{-2}\right)^2 = 2.5 J$$

26. (b) Constant power of car $P_0 = F.v = ma.v$

$$P_0 = m\frac{dv}{dt}.v$$

$P_0 dt = mvdv$ Integrating

$$P_0 t = \frac{mv^2}{2}$$

$$v = \sqrt{\frac{2P_0 t}{m}}$$

$\because P_0$, m and 2 are constant

$\therefore \quad v \propto \sqrt{t}$

27. (a) $x = 3t - 4t^2 + t^3$

$$\frac{dx}{dt} = 3 - 8t + 3t^2$$

$$\text{Acceleration} = \frac{d^2x}{dt^2} = -8 + 6t$$

Acceleration after 4 sec

$= -8 + 6 \times 4 = 16\ ms^{-2}$

Displacement in 4 sec

$= 3 \times 4 - 4 \times 4^2 + 4^3 = 12$ m

∴ Work = Force × displacement

= Mass × acc. × disp.

$= 3 \times 10^{-3} \times 16 \times 12 = 576$ mJ

28. (c) $K_i = \frac{1}{2} m_1 u_1^2$,

$$K_f = \frac{1}{2} m_1 v_1^2, v_1 = \frac{m_1 - m_2}{m_1 + m_2} u_1$$

Fractional loss

$$\frac{K_i - K_f}{K_i} = \frac{\frac{1}{2} m_1 u_1^2 - \frac{1}{2} m_1 v_1^2}{\frac{1}{2} m_1 u_1^2}$$

$$= 1 - \frac{v_1^2}{u_1^2} = 1 - \frac{(m_1 - m_2)^2}{(m_1 + m_2)^2} = \frac{4m_1 m_2}{(m_1 + m_2)^2}$$

$$(m_2 = m;\ m_1 = nm); \qquad = \frac{4n}{(1+n)^2}$$

Energy transfer is maximum when $K_f = 0$

$$\frac{4n}{(1+n)^2} = 1 \Rightarrow 4n = 1 + n^2 + 2n \Rightarrow n^2 + 1 - 2n = 0$$

$$(n-1)^2 = 0 \qquad n = 1 \text{ ie. } m_2 = m,\ m_1 = m$$

Transfer will be maximum when both masses are equal and one is at rest.

29. (a) For inelastic collision, linear momentum is conserved

$$\Rightarrow mv_1 = 2mv_2 \Rightarrow v_2 = \frac{v_1}{2}$$

Loss in K.E. = Gain in P.E.

$$= \frac{1}{2} mv_1^2 - \frac{1}{2}(2m)v_2^2 = 2mgh$$

$$\Rightarrow 4\ mgh = mv_1^2 - \frac{mv_1^2}{2} = \frac{mv_1^2}{2} = \frac{mv^2}{2}$$

$$\Rightarrow h = \frac{v^2}{8g}$$

30. **(c)** Volume of water to raise = 22380 l = 22380×10^{-3}m^3

$$P=\frac{mgh}{t}=\frac{V\rho gh}{t}\Rightarrow t=\frac{V\rho gh}{P}$$

$$t=\frac{22380\times10^{-3}\times10^{3}\times10\times10}{10\times746}=15\text{ min}$$

31. **(b)** $E=\frac{p^2}{2m}$

or, $E_1=\frac{p_1^2}{2m_1}, E_2=\frac{p_2^2}{2m_2}$

or, $m_1=\frac{p_1^2}{2E_1}, m_2=\frac{p_2^2}{2E_2}$

$m_1>m_2\Rightarrow\frac{m_1}{m_2}>1$

$\therefore \frac{p_1^2E_2}{E_1p_2^2}>1\Rightarrow\frac{E_2}{E_1}>1 \quad [\because p_1=p_2]$

or, $E_2>E_1$

32. **(d)** From, F = ma

$$a=\frac{F}{m}=\frac{0.1x}{10}=0.01x=V\frac{dV}{dx}$$

So, $\int_{V_1}^{V_2} VdV=\int_{20}^{30}\frac{x}{100}dx$

$$-\frac{V^2}{2}\Big|_{V_1}^{V_2}=\frac{x^2}{200}\Big|_{20}^{30}=\frac{30\times30}{200}-\frac{20\times20}{200}$$

$=4.5-2=2.5$

$=\frac{1}{2}m\left(V_2^2-V_1^2\right)=10\times2.5\text{ J}=-25\text{J}$

Final K.E.

$=\frac{1}{2}mV_2^2=\frac{1}{2}mV_1^2-25=\frac{1}{2}\times10\times10\times10-25$

$=500-25=475$ J

33. **(c)** Friction is a non-conservative force. Work done by a non-conservative force over a closed path is not zero.

34. **(b)** $F=\frac{12}{100}\times1000\times10\text{ N}=1200\text{ N}$

P = Fv = 1200 N × 15 ms^{-1} = 18 kW.

35. **(c)** When the ball is released from the top of tower then ratio of distances covered by the ball in first, second and third second

$h_I:h_{II}:h_{III}=1:3:5$: [Because $h_n \propto (2n-1)$]

$\therefore$ Ratio of work done $mgh_I:mgh_{II}:mgh_{III}=1:3:5$

36. **(c)** m_2 m_1

(B)→ v (A)

$u=0$

conservation of linear momentum along x-direction

$m_2v=m_1v_x \Rightarrow \frac{m_2v}{m_1}=v_x$

along y-direction

$m_2\times\frac{v}{2}=m_1v_y \Rightarrow v_y=\frac{m_2v}{2m_1}$

Note: Let A moves in the direction, which makes an angle θ with initial direction i.e.

$$\tan\theta=\frac{v_y}{v_x}=\frac{m_2v}{2m_1}\Big/\frac{m_2v}{m_1}$$

$$\tan\theta=\frac{1}{2}$$

$\Rightarrow \quad \theta=\tan^{-1}\left(\frac{1}{2}\right)$ to the x-axis.

37. **(b)** Let the block compress the spring by x before stopping.

Kinetic energy of the block = (P.E of compressed spring) + work done against friction.

$$\frac{1}{2}\times2\times(4)^2=\frac{1}{2}\times10{,}000\times x^2+15\times x$$

$10{,}000x^2+30x-32=0$

$\Rightarrow 5000x^2+15x-16=0$

$$\therefore\ x=\frac{-15\pm\sqrt{(15)^2-4\times(5000)(-16)}}{2\times5000}$$

= 0.055m = 5.5cm.

38. **(a)** Amount of water flowing per second from the pipe

$$=\frac{m}{time}=\frac{m}{\ell}\cdot\frac{\ell}{t}=\left(\frac{m}{\ell}\right)v$$

Power = K.E. of water flowing per second

$$=\frac{1}{2}\left(\frac{m}{\ell}\right)v\cdot v^2$$

$$=\frac{1}{2}\left(\frac{m}{\ell}\right)v^3$$

$$=\frac{1}{2}\times100\times8=400\ W$$

39. **(b)** Mass of over hanging chain $m'=\frac{4}{2}\times(0.6)$kg

Let at the surface PE = 0

C.M. of hanging part = 0.3 m below the table

$$U_i=-m'gx=-\frac{4}{2}\times0.6\times10\times0.30$$

$\Delta U=m'gx=3.6\text{J}$ = Work done in putting the entire chain on the table.

40. (d)

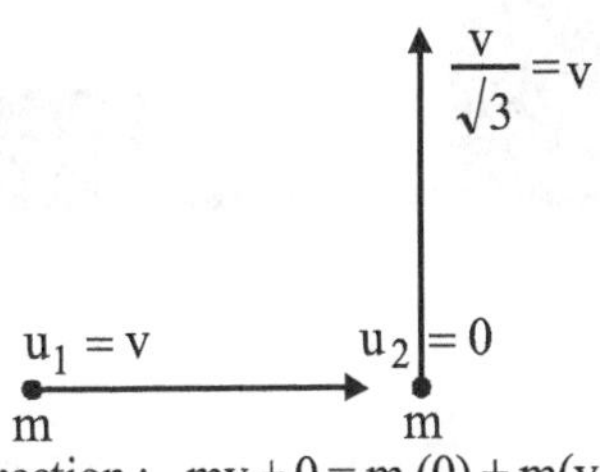

In x-direction : $mv + 0 = m(0) + m(v_2)_x$

In y-direction : $0 + 0 = m\left(\frac{v}{\sqrt{3}}\right) + m(v_2)_y$ is

$$\Rightarrow (v_2)_y = \frac{v}{\sqrt{3}} \text{ and } (v_2)_x = v$$

$$\therefore \; v_2 = \sqrt{\left(\frac{v}{\sqrt{3}}\right)^2 + v^2}$$

$$\Rightarrow \; v_2 = \sqrt{\frac{v^2}{3} + v^2} = v\sqrt{\frac{4}{3}} = \frac{2v}{\sqrt{3}}$$

Alternative method : In x-direction,

$mv = mv_1 \cos\theta$ ...(1)

where v_1 is the velocity of second mass

In y-direction,

$$0 = \frac{mv}{\sqrt{3}} - mv_1 \sin\theta$$

or $m_1 v_1 \sin\theta = \frac{mv}{\sqrt{3}}$ ...(2)

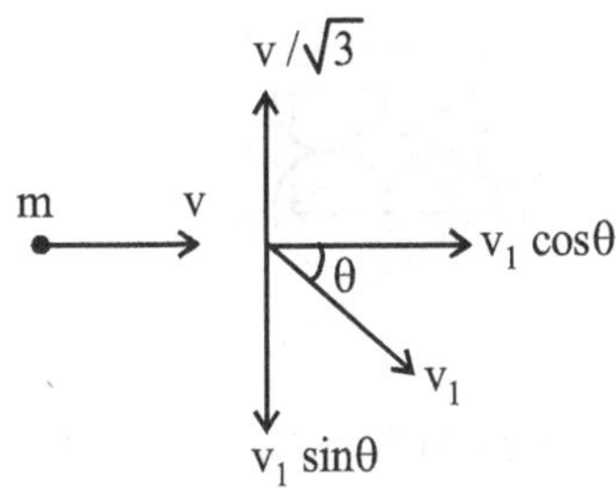

Squaring and adding eqns. (1) and (2)

$$v_1^2 = v^2 + \frac{v^2}{\sqrt{3}} \Rightarrow v_1 = \frac{2}{\sqrt{3}}v$$

41. (b)

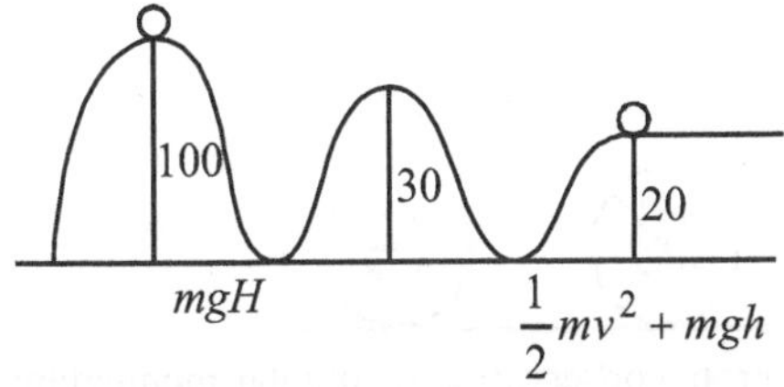

Using conservation of energy,

$$m(10 \times 100) = m\left(\frac{1}{2}v^2 + 10 \times 20\right)$$

or $\frac{1}{2}v^2 = 800$ or $v = \sqrt{1600} = 40$ m/s

42. (b)

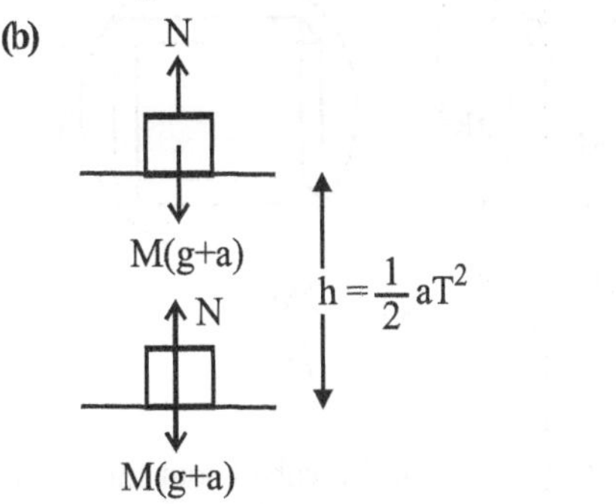

Work done by normal reaction

$$= Nh = M(g+a)\frac{1}{2}aT^2 = \frac{1}{2}M(g+a)\,aT^2$$

43. (c) Applying W-E theorem on the block for any compression x :

$$W_{ext} + W_g + W_{spring} = \Delta KE$$

$$\Rightarrow \; Fx + 0 - \frac{1}{2}Kx^2 = \frac{1}{2}mv^2$$

$\Rightarrow$ KE vs x is inverted parabola.

44. (d) $K.E. = \frac{1}{2}mv^2$

Further, $v^2 = u^2 + 2as = 0 + 2ad = 2ad$
$= 2(F/m)d$

Hence, $K.E. = \frac{1}{2}m \times 2(F/m)d = Fd$

or, K.E. acquired = Work done
$= F \times d =$ constant.

i.e., it is independent of mass m.

45. (a) Gravitational potential energy of ball gets converted into elastic potential energy of the spring.

$$mg(h+d) = \frac{1}{2}kd^2$$

Net work done $= mg(h+d) - \frac{1}{2}kd^2 = 0$

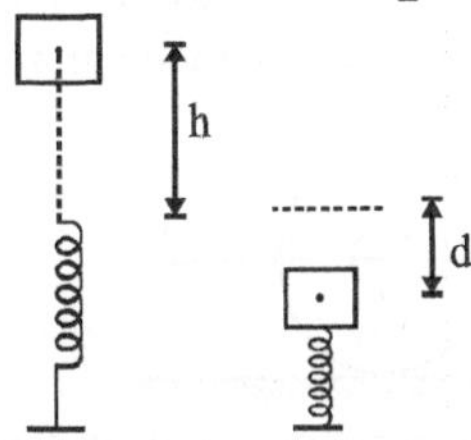

1. (a) Here $a = \frac{2}{\sqrt{3}}R$

Now, $\frac{M}{M'} = \frac{\frac{4}{3}\pi R^3}{a^3}$

$= \frac{\frac{4}{3}\pi R^3}{\left(\frac{2}{\sqrt{3}}R\right)^3} = \frac{\sqrt{3}}{2}\pi.\ \ M' = \frac{2M}{\sqrt{3}\pi}$

Moment of inertia of the cube about the given axis,

$$I = \frac{M'a^2}{6} = \frac{\frac{2M}{\sqrt{3}\pi} \times \left(\frac{2}{\sqrt{3}}R\right)^2}{6} = \frac{4MR^2}{9\sqrt{3}\pi}$$

2. (d) Initially centre of mass is at the centre. When sand is poured it will fall and again after a limit, centre of mass will rise.

3. (a) Does not shift as no external force acts. The centre of mass of the system continues its original path. It is only the internal forces which comes into play while breaking.

4. (d) Let the mass of loop P (radius = r) = m
So, the mass of loop Q (radius = nr) = nm

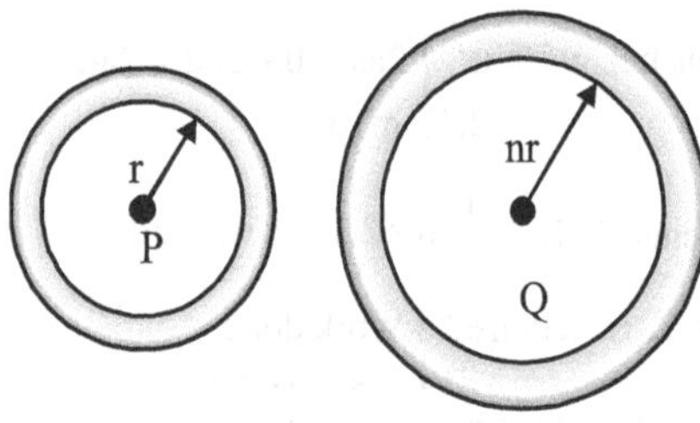

Moment of inertia of loop P, $I_P = mr^2$
Moment of inertia of loop Q, $I_Q = nm(nr)^2 = n^3mr^2$

$\therefore \frac{I_Q}{I_P} = n^3 = 8 \Rightarrow n = 2$

5. (d) When the ball is hit by a cue, the linear impulse imparted to the ball = change in momentum = mv_0

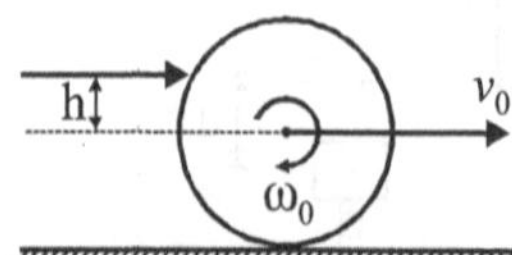

Angular momentum = Moment of momentum

$I\omega_0 = (mv_0)h$

$\frac{2}{5}mr^2\omega_0 = mv_0h$ or $\omega_0 = \frac{5v_0h}{2r^2}$

6. (d)

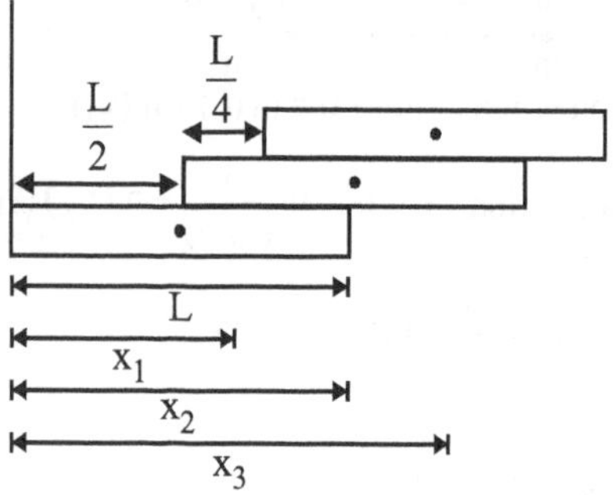

$x_1 = \frac{L}{2},\ x_2 = L,\ x_3 = \frac{5L}{4}$

$$\therefore X_{CM} = \frac{m_1x_1 + m_2x_2 + m_3x_3}{m_1 + m_2 + m_3}$$

$$= \frac{M \times \frac{L}{2} + M \times L + M \times \frac{5L}{4}}{M + M + M}$$

$$= \frac{\frac{11}{4}ML}{3M} = \frac{11L}{12}$$

7. (c)

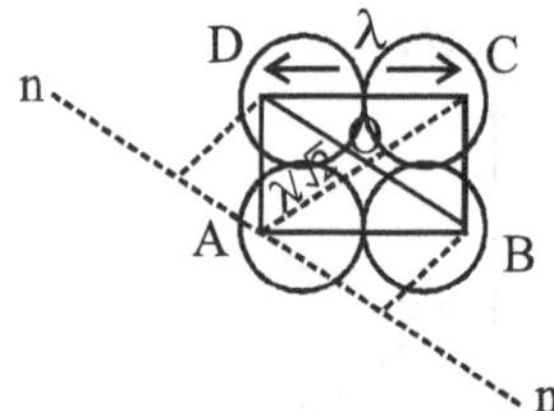

$I_{nn'}$ = $M.I$ due to the point mass at B + $M.I$ due to the point mass at D + $M.I$ due to the point mass at C.

$$I_{nn'} = 2 \times m\left(\frac{\ell}{\sqrt{2}}\right)^2 + m(\sqrt{2}\ell)^2$$

$$= m\ell^2 + 2m\ell^2 = 3m\ell^2$$

8. (c)

From conservation of angular momentum about any fix point on the surface,

$mr^2\omega_0 = 2mr^2\omega$

$\Rightarrow \omega = \omega_0/2 \Rightarrow v = \frac{\omega_0 r}{2}$ $[\because v = r\omega]$

9. (d) Initial position of cm $= \frac{m_2\ell}{m_1 + m_2}$

Also $x_{cm} = \frac{m_1\Delta x_1 + m_2\Delta x_2}{m_1+m_2} = \frac{m_1 v_0 t + 0}{m_1+m_2}$

$\therefore$ final position $= \frac{m_2\ell}{m_1+m_2} + \frac{m_1 v_0 t}{m_1+m_2}$

10. (a) Here, $L = 1.8\ \text{kg m}^2\ \text{s}^{-1}$, $M = 1.5$ kg, $\omega = 0.3\ \text{rad s}^{-1}$

Angular momentum, $L = I\omega$

$L = k^2 M\omega \qquad (\because I = MK^2)$

or $1.8 = k^2 \times 1.5 \times (0.3)$

$\Rightarrow k^2 = \frac{1.8}{1.5 \times 0.3} = 4$

$\Rightarrow k = 2$ m.

11. (b) $\vec{\tau} = \vec{r} \times \vec{F} \Rightarrow \vec{r}\cdot\vec{\tau} = 0 \qquad \vec{F}\cdot\vec{\tau} = 0$

Since, $\vec{\tau}$ is perpendicular to the plane of $\vec{r}$ and $\vec{F}$, hence the dot product of $\vec{\tau}$ with $\vec{r}$ and $\vec{F}$ is zero.

12. (c)

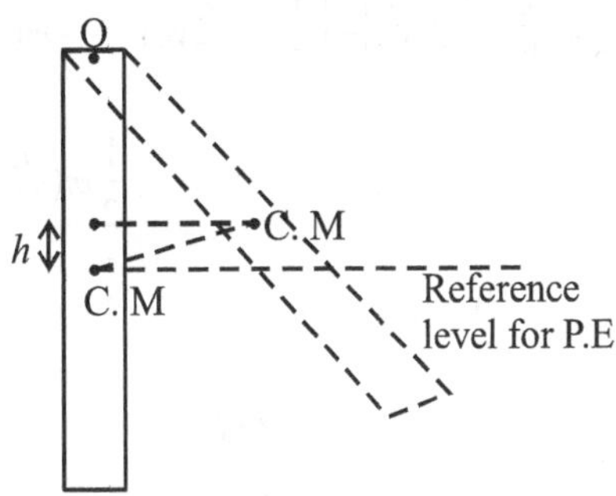

The moment of inertia of the rod about O is $\frac{1}{3}m\ell^2$.

The maximum angular speed of the rod is when the rod is instantaneously vertical. The energy of the rod in this condition is $\frac{1}{2}I\omega^2$ where I is the moment of inertia of the rod about O. When the rod is in its extreme portion, its angular velocity is zero momentarily. In this case, the energy of the rod is mgh where h is the maximum height to which the centre of mass (C.M) rises

$\therefore mgh = \frac{1}{2}I\omega^2 = \frac{1}{2}\left(\frac{1}{3}ml^2\right)\omega^2$

$\Rightarrow h = \frac{\ell^2\omega^2}{6g}$

13. (b)

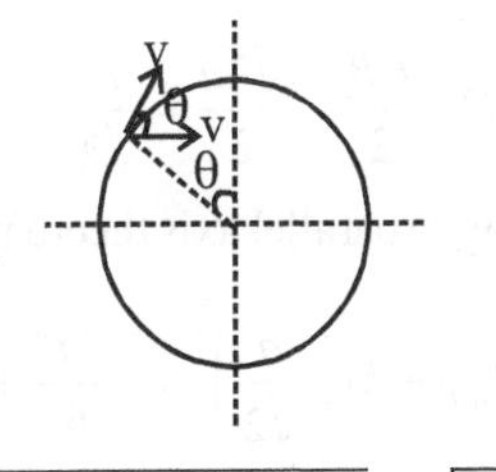

$v_R = \sqrt{v^2 + v^2 + 2v^2\cos\theta} = \sqrt{2v^2(1+\cos\theta)}$

$= 2v\cos\frac{\theta}{2}$

14. (b) $K.E_{rotational} = \frac{1}{2}I\omega^2$

$= \frac{1}{2}\frac{2}{5}\omega r^2 d^2 \quad \left(\because I_{\text{Solid sphere}} = \frac{2}{5}mr^2\right)$

$K.E_{translational} = \frac{1}{2}mv^2$

$\therefore \frac{K.E_{rotational}}{K.E_{translational}} = \frac{2}{5}$

Hence option (b) is correct

15. (c) Kinetic energy $_{(rotational)}\ K_R = \frac{1}{2}I\omega^2$

Kinetic energy $_{(translational)}\ K_T = \frac{1}{2}Mv^2$ $\qquad (v = R\omega)$

M.I.$_{(initial)}\ I_{ring} = MR^2$; $\omega_{initial} = \omega$

M.I.$_{(new)}\ I'_{(system)} = MR^2 + 2mR^2$

$\omega'_{(system)} = \frac{M\omega}{M+2m}$

Solving we get loss in K.E.

$= \frac{Mm}{(M+2m)}\omega^2 R^2$

16. (b) For no angular acceleration $\tau_{net} = 0$

$\Rightarrow F_1 \times 5 = F_2 \times 30$ (given $F_2 = 4$N) $\Rightarrow F_1 = 24$N

17. (c) For toppling $Mg\frac{L}{2} = F_1 \times h$

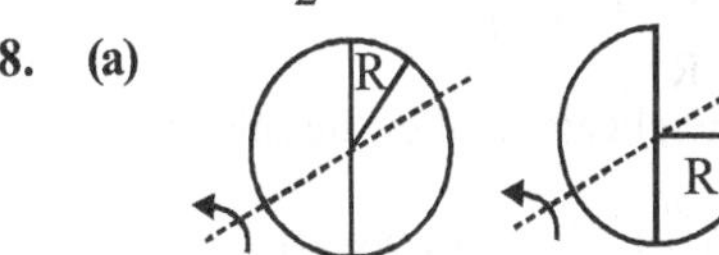

For sliding

$\mu Mg = F_2$

For sliding to occur first

$F_1 > F_2$

or $\frac{mgL}{2} > \mu Mg$ or $L > 2\mu h$

18. (a)

Moment of inertia of a ring about a given axis is $I = MR^2$

Mass of the remaining portion of the ring $= \frac{3M}{4}$

Moment of inertia of the remaining portion of the ring about a given axis is

$I' = \frac{3}{4}MR^2$

Given $I' = kMR^2$

$\therefore \quad k = 3/4$.

19. (b) Applying angular momentum conservation

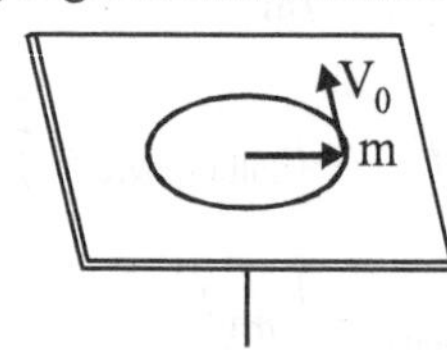

$$mV_0R_0 = (m)(V^1)\left(\frac{R_0}{2}\right)$$

$\therefore \quad v^1 = 2V_0$

Therefore, new KE $= \frac{1}{2}m(2V_0)^2 = 2mv_0^2$

20. (c) If rotation axis is passing through its middle point & is ⊥ to its plane, then moment of inertia about YY' is

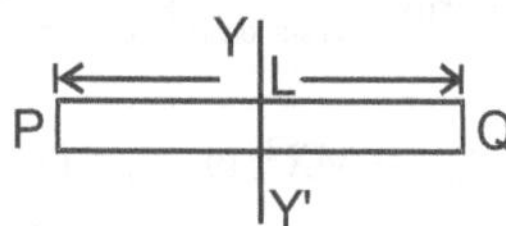

$I = \frac{ML^2}{12}$ where M = volume × density = (L×A)×ρ

so $I = \frac{L^3A\rho}{12}$

so rotational K.E $= \frac{1}{2}I\omega^2 = \frac{L^3A\rho\omega^2}{24}$

21. (c) If a body rolls on a horizontal surface, it possesses both translational and rotational kinetic energies. The net kinetic energy is given by

$$K_{net} = \frac{1}{2}mv^2\left(1+\frac{K^2}{R^2}\right),$$

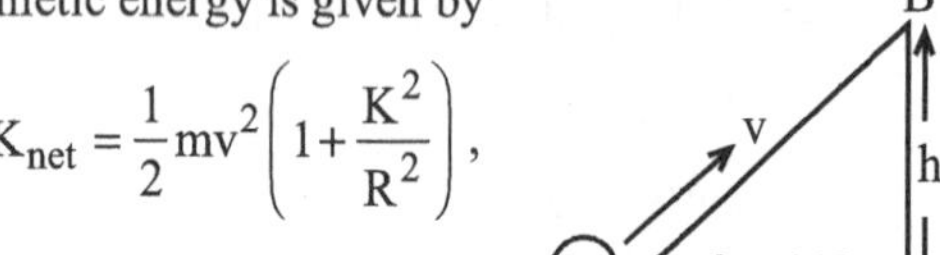

where K is the radius of gyration.
So from law of conservation of energy,

$$\frac{1}{2}mv^2\left(1+\frac{K^2}{R^2}\right) = mgh,$$

where h is the height attained by the sphere.

i.e., $\frac{1}{2}\times 2\times(10)^2\left(1+\frac{2}{5}\right) = 2\times 9.8\times h.$

i.e., $\frac{1}{2}\times 100\times\left(\frac{7}{5}\right) = 9.8h$

or $h = \frac{700}{98} = 7.1m$

22. (c) After collision velocity of COM of *A* becomes zero and that of *B* becomes equal to initial velocity of COM of *A*. But angular velocity of *A* remains unchanged as the two spheres are smooth.

23. (b) M.I. of disc $= \frac{1}{2}MR^2 = \frac{1}{2}M\left(\frac{M}{\pi t\rho}\right) = \frac{1}{2}\frac{M^2}{\pi t\rho}$

$$\left(As\ \rho = \frac{M}{\pi R^2 t}\ \text{Therefore}\ R^2 = \frac{M}{\pi t\rho}\right)$$

If mass and thickness are same then, $I \propto \frac{1}{\rho}$

$$\therefore \frac{I_1}{I_2} = \frac{\rho_2}{\rho_1} = \frac{3}{1}$$

24. (c)

When the system is released, heavier mass move downward and the lighter one upward. Thus, centre of mass will move towards the heavier mass with acceleration

$$a = \left(\frac{3m-m}{3m+m}\right)g = \frac{g}{2}$$

25. (c) $K = K_{ring} + K_{particles}$

$$= \left[\frac{1}{2}mv_0^2 + \frac{1}{2}I\omega^2\right] + \left[\frac{1}{2}m(\sqrt{2}v_0)^2 + \frac{1}{2}m(2v_0)^2 + \frac{1}{2}m(\sqrt{2}v_0)^2 + 0\right]$$

Also $\omega = \frac{v_0}{R}$, $I = mR^2$

$\therefore \quad K = 5mv_0^2$

26. (d) $I_{nn'} = \frac{1}{12}M(a^2+a^2) = \frac{Ma^2}{6}$

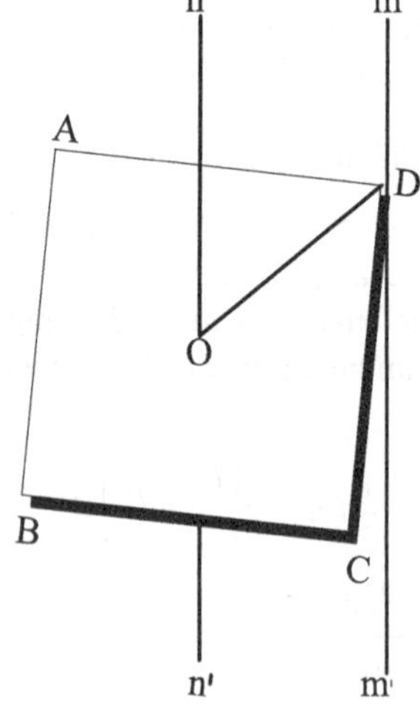

Also, $DO = \frac{DB}{2} = \frac{\sqrt{2}a}{2} = \frac{a}{\sqrt{2}}$

According to parallel axis theorem

$$I_{mm'} = I_{nn'} + M\left(\frac{a}{\sqrt{2}}\right)^2 = \frac{Ma^2}{6} + \frac{Ma^2}{2}$$

$$= \frac{Ma^2 + 3Ma^2}{6} = \frac{2}{3}Ma^2$$

27. (a) From law of conservation of angular momentum,

$I\omega = I'\omega'$

Given $I' = I/n$

$\therefore \quad \omega' = n\omega \quad$ or $\quad \omega' \propto n$

28. (b)

29. (d) Melting of ice produces water which will spread over larger distance away from the axis of rotation. This increases the moment of inertia so angular velocity decreases

30. (c) $I_p = \frac{mr^2}{2} + 2\left[\frac{mr^2}{2} + m(2r)^2\right] + \left[\frac{mr^2}{2} + m(2r)^2\right]$

$+\left[\frac{mr^2}{2} + (2r)^2\right] + 2\left[\frac{mr^2}{2} + m(2\sqrt{3}r)^2\right]$

$= \frac{111}{2}mr^2$

31. (d) $0 = \frac{m_1(-x_1) + m_2 x_2}{m_1 + m_2}$

$\therefore \quad m_1 x_1 = m_2 x_2$

$x_1 - d \quad x_2 - d'$

d d'

m_1 O m_2

Now, $0 = \frac{-m_1(x_1 - d) + m_2(x_2 - d)}{m_1 + m_2}$

$0 = m_1(d - x_1) + m_2(x_2 - d')$

$\Rightarrow 0 = m_1 d - m_1 x_1 + m_2 x_2 - m_2 d'$

$\therefore \; d' = \frac{m_1}{m_2} d$

32. (c) This is a torque problem. While the fulcrum can be placed anywhere, placing it at the far right end of the bar eliminated cable B from the calculation. There are now only two forces acting on the bar ; the weight that produces a counterclockwise rotation and the tension in cable A that produces a clockwise rotation. Since the bar is in equilibrium, these two torques must sum to zero.

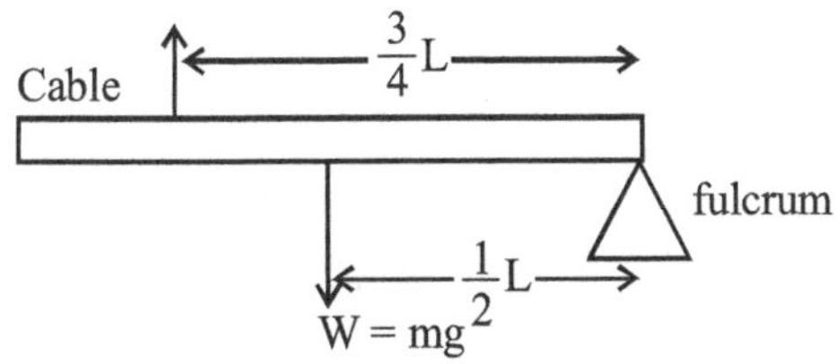

$\Sigma\tau = T_A(3/4L) - Mg(1/2L) = 0$

Therefore

$T_A = (MgL/2)/(3L/4) = (MgL/2)(4/3L) = 2Mg/3$

33. (b) Couple produces purely rotational motion.

34. (a)

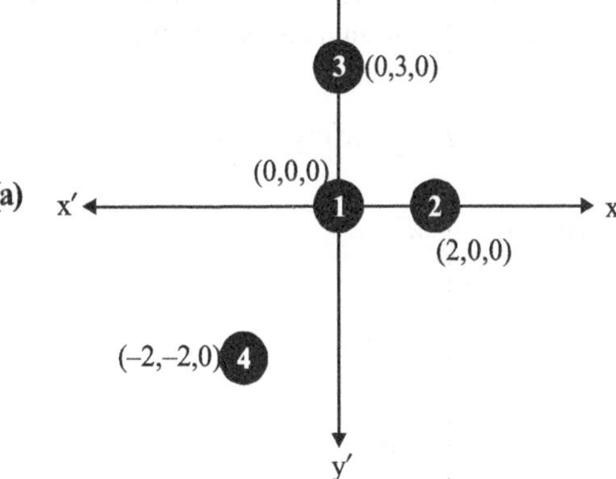

$I_1 = I_2 = 0$, because these particles are placed on x-axis

The M.I. of system about x-axis, $= I_1 + I_2 + I_3 + I_4$

$= 0 + 0 + 3 \times (3)^2 + 4 \times (-2)^2 = 27 + 16 = 43$ kg – m²

35. (b)

36. (b) $I = 1.2$ kg m², $E_r = 1500$ J,

$\alpha = 25$ rad/sec², $\omega_1 = 0$, t = ?

As $E_r = \frac{1}{2}I\omega^2$, $\omega = \sqrt{\frac{2E_r}{I}} = \sqrt{\frac{2 \times 1500}{1.2}} = 50$ rad / sec

From $\omega_2 = \omega_1 + \alpha t$

$50 = 0 + 25t$, $\quad \therefore \; t = 2$ seconds

37. (b) Since no external torque act on gymnast, so angular momentum ($L = I\omega$) is conserved. After pulling her arms & legs, the angular velocity increases but moment of inertia of gymnast, decreases in such a way that angular momentum remains constant.

38. (b) The *M.I.* about the axis of rotation is not constant as the perpendicular distance of the bead with the axis of rotation increases.

Also since no external torque is acting.

$\therefore \tau_{ext} = \frac{dL}{dt} \Rightarrow L = \text{constant} \Rightarrow I\omega = \text{constant}$

Since, *I* increases, ω decreases.

39. (c)

m

(2/π) r

r

z

Moment of inertia about z-axis, $I_z = mr^2$

(about centre of mass)

Applying parallel axes theorem,

$I_z = I_{cm} + mk^2$

$I_{cm} = I_z - m\left(\frac{2}{\pi}r\right)^2 = mr^2 - \frac{m4r^2}{\pi^2} = mr^2\left(1 - \frac{4}{\pi^2}\right)$

i.e., k = 4

40. (c) When two small spheres of mass m are attached gently, the external torque, about the axis of rotation, is zero and therefore the angular momentum about the axis of rotation is constant.

$\therefore I_1\omega_1 = I_2\omega_2 \Rightarrow \omega_2 = \frac{I_1}{I_2}\omega_1$

Here $I_1 = \frac{1}{2}MR^2$

and $I_2 = \frac{1}{2}MR^2 + 2mR^2$

$\therefore \omega_2 = \dfrac{\frac{1}{2}MR^2}{\frac{1}{2}MR^2 + 2mR^2} \times \omega_1 = \frac{M}{M+4m}\omega_1$

41. (b) $Tr = \frac{mr^2}{2}\alpha_1$ .... (1)

$Tr = \frac{mr^2}{2}\alpha$ .... (2)

$\alpha_1 = \alpha$ .... (3)

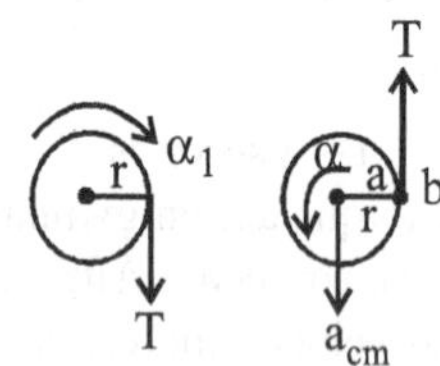

Acceleration of point b = acceleration of point a

$r\alpha_1 = a_{cm} - r\alpha$ .... (4)

Hence, $2r\alpha = a_{cm}$

42. (c) $X_{C.M.} = \frac{1\times0+2\times2+3\times0+4\times2+5\times1}{1+2+3+4+5}$

$= \frac{4+8+5}{15} = \frac{17}{15} = 1.1$

$Y_{C.M} = \frac{1\times0+2\times0+3\times2+4\times2+5\times1}{1+2+3+4+5}$

$= \frac{6+8+5}{15} = 1.3$

43. (d) $I_{AX} = m(AB)^2 + m(OC)^2 = m\ell^2 + m(\ell\cos 60°)^2$
$= m\ell^2 + m\ell^2/4 = 5/4\, m\ell^2$

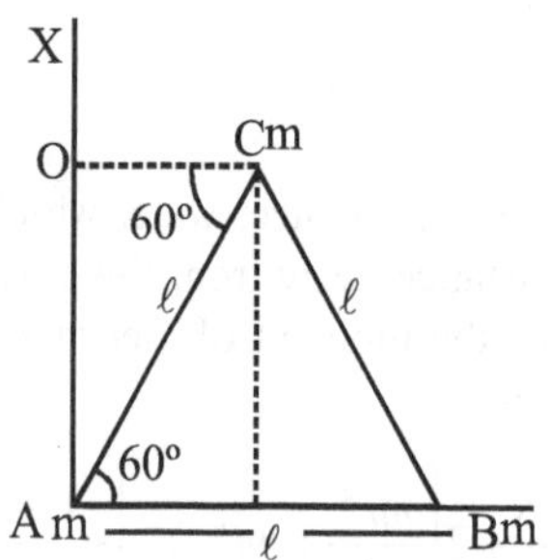

44. (c) Angle turned in three seconds, $\theta_{3s} = 2\pi \times 10 = 20\pi$ rad.

From $\theta = \omega_0 t + \frac{1}{2}\alpha t^2 \Rightarrow 20\pi = 0 + \frac{1}{2}\alpha \times (3)^2$

$\Rightarrow \alpha = \frac{40\pi}{9}$ rad/s^2

Now angle turned in 6 sec from the starting

$\theta_{6s} = \omega_0 t + \frac{1}{2}\alpha t^2 = 0 + \frac{1}{2}\times\left(\frac{40\pi}{9}\right)\times(6)^2 == 80\pi$ rad

$\therefore$ Angle turned between $t = 3$s to $t = 6$s

$\theta_{last\ 3s} = \theta_{6s} - \theta_{3s} = 80\pi - 20\pi = 60\pi$

Number of revolutions $= \frac{60\pi}{2\pi} = 30$.

45. (c) $a = \frac{f}{m} = \mu g$

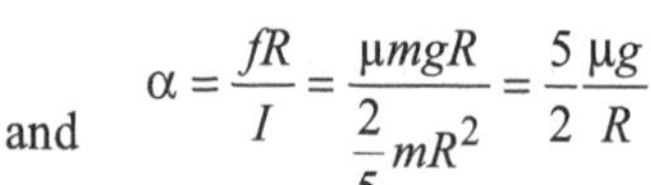

and $\alpha = \frac{fR}{I} = \frac{\mu mgR}{\frac{2}{5}mR^2} = \frac{5}{2}\frac{\mu g}{R}$

Now $v = 0 + at$

and $\omega' = \omega - \alpha t$

Also $\omega' = \frac{v}{R}$

After solving above equations, we get $\omega' = \frac{2\omega}{7}$

DAILY PRACTICE PROBLEMS	PHYSICS SOLUTIONS	DPP/CP07

1. **(a)** The escape velocity on the earth is defined as

$$v_e = \sqrt{2g_e R_e}$$

Where R_e & g_e are the radius & acceleration due to gravity of earth.
Now for planet $g_P=2g_e$, $R_P=R_e/4$

$$\text{So } v_P = \sqrt{2g_P R_P} = \sqrt{2\times 2g_e \times R_e/4} = \frac{v_e}{\sqrt{2}}$$

2. **(c)** Applying conservation of energy principle, we get

$$\frac{1}{2}mk^2v_e^2 - \frac{GMm}{R} = -\frac{GMm}{r}$$

$$\Rightarrow \frac{1}{2}mk^2\frac{2GM}{R} - \frac{GMm}{R} = -\frac{GMm}{r}$$

$$\Rightarrow \frac{k^2}{R} - \frac{1}{R} = -\frac{1}{r} \Rightarrow \frac{1}{r} = \frac{1}{R} - \frac{k^2}{R}$$

$$\Rightarrow \frac{1}{r} = \frac{1}{R}(1-k^2) \Rightarrow r = \frac{R}{1-k^2}$$

3. **(d)** The gravitational force due to the whole sphere at A point is

$F_1 = \frac{GM_e m_o}{(2R)^2}$, where m_0 is the assumed rest mass at point A.
In the second case, when we made a cavity of radius (R/2), then gravitational force at point A is

$$F_2 = \frac{GM_e m_o}{(R+R/2)^2} \qquad \therefore F_2/F_1 = 1/9$$

4. **(c)** According to Kepler's law of period $T^2 \propto R^3$

$$\frac{T_1^2}{T_2^2} = \frac{R_1^3}{R_2^3} = \frac{(6R)^3}{(3R)^3} = 8$$

$$\frac{24\times 24}{T_2^2} = 8$$

$$T_2^2 = \frac{24\times 24}{8} = 72 = 36\times 2$$

$$T_2 = 6\sqrt{2}$$

5. **(d)** Total energy $= -K.E = \frac{PE}{2}$

$$K.E = \frac{1}{2}mv^2$$

$$\therefore \text{ Total energy} = -\frac{1}{2}mv^2$$

6. **(a)** The force of attraction between sphere and shaded position $dF = GM\frac{\left(\frac{m}{l}dx\right)}{x^2}$

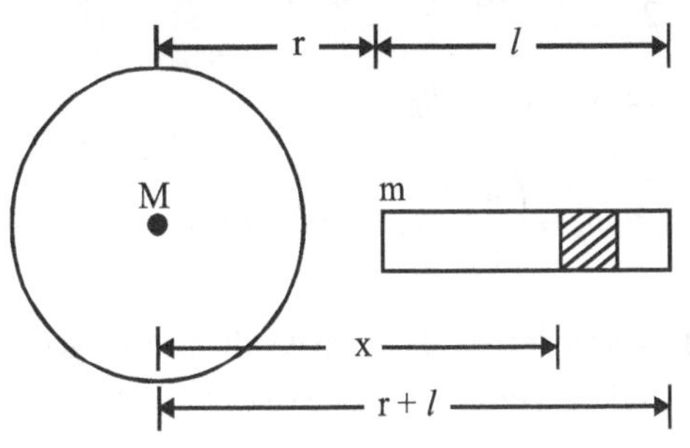

$$F = \int_r^{r+l}\frac{GMm}{lx^2}dx = \frac{GMm}{l}\int_r^{r+l}\frac{1}{x^2}dx$$

$$= \frac{GMm}{l}\int_r^{r+l}x^{-2}dx = \frac{GMm}{l}\left[\frac{x^{-2+1}}{-2+1}\right]_r^{r+l}$$

$$= -\frac{GMm}{l}\left[x^{-1}\right]_r^{r+l} = -\frac{GMm}{l}\left[\frac{1}{x}\right]_r^{r+l} = \frac{GMm}{r(r+l)}$$

7. **(b)** $F = \frac{k}{R} = \frac{Mv^2}{R}$. Hence $v \propto R^0$

8. **(d)** $\frac{mv^2}{(R+x)} = \frac{GmM}{(R+x)^2}$ also $g = \frac{GM}{R^2}$

$$\therefore \frac{mv^2}{(R+x)} = m\left(\frac{GM}{R^2}\right)\frac{R^2}{(R+x)^2}$$

$$\therefore \frac{mv^2}{(R+x)} = mg\frac{R^2}{(R+x)^2}$$

$$\therefore v^2 = \frac{gR^2}{R+x} \Rightarrow v = \left(\frac{gR^2}{R+x}\right)^{1/2}$$

9. **(b)** $v = \frac{3}{4}v_e$

$$K.E. = \frac{1}{2}mv^2 = \frac{1}{2}m\left(\frac{3}{4}v_e\right)^2 = \frac{9}{32}mv_e^2$$

$$= \frac{9}{32}m\left(\frac{2GM}{R}\right)$$

$$K.E. = \frac{9}{16}\frac{GMm}{R}\ ;\ P.E. = -\frac{GMm}{R}$$

$$\text{Total energy} = K.E. + P.E. = -\frac{7}{16}\frac{GMm}{R}$$

Let the height above the surface of earth be h, then

$$P.E. = -\frac{GMm}{h}$$

$$-\frac{7}{16}\frac{GMm}{R} = -\frac{GMm}{h} \quad \therefore\ h = \frac{16R}{7}$$

10. **(a)** When closer to the sun, velocity of planet will be greater. So time taken in covering a given area will be less.

11. **(c)** Applying conservation of total mechanical energy principle

$$\frac{1}{2}mv^2 = mg_A h_A = mg_B h_B$$

$$\Rightarrow g_A h_A = g_B h_B$$

$$\Rightarrow h_B = \left(\frac{g_A}{g_B}\right) h_A = 9 \times 2 = 18\,\text{m}$$

12. **(b)** Due to inertia of motion it will move tangentially to the original orbit with same velocity.

13. **(a)** $F \propto xM \times (1-x)M = xM^2(1-x)$

For maximum force, $\frac{dF}{dx} = 0$

$$\Rightarrow \frac{dF}{dx} = M^2 - 2xM^2 = 0 \Rightarrow x = 1/2$$

14. **(c)** Mass of the satellite = m and height of satellite from earth (h) = 6.4×10^6 m.

We know that gravitational potential energy of the satellite at height

$$h = -\frac{GM_e m}{R_e + h} = -\frac{gR_e^2 m}{2R_e} = -\frac{gR_e m}{2} = -0.5\,mgR_e$$

(where, $GM_e = gR_e^2$ and $h = R_e$)

15. **(d)** Acceleration due to gravity on earth's surface

$$g = G\frac{M}{R^2}$$

This implies that as radius decreases, the acceleration due to gravity increases.

$$\frac{\Delta g}{g} = -2\frac{\Delta R}{R} \quad \text{But } \frac{\Delta R}{R} = -1\%$$

('–' sign is due to shrinking of earth)

$$\therefore \frac{\Delta g}{g} = -2 \times (-1\%) = 2\%$$

16. **(a)** According to kepler's law of area

$$\frac{dA}{dt} = \frac{L}{2m}$$

For central forces, torque = 0

$\therefore$ L = constant

$$\therefore \frac{dA}{dt} = \text{constant}$$

17. **(b)** Potential energy of particle at the centre of square

$$= -4\left(\frac{GMm}{\frac{a}{\sqrt{2}}}\right)$$

$$\therefore -4\left(\frac{GMm}{a/\sqrt{2}}\right) + \frac{1}{2}mv^2 = 0 \Rightarrow v^2 = \frac{8\sqrt{2}\,GM}{a}$$

18. **(c)** The potential energy for a conservative force is defined as

$$F = \frac{-dU}{dr} \text{ or } U = -\int_{\infty}^{r} \vec{F}.d\vec{r} \qquad \ldots\ldots(i)$$

$$\text{or } U_r = \int_{\infty}^{r} \frac{GM_1M_2}{r^2}\,dr = \frac{-GM_1M_2}{r} \qquad \ldots\ldots(ii)$$

$(\because U_\infty = 0)$

If we bring the mass from the infinity to the centre of earth, then we obtain work, 'so it has negative (gravitational force do work on the object) sign & potential energy decreases. But if we bring the mass from the surface of earth to infinite, then we must do work against gravitational force & potential energy of the mass increases.

Now in equation (i) if $F = \frac{GM_1M_2}{r^{5/2}}$ instead of

$F = \frac{GM_1M_2}{r^2}$ then

$$U_r = \int_{\infty}^{r} \frac{GM_1M_2}{r^{5/2}}\,dr = \frac{-2}{3}\frac{GM_1M_2}{r^{3/2}}$$

$$\Rightarrow U_r \propto \frac{1}{r^{+3/2}}$$

19. **(b)** As we know, the minimum speed with which a body is projected so that it does not return back is called escape speed.

$$V_e = \sqrt{\frac{2GM}{r}} = \sqrt{\frac{2GM}{R+h}} = \sqrt{\frac{2GM}{4R}}$$

$$= \left(\frac{GM}{2R}\right)^{\frac{1}{2}} \quad (\because h = 3R)$$

20. **(a)** Acceleration due to gravity at a height h above the earth's surface is

$$g_h = g\left(1 - \frac{2h}{R}\right)$$

Acceleration due to gravity at a depth d below the earth's surface is

$$g_d = g\left(1 - \frac{d}{R}\right)$$

$$\text{Now, } \frac{g_h}{g_d} = \frac{\left(1 - \frac{2h}{R}\right)}{\left(1 - \frac{d}{R}\right)} = \frac{(R - 2h)}{(R - d)}$$

As h = 1 km, d = 1 km

$$\therefore \frac{g_h}{g_d} = \frac{R-2}{R-1}$$

21. (a) At the surface of earth, the value of $g = 9.8 m/sec^2$. If we go towards the centre of earth or we go above the surface of earth, then in both the cases the value of g decreases.

Hence $W_1 = mg_{mine}$, $W_2 = mg_{sea\ level}$, $W_3 = mg_{moun}$

So $W_1 < W_2 > W_3$ (g at the sea level = g at the suface of earth)

22. (d) Time period does not depend upon the mass of satellite

23. (a) $T = \frac{2\pi r}{v_0} = \frac{2\pi r}{(gR^2/r)^{1/2}} = \frac{2\pi r^{3/2}}{\sqrt{gR^2}} = \frac{2\pi}{\omega}$

Hence, $r^{3/2} = \frac{\sqrt{gR^2}}{\omega}$ or $r^3 = \frac{gR^2}{\omega^2}$

or, $r = (gR^2/\omega^2)^{1/3}$

24. (b) $g' = g - \omega^2 R\cos^2\lambda$

To make effective acceleration due to gravity zero at equator $\lambda = 0$ and $g' = 0$

$\therefore 0 = g - \omega^2 R \Rightarrow \omega = \sqrt{\frac{g}{R}} = \frac{1}{800}\frac{rad}{s}$

25. (a) $mg = 72$ N (body weight on the surface)

$$g = \frac{GM}{R^2}$$

At a height $H = \frac{R}{2}$,

$$g' = \frac{GM}{\left(R + \frac{R}{2}\right)^2} = \frac{4\ GM}{9\ R^2}$$

Body weight at height $H = \frac{R}{2}$,

$$mg' = m \times \frac{4\ GM}{9\ R^2}$$

$$= m \times \frac{4}{9} \times g = \frac{4}{9} mg$$

$$= \frac{4}{9} \times 72 = 32\text{ N}$$

26. (c) At a height h,

$$g' = g\frac{R^2}{(R+h)^2} \Rightarrow mg' = mg\left(\frac{R}{R+h}\right)^2$$

$$\Rightarrow W' = W\left(\frac{R}{R+h}\right)^2$$

Here, $h = R/2$

$$\therefore W' = \frac{4}{9}W$$

27. (c) Gravitational P.E. = m × gravitational potential

U = mV, so the graph of U will be same as that of V for a spherical shell.

28. (c) Applying the properties of ellipse, we have

$$\frac{2}{R} = \frac{1}{r_1} + \frac{1}{r_2} = \frac{r_1 + r_2}{r_1 r_2}$$

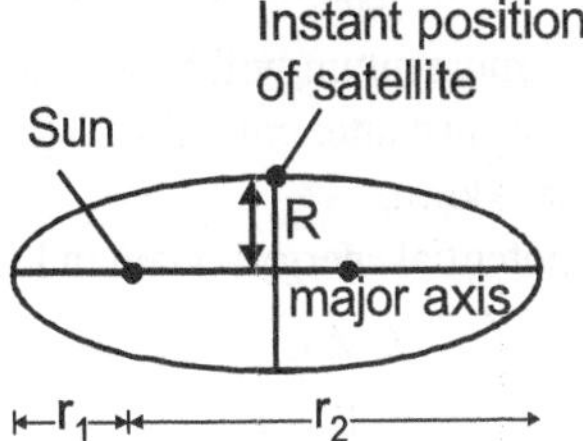

$$R = \frac{2 r_1 r_2}{r_1 + r_2}$$

29. (c) In a circular or elliptical orbital motion, torque is always acting parallel to displacement or velocity. So, angular momentum is conserved. In attractive field, potential energy is negative. Kinetic energy changes as velocity increase when distance is less. So, option (c) is correct.

30. (a) Here, $v = \sqrt{\frac{2GM}{R}}$ and $kv = \sqrt{\frac{2GM}{R+R}}$.

Solving $k = \frac{1}{\sqrt{2}}$

31. (a) $g = \frac{G(2M)}{(2R)^2} = \frac{GM}{2R^2}$

From $h = \frac{1}{2}gt^2$ $[\because U = 0]$

$$t = \sqrt{\frac{2h}{g}} = 2\sqrt{\frac{hR^2}{GM}}$$

32. (b) $\text{P.E.} = \int_{R_0}^{R} \frac{GMm}{r^2}dr = -GMm\left[\frac{1}{R} - \frac{1}{R_0}\right]$

The K.E. acquired by the body at the surface $= \frac{1}{2}mv^2$

$$\therefore \frac{1}{2}mv^2 = -GMm\left[\frac{1}{R} - \frac{1}{R_0}\right]$$

$$v = \sqrt{2GM\left(\frac{1}{R_0} - \frac{1}{R}\right)}$$

33. (b) $\frac{mv^2}{R} = \frac{k}{R^2}$ or $mv^2 = \frac{k}{R}$

Kinetic energy $= \frac{1}{2}mv^2 = \frac{k}{2R}$

In case of satellites P.E = – 2 K.E

and T.E = P. E + K. E

Total energy $= \frac{k}{2R} - \frac{k}{R} = -\frac{k}{2R}$

34. (d) Variation of g with altitude is,

$$g_h = g\left[1 - \frac{2h}{R}\right];$$

variation of g with depth is,

$$g_d = g\left[1-\frac{d}{R}\right]$$

Equating g_h and g_d, we get d = 2h

35. (a) The total momentum will be zero and hence velocity will be zero just after collisiion. The pull of earth will make it fall down.

36. (b) Loss in potential energy = Gain in kinetic energy

$$-\frac{GMm}{R}-\left(-\frac{3}{2}\frac{GMm}{R}\right)=\frac{1}{2}mv^2$$

$$\Rightarrow \frac{GMm}{2R}=\frac{1}{2}mv^2 \Rightarrow v=\sqrt{\frac{GM}{R}}=\sqrt{gR}$$

37. (d)

38. (b) $g \propto \frac{1}{R^2}$

R decreasing g increase hence, curve b represents correct variation.

39. (d) Angular momentum, L = Iω; moment of inertia of sphere along the axis passing through centre of mass,

$I=\frac{2}{5}MR^2$ and $\omega=\frac{2\pi}{T}$.

Putting these values, $L=\frac{4\pi MR^2}{5T}$

40. (c) $T=2\pi\sqrt{\frac{(R+h)^3}{GM}}$

$$T_1 = 2\pi\sqrt{\frac{R^3}{GM}},\quad T_2 = 2\pi\sqrt{\frac{(1.01R)^3}{GM}}$$

$$\frac{T_2-T_1}{T_1}\times 100 = 1.5\%$$

41. (a) The gravitational potential at the centre of uniform spherical shell is equal to the gravitational potential at the surface of shell i.e.,

$V=\frac{-GM}{a}$, where a is radius of spherical shell

Now, if the shell shrinks then its radius decrease then density increases, but mass is constant. so from above expression if a decreases, then V increases.

42. (b) $g' = g\left(1-\frac{d}{R}\right) \Rightarrow \frac{g}{n} = g\left(1-\frac{d}{R}\right)$

$$\Rightarrow d = \left(\frac{n-1}{n}\right)R$$

43. (d) $E_{earth} = E_{moon}$

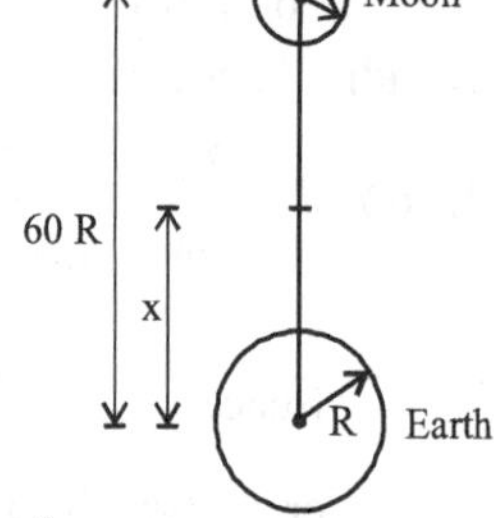

$$\Rightarrow \frac{GM}{x^2} = \frac{GM/81}{(60R-x)^2}$$

$$\Rightarrow \frac{1}{x} = \frac{1}{9(60R-x)}$$

$\Rightarrow$ x = 54 R from centre of earth.

44. (b) Acceleration due to gravity at lattitude' λ ' is given by

$$g_\lambda = g_e - R_e\omega^2\cos^2\lambda$$

At equator, $\lambda = 90° \Rightarrow \cos\lambda = \cos 90° = 0$

or $g_\lambda = g_e = g$ (as given in question)

At 30°, $g_{30} = g - R\omega^2\cos^2 30 = g - \frac{3}{4}R\omega^2$

or, $g - g_{30} = \frac{3}{4}R\omega^2$

45. (a) As we know,

Gravitational potential energy $= \frac{-GMm}{r}$

and orbital velocity, $v_0 = \sqrt{GM/R+h}$

$$E_f = \frac{1}{2}mv_0^2 - \frac{GMm}{3R} = \frac{1}{2}m\frac{GM}{3R} - \frac{GMm}{3R}$$

$$= \frac{GMm}{3R}\left(\frac{1}{2}-1\right) = \frac{-GMm}{6R}$$

$$E_i = \frac{-GMm}{R} + K$$

$$E_i = E_f$$

Therefore minimum required energy, $K = \frac{5GMm}{6R}$

1. (c) We know that Young's modulus

$$Y = \frac{F}{\pi r^2} \times \frac{L}{\ell}$$

Since Y, F are same for both the wires, we have,

$$\frac{1}{r_1^2}\frac{L_1}{\ell_1} = \frac{1}{r_2^2}\frac{L_2}{\ell_2}$$

or, $$\frac{\ell_1}{\ell_2} = \frac{r_2^2 \times L_1}{r_1^2 \times L_2} = \frac{(D_2/2)^2 \times L_1}{(D_1/2)^2 \times L_2}$$

or, $$\frac{\ell_1}{\ell_2} = \frac{D_2^2 \times L_1}{D_1^2 \times L_2} = \frac{D_2^2}{(2D_2)^2} \times \frac{L_2}{2L_2} = \frac{1}{8}$$

So, $\ell_1 : \ell_2 = 1 : 8$

2. (a) From the graph, it is clear that for the same value of load, elongation is maximum for wire *OA*. Hence *OA* is the thinnest wire among the four wires.

3. (b) Small amount of work done in extending the spring by *dx* is

$dW = kx\,dx$

$$\therefore W = k\int_{0.05}^{0.15} x\,dx$$

$$= \frac{800}{2}\left[(0.15)^2 - (0.05)^2\right]$$

$= 400\,[(0.15+0.05)(0.15-0.05)]$

$= 400 \times 0.2 \times 0.1 = 8\text{ J}$

4. (c) Using the usual expression for the Young's modulus, the force constant for the wire can be written as $k = \frac{F}{\Delta l} = \frac{YA}{L}$ where the symbols have their usual meanings. Now the two wires together will have an effective force constant $\left[\frac{k_1k_2}{k_1+k_2}\right]$. Substituting the corresponding lengths and the Young's moduli we get the answer.

5. (b) Compressibility of water,

$K = 45.4 \times 10^{-11}\text{ Pa}^{-1}$

density of water $P = 10^3\text{ kg/m}^3$

depth of ocean, $h = 2700$ m

We have to find $\frac{\Delta V}{V} = ?$

As we know, compressibility,

$$K = \frac{1}{B} = \frac{(\Delta V/V)}{P}\ (P = \rho gh)$$

So, $(\Delta V/V) = K\rho gh$

$= 45.4 \times 10^{11} \times 10^3 \times 10 \times 2700$

$= 1.2258 \times 10^{-2}$

6. (a) Young's modulus $Y = \frac{W}{A}.\frac{l}{\Delta l}$

$$\frac{W_1}{Y_1} = \frac{W_2}{Y_2}$$

[$\because$ A, l, Δl same for both brass and steel]

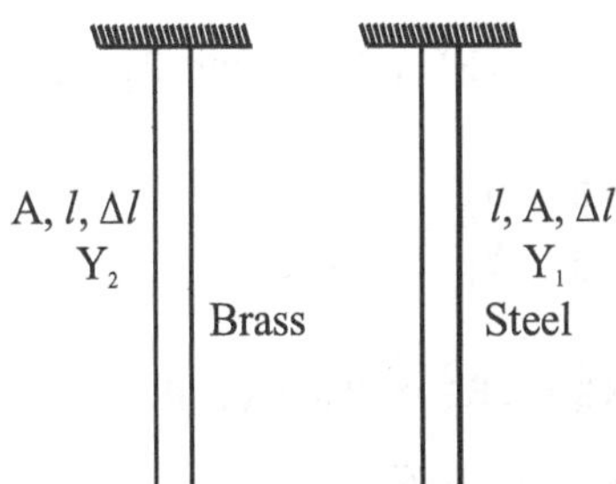

$$\frac{W_1}{W_2} = \frac{Y_1}{Y_2} = 2 \quad [Y_{steel}/Y_{brass} = 2 \text{ given}]$$

7. (b) Solids are least compressible whereas gases are highly compressible.

8. (d) $Y_c \times (\Delta L_c / L_c) = Y_s \times (\Delta L_s / L_s)$

$$\Rightarrow 1\times10^{11}\times\left(\frac{1\times10^{-3}}{1}\right) = 2\times10^{11}\times\left(\frac{\Delta L_s}{0.5}\right)$$

$$\therefore \Delta L_s = \frac{0.5\times10^{-3}}{2} = 0.25\text{ mm}$$

Therefore, total extension of the composite wire = $\Delta L_c + \Delta L_s$

$= 1\text{ mm} + 0.25\text{ mm} = 1.25\text{ mm}$

9. (b) Bulk modulus $B = \frac{-P}{(\Delta V/V)} = \frac{-PV}{\Delta V}$ ....(1)

and $\Delta V = \gamma V \Delta T = 3\alpha.V.T$ or $\frac{-V}{\Delta V} = \frac{1}{3\alpha.T.}$ ...(2)

From eqs. (1) and (2), $B = P/(3\alpha.T)$ or $T = \frac{P}{3\alpha B}$

10. (a) When same stress is applied at two different temperatures, the increase in length is more at higher temperature. Thus $T_1 > T_2$.

11. (c) According to questions,

$$\frac{\ell_s}{\ell_b} = a,\ \frac{r_s}{r_b} = b,\ \frac{y_s}{y_b} = c,\ \frac{\Delta\ell s}{\Delta\ell_b} = ?$$

As, $y = \frac{F\ell}{A\Delta\ell} \Rightarrow \Delta\ell = \frac{F\ell}{Ay}$

$$\Delta\ell_s = \frac{3mg\ell_s}{\pi r_s^2 . y_s} \quad [\because F_s = (M + 2M)g]$$

$$\Delta\ell_b = \frac{2Mg\ell_b}{\pi r_b^2 \cdot y_b} \quad [\because F_b = 2Mg]$$

$$\therefore \frac{\Delta\ell_s}{\Delta\ell_b} = \frac{\dfrac{3Mg\ell_s}{\pi r_s^2 . y_s}}{\dfrac{2Mg.\ell_b}{\pi r_b^2 . y_b}} = \frac{3a}{2b^2c}$$

12. (b) We know that $Y = FL/\pi r^2 \ell$ or $r^2 = FL/(Y\pi\ell)$

$$\therefore \quad R_B^2 = FL/(Y_B \pi \ell) \text{ and } R_S^2 = FL/(Y_S \pi \ell)$$

$$\text{or } \frac{R_B^2}{R_S^2} = \frac{Y_S}{Y_B} = \frac{2\times 10^{10}}{10^{10}} = 2$$

$$\text{or } R_B^2 = 2R_S^2 \text{ or } R_B = \sqrt{2}\, R_S$$

$$\therefore R_S = R_B/\sqrt{2}$$

13. (c) 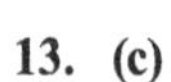

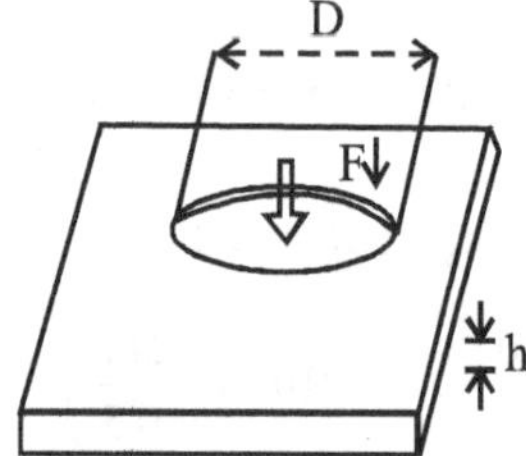

Shearing strain is created along the side surface of the punched disk. Note that the forces exerted on the disk are exerted along the circumference of the disk, and the total force exerted on its center only.

Let us assume that the shearing stress along the side surface of the disk is uniform, then

$$F = \int_{\text{surface}} dF_{max} = \int_{\text{surface}} \sigma_{max} dA = \sigma_{max} \int_{\text{surface}} dA$$

$$= \int \sigma_{max} . A = \sigma_{max} . 2\pi\left(\frac{D}{2}\right)h$$

$$= 3.5\times 10^8 \times \left(\frac{1}{2}\times 10^{-2}\right)\times 0.3\times 10^{-2}\times 2\pi$$

$$= 3.297\times 10^4 \simeq 3.3\times 10^4\,\text{N}$$

14. (d) Bulk modulus is given by, $k = \dfrac{F/A}{\Delta V/V}$

$$= \frac{mg}{A\left(\dfrac{\Delta V}{V}\right)} = \frac{h\rho g}{\left(\dfrac{\Delta V}{V}\right)}, \quad \left(\because \rho = \frac{m}{V}, V = A\times h\right)$$

Given, $h = 400$ m, $\dfrac{\Delta V}{V} = \dfrac{0.2}{100}$

and $\rho = 1\times 10^3$ kg/m^3

$$\therefore k = \frac{400\times 10^3 \times 9.8}{0.2/100} = 196\times 10^7\,\text{N m}^{-2}$$

$$k = 1.96\times 10^9\,\text{N m}^{-2}.$$

15. (a) $C_1 = \dfrac{\pi\eta(r_2^4 - r_1^4)}{2\ell}, C_2 = \dfrac{\pi\eta r^4}{2\ell}$

Initial volume = Final volume

$$\therefore \pi[r_2^2 - r_1^2]\ell\rho = \pi r^2 \ell\rho$$

$$\Rightarrow r^2 = r_2^2 - r_1^2 \Rightarrow r^2 = (r_2 + r_1)(r_2 - r_1)$$

$$\Rightarrow r^2 = (8.02 + 7.98)(8.02 - 7.98)$$

$$\Rightarrow r^2 = 16\times 0.04 = 0.64\text{ cm} \Rightarrow r = 0.8\text{ cm}$$

$$\therefore \frac{C_1}{C_2} = \frac{r_2^4 - r_1^4}{r^4} = \frac{[8.02]^4 - [7.98]^4}{[0.8]^4}$$

16. (c)

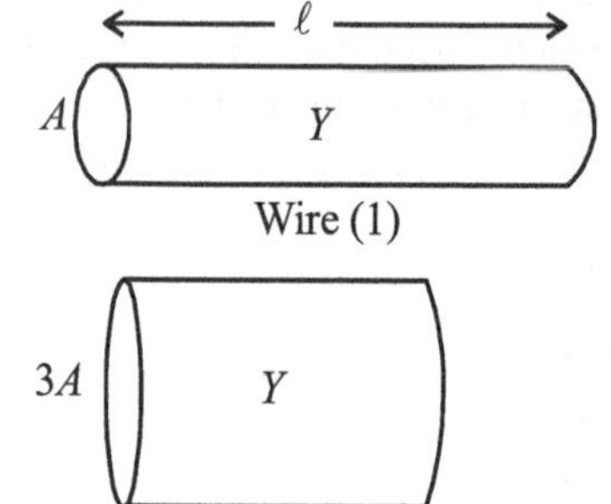

As shown in the figure, the wires will have the same Young's modulus (same material) and the length of the wire of area of cross-section $3A$ will be $\ell/3$ (same volume as wire 1).

For wire 1,

$$Y = \frac{F/A}{\Delta x/\ell} \quad \text{...(i)}$$

For wire 2,

$$Y = \frac{F'/3A}{\Delta x/(\ell/3)} \quad \text{...(ii)}$$

From (i) and (ii), $\dfrac{F}{A}\times\dfrac{\ell}{\Delta x} = \dfrac{F'}{3A}\times\dfrac{\ell}{3\Delta x} \Rightarrow F' = 9F$

17. (c) Poisson's ratio, $\sigma = \dfrac{\text{lateral strain }(\beta)}{\text{longitudinal strain }(\alpha)}$

For material like copper, $\sigma = 0.33$

And, $y = 3k(1 - 2\sigma)$

Also, $\dfrac{9}{y} = \dfrac{1}{k} + \dfrac{3}{n}$

$y = 2n(1 + \sigma)$

Hence, $n < y < k$

18. **(b)** Stress = 1 kg wt/mm^2 = 9.8 N/mm^2
= 9.8 × 10^6 N/m^2.

$$Y = 1\times10^{11} N/m^2, \quad \frac{\Delta\ell}{\ell}\times100 = ?$$

$$Y = \frac{\text{Stress}}{\text{Strain}} = \frac{\text{Stress}}{\Delta\ell/\ell}$$

$$\therefore \frac{\Delta\ell}{\ell} = \frac{\text{Stress}}{Y} = \frac{9.8\times10^6}{1\times10^{11}}$$

$$\frac{\Delta\ell}{\ell}\times100 = 9.8\times10^{-11}\times100\times10^6$$

$$= 9.8 \times 10^{-3} = 0.0098\ \%$$

19. **(d)**
$$W_1 = \frac{1}{2}kx^2$$
and
$$W_2 = \frac{1}{2}k(x+y)^2$$
$$\therefore \quad W = W_2 - W_1 = \frac{1}{2}k(x+y)^2 - \frac{1}{2}kx^2$$
$$= \frac{1}{2}ky(2x+y)$$

20. **(c)** Here, $k_Q = \frac{k_p}{2}$

According to Hooke's law

$\therefore \quad F_p = -k_p x_p$

$$F_Q = -k_Q x_Q \Rightarrow \frac{F_p}{F_Q} = \frac{k_p x_p}{k_Q x_Q}$$

$F_p = F_Q$ [Given]

$$\therefore \quad \frac{x_p}{x_Q} = \frac{k_Q}{k_p} \quad ...(i)$$

Energy stored in a spring is $U = \frac{1}{2}kx^2$

$$\therefore \quad \frac{U_p}{U_Q} = \frac{k_p x_p^2}{k_Q x_Q^2} = \frac{k_p}{k_Q}\times\frac{k_Q^2}{k_p^2} = \frac{1}{2} \quad \left[\because k_Q = \frac{k_p}{2}\right]$$

$$\Rightarrow \quad U_p = \frac{U_Q}{2} = \frac{E}{2} \quad [\because U_Q = E]$$

21. **(a)** Young's modulus $Y = \frac{\text{stress}}{\text{strain}}$

stress = Y × strain

Stress in steel wire = Applied pressure

Pressure = stress = Y × strain

Strain $= \frac{\Delta L}{L} = \alpha\,\Delta T$ (As length is constant)

$= 2\times10^{11}\times1.1\times10^{-5}\times100 = 2.2\times10^8$ Pa

22. **(b)** Let T be the tension in the ring, then

$$Y = \frac{T.2\pi r}{A.2\pi(R-r)} = \frac{Tr}{A(R-r)} \quad \therefore \quad T = \frac{YA(R-r)}{r}$$

23. **(a)** Ratio of radii $r_1 : r_2 = 1:2$

Ratio of area, $A_1 : A_2 = \pi r_1^2 : \pi r_2^2$

$A_1 : A_2 = 1 : 4$

Now, Stress$_1$: Stress$_2$ = 4 : 1

So, Strain$_1$: Strain$_2$ = 4 : 1

$$\therefore \quad \frac{l_1}{l_2} = \frac{4}{1} \Rightarrow 4l_2 = l_1 = 8$$

$\therefore \quad l_2 = 2$ mm

Increase in length of B is 2 mm.

24. **(b)** $\frac{dV}{V} = (1+2\sigma)\frac{dL}{L}$

$$\frac{dV}{V} = 2\times2\times10^{-3} = 4\times10^{-3}$$

$$\left[\because \sigma = 0.5 = \frac{1}{2}\right]$$

$\therefore$ Percentage change in volume = 4 × 10^{-1} = 0.4%

25. **(b)** $r\theta = \ell\phi \Rightarrow \phi = \frac{r\theta}{\ell} = \frac{6\text{mm}\times30^\circ}{1\text{m}} = 0.18^\circ$

26. **(d)** Bulk modulus $B = \frac{|-dp|}{\left|\left(\frac{dV}{V}\right)\right|}$

$\therefore$ Pressure, $dp = B\left(\frac{\Delta V}{V}\right)$

27. **(a)** Given: $F = 100$ kN $= 10^5$ N

$Y = 2\times10^{11}$ Nm^{-2}

$\ell_0 = 1.0$ m

radius r = 10 mm = 10^{-2} m

From formula, $Y = \frac{\text{Stress}}{\text{Strain}}$

$$\Rightarrow \quad \text{Strain} = \frac{\text{Stress}}{Y} = \frac{F}{AY}$$

$$= \frac{10^5}{\pi r^2 Y} = \frac{10^5}{3.14\times10^{-4}\times2\times10^{11}}$$

$$= \frac{1}{628}$$

Therefore % strain $= \frac{1}{628}\times100 = 0.16\%$

28. **(c)**

δ

For a beam, the depression at the centre is given by,

$$\delta = \left(\frac{fL}{4Ybd^3}\right)$$

[f, L, b, d are constants for a particular beam]

i.e. $\delta \propto \frac{1}{Y}$

29. (b) $K = \frac{F}{x} = \frac{4\times 9.8}{2\times 10^{-2}} = 19.6\times 10^{2}$

Work done $= \frac{1}{2}\times 19.6\times 10^{2}\times (0.05)^{2} = 2.45\,J$

30. (c) If ℓ is the original length of wire, then change in length of first wire, $\Delta\ell_1 = (\ell_1 - \ell)$

change in length of second wire, $\Delta\ell_2 = (\ell_2 - \ell)$

Now, $Y = \frac{T_1}{A}\times\frac{\ell}{\Delta\ell_1} = \frac{T_2}{A}\times\frac{\ell}{\Delta\ell_2}$

or $\frac{T_1}{\Delta\ell_1} = \frac{T_2}{\Delta\ell_2}$ or $\frac{T_1}{\ell_1-\ell} = \frac{T_2}{\ell_2-\ell}$

or $T_1\ell_2 - T_1\ell = T_2\ell_1 - \ell T_2$ or $\ell = \frac{T_2\ell_1 - T_1\ell_2}{T_2 - T_1}$

31. (a) $\delta = \frac{W\ell^3}{3YI}$, where W = load, ℓ = length of beam and I is geometrical moment of inertia for rectangular beam,

$I = \frac{bd^3}{12}$ where b = breadth and d = depth

For square beam b = d

$\therefore I_1 = \frac{b^4}{12}$

For a beam of circular cross-section, $I_2 = \left(\frac{\pi r^4}{4}\right)$

$\therefore \quad \delta_1 = \frac{W\ell^3\times 12}{3Yb^4} = \frac{4W\ell^3}{Yb^4}$ (for sq. cross section)

and $\delta_2 = \frac{W\ell^3}{3Y(\pi r^4/4)} = \frac{4W\ell^3}{3Y(\pi r^4)}$

(for circular cross-section)

Now $\frac{\delta_1}{\delta_2} = \frac{3\pi r^4}{b^4} = \frac{3\pi r^4}{(\pi r^2)^2} = \frac{3}{\pi}$

($\because b^2 = \pi r^2$ i.e., they have same cross-sectional area)

32. (a) Compressibility $= \frac{1}{\text{Bulk modulus}}$

As bulk modulus is least for ethanol (0.9) and maximum for mercury (25) among ehtanol, mercury and water.

Hence compression in volume $\frac{\Delta V}{V}$

Ethanol > Water > Mercury

33. (c) The given graph does not obey Hooke's law. and there is no well defined plastic region. So the graph represents elastomers.

34. (a) $U/\text{volume} = \frac{1}{2}Y\times \text{strain}^2 = 3600\ J\,m^{-3}$

[Strain $= 0.06\times 10^{-2}$]

35. (d) Potential energy per unit volume of the wire is given by :

$u = \frac{1}{2}\frac{(\text{Stress})^2}{\text{Young's modulus}} = \frac{1}{2}\frac{S^2}{Y}$

As stress, $S = \frac{\text{Force}}{\text{Area}}$

$\therefore \frac{S_1}{S_2} = \left(\frac{F_1}{F_2}\right)\left(\frac{A_2}{A_1}\right)$

As $F_1 = F_2$ (Given)

$\therefore \frac{S_1}{S_2} = \left(\frac{F_1}{F_2}\right)\left(\frac{A_2}{A_1}\right) = \left(\frac{A_2}{A_1}\right)$ ..(i)

The two wires are of the same material, therefore their Young's moduli will be same i.e., $Y_1 = Y_2$

$\therefore \frac{u_1}{u_2} = \left(\frac{S_1}{S_2}\right)^2 = \left(\frac{A_2}{A_1}\right)^2$

$= \left(\frac{\pi\left(\frac{d_2}{2}\right)^2}{\pi\left(\frac{d_1}{2}\right)^2}\right)^2 = \left[\left(\frac{d_2}{d_1}\right)^2\right]^2$

$= \left(\frac{d_2}{d_1}\right)^4 = \left(\frac{2}{1}\right)^4 = \frac{16}{1}$ $\left(\because \frac{d_1}{d_2} = \frac{1}{2}\ \text{(Given)}\right)$

36. (b) Using Hooke's law, $F = kx$ we can write

$4 = k(a - \ell_0)$ ...(i)

and $5 = k(b - \ell_0)$ ...(ii)

If ℓ be the length under tension 9N, then

$9 = k(\ell - \ell_0)$ ...(iii)

After solving above equations, we get

$\ell = (5b - 4a)$.

37. (b) $F = Y\times A\times\frac{l}{L} \Rightarrow F\propto r^2$ (Y, l and and L are constant)

If diameter is made four times then force required will be 16 times, i.e., 16×10^3 N

38. (c) Young's modulus of elasticity is

$Y = \frac{F/A}{\Delta L/L}$

$\therefore \Delta L = \frac{FL}{AY}$

So, $\Delta L \propto \frac{L}{A}$

$\therefore \frac{\Delta L_2}{\Delta L_1} = \frac{L_2}{L_1}\times\frac{A_1}{A_2} = \frac{2}{1}\times\frac{2}{1} = 4$

$\Delta L_2 = 4\times\Delta L_1 = 4\times 1 = 4$ cm

39. (a) $y = \frac{F/A}{\Delta l/l} = \frac{F}{A}.\frac{l}{\Delta l}$

$= \frac{20 \times 1}{10^{-6} \times 10^{-4}} = 2 \times 10^{11} \text{Nm}^{-2}$

40. (d) $K = \frac{\Delta P}{\Delta V/V} = \frac{h\rho g}{\Delta V/V} = \frac{200 \times 10^3 \times 10}{0.1/100} = 2 \times 10^9$

41. (d) Bulk Modulus = $\frac{dp}{\frac{dv}{v}}$

$dp = h\rho g = 200 \times 10^3 \times 9.8$

$\frac{dv}{v} = \frac{0.1}{100}$

Bulk modulus $= \frac{200 \times 10^3 \times 9.8}{0.1/100} = 19.6 \times 10^8 \text{N/m}^2$

42. (c)

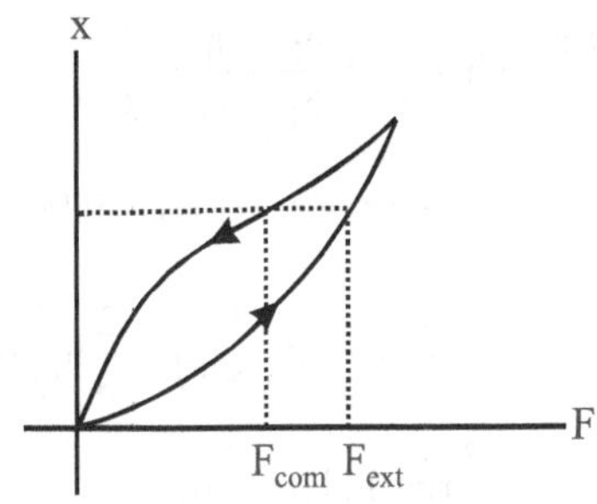

From the figure, it is clear that
$F_{com} < F_{ext}$.

43. (d) $\frac{\Delta r/r}{\Delta l/l} = 0.5 = \frac{1}{2}, \frac{\Delta r}{r} = \frac{1}{2}\frac{\Delta l}{l}$

44. (b) As $Y = \frac{\frac{F}{A}}{\frac{\Delta l}{l}} \Rightarrow \Delta l = \frac{Fl}{AY}$

But $V = Al$ so $A = \frac{V}{l}$

Therefore $\Delta l = \frac{Fl^2}{VY} \propto l^2$

Hence graph of Δl versus l^2 will give a straight line.

45. (b) $K = \frac{F}{x} = \frac{4 \times 9.8}{2 \times 10^{-2}} = 19.6 \times 10^2$

Work done $= \frac{1}{2} \times 19.6 \times 10^2 \times (0.05)^2 = 2.45 J$

1. **(a)** Bulk modulus,

$$B = -\frac{\Delta P}{\left(\frac{\Delta V}{V_0}\right)} \Rightarrow \Delta V = -V_0 \frac{\Delta P}{B}$$

or $V - V_0 = -V_0 \frac{\Delta P}{B}$ (Here V_0 = volume at the surface and V = volume at the depth)

or $V = V_0 - V_0 \frac{\Delta P}{B} \Rightarrow V = V_0\left(1 - \frac{\Delta P}{B}\right)$

$\therefore$ Density, $\rho' = \frac{m}{V} = \frac{m}{V_0\left(1 - \frac{\Delta P}{B}\right)}$

$$= \frac{m}{\frac{m}{\rho}\left(1 - \frac{nP_0 - P_0}{B}\right)} \quad (\because \Delta P = nP_0 - P_0)$$

$$\rho' = \frac{\rho B}{B - (n-1)P_0}$$

2. **(c)** Velocity of ball when it strikes the water surface

$v = \sqrt{2gh}$ ...(i)

Terminal velocity of ball inside the water

$v = \frac{2}{9} r^2 g \frac{(\rho - 1)}{\eta}$ ...(ii)

Equation (i) and (ii) we get $\sqrt{2gh} = \frac{2}{9}\frac{r^2 g}{\eta}(\rho - 1)$

$$\Rightarrow h = \frac{2}{81} r^4 \left(\frac{\rho - 1}{\eta}\right)^2 g$$

3. **(b)**

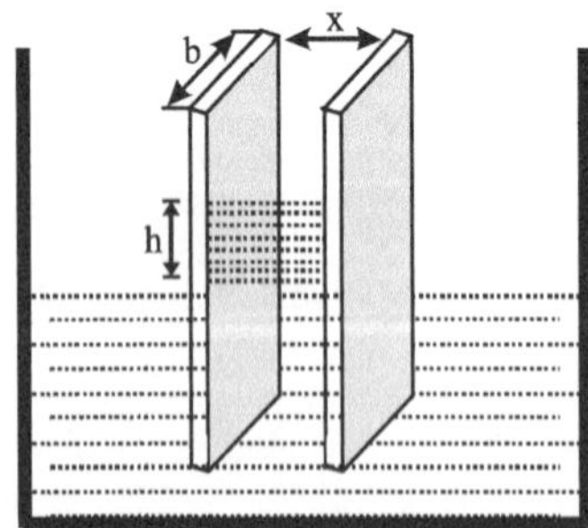

Let the width of each plate is b and due to surface tension liquid will rise upto height h then upward force due to surface tension.

= 2Tbcosθ ...(i)

Weight of the liquid rises in between the plates

= Vdg = (bxh)dg ...(ii)

Equating (i) and (ii) we get, 2T cosθ = xhdg

$$\therefore h = \frac{2T\cos\theta}{xdg}$$

4. **(b)** The theorem of continuity is valid.

$\therefore A_1 v_1 \rho = A_2 v_2 \rho$ as the density of the liquid can be taken as uniform.

$\therefore A_1 v_1 = A_2 v_2$

$\Rightarrow$ Smaller the area, greater the velocity.

5. **(c)** $P_a + \frac{1}{2}\rho_1 v_1^2 + 0 = P_a + \frac{1}{2}\rho_2 v_2^2 + (\rho_1 g h_1 + \rho_2 g h_2)$

As $v_2 << v_1$, $\therefore v_1 = \sqrt{2g(h_1 + h_2)\left(\frac{\rho_2}{\rho_1}\right)}$.

6. **(c)** $h = \frac{2T\cos\theta}{r\rho g} \Rightarrow h \propto \frac{1}{r} \Rightarrow \frac{h_2}{h_1} = \frac{r_1}{r_2} = \frac{2}{3}$

$$\left(\because r_1 = r,\ r_2 = r + 50\%\text{ of } r = \frac{3}{2}r\right)$$

New mass $m_2 = \pi r_2^2 h_2 \rho = \pi\left(\frac{3}{2}r_1\right)^2\left(\frac{2}{3}h_1\right)\rho$

$$= \frac{3}{2}\left(\pi r_1^2 h_1\right)\rho = \frac{3}{2}m$$

7. **(a)** When a body falls through a viscous liquid, its velocity increases due to gravity but after some time its velocity becomes uniform because of viscous force becoming equal to the gravitational force. Viscous force itself is a variable force which increases as velocity increases, so curve (a) represents the correct alternative.

8. **(c)** Sum of volumes of 2 smaller drops

= Volume of the bigger drop

$2.\frac{4}{3}\pi r^3 = \frac{4}{3}\pi R^3 \Rightarrow R = 2^{1/3} r$

Surface energy = $T.4\pi R^2$

$= T4\pi 2^{2/3} r^2 = T.2^{8/3}\pi r^2$.

9. **(c)** Angle of contact θ

$$\cos\theta = \frac{T_{SA} - T_{SL}}{T_{LA}}$$

when water is on a waxy or oily surface

$T_{SA} < T_{SL}$ cos θ is negative i.e., $90° < \theta < 180°$

i.e., angle of contact θ increases

And for $\theta > 90°$ liquid level in capillary tube fall. i.e., h decreases

10. **(a)** $F = 6\pi\eta r v$

$= 6 \times 3.14 \times (8 \times 10^{-5}) \times 0.03 \times 100$

$= 4.52 \times 10^{-3}$ dyne

11. **(a)** Velocity of water from hole = $v_1 = \sqrt{2gh}$

Velocity of water from hole B

$v_2 = \sqrt{2g(H_0 - h)}$

Time of reaching the ground from hole B

$t_1 = \sqrt{2(H_0 - h)/g}$

Time of reaching the ground from hole A

$t_2 = \sqrt{2h/g}$

12. (a) Fluid resistance is given by $R = \frac{8\eta L}{\pi r^4}$

When two capillary tubes of same size are joined in parallel, then equivalent fluid resistance is

$$R_S = R_1 + R_2 = \frac{8\eta L}{\pi R^4} + \frac{8\eta \times 2L}{\pi (2R)^4} = \left(\frac{8\eta L}{\pi R^4}\right) \times \frac{9}{8}$$

$$\text{Rate of flow} = \frac{P}{R_S} = \frac{\pi P R^4}{8\eta L} \times \frac{8}{9} = \frac{8}{9} X \quad \left[\text{as } X = \frac{\pi P R^4}{8\eta L}\right]$$

13. (b) The candle floats on the water with half its length above and below water level. Let its length be 10 cm with 5 cm below the surface and 5 cm above it. If its length is reduced to 8 cm, it will have 4 cm above water surface. So we see tip going down by 1 cm.

∴ rate of fall of tip = 1 cm/hour.

14. (a) Inside pressure must be $\frac{4T}{r}$ greater than outside pressure in bubble.

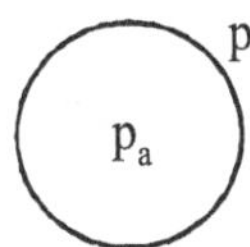

This excess pressure is provided by charge on bubble.

$$\frac{4T}{r} = \frac{\sigma^2}{2\varepsilon_0}; \quad \frac{4T}{r} = \frac{Q^2}{16\pi^2 r^4 \times 2\varepsilon_0} \quad \left[\sigma = \frac{Q}{4\pi r^2}\right]$$

$$Q = 8\pi r\sqrt{2rT\varepsilon_0}$$

15. (a) Because film tries to cover minimum surface area.

16. (d)

17. (a) The condition for terminal speed (v_t) is

Weight = Buoyant force + Viscous force

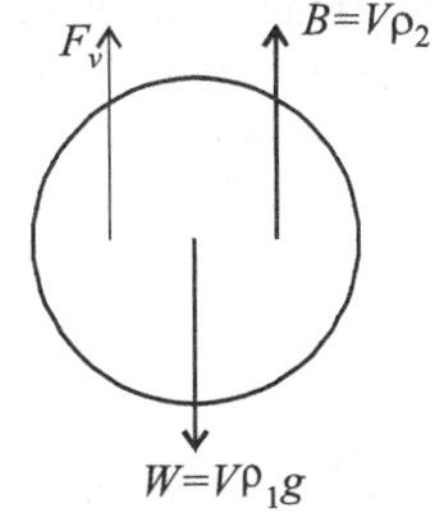

$$\therefore V\rho_1 g = V\rho_2 g + kv_t^2 \qquad \therefore v_t = \sqrt{\frac{Vg(\rho_1 - \rho_2)}{k}}$$

18. (d) According to Bernoulli's theorem, when velocity of liquid flow increases, the pressure decreases.

19. (c) Wetability of a surface by a liquid primarily depends on angle of contact between the surface and liquid. If angle of contact is acute liquids wet the solid and vice-versa.

20. (a) dv = 8 cm/s and dx = 0.1 cm

$$\text{Velocity gradient} = \frac{dv}{dx} = \frac{8}{0.1} = 80/s.$$

21. (a) Terminal velocities of rain drops are proportional to square of their radii.

Terminal velocity of a body is given by

$$v_T = \frac{2R^2}{9\eta}(d - \sigma)g. \text{ or, } V \propto R^2$$

22. (b)

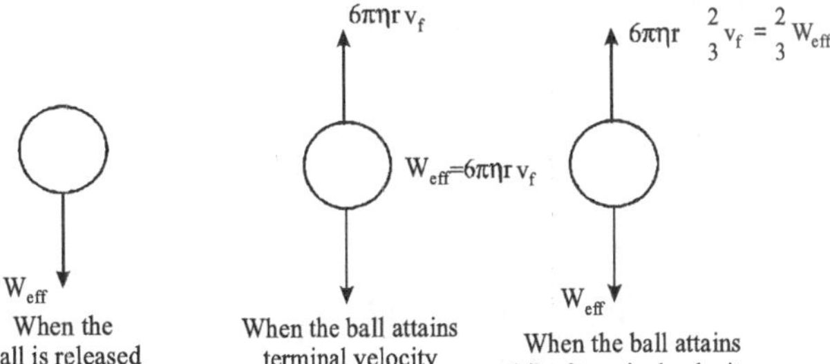

When the ball is just released, the net force on ball is W_{eff} (= mg – buoyant force)

The terminal velocity v_f of the ball is attained when net force on the ball is zero.

∴ Viscous force $6\pi\eta r\, v_f = W_{eff}$

When the ball acquires $\frac{2}{3}$rd of its maximum velocity v_f

the viscous force is $= \frac{2}{3} W_{eff}$

Hence net force is $W_{eff} - \frac{2}{3}W_{eff} = \frac{1}{3}W_{eff}$

∴ required acceleration is a/3

23. (a) $(2\pi r_1 + 2\pi r_2)\sigma = mg$

$$\left[2\pi \times \frac{8.7}{2} + 2\pi \times \frac{8.5}{2}\right]\sigma = 3.97 \times 980$$

$$\Rightarrow \sigma = 72 \text{ dyne cm}^{-1}$$

24. (b) Over a small temperature ranges, S.T. of water decreases linearly with rise of temperature.

25. (c) Volume of air bubble $V = \frac{4}{3}\pi r^3$

We get, $V \propto r^3$

If r is 2 times, V becomes 8 times at the surface of lake.

Pressure at the surface of lake is given by

$P_1 = 1$ atmosphere, $V_1 = 8V$

$\Rightarrow P_1 = Hdg$ where, d = density of water

Pressure at the bottom of lake,

P_2 = Pressure of atmosphere + Pressure of water

$P_2 = Hdg + hdg = (H + h)\, dg$ where, h = depth of lake

Let final volume, $V_2 = V$.

Because temperature is constant, hence from Boyle's law

$P_1V_1 = P_2V_2 \Rightarrow Hdg \times 8V = (H + h)dg \times V \Rightarrow h = 7H.$

26. (c)

27. (b) Volume of first piece of metal $= \frac{32}{8} = 4\,\text{cm}^3$

Upthrust = 4 gf

Effective weight = (32 – 4) gf = 28 gf

If m be the mass of second body, volume of second body is $\frac{m}{5}$

Now, $28 = m - \frac{m}{5} \Rightarrow m = 35$ g

28. (b) $g_{eff.} = 12\ \text{m/s}^2$, $= \frac{\rho_m}{\rho_w} = \frac{4}{10}$

$$a = \frac{V\rho_w \times 12 - V\rho_m \times 12}{V\rho_m} = 18\,\text{m/s}^2$$

$1 = \frac{1}{2} \times 18\,t^2$, $t = \frac{1}{3}$ s.

29. (b) Due to increase in velocity, pressure will be low above the surface of water.

30. (c) If ρ is the density of the ball and ρ′ that of the another ball, *m* for the balls are the same, but $r' = 2r$

$\therefore mg = 6\pi r\eta v$ (by Stoke's law)

or, $6\pi r\eta v = 6\pi 2r\eta v'$ So, $v' = \frac{v}{2}$

31. (d) As we know,

Pressure P = Vdg

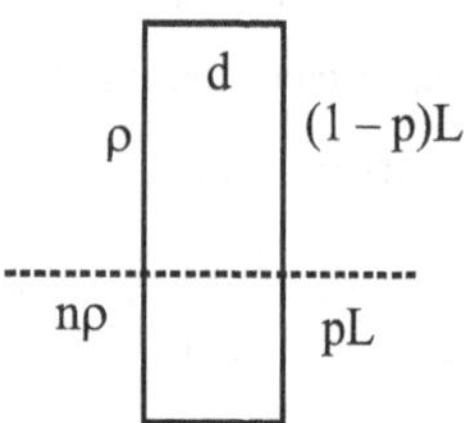

Here, LA d g = (pL) A (nρ)g + (1 – p)LA ρ g

$\Rightarrow d = (1 - p)\rho + pn\,\rho = [1 + (n - 1)p]\rho$

32. (d) At equilibrium, weight of the given block is balanced by force due to surface tension, i.e.,

2L. S = W

or $S = \frac{W}{2L} = \frac{1.5 \times 10^{-2}\,N}{2 \times 0.3\,m} = 0.025\,Nm^{-1}$

33. (a)

34. (d) $T_1 + T\cos(\pi - \theta) = T_2$

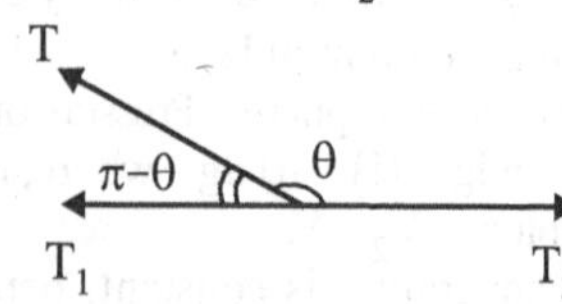

$\therefore \cos(\pi - \theta) = \frac{T_2 - T_1}{T}$

$\therefore -\cos\theta = \frac{T_2 - T_1}{T}$

$\therefore \cos\theta = \frac{T_1 - T_2}{T}$

35. (a) Let L = PQ = length of rod

$\therefore SP = SQ = \frac{L}{2}$

Weight of rod, W = A*l*ρg.

Acting at point S

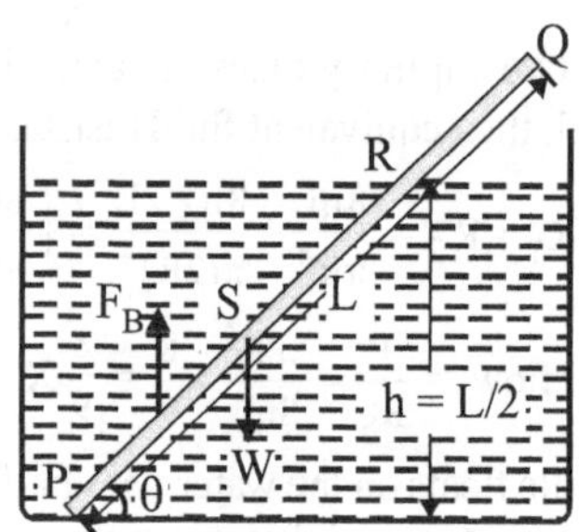

And force of buoyancy,

$F_B = Al\rho_0 g$. [l = PR]

Which acts at mid-point of PR.

For rotational equilibrium.

$$Al\rho_0 g \times \frac{\ell}{2}\cos\theta = AL\rho g \times \frac{L}{2}\cos\theta$$

$$\Rightarrow \frac{l^2}{L^2} = \frac{\rho}{\rho_0} \Rightarrow \frac{l}{L} = \sqrt{\frac{\rho}{\rho_0}}$$

From figure, $\sin\theta = \frac{h}{l} = \frac{L}{2l} = \frac{1}{2}\sqrt{\frac{\rho_0}{\rho}}$

36. (b) Terminal velocity, $v_0 = \frac{2\,r^2(\rho - \rho_0)g}{9\eta}$

$$= \frac{2 \times (2 \times 10^{-3})^2 \times (8 - 1.3) \times 10^3 \times 9.8}{9 \times 0.83} = 0.07\ \text{ms}^{-1}$$

37. (c) Work done = Surface tension × increase in area of the film

W = S × ΔA

Increase in area = Final area – initial area

$= 10 \times (0.5 + 0.1) - 10 \times 0.5 = 1\ \text{cm}^2$

$\therefore$ W = 72 × 2 × 1 = 144 erg

[∵ There are 2 free surfaces; ∴ ΔA = 2 × 1].

38. (b) Waterproofing agents are used so that the material does not get wet. This means angle of contact is obtuse.

39. (c)

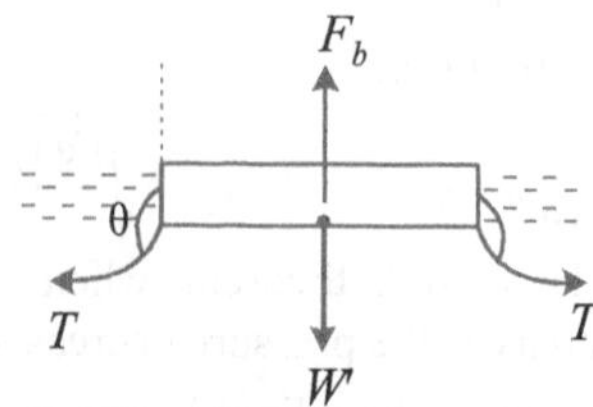

For floating disc, $F_{net} = 0$

or $F_b + 2\pi rT\cos\theta = W'$

or $W + 2\pi rT\cos\theta = W'$

40. **(b)**

41. **(d)** From the figure it is clear that liquid 1 floats on liquid 2. The lighter liquid floats over heavier liquid. Therefore we can conclude that $\rho_1 < \rho_2$

Also $\rho_3 < \rho_2$ otherwise the ball would have sink to the bottom of the jar.

Also $\rho_3 > \rho_1$ otherwise the ball would have floated in liquid 1. From the above discussion we conclude that

$$\rho_1 < \rho_3 < \rho_2.$$

42. **(a)** When the bubble gets detached,

Buoyant force = force due to surface tension

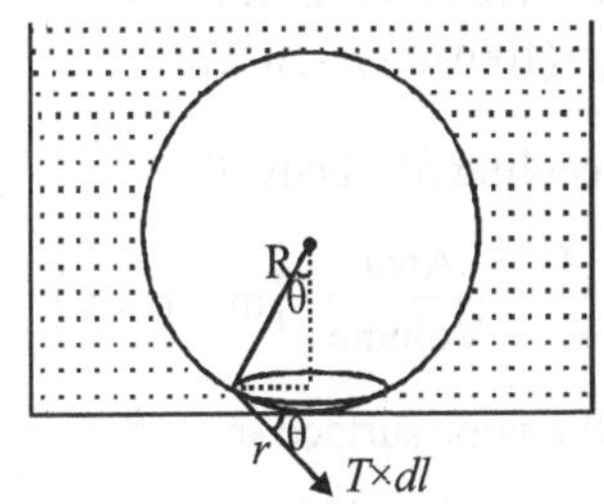

Force due to excess pressure = upthrust

Access pressure in air bubble = $\dfrac{2T}{R}$

$$\frac{2T}{R}(\pi r^2) = \frac{4\pi R^3}{3}\rho_w g$$

$$\Rightarrow \quad r^2 = \frac{2R^4 \rho_w g}{3T} \Rightarrow r = R^2\sqrt{\frac{2\rho_w g}{3T}}$$

43. **(b)** Bernoulli's theorem.

44. **(a)** Inflow rate of volume of the liquid = Outflow rate of volume of the liquid

$$\pi R^2 V = n\pi r^2 (v) \Rightarrow v = \frac{\pi R^2 V}{n\pi r^2} = \frac{VR^2}{nr^2}$$

45. **(c)** $T \times 2\pi r + mg = F_b$

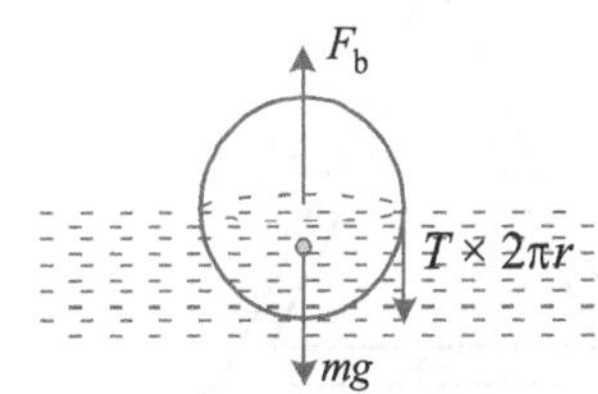

or $\quad T \times 2\pi r + \rho\frac{4}{3}\pi r^3 g = \left[\dfrac{\frac{4}{3}\pi r^3}{2}\right]\sigma g$

$$\therefore \quad r = \sqrt{\frac{3T}{g(2\rho - \sigma)}}$$

1. (a) $E = \frac{S}{S_0}\sigma T^4 = \frac{4\pi r^2}{4\pi R^2}\sigma T^4$

$= \sigma \frac{r^2}{R^2} T^4$

2. (c)

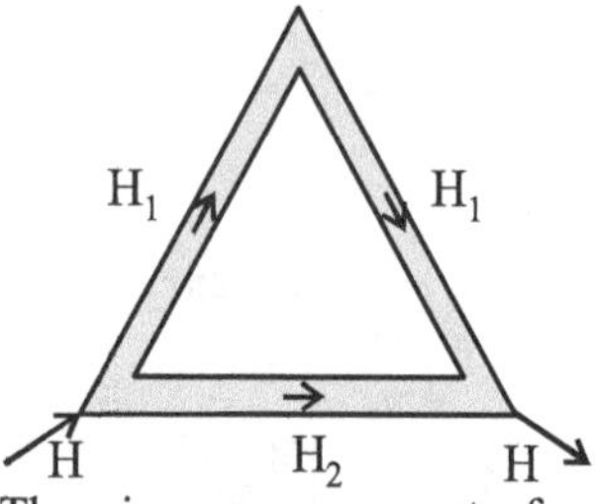

The given arrangement of rods can be redrawn as follows

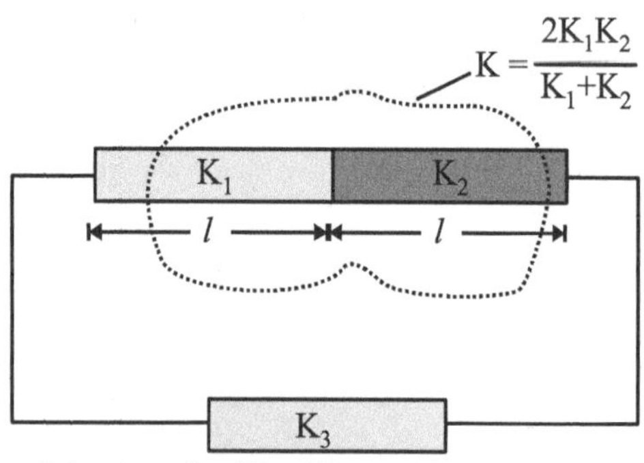

It is given that $H_1 = H_2$

$\Rightarrow \frac{KA(\theta_1 - \theta_2)}{2l} = \frac{K_3A(\theta_1 - \theta_2)}{l} \Rightarrow K_3 = \frac{K}{2} = \frac{K_1K_2}{K_1 + K_2}$

3. (c) $Q = mc\Delta T$

$Q = mc\,(T - T_0)$ ......(i)

$Q = Kt$ whereas K is heating rate

$\therefore$ from 50 to boiling temperature, T increases linearly.

At vaporization, equation is $Q = mL$

so, temperature remains constant till vaporisation is complete

After that, again Eqn (i) is followed and temperature increases linearly

4. (b) At constant temperature molar heat capacity

$C_T = \frac{\Delta Q}{n\Delta T}$

T is const. $\Rightarrow \Delta T = 0$

$\therefore \quad C_T = \frac{\Delta Q}{0} = \infty$

5. (d) Required work = energy released

Here, $Q = \int mc\, dT$

$= \int_{20}^{4} 0.1 \times 32 \times \left(\frac{T^3}{400^3}\right) dT \approx 0.002\,\text{kJ}.$

Therefore, required work = 0.002 kJ

6. (b) Since, e = a = 0.2

Since a = (1 – r – t) = 0.2 for the body B

Thus emissive power of B is given by,

$E = a\,E_b = (100)\,(0.2) = 20\ \text{W/m}^2$

7. (b) Rate of cooling of a body $R = \frac{\Delta\theta}{t} = \frac{A\varepsilon\sigma(T^4 - T_0^4)}{mc}$

$\Rightarrow R \propto \frac{A}{m} \propto \frac{\text{Area}}{\text{Volume}} \quad [m = \rho \times V]$

$\Rightarrow$ For the same surface area. $R \propto \frac{1}{\text{Volume}}$

$\because$ Volume of cube < Volume of sphere

$\Rightarrow R_{\text{cube}} > R_{\text{Sphere}}$ i.e., cube, cools down with faster rate.

8. (b) From question,

$\Delta\rho = (998 - 992)\ \text{kg/m}^3 = 6\ \text{kg/m}^3$

$\rho = \frac{998 + 992}{2}\ \text{kg/m}^3 = 995\ \text{kg/m}^3$

$\rho = \frac{m}{V}$

$\Rightarrow \frac{\Delta\rho}{\rho} = -\frac{\Delta V}{V} \quad \Rightarrow \left|\frac{\Delta\rho}{\rho}\right| = \left|\frac{\Delta V}{V}\right|$

$\therefore$ Coefficient of volume expansion of water,

$\frac{1}{V}\frac{\Delta V}{\Delta t} = \frac{1}{\rho}\frac{\Delta\rho}{\Delta t} = \frac{6}{995 \times 20} \approx 3 \times 10^{-4}\ /°\text{C}$

9. (a) $E = \frac{F/A}{\Delta l / l} = \frac{\text{stress}}{\text{strain}}$ where $\Delta\ell = (\ell' - \ell) = \ell\alpha t$ so $F = EA\alpha t$

10. (c)

11. (a) $F = Y\alpha t A$ or $F \propto \alpha$

($\because$ Y t A is same for both copper and iron)

or $F_C \propto \alpha_C$ and $F_I \propto \alpha_I$

$\therefore \quad \frac{F_C}{F_I} = \frac{3/2}{1} = \frac{3}{2}$

12. (c) According to question only one-quarter of the heat produced by falling piece of ice is absorbed in the melting of ice.

i.e., $\frac{mgh}{4} = mL$

$\Rightarrow \quad h = \frac{4L}{g} = \frac{4 \times 3.4 \times 10^5}{10} = 136\ \text{km}.$

13. (d) $W = W_1 - W_2 = mgh - mgh' = mg\,(h - h')$

$= 5 \times 10\ (20 - 0.2) = 5 \times 10 \times 19.8$

$= 5 \times 198 = 990$ joule

This energy is converted into heat when the ball strikes the earth. Heat produced is

$$Q = \frac{990}{4.2} \text{calorie}$$

$$\Delta T = \frac{Q}{mc} = \frac{99 \times 100}{42 \times 5000 \times 0.09} = \frac{11}{32}\,^{\circ}C$$

14. (b) Let the angle subtended by the arc formed be θ. Then

$$\theta = \frac{\ell}{r} \text{ or } \theta = \frac{\Delta\ell}{\Delta r} = \frac{\ell_2 - \ell_1}{r_1 - r_2}$$

$$\therefore\ \theta = \frac{\ell(\alpha_2 - \alpha_1)\Delta T}{t} \text{ or } \frac{\ell}{r} = \frac{\ell(\alpha_2 - \alpha_1)\Delta T}{t}$$

$$\text{So, } r = \frac{t}{(\alpha_2 - \alpha_1)\Delta T}$$

15. (d)

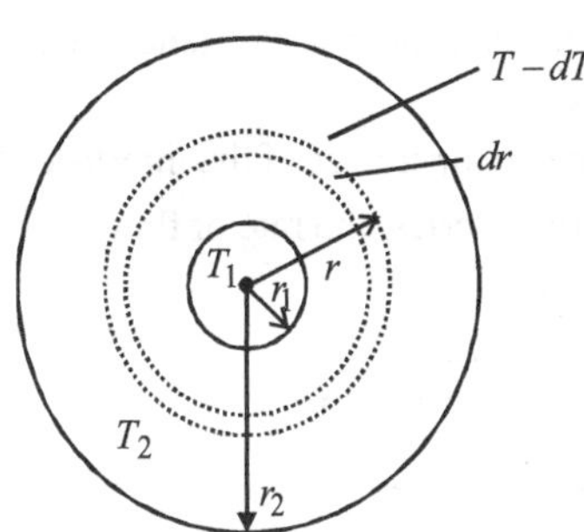

Consider a shell of thickness (dr) and of radius (r) and let the temperature of inner and outer surfaces of this shell be T and $(T - dT)$ respectively.

$\frac{dQ}{dt}$ = rate of flow of heat through it

$$= \frac{KA[(T - dT) - T]}{dr} = \frac{-KAdT}{dr}$$

$$= -4\pi K r^2 \frac{dT}{dr} \qquad (\because A = 4\pi r^2)$$

To measure the radial rate of heat flow, integration technique is used, since the area of the surface through which heat will flow is not constant.

$$\text{Then, } \left(\frac{dQ}{dt}\right)\int_{r_1}^{r_2} \frac{1}{r^2} dr = -4\pi K \int_{T_1}^{T_2} dT$$

$$\frac{dQ}{dt}\left[\frac{1}{r_1} - \frac{1}{r_2}\right] = -4\pi K [T_2 - T_1]$$

$$\text{or } \frac{dQ}{dt} = \frac{-4\pi K r_1 r_2 (T_2 - T_1)}{(r_2 - r_1)}$$

$$\therefore\ \frac{dQ}{dt} \propto \frac{r_1 r_2}{(r_2 - r_1)}$$

16. (a) According to Newton's law of cooling if temperature difference between body & surrounding is large, then rate of cooling is also fast hence curve A shows correct behaviour.

17. (a) Among glass, wood and metals, metals expand more for same rise in temperature.

18. (b) According to Wein's law $\lambda_m \propto \frac{1}{T}$ and from the figure $(\lambda_m)_1 < (\lambda_m)_3 < (\lambda_m)_2$ therefore $T_1 > T_3 > T_2$.

19. (a) Moment of inertia of a rod,

$$I = \frac{1}{12} ML^2$$

Differentiating w.r.t. to ΔL, we get

$$\frac{\Delta I}{\Delta L} = \frac{1}{12} \times 2ML$$

$$\Delta I = \frac{1}{12} 2ML\Delta L \qquad \therefore\ \frac{\Delta I}{I} = 2\frac{\Delta L}{L}$$

As we know, $\Delta L = L\alpha\Delta t$ or $\frac{\Delta L}{L} = \alpha\Delta t$

Substituting the value $\frac{\Delta L}{L}$, we get

$$\frac{\Delta I}{I} = 2\alpha\Delta t$$

20. (c) The lengths of each rod increases by the same amount

$$\therefore\ \Delta\ell_a = \Delta\ell_s \Rightarrow \ell_1\alpha_a t = \ell_2\alpha_s t$$

$$\Rightarrow \frac{\ell_2}{\ell_1} = \frac{\alpha_a}{\alpha_s} \Rightarrow \frac{\ell_2}{\ell_1} + 1 = \frac{\alpha_a}{\alpha_s} + 1$$

$$\Rightarrow \frac{\ell_2 + \ell_1}{\ell_1} = \frac{\alpha_a + \alpha_s}{\alpha_s} \Rightarrow \frac{\ell_1}{\ell_1 + \ell_2} = \frac{\alpha_s}{\alpha_a + \alpha_s}$$

21. (a) According to Kirchhoff law, good absorbers are good emitters. Since black spot is good absorber so it is also a good emitter & will be brighter than plate.

22. (a) From Wein's displacement law

$\lambda_m \times T = \text{constant}$

P – max. intensity is at violet

$\Rightarrow \lambda_m$ is minimum ⇒ temp maximum

R – max. intensity is at green

$\Rightarrow \lambda_m$ is moderate ⇒ temp moderate

Q – max. intensity is at red ⇒ λ_m is maximum ⇒ temp. minimum i.e., $T_p > T_R > T_Q$

23. (c)

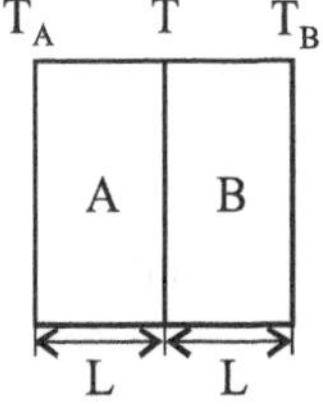

Let T be temperature of the junction

Here, $K_A = 2K_B$, $T - T_B = 50K$

At the steady state,

$$H_A = H_B$$

$$\therefore\ \frac{K_A A(T_A - T)}{L} = \frac{K_B A(T - T_B)}{L}$$

$2K_B(T_A - T) = K_B(T - T_B)$

$T_A - T = \frac{T - T_B}{2}$

$= \frac{50K}{2} = 25K$

24. **(b)**

25. **(a)** Suppose, height of liquid in each arm before rising the temperature is l.

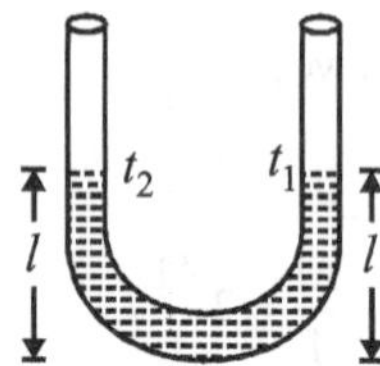
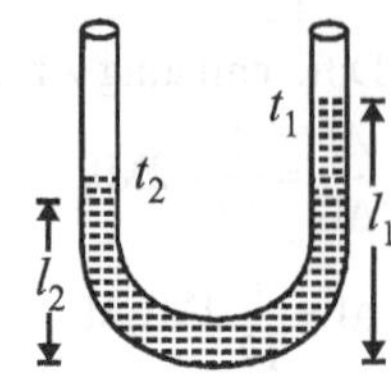

With temperature rise height of liquid in each arm increases *i.e.* $l_1 > l$ and $l_2 > l$

Also $l = \frac{l_1}{1+\gamma t_1} = \frac{l_2}{1+\gamma t_2}$

$\Rightarrow l_1 + \gamma l_1 t_2 = l_2 + \gamma l_2 t_1 \Rightarrow \gamma = \frac{l_1 - l_2}{l_2 t_1 - l_1 t_2}$.

26. **(a)** The rate of heat loss per unit area due to radiation $= \epsilon\sigma(T^4 - T_0^4)$
$= 0.6 \times 5.67 \times 10^{-8}[(400)^4 - (300)^4] = 595\ Jm^{-2}s^{-1}$.

27. **(d)** According to Wein's displacement law, product of wavelength belonging to maximum intensity and temperature is constant i.e., $\lambda_m T =$ constant.

28. **(c)** According to Newton's law of cooling, the temperature goes on decreasing with time non-linearly.

29. **(d)** $t \propto \frac{\ell}{A}, t' \propto \frac{2\ell}{A/2}$

$\frac{t'}{t} = 4\frac{\ell/A}{\ell/A}$

$t' = 4 \times t$

$3/t' = 48s$

30. **(a)** In series, equivalent thermal conductivity

$K_{eq} = \frac{2K_1K_2}{K_1 + K_2}$

or, $K_{eq} = \frac{2 \times K \times 2K}{K + 2K} = \frac{4}{3}K$

31. **(c)** $Q = -KA\left(\frac{d\theta}{dx}\right) \times t$

32. **(c)** Using Wein's law, $\lambda_m T =$ constant

$\lambda_1 T_1 = \lambda_2 T_2$

$\lambda_2 = \lambda_1 \frac{T_1}{T_2}$

$\lambda_2 = \frac{\lambda_0 T}{2T} = \frac{\lambda_0}{2}$

33. **(b)** Total power radiated by Sun $= \sigma T^4 \times 4\pi R^2$

The intensity of power at earth's surface

$= \frac{\sigma T^4 \times 4\pi R^2}{4\pi r^2}$

Total power received by Earth

$= \frac{\sigma T^4 R^2}{r^2}(\pi r_0^2)$

34. **(c)** The upthrust is given by $\frac{4}{3}\pi R_t^3 \rho g$

Here $R_t^3 = R_0^3(1 + \gamma_m t)$ and $\rho_t = \rho_0/(1 + \gamma_a t)$

So, the upthrust at t°C is given by

$= \frac{4}{3}\pi R_0^3(1 + \gamma_m t) \times \{\rho_0/(1 + \gamma_a t)\} g$

As $\gamma_m < \gamma_a$, hence upthrust at t°C < upthrust at 0°C

So, the upthrust is decreased. Hence weight in liquid gets increased.

35. **(d)** Let T be the temperature of the interface. As the two sections are in series, the rate of flow of heat in them will be equal.

$T_1 \quad \ell_1 \qquad \ell_2 \quad T_2$

$K_1 \qquad K_2$

$\therefore \frac{K_1 A(T_1 - T)}{\ell_1} = \frac{K_2 A(T - T_2)}{\ell_2}$,

where A is the area of cross-section.

or, $K_1 A(T_1 - T)\ell_2 = K_2 A(T - T_2)\ell_1$

or, $K_1 T_1 \ell_2 - K_1 T \ell_2 = K_2 T \ell_1 - K_2 T_2 \ell_1$

or, $(K_2\ell_1 + K_1\ell_2)T = K_1 T_1 \ell_2 + K_2 T_2 \ell_1$

$\therefore T = \frac{K_1 T_1 \ell_2 + K_2 T_2 \ell_1}{K_2 \ell_1 + K_1 \ell_2} = \frac{K_1 \ell_2 T_1 + K_2 \ell_1 T_2}{K_1 \ell_2 + K_2 \ell_1}$.

36. **(d)** Radius of small sphere = r

Thickness of small sphere = t

Radius of bigger sphere = 2r

Thickness of bigger sphere = t/4

Mass of ice melted = (volume of sphere) × (density of ice)

Let K_1 and K_2 be the thermal conductivities of larger and smaller sphere.

For bigger sphere,

$\frac{K_1 4\pi(2r)^2 \times 100}{t/4} = \frac{\frac{4}{3}\pi(2r)^3 \rho L}{25 \times 60}$

For smaller sphere,

$\frac{K_2 \times 4\pi r^2 \times 100}{t} = \frac{\frac{4}{3}\pi r^3 \rho L}{16 \times 60}$

$\therefore \frac{K_1}{K_2} = \frac{8}{25}$

37. **(b)** According to Wein's displacement law,

$\lambda_m T = 2.88 \times 10^{-3}$

When T = 2000 K,

$\lambda_m (2000) = 2.88 \times 10^{-3}$ ....(1)

When T = 3000 K,

$\lambda'_m (3000) = 2.88 \times 10^{-3}$ ....(2)

Dividing (1) by (2),

$$\frac{2}{3}\frac{\lambda_m}{\lambda'_m} = 1 \Rightarrow \frac{\lambda_m}{\lambda'_m} = \frac{3}{2} \Rightarrow \lambda'_m = \frac{2}{3}\lambda_m$$

38. **(c)** AB represents latent heat of fusion

$Q_1 = mL_F$

Here, $L_F \propto$ length of line AB

CD represents latent heat of vaporization $Q_2 = mL_V$

Here, $L_V \propto$ length of line CD

$\therefore\ Q_2 = 2Q_1$ $[\because$ As CD = 2AB]

39. **(b)** Let the final temperature be T

Heat gained by ice $= mL + m \times s \times (T - 0)$

$= 10 \times 80 + 10 \times 1 \times T$

Heat lost by water $= 55 \times 1 \times (40 - T)$

By using law of calorimetery,

$800 + 10\,T = 55 \times (40 - T)$

$\Rightarrow T = 21.54°C = 22°C$

40. **(a)** According to Newton's law of cooling,

$$\frac{\theta_1 - \theta_2}{t} = K\left[\frac{\theta_1 + \theta_2}{2} - \theta_0\right]$$

where θ_0 is the surrounding temperature.

$$\therefore\ \frac{60 - 40}{7} = K\left(\frac{60 + 40}{2} - 10\right)$$

$$\Rightarrow\ \frac{20}{7} = 40K \Rightarrow K = \frac{1}{14}$$

$$\therefore\ \frac{40 - 28}{t} = K\left[\frac{40 + 28}{2} - 10\right] \Rightarrow \frac{12}{t} = 24K$$

$$\text{or } t = \frac{12}{24K} = \frac{12 \times 14}{24} = 7 \text{ min}$$

41. **(d)** $\frac{Q}{t} = K_1 A_1 \frac{d\theta}{dx} = K_2 A_2 \frac{d\theta}{dx}$

42. **(c)** Since specific heat = 0.6 $kcal/g \times °C$ = 0.6 $cal/g \times °C$

From graph it is clear that in a minute, the temperature is raised from 0°C to 50°C.

$\Rightarrow$ Heat required for a minute = $50 \times 0.6 \times 50 = 1500$ *cal.*

Also from graph, Boiling point of wax is 200°C.

43. **(b)** Temperature of B will be higher because, due to expansion centre of mass B will come down same heat is supplied but in B, potential energy is decreased therefore internal energy gain will be more.

44. **(d)** According to the principle of calorimetry.

Heat lost = Heat gained

$mL_v + ms_w\Delta\theta = m_w s_w \Delta\theta$

$\Rightarrow m \times 540 + m \times 1 \times (100 - 80)$

$= 20 \times 1 \times (80 - 10)$

$\Rightarrow m = 2.5$ g

Therefore total mass of water at 80°C

$= (20 + 2.5)$ g $= 22.5$ g

45. **(a)** Initial rate of loss of heat $= \frac{\sigma T^4 \times A_1 \times e}{\sigma T^4 \times A_2 \times e} = \frac{R_1^2}{R_2^2}$

1. **(a)** $U = a + bPV$ ......(1)

In adiabatic change,

$$dU = -dW = \frac{nR}{\gamma - 1}(T_2 - T_1) = \frac{nR}{\gamma - 1}(dT)$$

$$\Rightarrow U = \int dU = \frac{nR}{\gamma - 1}\int dT$$

or $U = \left(\frac{nR}{\gamma - 1}\right)T + a = \frac{PV}{\gamma - 1} + a$ ......(2)

where a is the constant of integration.

Comparing (1) and (2), we get

$$b = \frac{1}{\gamma - 1} \Rightarrow \gamma = \frac{b+1}{b}.$$

2. **(d)** For path ab : $(\Delta U)_{ab} = 7000\,J$

By using $\Delta U = \mu C_V \Delta T$

$$7000 = \mu \times \frac{5}{2}R \times 700 \Rightarrow \mu = 0.48$$

For path ca :

$(\Delta Q)_{ca} = (\Delta U)_{ca} + (\Delta W)_{ca}$ ...(i)

$\because (\Delta U)_{ab} + (\Delta U)_{bc} + (\Delta U)_{ca} = 0$

$\because 7000 + 0 + (\Delta U)_{ca} = 0 \Rightarrow (\Delta U)_{ca} = -7000\,J$ ...(ii)

Also $(\Delta W)_{ca} = P_1(V_1 - V_2) = \mu R(T_1 - T_2)$

$= 0.48 \times 8.31 \times (300 - 1000) = -2792.16\text{J}$ ...(iii)

On solving equations (i), (ii) and (iii)

$(\Delta Q)_{ca} = -7000 - 2792.16 = -9792.16\,J \approx -9800\,J$

3. **(c)** The efficiency (η) of a Carnot engine and the coefficient of performance (β) of a refrigerator are related as

$\beta = \frac{1-\eta}{\eta}$ Here, $\eta = \frac{1}{10}$

$$\therefore \beta = \frac{1 - \frac{1}{10}}{\left(\frac{1}{10}\right)} = 9.$$

Also, Coefficient of performance (β) is given by $\beta = \frac{Q_2}{W}$,

where Q_2 is the energy absorbed from the reservoir.

or, $9 = \frac{Q_2}{10}$ $\therefore Q_2 = 90\,\text{J}$.

4. **(b)** According to first law of thermodynamics,

$\Delta Q = \Delta U + \Delta W$

ΔQ = heat absorbed by gas

ΔW = work done by gas.

$-20\text{J} = \Delta U - 8\text{J}$

$\Delta U = -12\text{J} = U_{\text{Final}} - U_{\text{initial}}$

$U_{\text{initial}} = 30\text{J}$.

$U_{\text{Final}} = 30 - 12 = 18\,\text{J}$.

5. **(a)** The volume on both sides will be so adjusted that the original pressure × volume is kept constant as the piston moves slowly (isothermal change)

P5V = P'V' .... (1)

10PV = P'V" .... (2)

From (1) and (2), V" = 2V'

and from V' + V" = 6V

V' = 2V, V" = 4V

6. **(a)** $W_{AB} = 0$, $W_{BC} = P\Delta V = nR\Delta T = -nRT_0$

$$W_{CA} = nRT\ell n\frac{V_f}{V_i} = nR(2T_0)\ell n2$$

$$Q_{BC} = nC_p\Delta T = \left(\frac{nR\gamma}{\gamma - 1}\right)T_0$$

Efficiency, $\eta = \frac{W}{Q} = \left[\frac{2\ell n2 - 1}{\gamma/(\gamma - 1)}\right]$

7. **(c)** $T_1 = 273 + 27 = 300\text{K}$

$T_2 = 273 + 927 = 1200\text{K}$

For adiabatic process,

$P^{1-\gamma}\,T^{\gamma} = \text{constant}$

$\Rightarrow P_1^{1-\gamma}\,T_1^{\gamma} = P_2^{1-\gamma}\,T_2^{\gamma}$

$$\Rightarrow \left(\frac{P_2}{P_1}\right)^{1-\gamma} = \left(\frac{T_1}{T_2}\right)^{\gamma} \quad \Rightarrow \left(\frac{P_1}{T_2}\right)^{1-\gamma} = \left(\frac{T_2}{T_1}\right)^{\gamma}$$

$$\left(\frac{P_1}{P_2}\right)^{1-1.4} = \left(\frac{1200}{300}\right)^{1.4} \quad \Rightarrow \left(\frac{P_1}{P_2}\right)^{-0.4} = (4)^{1.4}$$

$$\left(\frac{P_2}{P_1}\right)^{0.4} = 4^{1.4}$$

$$P_2 = P_1\,4^{\left(\frac{1.4}{0.4}\right)} = P_1\,4^{\left(\frac{7}{2}\right)}$$

$= P_1\,(2^7) = 2 \times 128 = 256$ atm

8. **(b)**

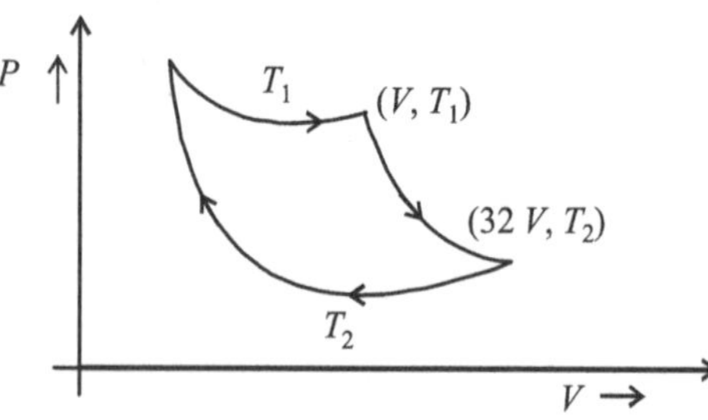

We have, $TV^{\gamma-1}$ = constant

$\Rightarrow T_1V^{\gamma-1} = T_2(32V)^{\gamma-1}$

$\Rightarrow T_1 = (32)^{\gamma-1}.T_2$

For diatomic gas, $\gamma = \frac{7}{5}$

$\therefore \gamma - 1 = \frac{2}{5}$

$\therefore T_1 = (32)^{\frac{2}{5}} . T_2 \Rightarrow T_1 = 4T_2$

Now, efficiency $= 1 - \frac{T_2}{T_1}$

$= 1 - \frac{T_2}{4T_2} = 1 - \frac{1}{4} = \frac{3}{4} = 0.75.$

9. (d) Isobaric compression is represented by curve AO
Work done = area under AD
$= 2 \times 10^2 \times (3-1)$
$= 4 \times 10^2 = 400$ J.

10. (a) As $P \propto \frac{1}{V^{1.5}}$, So $PV^{1.5}$ = constant

$\therefore \gamma = 1.5$ ($\because$ Process is adiabatic)

As we know, $\frac{C_p}{C_v} = \gamma \qquad \therefore \frac{C_p}{C_v} = 1.5$

11. (a) $W = \frac{nR\Delta T}{1-\gamma} \Rightarrow -146000 = \frac{1000 \times 8.3 \times 7}{1-\gamma}$

or $1 - \gamma = -\frac{58.1}{146} \Rightarrow \gamma = 1 + \frac{58.1}{146} = 1.4$

Hence the gas is diatomic.

12. (a) As, $P = \frac{1}{3}\left(\frac{U}{V}\right)$

But $\frac{U}{V} = KT^4$

So, $P = \frac{1}{3}KT^4$

or $\frac{uRT}{V} = \frac{1}{3}KT^4$ [As $PV = u RT$]

$\frac{4}{3}\pi R^3 T^3$ = constant

Therefore, $T \propto \frac{1}{R}$

13. (c) Heat required to change the temperature of vessel by a small amount dT

$-dQ = mCdT$

Total heat required

$$-Q = m\int_{20}^{4} 32\left(\frac{T}{400}\right)^3 dT = \frac{100 \times 10^{-3} \times 32}{(400)^3}\left[\frac{T^4}{4}\right]_{20}^{4}$$

$\Rightarrow Q = 0.001996$ kJ

Work done required to maintain the temperature of sink to T_2

$$W = Q_1 - Q_2 = \frac{Q_1 - Q_2}{Q_2} Q_2 = \left(\frac{T_1}{T_2} - 1\right) Q_2$$

$$\Rightarrow W = \left(\frac{T_1 - T_2}{T_2}\right) Q_2$$

For $T_2 = 20$ K

$W_1 = \frac{300 - 20}{20} \times 0.001996 = 0.028$ kJ

For $T_2 = 4$ K

$W_2 = \frac{300 - 4}{4} \times 0.001996 = 0.148$ kJ

As temperature is changing from 20k to 4 k, work done required will be more than W_1 but less than W_2.

14. (a) Initially
$V_1 = 5.6\ell, T_1 = 273K, P_1 = 1$ atm,

$\gamma = \frac{5}{3}$ (For monatomic gas)

The number of moles of gas is

$n = \frac{5.6\ell}{22.4\ell} = \frac{1}{4}$

Finally (after adiabatic compression)

$V_2 = 0.7\ell$

For adiabatic compression

$T_1 V_1^{\gamma-1} = T_2 V_2^{\gamma-1}$

$\therefore T_2 = T_1\left(\frac{V_1}{V_2}\right)^{\gamma-1} = T_1\left(\frac{5.6}{0.7}\right)^{\frac{5}{3}-1}$

$= T_1(8)^{2/3} = 4T_1$

We know that work done in adiabatic process is

$W = \frac{nR\Delta T}{\gamma - 1} = \frac{9}{8}RT_1$

15. (c) Curve A, B shows expansion. For expansion of a gas,
$W_{isothermal} > W_{adiabatic}$
$P_{isothermal} > P_{adiabatic}$
$T_{isothermal} > T_{adiabatic}$
$\Rightarrow$ Slope of curve for isothermal change < slope of curve for adiabatic change.
So, curve B shows isothermal change and curve A shows adiabatic change.

16. (a) $PV^{3/2} = K, \qquad \log P + \frac{3}{2}\log V = \log K$

$\frac{\Delta P}{P} + \frac{3}{2}\frac{\Delta V}{V} = 0$

$\frac{\Delta V}{V} = -\frac{2}{3}\frac{\Delta P}{P}$ or $\frac{\Delta V}{V} = \left(-\frac{2}{3}\right)\left(\frac{2}{3}\right) = -\frac{4}{9}$

17. (a) Initially the efficiency of the engine was $\frac{1}{6}$ which increases to $\frac{1}{3}$ when the sink temperature reduces by 62° C.

$\eta = \frac{1}{6} = 1 - \frac{T_2}{T_1}$, when T_2 = sink temperature

T_1 = source temperature

$\Rightarrow T_2 = \frac{5}{6} T_1$

Secondly,

$$\frac{1}{3} = 1 - \frac{T_2 - 62}{T_1} = 1 - \frac{T_2}{T_1} + \frac{62}{T_1} = 1 - \frac{5}{6} + \frac{62}{T_1}$$

or, $T_1 = 62 \times 6 = 372K = 372 - 273 = 99°C$

& $T_2 = \frac{5}{6} \times 372 = 310 K = 310 - 273 = 37°C$

18. (b) For an adiabatic process, the temperature-volume relationship is

$$T_1 V_1^{\gamma-1} = T_2 V_2^{\gamma-1} \Rightarrow T_1 = T_2 \left(\frac{V_2}{V_1}\right)^{\gamma-1}$$

Here $\gamma = 1.4$ (for diatomic gas). $V_2 = \frac{V_1}{32}, T_1 = T_i, T_2 = aT_i$

$\therefore T_i = aT_i \left[\frac{1}{32}\right]^{1.4-1} \quad \therefore T_i = aT_i \left[\frac{1}{2^5}\right]^{0.4} = \frac{aT_i}{4}$

$\therefore a = 4$

19. (b) In the first-case adiabatic change,

$\Delta Q = 0, \Delta W = -35$ J

From 1st law of thermodynamics,

$\Delta Q = \Delta U + \Delta W$,

or $0 = \Delta U - 35$

$\therefore \Delta U = 35$ J

In the second case

$\Delta Q = 12 \text{ cal} = 12 \times 4.2 \text{ J} = 50.4$ J

$\Delta W = \Delta Q - \Delta U = 50.4 - 35 = 15.4$ J

20. (a) For an adiabatic change PV^γ = constant

$P_1 V_1^\gamma = P_2 V_2^\gamma$

As molar specific heat of gas at constant volume

$$C_v = \frac{3}{2} R$$

$$C_P = C_V + R = \frac{3}{2} R + R = \frac{5}{2} R;$$

$$\gamma = \frac{C_P}{C_V} = \frac{(5/2)R}{(3/2)R} = \frac{5}{3}$$

$\therefore$ From eq^n. (1)

$$P_2 = \left(\frac{V_1}{V_2}\right)^\gamma P_1 = \left(\frac{6}{2}\right)^{5/3} \times 10^5 N/m^2$$

$= (3)^{5/3} \times 10^5 = 6.19 \times 10^5 N/m^2$

Work done

$$= \frac{1}{1-(5/3)} [6.19 \times 10^5 \times 2 \times 10^{-3} - 10^{-5} \times 6 \times 10^{-3}]$$

$$= -\left[\frac{2 \times 10^2 \times 3}{2} (6.19 - 3)\right]$$

$= -3 \times 10^2 \times 3.19 = -957$ joules

[–ve sign shows external work done on the gas]

21. (b) Efficiency of Carnot engine, $\eta = 1 - \frac{T_2}{T_1}$

where T_1 and T_2 be the temperature of source and sink respectively.

$$\therefore \frac{T_2}{T_1} = 1 - \eta = 1 - \frac{40}{100} = \frac{60}{100} = \frac{3}{5} \quad (\because \eta = 40\%)$$

$$T_2 = \frac{3}{5} T_1 = \frac{3}{5} \times 500 \text{ K} = 300 \text{ K} \quad ...(i)$$

$(\because T_1 = 500 K)$

Let T_1' be the temperature of the source for the same sink temperature when efficiency $\eta' = 50\%$

$$\therefore \frac{T_2}{T_1'} = 1 - \eta' = 1 - \frac{50}{100} = \frac{1}{2}$$

$T_1' = 2T_2 = 2 \times 300 \text{ K} = 600$ K (Using eq. (i))

22. (b) $dW = P \Delta V = 1.01 \times 10^5 [1671 - 1] \times 10^{-6}$ Joule

$= \frac{1.01 \times 167}{4.2}$ cal.

$= 40$ cal. nearly

$\Delta Q = mL = 1 \times 540$,

$\Delta Q = \Delta W + \Delta U$

or $\Delta U = 540 - 40 = 500$ cal.

23. (c) The temperature remains unchanged therefore

$U_f = U_i$.

Also, $\Delta Q = \Delta W$.

In the first step which is isochoric, $\Delta W = 0$.

In second step, pressure $= \frac{P}{n}$. Volume V is increased from V to nV.

$$\therefore \quad W = \frac{P}{n}(nV - V)$$

$$= PV\left(\frac{n-1}{n}\right)$$

$$= RT(1 - n^{-1})$$

24. (b) Efficiency of carnot engine

$n = 1 - \frac{T_2}{T_1}$ i.e., $\frac{1}{10} = 1 - \frac{T_2}{T_1}$

$$\Rightarrow \frac{T_2}{T_1} = 1 - \frac{1}{10} = \frac{9}{10} \Rightarrow \frac{T_1}{T_2} = \frac{10}{9}$$

$$\therefore \quad w = Q_2\left(\frac{T_1}{T_2} - 1\right)$$

i.e., $10 = Q_2\left(\frac{10}{9} - 1\right)$ $10 = Q_2\left(\frac{1}{9}\right)$

$\Rightarrow Q_2 = 90$J

So, 90 J heat is absorbed at lower temperature.

25. (a) $V_1 = 1\ell = 1000\ cm^3$, $P_1 = 72$ cm of Hg.
$V_2 = 900\ cm^3$, $P_2 = ?$
$\because$ The process is isothermal
$\therefore P_1V_1 = P_2V_2$
$72 \times 1000 = P_2 \times 900$
$P_2 = 80$ cm of Hg
$\therefore$ Stress $= P_2 - P_1 = 80 - 72 = 8$ cm of Hg.

26. (a) Process A $\rightarrow$ B occurs at constant pressure Hence work done in this process is

$$W_{AB} = PdV = P(V_2 - V_1)$$
$$= 10 \times (2-1) = 10\ J$$

Process B $\rightarrow$ C, occurs at constant volume.

Hence, $W_{BC} = 0$

Given : Q = 5 J therefore, total work done is $W_1 = 5\ J$

($\because \Delta U = 0$ in a cyclic process)

Therefore, we have

$$W_1 = W_{AB} + W_{BC} + W_{CA}$$

or $5J = 10\ J + 0 + W_{CA}$

$\therefore\ W_{CA} = -5$ joule

27. (d) We know that in adiabatic process,
PV^γ = constant ....(1)
From ideal gas equation, we know that
$PV = nRT$

$$V = \frac{nRT}{P} \quad ....(2)$$

Puttingt the value from equation (2) in equation (1),

$$P\left(\frac{nRT}{P}\right)^\gamma = \text{constant}$$

$P^{(1-\gamma)}\ T^\gamma$ = constant

28. (b) According to first law of thermodynamics,
$\Delta Q = \Delta U + \Delta W$
ΔQ = heat absorbed by gas
ΔW = work done by gas.
$-20J = \Delta U - 8J$
$\Delta U = -12J = U_{Final} - U_{initial}$
$U_{initial} = 30J$.
$U_{Final} = 30 - 12 = 18J$.

29. (d) Coefficient of performance,

$$Cop = \frac{T_2}{T_1 - T_2}$$

$$5 = \frac{273-20}{T_1-(273-20)} = \frac{253}{T_1 - 253}$$

$5T_1 - (5 \times 253) = 253$
$5T_1 = 253 + (5 \times 253) = 1518$

$$\therefore T_1 = \frac{1518}{5} = 303.6$$

or, $T_1 = 303.6 - 273 = 30.6 \cong 31°C$

30. (a) $W = \dfrac{nRdT}{\gamma - 1}$ γ is minimum for a polyatomic gas

Hence, W is greatest for polyatomic gas

31. (b) Heat is extracted from the source in path DA and AB is

$$\Delta Q = \frac{3}{2}R\left(\frac{P_0V_0}{R}\right) + \frac{5}{2}R\left(\frac{2P_0V_0}{R}\right)$$

$$\Rightarrow \frac{3}{2}P_0V_0 + \frac{5}{2}2P_0V_0 = \left(\frac{13}{2}\right)P_0V_0$$

32. (d) Isochoric proceess $dV = 0$

$W = 0$ proceess 1

Isobaric : $W = P\Delta V = nR\Delta T$

Adiabatic $|W| = \dfrac{nR\Delta T}{\gamma - 1}$ $0 < \gamma - 1 < 1$

As workdone in case of adiabatic process is more so process 3 is adiabatic and process 2 is isobaric

33. (c) For adiabatic process, dU = – 100 J

which remains same for other processes also.

Let C be the heat capacity of 2nd process then

$-(C)\,5 = dU + dW$

$= -100 + 25 = -75$

$\therefore\ C = 15$ J/K

34. (c) Coefficient of performance of a refrigerator,

$$\beta = \frac{Q_2}{W} = \frac{T_2}{T_1 - T_2} \qquad \text{(Where } Q_2 \text{ is heat removed)}$$

Given: $T_2 = 4°C = 4 + 273 = 277$ k
$T_1 = 30°C = 30 + 273 = 303$ k

$$\therefore\ \beta = \frac{600 \times 4.2}{W} = \frac{277}{303 - 277}$$

$\Rightarrow\ W = 236.5$ joule

Power $P = \dfrac{W}{t} = \dfrac{236.5\text{joule}}{1\sec} = 236.5$ watt.

35. (d) $dU = dQ - dW = (8 \times 10^5 - 6.5 \times 10^5) = 1.5 \times 10^5\ J$

$dW = dQ - dU == 10^5 - 1.5 \times 10^5 = -0.5 \times 10^5\ J$

– ve sign indicates that work done on the gas is 0.5×10^5 J.

36. (d) In cyclic process, change in total internal energy is zero.

$\Delta U_{cyclic} = 0$

$$\Delta U_{BC} = nC_v\Delta T = 1 \times \frac{5R}{2}\Delta T$$

Where, C_V = molar specific heat at constant volume.
For BC, $\Delta T = -200$ K

$\therefore \quad \Delta U_{BC} = -500R$

37. **(d)** Efficiency of engine A, $\eta_1 = 1 - \dfrac{T}{T_1}$,

Efficiency of engine B, $\eta_2 = 1 - \dfrac{T_2}{T}$

Here, $\eta_1 = \eta_2$

$\therefore \dfrac{T}{T_1} = \dfrac{T_2}{T} \Rightarrow T = \sqrt{T_1 T_2}$

38. **(b)** In the first process W is + ve as ΔV is positive, in the second process W is – ve as ΔV is – ve and area under the curve of second process is more

$\therefore$ Net Work < 0 and also $P_3 > P_1$.

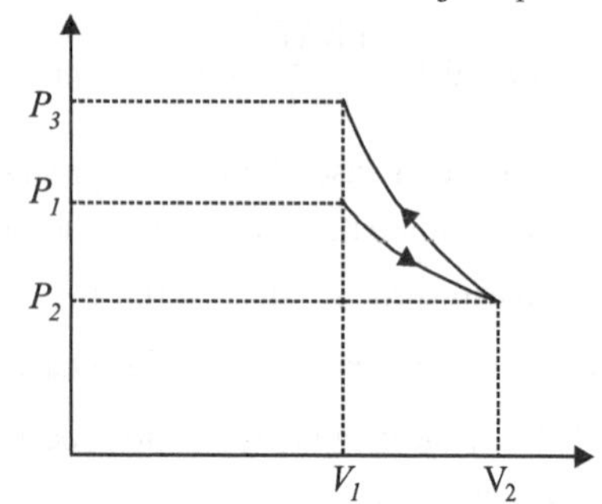

39. **(b)** Internal energy and entropy are state function, they do not depend upon path but on the state.

40. **(d)** 1st process is isothermal expansion which is only correct shown in option (d)

2nd process is isobaric compression which is correctly shown in option (d)

41. **(c)** $P_1 V_1^{\gamma} = P_2 V_2^{\gamma}$ (Adiabatic change)

$$P_2 = P_1\left(\frac{V_1}{V_2}\right)^{\gamma} = P_1\left(\frac{V_1}{V_1/3}\right)^{\gamma} = P_2(3)^{\gamma}$$

42. **(b)** W_{ext} = negative of area with volume-axis
W(adiabatic) > W(isothermal)

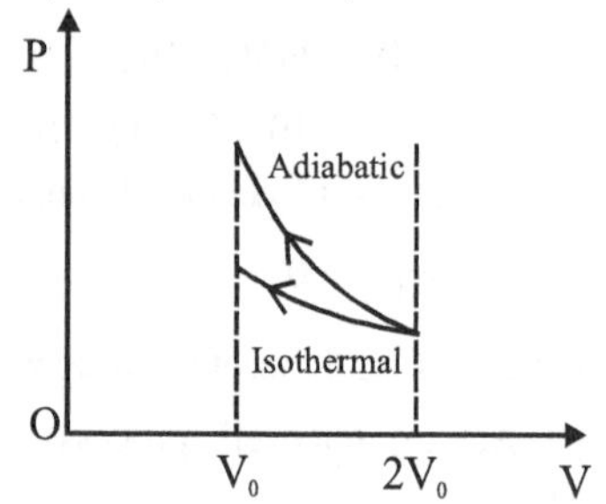

43. **(a)** Initial and final condition is same for all process

$\Delta U_1 = \Delta U_2 = \Delta U_3$

from first law of thermodynamics

$\Delta Q = \Delta U + \Delta W$

Work done

$\Delta W_1 > \Delta W_2 > \Delta W_3$ (Area of P.V. graph)

So $\Delta Q_1 > \Delta Q_2 > \Delta Q_3$

44. **(c)** $W = \dfrac{\pi r_1 r_2}{2} = \dfrac{\pi \times 1 \times 1}{2}$

$= \pi/2$ J

45. **(b)** Differentiate PV = constant w.r.t V

$$\Rightarrow P\Delta V + V\Delta P = 0 \Rightarrow \frac{\Delta P}{P} = -\frac{\Delta V}{V}$$

1. (d) Molar mass of the gas = 4g/mol
Speed of any quantity x

$V=\sqrt{\frac{\gamma RT}{m}} \Rightarrow 952=\sqrt{\frac{\gamma\times 3.3\times 273}{4\times 10^{-3}}}$

$\Rightarrow \gamma=1.6=\frac{16}{10}=\frac{8}{5}$

Also, $\gamma=\frac{C_P}{C_V}=\frac{8}{5}$

So, $C_P=\frac{8\times 5}{5}=8JK^{-1}mol^{-1}$

2. (d) Since v_{rms} is doubled by increasing the temp. so by

$v_{rms}=\sqrt{\frac{3KT}{m}}$, the temp. increase by four times.

Now for constant pressure $\frac{V_1}{T_1}=\frac{V_2}{T_2}$

$V_1=V, T_1=T°K, T_2=4T°K, V_2=?$

$V_2=4V$

3. (a) For mixture of gas, $C_v=\frac{n_1C_{v_1}+n_2C_{v_2}}{n_1+n_2}$

$$=\frac{4\times\frac{3}{2}R+\frac{1}{2}\times\frac{5}{2}R}{\left(4+\frac{1}{2}\right)}=\frac{6R+\frac{5}{4}R}{\frac{9}{2}}=\frac{29R\times 2}{9\times 4}=\frac{29R}{18}$$

and $C_p=\frac{n_1C_{p_1}+n_2C_{p_2}}{(n_1+n_2)}=\frac{4\times\frac{5R}{2}+\frac{1}{2}\times\frac{7R}{2}}{\left(4+\frac{1}{2}\right)}$

$$=\frac{10R+\frac{7}{4}R}{\frac{9}{2}}=\frac{47R}{18}$$

$$\therefore \frac{C_p}{C_v}=\frac{47R}{18}\times\frac{18}{29R}=1.62$$

4. (b) $\frac{P_1V_1}{T_1}=\frac{P_2V_2}{T_2}$ Here, $P_1=200kPa$

$T_1=22°C=295K$, $T_2=42°C=315K$

$V_2=V_1+\frac{2}{100}V_1=1.02V_1$

$\therefore\ P_2=\frac{200\times 315V_1}{295\times 1.02V_1}=209.37kPa$

5. (c) By kinetic theory of gases,
rms (root mean square) velocity of gas

$V_{rms}=\sqrt{\frac{3RT}{M}}=\sqrt{\frac{3KT}{m}}$

Where M and m are mol. wt and mass of gas respectively

$V_{rms}=\sqrt{\frac{3\times 1.38\times 10^{-23}\times 273}{5\times 10^{-17}}}$

$=1.5\times 10^{-2}\,m/s=1.5\,cm/s$

6. (d) $C_p=\frac{5}{2}R$ and $C_v=\frac{3}{2}R$

We know that $Q_v=nC_v\Delta T$ and $Q_p=nC_p\Delta T$

$\Rightarrow \frac{Q_v}{Q_p}=\frac{3}{5}$.

Given $Q_p=207\,J \Rightarrow Q_v\cong 124$ J

7. (c) On giving same amount of heat at constant pressure, there is change in temperature for mono, dia and polyatomic gas.

$(\Delta Q)_P=\mu C_p\Delta T\left(\mu=\frac{\text{No. of molecules}}{\text{Avogadro's no.}}\right)$

or $\Delta T\propto\frac{1}{\text{no. of molecules}}$

8. (a) $C=C_{v\,mix}+\frac{R}{1-n}$ .....(1)

Now, $C_{vmix}=\frac{n_1C_{v1}+n_2C_{v2}}{n_1+n_2}$

$=\frac{1\times\frac{3R}{2}+2\times\frac{5R}{2}}{1+2}=\frac{13R}{6}$

From (1), $3R=\frac{13R}{6}+\frac{R}{1-n}\Rightarrow n=-\frac{1}{5}$

9. (a) $P=\frac{1}{3}\rho\bar{\upsilon}^2=\frac{1}{3}\times(6\times 10^{-2})\times(500)^2$

$=5\times 10^3\,N/m^2$

10. (d) $v_{rms}=\sqrt{\frac{3kT}{m}}$ ....(1)

$v'_{rms}=\sqrt{\frac{3k\times 3T}{m}}$ ....(2)

Equ. (2) is dividing by equ. (1)

$$\frac{v'_{rms}}{v_{rms}}=\sqrt{\frac{3k.3T.m}{m.3kT}}=\sqrt{3}$$

$$v'_{rms}=\sqrt{3}v_{rms}$$

11. (a) $\tau=\frac{1}{\sqrt{2}\pi d^2\left(\frac{N}{V}\right)\sqrt{\frac{3RT}{M}}}$

$\tau\propto\frac{V}{\sqrt{T}}$

As, $TV^{\gamma-1}=K$

So, $\tau\propto V^{\gamma+1/2}$

Therefore, $q=\frac{\gamma+1}{2}$

12. (a) Volume$=\frac{\text{mass}}{\text{density}}=\frac{1}{4}m^3$

$K.E=\frac{5}{2}PV=\frac{5}{2}\times8\times10^4\times\frac{1}{4}=5\times10^4 J$

13. (c) As no heat is lost,
Loss of kinetic energy = gain of internal energy of gas

$\frac{1}{2}mv^2=nC_V\Delta T\Rightarrow\frac{1}{2}mv^2=\frac{m}{M}\cdot\frac{R}{\gamma-1}\Delta T$

$\Rightarrow\ \Delta T=\frac{Mv^2(\gamma-1)}{2R}K$

14. (b) From graph, $T^2V=$ const. .....(1)
As we know that $TV^{\gamma-1}=$ const

$\Rightarrow VT^{\frac{1}{\gamma-1}}=$const. ....(2)

On comparing (1) and (2), we get
$\Rightarrow\gamma=3/2$

Also $v_{rms}=\sqrt{\frac{3P}{\rho}}$ and $v_{sound}=\sqrt{\frac{P\gamma}{\rho}}$

$\Rightarrow\ \frac{v_{rms}}{v_{sound}}=\sqrt{\frac{3}{\gamma}}=\sqrt{2}$

15. (a) $W=\frac{nR\Delta T}{1-\gamma}\Rightarrow-146000=\frac{1000\times8.3\times7}{1-\gamma}$

or $1-\gamma=-\frac{58.1}{146}\Rightarrow\gamma=1+\frac{58.1}{146}=1.4$

Hence the gas is diatomic.

16. (a) $C_{rms}=\sqrt{\frac{3RT}{M}}$

M is molecular wt.

$=\sqrt{\frac{3R(273+47)}{16}}=\sqrt{\frac{3RT}{2}}$

$\Rightarrow T=40$ K.

17. (a) When temperature is same according to kinetic theory of gases, kinetic energy of molecules will be same.

$K.E.=\frac{1}{2}\times32\times\left(\frac{1}{2}\right)^2=\frac{1}{2}\times2\times v^2$

RMS velocity of hydrogen molecules = 2 km/sec.

18. (a) Let T be the temperature of the mixture, then
$U=U_1+U_2$

$\Rightarrow\ \frac{f}{2}(n_1+n_2)RT$

$=\frac{f}{2}(n_1)(R)(T_0)+\frac{f}{2}(n_2)(R)(2T_0)$

$\Rightarrow (2+4)T=2T_0+8T_0\ (\because\ n_1=2, n_2=4)$

$\therefore\ T=\frac{5}{3}T_0$

19. (b) Coefficient of volume expansion at constant pressure is $\frac{1}{273}$ for all gases. The average transnational K.E. is same for molecules of all gases and for each molecules it is $\frac{3}{2}kT$

Mean free path $\lambda=\frac{kT}{\sqrt{2}\,\pi d^2P}$ (as P decreases, λ increases)

20. (c) For a given pressure, volume will be more if temperature is more (Charle's law)

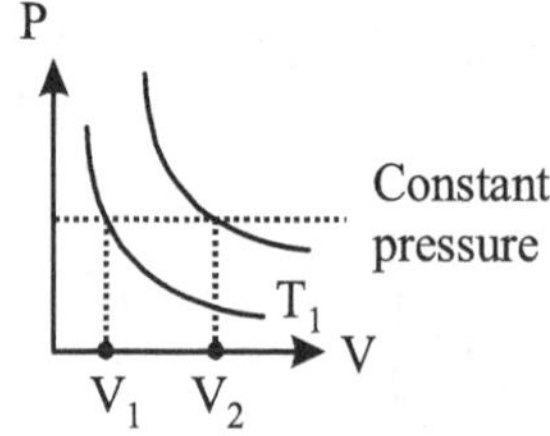

From the graph it is clear that $V_2>V_1\Rightarrow T_2>T_1$

21. (a) $\frac{c_2}{c_1}=\sqrt{\frac{400}{300}}=\frac{2}{\sqrt{3}}\Rightarrow c_2=\frac{2}{\sqrt{3}}\times200=\frac{400}{\sqrt{3}}\ ms^{-1}$

22. (c) $\frac{PV}{T}=nR=\left(\frac{m}{M}\right)R$ or $\frac{PV}{T}=\left(\frac{R}{M}\right)m$

i.e., $\frac{PV}{T}$ versus m graph is straight line passing through origin with slope R/M, i.e. the slope depends on molecular mass of the gas M and is different for different gases.

23. (d) Molecule number ratio is $H_2:O_2=\frac{2}{3}:\frac{1}{3}$.

That gives $(c_{rms})^2=16\left(\frac{2}{3}\right)+1\left(\frac{1}{3}\right)$ times the value for O_2.

24. (a) According to Vander Waal's equation

$$P=\frac{nRT}{V-n\beta}-\frac{\alpha n^2}{V^2}$$

Work done,

$$W=\int_{V_1}^{V_2} PdV = nRT\int_{V_1}^{V_2}\frac{dV}{V-n\beta}-\alpha n^2\int_{V_1}^{V_2}\frac{dV}{V^2}$$

$$= nRT\left[\log_e(V-n\beta)\right]_{V_1}^{V_2}+\alpha n^2\left[\frac{1}{V}\right]_{V_1}^{V_2}$$

$$= nRT\log_e\left(\frac{V_2-n\beta}{V_1-n\beta}\right)+\alpha n^2\left[\frac{V_1-V_2}{V_1V_2}\right]$$

25. **(a)** From PV = nRT

$$P_A=\frac{\rho_A M_A}{RT} \text{ and } P_B=\frac{\rho_B M_B}{RT}$$

From question,

$$\frac{P_A}{P_B}=\frac{\rho_A}{\rho_B}\frac{M_A}{M_B}=2\frac{M_A}{M_B}=\frac{3}{2}$$

So, $\frac{M_A}{M_B}=\frac{3}{4}$

26. **(b)** $(C_p)_{mix}=\frac{\mu_1 C_{p_1}+\mu_2 C_{p_2}}{\mu_1+\mu_2}$

$(C_{p_1}(He)=\frac{5}{2}R \text{ and } C_{p_2}(H_2)=\frac{7}{2}R)$

$$(C_p)_{mix}==\frac{1\times\frac{5}{2}R+1\times\frac{7}{2}R}{1+1}\ 3R=3\times 2=6 \text{ cal/mol.°C}$$

∴ Amount of heat needed to raise the temperature from 0°C to 100°C

$$(\Delta Q)_p=\mu C_p\Delta T=2\times 6\times 100=1200\,cal$$

27. **(a)** 1 mole = 22.4 L at S.T.P.

$$\frac{4.5g}{18g}=22.4\times\frac{4.5}{18}=5.6\,L$$

28. **(d)** Internal energy of 2 moles of oxygen

$$U_{O_2}=\mu\left(\frac{5}{2}RT\right)=2.\frac{5}{2}RT=5RT$$

Internal energy of 4 moles of Argon.

$$U_{Ar}=\mu\left(\frac{3}{2}RT\right)=4.\frac{3}{2}RT=6RT$$

∴ Total internal energy

$$U=U_{O_2}+U_{Ar}=11RT$$

29. **(d)** $PV=\frac{m}{M}RT$

Initially, $PV=\frac{6}{M}R\times 500$

Finally, $\frac{P}{2}V=\frac{(6-x)}{M}R\times 300$ (if x g gas leaks out)

Hence, $2=\frac{6}{6-x}\times\frac{5}{3}$ ∴ x = 1 gram

30. **(b)** $\gamma=1+\frac{2}{f},\Rightarrow\gamma-1=\frac{2}{f}\Rightarrow\frac{f}{2}=\frac{1}{\gamma-1}\Rightarrow f=\frac{2}{\gamma-1}$

31. **(a)**

32. **(a)** Number of moles of first gas $=\frac{n_1}{N_A}$

Number of moles of second gas $=\frac{n_2}{N_A}$

Number of moles of third gas $=\frac{n_3}{N_A}$

If there is no loss of energy then

$P_1V_1+P_2V_2+P_3V_3=PV$

$$\frac{n_1}{N_A}RT_1+\frac{n_2}{N_A}RT_2+\frac{n_3}{N_A}RT_3=\frac{n_1+n_2+n_3}{N_A}RT_{mix}$$

$$\Rightarrow T_{mix}=\frac{n_1T_1+n_2T_2+n_3T_3}{n_1+n_2+n_3}$$

33. **(d)** Since it hits the plane wall parallel to $y\,z$ – plane and it rebounds with same velocity, its y and z components of velocity do not change, but the x-component reverses the sign.

∴ Velocity after collision is $(-v_x, v_y$ and $v_z)$.

The change in momentum is

$$-mv_x-mv_x=-2mv_x$$

34. **(c)** V and T will be same for both gases.

$P_1V=\mu_1RT$ and $P_2V=\mu_2RT$

$(P_1/P_2)=\frac{5}{3}$ ∴ $\left(\frac{\mu_1}{\mu_2}\right)=\frac{5}{3}$

By definition, $\mu_1=\frac{N_1}{N_A}$ and $\mu_2=\frac{N_2}{N_A}$

∴ $\frac{N_1}{N_2}=\frac{\mu_1}{\mu_2}=\frac{5}{3}$

35. **(d)** $E=\frac{3}{2}\times 300$; $E'=\frac{3}{2}R(600)=2E=2\times 6.21\times 10^{-21}$

$= 12.42\times 10^{-21}$ J.

$$v_{rms}=\sqrt{\frac{3R\times 300}{M}};\ v'_{rms}=\sqrt{\frac{3R\times 600}{M}}=\sqrt{2}\,v_{rms}$$

$= 684.44$ m/s

36. **(d)** Let the mass of the gas be m.

At a fixed temperature and pressure, volume is fixed.

Density of the gas, $\rho=\frac{m}{V}$

Now $\frac{\rho}{P}=\frac{m}{PV}=\frac{m}{nRT}$

$\Rightarrow\frac{m}{nRT}=x$ (By question)

$\Rightarrow xT=\text{constant}\Rightarrow x_1T_1=x_2T_2$

$$\Rightarrow x_2\Rightarrow\frac{x_1T_1}{T_2}=\frac{283}{383}x\quad\left[\begin{array}{ccc}\therefore & & \\ T_1 & = & 283K\\ T_2 & = & 383K\end{array}\right]$$

37. **(h)** $F=\frac{dU}{dr}=-\frac{d}{dr}\left[\frac{M}{r^3}-\frac{N}{R^{12}}\right]=-\left[\frac{-6M}{r^2}+\frac{12N}{r^{13}}\right]$

In equilibrium position, $F = 0$

$\therefore \frac{6M}{r^2} - \frac{12N}{r^{13}} = 0$ or, $r^6 = \frac{2N}{M}$

$\therefore$ Potential energy at equilibrium position

$$U = \frac{M}{(2N/M)} = \frac{N}{(2N/M)^2} = \frac{M^2}{2N} - \frac{M^2}{4N} = \frac{M^2}{4N}$$

38. (a) Pressure of the gas will not be affected by motion of the system, hence by

$$v_{rms} = \sqrt{\frac{3P}{\rho}} \Rightarrow \bar{c}^2 = \frac{3P}{\rho} \Rightarrow P = \frac{1}{3}\rho\bar{c}^2$$

39. (b) Mean free path in a gas is 100 times the interatomic distance.

40. (a) $v_{rms} = \left[\frac{(2)^2 + (3)^2 + (4)^2 + (5)^2}{4}\right]^{1/2} = \sqrt{\left[\frac{54}{4}\right]}$

41. (c) If a gas is heated at constant volume then no work is done. The heat supplied is given by

$dQ = nC_v dT$

But $C_v = \frac{f}{2}R$ where f is the degree of freedom of the gas

$$\therefore dQ = \frac{nfRdT}{2}$$

$$= \frac{2 \times 3 \times R \times (373 - 273)}{2} = 300\,R$$

42. (b) For 1 molecule of a gas, $V_{rms} = \sqrt{\frac{3KT}{m}}$

where m is the mass of one molecule

For N molecule of a gas, $V_1 = \sqrt{\frac{3KT \times N}{m}}$

For 2N molecule of a gas $V_2 = \sqrt{\frac{3KT \times 2N}{(2m)}}$

$$\therefore \frac{V_1}{V_2} = 1$$

43. (c)

44. (c) P-V diagram of the gas is a straight line passing through origin. Hence $P \propto V$ or PV^{-1} = constant

Molar heat capacity in the process PV^x = constant

$C = \frac{R}{\gamma - 1} + \frac{R}{1 - x}$; Here $\gamma = 1.4$ (For diatomic gas)

$$\Rightarrow C = \frac{R}{1.4 - 1} + \frac{R}{1 + 1} \Rightarrow C = 3R$$

45. (c) Given

$C_P - C_V = 5000$ J/mole°C ....(i)

$\frac{C_P}{C_V} = 1.6$ ....(ii)

From Equation (i) & (ii),

$$\Rightarrow \frac{C_P}{C_V} - \frac{C_V}{C_V} = \frac{5000}{C_V}$$

$$\Rightarrow 1.6 - 1 = \frac{5000}{C_V}$$

$$\Rightarrow C_V = \frac{5000}{0.6} = 8.33 \times 10^3$$

Hence $C_P = 1.6\,C_V = 1.6 \times 8.33 \times 10^3$

$C_P = 1.33 \times 10^4$

1. (a) For an SHM, the acceleration $a = -\omega^2 x$ where ω^2 is a constant. Therefore, $\frac{a}{x}$ is a constant. The time period T is also constant. Therefore, $\frac{aT}{x}$ is a constant.

2. (b) $t_1 = 2\pi\sqrt{\frac{m}{k_1}}$ or $t_1^2 = \frac{4\pi^2 m}{k_1}$ or $k_1 = \frac{4\pi^2 m}{t_1^2}$

Similarly, $k_2 = \frac{4\pi^2 m}{t_2^2}$ and $(k_1 + k_2) = \frac{4\pi^2 m}{t_0^2}$

$\therefore \frac{4\pi^2 m}{t_0^2} = \frac{4\pi^2 m}{t_1^2} + \frac{4\pi^2 m}{t_2^2}$ or $\frac{1}{t_0^2} = \frac{1}{t_1^2} + \frac{1}{t_2^2}$

3. (b)

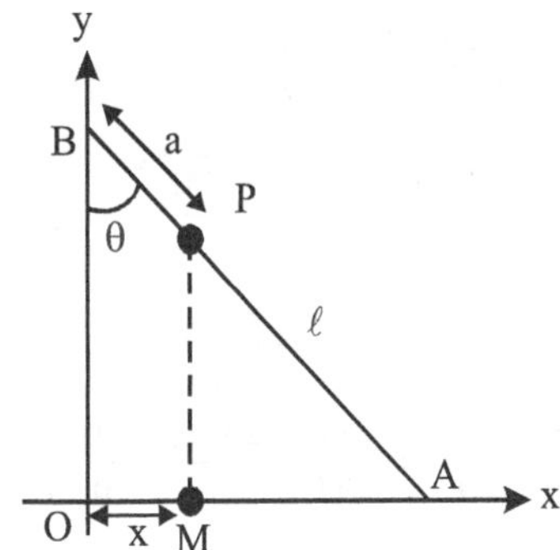

$\frac{d\theta}{dt} = 2 \therefore \theta = 2t$

Let BP = a, $\therefore$ x = OM = a sin θ = a sin (2t)
Hence M executes SHM within the given time period and its acceleration is opposite to x that means towards left.

4. (b) The kinetic energy of a particle executing S.H.M. is given by

$K = \frac{1}{2} m a^2 \omega^2 \sin^2 \omega t$

where, m = mass of particle
a = amplitude
ω = angular frequency
t = time

Now, average K.E. $= \langle K \rangle = \langle \frac{1}{2} m\omega^2 a^2 \sin^2 \omega t \rangle$

$= \frac{1}{2} m\omega^2 a^2 \langle \sin^2 \omega t \rangle$

$= \frac{1}{2} m\omega^2 a^2 \left(\frac{1}{2}\right) \quad \left(\because \langle \sin^2 \theta \rangle = \frac{1}{2}\right)$

$= \frac{1}{4} m\omega^2 a^2 = \frac{1}{4} m a^2 (2\pi\nu)^2 \quad (\because \omega = 2\pi\nu)$

or, $\langle K \rangle = \pi^2 m a^2 \nu^2$

5. (a) We know that $T = 2\pi\sqrt{\frac{M}{k}}$

From first case, $2 = 2\pi\sqrt{\frac{M}{k}}$ .... (1)

In second case, $4 = 2\pi\sqrt{\frac{M+2}{k}}$ .... (2)

From eq. (1) and eq. (2)

$\frac{4}{2} = \sqrt{\frac{M+2}{M}} \Rightarrow 4 = 1 + \frac{2}{M}$

$\frac{2}{M} = 3 \Rightarrow M = \frac{2}{3}$ kg

6. (b) Amplitude of a damped oscillator at any instant t is given by
$A = A_0 e^{-bt/2m}$
where A_0 is the original amplitude
From question,

When t = 2 s, $A = \frac{A_0}{3}$

$\therefore \frac{A_0}{3} = A_0 e^{-2b/2m}$

or, $\frac{1}{3} = e^{-b/m}$ ...(i)

When t = 6 s, $A = \frac{A_0}{n}$

$\therefore \frac{A_0}{n} = A_0 e^{-6b/2m}$

or, $\frac{1}{n} = e^{-3b/m} = (e^{-b/m})^3$

or, $\frac{1}{n} = \left(\frac{1}{3}\right)^3$ (Using eq. (i))

$\therefore n = 3^3$

7. (c) Acceleration due to gravity at a depth 'd' is given by,

$g' = g\left(1 - \frac{d}{R}\right) = g\left(\frac{R-d}{R}\right) = \frac{g}{R} y$

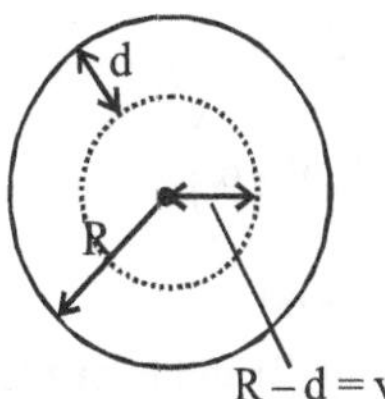

or acceleration ∝ displacement
which is the condition for SHM. So, body will oscillate simple harmonically in tunnel.

8. (c) Time to complete 1/4th oscillation is $\frac{T}{4}$ s. Time to complete $\frac{1}{8}$th vibration from extreme position is obtained from

$$y=\frac{a}{2}=a\cos\omega t=a\cos\frac{2\pi}{T}t \text{ or } t=\frac{T}{6}\text{s}$$

So time to complete 3/8th oscillation

$$=\frac{T}{4}+\frac{T}{6}=\frac{5T}{12}$$

9. (a) As we know, kinetic energy $=\frac{1}{2}m\omega^2(A^2-x^2)$

Potential energy $=\frac{1}{2}m\omega^2x^2$

$$\therefore \frac{\frac{1}{2}m\omega^2(A^2-x^2)}{\frac{1}{2}m\omega^2x^2}=\frac{1}{4}\Rightarrow\frac{A^2-x^2}{x^2}=\frac{1}{4}$$

$$4A^2-4x^2=x^2 \Rightarrow x^2=\frac{4}{5}A^2 \quad \therefore x=\frac{2}{\sqrt{5}}A.$$

10. (d) $T=2\pi\sqrt{\frac{\ell}{g}}$ and $T'=2\pi\sqrt{\frac{1.21\ell}{g}}$

$(\because \ell'=\ell+21\%\text{ of }\ell)$

$$\%\text{ increase}=\frac{T'-T}{T}\times 100$$

$$=\frac{\sqrt{1.21\ell}-\sqrt{\ell}}{\sqrt{\ell}}\times100=\left(\sqrt{1.21}-\sqrt{1}\right)\times 100$$

$$=(1.1-1)\times 100=10\%$$

11. (d) $T=2\pi\sqrt{\frac{m}{k}}$

When a spring is cut into n parts
Spring constant for each part = nk
Here, $n=4$

$$T_1=2\pi\sqrt{\frac{m}{4k}}=\frac{T}{2}$$

12. (d) $x=A\cos(\omega t+\delta)$

$$y=A\cos(\omega t+\alpha) \quad ....(1)$$

When $\delta=\alpha+\frac{\pi}{2}$

$$x=A\cos\left(\frac{\pi}{2}+\omega t+\alpha\right)$$

$$x=-A\sin(\omega t+\alpha) \quad ....(2)$$

Squaring (1) and (2) and then adding
$x^2+y^2=A^2[\cos^2(\omega t+\alpha)+\sin^2(\omega t+\alpha)]$
or $x^2+y^2=A^2$, which is the equation of a circle. The present motion is anticlockwise.

13. (a) Here,
$x=x_0\cos(\omega t-\pi/4)$

$\therefore$ Velocity, $v=\frac{dx}{dt}=-x_0\omega\sin\left(\omega t-\frac{\pi}{4}\right)$

Acceleration,

$$a=\frac{dv}{dt}=-x_0\omega^2\cos\left(\omega t-\frac{\pi}{4}\right)$$

$$=x_0\omega^2\cos\left[\pi+\left(\omega t-\frac{\pi}{4}\right)\right]$$

$$=x_0\omega^2\cos\left(\omega t+\frac{3\pi}{4}\right) \quad ...(1)$$

Acceleration, $a=A\cos(\omega t+\delta)$ ...(2)
Comparing the two equations, we get

$A=x_0\omega^2$ and $\delta=\frac{3\pi}{4}$.

14. (c) $T=2\pi\sqrt{\frac{M}{k}}$

$$T'=2\pi\sqrt{\frac{M+m}{k}}=\frac{5T}{3}$$

$$\therefore 2\pi\sqrt{\frac{M+m}{k}}=\frac{5}{3}\times 2\pi\sqrt{\frac{M}{k}}$$

$$M+m=\frac{25}{9}\times M$$

$$1+\frac{m}{M}=\frac{25}{9} \Rightarrow \frac{m}{M}=\frac{25}{9}-1=\frac{16}{9}$$

15. (d) A = 0.05 m, y = 0.01 m
Acceleration, a = 1.0 m/s²
We have, $a=-\omega^2y$ or $|a|=\omega^2y$

$$\Rightarrow 1.0=\omega^2\times 0.01$$

$$\therefore \omega^2=\frac{1.0}{0.01}=100\Rightarrow\omega=10$$

Now, time period, $T=\frac{2\pi}{\omega}=\frac{2\pi}{10}=\frac{\pi}{5}$ sec.

16. (c) We have, $U+K=E$
where, U = potential energy, K = Kinetic energy, E = Total energy.
Also, we know that, in S.H.M., when potential energy is maximum, K.E. is zero and vice-versa.

$\therefore U_{max}+0=E \Rightarrow U_{max}=E$

Further,

$$K.E.=\frac{1}{2}m\omega^2a^2\cos^2\omega t$$

But by question, $K.E.=K_0\cos^2\omega t$

$$\therefore K_0=\frac{1}{2}m\omega^2a^2$$

Hence, total energy, $E=\frac{1}{2}m\omega^2a^2=K_0$

$\therefore U_{max}=K_0$ & $E=K_0$.

17. (b) Distance covered by lift is given by
$y = t^2$
$\therefore$ Acceleration of lift upwards

$$= \frac{d^2y}{dt^2} = \frac{d}{dt}(2t) = 2\ m/s^2 = \frac{g}{5}$$

Now, $T = 2\pi\sqrt{\frac{\ell}{g}}$

$$T' = 2\pi\sqrt{\frac{\ell}{g+\frac{g}{5}}} = 2\pi\sqrt{\frac{\ell}{\frac{6}{5}g}} = \sqrt{\frac{5}{6}}T.$$

18. (d) In simple harmonic motion, starting from rest,
At $t = 0, x = A$
$x = A\cos\omega t$ ..... (i)
When $t = \tau, x = A - a$
When $t = 2\tau, x = A - 3a$
From equation (i)
$A - a = A\cos\omega\tau$ ......(ii)
$A - 3a = A\cos 2\omega\tau$ ....(iii)
As $\cos 2\omega\tau = 2\cos^2\omega\tau - 1$ ...(iv)
From equation (ii), (iii) and (iv)

$$\frac{A-3a}{A} = 2\left(\frac{A-a}{A}\right)^2 - 1$$

$$\Rightarrow \frac{A-3a}{A} = \frac{2A^2 + 2a^2 - 4Aa - A^2}{A^2}$$

$\Rightarrow A^2 - 3aA = A^2 + 2a^2 - 4Aa$
$\Rightarrow 2a^2 = aA$
$\Rightarrow A = 2a$

$$\Rightarrow \frac{a}{A} = \frac{1}{2}$$

Now, $A - a = A\cos\omega\tau$

$$\Rightarrow \cos\omega\tau = \frac{A-a}{A}$$

$$\Rightarrow \cos\omega\tau = \frac{1}{2} \text{ or } \frac{2\pi}{T}\tau = \frac{\pi}{3}$$

$\Rightarrow T = 6\tau$

19. (b) $v_1 = \frac{dy_1}{dt} = 0.1\times 100\pi\cos\left(100\pi t + \frac{\pi}{3}\right)$

$$v_2 = \frac{dy_2}{dt} = -0.1\pi\sin\pi t = 0.1\pi\cos\left(\pi t + \frac{\pi}{2}\right)$$

$$\therefore \text{Phase diff.} = \phi_1 - \phi_2 = \frac{\pi}{3} - \frac{\pi}{2} = \frac{2\pi - 3\pi}{6} = -\frac{\pi}{6}$$

20. (c) $f_A = \frac{1}{2\pi}\sqrt{\frac{g}{L_A}}$ and $f_B = \frac{f_A}{2} = \frac{1}{2\pi}\sqrt{\frac{g}{L_B}}$

$$\therefore \frac{f_A}{f_{A/2}} = \frac{1}{2\pi}\sqrt{\frac{g}{L_A}} \times 2\pi\sqrt{\frac{L_B}{g}} \Rightarrow 2 = \sqrt{\frac{L_B}{L_A}} \Rightarrow 4 = \frac{L_B}{L_A},$$

regardless of mass

21. (d)

22. (c) Here all the three springs are connected in parallel to mass m. Hence equivalent spring constant
$k = K + K + 2K = 4K$.

23. (a) $K.E. = \frac{1}{2}m\omega^2(a^2 - x^2)$

When $x = 0$, K.E is maximum and is equal to $\frac{1}{2}m\omega^2 a^2$.

24. (d) $y = 5\sin(\pi t + 4\pi)$, comparing it with standard equation

$$y = a\sin(\omega t + \phi) = a\sin\left(\frac{2\pi t}{T} + \phi\right)$$

$$a = 5m \text{ and } \frac{2\pi t}{T} = \pi t \Rightarrow T = 2\sec.$$

25. (b) $T = 2\pi\sqrt{\frac{\text{displacement}}{\text{acceleration}}} = 2\pi\sqrt{\frac{x}{zx}} = 2\pi/\sqrt{z}$

26. (b) Here, $x = 2\times 10^{-2}\cos\pi t$
Speed is given by

$$v = \frac{dx}{dt} = 2\times 10^{-2}\,\pi\sin\pi t$$

For the first time, the speed to be maximum,

$$\sin\pi t = 1 \text{ or, } \sin\pi t = \sin\frac{\pi}{2}$$

$$\Rightarrow \pi t = \frac{\pi}{2} \text{ or, } t = \frac{1}{2} = 0.5\text{ sec.}$$

27. (d) $T = 2\pi\sqrt{\frac{R}{g}} = 2\pi\sqrt{\frac{64\times 10^6}{9.8}} = 2\times\frac{22}{7}\times\frac{8\times 10^3}{7\times\sqrt{2}}$

$$= \frac{\sqrt{2}\times 22\times 8\times 1000}{49\times 60}\text{ min} = 84.6\text{ min}$$

28. (a) $x = 3\sin 2t + 4\cos 2t$. From given equation

$$a_1 = 3,\ a_2 = 4, \text{ and } \phi = \frac{\pi}{2}$$

$$\therefore a = \sqrt{a_1^2 + a_2^2} = \sqrt{3^2 + 4^2} = 5$$

$$\Rightarrow v_{max} = a\omega = 5\times 2 = 10$$

29. (d) Slope of F - x curve $= -k = -\frac{80}{0.2} \Rightarrow k = 400$ N/m,

Time period, $T = 2\pi\sqrt{\frac{m}{k}} = 0.0314$ sec.

30. (b) Phase change π in 50 oscillations.
Phase change 2π in 100 oscillations.
So frequency different ~ 1 in 100.

31. (a) $T = 2\pi\sqrt{\frac{\ell}{g}}$; $2 = 2\pi\sqrt{\frac{\ell}{g}} = 2\pi\sqrt{\frac{\ell'}{(g/6)}}$

Time period will remain constant if on moon,
$\ell' = \ell/6 = 1/6$ m

32. (b) Let k be the force constant of spring of length l_2. Since $l_1 = n\,l_2$, where n is an integer, so the spring is made of $(n+1)$ equal parts in length each of length l_2.

$\therefore \quad \frac{1}{K} = \frac{(n+1)}{k}$ or $k = (n+1)K$

The spring of length l_1 (= n l_2) will be equivalent to n springs connected in series where spring constant

$k' = \frac{k}{n} = (n+1)K/n$ & spring constant of length ℓ_2 is K(n+1).

33. (c) Given

$y = 0.2 \sin(10\pi t + 1.5\pi)\cos(10\pi t + 1.5\pi)$

We know that 2 sinA cos A = sin2A, we get

$y = 0.1 \sin 2\ (10\pi t + 1.5\pi) = 0.1 \sin\ (20\pi t + 3\pi)$

On comparing with wave equation

$y = a \sin(\omega t + \phi)$ we get

$\omega = 20\pi$

$$T = \frac{2\pi}{\omega} = \frac{2\pi}{20\pi} = \frac{1}{10}\text{sec.} = 0.1\text{sec.}$$

34. (a) The displacement of a particle in S.H.M. is given by

$y = a \sin(\omega t + \phi)$

$$\text{velocity} = \frac{dy}{dt} = \omega a \cos(\omega t + \phi)$$

The velocity is maximum when the particle passes through the mean position i.e.,

$$\left(\frac{dy}{dt}\right)_{max} = \omega a$$

The kinetic energy at this instant is given by

$$\frac{1}{2}m\left(\frac{dy}{dt}\right)^2_{max} = \frac{1}{2}m\omega^2 a^2 = 8 \times 10^{-3}\text{ joule}$$

or $\quad \frac{1}{2} \times (0.1)\,\omega^2 \times (0.1)^2 = 8 \times 10^{-3}$

Solving we get $\quad \omega = \pm 4$

Substituting the values of a, ω and φ in the equation of S.H.M., we get

$y = 0.1 \sin(\pm 4t + \pi/4)$ metre.

35. (d) $K.E = \frac{1}{2}k(A^2 - d^2)$

and $P.E. = \frac{1}{2}kd^2$

At mean position $d = 0$. At extrement positions $d = A$

36. (b) $y = 3\sin\frac{\pi}{2}(50t - x)$

$y = 3\sin\left(25\pi t - \frac{\pi}{2}x\right)$ on comparing with the standard wave equation

$y = a \sin(\omega t - kx)$

$$\text{Wave velocity } v = \frac{\omega}{k} = \frac{25\pi}{\pi/2} = 50\text{ m/sec.}$$

The velocity of particle

$$v_p = \frac{\partial y}{\partial t} = 75\pi\cos\left(25\pi t - \frac{\pi}{2}x\right)$$

$v_{p\,max} = 75\pi$

then $\frac{v_{p\,max}}{v} = \frac{75\pi}{50} = \frac{3\pi}{2}$

37. (b) The equivalent situation is a series combination of two springs of spring constants k and 2k.

If k' is the equivalent spring constant, then

$$k' = \frac{(k)(2k)}{3k} = \frac{2k}{3}$$

$$\Rightarrow T = 2\pi\sqrt{\frac{3m}{2k}}$$

38. (a) Time period of simple pendulum $T = 2\pi\sqrt{\left(\frac{l}{g}\right)} \propto \sqrt{l}$

where l is effective length.

[i.e distance between centre of suspension and centre of gravity of bob]

Initially, centre of gravity is at the centre of sphere. When water leaks the centre of gravity goes down until it is half filled; then it begins to go up and finally it again goes at the centre. That is effective length first increases and then decreases. As $T \propto \sqrt{l}$, so time period first increases and then decreases.

39. (d) At $t = 0, x = 5 = \frac{A}{2}$

$\Rightarrow$ Initial phase, $\phi = 30° = \frac{\pi}{6}$

$\Rightarrow x = A\sin(\omega t + \phi)$

$$= 10\sin\left(\frac{2\pi}{T}t + \frac{\pi}{6}\right) = 10\sin\left(\pi t + \frac{\pi}{6}\right)$$

40. (b) For block A to move in S.H.M.

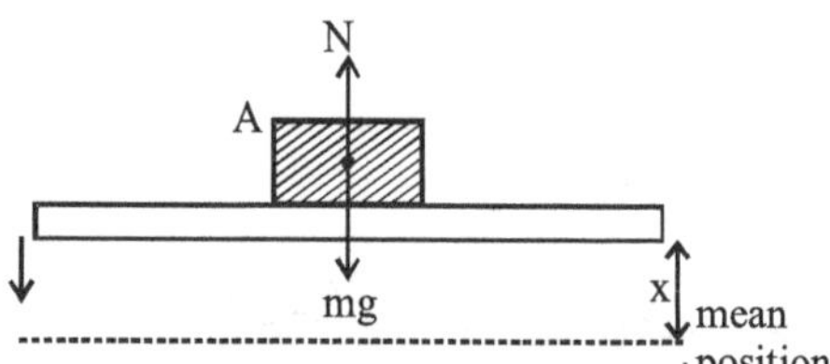

$mg - N = m\omega^2 x$

where x is the distance from mean position

For block to leave contact N = 0

$$\Rightarrow mg = m\omega^2 x \Rightarrow x = \frac{g}{\omega^2}$$

41. (a) $t = 2\pi\sqrt{\frac{\ell}{g_{eff}}}$; $t_0 = 2\pi\sqrt{\frac{\ell}{g}}$

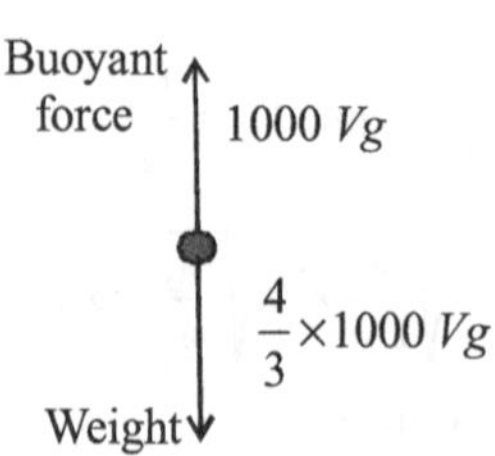

Net force $=\left(\frac{4}{3}-1\right)\times 1000\,Vg=\frac{1000}{3}Vg$

$$g_{eff}=\frac{1000\,Vg}{3\times\frac{4}{3}\times 1000V}=\frac{g}{4}$$

$$\therefore t=2\pi\sqrt{\frac{\ell}{g/4}}$$

$t=2t_0$

42. **(a)** K.E. of a body undergoing SHM is given by,

$$K.E.=\frac{1}{2}ma^2\omega^2\cos^2\omega t$$

$$T.E.=\frac{1}{2}ma^2\omega^2$$

Given K.E. = 0.75 T.E.

$$\Rightarrow 0.75=\cos^2\omega t\Rightarrow \omega t=\frac{\pi}{6}$$

$$\Rightarrow t=\frac{\pi}{6\times\omega}\Rightarrow t=\frac{\pi\times 2}{6\times 2\pi}\Rightarrow t=\frac{1}{6}\,s$$

43. **(a)** Under the action of first force, $F_1=m\omega_1^2 y$

Under the action of second force,

$F_2=m\omega_2^2 y$

Under the action of resultant force,

$F_1+F_2=m\omega^2 y$

$$\Rightarrow m\omega^2 y=m\omega_1^2 y+m\omega_2^2 y$$

$$\Rightarrow \omega^2=\omega_1^2+\omega_2^2$$

$$\Rightarrow\left(\frac{2\pi}{T}\right)^2=\left(\frac{2\pi}{T_1}\right)^2+\left(\frac{2\pi}{T_2}\right)^2$$

$$\Rightarrow T=\sqrt{\frac{T_1^2T_2^2}{T_1^2+T_2^2}}=\sqrt{\frac{\left(\frac{4}{5}\right)^2\cdot\left(\frac{3}{5}\right)^2}{\left(\frac{4}{5}\right)^2+\left(\frac{3}{5}\right)^2}}=\frac{12}{25}.$$

44. **(d)** $F=-bV$, b depends on all the three i.e, shape and size of he block and viscosity of the medium.

45. **(c)** When the bob moves from maximum angular displacement θ to mean position, then the loss of gravitational potential energy is mgh where $h=l(1-\cos\theta)$

DAILY PRACTICE PROBLEMS

PHYSICS SOLUTIONS

DPP/CP14

1. (a) ℓ_1 ▽ ℓ_2 ▽ ℓ_3 ← 110 cm →

$n_1 : n_2 : n_3 = 3 : 2 : 1$

$n \propto \frac{1}{\ell}$

$\ell_1 : \ell_2 : \ell_3 = \frac{1}{3} : \frac{1}{2} : \frac{1}{1} = 2 : 3 : 6$

$\ell_1 + \ell_2 + \ell_3 = 110$

$\Rightarrow 2x + 3x + 6x = 110 \Rightarrow x = 10$

$\therefore$ The two bridges should be set at 2x i.e, 20 cm from one end and 6x i.e, 60 cm from the other end.

2. (a) Equation of the harmonic progressive wave given by: $y = a \sin 2\pi(bt - cx)$.
Here $\upsilon = b$

$k = \frac{2\pi}{\lambda} = 2\pi c$ take, $\frac{1}{\lambda} = c$

$\therefore$ Velocity of the wave $= \upsilon\lambda = b\frac{1}{c} = \frac{b}{c}$

$\frac{dy}{dt} = a\,2\pi b \cos 2\pi(bt - cx) = a\omega \cos(\omega t - kx)$

Maximum particle velocity $= a\omega = a2\pi b = 2\pi ab$

given this is $2 \times \frac{b}{c}$ i.e. $2\pi a = \frac{2}{c}$ or $c = \frac{1}{\pi a}$

3. (c) $y = 0.25 \sin(10\pi x - 2\pi t)$
Comparing this equation with the standard wave equation
$y = a \sin(kx - \omega t)$
We get, $k = 10\pi$

$\Rightarrow \frac{2\pi}{\lambda} = 10\pi \Rightarrow \lambda = 0.2$ m

And $\omega = 2\pi$ or, $2\pi v = 2\pi \Rightarrow v = 1$Hz.
The sign inside the bracket is negative, hence the wave travels in + ve x- direction.

4. (b) Amplitude of reflected wave $= \frac{2}{3} \times 0.9 = 0.6$

It would travel along negative direction of x-axis, and on reflection at a rigid support, there occurs a phase change of π.

5. (c) Velocity of source $= 18$ km h^{-1} $= 5$ m s^{-1}
(i) S moves towards listener (v_S)
(ii) listener moves towards source (v_L)

$v' = \frac{v + v_L}{v - v_S} v = 280$ Hz, Beats $= v' - v = 8$.

6. (d) Third overtone has a frequency 7 n, which means $L = \frac{7\lambda}{4}$ = three full loops + one half loop, which would make four nodes and four antinodes.

7. (c) Comparing it with $y(x, t) = A\cos(\omega t + \pi/2)\cos kx$.
If $kx = \pi/2$, a node occurs ; $\therefore$ $10\pi x = \pi/2 \Rightarrow x = 0.05$ m
If $kx = \pi$, an antinode occurs $\Rightarrow 10\pi x = \pi$
$\Rightarrow x = 0.1$ m

Also speed of wave $\omega/k = \frac{50\pi}{10\pi} = 5$m/s and

$\lambda = 2\pi/k = 2\pi/10\pi = 0.2$ m

8. (a) $\ell_1 + x = \frac{\lambda}{4} = 22.7$ equation (1)

$\ell_2 + x = \frac{3\lambda}{4} = 70.2$ equation (2)

$l_3 + x = \frac{5\lambda}{4}$ equation (3)

From equation (1) and (2)

$x = \frac{\ell_2 - 3\ell_1}{2} = \frac{70.2 - 68.1}{2} = \frac{2.1}{2} = 1.05$ cm

From equation (2) and (3) $\frac{\ell_3 + x}{\ell_1 + x} = 5$

$\ell_3 = 5\ell_1 + 4x = 5 \times 22.7 + 4 \times 1.05 = 117.7$ cm

9. (d)

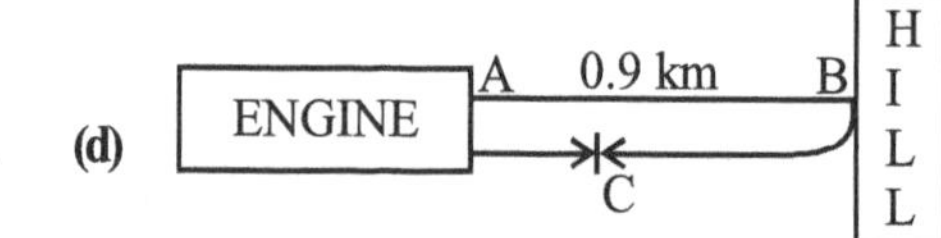

Let after 5 sec engine at point C

$t = \frac{AB}{330} + \frac{BC}{330}$

$5 = \frac{0.9 \times 1000}{330} + \frac{BC}{330}$

$\therefore$ BC = 750 m
Distance travelled by engine in 5 sec
= 900 m – 750 m = 150 m
Therefore velocity of engine

$= \frac{150\text{m}}{5\text{sec}} = 30$ m/s

10. (a) For fundamental mode,

$f = \frac{1}{2\ell}\sqrt{\frac{T}{\mu}}$

Taking logarithm on both sides, we get

$$\log f = \log\left(\frac{1}{2\ell}\right) + \log\left(\sqrt{\frac{T}{\mu}}\right)$$

$$= \log\left(\frac{1}{2\ell}\right) + \frac{1}{2}\log\left(\frac{T}{\mu}\right)$$

$$\text{or } \log f = \log\left(\frac{1}{2\ell}\right) + \frac{1}{2}[\log T - \log \mu]$$

Differentiating both sides, we get

$$\frac{df}{f} = \frac{1}{2}\frac{dT}{T} \text{ (as } \ell \text{ and } \mu \text{ are constants)}$$

$$\Rightarrow \frac{dT}{T} = 2 \times \frac{df}{f}$$

Here df = 6

f = 600 Hz

$$\therefore \frac{dT}{T} = \frac{2\times 6}{600} = 0.02$$

11. (b) Frequency received by listener from the rear source,

$$n' = \frac{v-u}{v}\times n = \frac{v-u}{v}\times\frac{v}{\lambda} = \frac{v-u}{\lambda}$$

Frequency received by listener from the front source,

$$n'' = \frac{v+u}{v}\times\frac{v}{\lambda} = \frac{v+u}{\lambda}$$

No. of beats = $n'' - n'$

$$= \frac{v+u}{\lambda} - \frac{v-u}{\lambda} = \frac{v+u-v+u}{\lambda} = \frac{2u}{\lambda}$$

12. (c) $n' = n\left[\frac{v+v_0}{v}\right] = n\left[\frac{v+\frac{v}{5}}{v}\right] = n\left[\frac{6}{5}\right]$

$$\frac{n'}{n} = \frac{6}{5}; \frac{n'-n}{n} = \frac{6-5}{5}\times 100 = 20\%$$

13. (c) In a closed organ pipe the fundamental frequency is

$$\upsilon = \frac{\upsilon}{4L}$$

$$\upsilon = \frac{320 \text{ ms}^{-1}}{4\times 1 \text{ m}} = 80 \text{ Hz}$$

In a closed organ pipe only odd harmonics are present. So, it can resonate with 80 Hz, 240 Hz, 400 Hz, 560 Hz.

14. (b)

f (S) ⟶ 30 ms^{-1} ⇢ |f' (O) f'' (O) ⟶ 30 ms^{-1} ⟵ |f' (S)

f' is the apparent frequency received by an observer at the hill. f'' is the frequency of the reflected sound as heard by driver.

$$f' = \frac{v}{v-30}f,$$

$$f'' = \frac{v+30}{v}f' = \frac{v+30}{v-30}f = \frac{360}{300}\times 600$$

= 720 Hz

15. (d) Figure(a) represents a harmonic wave of frequency 7.0 Hz, figure (b) represents a harmonic wave of frequency 5.0 Hz. Therefore beat frequency $v_s = 7 - 5 = 2.0$ Hz.

16. (a) $\upsilon \propto \sqrt{T}$

17. (b) In fundamental mode,

$$\frac{\lambda}{2} = l \Rightarrow \lambda = 2l$$

$$\therefore f = \frac{v}{\lambda} = \frac{v}{2l} \quad ....(1)$$

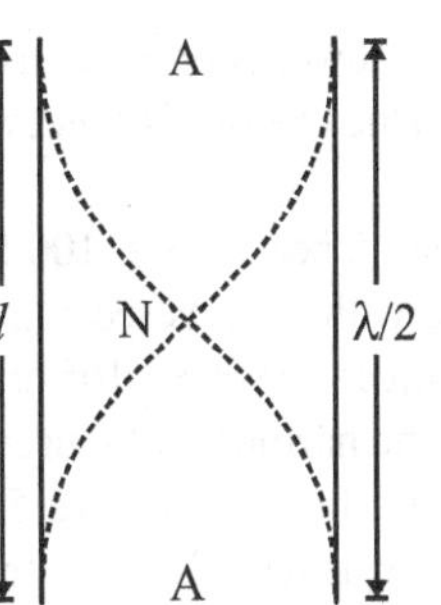

Fundamental mode

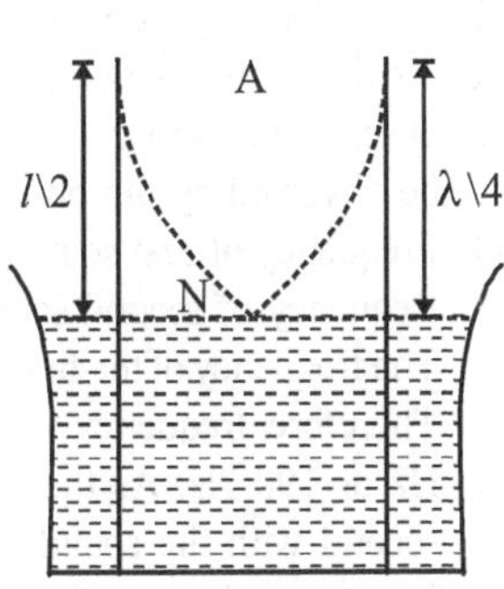

Half length dipped in water

In half length dipped in water mode,

$$\frac{l}{2} = \frac{\lambda}{4} \Rightarrow \lambda = 2l$$

$$\therefore f' = \frac{v}{\lambda} = \frac{v}{2l} = f$$

18. (a) Given wave equation is $y(x,t)$

$$= e^{\left(-ax^2+bt^2+2\sqrt{ab}\,xt\right)}$$

$$= e^{-[(\sqrt{a}x)^2+(\sqrt{b}t)^2+2\sqrt{a}x.\sqrt{b}t]}$$

$$= e^{-(\sqrt{a}x+\sqrt{b}t)^2}$$

$$= e^{-\left(x+\sqrt{\frac{b}{a}}t\right)^2}$$

It is a function of type $y = f(x+vt)$

$\Rightarrow$ Speed of wave = $\sqrt{\frac{b}{a}}$

19. (c) Particle velocity

$$v = \frac{d}{dt}\left[x_0 \sin 2\pi\left(nt - \frac{x}{\lambda}\right)\right]$$

$$= 2\pi n x_0 \cos 2\pi\left(nt - \frac{x}{\lambda}\right)$$

$\therefore$ Maximum particle velocity = $2\pi n x_0$

Wave velocity = $\frac{\lambda}{T} = n\lambda$

Given, $2\pi n x_0 = 4n\lambda \Rightarrow \lambda = \frac{2\pi n x_0}{4n} = \frac{\pi x_0}{2}$

20. (d) As number of beats/sec = diff. in frequencies has to be less than 10, therefore $0 < (n_1 - n_2) < 10$

21. (c) Length of pipe = 85 cm = 0.85m
Frequency of oscillations of air column in closed organ pipe is given by,

$$f=\frac{(2n-1)\upsilon}{4L}$$

$$f=\frac{(2n-1)\upsilon}{4L}\le 1250$$

$$\Rightarrow \frac{(2n-1)\times 340}{0.85\times 4}\le 1250$$

$$\Rightarrow 2n-1\le 12.5\approx 6$$

22. (b) As the source is not moving towards or away from the observer in a straight line, so the Doppler's effect will not be observed by the observer.

23. (c) Frequency of first source with 5 beats/ sec = 100 Hz and frequency of second source with 5 beats/sec = 205 Hz. The frequency of the first source = 100 ± 5 = 105 or 95 Hz. Therefore, frequency of second harmonic of source = 210 Hz or 190 Hz. As the second harmonic gives 5 beats/ second with the sound of frequency 205 Hz, therefore, frequency of second harmonic source should be 210 Hz or frequency of source = 105 Hz.

24. (c) Pressure change will be minimum at both ends. In fact, pressure variation is maximum at $\ell/2$ because the displacement node is pressure antinode.

25. (a) $n_{Last}=n_{First}+(N-1)x$
$2n=n+(41-1)\times 5$

$\Rightarrow n_{First}=200$ Hz and $n_{Last}=400$ Hz

26. (b) Here, T = 0.05 sec, v = 300 ms^{-1}.

Now $\lambda=\frac{v}{\nu}=vT=(300\times 0.05)$m

or, $\lambda=15$ m
Phase of the point at 10 m from the source

$$=\frac{2\pi}{\lambda}\times x=\frac{2\pi}{15}\times 10=\frac{4\pi}{3}\text{rad}$$

Phase of the point at 15 m from the source

$$\frac{2\pi}{\lambda}\times x=\frac{2\pi}{15}\times 15=2\pi\,\text{rad}$$

∴ The phase difference between the points

$$=2\pi-\frac{4\pi}{3}=\frac{2\pi}{3}\text{rad}$$

27. (a) We have, $L_1=10\log\left(\frac{I_1}{I_0}\right)$; $L_2=10\log\left(\frac{I_2}{I_0}\right)$

$$\therefore L_1-L_2=10\log\left(\frac{I_1}{I_0}\right)-10\log\left(\frac{I_2}{I_0}\right)$$

or, $\Delta L=10\log\left(\frac{I_1}{I_0}\times\frac{I_0}{I_2}\right)$ or, $\Delta L=10\log\left(\frac{I_1}{I_2}\right)$

or, $20=10\log\left(\frac{I_1}{I_2}\right)$ or, $2=\log\left(\frac{I_1}{I_2}\right)$

or, $\frac{I_1}{I_2}=10^2$ or, $I_2=\frac{I_1}{100}$.

⇒ Intensity decreases by a factor 100.

28. (a) $y(x,t)=0.005\cos(\alpha x-\beta t)$ (Given)
Comparing it with the standard equation of wave
$y(x,t)=a\cos(kx-\omega t)$ we get
$k=\alpha$ and $\omega=\beta$

But $k=\frac{2\pi}{\lambda}$ and $\omega=\frac{2\pi}{T}$

$\Rightarrow \frac{2\pi}{\lambda}=\alpha$ and $\frac{2\pi}{T}=\beta$

Given that $\lambda=0.08$ m and $T=2.0$s

$\therefore \alpha=\frac{2\pi}{0.08}=25\pi$ and $\beta=\frac{2\pi}{2}=\pi$

29. (b) Equation is of stationary wave. Comparing with the standard equation

$$y=2A\sin\left(\frac{2\pi}{T}\right)t\,\cos\left(\frac{2\pi}{\lambda}\right)x$$

$$\frac{2\pi}{\lambda}=4.5 \text{ or } \lambda=\frac{2\pi}{4.5}=1.4\text{m}$$

30. (b) Waves are kind of disturbances which moves from one place to another without the actual physical transfer of matter of the medium as a whole. The particles of the medium only oscillate but do not travel from one place to another.
Waves transport energy and the pattern of disturbance has information that propagate from one point to another. Here, wave pattern propagates.
All our communication essentially depend on transmission of signals through the waves.

31. (b) $\frac{v}{4\ell_1}=\frac{3v}{2\ell_2}$, $\therefore \frac{\ell_1}{\ell_2}=\frac{1}{6}$

32. (c) $\omega_1=600\pi$, $\omega_2=604\pi$,
$f_1=300$ Hz, $f_2=302$ Hz
Beat frequency, $f_2-f_1=2$ Hz
⇒ number of beats in three seconds = 6

33. (b) Given $f_A=1800$Hz
$v_t=v$
$f_B=2150$ Hz
Reflected wave frequency received by A, $f_A'=?$
Applying doppler's effect of sound,

$$f'=\frac{v_s f}{v_s-v_t}$$

here, $v_t=v_s\left(1-\frac{f_A}{f_B}\right)$

$$=343\left(1-\frac{1800}{2150}\right)$$

$v_t=55.8372$ m/s

Now, for the reflected wave,

$$\therefore \quad f_A' = \left(\frac{v_s + v_t}{v_s - v_t}\right) f_A$$

$$= \left(\frac{343 + 55.83}{343 - 55.83}\right) \times 1800$$

$= 2499.44 \approx 2500\text{Hz}$

34. (a) Standing waves are produced when two waves propagate in opposite direction

As z_1 & z_2 are propagating in +ve x-axis & –ve x-axis

so, $z_1 + z_2$ will represent a standing wave.

35. (d) Load supported by sonometer wire = 4 kg

Tension in sonometer wire = 4 g

If μ = mass per unit length

then frequency $\upsilon = \frac{1}{2l}\sqrt{\frac{T}{\mu}}$

$$\Rightarrow 416 = \frac{1}{2l}\sqrt{\frac{4g}{\mu}}$$

When length is doubled, i.e., $l' = 2l$

Let new load = L

As, $\upsilon' = \upsilon$

$$\therefore \frac{1}{2l'}\sqrt{\frac{Lg}{\mu}} = \frac{1}{2l}\sqrt{\frac{4g}{\mu}}$$

$$\Rightarrow \frac{1}{4l}\sqrt{\frac{Lg}{\mu}} = \frac{1}{2l}\sqrt{\frac{4g}{\mu}}$$

$\Rightarrow \sqrt{L} = 2 \times 2 \Rightarrow L = 16\,\text{kg}$

36. (b) Let the string vibrates in p loops, wavelength of the pth mode of vibration is given by

$$\lambda_p = \frac{2l}{p}$$

Given, $y = 2\sin\left(\frac{4\pi x}{15}\right)\cos(96\pi t)$

or $y = 2\left[\sin\left(\frac{4\pi x}{15} + 96\pi t\right) + \sin\left(\frac{4\pi x}{15} - 96\pi t\right)\right]$

Comparing it with standard equation, we get

$$\upsilon = \frac{96\pi}{2\pi} = 48 \text{ Hz and } k = \frac{4\pi}{15}$$

$$\frac{1}{48} = \frac{2 \times 60}{p} \times \frac{4\pi}{15 \times 96\pi}$$

$\Rightarrow p = 16.$

37. (b) [Figure: string with three loops labelled l_1, l_2, l_3]

$$n = \frac{1}{2l}\sqrt{\frac{T}{m}}$$

or, $n \propto \frac{1}{l}$ or nl = constant, K

$\therefore n_1 l_1 = K,$

$n_2 l_2 = K, n_3 l_3 = K$

Also, $l = l_1 + l_2 + l_3$

or, $\frac{K}{n} = \frac{K}{n_1} + \frac{K}{n_2} + \frac{K}{n_3}$

or, $\frac{1}{n} = \frac{1}{n_1} + \frac{1}{n_2} + \frac{1}{n_3}$

38. (a) Time taken for two syllables $t = \frac{2}{5}$ sec.

$x + x = v \times t = 330 \times \frac{2}{5} \quad \therefore x = 66\,\text{m}$

39. (a) Velocity of sound $= \sqrt{\frac{\gamma RT}{M}}$

When water vapour are represent in air average molecular weight of air decreases and hence velocity increases.

40. (b) $\frac{I_{max}}{I_{min}} = \frac{(a+b)^2}{(a-b)^2} = 49 \qquad \therefore \frac{a+b}{a-b} = 7$

$7a - 7b = a + b$ or $6a = 8b$ or $\frac{a}{b} = \frac{8}{6} = \frac{4}{3}$

41. (a) By the concept of accoustic, the observer and source are moving towards each other, each with a velocity of 18 m s^{-1}.

$\therefore v' = \frac{330 + 18}{330 - 18} \times 1000 \approx 1115$ Hz

42. (c) Compare the given equation with standard form

$$y = r\sin\left[\frac{2\pi x}{\lambda} - \frac{2\pi t}{T}\right]$$

$\frac{2\pi}{\lambda} = 3, \lambda = \frac{2\pi}{3}$ and $\frac{2\pi}{T} = 15$

$$T = \frac{2\pi}{15}$$

Speed of propagation, $v = \frac{\lambda}{T} = \frac{2\pi/3}{2\pi/15} = 5$

43. (c) The contrast will be maximum, when $I_1 = I_2$ i.e. $a = b$. In that event, $I_{min} = (a - b)^2 = 0$, where a and b are the amplitudes of interfering waves.

44. (b) Fundamental frequency of closed organ pipe

$$V_c = \frac{V}{4l_c}$$

Fundamental frequency of open organ pipe

$$V_0 = \frac{V}{2l_0}$$

Second overtone frequency of open organ pipe = $\frac{3V}{2l_0}$

From question,

$$\frac{V}{4l_c} = \frac{3V}{2l_0}$$

$\Rightarrow$ $l_0 = 6l_c = 6 \times 20 = 120$ cm

45. **(b)** $y = 60\cos(180t - 6x)$ ....(1)

$$\omega = 180, k = 6 \Rightarrow \frac{2\pi}{\lambda} = 6$$

$$v = \frac{\omega}{k} = \frac{2\pi}{T} \times \frac{\lambda}{2\pi} = \frac{180}{6} = 30 \text{ m/s}$$

Differentiating (1) w.r.t. t,

$$v = \frac{dy}{dt} = -60 \times 180 \sin(180t - 6x)$$

$v_{max} = 60 \times 180$ μm/s

$= 10800$ μm/s $= 0.0108$ m/s

$$\frac{v_{max}}{v} = \frac{0.0108}{30} = 3.6 \times 10^{-4}$$

DAILY PRACTICE PROBLEMS — PHYSICS SOLUTIONS — DPP/CP15

1. (a) Electric field intensity at the centre of the disc.

$E = \frac{\sigma}{2\epsilon_0}$ (given)

Electric field along the axis at any distance x from the centre of the disc

$$E' = \frac{\sigma}{2\epsilon_0}\left(1 - \frac{x}{\sqrt{x^2+R^2}}\right)$$

From question, x = R (radius of disc)

$$\therefore E' = \frac{\sigma}{2\epsilon_0}\left(1 - \frac{R}{\sqrt{R^2+R^2}}\right)$$

$$= \frac{\sigma}{2\epsilon_0}\left(\frac{\sqrt{2}R - R}{\sqrt{2}R}\right)$$

$$= \frac{4}{14}E$$

$\therefore$ % reduction in the value of electric field

$$= \frac{\left(E - \frac{4}{14}E\right)\times 100}{E} = \frac{1000}{14}\% \simeq 70.7\%$$

2. (a) Surface charge density $(\sigma) = \frac{\text{Charge}}{\text{Surface area}}$

So $\sigma_{inner} = \frac{-2Q}{4\pi b^2}$

and $\sigma_{Outer} = \frac{Q}{4\pi c^2}$

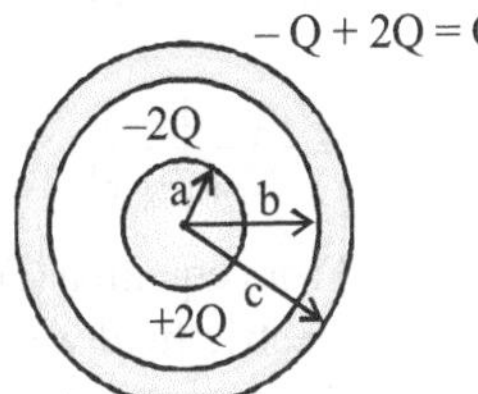

3. (a) Initial force between the two spheres carrying charge (say q) is

$F = \frac{1}{4\pi\varepsilon_0}\frac{q^2}{r^2}$ (r is the distance between them)

Further when an uncharged sphere is kept in touch with the sphere of charge q, the net charge on both become $\frac{q+0}{2} = \frac{q}{2}$. Force on the 3rd charge, when placed in center of the 1st two

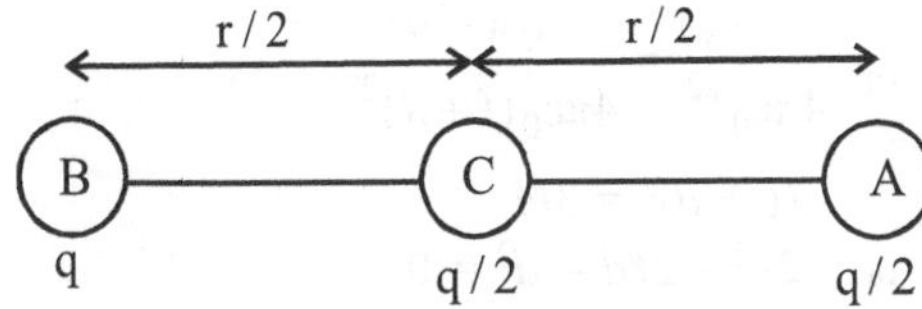

$$F_3 = \frac{1}{4\pi\varepsilon_0}\frac{q\left(\frac{q}{2}\right)}{\left(\frac{r}{2}\right)^2} - \frac{1}{4\pi\varepsilon_0}\frac{\left(\frac{q}{2}\right)^2}{\left(\frac{r}{2}\right)^2}$$

$$= \frac{1}{4\pi\varepsilon_0}\frac{q^2}{r^2}[2-1] = F$$

4. (d) Since electric field $\vec{E}$ decreases inside water, therefore flux $\phi = \vec{E}.\vec{A}$ also decreases.

5. (d) Unit positive charge at O will be repelled equally by three charges at the three corners of triangle. By symmetry, resultant $\vec{E}$ at O would be zero.

6. (c) When a dipole is placed in a uniform electric field, two equal and opposite forces act on it. Therefore, a torque acts which rotates the dipole.

7. (d) Force acting on the charged particle due to electric field $= q\vec{E}$

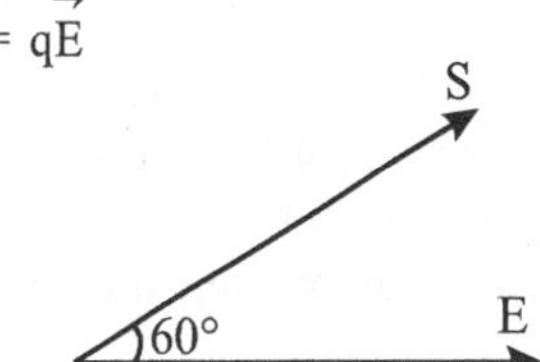

work done in moving through distance S,

$$W = q\vec{E}.\vec{S} = (qE)\times S\times\cos\theta$$

$$\therefore\ 10\,J = (0.5\,C)\times E\times 2\cos 60°$$

$$E = 10\times 2 = 20\ NC^{-1} = 20\ Vm^{-1}$$

8. (c) The charged sphere is a conductor. Therefore the field inside is zero and outside it is proportional to $1/r^2$

9. (d) Since $\phi_{total} = \phi_A + \phi_B + \phi_C = \frac{q}{\varepsilon_0}$,

where q is the total charge.

As shown in the figure, flux associated with the curved surface B is $\phi = \phi_B$

Let us assume flux linked with the plane surfaces A and C be

$\phi_A = \phi_C = \phi'$

Therefore,

$$\frac{q}{\varepsilon_0} = 2\phi' + \phi_B = 2\phi' + \phi$$

$$\Rightarrow \phi' = \frac{1}{2}\left(\frac{q}{\varepsilon_0} - \phi\right)$$

10. (b) We have $E_a = \frac{2kp}{r^3}$ and $E_e = \frac{kp}{r^3}$; $\therefore\ E_a = 2E_e$

11. (c) Electric lines of force due to a positive charge is spherically symmetric.

All the charges are positive and equal in magnitude. So repulsion takes place. Due to which no lines of force are present inside the equilateral triangle and the resulting lines of force obtained as shown:

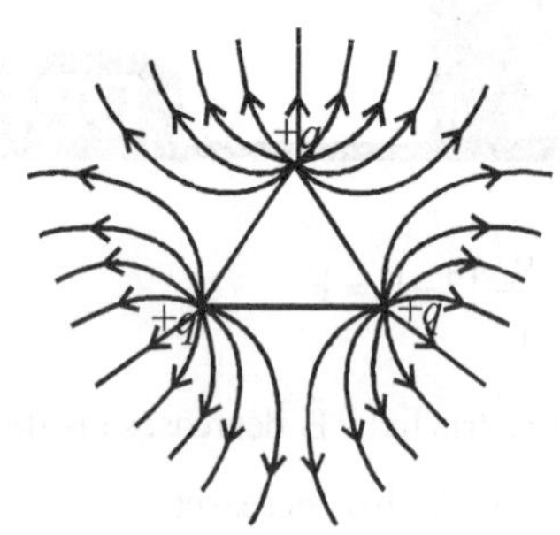

12. (a) ←a→←a→

Q_1 Q_2 Q_3

$Q_2 = -Q_3 = Q$

Force on Q_3 due to Q_2 + Force on Q_3 due to $Q_1 = 0$.

$$\frac{1}{4\pi \in_0}\left(\frac{-Q^2}{a^2}\right) + \frac{1}{4\pi \in_0}\frac{Q_1 Q}{4a^2} = 0 \Rightarrow Q_1 = 4Q_3$$

13. (b) Charge per cm length of the wire = qC

$\therefore$ Charge per metre of the wire = 100q C

According to Gauss's law,

Total electric flux passing through the cylindrical surface

$$\phi = \frac{q_{enclosed}}{\varepsilon_0} = \frac{100q}{\varepsilon_0}$$

1 m

50 cm

14. (c) $T_0 = 2\pi\sqrt{\frac{L}{g}}$

When the plates are charged, the net acceleration is,

$g' = g + a$

$$g' = g + \frac{qE}{m} \qquad \left(a = \frac{qE}{m}\right)$$

$$\therefore \quad T = 2\pi\sqrt{\frac{L}{g + \frac{qE}{m}}}$$

$$\therefore \quad \frac{T}{T_0} = \left(\frac{g}{g + \frac{qE}{m}}\right)^{1/2}$$

15. (c) Charges (q) = 2×10^{-6} C, Distance (d) = 3 cm = 3×10^{-2} m and electric field (E) = 2×10^5 N/C. Torque (τ) = q.d.

E = $(2 \times 10^{-6}) \times (3 \times 10^{-2}) \times (2 \times 10^5)$

= 12×10^{-3} N–m.

16. (b) Net flux emmited from a spherical surface of radius a according to Gauss's theorem

$$\phi_{net} = \frac{q_{in}}{\varepsilon_0}$$

or, $(Aa)(4\pi a^2) = \frac{q_{in}}{\varepsilon_0}$

So, $q_{in} = 4\pi\varepsilon_0 A a^3$

17. (a) Since lines of force starts from A and ends at B, so A is +ve and B is –ve. Lines of forces are more crowded near A, so A > B.

18. (c) Net force on each of the charge due to the other charges is zero. However, disturbance in any direction other than along the line on which the charges lie, will not make the charges return.

19. (c)

20. (d) For distances far away from centre of dipole

$$E_{axis} = E_a = \frac{1}{4\pi\varepsilon_0}\frac{2p}{r^3}$$

$$E_{equa} = E_e = \frac{1}{4\pi\varepsilon_0} - \frac{p}{r^3}$$

$$\frac{d}{dr}(E_a) = \frac{1}{4\pi\varepsilon_0} 2p \frac{d}{dr}(r^{-3})$$

$$= -6 \cdot \frac{1}{4\pi\varepsilon_0}\frac{p}{r^4} \quad \text{...(i)}$$

$$\frac{d}{dr}(E_e) = \frac{1}{4\pi\varepsilon_0} p \frac{d}{dr}(r^{-3})$$

$$= -3\frac{1}{4\pi\varepsilon_0}\frac{p}{r^4} \quad \text{...(ii)}$$

From equation (i) and (ii) the magnitude of change in electric field w.r.t. distance is more in case of axis of dipole as compared to equatorial plane.

21. (b)

$2q$ q $-3q$

P A B

←ℓ→←d→

Let a charge $2q$ be placed at P, at a distance I from A where charge q is placed, as shown in figure. The charge $2q$ will not experience any force, when force, when force of repulsion on it due to q is balanced by force of attraction on it due to $-3q$ at B where AB = d

$$\text{or} \quad \frac{(2q)(q)}{4\pi\varepsilon_0 \ell^2} = \frac{(2q)(-3q)}{4\pi\varepsilon_0(\ell + d)^2}$$

$(\ell + d)^2 = 3\ell^2$

or $2\ell^2 - 2\ell d - d^2 = 0$

$$\therefore \quad \ell = \frac{2d \pm \sqrt{4d^2 + 2d^2}}{4} = \frac{d}{2} \pm \frac{\sqrt{3}d}{2}$$

$$\ell = \frac{d + \sqrt{3}d}{2}$$

22. (d) Let F be the force between Q and Q. The force between q and Q should be attractive for net force on Q to be zero. Let F' be the force between Q and q. The resultant of F' and F' is R. For equilibrium

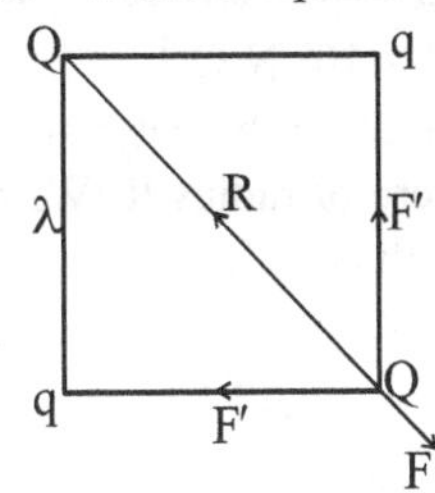

$\vec{R}+\vec{F}=0 \qquad \sqrt{2}F'=-F$

$$\sqrt{2}\times k\frac{Qq}{\ell^2}=-k\frac{Q^2}{(\sqrt{2}\ell)^2} \Rightarrow \frac{Q}{q}=-2\sqrt{2}$$

23. (b) Nuclear force binds the protons and neutrons in the nucleus of an atom.

24. (c) Net downward force on the drop $=\frac{4}{3}\pi r^3(\rho-\rho_0)g$

For equilibrium, electric force must be upwards i.e. charge on the drop is positive.

$$neE=\frac{4}{3}\pi r^3(\rho-\rho_0)g \text{ i.e. } n=\frac{4\pi r^3(\rho-\rho_0)g}{3eE}$$

25. (d) Electric flux, $\phi=EA\cos\theta$,

where θ = angle between E and normal to the surface.

Here $\theta=\frac{\pi}{2}$

$\Rightarrow \phi=0$

26. (a) Potential energy of an electric dipole in an electric field is, $U=-\vec{p}.\vec{E}$ i.e. $U=-pE\cos\theta$

For minimum U, $\theta=0°$

$\Rightarrow U_{min}=-pE\cos 0=-pE$

27. (c) Milikan demonstrated the quantisation of charge experimentally. Charge on electron $=-e=-1.6\times10^{-19}$ C. Addition of charge can occur in integral multiples of e.

28. (c) Let n be the number of electrons missing.

$$F=\frac{1}{4\pi\varepsilon_0}\cdot\frac{q^2}{d^2}$$

$$\Rightarrow q=\sqrt{4\pi\varepsilon_0 d^2F}=ne$$

$$\therefore n=\sqrt{\frac{4\pi\varepsilon_0 Fd^2}{e^2}}$$

29. (b) $F=\frac{1}{4\pi\varepsilon_0}\frac{(4q)(-4q)}{r^2}$

when C is touched with A, then charge on A & C each = 2q after that C is touched with B, charge on

$$B=\frac{2q+(-4q)}{2}=-q$$

Now, force $F'=\frac{1}{4\pi\varepsilon_0}\frac{(2q)(-q)}{r^2} \Rightarrow F'=\frac{F}{8}$

30. (a) $-eE=mg$

$$\vec{E}=-\frac{9.1\times10^{-31}\times10}{1.6\times10^{-19}}=-5.6\times10^{-11}\text{ N/C}$$

31. (d) Torque, $\vec{\tau}=\vec{p}\times\vec{E}=pE\sin\theta$

$4=p\times2\times10^5\times\sin30°$

or, $p=\frac{4}{2\times10^5\times\sin30°}=4\times10^{-5}$ Cm

Dipole moment, $p=q\times l$

$$q=\frac{p}{l}=\frac{4\times10^{-5}}{0.02}=2\times10^{-3}\text{C}=2\text{mC}$$

32. (c) K.E. = Force × distance = qE.y

33. (d) Charge (q) = 0.2 C; Distance (d) = 2 m; Angle $\theta=60°$ and Work done (W) = 4J.

Work done in moving the charge (W)
$= F.d\cos\theta = qEd\cos\theta$

or, $E=\frac{W}{qd\cos\theta}=\frac{4}{0.2\times2\times\cos60°}=\frac{4}{0.4\times0.5}=20$ N/C.

34. (a) $\phi=\vec{E}.\vec{A}=4\hat{i}.(2\hat{i}+3\hat{j})=8$ V-m

35. (d) The dipole is placed in a non-uniform field, therefore a force as well as a couple acts on it. The force on the negative charge is more ($F\propto E$) and is directed along negative x-axis. Thus the dipole moves along negative x-axis and rotates in an anticlockwise direction.

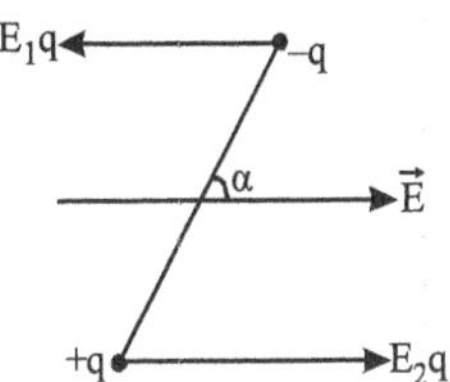

36. (b) Flux $=\vec{E}.\vec{A}$.

$\vec{E}$ is electric field vector & $\vec{A}$ is area vector.

Here, angle between $\vec{E}$ & $\vec{A}$ is 90°.

So, $\vec{E}.\vec{A}=0$; Flux = 0

37. (c) The charge on disc A is 10^{-6} μC. The charge on disc B is 10×10^{-6} μC. The total charge on both = 11 μC. When touched, this charge will be distributed equally i.e. 5.5 μC on each disc.

38. (a) According to Gauss's law total electric flux through a closed surface is $\frac{1}{\varepsilon_0}$ times the total charge inside that surface.

Electric flux, $\phi_E=\frac{q}{\varepsilon_0}$

Charge on α-particle = 2e

$$\phi_E = \frac{2e}{\varepsilon_0}$$

39. (b) It is possible to create or destroy charged particles but it is not possible to create or destroy net charge. The charge of an isolated system is conserved.

40. (a) The flux is zero according to Gauss' Law because it is a open surface which enclosed a charge q.

41. (d) By Gauss law, we know that

$\phi = \frac{q}{\varepsilon_0}$ Here, Net electric flux, $\phi = \phi_2 - \phi_1$

$$= 9 \times 10^6 - 6 \times 10^6 = \frac{q}{\varepsilon_0} \Rightarrow q = 3 \times 10^6 \times \varepsilon_0.$$

42. (d) They will not experience any force if $|\vec{F}_G| = |\vec{F}_e|$

$$\Rightarrow G\frac{m^2}{(16 \times 10^{-2})^2} = \frac{1}{4\pi\varepsilon_0} \cdot \frac{q^2}{(16 \times 10^{-2})^2} \Rightarrow \frac{q}{m} = \sqrt{4\pi\varepsilon_0 G}$$

43. (c) Here, $\ell = 2.4\ m$, $r = 4.6\ mm = 4.6 \times 10^{-3}\ m$

$q = -4.2 \times 10^{-7}$ C

Linear charge density, $\lambda = \frac{q}{\ell}$

$$= \frac{-4.2 \times 10^{-7}}{2.4} = -1.75 \times 10^{-7}\ \text{C m}^{-1}$$

Electric field, $E = \frac{\lambda}{2\pi\varepsilon_0 r}$

$$= \frac{-1.75 \times 10^{-7}}{2 \times 3.14 \times 8.854 \times 10^{-12} \times 4.6 \times 10^{-3}}$$

$= -6.7 \times 10^5\ \text{N C}^{-1}$

44. (a) Charge resides on the outer surface of a conducting hollow sphere of radius R. We consider a spherical surface of radius r < R.

By Gauss theorem

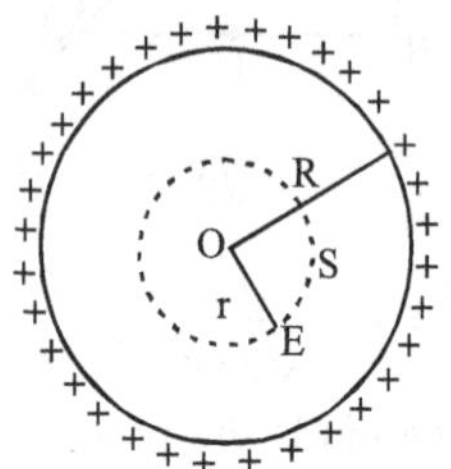

$$\oint \vec{E} \cdot d\vec{s} = \frac{1}{\varepsilon_0} \times \text{charge enclosed or } E \times 4\pi r^2 = \frac{1}{\varepsilon_0} \times 0$$

$\Rightarrow E = 0$

i.e electric field inside a hollow sphere is zero.

45. (c) By Gauss's theorem, $\phi = \frac{Q_{in}}{\varepsilon_0}$

Thus, the net flux depends only on the charge enclosed by the surface. Hence, there will be no effect on the net flux if the radius of the surface is doubled.

DAILY PRACTICE PROBLEMS

PHYSICS SOLUTIONS

DPP/CP16

1. (d) As volume remains constant, therefore,

$\frac{4}{3}\pi R^3 = n\times\frac{4}{3}\pi r^3 \quad \therefore \quad R = n^{1/3}r.$

New potential $= V' = \frac{nq}{4\pi\varepsilon_o R} = \frac{nq}{4\pi\varepsilon_o(n^{1/3}r)}$

$$= n^{2/3}\frac{q}{4\pi\varepsilon_o r} = n^{2/3}V.$$

2. (a) $C_a = \frac{\epsilon_0 A}{d}$ and $C_b = \frac{\epsilon_0 A}{\frac{d}{2}+\frac{d}{2K}} = \frac{2\epsilon_0 A(1+K)}{d}$

and $C_c = \frac{\epsilon_0 \frac{A}{2}}{d} + \frac{\epsilon_0 \frac{A}{2}K}{d} = \frac{\epsilon_0 A}{2d}(1+K)$

or $C_b = \frac{\epsilon_0 A}{d}2(1+K) > C_a$

or $C_c = \frac{\epsilon_0 A}{d}\frac{1+K}{2} > C_a \quad \therefore C_b$ and $C_c > C_a$

3. (c) Charges reside only on the outer surface of a conductor with cavity.

4. (b) In oil, C becomes twice, V becomes half. Therefore, E = V/d becomes half.

5. (d) 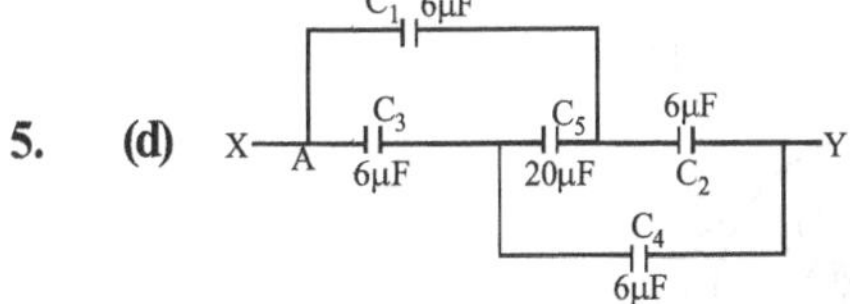

Equivalent circuit

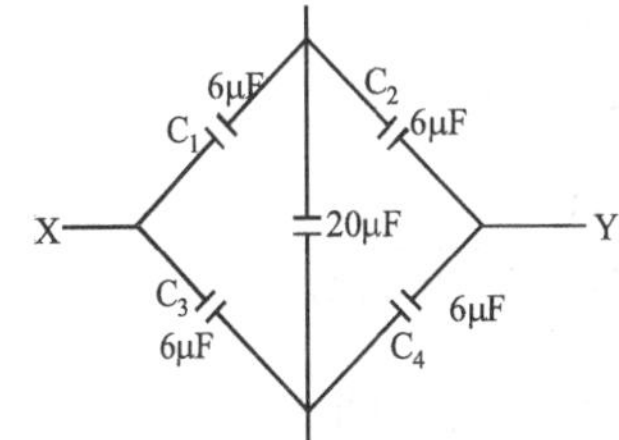

Here, $\frac{C_1}{C_3} = \frac{C_2}{C_4}$

Hence, no charge will flow through 20μF

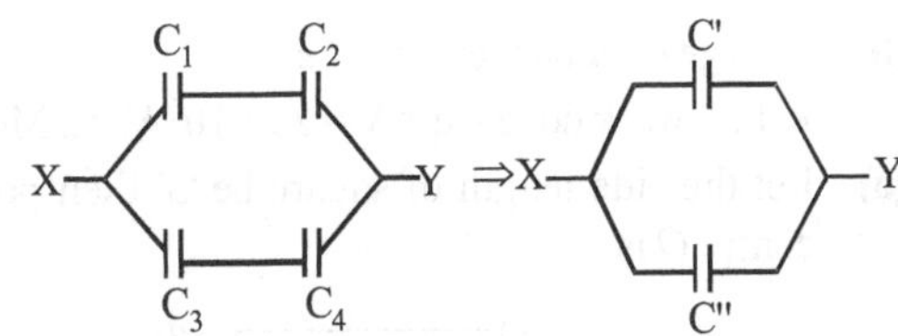

C_1 and C_2 are in series, also C_3 and C_4 are in series.

Hence, $C' = 3\ \mu F$, $C'' = 3\ \mu F$

C' and C'' are in parallel.

Hence net capacitance $= C' + C'' = 3 + 3 = 6\ \mu F$

6. (c) 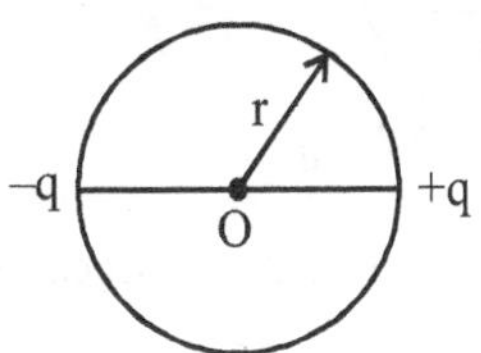

$\frac{mv^2}{r} = \frac{kq^2}{(2r)^2}$; $mv^2 = \frac{kq^2}{4r}$

Kinetic energy of each particle

$= \frac{1}{2}mv^2 = \frac{kq^2}{8r}$

7. (b) It consists of two capacitors in parallel, therefore, the total capacitance is $= \frac{2\epsilon_0 A}{d}$

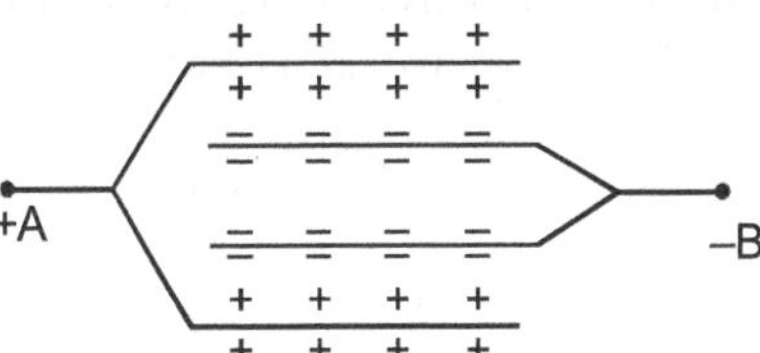

(The plates of B, having negative charge do not constitute a capacitor).

8. (a) The potential energy of a charged capacitor is given by $U = \frac{Q^2}{2C}$.

If a dielectric slab is inserted between the plates, the energy is given by $\frac{Q^2}{2KC}$, where K is the dielectric constant.

Again, when the dielectric slab is removed slowly its energy increases to initial potential energy. Thus, work done is zero.

9. (a) As $x = t\left(1-\frac{1}{K}\right)$, where x is the addition distance of plate, to restore the capacity of original value.

$\therefore \quad 3.5\times10^{-5} = 4\times10^{-5}\left(1-\frac{1}{K}\right).$

Solving, we get, K = 8.

10. (b) At. equipotential surface, the potential is same at any point i.e., $V_A = V_B$ as shown in figure. Hence no work is required to move unit change from one point to another i.e.,

$$V_A - V_B = \frac{W}{\text{unit charge}} = 0 \Rightarrow W = 0$$

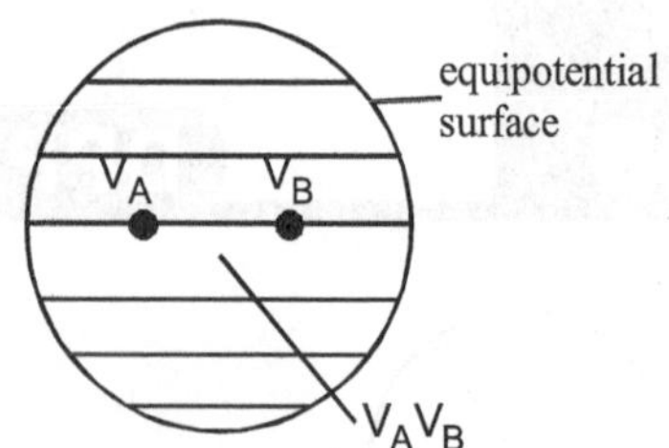

11. **(b)**
(i) Electrostatic field is zero inside a charged conductor or neutral conductor.
(ii) Electrostatic field at the surface of a charged conductor must be normal to the surface at every point.
(iii) There is no net charge at any point inside the conductor and any excess charge must reside at the surface.
(iv) Electrostatic potential is constant throughout the volume of the conductor and has the same value (as insde) on its surface.
(v) Electric field at the surface of a charged conductor is

$$\vec{E} = \frac{\sigma}{\varepsilon_0}\hat{n}$$

12. **(b)** In shell, q charge is uniformly distributed over its surface, it behaves as a conductor.

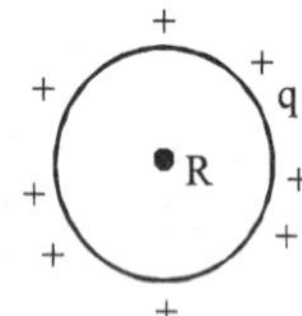

V= potential at surface $= \frac{q}{4\pi\varepsilon_0 R}$ and inside

$$V = \frac{q}{4\pi\varepsilon_0 R}$$

Because of this it behaves as an equipotential surface.

13. **(c)** Volume of big drop = 1000 × volume of each small drop

$$\frac{4}{3}\pi R^3 = 1000 \times \frac{4}{3}\pi r^3 \Rightarrow R = 10r$$

$$\because \quad V = \frac{kq}{r} \text{ and } V' = \frac{kq}{R} \times 1000$$

Total charge on one small droplet is q and on the big drop is 1000q.

$$\Rightarrow \frac{V'}{V} = \frac{1000r}{R} = \frac{1000}{10} = 100$$

$$\therefore \quad V' = 100V$$

14. **(c)** In a round trip, displacement is zero. Hence, work done is zero.

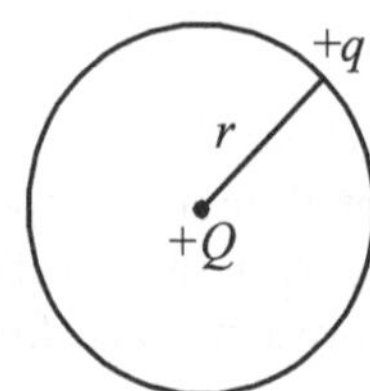

15. **(b)** The two capacitors are in parallel so

$$C = \frac{\varepsilon_0 A}{t \times 2}(k_1 + k_2)$$

16. **(d)** In equilibrium, $F = qE = (ne)\frac{V}{d} = mg$

$$n = \frac{mg\,d}{eV} = \frac{1.96 \times 10^{-15} \times 9.8 \times 0.02}{1.6 \times 10^{-19} \times 800} = 3$$

17. **(a)** $E = \frac{1}{2}CV^2 = \frac{1}{2} \times 1 \times 10^{-6} \times (4000)^2 = 8\,J.$

18. **(d)** As we know,

$$\text{Common potential} = \frac{\text{Total charge}}{\text{Total capacity}}$$

$Q_1 = C_0V_1, Q_2 = 0$, therefore

$$V_2 = \frac{C_0V_1 + 0}{C_0 + kC_0} = \frac{V_1}{1+k}$$

$$1 + k = \frac{V_1}{V_2} \text{ or } k = \frac{V_1}{V_2} - 1 = \frac{V_1 - V_2}{V_2}$$

19. **(b)** Potential difference across the branch de is 6 V. Net capacitance of de branch is 2.1 μF
So, q = CV
$\Rightarrow q = 2.1 \times 6\,\mu C$
$\Rightarrow q = 12.6\,\mu C$
Potential across 3 μF capacitance is

$$V = \frac{12.6}{3} = 4.2 \text{ volt}$$

Potential across 2 and 5 combination in parallel is 6 – 4.2 = 1.8 V
So, q' = (1.8) (5) = 9 μC

20. **(c)**

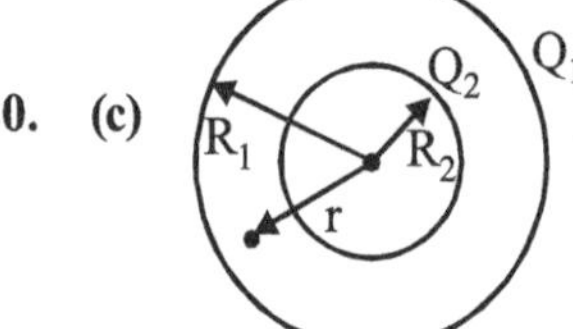

$$V_r = \frac{Q_2}{4\pi\varepsilon_0 r} + \frac{Q_1}{4\pi\varepsilon_0 R_1}$$

$$V_r = \frac{1}{4\pi\varepsilon_0}\left(\frac{Q_2}{r} + \frac{Q_1}{R_1}\right)$$

21. **(d)** $U = \frac{1}{2}QV$ = Area of triangle OAB

22. **(b)** Charge on α particle, q = 2 e.
K.E. = work done = q × V = 2e × 10^6 V = 2 MeV.

23. **(a)** Let the side length of square be '*a*' then potential at centre *O* is

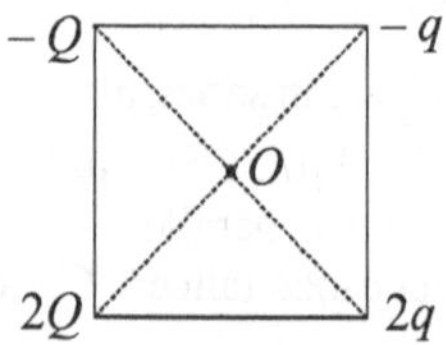

$$V=\frac{k(-Q)}{\left(\frac{a}{\sqrt{2}}\right)}+\frac{k(-q)}{\frac{a}{\sqrt{2}}}+\frac{k(2q)}{\frac{a}{\sqrt{2}}}+\frac{k(2Q)}{\frac{a}{\sqrt{2}}}=0$$

(Given)

$$=-Q-q+2q+2Q=0=Q+q=0$$
$$=Q=-q$$

24. (b) $C_0=\frac{k\in_0 A}{d}$

$$C=\frac{k\in_0 2}{3d}+\frac{2k\in_0 A}{3d}=\frac{4}{3}\frac{k\in_0 A}{d}$$

$$\therefore \frac{C}{C_0}=\frac{\frac{4}{3}\frac{k\in_0 A}{d}}{\frac{k\in_0 A}{d}}=\frac{4}{3}$$

25. (b) Work done = Change in energy

$$=\frac{1}{2}\left(C+\frac{C}{2}\right)V^2=\frac{1}{2}\left(\frac{3C}{2}\right)V^2=\frac{3CV^2}{4}$$

26. (a) Potential at B, V_B is maximum

$$V_B>V_C>V_A$$

As in the direction of electric field potential decreases.

27. (a) The equivalent circuit diagram as shown in the figure.

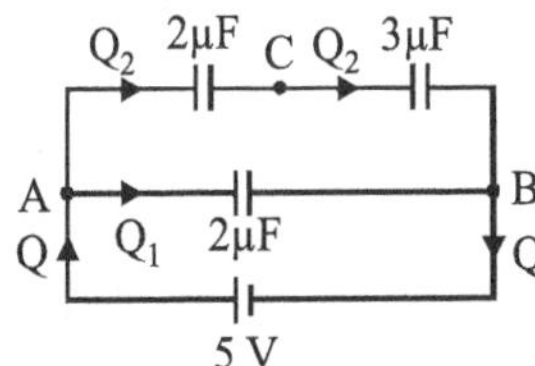

The equivalent capacitance between A and B is

$$C_{eq}=\frac{2\mu F\times 3\mu F}{2\mu F+3\mu F}+2\mu F=\frac{16}{5}\mu F$$

Total charge of the given circuit is

$$Q=\frac{16}{5}\mu F\times 5V=16\mu C$$

$$Q_1=(2\mu F)\times 5V=10\mu C$$

$\therefore$ $Q_2=Q-Q_1=16\ \mu C-10\ \mu C=6\ \mu C$

$\therefore$ Voltage between B and C is

$$V_{BC}=\frac{Q_2}{3\mu F}=\frac{6\mu C}{3\mu F}=2\,V$$

28. (d) Electric field

$$E=\frac{\sigma}{\varepsilon}=\frac{Q}{A\varepsilon}$$

ε of kerosine oil is more than that of air.

As ε increases, E decreases.

29. (c) When a battery across the plates of capacitor is disconnected and dielectric slab is placed in between the plates, then

(i) capacity C increases

(ii) charge q remains unchanged

(iii) potential V decreases

(iv) energy E decreases

30. (b) Electric lines of force are always perpendicular to an equipotential surface.

31. (c) All the charge given to inner sphere will pass on to the outer one. So capacitance that of outer one is $4\pi\in_0 b$.

32. (a) In Ist case when capacitor C attached with battery charged with the energy.

U_1= U (stored energy on capacitor).

In IInd case after disconnect of battery similar capacitor is attached in parallel with Ist capacitor then

C_{eq}.= C' = 2C.

$$\text{Now, }\frac{U_1}{U_2}=\frac{\frac{1}{2}\frac{q^2}{C}}{\frac{1}{2}\frac{q^2}{C'}}=\frac{C'}{C}=\frac{2C}{C}\qquad(\because C'=2C)$$

$$U_2=\frac{U}{2}$$

33. (a) Here we have to findout the shape of equipotential surface, these surface are perpendicular to the field lines, so there must be electric field which can not be without charge.

So, the collection of charges, whose total sum is not zero, with regard to great distance can be considered as a point charge. The equipotentials due to point charge are spherical in shape as electric potential due to point charge q is given by

$$V=K_e\frac{q}{r}$$

This suggest that electric potentials due to point charge is same for all equidistant points. The locus of these equidistant points which are at same potential, form spherical surface.

The lines of field from point charge are radial. So the equipotential surface perpendicular to field lines from a sphere.

34. (c) Equipotential surfaces are normal to the electric field lines. The following figure shows the equipotential surfaces along with electric field lines for a system of two positive charges.

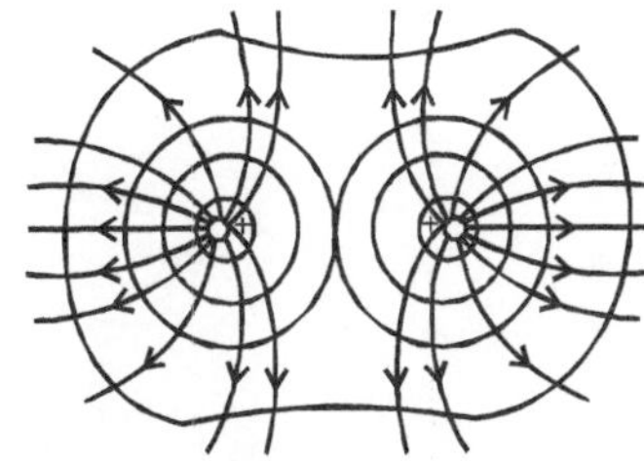

35. (c) Let plate A plate B be carrying charges Q_1 and Q_2 respectively. When they are brought closer, they induce equal and opposite charges on each other i.e. $-Q_2$ on plate A and $-Q_1$ on plate B. Therefore, net charge on plate A = $Q_1 - Q_2$ and net charge on plate B = $-(Q_1 - Q_2)$, so the charge on the capacitor = $Q_1 - Q_2$.

$\therefore$ Potential different between the plates

$$V = \frac{Q_1 - Q_2}{C}$$

36. **(d)** $C = \frac{\epsilon_0 A}{d}$

$A \to$ common area, Here $A = A_1$

37. **(a)** The equivalent circuit is shown in figure.

$C_{AB} = 3\mu F$.

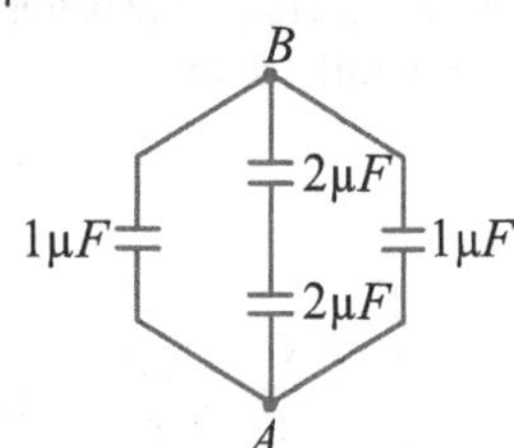

38. **(c)** If we increase the distance between the plates its capacity decreases resulting in higher potential as we know $Q = CV$. Since Q is constant (battery has been disconnected), on decreasing C, V will increase.

39. **(a)** $V = V_1 + V_2 + V_3 = \frac{1}{4\pi \epsilon_0} \cdot \frac{Q}{R} + \frac{1}{4\pi \epsilon_0}\left(\frac{-2Q}{R}\right)$

$$+ \frac{1}{4\pi \epsilon_0}\left(\frac{3Q}{R}\right) = \frac{1}{4\pi \epsilon_0}\left(\frac{2Q}{R}\right)$$

40. **(a)** As we know, $E = -\frac{dV}{dx}$

Potential at the point x = 2m, y = 2m is given by :

$$\int_0^V dV = -\int_0^{2,2} (25dx + 30dy)$$

on solving we get,

V = – 110 volt.

41. **(d)** On the equipotential surface, electric field is normal to the charged surface (where potential exists) so that no work will be done.

42. **(b)**

43. **(b)** Potential at the centre of the triangle,

$$V = \frac{\Sigma q}{4\pi\varepsilon_0 r} = \frac{2q - q - q}{4\pi\varepsilon_0 r} = 0$$

Obviously, $E \neq 0$

44. **(a)** Whenever a charge (+50 nC) is kept inside a hollow metallic spherical shell, it induces an equal and opposite charge on the inner surface and an equal and same type of charges on the outer surface.

$\therefore$ Inside, induced charge is – 50 nC and outside, +50 nC – 150 nC already present.

45. **(b)** In parallel, potential is same, say V

$$\frac{Q_1}{Q_2} = \frac{C_1 V}{C_2 V} = \frac{C_1}{C_2}$$

1. (b) $V = IR = (neAv_d)\rho\frac{\ell}{A}$

$\therefore \quad \rho = \frac{V}{V_d l n e}$

Here V = potential difference

l = length of wire

n = no. of electrons per unit volume of conductor.

e = no. of electrons

Placing the value of above parameters we get resistivity

$$\rho = \frac{5}{8\times10^{28}\times1.6\times10^{-19}\times2.5\times10^{-4}\times0.1}$$

$= 1.6\times10^{-5}\Omega m$

2. (d) From the curve it is clear that slopes at points A, B, C, D have following order $A > B > C > D$.

And also resistance at any point equals to slope of the V-i curve.

So order of resistance at three points will be

$R_A > R_B > R_C > R_D$

3. (d) From the principle of potentiometer, $V \propto l$

$\Rightarrow \frac{V}{E} = \frac{l}{L}$; where

V = emf of battery, E = emf of standard cell.

L = length of potentiometer wire

$V = \frac{El}{L} = \frac{30E}{100}$

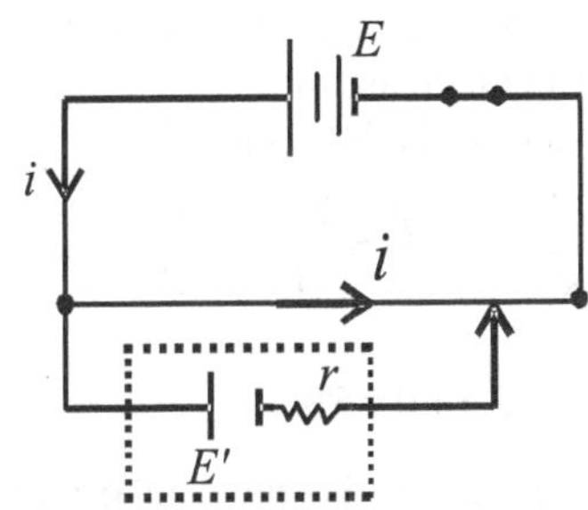

NOTE In this arrangement, the internal resistance of the battery E does not play any role as current is not passing through the battery.

4. (d) $R = \frac{\rho l}{\pi r^2}$. But $m = \pi r^2 ld \therefore \pi r^2 = \frac{m}{ld}$

$\therefore R = \frac{\rho l^2 d}{m}$, $R_1 = \frac{\rho l_1^2 d}{m_1}$, $R_2 = \frac{\rho l_2^2 d}{m_2}$

$R_3 = \frac{\rho l_3^2 d}{m_3}$

$R_1 : R_2 : R_3 = \frac{l_1^2}{m_1} : \frac{l_2^2}{m_2} : \frac{l_3^2}{m_3}$

$R_1 : R_2 : R_3 = \frac{25}{1} : \frac{9}{3} : \frac{1}{5} = 125 : 15 : 1$

5. (c) In series, $R_s = nR$

In parallel, $\frac{1}{R_p} = \frac{1}{R} + \frac{1}{R} + \ldots n$ terms

$\therefore R_s/R_p = n^2/1 = n^2$

6. (a) Efficiency is given by $\eta = \frac{\text{output}}{\text{input}}$

$= \frac{5\times15\times14}{10\times8\times15} = 0.875$ or 87.5%

7. (b) According to the condition of balancing

$\frac{55}{20} = \frac{R}{80} \Rightarrow R = 220\Omega$

8. (a) $J = \sigma E \Rightarrow J\rho = E$

J is current density, E is electric field

so B = ρ = resistivity.

9. (d) Kirchhoff's first law is based on conservation of charge and Kirchhoff's second law is based on conservation of energy.

10. (c) $R = \frac{\rho\ell}{A}$

When wire is cut into 4 pieces and connected in parallel.

$R_{eff.} = \frac{R}{16} \Rightarrow P_C = 16P$

$P_A : P_B : P_C : P_D = \frac{V^2}{R} : \frac{V^2}{R/4} : \frac{V^2}{R/16} : \frac{V^2}{R/2}$

11. (b) $S = \frac{I_g R}{nI_g - I_g} \Rightarrow S = \frac{I_g}{(n-1)I_g} R$

12. (d) Resistance of a conductor, $R = \frac{m}{ne^2\tau}\frac{l}{A}$

As the temperature increases, the relaxation time τ decreases because the number of collisions of electrons per second increases due to increase in thermal energy of electrons.

13. (b)

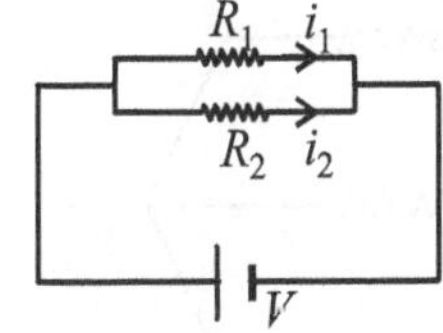

$R_1 = \frac{\rho \ell_1}{\pi r_1^2}$; $R_2 = \frac{\rho \ell_2}{\pi r_2^2}$

$i_1 R_1 = i_2 R_2$ (same potential difference)

$\therefore \frac{i_1}{i_2} = \frac{R_2}{R_1} = \frac{\ell_2}{\ell_1} \times \frac{r_1^2}{r_2^2} = \frac{3}{4} \times \frac{4}{9} = \frac{1}{3}$

14. **(c)** $\frac{R_1}{R_2} = \frac{\ell_1}{\ell_2}$ where $\ell_2 = 100 - \ell_1$

In the first case $\frac{X}{Y} = \frac{20}{80}$

In the second case

$\frac{4X}{Y} = \frac{\ell}{100 - \ell} \Rightarrow \ell = 50$

15. **(c)** Before connecting E, the circuit diagram is

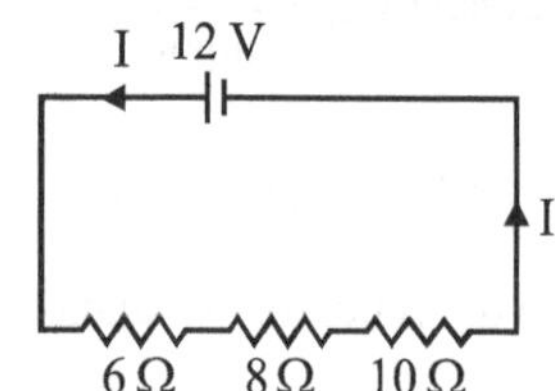

Then, $R_{eq} = 6\,\Omega + 8\,\Omega + 10\,\Omega = 24\,\Omega$

Current in the 8 Ω resistance, $I = \frac{12V}{24\Omega} = \frac{1}{2} A$

After connecting E, the current through 8 Ω is

$I = \frac{1}{2} A$

$\therefore \quad E = \frac{1}{2} A \times 8\Omega = 4V$

16. **(d)** By junction rule at point B

$-I + 1A + 2A = 0$

So, $I = 3A$

By Loop rule,

$-3 \times 2 - 1 \times 1 - E + 12 = 0$

$E = 5V$

17. **(d)** Resistance of bulb $R_b = \frac{(1.5)^2}{4.5} = 0.5\Omega$

Current drawn from battery $= \frac{E}{2.67 + 0.33} = \frac{E}{3}$

Share of bulb $= \frac{2}{3} \times \frac{E}{3} = \frac{2E}{9}$

$\therefore \left(\frac{2E}{9}\right)^2 \times 0.5 = 4.5$ or $E = 13.5 V$.

18. **(d)** The equivalent circuit is given below :

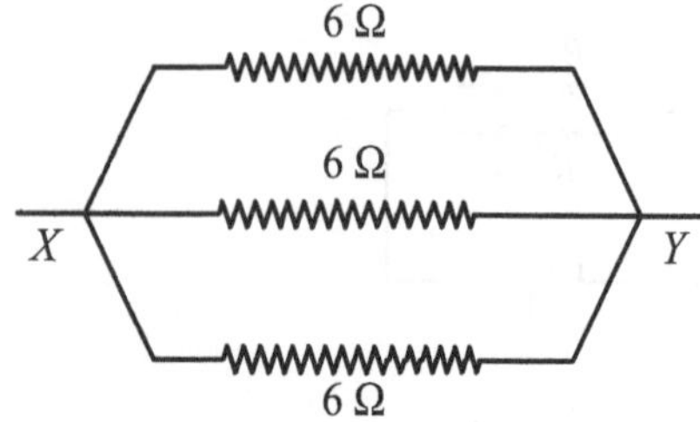

The equivalent resistance is given by

$\frac{1}{R} = \frac{1}{6} + \frac{1}{6} + \frac{1}{6} = \frac{3}{6} = \frac{1}{2}$

$\Rightarrow R_{eq} = 2\Omega$

19. **(a)** Since average drift velocity $= \frac{1}{2} \frac{eE}{m} \times (\tau)$

Now $I = NeA \times$ (avg. drift velocity)

$= \frac{Ne^2 AE}{2m\ell} \times \tau = \frac{Ne^2 AV}{2m\ell} \times \tau$

$R = \frac{V}{I} = \frac{2m\ell}{Ne^2 \tau A}$, where N is electron density.

20. **(c)** The current through the resistance R

$I = \left(\frac{\varepsilon}{R + r}\right)$

The potential difference across R

$V = IR = \left(\frac{\varepsilon}{R + r}\right) R$

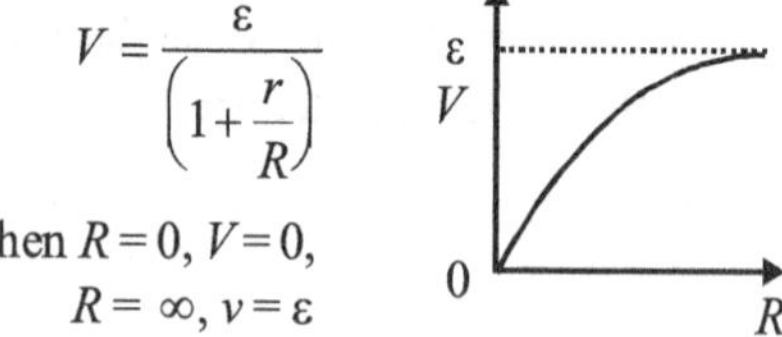

$V = \frac{\varepsilon}{\left(1 + \frac{r}{R}\right)}$

when $R = 0$, $V = 0$,

$R = \infty$, $v = \varepsilon$

Thus V increases as R increases upto certain limit, but it does not increase further.

21. **(c)** Resistance of bulb is constant

$P = \frac{V^2}{R} \Rightarrow \frac{\Delta p}{p} = \frac{2\Delta V}{V} + \frac{\Delta R}{R}$

$\frac{\Delta p}{p} = 2 \times 2.5 + 0 = 5\%$

22. **(a)** Potential gradient = Potential fall per unit length. In this case resistance of unit length.

$R = \frac{\rho l}{A} = \frac{10^{-7} \times 1}{10^{-6}} = 10^{-1}\Omega$

Potential fall across R is

$V = I.R = 0.1 \times 10^{-1} = 0.01$ volt/m.

$= 10^{-2}$ volt / m

23. **(d)** $R_1 + R_2 =$ Constant, R_1 will increase, R_2 will decrease.

$R_1 \alpha \Delta T - R\beta \Delta T = 0 \Rightarrow R_1 \alpha \Delta T = R_2 \beta \Delta T$

$\therefore \frac{R_1}{R_2} = \frac{\beta}{\alpha}$

24. (d) Given : Number of cells, $n = 5$, emf of each cell $= E$
Internal resistance of each cell $= r$
In series, current through resistance R

$$I = \frac{nE}{nr + R} = \frac{5E}{5r + R}$$

In parallel, current through resistance R

$$I' = \frac{E}{\frac{r}{n} + R} = \frac{nE}{r + nR} = \frac{5E}{r + 5R}$$

According to question, $I = I'$

$$\therefore \frac{5E}{5r + 5R} = \frac{5E}{r + 5R} \Rightarrow 5r + R = r + 5R$$

or $R = r \quad \therefore \frac{R}{r} = 1$

25. (d) The total volume remains the same before and after stretching.

Therefore $A \times \ell = A' \times \ell'$

Here $\ell' = 2\ell$

$$\therefore A' = \frac{A \times \ell}{\ell'} = \frac{A \times \ell}{2\ell} = \frac{A}{2}$$

Percentage change in resistance

$$= \frac{R_f - R_i}{R_i} \times 100 = \frac{\rho\left(\frac{\ell'}{A'} - \frac{\ell}{A}\right)}{\rho\frac{\ell}{A}} \times 100$$

$$= \left[\left(\frac{\ell'}{A'} \times \frac{A}{\ell}\right) - 1\right] \times 100 = \left[\left(\frac{2\ell}{A/2} \times \frac{A}{\ell}\right) - 1\right] \times 100$$

$= 300\%$

26. (a) Pot. gradient $= 0.2\text{mV/cm}$

$$= \frac{0.2 \times 10^{-3}}{10^{-2}} = 2 \times 10^{-2}\text{V/m}$$

Emf of cell $= 2 \times 10^{-2} \times 1\text{m} = 2 \times 10^{-2}\text{V} = 0.02\text{ V}$
As per the condition of potentiometer
$0.02\,(R + 490) = 2\,(R)$ or $1.98\,R = 9.8$

$$\Rightarrow \quad R = \frac{9.8}{1.98} = 4.9\ \Omega$$

27. (d)

Applying Kirchhoff's rule in loop **abcfa**
$\varepsilon_1 - (i_1 + i_2)R - i_1 r_1 = 0.$

28. (c) Total power consumed by electrical appliances in the building, $P_{total} = 2500\text{W}$

Watt = Volt × ampere

$\Rightarrow \quad 2500 = V \times I \Rightarrow 2500 = 220\,I$

$$\Rightarrow \quad I = \frac{2500}{220} = 11.36 \approx 12\text{A}$$

(Minimum capacity of main fuse)

29. (a)

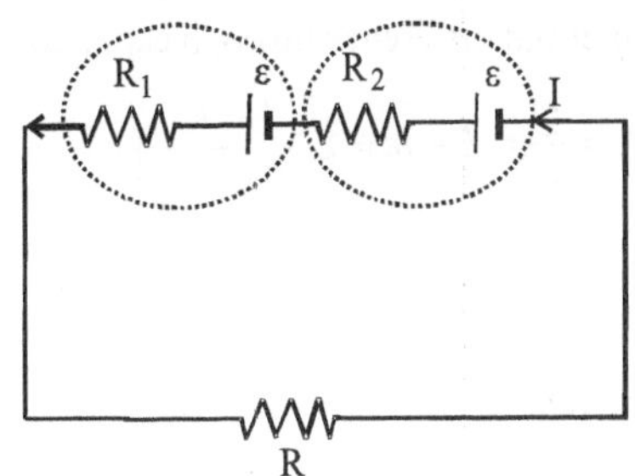

$$I = \frac{2\varepsilon}{R + R_1 + R_2}$$

Potential difference across second cell

$= V = \varepsilon - iR_2 = 0$

$$\varepsilon - \frac{2\varepsilon}{R + R_1 + R_2}.R_2 = 0$$

$R + R_1 + R_2 - 2R_2 = 0$

$R + R_1 - R_2 = 0$

$\therefore R = R_2 - R_1$

30. (c)

Resistance of the series combination,
$S = R_1 + R_2$
Resistance of the parallel combination,

$$P = \frac{R_1 R_2}{R_1 + R_2}$$

$$S = nP \Rightarrow R_1 + R_2 = \frac{n(R_1 R_2)}{(R_1 + R_2)}$$

$\Rightarrow (R_1 + R_2)^2 = nR_1R_2$
Minimum value of n is 4 for that

$(R_1 + R_2)^2 = 4R_1R_2 \Rightarrow (R_1 - R_2)^2 = 0$

31. (c) To convert a galvanometer into a voltmeter we connect a high resistance in series with the galvanometer. The same procedure needs to be done if ammeter is to be used as a voltmeter.

32. (c) Given, emf of cell E = 200 V
Internal resistance of cells = 1 Ω
D. C. main supply voltage V = 220 V
External resistance R = ?

$$r = \left(\frac{E - V}{V}\right)R$$

$$1 = \left(\frac{20}{220}\right) \times R$$

$\therefore \quad R = 11\ \Omega.$

33. (a) In steady state, flow fo current through capacitor will be zero.

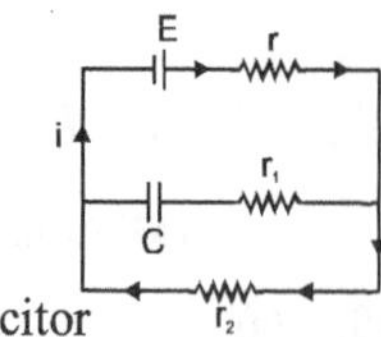

Current through the circuit,

$$i = \frac{E}{r + r_2}$$

Potential difference through capacitor

$$V_c = \frac{Q}{C} = E - ir = E - \left(\frac{E}{r + r_2}\right) r$$

$$\therefore \quad Q = CE \frac{r_2}{r + r_2}$$

34. (c) $i = neAV_d$ and $V_d \propto \sqrt{E}$ (Given)

or, $i \propto \sqrt{E}$

$i^2 \propto E$

$i^2 \propto V$

Hence graph (c) correctly dipicts the *V-I* graph for a wire made of such type of material.

35. (b) Current, $I = (2.9 \times 10^{18} + 1.2 \times 10^{18}) \times 1.6 \times 10^{-19}$

$= 0.66$A towards right.

36. (a) Copper rod and iron rod are joined in series.

$$\therefore R = R_{Cu} + R_{Fe} = (\rho_1 + \rho_2) \frac{\ell}{A}$$

$$\left(\because R = \rho \frac{\ell}{A}\right)$$

From ohm's law $V = RI$

$= (1.7 \times 10^{-6} \times 10^{-2} + 10^{-5} \times 10^{-2}) \div 0.01 \times 10^{-4}$ volt

$= 0.117$ volt $(\because I = 1A)$

37. (d) $I = \frac{E}{R + r}$, Internal resistance (r) is zero, $I = \frac{E}{R}$ = constant.

38. (b) $R_t = R_0(1 + \alpha t)$

Initially, $R_0(1 + 30\alpha) = 10\,\Omega$

Finally, $R_0(1 + \alpha t) = 11\,\Omega$

$$\therefore \frac{11}{10} = \frac{1 + \alpha t}{1 + 30\alpha}$$

or, $10 + (10 \times 0.002 \times t) = 11 + 330 \times 0.002$

or, $0.02t = 1 + 0.66 = 1.066$ or $t = \frac{1.66}{0.02} = 83°C$.

39. (b) As $P = I^2 R$, so $P_1 = (1.01\ I)^2 R = 1.02 I^2 R = 1.02 P$.

It means % increase in power

$$= \left(\frac{P_1}{P} - 1\right) \times 100 = 2\%.$$

40. (b) Let I_1 be the current throug 5 Ω resistance, I_2 through (6 + 9) Ω resistance. Then as per question,

$I_1^2 \times 5 = 20$ or, $I_1 = 2A$.

Potential difference across C and D $= 2 \times 5 = 10V$

Current $I_2 = \frac{10}{6+9} = \frac{2}{3}A$.

Heat produced per second in 2 Ω

$$= I^2 R \left(\frac{8}{3}\right)^2 \times 2 = 14.2\,\text{cal/s}.$$

41. (b) $\frac{P}{Q} = \frac{R}{S}$ where $S = \frac{S_1 S_2}{S_1 + S_2}$

42. (c) $R = \frac{\rho \ell_1}{A_1}$, now $\ell_2 = 2\ell_1$

$A_2 = \pi(r_2)^2 = \pi(2r_1)^2 = 4\pi r_1^2 = 4A_1$

$$\therefore \quad R_2 = \frac{\rho(2\ell_1)}{4A_1} = \frac{\rho \ell_1}{2A_1} = \frac{R}{2}$$

∴ Resistance is halved, but specific resistance remains the same.

43. (d) $E = V + Ir$

$V = 12 - 3 = 9$ volt

44. (c) $I = neAV_d$

$V_d = \frac{I}{neA} = 5 \times 10^{-3}$ m/sec

45. (d) Since due to wrong connection of each cell the total emf reduced to 2ε then for wrong connection of three cells the total emf will reduced to $(n\varepsilon - 6\varepsilon)$ whereas the total or equivalent resistance of cell combination will be *nr.*

DAILY PRACTICE PROBLEMS

PHYSICS SOLUTIONS

DPP/CP18

1. (a) At a distance x consider small element of width dx.
Magnetic moment of the small element is

$$dm = \frac{\left(\frac{q}{\ell}dx\right)\omega}{2\pi}.\pi x^2$$

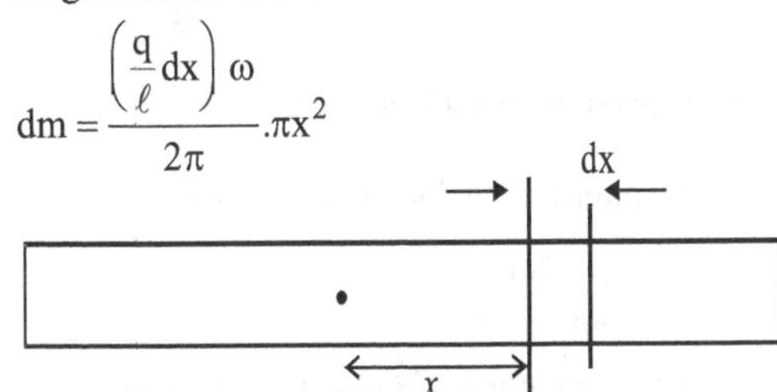

$$M = \int_{-\ell/2}^{\ell/2} \frac{q\omega}{2\ell} x^2 dx \;;\; M = \frac{q\omega\ell^2}{24} = \frac{q\pi f\ell^2}{12}$$

2. (d) The straight part will not contribute magnetic field at the centre of the semicircle because every element of the straight part will be 0° or 180° with the line joining the centre and the element

Due to circular portion, the field is $\frac{1}{2}\frac{\mu_0 i}{2r} = \frac{\mu_0 i}{4r}$

Hence total field at O = $\frac{\mu_0 i}{4r}$ tesla

3. (d) Torque on the solenoid is given by
$\tau = MB\sin\theta$
where θ is the angle between the magnetic field and the axis of solenoid.
$M = niA$
$\therefore \tau = niAB\sin 30°$
$= 2000 \times 2 \times 1.5 \times 10^{-4} \times 5 \times 10^{-2} \times \frac{1}{2}$
$= 1.5 \times 10^{-2}\, N-m$

4. (c) Time period of cyclotron is

$$T = \frac{1}{\upsilon} = \frac{2\pi m}{eB};\; B = \frac{2\pi m}{e}\upsilon;\; R = \frac{m\upsilon}{eB} = \frac{p}{eB}$$

$$\Rightarrow \quad p = eBR = e \times \frac{2\pi m\upsilon}{e} R = 2\pi m\upsilon R$$

$$K.E. = \frac{p^2}{2m} = \frac{(2\pi m\upsilon R)^2}{2m} = 2\pi^2 m\upsilon^2 R^2$$

5. (a) $R_g = 50\Omega, I_g = 25 \times 4 \times 10^{-4}\Omega = 10^{-2}$ A
Range of V = 25 volts
$V = I_g(R_e + R_g)$

$$\therefore R_e = \frac{V}{I_g} - R_g = 2450\Omega$$

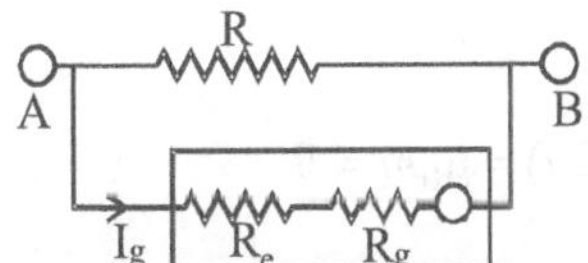

6. (c) $B_{axis} = \left(\frac{\mu_0 NI}{2x^3}\right)R^2$

$B \propto R^2$
So, when radius is doubled, magnetic field becomes four times.

7. (c) When a charged particle enters a transverse magnetic field it traverse a circular path. Its kinetic energy remains constant.

8. (c) K.E. of electron = 10 eV

$$\Rightarrow \frac{1}{2}mv^2 = 10 \text{ eV}$$

$$\Rightarrow \frac{1}{2}(9.1 \times 10^{-31})v^2 = 10 \times 1.6 \times 10^{-19}$$

$$\Rightarrow v^2 = \frac{2 \times 10 \times 1.6 \times 10^{-19}}{9.1 \times 10^{-31}}$$

$\Rightarrow v^2 = 3.52 \times 10^{12} \Rightarrow v = 1.88 \times 10^6$ m
Also we know that for circular motion

$$\frac{mv^2}{r} = Bev \Rightarrow r = \frac{mv}{Be} = 11 \text{ cm}$$

9. (d) No magnetic force acts on the electron and force due to electric field will act opposite to its initial direction of motion. Hence its velocity decreases in magnitude.

10. (b) $\frac{mv^2}{r} = qvB \Rightarrow r = \frac{mv}{qB}$

$$\Rightarrow r_p = \frac{m_p v_p}{q_p B};$$

$$r_d = \frac{m_d v_d}{q_d B};\; r_\alpha = \frac{m_\alpha v_\alpha}{q_\alpha B}$$

$m_\alpha = 4m_p,\; m_d = 2m_p$

$q_\alpha = 2q_p,\; q_d = q_p$
From the problem

$$E_p = E_d = E_\alpha = \frac{1}{2}m_p v_p^2$$

$$= \frac{1}{2}m_d v_d^2 = \frac{1}{2}m_\alpha v_\alpha^2$$

$$\Rightarrow v_p^2 = 2v_d^2 = 4mv_2^2$$

Thus we have, $r_\alpha = r_p < r_d$

11. (c) Resistance of Galvanometer,

$$G = \frac{\text{Current sensitivity}}{\text{Voltage sensitivity}} \Rightarrow G = \frac{10}{2} = 5\Omega$$

Here i_g = Full scale deflection current

$= \frac{150}{10} = 15\text{ mA}$

V = voltage to be measured = 150 volts
(such that each division reads 1 volt)

$\Rightarrow R = \frac{150}{15\times10^{-3}} - 5 = 9995\Omega$

12. (d) Magnetic field at the centre of the current loop is

$$B = \frac{\mu_0 2\pi I}{4\pi R}$$

$$\text{or, } B = \frac{\mu_0 2\pi q\upsilon}{4\pi R},\quad R = \frac{\mu_0 2\pi q\upsilon}{4\pi B}$$

Substituting the given values, we get

$$R = \frac{4\pi\times10^{-7}\times2\pi\times2\times10^{-6}\times6.25\times10^{12}}{4\pi\times6.28} = 1.25\text{ m}$$

13. (b) Here, $\vec{E}$ and $\vec{B}$ are perpendicular to each other and the velocity $\vec{v}$ does not change; therefore

$qE = qvB \Rightarrow v = \frac{E}{B}$

Also,

$$\left|\frac{\vec{E}\times\vec{B}}{B^2}\right| = \frac{E\ B\sin\theta}{B^2} = \frac{E\ B\ \sin 90°}{B^2} = \frac{E}{B} = |\vec{v}| = v$$

14. (b) The force on the two arms parallel to the field is zero.

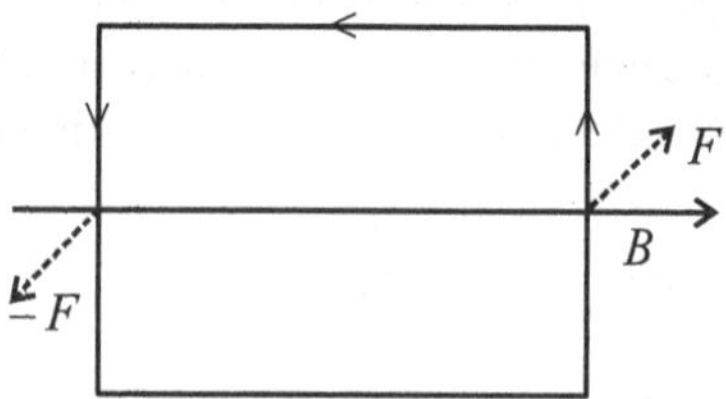

15. (d) Magnetic field at a point on the axis of a current carrying wire is always zero.

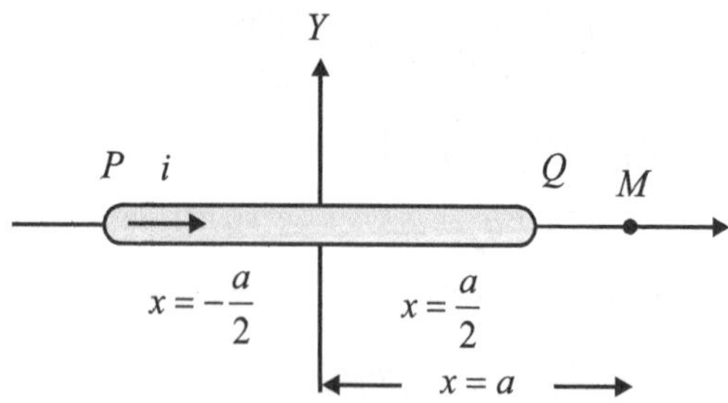

16. (b) Current carrying conductors will attract each other, while electron beams will repel each other.

17. (c) To keep the main current in the circuit unchanged, the resistance of the galvanometer should be equal to the net resistance.

$$\therefore G = \left(\frac{GS}{G+S}\right) + S'$$

$$\Rightarrow G - \frac{GS}{G+S} = S'$$

$$\therefore S' = \frac{G^2}{G+S}.$$

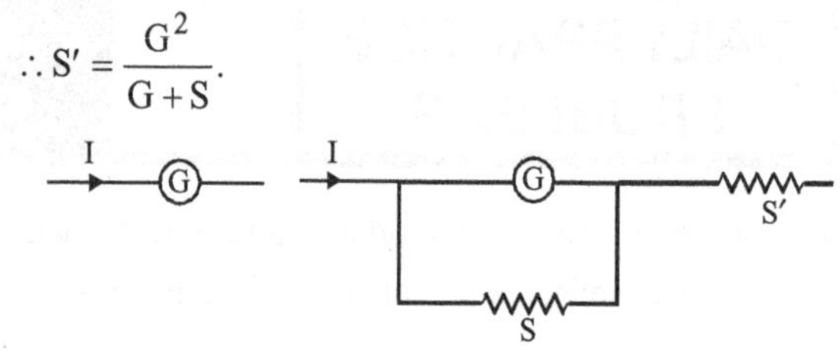

18. (d) Current in a small element, $dI = \frac{d\theta}{\pi} I$

Magnetic field due to the element

$$dB = \frac{\mu_0}{4\pi}\frac{2dI}{R}$$

The component dB cos θ, of the field is cancelled by another opposite component.

Therefore,

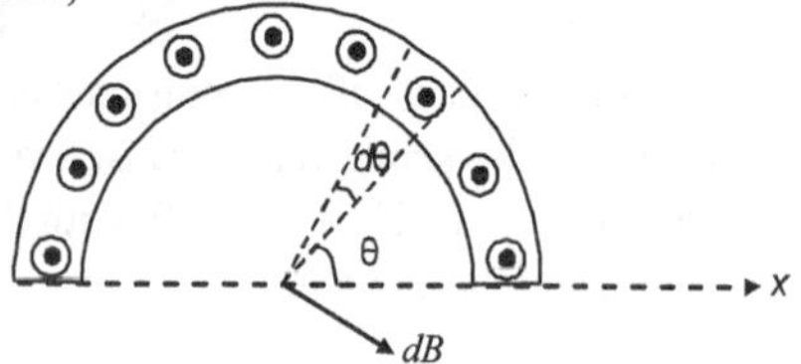

$$B_{net} = \int dB\sin\theta = \frac{\mu_0 I}{2\pi^2 R}\int_0^{\pi}\sin\theta d\theta = \frac{\mu_0 I}{\pi^2 R}$$

19. (a)

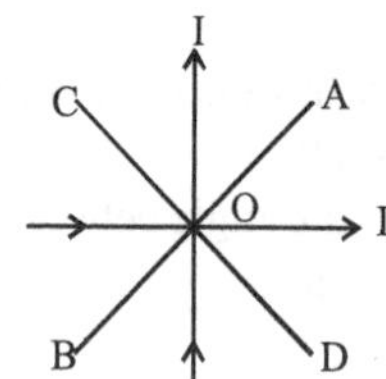

Net magnetic field on AB is zero because magnetic field due to both current carrying wires is equal in magnitude but opposite in direction.

20. (b) According to the figure the magnitude of force on the segment QM is $F_3 - F_1$ and PM is F_2.

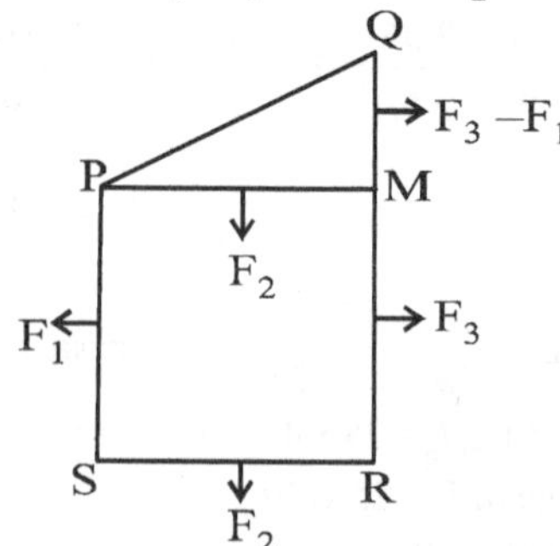

Therefore, the magnitude of the force on segment PQ is $\sqrt{(F_3 - F_1)^2 + F_2^2}$

21. (c) $B = \mu_0 ni$

$$B_1 = (\mu_0)\left(\frac{n}{2}\right)(2\ i) = \mu_0 ni = B$$

$$\Rightarrow B_1 = B$$

22. (c) The angular momentum L of the particle is given by $L = mr^2\omega$ where $\omega = 2\pi n$.

$\therefore$ Frequency $n = \frac{\omega}{2\pi}$; Further $i = q \times n = \frac{\omega q}{2\pi}$

Magnetic moment, $M = iA = \frac{\omega q}{2\pi} \times \pi r^2$;

$\therefore M = \frac{\omega q r^2}{2}$ So, $\frac{M}{L} = \frac{\omega q r^2}{2mr^2\omega} = \frac{q}{2m}$

23. (c) A current loop in a magnetic field is in equilibrium in two orientations one is stable and another unstable.

$\because \quad \vec{\tau} = \vec{M} \times \vec{B} = MB\sin\theta$

If $\theta = 0° \Rightarrow \tau = 0$ (stable)

If $\theta = \pi \Rightarrow \tau = 0$ (unstable)

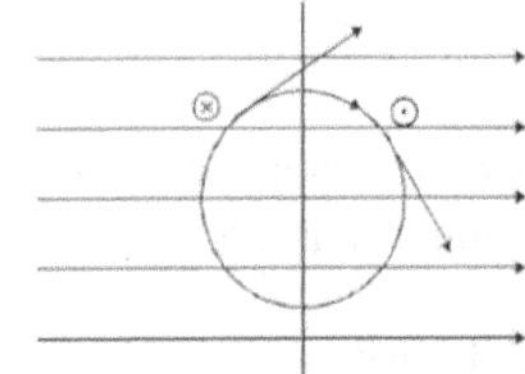

Do not experience a torque in some orientations Hence option (c) is correct.

24. (c) $B = \frac{\mu_0}{4\pi}\frac{2i_2}{(r/2)} - \frac{\mu_0}{4\pi}\frac{2i_1}{(r/2)} = \frac{\mu_0}{4\pi}\frac{4}{r}(i_2 - i_1)$

$= \frac{\mu_0}{4\pi}\frac{4}{5}(5 - 2.5) = \frac{\mu_0}{2\pi}$.

25. (a) The direction of $\vec{B}$ is along $(-\hat{k})$

$\therefore$ The magnetic force

$\vec{F} = Q(\vec{v} \times \vec{B}) = Q(v\hat{i}) \times B(-\hat{k}) = QvB\hat{j}$

$\Rightarrow \vec{F}$ is along OY.

26. (a) According to Ampere's circuit law

$\oint \vec{B}.d\vec{l} = \mu_0 I_{enclosed} = \mu_0(2A - 1A) = \mu_0$

27. (a) We know that the magnetic field produced by a current carrying circular coil of radius r at its centre is

$B = \frac{\mu_0}{4\pi}\frac{I}{r} \times 2\pi$

Here $B_A = \frac{\mu_0}{4\pi}\frac{I}{R} \times 2\pi$

and $B_B = \frac{\mu_0}{4\pi}\frac{2I}{2R} \times 2\pi$

$\Rightarrow \frac{B_A}{B_B} = 1$

28. (b) When a charged particle enters a magnetic field at a direction perpendicular to the direction of motion, the path of the motion is circular. In circular motion the direction of velocity changes at every point (the magnitude remains constant).

Therefore, the tangential momentum will change at every point. But kinetic energy will remain constant as it is given by $\frac{1}{2}mv^2$ and v^2 is the square of the magnitude of velocity which does not change.

29. (a) $I = 50\,k$; $I_g = 20k$, where k is the figure of merit of galvanometer; $S = I_g R_g (I - I_g)$; so $12 = \frac{20k.R_g}{(50k - 20k)}$

On solving we get $R_g = 18\,\Omega$.

30. (d) Let I be current and l be the length of the wire.

For Ist case : $B = \frac{\mu_0 In}{2r} = \frac{\mu_0 I \times \pi}{l}$ where $2\pi r = l$

and n = 1

For IInd case : $l = n(2\pi r') \Rightarrow r' = \frac{l}{2n\pi}$

$$B' = \frac{\mu_0 nI}{2r'} = \frac{\mu_0 nI}{2\frac{l}{2n\pi}} = \frac{n^2\mu_0\pi I}{l} = n^2 B$$

31. (c) $B = \frac{\mu_0 i\, a^2}{2(x^2 + a^2)^{3/2}}$

$$B' = \frac{\mu_0 i}{2a} = \frac{\mu_0 i\, a^2}{2a(x^2 + a^2)^{3/2}}\left(\frac{(x^2 + a^2)^{3/2}}{a^2}\right)$$

$$B' = \frac{B.(x^2 + a^2)^{3/2}}{a^3}$$

Put $x = 4$ & $a = 3 \Rightarrow B' = \frac{54(5^3)}{3 \times 3 \times 3} = 250\mu T$

32. (a) The force acting on a charged particle in magnetic field is given by

$F = q(\vec{v} \times \vec{B})$ or $F = qvB\sin\theta$

when angle between v and B is 180°,

$F = 0$

33. (b) The force acting on electron will be perpendicular to the direction of velocity till the electron remains in the magnetic field. So the electron will follow the path as given.

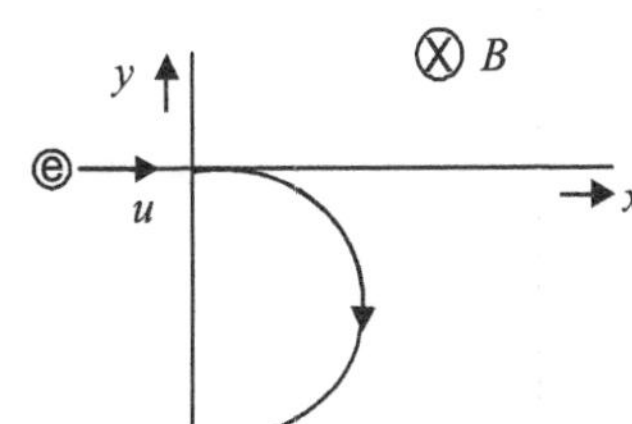

34. (c) A voltmeter is a high resistance galvanometer and is connected in parallel to circuit and ammeter is a low resitance galvanometer so if we connect high resistance in series with ammeter its resistance will be much high.

35. (d) Here, the wire does not produce any magnetic field at O because the conductor lies on the line of O. Also, the loop does not produce magnetic field at O.

36. (d) Magnetic field between the plates in this case is zero.

37. (d) For a given perimeter the area of circle is maximum. So magnetic moment of (S) is greatest.

38. (a) Lorentz force, $\vec{F} = q\{\vec{E} + (\vec{v} \times \vec{B})\}$

$$\vec{v} \times \vec{B} = \begin{vmatrix} \hat{i} & \hat{j} & \hat{k} \\ 1 & 2 & 0 \\ 5 & 3 & 4 \end{vmatrix} = 8\hat{i} - 4\hat{j} - 7\hat{k}$$

$$\vec{F} = 1(2\hat{i} - 3\hat{j} + 8\hat{i} - 4\hat{j} - 7\hat{k}) = (10\hat{i} - 7\hat{j} - 7\hat{k})$$

39. (b) Here, $R_g = 100\,\Omega$; $I_g = 10^{-5}$ A; I = 1A; S = ?

$$S = \frac{I_g R_g}{I - I_g} = \frac{10^{-5} \times 100}{1 - 10^{-5}} = 10^{-3}\,\Omega \text{ in parallel}$$

40. (d)

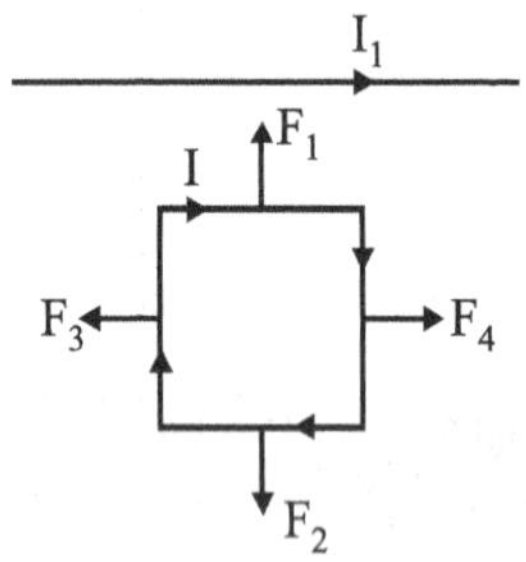

I_1, I, F_1, F_2, F_3, F_4

$F_1 > F_2$ as $F \propto \frac{1}{d}$, and F_3 and F_4 are equal and opposite.

Hence, the net attraction force will be towards the conductor.

41. (c) $\vec{F}_1 = \vec{F}_2 = 0$

because of action and reaction pair

42. (c) As electron move with constant velocity without deflection. Hence, force due to magnetic field is equal and opposite to force due to electric field.

$$qvB = qE \Rightarrow v = \frac{E}{B} = \frac{20}{0.5} = 40 \text{ m/s}$$

43. (c)

44. (b) To measure AC voltage across a resistance a moving coil galvanometer is used.

45. (c) As $\vec{F} = q\vec{V}\vec{B}\sin\theta$

F is zero for sin 0° or sin 180° and is non-zero for angle between $\vec{V}$ and $\vec{B}$ any value other than zero and 180°.

DAILY PRACTICE PROBLEMS — PHYSICS SOLUTIONS — DPP/CP19

1. (b) As the axes are perpendicular, mid point lies on axial line of one magnet and on equatorial line of other magnet.

$$\therefore\ B_1 = \frac{\mu_0}{4\pi}\frac{2M}{d^3} = \frac{10^{-7}\times 2\times 1}{1^3} = 2\times 10^{-7}$$

and $B_2 = \frac{\mu_0}{4\pi}\frac{M}{d^3} = 10^{-7}$

$\therefore$ Resultant field $= \sqrt{B_1^2 + B_2^2} = \sqrt{5}\times 10^{-7}\,T$

2. (c) Initial magnetic moment of each magnet $= m \times \ell$.
As is clear from Fig., S_1 and N_2 neutralize each other.
Effective distance between

N_1 and $S_2 = \sqrt{\ell^2 + \ell^2} = \ell\sqrt{2}$

$\therefore\ M' = m\ell\sqrt{2}$.

3. (d) As shown in the figure, the magnetic lines of force are directed from south to north inside a bar magnet.

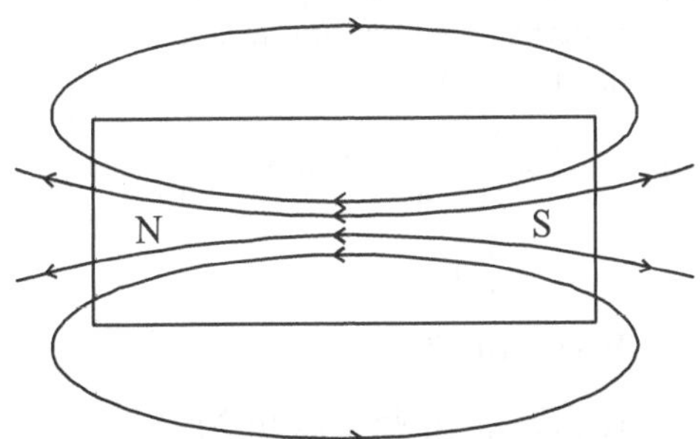

4. (b) For a diamagnetic material, the value of μ_r is less than one. For any material, the value of ϵ_r is always greater than 1.

5. (a) The time period of oscillation of a freely suspended magnet is given by

$$T = 2\pi\sqrt{\frac{I}{MH}}$$

Thus, $\frac{T}{T'} = \frac{2\pi\sqrt{\frac{I}{MH}}}{2\pi\sqrt{\frac{I}{MH'}}}$

Given, $T = 4$ sec, $T' = 2$ sec,

So, $\frac{4}{2} = \sqrt{\frac{H'}{H}}$

or $\sqrt{\frac{H'}{H}} = 2$

or $H' = 4H$

6. (b) Diamagnetic materials are repelled in an external magnetic field.
Bar *B* represents diamagnetic materials.

7. (a) The temperature above which a ferromagnetic substance becomes paramagnetic is called Curie's temperature.

8. (a) Iron is ferromagnetic.

9. (b) $\tau = MB\sin\theta$

$\tau = iAB\sin 90°$

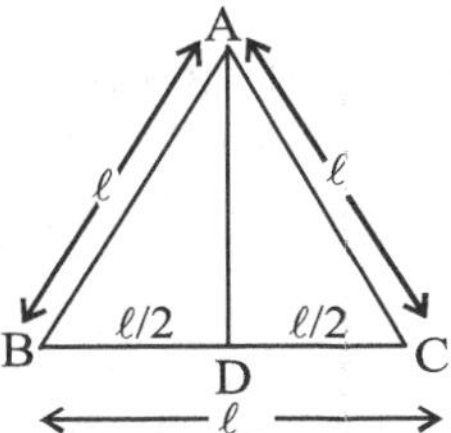

$\therefore\ A = \frac{\tau}{iB}$

Also, A = 1/2 (BC) (AD)

But $\frac{1}{2}(BC)(AD) = \frac{1}{2}(l)\sqrt{l^2 - \left(\frac{l}{2}\right)^2} = \frac{\sqrt{3}}{4}l^2$

$$\Rightarrow\quad \frac{\sqrt{3}}{4}(l)^2 = \frac{\tau}{Bi}$$

$$\therefore\quad l = 2\left(\frac{\tau}{\sqrt{3}\,B.i}\right)^{\frac{1}{2}}$$

10. (a) $M = 60\,Am^2$

$\vec{\tau} = 1.2\times 10^{-3}$ Nm, $B_H = 40\times 10^{-6}$ Wb/m^2

$\vec{\tau} = \vec{M}\times\vec{B}_H \Rightarrow \tau = MB_H\sin\theta$

$\Rightarrow 1.2\times 10^{-3} = 60\times 40\times 10^{-6}\sin\theta$

$\Rightarrow \sin\theta = \frac{1.2\times 10^{-3}}{60\times 40\times 10^{-6}} = \frac{1}{2} = \sin 30°$

$\Rightarrow \theta = 30°$

11. (b) Electro magnet should be amenable to magnetisation and demagnetization.
$\therefore$ retentivity and coercivity should be low.

12. (b) $T = 2\pi\sqrt{\frac{I}{M\times B}} = 2\pi\sqrt{\frac{I}{MB}}$ where $I = \frac{1}{12}m\ell^2$

When the magnet is cut into three pieces the pole strength will remain the same and

M.I. $(I') = \frac{1}{12}\left(\frac{m}{3}\right)\left(\frac{\ell}{3}\right)^2\times 3 = \frac{I}{9}$

We have, Magnetic moment (M)
= Pole strength $(m)\times\ell$
$\therefore$ New magnetic moment,

$M' = m\times\left(\frac{\ell}{3}\right)\times 3 = m\ell = M$

$\therefore\ T' = \frac{T}{\sqrt{9}} = \frac{2}{3}s$.

13. (b) Graph [A] is for material used for making permanent magnets (high coercivity)
Graph [B] is for making electromagnets and transformers.

14. **(a)** The earth's core is hot and molten. Hence, convective current in earth's core is responsible for it's magnetic field.

15. **(b)** $T \propto \frac{1}{\sqrt{H}} \Rightarrow \frac{T_1}{T_2} = \sqrt{\frac{H_2}{H_1}} \Rightarrow \frac{2}{1} = \sqrt{\frac{H+F}{H}}$

$\Rightarrow F = 3H \, or \frac{H}{F} = \frac{1}{3}$

16. **(a)** H = B cos θ, V = B sin θ
Here B = earth's magnetic field
θ = angle of dip = 90° at north pole
⇒ H = B cos 90° = 0
V = B sin 90° = B
⇒ V >> H

17. **(d)** Initially for circular coil $L = 2\pi r$ and $M = i \times \pi r^2$

$= i \times \pi \left(\frac{L}{2\pi}\right)^2 = \frac{iL^2}{4\pi}$ ...(i)

Finally for square coil side $a = \frac{L}{4}$ and

$M' = i \times \left(\frac{L}{4}\right)^2 = \frac{iL^2}{16}$ ...(ii)

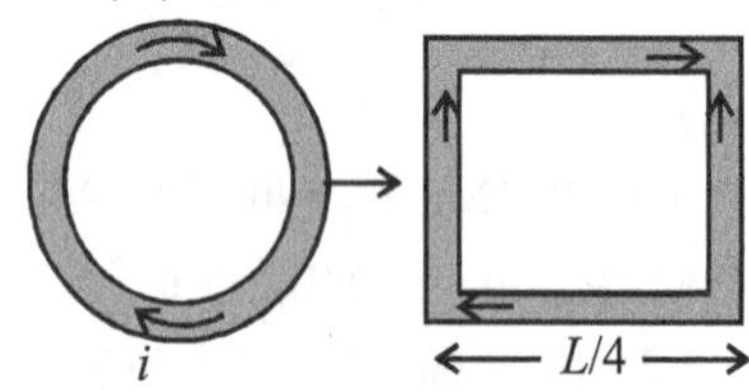

Solving equation (i) and (ii) $M' = \frac{\pi M}{4}$

18. **(b)** $FL = MB$ (= Torque) $\Rightarrow L = \frac{MB}{F}$

19. **(a)** $\chi_d < \chi_p < \chi_f$
For diamagnetic substance χ_d is small and negative (10^{-5})
For paramagnetic substances χ_p is small and positive (10^{-3} to 10^{-5})
For ferromagnetic substanes χ_f is very large (10^3 to 10^5)

20. **(b)** $B = \mu_0 \mu_r H \Rightarrow \mu_r \propto \frac{B}{H}$ = slope of B-H curve

According to the given graph, slope of the graph is highest at point Q.

21. **(d)** On increasing the temperature by 700°C, the magnetic needle is demagnetised. Therefore, the needle stops vibrating.

22. **(b)** $\tau = MB \sin \theta$ $(\theta = 90°)$

$\tau = MB \Rightarrow \frac{B_1}{B_2} = \frac{\tau_1}{\tau_2}$ (since magnetic moment is same)

23. **(a)** Magnetic field due to a bar magnet in the broad-side on position is given by

$$B = \frac{\mu_0}{4\pi} \frac{M}{\left[r^2 + \frac{\ell^2}{4}\right]^{3/2}} \; ; M = m\ell.$$

After substituting the values and simplifying we get $B = 6 \times 1^{-5}$ A-m

24. **(c)** Initially magnetic moment of system

$M_1 = \sqrt{M^2 + M^2} = \sqrt{2}M$ and moment of inertia

$I_1 = I + I = 2I$.

Finally when one of the magnet is removed then

$M_2 = M$ and $I_2 = I$

So, $T = 2\pi \sqrt{\frac{I}{M B_H}}$

$\frac{T_1}{T_2} = \sqrt{\frac{I_1}{I_2} \times \frac{M_2}{M_1}} = \sqrt{\frac{2I}{I} \times \frac{M}{\sqrt{2}M}}$

$\Rightarrow T_2 = \frac{2^{5/4}}{2^{1/4}} = 2 \sec$

25. **(d)** A magnetic needle kept in non uniform magnetic field experience a force and torque due to unequal forces acting on poles.

26. **(d)** $\tan \delta = \frac{V}{H} = \frac{3}{4}$ $\left[\because \tan 37° = \frac{3}{4}\right]$

$\therefore \; V = \frac{3}{4} H$

$V = 6 \times 10^{-5}$ T

$H = \frac{4}{3} \times 6 \times 10^{-5} T = 8 \times 10^{-5} T$

$\therefore \; B_{total} = \sqrt{V^2 + H^2} = \sqrt{(36+64)} \times 10^{-5}$
$= 10 \times 10^{-5} = 10^{-4}$T.

27. **(b)** Ferromagnetic substance has magnetic domains whereas paramagnetic substances have magnetic dipoles which get attracted to a magnetic field.
Diamagnetic substances do not have magnetic dipole but in the presence of external magnetic field due to their orbital motion of electrons these substances are repelled.

28. **(d)** PQ_6 corresponds to the lowest potential energy among all the configurations shown.

29. **(a)** $\tan \theta = \frac{V}{H}, \tan \theta' = \frac{V}{H \cos x}; \; \frac{\tan \theta'}{\tan \theta} = \frac{1}{\cos x}$

30. **(d)** In series, same current flows through two tangent galvanometers.

31. **(c)** Net magnetic dipole moment = $2 M \cos \frac{\theta}{2}$

As value of $\cos \frac{\theta}{2}$ is maximum in case (c) hence net magnetic dipole moment is maximum for option (c).

32. (b) Since magnetic field is in vertical direction and needle is free to totate in horizontal plane only so magnetic force cannot rotate the needle in horizontal plane so needle can stay in any position.

33. (b) Work done in rotating the magnetic dipole from position $\theta_1 = 0°$ to $\theta_2 = 180°$

$\because W = MB(\cos\theta_1 - \cos\theta_2)$

$\therefore W = MB(\cos 0° - \cos 180°) = 2MB$

34. (b) The time period of a bar magnet in a magnetic field is given by.

$$T = 2\pi\sqrt{\frac{I}{MB}};$$

Here, I = moment of inertia $\propto m$, M = moment of magnet, B = magnetic field.

$T \propto \sqrt{I} \propto \sqrt{m}$; so, T becomes twice as mass becomes four times

35. (a) Given, $B = 4 \times 10^{-5}$ T

$R_E = 6.4 \times 10^6$ m

Dipole moment of the earth M = ?

$$B = \frac{\mu_0}{4\pi}\frac{M}{d^3}$$

$$4\times10^{-5} = \frac{4\pi\times10^{-7}\times M}{4\pi\times\left(6.4\times10^6\right)^3}$$

$\therefore \quad M \cong 10^{23}$ Am^2

36. (b) From $\mu_r = 1 + \chi_m$;

Magnetic suscaptibility, $\chi_m = \mu_r - 1$

$\chi_m = 0.075 - 1 = -0.925.$

37. (d) $\delta_1 = 40°, \delta_2 = 30°, \delta = ?$

$$\cot\delta = \sqrt{\cot^2\delta_1 + \cot^2\delta_2} = \sqrt{\cot^2 40° + \cot^2 30°}$$

$$\cot\delta = \sqrt{1.19^2 + 3} = 2.1$$

$\therefore \delta = 25°$ i.e. $\delta < 40°$.

38. (d) In magnetic dipole

Force $\propto \frac{1}{r^4}$

In the given question,

Force $\propto x^{-n}$

Hence, n = 4

39. (b) According to Curie's law, $\chi_m = \frac{\mu_0 C}{T}$

where C is Curie constant, T = temperature

$$\therefore \chi_m \alpha \frac{1}{T}$$

$$\frac{\chi_{m_1}}{\chi_{m_2}} = \frac{T_2}{T_1} = \frac{273+333}{273+30} = \frac{606}{303} = 2$$

$$\therefore \chi_{m_2} = \chi_{m_1}/2 = 0.5\chi_{m_1} = 0.5\chi.(\because \chi_{m_1} = \chi)$$

40. (a) We know that $\mu_r = 1 + x$

$= 1 + 5500 = 5501$

$\therefore \quad \mu = \mu_r\mu_0 = (5501)\times(4\pi\times10^{-7})$

$= 6.9\times10^{-3}$

41. (a) We know that $T_1 = 2\pi\sqrt{\frac{T}{MB_{H_1}}}$ ...(i)

Where $B_{H_1} = 24\times10^{-6}$ T

The magnetic field produced by, wire

$$B = \frac{\mu_0}{2\pi}\cdot\frac{i}{r}$$

$$= (2\times10^{-7})\times\frac{(18)}{0.20}$$

$$= 1.8\times10^{-6} \text{ T}$$

Now $B_{H_2} = B_{H_1} + B = 42\times10^{-6}$ T

$$T_2 = 2\pi\sqrt{\frac{I}{MBH_2}} \quad ...(ii)$$

Using equations (i) and (ii), and substituting the values, we get

$T_2 = 0.076$ s

42. (a) As $BI = \mu_0 MI_M = \mu_0(I + I_M)$

Here, $I = 0$

Then $\mu_0 MI = \mu_0(I_M)$

$\Rightarrow \quad I_M = MI = 10^5$ A

43. (a) Given $M = 8\times10^{22}$ Am^2

$d = R_e = 6.4\times10^6$m

Earth's magnetic field, $B = \frac{\mu_0}{4\pi}\cdot\frac{2M}{d^3}$

$$= \frac{4\pi\times10^{-7}}{4\pi}\times\frac{2\times8\times10^{22}}{(6.4\times10^6)^3}$$

$\cong 0.6$ Gauss

44. (c) Magnetic field in solenoid $B = \mu_0 n i$

$$\Rightarrow \frac{B}{\mu_0} = ni$$

(Where n = number of turns per unit length)

$$\Rightarrow \frac{B}{\mu_0} = \frac{Ni}{L}$$

$$\Rightarrow 3\times10^3 = \frac{100i}{10\times10^{-2}}$$

$\Rightarrow i = 3$A

45. (a) As length of each part also becomes half, therefore magnetic moment M = pole strength × length

$\Rightarrow \frac{1}{2}\times\frac{1}{2} = \frac{1}{4}$th i.e. M/4.

1. (b) Induced emf produced between the centre and a point on the disc is given by

$$e = \frac{1}{2}\omega BR^2$$

Putting the values,

$\omega = 60$ rad/s, $B = 0.05$ Wb/m^2

and $R = 100$ cm $= 1$m

We get $e = \frac{1}{2} \times 60 \times 0.05 \times (1)^2 = 1.5V$

2. (a) According to Faraday's law of electromagnetic induction, $\varepsilon = \frac{d\phi}{dt}$

Also, $\varepsilon = iR$

$\therefore \quad iR = \frac{d\phi}{dt} \quad \Rightarrow \quad \int d\phi = R\int idt$

Magnitude of change in flux $(d\phi) = R \times$ area under current vs time graph

or, $d\phi = 100 \times \frac{1}{2} \times \frac{1}{2} \times 10 = 250$ Wb

3. (a) If a wire, ℓ meter in length, moves perpendicular to a magnetic field of B weber/meter^2 with a velocity of v meter/second, then the e.m.f. induced in the wire is given by

$V = B v\ell$ volt.

Here, $B = 0.30 \times 10^{-4}$ weber/meter^2,

$v = 5.0$ meter/second and $\ell = 10$ meter.

$\therefore \quad B = 0.30 \times 10^{-4} \times 5.0 \times 10 = 0.0015$ volt.

4. (d) The magnetic field is increasing in the downward direction. Therefore, according to Lenz's law, the current I_1 will flow in the direction ab and I_2 in the direction dc.

5. (a) Self inductance of a solenoid,

$$L = \frac{\mu_0 N^2 A}{l} = \frac{\mu_0 N^2 \pi r^2}{l}$$

$$\therefore \quad \frac{L_1}{L_2} = \left(\frac{r_1}{r_2}\right)^2 \left(\frac{l_2}{l_1}\right) \qquad [\because N_1 = N_2]$$

Here, $\frac{l_1}{l_2} = \frac{1}{2}, \frac{r_1}{r_2} = \frac{1}{2}$

$$\therefore \quad \frac{L_1}{L_2} = \left(\frac{1}{2}\right)^2 \left(\frac{2}{1}\right) = \frac{1}{2}$$

6. (b) $\ell = 1$m, $\omega = 5$ rad/s, $B = 0.2 \times 10^{-4} T$

$$\varepsilon = \frac{B\omega\ell}{2} = \frac{0.2 \times 10^{-4} \times 5 \times 1}{2} = 50\mu V$$

7. (c)

8. (c) Emf induced in side 1 of frame $e_1 = B_1 V\ell$

$$B_1 = \frac{\mu_0 I}{2\pi(x - a/2)}$$

Emf induced in side 2 of frame $e_2 = B_2 V\ell$

$$B_2 = \frac{\mu_0 I}{2\pi(x + a/2)}$$

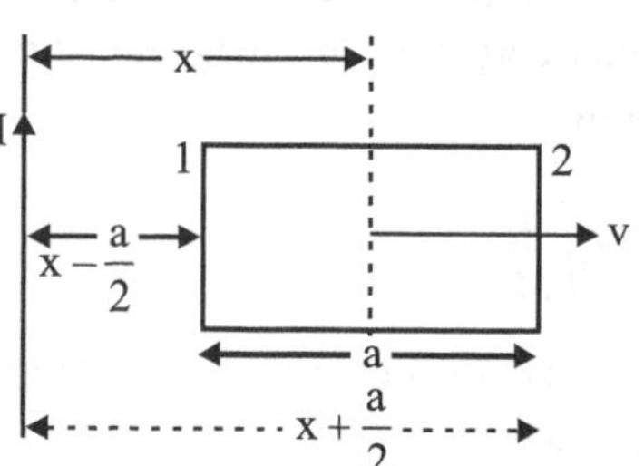

Emf induced in square frame

$e = B_1 V\ell - B_2 V\ell$

$$= \frac{\mu_0 I}{2\pi(x - a/2)}\ell v - \frac{\mu_0 I}{2\pi(x + a/2)}\ell v$$

or, $e \propto \frac{1}{(2x - a)(2x + a)}$

9. (a) When a north pole of a bar magnet moves towards the coil, the induced current in the coil flows in a direction such that the coil presents its north pole to the bar magnet as shown in figure (a). Therefore, the induced current flows in the coil in the anticlockwise direction. When a north pole of a bar magnet moves away from the coil, the induced current in the coil flows in a direction such that the coil presents its south pole to the bar magnet as shown in figure (b).

(a)

(b)

Therefore induced current flows in the coil in the clockwise direction.

10. (d) Given : $\phi = 4t^2 + 2t + 1$ wb

$$\therefore \quad \frac{d\phi}{dt} = \frac{d}{dt}(4t^2 + 2t + 1) = 8t + 2 = |\varepsilon|$$

Induced current, $I = \frac{|\varepsilon|}{R} = \frac{8t + 2}{10\Omega} = \frac{8t + 2}{10}$A

At t = 1 s,

$$I = \frac{8 \times 1 + 2}{10}A = 1A$$

11. (d) $M = \frac{\mu_0 N_1 N_2 A}{\ell}$

$$= \frac{4\pi \times 10^{-7} \times 300 \times 400 \times 100 \times 10^{-4}}{0.2}$$

$M=\frac{\mu_0 N_1 N_2 A}{\ell}$

$=2.4\pi\times10^{-4}$H

12. (d) $e=-\frac{\Delta\phi}{\Delta t}=\frac{-\Delta(LI)}{\Delta t}=-L\frac{\Delta I}{\Delta t}$

$\therefore |e|=L\frac{\Delta I}{\Delta t}\Rightarrow 8=L\times\frac{4}{0.05}$

$\Rightarrow L=\frac{8\times0.05}{4}=0.1\text{H}$

13. (c) Total number of turns in the solenoid, N = 500
Current, I = 2A.
Magnetic flux linked with each turn
$=4\times10^{-3}$ Wb

As, $\phi=LI$ or $N\phi=LI\Rightarrow L=\frac{N\phi}{1}$

$=\frac{500\times4\times10^{-3}}{2}$ henry = 1 H.

14. (d) Electric field will be induced, as ABCD moves, in both *AD* and *BC*. The metallic square loop moves in its own plane with velocity *v*. A uniform magnetic field is imposed perpendicular to the plane of the square loop. *AD* and *BC* are perpendicular to the velocity as well as perpendicular to applied field so an emf is induced in both, this will cause electric fields in both.

A B C D v

15. (d) E.M.F. generated, $e=-\frac{d\phi}{dt}=-\frac{d(N\vec{B}.\vec{A})}{dt}$

$=-N\frac{d}{dt}(BA\cos\omega t)=NBA\omega\sin\omega t$

$\Rightarrow e_{max}=NBA\omega$

16. (c) $L=2\text{mH}, i=t^2e^{-t}$

$E=-L\frac{di}{dt}=-L[-t^2e^{-t}+2te^{-t}]$

when $E=0$,

$-e^{-t}t^2+2te^{-t}=0$

or, $2t\,e^{-t}=e^{-t}t^2$

$\Rightarrow t=2$ sec.

17. (b)

18. (b) $\frac{d\phi}{dt}=\frac{(W_2-W_1)}{t}$ $R_{tot}=(R+4R)\Omega=5R\ \Omega$

$i=\frac{nd\varphi}{R_{tot}dt}=\frac{-n(W_2-W_1)}{5Rt}$.

($\because W_2$ & W_1 are magnetic flux)

19. (b) The individual emf produced in the coil $e=\frac{-d\phi}{dt}$

$\therefore$ The current induced will be $i=\frac{|e|}{R}\Rightarrow i=\frac{1}{R}\frac{d\phi}{dt}$

But $i=\frac{dq}{dt}\Rightarrow\frac{dq}{dt}=\frac{1}{R}\frac{d\phi}{dt}\Rightarrow\int dq=\frac{1}{R}\int d\phi\Rightarrow q=\frac{BA}{R}$

20. (c) Induced emf $=vB_H l=1.5\times5\times10^{-5}\times2$
$=15\times10^{-5}$
$=0.15$mV

21. (b) $\varepsilon=\frac{d\phi}{dt}=nA\frac{dB}{dt}$

$\therefore\ \varepsilon=10\times(10\times10^{-4})(10^4)$ [10^8 Gauss/sec=10^4 T/s]
$=100$ V.
$I=(\varepsilon/R)=(100/20)=5$amp.

22. (a)

W E

$\varepsilon_{ind}=Bv\ell$

$=0.3\times10^{-4}\times5\times20$

$=3\times10^{-3}$ V = 3 mV.

23. (d) The self inductance of a long solenoid is given by

$L=\mu_r\mu_0 n^2 Al$

Self inductance of a long solenoid is independent of the current flowing through it.

24. (d) Here, induced e.m.f.

ω 2λ λ x dx

$e=\int_{2\ell}^{3\ell}(\omega x)Bdx$

$=B\omega\frac{[(3\ell)^2-(2\ell)^2]}{2}=\frac{5B\ell^2\omega}{2}$

25. (b)

26. (b) Induced e.m.f. in the ring opposes the motion of the magnet.

27. (a)

28. (b)

29. (d) Magnetic flux, $\phi_B=BA\cos\theta$
Induced emf, $\varepsilon=BA\sin\theta$
Here, $\theta=0°$
$\therefore$ Magnetic flux is maximum and induced emf is zero.

30. (c) e.m.f. induced $=\frac{1}{2}BR^2\omega=\frac{1}{2}BR^2(2\pi n)$

$=\frac{1}{2}\times(0.1)\times(0.1)^2\times2\pi\times10=(0.1)^2\ \pi$ volts

31. (a) $E=\frac{d}{dt}(NMI)\Rightarrow E=NM\frac{dI}{dt}\Rightarrow E=\frac{NMI}{t}$

emf induced per unit turn $=\frac{E}{N}=\frac{MI}{t}$

32. **(d)** According to Lenz's law, when switch is closed, the flux in the loop increases out of plane of paper, so induced current will be clockwise.

33. **(a)** Since $\varepsilon = -\frac{Nd\phi}{dt}$ if $\frac{d\phi}{dt}$ is fast, so ε is large.

34. **(d)** The e.m.f. is induced when there is change of flux. As in this case there is no change of flux, hence no e.m.f. will be induced in the wire.

35. **(b)** Given, $B = 0.01\ T, A = \pi R^2 = \pi \times (1\ m)^2 = \pi m^2$

$\omega = 100\ rads^{-1}$

$\therefore$ The maximum induced emf $\varepsilon_{max} = BA\omega$

$= 0.01 \times \pi \times 100\ V = \pi V$

36. **(b)** $e = \frac{-(\phi_2 - \phi_1)}{t} = \frac{-(0 - NBA)}{t} = \frac{NBA}{t}$

$$t = \frac{NBA}{e} = \frac{50 \times 2 \times 10^{-2} \times 10^{-2}}{0.1} = 0.1\ s$$

37. **(c)** $\frac{\Delta\phi}{\Delta t} = \varepsilon = iR \Rightarrow \Delta\phi = (i\Delta t)R = QR$

$$\Rightarrow Q = \frac{\Delta\phi}{R}$$

38. **(d)** $\phi = BA\cos\theta = 2.0 \times 0.5 \times \cos 60°$

$$= \frac{2.0 \times 0.5}{2} = 0.5\ \text{weber.}$$

39. **(a)** $\xi = \frac{W}{Q} \Rightarrow V = \frac{W}{Q} \Rightarrow W = QV$

40. **(b)** Mutual inductance depends on the relative position and orientation of the two coils.

41. **(c)**

42. **(a)** As the magnetic field increases, its flux also increases into the page and so induced current in bigger loop will be anticlockwise. i.e., from D to C in bigger loop and then from B to A in smaller loop.

43. **(c)** As I increases, ϕ increases

$\therefore$ I_i is such that it opposes the increases in ϕ.

Hence, ϕ decreases (By Right Hand Rule). The induced current will be counterclockwise.

44. **(d)** According to Faraday's law of electromagnetic induction,

Induced emf, $e = \frac{Ldi}{dt}$

$$50 = L\left(\frac{5-2}{0.1\sec}\right)$$

$$\Rightarrow\ L = \frac{50 \times 0.1}{3} = \frac{5}{3} = 1.67\ H$$

45. **(d)** Mutual inductance between two coil in the same plane with their centers coinciding is given by

$$M = \frac{\mu_0}{4\pi}\left(\frac{2\pi^2 R_2^2 N_1 N_2}{R_1}\right) \textit{henry.}$$

1. (c) Across resistor, $I = \frac{V}{R} = \frac{100}{1000} = 0.1\,A$

At resonance,

$$X_L = X_C = \frac{1}{\omega C} = \frac{1}{200 \times 2 \times 10^{-6}} = 2500$$

Voltage across L is

$$I X_L = 0.1 \times 2500 = 250\,\text{V}$$

2. (a) The phase angle between voltage V and current I is $\pi/2$. Therefore, power factor $\cos\phi = \cos(\pi/2) = 0$. Hence the power consumed is zero.

3. (c) The current drawn by inductor and capacitor will be in opposite phase. Hence net current drawn from generator $= I_L - I_C = 0.9 - 0.4 = 0.5$ amp.

4. (c) Capacitive reactance, $X = \frac{1}{\omega C} = \frac{1}{2\pi f C}$

$$\Rightarrow X \propto \frac{1}{f\,C}$$

$$\therefore \frac{X'}{X} = \frac{f}{f'} \times \frac{C}{C'} = \frac{f}{2f} \times \frac{C}{2C} = \frac{1}{4} \Rightarrow X' = \frac{X}{4}$$

5. (a) The charging of inductance given by,

$$i = i_0\left(1 - e^{-\frac{Rt}{L}}\right)$$

$$\frac{i_0}{2} = i_0\left(1 - e^{-\frac{Rt}{L}}\right) \Rightarrow e^{-\frac{Rt}{L}} = \frac{1}{2}$$

Taking log on both the sides,

$$-\frac{Rt}{L} = \log 1 - \log 2$$

$$\Rightarrow t = \frac{L}{R}\log 2 = \frac{300 \times 10^{-3}}{2} \times 0.69$$

$$\Rightarrow t = 0.1 \text{ sec.}$$

6. (c) $\tan\phi = \frac{\omega L}{R} = \frac{X_L}{R}$

Given $\phi = 45°$. Hence $X_L = R$.

7. (d)

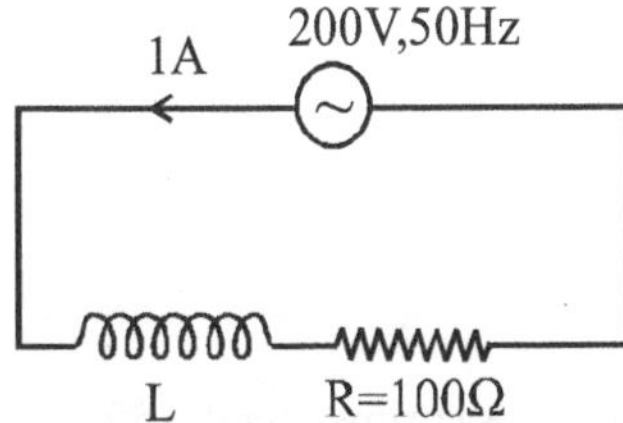

From the rating of the bulb, the resistance of the bulb can be calculated.

$$R = \frac{V_{rms}^2}{P} = 100\Omega$$

For the bulb to be operated at its rated value the rms current through it should be 1A

Also, $I_{rms} = \frac{V_{rms}}{Z}$

$$\therefore \quad 1 = \frac{200}{\sqrt{100^2 + (2\pi 50.L)^2}}$$

$$L = \frac{\sqrt{3}}{\pi}\text{H}$$

8. (a) At angular frequency ω, the current in RC circuit is given by

$$i_{max} = \frac{V_{max}}{\sqrt{R^2 + \left(\frac{1}{\omega C}\right)^2}} \qquad ....(i)$$

$$\text{Also } \frac{i_{rms}}{2} = \frac{V_{rms}}{\sqrt{R^2 + \left(\frac{1}{\frac{\omega}{3}C}\right)^2}} = \frac{V_{max}}{\sqrt{R^2 + \frac{9}{\omega^2 C^2}}} \qquad .....(ii)$$

From equation (i) and (ii) we get

$$3R^2 = \frac{5}{\omega^2 C^2} \Rightarrow \frac{\frac{1}{\omega C}}{R} = \sqrt{\frac{3}{5}} \Rightarrow \frac{X_C}{R} = \sqrt{\frac{3}{5}}$$

9. (c)

10. (c) Growth in current in LR_2 branch when switch is closed is given by

$$i = \frac{E}{R_2}[1 - e^{-R_2 t/L}]$$

$$\Rightarrow \frac{di}{dt} = \frac{E}{R_2}.\frac{R_2}{L}.e^{-R_2 t/L} = \frac{E}{L}e^{-\frac{R_2 t}{L}}$$

Hence, potential drop across L

$$= \left(\frac{E}{L}e^{-R_2 t/L}\right)L = Ee^{-R_2 t/L}$$

$$= 12e^{-\frac{2t}{400 \times 10^{-3}}} = 12e^{-5t}\text{V}$$

11. (a) $I = I_o\left(1 - e^{-\frac{R}{L}t}\right)$

(When current is in growth in LR circuit)

$$= \frac{E}{R}\left(1 - e^{-\frac{R}{L}t}\right) = \frac{5}{5}\left(1 - e^{-\frac{5}{10} \times 2}\right)$$

$$= (1 - e^{-1})$$

12. (d) Power, $P = I_{r.m.s} \times V_{r.m.s} \times \cos\phi$

In the given problem, the phase difference between voltage and current is $\pi/2$. Hence

$$P = I_{r.m.s} \times V_{r.m.s} \times \cos(\pi/2) = 0.$$

13. (c) When the capacitor is completely charged, the total energy in the LC circuit is with the capacitor and that energy is $E=\frac{1}{2}\frac{Q^2}{C}$

When half energy is with the capacitor in the form of electric field between the plates of the capacitor we get

$\frac{E}{2}=\frac{1}{2}\frac{Q'^2}{C}$ where Q' is the charge on one plate of the capacitor

$$\therefore \frac{1}{2}\times\frac{1}{2}\frac{Q^2}{C}=\frac{1}{2}\frac{Q'^2}{C} \Rightarrow Q'=\frac{Q}{\sqrt{2}}$$

14. (a) Energy stored in magnetic field $=\frac{1}{2}Li^2$

Energy stored in electric field $=\frac{1}{2}\frac{q^2}{C}$

$$\therefore \frac{1}{2}Li^2=\frac{1}{2}\frac{q^2}{C}$$

Also $q=q_0\cos\omega t$ and $\omega=\frac{1}{\sqrt{LC}}$

On solving $t=\frac{\pi}{4}\sqrt{LC}$

15. (c) The circuit will have inductive nature if

$$\omega>\frac{1}{\sqrt{LC}}\left(\omega L>\frac{1}{\sqrt{LC}}\right)$$

Hence (a) is false. Also if circuit has inductive nature the current will lag behind voltage. Hence (d) is also false.

If $\omega=\frac{1}{\sqrt{LC}}\left(\omega L=\frac{1}{\omega C}\right)$ the circuit will have resistance nature. Hence (b) is false.

Power factor

$$\cos\phi=\frac{R}{\sqrt{R^2+\left(\omega L-\frac{1}{\omega C}\right)^2}}=1 \text{ if } \omega L=\frac{1}{\omega C}$$

16. (b) $V_{rms}=\sqrt{\frac{(T/2)V_0^2+0}{T}}=\frac{V_0}{\sqrt{2}}$.

17. (d) Option (d) is false because the reason why the voltage leads the current is because $\frac{1}{C\omega}>L\omega$ and if the voltage lags, the inductive reactance is greater than the capacitive reactance.

18. (b) $P=\frac{1}{2}V_0 i_0\cos\phi \Rightarrow P=P_{peak}.\cos\phi$

$$\Rightarrow \frac{1}{2}(P_{peak})=P_{peak}\cos\phi \Rightarrow \cos\phi=\frac{1}{2}\Rightarrow\phi=\frac{\pi}{3}$$

19. (c) $\eta=\frac{E_sI_s}{E_pI_p}$ $\therefore \eta=\frac{110\times 9}{220\times 5}=0.9\times 100\%=90\%$

20. (d) $V=\frac{V_0}{T/4}t \Rightarrow V=\frac{4V_0}{T}t$

$$\Rightarrow V_{rms}=\sqrt{<V^2>}=\frac{4V_0}{T}\sqrt{<t^2>}=\frac{4V_0}{T}\left\{\frac{\int_0^{T/4}t^2dt}{\int_0^{T/4}dt}\right\}^{1/2}=\frac{V_0}{\sqrt{3}}$$

21. (a) $L=10\,mHz=10^{-2}Hz$

$f=1MHz=10^6Hz$

$$f=\frac{1}{2\pi\sqrt{LC}}$$

$$f^2=\frac{1}{4\pi^2LC}$$

$$\Rightarrow C=\frac{1}{4\pi^2f^2L}=\frac{1}{4\times 10\times 10^{-2}\times 10^{12}}=\frac{10^{-12}}{0.4}=2.5\,pF$$

22. (d) Current is maximum when $X_L=X_C$

$$\Rightarrow \omega L=\frac{1}{\omega C}\Rightarrow \omega=\frac{1}{\sqrt{LC}}=\frac{1}{\sqrt{0.5\times 8\times 10^{-6}}}$$

$$=\frac{1}{2\times 10^{-3}}=500\text{ rad/s.}$$

23. (a) If $\omega=50\times 2\pi$ then $\omega L=20\Omega$

If $\omega'=100\times 2\pi$ then $\omega'L=40\Omega$

Current flowing in the coil is

$$I=\frac{200}{Z}=\frac{200}{\sqrt{R^2+(\omega'L)^2}}=\frac{200}{\sqrt{(30)^2+(40)^2}}$$

$I=4A$.

24. (a) At resonance impedance is minimum ($\therefore X_L=X_C$) current is maximum, because V_L and V_C are equal in magnitude

$\therefore V_{LC}=V_L-V_C=0$

Hence, voltmeter V_2 read 0 volt.

25. (b)

26. (d) As $E_pI_p=P_i$ $\therefore I_p=\frac{P_i}{E_p}=\frac{4000}{100}=40\,A$.

27. (c) The phase angle is given by

$$\tan\phi=\frac{\omega L}{R}=\frac{2\pi\times 50\times 0.21}{12}=5.5$$

$$\phi=\tan^{-1}5.5=80^\circ$$

28. (c)

29. (a) Since $\frac{V_s}{V_p}=\frac{N_s}{N_p}$

Where

N_s = No. of turns across primary coil = 50

N_p = No. of turns across secondary coil = 1500

and $V_p=\frac{d\phi}{dt}=\frac{d}{dt}(\phi_0+4t)=4$

$$\Rightarrow V_s=\frac{1500}{50}\times 4=120\text{ V}$$

30. (a) $\frac{E_s}{E_p}=\frac{n_s}{n_p}$ or $E_s=E_p\times\left(\frac{n_s}{n_p}\right)$

$\therefore\ E_s=120\times\left(\frac{200}{100}\right)=240\,V$

$\frac{I_p}{I_s}=\frac{n_s}{n_p}$ or $I_s=I_p\left(\frac{n_p}{n_s}\right)$ $\therefore\ I_s=10\left(\frac{100}{200}\right)=5$ amp

31. (b) $R\uparrow\Rightarrow I\downarrow\Rightarrow\frac{dI}{dt}\to(-ve)\to e=(+ve)$

$\left[\text{As } e=-L\frac{dI}{dt}\right]$

Supporting $\to I_{net}\uparrow$

32. (b) We know that, $i=i_0(1-e^{-t/\tau})$

or $\frac{3}{4}i_0=i_0(1-e^{-4/\tau})$

or $e^{-4/\tau}=\frac{1}{4}$

or $e^{4/\tau}=4$

$\therefore\ \frac{4}{\tau}=\ln 4$

or $\tau=\frac{2}{\ln 2}s$

33. (d) At resonance, $\omega L=\frac{1}{\omega C}$. The circuit behaves as if it contains R only. So, phase difference = 0

At resonance, impedance is minimum $Z_{min}=R$ and current is maximum, given by

$$I_{max}=\frac{E}{Z_{min}}=\frac{E}{R}$$

It is interesting to note that before resonance the current leads the applied emf, at resonance it is in phase, and after resonance it lags behind the emf. LCR series circuit is also called as acceptor circuit and parallel LCR circuit is called rejector circuit.

34. (d) Condition for which the current is maximum in a series LCR circuit is,

$\omega=\frac{1}{\sqrt{LC}}$

$1000=\frac{1}{\sqrt{L(10\times10^{-6})}}$

$\Rightarrow =L=100\,mH$

35. (c) When a circuit is broken, the induced e.m.f. is largest. So the answer is (c).

36. (b) $V=50\times2\sin 100\pi\cos 100\pi t=50\sin 200\pi t$

$\Rightarrow V_0=50\,Volts$ and $\nu=100\,Hz$

37. (d) $V=-L\frac{di}{dt}$

Here $\frac{di}{dt}$ is + ve for $\frac{T}{2}$ time and

$\frac{di}{dt}$ is – ve for next $\frac{T}{2}$ time

38. (b) $V=V_0\sin\omega t$

Voltage in r.m.s. value

$V_0=\sqrt{2}\times234\,V=331$ volt

and $\omega t=2\pi nt=2\pi\times50\times t=100\pi t$

Thus, the equation of line voltage is given by

$V=331\sin(100\pi t)$

39. (a) The resistance in the middle plays no part in the charging process of C, as it does not alter either the potential difference across the RC combination or the current through it.

40. (a) Here, $C=100\,\mu F=100\times10^{-6}F, R=40\,\Omega$, $V_{rms}=110\,V, f=60\,Hz$

Peak voltage,

$$V_0=\sqrt{2}\,.\,V_{rms}=100\sqrt{2}=155.54\,V$$

Circuit impedance,

$$Z=\sqrt{R^2+\frac{1}{\omega^2C^2}}$$

$$=\sqrt{40^2+\frac{1}{(2\times\pi\times60\times100\times10^{-6})^2}}$$

$$=\sqrt{1600+703.60}=\sqrt{2303.60}=48\,\Omega$$

hence, maximum current in coil,

$$I_0=\frac{V_0}{Z}=\frac{155.54}{48}=3.24\,A$$

41. (a) Laminated core provide less area of cross-section for the current to flow. Because of this, resistance of the core increases and current decreases thereby decreasing the eddy current losses.

42. (b) $V=200V;\ r=10\Omega$

$R'=10+100\Omega=110\Omega$

$I=\frac{V}{R'}=\frac{220}{100}=2A$

$P=I^2R=4\times100=400\,W$

43. (b) For step-down transformer,

$\frac{V_P}{V_S}=\frac{I_S}{I_P}$ $\because V_P>V_S$ $\therefore I_S>I_P$

44. (b) At $t=0$, no current will flow through L and R_1

$\therefore$ Current through battery $=\frac{V}{R_2}$

At $t=\infty$,

effective resistance, $R_{eff}=\frac{R_1R_2}{R_1+R_2}$

$\therefore$ Current through battery $=\frac{V}{R_{eff}}$

$=\frac{V(R_1+R_2)}{R_1R_2}$

45. (d) These three inductors are connected in parallel. The equivalent inductance L_p is given by

$$\frac{1}{L_p}=\frac{1}{L_1}+\frac{1}{L_2}+\frac{1}{L_3}=\frac{1}{3}+\frac{1}{3}+\frac{1}{3}=\frac{3}{3}=1$$

$\therefore L_p=1$

1. (b) ∵ The E.M. wave are transverse in nature i.e.,

$$= \frac{\vec{k}\times\vec{E}}{\mu\omega} = \vec{H} \quad \ldots(i)$$

where $\vec{H} = \frac{\vec{B}}{\mu}$

and $\frac{\vec{k}\times\vec{H}}{\omega\varepsilon} = -\vec{E} \quad \ldots(ii)$

$\vec{k}$ is ⊥ $\vec{H}$ and $\vec{k}$ is also ⊥ to $\vec{E}$

or In other words $\vec{X} \parallel \vec{E}$ and $\vec{k} \parallel \vec{E}\times\vec{B}$

2. (a) $E_{rms} = 720$

The average total energy density

$= \frac{1}{2}\epsilon_0 E_0^2 = \frac{1}{2}\epsilon_0 [\sqrt{2}E_{rms}]^2 = \epsilon_0 E_{rms}^2$

$= 8.85\times10^{-12}\times(720)^2$

$= 4.58\times10^{-6}\ J/m^3$

3. (d) $I_d = 1\ mA = 10^{-3}$ A

$C = 2\mu F = 2\times10^{-6}$ F

$$I_D = I_C = \frac{d}{dt}(CV) = C\frac{dV}{dt}$$

Therefore, $\frac{dV}{dt} = \frac{I_D}{C} = \frac{10^{-3}}{2\times10^{-6}} = 500\ Vs^{-1}$

Therefore, applying a varying potential difference of 500 V s^{-1} would produce a displacement current of desired value.

4. (c) $E_0 = CB_0$ and $C = \frac{1}{\sqrt{\mu_0\varepsilon_0}}$

Electric energy density $= \frac{1}{2}\varepsilon_0 E_0{}^2 = \mu_E$

Magnetic energy density $= \frac{1}{2}\frac{Bo^2}{\mu_0} = \mu_B$

Thus, $\mu_E = \mu_B$

Energy is equally divided between electric and magnetic field

5. (c) In an electromagnetic wave electric field and magnetic field are perpendicular to the direction of propagation of wave. The vector equation for the electric field is

$$\vec{E} = E_0\cos\left(\omega t - \frac{2\pi}{\lambda}y\right)\hat{z}$$

6. (a) Frequency remains constant during refraction

$$v_{med} = \frac{1}{\sqrt{\mu_0\ \epsilon_0 \times 4}} = \frac{c}{2}$$

$$\frac{\lambda_{med}}{\lambda_{air}} = \frac{v_{med}}{v_{air}} = \frac{c/2}{c} = \frac{1}{2}$$

∴ wavelength is halved and frequency remains unchanged.

7. (b) For electromagnetic waves we know that,

$$\frac{E}{B} = c$$

$$\therefore \frac{9\times10^{-4}}{B} = 3\times10^8\ ms^{-1}$$

$B = 3\times10^{-12}$ T.

8. (b) Here, $k = \frac{2\pi}{\lambda}$, $\omega = 2\pi\upsilon$

$$\therefore \frac{k}{\omega} = \frac{2\pi/\lambda}{2\pi\upsilon} = \frac{1}{\pi\upsilon} = \frac{1}{c} \qquad (\because c = \upsilon\lambda)$$

where c is the speed of electromagnetic wave in vacuum. It is a constant whose value is $3\times10^6\ m\ s^{-1}$

9. (a) E.M. wave always propagates in a direction perpendicular to both electric and magnetic fields. So, electric and magnetic fields should be along + X– and + Y–directions respectively. Therefore, option (a) is the correct option.

10. (d) $\frac{1}{2}\varepsilon_0 E_0^2$ is electric energy density.

$\frac{B^2}{2\mu_0}$ is magnetic energy density.

So, total energy $= \frac{1}{2}\varepsilon_0 E_0^2 + \frac{B_0^2}{2\mu_0}$

11. (c) Incident momentum, $p = \frac{E}{c}$

For perfectly reflecting surface with normal incidence

$$\Delta p = 2p = \frac{2E}{c}$$

$$F = \frac{\Delta p}{\Delta t} = \frac{2E}{ct}$$

$$P = \frac{F}{A} = \frac{2E}{ctA}$$

12. (a) E_x and B_y would generate a plane EM wave travelling in z-direction, $\vec{E}$, $\vec{B}$ and $\vec{k}$ from a right handed system $\vec{k}$ is along z-axis. As $\hat{i}\times\hat{j} = \hat{k}$

$\Rightarrow E_x\hat{i}\times B_y\hat{j} = C\hat{k}$ i.e., E is along x-axis and B is along y-axis.

13. (b) From question,

$B_0 = 20\ nT = 20\times10^{-9}$T

(∵ velocity of light in vacuum $C = 3\times10^8\ ms^{-1}$)

$$\vec{E}_0 = \vec{B}_0\times\vec{C}$$

$$|\vec{E}_0| = |\vec{B}|.|\vec{C}| = 20\times10^{-9}\times3\times10^8$$

$= 6$ V/m.

14. **(c)** Microwave oven acts on the principle of giving vibrational energy to water molecules.

15. **(a)** Displacement current is set up in a region where the electric field is changing with time.

16. **(a)** On comparing the given equation to

$\vec{E} = a_0\hat{i}\cos(\omega t - kz)$

$\omega = 6\times10^{8z}$,

$k = \frac{2\pi}{r} = \frac{\omega}{c}$

$k = \frac{\omega}{c} = \frac{6\times10^8}{3\times10^8} = 2\,\text{m}^{-1}$

17. **(b)** Displacement current, I_D = conduction current, I_C

$\therefore \frac{dq}{dt} = \frac{d}{dt}[q_0\cos 2\pi\upsilon t] = -q_0 2\pi\upsilon \sin 2\pi\upsilon t$

18. **(a)** Momentum of light falling on reflecting surface $p = \frac{E}{C}$

As surface is perfectly reflecting so momentum reflect

$p^1 = -\frac{E}{C}$

$\frac{E}{C}$ ↓ ↑ $\frac{E}{C} = P$

So, momentum transferred

$= P - P^1 = \frac{E}{C} - \left(-\frac{E}{C}\right) = \frac{2E}{C}$

19. **(d)** (1) Infrared rays are used to treat muscular strain because these are heat rays.

(2) Radio waves are used for broadcasting because these waves have very long wavelength ranging from few centimeters to few hundred kilometers

(3) X-rays are used to detect fracture of bones because they have high penetrating power but they can't penetrate through denser medium like dones.

(4) Ultraviolet rays are absorbed by ozone of the atmosphere.

20. **(d)** Given: $\vec{B} = 1.2\times10^{-8}\hat{k}\text{T}$

$\vec{E} = ?$

From formula,

$E = Bc = (1.2\times10^{-8}\text{T})(3\times10^8\text{ms}^{-1}) = 3.6\text{ Vm}^{-1}$

$\vec{B}$ is along Z-direction and the wave propagates along X-direction. Therefore $\vec{E}$ should be along Y-direction.

Thus, $\vec{E} = 3.6\hat{j}\text{ Vm}^{-1}$

21. **(d)**

22. **(a)** $E = \frac{hc}{\lambda} \Rightarrow \lambda = \frac{hc}{E}$

$\Rightarrow \lambda = \frac{6.6\times10^{-34}\times3\times10^8}{11\times1000\times1.6\times10^{-19}} = 12.4\text{ Å}$

Increasing order of frequency →

x-rays u-v rays visible Infrared

23. **(a)** Velocity of light

$C = \frac{E}{B} \Rightarrow B = \frac{E}{C} = \frac{9.3}{3\times10^8} = 3.1\times10^{-8}\text{T}$

24. **(b)** The average energy stored in the electric field

$U_E = \frac{1}{2}\varepsilon_0 E^2$

The average energy stored in the magnetic field = U_B =

$\frac{1}{2}\frac{B^2}{\mu_0}$,

According to conservation of energy $U_E = U_B$

$\varepsilon_0\mu_0 = \frac{B^2}{E^2}$

$\frac{B}{E} = \sqrt{\varepsilon_0\mu_0} = \frac{1}{c}$

25. **(b)** EM waves carry momentum and hence can exert pressure on surfaces. They also transfer energy to the surface so $p \neq 0$ and $E \neq 0$.

26. **(b)**

27. **(a)** The decreasing order of the wavelengths is as given below:

microwave, infrared, ultraviolet, gamma rays.

28. **(b)** Infrared causes heating effect.

29. **(c)** Speed of EM waves in vacuum $= \frac{1}{\sqrt{\mu_0\,\epsilon_0}}$ = constant

30. **(b)**

31. **(a)** $I_D = \varepsilon_o\, d\phi_E/dt$.

32. **(d)** $\nu_{\gamma\text{-rays}} > \nu_{\text{visible radiation}} > \nu_{\text{infrared}} > \nu_{\text{Radio waves}}$

33. **(b)**

34. **(d)** Wave is uv rays.

35. **(b)**

36. **(c)** Electromagnetic waves are the combination of mutually perpendicular electric and magnetic fields.

37. **(b)** Audible waves are not electromagnetic wave.

38. **(b)** Wave impedance $= Z = \sqrt{\frac{\mu_0}{\varepsilon_0}} = 376.6\ \Omega$

39. **(b)**

40. **(b)** Depends on the magnitude of frequency

41. **(d)** $B_0 = \frac{E_0}{c} = \frac{9\times10^3}{3\times10^8} = 3\times10^{-5}\text{T}$.

42. **(d)** To generate electromagnetic waves we need accelerating charge particle.

43. **(d)**

44. **(d)** For an E.M. wave power is transmitted in a direction perpendicular to both the fields.

45. **(c)** Speed of light of vacuum $c = \frac{1}{\sqrt{\mu_0\varepsilon_0}}$ and in another

medium $v = \frac{1}{\sqrt{\mu\varepsilon}}$

$\therefore \frac{c}{v} = \sqrt{\frac{\mu\varepsilon}{\mu_0\varepsilon_0}} = \sqrt{\mu_r K} \Rightarrow v = \frac{c}{\sqrt{\mu_r K}}$.

DAILY PRACTICE PROBLEMS — PHYSICS SOLUTIONS — DPP/CP23

1. **(b)** $\frac{1}{f_R} = (1.5-1)\left(\frac{1}{R_1} - \frac{1}{R_2}\right)$

$\frac{1}{f_V} = (1.45-1)\left(\frac{1}{R_1} - \frac{1}{R_2}\right)$

$\frac{f_V}{f_R} = \frac{0.5}{0.45} = \frac{10}{9}$

$f_R = \frac{9}{10} f_V = \frac{9}{10} \times 20\text{ cm} = 18\text{ cm}.$

2. **(a)** We have,

$$\mu = \frac{\sin\left(\frac{A+\delta_m}{2}\right)}{\sin\left(\frac{A}{2}\right)}$$

$$\Rightarrow \cot\frac{A}{2} = \frac{\sin\left(\frac{A+\delta_m}{2}\right)}{\sin\left(\frac{A}{2}\right)}$$

$$\text{or } \sin\frac{A}{2} \cdot \cot\frac{A}{2} = \sin\left(\frac{A+\delta_m}{2}\right)$$

$$\text{or } \sin\frac{A}{2} \cdot \frac{\cos\frac{A}{2}}{\sin\frac{A}{2}} = \sin\left(\frac{A+\delta_m}{2}\right)$$

$$\text{or } \cos\frac{A}{2} = \cos\left[\frac{\pi}{2} - \left(\frac{A+\delta_m}{2}\right)\right]$$

$$\Rightarrow \frac{A}{2} = \frac{\pi}{2} - \left(\frac{A+\delta_m}{2}\right)$$

or $A = \pi - A - \delta_m \Rightarrow \delta_m = \pi - 2A$.

3. **(a)** Let the distance between the lenses be d.

Then, equivalent power is

$P = P_1 + P_2 - d P_1 P_2$

Given $P_1 = P_2 = +5\text{ D}$

$\therefore\ P = (10 - 25d)\text{ D}$

For P to be –ve,

$10 - 25d < 0 \Rightarrow d > \frac{2}{5}\text{m}$

or, $d > 0.4\text{ m}$ or $d > 40\text{ cm}$

4. **(b)** ${}^a\mu_g = \frac{\sin 60°}{\sin 35°}$ ... (i)

${}^a\mu_w = \frac{\sin 60°}{\sin 41°}$ ... (ii)

${}^w\mu_g = \frac{\sin 41°}{\sin\theta}$ ... (iii)

${}^a\mu_w \times {}^w\mu_g = {}^a\mu_g$

$$\frac{\sin 60°}{\sin 41°} \times \frac{\sin 41°}{\sin\theta} = \frac{\sin 60°}{\sin 35°}$$

(Using (i), (ii) and (iii))

$= \sin\theta = \sin 35° \qquad \theta = 35°$

5. **(d)**

r

$h = 12$ cm

C

$$\sin C = \frac{1}{\mu} = \frac{1}{4/3} = \frac{3}{4}.$$

Now $r = h\tan C$

$$= 12 \times \frac{3}{\sqrt{7}} = \frac{36}{\sqrt{7}}\text{ cm}$$

6. **(a)** $\frac{1}{f} = (\mu - 1)\left(\frac{1}{R_1} - \frac{1}{R_2}\right)$

According to Cauchy relation

$\mu = A + \frac{B}{\lambda^2} + \frac{C}{\lambda^4} \cdots$ Hence $f \propto \lambda$.

Hence, red light having maximum wavelength has maximum focal length.

$\therefore$ $f_v < f_r$ and also $F_v > F_r$ as focal length is negative for a concave lens.

7. **(c)** To minimise spherical aberration in a lens, the total deviation should be equally distributed over the two surfaces.

8. **(d)**

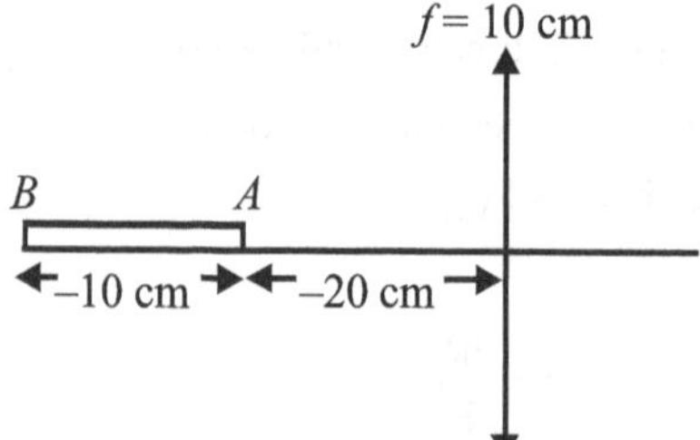

The focal length of the mirror

$$-\frac{1}{f} = \frac{1}{v} + \frac{1}{u}$$

For A end of the rod the image distance

When $u_1 = -20$ cm

$$\Rightarrow \frac{-1}{10} = \frac{1}{v_1} - \frac{1}{20}$$

$$\frac{1}{v_1}=\frac{-1}{10}+\frac{1}{20}=\frac{-2+1}{20}$$

$v_1=-20$ cm

For when $u_2=-30$ cm

$$\frac{1}{f}=\frac{1}{v_2}-\frac{1}{30}$$

$$\frac{1}{v_2}=\frac{-1}{10}+\frac{1}{30}=\frac{-30+10}{300}=\frac{-20}{300}$$

$v_2=-15$ cm

$L=v_2-v_1=-15-(-20)$

$L=5$ cm

9. (a) Magnification

$$=\frac{f_0}{f_e}=\frac{\text{Angle subtended by final image on the eye}}{\text{Angle subtended by the object on eye (or objective)}}$$

$$\Rightarrow \frac{0.3\text{m}}{3\text{cm}}=\frac{\beta}{0.5°}\Rightarrow\frac{30\text{ cm}}{3\text{cm}}=\frac{\beta}{0.5°}$$

$\Rightarrow \beta=5°$

10. (b) Due to difference in refractive indices images obtained will be two. Two media will form images at two different points due to difference in focal lengths.

11. (c) For reading purposes :

u = – 25 cm, v = – 50 cm, f = ?

$$\frac{1}{f}=\frac{1}{v}-\frac{1}{u}=-\frac{1}{50}+\frac{1}{25}=\frac{1}{50};$$

$$P=\frac{100}{f}=+2\,D$$

For distant vision, f' = distance of far point = –3 m

$$P=\frac{1}{f'}=-\frac{1}{3}D=-0.33\ D$$

12. (a) Clearly,

$i+r+90°=180°$

$\Rightarrow\ i+r=90°$ ...(i)

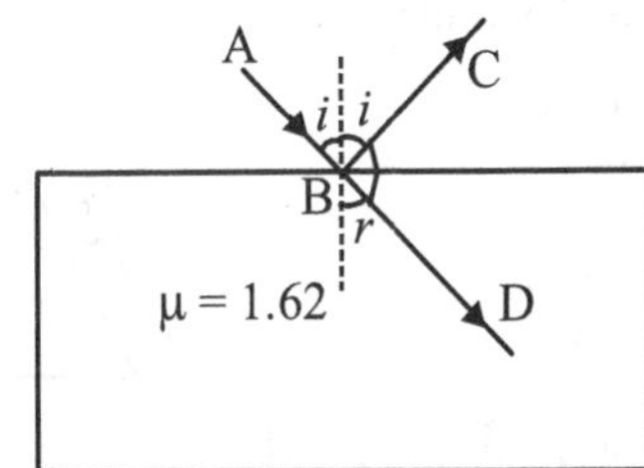

Now, $\frac{\sin i}{\sin r}=\mu$

$\Rightarrow \frac{\sin i}{\sin(90°-i)}=\mu$, from (1)

or $\frac{\sin i}{\cos i}=\mu\Rightarrow\tan i=\mu$

or $i=\tan^{-1}(\mu)$ i.e., $i=\tan^{-1}(1.62)$

13. (b) $f_0=100$ cm, $f_e=5$ cm

When final image is formed at least distance of distinct vision (d), then

$$M=\frac{f_0}{f_e}\left(1+\frac{f_e}{d}\right)=\frac{100}{5}\left(1+\frac{5}{25}\right)\quad[\because D=25\text{ cm}]$$

$$M=20\times\frac{6}{5}=24$$

14. (b) Secondary rainbow is formed by rays undergoing internal reflection twice inside the drop.

15. (b) $\tan 45°=\frac{h}{60}\Rightarrow h=60m$

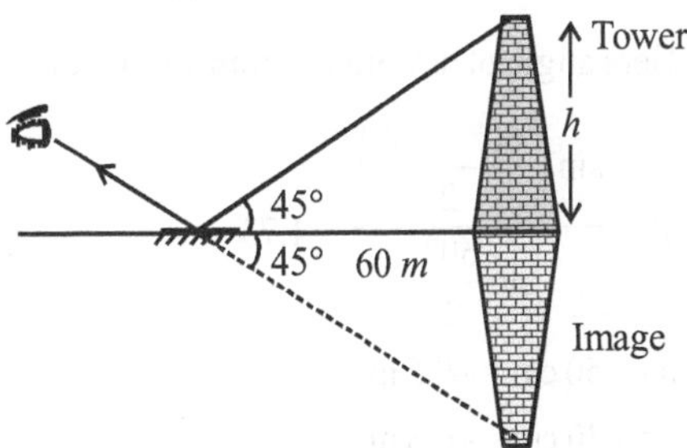

16. (c) Using, $\frac{\mu}{v}-\frac{1}{u}=\frac{\mu-1}{R}$

or $\frac{2}{v}-\frac{1}{\infty}=\frac{2-1}{R}$

$\therefore\ v=2R$

17. (a) ${}_a n_\ell=1.6,\ {}_a n_w=1.33$

f = 20 cm

We have,

$$\frac{1}{f}=({}_a n_\ell-1)\left(\frac{1}{R_1}-\frac{1}{R_2}\right)$$

$$\frac{1}{20}=(1.6-1)\left(\frac{1}{R_1}-\frac{1}{R_2}\right)\quad .....(1)$$

Also, $\frac{1}{f'}=({}_w n_\ell-1)\left(\frac{1}{R_1}-\frac{1}{R_2}\right)$

$$=\left(\frac{{}_a n_\ell}{{}_a n_w}-1\right)\left(\frac{1}{R_1}-\frac{1}{R_2}\right)$$

$$\frac{1}{f'}=\left(\frac{1.6}{1.33}-1\right)\left(\frac{1}{R_1}-\frac{1}{R_2}\right)\quad .....(2)$$

Dividing equation (1) by (2)

$$\Rightarrow\frac{f'}{20}=\frac{0.6}{(1.2-1)}$$

$$f'=\frac{0.6\times 20}{0.2}=60\text{ cm.}$$

Hence it's focal length is three times longer than in air.

18. **(a)** $m = \frac{v_0}{|u_0|}\left(1+\frac{d}{f_e}\right) = \frac{20}{5}\left(1+\frac{20}{10}\right)$

$= 4\left(\frac{10+20}{10}\right) = \frac{4\times 30}{10} = 12$

19. **(a)** Given $i = 60°$

$A = \delta = e$

$\delta = i + e - A \Rightarrow \delta = i \quad (\because e = A)$

$$\mu = \frac{\sin\left(\frac{A+\delta_m}{2}\right)}{\sin\frac{A}{2}}$$

Here angle of deviation is min. ($\because$ i = e)

$$\mu = \frac{\sin\left(\frac{60°+60°}{2}\right)}{\sin\frac{60°}{2}} = 1.73$$

20. **(b)** $u = -50\text{ cm} = -0.5\text{ m}$

$v = -30\text{ cm} = -0.3\text{ m}$

$P = \frac{1}{f} = \frac{1}{v} - \frac{1}{u} = \frac{-1}{0.3} + \frac{1}{0.5} = \frac{-0.2}{0.15} = -1.33\text{ D}$.

21. **(b)** Object distance $u = -40$ cm

Focal length $f = -20$ cm

According to mirror formula

$\frac{1}{u} + \frac{1}{v} = \frac{1}{f}$ or $\frac{1}{v} = \frac{1}{f} - \frac{1}{u}$

or $\frac{1}{v} + \frac{1}{-20} - \frac{1}{(-40)} = \frac{1}{-20} + \frac{1}{40}$

$\frac{1}{v} = \frac{-2+1}{40} = -\frac{1}{40}$ or $v = -40$ cm.

Negative sign shows that image is infront of concave mirror. The image is real.

Magnification, $m = \frac{-v}{u} = -\frac{(-40)}{(-40)} = -1$

The image is of the same size and inverted.

22. **(a)**

Oil μ_1

Water μ_2

As refractive index, $\mu = \frac{\text{Real depth}}{\text{Apparent depth}}$

$\therefore$ Apparent depth of the vessel when viewed from above is

$d_{\text{apparent}} = \frac{x}{2\mu_1} + \frac{x}{2\mu_2} = \frac{x}{2}\left(\frac{1}{\mu_1} + \frac{1}{\mu_2}\right)$

$= \frac{x}{2}\left(\frac{\mu_2+\mu_1}{\mu_1\mu_2}\right) = \frac{x(\mu_1+\mu_2)}{2\mu_1\mu_2}$

23. **(d)** As $r_1 < i_1$ i.e., the incident ray bends towards the normal $\Rightarrow$ medium 2 is denser than medium 1.

Or $r_2 < i_1 \Rightarrow$ medium 3 is denser than medium 1.

Also, $r_2 > r_1 \Rightarrow$ medium 2 is denser than medium 3.

24. **(d)** Here, $v_A = 1.8 \times 10^8\text{ m s}^{-1}$

$v_B = 2.4 \times 10^8\text{ m s}^{-1}$

Light travels slower in denser medium. Hence medium A is a denser medium and medium B is a rarer medium. Here, Light travels from medium A to medium B. Let C be the critical angle between them.

$\therefore \quad \sin C = {}^{A}\mu_B = \frac{1}{{}^{B}\mu_A}$

Refractive index of medium B w.r.t. to medium A is

${}^{A}\mu_B = \frac{\text{Velocity of light in medium } A}{\text{Velocity of light in medium } B} = \frac{v_A}{v_B}$

$\therefore \quad \sin C = \frac{v_A}{v_B} = \frac{1.8\times10^8}{2.4\times10^8} = \frac{3}{4}$ or $C = \sin^{-1}\left(\frac{3}{4}\right)$

25. **(a)** For a thin prism, $D = (\mu - 1)A$

Since $\lambda_b < \lambda_r \Rightarrow \mu_r < \mu_b \Rightarrow D_1 < D_2$

26. **(b)** Difference between apparent and real depth of a pond is due to the refraction of light, not due to the total internal reflection. Other three phenomena are due to the total internal reflection.

27. **(b)** Using the lens formula $\frac{1}{f} = \frac{1}{v} - \frac{1}{u}$

Given $v = d$, for equal size image $|v| = |u| = d$

By sign convention $u = -d$

$\therefore \quad \frac{1}{f} = \frac{1}{d} + \frac{1}{d}$ or $f = \frac{d}{2}$

28. **(a)** Due to covering the reflection from lower part is not there so it makes the image less bright.

29. **(b)** From the fig.

Angle of deviation,

$\delta = i + e - A$

Here, $e = i$

and $e = \frac{3}{4}A$

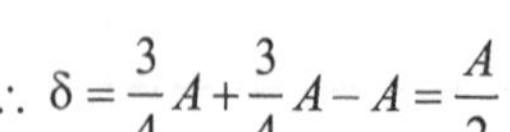

$\therefore \delta = \frac{3}{4}A + \frac{3}{4}A - A = \frac{A}{2}$

For equilateral prism, $A = 60°$

$\therefore \delta = \frac{60°}{2} = 30°$

30. **(a)** Power of lens, P (in dioptre)

$$= \frac{100}{\text{focal length f (in cm)}}$$

$$\therefore \quad f = \frac{100}{10} = 10 \text{ cm}$$

By lens maker's formula, $\frac{1}{f} = (\mu - 1)\left(\frac{1}{R_1} - \frac{1}{R_2}\right)$

For biconvex lens, $R_1 = +R$, and $R_2 = -R$

$$\therefore \quad \frac{1}{f} = (\mu - 1)\left(\frac{1}{R} + \frac{1}{R}\right)$$

$$\frac{1}{f} = (\mu - 1)\left(\frac{2}{R}\right)$$

$$\frac{1}{10} = (\mu - 1)\left(\frac{2}{10}\right)$$

$$(\mu - 1) = \frac{1}{2} \text{ or } \mu = \frac{1}{2} + 1 = \frac{3}{2}$$

31. (d) In the later case microscope will be focussed for O'. So, it is required to be lifted by distance OO'.
OO' = real depth of O – apparent depth of O.

Image → •O' ... O

$$= 3 - \frac{3}{1.5} \quad \left[\mu = \frac{\text{real depth}}{\text{apparent depth}}\right]$$

$$= 3\left[\frac{1.5 - 1}{1.5}\right] = \frac{3 \times .5}{1.5} = 1 \text{ cm}$$

32. (d) The cause of chromatic aberration is that lens focusses different colours at different points.

33. (c) For the prism as the angle of incidence (i) increases, the angle of deviation (δ) first decreases goes to minimum value and then increases.

34. (d) $d_A : d_B = 6 : 4$

$\because$ Time taken $\propto$ thickness

and time taken $\propto \frac{1}{\text{velocity}}$

$\therefore$ Thickness $\propto \frac{1}{\text{velocity}}$

$$\therefore \quad \frac{d_A}{d_B} = \frac{v_B}{v_A}$$

Also, $\mu = \frac{c}{v} \quad \therefore \frac{\mu_A}{\mu_B} = \frac{v_B}{v_A}$

$$\therefore \quad \frac{d_A}{d_B} = \frac{\mu_A}{\mu_B} = \frac{6}{4} = \frac{3}{2} = 1.5$$

$$\therefore \quad {}_B\mu_A = 1.5$$

35. (b) Since $\frac{\text{Apparent depth}}{\text{Real depth}} = \frac{1}{\mu}$

$\Rightarrow$ Apparent depth = d/μ

So mark raised up = Real depth – Apparent depth

$$= d - \frac{d}{\mu} = d\left(1 - \frac{1}{\mu}\right) = \left(\frac{\mu - 1}{\mu}\right)d$$

36. (b) Dispersive power of a prism $\omega = \frac{\mu_V - \mu_R}{\mu_y - 1} = \frac{d\mu}{\mu - 1}$,

where $\mu = \mu_y = \frac{\mu_V + \mu_R}{2}$

37. (a) Considering refraction at the curved surface,

$u = -20, \ \mu_2 = 1$

$\mu_1 = 3/2, \ R = +20$

Applying $\frac{\mu_2}{v} - \frac{\mu_1}{u} = \frac{\mu_2 - \mu_1}{R}$

$$\Rightarrow \frac{1}{v} - \frac{3/2}{-20} = \frac{1 - 3/2}{20} \Rightarrow v = -10$$

i.e., 10 cm below the curved surface or 10 cm above the actual position of flower.

38. (b) When $\theta = 90°$ then $\frac{360}{\theta} = \frac{360}{90} = 4$

is an even number. The number of images formed is given by

$$n = \frac{360}{\theta} - 1 = \frac{360}{90} - 1 = 4 - 1 = 3$$

39. (b) The critical angle of incidence is that angle at which angle of refraction is 90°.

n_1 / n_2 / i_c

$\sin i_c = \frac{n_1}{n_2}$ where $n_2 > n_1$

As, refractive index = $\frac{\text{velocity (air)}}{\text{velocity (medium)}}$

$$\therefore \sin i_c = \frac{2.2 \times 10^8 \text{ m/sec}}{2.4 \times 10^8 \text{ m/sec}} = \frac{11}{12}$$

$$\Rightarrow i_C = \sin^{-1}\left(\frac{11}{12}\right)$$

40. (b) $\frac{P_a}{P_l} = \frac{\left(\frac{\mu_g}{\mu_a} - 1\right)}{\left(\frac{\mu_g}{\mu_l} - 1\right)} = \frac{+5}{-100/100} = -5$

$$-5\left(\frac{\mu_g}{\mu_l} - 1\right) = \frac{\mu_g}{\mu_a} - 1$$

$$\frac{1.5}{\mu_1}-1=\frac{-1}{5}(1.5-1)=-0.1;\ \mu_1=\frac{1.5}{0.9}=\frac{5}{3}$$

41. (d) $\sin C=\frac{1}{\mu}=\frac{1}{\sqrt{2}}$ $\quad\therefore C=\sin^{-1}\left(\frac{1}{\sqrt{2}}\right)=45^\circ$

Now $\frac{\sin C}{\sin r}=\frac{1}{\mu}$ or $\frac{\sin 45^\circ}{\sin r}=\frac{1}{\sqrt{2}}$

$\sin r=1$ or $r=90^\circ$

42. (d)

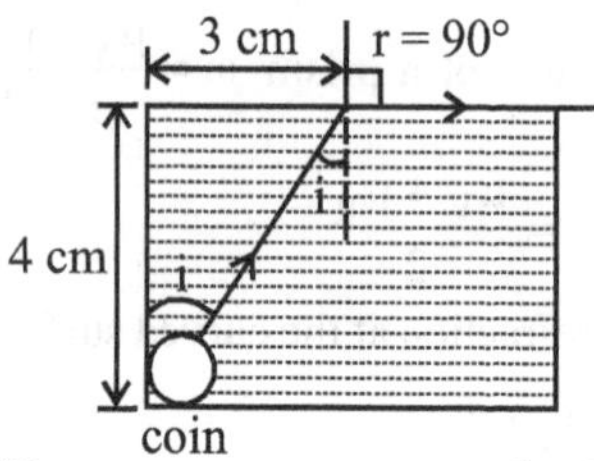

Hypotenuse comes out to be 5 cm.

Since, $\frac{1}{\mu}=\frac{\sin i}{\sin 90^\circ}$

$\mu=\frac{1}{\sin i}=\frac{5}{3}$

Speed, $v=\frac{c}{\mu}=\frac{3\times10^8}{5/3}=1.8\times10^8$ m/s

43. (a)

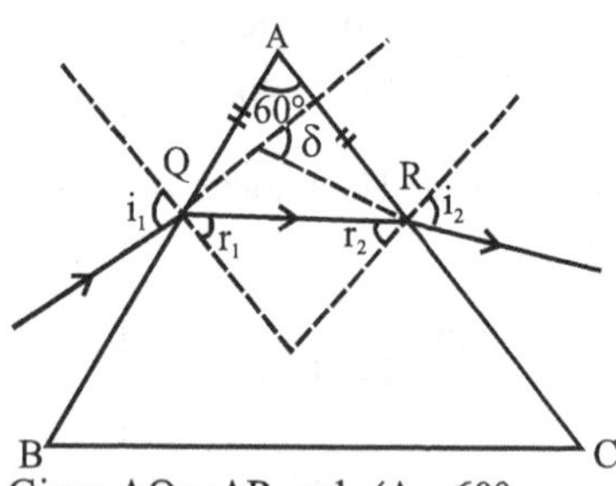

Given AQ = AR and $\angle A=60^\circ$

$\therefore\ \angle AQR=\angle ARQ=60^\circ$

$\therefore\ r_1=r_2=30^\circ$

Applying Snell's law on face AB.

$\sin i_1=\mu\sin r_1$

$\Rightarrow\ \sin i_1=\sqrt{3}\sin 30^\circ=\sqrt{3}\times\frac{1}{2}=\frac{\sqrt{3}}{2}$

$\therefore\ i_1=60^\circ$

Similarly, $i_2=60^\circ$

In a prism, deviation

$\delta=i_1+i_2-A=60^\circ+60^\circ-60^\circ=60^\circ$

44. (a) $\frac{1}{f}=\left(\frac{\mu_g}{\mu_m}-1\right)\left(\frac{1}{R_1}-\frac{1}{R_2}\right)$

If $\mu_g=\mu_m$, then $\frac{1}{f}=(1-1)\left(\frac{1}{R_1}-\frac{1}{R_2}\right)$

$\Rightarrow\ \frac{1}{f}=0$

$$\boxed{f=\frac{1}{0}=\infty}$$

This implies that the liquid must have refractive index equal to glass.

45. (b) Minimum deviation of the prism when it is dipped in water $=\delta_m'=({}_w\mu_g-1)A$

$$=\left(\frac{{}_a\mu_g}{{}_a\mu_\omega}-1\right)A=\left(\frac{\frac{3}{2}}{\frac{4}{3}}-1\right)A=\frac{1}{8}A$$

Minimum deviation of the prism with respect to air

$$=\delta_m=(\mu-1)A=\left(\frac{3}{2}-1\right)A=\frac{1}{2}A$$

$$\frac{\delta_m'}{\delta_m}=\frac{\frac{1}{8}A}{\frac{1}{2}A}=\frac{1}{4}$$

1. (b) For path difference λ, phase

$$\text{difference} = 2\pi \left(Q = \frac{2\pi}{\lambda} x = \frac{2\pi}{\lambda} . \lambda = 2\pi \right)$$

$\Rightarrow I = I_0 + I_0 + 2I_0 \cos 2\pi$

$\Rightarrow I = 4I_0 \qquad (\therefore \cos 2\pi = 1)$

For $x = \frac{\lambda}{4}$, phase difference $= \frac{\pi}{2}$

$$\therefore I' = I_1 + I_2 + 2\sqrt{I_1}\sqrt{I_2} \cos\frac{\pi}{2}$$

If $I_1 = I_2 = I_0$ then $I' = 2I_0 = 2.\frac{I}{4} = \frac{I}{2}$

2. (c) Here Angle of incidence, i = 57
$\tan 57° = 1.54$
$u_{glass} = \tan i$
It means, Here Brewster's law is followed and the reflected ray is completely polarised.
Now, when reflected ray is analysed through a polaroid then intensity of light is given by malus law.
i.e. $I = I_0 \cos^2\theta$
on rotating polaroid 'θ' changes. Due to which intensity first decreases and then increases.

3. (b) Let nth fringe of 2500 Å coincide with (n – 2)th fringe of 3500Å.
∴ $3500 (n - 2) = 2500 \times n$
$1000\, n = 7000, n = 7$
∴ 7th order fringe of 1st source will coincide with 5th order fringe of 2nd source.

4. (a) When incident wavefronts passes through a prism, then lower portion of wavefront (B) is delayed resulting in a tilt. So, time taken by light to reach *A'* from *A* is equal to the time taken to reach *B'* from *B*.

5. (b) ${}^a\mu_g = \tan \theta_P$ where θ_P = polarising angle.
or, ${}^a\mu_g = \tan 60°$

or, $\frac{c}{v_g} = \sqrt{3}$

or, $v_g = \frac{c}{\sqrt{3}} = \frac{3 \times 10^8}{\sqrt{3}} = \sqrt{3} \times 10^8 \text{ms}^{-1}$

6. (c) Angular width $= \frac{\lambda}{d} = 10^{-3}$ (given)

∴ No. of fringes within 0.12° will be

$$n = \frac{0.12 \times 2\pi}{360 \times 10^{-3}} = [2.09]$$

∴ The number of bright spots will be two.

7. (c) Here $A^2 = a_1^2 + a_2^2 + 2a_1a_2 \cos\delta$

∵ $a_1 = a_2 = a$

$$\therefore \quad A^2 = 2a^2(1 + \cos\delta) = 2a^2\left(1 + 2\cos^2\frac{\delta}{2} - 1\right)$$

$\Rightarrow A^2 \propto \cos^2\frac{\delta}{2}$

Now, $I \propto A^2 \quad \therefore I \propto A^2 \propto \cos^2\frac{\delta}{2}$

$\therefore I \propto \cos^2\frac{\delta}{2}$.

8. (b) $\Delta x_1 = (\mu_1 - 1)t = (1.5 - 1)t = 0.5t$

and $\Delta x_2 = (\mu_2 - 1) \times 2t = \left(\frac{4}{3} - 1\right) \times 2t = \frac{2}{3}t$.

As $\Delta x_2 > \Delta x_1$, so shift will be along –ve *y*-axis.

9. (d) Given: D = 2m; d = 1 mm = 1×10^{-3} m
$\lambda = 600$ nm $= 600 \times 10^{-6}$ m
Width of central bright fringe (= 2β)

$$= \frac{2\lambda D}{d} = \frac{2 \times 600 \times 10^{-6} \times 2}{1 \times 10^{-3}} \text{m} = 2.4 \text{mm}$$

10. (d) Conditions for diffraction minima are
Path diff. $\Delta x = n\lambda$ and Phase diff. $\delta\phi = 2n\pi$
Path diff. $= n\lambda = 2\lambda$
Phase diff. $= 2n\pi = 4\pi \quad (\because n = 2)$

11. (c) When the wavelength of light used is comparable with the separation between two points, the image of the object will be a ϕ diffraction pattern whose size will be

$$\theta = \frac{1.22\lambda}{D}$$

where λ = wavelength of light used
D = diameter of the objective
Two objects whose images are closer than this distance, will not be resolved.

12. (d) The waves reflected from the top layer of oil interfere with the wave train reflected from the lower surface of thin oil film producing light and dark coloured pattern.

13. (d) Phase difference, $\phi = \frac{2\pi}{\lambda} \times$ Path difference

$$\phi = \frac{2\pi}{\lambda} \times \frac{\lambda}{6} = \frac{\pi}{3} = 60°$$

As, $I = I_{max} \cos^2\frac{\phi}{2}$

$$I = I_0 \cos^2\frac{60°}{2} = I_0 \times \left(\frac{\sqrt{3}}{2}\right)^2 = \frac{3}{4}I_0 \quad \frac{I}{I_0} = \frac{3}{4}$$

14. (b)

15. (a) We know that for maxima

$$b\sin\theta = (2n+1)\frac{\lambda}{2}$$

or $\sin\theta = \frac{2n+1}{2}\left(\frac{\lambda}{b}\right)$

So on decreasing the slit width, 'b', keeping λ same, sin θ and hence θ increases.

16. (b) If the angular limit of resolution of human eye is R then

$$R = \frac{1.22\lambda}{a} = \frac{1.22\times5\times10^{-7}}{2\times10^{-3}}\text{ rad}$$

$$= \frac{1.22\times5\times10^{-7}}{2\times10^{-3}}\times\frac{180}{\pi}\times60\text{ minute} = 1\text{ minute}$$

17. (c) $\mu = \tan i$

$\Rightarrow i = \tan^{-1}(\mu) = \tan^{-1}(\sqrt{3}) = 60°$.

18. (d) Order of the fringe can be counted on either side of the central maximum. For example, no. 3 is first order bright fringe.

$$\Delta X_C = \lambda,\ \Delta X_A = \frac{\lambda}{2}$$

$$\Delta X_C - \Delta X_A = \frac{\lambda}{2} = 300\text{nm}$$

19. (a) For a circularly polarised light electric field remains constant with time.

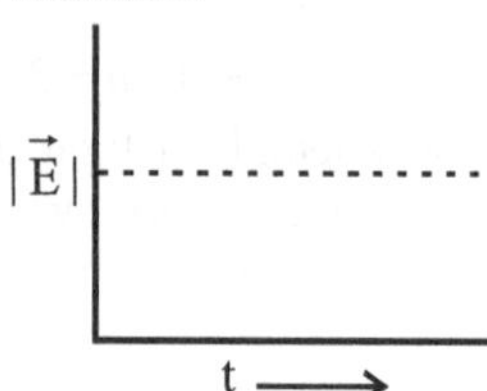

20. (b) $\beta' = \frac{\beta}{\mu} = \frac{0.133}{1.33} = 0.1$ cm

21. (b)

22. (d) Let λ be wavelength of monochromatic light incident on slit S, then angular distance between two consecutive fringes, that is the angular fringe width is

$$\theta = \frac{\lambda}{d}$$

where d is distance between coherent sources.

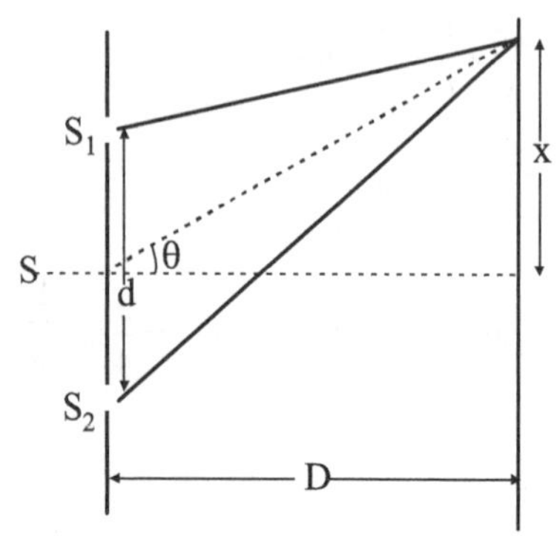

Give, $\frac{\Delta\theta}{\theta} = \frac{10}{100}$

So, from eq. (1),

$$\frac{\Delta\lambda}{\lambda} = \frac{\Delta\theta}{\theta} = \frac{10}{100} = 0.1$$

$\Rightarrow \Delta\lambda = 0.1\lambda = 0.1\times5890\text{Å} = 589\text{Å}$ (increases)

Note : Since, $\theta \propto \lambda$, as θ increases, λ increases.

23. (c) $\sin\theta = \frac{\lambda}{d} = \frac{589\times10^{-9}}{0.589\times10^{-3}} = 10^{-3} = \frac{1}{1000} = 0.001$

24. (c) In Fraunhoffer diffraction, for minimum intensity,

$$\Delta x = m\frac{\lambda}{2}$$

For first minimum, m = 1

$$\therefore\ \Delta x = \frac{\lambda}{2}$$

25. (d) Optical path difference

$\Delta x = (\mu_2 - \mu_1)t$.

26. (b) Separation between slits are (r_1=) 16 cm and (r_2=) 9 cm. Actual distance of separation

$$= \sqrt{r_1 r_2} = \sqrt{16\times9} = 12\text{cm}$$

27. (b) $\phi = \frac{\pi}{3}, a_1 = 4, a_2 = 3$

So, $A = \sqrt{a_1^2 + a_2^2 + 2a_1a_2\cos\phi} \approx 6$

28. (c) $\beta = \frac{D\lambda}{d}$, where D is the distance between the slits & screen and d is the separation between the slits.

$$\beta' = \frac{2D\lambda}{d/2} = \frac{4D\lambda}{d} = 4\beta$$

29. (b) For first minima at P

$AP - BP = \lambda$

$$AP - MP = \frac{\lambda}{2}$$

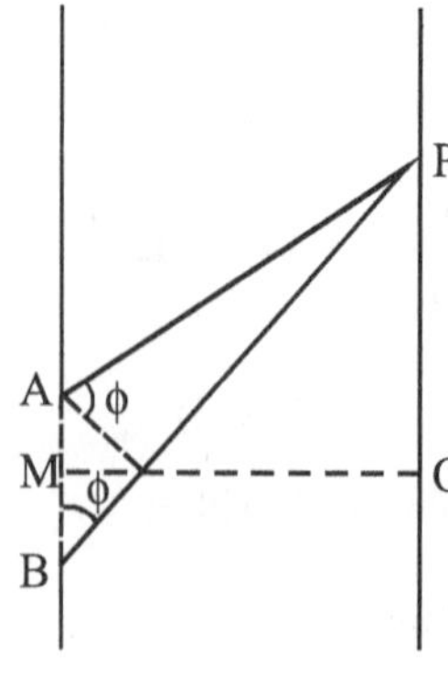

So phase difference, $\phi = \frac{2\pi}{\lambda}\times\frac{\lambda}{2} = \pi$ radian

30. **(a)** Shift $= \frac{D}{d}(\mu - 1)\, t = \frac{\beta}{\lambda}(\mu - 1)\, t = 4\beta$

$$t = \frac{4\lambda}{\mu - 1} = \frac{4 \times 6000 \times 10^{-10}}{1.5 - 1} = 4.8\ \mu m$$

31. **(c)** The fringe width is given by, $\beta = \frac{\lambda D}{d}$

The angular width of fringe is given by

$$\frac{d}{D} = \frac{\lambda}{\beta} = \frac{6 \times 10^{-7}}{0.12 \times 10^{-3}} = 5 \times 10^{-3}\ \text{rad.}$$

32. **(c)** Distance of nth maxima, $x = n\lambda \frac{D}{d} \propto \lambda$

As $\lambda_b < \lambda_g \quad \therefore\ x_{blue} < x_{green}$

33. **(a)** As $\beta = \frac{\lambda D}{d}$ and $\lambda_b < \lambda_y$,

$\therefore$ fringe width β will decrease

34. **(c)** At Brewster's angle, only the reflected light is plane polarised, but transmitted light is partially polarised.

35. **(b)** $\Delta x_{max} = 2\lambda$.

So there are five maxima.

These are for $\Delta x = 0, \pm\lambda, \pm 2\lambda$.

36. **(c)** The nearest white spot will be at P, the central maxima.

$$\therefore\ y = \frac{2d}{3} - \frac{d}{2} = \frac{d}{6}$$

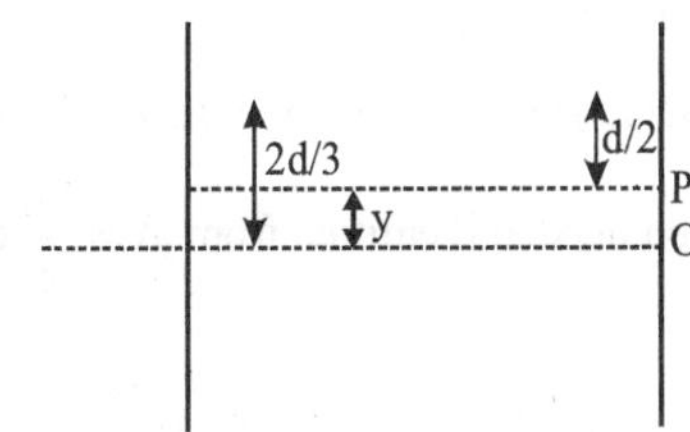

37. **(d)** : Resultant amplitude,

$$A = \sqrt{(A_1)^2 + (A_2)^2 + 2A_1A_2\cos\theta}$$

Here, $A_1 = A_2 = 1$ cm, $\phi = 3\pi$ rad

$$\therefore\ A = \sqrt{1^2 + 1^2 + 2 \times 1 \times 1 \times \cos 3\pi}$$

$$= \sqrt{2 + 2 \times (-1)} = 0$$

38. **(c)**

39. **(c)**

40. **(c)**

41. **(b)** (Light bends upwards)

Plane WF

μ increases

Vel decreases

Refracted WF

42. **(a)** $I = I_0\left(\frac{\sin\phi}{\phi}\right)^2$ and $\phi = \frac{\pi}{\lambda}(b\sin\theta)$

When the slit width is doubled, the amplitude of the wave at the centre of the screen is doubled, so the intensity at the centre is increased by a factor 4.

43. **(a)**

44. **(a)** Where n is equivalent number of fringe by which the centre fringe is shifted due to mica sheet

$$\lambda = \frac{(\mu - 1)t}{n} = \frac{(1.5 - 1)6 \times 10^{-6}}{5}$$

$$= 6 \times 10^{-7}\ m = 6000\ \text{Å}$$

45. **(c)** Suppose intensity of unpolarised light = 100.

$\therefore$ Intensity of polarised light from first nicol prism

$$= \frac{I_0}{2} = \frac{1}{2} \times 100 = 50$$

According to law of Malus,

$$I = I_0\cos^2\theta = 50\ (\cos 60°)^2 = 50 \times \left(\frac{1}{2}\right)^2 = 12.5$$

1. (d) Wavelength of particle $(\lambda_1) = \frac{h}{mv} = \frac{h}{(1\times10^{-3})\times v}$

where v is the velocity of the particle.
Wavelength of electron

$$(\lambda_2) = \frac{h}{(9.1\times10^{-31})\times(3\times10^6)}$$

But $\lambda_1 = \lambda_2$

$$\therefore \frac{h}{(1\times10^{-3})\times v} = \frac{h}{(9.1\times10^{-31})\times(3\times10^6)}$$

$$\Rightarrow v = \frac{9.1\times10^{-31}\times3\times10^6}{10^{-3}}$$

$$= 2.73\times10^{-21}\,ms^{-1}$$

2. (a) For electron De-Broglie wavelength,

$$\lambda_e = \frac{h}{\sqrt{2mE}}$$

For photon $E = pc$

$\Rightarrow$ De-Broglie wavelength, $\lambda_{Ph} = \frac{hc}{E}$

$$\therefore \frac{\lambda_e}{\lambda_{Ph}} = \frac{h}{\sqrt{2mE}}\times\frac{E}{hc} = \left(\frac{E}{2m}\right)^{1/2}\frac{1}{c}$$

3. (d) The electron ejected with maximum speed v_{max} are stopped by electric field $E = 4N/C$ after travelling a distance $d = 1m$

$$\frac{1}{2}mv_{max}^2 = eEd = 4eV$$

The energy of incident photon $= \frac{1240}{200} = 6.2$ eV

From equation of photo electric effect

$$\frac{1}{2}mv_{max}^2 = h\nu - \phi_0$$

$$\therefore \phi_0 = 6.2 - 4 = 2.2\text{ eV}$$

4. (b) $\lambda_{min} = 1$ Å (given)

$$\because \lambda_{min} = \frac{1240}{E}\text{ (eV) (nm)}$$

$$\text{Thus, } E = \frac{1240(\text{eV})(\text{nm})}{0.01(\text{nm})} = 12400\text{ eV}$$

$E = 12.4$ KeV

5. (a) The maximum kinetic energy of an electron accelerated through a potential difference of V volt is $\frac{1}{2}mv^2 = eV$

$\therefore$ maximum velocity $v = \sqrt{\frac{2eV}{m}}$

$$v = \sqrt{\frac{2\times1.6\times10^{-19}\times15000}{9.1\times10^{-31}}}$$

$v = 7.26\times10^7$ m/s

6. (b) Photoelectrons are emitted in A alone. Energy of electron needed if emitted from $A = \frac{h\upsilon}{e}$eV

$$\therefore E_A = \frac{(6.6\times10^{-34})\times(1.8\times10^{14})}{1.6\times10^{-19}} = 0.74\text{ eV}$$

$$E_B = \frac{(6.6\times10^{-34})\times(2.2\times10^{14})}{1.6\times10^{-19}} = 0.91\text{ eV}$$

Incident energy 0.825 eV is greater than E_A (0.74 eV) but less than E_B (0.91 eV).

7. (a) According to relation, $E = \frac{1}{2}mv^2$

$$\sqrt{\frac{2E}{m}} = v$$

$$\lambda = \frac{h}{\sqrt{2mE}}$$

Because $m_1 < m_3 < m_2$

So for same λ, $E_1 > E_3 > E_2$.

8. (a) Emission of electron from a substance under the action of light is photoelectric effect. Light must be at a sufficiently high frequency. It may be visible light, U.V, X-rays. So U.V. cause electron emission.

9. (b) $\lambda_0 = \frac{c}{v_0} = \frac{3\times10^8}{5\times10^{14}} = 6\times10^{-7}\text{m} = 6000\text{Å}$

10. (c) $\lambda = \frac{h}{mv}$, $v = \frac{m_0}{\sqrt{1-\left(\frac{v}{c}\right)^2}}$, $v \to c$, $m \to \infty$

hence, $\lambda \to 0$.

11. (a) Give that, only 25% of 200W converter electrical energy into light of yellow colour

$$\left(\frac{hc}{\lambda}\right)\times N = 200\times\frac{25}{100}$$

Where N is the No. of photons emitted per second, h is planck's constant and c is speed of light.

$$N = \frac{200\times25}{100}\times\frac{\lambda}{hc}$$

$$= \frac{200\times25\times0.6\times10^{-6}}{100\times6.2\times10^{-34}\times3\times10^8} = 1.5\times10^{20}$$

12. (d) For photon $E = h\nu$

$$E = \frac{hc}{\lambda} \Rightarrow \lambda_2 = \frac{hc}{E} \quad \text{...(i)}$$

for proton $E = \frac{1}{2}m_p v_p^2$

$$E = \frac{1}{2}\frac{m_p^2 v_p^2}{m} \Rightarrow p = \sqrt{2mE}$$

From De Broglie Eqn.

$$p = \frac{h}{\lambda_1} \Rightarrow \lambda_1 = \frac{h}{p} = \frac{h}{\sqrt{2mE}} \quad \text{...(ii)}$$

$$\frac{\lambda_2}{\lambda_1} = \frac{hc}{E\times\frac{h}{\sqrt{2mE}}} \propto E^{-1/2}$$

13. (a) $h\nu = W_0 + E_k = 3.5 + 1.2 = 4.7\text{ eV}$

14. (a) $\phi = 6.2\text{ eV} = 6.2\times1.6\times10^{-19}\text{ J}$

$V = 5$ volt

$$\frac{hc}{\lambda} - \phi = eV_0$$

$$\Rightarrow \lambda = \frac{hc}{\phi + eV_0}$$

$$= \frac{6.6\times10^{-34}\times3\times10^8}{1.6\times10^{-19}(6.2+5)} \approx 10^{-7}\text{ m}$$

This range lies in ultra violet range.

15. (c) Applying Einstein's formula for photo-electricity

$$h\nu = \phi + \frac{1}{2}mv^2; \qquad h\nu = \phi + K$$

$$\phi = h\nu - K$$

If we use 2ν frequency then let the kinetic energy becomes K'

So, $h.2\nu = \phi + K'$

$$2h\nu = h\nu - K + K'$$

$$K' = h\nu + K$$

16. (a) $\because \lambda_0 = \frac{hc}{\phi}$

$$\therefore (\lambda_0)_{sodium} = \frac{6.6\times10^{-34}\times3\times10^8}{2\times1.6\times10^{-19}} = 6188\text{ Å}$$

$$\because \lambda_0 \propto \frac{1}{\phi} \Rightarrow \frac{(\lambda_0)_{sodium}}{(\lambda_0)_{copper}} = \frac{(\phi)_{copper}}{(\phi)_{sodium}}$$

$$\Rightarrow (\lambda_0)_{copper} = \frac{2}{4}\times6188 = 3094\text{ Å}$$

To eject photo-electrons from sodium the longest wavelength is 6188 Å and that for copper is 3094 Å.

Hence for light of wavelength 4000 Å, sodium is suitable.

17. (c) $\frac{1}{2}mv^2 = \frac{hc}{\lambda} - \phi \Rightarrow v = \sqrt{\frac{2(hc - \lambda\phi)}{\lambda m}}$

18. (d) de-Broglie wavelength,

$$\lambda = \frac{h}{p} = \frac{h}{\sqrt{2.m.(K.E)}}$$

$$\therefore \lambda \propto \frac{1}{\sqrt{K.E}}$$

If K.E is doubled, wavelength becomes $\frac{\lambda}{\sqrt{2}}$

19. (b) $\frac{1}{2}mv_1^2 = 2W_0 - W_0 = W_0$ and

$$\frac{1}{2}mv_2^2 = 10W_0 - W_0 = 9W_0$$

$$\therefore \frac{v_1}{v_2} = \sqrt{\frac{W_0}{9W_0}} = \frac{1}{3}$$

20. (a) The work function has no effect on photoelectric current so long as $h\nu > W_0$. The photoelectric current is proportional to the intensity of incident light. Since there is no change in the intensity of light, hence $I_1 = I_2$.

21. (c) $n \to 2-1$

$E = 10.2\text{ eV}$

$kE = E - \phi$

$Q = 10.20 - 3.57$

$h\upsilon_0 = 6.63\text{ eV}$

$$\upsilon_0 = \frac{6.63 \times 1.6 \times 10^{-19}}{6.67 \times 10^{-34}} = 1.6 \times 10^{15}\text{ Hz}$$

22. **(d)** $h\nu = W + \frac{1}{2}mv^2$ or $\frac{hc}{\lambda} = W + \frac{1}{2}mv^2$

Here $\lambda = 3000$ Å $= 3000 \times 10^{-10}$ m

and $W = 1$ eV $= 1.6 \times 10^{-19}$ joule

$$\therefore \frac{(6.6\times10^{-34})(3\times10^{8})}{3000\times10^{-10}}$$

$$= (1.6\times10^{-19}) + \frac{1}{2}\times(9.1\times10^{-31})v^2$$

Solving we get, $v \cong 10^6$ m/s

23. **(b)** According to Einstein's photoelectric equation, $h\nu = \phi_0 + K_{max}$

We have

$h\nu = \phi_0 + 0.5$ ...(i)

and $1.2h\nu = \phi_0 + 0.8$ ...(ii)

Therefore, from above two equations $\phi_0 = 1.0$ eV.

24. **(c)** $\lambda_{max.} = \frac{2d\sin\theta}{n_{min.}} = \frac{2\times15\times\sin 90°}{1} = 30$Å

25. **(d)** $W_0 = h\nu_1 - eV_1$

$= h\,\nu_2 - e\,V_2$

$eV_2 = h(\nu_2 - \nu_1) + eV_1$

$$V_2 = \frac{h(n_2 - n_1)}{e} + V_1$$

26. **(b)** $KE_{max} = h\nu - \phi$

$1eV = h\nu - 1.9eV \Rightarrow h\nu = 2.9$ eV

Now threshold wavelength (maximum wavelength), $\lambda_0 = \frac{hc}{E}$

$$\Rightarrow \lambda_0 = \frac{6.6\times10^{-34}\times3\times10^{8}}{1.9\times1.6\times10^{-19}} = 6513 \text{ Å}$$

And threshold frequency

$$\nu_0 = \frac{c}{\lambda_0} = \frac{3\times10^8}{6513\times10^{-10}} = 4.6\times10^{14}\text{Hz}$$

27. **(b)** $E = W_0 + K_{max}$ ...(i)

$\Rightarrow hf = W_A + K_A$ ...(ii)

and $2hf = W_B + K_B = 2W_A + K_B \left(\because \frac{W_A}{W_B} = \frac{1}{2}\right)$

Dividing equation (i) by (ii)

$$\frac{1}{2} = \frac{W_A + K_A}{2W_A + K_B} \Rightarrow \frac{K_A}{K_B} = \frac{1}{2}$$

28. **(d)** $h\upsilon - h\upsilon_0 = E_K$, according to photoelectric equation, when $\upsilon = \upsilon_0$, $E_K = 0$.

Graph (d) represents $E_K - \upsilon$ relationship.

29. **(d)** $K_{max} = \frac{hc}{\lambda} - W = \frac{hc}{\lambda} - 5.01$

$$= \frac{12375}{\lambda(\text{in Å})} - 5.01$$

$$= \frac{12375}{2000} - 5.01 = 6.1875 - 5.01 = 1.17775$$

$\simeq 1.2$ V

30. **(b)**

31. **(b)** $\lambda \propto \frac{1}{\sqrt{V}}$

$$\Rightarrow \frac{\lambda_1}{\lambda_2} = \sqrt{\frac{V_2}{V_1}} = \sqrt{\frac{100\text{keV}}{25\text{keV}}} = 2$$

$$\Rightarrow \lambda_2 = \frac{\lambda_1}{2}$$

32. **(a)** In the Davisson and Germer experiment, the velocity of electrons emitted from the electron gun can be increased by increasing the potential difference between the anode and filament.

33. **(b)** According to Einsten's photoelectric effect, the K.E. of the radiated electrons

$K.E_{max} = E - W$

$\frac{1}{2}mv_1^2 = (1 - 0.5)$ eV $= 0.5$ eV

$\frac{1}{2}mv_2^2 = (2.5 - 0.5)$ eV $= 2$ eV

$$\frac{v_1}{v_2} = \sqrt{\frac{0.5}{2}} = \frac{1}{\sqrt{4}} = 1/2$$

34. **(b)** By using $h\nu - h\nu_0 = K_{max}$

$\Rightarrow h(\nu_1 - \nu_0) = K_1$ ..... (i)

And $h(\nu_2 - \nu_0) = K_2$ ..... (ii)

$\Rightarrow \frac{\nu_1 - \nu_0}{\nu_2 - \nu_0} = \frac{K_1}{K_2} = \frac{1}{K}$, Hence $\nu_0 = \frac{k\nu_1 - \nu_2}{K-1}$.

35. **(b)** Cathode rays get deflected in the electric field.

36. **(c)** As we know

$$\lambda \propto \frac{1}{\sqrt{V}}$$

$$\therefore \frac{1}{\sqrt{100}} : \frac{1}{\sqrt{200}} : \frac{1}{\sqrt{300}} = 1 : \frac{1}{\sqrt{2}} : \frac{1}{\sqrt{3}}$$

37. **(d)** Number of emitted electrons N_E

$\propto$ Intensity

$$\propto \frac{1}{(\text{Distance})^2}$$

Therefore, as distance is doubled, N_E decreases by (1/4) times.

38. (d) Photoelectrons are emitted if the frequency of incident light is greater than the threshold frequency.

39. (a) K.E. = $h\nu - h\nu_{th} = eV_0$ (V_0 = cut off voltage)

$$\Rightarrow V_0 = \frac{h}{e}(8.2\times10^{14} - 3.3\times10^{14})$$

$$= \frac{6.6\times10^{-34}\times4.9\times10^{14}}{1.6\times10^{-19}} \approx 2V.$$

40. (d) $\frac{hc}{\lambda} - \phi = eV_0$

$$v_0 = \frac{hc}{e\lambda} - \frac{\phi}{e}$$

For metal A For metal B

$$\frac{\phi_A}{hc} = \frac{1}{\lambda} \qquad \frac{\phi_B}{hc} = \frac{1}{\lambda}$$

As the value of $\frac{1}{\lambda}$ (increasing and decreasing) is not specified hence we cannot say that which metal has comparatively greater or lesser work function (ϕ).

41. (c)

42. (d) Potential difference = 100 V

K.E. acquired by electron = e (100)

$$\frac{1}{2}mv^2 = e(100) \Rightarrow v = \sqrt{\frac{2e(100)}{m}}$$

According to de Broglie's concept

$$\lambda = \frac{h}{mv}$$

$$\Rightarrow \lambda = \frac{h}{m\sqrt{\frac{2e(100)}{m}}}$$

$$= \frac{h}{\sqrt{2me(100)}} = 1.2\times10^{-10} = 1.2Å$$

43. (d) Since $p = nh\nu$

$$\Rightarrow n = \frac{p}{h\nu} = \frac{2\times10^{-3}}{6.6\times10^{-34}\times6\times10^{14}} = 5\times10^{15}$$

44. (a) From formula

$$\lambda = \frac{h}{\sqrt{2mKT}}$$

$$= \frac{6.63\times10^{-34}}{\sqrt{2\times1.67\times10^{-27}\times1.38\times10^{-23}T}} m$$

[By placing value of h, m and k)

$$= \frac{30.8}{\sqrt{T}} Å$$

45. (c) The photoelectric equation

$K_{max} = h\nu - \phi_0$

Explains that the intensity of incident radiation will increase photocurrent only beyond the threshold frequency.

1. (b) P.E. $=\frac{-Ze^2}{4\pi\varepsilon_0 r}$. Negative sign indicates that revolving electron is bound to the positive nucleus.
So, it decreases with increase in radii of orbit.

2. (b) $E = Rhc\left[\frac{1}{n_1^2}-\frac{1}{n_2^2}\right]$
E will be maximum for the transition for which $\left[\frac{1}{n_1^2}-\frac{1}{n_2^2}\right]$ is maximum. Here n_2 is the higher energy level.

Clearly, $\left[\frac{1}{n_1^2}-\frac{1}{n_2^2}\right]$ is maximum for the third transition, i.e. $2 \to 1$. I transition represents the absorption of energy.

3. (a) Number of emission spectral lines
$$N=\frac{n(n-1)}{2}$$
$$\therefore 3=\frac{n_1(n_1-1)}{2}, \text{ in first case.}$$
Or $n_1^2-n_1-6=0$ or $(n_1-3)(n_1+2)=0$
Take positive root.
$\therefore n_1=3$
Again, $6=\frac{n_2(n_2-1)}{2}$, in second case.
Or $n_2^2-n_2-12=0$ or $(n_2-4)(n_2+3)=0$.
Take positive root, or $n_2=4$
Now velocity of electron $\upsilon=\frac{2\pi KZe^2}{nh}$
$$\therefore \frac{\upsilon_1}{\upsilon_2}=\frac{n_2}{n_1}=\frac{4}{3}.$$

4. (c) $N \propto \frac{1}{\sin^4\theta/2}$; $\frac{N_2}{N_1}=\frac{\sin^4(\theta_1/2)}{\sin^4(\theta_2/2)}$
or $\frac{N_2}{5\times10^6}=\frac{\sin^4(60°/2)}{\sin^4(120°/2)}$
or $\frac{N_2}{5\times10^6}=\frac{\sin^4 30°}{\sin^4 60°}$
or $N_2=5\times10^6\times\left(\frac{1}{2}\right)^4\left(\frac{2}{\sqrt{3}}\right)^4=\frac{5}{9}\times10^6$

5. (c) Magnetic moment of the hydrogen atom, when the electron is in n^{th} excited state, i.e., $n'=(n+1)$
As magnetic moment $M_n=I_nA=i_n(\pi r_n^2)$
$$i_n=eV_n=\frac{mz^2e^5}{4\varepsilon_0^2n^3h^3}$$
$$r_n=\frac{n^2h^2}{4\pi^2kzme^2}\left(k=\frac{1}{4\pi\epsilon_0}\right)$$
Solving we get magnetic moment of the hydrogen atom for n^{th} excited state
$$M_{n'}=\left(\frac{e}{2m}\right)\frac{nh}{2\pi}$$

6. (a) $E=\frac{hc}{\lambda}\Rightarrow\lambda=\frac{hc}{E}=\frac{6.62\times10^{-34}\times3\times10^8}{12.5\times1.6\times10^{-19}}$
$=993$ A°
$$\frac{1}{\lambda}=R\left(\frac{1}{n_1^2}-\frac{1}{n_2^2}\right)$$
(where Rydberg constant, $R=1.097\times10^7$)
or, $\frac{1}{993\times10^{-10}}=1.097\times10^7\left(\frac{1}{1^2}-\frac{1}{n_2^2}\right)$
Solving we get $n_2=3$
Spectral lines
Total number of spectral lines = 3
Two lines in Lyman series for $n_1=1, n_2=2$ and $n_1=1, n_2=3$ and one in Balmer series for $n_1=2, n_2=3$

n = 3
n = 2 — Balmer
n = 1 — Lyman, Lyman

7. (b) $l=\frac{nh}{2\pi}, |E|\propto Z^2/n^2; n=3$
$\Rightarrow l_H=l_{Li}$ and $|E_H|<|E_{Li}|$

8. (b) $r\propto n^2$
$$\therefore \frac{\text{radius of final state}}{\text{radius of initial state}}=n^2$$
$$\frac{21.2\times10^{-11}}{5.3\times10^{-11}}=n^2$$
$\therefore$ $n^2=4$ or $n=2$

9. (a) $R=\frac{R_0n^2}{Z}$
Radius in ground state $=\frac{R_0}{Z}$
Radius in first excited state $=\frac{R_0\times4}{Z}$ $(\because n=2)$
Hence, radius of first excited state is four times the radius in ground state.

10. (a) Speed of electron in nth orbit

$$V_n = \frac{2\pi KZe^2}{nh}$$

$$V = (2.19 \times 10^6 \text{ m/s}) \frac{Z}{n}$$

$$V = (2.19 \times 10^6) \frac{2}{3} \text{ (Z=2 \& n=3)}$$

$$V = 1.46 \times 10^6 \text{ m/s}$$

11. (b) $KE_{max} = 10\text{eV}$

$\phi = 2.75 \text{ eV}$

Total incident energy

$E = \phi + KE_{max} = 12.75 \text{ eV}$

$\therefore$ Energy is released when electron jumps from the excited state n to the ground state.

$\because E_4 - E_1 = \{-0.85 - (-13.6) \text{ ev}\}$
$= 12.75\text{eV}$

$\therefore$ value of n = 4

12. (a) As the electron comes nearer to the nucleus the potential energy decreases

$$\left(\because \frac{-k.Ze^2}{r} = \text{P.E. and } r \text{ decreases}\right)$$

The K.E. will increase $\left[\because \text{K.E.} = \frac{1}{2}|\text{P.E.}| = \frac{1}{2}\frac{kZe^2}{r}\right]$

The total energy decreases $\left[\text{T.E.} = -\frac{1}{2}\frac{kZe^2}{r}\right]$

13. (d) When one e^- is removed from neutral helium atom, it becomes a one e^- species.

For one e^- species we know

$$E_n = \frac{-13.6Z^2}{n^2} \text{ eV/atom}$$

For helium ion, $Z = 2$ and for first orbit $n = 1$.

$$\therefore \quad E_1 = \frac{-13.6}{(1)^2} \times 2^2 = -54.4 \text{ eV}$$

$\therefore$ Energy required to remove this $e^- = +54.4$ eV

$\therefore$ Total energy required = 54.4 + 24.6 = 79 eV

14. (b) For 2nd line of Balmer series in hydrogen spectrum

$$\frac{1}{\lambda} = R(1)\left(\frac{1}{2^2} - \frac{1}{4^2}\right) = \frac{3}{16}R$$

For Li^{2+} $\left[\frac{1}{\lambda} = R \times 9\left(\frac{1}{x^2} - \frac{1}{12^2}\right) = \frac{3R}{16}\right]$

which is satisfied by $n = 12 \rightarrow n = 6$.

15. (d) For an atom following Bohr's model, the radius is given by

$r_m = \frac{r_0 m^2}{Z}$ where r_0 = Bohr's radius and m = orbit number.

For *Fm*, $m = 5$ (Fifth orbit in which the outermost electron is present)

$$\therefore \quad r_m = \frac{r_0 5^2}{100} = nr_0 \text{ (given)} \Rightarrow n = \frac{1}{4}$$

16. (a) Energy of electron in n^{th} orbit is

$$E_n = -(\text{Rch})\frac{Z^2}{n^2} = -54.4 \text{ eV}$$

For He^+ is ground state

$$E_1 = -(\text{Rch})\frac{(2)^2}{(1)^2} = -54.4 \Rightarrow \text{Rch} = 13.6$$

$\therefore$ For Li^{++} in first excited state (n = 2)

$$E' = -13.6 \times \frac{(3)^2}{(2)^2} = -30.6 \text{ eV}$$

17. (a) Angular momentum = mrv = J

$$\therefore \quad v = \frac{J}{mr}$$

$$\text{K. E. of electron} = \frac{1}{2}mv^2 = \frac{1}{2}m\left(\frac{J}{mr}\right)^2$$

$$= \frac{J^2}{2mr^2}$$

18. (b) When $F = \frac{k}{r}$ = centripetal force, then $\frac{k}{r} = \frac{mv^2}{r}$

$\Rightarrow mv^2$ = constant $\Rightarrow$ kinetic energy is constant

$\Rightarrow T$ is independent of n.

19. (b) $\frac{1}{\lambda'} = \frac{1}{\lambda}\sqrt{\frac{c-v}{c+v}}$

Here, $\lambda' = 706$ nm, $\lambda = 656$ nm

$$\therefore \quad \frac{c-v}{c+v} = \left(\frac{\lambda}{\lambda'}\right)^2 = \left(\frac{656}{706}\right)^2 = 0.86$$

$$\Rightarrow \quad \frac{v}{c} = \frac{0.14}{1.86}$$

$$\Rightarrow \quad v = 0.075 \times 3 \times 10^8 = 2.25 \times 10^7 \text{m/s}$$

20. (d) $\because \quad B = \frac{\mu_0 I}{2r}$ and $I = \frac{e}{T}$

$$B = \frac{\mu_0 e}{2rT} \quad [r \propto n^2, T \propto n^5]; \quad B \propto \frac{1}{n^5}$$

21. (a) 53 electrons in iodine atom are distributed as 2, 8, 18, 18, 7

$\therefore n = 5$

$$r_n = (0.53 \times 10^{-10})\frac{n^2}{Z}$$

$$= \frac{0.53 \times 10^{-10} \times 5^2}{53} = 2.5 \times 10^{-11} \text{m}$$

22. (a) At closest distance of approach, the kinetic energy of the particle will convert completely into electrostatic potential energy.

Kinetic energy K.E. $= \frac{1}{2}mv^2$

Potential energy P.E. $= \frac{KQq}{r}$

$\frac{1}{2}mv^2 = \frac{KQq}{r} \Rightarrow r \propto \frac{1}{m}$

23. (c) $\frac{n(n-1)}{2} = 6$

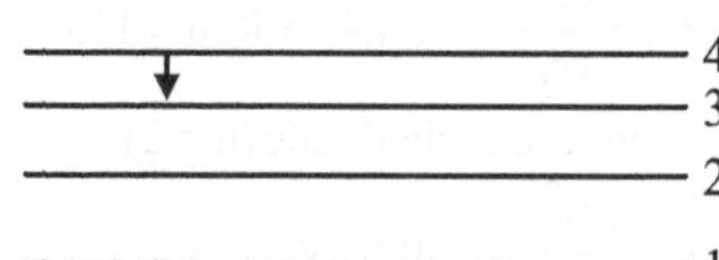

$n^2 - n - 12 = 0$

$(n-4)(n+3) = 0$ or $n = 4$

24. (c) The wavelength of spectrum is given by

$\frac{1}{\lambda} = Rz^2\left(\frac{1}{n_1^2} - \frac{1}{n_2^2}\right)$ where $R = \frac{1.097 \times 10^7}{1 + \frac{m}{M}}$

where m = mass of electron

M = mass of nucleus.

For different M, R is different and therefore λ is different.

25. (a) $\because T \propto n^3$

$Tn_1 = 8\,Tn_2$ (given)

Hence, $n_1 = 2n_2$

26. (d) $\Delta E = h\nu$

$\nu = \frac{\Delta E}{h} = k\left[\frac{1}{(n-1)^2} - \frac{1}{n^2}\right] = \frac{k(2n-1)}{n^2(n-1)^2}$

$\approx \frac{2k}{n^3}$ or $\nu \propto \frac{1}{n^3}$

27. (c) A spectrum is observed, when light coming directly from a source is examined with a spectroscope. Therefore spectrum obtained from a sodium vapour lamp is emission spectrum.

28. (a) Energy of ground state 13.6 eV

Energy of first excited state

$= -\frac{13.6}{4} = -3.4$ eV

Energy of second excited state

$= -\frac{13.6}{9} = -1.5$ eV

Difference between ground state and 2nd excited state $= 13.6 - 1.5 = 12.1$ eV

So, electron can be excited upto 3rd orbit

No. of possible transition

$1 \to 2, 1 \to 3, 2 \to 3$

So, three lines are possible.

29. (b) In Bohr's model, angular momentum is quantised i.e

$\ell = n\left(\frac{h}{2\pi}\right)$

30. (b) The smallest frequency and largest wavelength in ultraviolet region will be for transition of electron from orbit 2 to orbit 1.

$\therefore \frac{1}{\lambda} = R\left(\frac{1}{n_1^2} - \frac{1}{n_2^2}\right)$

$\Rightarrow \frac{1}{122 \times 10^{-9} m} = R\left[\frac{1}{1^2} - \frac{1}{2^2}\right] = R\left[1 - \frac{1}{4}\right] = \frac{3R}{4}$

$\Rightarrow R = \frac{4}{3 \times 122 \times 10^{-9}} m^{-1}$

The highest frequency and smallest wavelength for infrared region will be for transition of electron from ∞ to 3rd orbit.

$\therefore \frac{1}{\lambda} = R\left(\frac{1}{n_1^2} - \frac{1}{n_2^2}\right) \Rightarrow \frac{1}{\lambda} = \frac{4}{3 \times 122 \times 10^{-9}}\left(\frac{1}{3^2} - \frac{1}{\infty}\right)$

$\therefore \lambda = \frac{3 \times 122 \times 9 \times 10^{-9}}{4} = 823.5$nm

31. (c) $\frac{1}{\lambda} = R\left(\frac{1}{n_1^2} - \frac{1}{n_2^2}\right)$ where R = Rydberg constant

$\frac{1}{\lambda_{32}} = \left(\frac{1}{4} - \frac{1}{9}\right) = \frac{5}{36}$

$\Rightarrow \lambda_{32} = \frac{36}{5}$

Similarly solving for λ_{31} and λ_{21}

$\lambda_{31} = \frac{9}{8}$ and $\lambda_{21} = \frac{4}{3}$

$\therefore \frac{\lambda_{32}}{\lambda_{31}} = 6.4$ and $\frac{\lambda_{21}}{\lambda_{31}} \simeq 1.2$

32. (d) $b = \frac{Ze^2 \cot\left(\frac{\theta}{2}\right)}{4\pi \epsilon_0 k_i} = 0 \Rightarrow \cot\left(\frac{\theta}{2}\right) = 0$

$\Rightarrow \frac{\theta}{2} = 90°$ or $\theta = 180°$

33. (a) Speed of electron in nth orbit

$V_n = \frac{2\pi KZe^2}{nh}$

$V = (2.19 \times 10^6 \text{m/s}) \frac{Z}{n}$

$V = (2.19 \times 10^6) \frac{2}{3}$ (Z = 2 & n = 3)

$V = 1.46 \times 10^6$ m/s

34. (d) $E = E_4 - E_3$

$= -\frac{13.6}{4^2} - \left(-\frac{13.6}{3^2}\right) = -0.85 + 1.51$

$= 0.66$ eV

35. (d) $\because$ The frequency of the transition $\nu \propto \frac{1}{n^2}$, when $n = 1, 2, 3$.

36. (c) According to Bohr's theory, the wave number of the last line of the Balmer series in hydrogen spectrum,

For hydrogen atom z = 1

$$\frac{1}{\lambda} = RZ^2\left(\frac{1}{n_2^2} - \frac{1}{n_1^2}\right)$$

$$= 10^7 \times 1^2\left(\frac{1}{2^2} - \frac{1}{\infty^2}\right)$$

$\Rightarrow$ wave number $\frac{1}{\lambda} = 0.25 \times 10^7 \text{ m}^{-1}$

37. (a) Velocity of electron in n^{th} orbit of hydrogen atom is given by :

$$V_n = \frac{2\pi KZe^2}{nh}$$

Substituting the values we get,

$$V_n = \frac{2.2\times10^6}{n}\text{m/s} \quad \text{or} \quad V_n \propto \frac{1}{n}$$

As principal quantum number increases, velocity decreases.

38. (c) $\frac{1}{\lambda} = R\left[\frac{1}{n_1^2} - \frac{1}{n_2^2}\right]$

$$\Rightarrow \frac{1}{970.6\times10^{-10}} = 1.097\times10^7\left[\frac{1}{1^2} - \frac{1}{n_2^2}\right] \Rightarrow n_2 = 4$$

$\therefore$ Number of emission line $N = \frac{n(n-1)}{2} = \frac{4\times3}{2} = 6$

39. (a) We have $E_n = \frac{-2\pi^2 mK^2Z^2e^4}{n^2h^2}$. For helium $Z=2$. Hence requisite answer is $4E_n$

40. (c) As α-particles are doubly ionised helium He^{++} i.e. Positively charged and nucleus is also positively charged and we know that like charges repel each other.

41. (b) $\bar{v} = R\left(\frac{1}{n_1^2} - \frac{1}{n_2^2}\right)$, where $n_1 = 2, n_2 = 4$

$$\bar{v} = R\left(\frac{1}{4} - \frac{1}{16}\right)$$

$$\frac{1}{\lambda} = R\left(\frac{12}{4\times16}\right) \quad \Rightarrow \quad \lambda = \frac{16}{3R}$$

42. (a) The kinetic energy of the projectile is given by

$$\frac{1}{2}mv^2 = \frac{Ze(2e)}{4\pi\varepsilon_0 r_0} = \frac{Z_1 Z_2}{4\pi\varepsilon_0 r_0}$$

Thus energy of the projectile is directly proportional to Z_1, Z_2

43. (a) We know that $\frac{1}{\lambda} = RZ^2\left[\frac{1}{n_1^2} - \frac{1}{n_2^2}\right]$

The wave length of first spectral line in the Balmer series of hydrogen atom is 6561Å. Here $n_2 = 3$ and $n_1 = 2$

$$\therefore \frac{1}{6561} = R(1)^2\left(\frac{1}{4} - \frac{1}{9}\right) = \frac{5R}{36} \quad \text{...(i)}$$

For the second spectral line in the Balmer series of singly ionised helium ion $n_2 = 4$ and $n_1 = 2$; $Z = 2$

$$\therefore \frac{1}{\lambda} = R(2)^2\left[\frac{1}{4} - \frac{1}{16}\right] = \frac{3R}{4} \quad \text{...(ii)}$$

Dividing equation (i) and equation (ii) we get

$$\frac{\lambda}{6561} = \frac{5R}{36}\times\frac{4}{3R} = \frac{5}{27}$$

$\therefore \lambda = 1215$ Å

44. (a) For Lyman series

$$\upsilon = R_C\left[\frac{1}{1^2} - \frac{1}{n^2}\right]$$

where n = 2, 3, 4, ...

For the series limit of Lyman series, $n = \infty$

$$\therefore \upsilon_1 = R_C\left[\frac{1}{1^2} - \frac{1}{\infty^2}\right] = R_C \quad \text{...(i)}$$

For the first line of Lyman series, $n = 2$

$$\therefore \upsilon_2 = R_C\left[\frac{1}{1^2} - \frac{1}{2^2}\right] = \frac{3}{4}R_C \quad \text{...(ii)}$$

For Balmer series

$$\upsilon = R_C\left[\frac{1}{2^2} - \frac{1}{n^2}\right]$$

where n = 3, 4, 5 .....

For the series limit of Balmer series, $n = \infty$

$$\therefore \upsilon_3 = R_C\left[\frac{1}{2^2} - \frac{1}{\infty^2}\right] = \frac{R_C}{4} \quad \text{...(iii)}$$

From equation (i), (ii) and (iii), we get

$\upsilon_1 = \upsilon_2 + \upsilon_3$ or $\upsilon_1 - \upsilon_2 = \upsilon_3$

45. (d) As $r \propto \frac{1}{m}$ $\therefore r_0' = \frac{1}{2}r_0$

As $E \propto m$ $\therefore E_0' = 2(-13.6) = -27.2$ eV

DAILY PRACTICE PROBLEMS

PHYSICS SOLUTIONS

DPP/CP27

1. **(b)** B.E. $= 0.042 \times 931 \simeq 42$ MeV

Number of nucleons in ${}^{7}_{3}Li$ is 7.

$\therefore$ B.E./ nucleon $= \frac{42}{7} = 6\text{ MeV} \simeq 5.6\text{ MeV}$

2. **(d)** ${}^{A}_{Z}X \longrightarrow {}^{A}_{Z+1}Y : \beta,\ {}^{A}_{Z+1}Y \longrightarrow {}^{A-4}_{Z-1}B^{*} : \alpha$

${}^{A-4}_{Z-1}B^{*} \longrightarrow {}^{A-4}_{Z-1}B : \gamma$

(β, α, γ) ($\because \beta = {}^{0}_{-1}e$, $\alpha = {}^{4}_{2}He$, mass number and charge number of a nucleus remains unchanged during γ decay)

3. **(c)** The radius of the nuclears is directly proportional to cube root of atomic number i.e. $R \propto A^{1/3}$

$\Rightarrow$ $R = R_0\ A^{1/3}$, where R_0 is a constant of proportionality

$$\frac{R_2}{R_1} = \left(\frac{A_2}{A_1}\right)^{1/3} \left(\frac{64}{27}\right)^{1/3} = \frac{4}{3}$$

where R_1 = the radius of ${}^{27}Al$, and A_1 = Atomic mass number of Al

R_2 = the radius of ${}^{64}Cu$ and A_2 = Atomic mass number of C4

$$R_2 = 3.6 \times \frac{4}{3} = 4.8\text{m}$$

4. **(c)** Nuclear forces are short range attractive forces which balance the repulsive forces between the protons inside the nucleus.

5. **(a)** $\lambda = \frac{1}{t}\log_e \frac{A_o}{A} = \frac{1}{5}\log_e \frac{5000}{1250}$

$$= \frac{2}{5}\log_e 2 = 0.4\log_e 2$$

6. **(d)** Radioactivity at T_1, $R_1 = \lambda N_1$

Radioactivity at T_2, $R_2 = \lambda N_2$

$\therefore$ Number of atoms decayed in time

$(T_1 - T_2) = (N_1 - N_2)$

$$= \frac{(R_1 - R_2)}{\lambda} = \frac{(R_1 - R_2)T}{0.693} \propto (R_1 - R_2)T$$

7. **(c)** ${}^{2}_{1}H$ and ${}^{3}_{1}H$ requires a and b amount of energies for their nucleons to be separated.

${}^{4}_{2}He$ releases c amount of energy in its formation i.e., in assembling the nucleons as nucleus.

Hence, Energy released $= c - (a + b) = c - a - b$

8. **(a)** Mass defect $= ZM_p + (A - Z)M_n - M(A,Z)$

or, $\frac{B.E.}{c^2} = ZM_p + (A-Z)\,M_n - M(A,Z)$

$\therefore M(A, Z) = ZM_p + (A-Z)M_n - \frac{B.E.}{c^2}$

9. **(d)**

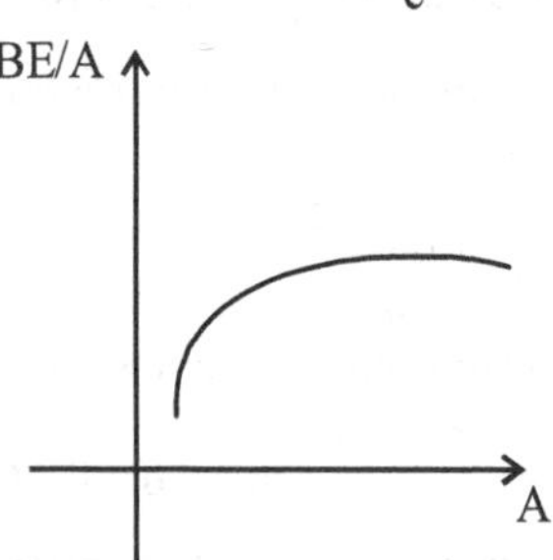

From the graph of BE/A versus mass number A it is clear that, BE/A first increases and then decreases with increase in mass number.

10. **(c)** The range of energy of β-particles is from zero to some maximum value.

11. **(a)** ${}_{72}A^{180} \xrightarrow{\alpha} {}_{70}A_1^{176} \xrightarrow{\beta} {}_{71}A_2^{176}$

$$\xrightarrow{\alpha} {}_{69}A_3^{172} \xrightarrow{\gamma} {}_{69}A_4^{172}$$

12. **(c)** $\frac{dN}{dt} = KN$

$9750 = KN_0$ .... (1)

$975 = KN$ ..... (2)

Dividing (1) by (2)

$$\frac{N}{N_0} = \frac{1}{10}$$

$$K = \frac{2.303}{t}\log\frac{N_0}{N} = \frac{2.303}{5}\log 10$$

$= 0.4606 = 0.461$ per minute

13. **(d)**

14. **(d)** Extremely high temps needed for fusion make K.E. large enough to overcome repulsion between nuclei.

15. **(c)** Binding energy

$= [ZM_P + (A - Z)M_N - M]c^2$

$= [8M_P + (17 - 8)M_N - M]c^2$

$= [8M_P + 9M_N - M]c^2$

$= [8M_P + 9M_N - M_o]c^2$

16. **(c)** In this reaction mass is not conserved.

17. **(a)** $T_{1/2} = \frac{\ln 2}{\lambda} \therefore \lambda = \frac{\ln 2}{T_{1/2}}$

$$\Rightarrow \lambda_A = \frac{\ln 2}{T_A}, \lambda_B = \frac{\ln 2}{T_B} \Rightarrow \frac{\lambda_A}{\lambda_B} = \frac{T_B}{T_A}.$$

18. (d) $N_1 = N_0 e^{-10\lambda t}$, $N_2 = N_0 e^{-\lambda t}$

$\frac{N_1}{N_2} = e^{-9\lambda t} = e^{-1}$; $9\lambda t = 1 \Rightarrow t = \frac{1}{9\lambda}$

19. (d) Let at time t_1 & t_2, number of particles be N_1 & N_2. So,

$R_1 = \frac{dN_1}{dt} = -\lambda N_1;\quad R_2 = \frac{dN_2}{dt} = -\lambda N_2$

$\frac{R_1}{R_2} = \frac{\lambda N_1}{\lambda N_2} = \frac{N_1}{N_1 e^{-\lambda(t_2-t_1)}} = e^{\lambda(t_2-t_1)}$

$R_1 = R_2 e^{\lambda(t_2-t_1)} = R_2 e^{-\lambda(t_1-t_2)}$

20. (c) Average life of the nuclei is

$t_{av} = \frac{1}{\lambda}$ ....(i)

Half life of the nuclei

$t_{1/2} = \frac{0.693}{\lambda}$ ....(ii)

from (i) and (ii)

$t_{av} = \frac{t_{1/2}}{0.693}$

21. (d) Nuclear force is not the same between any two nucleons.

22. (a)

23. (a) $P = n\left(\frac{E}{t}\right) \Rightarrow 1000 = \frac{n \times 200 \times 10^6 \times 1.6 \times 10^{-19}}{t}$

$\Rightarrow \frac{n}{t} = 3.125 \times 10^{13}$.

24. (c) Binding energy per nucleon for fission products is higher relative to Binding energy per nucleon for parent nucleus, i.e., more masses are lost and are obtained as kinetic energy of fission products. So, the given ratio <1.

25. (b) We have $K_\alpha = \frac{m_y}{m_y + m_\alpha}.Q$

$\Rightarrow K_\alpha = \frac{A-4}{A}.Q \Rightarrow 48 = \frac{A-4}{A}.50 \Rightarrow A = 100$

26. (b) Using the relation for mean life.

Given : $t = 2\tau = 2\left(\frac{1}{\lambda}\right) \quad \left(\because \tau = \frac{1}{\lambda}\right)$

Then from $M = M_0 e^{-\lambda t} = 10e^{-\lambda \times \frac{2}{\lambda}}$

$= 10\left(\frac{1}{e}\right)^2 = 1.35\text{g}$

27. (d) Because radioactivity is a spontaneous phenomenon.

28. (a) α-particle = ${}_2He^4$. It contains 2 p and 2 n. As some mass is converted into B.E., therefore, mass of α particle is slightly less than the sum of the masses of 2 *p* and 2 *n*.

29. (c) $T_{av} = \frac{T_\alpha T_\beta}{T_\alpha + T_\beta}$

If α and *B* are emitted simultaneously.

30. (a) Due to irradiation of α-rays on end A will make it (positive) and irradiation of β-rays on end B will make it (negative) hence current will flow from A to B (or from positive to negative).

31. (b) Momentum

$Mu = \frac{E}{c} = \frac{h\nu}{c}$

Recoil energy

$\frac{1}{2}Mu^2 = \frac{1}{2}\frac{M^2u^2}{M} = \frac{1}{2M}\left(\frac{h\nu}{c}\right)^2$

$= \frac{h^2\nu^2}{2Mc^2}$

32. (c) No. of nuclide at time t is given by $N = N_0 e^{-\lambda t}$

Where N_0 = initial nuclide

This equation is equivalent to $y = ae^{-kx}$

Thus correct graph is

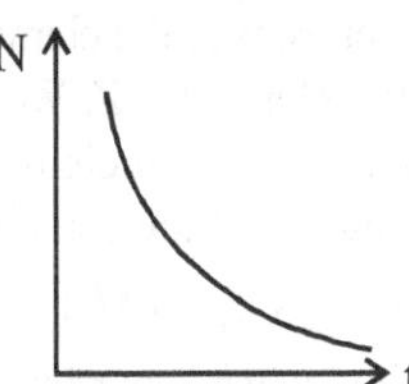

33. (b) By conservation of energy,

$(M + \Delta m)c^2 = \frac{2.M}{2}c^2 + \frac{1}{2}.\frac{2M}{2}v^2$,

where v is the speed of the daughter nuclei

$\Rightarrow \Delta mc^2 = \frac{M}{2}v^2 \qquad \therefore v = c\sqrt{\frac{2\Delta m}{M}}$

34. (a) Suppose that,

The number of ${}^{10}B$ type atoms = x

and the number of ${}^{11}B$ type atoms = y

Weight of ${}^{10}B$ type atoms = $10x$

Weight of ${}^{11}B$ type atoms = $11y$

Total number of atoms = $x + y$

$\therefore$ Atomic weight $= \frac{10x + 11y}{x + y} = 10.81$

$\Rightarrow 10x + 11y = 10.81x + 10.81y$

$\Rightarrow 0.81x = 0.19y \Rightarrow \frac{x}{y} = \frac{19}{81}$

35. (b) Applying law of conservation of momentum,

$m_1 v_1 = m_2 v_2$

$\frac{v_1}{v_2} = \frac{m_2}{m_1}$

As $m = \frac{4}{3}\pi r^3 \rho \Rightarrow m \propto r^3$

Hence, $\frac{m_2}{m_1}=\frac{r_2^3}{r_1^3}$

$\therefore \frac{v_1}{v_2}=\frac{r_2^3}{r_1^3} \Rightarrow \frac{r_2}{r_1}=\left(\frac{1}{2}\right)^{\frac{1}{3}}$

36. **(d)** In an explosion a body breaks up into two pieces of unequal masses both part will have numerically equal momentum and lighter part will have more velocity.

$U \rightarrow Th + He$

$KE_{Th}=\frac{p^2}{2m_{Th}}, KE_{He}=\frac{p^2}{2m_{He}}$

since m_{He} is less so KE_{He} will be more.

37. **(a)** As we know, $R=R_0(A)^{1/3}$

where A = mass number

$R_{Al}=R_0(27)^{1/3}=3R_0$

$R_{Te}=R_0(125)^{1/3}=5R_0=\frac{5}{3}R_{Al}$

38. **(a)** Given : Mass of neutron = M_n

Mass of proton = M_p; Atomic mass of the element = M; Number of neutrons in the element = N and number of protons in the element = Z. We know that the atomic mass (M) of any stable nucleus is always less than the sum of the masses of the constituent particles.

Therefore, $M < [NM_n + ZM_p]$.

X is a neutrino, when β-particle is emitted.

39. **(a)** Activity decreases

5000 dps to 2500 dps in 150 days

$\therefore$ Half life period $T_{1/2}=150$ days

$\therefore$ 300 days $=2T_{1/2}$

Therefore, initial activity $=5000\times 2T_{1/2}=5000\times 2\times 2$ $=20000$ dps

40. **(b)** The order of density of uranium nucleus is 10^{17} kg/m^2.

41. **(b)**

42. **(a)** $B.E_H=\frac{2.22}{2}=1.11$

$B.E_{He}=\frac{28.3}{4}=7.08$

$B.E_{Fe}=\frac{492}{56}=8.78=$ maximum

$B.E_U=\frac{1786}{235}=7.6$

$^{56}_{26}Fe$ is most stable as it has maximum binding energy per nucleon.

43. **(d)** Neutrons can't be deflected by a magnetic field.

44. **(b)** $_{-1}e^0$ is known as β-particle & $\bar{\nu}$ is known as antineutrino. Since in this reaction $\bar{\nu}$ is emitted with $_{-1}e^0$ (β-particle or electron), so it is known as β-decay.

45. **(a)** Given, $\lambda_A=8\lambda, \lambda_B=\lambda$

$N_B=\frac{N_A}{e}$

$\Rightarrow N_o e^{-\lambda_B t}=N_o\frac{e^{-\lambda_A t}}{e}$

$e^{-\lambda t}=e^{-8\lambda t}e^{-1}$

$e^{-\lambda t}=e^{-8\lambda t-1}$

Comparing both side powers

$-\lambda t=-8\lambda t-1$

$-1=7\lambda t$

$t=-\frac{1}{7\lambda}$

The best possible answer is $t=\frac{1}{7\lambda}$

1. (d) $\Delta I_E = 8.0\,mA$
$\Delta I_C = 7.9\,mA$
$$\alpha = \frac{\Delta I_C}{\Delta I_E} = \frac{7.9}{8.0} = 0.9875 \simeq 0.99$$
Also, $\beta = \frac{\alpha}{1-\alpha} = \frac{0.9875}{(1-0.9875)} = 79$

2. (d) Here, $n_i = 10^{16}\,m^{-3}$, $n_h = 5 \times 10^{22}\,m^{-3}$
As $n_e n_h = n_i^2$
$$\therefore\ n_e = \frac{n_i^2}{n_h} = \frac{(10^{16} m^{-3})^2}{5\times10^{22} m^{-3}} = 2\times10^9\,m^{-3}$$

3. (d) Energy band gap range is given by,
$$E_g = \frac{hc}{\lambda}$$
For visible region $\lambda = (4 \times 10^{-7} \sim 7 \times 10^{-7})$m
$$E_g = \frac{6.6\times10^{-34}\times3\times10^8}{7\times10^{-7}} = \frac{19.8\times10^{-26}}{7\times10^{-7}} = \frac{2.8\times10^{-19}}{1.6\times10^{-19}}$$
$E_g = 1.75\,eV$

4. (b) Voltage gain $= \beta \times$ Impedance gain
$$50 = \beta \times \frac{200}{100} = 2\beta \Rightarrow \beta = 25$$
and power gain $= \beta^2 \times \frac{200}{100} = 1250.$

5. (b) When either of A or B is 1 i.e. closed then lamp will glow.
In this case, Truth table

Inputs		Output
A	B	Y
0	0	0
0	1	1
1	0	1
1	1	1

This represents OR gate.

6. (c) In p-region of p-n junction
holes concentration > electrons concentration and in n-region electrons concentration > holes concentration.

7. (c) Peak value of rectified output voltage
= peak value of input voltage – barrier voltage
$= 2 - 0.7 = 1.3$ V.

8. (a) Current gain $(\alpha) = 0.96$
$I_e = 7.2\,mA$
$$\frac{I_c}{I_e} = \alpha = 0.96$$
$I_c = 0.96 \times 7.2\,mA = 6.91\,mA$
$I_e = I_c + I_b$
$\Rightarrow I_b = I_e - I_c = 7.2 - 6.91 = 0.29\,mA$

9. (c) No. of electrons reaching the collector,
$$n_C = \frac{96}{100}\times10^{10} = 0.96\times10^{10}$$
Emitter current, $I_E = \frac{n_E \times e}{t}$
Collector current, $I_C = \frac{n_C \times e}{t}$
$\therefore$ Current transfer ratio,
$$\alpha = \frac{I_C}{I_E} = \frac{n_C}{n_E} = \frac{0.96\times10^{10}}{10^{10}} = 0.96$$

10. (c) Here diode is forward biased with voltage $= 2 - 0 = 2$ V.
$V_B = V_{knee} + IR$
$2 = 0.7 + I \times 200$
($\therefore$ Total resistance $= 180 + 20 = 200\Omega$)
$$\therefore\ I = \frac{1.3}{200} = 6.5\,mA$$

11. (b) I $\rightarrow$ ON
II $\rightarrow$ OFF
In IInd state it is used as a amplifier it is active region.

12. (b) In half wave rectifier only half of the wave is rectified.

13. (c) $V' = V + IR = 0.5 + 0.1 \times 20 = 2.5$ V

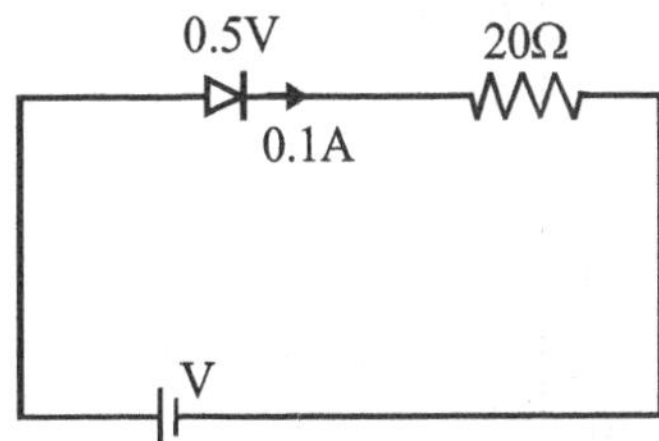

14. (b) $$\frac{V_o}{V_{in}} = \frac{R_o}{R_{in}}\times\beta = \frac{5\times10^3\times62}{500} = 10\times62 = 620$$
$V_o = 620 \times V_{in} = 620 \times 0.01 = 6.2$ V
$\therefore V_o = 6.2$ volt.

15. (b) Conductivity $\sigma = n_i e \mu_e = 10^{17} \times (1.6\times10^{-19}) \times 3800$
$= 60.8$ mho/cm

16. **(d)** Negative feedback is applied to reduce the output voltage of an amplifier. If there is no negative feedback, the value of output voltage could be very high. In the options given, the maximum value of voltage gain is 100. Hence it is the correct option.

17. **(a)** In the given system all four gate is NOR gate

Truth Table

A	B	$(y' = \overline{A+B})$	$y'' = \overline{(A+y')}$	$y''' = \overline{(A+y'')}$	$y = \overline{y''+y'''}$
0	0	1	0	0	1
0	1	0	1	0	0
1	0	0	0	1	0
1	1	0	0	0	1

i.e.,

A	B	y
0	0	1
0	1	0
1	0	0
1	1	1

18. **(a)** Conductivity, $\sigma = \frac{1}{\rho} = e(n_e\mu_e + n_h\mu_h)$

$2.13 = 1.6 \times 10^{-19}(0.38 + 0.18)\, n_i$

(Since in intrinsic semi-conductor, $n_e = n_h = n_i$)

$\therefore$ density of charge carriers, n_i

$$= \frac{2.13}{1.6 \times 10^{-19} \times 0.56} = 2.37 \times 10^{19}\,\text{m}^{-3}$$

19. **(d)** Here $Y = \overline{(\overline{A} + \overline{B})} = \overline{\overline{A.B}} = A \cdot B$. Thus, it is an AND gate for which truth table is

A	*B*	*Y*
0	0	0
0	1	0
1	0	0
1	1	1

20. **(d)** $n_i^2 = n_e n_h$

$(1.5 \times 10^{16})^2 = n_e (4.5 \times 10^{22})$

$\Rightarrow \quad n_e = 0.5 \times 10^{10}$

or $\quad n_e = 5 \times 10^9$

Given $\quad n_h = 4.5 \times 10^{22}$

$\Rightarrow n_h >> n_e$

$\therefore$ Semiconductor is p-type and

$n_e = 5 \times 10^9\,\text{m}^{-3}$.

21. **(c)** In *n*-type semiconductors, electrons are the majority charge carriers.

22. **(d)** For semiconductor, $n = AT^{3/2} e^{-\frac{E_g}{2KT}}$;

so $n \propto T^{3/2}$

23. **(d)** When PN junction diode is forward biased both depletion layer width W and barrier height V_0 decrease and current due to molarity carrier increases.

24. **(b)** D_2 is forward biased whereas D_1 is reversed biased. So effective resistance of the circuit

$R = 4 + 2 = 6\Omega$

$\therefore i = \frac{12}{6} = 2$ A.

25. **(d)** In common emitter configuration current gain

$$A_i = \frac{-hf_e}{1 + h_{oe}R_L} = \frac{-50}{1 + 25 \times 10^{-6} \times 1 \times 10^3} = -48.78$$

26. **(c)**

27. **(b)** It is a *p-n-p* transistor with *R* as base.

28. **(c)** Here P-N junction diode rectifies half of the ac wave i.e., acts as half wave rectifier. During + ve half cycle Diode $\rightarrow$ forward biased output across will be

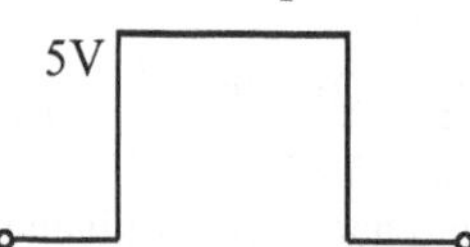

During –ve half cycle Diode $\rightarrow$ reverse biased output will not obtained.

29. **(d)** Due to heating, when a free electron is produced then simultaneously a hole is also produced.

30. **(b)** $I = nA\,ev_d$ or $I \propto nv_d$

$$\therefore \quad \frac{I_e}{I_h} = \frac{n_e v_e}{n_h v_h} \text{ or } \frac{n_e}{n_h} = \frac{I_e}{I_h} \times \frac{v_h}{v_e} = \frac{7}{4} \times \frac{4}{5} = \frac{7}{5}$$

31. **(c)** Electronic configuration of ^{6}C

$^6\text{C} = 1s^2, 2s^2\, 2p^2$

The electronic configuration of ^{14}Si

$^{14}\text{Si} = 1s^2, 2s^2\, 2p^6, 3s^2\, 3p^2$

As they are away from Nucleus, so effect of nucleus is low for Si even for Sn and Pb are almost mettalic.

32. **(d)** V_1 —▷|— /\/\/\ — V_2

In forward bias, $V_1 > V_2$ i.e., in figure (d) p-type semiconductor is at higher potential w.r.t. n-type semiconductor.

33. **(a)** A positive feed back from output to input in an amplifier provides oscillations of constant amplitude.

34. **(b)** The power gain in case of CE amplifier,

Power gain = β^2 × Resistance gain

$= \beta^2 \times \frac{R_o}{R_i}$

$= (10)^2 \times 5 = 500$.

35. **(c)** **Given :** Voltage gain $A_V = 150$

$V_i = 2\cos\left(15t + \frac{\pi}{3}\right); V_0 = ?$

For CE transistor phase difference between input and output signal is $\pi = 180°$

Using formula, $A_V = \frac{V_0}{V_i}$

$\Rightarrow V_0 = A_V \times V_i$

$= 150 \times 2\cos\left(15t + \frac{\pi}{3}\right)$

or $V_0 = 300\cos\left(15t + \frac{\pi}{3} + \pi\right)$

$V_0 = 300\cos\left(15t + \frac{4}{3}\pi\right)$

36. **(a)** To use a transistor as an amplifier the emitter base junction is forward biased while the collector base junction is reverse biased.

37. **(d)** Copper is a conductor, so its resistance decreases on decreasing temperature as thermal agitation decreases; whereas germanium is semiconductor therefore on decreasing temperature resistance increases.

38. **(b)** In forward biasing, the diode conducts. For ideal junction diode, the forward resistance is zero; therefore, entire applied voltage occurs across external resistance R i.e., there occurs no potential drop, so potential across R is V in forward biased.

39. **(a)** Current gain $(\alpha) = 0.96$

$I_e = 7.2\,\text{mA}$

$\frac{I_c}{I_e} = \alpha = 0.96$

$I_c = 0.96 \times 7.2\,\text{mA} = 6.91\,\text{mA}$

$I_e = I_c + I_b$

$\Rightarrow I_b = I_e - I_c = 7.2 - 6.91 = 0.29\,\text{mA}$

40. **(d)**

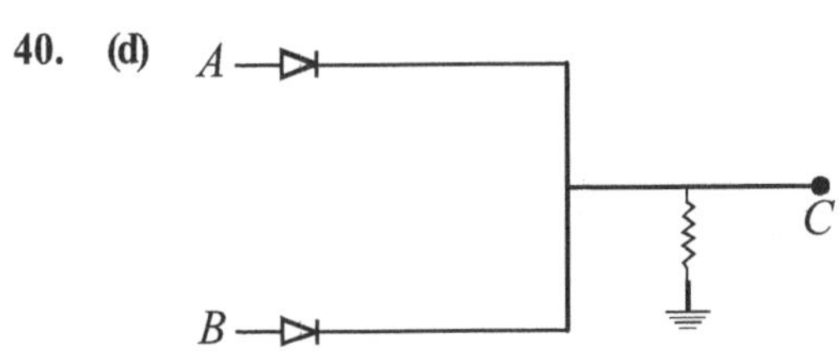

The truth table for the above logic gate is :

A	B	C
1	1	1
1	0	1
0	1	1
0	0	0

This truth table follows the boolean algebra $C = A + B$ which is for OR gate

41. **(b)** $R = \frac{\Delta V}{\Delta I} = \frac{2.1 - 2}{(800 - 400) \times 10^{-3}} = \frac{1}{4} = 0.25\,\Omega$

42. **(c)**

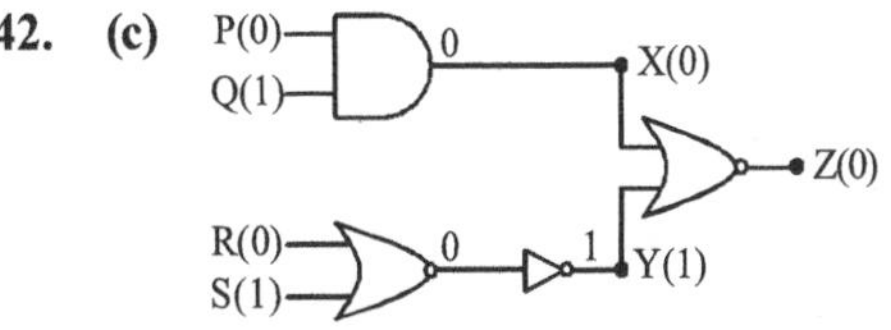

43. **(b)**

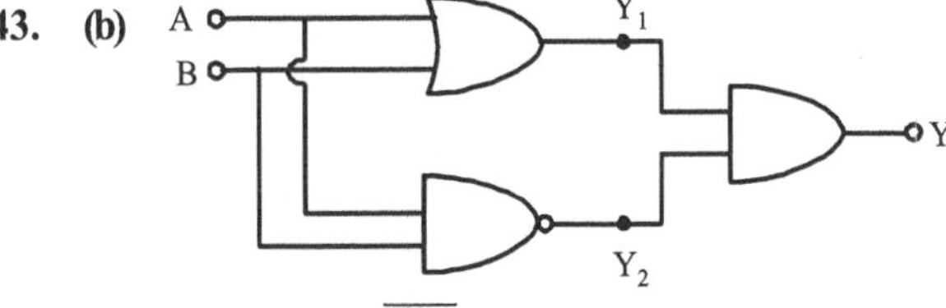

$Y_1 = A + B,\ Y_2 = \overline{A.B}$

$Y = (A + B)\bullet\overline{AB} = A\bullet\overline{A} + A\bullet\overline{B} + B\bullet\overline{A} + B\bullet\overline{B}$

$= 0 + A\bullet\overline{B} + B\bullet\overline{A} + 0 = A\bullet\overline{B} + B\bullet\overline{A}$ (XOR gate)

44. **(b)** $E_g = 2.0\,\text{eV} = 2 \times 1.6 \times 10^{-19}\,\text{J}$

$E_g = h\nu$

$\therefore\ \nu = \frac{E_g}{h} = \frac{2 \times 1.6 \times 10^{-19}\,\text{J}}{6.62 \times 10^{-34}\,\text{Js}}$

$= 0.4833 \times 10^{15}\,\text{s}^{-1} = 4.833 \times 10^{14}\,\text{Hz}$

$\simeq 5 \times 10^{14}\,\text{Hz}$

45. **(d)** The average value of output direct current in a full wave rectifier = (average value of current over a cycle)

$= (2I_0/\pi) = \frac{2I_0}{\pi}$